Y0-BYB-333

BIOLOGICAL REACTIVE INTERMEDIATES IV

Molecular and Cellular Effects and Their Impact on Human Health

ADVANCES IN EXPERIMENTAL MEDICINE AND BIOLOGY

Editorial Board:

NATHAN BACK, *State University of New York at Buffalo*

IRUN R. COHEN, *The Weizmann Institute of Science*

DAVID KRITCHEVSKY, *Wistar Institute*

ABEL LAJTHA, *N.S. Kline Institute for Psychiatric Research*

RODOLFO PAOLETTI, *University of Milan*

Recent Volumes in this Series

Volume 276
CORONAVIRUSES AND THEIR DISEASES
Edited by David Cavanagh and T. David K. Brown

Volume 277
OXYGEN TRANSPORT TO TISSUE XII
Edited by Johannes Piiper, Thomas K. Goldstick, and Michael Meyer

Volume 278
IMMUNOBIOLOGY AND PROPHYLAXIS OF HUMAN HERPESVIRUS INFECTIONS
Edited by Carlos Lopez, Ryoichi Mori, Bernard Roizman, and Richard J. Whitley

Volume 279
BIOCHEMISTRY, MOLECULAR BIOLOGY, AND PHYSIOLOGY OF PHOSPHOLIPASE A_2 AND ITS REGULATORY FACTORS
Edited by Anil B. Mukherjee

Volume 280
MYOBLAST TRANSFER THERAPY
Edited by Robert C. Griggs and George Karpati

Volume 281
FIBRINOGEN, THROMBOSIS, COAGULATION, AND FIBRINOLYSIS
Edited by Chung Yuan Liu and Shu Chien

Volume 282
NEW DIRECTIONS IN UNDERSTANDING DEMENTIA AND ALZHEIMER'S DISEASE
Edited by Taher Zandi and Richard J. Ham

Volume 283
BIOLOGICAL REACTIVE INTERMEDIATES IV: Molecular and Cellular Effects and Their Impact on Human Health
Edited by Charlotte M. Witmer, Robert R. Snyder, David J. Jollow, George F. Kalf, James J. Kocsis, and I. Glenn Sipes

---

A Continuation Order Plan is available for this series. A continuation order will bring delivery of each new volume immediately upon publication. Volumes are billed only upon actual shipment. For further information please contact the publisher.

[International Symposium on Biological Reactive Intermediates (1990: Tucson, Ariz.)

BIOLOGICAL REACTIVE INTERMEDIATES IV

Molecular and Cellular Effects and Their Impact on Human Health

Edited by

Charlotte M. Witmer and Robert R. Snyder
Rutgers University
Piscataway, New Jersey

David J. Jollow
Medical University of South Carolina
Charleston, South Carolina

George F. Kalf and James J. Kocsis
Thomas Jefferson University
Philadelphia, Pennsylvania

and

I. Glenn Sipes
University of Arizona
Tucson, Arizona

RA1220
I58
1990

PLENUM PRESS • NEW YORK AND LONDON

Library of Congress Cataloging in Publication Data

International Symposium on Biological Reactive Intermediates (4th: 1990: Tucson, Ariz.)
Biological reactive intermediates IV: molecular and cellular effects and their impact on human health / edited by Charlotte M. Witmer . . . [et al.].
p. cm.—(Advances in experimental medicine and biology: v. 283)
"Proceedings of the Fourth International Symposium on Biological Reactive Intermediates, held January 14–17, 1990, in Tucson, Arizona"—T.p. verso.
Includes bibliographical references and index.
ISBN 0-306-43737-6
1. Biotransformation (Metabolism)—Congresses. 2. Biochemical toxicology—Congresses. I. Witmer, Charlotte M. II. Title. III. Title: Biological reactive intermediates 4. IV. Series.
[DNLM: 1. Biotransformation—congresses. 2. Toxicology—congresses. W1 AD559 v. 283 / QV 600 I63b 1990]
RA1220.I58 1990
615.9—dc20
DNLM/DLC 90-14326
for Library of Congress CIP

Proceedings of the Fourth International Symposium on
Biological Reactive Intermediates, held January 14–17, 1990,
in Tucson, Arizona

© 1991 Plenum Press, New York
A Division of Plenum Publishing Corporation
233 Spring Street, New York, N.Y. 10013

All rights reserved

No part of this book may be reproduced, stored in a retrieval system, or transmitted in any form or by any means, electronic, mechanical, photocopying, microfilming, recording, or otherwise, without written permission from the Publisher

Printed in the United States of America

PREFACE

The finding that chemicals can be metabolically activated to yield reactive chemical species capable of covalently binding to cellular macromolecules and the concept that these reactions could initiate toxicological and carcinogenic events stimulated a meeting by a small group of toxicologists at the University of Turku, in Finland, in 1975 (Jollow et al., 1977). The growing interest in this field of research led to subsequent symposia at the University of Surrey, in England in 1980 (Snyder et al., 1982), and the University of Maryland in the U.S.A. in 1985 (Kocsis et al., 1986). The Fourth International Symposium on Biological Reactive Intermediates was hosted by the Center for Toxicology at the University of Arizona and convened in Tucson, Arizona, January 14-17, 1990. Over 300 people attended. There were 60 platform presentations by invited speakers, and 96 volunteer communications in the form of posters were offered. These meetings have grown from a small group of scientists working in closely related areas to a major international series of symposia which convene every five years to review, and place in context, the latest advances in our understanding of the formation, fate and consequences of biological reactive intermediates.

The Organizing Committee: Allan H. Conney, Robert Snyder (Co-chairman), and Charlotte M. Witmer (Rutgers University, Piscataway, NJ), David J. Jollow Co-chairman) (Medical University, South Carolina, Charleston, SC), I. Glenn Sipes (Co-chairman) (University of Arizona, Tucson, AZ), James J. Kocsis and George F. Kalf (Thomas Jefferson University, Philadelphia, PA), and Donald J. Reed (Oregon State University), developed the program on the basis of a series of communications with the members of the International Program Committee: M.W. Anders (University of Tubingen), A. Jay Gandolfi (University of Arizona), G. Gordon Gibson (University of Surrey), James R. Gillette (National Institutes of Health), Doyle G. Graham (Duke University), Ronald W. Estabrook (University of Texas), Helmut Greim (GSF, Munich), James Halpert (University of Arizona), Jack Hinson (National Center for Toxicological Research), Daniel C. Liebler (University of Arizona), Lance Pohl (National Institutes of Health), John B. Schenkman (University of Connecticut). The result was a program which encompassed a broad range of areas related to the field of biological reactive intermediates, the role of reactive oxygen, carcinogenesis, the application of new techniques in molecular biology to study biological reactive intermediates, the use of information relating to biological reactive intermediates as biomarkers of exposure, and the study of biological reactive intermediates in humans.

These meetings are not sponsored by a major society. It falls upon the organizing committee to secure funding for the symposium. The committee expresses their gratitude to the following organizations for contributing to the support of the symposium. Supporters among government agencies included the National Institute for Environmental Health Sciences, the U.S. Environmental Protection Agency, the Agency for Toxic Substances and Disease Registry, the U.S. Army, and the U.S. Air Force. In addition, industrial support came from Beckman Instruments, the Burroughs Wellcome Company, the Dow Chemical Company, Eli Lilly and Company, Hoffmann-LaRoche, Inc., Glaxo, Inc., ICI Americas, Inc., G.D. Searle and Company, Merck Sharp and Dohme Research Laboratories, the Proctor and Gamble Company, Smith Kline Beecham, Sterling Winthrop Research Institute, and the Upjohn Company. Support from academic institutions came from the University of Arizona Center for

Toxicology , and from the Environmental and Occupational Health Sciences Institute and the Joint Graduate Program in Toxicology of Rutgers University and The University of Medicine and Dentistry of New Jersey/Robert Wood Johnson Medical School.

It is not possible to mount a major international symposium without the dedicated effort of session chairpeople. The session chairs contributed programmatic ideas prior to the meeting and kept the program on track during the meeting. We express our gratitude to Ronald W. Estabrook (University of Texas), Steven D. Aust (Utah State University), Robert P. Hanzlik (Kansas State University), A. Jay Gandolfi (University of Arizona), Allan H. Conney (Rutgers University), G. Tim Bowden (University of Arizona), Dietrich W. Henschler (University of Wurzburg), Hermann Kappus (Free University of Berlin), Donald S. Davies (University of Arizona), Peter Farmer (Medical Research Council), Roger O. McClellan (Chemical Industry Institute of Toxicology).

Professor Elizabeth C. Miller, who played an important role in the previous Biological Reactive Intermediates Symposia, died in the time between the third and fourth in this series of symposia. She, along with her husband and collaborator, Professor James A. Miller, reported the original observation of the covalent binding phenomenon. Together, they helped open up a whole new area of investigation for two generations of scientists. The organizing committee wishes to pay tribute to the memory of Betty Miller for her fundamental scientific contributions and for the enthusiasm with which she participated in these conferences.

Critical to the success of these symposia has been the selection of an appropriate venue and this year we were particularly fortunate in having asked the Center for Toxicology at the University of Arizona, directed by Professor I. Glenn Sipes to act as host. The stimulating scientific spirit, the hospitality, and the excellent facilities located against the background of the great southwestern desert created an ideal locale for these deliberations. We are forever indebted to Professor Sipes and the local organizing committee (Professors Daniel C. Liebler, A. Jay Gandolfi and James R. Halpert) for providing the environment in which our colleagues from around the world could gather, communicate and learn together.

Without doubt, the people who worked hardest to insure the success of the conference were the administrators and secretarial staff including Sena Taylor, Charleen Prytula, Kathy Sousa, Pamela Murray, Bert Sanchez and Judith Stanfield of the University of Arizona and Catherine Raymore and Bernadine Chmielowicz of Rutgers University.

The field of biological reactive intermediates has grown to include toxicologists, biochemists, oncologists, pathologists, chemists, physicians, epidemiologists, geneticists, molecular biologists and scientists in an increasing circle of related disciplines. The need for interdisciplinary approaches to the solution of basic biomedical problems has of necessity lead to the dissolution of boundaries between the classical disciplines. The benefit to the individual scientist is the opportunity to expand one's horizons and to grow intellectually. The benefit to man is that we can more readily approach solutions to key problems in human health. The meeting in Arizona demonstrated once again that our colleagues from around the world can join together with the common goal of understanding chemically induced human disease. Of equal importance was that once again we demonstrated that this could be accomplished in an atmosphere of collegiality, appreciation of the talents of our fellow scientists, and the comradeship which attracts us to meet together on a regular basis to continue our discussion.

Charlotte M. Witmer
Robert Snyder
David J. Jollow
George F. Kalf
James J. Kocsis
I. Glenn Sipes

CONTENTS

SESSIONS I - VI

SHORT COMMUNICATIONS

SESSION VII - XII

SHORT COMMUNICATIONS

CYTOCHROME P-450 OXIDATIONS AND THE GENERATION OF BIOLOGICALLY REACTIVE INTERMEDIATES

F. Peter Guengerich,* Tsutomu Shimada,* Arnaud Bondon,* and Timothy L. Macdonald†

Department of Biochemistry* and Center in Molecular Toxicology, Vanderbilt University School of Medicine, Nashville, Tennessee 37232
Department of Chemistry†, University of Virginia, Charlottesville, Virginia 22901

INTRODUCTION

Cytochrome P-450 (P-450) enzymes are involved in the oxidation of many steroids, eicosanoids, pesticides, drugs, and carcinogens. Considerable evidence has been accrued over the years to support the view that the majority of chemical carcinogens require bioactivation in order to elicit tumor initiation, and many toxic chemicals other than carcinogens also require bioactivation. The P-450 enzymes are probably involved to a greater extent than any other enzymes in the generation of the biological reactive intermediates involved in such toxicities (for reviews see Nelson, 1982; Guengerich and Liebler, 1985; Nelson and Harvison, 1987; Kadlubar and Hammons, 1987; Guengerich, 1988). Thus, a proper knowledge of these enzymes and the chemistry involved in catalysis is requisite for a rational understanding of schemes of toxicity and carcinogenesis.

Today ample evidence supports the view that many P-450 genes exist in each species (30-50?) and many of these are expressed simultaneously to give a mixture of P-450 proteins in the liver or another tissue. The P-450s have now been classified on the basis of their primary sequences, when available (Nebert et al., 1989). The sequences (almost always derived from cDNA analysis) provide a rather logical way of classifying the P-450s and the use of such nomenclature in the literature is recommended. However, attention is called to the dramatic differences among some of the closely related P-450s as regards their catalytic specificity (e.g., P-450 IIC and IIIA families) and regulation of tissue- and inducer-specific expression (e.g., P-450 IIIA and IVA families; see, for instance, Umbenhauer et al., 1987 and Kaminsky et al., 1984). Indeed, recent studies involving the sequencing of natural mutants and site-directed mutagenesis have shown the dramatic effects that alteration of even a single residue can have on the catalytic specificity of a P-450 (Lindberg and Negishi, 1989; Kronbach et al., 1989). These considerations are important as they relate to efforts to extrapolate conclusions about the catalytic specificity of individual P-450s to their orthologs in other species, particularly humans.

In recent years the purification and characterization of human P-450s has become possible, and much information is now available regarding the human P-450s (see Guengerich, 1989a). Obviously knowledge about the catalytic specificities of the human P-450s regarding the generation of biological reactive intermediates is of importance in the proper assessment of risk. Recently we have addressed this problem and, on the basis of *in vitro* studies, concluded that three human P-450 enzymes play the most prominent roles in the bioactivation of most of the stronger pro-mutagens and -carcinogens studied to date (Table 1).

Ultimately the significance of these findings will have to be evaluated in in vivo settings. Another focus of our work has been the development of appropriate non-invasive assays to address the hypothesis that differences in the levels of particular P-450s in humans can actually contribute to variations in risk from chemicals in the environment.

While much remains to be learned about the roles of particular structural elements of the P-450 proteins in influencing catalytic specificity, a review of the mechanisms P-450 enzymes employ in catalysis is in order. The remainder of this article will be devoted to that task, with recent examples presented of how knowledge of catalytic events can be used to rationalize events in biotransformation and predict reactions with new compounds.

RESULTS AND DISCUSSION

General P-450 Mechanism

The oxidative reactions catalyzed by P-450 enzymes can be understood in the context of a general scheme involving formation of a perferryl complex and subsequent odd-electron reactions (Fig. 1). That is, the formal $(FeO)^{3+}$ species abstracts a hydrogen atom or a p- or non-bonded electron to form a transient reaction pair, which quickly ($k \sim 10^8$ s^{-1}) collapses to give the products (this collapse is often termed "oxygen rebound"). With this general scheme it is possible to explain carbon hydroxylation, epoxidation and related reactions (oxidative group migration and porphyrin N-alkylation), heteroatom release, and heteroatom oxygenation.

Table 1. Pro-carcinogens Activated by Human P-450 Enzymes

P-450 IIE1 (P-450$_j$)
N-Nitroso-*N*-benzyl-*N*-methylamine
N-Nitroso-*N*-butyl-*N*-methylamine
N-Nitroso-*N*,*N*-diethylamine
N-Nitroso-*N*,*N*-dimethylamine
P-450 IA2 (P-450$_{PA}$)
2-Acetylaminofluorene
2-Aminoanthracene
4-Aminobiphenyl
2-Amino-3,5-dimethylimidazo[4,5-*f*]quinoline (MeIQ)
2-Amino-3,8-dimethylimidazo[4,5-*f*]quinoxaline (MeIQx)
2-Aminodipyrido[1,2-*a*:3',2'-d]imidazole (Glu P-2)
2-Aminofluorene
2-Amino-6-methyldipyrido[1,2-*a*:3',2'd]imidazole (Glu P-1)
2-Amino-3-methylimidazo[4,5-*f*]quinoline (IQ)
3-Amino-1-methyl-5*H*-pyrido[4,3-*b*]indole (Trp P-2)
P-450 IIIA4 (P-450$_{NF}$)
Aflatoxin B_1
Aflatoxin G_1
6-Aminochrysene
Benzo(*b*)fluoranthene-9,10-diol
Benzo(*a*)pyrene-7,8-diol (+ and -)
tris-(2,3-Dibromopropyl)phosphate
7,12-Dimethylbenz(*a*)anthracene-3,4-diol
Sterigmatocystin

From Yoo et al., 1988; Shimada and Guengerich, 1989; and Shimada et al., 1989a, 1989b.

Much has been written elsewhere about carbon hydroxylation, and the reaction appears to be fairly well understood (Groves et al., 1978; Ortiz de Montellano, 1986). Formal dehydrogenation of alkanes sometimes accompanies hydroxylation, and this reaction can be understood in terms of the same initial step as postulated for carbon hydroxylation (hydrogen atom abstraction) followed by the formal transfer of a proton and electron rather than oxygen rebound (Ortiz de Montellano, 1989). Epoxidation and related reactions involving p systems have been considered elsewhere (Ortiz de Montellano et al., 1982; Liebler and Guengerich, 1983; Guengerich and Macdonald, 1984; Groves et al., 1986; Groves and Watanabe, 1986; Ortiz de Montellano, 1986; Garrison and Bruice, 1989; Guengerich, 1990a). While a cation Fe^{3+} complex is shown in Fig. 1, such an intermediate may well not explain all of the reactions that are seen and other intermediates such as p charge transfer complexes, radical-Fe^{4+} species, and metallo-oxetanes have been considered in the cases of the enzymes and biomimetic models, and the possibility exists that different intermediates may be used depending upon the fine structure of the model or enzyme (Yamaguchi et al., 1986; Traylor and Miksztal, 1989; Guengerich and Macdonald, 1990).

N-Dealkylation and -Oxygenation

Considerable evidence has now been gathered to support the view that these oxidation reactions proceed via initial one-electron oxidation at the nitrogen. This evidence includes the following points: (1) N-oxides are generally rather stable

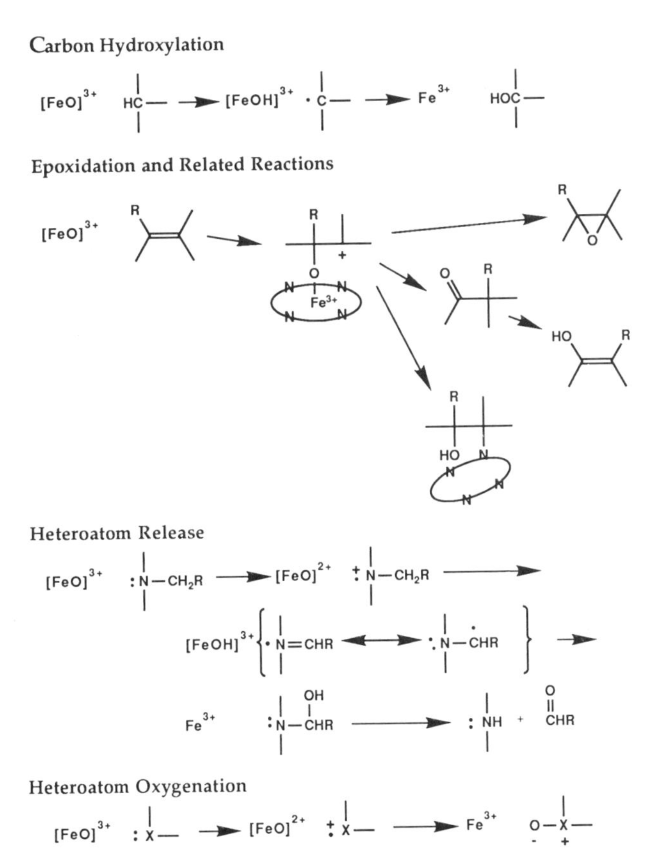

Figure 1. Generalized scheme of P-450-catalyzed oxidation reactions (Guengerich and Macdonald, 1984, 1990).

entities and cannot be considered obligatory intermediates in N-dealkylation (Guengerich, 1984); indeed, some substrates are oxidized to give both stable N-oxides and products derived by dealkylation paths (Williams et al., 1989). (2) Reactions involving strained cycloalkylamines gives rise to rearranged products expected for pathways originating with aminium radicals (Guengerich et al., 1984). For instance, a cyclobutylamine is oxidized to a pyrroline (Fig. 2) (Bondon et al., 1989). (3) P-450-catalyzed oxidation of 4-alkyl 1,4-dihydropyridines results in the release of alkyl radicals, which can either be trapped or alkylate the porphyrin (Augusto et al., 1982). The ability of different 4-substituents to leave is a function of radical stability (Lee et al., 1988). The proposed analogy of dihydropyridine oxidation to N-dealkylation is depicted in Fig. 3, with similar steps being involved (the exception is the oxygen transfer itself). (4) P-450-catalyzed carbon hydroxylation is accompanied by relatively high intrinsic kinetic hydrogen isotope effects (Groves et al., 1978) but N-dealkylation is not (Miwa et al., 1983), as in the case of electrochemical and chemical reactions which are known to proceed via aminium radicals (Shono et al., 1982). Further, low deuterium and tritium isotope effects have been found to be associated with the oxidation of 1,4-dihydropyridines such as nifedipine (Guengerich and Böcker, 1988; Guengerich, 1990b). (5) When some p-substituted N,N-dimethylanines were examined for rates of N-demethylation by P-450 IIB1, a value of $r = -0.7$ was found and is consistent with the view that a positively-charged intermediate exists (Burka et al., 1985). Further analysis of the same type of data using models based on Marcus equations suggest values of ~+1.8 V for $E_{1/2}$, the formal oxidation-reduction potential of the FeO^{3+} complex in P-450 (for the FeO^{3+}/FeO^{2+} couple) and 23 kcal mol^{-1} for l, a complex self-exchange parameter (Macdonald et al., 1989). Further, it is of interest to point out that in such a Marcus model the effective $E_{1/2}$ of the P-450 is thought to be heavily influenced by the dielectric constant of the protein interior and the distance between the atoms participating in such an outer sphere electron transfer process, and both of these parameters are probably heavily influenced by the protein structure (Macdonald et al., 1989).

N-Dealkylation and -oxygenation are both thought to proceed from an initial animium radical intermediate, as depicted in Figure 4. If the $E_{1/2}$ for electron transfer is unfavorable (lower path of Fig. 4), then hydrogen atom abstraction from the substrate may be the only course available (and such a process requires a closer distance between the FeO^{3+} species and the oxidizable site of the substrate than does electron transfer). If the aminium radical is formed, partitioning between the dealkylation and oxygenation routes is dependent upon the relative values of the two rate constants, $k_{rebound\ (N)}$ and k_{deprot}, which, of course, cannot be measured directly. Rates of a-deprotonation of aminium radicals have been long considered to be very rapid, but this view may be misleading in that most of the purely chemical studies have been done in the presence of free amines, which serve to help abstract the a-protons (see Hammerich and Parker, 1984; Dinnocenzo and Banach, 1989).

It is possible that these rates may be slower than originally thought, and greater stability of an aminium radical favors N-oxygenation, for oxygen rebound to an aminium radical would be expected to be as fast as to a carbon radical ($k \sim 10^8\ s^{-1}$). In most cases of nitrogen oxidation by P-450, N-dealkylation is favored. The main exceptions are primary arylamines (devoid of a-protons, of course), compounds in which the nitrogen is at a bridgehead and Bredt's rule applies (e.g., quinidine), and amines in which a neighboring atom inductively stabilizes the aminium radical (e.g.,

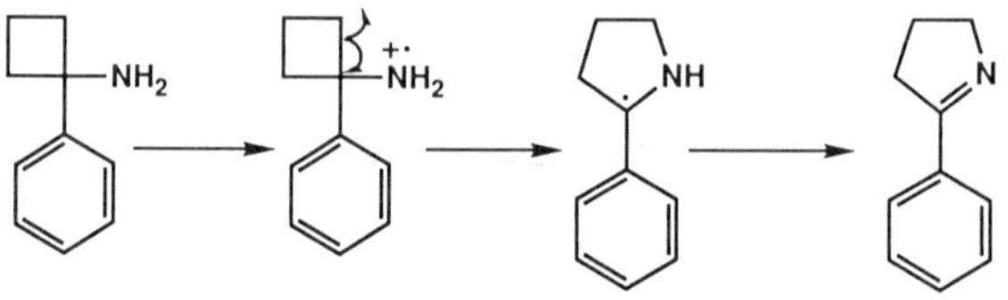

Figure 2. P-450-catalyzed ring expansion of 1-phenylcyclobutylamine (Bondon et al., 1989).

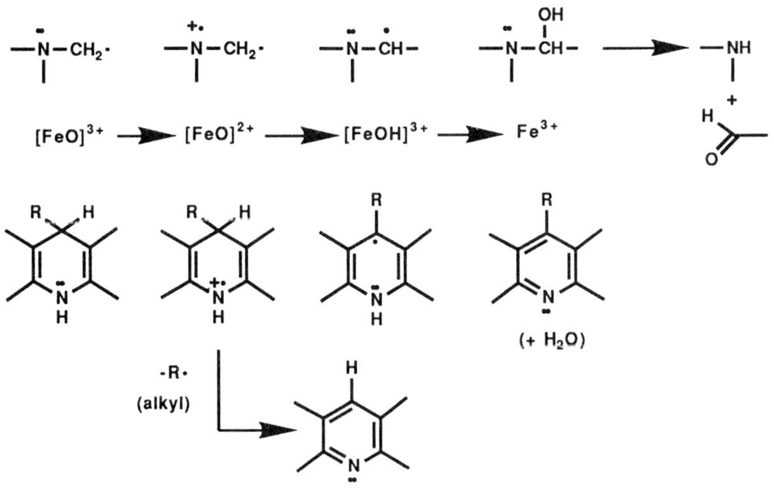

Figure 3. N-Dealkylation and 1,4-dihydropyridine oxidation catalyzed by P-450 (Augusto et al., 1982; Böcker and Guengerich, 1986; Lee et al., 1988; Guengerich and Böcker, 1988; Guengerich, 1990a).

azoprocarbazine) (Guengerich and Macdonald, 1984). Recently several other instances in which P-450s catalyze N-oxygenation have been clearly documented. N-(1-Phenylcyclobutyl)-benzylamine is oxidized to a hydroxylamine and then on to the nitrone (Bondon et al., 1989), methamphetamine is oxidized to a hydroxylamine (Baba et al., 1988), and the pyrrolizidine alkaloid senecionine is converted to the N-oxide (Williams et al., 1989). Thus, in some P-450 proteins the conditions are favorable for N-oxygenation of certain substrates, and further studies will be required to elucidate the details of why this is the case.

Peroxidases also catalyze the oxidative N-dealkylation of substituted amines. However, the mechanism is probably different in most cases. The reaction is probably initiated by 1-electron oxidation and aminium radicals can be detected using EPR methods (Griffin and Ting, 1978; Van der Zee et al., 1989). However, it does not appear that most peroxidases are capable of carrying out the oxygen rebound reaction in the formal FeO^{2+}/-NR- pair. This may be because of steric restriction in the protein, and Ortiz de Montellano (1987) has postulated that in horseradish peroxidase the mode of electron flow is through the meso edge of the heme. Sometimes P-450s

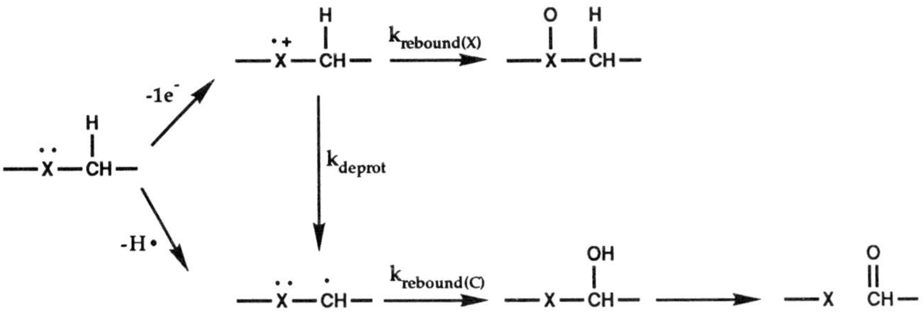

Figure 4. General scheme for heteroatom oxygenation and dealkylation (Guengerich and Macdonald, 1984, 1990).

act in a manner more akin to the peroxidases--for instance, in the oxidation of 1,4-dihydropyridines there is no evidence that oxygen is transferred (Fig. 3). Miwa et al (1983) observed that the kinetic deuterium isotope effects associated with peroxidase-catalyzed N-dealkylation reactions are relatively high--although such results may appear to indicate hydrogen atom abstraction (Miwa et al., 1983), Van der Zee et al. (1989) have argued in favor of initial oxidation by 1-electron transfer and, if the meso edge of the porphyrin is actually involved in the transfer process, then it would seem unlikely that a hydrogen atom would be abstracted directly. In the oxidation of the 1,4-dihydropyridine nifedipine, both P-450s and horseradish peroxidase showed low kinetic hydrogen isotope effects (Guengerich, 1990b). Why peroxidases show such high kinetic deuterium isotope effects is yet unclear and will require further investigation.

What is the significance of the partitioning between N-dealkylation and N-oxygenation? Understanding these processes is important because in some cases the two pathways can lead to quite different products. For instance, a single P-450 can oxidize the pyrrolizidine alkaloid senecionine to either the N-oxide (N-oxygenation) or a pyrrole (dealkylation and rearrangement) (Fig. 5) (Williams et al., 1989). The pyrrole is probably the ultimate toxic species and the N-oxide should be considered a detoxication product. N-Oxygenation converts arylamines to dangerous hydroxylamines (Kadlubar and Hammons, 1987)YN-dealkylation of an alkyl-substituted arylamine can be considered a detoxication reaction in that it renders the compound unable to be hydroxylated by the flavin-containing monooxygenase (see Ziegler et al., 1988). Numerous other examples can be considered.

Halogen Oxygenation

Many seemingly unusual reactions catalyzed by P-450s can be rationalized using general schemes of the principles depicted in Figure 1. For instance, Ullrich has characterized the novel rearrangements of prostaglandins to thromboxanes and prostacyclins in terms of such chemistry, although the oxidation of the iron atom is an internal process (Hecker and Ullrich, 1989). The dehydrogenation of alkanes (Nagata et al., 1986; Rettie et al., 1987; Ortiz de Montellano, 1989) and the oxidative cleavage of carboxylic acid esters (Guengerich, 1987; Guengerich et al., 1988) can also be understood in the context of hydrogen atom abstraction and subsequent transfers. The high effective $E_{1/2}$ of the P-450 FeO^{3+} species (Macdonald et al., 1989) suggests that other reactions not generally considered may also be catalyzed by P-450s. In this

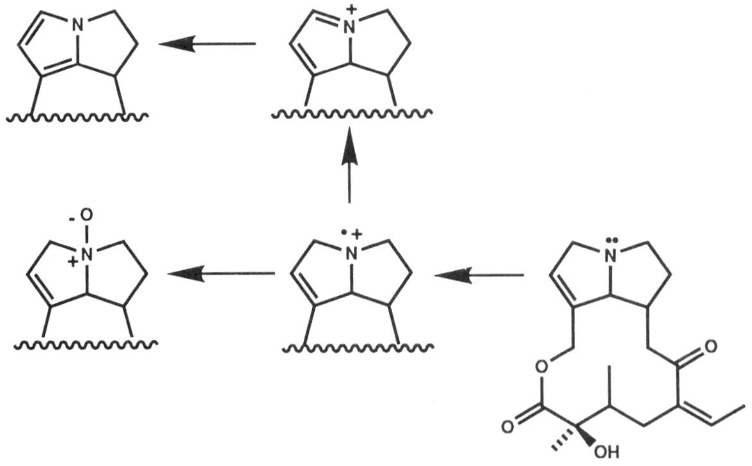

Figure 5. Oxidation of the pyrrolizidine alkaloid senecionine by P-450 (Williams et al., 1989).

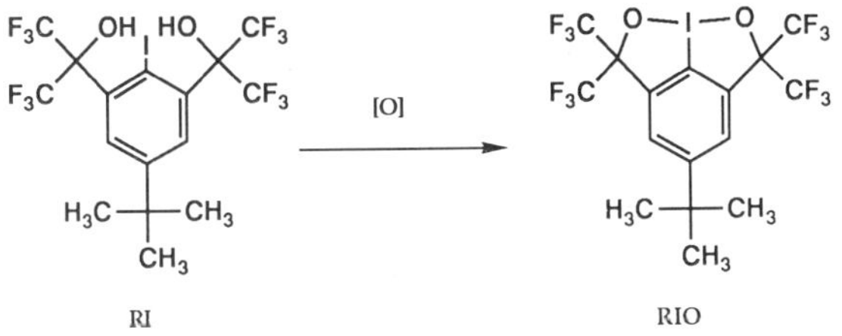

RI

RIO

Figure 6. P-450-catalyzed iodine oxidation (Guengerich, 1989b).

regard we considered the oxidation of halides. Previously we had demonstrated that P-450 IIB1 could catalyze the transfer of oxygen from iodosylbenzene to iodobenzene (Burka et al., 1980), but it was not possible to demonstrate the NADPH-dependent oxidation of a halide directly.

The compound 10-tert-butyl-3,3,6,6-tetrakis(trifluoromethyl)-4,5,6-benzo-1-ioda-7,8-dioxabicyclo[3.3.1]octane (RIO) is relatively stable, unlike iodosylbenzene, and can be readily quantified by high performance liquid chromatography. P-450 IIB1 was shown to oxidize 4-tert-butyl-2,6-bis[1-hydroxyl-(trifluorylmethyl)-2,2,2-trifluoroethyl]iodobenzene (RI) to RIO, and the UV spectrum of the product provided evidence that it had actually been formed (Guengerich, 1989b) (Figure 6).

The same reaction could be demonstrated using horseradish peroxidase or model metalloporphyrins. Such an oxidation was not detectable with the corresponding bromide, at least not at a rate exceeding 1% that observed with the iodide. Such a result is in line with the difference in $E_{1/2}$ values for 1-electron oxidation of these halides (Guengerich, 1989b).

Haloso compounds can oxidize protein sulfhydryls non-enzymatically. Thus both iodosylbenzene and RIO were able to oxidize the sulfhydryl groups in a mixture of P-450 IIB1 and NADPH-P-450 reductase (Table 2).

When iodobenzene was incubated with the mixture of the two proteins and NADPH, about one quarter of the protein sulfhydryl groups were oxidized (Table 2). This oxidation could be the result of formation of iodosylbenzene and sulfhydryl oxidation or of oxidation to moieties such as the epoxide or catechol, which have been suggested to be biological reactive intermediates formed from bromobenzene (Lau et al., 1984) and could deplete sulfhydryl compounds by conjugation (instead of oxidation). However, other iodides which would not easily be transformed to such Michael acceptors were also found to decrease the number of free sulfhydryl groups, to about the same extent (Table 2). The decrease in thiol groups was not seen in similar incubations with the bromide analog of RI or ethyl bromide, so we do not feel that direct oxygenation is a general event in the oxidation of bromides (Guengerich, 1989b).

CONCLUSIONS

Many P-450s exist and each appears to have its own elements of catalytic specificity and regulation. The diverse reactions can be rationalized in terms of a relatively general chemistry, and the differences seen in catalysis are attributed to the apoprotein structures. Many seemingly unusual reactions can be understood in context of the general chemistry, and further elucidation of chemical details within the enzymes is necessary for a full appreciation of how P-450s activate and detoxify potentially toxic chemicals.

Table 2. Loss of Protein Sulfhydryl Groups Accompanying the Incubation of Halides with P-450 and NADPH-P-450 Reductase

Additions	Residual free sulfhydryls	Percent control
	μM	%
---	33	100
PhIO [a]	6	17 [b]
RIO [a]	26	80 [b]
RBrO [a]	19	60 [b]
NADPH	26	100
RI, NADPH	19	75 [c]
Iodobenzene, NADPH	20	76 [c]
7-Iodo-n-heptane, NADPH	16	63 [c]
C_2H_5I, NADPH	22	86 [c]

All incubations (V = 1.0 ml) included 2 mM P-450 IIB1, 4 mM NADPH-P-450 reductase, 2 mg catalase ml^{-1}, 45 mM L-a-dilauroyl-sn-glycero-3-phosphocholine, and 50 mM Tris-HCl (pH 7.5) and were carried out at 37BC for 45 min. Sulfhydryl groups (means of duplicate experiments, < 10% difference) were determined using 5,5'-bis(dithiobenzoic acid) (Guengerich, 1989b).

[a] Added at 0.1 mM.

[b] With reference to incubation without NADPH or other additions.

[c] With reference to incubation including only NADPH.

REFERENCES

Augusto, O., Beilan, H. S., and Ortiz de Montellano, P. R. (1982). The catalytic mechanism of cytochrome P-450. Spin-trapping evidence for one-electron substrate oxidation. *J. Biol. Chem.* **257,** 11288-11295.

Baba, T., Yamada, H., Oguri, K., and Yoshimura, H. (1988). Participation of cytochrome P-450 isozymes in N-demethylation, N-hydroxylation and aromatic hydroxylation of methamphetamine. *Xenobiotica* **18,** 475-484.

Bondon, A., Macdonald, T. L., Harris, T. M., and Guengerich, F. P. (1989). Oxidation of cyclobutylamines by cytochrome P-450: Mechanism-based inactivation, adduct formation, ring expansion, and nitrone formation. *J. Biol. Chem.* **264,** 1988-1997.

Böcker, R. H., and Guengerich, F. P. (1986). Oxidation of 4-aryl- and 4-alkyl-substituted 2,6-dimethyl-3,5-bis-(alkoxycarbonyl)-1,4-dihydropyridines by human liver microsomes and immunochemical evidence for the involvement of a form of cytochrome P-450. *J. Med. Chem.* **29,** 1596-1603.

Burka, L. T., Guengerich, F. P., Willard, R. J., and Macdonald, T. L. (1985). Mechanism of cytochrome P-450 catalysis. Mechanism of N-dealkylation and amine oxide deoxygenation. *J. Am. Chem. Soc.* **107,** 2549-2551.

Burka, L. T., Thorsen, A., and Guengerich, F. P. (1980). Enzymatic monooxygenation of halogen atoms: Cytochrome P-450-catalyzed oxidation of iodobenzene by iodosobenzene. *J. Am. Chem. Soc.* **102,** 7615-7616.

Dinnocenzo, J. P., and Banach, T. E. (1989). Deprotonation of tertiary amine cation radicals. A direct experimental approach. *J. Am. Chem. Soc.* **111,** 8646-8653.

Garrison, J. M., and Bruice, T. C. (1989). Intermediates in the epoxidation of alkenes by cytochrome P-450 models. 3. Mechanism of oxygen transfer from substyituted oxochromium (V) porphyrins to olefinic substrates. *J. Am. Chem. Soc.* **111,** 191-198.

Griffin, B. W., and Ting, P. L. (1978). Mechanism of N-demethylation of aminopyrine by hydrogen peroxide catalyzed by horseradish peroxidase, not myoglobin, and protohemin. *Biochemistry* **17,** 2206-2211.

Groves, J. T., and Watanabe, Y. (1986). On the mechanism of olefin epoxidation by oxo-iron porphyrins. Direct observation of an intermediate. *J. Am. Chem. Soc.* **108,** 507-508.

Groves, J. T., Avaria-Neisser, G. E., Fish, K. M., Imachi, M., and Kuczkowski, R. L. (1986). Hydrogen-deuterium exchange during propylene oxidation by cytochrome P-450. *J. Am. Chem. Soc.* **108,** 3837-3838.
Groves, J. T., McClusky, G. A., White, R. E., and Coon, M. J. (1978). Aliphatic hydroxylation by highly purified liver microsomal cytochrome P-450. Evidence for a carbon radical intermediate. *Biochem. Biophys. Res. Commun.* **76,** 541-549.
Guengerich, F. P. (1984). Oxidation of sparteines by cytochrome P-450: Evidence against the formation of N-oxides. J. *Med. Chem.* **27,** 1101-1103.
Guengerich, F. P. (1987). Oxidative cleavage of carboxylic esters by cytochrome P-450. *J. Biol. Chem.* **262,** 8459-8462.
Guengerich, F. P. (1988). Roles of cytochrome P-450 enzymes in chemical carcinogenesis and cancer chemotherapy. *Cancer Res.* **48,** 2946-2954.
Guengerich, F. P. (1989a). Biochemical characterization of human cytochrome P-450 enzymes. *Ann. Rev. Pharmacol. Toxicol.* **29,** 241-264.
Guengerich, F. P. (1989b). Oxidation of halogenated compounds by metalloporphyrins, peroxidases, and cytochrome P-450. *J. Biol. Chem.* **264,** 17198-17205.
Guengerich, F. P. (1990a). Enzymatic oxidation of xenobiotic chemicals. *CRC Crit. Rev. Biochem.*, in press.
Guengerich, F. P. (1990b). Low kinetic hydrogen isotope effects in the oxidation of 1,4-dihydro-2,6-dimethyl-4-(2-nitrophenyl)-3,5-pyridinedicarboxylic acid dimethyl ester (nifedipine) by cytochrome P-450 enzymes are consistent with an electron-proton-electron transfer mechanism. *Chem. Res. Toxicol.*, in press.
Guengerich, F. P., and Böcker, R. H. (1988). Cytochrome P-450-catalyzed dehydrogenation of 1,4-dihydropyridines. *J. Biol. Chem.* **263,** 8168-8175.
Guengerich, F. P., and Liebler, D. C. (1985). Enzymatic activation of chemicals to toxic metabolites. *CRC Crit. Rev. Toxicol.* **14,** 259-307.
Guengerich, F. P., and Macdonald, T. L. (1984). Chemical mechanisms of catalysis by cytochromes P-450: A unified view. *Acct. Chem. Res.* **17,** 9-16.
Guengerich, F. P., and Macdonald, T. L. (1990). Catalytic mechanism of cytochrome P-450. *FASEB J.*, in press.
Guengerich, F. P., Peterson, L. A., and Böcker, R. H. (1988). Cytochrome P-450-catalyzed hydroxylation and carboxylic acid ester cleavage of Hantzsch pyridine esters. *J. Biol. Chem.* **263,** 8176-8183.
Guengerich, F. P., Willard, R. J., Shea, J. P., Richards, L. E., and Macdonald, T. L. (1984). Mechanism-based inactivation of cytochrome P-450 by heteroatom-substituted cyclopropanes and formation of ring-opened products. *J. Am. Chem. Soc.* **106,** 6446-6447.
Hammerich, O., and Parker, V. D. (1984). Kinetics and mechanisms of reaction of organic cation radicals in solution. *Adv. Phys. Org. Chem.* **20,** 55-189.
Hecker, M., and Ullrich, V. (1989). On the mechanism of prostacyclin and thromboxane A_2 biosynthesis. *J. Biol. Chem.* **264,** 141-150.
Kadlubar, F. F., and Hammons, G. J. (1987). The role of cytochrome P-450 in the metabolism of chemical carcinogens. In *Mammalian Cytochromes* P-450 (F. P. Guengerich, Ed.), Vol. II, pp. 81-130, CRC Press, Boca Raton, Florida, USA.
Kaminsky, L. S., Dannan, G. A., and Guengerich, F. P. (1984). Composition of cytochrome P-450 isozymes from hepatic microsomes of C57BL/6 and DBA/2 mice assessed by warfarin metabolism, immunoinhibition, and immunoelectrophoresis with anti-(rat cytochrome P-450). *Eur. J. Biochem.* **141,** 141-148.
Kronbach, T., Larabee, T. M., and Johnson, F. F. (1989). Hybrid cytochromes P450 identify a substrate binding domain in P450 IIC5 and P450 IIC4. *Proc. Natl. Acad. Sci., USA* **86,** 8262-8265.
Lau, S. S., Monks, T. J., and Gillette, J. R. (1984). Multiple reactive metabolites derived from bromobenzene. *Drug Metab. Disp.* **12,** 291-296.
Lee, J. S., Jacobsen, N. E., and Ortiz de Montellano, P. R. (1988). 4-Alkyl radical extrusion in the cytochrome P-450-catalyzed oxidation of 4-alkyl-1,4-dihydropyridines. *Biochemistry* **27,** 7703-7710.
Liebler, D. C., and Guengerich, F. P. (1983). Olefin oxidation by cytochrome P-450: Evidence for group migration in catalytic intermediates formed with vinylidene chloride and trans-1-phenyl-1-butene. *Biochemistry* **22,** 5482-5489.
Lindberg, R-L. P., and Negishi, M. (1989). Alteration of mouse cytochrome $P450_{coh}$

substrate specificity by mutation of a single amino-acid residue. *Nature* (London) **339,** 632-634.
Macdonald, T. L., Gutheim, W. G., Martin, R. B., and Guengerich, F. P. (1989). Oxidation of substituted N,N-dimethylanilines by cytochrome P-450: Estimation of the effective oxidation-reduction potential of cytochrome P-450. *Biochemistry* **28,** 2071-2077.
Miwa, G. T., Walsh, J. S., Kedderis, G. L. and Hollenberg, P. F. (1983). The use of intramolecular isotope effects to distinguish between deprotonation and hydrogen atom abstraction mechanisms in cytochrome P-450 and peroxidase-catalyzed N-demethylation reactions. *J. Biol. Chem.* **258,** 14445-14449.
Nagata, K., Liberato, D. J., Gillette, J. R., and Sasame, H. A. (1986). An unusual metabolite of testosterone. 17b-Hydroxy-4,6-androstadiene-3-one. *Drug Metab. Disp.* **14,** 559-565.
Nebert, D. W., Nelson, D. R., Adesnik, M., Coon, M. J., Estabrook, R. W., Gonzalez, F. J., Guengerich, F. P., Gunsalus, I. C., Johnson, E. F., Kemper, B., Levin, W., Phillips, I. R., Sato, R., and Waterman, M. R. (1989). The P450 superfamily: Update on listing of all genes and recommended nomenclature of the chromosomal loci. *DNA* **8,** 1-13.
Nelson, S. D. (1982). Metabolic activation and drug toxicity. *J. Med. Chem.* **25,** 753-765.
Nelson, S. D. and Harvison, P. J. (1987). Roles of cytochromes P-450 in chemically induced cytotoxicity. In *Mammalian Cytochromes P-450* (F. P. Guengerich, Ed.), Vol. II, pp. 19-79, CRC Press, Boca Raton, Florida, USA.
Ortiz de Montellano, P. R. (1986). Oxygen activation and transfer. In *Cytochrome P-450* (P. R. Ortiz de Montellano, Ed.), pp. 217-271, Plenum Press, New York, USA.
Ortiz de Montellano, P. R. (1987). Control of the catalytic activity of prosthetic heme by the structure of hemoproteins. *Acct. Chem. Res.* **20,** 289-294.
Ortiz de Montellano, P. R. (1989). Cytochrome P-450 catalysis: radical intermediates and dehydrogenation intermediates. *Trends Pharmacol. Sci.* **10,** 354-359.
Ortiz de Montellano, P. R., Kunze, K. L., Beilan, H. S., and Wheeler, C. (1982). Destruction of cytochrome P-450 by vinyl fluoride, fluroxene, and acetylene. Evidence for a radical intermediate in olefin oxidation. *Biochemistry* **21,** 1331-1339.
Rettie, A. E., Rettenmeier, A. W., Howald, W. N., and Baillie, T. A. (1987) Cytochrome P-450 catalyzed formation of D^4-VPA, a toxic metabolite of valproic acid. *Science* (Washington, D. C.) **235,** 890-893.
Shimada, T., and Guengerich, F. P. (1989). Evidence for cytochrome P-450_{NF}, the nifedipine oxidase, being the principal enzyme involved in the bioactivation of aflatoxins in human liver. *Proc. Natl. Acad. Sci. U.S.A.* **86,** 462-465.
Shimada, T., Iwasaki, M., Martin, M. V., and Guengerich, F. P. (1989a). Human liver microsomal cytochrome P-450 enzymes involved in the bioactivation of procarcinogens detected by umu gene response in Salmonella typhimurium TA1535/pSK1002. *Cancer Res.* **49,** 3218-3228.
Shimada, T., Martin, M. V., Pruess-Schwartz, D., Marnett, L. J., and Guengerich, F. P. (1989b). Roles of individual forms of human cytochrome P-450 enzymes in the bioactivation of benzo(a)pyrene, 7,8-dihydroxy-7,8-dihydrobenzo(a)pyrene, and other dihydrodiol derivatives of polycyclic aromatic hydrocarbons. *Cancer Res.* **49,** 6304-6312.
Shono, T., Toda, T., and Oshino, N. (1982). Electron transfer from nitrogen in microsomal oxidation of amine and amide. Simulation of microsomal oxidation by anodic oxidation. *J. Am. Chem. Soc.* **104,** 2639-2641.
Traylor, T. G., and Miksztal, A. R. (1989). Alkene epoxidations catalyzed by iron(III), manganese(III), and chromium(III) porphyrins. Effects of metal and porphyrin substituents on selectivity and regiochemistry of epoxidation. *J. Am. Chem. Soc.* **111,** 7443-7448.
Umbenhauer, D. R., Martin, M. V., Lloyd, R. S., and Guengerich, F. P. (1987). Cloning and sequence determination of a complementary DNA related to human liver microsomal cytochrome P-450 S-mephenytoin 4-hydroxylase. *Biochemistry* **26,** 1094-1099.
Van der Zee, J., Duling, D. R., Mason, R. P., and Eling, T. E. (1989). The oxidation of N-substituted aromatic amines by horseradish peroxidase. *J. Biol. Chem.* **264,** 19828-19836.

Williams, D. E., Reed, R. L., Kedzierski, B., Guengerich, F. P., and Buhler, D. C. (1989). Bioactivation and detoxication of the pyrrolizidine alkaloid senecionine by cytochrome P-450 isozymes in rat liver. *Drug Metab. Disp.* **17,** 387-392.

Yamaguchi, K., Takahara, Y., and Fueno, T. (1986). Ab-initio molecular orbital studies of structure and reactivity of transition metal-oxo compounds. In *Applied Quantum Chemistry* (V. H. Smith, Jr., Ed.), pp. 155-184, D. Reidel Publishing, New York, USA.

Yoo, J-S. H., Guengerich, F. P., and Yang, C. S. (1988). Metabolism of N-nitrosodialkylamines in human liver microsomes. *Cancer Res.* **88,** 1499-1504.

Ziegler, D. M., Ansher, S. S., Nagata, T., Kadlubar, F. F., and Jakoby, W. B. (1988). N-methylation: Potential mechanism for metabolic activation of carcinogenic primary arylamines. *Proc. Natl. Acad. Sci. USA* **85,** 2514-2517.

FORMATION OF REACTIVE INTERMEDIATES BY PHASE II ENZYMES:

GLUTATHIONE-DEPENDENT BIOACTIVATION REACTIONS

Spyridon Vamvakas and M. W. Anders

Department of Pharmacology
University of Rochester
601 Elmwood Avenue
Rochester, NY 14642

INTRODUCTION

Xenobiotic metabolism serves several important functions in the body: The metabolism of drugs to pharmacologically inactive metabolites is the major mechanism for the termination of drug action. Metabolism may quantitatively alter drug action and may also convert pharmacologically inactive prodrugs to their pharmacologically active metabolites. In addition, metabolism of xenobiotics to polar, readily excretable metabolites is a major detoxication mechanism. Although the metabolism of drugs and chemicals is usually beneficial, it is now well known that the toxicity of most organic chemicals is associated with their enzymatic conversion to toxic metabolites, a process termed bioactivation (Anders, 1985).

The metabolism of foreign compounds can be grouped into two phases (Williams, 1959): Phase I metabolism involves oxidations, reductions, and hydrolyses. These reactions may be regarded as functionalization reactions that prepare compounds for Phase II metabolism and may serve to detoxify or bioactivate xenobiotics (Anders, 1985; Guengerich and Liebler, 1985). Phase II reactions are synthetic reactions and include, for example, glucuronide, sulfate, glutathione, and amino acid conjugation as well as *N*-acetylation; Phase II metabolism has long been regarded as being solely a detoxication mechanism. Recent studies have elaborated Phase II enzyme-catalyzed bioactivation reactions. For example, sulfate and, perhaps, glucuronide conjugate formation is a bioactivation reaction for *N*-hydroxyacetylaminofluorene (Hanna and Banks, 1985), and acyl glucuronide formation may also be a bioactivation reaction (Faed, 1984). The important Phase II pathway of mercapturic acid formation, which includes glutathione *S*-conjugate formation, metabolism of the glutathione conjugate to the cysteine *S*-conjugate, and *N*-acetylation of the cysteine conjugate, has also been regarded as a detoxication mechanism. Recent studies show, however, that glutathione conjugate formation is the initial step in the bioactivation of several groups of xenobiotics to metabolites that are toxic, mutagenic, and carcinogenic (Anders, 1988; Anders et al., 1988; Dekant et al., 1988a; 1989).

The objective of the review is to discuss glutathione-dependent bioactivation mechanisms. Briefly, there are four types of glutathione-dependent bioactivation mechanisms: Direct-acting glutathione conjugates, cysteine conjugate β-lyase-dependent bioactivation, glutathione *S*-conjugates as transport forms, and glutathione-dependent release of toxic agents. Examples will be used to illustrate each type.

Biological Reactive Intermediates IV, Edited by C.M. Witmer *et al.*
Plenum Press, New York, 1990

RESULTS AND DISCUSSION: Glutathione-Dependent Bioactivation

Direct-acting glutathione conjugates

This type of glutathione-dependent bioactivation reactions involves the formation of glutathione conjugates that are toxic without further enzymatic intervention, although nonenzymatic reactions may occur after conjugate formation. The involvement of direct-acting glutathione conjugates in genotoxic events *in vitro* and *in vivo* has been extensively studied with ethylene dibromide, which has been used as an agricultural fumigant and as a lead scavenger in gasoline; because of its toxicity, its use is now curtailed.

Ethylene dibromide is metabolized by two competing pathways: One, microsomal cytochromes P-450 convert ethylene dibromide to bromoacetaldehyde, which may be conjugated with glutathione and reduced to form *S*-(2-hydroxyethyl)glutathione (Nachtomi, 1970). Two, conjugation of ethylene dibromide with glutathione yields *S*-(2-bromoethyl)glutathione, which may rearrange to form a reactive episulfonium ion (Fig. 1). The episulfonium ion may react with water to give *S*-(2-hydroxyethyl)glutathione (Hill et al., 1978). The ratio of the oxidative and conjugative pathways in rats is about 4:1 (van Bladeren et al., 1981). Further metabolism of the glutathione conjugate affords the mercapturic acid *S*-(2-hydroxyethyl)-L-cysteine, which is the major urinary metabolite of ethylene dibromide.

Both bromoacetaldehyde and the episulfonium ion of *S*-(2-bromoethyl)glutathione are reactive metabolites that may be responsible for the observed mutagenic and carcinogenic effects of ethylene dibromide, and several studies have focused on establishing the relative roles of the oxidative (Phase I) and conjugative (Phase II) pathways. The extrahepatic toxicity of ethylene dibromide is associated with its glutathione-dependent metabolism (Sipes et al., 1986). The evidence also indicates that the genotoxic effects of ethylene dibromide are attributable to glutathione conjugation. Both the bacterial mutagenicity and the in vitro binding of ethylene dibromide metabolites to DNA depend upon the presence of both cytosolic glutathione *S*-transferases and glutathione and are not blocked by inhibitors of cytochromes P-450. Also the rate of DNA strand breakage is enhanced by deuterium substitution, which reduces the oxidative metabolism of ethylene dibromide and favors conjugation with glutathione. The role of the latter pathway as the primary source of genotoxic metabolites from ethylene dibromide was supported by DNA repair experiments *in vivo* and *in vitro* (Working et al., 1986); these experiments demonstrated that the microsomal oxidation of ethylene dibromide is a detoxifying reaction in terms of ethylene dibromide-induced genotoxicity, because inhibitors of the cytochromes P-450-dependent oxidation in the liver lead to increased tissue-bound metabolites and to increased genotoxicity. Finally with [1,2-^{14}C]ethylene dibromide, more than 90% of the radioactivity bound to DNA was identified as *S*-[2-(N^7-guanyl)ethyl]glutathione (Guengerich et al., 1987).

Ethylene dichloride, which is used as a lead scavenger in gasoline, as an industrial solvent, and as a grain fumigant, is hepatotoxic, nephrotoxic, and mutagenic and produces tumors in mice. Ethylene dichloride is bioactivated by the same mechanism as ethylene dibromide. Oxidative metabolism of ethylene dichloride does not contribute to bacterial mutagenesis or to DNA binding of ethylene dichloride metabolites (Banerjee et al., 1980; Guengerich et al., 1980; Inskeep et al., 1986; Rannung et al., 1978). Glutathione conjugation serves to activate ethylene dichloride (Fig. 1). *S*-(2-Chloroethyl)-DL-cysteine, a putative metabolite of ethylene dichloride, has been implicated as the reactive intermediate involved in ethylene dichloride-induced nephrotoxicity (Elfarra et al., 1985) and mutagenicity (Rannung et al., 1978). The role for an episulfonium ion intermediate in the toxicity of *S*-(2-chloroethyl)-DL-cysteine *in vivo* and *in vitro* was supported by studies with the analogs *S*-(3-chloropropyl)-, *S*-ethyl-, and *S*-(2-hydroxyethyl)-DL-cysteine; *S*-(2-Chloroethyl)-DL-cysteine, the only compound capable of forming an episulfonium ion, was cytotoxic, whereas equimolar concentrations of the other analogs were not cytotoxic (Webb et al., 1987). Further evidence for an episulfonium ion intermediate was provided by

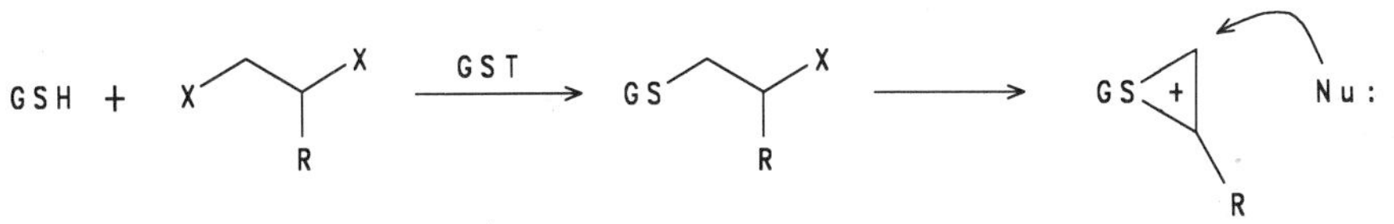

GSH + X–CH₂–CH(R)–X —GST→ GS–CH₂–CH(R)–X ———→ GS⁺(episulfonium, R) ← Nu:

GSH = glutathione
GST = glutathione S-transferase
Ethylene dibromide, X = Br, R = H
Ethylene dichloride, X = Cl, R = H
Dibromochloropropane, X = Br, R = CH_2Cl
Nu: = tissue nucleophile (protein, DNA), thiol, H_2O

Figure 1. Glutathione-dependent bioactivation of vicinal dihaloalkanes.

structural confirmation of the glutathione adduct of *S*-(2-chloroethyl)-DL-cysteine as *S*-[2-(DL-cysteinyl)ethyl]glutathione by NMR and mass spectrometry (Webb et al., 1987).

1,2-Dibromo-3-chloropropane, a potent nematocide, was formerly used as a soil fumigant. 1,2-Dibromo-3-chloropropane causes male infertility in humans, produces renal and testicular damage in rats, is mutagenic in bacteria, and is carcinogenic in rodents. A cytochromes P-450-dependent or a glutathione *S*-transferase-dependent pathway (Fig. 1), or both, may be involved in the bioactivation of 1,2-dibromo-3-chloropropane. In contrast to 1,2-dihaloethanes, the mutagenic activity of 1,2-dibromo-3- chloropropane in bacteria is associated with oxidation by cytochromes P-450; 2-bromoacrolein has been identified as a mutagenic metabolite (Miller et al., 1986; Omichinski et al., 1988). Reduction of oxidative metabolism by perdeuteration or inclusion of glutathione and glutathione *S*-transferases in the activating system decreases the bacterial mutagenicity of DBCP. In contrast, deuterium substitution has no influence on *in vivo* DNA damage or on the kidney and testicular toxicity produced by 1,2-dibromo-3-chloropropane. Also, urinary metabolites of 1,2-dibromo-3-chloropropane are largely derived from glutathione conjugates (Jones et al., 1979), and the concentrations of [2H5]1,2-dibromo-3-chloropropane and of 1,2-dibromo-3-chloropropane in plasma, liver, kidney, or testes are similar, indicating that oxidative metabolism is not a major pathway of 1,2-dibromo-3-chloropropane elimination and is not responsible for the organotropic effects in mammals. Evidence for the role of the glutathione conjugation pathway in the genotoxic effects of 1,2-dibromo-3-chloropropane has been obtained *in vitro*: Incubation of liver or testicular cells with diethyl maleate or buthionine sulfoximine to lower cellular glutathione concentrations decreased 1,2-dibromo-3-chloropropane-induced DNA damage, but *in vivo* treatment of rats with diethyl maleate or buthionine sulfoximine did not alter 1,2-dibromo-3-chloropropane-induced renal necrosis or DNA damage, perhaps because of insufficient depletion of tissue glutathione concentrations (Låg et al., 1989).

The glutathione-dependent bioactivation on nonhalogenated compounds has also been reported: The thiol-dependent binding of 1-methyl-4-phenyl-5-nitrosoimidazole, a reduced metabolite of 1-methyl-4-phenyl-5-nitroimidazole, to DNA has been reported (Ehlhardt and Goldman, 1989). These authors speculate that the 5-nitrosoimidazole reacts with a thiol to yield an adduct (R-N=0 + R'SH R'-S-N(OH)-R) that may lose water to give a 2-(*N*-hydroxy)-iminoimidazole that reacts with DNA. The glutathione conjugate of 3-nitroso-1-methyl-5*H*-pyrido[4,3-b]indole is apparently directly mutagenic but the structure of the adduct has not been determined (Saito et al., 1983).

Cysteine conjugate β-lyase-dependent activation of glutathione S-conjugates

Several haloalkenes are selective kidney toxins and carcinogens. Hexachlorobutadiene is nephrotoxic and nephrocarcinogenic (Kociba et al., 1977; Lock and Ishmael, 1979), but bioactivation by cytochromes P-450 is not involved (Wolf et al., 1984). Chlorotrifluoroethene is a potent nephrotoxin whose organ-selective effects are not associated with inorganic fluoride release (Potter et al., 1981), as would be expected if cytochromes P-450-dependent metabolism was involved. These observations have lead investigators to seek an alternative bioactivation pathway for nephrotoxic haloalkenes. Important clues for the involvement of glutathione in the organotropic effects were provided by the findings that *S*-(1,2-dichlorovinyl)-L-cysteine, derived from trichloroethene, is nephrotoxic in all species studied (Terracini and Parker, 1965) and that the glutathione conjugates of hexachlorobutadiene, chlorotrifluoroethene and tetrafluoroethene are also nephrotoxic (Dohn et al., 1985; Odum and Green, 1984; Wolf et al., 1984). These observations lead to the assertion of a hypothesis to explain the nephrotoxicity of haloalkenes (Elfarra and Anders, 1984): The organ-selective toxicity is due to hepatic glutathione *S*-conjugate formation, translocation of the glutathione *S*-conjugate (or metabolites thereof) to the kidney, and bioactivation by cysteine *S*-conjugate β-lyase (β-lyase) (Fig. 2, 3). Evidence supporting this hypothesis has been obtained from both *in vivo* and *in vitro* studies.

The chloroalkenes hexachlorobutadiene, 1,1,2-trichloro-3,3,3-trifluoro-1-propene, and tetrachloroethene are metabolized by soluble and microsomal glutathione *S*-transferases from rat liver by an addition-elimination reaction to give *S*-(1,2,3,4,4-pentachlorobutadienyl)glutathione, *S*-(1,2-dichloro-3,3,3-trifluoro-1-propenyl)-glutathione and *S*-(1,2,2-trichlorovinyl)-glutathione respectively (Dekant et al., 1987a; Vamvakas et al. 1989a; Wolf et al., 1984) (Fig. 2). In contrast to the vinylic *S*-conjugates formed from chloroalkenes, conjugation of fluoroalkenes with glutathione results in the formation of *S*- (fluoroalkyl)glutathione conjugates; *S*-(2-chloro-1,1,2-trifluoroethyl)-glutathione and *S*-(1,1,2,2-tetrafluoroethyl)-glutathione are formed from chlorotrifluoroethene and tetrafluoroethene respectively (Dohn and Anders, 1982; Odum and Green, 1984) (Fig. 2). The hepatic microsomal glutathione *S*-transferase exhibit a 2- to 10-fold higher activity toward haloalkenes compared with the cytosolic transferases, perhaps due to the preferential distribution of the lipophilic haloalkenes into the lipid membranes of the endoplasmic reticulum. The observation that the cytosolic transferases catalyze the nonstereoselective addition of glutathione to chlorotrifluoroethene and that the reaction catalyzed by the microsomal transferase is stereoselective enabled the determination of the relative contributions of the microsomal and cytosolic transferases. When isolated rat hepatocytes were incubated with chlorotrifluoroethene, the contribution of the microsomal glutathione transferases to glutathione conjugate formation was about 75% (S. Hargus, M. Fitzsimmons, and M. W. Anders, this volume). Nonenzymatic conjugation with glutathione is observed only with the highly nephrotoxic alkyne dichloroacetylene (Kanhai et al., 1990).

Hepato-renal transport and metabolism of the glutathione *S*- conjugates to the cysteine *S*-conjugates are required for the expression of toxicity (Anders et al., 1988; Dekant et al., 1988a). Cysteine *S*-conjugate formation is catalyzed by γ-glutamyl transpeptidase and dipeptidases, which are concentrated in the kidney in mammals. The organ distribution of glutathione conjugate processing enzymes together with the ability of the kidney to transport and concentrate *S*-conjugates are probably involved in the organ-selective effects of the haloalkenes. Indeed the nephrotoxicity of cysteine *S*-conjugates can be blocked by probenecid (Elfarra et al., 1986; Lock and Ishmael, 1985).

Cysteine *S*-conjugates may be acetylated to form the corresponding mercapturic acids, which are excreted in the urine. Alternatively, cysteine *S*-conjugates may be metabolized by the pyridoxal phosphate-dependent β-lyase, which is located in proximal tubule cells, to unstable thiols that yield electrophilic products whose interaction with macromolecules is associated with the acute toxicity and nephrocarcinogenicity (Fig. 3). Attempts to identify the nature of these reactive intermediates have been successful. As shown in Figure 3, thioketenes (from halovinyl *S*-conjugates) and thionoacyl fluorides (from fluoroalkyl) *S*-conjugates are presently thought to be the ultimate reactive metabolites formed from haloalkene-derived *S*-conjugates (Dekant et al., 1987b; 1988b; 1988c). Both thionoacyl fluorides and thioketenes are potent acylating agents that react with nucleophiles.

After giving hexachlorobutadiene or tetrachloroethene to rats, glutathione S-conjugates identical to those formed in liver subcellular fractions are found in the bile (Nash et al., 1984; Odum and Green, 1987), and the corresponding mercapturic acids of several haloalkenes were identified in urine (Dekant et al., 1986a; 1986b; Reichert and Schutz, 1986).

The glutathione and cysteine S-conjugates derived from the nephrotoxic haloalkenes, hexachlorobutadiene, chlorotrifluoroethene, and trichloroethene are nephrotoxic in rats (Anders et al., 1986; Dohn et al., 1985; Ishmael and Lock, 1986) and are cytotoxic in freshly isolated kidney cells and in renal cells in culture (Dohn et al., 1985; Jones et al., 1986; Lash and Anders, 1986; Stevens et al., 1986). In agreement with the proposed mechanism (Fig. 2, 3) inhibition of γ-glutamyltransferase, of dipeptidases, or of β-lyase by acivicin, by 1,10-phenanthroline and phenylalanylglycine, or by aminooxyacetic acid, respectively, blocks the toxic effects both *in vivo* and *in vitro*. Furthermore the α-methyl analogues of S- (1,2-dichlorovinyl)-L-cysteine and S-(2-chloro-1,1,2-trifluoroethyl)-L-cysteine, which cannot be metabolized by β-lyase, are not cytotoxic (Dohn et al., 1985; Elfarra et al., 1986). The role of thionoacyl halides and thioketenes in S- conjugate-induced toxicity has been confirmed by structure-activity studies: Only S-conjugates that form acylating agents are cytotoxic in kidney cells and mutagenic in bacteria (Vamvakas et al., 1988a; 1989b). Mitochondrial dysfunctions is an early central event in the cytotoxicity of S-conjugates (Lash and Anders, 1987). Recent studies with fluorescence digital imaging microscopy indicate that alterations in the mitochondrial Ca2+ homeostasis is an early and specific event in kidney cells treated with S-(1,2-dichlorovinyl)-L-cysteine (Vamvakas et al., submitted).

Hexachlorobutadiene and tetrachloroethene are mutagenic in the Ames test under conditions favoring glutathione conjugation (Vamvakas et al., 1988b; 1989c). Moreover, β-lyase-dependent mutagenicity in bacteria (Green and Odum 1985), induction of DNA repair in renal cells in culture (Vamvakas et al., 1988c; 1989d), and DNA damage in rabbit renal tubule both *in vivo* and *in vitro* (Jaffe et al. 1985) is observed with glutathione and cysteine S- conjugates derived from chloroalkenes. In contrast to chloroalkenyl S-conjugates, fluoroalkyl S-conjugates are cytotoxic but not genotoxic (Green and Odum 1985; Vamvakas et al., 1989d); the reason for this different behavior has not been elucidated yet.

Glutathione conjugates as transport forms

A. Bromohydroquinones. Bromobenzene is nephrotoxic in rats, and 2-bromohydroquinone has been implicated as a toxic metabolite of bromobenzene (Lau et al., 1984). Subsequent experiments showed that glutathione conjugates of 2-bromohydroquinone, including 2-bromo-3- (glutathion-S-yl)hydroquinone and 2-bromo-(diglutathion-S- yl)hydroquinone, are formed as hepatic metabolites of 2-bromohydroquinone (Monks et al., 1985) (Fig. 4). The nephrotoxicity of 2-bromo-(diglutathion-S-yl)hydroquinone and of glutathione conjugates of 1,4-benzoquinone is blocked by acivicin, an irreversible inhibitor of γ-glutamyltransferase (Lau et al., 1988; Monks et al., 1988). These data indicate that the selective nephrotoxicity of glutathione conjugates of hydroquinones is associated with their selective uptake into renal cells after metabolism to the corresponding cysteine conjugates by γ-glutamyltransferase and dipeptidases. The detailed mechanism by which glutathione conjugates of quinones produce nephrotoxicity in rats is still not understood; bromohydroquinone itself is toxic in rabbit renal proximal tubules and its toxicity in this preparation does not involve glutathione conjugate formation (Schnellmann et al., 1989).

B. Isothiocyanates. Organic isothiocyanates are important natural products that are released enzymatically from precursor glucosinolates present in foods; allyl isothiocyanate is a suspected bladder carcinogen, and, in contrast, benzyl isothiocyanate has anticarcinogenic properties. The glutathione and cysteine conjugates of allyl and benzyl isothiocyanates, S-(N-allylthiocarbamoyl)glutathione, S-(N-allylthiocarbamoyl)-L-cysteine, S-(N-benzylthiocarbamoyl)glutathione, and S-(N-benzylthiocarbamoyl)-L-cysteine, respectively, are cytotoxic in RL-4 rat liver cell suspensions (Bruggeman et al., 1986; Temmink et al., 1986). Because the reaction

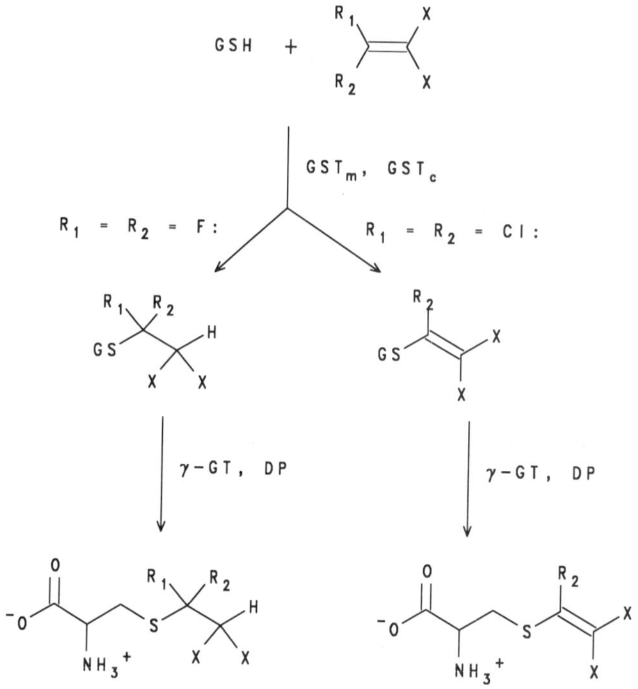

GSH = glutathione
X = Cl, F, CF_3, CCl = CCl_2
GST = glutathione S-transferase
γ-GT = γ-glutamyltransferase
DP = dipeptidases

Figure 2. Biosynthesis of *S*-(haloalkyl)- and *S*-(haloalkenyl)glutathione conjugates and metabolism to the corresponding cysteine *S*-conjugates.

between thiols and organic isothiocyanates is readily reversible, these workers postulated that glutathione and cysteine may serve as transport forms for the cytotoxic isothiocyanates; the finding that the glutathione conjugates of isothiocyanates, which presumably do not enter cells, are cytotoxic may indicate that isothiocyanate release from thiol conjugates occurs at the plasma membrane.

C. Furazolidone metabolites. The microsomal metabolism of furazolidone (*N*-(5-nitro-2- furfurylidene)-3-amino-2-oxazolidone) yields the acrylonitrile derivative *N*-(4-cyano-2-oxo-3-butenylidene)-3-amino-2-oxazolidone as a metabolite (Vroomen et al., 1988). This metabolite reacts reversibly with glutathione and with protein sulfhydryl groups, and the glutathione conjugate may, therefore, act as a transport form of the reactive metabolite.

Glutathione-dependent release of toxic metabolites

A Phase II enzyme-catalyzed pathway for dichloromethane metabolism has been elucidated that involves glutathione *S*-transferase-catalyzed formation of *S*-chloromethylglutathione followed by nonenzymatic conversion to *S*-hydroxymethyl-glutathione; this hemimercaptal of glutathione and formaldehyde eliminates formaldehyde and regenerates glutathione (Ahmed and Anders, 1978). The significance of the Phase II metabolism of dichloromethane is that the development of lung and

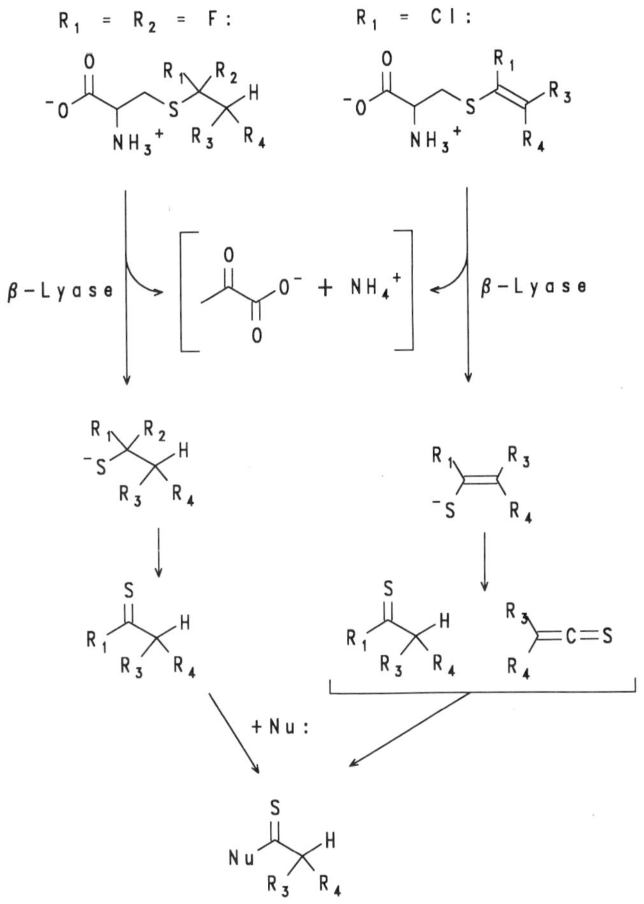

Figure 3. Cysteine conjugate β-lyase-dependent metabolism of *S*-(haloalkyl)- and *S*-(haloalkenyl)-L-cysteine conjugates. R3, R4 = Cl, F, CF3, CCl=CCl2; Nu: = tissue nucleophile (protein, DNA, thiol).

liver tumors seen after giving dichloromethane to animals is associated with the glutathione *S*-transferase-catalyzed metabolism of dichloromethane (Andersen et al., 1987; Reitz et al., 1989).

The glutathione *S*-transferase-catalyzed release of hydrogen cyanide from organic thiocyanates is a prototypical example of the Phase II-dependent release of a stable, but toxic, metabolite. Organic thiocyanates, which are used as pesticides, are metabolized to hydrogen cyanide, glutathione disulfide, and a thiol by the glutathione *S*-transferases (Ohkawa and Casida, 1971).

The reaction of thiols, including glutathione, with *N*-methyl- *N*'-nitro-*N*-nitrosoguanidine releases methyldiazohydroxide (Lawley and Thatcher, 1970), the same reactive intermediate formed by the microsomal α-hydroxylation of *N*-nitrosodimethylamine.

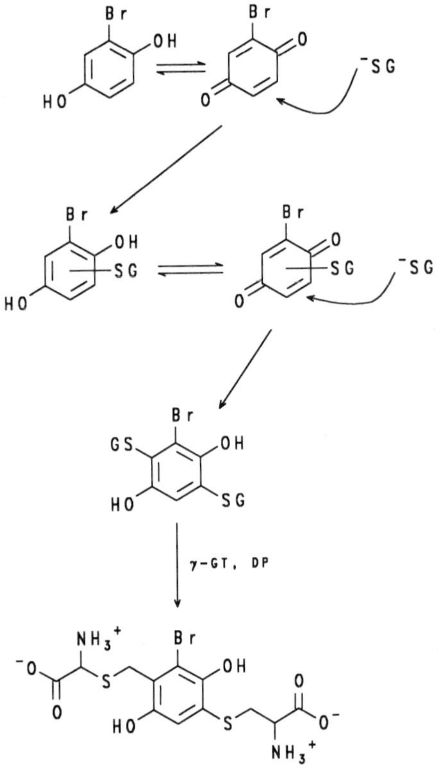

GSH = glutathione
γ-GT = γ-glutamyltransferase
DP = dipeptidases

Figure 4. Glutathione-dependent bioactivation of bromohydroquinone.

CONCLUSIONS

In contrast to the previously held view that glutathione-dependent metabolism serves only to detoxify xenobiotics, the information presented in this brief review indicates that glutathione conjugation reactions serve as a bioactivation mechanism for several groups of compounds. These reactions are involved in the organotropic toxicity as well as in the mutagenicity and carcinogenicity of some xenobiotics.

It is likely that additional mechanisms for the glutathione-dependent bioactivation of xenobiotics will be elaborated in the future. Indeed, Huxtable et al. (This volume) reported that the pneumotoxicity of certain pyrrolizidine alkaloids may be associated with the formation of 7-glutathionyl dehydroretronecine.

ACKNOWLEDGEMENTS

Work in the authors' laboratory was supported by NIEHS grant ES-03127; S.V. was supported by Boehringer-Ingelheim.

REFERENCES

Ahmed, A.E. and Anders, M.W. (1978). Metabolism of dihalomethanes to formaldehyde and inorganic halide. II. Studies on the mechanism of the reaction. *Biochem. Pharmacol.* **27,** 2021-2025.

Anders, M.W., ed. (1985). *Bioactivation of Foreign Compounds,* Academic Press, Orlando.

Anders, M.W. (1988). Glutathione-dependent toxicity: Biosynthesis and bioactivation of cytotoxic *S*-conjugates. *ISI Atlas of Science: Pharmacology* **2,** 99-104.

Anders, M.W., Lash, L.H., and Elfarra, A.A. (1986). Nephrotoxic amino acid and glutathione *S*-conjugates: Formation and renal activation. *Adv. Exp. Med. Biol.* **197,** 443-455.

Anders, M.W., Lash, L.H., Dekant, W., Elfarra, A.A., and Dohn, D.R. (1988). Biosynthesis and biotransformation of glutathione *S*-conjugates to toxic metabolites. *CRC Crit. Rev. Toxicol.* **18,** 311-341.

Andersen, M.E., Clewell, H.J., III, Gargas, M.L., Smith, F.A., and Reitz, R.H. (1987). Physiologically based pharmacokinetics and the risk assessment process for methylene chloride. *Toxicol. Appl. Pharmacol.* **87,** 185-205.

Banerjee, S., Van Duuren, B.L., and Oruambo, F.J. (1980). Microsome mediated covalent binding of 1,2-dichloroethane to lung microsomal proteins and salmon sperm DNA. *Cancer Res.* **40,** 2170- 2173.

Bruggeman, I.M., Temmink, J.H.M., and van Bladeren, P.J. (1986). Glutathione- and cysteine-mediated cytotoxicity of allyl and benzyl isothiocyanate. *Toxicol. Appl. Pharmacol.* **83,** 349-359.

Dekant, W., Metzler, M., and Henschler, D. (1986a). Identification of *S*-1,2,2-trichlorovinyl-*N*-acetylcysteine as a urinary metabolite of tetrachloroethylene: Bioactivation through glutathione conjugation as a possible explanation of its nephrocarcinogenicity. *J. Biochem. Toxicol.* **1,** 57-72.

Dekant, W., Metzler, M., and Henschler, D. (1986b). Identification of S-1,2-dichlorovinyl-N-acetyl-cysteine as a urinary metabolite of trichloroethylene: A possible explanation for its nephrocarcinogenicity in male rats. *Biochem. Pharmacol.* **35,** 2455-2458.

Dekant, W., Martens, G., Vamvakas, S., Metzler, M., and Henschler, D. (1987a). Bioactivation of tetrachloroethylene: Role of glutathione *S*-transferase-catalyzed conjugation versus cytochrome P-450-dependent alkylation. *Drug Metab. Dispos.* **15,** 702-709.

Dekant, W., Lash, L.H., and Anders, M.W. (1987b). Bioactivation mechanism of the cytotoxic and nephrotoxic *S*-conjugate *S*-(2-chloro-1,1,2-trifluoroethyl)-L-cysteine. *Proc. Natl. Acad. Sci. USA* **84,** 7443-7447.

Dekant, W., Lash, L.H., and Anders, M.W. (1988a). Fate of glutathione conjugates and bioactivation of cysteine *S*- conjugates by cysteine conjugate β-lyase. In *Glutathione Conjugation: Its Mechanism and Biological Significance* (H. Sies and B. Ketterer, Eds.), pp. 415-447. Academic Press, Orlando.

Dekant, W., Berthold, K., Vamvakas, S., Henschler, D., and Anders, M.W. (1988b). Thioacylating intermediates as metabolites of *S*-(1,2-dichlorovinyl)-L-cysteine and *S*-(1,2,2-trichlorovinyl)-L-cysteine formed by cysteine conjugate β-lyase. *Chem. Res. Toxicol.* **1,** 175-178.

Dekant, W., Berthold, K., Vamvakas, S., and Henschler, D. (1988c). Thioacylating agents as ultimate intermediates in the β-lyase catalyzed metabolism of *S*-(pentachlorobutadienyl)-L- cysteine. *Chem.-Biol. Interact.* **67,** 139-148.

Dekant, W., Vamvakas, S., and Anders, M.W. (1989). Bioactivation of nephrotoxic haloalkenes by glutathione conjugation: Formation of toxic and mutagenic intermediates by cysteine conjugate β-lyase. *Drug Metab. Rev.* **20,** 43-83.

Dohn, D.R., and Anders, M.W. (1982). The enzymatic reaction of chlorotrifluoroethylene with glutathione. *Biochem. Biophys. Res. Commun.* **109,** 1339-1345.

Dohn, D.R., Leininger, J.R., Lash, L.H., Quebbemann, A.J., and Anders, M.W. (1985). Nephrotoxicity of *S*-(2-chloro-1,1,2-trifluoroethyl)glutathione and *S*-(2-chloro-1,1,2- trifluoroethyl)-L-cysteine, the glutathione and cysteine conjugates of chlorotrifluoroethene. *J. Pharmacol. Exp. Ther.* **235,** 851-857.

Ehlhardt, W.J., and Goldman, P. (1989). Thiol-mediated incorporation of radiolabel from 1-[14]-methyl-4-phenyl-5-nitroimidazole into DNA. *Biochem. Pharmacol.* **38,** 1175-1180.

Elfarra, A.A., and Anders, M.W. (1984). Renal processing of glutathione conjugates: Role in nephrotoxicity. *Biochem. Pharmacol.* **33,** 3729-3732.

Elfarra, A.A., Baggs, R.B., and Anders M.W. (1985). Structure-nephrotoxicity ralationships of S-(2-chlororethyl)-DL-cysteine and analogs: Role for an episulfonium ion. *J. Pharmacol. Exp. Therap.* **233,** 2, 512-516.

Elfarra, A.A., Jakobson, I., and Anders, M.W. (1986). Mechanism of *S*-(1,2-dichlorovinyl)glutathione-induced nephrotoxicity. *Biochem. Pharmacol.* **35,** 283-288.

Faed, E. M. (1984). Properties of acyl glucuronides: Implications for studies of the pharmacokinetics and metabolism of acidic drugs. *Drug Metab. Rev.* **15,** 1213-1249.

Green, T., and Odum, J. (1985). Structure/activity studies of the nephrotoxic and mutagenic action of cysteine conjugates of chloro- and fluoroalkenes. *Chem. Biol. Interact.* **54,** 15-31.

Guengerich, F.P., Crawford, W.M. Jr., Domoraddzki, J.Y., Macdonald, T.L., and Watanabe, P.G. (1980). In vitro activation of 1,2-dichloroethane by microsomal and cytosolic enzymes. *Toxicol. Appl. Pharmacol.* **55,** 303-317.

Guengerich, F.P. and Liebler, D.C. (1985). Enzymatic activation of chemicals to toxic metabolites. *CRC Crit. Rev. Toxicol.* **14,** 259-307.

Guengerich, F.P., Peterson, L.A., Cmarik, J.L., Koga, N., and Inskeep, P.B. (1987). Activation of dihaloalkanes by glutathione conjugation and formation of DNA adducts. *Environ. Health. Perspect.* **76,** 15-18.

Hanna, P.E. and Banks, R.B. (1985). Arylhydroxylamines and arylhydroxamic acids: Conjugation reactions. In *Bioactivation of Foreign Compounds* (M.W. Anders, Ed.), pp. 375-402. Academic Press, Orlando.

Hill, D.L., Shih, T.-W., Johnston, T.P., and Struck, R.F. (1978). Macromolecular binding and metabolism of the carcinogen 1,2-dibromoethane. *Cancer Res.* **38,** 2438-2442.

Ishmael, J., and Lock, E.A. (1986). Nephrotoxicity of hexachlorobutadiene and its glutathione-derived conjugates. *Toxicol. Pathol.* **14,** 258-262.

Inskeep, P.B., Koga, N., Cmarik, J.L., and Guengerich, F.P. (1986). Covalent binding of 1,2-dihaloalkanes to DNA and stability of the major DNA adduct *S*-[2-(N^7-guanyl)ethyl]glutathione. *Cancer Res.* **46,** 2839-2844.

Jaffe, D.R., Hassall, C.D., Gandolfi, A.J., and Brendel, K. (1985). Production of DNA single strand breaks in rabbit renal tissue after exposure to 1,2-dichlorovinylcysteine. *Toxicology* **35,** 25-33.

Jones, A.R., Fakhouri, G., and Gadiel, P. (1979). The metabolism of the soil fumigant of 1,2-dibromo-3-chloropropane in the rat. *Experentia* **35,** 1432-1434.

Jones, T.W., Gerdes, R.G., Ormstad, K., and Orrenius, S. (1985). The formation of both a mono- and bis-substituted glutathione conjugate of hexachlorobutadiene by isolated hepatocytes and following *in vivo* administration to the rat. *Chem. Biol. Interact.* **56,** 251-267.

Jones, T.W., Wallin, A., Thor, H., Gerdes, R.G., Ormstad, K., and Orrenius, S. (1986). The mechanism of pentachlorobutadienyl-glutathione nephrotoxicity studied with isolated rat renal epithelial cells. *Arch. Biochem. Biophys.* **251,** 504-513.

Kanhai, W., Dekant, W., and Henschler, D. (1989). Metabolism of the nephrotoxin dichloroacetylene by glutathione conjugation. *Chem. Res. Toxicol.* **2,** 51-561.

Kociba, R.J., Keyes, D.G., Jersey, G.C., Ballard, J.J., Dittenber, D.A., Quast, J.F., Wade, C.E., Humiston, C.G., and Schwetz, B.A. (1977). Results of a two year chronic toxicity study with hexachlorobutadiene in rats. *Am. Ind. Hyg. Assoc. J.* **38,** 589-602.

Låg, M., Omichinski, J.G., Søderlund, E.J., Brunborg, G., Holme, J.A., Dahl, E.J., Nelson, S.D., and Dybing, E. (1989). Role of P-450 activity and glutathione levels in 1,2-dibromo-3-chloropropane tissue distribution, renal necrosis and in vivo DNA damage. *Toxicology* **56,** 273-288.

Lash, L. H., and Anders, M. W. (1986). Cytotoxicity of *S*-(1,2-dichlorovinyl)-glutathione and *S*-(1,2-dichlorovinyl)-L-cysteine in isolated rat kidney cells. *J. Biol. Chem.* **261,** 13076-13081.

Lash, L.H., and Anders, M.W. (1987). Mechanism of *S*-(1,2-dichlorovinyl)-L-cysteine- and *S*-(1,2-dichlorovinyl)- L-homocysteine-induced renal mitochondrial toxicity. *Mol. Pharmacol.* **32,** 549-556.

Lau, S.S., Monks, T.J., and Gillette, J.R. (1984). Identification of 2-bromohydro-

quinone as a metabolite of bromobenzene and o-bromophenol: Implications for bromobenzene- induced nephrotoxicity. *J. Pharmacol. Exp. Ther.* **230,** 360-366.
Lau, S.S., Hill, B.A., Highet, R.J., and Monks, T.J. (1988). Sequential oxidation and glutathione addition to 1,4-benzoquinone: Correlation of toxicity with increased glutathione substitution. *Mol. Pharmacol.* **34,** 829-836.
Lawley, P.D., and Thatcher, C.J. (1970). Methylation of deoxyribonucleic acid in cultured mammalian cells by N-methyl- N'-nitro-N-nitrosoguanidine. *Biochem. J.* **116,** 693-707.
Lock, E.A., and Ishmael, J. (1979). The acute toxic effects of hexachloro-1:3-butadiene on the rat kidney. *Arch. Toxicol.* **43,** 47-57.
Lock, E.A., and Ishmael, J. (1985). Effect of the organic acid transport inhibitor probenecid on renal cortical uptake and proximal tubular toxicity of hexachloro-1,3-butadiene and its conjugates. *Toxicol. Appl. Pharmacol.* **81,** 32-42.
Miller, G.M., Brabec, M.J., and Kulkarni, A.P. (1986). Mutagen activation of 1,2-dibromo-3-chloropropane by cytosolic glutathione S-transferases and microsomal enzymes. *J. Toxicol. Environ. Health.* **19,** 503-518.
Monks, T.J., Lau, S.S., Highet, R.J., and Gillette, J.R. (1985). Glutathione conjugates of 2-bromohydroquinone are nephrotoxic. *Drug Metab. Dispos.* **13,** 553-559.
Monks, T.J., Highet, R.J., and Lau, S.S. (1988). 2-Bromo- (diglutathion-*S*-yl)hydroquinone nephrotoxicity: Physiological, biochemical, and electrochemical determinants. *Mol. Pharmacol.* **34,** 492-500.
Nachtomi, E. (1970). The metabolism of ethylene dibromide in the rat. The enzymic reaction with glutathione *in vitro* and *in vivo*. *Biochem. Pharmacol.* **19,** 2853-2860.
Nash, J. A., King, L. H., Lock, E. A., and Green, T. (1984). The metabolism and disposition of hexachloro-1:3-butadiene in the rat and its relevance to nephrotoxicity. *Toxicol. Appl. Pharmacol.* **73,** 124-137.
Odum, J., and Green, T. (1984). The metabolism and nephrotoxicity of tetrafluoroethylene in the rat. *Toxicol. Appl. Pharmacol.* **76,** 306-318.
Odum, J., and Green, T. (1987). Perchloroethylene metabolism by the glutathione conjugation pathway. *Toxicologist* **7,** 269.
Ohkawa, H. and Casida, J.E. (1971). Glutathione S-transferases liberate hydrogen cyanide from organic thiocyanates. *Biochem. Pharmacol.* **20,** 1708-1711.
Omichinski, J.G., Soderlund, E.J., Dybing, E., Pearson, P.G., and Nelson, S.D. (1988). Detection and mechanism of formation of the potent direct acting mutagen 2-bromoacrolein from 1,2- dibromo-3-chloropopane. *Toxicol. Appl. Pharmacol.* **92,** 286-292.
Potter, C.L., Gandolfi, A.J., Nagle, R., and Clayton, J.W. (1981). Effects of inhaled chlorotrifluoroethylene and hexafluoropropene on the rat kidney. *Toxicol. Appl. Pharmacol.* **59,** 431-440.
Rannung, U., Sundvall, A., and Ramel, C. (1978). The mutagenic effect of 1,2-dichloroethane on *Salmonella typhimurium*. I. Activation through conjugation with glutathione in vitro. *Chem.- Biol. Interact.* **20,** 1-16.
Reichert, D., and Schutz, S. (1986). Mercapturic acid formation is an activation and intermediary step in the metabolism of hexachlorobutadiene. *Biochem. Pharmacol.* **35,** 1271-1275.
Reitz, R.H., Mendrala, A.L., and Guengerich, F.P. (1989). *In vitro* metabolism of methylene chloride in human and animal tissues: Use in physiologically based pharmacokinetic models. *Toxicol. Appl. Pharmacol.* **97,** 230-246.
Saito, K., Yamazoe, Y., Kamataki, T., and Kato, R. (1983). Activation and detoxication of N-hydroxy-Trp-P-2 by glutathione and glutathione transferases. *Carcinogenesis* **4,** 1551-1557.
Schnellmann, R.G., Monks, T.J., Mandel, L.J., and Lau, S.S. (1989). 2-Bromohydroquinone-induced toxicity to rabbit renal proximal tubules: The role of biotransformation, glutathione, and covalent binding. *Toxicol. Appl. Pharmacol.* **99,** 19-27.
Sipes, I.G., Wiersma, D.A., and Amstrong, D.J. (1986). The role of glutathione in the toxicity of xenobiotic compounds: metabolic activation of 1,2-dibromoethane by glutathione. *Adv. Exp. Med. Biol.* **197,** 457-464.
Stevens, J., Hayden, P., and Taylor, G. (1986). The role of glutathione metabolism and cysteine conjugate β-lyase in the mechanism of *S*-cysteine conjugate toxicity in LLC-PK1 cells. *J. Biol. Chem.* **261,** 3325-3332.

Temmink, J.H.M., Bruggeman, I.M., and van Bladeren, P.J. (1986). Cytomorphological changes in liver cells exposed to allyl and benzyl isothiocyanate and their cysteine and glutathione conjugates. *Arch. Toxicol.* **59,** 103-110.

Terracini, B., and Parker, V.H. (1965). A pathological study on the toxicity of *S*-dichlorovinyl-L-cysteine. *Food Cosmet. Toxicol.* **3,** 67-74.

Vamvakas, S., Berthold, K., Dekant, W., and Henschler, D. (1988a). Bacterial cysteine conjugate β-lyase and the metabolism of cysteine *S*-conjugates: Structural requirements for the cleavage of *S*-conjugates and the formation of reactive intermediates. *Chem. Biol. Interact.* **65,** 59-71.

Vamvakas, S., Kordowitch, F.J., Dekant, W., Neudecker, T., and Henschler, D. (1988b). Mutagenicity of hexachloro-1,3- butadiene and its *S*-conjugates in the Ames test: Role of activation by the mercapturic acid pathway in its nephrocarcinogenicity. *Carcinogenesis* **9,** 907-210.

Vamvakas, S., Dekant, W., and Henschler, D. (1988c). Genotoxicity of haloalkene and haloalkane glutathione *S*- conjugates in a cultured line of porcine kidney cells. *Toxicol. In Vitro* **3,** 151-156.

Vamvakas, S., Kremling, E., and Dekant, W. (1989a). Metabolic activation of the nephrotoxic haloalkane 1,1,2-trichloro-3,3,3-trifluoro-1-propene by glutathione conjugation. *Biochem. Pharamcol.* **38,** 2297-2304.

Vamvakas, S., Köchling, A., and Dekant, W. (1989b). Cytotoxicity of cysteine *S*-conjugates: Structure activity relationships. *Chem.-Biol. Interact.* **71,** 79-90.

Vamvakas, S., Hergenhof, M., Dekant, W., and Henschler, D. (1989c). Mutagenicity of tetrachloroethylene in the Ames-test - Metabolic activation by conjugation with glutathione. *J. Biochem.-Toxicol.* **4,** 21-28.

Vamvakas, S., Dekant, W., and Henschler, D. (1989d). Genotoxicity of cysteine *S*-conjugates derived from halogenated alkenes and alkanes in a cultured line of porcine kidney cells (LLC-PK1). *Mutat. Res.* **222,** 329-335.

Vamvakas, S., Sharma, V.K., Sheu, S-S., and Anders, M.W. Perturbations of intracellular calcium distribution in kidney cells by nephrotoxic haloalkenyl cysteine *S*-conjugates studied with fluroescence digital imaging microscopy. (Submitted).

van Bladeren , P.J., Breimer, D.D., van Huijgevoort, J.A.T.C.M., Vermeulen, N.P.E., and van der Gen, A. (1981). The metabolic formation of N-acetyl-S-2-hydroxyethyl-L-cysteine from tetradeutero-1,2-dibromoethane. Relative importance of oxidation and glutathione conjugation *in vivo*. *Biochem. Pharmacol.* **30,** 2499-2502.

Vroomen, L.H.M., Berghmans, M.C.J., Groten, J.P., Koemen, J.H., and van Bladeren, P.J. (1988). Reversible interaction of a reactive intermediate derived from furazolidone with glutathione and protein. *Toxicol. Appl. Pharmacol.* **95,** 53-60.

Webb. W.W., Elfarra, A.A., Webster, K.D., Thom, R.E., and Anders, M.W. (1987). Role for an episulfonium ion in *S*-(2-chloroethyl)-DL-cysteine-induced cytotoxicity and its reaction with glutathione. *Biochemistry* **26,** 3017-3023.

Williams, R.T. (1959). *Detoxication Mechanisms*, 2d ed., pp. 734-740. Chapman and Hall, London.

Wolf, C.R., Berry, P.N., Nash, J.A., Green, T., and Lock, E.A. (1984). The role of microsomal and cytosolic glutathione-*S*-transferases in the conjugation of hexachloro-1:3-butadiene and its possible relevance to toxicity. *J. Pharmacol. Exp. Ther.* **228,** 202-208.

Working, P.K., Smith-Oliver, T., White, R.D., and Butterworth, B.E. (1986). Induction of DNA repair in rat spermatocytes and hepatocytes by 1,2-dibromoethane: The role of glutathione conjugation. *Carcinogenesis* **7,** 467-472.

ROLE OF THE WELL-KNOWN BASIC AND RECENTLY DISCOVERED ACIDIC GLUTATHIONE S-TRANSFERASES IN THE CONTROL OF GENOTOXIC METABOLITES

Franz Oesch, Ingolf Gath, Takashi Igarashi, Hansruedi Glatt, Barbara Oesch-Bartlomowicz, and Helmut Thomas

Institute of Toxicology
University of Mainz
Obere Zahlbacher Strasse 67
D-6500 Mainz, Federal Republic of Germany

INTRODUCTION

The discovery of glutathione S-transferases (GSTs; E.C. 2.5.1.18) active in the metabolism of carcinogens dates back to Booth et al. (1961). GSTs were initially believed to serve as intracellular transport proteins for endogenous compounds with limited solubility in water, thereby acting as an intracellular equivalent to albumin. In this assumed capacity of reversible binding and transport of various ligands, the corresponding protein was named ligandin (Litwack et al., 1971). Following the discovery of abundant GST occurrence in most forms of arerobic life including plants, and the GST-catalysed conjugation of a wide variety of electrophilic substrates with glutathione, GSTs are now generally considered to play a crucial role in the detoxification of foreign compounds (for reviews see Mannervik, 1985; Ketterer, 1988; Mannervik and Danielson, 1988; Sies and Ketterer, 1988). GSTs are also believed to provide cellular protection by covalent binding of reactive electrophiles to the enzyme itself resulting in immobilization and inactivation of the compound.

Individual GST isoenzymes are differently expressed in different species and tissues (Igarashi et al., 1983, 1986) and may be altered by enzyme inducers. Dramatic ontogenic and organ specific changes in the expression of GSTs have been observed (Fryer et al., 1986; Faulder et al., 1987), and sex-related differences in the expression of specific GST isoenzymes in rat and mouse liver have been described (Igarashi et al., 1985; Hatayama et al., 1986). Attention has focused recently on the high level expression of certain isoenzymes in mammalian tumors, their possible clinical use a sdiagnostic markers for neoplasia, and as factors responsible for the development of drug resistance (Sato, 1989).

Studies on the structure of GST genes and control of expression are expected to provide deeper insight into GST multiplicity, regulation and function (Telakowski-Hopkings et al., 1988).

We have concentrated mainly on the GST-mediated detoxification of epoxides and epoxide hydrolase-resistant diol epoxides on polycyclic aromatic hydrocarbons, and the characterization of the responsible GST isoenzymes.

Biological Reactive Intermediates IV, Edited by C.M. Witmer *et al.*
Plenum Press, New York, 1990

RESULTS AND DISCUSSION

Nomenclature and Properties of Glutathione S-Transferases

GSTs have been most extensively investigated in rat, mouse and human tissues, and, with the exception of microsomal GST, been found to constitute functional homo- and heterodimers (Mannervik and Danielson, 1988). In the rat at least eight different subunites have been characterized, although more are known to exist, constituting 12 different isoenzymes. Several nomenclatures are used, the most favoured designating each subunit with an arabic numeral based on the chronological order of its characterization (Table 1).

For subunits 1, 2, 3, 4, and 7 complete cDNA sequences and deduced primary structures are knwon (Pickett et al., 1984; Telakowski-Hopkins et al., 1986; Ding et al., 1985; Ding et al., 1986; Suguoka et al., 1985). These data show 69% identity in full length DNA sequences between subunits 1 and 2, and 77% identity between subunits 3 and 4. Very little identity exists between the group comprising subunits 1 and 2, as compared with the group comprising subunits 3 and 4, and with subunit 7, indicating that these three categories belong to separate multigene families. correspondingly, and supported by complete or partial amino acid sequences, immunological crossreactivity as well as enzymatic properties of the corresponding homodimers, the subunits given in Table 1 have been assigned to three families, Alpha (subunits 1, 2 and

Table 1. Nomenclature for the Cytosolic Rat Glutathione Transfersases

New Nomenclature	Previous Nomenclature		Alternative subunit Designations
Glutatione transferase 1-1	B (ligandin)	L_2	Ya
Glutatione transferase 1-2		BL	Ya, Yc
Glutatione transferase 2-2	AA	B_2	Yc
Glutatione transferase 3-3	A	A_2	Yb1
Glutatione transferase 3-4	C	AC	Ybl, Yb2
Glutatione transferase 3-6	P		
Glutatione transferase 4-4	D (?)	C_2	Yb2
Glutatione transferase 4-6	S		
Glutatione transferase 5-5	E		
Glutatione transferase 6-6	M_T		Yn
Glutatione transferase 7-7	P		Yp or Yf
Glutatione transferase 8-8	K		Yk

Modified and updated from Jacoby et al., 1984.

8), Mu (subunits 3, 4 and 6), and Pi (subunit 7). Subunit 5 has yet to be classified (Mannervik and Danielson, 1988). Evidence has been obtained only for the formation of heterodimers from subunits within a multigene family with the resulting dimer reflecting the functional properties of its constituting subunits (Table 2). Comparison of orthologous isoenzymes from various species has shown, that the classification into classes Alpha, Mu, and Pi is species-independent.

GSTs of a certain class possess diagnostic substrate specificities, which have proven useful for the identification and classification of newly discovered forms (Table 3). Class Alpha GSTs and the as yet unassigned GST 5-5 for example are highly active with organic hydroperoxides such as cumene hydroperoxide, class Mu transferases conjugate *trans*-stilbene oxide with a comparatively high efficiency, while ethacrynic acid is a marker substrate for class Pi GSTs (Table 3). From the substrates listed in Table 3, 1-chloro-2,4-dinitrobenzene (CDNB) is the most commonly used to determine the overall activity of unfractionated GSTs. It is well utilized by all rat GSTs except subunit 5, which, however, is usually present in small amounts. The other substrates are chosen, because they are fairly selective for individual subunits.

Mutagenic, Carcinogenic and Cytotoxic Substrates of Glutathione S-Trnasferases

A recently discovered Alpha class acidic isoenzyme from rat liver GST 8-8 is highly reactive in detoxifying cytotoxic 4-hydroxyalkenals (Jensson et al., 1986). Among all other GSTs of rat liver investigated so far, the next most active (the near neutral Mu class isoenzyme 4-4) exhibits a specific activity towards 4-hydroxynonenal (4-HNE) which is more than 20-fold lower. This suggests an important role of the mainly acidic GSTs in the detoxification of hydroxyalkenals, which are major lipid peroxidation products (Alin et al., 1985).

Aflatoxin B_1-8,9-oxide is a major metabolite of aflatoxin B_1 and a potent hepatocarcinogen in the rat. It is conjugated, although at the low rate of 1 nmol/min/mg, exclusively by the subunits 1 and 2 (Coles et al., 1985). The important point of this reaction is its 4- to 5-fold acceleration following the induction of subunits 1 for example by ethoxyquin. Ethoxyquin is one of a wide range of chemicals which induce GSTs in the rat and include polycyclic aromatic hydrocarbons, chlorinated biphenyls, certain antioxidants and *trans*-stilbene oxide. Most of these inducers are selective for subunits 1 and 3, and influence the detoxification of substrates for the corresponding enzymes (Igarashi et al., 1987; Gregus et al., 1989).

Various electrophilic metabolites including "K-region" epoxides and "bay-region" dihydrodiol epoxides are formed from polycyclic aromatic hydrocarbons, some of which are powerful mutagens and initiators of carcinogenesis (Thakker et al., 1985). These epoxides have been recognized as another important group of substrates primarily for near neutral and acidic GSTs. *Anti*-benzo(*a*)-pyrene-7,8-diol-9,10-epoxide (BPDE), an ultimate carcinogen derived from benzo(*a*)pyrene, was shown to be metabolized by all rat GST isoenzymes investigated, and best by GST 7-7 and GST 4-4 (Jernström et al., 1985; Robertson et al, 1986). There was a considerable difference in the catalytic efficiency of 0.287 s^{-1} μM^{-1} for GST 7-7 compared to 0.046 s^{-1} μM^{-1} for GST 4-4 as determined from isoenzymes of rat lung (Robertson et al., 1986). Rat liver GST 4-4 metabolized BPDE at a rate of 357 nmol/min/mg about 2- and 3.4-times more efficiently than GST 3-4 and GST 1-1 (Jernström et al., 1985). Similar to the rat GSTs, the acidic human placental GST π metabolized BPDE with an apparent Vmax of 825 nmol/min/mg more efficently than the near neutral and basic human liver GSTs μ and α-E with apparent Vmax values of 570 nmol/min/mg and 38 nmol/min/mg, respectively (Robertson et al., 1986). The rat GSTs 4-4 and 7-7 as well as the human GST π express high selectivity for the (+)-enantiomer of BPDE which is fortunate with regard to the preferential metabolic formation of this ultimately carcinogenic intermediate. The same stereoselectivity is shown by GST 4-4 towards chrysene and benz(*a*)anthracene bay-region diol epoxides (Robertson and Jernström, 1986).

Table 2. Physicochemical Characteristics of Rat Glutathione Transferases

Isoenzyme	Class	Apparent subunit MW (kdalton)[a]	Subunit MW[b]	No. of amino acids per subunit[b]	Isoelectric point
1-1	Alpha	25	25 434	221	10
1-2	Alpha	25 + 28	-	-	9.9
2-2	Alpha	28	25 209	220	9.8
3-3	Mu	26.5	25 806	217	8.9
3-4	Mu	26.5	-	-	8
3-6	Mu	26.5 + 26	-	-	7.4
4-4	Mu	26.5	25 592	217	6.9
4-6	Mu	26.5 + 26	-	-	6.1
5-5	-[c]	26.5	-	-	7.3
6-6[d]	Mu	26	-	-	5.8
7-7	Pi	24	23 307	209	7.0
8-8	Alpha	24.5	-	-	6.0
Microsomal	-	17	17 237	154	10.1

Data from Mannervik and Danielson, 1988, with permission.

[a] Relative values estimated by sodium dodecyl sulfate/polyacrylamide gel electrophoresis.
[b] N-terminal methionine residue not included.
[c] Not yet classified.
[d] Recent work suggests that the major testicular enzyme, designated transferase 6-6, is a heterodimer.

Table 3. Specific Activities (μMOL/MIN/MG) of Rat Glutathione Transferases

Class:	Alpha			Mu			Pi	Unknown	Microsomal
Substrate Enzyme:	1-1	2-2	8-8	3-3	4-4	6-6[a]	7-7	5-5	(activated)
1-Chloro-2,4-dinitrobenzene	50	17	10	58	17	190	24	<0.15	30
1,2-Dichloro-4-nitrobenzene	<0.04	<0.04	0.12	5.3	0.18	2.85	0.048	Nil	<0.06
Bromosulfophthalein	<0.01	<0.01	-	0.94	0.04	-	0.01[b]	-	<0.01
Ethacrynic acid	0.08	1.24	7.0	0.08	0.62	0.057	3.84	<Nil	<0.01
trans-4-Phenyl-3-buten-2-one	<0.004	<0.004	0.10	0.05	1.18	0.019	0.22	<0.001	0.001
4-Hydroxynonenal	2.6	0.67	170	2.7	6.9	-	-	-	-
Leukotriene A_4[c]	0.002	0.0005	-	0.002	0.077	-	-	-	-
1,2-Epoxy-3-(p-nitrophenoxy)-propane	<0.1	<0.1	-	0.53	1.37	-	-	25.5	<0.01
trans-Stilbene oxide[d]	0.001	0.003	0.033	0.10	2.0	0.13	0.005	-	-
Benzo(a)pyrene 7,8-diol-9,10-oxide	-	0.006	0.18	0.012	0.68	-	5.5	-	-
Cumene hydroperoxide	3.1	7.9	1.10	0.35	0.72	0.19	0.048	12.5	0.8
H_2O_2	<0.01	<0.01	<0.01	<0.01	<0.01	<0.01	<0.01	-	<0.04
Δ5-Androstene-3,17-dione	4.2	0.36	-	0.02	0.002	-	-	-	-
p-Nitrophenyl acetate	0.79	0.20	-	1.01	0.28	0.19	-	-	-

Data from Mannervik and Danielson, 1988, with permission.

a Unpublished work by Guthenberg, C. and Mannervik, B.
b Unpublished work by Tahir, M.K. and Mannervik, B.
c Unpublished work by Örning, L., Söderström, M., Hammarström, S., and Mannervik, B.
d Unpublished work by Seidegård, J., Danielson, U.H., and Mannervik, B.
\- not detected

Table 4. Substrate Specificity of Glutathione S-Transferase X Compared with Those of Glutathione S-Transferases C and B

	Specific activity (nmol/min/mg)		
Substrate	Transferase X	Transferase C (3-4)	Transferase B (1-2)
1-Chloro-2,4-dinitrobenzene	21 500	22 500 (10 000)	16 000 (11 000)
1,2-Dichloro-4-nitrobenzene	2 700	1 000 (2 000)	4 (3)
trans-4-Phenylbut-3-en-2-one	700	470 (400)	< 5 (1)
Menaphthyl sulphate	0[a]	-	6 (4)
1,2-Epoxy-3-(p-nitrophenoxy)-propane	360	280 (0)[a]	- (0)[a]
2-(4-Nitrophenyl)-ethyl bromide	180	45	-
p-Nitrobenzyl chloride	3 300	(10 200)	(100)

Data from Friedberg *et al.*, 1983, with permission.

[a] A value of zero does not necessarily mean that the enzyme is incapable of catalysing the reaction, but rather that the reaction rate was not significantly greater than the spontaneous rate in trials with reasonable amounts of enzyme (> 1 mg of enzyme/assay in our experiments).

Numbers in parentheses give other literature values (Jakoby *et al.*, 1976). Apparently homogeneous preparations of glutathione S-transferases X, C and B were used throughout. Assays were performed as described previously (Habig *et al.*, 1974; Gilham, 1971). Assays were performed in triplicate. Deviations from the mean were less than 10 %.

Table 5. Quantitative Comparison of the Inactivation of Benz(*A*)Anthracene 5,6-Oxide and Benz(*A*)Anthracene-8,9-Diol-10,11-Oxide by Different Enzymes

Enzyme	Enzyme concentration in liver[a] (µg/mg tissue)	Amounts of enzyme required for a 50 % reduction in mutagenicity			
		BA 5,6-oxide		BA-8,9-diol-10,11-oxide	
		µg/incubation	mg liver equivalents	µg/incubation	mg liver equivalents
Experiment shown in Figure 1					
Glutathione transferase A (3-3)	0.5	0.2	0.4	100	200
Glutathione transferase C (3-4)	1.1	0.02	0.02	20	20
Glutathione transferase X	0.25	<<0.01[b]	<<0.06	8	30
Experiment shown in Figure 2					
Glutathione transferase A (3-3)	0.5	0.11	0.2	110	200
Glutathione transferase B (1-2)	2.2	0.5	0.2	130	60
Glutathione transferase C (3-4)	1.1	0.02	0.017	30	20
Glutathione transferase X	0.25	0.003	0.011	6	20
Experiments using other enzymes[c]					
Major microsomal epoxide hydrolase (mEH_b)	0.5	0.7	1.3	Inactive(>>170)	>>300
Major cytosolic epoxide hydrolase (cEH_{TSO})	0.16	4	30	Inactive(>> 30)	>>200
Dihydrodiol dehydrogenase	0.45	Inactive(>>3)	>>1000	70	170

Data from Glatt *et al.*, 1983, with permission.

a Values refer to untreated, adult males of the species from which the enzyme was purified.

b Determinable only as an upper limit from the experiment.

c Data taken from Glatt *et al.*, 1982. These experiments were performed in the same way as those using glutathione transferases.

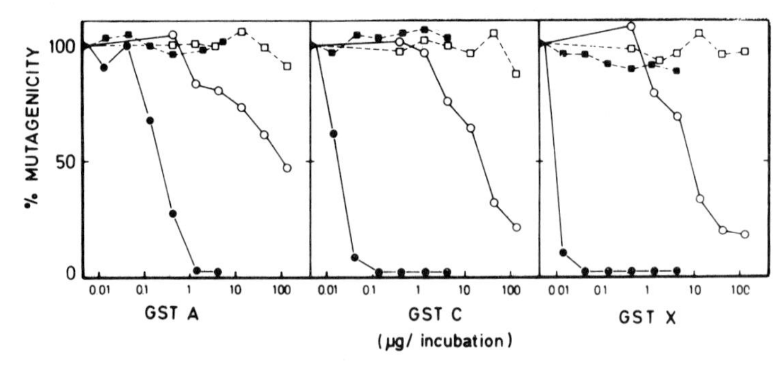

Figure 1. Effect of different forms of glutathione transferase (GST), purified from untreated rats, on the mutagenicity of BA 5,6-oxide in the presence (●) and absence (o) of glutathione (2 mM) and of BA-8,9-diol,10,11-oxide in the presence (▪) and absence (□) of glutathione (2 mM). The numbers of mutants above solvent control induced by 1 μg of BA 5,6-oxide or by 3 μg of BA-8,9-diol-10,11-oxide in the presence of various amounts of the purified glutathione transferases are expressed as percentage of the corresponding value obtained in the absence of enzyme. The absolute numbers of colonies obtained in the absence of enzyme were 74 and 85 for the solvent control, 730 and 1040 for BA5,6-oxide, and 1520 and 1540 for BA-8,9-diol-10,11-oxide, in the presence and absence of glutathione, respectively. Triplicate incubations were performed. The fariation in the numbers of colonies on replicate plates was less than 10%. Data from Glatt et al. (1983), with permission.

The prominent contribution of near neutral and acidic GSTs to the detoxification of polycyclic aromatic epoxides became obvious with the discovery of a weakly acidic rat liver GST termed GST-X (Friedberg et al., 1983). GST-X was identified as a homodimer with a subunit molecular weight of 23,500 Da, an isoelectric point of 6.9, and a particularly high turnover of 1,2-dichloro-4-nitrobenzene in comparison to other Alpha and Mu class GSTs (Table 4). On a molar basis this isoenzyme was at least 6.7- and 5-fold more efficient in the conjugation of the mutagenic K-region epoxide benz)*a*)anthracene-5,6-oxide (BA-5,6-oxide) and the non-bay-region diol epoxide r-8,t-9-dihydroxy-t-10,11-oxy-8,9,10,11-tetrahydrobenz(*a*)-anthracene (BA-8,9-diol-10,11-oxide) when compared with the major rat liver GSTs C (3-4), B (1-2) and A (3-3) (Table 5; Figures 1 and 2). The relative efficiency of all four investigated GSTs was similar with both substrates in spite of the structural differences in the epoxides. The inactivation of the diol epoxide required about 1000-fold higher concentrations of GSTs than the inactivation of the K-region epoxide. An even stronger preference for K-region epoxides, as compared to diol-epoxides, was observed for the major microsomal epoxide hydrolase mEH_b (Glatt et al., 1983).

Glutathione S-Transferase X and Low Abundance Class MU Isoenzymes

Characterization of GST-X by fast atom bombardement mass spectroscopy of tryptic peptides revealed clear differences from GST 4-4, but also a number of structural homologies to this isoenzyme, from which a close relationship of GST-X to the Mu class GSTs could be deduced.

Recently evidence was obtained by chromatofocussing of S-hexylglutathione-Sepharose 6B purified GSTs for at least seven near neutral to acidic isoenzymes in rat liver (Milbert, 1986). These newly discovered isoenzymes also appeared to be mainly members of the Mu class GSTs. Therefore it was assumed that they may have equal or even higher capacities for the detoxification of polycyclic aromatic epoxides or lipid peroxidation products than the rat GSTs so far characterized.

In order to investigate the properties of these isoenzymes and their role in the cellular protection against mutagenic, carcinogenic and cytotoxic intermediates of endogenous and foreign compounds, we designed a method for the efficient purification particularly of acidic rat liver GSTs. This method includes chromatography of the cytosol on QA-cellulose at pH 8.5, affinity chromatography on GSH-Sepharose and chromatofocussing between pH 7.0 and 4.0. Throughout the purification GST activity was monitored with CDNB and 4-HNE as substrates, the latter acting as a marker substrate specifically for acidic GSTs. Chromatofocussing resolved three major GST containing fractions (CV I- CF III) with isoelectric points at 6.8, 6.2, and 5.8 (Figure 3). CF I and CF II were further purified on hydroxylapatite to yield a total of four near neutral to acidic GST isoenzymes, HA I-a (Y3-Y4), HA I-b (Y5-Y5), HA II (Y1-Y1) and CF III (mainly Y2-Y2 and a trace of Y6-Y6), consisting of six different subunits which were tentatively termed Y1 to Y6 (Table 6; Figure 4). The apparent molecular weight of these subunits ranged between 25,500 Da and 27,000 Da.

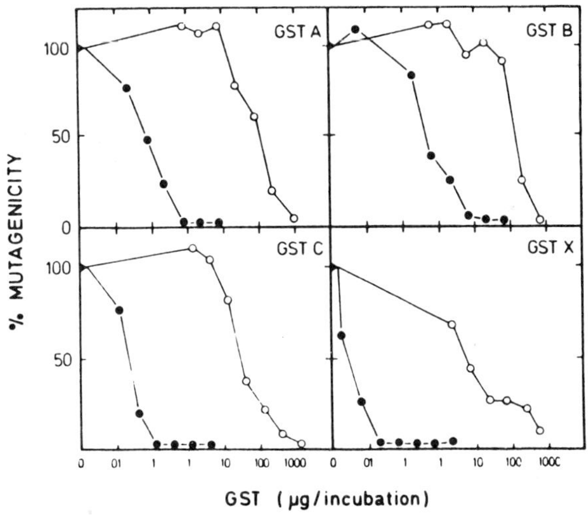

Figure 2. Effect of different forms of glutathione transferases (GST), purified from Aroclor 1254-treated rats, on the mutagenicity of BA 5,6-oxide (•) and BA-8,9-diol-10,11-oxide (o). The experiment was conducted as described in the legend to Figure 1 except that the enzymes were purified from Aroclor 1254-treated rats and they were tested only in the presence of glutathione (2 mM). The absolute numbers of colonies obtained in the absence of enzyme were 109, 405, and 1270 for the solvent controls, for incubations with 1 µg BA 5,6-oxide, and for incubations with 2.5 µg BA-8,9-diol-10,11-oxide, respectively. Incubations were performed in triplicate, and the variation in the nubmers of colonies on replicate plates was less than 15%. Data from Glat et al. (1983), with permission.

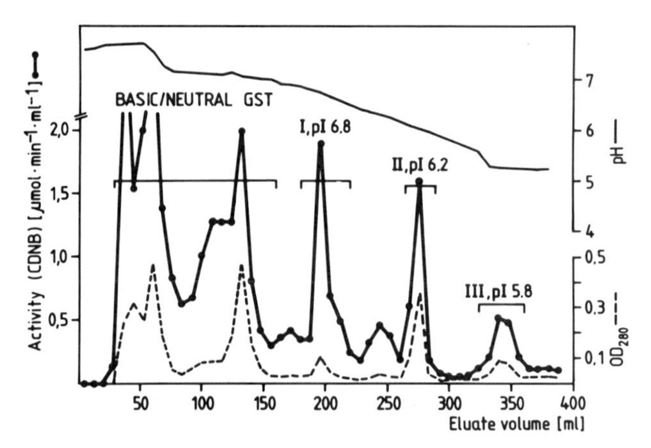

Figure 3. Chromatofocussing of GSTs between pH 7-4. GSH-Sepharose 6B eluates were concentrated and dialyzed against 25 mM imidazole/HCl buffer (pH 7.4). Chromatofocussing was performed on a PBE 74 column.

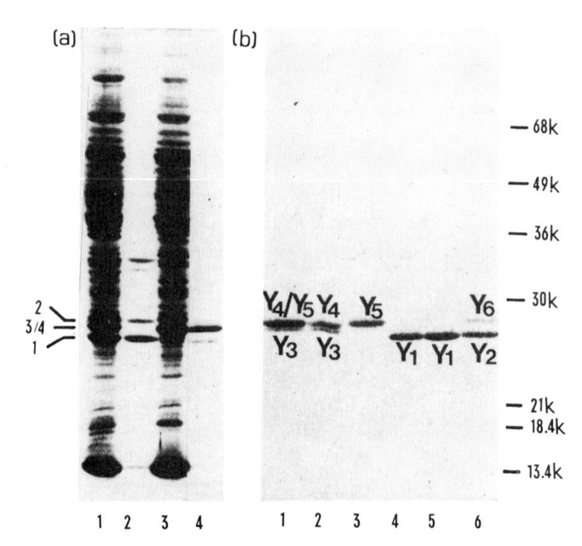

Figure 4. SDS/polyacrylamide-gel electrophoresis of (a) crude and (b) purified acidic GSTs of rat liver. GSTs were electrophoresed in a 12.5% SDS/polyacrylamide gel and protein bands were vasualized by Coomassie Brillant Blue R 250.

a. Lanes 1: cytosol; lane 2: flow through of QA-52 cellulose; lane 3: QA-52 cellulose eluate; lane 4: GSH affinity eluate. Subunits apparently corresponding to characterized rat liver GSTs are marked with arabic numberals (see Table 1).

b. Purified GSTs after chromatofocussing (CF) or chromatofocussing and hydroxylapatite chromatography (HA). Lane 1: CF I; lane 2: HA I-a; lane 3: HA I-b; lane 4: CF II; lane 5: HA II; lane 6: CF III.

Table 6. Purification of Rat Liver Acidic Glutathione S-Transferases

	Volume (ml)	Protein (mg)	CDNB[a] Specific activity (μmol/min x mg)	CDNB[a] Total activity (μmol/min)	4-HNE[b] Specific activity (μmol/min x mg)	4-HNE[b] Total activity (μmol/min)	Ratio 4-HNE/CDNB
Cytosol	192	2937.6	2.44	7181(100%)	1.21	3562(100%)	0.50
QA-cellulose eluate	375	1237.5	0.71	879(12.2%)	2.92	3614(101.5%)	4.11
GSH affinity eluate	158	45.8	12.07	552(7.7%)	52.41	2401(67.4%)	4.35
CF (basic/neutral forms)	130	23.1	14.27	330(4.6%)	2.53	59(1.7%)	0.18
CF I (pI 6.8)	43.5	2.3	15.66	36(0.5%)	6.79	16(0.5%)	0.44
CF II (pI 6.2)	28	5.5	4.36	24(0.3%)	121.23	662(18.6%)	27.58
CF III (pI 5.8)	40	5.3	2.41	13(0.2%)	64.14	341(9.6%)	26.23
HA I-a	23	0.9	22.75	21(0.3%)	4.50	4.2(0.1%)	0.14
HA I-b	14	0.4	12.00	4(0.06%)	0.80	0.2(0.01%)	0.21
HA II	16.5	3.1	4.61	13.8(0.2%)	185.44	550.8(15.5%)	35.59

[a] CDNB: 1-chloro-2,4-dinitrobenzene as substrate

[b] 4-HNE: 4-hydroxynon-2-enal as substrate

Table 7. Specific Activities of the New Acidic Rat Liver Glutathione S-Transferases

Subunit (MW x 10^{-3}) / Substrate	HA I-a Y_3 (26.5) Y_4 (27.0)	HA I-b Y_5 (27.0)	HA II Y_1 (25.5)	CF III Y_2 (25.5) Y_6 (27.0)
1-Chloro-2,4-dinitro-benzene	13.08[a]	2.84	5.03	8.16
1,2-Dichloro-4-nitrobenzene	1.23	bd	bd	0.27
4-Hydroxynon-2-enal	2.32	2.57	114.36	99.29
Ethacrynic acid	0.42	0.91	3.64	3.73
Δ^5-Androsten-3,17-dione	bd	bd	0.02	0.03
trans-4-Phenyl-3-buten-2-one	0.34	0.07	bd	0.03
1,2-Epoxy-3-(p-nitro-phenoxy)propane	bd	bd	bd	bd
Cumene hydroperoxide	0.10	bd	0.23	0.19

[a] Activities in $\mu mol \cdot min^{-1} \cdot mg^{-1}$.
bd: Below detection.

Subunits Y1 and Y2 were shown by strong immunological crossreactivity with antiserum against subunit 8 to be members of the class Alpha GSTs. Subunits Y3 to Y6 were identified by Western blotting as belonging to the class Mu subunits. Ha II (Y1-Y1), based on its isoelectric point and crossreactivity with GST 8-8, is closely related to, if not identical with, GST 8-8. CF III (Y2-Y2), although immunologically closely related to the Alpha class subunit 8 differs in its isoelectric point and enzymatic properties from GST 8-8 and appears to be a new isoenzyme. Subunit Y3 expressed similarity to GST subunit 6, which has been characterized from rat brain and testis, and suggests the presence of a 6-Y4 heterodimer (HA I-a) in rat liver. Subsrate specificities of the new isoenzymes tested with a number of model compounds are rather undiagnostic with the exception of the high turnover of 4-HNE with CF III (Table 7).

In summary, we have identified and purified at least three novel acidic GSTs which appear to be minor forms representing between 0.06% (HA I-b) and 0.3% (HA I-a) of total rat liver CDNB-conjugating GST activity. Their efficiency in the metabolism of polycyclic aromatic epoxides is at present studied.

CONCLUSIONS

Glutathione S-transferases constitute a superfamily of dimeric proteins, the subunits of which may be assigned with respect to structure homologies, enzymatic, immunological and functional properties to three gene families, Alpha, Mu, and Pi. The main function of these enzymes appears to be the protection against mutagenic, carcinogenic and cytotoxic electrophilic metabolites formed from endogenous and exogenous compounds. Mu and Pi class isoenzymes containing the subunits 4 and 7 have proven particularly efficient in the detoxification of K-region and bay-region diol epoxides from polycyclic aromatic hydrocarbons. The Alpha class GST 8-8 was by far the most effective in conjugating the highly mutagenic and highly cytotoxic lipid peroxidation product 4-hydroxynonenal. These findings provide evidence for an important role of near neutral and acidic GSTs in the control of genotoxic metabolites. Four acidic GSTs consisting of 6 individual subunits were purified by a newly developed method. From their close relationship to the Mu class GSTs, which are very potent in the conjugation of polycyclic aromatic epoxides it is concluded, that they, although minor forms in rat liver, may be of importance in the protection against mutagenic, carcinogenic and cytotoxic metabolites of various xenobiotics.

ACKNOWLEDGEMENTS

The support by the Deutsche Forschungsgemeinschaft (SFB 302), the kind gifts of 4-HNE (H. Esterbauer, University of Graz) and antibodies against GST 6-6 (K. Sato, Hirosaki University) are greatfully acknowledged. T.I. is the recipient of a Humboldt Foundation postdoctoral fellowship.

REFERENCES

Alin, P., Danielson, H., and Mannervik, B. (1985). 4-Hydroxy-alk-2-enals are substrates for glutathione transferase. *FEBS Lett.* **179**, 267-270.

Booth, J., Boyland, E. and Sims, P. (1961). An enzyme from rat liver catalysing conjugations with glutathione. *Biochem. J.* **79**, 516-526.

Coles, B., Meyer, D.J., Ketterer, B., Stanton, C.A., and Garner, R.C. (1985). Studies on the detoxication of microsomally-activated afla-toxin B1 by glutathione and glutathione S-transferases *in vitro*. *Carcinogenesis* **6**, 693-697.

Ding, G.J.F., Ding, V.D.H., Rodkey, J.A., Bennett, C.D., Lu, A.Y.H., and Pickett, C.B. (1986). Rat liver glutathione S-transferases. DNA sequence analysis of a Yb2 cDNA clone and regulation of the Yb1 and Yb2 mRNAs by phenobarbital. *J. Biol. Chem.* **261**, 7952-7957.

Ding, G.J.F., Lu, A.Y.H., and Pickett, C.B. (1985). Rat liver glutathione S-transferases. Nucleotide sequence analysis of a Yb1 cDNA clone and prediction of the complete amino acid sequence of the Yb1 subunit. *J. Biol. Chem.* **260**, 13268-13271.

Faulder, C.G., Hirrell, P.A., Hume, R., and Strange, R.C. (1987). Studies of the development of basic, neutral and acidic isoenzymes of glutathione S-transferases in human liver, adrenal, kidney and spleen. *Biochem. J.* **241**, 221-228.

Friedberg, T., Milbert, U., Bentley, P., Guenthner, T.M. and Oesch, F. (1983). Purification and characterization of a new cytosolic glutathione S-transferase (glutathione S-transferase X) from rat liver. *Biochem. J.* **215**, 617-625.

Fryer, A.A., Hume, R., and Strange, R.C. (1986). The development of glutathione S-transferase and glutathione peroxidase activities in human lung. *Biochim. Biophys. Acta.* **883**, 448.

Gilham, B. (1971). The reaction of aralkyl sulphate esters with glutathione catalysed by rat liver preparations. *Biochem. J.* **121**, 667-672.

Glatt, H.R., Cooper, C.S., Grover, P.L., Sims, P., Bentley, P., Merdes, I., Waechter, F., Vogel, K., Guenthner, T.M. and Oesch, F. (1982). Inactivation of a diol-epoxide by dihydrodiol dehydrogenase, but not by two epoxide hydrolases. *Science* **215**, 1507-1509.

Glatt, H.R., Friedberg, T., Grover, P.L., Sims, P. and Oesch, F. (1983). Inactivation of a diol epoxide and a K-region epoxide with high efficiency by glutathione transferase X. *Cancer Res.* **43**, 5713-5717.

Gregus, Z., Madhu, C., and Klaasen, C.D. (1989). Inducibility of glutathione S-transerases in hamsters. *Cancer Lett.* **44**, 89-94.

Habig, W.H., Pabst, M.J. and Jakoby, W.B. (1974). Glutathione S-transferases. The first enzymatic step in mercapturic acid formation. *J. Biol. Chem.* **249**, 7130-7139.

Hatayama, I., Satoh, K., and Sato, K. (1986). Development and hormonal regulation of

the major form of hepatic glutathione S-transferase in male mice. *Biochem. Biophys. Res. Comun.* **140**, 581-588.
Igarashi, T., Irokawa, N., Ono, S., Ohmori, S., Ueno, K., and Kitagawa, H. (1987). Difference in the effects of phenobarbital and 3-methylchol-anthrene treatment on subunit composition of hepatic glutathione S-transferases in male and female rats. *Xenobiotica* **17**, 127-137.
Igarashi, T., Satoh, T., Ueno, K., and Kitagawa, H. (1983). Species difference in glutathione level and glutathione related enzyme activities in rats, mice, guinea pigs and hamsters. *J. Pharm. Dyn.* **6**, 941-949.
Igarashi, T., Satoh, T., Iwashita, K., Ono, S., Ueno, K., and Kitagawa, H. (1985). Sex difference in subunit composition of hepatic glutathione S-transferase in rats. *J. Biochem.* **98**, 117-123.
Igarashi, T., Tomihari, N., Ohmori, S., Ueno, K., Kitagawa, H., and Satoh, T. (1986). Comparison of glutathione S-transferases in mouse, guinea pig, rabbit and hamster liver cytosol to those in rat liver. *Biochem. International* **13**, 641-648.
Jakoby, W.B., Habig, W.H., Keen, J.N., Ketley, J.N. and Pabst, M.J. (1976). Glutathione S-transferases: catalytical aspects. In: *Glutathione: Metabolism and Function*, Arias, I.M. and Jakoby, W.B. (eds): New York, *Raven Press*, pp. 189-201.
Jakoby, W.B., Ketterer, B. and Mannervik, B. (1984). Glutathione transferases: nomenclature. *Biochem. Pharmacol.* **33**, 2539-2540.
Jensson, H., Guthenberg, C., alin, P., and Mannervik, B. (1986). Rat glutathione transferase 8-8, an enzyme efficiently detoxifying 4-hydroxyalk-2-enals. *FEBS Lett.* **203**, 207-209.
Jernström, B., Martinez, M., Meyer, D.J., and Ketterer, B. (1985). Glutathione conjugation of the carcinogenic and mutagenic electrophile (+)-7β,8α-dihydroxy-9α,10α-oxy-7,8,9,10-tetrahydrobenzo(*a*)pyrene catalyzed by purified rat liver glutathione transerases. *Carcinogenesis* **6**, 85-89.
Ketterer, B. (1988). Protective role of glutathione and glutathione transferases in mutagenesis and carcinogenesis. *Mutat. Res.* **202**, 343-361.
Litwack, G., Ketterer, B., and Arias, I.M. (1971). Ligandin: a hepatic protein which binds steroids, bilirubin, carcinogens and a number of organic anions. *Nature* (London) **234**, 466-467.
Mannervik, B. (1985). The isoenzymes of glutathione transferase. *Adv. Enzymol. Rel. Areas Mol. Biol.* **57**, 357-417.
Mannervik, B., and danielson, U.H. (1988). Glutathione transferases - structure and catalytic activity. *CRC Crit. Rev. Biochem.* **23**, 283-337.
Milbert, U. (1986). Ph.D. - Thesis, University of Mainz.
Pickett, C.B., Telakowski-Hopkins, C.A., Ding, G.J., argenbright, L., and Lu, A.Y.H. (1984). Rat liver glutathione S-transferases. Complete nucleotide sequence of a glutathione S-transferase mRNA and the regulation of the Ya, Yb, and Yc mRNA by 3-methylcholanthrene and phenobarbital. *J. Biol. Chem.* **259**, 5182-5188.
Robertson, I.G.C., Guthenberg, C., Mannervik, B. and Jernström, B. (1986). Differences in stereoselectivity and catalytic efficiency of three human glutathione transferases in the conjugation of glutathione with 7β,8α-dihydroxy-9α,10α-oxy-7,8,9,10-tetrahydrobenzo(*a*)pyrene. *Cancer Res.* **46**, 2220-2224.
Robertson, I.G.C., Jensson, H., Mannervik, B. and Jernström, B. (1986). Glutathione transferases in rat lung: the presence of transferase 7-7, highly efficient in the conjugation of glutathione with the carcinogenic (+)-7β,8α-dihydroxy-9α,10α-oxy-7,8,9,10-tetrahydrobenzo(*a*)pyrene. *Carcinogenesis* **7**, 295-299.
Robertson, I.G.C. Jernström, B. (1986). The enzymatic conjugation of glutathione with bay-region diol-epoxides of benzo(*a*)pyrene, benz(*a*)anthracene and chrysene. *Carcinogenesis* **7**, 1633-1636.
Sato, K. (1989). Glutathione transferases as markers of preneoplasia and neoplasia. *Adv. Cancer Res.* **52**, 205-255.
Sies, H., and Ketterer, B. (1988) (Eds.). Glutathione conjugation: mechanisms and biological significance. Academic Press, New York.
Suguoka, Y., Kano, T., Okuda, A., Sakai, M., Kitagawa, T., and Muramatsu, M. (1985). Cloning and the nucleotide sequence of rat glutathione S-transferase P cDNA. *Nucleic Acids Res.* **13**, 6049-6057.
Telakowski-Hopkins, C.A., King, R.G., and Pickett, C.B. (1988). Glutathione S-

transferase Ya subunit gene: identification of regulatory elements required for basal level and inducible expression. *Proc. Natl. Acad. Sci. USA* **85**, 1000-1004.

Telakowski-Hopkins, C.A., Rothkopf, G.S., and Pickett, C.B. (1986). Structural analysis of a rat liver glutathione S-transferase Ya gene. *Proc. Natl. Acad. Sci. USA* **83**, 9393-9397.

Thakker, D.R., Yagi, H., Levin, W., Wood, A.W., Conney, A.H., and Jerina, D.M. (1985). Polycyclic aromatic hydrocarbons: metabolic activation to ultimate carcinogens. In: *Bioactivation of Foreign Compounds* (M.W. Anders, Ed.) pp. 177-242. New York, Academic Press.

BIOACTIVATION OF XENOBIOTICS BY FLAVIN-CONTAINING MONOOXYGENASES

Daniel M. Ziegler

Clayton Foundation Biochemical Institute and Department of Chemistry
The University of Texas at Austin
Austin, TX 78712

INTRODUCTION

Flavoproteins are ubiquitous in nature but the ones referred to as flavin-containing monooxygenases, or simply by the acronym FMO, share a catalytic mechanism distinctly different from all other known oxidases or monooxygenases bearing flavin, heme or other redox active prosthetic groups. Like other mammalian monooxygenases, FMO's require NADPH and oxygen as cosubstrates for the oxygenation of the third substrate but they differ in that the third substrate is not required for the generation of the enzyme bound oxygenating intermediate. Kinetic studies on mechanism (Poulsen and Ziegler, 1979, Beaty and Ballou, 1981a, 1981b) have shown that the xenobiotic substrate is not required for flavin reduction by NADPH nor for reoxidation of dihydroflavin by molecular oxygen. The latter reaction produces the 4a-hydroperoxyflavin which, in FMO, is stabilized by the protein microenvironment around the prosthetic group. The enzyme is apparently present within the cell in this form and any soft nucleophile that can gain access to the enzyme-bound oxygenating intermediate will be oxidized. Precise fit of substrate to enzyme is not necessary, and FMO catalyzes at the same maximum velocity the oxidation of compounds that possess few, if any structural features in common (Ziegler, 1988). These flavoproteins apparently discriminate between physiologically essential and xenobiotic soft nucleophiles by excluding the former rather than by selectively binding the latter. This property is largely responsible for the exceptionally broad specificity of these enzymes. Steric parameters controlling access of nucleophiles to the hydroperoxyflavin apparently differ in various forms of FMO.

Species and tissue specific forms of these flavoproteins have been described (Williams et al., 1984a, 1984b, Tynes et al., 1985, Tynes and Philpot, 1987). Species differences in hepatic FMO appear quantitative rather than qualitative, whereas tissue specific forms present in some species are distinctly different enzymes. For instance, the hepatic and pulmonary FMOs in rabbits exhibit distinct but overlapping substrate specificities (Williams, 1984a, Tynes et al., 1985) and are apparently different gene products. The following discussion is therefore largely restricted to bioactivation reactions catalyzed by the hepatic form of FMO, but since this form is present in the major organs of entry in most species (Dannon and Guengerich, 1982,_Tynes and Philpot, 1987) bioactivation reactions determined with hepatic FMO can be, in general, extrapolated to other tissues.

While the oxygenation of most xenobiotic substrates catalyzed by FMO leads to less reactive metabolites, a few are converted to more reactive intermediates and the following sections will focus primarily on compounds bearing functional groups that are bioactivated primarily or exclusively by the mammalian microsomal hepatic FMO.

Functional Groups Bearing Nitrogen

Of the various organic nitrogen substrates for FMO, only secondary arylamines (aromatic amines) and 1,1-disubstituted hydrazines are N-oxygenated to more reactive intermediates. The contribution of FMO to the N-oxygenation of carcinogenic primary arylamines is less certain, although fully enzymic routes for the bioactivation of these amines are known and a discussion of this pathway is included herein. However, the extent to which pathways deduced from studies with enzymes occur *in vivo* is not known. The latter statement is also true for the N-hydroxylation of primary alkylamines catalyzed exclusively by rabbit lung FMO (Williams et al., 1985, Tynes et al., 1985). While primary hydroxylamines are chemically more reactive than parent amines, their facile further oxidation to oximes also catalyzed by FMO (Poulsen et al., 1985) would limit their concentration *in vivo*. N-Oxygenation of primary alkylamines to oximes by this route is most likely a pathway for detoxication rather than intoxication.

Primary Arylamines

None of the mammalian FMO's described to date catalyze facile direct N-hydroxylation of most primary arylamines. The liver FMO apparently can catalyze, at very slow rates, the N-oxidation of 2-aminofluorene to its carcinogenic 2-aminofluorene hydroxylamine (Fredrick et al., 1982) and of 1-aminopyrene and 2-aminoanthracene to *Salmonella tryphinaurium* mutagens (Pelroy and Gandolfi, 1980). However the rates are quite slow and it appears unlikely that direct N-hydroxylation of primary arylamines contributes substantially to bioactivation of primary arylamines in the intact animal. On the other hand, primary arylamines that are N-methylated by S-adenosylmethionine *in vivo* (reaction 1) become excellent substrates for FMO.

$$ArNH_2 \xrightarrow{Adomet} ArNHCH_3 \qquad (1)$$

Kinetic constants for the N-methylation of four primary alkylamines catalyzed by purified rabbit liver enzymes (Table 1) indicate that the two carcinogenic amines are methylated whereas the two that are not carcinogenic in adult animals show virtually no activity with either transferase A or B. Although N-methylated arylamines are no more reactive chemically than the parent amine, they are bioactivated via FMO as illustrated by examples described in the next section.

Secondary Arylamines

The porcine liver FMO catalyzes the oxidation of all N-methylarylamines to nitrones through intermediate hydroxylamines as indicated by reactions 2 and 3.

$$ArNHCH_3 \xrightarrow[O_2]{NADPH} ArN(OH)CH_3 \qquad (2)$$

$$ArN(OH)\text{-}CH_3 \xrightarrow[O_2]{NADPH} ArN^+(O^-)=CH_2 \qquad (3)$$

Table 1. Rabbit Liver Amine N-Methyl Transferases: Arylamine Substrates

Arylamines	kcat/Km x 10^{-3} Transferase A	Transferase B
Benzidine	15	40
4-Aminobiphenyl	30	52
2-Aminobiophenyl	> 0.4	> 0.5
4-aminoazobenzene	0.0	0.0

Kinetic constants calculated from data taken from Ziegler et al. (1988).

Table 2. N-Methylarylamine Substrates for FMO

N-Methylarylamine	kcat/Km 10^{-3}
N-Methylaniline	100
N-Methylbenzidine	52
N,N'-Dimethylbenzidine	161
N-Methyl-4-aminobiphenyl	72

Kinetic constants calculated from data taken from Ziegler et al. (1988).

The oxidative N-demethylation of N-methylarylamines by reactions 2-3 followed by 4 has been known for some time (Prough and Ziegler, 1977), and as indicated (reaction 4) the primary arylhydroxlamines are always the other product. The studies

$$ArN^{+}(O^{-})=CH_2 \xrightarrow{H_2O} ArNHOH + CH_2O \quad (4)$$

of Kadlubar et al. (1976) have shown that FMO catalyzes the N-oxidation of N-methyl-4-aminoazobenzene to its hydroxylamine derivative via reactions 2-4 and FMO is the primary agent catalyzing the critical first step in the bioactivation of this carcinogen. The contribution of FMO to the bioactivation of other N-methylarylamines administered directly or of those formed by N-methylation is, therefore, a distinct possibility.

Hydrazines

The oxidation of 1,1-disubstituted hydrazines is catalyzed exclusively by FMO in isolated liver microsomes (Prough and Moloney, 1985). This includes both acyclic (eg. 1,1-dimethylhydrazine) and cyclic (eg. N-aminopiperidine) hydrazines. The studies of Prough et al. (1981) indicate that acyclic hydrazines are oxidized to N-oxides (equation 5). The hydrazine oxides are quite unstable and dehydrate to the unsymmetrical diazine as indicated in equation 6.

The intermediate diazine is in equilibrium with the far more stable hydrazone which readily hydrolyzes to the monosubstituted hydrazine and aldehyde (equation 7). While hydrazones produced by the N-oxidation of 1,1-disubstituted hydrazines are no

$$R_2N\text{-}NH_2 \xrightarrow[O_2]{NADPH} R_2N^{+}(O^{-})\text{-}NH_2 \quad (5)$$

$$R_2N^{+}(O^{-})\text{-}NH_2 \xrightarrow[H^+]{H_2O} R_2N^{+}\text{=}NH \underset{}{\overset{H^+}{\rightleftharpoons}} RNHN\text{=}R \quad (6)$$

$$RNHN\text{=}R \xrightarrow{H_2O} RNHNH_2 + R\text{=}O \quad (7)$$

more reactive chemically than the parent hydrazine, the further oxidation of the hydrolysis product (the monosubstituted hydrazine), catalyzed at least in part by FMO, does produce alkyl radicals (Prough et al., 1981) presumably by decomposition of the monosubstituted diazine. To what extent FMO contributes to the bioactivation of 1,1-disubstituted hydrazines is not known, although the enzymology suggests a potential role.

Functional Groups Bearing Sulfur

Most functional groups bearing sulfur are quite nucleophilic and, with few exceptions, are excellent substrates for FMO. However the oxidation of only some groups bearing this element are uniformly converted to more reactive intermediates and the following examples will be restricted to thiocarbamides, thioamides and aminothiols that are known to be bioactivated primarily by oxidations catalyzed by mammalian FMO.

Thiocarbamides

At toxic doses, thiocarbamides (thioureas) produced pulmonary edema and pleural effusion in susceptible animals (Richter, 1952). While the nature of the ultimate toxic metabolite is not known with certainty, the initial steps in the metabolism of thiocarbamides are catalyzed exclusively by FMO (Poulsen et al., 1979). FMO catalyzes the oxidation of thiourea and thiourea derivatives to formamidine sulfinic acids through intermediate sulfenic acids as indicated in equations 8 and 9.

$$R_2N\text{-}C(SH)\text{-}NH_2 \xrightarrow[O_2]{NADPH} RN\text{=}C(SOH)\text{-}NH_2 \quad (8)$$

$$RN\text{=}C(SOH)\text{-}NH_2 \xrightarrow{NADPH} RN\text{=}C(SO_2H)\text{-}NH_2 \quad (9)$$

Both the intermediate and final enzymically generated oxidation products are far more reactive than the parent thiourea, and they also differ considerably in their chemical and biological properties.

The intermediate formamidine sulfenic acids readily oxidize glutathione (GSH) to the mixed disulfides which, as shown equation 10 can react with GSH yielding GSSG and the parent thiourea.

$$\text{RN=C(SOH)-NH}_2 + \text{GSH} \xrightarrow{H_2O} \text{RN=C(SSG)-NH}_2 \xrightarrow{GSH} \text{RN=C-NH}_2 + \text{GSSG} \quad (10)$$

Regeneration of the parent thiourea can establish a futile cycle that oxidizes GSH to GSSG at the expense of NADPH and molecular oxygen. The operation of such a thiocarbamide-dependent futile cycle has been demonstrated with microsomes, and purified FMO (Poulsen et al., 1979) as well as in perfused rat liver, and in intact animals (Kreiter et al., 1984). Futile cycling will persist until the intermediate sulfenate is further oxidized to the sulfinate (equation 9). However kinetic constants for the second oxidation are not favored (Table 3), and it is quite likely that at toxic doses of thiocarbamides significant formation of the formamidine sulfinic acids will not take place until GSH has decreased substantially. At this point oxidation of other cellular thiols, including protein thiols, by the sulfenic acid can occur which would destroy the integrity of intracellular and membrane proteins bearing exposed thiols.

Table 3. Thiocarbamide and Mercaptoimidazole Substrates for FMO

Compound	kcat/km x 10^{-3}
Thiourea	1,500
Ethylenethiourea	1,030
Phenylthiourea	6,400
Thiocarbanilide	5,300
Methimazole	5,300
2-Mercaptobenzimidazole	2,900
Formamidine sulfenate	230
Ethyleneformamidine sulfenate	630
Diphenylformamidine sulfenate	4,600
Methimazole sulfenate	1,500
2-Mercaptobenzimidazole sulfenate	1,400

Kinetic constants based on data taken from Poulsen et al. (1979).

The first oxidation product (the formamidine sulfenic acids) are probably the reactive metabolites responsible for acute toxicity of thiocarbamides and of the closely related mercaptoimidazoles (Table 3) which are also oxidized to sulfenic acids via FMO. The role of FMO in the acute toxicity of thiocarbamides is also consistent with the studies of Smith and Williams (1961) on differences in toxicity of phenylthiourea and thiocarbanilide in rabbits. The latter is considerably less toxic in this species and recent studies (Nagata et al., 1990) have shown that, unlike phenylthiourea, thiocarbanilide is not a substrate for the purified rabbit lung FMO.

On the other hand, the sulfenic acid metabolites are probably not responsible for all of the biological effects of thiocarbamides. The second oxidation products (the formamidine sulfinic acids) are also quite reactive and are the more likely candidates for the carcinogenicity of low chronic doses of some members of this class of compounds.

Formamidine sulfinic acids are moderate electrophiles sufficiently reactive to alkylate amino groups of protein and nucleic acids as illustrated by equation 11.

$$\text{RN=C(SO}_2\text{H)-NH}_2 + \text{Prot-NH}_2 \longrightarrow \text{RN=C(HN-Prot)-N} + \text{H}_2\text{SO}_3 \quad (11)$$

Although the alkylation of DNA by formamidine sulfinic acids has not been tested, such a reaction is consistent with the known chemistry of these metabolites (Sterling, 1974) and the stable adducts formed could account for the carcinogencity of thiourea (Fitzhugh and Nelson, 1948).

Thioamides

Relatively simple compounds bearing a thioamide group are quite toxic and some are also carcinogenic, which suggests that their toxicity is due to the thioamide moiety. One of the simplest members of this series (thioacetamide) produces liver tumors upon chronic administration to rodents (Fitzhugh and Nelson, 1948), and at acute doses causes hepatic necrosis and damage to kidney tubules (Barker and Smukler, 1972, 1973).

The studies of Hanzlik and his associates clearly demonstrated that the deleterious properties of the thioamides is due to their oxidation to sulfenes through intermediate sulfoxides (Hanzlik et al., 1980, Cashman and Hanzlik, 1981) (equations 12 and 13).

$$\text{R-}\overset{\overset{\text{S}}{\|}}{\text{C}}\text{-NH}_2 \xrightarrow[\text{O}_2]{\text{NADPH}} \text{R-}\overset{\overset{\text{S=O}}{\|}}{\text{C}}\text{-NH}_2 \qquad (12)$$

$$\text{R-}\overset{\overset{\text{S=O}}{\|}}{\text{C}}\text{-NH}_2 \xrightarrow[\text{O}_2]{\text{NADPH}} \text{R-}\overset{\overset{\text{SO}_2\text{H}}{|}}{\text{C}}\text{=NH}_2 \qquad (13)$$

The oxidation of thioamides to thioamide sulfoxides (equation 12) is catalyzed almost exclusively by FMO. FMO can also catalyze the oxidation of sulfoxides to the dioxygenated products (equation 13), but the concentrations required to saturate the enzyme are quite high (Cashman and Hanzlik, 1981). While the biochemical mechanism for the second oxidative step has not been completely resolved, the studies of Chieli and Malvaldi (1984, 1985) have shown that FMO plays an essential role in the bioactivation of thioacetamide and thiobenzamide in the rat. Alternate substrates for FMO capable of competitively inhibiting the sulfoxidation of the thioamides reduced their hepatotoxicity, whereas selective inhibitors for P-450-dependent monooxygenases had no effect.

Aminothiols

The aminothiol, cysteamine, and a few structural analogs produce extensive oxidation of tissue thiols at doses used to protect animals against the lethal effects of ionizing radiation (Bacq, 1975). The aminothiols also produce extensive ulceration of stomach and duodenal mucosa (Selye and Szabo, 1973) and as little as 30 mg/kg cysteamine depletes tissue of prolactin (Millard et al., 1982), and somatostatin in the central nervous system (Sagar et al., 1982). Whether the depletion of prolactin and somatostatin is related to action of aminothiols on tissue thiols is not known, but the synthesis and transport of extracellular peptides may well depend on the integrity of the cellular thiol:disulfide balance.

On the other hand, there is little doubt that the toxicity of aminothiols is due largely to their rapid oxidation to disulfides catalyzed exclusively by FMO (Ziegler et al., 1983). As indicated by the kinetic constants in Table 4, the toxic radioprotective aminothiols are all excellent substrates for FMO. But unlike most other organic sulfur substrates for FMO, aminothiols are oxidized only to disulfides (equation 14). Further oxidation of the 2-aminodisulfides does not occur since they are present in solution largely as dications and dications are not substrates for any of the mammalian FMO's (Ziegler, 1988).

Table 4. Aminothiol Substrates for the Porcine Liver FMO

Compound	kcat/Km x 10^{-3}
Cysteamine	290
N,N-Dimethylcysteamine	290
Piperazinylethanethiol	60
Guanidinoethanethiol	50

Kinetic constants based on data from Ziegler et al. (1983).

$$2 \; \rangle N^{+}HCH_2CH_2\text{-}SH \xrightarrow[O_2]{NADPH} \rangle N^{+}HCH_2CH_2\text{-}SS\text{-}CH_2CH_2N \; H^{+} \langle \qquad (14)$$

The dicationic disulfides produced by the oxidation of aminothiols via FMO are excellent thiol oxidants capable of oxidizing both GSH and protein thiols. The oxidation of GSH by the metabolically generated dicationic disulfide regenerates the aminothiol along with GSSG (equation 15).

$$(\rangle N^{+}HCH_2CH_2S\text{-})_2 + 2GSH \rightleftharpoons 2 \; \rangle N^{+}HCH_2CH_2NH \; + \; GSSG \qquad (15)$$

Reoxidation of the aminothiol (equation 14) establishes a futile cycle that catalyzes the oxidation of GSH to GSSG at the expense of NADPH and oxygen. The uncontrolled oxidation of GSH by this mechanism has been demonstrated CCinDD CCvitroDD with purified FMO (Poulsen and Ziegler, 1977) and is undoubtedly the basis for the dramatic increase in the efflux of GSSG into bile in rats treated with cysteamine (Lauterburg and Mitchell, 1981).

The dicationic disulfides are also excellent protein thiol oxidants and readily form mixed disulfides with proteins bearing accessible thiols as indicated by equation 16.

$$(\rangle N^{+}HCH_2CH_2S\text{-})_2 + Prot\text{-}SH \rightleftharpoons Prot\text{-}S\text{-}SCH_2CH_2N^{+} \; H \langle \qquad (16)$$

The oxidation of cellular protein thiols is known to disrupt a number of metabolic systems (Kosower and Kosower, 1978) and the rapid oxidation of aminothiols to their disulfides appears responsible for most, if not all, the deleterious effects of acute doses of these radioprotective agents including the dramatic loss of liver glycogen described by Bacq and Fischer (1953) almost 40 years ago. The oxidation of aminothiols to the toxic disulfides is apparently catalyzed exclusively by FMO and this is perhaps the most clearly documented bioactivation of a class of xenobiotics by FMO.

CONCLUDING REMARKS

The catalytic mechanism of the flavin-containing monooxygenases suggests that these enzymes have evolved to catalyze oxidative detoxication of structurally diverse soft nucleophiles so abundant in food derived from plants (Liener, 1980). Without exception, aqueous extracts of plants commonly consumed by humans contain substrates for the mammalian hepatic FMO (unpublished studies - this laboratory). While the nature of these compounds has not been positively identified, various organic sulfur compounds derived from glucosinolates (Tookey et al., 1980) or endogenous alkaloids (Lovenberg, 1973) are the most likely candidates. Inspection of the structures of xenobiotics prevalent in food from plants suggests that most of the soft nucleophiles would be oxidized to less reactive derivatives, but a few may be bioactivated via FMO. For instance, goitrin present in processed cabbage and allyldisulfide, a common constituent of *Brassica* plants, are substrates for FMO and both are capable of initiating futile metabolic cycles similar to those described above with thiocarbamides and aminothiols.

The potential bioactivation of constituents so common in plant food consumed by humans is somewhat enigmatic because it appears to contradict speculations regarding the role of FMO in the detoxication of naturally occurring xenobiotic soft nucleophiles. However, this dilemma may be more apparent than real for the following reasons. Organic sulfur compounds oxidized to sulfenic acids preferentially react with glutathione and the first product is always the corresponding glutathione-xenobiotic mixed disulfide. While the mixed disulfides can be reduced by GSH they are also excreted into bile. The latter pathway may be a significant route for detoxication and elimination of small amounts of sulfur xenobiotics oxidized to sulfenic acids via FMO. Metabolic cycling resulting in excessive loss of tissue glutathione and consequent oxidation of protein thiols probably requires much higher concentrations of the toxins. The subversion of a pathway designed for detoxication to intoxication may, therefore, depend largely upon the amount of toxin consumed.

REFERENCES

Bacq, A. M. (1975). In *Sulfur Containing Radioprotective Agents* (A. M. Bacq, ed.) p. 8, Pergammon Press, Oxford.

Barker, E. A. and Smukler, E. A. (1972). Altered Microsome Function During Acute Thioacetamide Poisoning. *Mol. Pharmacol.* **8**, 318-326.

Barker, E. A. and Smukler, E. A. (1973). Non-hepatic Thioacetamide Injury: I Thymic Corticol Necrosis. *Am. J. Pathol.* **71**, 409-418.

Beaty, N. S. and Ballou, D. P. (1981a). The Reductive Half-Reaction of Liver Microsomal FAD-Containing Monooxygenase. *J. Biol. Chem.* **256**, 4611-4618.

Beaty, N. S. and Ballou, D. P. (1981b). The Oxidative Half-Reaction of Liver Microsomal FAD-Containing Monooxygenase. *J. Biol. Chem.* **256**, 4619-4625.

Cashman, J. R. and Hanzlik, R. P. (1981). Microsomal Oxidation of Thiobenzamide. A Photometric Assay for the Flavin-Containing Monooxygenase. *Biophys. Res. Commun.* **98**, 147-153.

Chieli, E. and Malvaldi, G. (1984). Role of the Microsomal FAD-Containing Monooxygenase in the Liver Toxicity of Thioacetamide S-Oxide. *Toxicology* **31**, 41-52.

Chieli, E. and Malvaldi, G. (1985). Role of the P-450 and FAD-Containing Monooxygenases in the Bioactivation of Thioacetamide, Thiobenzamides and Their Sulfoxides. *Biochem. Pharmacol.* **34**, 395-396.

Dannan, G. A. and Guengerich, F. P. (1982). Immunochemical Comparison and Quantitation of Microsomal Flavin-Containing Monooxygenase in Various Hog, Mouse, Rat, Rabbit, Dog and Human Tissues. *Mol. Pharmacol.* **22**, 787-794.

Fitzhugh, O. A. and Nelson, A. A. (1948). Liver Tumors in Rats Fed Thiourea or Thioacetamide. *Science* **108**, 626-628.

Frederick, C. B., Mays, J. B., Ziegler, D. M., Guengerich, F. P. and Kadlubar, F. F. (1982). Cytochrome P-450 and Flavin-Containing Monooxygenase-Catalyzed Formation of the Carcinogen N-Hydroxy-2-Aminofluorene and Its Covalent Binding to Nuclear DNA. *Cancer Res.* **42**, 2671-2677.

Hanzlik, R. P., Cashman, J. P. and Traiger, G. J. (1980). Relative Hepatotoxicity of Thiobenzamides and Thiobenzamide-S-Oxides in the Rat. *Toxicol. Appl. Pharmacol.* **55**, 260-272.

Kreiter, P. A., Ziegler, D. M., Hill, K. A. and Burk, R. F. (1984). Increased Biliary GSSG Efflux from Rat Livers Perfused with Thiocarbamide Substrates for the Flavin-Containing Monooxygenase. *Mol. Pharmacol.* **26**, 122-127.

Liener, J. E. (1980). *Toxic Constituents of Plant Foodstuff* (J. E. Liener, ed.) 2nd edition, Academic Press Inc., New York.

Lovenberg, W. (1973). Some Vaso-and Psychoactive Substances in Food: Amines, Stimulants, Depressants and Hallucinogens in *Toxicants Occurring Naturally in Foods* (F. M. Strong, L. Atkin, J. M. Coon, D. W. Fassett, B. J. Wilson and I. A. Wolff) pp. 170-188, National Academy of Sciences, Washington, D.C.

Millard, W. J., Sagar, S. M., Landis, D. M. D., Martin, J. B. and Badger, T. M. (1982). Cysteamine: A Potent and Specific Depletor of Pituitary Prolactin. *Science* **217**, 452-0454.

Nagata, T., Williams, D. E. and Ziegler, D. M. (1990). Substrate Specificities of Rabbit Lung and Porcine Liver Flavin-Containing Monooxygenases: Differences Due to Substrate Size. *Chem. Res. Toxicol.* (submitted).

Pelroy, R. A., Gandolfi, A. J. (1980). Use of a Mixed Function Amine Oxidase for Metabolic Activation in the Ames/Salmonella Assay System. *Mutat. Res.* **72**, 329-334.

Poulsen, L. L., Hyslop, R. M. and Ziegler, D. M. (1979). S-Oxygenation of N-Substituted Thioureas Catalyzed by the Liver Microsomal FAD-Containing Monooxygenase. *Arch. Biochem. Biophys.* **198**, 78-98.

Poulsen, L. L., Taylor, K., Williams, D. E., Masters, B. S. S. and Ziegler, D. M. (1986). Substrate Specificity of the Rabbit Lung Flavin-Containing Monooxygenase for Amines: Oxidation Products of Primary Alkylamines. *Mol. Pharmacol.* **30**, 680-685.

Poulsen, L. L. and Ziegler, D. M. (1977). Microsomal Mixed-Function Oxidase-Dependent Renaturation of Reduced Ribonuclease. *Arch. Biochem. Biophys.* **183**, 563-570.

Poulsen, L. L. and Ziegler, D. M. (1979). The Liver Microsomal FAD-Containing Monooxygenases: Spectral Characterization and Kinetic Studies. *J. Biol. Chem.* **254**, 6449-6455.

Prough, R. A. and Moloney, S. J. (1985). "Hydrazines" in *Bioactivation of Foreign Compounds* (M. M. Anders, ed.) pp. 433-446, Academic Press Inc., New York.

Prough, R. A., Freeman, P. C. and Hines, R. N. (1981). The Oxidation of Hydrazine Derivatives Catalyzed by the Purified Microsomal FAD-Containing Monooxygenase. *J. Biol. Chem.* **256**, 4178-4184.

Prough, R. A. and Ziegler, D. M. (1977). The Relative Participation of Liver Microsomal Amine Oxidase and Cytochrome P-450 in N-Demethylation Reactions. *Arch. Biochem. Biophysic.* **180**, 363-373.

Richter, C. P. (1952). The Physiology and Cytology of Pulmonary Edema and Pleural Effusion Produced in Rats by Alpha-naphthyl Thiourea. *J. Thoracic Cardiovac. Surg.* **23**, 66-91.

Sagar, S. M., Landry, D., Millard, W. J., Badger, T. M., Arnold, M. A. and Martin, J. B. (1982). Depletion of Somatostatin-Like Immunoreactivity in the Rat Central Nervous System by Cysteamine. *J. Neuroscience* **2**, 225-231.

Selye, H. and Szabo, S. (1973). Experimental Model for Production of Perforating Duodenal Ulcers by Cysteamine in the Rat. *Nature* **244**, 458-459.

Smith, R. L. and Williams, R. T. (1961). The Metabolism of Arylthioureas I. The Metabolism of 1,3-Diphenyl-2-thiourea (Thiocarbanilide) and Its Derivatives. *J. Med. Pharma. Chem.* **4**, 97-107.

Sterling, C. J. M. (1974). The Sulfinic Acids and Their Derivatives. *Int. J. Sulfur Chem.* **6**, 277-316.

Tynes, R. E. and Philpot, R. M. (1987). Tissue and Species-Dependent Expression of Multiple Forms of Mammalian Microsomal Flavin-Containing Monooxygenase. *Mol. Pharmacol.* **31**, 569-574.

Tynes, R. E., Sabourin, P. J. and Hodgson, E. (1985). Identification of Distinct Hepatic and Pulmonary Forms of Microsomal Flavin-Containing Monooxygenase in the Mouse and Rabbit. *Biochem. Biophy. Res. Commun.* **126**, 1069-1075.

Williams, D. E., Hale, S. E., Meurhoff, A. S. and Masters, B. S. S. (1984a). Rabbit Lung

Flavin-Containing Monooxygenase: Purification, Characterization and Induction During Pregnancy. *Mol. Pharmacol.* **28**, 381-390.

Williams, D. E., Ziegler, D. M., Nordin, D. J., Hale, S. E. and Masters, B. S. S. (1984b). Rabbit Lung Flavin-Containing Monooxygenase is Immunochemically and Catalytically Distinct from the Liver Enzyme. *Biochem. Biophys. Res. Commun.* **125**, 116-122.

Ziegler, D. M. (1988). Flavin-Containing Monooxygenases: Catalytic Mechanism and Substrate Specificities. *Drug Meta. Revs.* **19**, 1-32.

Ziegler, D. M., Ansher, S. S., Nagata, T., Kadlubar, F. F. and Jakoby, W. B. (1988). N-Methylation: Potential Mechanism for Metabolic Activation of Carcinogenic Primary Arylamines. *Proc. Natl. Acad. Sci. USA* **85**, 2514-2517.

Ziegler, D. M., Poulsen, L. L. and Richerson, R. B. (1983). Oxidative Metabolism of Sulfur-Containing Radioprotective Agents in *Radioprotective and Anticarcinogens* (D. F. Nygaard and M. G. Simic, eds.) pp. 191-202, Academic Press Inc., New York.

FORMATION OF BIOLOGICAL REACTIVE INTERMEDIATES BY PEROXIDASES: HALIDE MEDIATED ACETAMINOPHEN OXIDATION AND CYTOTOXICITY

Peter J. O'Brien, Sumsullah Khan and Samuel D. Jatoe

Faculty of Pharmacy
University of Toronto
Toronto, Ontario, Canada, M5S 2S2

INTRODUCTION

The physiological role of heme containing peroxidases in circulating or tissue infiltrating white cells, polymorphonuclear leukocytes, monocytes and eosinophils is to enable these cells to carry out their function of killing invading bacteria, viruses, parasites, protozoa, etc. or tumor cells. Another role is to inactivate regulators released into the blood stream such as estradiol, leukotrienes and chemotactic factors. The peroxidases in these cells are soluble but are located in granules or lysosomes. They are released into phagocytic vacuoles and/or from the cell when the cells are activated by invading microorganisms. Lactoperoxidase secreted into body fluids from the mammary gland, Zymbal gland, salivary gland, lacrimal gland and Harderian glands, etc may also play a role in killing invading organisms in body fluids. Peroxidases located in the rough endoplasmic reticular membrane and nuclear envelope of other cells are not released from the cell and are probably more concerned with the synthesis of regulators (eg. thyroxine by thyroid peroxidase, prostaglandin by prostaglandin synthetase) or the inactivation of regulators (eg. oestradiol by uterine peroxidase). The peroxidase of resident peritoneal macrophages or Kupffer cells of the liver are also located in the endoplasmic reticulum but are not discharged to phagocytic vacuoles so that their function may not be antimicrobial. Intestinal peroxidase and another uterine peroxidase have been attributed to infiltrated eosinophils whereas spleen peroxidase has been attributed to infiltrated monocytes (Banerjee, 1988). Peroxidases are therefore ubiquitous in tissues either as a result of endogenous activity or as a result of infiltrated white blood cells.

Peroxidases are thought to play a minor role in the metabolism of most drugs or xenobiotics which have largely been attributed to cytochrome P-450 dependent mixed function oxidases and phase II enzyme systems. Recently however, butylated hydroxyanisole fed to rats was found to undergo dimerization attributed to intestinal peroxidase (Guarna et al., 1983). Furthermore, considerable evidence has accumulated suggesting that peroxidases or prostaglandin synthase play a major role in the activation of some xenobiotics to reactive intermediates (reviewed O'Brien, 1984, 1985, 1988 a,b; Meunier, 1987; Marnett and Eling, 1983; Boyd and Eling, 1985; Josephy, 1988). This has received particular attention with respect to the arylamine induced tumor formation in nonhepatic tissues such as bladder and the peroxidase rich secretory Zymbal or Harderian glands. It has been estimated that 20% of the DNA adducts formed in the urothelium of dogs treated with the bladder carcinogen, 2-naphthylamine are estimated to be of peroxidative origin (Yamazoe et al., 1985). Similar results probably exist for the bladder carcinogens benzidine (O'Brien et al., 1985; Yamazoe et al., 1989) or aminofluorene (Krauss et al., 1989).

Biological Reactive Intermediates IV, Edited by C.M. Witmer *et al.*
Plenum Press, New York, 1990

The antimicrobial and antitumor activity of peroxidases is believed to be mediated by halogenating oxidants formed by the peroxidase catalyzed oxidation of physiological halides (Weiss, 1986).

$$H_2O_2 + X^- + H^+ \rightarrow HOX + H_2O$$
$$(X^- = Cl^-, Br^-, I^- \text{ and } SCN^-)$$

The plasma contains approximately 0.1M chloride, 0.1 mM bromide, 1μM iodide and 0.12mM thiocyanate. With neutrophils, chlorination dominates at a rate of approx. 100 nmol HOCl. formed per 10^6 cells hr^{-1} (Grisham et al., 1981; Thomas et al., 1983). Bromination with neutrophils only becomes significant at bromide concentrations (>1mM) whereas eosinophils brominate at physiological bromide concentrations at a rate of approx. 40 nmol hypohalite formed per 10^6 cells. hr^{-1} (Mayeno et al., 1989). This halide specificity reflects the properties of myeloperoxidase found in neutrophils or monocytes versus eosinophil peroxidase found in eosinophils. The hypohalites formed and released are believed to be responsible for the antimicrobial function of the neutrophil, monocyte or eosinophil. It is ironic that we treat our municipal sewage and other waste waters with the same chlorine derived halogenating oxidants used physiologically against microorganisms.

Whilst the halides seem to play a major role in the antimicrobial activity and inactivation of regulators catalysed by mammalian peroxidases, the role of halides in xenobiotic activation remains to be explored. Activated polymorphonuclear leukocytes have been shown to catalyse the oxidation or binding of nucleic acids of some antithyroid drugs, carcinogenic arylamines, nonsteroidal anti-inflammatory drugs, procainamide and acetaminophen (Tsuruta et al., 1985; Corbett et al., 1989; O'Brien 1988; Uetrecht, 1988, 1989). However, the role of halides in the leukocyte catalysed oxidations of these drugs has not been investigated. Although chlorination of traps by leukocytes have been reported (Foote et al., 1983; O'Brien, 1988), no chlorination of these drugs has been reported, presumably because any intermediates involved, eg. N-chloramines, react with neutrophil components. In the following, the effect of halides on the mammalian peroxidase catalysed oxidation of drugs to cytotoxic reactive metabolites has been investigated. Acetaminophen was the drug selected for study.

Acetaminophen is a widely used over the counter analgesic/antipyretic drug which can cause fatal liver necrosis if taken in a large overdose (Prescott, 1983). The major biological reactive metabolite believed to be responsible for the hepatotoxicity is N-acetyl-p-benzoquinoneimine (Dahlin et al., 1984). The latter may initiate cytotoxicity by covalent binding to and oxidation of cysteinyl thiol groups in essential proteins. Cytochrome P-450 is believed to catalyse the oxidation of acetaminophen to quinoneimine although in vitro experiments with microsomal: NADPH systems or reconstituted mixed function oxidase systems give a maximum conversion of 0.5-2% even when GSH is used to trap the quinoneimine and the most active β-naphthoflavone induced cytochrome P-450 is used (Potter and Hinson, 1987a; Harrison et al., 1988). Presumably the low yield formed reflects the small amount of acetaminophen was oxidised. Peroxidase-H_2O_2 systems (Potter and Hinson, 1987a) also formed little quinoneimine even though most of the acetaminophen was oxidised. This was largely because most of the acetaminophen was oxidised by a one-electron oxidation to an oxygen centered phenoxyl free radical (Mason and Fischer, 1986) which underwent far more dimerization and polymerization than disproportionation to N-acetyl-p-benzoquinoneimine and acetaminophen (Potter and Hinson, 1987a). The low yield of GSH conjugate formed also reflects futile redox cycling of the phenoxy radicals by GSH resulting in extensive GSH oxidation (Ross et al., 1985). Little quinoneimine was also formed with prostaglandin H synthase and arachidonate even though both a one- and two -electron oxidation of acetaminophen occurred (Potter and Hinson, 1987b). The analgesic, antipyretic action of acetaminophen has been explained by the inhibition of prostaglandin synthetase that occurs when acetaminophen undergoes cooxidation at low peroxide tone (Hanel and Lands, 1982). Activation of acetaminophen by prostaglandin synthetase located in the renal papilla has also been implicated in the chronic analgesic nephropathy associated with the chronic abuse of analgesic mixtures that contain acetaminophen (Moldeus and Rahimtula, 1980; Mohandas et al., 1981; Moldeus et al., 1982; Boyd and Eling, 1981).

The research described in the following shows that acetaminophen is highly toxic to isolated hepatocytes when oxidised by mammalian peroxidases in the presence of physiological concentrations of chloride and bromide as a result of stoichiometric oxidation to quinoneimine. Acetaminophen is not toxic when oxidised by peroxidases in the absence of halides.

METHODS

Myeloperoxidase (human leukocyte) (MPO) was supplied by Alpha Therapeutics Corporation (Los Angeles, CA). Horseradish peroxidase type VI (HRP), sodium bromide, H_2O_2 and hypochlorite were purchased from Sigma Chemical Co. (St. Louis, MO). Acetaminophen was obtained from Aldrich Chemical Co. (Milwaukee, WI). N-Acetylbenzoquinoneimine was obtained from Dalton Chemicals, Toronto. 3-Chloroacetaminophen and 3,5-dichloroacetaminophen were synthesized by reducing 2-chloro-4-nitrophenol and 2,6-dichloro-4-nitrophenol to the corresponding aminophenol followed by acetylating the amino group as described (Fernando et al., 1980).

Male Sprague-Dawley rats (200-250 g body weight) were used throughout the experiments. Hepatocytes were isolated by collagenase perfusion of the liver as described (Moldeus et al., 1978). Isolated hepatocytes (10^6 cells per ml) were incubated in 10ml Krebs-Henseleit buffer, pH 7.4, in rotating round bottom flasks at 37°C under a carbogen atmosphere. Cell viability was estimated by the Trypan Blue exclusion method. Where indicated, the reaction mixture was preincubated before addition to the isolated hepatocytes. However normally, unless otherwise shown, H_2O_2 (0.1 mM) was added last to the incubation mixture containing hepatocytes (10^6 cells/ml), acetaminophen, horseradish peroxidase type VI (10 µg/ml) or myeloperoxidase (1 unit/ml), and where indicated bromide or chloride. The spectral changes associated with a standard incubation mixture without hepatocytes were recorded (200-500 nm) on a Shimadzu UV 240 spectrophotometer at various times.

The oxidation of acetaminophen was carried out by reacting acetaminophen (100µM) with HOCl (100µM) in the presence or absence of KBr(100µM) or KI(10µM) in distilled water (1ml) at room temperature. After about 5 minutes, aliquots (10µl) were analyzed with a Shimadzu LC-6A HPLC system. The mixtures were separated on a Waters µBondapak C_{18} column (25cm x 46mm) using a methanol/water gradient of 10% to 100% over 25 minutes and UV detection at 260nm.

To isolate compounds for mass spectrometry, acetaminophen (2mM) was reacted with HOCl (2mM) over 10 minutes in distilled water at room temperature. The mixture was extracted twice with 3 volumes of diethyl ether and the organic phase was evaporated to dryness. The residue was dissolved in methanol (2ml) and purified by preparative HPLC by injecting aliquots (100µl) on a Whatman Partisil M9 10/50 ODS column. The sample was eluted with acetonitrile/water mixture at a flow rate of 3 ml/min using a gradient of 20% acetonitrile to 50% acetonitrile over 20 minutes. Isolation of compounds was achieved by collecting the column effluent corresponding to the peaks of interest. The acetonitrile in each collected fraction was evaporated under nitrogen gas and the remaining aqueous phase was frozen in liquid nitrogen and lyophilized. The dry powder was dissolved in methanol for mass spectral determination. Mass spectra were obtained with a VG Zab -SE Mass Spectrometer operated in the electron ionization mode at 70eV with a source temperature of 50-300°C and a trap current of 100µA.

RESULTS

As shown in Table 1, acetaminophen was not toxic to isolated rat hepatocytes even at a concentration of 20mM. Furthermore, acetaminophen (1mM) was not cytotoxic even when oxidized in the presence of hepatocytes with H_2O_2 and horseradish peroxidase to phenoxy radicals, dimers and polymers. Preincubation of acetaminophen (5 mM) with H_2O_2 (5 mM) and horseradish peroxidase for 5 minutes before addition to hepatocytes was also not cytotoxic indicating the dimeric and polymeric products were not cytotoxic even at high concentrations. Acetaminophen was also not cytotoxic when oxidized by H_2O_2 and myeloperoxidase. However, the acetaminophen oxidation products were highly cytotoxic if oxidised by H_2O_2 and myeloperoxidase in the presence of bromide. Acetaminophen also became toxic when oxidized by hypochlorite particularly in the presence of physiological concentrations of bromide. It can also be seen that control hepatocytes were not affected by these concentrations of hypochlorite with or without bromide in the absence of acetaminophen even at higher concentrations than were used with acetaminophen above.

The identity of the cytotoxic intermediates and products formed from acetaminophen by H_2O_2 and myeloperoxidase with physiological concentrations of bromide was determined spectrally. As shown in Figure 1a, N-acetylbenzoquinoneimine (at 263mμ) was the principal product formed. With chloride present instead of bromide less oxidation occurred and the N-acetylbenzoquinoneimine formed was converted to benzoquinone (Figure 1b).

The acetaminophen oxidation products formed when acetaminophen was oxidised by 1-5 equivalents of hypochlorite determined by HPLC chromatography are shown in Figure 2a-c. Identification of the peaks was obtained by running synthetic standards. In addition the various HPLC fractions were collected and mass spectrometry was used to confirm their identification. As shown in Figure 3, the EI mass spectrum of peak 5

Table 1. Activation by Hypohalites Versus Peroxidase

Additions	Cytotoxicity			
Incubation time (mins)	60'	120'	180'	240'
None	15 ± 2	18 ± 2	19 ± 2	20 ± 2
Acetaminophen (20 mM)	17 ± 3	18 ± 2	19 ± 3	19 ± 3
Acetaminophen (1 mM) +HRP + H_2O_2 (1 mM)	18 ± 3	21 ± 2	22 ± 3	22 ± 4
Radical coupling products: (5 mM)	15 ± 4	18 ± 4	19 ± 3	20 ± 4
Acetaminophen (0.5 mM) + HOC1 (0.3 mM)	21 ± 2	23 ± 3	25 ± 3	26 ± 3
Acetaminophen (0.3 mM) + HOC1 (0.3 mM) + KBr (0.3 mM)	38 ± 3	45 ± 4	66 ± 5	92 ± 4
Acetaminophen (0.5 mM) + MPO + H_2O_2 (0.5 mM) + KBr (0.5 mM)	33 ± 3	48 ± 5	81 ± 8	92 ± 4
N-acetylbenzoquinoneimine (0.3 mM)	37 ± 3	47 ± 4	63 ± 6	96 ± 4
Benzoquinone (0.15 mM)	42 ± 4	56 ± 5	76 ± 7	96 ± 4
HOC1 (1 mM)	15 ± 3	18 ± 3	19 ± 3	20 ± 3
HOC1 (0.6 mM) + KBr (0.6 mM)	15 ± 3	18 ± 3	17 ± 3	20 ± 4

Reaction conditions: Isolated hepatocytes (10^6 cells/ml) in Krebs-Henseleit buffer. Results (mean ± SE) are the average of three determinations with different batches of cells.

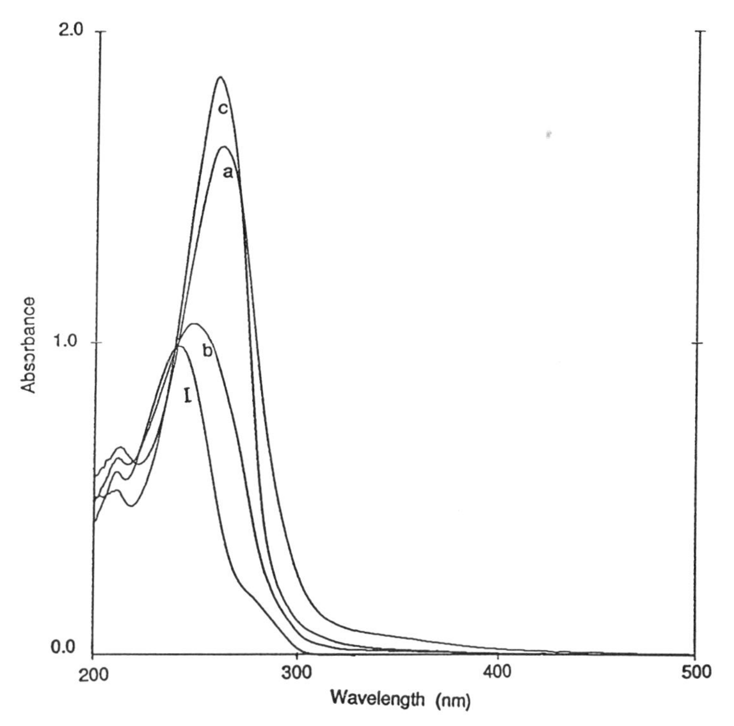

Figure 1. Oxidation of acetaminophen (I). Reaction conditions: acetaminophen (0.1mM) and (a) potassium bromide (0.1mM) MPO (1 unit/ml) and H_2O_2 (0.2mM); (b) potassium chloride (0.1M), MPO (1 unit/ml) and H_2O_2 (0.2mM); (c) potassium bromide (0.1mM) and hypochlorite (0.1mM) in 3.0 ml of 50mM acetate buffer, pH 5.0 at 20°C. The spectra were recorded 1-2 minutes after the addition of either H_2O_2 or hypochlorite.

was found to be identical to that of synthetic 3,5-dichloroacetaminophen. The base ion peak in both spectra was the [M-42] at m/z 177 which originated from the molecular ion M^+ at m/z 219. About 50% of the acetaminophen was oxidised and chlorinated by 1 equivalent of hypochlorite to benzoquinone, N-acetyl-p-benzoquinoneimine, 3-chloroacetaminophen and a lesser amount of 3,5-dichloroacetaminophen. Two equivalents of hypochlorite were required to oxidize and chlorinate about 90% of the acetaminophen and more benzoquinone and ring chlorinated products were formed. With 5 equivalents of hypochlorite, however most of the acetaminophen, had been converted to 3,5 dichloroacetaminophen. The addition of ascorbate to the reaction mixture before separating the products by HPLC resulted in the reduction of N-acetyl-p-benzoquinoneimine to acetaminophen, the disappearance of the benzoquinone peak and no change in the chloro-and dichloroacetaminophen peaks.

The presence of one equivalent of bromide, however, increased the yield of N-acetyl-p-benzoquinoneimine formed from 17% to 80% when acetaminophen was oxidized with one equivalent of HOCl. This yield was determined by HPLC chromatography (Figure 2d) and UV absorption spectroscopy (Figure 1c). When iodide was used instead of bromide a nearly stoichiometric conversion to N-acetylbenzoquinoneimine was obtained. The effect of halide concentration on the yield of quinoneimine when acetaminophen is oxidized by HOCl is shown in Figure 4.

Chloride was without effect whereas iodide was so effective that only 0.25 equivalents were required. The latter stoichiometry suggests that iodide recycling occurs.

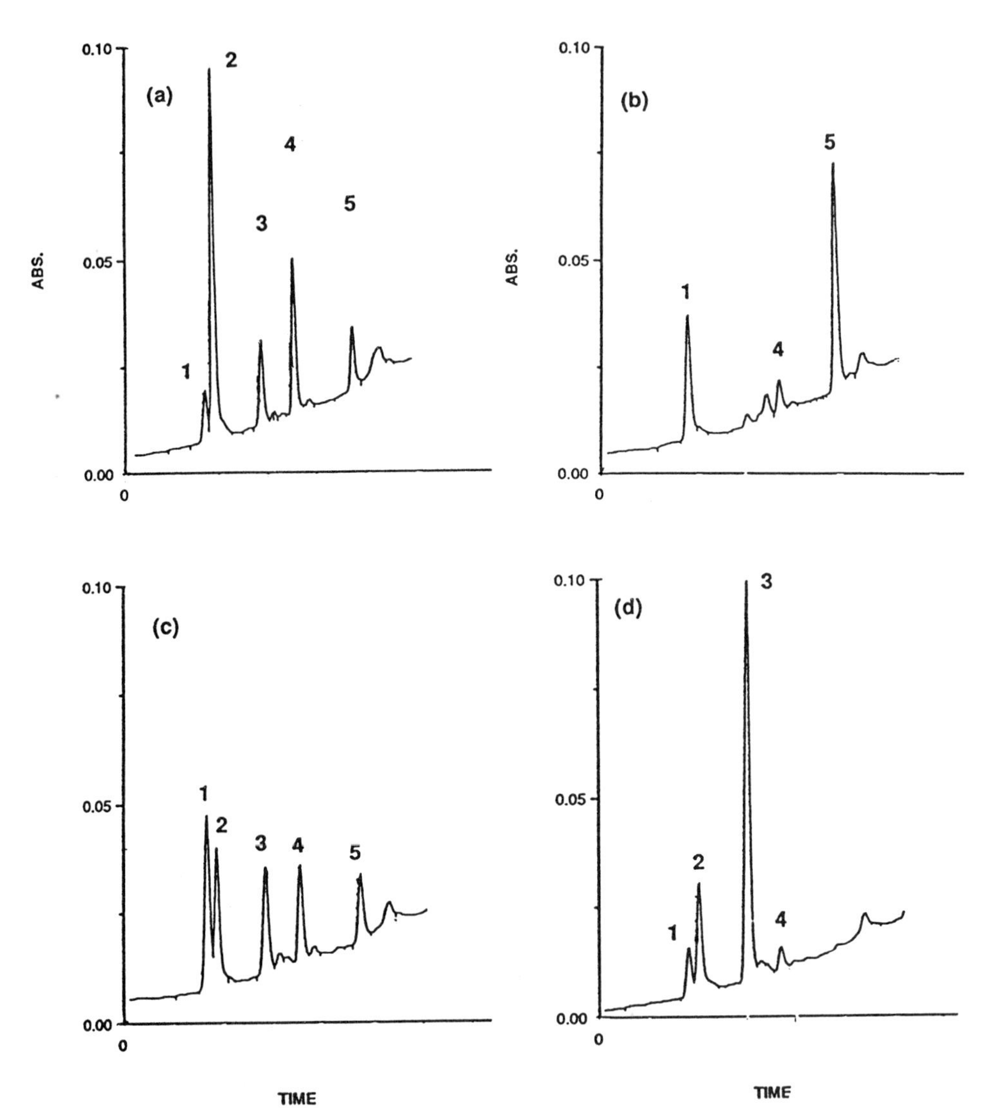

Figure 2. HPLC profiles of acetaminophen oxidation products by HOCl (a-c) or the HOCl/Br^- system (d).

Identification of Peaks:
1. benzoquinone (10.4 min)
2. acetaminophen (10.9 min)
3. N-acetyl-p-benzoquinoneimine (12.9 min)
4. chloroacetaminophen (14.3 min)
5. dichloroacetaminophen (16.7 min)

Chromatographic conditions: as described in Methods.

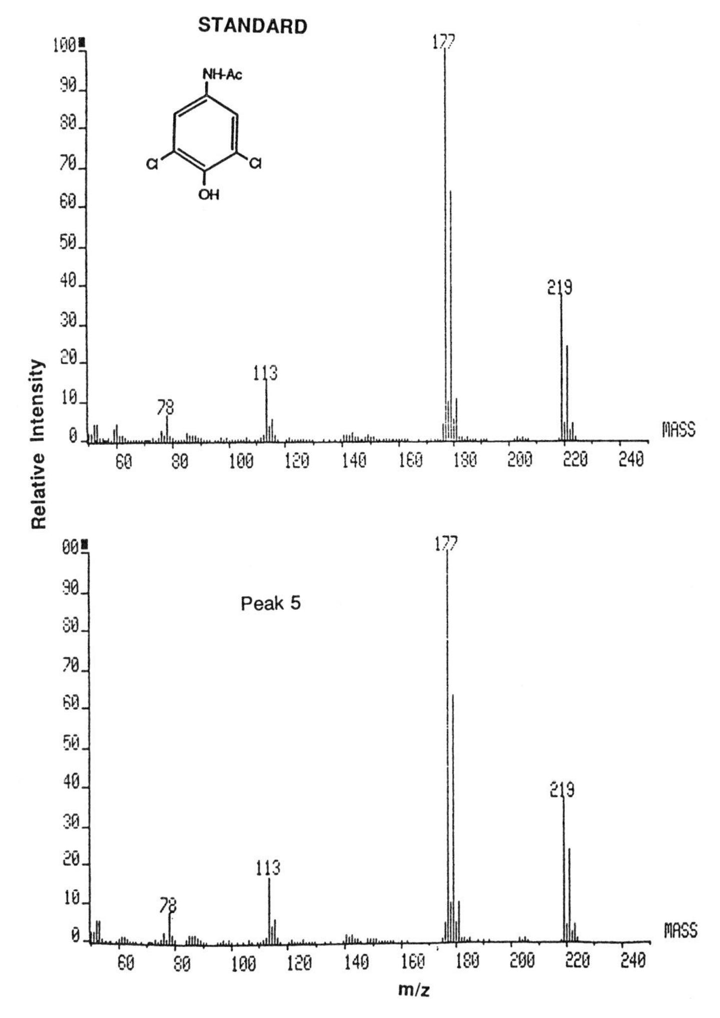

Figure 3. Mass spectra of 3,5-dichloroacetamenophen and HPLC peak 5 isolated from the acetaminophen + HOCl reaction mixture. Mass spectra were obtained as described in Methods.

As shown in Table 1, cytotoxicity was also most marked when acetaminophen was treated with hypochlorite + bromide or iodide than with hypochlorite alone. Cytotoxicity of the acetaminophen + hypochlorite + bromide or iodide system could also be accounted for by N-acetyl-p-benzoquinoneimine formation whereas cytotoxicity of the acetaminophen + hypochlorite system could be accounted for by benzoquinone formation.

DISCUSSION

The lack of cytotoxicity when rat hepatocytes are incubated with high concentrations of acetaminophen indicate that cytochrome P-450 dependent mixed function oxidase activity in hepatocytes is not sufficient to cause acetaminophen

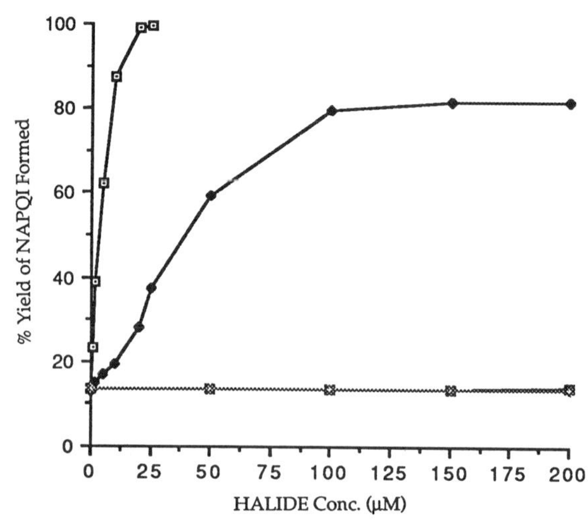

Figure 4. Effect of various concentrations of halides on the formation of N-acetylbenzoquinoneimine. Reaction conditions: acetaminophen (0.1 mM), hypochlorite (0.1 mM) and halides (0-0.2mM) in 3.0 ml phosphate buffer, PH 7.4 at 20°C. The formation of N-acetylbenzoquinoneimine was monitored spectrophotometrically at 263 nm. ( ) HOCl; ( ) HOCl + Cl^-; ( ) HOCl + Br^- and ( ) HOCl + I^-.

toxicity over the five hour incubation period studied. Other investigators have shown that isolated hepatocytes can be susceptible to acetaminophen if prepared from rats in which hepatic P-450 isozymes have been induced by pretreatment of the animals with 3-methylcholanthrene or phenobarbital (Porubek et al., 1987).

Hydrogen peroxide and myeloperoxidase or horseradish peroxidase in the absence of chloride or bromide did not oxidise acetaminophen to cytotoxic products indicating that phenoxy radicals (Mason and Fischer, 1986) or radical coupling products (Potter and Hinson, 1987b) were not cytotoxic. Thus no cytotoxicity was observed if the hepatocytes were present during the peroxidase catalysed oxidation of acetaminophen (1mM) even though most of the acetaminophen was oxidised. The stable radical coupling products dimers or "melanin" polymers were also not cytotoxic. For this experiment, acetaminophen (5mM) was oxidised with H_2O_2 (5mM) and horseradish peroxidase 5 minutes before addition to hepatocytes.

Marked cytotoxicity occurred when acetaminophen was oxidised by myeloperoxidase, H_2O_2 and bromide but not chloride. The oxidation of acetaminophen to N-acetylbenzoquinoneimine when the peroxidase oxidising system contains physiological concentrations of bromide can be attributed to a change from a free radical mechanism involving a one electron oxidation of acetaminophen to a non-radical peroxidase catalysed two electron oxidation of acetaminophen by hypobromite formed by the peroxidase catalysed oxidation of bromide (Fig. 5). Evidence for this is that methionine (1mM) a hypohalite trap prevented the acetaminophen oxidation and cytotoxicity by the peroxidase, H_2O_2, bromide system. Recently a similar change in mechanism from a one electron oxidation mechanism to a two electron oxidation mechanism was reported when chloride or bromide was added to a myeloperoxidase-

H_2O_2-N-hydroxy-N-(2-fluorenyl)acetamide system. The oxidative cleavage of the N-acetyl group observed was much faster with bromide than chloride (Ritter and Malejka-Giganti, 1989). We have also obtained a similar change in peroxidase mechanism for the lactoperoxidase or myeloperoxidase catalysed oxidation of N-hydroxy-phenacetin or p-phenetidine when halides are present (O'Brien, 1990a,b).

Evidence on the halogenating mechanisms involved were obtained from studies of the reaction between acetaminophen and hypochlorite. One equivalent of hypochlorite oxidised and chlorinated about 50% of the acetaminophen at pH 7.4 to N-acetyl-p-benzoquinoneimine, benzoquinone and the ring chlorinated products, 3-chloroacetaminophen and a lesser amount of 3,5-dichloroacetaminophen. However addition of 1 equivalent of bromide or 0.25 equivalents of iodide resulted in the oxidation of 80-95% respectively of the acetaminophen to N-acetyl-p-benzoquinoneimine without the formation of benzoquinone or ring halogenated products. Cytotoxicity correlated with oxidation to N-acetyl-p-benzoquinoneimine as cytotoxicity was much more marked when acetaminophen was treated with hypochlorite + bromide or iodide than with hypochlorite alone. Cytotoxicity of the acetaminophen + hypochlorite + bromide system was also similar to that found for N-acetyl-p-benzoquinoneimine whereas the cytotoxicity of the acetaminophen + hypochlorite system appears to be due to the formation of benzoquinone. The other products $3,5(Cl)_2$ acetaminophen and 3Cl acetaminophen were not toxic to hepatocytes at 1mM (results not shown) and sufficient benzoquinone was formed to account for the cytotoxicity. The requirement of only 0.25 equivalent of iodide for N-acetyl-p-benzoquinoneimine ($Ph = NCOCH_3$) formation from acetaminophen ($Ph.NHCOCH_3$) is particularly interesting and suggests that active halide recycling occurs as described by the two following reactions:

$$HOCl + I^- \rightarrow HOI + Cl^-$$
$$Ph.NHCOCH_3 + HOI \rightarrow Ph = NCOCH_3 + I^- + H_2O$$

Hydroquinones are also oxidised to benzoquinones by hypochlorite without ring chlorination (Rice and Gomez-Taylor, 1986). However although the effectiveness of hypohalites at oxidising acetaminophen is in the order HOI > HOBr > HOCl, the order of oxidation potentials at 25°C is HOCl (1.49) > HOBr (1.33) > HOI (0.99) (Rice and Gomez-Taylor, 1986). The order of oxidative effectiveness therefore corrrelates with the order of nucleophilicity ie. HOI > HOBr > HOCl (March, 1988) and thus N-halogenating ability. The following two reactions may therefore be involved.

$$Ph.NHCOCH_3 + HOBr \rightarrow Ph.N(Br)COCH_3 + H_2O$$
$$Ph.N(Br)COCH_3 \rightarrow Ph = N.COCH_3 + Br^- + H^+$$

Halide ions are good leaving groups in the latter reaction because they are very weak bases. The leaving group ability of halides is also in the order $I^- > Br^- > Cl^-$ (March, 1988) and would therefore explain the halide specificity of acetaminophen oxidation.

Recently Harvison et al (1988) have proposed that cytochrome P-450 oxidises the amide nitrogen of acetaminophen by a one-electron abstraction followed by rapid recombination of the radical pair to form a cytochrome P-450 hemesite-bound hydroxamic acid which dehydrates to generate N-acetyl-p-benzoquinoneimine. Peroxidases however probably catalyse a one electron oxidation of the phenolic oxygen to form a phenoxy radical which dimerises or to a lesser degree disproportionates to N-acetyl-p-benzoquinoneimine (Potter and Hinson 1987a) as shown in Figure 5a. In Figure 5b a mechanism for the oxidation of acetaminophen by hypobromite is outlined. The first step involves an attack on the acetaminophen amide nitrogen by hypobromite to form an unstable intermediate N-bromoacetaminophen. This then eliminates HBr to give N-acetyl-p-benzoquinoneimine as the major product. A mechanism for the chlorination of acetaminophen by hypochlorite observed in the absence of bromide or iodide is outlined in Figure 5c. No ring chlorination occurred when hypochlorite was added to N-acetyl-p-benzoquinoneimine. It is concluded that ring chlorination occurred either by an intramolecular rearrangement of N-chloroacetaminophen and/or by direct ring chlorination as occurs with phenol (Rice and Gomez-Taylor, 1986). Benzoquinone may be formed by hydrolysis of N-acetyl-p-benzoquinoneimine with elimination of acetamide via an ipso attack by water to form an intermediate carbinolamide (Novak et al., 1986). However, a much faster conversion by hypobromite would be expected if hypochlorite was acting as a nucleophile. Why N-acetyl-p-benzoquinoneimine is more stable with hypobromite than hypochlorite is not clear.

Peroxidases in the liver seem to be located in the endoplasmic reticulum and nuclear envelope of the phagocytic reticuloendothelial cells (Kupffer cells) which line the bile ducts in the liver (Fahimi et al., 1976) and form as much as 14% of the liver. These non-parenchymal cells play an essential role as a selective filter to rapidly remove blood-borne abnormal particulate matter and gut derived antigens (eg. endotoxin from the portal circulation), primarily through the process of phagocytosis. Kupffer cells contain 30 times higher peroxidase specific activity than the hepatocytes (Van Berkel, 1974) and could therefore play a role in the peroxidase catalysed activation of acetaminophen. The reduced spectra of this peroxidase is also similar to myeloperoxidase (Fahimi et al., 1976) which could also indicate that the peroxidase can form hypochlorite although it is not known whether the peroxidase is released from Kupffer cells.

Mononuclear phagocytes however have the ability to release peroxidases and can be recruited into the liver and activated by bacterial, virus or parasitic infections; lipopolysacchloride endotoxins; or by autoimmune responses to liver antigens (Ferluga and Allison, 1978). This may arise because damaged cells in the liver release chemotactic and activating factors towards polymorphonuclear leukocytes, Kupffer cells and monocytes (Laskin et al., 1986), particularly if sensitised by previous infections. In response to these inflammatory stimuli, the peroxidase containing mononuclear phagocytes migrate into the liver from the blood and bone marrow and become activated. These cells have more peroxidase than resident Kupffer cells. They are also 10-15 times more phagocytic when activated, and release 30% more hydrogen peroxide and other activated oxygen species than resident Kupffer cells (Pilaro and Laskin, 1986; Laskin et al, 1988). They may also promote hepatic damage e.g., hepatitis through the release of toxic secretory products (Ferluga and Allison, 1978). Laskin et al. 1986, have proposed that the release of activated oxygen species contributes to hepatic injury, tumor promotion and carcinogenesis.

Treatments of rats with acetaminophen results in the accumulation of activated mononuclear phagocytes in the centrilobular regions of the liver within 24 hours in the absence of necrosis. Levels of migration of mononuclear phagocytes were 4-7 times greater than the migration of resident Kupffer cells. Furthermore, cultured hepatocytes treated with acetaminophen released chemotactic and activating factors towards Kupffer cells and monocytes (Laskin et al., 1986). In mice, Jaeschke and Mitchell, (1988), have found a rapid 15-fold increase in neutrophil accumulation between 1.5 and 3 hours post-dose i.e., after glutathione depletion and macromolecule alkylation. They suggested this caused necrosis as a result of plugging the hepatic microvasculature which resulted in ischemic infarction. A variety of other agents including ethanol, phenobarbital, carbon tetrachloride, halothane, phenytoin, and para-aminosalicylate also induce the accumulation of mononuclear phagocytes in the periportal region but unlike acetaminophen is concomitant with the induction of zone-specific injury or hepatitis. Phenobarbital or endotoxin also induces the accumulation of activated mononuclear phagocytes in the liver (Laskin et al., 1988; Pilaro and Laskin, 1986). The increased susceptibility of alcoholics to acetaminophen toxicity (McClain et al., 1980; Licht et al., 1980) has been attributed to decreased hepatic glutathione levels (Lauterburg and Veley, 1988) or to the induction of a P-450 isoenzyme that activates acetaminophen (Sato et al., 1981). However, distinctive cytological features of alcoholic hepatitis is the infiltration of polymorphonuclear leukocytes associated with necrotic and Mallory-body-containing liver cells (MacSween, 1981). Infiltrated or recruited mononuclear phagocytes could be highly effective at causing liver necrosis as a result of causing the oxidation of acetaminophen by hypohalites.

Recently eosinophils have been shown to preferentially utilize bromide to generate active bromine in the presence of at least a 1000-fold excess of chloride

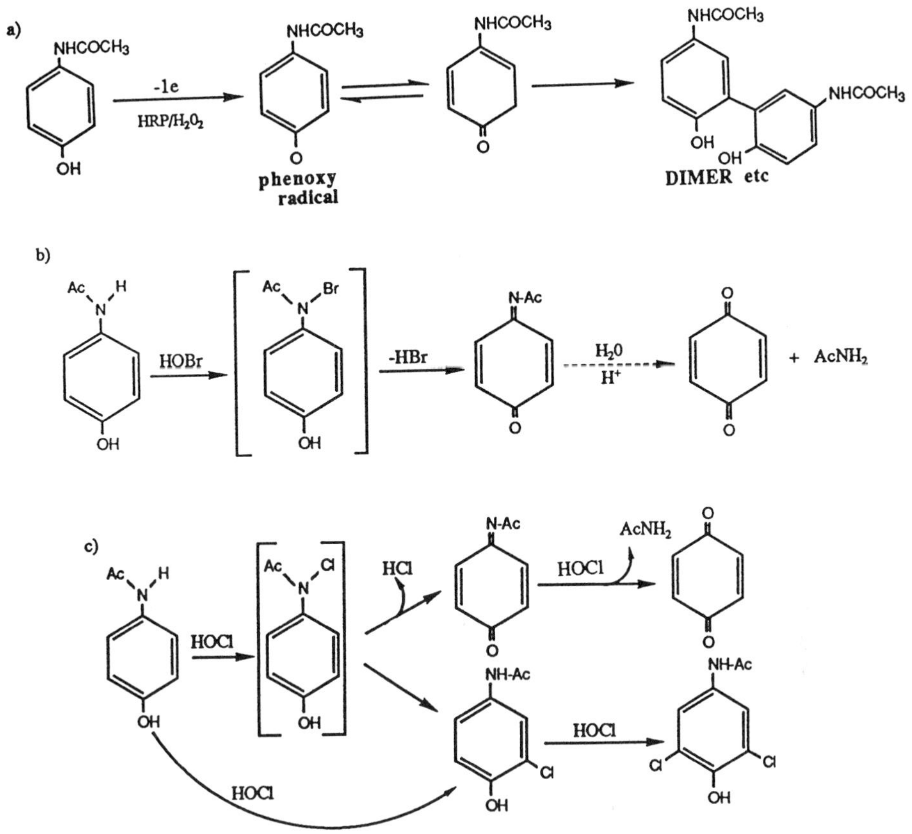

Figure 5. Acetaminophen activation mechanism(s): (a) -1e oxidation; (b) HOBr and (c) HOCl.

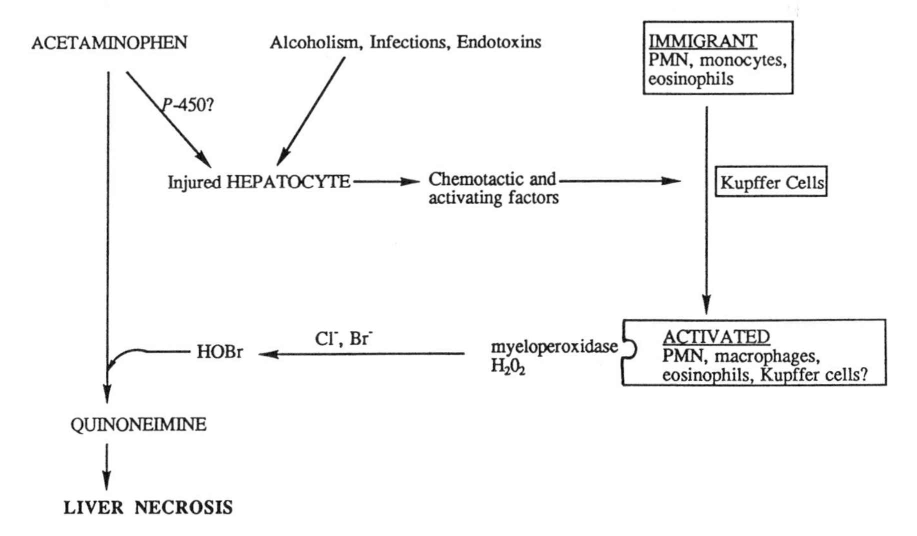

Figure 6. A Peroxidase mechanism for acetaminophen hepatotoxicity involving parenchyman and non-parenchyman cells.

(Mayeno et al., 1989). Myeloperoxidase has different properties so that polymorphonuclear leukocytes do not preferentially utilise bromide. However, the results presented here suggest that eosinophils or polymorphonuclear leukocytes by generating hypochlorite or hypobromite in the presence of physiological concentrations of chloride and bromide could oxidise acetaminophen stoichiometrically to the highly cytotoxic N-acetylbenzoquinoneimine. In Figure 6 a mechanism for acetaminophen hepatotoxicity involving oxidation-halogenation by activated infiltrated polymorphonuclear leukocytes, macrophages or eosinophils is outlined. The mechanism would predict an enhanced susceptibility of the liver to acetaminophen induced necrosis in alcoholism, viral hepatitis or parasitic infection, etc. Whether the peroxidases of nonparenchymal cells or infiltrating white cells also play a role in acetaminophen induced necrosis of the normal liver however is much more speculative.

ACKNOWLEDGEMENTS

This research was supported by the Natural Science and Engineering Council of Canada.

REFERENCES

Banerjee, R.K. (1988). Membrane peroxidases. *Molec. Cell. Biochem.* **83**, 105-128.

Boyd, J.A. and Eling, T.E. (1981). Prostaglandin endoperoxide synthetase dependent cooxidation of acetaminophen to intermediates which covalently bind to rabbit renal medullary microsomes. *J. Pharmacol. Exp. Ther* **219**, 659-664.

Boyd, J.A. and Eling, T.E. (1985). Metabolism of aromatic amines by prostaglandin H sunthase. *Environ. Health Perspect.* **64**, 45-51.

Corbett, M.D. and Corbett, B.R. (1988). Metabolic activation and nucleic acid binding of acetaminophen and related amines during the respiratory burst of human gramlocytes. *Chem. Res. Toxicol.* **2**, 260-267.

Dahlin, D.C., Miwa, G.T., Lu, A.Y.H. and Nelson, S.D. (1984). N-acetyl-p-benzoquinoneimine: a cytochrome P-450 mediated oxidation product of acetaminophen. *Proc. Natl. Acad. Sci.* **81**, 1327-1331.

Fahimi, H.D., Gray, B.A. and Herzog, V.K. (1976). Cytochemical localization of catalase and peroxidase in sinusoidal cells of rat liver. *Laboratory Investigation* **34**, 192-201.

Fernando, C.R., Calder, I.C. and Ham, K.N. (1980). Studies on the mechanism of toxicity of acetaminophen. Synthesis and reactions of N-acetyl-2,6-dimethyl-p-benzoquinone imines. *J. Medic. Chem.* **23**, 1153-1158.

Ferluga, J. and Allison, A. (1978). Role of mononuclear infiltrating cells in pathogenetics of hepatitis. *Lancet* **2**, 610-611.

Foote, C.S., Goyne, T.E. and Lehrer, R.I. (1983). Assessment of chlorination by human neutrophils. *Nature* **301**, 715-716.

Grisham, M.B., Jefferson, M.M., Melton, D.F. and Thomas, E.L. (1984). Chlorination of endogenous amines by isolated neutrophils. *J. Biol. Chem.* **259**, 10404-10412.

Guarna, A., Corte, L.D., Giovannini, M.G., De Sarlo, F. and Sgaragli, G. (1983). A dimer metabolite of 2t-butyl-4-methoxyphenol in the rat. *Drug Metab. Dispos.* **11**, 581-584.

Hanel, A.M. and Lands, W.E.M. (1982). Modification of anti-inflammatory drug effectiveness by ambient lipid peroxides. *Biochem. Pharmacol.* **31**, 3307-3311.

Harvison, P.J., Guengerich, F.P., Rashed, M.S. and Nelson, S.D. (1988). Cytochrome P-450 isozyme selectivity in the oxidation of acetaminophen. *Chem. Res. Toxicol.* **1**, 47-52.

Jaeschke, H. and Mitchell, J.R. (1988). Acetaminophen toxicity. *New Engl. J. Med.* **319**, 1601-1602.

Josephy, D. (1988). Activating aromatic amines by prostaglandin synthetose. *Free Rad. Biol. Med.* **6**, 533-542.

Krauss, R.S., Angerman-Stewart, J., Eling, T.E., Dooley, K.L. and Kadlubar, F.F. (1989). The formation of 2-aminofluorene-DNA adducts in vivo: evidence for peroxidase-mediated activation. *J. Biochem. Toxicol.* **4**, 111-117.

Laskin, D.L., Robertson, F.M., Pilaro, A.M. and Laskin, J.D. (1988). Activation of liver macrophages following phenobarbital treatment of rats. *Hepatology* **8**, 1051-1055.

Laskin, D.L., Pilaro, A.M. and Sungchul, J. (1986). Potential role of activated macrophages in acetaminophen hepatotoxicity. *Toxicol. Appl. Pharmacol.* **86**, 216-226

Lauterburg, B.H. and Velez, M.E. (1988). Glutathione deficiency in alcoholics: risk factor for paracetamol hepatotoxicity. *Gut.* **29**, 1153-1157.

Licht, H., Seeff, I.B. and Zimmermann, H.J. (1980). Apparent potentiation of acetaminophen hepatotoxicity by alcohol. *Ann. Intern. Med.* **92**, 511.

Mason, R.P. and Fischer, V. (1986). Free radicals of acetaminophen: their subsequent reactions and toxicological significance. *Federation Proc.* **45**, 2493-2499.

McClain, C.J., Kromhout, J.P., Peterson, F.J. and Holtzman, J.L. (1980). Potentiation of paracetamol hepatotoxicity by alcohol. *JAMA* **244**, 251-253.

MacSween, R.N.M. (1981). Alcoholic liver disease: morphological manifestations: review by an international group. *Lancet* **1**, 707.

Marnett, L.J. and Eling, T.E. (1983). Cooxidation during prostaglandin biosynthesis: a pathway for the metabolic activation of xenobiotics. *Rev. Biochem. Toxicol.* **5**, 135-172.

Mayeno, A.N., Curran, A.J., Roberts, R.L. and Foote, C.S. (1989). Eosinophils preferentially use bromide to generate halogenating agents. *J. Biol. Chem.* **264**, 5660.

Meunier, B. (1987). Horseradish peroxidase: a useful tool for modeling the extra-hepatic biooxidation of oxogens. *Biochimie* **69**, 3-9.

Mohandas, J., Duggin, G.G., Horvath, J.S. and Tiller, D.J. (1981). Metabolic oxidation of acetaminophen mediated by cytochrome P-450 mixed function oxidase and prostaglandin endoperoxide synthetase in rabbit kidney. *Toxicol. Appl. Pharmacol.* **61**, 252-259.

Moldeus, P. and Rahimtula, A. (1980). Metabolism of paracetamol to a glutathione conjugate catalysed by prostaglandin synthetase. Biochem. Biophys. *Res. Comm.* **96**, 469-475.

Moldeus, P., Hogberg, J. and Orrenius, S. (1978). Isolation and use of liver cells. *Methods Enzymol.* **52**, 60-66.

Moldeus, P., Andersson, B., Rahimtula, A. and Berggren, M. (1982). Prostaglandin synthetase catalysed activation of paracetamol. *Biochem. Pharmacol.* **31**, 1363-1370.

O'Brien, P.J. (1984). Multiple mechanisms for the metabolic activation of carcinogenic arylamines. *Free Radicals in Biology* VI (Pryor, W.A., ed), pp. 289-322, Academic Press.

O'Brien, P.J. (1985). Free-Radical Mediated DNA binding. *Env. Health Perspect.* **64**, 219-232.

O'Brien, P.J. (1988a). Radical formation during the peroxidase catalyzed metabolism of carcinogens and xenobiotics. *Free Radicals Biol. Med.* **4**, 169-183.

O'Brien, P.J. (1988b). Oxidants formed by the respiratory burst: their physiological role and their involvement in the oxidative metabolism and activation of drugs. The Respiratory Burst and Its Physiological Significance (Sbarra, A.J., and Strauss, R.R., eds) pp. 203-232, Plenum Press.

O'Brien, P.J. (1990a). Activation of xenobiotics by hypohalites or peroxidases, H_2O_2 and halide in Biological Oxidation Systems. Ed. Reddy, C.C., Hamilton, G.A. and Madyastha, K.M. Academic Press, N.Y.

O'Brien, P.J., Gregory, B., Fanney, R., Davison, W., Rahintula, A.D. and Tsuruta, Y. (1985). Microsomes and Drug Oxidations (Boobis, A.R., Caldwell, J., de Matteis, F., and Elcombe, C.R., eds) pp. 100-112, Taylor and Francis Ltd., London.

O'Brien, P.J., Jatoe, S., McGirr, L. G., Khan, S. (1990b). Molecular activtion mechanisms involved in arylamine cytotoxicity: peroxidase products in "N-oxidation of Drugs", Biochemistry, Pharmacology and Toxicology, Ed. Hlavica, P., Damani, L.A., and Gorrod, J.W., Chapman and Hall (in press).

Pilaro, A.M. and Laskin, D.L. (1986). Accumulation of activated mononuclear phagocytes in the liver following lipopolysaccharide treatment of rats. *J. Leukocyte Biology* **40**, 29-41.

Porubek, D.J., Rundgren, M., Harrison, P.J., Nelson, S.D. and Moldeus, P. (1987). Investigation of mechanisms of acetaminophen toxicity in isolated rat hepatocytes with acetaminophen analogues. *Molec. Pharmacol.* **31**, 647-653.

Potter, D.W. and Hinson, J.A. (1987a). Mechanisms of acetaminophen oxidation to N-acetyl-p-benzoquinoneimine by horseradish peroxidase and cytochrome P-450. *J. Biol. Chem.* **262**, 966-973.

Potter, D.W. and Hinson, J.A. (1987b). The one and two-electron oxidation of acetaminophen catalysed by prostaglandin H synthase. *J. Biol. Chem.* **262**, 974-980.

Prescott, L.G. (1983). Paracetamol Overdosage. *Drugs* **25**, 290-314.

Ritter, C.L. and Malejka-Giganti, D. (1989). Oxidations of the carcinogen N-hydroxy-N-(2-fluorenyl) acetamide by enzymatically or chemically generated oxidants of chloride and bromide. *Chem. Res. Toxicol.* **2**, 325-333.

Ross, D., Norbeck, K. and Moldeus, P. (1985). The generation and subsequent fate of glutathionyl radicals in biological systems. *J. Biol. Chem.* **260**, 15028-15032.

Rice, R.G. and Gomez-Taylor, M. (1986). Occurrence of by-products of strong oxidants reacting with drinking water contaminants. *Env. Health Perspectives* **69**, 31-44.

Sato, C., Matsuda, Y. and Lieber, C.S. (1981). Increased hepatotoxicity of acetaminophen after chronic ethanol consumption in the rat. *Gastroenterology* **80**, 140-148.

Thomas, E.L., Grisham, M.B. and Jefferson, M.M. (1983). Myeloperoxidase dependent effect of amines on functions of isolated neutrophils. *J. Clin. Invest.* **72**, 441-453.

Tsuruta, Y., Subrahmanyam, V.V., Marshall, W. and O'Brien, P.J. (1985). Peroxidase mediated irreversible binding of arylamine carcinogens to DNA intact polymorphonuclear leukocytes activated by a tumour promoter. *Chem.-Biol. Interacn.* **53**, 25-35.

Uetrecht, J.P. (1988). Drug-induced agranulocytosis and other effects mediated by peroxidases during the respiratory burst. The Respiratory Burst and Its Physiological Significance (Sbarra, A.J., and Strauss, R.R., eds) pp. 233-244, Plenum Press.

Uetrecht, J.P. (1989). Idiosyncratic Drug Reactions: possible role of reactive metabolites generated by leukocytes. *Pharmaceutical Research* **6**, 265-273.

Van Berkel, T.J. (1974). Difference spectra, catalase- and peroxidase activities of isolated parenchymal and non-parenchymal cells from rat liver. *Biochem. biophys. Res. Comm.* **61**, 204-209.

Weiss, S.J. (1986). Chlorinated oxidants generated by leukocytes. *Adv. in Free Radical Biology and Medicine* **2**, 91-106.

Weiss, S.J., Test, S.T., Eckmann, C.M., Roos, D. and Regiane, D. (1986). Brominating oxidants generated by human eosinophils. *Science* **234**, 200-203.

Yamazoe, Y., Miller, D.W., Weiss, C.C., Dooley, K.L., Zenser, T.V., Beland, F.A. and Kadlubar, F.F. (1985). DNA adducts formed by ring-oxidation of the carcinogen 2-naphthylamine with prostaglandin H synthase in vitro and in dog urothelium *in vivo*. *Carcinogenesis*, **6**, 1379-1387.

Yamazoe, Y., Zenser, T.V., Miller, D.W. and Kadlubar, F.F. (1989). The benzidine: DNA adduct formed by peroxidase and H_2O_2. *Carcinogenesis* **9**, 1635-1641.

PEROXYL FREE RADICALS: BIOLOGICAL REACTIVE INTERMEDIATES PRODUCED DURING LIPID OXIDATION

Lawrence J. Marnett

A.B. Hancock, Jr. Memorial Laboratory for Cancer Research
Center in Molecular Toxicology
Vanderbilt University School of Medicine
Nashville, TN 37232

INTRODUCTION

Peroxyl radicals are intermediates in the generation of hydroperoxides and are products of their metabolism. Peroxyl radicals possess a unique combination of reactivity and selectivity that enables them to diffuse to remote cellular loci and to oxidize cellular constituents to intermediates that may play a role in toxicity and carcinogenicity. Because of the widespread occurrence of polyunsaturated fatty acid hydroperoxides in the plant and animal kingdom, peroxyl radicals must be considered biological reactive intermediates of considerable toxicological importance.

Several pathways exist for the production of hydroperoxides from polyunsaturated fatty acids in biological systems. Lipoxygenases and cyclooxygenase oxygenate arachidonic acid to peroxy intermediates that are converted to leukotrienes, prostaglandins, and thromboxane *inter alia* (Pace-Asciak, C.R. et al., 1983). Oxidative stress triggers lipid peroxidation and generation of phospholipid hydroperoxides (Porter, N.A., 1986). Polyunsaturated fatty acid hydroperoxides are mild chemical oxidants that are converted by metal complexes and metalloproteins to a variety of potent oxidizing agents (Kochi, J.K., 1978). The principal metal that reacts with hydroperoxides in biological systems is iron and the major pathway of its reaction with hydroperoxides is reductive. One-electron reduction produces alkoxyl radicals and a one-electron oxidized metal whereas two-electron oxidation produces an alkoxide ion and a two-electron oxidized metal center. These reactions are illustrated for heme complexes in equations 1 and 2. Alkoxyl radicals and ferryl-oxo complexes **1** and **2** are extremely reactive and exhibit half-lives on the order of μsec (Pryor, W.A., 1986). In the absence of any stabilizing factors, it is unlikely these species diffuse more than a few molecular diameters before reacting with an oxidizable cellular constituent.

eq 1 $$Fe^{3+} + ROOH \longrightarrow {}^{+\bullet}Fe^{4+}{=}O + ROH \quad (\mathbf{1})$$

eq 2 $$Fe^{3+} + ROOH \longrightarrow Fe^{4+}{=}O + RO^{\bullet} + H^{+} \quad (\mathbf{2})$$

Biological Reactive Intermediates IV, Edited by C.M. Witmer *et al.*
Plenum Press, New York, 1990

Alkoxyl radicals and ferryl-oxo complexes oxidize hydroperoxides to peroxyl radicals with unexpectedly high efficiency (eq. 3 and 4) (Traylor, T.G. et al., 1987). In addition, alkoxyl radicals derived from polyunsaturated fatty acid hydroperoxides undergo intramolecular cyclization and coupling to oxygen to produce peroxyl radicals (eq. 5) (Dix, T.A., et al., 1985). Finally, as mentioned above, peroxyl radicals are intermediates in lipid peroxidation, a process triggered by a variety of endogenous and exogenous stimuli (eq. 6) (Porter, N.A., 1986). In contrast to alkoxyl radicals and ferryl-oxo complexes, peroxyl radicals have half-lives up to sec which enables them to diffuse to regions of the cell remote from the site of their generation (Pryor, W.A., 1986). Characteristic reactions of peroxyl radicals include H-abstraction, bimolecular coupling to form tetroxides, and oxygenation of isolated double bonds (Ingold, K., 1969). It is the latter reaction that is of toxicological interest.

eq 3 $$ROOH + \bullet OR \longrightarrow ROO\bullet + HOR$$

eq 4 $$ROOH + Fe^{4+}{=}O \longrightarrow ROO\bullet + Fe^{3+}\text{-}OH$$

eq 5

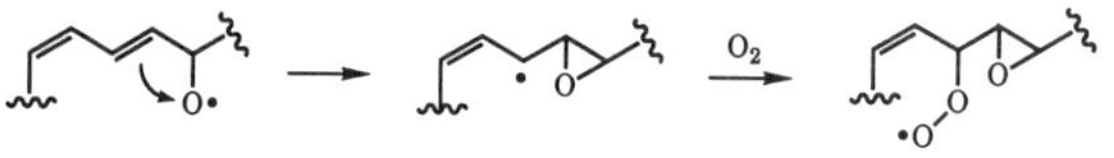

eq 6

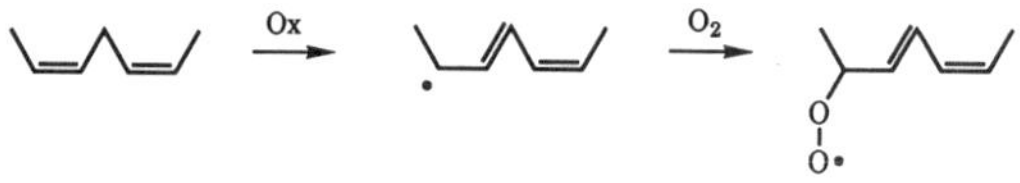

Epoxidation is a key reaction in the metabolic activation of several toxins and carcinogens (Sims, P., et al., 1974). For example, benzo[a]pyrene (BP), a widespread environmental pollutant, requires metabolic activation to exert its mutagenic and carcinogenic effects (Dipple, A., et al., 1984). BP or its metabolites are activated by cytochrome P-450 enzymes and by oxidants generated in metal-hydroperoxide reactions. Cytochrome P-450 oxygenates BP to several derivatives including 7,8-epoxy-7,8-dihydrobenzo[a]pyrene (BP-7,8-oxide) which may be enzymatically hydrated to 7,8-dihydroxy-7,8-dihydrobenzo[a]pyrene (BP-7,8-diol). The latter is further oxidized to 7,8-dihydroxy-9,10-epoxy-7,8,9,10-tetrahydrobenzo[a]pyrene (BPDE), the ultimate carcinogenic form of BP (Dipple, A., et al., 1984;Pelkonen, O., et al., 1982). Metal-hydroperoxide reactions oxidize BP to quinones and BP-7,8-diol to BPDE (Marnett, L.J. et al., 1977; Sivarajah, K. et al, 1979). The principal oxidants generated during metal-hydroperoxide reactions appear to be peroxyl radicals (ROO·) (Dix, T.A. et al., 1985). Peroxyl radical-dependent oxidations are distinct from cytochrome P-450-catalyzed oxidations and may constitute a complementary pathway of carcinogen activation (Marnett, L.J., 1987).

In mouse skin, the major DNA adducts formed after topical application of BP are formed from BPDE (Koreeda, M. et al., 1978). Thus, metabolic activation of BP in skin occurs by pathways that generate BP-7,8-oxide, hydrate it to BP-7,8-diol, and epoxidize BP-7,8-diol to BPDE. Oxidation of BP to BP-7,8,-oxide can only be catalyzed by cytochrome P-450 because no other mammalian enzymes epoxidize polycyclic aromatic hydrocarbons. This step is not effected by peroxyl free radicals as they oxidize BP to quinones (Marnett, L.J. et al., 1977). The relative contributions of cytochrome P-450 and peroxyl radicals to epoxidation of polycyclic hydrocarbon dihydrodiols (e.g. BP-7,8-diol → BPDE) is under active investigation.

RESULTS AND DISCUSSION

The stereoselectivity of epoxidation of the enantiomers of BP-7,8-diol provides an opportunity to estimate the importance of these two complementary pathways of metabolic activation. The major determinant of cytochrome P-450 epoxidation is the orientation of the pyrene ring at the active site of the enzyme, whereas the major determinant of peroxyl radical epoxidation stereochemistry is hydrogen bonding to the allylic hydroxyl group. (-)-BP-7,8-diol is epoxidized mainly to (+)-*anti*-BPDE by both systems (Panthananickal, A. et al., 1981). In contrast, (+)-BP-7,8-diol is epoxidized to (+)-*syn*-BPDE by cytochrome P-450, but to (-)-*anti*-BPDE by peroxyl-radicals (Figure 1) (Deutsch, J. et al., 1978; Thakker, D.R. et al., 1977). Small amounts of (+)-*syn*-BPDE are also formed from (+)-BP-7,8-diol by peroxyl radicals (Panthananickal, A. et al., 1981). The distinct stereochemistry of (+)-BP-7,8-diol oxidation has been exploited to demonstrate that peroxyl radicals contribute to the oxidation of (+)-BP-7,8-diol by crude tissue preparations, cultured fibroblasts (Boyd, J.A. et al., 1982), hamster tracheal explants (Reed, G.A. et al., 1984), and freshly isolated mouse skin keratinocytes (Eling, T. et al., 1986).

The role of peroxyl radicals and cytochrome P-450 in the metabolic activation of the (+)-enantiomer of 7,8-dihydroxy-7,8-dihydrobenzo[a]pyrene [(+)-BP-7,8-diol] was investigated in the epidermis of CD-1 mice (Pruess-Schwartz, et al., 1989). In skin homogenates from untreated or acetone-pretreated animals [7-^{14}C]-(+)-BP-7,8-diol (20 μM) was metabolized primarily to (-)*anti*-BPDE as detected by HPLC of the stable tetraol hydrolysis products (Figure 2). The amounts of *anti*-BPDE-tetraols increased with the length of time of incubation (0-90 min). Only small amounts of (+)-*syn*-BPDE-tetraols were detected. Epoxidation was not dependent upon NADPH. The addition of butaylated hydroxyanisole, a free radical scavenger, decreased the formation of both *anti*- and *syn*-BPDE-tetraols (I_{50} < 1 μM).

In epidermal homogenates from animals pretreated with β-naphthoflavone (β-NF, an inducer of cytochrome P-450_c), (+)-BP-7,8-diol was metabolized almost exclusively to (+)-*syn*-BPDE (Figure 2) (Pruess-Schwartz, et al., 1989). The amounts of *syn*-BPDE-tetraols also increased with time of incubation with only small amounts of *anti*-BPDE-tetraols being detected. Epoxidation was NADPH-dependent and was not inhibited by the addition of BHA. The addition of α-naphthoflavone (an inhibitor of cytochrome P-450) inhibited *syn*-BPDE-tetraol formation (I_{50} ~ 2.5 μM).

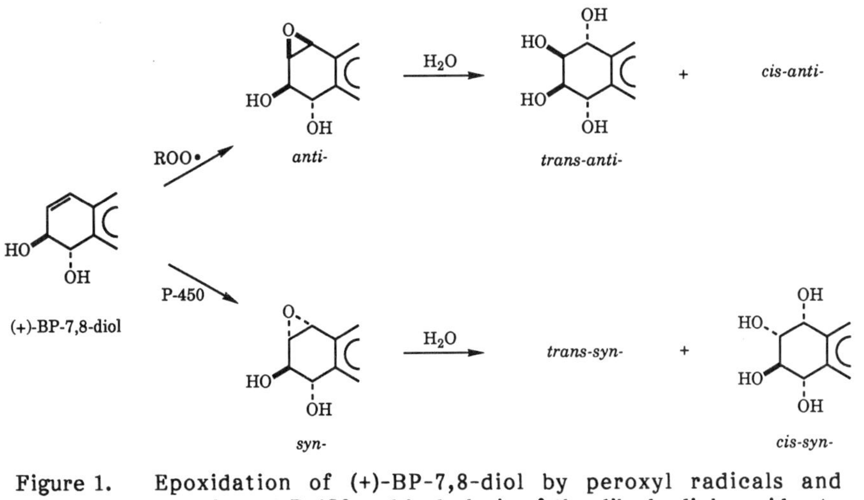

Figure 1. Epoxidation of (+)-BP-7,8-diol by peroxyl radicals and cytochrome P-450 and hydrolysis of the dihydrodiolepoxides to tetrahydrotetraols. Only the major hydrolysis products are shown. The minor products are named but not drawn.

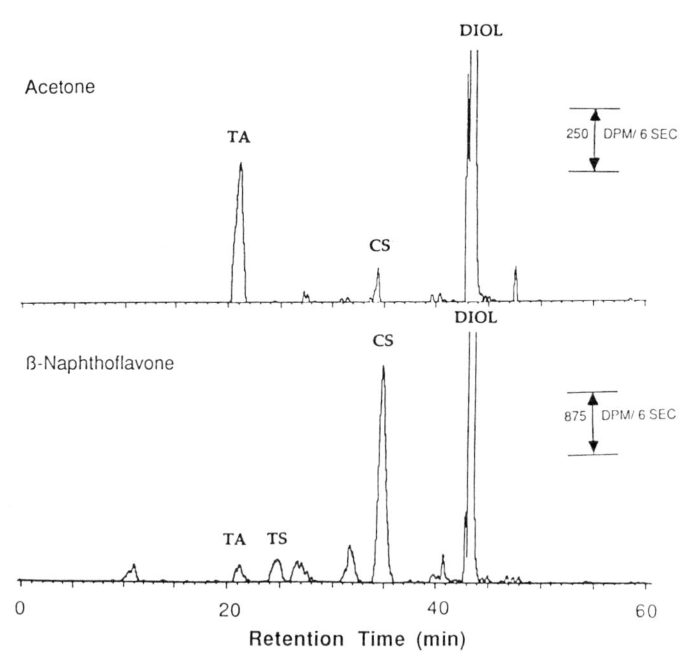

Figure 2. Reverse-phase HPLC profiles of the tetraol hydrolysis products formed after (+)-BP-7,8-diol epoxidation by mouse skin homogenates. [^{14}C]-(+)-BP-7,8-diol (20 μM) was incubated for 90 min at 37°C with mouse epidermal homogenates prepared from mice that had been treated with acetone (upper panel) or 350 nmol β-NF in acetone (lower panel) 24 hr prior to sacrifice. Peaks were identified by cochromatography with authentic standards.

The results of these experiments suggest that two separate pathways exist in mouse skin homogenates for epoxidation of BP-7,8-diol to dihydrodiolepoxides. In untreated mouse skin, the major oxidizing agent(s) appears to be peroxyl radicals generated by an unknown mechanism. Treatment of the animals with inducers that act at the A_h locus remodels cellular metabolism so that cytochrome P-450 is the major oxidizing agent. To investigate the contribution of peroxyl radicals and cytochrome P-450 to the metabolic activation of (+)-BP-7,8-diol in an intact biological system, [1,3-^{3}H]-(+)-BP-7,8-diol (200 nmol/mouse, 50 μCi/mouse) was applied topically to mice and the hydrocarbon-modified deoxyribonucleosides were isolated and analyzed. After 3 hr of exposure, both (-)-*anti*-BPDE- and (+)-*syn*-BPDE-modified deoxyribonucleosides were detected as determined by comparison of elution volume with known standards (Pruess-Schwartz, D. et al., 1989). In mice that received acetone pretreatment, similar amounts of (-)-*anti*-BPDE-dGuo and (+)-*syn*-BPDE-dGuo were formed (Figure 3). The amounts of these adducts varied in two separate experiments from 0.1 to 0.3 pmol (-)*anti*-BPDE-dGuo and 0.22 to 0.25 pmol (+)-*syn*-BPDE-dGuo, giving an *anti/syn* ratio of 0.6 to 1.2. Similar proportions of these adducts were formed in mice that received no pretreatment (data not shown). In mice that were pretreated with β-NF, (+)-*syn*-BPDE-dGuo was the major DNA adduct formed, but small amounts of (-)*anti*-BPDE-dGuo were also detected (0.2 pmol of (-)-*anti*-BPDE-dGuo and 1.8 pmol of (+)-*syn*-BPDE-dGuo. The *anti/syn* ratio was 0.1.

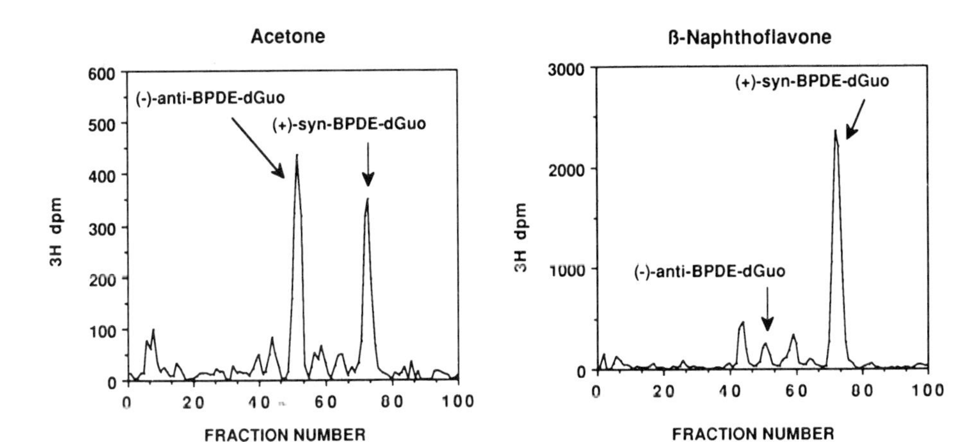

Figure 3. Reverse-phase HPLC profiles of the (+)-BP-7,8-diol-deoxyribonucleoside adducts formed in mouse epidermis *in vivo*. [^{3}H]-(+)-BP-7,8-diol (200 nmol) was applied topically to groups of 10 mice that had received either 100 µl acetone/animal or 350 nmol β-NF in 100 ml acetone/animal before diol application. After 3 hr, the mice were sacrificed and the (+)-BP-7,8-diol-modified deoxyribonucleosides were isolated and analyzed. Acetone-pretreated mice, 4.5 pmol (+)-BP-7,8-diol bound/mg DNA; β-NF-mice, 11 pmol (+)-BP-7,8-diol bound/mg DNA.

The results of the study of deoxynucleoside adducts produced in mouse skin *in vivo* are consistent with the results of the investigations of tetraol hydrolysis products generated *in vitro* and imply a significant role for peroxyl radicals in the epoxidation of BP-7,8-diol. This illustrates the potential importance of peroxyl radicals as biological reactive intermediates and the sensitivity of metabolic activation pathways to pharmacological manipulation. Studies are underway in our laboratory using BP-7,8-diol to quantitate the production of peroxyl radicals *in vivo* and in cellular systems following treatment with a variety of pharmacological stimuli.

ACKNOWLEDGEMENTS

This work was supported by a research grant from the National Institutes of Health (CA47479).

REFERENCES

Boyd, J.A., Barrett, J.C. and Eling, T.E. (1982). Prostaglandin endoperoxide synthetase dependent cooxidation of (±)-trans-7,8-dihydroxy-7,8-dihydrobenzo[a]pyrene in C3H10T1/2 clone 8 cells. *Cancer Res.* **42**, 2828-2832.

Deutsch, J., Leutz, J.C., Yang, S.K., Gelboin, H.V., Chang, R.L., Vatsis, K.P. and Coon, M.J. (1978). Regio- and stereoselectivity of various forms of cytochrome P-450 in the metabolsim of benzo[a]pyrene and (±)-7,8-dihydroxy-7,8-dihydrobenzo[a]pyrene as shown by product formation and binding to DNA. *Proc. Natl. Acad. Sci. USA* **75**, 3123-3127.

Dipple, A., Moschel, R.C. and Bigger, C.A. H. (1984). Polycyclic aromatic carcinogens. In *Chemical Carcinogens*, Second edition (Searle, C.E., Ed.) pp 41-163, American Chemical Society, Washington, D.C.

Dix, T.A. and Marnett, L.J. (1985). Conversion of linoleic acid hydroperoxide to

hydroxy, keto, epoxyhydroxy, and trihydroxy fatty acids by hematin. *J. Biol. Chem.* **260**, 5351-5367.
Dix, T.A., Fontana, R., Panthani, A. and Marnett, L.J. (1985). Hematin-catalyzed epoxidation of 7,8-dihydroxy-7,8-dihydrogenzo[a]pyrene)BP-7,8-diol) by polyunsaturated fatty acid hydroperoxides. *J. Biol. Chem.* **260**, 5358-5365.
Eling, T., Curtis, J., Battista, J. and Marnett, L.J. (1986). Oxidation of (+)-7,8-dihydroxy-7,8-dihydrobenzo[a]pyrene by mouse keratinocytes: evidence for peroxyl radical- and monoxygenase-dependent metabolism. *Carcinogenesis* **7**, 1957-1963.
Ingold, K. (1969). Peroxy radicals. *Acc. Chem. Res.* **2**, 1-19.
Kochi, J.K. (1978). In *Organometallic Mechanisms and Catalysis*, pp 50-83, Academic Press, New York.
Koreeda, M., Moore, P.D., Wislocki, P.G., Levin, W., Conney, A.H., Yagi, H. and Jerina, D.M. (1978). Binding of benzo[a]pyrene 7,8-diol-9,10-epoxide to DNA, RNA, and protein of mouse skin occurs with high stereoselectivity. *Science* **199**, 778-781.
Marnett, L.J., Reed, G.A. and Johnson, J.T. (1977). Prostaglandin synthetase-dependent benzo[a]pyrene oxidation: products of the oxidation and inhibition of their formation by antioxidants. *Biochem. Biophys. Res. Comm.* **79**, 569-576.
Marnett, L.J., Johnson, J.T. and Bienkowski, M.J. (1979). Arachidonic acid-dependent metabolism of 7,8-dihydroxy-7,8-dihydrobenzo[a]pyrene by ram seminal vesicles. *FEBS Letts.* **106**, 13-16.
Marnett, L.J. (1987). Peroxyl free radicals: potential mediators of tumor initiation and promotion. *Carcinogenesis* **8**, 1365-1373.
Pace-Asciak, C.R. and Smith, W.L. (1983). Enzymes in the biosynthesis and catabolism of the eicosanoids: prostaglandins, thromboxanes, leukotrienes and hydroxy fatty acids. In *The Enzymes*, Vol. XVI (Boyer, Paul D., Ed.) pp 544-603, Academic Press, New York.
Panthananickal, A. and Marnett, L.J. (1981). Arachidonic acid-dependent metabolism of (±)-7,8-dihydroxy-7,8-dihydrobenzo[a]pyrene to polyguanylic acid-binding derivatives. *Chem. Biol. Interactions* **33**, 239-252.
Pelkonen, O. and Nebert, D.W. (1982). Metabolism of polycyclic aromatic hydrocarbons: etiological role in carcinogenesis. *Pharmacol. Rev.* **34**, 189-222.
Porter, N.A. (1986). Mechanisms for the autoxidation of polyunsaturated lipids. *Acc. Chem. Res.* **19**, 262-268.
Pruess-Schwartz, D., Nimesheim, A. and Marnett, L.J. (1989). Peroxyl radical- and cytochrome P-450-dependent metabolic activation of (+)-7,8-dihydroxy-7,8-dihydrobenzo[a]pyrene in mouse skin *in vitro* and *in vivo*. *Cancer Res.* **49**, 1732-1737.
Pryor, W.A. (1986). Oxy-radicals and related species: their formation, lifetimes and reactions. *Ann. Rev. Physiol.* **48**, 657-667.
Reed, G.A., Grafstrom, R.C., Krauss, R.S., Autrup, H. and Eling, T.E. (1984). Prostaglandin H synthase-dependent co-oxygenation of (±)-7,8-dihydroxy-7,8-dihydrobenzo[a]pyrene in hamster trachea and human bronchus explants. *Carcinogenesis* **5**, 955-960.
Sims, P. and Grover, P.L. (1974). Epoxides in polycyclic aromatic hydrocarbon metabolism and carcinogenesis. *Adv. Cancer Res.* **20**, 165-274.
Sivarajah, K., Mukhtar, H. and Eling, T.E. (1979). Arachidonic acid-dependent metabolism of (±)-trans-7,8-dihydroxy-7,8-dihydrobenzo[a]pyrene (BP-7,8-diol) to 7,10/8,9-tetrols. *FEBS Letts.* **106**, 17-20.
Thakker, D.R., Yagi, H., Akagi, H., Koreeda, M., Lu, A.Y.H., Levin, W., Wood, A.W., Conney, A.H. and Jerina, D.M. (1977). Stereoselective metabolism of benzo[a]pyrene and benzo[a]pyrene-7,8-dihydrodiol to diol-epoxides. *Chem. Biol. Interactions* **16**, 281-300.
Traylor, T.G. and Xu, F. (1987). A biomimetic model for catalase: the mechanisms of reaction of hydrogen peroxide and hydroperoxides with iron(III) porphyrins. *J. Am. Chem. Soc.* **109**, 6201-6202.

BIOLOGICAL SIGNIFICANCE OF ACTIVE OXYGEN SPECIES: IN VITRO STUDIES ON SINGLET OXYGEN-INDUCED DNA DAMAGE AND ON THE SINGLET OXYGEN QUENCHING ABILITY OF CAROTENOIDS, TOCOPHEROLS AND THIOLS

Paolo Di Mascio, Stephan P. Kaiser, Thomas, P.A. Devasagayam and Helmut Sies

Institut für Physiologische Chemie I
University of Düsseldorf
Moorenstrasse 5,
D-4000-Düsseldorf, West Germany

INTRODUCTION

Reactive oxygen species have been in focus for some time now, and there are several reviews on this topic. since our contribution to Biological Reactive Intermediates III (Wefers and Sies, 1986), work from our laboratory has been presented comprehensively (Sies, 1986, 1988; Ishikawa and Sies, 1989). Further, the biochemistry of oxygen toxicity has been presented (Cadenas, 1989).

This chapter emphasizes recent work on singlet molecular oxygen, one of the reactive oxygen species which is not a radical. The advances in generating and detecting singlet oxygen permitted a more close examination of the biological antioxidant capacity against singlet oxygen.

A more general introduction into the field of reactive oxygen species has been given by Pryor (1986). One aspect that deserves particular attention is the range of lifetimes of the various species, since a short lifetime dictates a short diffusion pathlength, and therefore the range of targets that can be reached following generation of the reactive species. It is well-known that the fastest decay time applies for the hydroxyl radical, being of the order of nanoseconds, whereas for singlet oxygen the halftimes are in the microsecond range. This means that there is a considerable diffusion of singlet oxygen once it is generated, determined to be in the range of 100 Angstrom (see Sies, 1986 for further information). Biologically, singlet molecular oxygen may be generated by photochemical reactions through transfer of excitation energy from a suitable excited triplet state sensitizer (photoexcitation), and also by dark reactions (chemiexcitation) which include enzymatic reactions or radical interactions (Murphy and Sies, 1990).

RESULTS AND DISCUSSION

Generation and Detection of Singlet Molecular Oxygen

The generation of singlet molecular oxygen was performed in three different ways. The generation by microwave discharge was performed according to Wefers et al. (1987), and the method of Midden and Wang (1983) was employed for the generation by photosensitization with Rose Bengal immobilized on a glass plate. The third method, employed most widely in our studies, is the chemical generation of singlet oxygen using the thermal decomposition of the endoperoxide of naphthalene dipropionate, described in detail (Di Mascio and Sies, 1989), and based on previous literature (Saito et al., 1983; Aubry, 1985).

The detection was via light emission, either in the monomol emission at 1270 nm, or in the dimol emission bands at 634 nm and 703 nm. The detectors were a liquid nitrogen-cooled germanium diode, or a cooled (-25°C) red-sensitive photomultiplier used for single-photon counting, respectively. The chemical detection was via chemical trapping with the water-soluble salt of anthracene-9,10-diyldiethyl disulfate. For details, see Di Mascio and Sies (1989).

Singlet Oxygen-dependent Loss of Transforming Activity (Di Mascio et al., 1989a)

Exposure of single-stranded bacteriophage M13 DNA to singlet oxygen led to a decrease of tranforming activity as a function of time. Loss of transforming activity was about doubled when water in the buffer was replaced by deuterium oxide, increasing the lifetime and, consequently, the diffusion pathlength of the singlet oxygen generated. When Rose Bengal was used as a generation source, the controls without Rose Bengal gave only minute effects. Likewise, when methionine was present, the scavenging of singlet oxygen provided protection of the DNA.

Single-stranded DNA was more susceptible to singlet oxygen than double-stranded DNA indicated by a more pronounced loss of the surviving fraction versus the singlet oxygen generation, studied in experiments with a bacteriophage (Di Mascio et al., 1989a).

Single-strand Break Formation upon Exposure to Singlet Oxygen

Single-strand break formation of pBR322 after exposure to singlet oxygen generated by microwave discharge was detected as an increase in the open circular form at the expense of the closed circular supercoiled form, measured after electrophoresis with subsequent ethidium bromide staining (Wefers et al, 1987; Di Mascio et al., 1989a). Like the effects on the transforming activity, deuterium oxide increased strand-break formation, whereas methionine was protective and methionine sulfone as the oxidation product of methionine did not show a protective effect.

To assess the time course, sodium azide (10 mM) was added in incubations of pBR322 at various time points to quench singlet oxygen. It was seen (Di Mascio et al., 1989a) that the reaction yielding DNA damage was half-completed at 11 min, and after 1 h more than 90% of the reaction had occurred. This method allowed us to determine the average rate of strand-break formation in a certain time interval. Over the same time inverval, we calculated the singlet oxygen generation rate (Di Mascio and Sies, 1989), and this steady state analysis reveals a complex relationship between strand-break formation rate and the singlet oxygen production rate (Figure 1). The efficiency of singlet oxygen to yield single-strand breaks corresponds to the slope in Figure 1, and it increases with increasing rates of singlet oxygen formation. In a rough calculation, the following was obtained (Di Mascio et al., 1989a). The concentration of pBR322 in the incubation mixtrue was 15 µg/ml or 5.2 nM. At 5 µM singlet oxygen per min the efficiency calculated by the slope was 1.5 µM open circular DNA per mol singlet oxygen generated. At 30 µM singlet oxygen generated per min, the efficiency is about 4 times higher, 6.1 µM open circular DNA per mol singlet oxygen generated. In the latter case the steady state concentration of singlet oxygen is 24,8 x 10^{-12}M.

A second-order mechanism is suggested, as the rate of single-strand breaks is proportional to the square of the singlet oxygen production rate (inset in Figure 1). This is in line with results obtained by Cadet et al. (1983), who investigated the attachment of singlet oxygen on deoxyguanosine and found that in the pathway leading to cyanuric acid the consecutive addition of two molecules of singlet oxygen is required.

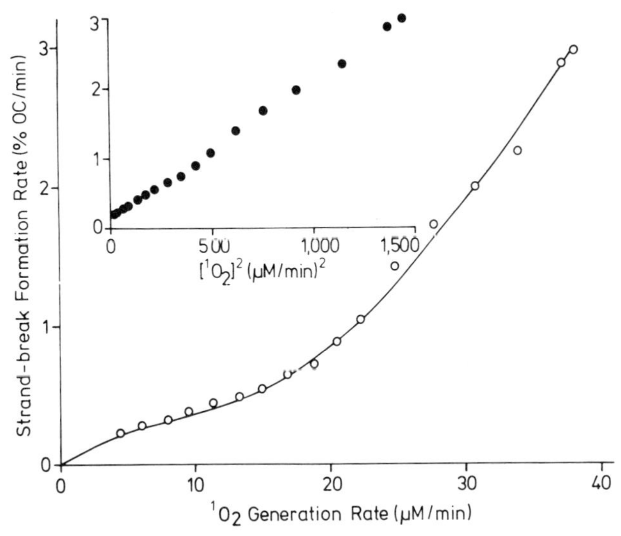

Figure 1. Rate of strand-break formation in pBR322 DNA as a function of $^1O_2$ generation rate. Inset: Strand-break-formation rate plotted vs. square of $^1O_2$ production rate. correlation coefficient, r = 0.998 (from Di Mascio et al., 1989a).

Recent work by Epe et al. (1988 and 1989) underscores the capability of singlet oxygen to inflict DNA damage by causing single-strand breaks. Work from our group was extended to a shuttle-vector system (Di Mascio et al., 1990). With these *in vitro* results on singlet oxygen, we became more interested in potential singlet oxygen scavenging activities, and these were studied also in a simple *in vtiro* system.

Ranking Bioloigcal Singlet Oxygen Quenchers (see Di Mascio et al., 1989b)

We investigated the relative quenching ability of various naturally occurring carotenoids and compared them with alpha-tocopherol and bile pigments, using the thermodissociable endoperoxide of naphthylidene dipropionate, as mentioned above.

Background: There is increasing interest in the role of diet and nutrition in the pathogenesis and possible prevention of cancer (Ames, 1983), and the question has been raised whether beta-carotene might have anticarcinogenic properites independent of its provitamin A activity. An inverse relationship between beta-carotene intake and the incidence of certain types of cancer such as lung and gastrointestinal tract cancer has been observed (Peto et al., 1981). Animal experiments have revealed anticarcinogenic properties of carotenoids (Mathews-Roth, 1985). As many as 20 regularly occurring carotenoids have been reported in plasma (Thompson et al., 1985). The physiological functions of carotenoids may be highly specialized as indicated, for instance, by the presence of zeaxanthin and lutein in the macula area of the retina in the virtual absence of beta-carotene (Handelman et al., 1988). The ability of beta-carotene in biological systems has been attributed to its ability to physically quench singlet oxygen, first described by Foote and Denny (1968). However, this carotenoid also reacts rapidly with organic radicals such as the trichloromethyl peroxyl radical (Packer et al., 1981), and it exerts interesting types of behaviour as an antioxidant (Burton and Ingold, 1984). A recent compilation of information on carotenoids is given by Krinsky (1990).

Our own recent work on ranking of the different carotenoids (Figure 2) showed that lycopene is the most efficient biological carotenoid singlet oxygen quencher (Di Mascio et al., 1989b). The calculated k_q value of this open-chain isomer of beta-carotene was 31 x 109 M^{-1} s^{-1}, more than double that of beta-carotene. Preliminary

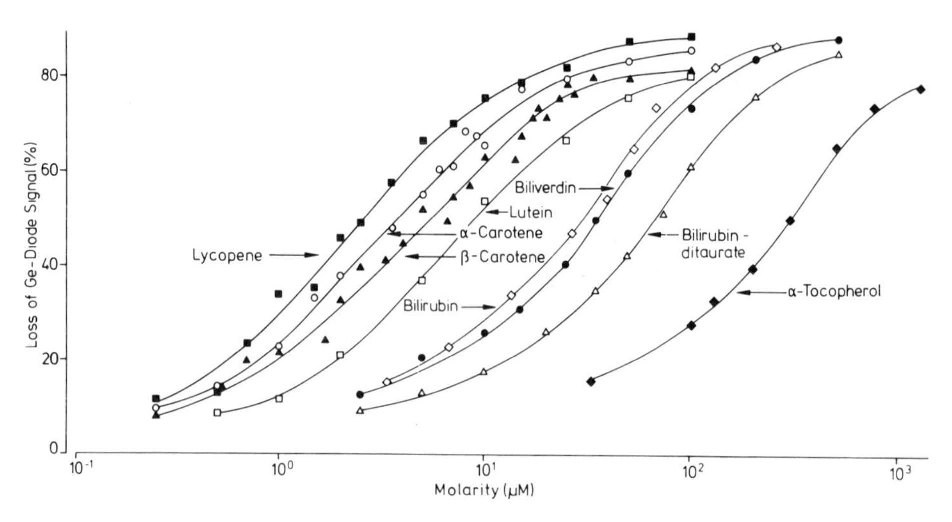

Figure 2. Dependence of loss of germanium-diode signal on the concentration of carotenoids, xanthophylls, α-tocopherol and bile pigments (from Di Mascio et al.., 1989b).

results indicate that the carotenoids phytoene and phytofluene, derived from plants, have much lower singlet oxygen quenching abilities.

Other classes of compounds, e.g. bilirubin, tocopherols (Kaiser et al., 1990), and thiols (Kaiser et al., 1989; Devasagayam et al., 1990), were less active, but these may also be biologically important in singlet oxygen quenching because of their higher concentration and/or different subcellular location in biological targets, besides solubility characterisitcs (see Table 1).

While the role of carotenoids in protecting plants against the photosensitisation by their own chlorophyll as well as in treatment of patients with photosensitivity diseases is established, the mechanism by which beta-carotene exerts a protective function against cancer remains unknown. However there are several lines of evidence which suggest that the generation of reactive oxygen species may play an important role in the development of cancer (Cerutti 1985).

The present work emphasizes that attention should be extended from beta-carotene to lycopene and other carotenoids. Studies on the relationship between plasma beta-carotene and cancer incidence have focused on beta-carotene (Peto et al., 1981; Gey et al., 1987; Cavina et al., 1988; Ziegler 1989). However, beta-carotene comprises only 15-30% of the total carotenoids. Lycopene has a plasma concentration slightly higher than beta-carotene, but there is a large variation of plasma levels and carotenoid patterns. Both these carotenoids were found in low-density lipoproteins (Esterbauer et al., 1989) and therefore could exert functions in the prevention of atherosclerosis.

These remarks on our recent interests in reactive oxygen species only represent a small segment of current attention of these important byproducts of aerobic life. One new insight seems to be that, in addition to being used in cytotoxicity, reactive oxygen species may also be utilized as signal molecules. Our recent work on the release of reactive oxygen species by human fibroblasts in response to interleukin-1 or tumor necrosis factor-alpha (Meier et al., 1989) points in that direction.

Table 1. Singlet Oxygen Quenching Constants, Chemical Reaction Rate Constants and Content in Mammalian Tissues of the Carotenoids, Bile Pigments, Tocopherols, Thiols and Related Compounds.

Compound	$k_q + k_r$	k_r	Contents in tissues
	10^6 $M^{-1}s^{-1}$		(μmol/g or μmol/l)
Lycopene[a]	31,000		0.5 - 1, plasma
γ-Carotene[a]	25,000		
Canthaxanthin[a]	21,000		
α-Carotene[a]	19,000		0.5 - 1, plasma
β-Carotene[a]	14,000		0.3 - 0.6, plasma
Bixin[a]	14,000		
Zeaxanthin[a]	10,000		
Lutein[a]	8,000		
Bilirubin[a]	3,200		5 - 20, plasma
Biliverdin[a]	2,300		
L-Dithiothreitol[b]	977	3.5	
α-Tocopherol[c]	280	3.6	15 - 31, plasma
β-Tocopherol[c]	270	0.2	
Cysteine[b]	247	53	30 - 100, tissues
γ-Tocopherol[c]	230	2.8	
N-Acetyl-L-cysteine[b]	190	0.7	
β,β-Dimethylcysteine[b]	167	3.6	
δ-Tocopherol[c]	160	1.7	
Cysteamine[b]	41.1	8.4	3 - 13, tissues
Captopril[b]	4.0		
Clutathione[b]	2.2	1.3	500-10,000,.tissues

The $k_q + k_r$ values, were obtained from Stern-Volmer plots. k_r values were obtained by measuring the thiol or tocopherol depletion by $NDPO_2$-generated $^1O_2$. Data from, a) Di Mascio et al., 1989b; b) Devasagayam et al., 1990; and c) Kaiser et al., 1990.

ACKNOWLEDGEMENTS

Helpful discussions with the colleagues mentioned in the references are gratefully acknowledged. Our studies were supported by the National Foundation for Cancer Research, Bethesda, Kernforchungsanlage Jülich GmbH and by Deutsche Forschungsgemeinschaft, Bonn.

REFERENCES

Ames, B.N. (1983). Dietary carcinogens and anticarcinogens. *Science* **221,** 1256-1264.

Aubry, J.M. (1985). Search for singlet oxygen in the decomposition of hydrogen peroxide by mineral compounds in aqueous solutions. *J. Am. Chem. Soc.* **103,** 7218-7224.

Burton, G.W. and Ingold, K.U. (1984). β-Carotene: An unusual type of lipid antioxidant. *Science* **224,** 569-573.

Cadenas, E. (1989). Biochemistry of oxygen toxicity. *Annu. Rev. Biochem.* **58,** 79-110.

Cadet, J., Decarroz, C., Wang, S.Y. and Midden, W.R. (1983). Mechanisms and products of photosensitised degradation of nucleic acids and related model compounds. *Isr. J. Cehm.* **23,** 420-429.

Cavina, G., Gallinella, B., Porra, R., Pecora, P. and Saraci, C. (1988). Carotenoids, retinoids and alpha-tocopherol in human serum - identification and determination by reversed-phase HPLC. *J. Pharm. Biomed. Anal.* **6,** 259-269.

Cerutti, P.A. (1985). Prooxidant state and tumour promotion. *Science* **221,** 1256-1264.

Devasagayam, T.P.A., Di Mascio, P., Kaiser, S. and Sies, H. (1990). Activity of thiols as singlet molecular oxygen quenchers. *J. Photochem. Photobiol.* (submitted).
Di Mascio, P. and Sies, H. (1989). Quantification of singlet oxygen generated by thermolysis of 3,3'-(1,4-naphthylidene) dipropionate. Monomol and dimol photoemission and the effects of 1,4-diazabicyclo (2.2.2) octane. *J. Am. Chem. Soc.* **111**, 2909-2914.
Di Mascio, P., Wefers, H., Do-Thi, H-P., Lafleur, M.V.M. and Sies, H. (1989a). Singlet molecular oxygen causes loss of biological activity in plasmid and bacteriophage DNA and induces single-strand breaks. *Biochim. Biophys. Acta* **1007**, 151-157.
Di Mascio, P., Kaiser, S. and Sies, H. (1989b). Lycopene as the most efficient biological carotenoid singlet oxygen quencher. *Arch. Biochem. Biophys.* **274**, 532-538.
Di Mascio, P., Mench, C.F.M., Nigro, R.G., Sarasin, A. and Sies, H. (1990). Singlet molecular oxygen induced mutagenicity in a mammalain SV40-based shuttle vector. *Photochem. Photobiol.*
Epe, B., Mützel, P. and Adam, W. (1988). DNA damage by oxygen radicals and excited state species: A comparative study using enzymatic probes *in vitro*. *Chem. Biol. Interact.* **67**, 149-165.
Epe, B., Hegler, J. and Wild, D. (1989). Singlet oxygen as the ultimately reactive species in Salmonella typhimurium DNA damage induced by methylene blue/visible light. *Carcinogenesis* **10**, 2019-2024.
Esterbauer, H., Striegl, G., Puhl, H. and Rotheneder, M. (1989). Continuous monitoring of *in vitro* oxidation of human low density lipoprotein. *Free Rad. Res. Commun.* **6**, 67-75.
Foote, C.S. and Denny, r.W. (1968). Chemistry of singlet oxygen VIII. Quenching by β-crotene. *J. Am. Chem. Soc.* **90**, 6233-6235.
Gey, K.F., Brubacher, G.B. and Stähelin, H.B. (1987). Plasma levels of antioxidant vitamins in relation to ischemic heart disease and cancer. *Am. J. Clin. Nutr.* **45**, 1368-1377.
Handelman, G.J., Dratz, E.A., Reay, C.C. and van Kuijk, F.J.G.M. (1988). Carotenoids in the human macula and whole retina. *Invest. Ophthamol. Vis. Sci.* **29**, 850-855.
Ishikawa, T. and Sies, H. (1989). Glutathione as an antioxidant: Toxicological aspects. In *Glutathione: Chemical, Biochemical and Medical Aspects*, Part B. (D. Dolphin, R. Poulson and D. Avramovic, Eds.) pp. 86-109.
Kaiser, S., Di Mascio, P. and Sies, H. (1989). Lipoat and Singulettsauerstoff. In *Thioctsäure* (H.O. Borbe and H. Ulrich, Eds.) pp. 69-76, pmi Verlag GmbH, Frankfurt.
Kaiser, S., Di Mascio, P., Murphy, M.E. and Sies, H. (1990). Physical and chemical scavenging of singlet oxygen by tocopherols. *Arch. Biochem. Biophys.* **276**, XXX-XXX.
Krinsky, N.I. (1990). antioxidant functions of carotenoids. *Free Rad. Biol. Med.* **7**, XXX-XXX.
Mathews-Roth, M.M. (1985). Carotenoids and cancer prevention-experimental and epidemiological studes. *Pure Appl. Chem.*, **57**, 717-722.
Meier, B., Radeke, H.H., Selle, S., Younes, M., Sies, H., Resch, K. and Habermehl, G.G. (1989). Human fibroblasts release reactive xoygen species in response to interleukin-1 or tumour necrosis factor-alpha. *Biochem. J.* **263**, 539-545.
Midden, W.R. and Wang, S.Y. (1983). Singlet oxygen generation for solution kinetics: clean and simple. *J. Am. Chem. Soc.* **105**, 4129-4135.
Murphy, M.E. and Sies, H. (1990). Visible-range low-level chemiluminescence in biological systems. *Meth. Enzymol.* (in press).
Packer, J.E., Mahood, J.S., Mora-Arellano, O., Slater, T.F., Wilson, R.L. and Wolfenden, B.S. (1981). Free radical and singlet oxygen scavengers. Reaction of a peroxy-radical with betacarotene, diphenyl furan and 1,4-diazabicyclo (2.2.2)-octane. *Biochem. Biophys. Res. Commun.* **98**, 901-906.
Peto, R., Doll, R., Buckley, J.D. and Sporn, M.B. (1981). Can dietary beta-carotene materially reduce human cancer rates? *Nature (London)* **290**, 201-208.
Pryor, W.A. (1986). Oxy-radicals and related species: Their formation, lifetimes, and ractions. *Annu. Rev. Physiol.* **48**, 657-667.
Saito, I., Matsuura, T. and Inoue, K. (1983). Formation of superoxide ion via one-electron transfer from electron donors to singlet oxygen. *J. Am. Chem. Soc.* **105**, 3200-3206.

Saito, I., Nagata, R., Nakagawa, H., Moriyama, H., Matsuura, T. and Inoue, K. (1987). New singlet oxygen source and trapping reagent for peroxide intermediates. *Free Rad. Res. Commun.* **2**, 327-336.

Sies, H. (1986). Biochemistry of oxidative stress. *Angew. Chem. (Int. Ed. Engl.)* **25**, 1058-1071.

Sies, H. (1988). Oxidative stress: Quinone redox cycling. *ISI Atlas of Science: Biochemistry* **1**, 109-114.

Thompson, J.N., Duval, S. and Verdier, P. (1985). Investigations of carotenoids in human blood using high performance liquid chromatography. *J. Micronutr. Anal.* **1**, 81-91.

Wefers, H. and Sies, H. (1986). Reactive oxygen species formed *in vitro* and in cells: role of thiols (GSH). Model studies with xanthine and xanthine oxidase and horseradish peroxidase. In *Biological Reactive Intermediates III* (J.J. Kocsis, D.J. Jollow, C.M. Witmer, J.O. Nelson and R. Snyder, Eds.) pp. 505-512, Plenum Publishing Corporation.

Wefers, H., Schulte-Frohlinde, D. and Sies, H. (1987). Loss of transforming activity of plasmid DNA (pBR322). In *E. coli* caused by singlet molecular oxygen. *FEBS Lett.* **211**, 49-52.

Ziegler, R.G. (1989). A review of epidemiologic evidence that carotenoids reduce the risk of cancer. *J. Nutr.* **119**, 116-122.

PHYSIOLOGICAL AND TOXICOLOGICAL ROLES OF HYDROPEROXIDES

V. Ullrich, M. Schatz, and F. Meisch

Faculty of Biology
University of Konstanz
Federal Republic of Germany

INTRODUCTION

In polymorphonuclear leukocytes (PMN) arachidonic acid is converted both to (5S)-5-hydroperoxy-(E,Z,Z,Z)-6,8,11,14-eicosatetraenoic acid (5-HPETE) and to (5S,6S)-5(6)-oxido-(E,E,Z,Z)-7,9,11,14-eicosatetraenoic acid (leukotriene A_4, LTA_4) by the action of 5-lipoxygenase. Furthermore, the specific dioxygenation product 5-HPETE is rapidly metabolized into (5S)-5-hydroxy-(E,Z,Z,Z)-6,8,11,14-eicosatetraenoic acid (5-HETE), presumably by the action of a peroxidase on 5-HPETE. In a further enzymatic step, LTA_4 is hydrolysed to give (5S,12R)-5,12-dihydroxy-(Z,E,E,Z)-6,8,10,14-eicosatetraenoic acid (leukotriene B_4, LTB_4), which acts as a potent chemoattractant and thus causes further accumulation of circulating PMN (Samuelsson et al., 1987).

In addition to the 5-lipoxygenase activity, neutrophils and mainly eosinophils from several species have been shown to contain a 15-lipoxygenase. The (15S)-15-hydroperoxy-(Z,Z,Z,E)-5,8,11,13-eicosatetraenoic acid (15-HPETE) is the product of the action of 15-lipoxygenase on arachidonic acid. Like 5-HPETE the 15-HPETE is rapidly reduced to (15S)-15-hydroxy-(Z,Z,Z,E)-5,8,11,13-eicosatetraenoic acid (15-HETE). The 15-lipoxygenase of human neutrophils has a much lower affinity for both linoleic acid and arachidonic acid (K_m of 77 and 63 μM, respectively) than the 5-lipoxygenase for arachidonic acid (K_m of 12.2 μM) (Borgeat, 1989). The biological significance of products from 15-lipoxygenase seems to be connected with the synthesis of the bioactive compounds 5S,6R,15S-trihydroxy-7,9,13-trans-11-cis-eicosatetraenoic acid (lipoxin A, LXA) and 5S,14R,15S-trihydroxy-6,10,12-trans-8-cis-eicosatetraenoic acid (lipoxin B, LXB) (Serhan et al., 1984).

The third mammalian lipoxygenase catalyses the insertion of molecular oxygen into arachidonic acid at position 12 to give (12S)-12-hydroperoxy-5,8,10,14-(Z,Z,E,Z)-eicosatetraenoic acid (12-HPETE). The human 12-lipoxygenase is present in platelets, but the physiological function of 12-HPETE and its enzymatic reduction product, (12S)-12-hydroxy-5,8,10,14-(Z,Z,E,Z)-eicosatetraenoic acid (12-HETE), is still unclear (Smith, 1989).

Ca^{2+} ions have been shown to be essential for purified 5-lipoxygenase activity and ATP further enhances this stimulation. As with other lipoxygenases, an addition of fatty acid hydroperoxides results in a significant activation, which favours the existence of regulatory redox mechanisms for 5-lipoxygenase activation. Furthermore, the purified enzyme requires the presence of two high molecular weight cytosolic factors and one membrane-bound factor for full activation (Rouzer et al., 1985;

Biological Reactive Intermediates IV, Edited by C.M. Witmer *et al.*
Plenum Press, New York, 1990

Rouzer et al., 1986), which points out the complexity of the regulatory mechanisms of this enzymatic activity. These studies have also demonstrated that the 5-lipoxygenase and the LTA4 synthase activities reside in a single protein, as indicated by the copurification of the two activities, the same stability, and the same requirements for activation.

Our studies have clearly established the importance of the GSH status in the neutrophil for activation of the enzyme (Hatzelmann et al., 1987), an observation likely related to the requirement of the enzyme for lipid hydroperoxides. We recently demonstrated that preincubation of human PMN with GSH-depleting agents prior to addition of arachidonic acid resulted in the stimulation of 5-lipoxygenase activity. For one of these agents, Dnp-Cl, we could exclude an involvement of enhanced intracellular Ca^{2+} levels, so that the underlying mechanism seemed to imply a different regulatory principle (Hatzelmann et al., 1989).

Studies on the mechanism of activation of lipoxygenases by hydroperoxides using the soybean enzyme have shown that these compounds oxidize the non-haem iron of the protein from the ferrous to the ferric form (Schewe et al., 1986). Other experiments show that ferric lipoxygenase can easily be inactivated by reduction, suggesting that changes in the oxidation state of the enzyme may be important for the expression in vivo of lipoxygenase activity (Kemal et al., 1987). Soybean lipoxygenase Type I is known to produce (13S)-13-hydroperoxy-(Z,E)-9,11-octadecadienoic acid (13-HPODE) from its natural substrate linoleic acid, but also 15-HPETE from arachidonic acid. The formation of these products can easily be measured, always together with their reduction products (13S)-13-hydroxy-(Z,E)-9,11-octadecadienoic acid (13-HODE) and 15-HETE, at 237 nm in a spectrophotometer.

Kinetic studies on prostaglandin H synthase indicated that the concentration of hydroperoxide needed for full cyclooxygenase (also a dioxygenase) activity is much less than that which gives 50 percent effectiveness with the peroxidase (Kulmacz et al., 1983). The hydroperoxide "tonus" was calculated to be 800 nM.

To prove whether these findings are due to the redox-state of the lipoxygenase, it was necessary to define a model system independent of other cellular factors. We found that an enzyme assay containing soybean lipoxygenase Type I and bovine erythrocyte glutathione peroxidase was able to simulate these effects.

MATERIALS AND METHODS

Chemicals

Prostaglandin B_2, Dnp-Cl (1-chloro-2,4-dinitrobenzene), GSH (reduced glutathione), isoluminol (6-amino-2,3-dihydro-1,4-phthalazinedione), microperoxidase (MP-11), lipoxygenase from soybean type I, glutathione peroxidase from bovine erythrocytes were purchased from Sigma Chemical Co. (Deisenhofen, FRG). Arachidonic acid and linoleic acid were obtained from Larodan (Malmö, Sweden). Chelex 100 analytical grade chelating resin, 100-200 mesh, was purchased from BIO-RAD (Munich, FRG). Boric acid "suprapur" and sodium hydroxide monohydrate "suprapur" were purchased from Merck (Darmstadt, FRG). All other solvents and chemicals utilized were of HPLC or analytical grade from Merck.

Isolation of neutrophils, reverse-phase HPLC analysis, incubation conditions and extraction of the metabolites were as described by Hatzelmann et al. (1987). Experiments in human PMN supernatant were performed as described by Hatzelmann et al. (1989).

10,000 xg supernatants were prepared in an Eppendorf centrifuge (5 min at room temperature), 265,000 xg supernatants in a TL-100 ultracentrifuge from Beckmann (Frankfurt, FRG) (30 min at 4°C).

Equipment for hydroperoxide detection

All HPLC analytical columns (Nucleosil ODS silica prepacked columns; 250x4.6 mm, 5 μm particle size) and guard columns (10x4.6 mm) were purchased from Bischoff (Leonberg, FRG). The HPLC system including Rheodyne injector (Cotati, CA, U.S.A.), model 2150 constant flow pump for HPLC eluant from LKB Pharmacia (Freiburg, FRG), model 600/200 constant flow pump for chemiluminescence solution from Kontron (Eching, FRG), mixing coil, UV- detector Uvicon 740 LC from Kontron with writer Servogor Z 10 from BBC Goerz (Austria), radioactivity monitor (Ramona) equipped with 650 μl flow cell from Raytest Isotopenmeßgeräte (Straubenhardt, FRG) and D-2500 Chromato-integrator (Merck) was configured as described by Yamamoto et al. (1987) and Yamamoto (1988). The radioactivity monitor was used as a photomultiplier, time constant set to 2 s. We found a flow rate ratio (HPLC eluant / chemiluminescence solution) of 1.0 to give the optimal signal / noise ratio.

The Millipore Milli-Q water purification system was from Waters Chromatographie (Königsbach-Stein, FRG).

HPLC eluant was identical with solvent B from Hatzelmann et al. (1987). Chemiluminescence eluant constituted of 500 mM sodium borate buffer, pH 10.0, containing 1 mM isoluminol and 2 μM microperoxidase/methanol (710/290 v/v).

To measure hydroperoxide levels, the conditions described were altered, as 9-HPETE was used as an internal standard. One GSH concentration slightly above the activation threshold (control) and one GSH concentration slightly below were chosen to measure the hydroperoxide concentration during the first 5 minutes of the lipoxygenase reaction.

Results are expressed as the means (+/-SEM) of a minimum of 3 experiments.

RESULTS AND DISCUSSION

We already concluded that in human PMN a defined threshold level of hydroperoxides is necessary for activation of 5-lipoxygenase (Hatzelmann et al., 1989). Figure 1 shows, that in human PMN supernatant, a certain concentration of glutathione peroxidase almost completely inhibits 5-lipoxygenase activity in the presence of Ca^{2+} and ATP. But, as can be seen in Figure 1, a resting basal activity could not be inhibited even with higher glutathione peroxidase concentrations. This result shows, that in human PMN supernatant a part of the enzyme always stays in the active form by mechanisms yet to be elucidated.

We found that an enzyme assay containing soybean lipoxygenase Type I (formally a 15-lipoxygenase) and bovine erythrocyte glutathione peroxidase was able to simulate the dependence on various substances described for various cells and homogenates. Incubation of defined amounts of these two enzymes with GSH in varying concentrations and a fixed amount of linoleic acid resulted in a threshold of lipoxygenase action (i.e. production of 13-HPODE). This threshold was dependent on the enzyme concentrations (Figure 2). As can be seen in Figure 2, preincubation with 50 μM GSH was sufficient to delete the lipoxygenase reaction, whereas in the case of tenfold higher enzyme concentrations, an amount of 250 μM GSH was necessary (data not shown).

To measure the corresponding hydroperoxide level in this enzyme assay it is necessary to deplete the system of bivalent transition metal ions, since they are responsible for the destruction of the hydroperoxides to be measured. This is achieved by using "suprapur"-grade chemicals for the buffers and preparing all reagents in bidest water, that was run over a "Milli-Q"-column prior to use. A comparison of the experiments run with transition-metal rich and transition-metal poor buffers, respectively, shows a marked decrease in the need for GSH to delete the lipoxygenase reaction (Figure 3).

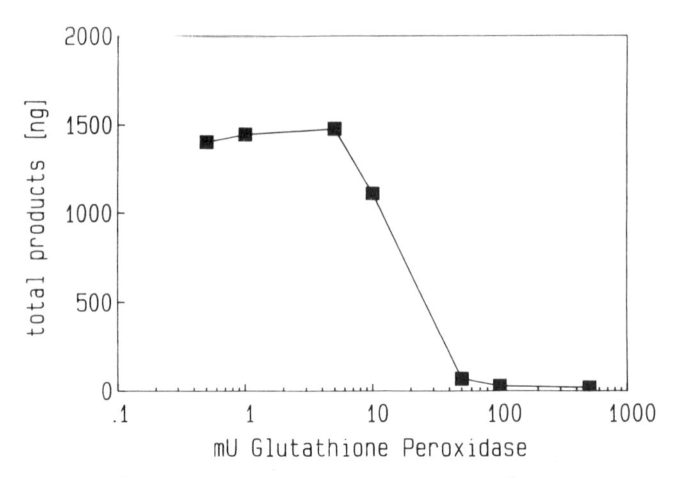

Figure 1. $5x10^6$ cells preincubated with 1 mM Ca^{2+}, 0.1 mM GSH, 5 mM ATP, and varying concentrations of glutathione peroxidase at 37°C for 5 min. Incubated with 10 µM arachidonic acid for 10 min in a total volume of 1 ml.

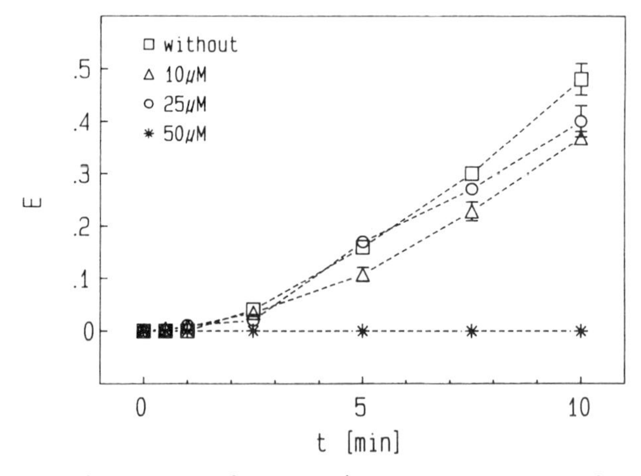

Figure 2. Preincubation of 0.64 µg/ml glutathione peroxidase, 84.5 µM linoleic acid, and various concentrations of GSH at 25°C for 5 min. Incubation with 0.5 µg/ml lipoxygenase in a total volume of 1 ml.

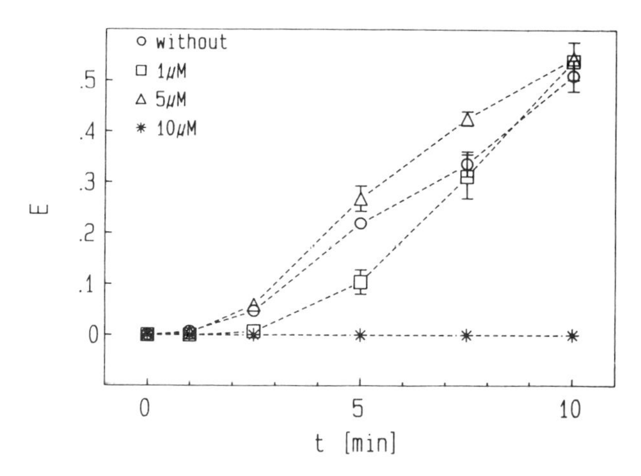

Figure 3. Preincubation of 0.64 μg/ml glutathione peroxidase, 84.5 μM linoleic acid, and various concentrations of GSH at 25°C for 5 min. Incubation with 0.5 μg/ml lipoxygenase in a total volume of 1 ml.

Figure 3 shows that after the treatment described above a GSH concentration of 10 μM is sufficient to delete the peroxidase reaction, whereas in the case of tenfold higher enzyme concentrations 150 μM are necessary (data not shown). This gives rise to the assumption that transition metals are indeed responsible for alterations in the redox state of the lipoxygenase in the model system. The corresponding hydroperoxide levels were measured to be 2.5 nM (+/- 0.3) and 13.9 nM (+/- 0.7), respectively. Unfortunately, these results show no linear relationship between enzyme concentrations, GSH level and hydroperoxide "tonus" in the model system.

Riendeau and coworkers (1989) could demonstrate in the high speed supernatant fraction from rat PMN that the amount of 5-HPETE required to activate the 5-lipoxygenase was directly related to the concentration of thiol and that maximal product formation could be obtained in the absence of 5-HPETE when reducing agents were also omitted from the reaction mixture. Their results also provide evidence that several components of the supernatant fraction modulate the level of 5-lipoxygenase activity measured in the absence of exogenous peroxides: the requirement for hydropeoxides could be replaced by ATP, other nucleotide enzyme cofactors, iron-II (Riendeau et al., 1989) and GTPγS (Denis et al., 1989). Our enzyme assay shows the same dependence on the concentration of thiols (GSH) and the addition of Fe-II and GTPγS (Figures 4 and 5). As can be seen in Figure 4, a concentration of 10 μM Fe-II is able to restore lipoxygenase activity, whereas Fe-III has no effect even at concentrations of 100 μM. Figure 5 shows the same result for GTPγS, which also starts reacting at a concentration of 10 μM.

NADPH has no effect in this model system, which is understandable because none of the enzymes used is dependent on this cofactor. The results of Riendeau and coworkers are probably due to the involvement of glutathione reductase in their supernatant. For us it was necessary to transfer these results to human PMN. Therefore the supernatants of human PMN were run over a Chelex 100 column prior to incubation. But, as can be seen in Figures 6 and 7, a basal level of hydroperoxides (basicly 12-HPETE) could not be reduced in order to elucidate the hydroperoxide "tonus" of 5-lipoxygenase in PMN.

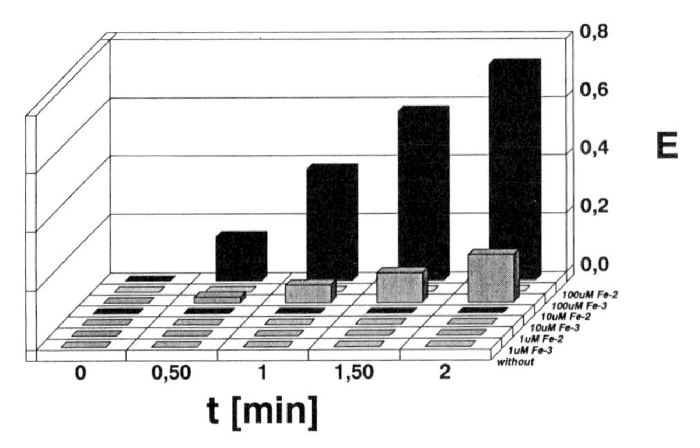

Figure 4. Preincubation of 6.4 µg/ml glutathione peroxidase, 84.5 µM linoleic acid, 150 µM GSH, and various concentrations of Fe-II or Fe-III at 25°C for 5 min. Incubation with 5 µg/ml lipoxygenase in a total volume of 1 ml.

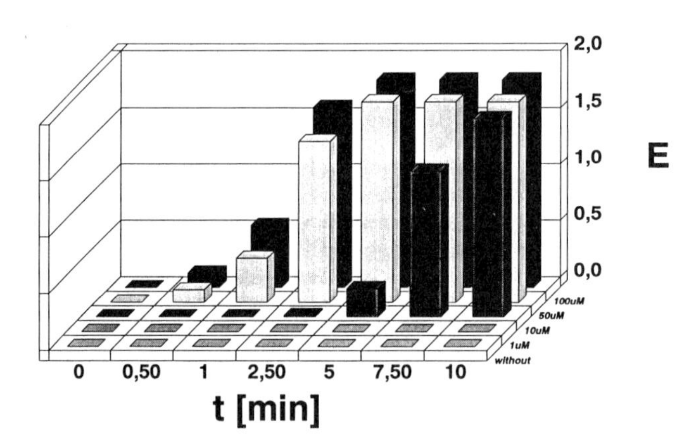

Figure 5. Preincubation of 6.4 µg/ml glutathione peroxidase, 84.5 µM linoleic acid, 150 µM GSH, and various concentrations of GTPγS at 25°C for 5 min. Incubation with 5 µg/ml lipoxygenase in a total volume of 1 ml.

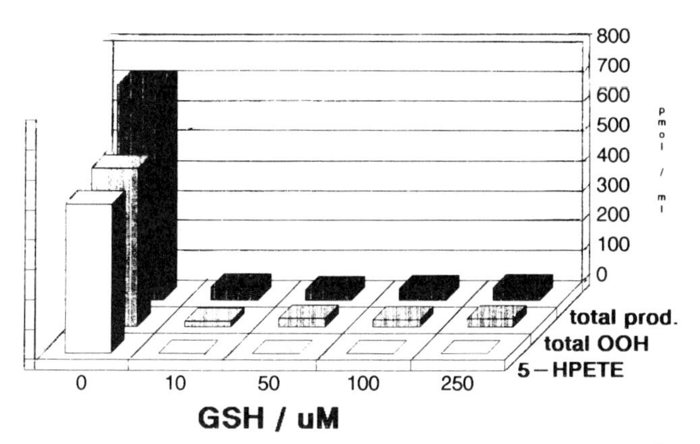

Figure 6. Preincubation of the 10,000 xg supernatant of $5x10^6$ cells/ml with 1 u/ml glutathione peroxidase, 0.5 mM Ca^{2+}, 10 µM ETYA, and various concentrations of GSH at 37°C for 5 min. Incubation with 20 µM arachidonic acid for 30 s in a total volume of 1 ml.

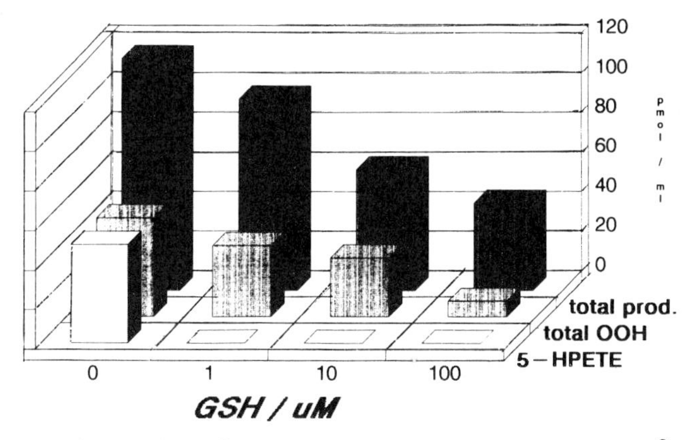

Figure 7. Preincubation of the 265,000 xg supernatant of $5x10^6$ cells/ml with 0.1 u/ml glutathione peroxidase, 0.5 mM Ca^{2+}, 10 µM ETYA, and various concentrations of GSH at 37°C for 5 min. Incubation with 20 µM arachidonic acid for 30 s in a total volume of 1 ml.

We conclude, that our PMN preparation still contains hydroperoxides other than 5-HPETE which may potentially keep 5-lipoxygenase in its active form.

REFERENCES

Borgeat, P. (1989). Biochemistry of the lipoxygenase pathways in neutrophils. *Can. J. Physiol. Pharmacol.* **67**, 936-942.

Denis, D., Choo, L.Y., and Riendeau, D. (1989). Activation of 5-lipoxygenase by guanosine 5'-O-(3-thiotriphosphate) and other nucleoside phosphorothioates: Redox properties of thionucleotide analogs. *Archives of Biochemistry and Biophysics* **273**, 592-596.

Hatzelmann, A., and Ullrich, V. (1987). Regulation of 5-lipoxygenase activity by the glutathione status in human polymorphonuclear leukocytes. *Eur. J. Biochem.* **169**, 175-184.

Hatzelmann, A., Schatz, M., and Ullrich, V. (1989). Involvement of glutathione peroxidase activity in the stimulation of 5-lipoxygenase activity by glutathione-depleting agents in human polymorphonuclear leukocytes. *Eur. J. Biochem.* **180**, 527-533.

Kemal, C., Louis-Flamberg, P., Krupinski-Olsen, R., and Shorter, A.L. (1987). Reductive inactivation of soybean lipoxygenase 1 by catechols: A possible mechanism for regulation of lipoxygenase activity. *Biochemistry* **26**, 7064-7072.

Kulmacz, R.J., and Lands, W.E.M. (1983). Requirements for hydroperoxide by the cyclooxygenase and peroxidase activities of prostaglandin H synthase. *Prostaglandins* **25**, No.4, 531-540.

Riendeau, D., Denis, D., Choo, L.Y., and Nathaniel, D.J. (1989). Stimulation of 5-lipoxygenase activity under conditions which promote lipid peroxidation. *Biochem. J.* **263**, 565-572.

Rouzer, C.A., and Samuelsson,B. (1985). On the nature of the 5-lipoxygenase reaction in human leukocytes: Enzyme purification and requirement for multiple stimulatory factors. *Proc. Natl. Acad. Sci. U.S.A.* **82**, 6040-6044.

Rouzer, C.A., Matsumoto, T., and Samuelsson, B. (1986). Single protein from human leukocytes possesses 5-lipoxygenase and leukotriene A4 synthase activities. *Proc. Natl. Acad. Sci. U.S.A.* **83**, 857-861.

Samuelsson, B., Dahlen, S.-E., Lindgren, J.A., Rouzer, C.A., and Serhan, C.N. (1987). Leukotrienes and lipoxins: Structures, biosynthesis, and biological effects. *Science* **237**, 1171-1176.

Schewe, T., Rapoport, S.M., and K?hn, H. (1986). Enzymology and physiology of reticulocyte lipoxygenase: Comparison with other lipoxygenases. In *Advances in Enzymology* (Meister, Ed.), Vol. 58, pp. 191-272. John Wiley and Sons.

Serhan, C.N., Hamberg, M., and Samuelsson, B. (1984). Lipoxins: Novel series of biologically active compounds formed from arachidonic acid in human leukocytes. *Proc. Natl. Acad. Sci. U.S.A.* **81**, 5335-5339.

Smith, W.L. (1989). The eicosanoids and their biochemical mechanisms of action. *Biochem. J.* **259**, 315-324.

Yamamoto, Y., Brodsky, M.H., Baker, J.C., and Ames, B.N. (1987). Detection and characterization of lipid hydroperoxides at picomole levels by High-Performance Liquid Chromatography. *Analytical Biochemistry* **160**, 7-13.

Yamamoto, Y. (1988). Analysis of lipid peroxidation by HPLC. In *Medical, Biochemical and Chemical Aspects of Free Radicals* (O. Hayaishi, E. Niki, M. Kondo, T. Yoshikawa, Eds.), pp. 797-804. Elsevier Science Publishers, B.V., Amsterdam.

OVERVIEW: THEORETICAL ASPECTS OF ISOTOPE EFFECTS ON THE PATTERN OF METABOLITES FORMED BY CYTOCHROME P-450

James R. Gillette and Kenneth Korzekwa

Laboratory of Chemical Pharmacology
National Heart, Lung, and Blood Institute
Bethesda, MD 20892

INTRODUCTION

Measurements of the effects of substitutions of deuterium at specific positions of substrates on the kinetics of metabolism of substrates by enzymes have provided valuable information on the mechanisms of enzymes. Most of the theories that permit interpretation of these measurements are based on the following assumptions: 1) Deuterium is substituted on only one position. 2) The isotope effect is on only one rate constant, namely the one which is associated with the breaking of a C-H Bond. 3) The enzyme forms only one product from the substrate. These assumptions thus imply that isotope effects on the reversible binding of the substrate in the active site of the enzyme are negligible. Secondary isotope effects also are usually assumed to be negligible. Northrop (1977) has pointed out that when these assumptions are valid, the apparent isotope effects manifested by the mechanisms of most enzymes may be shown to fit the following relationships:

$$^{D}(V/K) = \frac{(V_{max}/K_m)_H}{(V_{max}/K_m)_D} = \frac{C + k_H/k_D}{C + 1} \qquad (1)$$

where C is the "Commitment to Catalysis".

$$^{D}V = \frac{(V_{max})_H}{(V_{max})_D} = \frac{R + k_H/k_D}{R + 1} \qquad (2)$$

where R is the "Ratio of Catalysis".

Both C and R are collections of rate constants unique to the mechanism of the enzyme, and "mask" the "intrinsic isotope effect", k_H/k_D. Because most investigators of cytochrome P-450 enzymes have focused on the estimation of the "intrinsic isotope effects" and because R is usually (but not invariably) smaller than C, they have frequently restricted their studies to measurements of ^{D}V. By doing so, however, they have failed to obtain all of the information that is necessary to elucidate the mechanism of the enzyme. Indeed, they cannot even predict whether the isotope effects would be manifested at the low concentrations of the substrate usually obtained *in vivo*.

When investigators have studied isotope effects of the metabolism of compounds by the cytochrome P-450 enzymes, they frequently try to interpret their results in

Biological Reactive Intermediates IV, Edited by C.M. Witmer *et al.*
Plenum Press, New York, 1990

terms of very simple mechanisms. But the relevance of these simple mechanisms to the mechanisms of cytochrome P-450 enzymes is questionable.

Single Product Mechanisms

In the current view of the mechanisms of aliphatic hydroxylation by cytochrome P-450 enzymes, the enzyme first combines with the substrate to form a complex, (ES), that undergoes a one electron reduction followed by oxygenation and the introduction of second electron to form a dioxygenated complex, (EO_2S), which in turn undergoes rearrangement to form an oxenoid complex, (EOS), and water. The oxenoid complex then abstracts a hydrogen (presumably the isotopically sensitive step) to form a C-centered radical which then reacts with heme-OH radical to form the enzyme-product complex, which then dissociates to the product and the free enzyme.

In this mechanism (Figure 1A), the rearrangement of (EO_2S) to (EOS) plus water is irreversible. But as has been pointed out by Northrop (1975) when a single product is formed and when the isotopically sensitive step is preceded by an irreversible step, the equation for V_{max}/K_m will not contain the isotopically sensitive rate constant, and therefore the apparent isotope effect on V_{max}/K_m should be 1.0, i.e. the value of C in the Northrop's equation will be infinite. What happens in this situation is that at low substrate concentrations the concentration of the (EOS) complex is increased to such an extent that the rate of metabolism of the deuterated form will equal that of the protonated form, i.e. $(EOS_D)\ k_{46D} = (EOS_H)k_{46H}$.

There still may be an apparent isotope effect on V_{max} for the formation of a single product in such mechanisms (Figure 1A). But it follows that when an isotope effect on V_{max} is observed in these mechanisms, the magnitude of the apparent isotope effect on V_{max} should equal the magnitude of the isotope effect on K_m. Thus the approach to saturation of the enzyme will occur at lower concentrations of the deuterated substrate than at those of the protonated substrate.

It follows from these basic concepts that when an isotope effect on V_{max}/K_m is observed, the mechanism of metabolism by a cytochrome P-450 cannot be that shown in Figure 1A. Instead, there must be another pathway emanating from (EOS), the most likely of which is the reduction of (EOS) to (ES) with the formation of water (Figure 1B). When water is formed from (EOS) (that is the system is uncoupled) the concentration of the (EOS_D) complex at low concentrations of the deuterated substrate will be greater than that of (EOS_H) and thus there will be an increase in the rate of water formation. But the concentration of the (EOS_D) complex with the deuterated substrate will not reach a concentration that would permit the rate of metabolism of the deuterated form to equal the rate of metabolism of the protonated form. One might think that the formation of hydrogen peroxide would in itself also causes an apparent isotope effect on V_{max}/K_m, but it cannot because the hydrogen peroxide is formed at a step preceding the irreversible step that forms the immediate precursor of the isotope sensitive step. Nevertheless, the formation of hydrogen peroxide can potentiate the unmasking effects of water formation when the rate constant for the formation of the (EO_2S) complex (k_{23} in Figure 1) and the rate constant for the dissociation of (ES) to the substrate and the free enzyme (k_{21}) are either of the same order of magnitude or $k_{23} >> k_{21}$.

It becomes evident that the occurrence of isotope effects on the V_{max} of single metabolite reactions does not provide sufficient evidence to permit extrapolations to the low concentrations of the substrate that are frequently found *in vivo*. If water is not formed from (EOS) the pharmacokinetics of deuterated substrates may be virtually identical to those of protonated substrates. If water is formed from (EOS), however, a decrease in the clearance of formation of the metabolite from the deuterated form may be less than that from the protonated form. Only by measuring the isotope effects on both V_{max}/K_m and V_{max} can we begin to make such extrapolations possible.

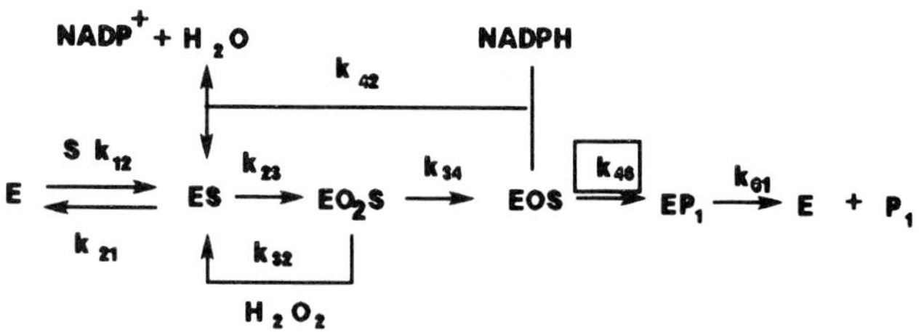

$$
E \underset{k_{21}}{\overset{S\ k_{12}}{\rightleftharpoons}} ES \xrightarrow{k_{23}} EO_2S \xrightarrow{k_{34}} EOS \xrightarrow{\boxed{k_{46}}} EP_1 \xrightarrow{k_{61}} E + P_1
$$

(ES ← k_{42} ← EOS: $NADP^+ + H_2O$ / NADPH; ES ← k_{32} ← EO_2S: H_2O_2)

k46 IS THE ISOTOPICALLY SENSITIVE RATE CONSTANT.

A. NO WATER FORMATION; $k_{42} = 0$.

$$^{D}(V/K) = 1$$

$$^{D}(V) = \frac{k46H\ (1/k23 + 1/k34 + 1/k61) + \boxed{k46H/k46D}}{k46H\ (1/k23 + 1/k34 + 1/k61) + 1}$$

B. WATER FORMATION; $k_{42} > 0$.

$$^{D}(V/K) = \frac{\dfrac{k46H((k23 + k21)k34 + k21\ k32)}{k21(k34 + k32)k42} + \boxed{k46H/k46D}}{\dfrac{k46H((k23 + k21)k34 + k21\ k32)}{k21(k34 + k32)k42} + 1}$$

$$^{D}(V) = \frac{\dfrac{k46H(k34 + k32 + k23 + k23\ k34/k61)}{(k34 + k32 + k23)\ k42 + k23\ k34} + \boxed{k46H/k46D}}{\dfrac{k46H(k34 + k32 + k23 + k23\ k34/k61)}{(k34 + k32 + k23)\ k42 + k23\ k34} + 1}$$

MODEL A

Figure 1. Single Metabolite Mechanisms

Multimetabolite Mechanisms

A cytochrome P-450 enzyme frequently may convert a substrate to several different metabolites, but the kinetic mechanisms by which the various metabolites are formed may differ not only with the enzyme and the substrate, but also with the metabolite. Studies of isotope effects on various parameters, however, can greatly aid the elucidation of different mechanisms. Studying isotope effects on the formation of several metabolites also permits us to study the apparent isotope effects on individual metabolites formed by either isotopically sensitive or isotopically insensitive pathways, total metabolism, and the ratio of the rates of formation of the metabolites. It is important, therefore, that investigators clearly define the parameter for which the isotope effect is calculated.

We have derived kinetic equations for three plausible mechanisms by which several metabolites may be formed from a substrate by a single enzyme. In deriving these equations we have assumed that k_{46} is the only isotopically sensitive rate constant and that secondary isotope effects are negligible.

Rapid-equilibrium Mechanism (Model A (Branched)

In this mechanism (Figure 2), a substrate may become reversibly bound in the active site of the cytochrome P-450 in several different orientations, but the different orientations are interconverted within the active site so rapidly that a virtual equilibrium always exists. At no time, however, does the substrate leave the active site as the substrate after the formation of the (EO_2S) complexes. In this mechanism, the conversion of the (ES) complexes to (EO_2S) complexes and then to the (EOS) complexes may be represented by single rate constants. A portion of the (EOS) complex is converted to one enzyme metabolite complex (EP1) by an isotopically sensitive step, while another portion of the (EOS) complex is converted to another enzyme-metabolite complex (EP2) by an isotopically insensitive step.

When the (EOS) complexes are not reduced to (ES) and water (Figure 2A), the equation for $^D(V/K)$ for the formation of P_1 may be rearranged to the Northrop equation. Notice that the intrinsic isotope effect term in the equation is (k_{46H}/k_{46D}), indicating that the isotope effect will be normal, i.e. the formation of P_1 from the protonated substrate will be less than that from the deuterated form of the substrate. A similar Northrop equation may be derived for the formation of P_2. But notice that in the equation for $^D(V/K)_{P2}$ the intrinsic isotope effect term is (k_{46D}/k_{46H}) indicating that there will be an apparent inverse isotope effect on the formation of P_2, that is the P_2 will be formed more rapidly from the deuterated form of the substrate than from the protonated form.

Since in this mechanism the formation of both (EP_1) and (EP_2) originate from the same intermediate (EOS), the ratio of the rates of formation will be independent of the concentration of (EOS) and therefore will be the same at any substrate concentration. The ratio of the products thus equals the ratio of the rate constants. According to this mechanism the isotope effect on the ratio of the products should therefore equal the "intrinsic isotope effect". In fact when this mechanism is valid $^D(P_1/P_2)$ may be substituted into the Northrop equations to estimate R or C.

The rate equation for $(V_{max(P1)} + V_{max(P2)})/K_m$ does not contain k_{46} when no water is formed from (EOS). Thus the isotope effect on $(V_m/K_m)_{total}$ for total metabolism should be 1.0. According to this mechanism, therefore, there should be isotope effects of the relative rates of formation of the metabolites, but no difference in the disappearance of the substrate at the low concentrations of the deuterated and protonated forms of the substrate that frequently occur *in vivo*.

The finding of a normal apparent isotope effect on $(V_{max(P1)} + V_{max(P2)})/K_m$ therefore, implies either the presence of an unidentified metabolite or the reduction of (EOS) to (ES) and water.

Parallel Pathway Mechanism (Model B)

In this model (Figure 3), the substrate combines with the enzyme in two (or more) distinctly different orientations in the active site, but there is no interconversion between the orientations and thus two (or more) distinctly different (ES) complexes are formed. There also is no interconversion between the orientations in any of the other intermediates, i.e., between (EO_2S_1) and (EO_2S_2) and between (EOS_1) and (EOS_2).

When (EOS) is not reduced to (ES) and water (Figure 3A), the isotope effects on (V_{max}/K_m) for both P_1 and P_2 will be 1.0. (EOS_{D1}) will be increased to the extent that $(EOS_{D1})k_{46D} = (EOS_{H1})k_{46H}$. Moreover, the isotope effects on V_{max} for P_2 will be normal and equal to that of P_1. The tendency of the (EOS_{D1}) to increase will cause a decrease in (EOS_{D2}) under saturating conditions. Consequently, the isotope effect

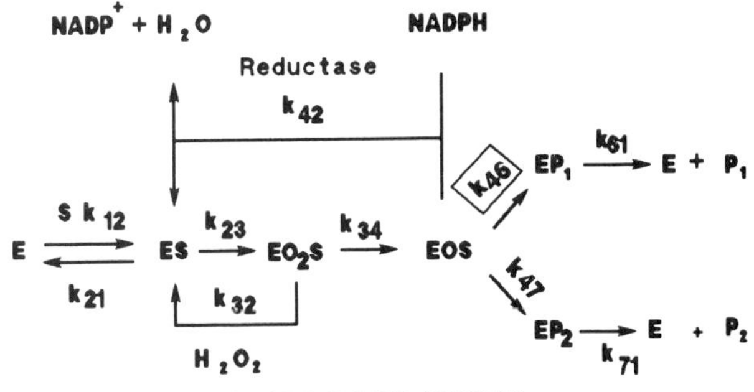

k46 IS THE ISOTOPICALLY SENSITIVE RATE CONSTANT.

A. NO WATER FORMATION; k42 = 0.

$$^{D}\left(\frac{V(1)}{K}\right) = \frac{k46H/k47 + \boxed{k46H/k46D}}{k46H/k47 + 1} \qquad ^{D}\left(\frac{V(1) + V(2)}{K}\right) = 1$$

$$^{D}\left(\frac{V(2)}{K}\right) = \frac{k47/k46H + \boxed{k46D/k46H}}{k47/k46H + 1} \qquad ^{D}(P1/P2) = \boxed{k46H/k46D}$$

B. WATER FORMATION; k42 > 0.

$$^{D}\left(\frac{V(1)}{K}\right) = \frac{\dfrac{k46H\ (k23\ k34 + A)}{k23\ k34\ k47 + A(k47 + k42)} + \boxed{\dfrac{k46H}{k46D}}}{\dfrac{k46H\ (k23\ k34 + A)}{k23\ k34\ k47 + A(k47 + k42)} + 1}$$

$$^{D}\left(\frac{V(2)}{K}\right) = \frac{\dfrac{k23\ k34\ k47 + A(k47 + k42)}{k46H\ (k23\ k34 + A)} + \boxed{\dfrac{k46D}{k46H}}}{\dfrac{k23\ k34\ k47 + A(k47 + k42)}{k46H\ (k23\ k34 + A)} + 1}$$

$$^{D}\left(\frac{V(1) + V(2)}{K}\right) = \frac{\dfrac{(k46H + k47)\ (k23\ k34 + A)}{A\ k42} + \boxed{\dfrac{(k46H + k47)}{(k46D + k47)}}}{\dfrac{(k46H + k47)\ (k23\ k34 + A)}{A\ k42} + 1}$$

$$^{D}(P1/P2) = \boxed{k46H/k46D} \qquad A = k21\ (k34 + k32)$$

MODEL A (BRANCHED)

Figure 2. Rapid Equilibrium Mechanisms

on (P_1/P_2) will be 1.0, that is the ratio of the products (P_1/P_2) obtained from the deuterated form of the substrate will be the same as that obtained from the protonated form.

When (EOS_1) is reduced to (ES_1) and water, the apparent isotope effect on the (V_m/K_m) for P_1 will be normal, but the apparent isotope effect on P_2 will still be 1.0, even when water is also formed from (EOS_2) (Figure 3B). By contrast, the isotope effect on V_{max} for P_1 will be normal, but that from P_2 may be either normal or inverse (Equation not shown).

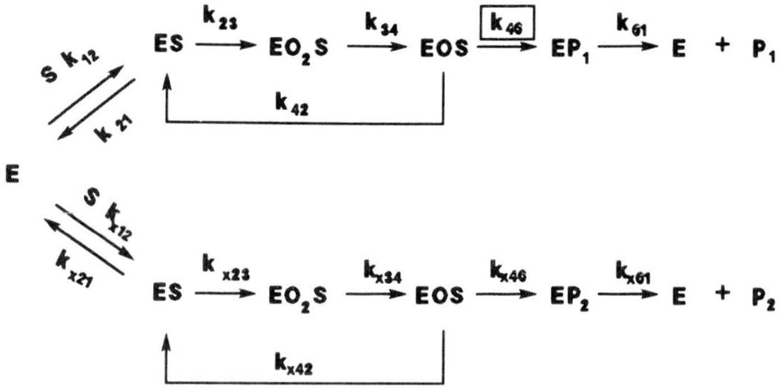

$$E \underset{k_{21}}{\overset{S\,k_{12}}{\rightleftharpoons}} ES \xrightarrow{k_{23}} EO_2S \xrightarrow{k_{34}} EOS \xrightarrow{\boxed{k_{46}}} EP_1 \xrightarrow{k_{61}} E + P_1$$

$$EOS \xrightarrow{k_{42}} ES$$

$$E \underset{k_{x21}}{\overset{S\,k_{x12}}{\rightleftharpoons}} ES \xrightarrow{k_{x23}} EO_2S \xrightarrow{k_{x34}} EOS \xrightarrow{k_{x46}} EP_2 \xrightarrow{k_{x61}} E + P_2$$

$$EOS \xrightarrow{k_{x42}} ES$$

k46 IS THE ISOTOPICALLY SENSITIVE RATE CONSTANT.

A. NO WATER FORMTION; k42 = 0

^{D}(V/K) SHOULD BE 1.0 FOR BOTH P1 AND P2.

^{D}(V) OF P1 EQUALS THAT OF P2. BOTH ARE NORMAL.

^{D} (P1/P2) SHOULD BE 1.0

B. WATER FORMATION; k42 > 0

^{D}(V/K) SHOULD BE NORMAL FOR P1.

^{D}(V/K) SHOULD BE 1.0 FOR P2.

^{D}(P1/P2) SHOULD BE NORMAL BUT MASKED.

MODEL B

Figure 3. Parallel Pathway Mechanisms

The equations thus reveal several ways by which the parallel pathway mechanism may be distinguished from the rapid equilibrium mechanism. The most obvious difference is that in the rapid equilibration mechanism the $^{D}(V/K)$ for the formation of P_2 is inverse but in the parallel pathway mechanism $^{D}(V/K)$ is 1.0.

Dissociative Mechanism (Model C)

This model was suggested by Harada *et al.* (1984). It is the same as the parallel pathway model, except that (EOS_1) and (EOS_2) may dissociate to (EO) and substrate, and the substrate may recombine with EO in either orientation. Because (EOS_1) may be converted to (EOS_2) by this process, the $^{D}(V/K)$ and $^{D}(V)$ for the formation P_1 may be normal and that for the formation of P_2 may be inverse when no water is formed from (EOS_1), (EOS_2) or EO. In this mechanism, however, the $^{D}(P_1/P_2)$ is no longer the "intrinsic isotope effect", k_{46H}/k_{46D}. Instead, it is masked by a term that depends

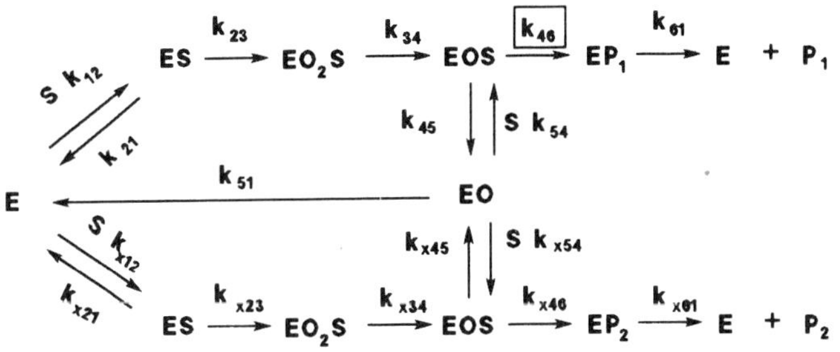

$$
\begin{array}{l}
ES \xrightarrow{k_{23}} EO_2S \xrightarrow{k_{34}} EOS \xrightarrow{\boxed{k_{46}}} EP_1 \xrightarrow{k_{61}} E + P_1 \\
E \underset{k_{21}}{\overset{S\,k_{12}}{\rightleftharpoons}} ES \qquad EOS \underset{S\,k_{54}}{\overset{k_{45}}{\rightleftharpoons}} EO \qquad EO \xrightarrow{k_{51}} E \\
EO \underset{k_{x45}}{\overset{S\,k_{x54}}{\rightleftharpoons}} EOS \qquad E \underset{k_{x21}}{\overset{S\,k_{x12}}{\rightleftharpoons}} ES \\
ES \xrightarrow{k_{x23}} EO_2S \xrightarrow{k_{x34}} EOS \xrightarrow{k_{x46}} EP_2 \xrightarrow{k_{x61}} E + P_2
\end{array}
$$

k46 IS THE ISOTOPICALLY SENSITIVE RATE CONSTANT.

A. NO WATER FORMATION; k51 = 0.

^{D}(V/K) FOR P1 SHOULD BE NORMAL BUT MASKED.

^{D}(V/K) FOR P2 SHOULD BE INVERSE BUT MASKED.

(k46H/k46D) SHOULD BE UNMASKED TO A GREATER EXTENT IN COMPETITIVE EXPERIMENTS THAN IN NONCOMPETITIVE EXPERIMENTS.

^{D}(P1/P2) SHOULD BE NORMAL BUT DOES NOT = (k46H/k46D).

B. WATER FORMATION; k51 > 0.

LINEWEAVER-BURK PLOTS SHOULD BE CURVED.

(k46H/k46D) SHOULD BE UNMASKED TO A GREATER EXTENT IN COMPETITIVE EXPERIMENTS THAN IN NONCOMPETITIVE EXPERIMENTS.

MODEL C

Figure 4. Dissociative Mechanisms

predominantly on the dissociation rate constant, k_{45}. As the values of k_{45} are increased, the masking term decreases and $^{D}(P_1/P_2)$ approaches the intrinsic isotope effect.

When water is formed from EO, the denominators of the equations for the rates of formation of P_1 and P_2 become binomial, which implies that the various kinds of Lineweaver-Burk plots should give curved rather than straight lines.

Whether or not water is formed from (EO), however, the dissociative model can be distinguished from the rapid equilibration mechanism by comparing the apparent isotope effects at low substrate concentrations obtained in noncompetitive experiments (the protonated and deuterated forms of the substrate are incubated separately) with those obtained in competitive experiments (the protonated and deuterated forms are incubated together). In the dissociative model the apparent isotope effect for P_1 at low substrate concentrations should be greater in the competitive experiments than in the noncompetitive experiments. This occurs because the presence of the protonated form offers other pathways emanating from (EOS_1)

that would not exist in the noncompetitive experiment. By contrast, in the rapid equilibration mechanism the $^{D}(V/K)$ values should be the same for the competitive and noncompetitive experiments.

DISCUSSION

From the insights gained from the equations for the various mechanisms, it becomes obvious that at the low concentrations of drugs that are frequently obtained *in vivo*, the isotope effects either for the total metabolism of the substrate or the formation of biologically reactive metabolites cannot be predicted solely by studies of $^{D}(V)$ for the formation of metabolites by cytochrome P-450 enzymes. Instead $^{D}(V/K)$ values must also be obtained . Moreover, even when $^{D}(V/K)$ values are obtained for the formation of metabolites by isotopically sensitive pathways, it would still be necessary to determine the $^{D}((V_{(P1)} + V_{(P2)})/K)$ values to determine whether isotope effects on the metabolic clearance of the substrate would be expected.

A complete analysis of isotope effects on several different parameters including the $^{D}(V/K)$ and $^{D}(V)$ for both the formation of the individual metabolites and various combinations of the different metabolites together with the $^{D}(P1/P2)$ values can provide valuable information not only for predicting the isotope effects *in vivo*, but also for elucidating different kinetic mechanisms by which several metabolites may be formed from a substrate by a single cytochrome P-450 enzyme.

REFERENCES

Harada, N., Miwa, G.T., Walsh, J.R. and Lu, A.Y.H. (1984). Kinetic isotope effects on cytochrome P-450-catalyzed oxidation reactions. Evidence for the irreversible formation of an activated oxygen intermediate of cytochrome P-448. *J. Biol. Chem.* **259**, 3005-3010.

Northrop, D.B. (1975). Steady-state analysis of kinetic isotope effects in enzymic reactions. *Biochemistry* **14**, 2644-2651.

Northrop, D.B. (1977). Determining the absolute magnitude of hydrogen isotope effects. In *Isotope Effects on Enzyme Catalyzed Reactions.* (W.W. Cleland, M.H. O'Leary and D.B. Northrop, Eds.), pp. 122-151, University Park Press, Baltimore.

S-THIOLATION OF PROTEIN SULFHYDRYLS

James A. Thomas, Eun-Mi Park, Yuh-Cherng Chai, Robert Brooks, Kazuhito Rokutan+, and Richard B. Johnston, Jr.*

Department of Biochemistry and Biophysics, Iowa State University
Ames, IA, 50011
Department of Preventive Medicine, Kyoto Prefectural+
University, Kyoto, JN., and
Department of Pediatrics*
University of Pennsylvania, School of Medicine
Philadelphia, PA

INTRODUCTION

Many cellular proteins have highly reactive sulfhydryls that are especially prone to modification during oxidative stress. Proteins with at least one "reactive" sulfhydryl are prevelant in cytoplasm, membranes, and nuclei, and the concentration of these reactive sites is impressive, reaching or exceeding the concentration of glutathione in many cells. Most proteins contain a single reactive sulfhydryl and are unlikely to form protein-protein disulfides under oxidizing conditions. Therefore, the most likely modification of these single reactive sulfhydryls by oxidative stress is formation of a mixed-disulfide with a low molecular weight cellular thiol such as glutathione, a modification we have termed protein S-thiolation (Ziegler, D.M., 1985 and Grimm, et.al., 1985). (If the modification is known to involve glutathione, the term S-glutathiolation is more descriptive.) This modification is metabolically labile and the rapid "dethiolation" of these proteins by several reductive processes (Park and Thomas, 1989) prevents lasting damage to the protein. Thus, the dynamic modification of proteins by S-thiolation/dethiolation processes represents one of the more important cellular mechanisms for protection against oxidative stress.

This paper describes experiments designed to establish the importance of this modification in a number of different cultured cell types by identifying some of the most abundant proteins that are modified, and by studying the metabolic processes that initiate protein S-thiolation in intact cells. These studies are interpreted in light of model studies with purified proteins that help define mechanisms of both S-thiolation and dethiolation. In this report the S-thiolation of proteins in heart cells, liver cells, macrophages, and 3T3-L1 cells will be discussed.

S-thiolation in Cells; Methodology

Cells were cultured by standard methods and immediately before experiments the culture medium was replaced with medium containing cycloheximide sufficient to inhibit protein synthesis. The minimum concentration of cycloheximide was determined in a preliminary experiment. After 15 to 30 min ^{35}S-cystine was added to those cell types that used only cysteine for glutathione synthesis, i.e., cardiac cells, 3T3-L1 cells, and macrophages. Liver cells used both cysteine and methionine for

Biological Reactive Intermediates IV, Edited by C.M. Witmer *et al.*
Plenum Press, New York, 1990

glutathione synthesis and they were incubated with Tran^{35}S-label (ICN, Inc.), a protein hydrolysate that contains both labeled methionine and cysteine. All cells were incubated with the isotope for 3 to 4 hours, during which time there was no incorporation of ^{35}S into proteins while approximately 1 to 5 % of the added label was incorporated into the glutathione. The specific activity of the glutathione pool in the cells needed to be at least 10^5 cpm/nmol to observe protein modifications by autoradiography.

The cells were immediately used for experiments in which various agents were used to either generate an oxidative stress (t-butylhydroperoxide, menadione, phorbol myristate acetate, or zymosan) or oxidize cellular sulfhydryls (diamide). At appropriate times after these additions, cells were rapidly washed and lysed by addition of hypotonic buffer (pH 7.0) containing 40 mM N-ethylmaleimide at room temperature. Cell debris was removed from this extract by a rapid low speed sedimentation step and an appropriate sample buffer containing additional N-ethylmaleimide was added for SDS-page analysis. These samples were prepared without any sulfhydryl reducing agent except in those cases where the purpose was to demonstrate that all radioactivity in the protein bands was lost on reduction by these agents. Samples were heated and equal amounts of protein were added to each sample well in the gel. Membrane fractions were prepared by a modification of published procedures (Cuatraecasas and Parikh, 1974). Basically, membrane fractions were isolated by centrifugation in an Airfuge, and after several washing steps, proteins were solubilized with 1% Triton X-100 at pH 7.4. The solubilized proteins were then treated with SDS-containing sample buffer in preparation for SDS-page analysis in the manner described for cytosolic proteins.

Mechanisms of S-Thiolation/dethiolation

In model studies on the effect of oxidative stress on proteins, we have used proteins that are known to be S-thiolated in intact cells, i.e., glycogen phosphorylase b, creatine kinase, and a 30 kDa protein from rat liver that has yet to be identified. Each is a major protein in the cytoplasm of these cells, representing between 1 and 7% of the soluble protein. Thus, these proteins represent a large pool of reactive protein sulfhydryl groups in the cytoplasmic fraction.

The mechanisms for the formation of S-glutathiolated derivatives (a mixed-disulfide with glutathione) of these proteins has been studied in some detail. It was early discovered that S-glutathiolation occurred rapidly without catalysis (Ziegler, D.M., 1985) and that partially reduced oxygen species (ROS) could cause protein S-glutathiolation without appreciable oxidation of reduced glutathione under conditions that simulated those that might be present in cell cytoplasm (Park and Thomas, 1988). In addition, the rate of S-glutathiolation of proteins by thiol/disulfide exchange with glutathione disulfide was appreciable only when the ratio of glutathione disulfide to reduced glutathione was high. These facts suggested that two different mechanisms for S-glutathiolation of proteins might be important under different conditions in intact cells. Thus, under conditions in which ROS are generated at low rates, direct reaction of ROS with protein sulfhydryls as shown in Figure 1A may predominate, while at the same time ROS that react with superoxide dismutase, and glutathione peroxidase produce small amounts of glutathione disulfide. The relative rates of protein S-thiolation and glutathione disulfide formation in intact cells are difficult to ascertain, because both processes are rapidly reversible. It seems likely that direct reaction of ROS with certain protein sulfhydryls might compete favorably with the enzymatic formation of glutathione disulfide. A second mechanism of protein S-thiolation may come into play when ROS are generated at higher rates, and enough glutathione disulfide is generated to participate in protein S-glutathiolation by thiol/disulfide exchange (Figure 1B).

These ideas lead to the hypothesis that protein S-thiolation may represent a major mechanism for detoxification of ROS in intact cells. Therefore, the dethiolation of these proteins represents an important component of the ability of cells to detoxify reactive ROS that are generated during oxidative stress. Continuing

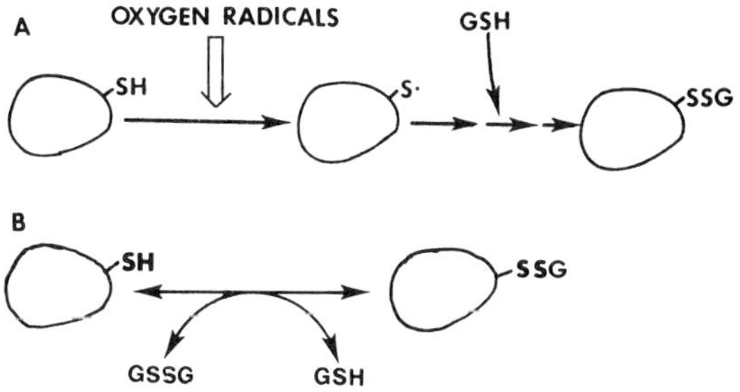

A OXYGEN RADICALS GSH
SH S· SSG

B
SH SSG
GSSG GSH

Figure 1. Mechanisms of protein S-thiolation. A. ROS-INITIATED S-THIOLATION: Direct action of ROS on proteins may result in thiyl radical intermediates (Ornerod and Singh, 1966) which can rapidly react with the large pool of reduced glutathione in intact cells. In cells with depleted glutathione pools, oxygen may compete for the thiyl radical intermediate resulting in irreversibly oxidized protein sulfhydryls (Park, and Thomas, 1988). B. THIOL/DISULFIDE EXCHANGE: The concentration of low molecular weight disulfides may determine the rate of thiol/disulfide exchange with specific protein sulfhydryls (Gilbert, H.F., 1984). It is likely that glutathione disulfide, cystine, and cystamine are all candidates for such reactions if adequate amounts of the disulfide form are generated in cells (Miller, et.al., 1990).

work on the mechanisms of protein dethiolation has shown that at least three different mechanisms may participate in this process (Park and Thomas, 1989). They include the three NADPH-dependent dethiolation mechanisms summarized in Figure 2. The first uses a thioredoxin-like protein (TLP) for dethiolation (Figure 2A), a second uses glutathione as an intermediate for reduction of a glutaredoxin-like protein (GLP) (Figure 2B), and a third reduces proteins directly with glutathione (Figure 2C). Our studies showed that creatine kinase is probably dethiolated directly by glutathione, while both glycogen phosphorylase and rat liver 30 kDa protein are only dethiolated appreciably by either of the first two mechanisms. Neither protein substrate showed a strong preference for TLP or GLP-dependent dethiolases. In recent work not shown here we have developed new assay methods for dethiolases and it is clear that most normal cells contain 3 to 5 times more TLP-dependent dethiolase tham GLP-dependent dethiolase. We have found only one cell type where GLP-dependent activity is absent. There are likely to be other important dethiolases, since protein disulfide isomerase is largely a membrane-associated enzyme with dethiolase-like activity (Freedman, R.B., 1989) and other thioltransferases have been described that may be active as dethiolases (Mannervik, etal., 1983, and Dotan and Shechter, 1987). In addition both cysteamine and cysteine may participate in protein dethiolation as well.

S-Thiolation of Proteins in Cells

Agents Used to Stimulate Protein S-Thiolation in Cells: Our experiments have emphasized identification of major S-thiolated proteins in cultured cells by using several different techniques for producing oxidative stress. Diamide has been used extensively because it is known to chemically oxidize sulfhydryl groups to the disulfide in a very specific manner, acting on both membranes and intracellular proteins. It is clear that protein sulfhydryls and low molecular weight thiols are both targets for this reagent, making it useful to identify many of the proteins that might become S-thiolated during oxidative stress. This reagent is not as selective in its action as agents that generate oxyradicals because oxyradicals may be generated at specific cellular sites, reacting quickly with neighboring proteins. t-Butylhydroperoxide (t-BuOOH) is a second agent that is used extensively. It can function as an effective substrate for intracellular glutathione peroxidase, causing extensive formation of

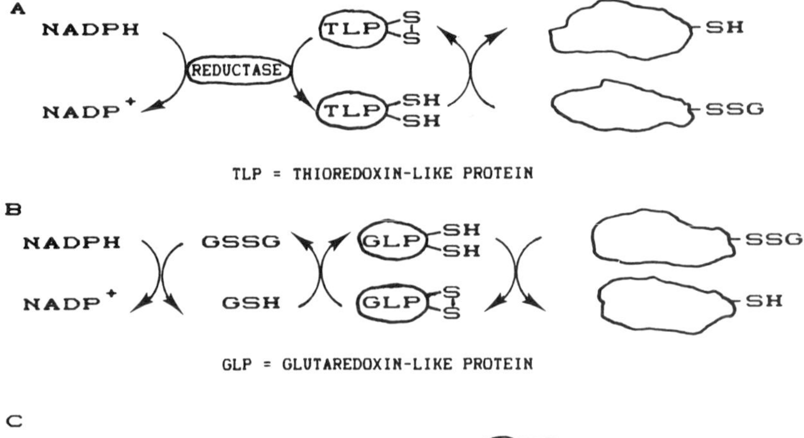

Figure 2. Mechanisms of Dethiolation. A. TLP-DEPENDENT DETHIOLASE: Dethiolation by a low molecular weight protein with properties similar to thioredoxin, i.e., thioredoxin-like protein (TLP). This is a two protein system that requires an NADPH-dependent reductase and TLP. B. GSH-DEPENDENT DETHIOLATION: Dethiolation by a low molecular weight protein with properties similar to glutaredoxin, i.e., glutaredoxin-like protein (GLP). This mechanism requires glutathione reductase and GLP. C. GLP-DEPENDENT DETHIOLASE: Direct reduction of S-thiolated proteins by reduced glutathione. Glutathione reductase is also an important component of this mechanism.

glutathione disulfide. This reagent may also generate radical species in treated cells and therefore a single mode of action as a substrate of glutathione peroxidase is not indicated. Menadione has been used to generate radicals inside cells by the redox cycling mechanism, but again it may cause other reactions that do not generate oxyradicals directly. Agents used to stimulate macrophages in our experiments are known to stimulate the NADPH oxidase of these cells to produce massive amounts of superoxide anion. Superoxide anion may be found both inside and outside the stimulated cells.

Cardiac Cells

We compared the effect of diamide and t-BuOOH on the glutathione pool and on protein S-thiolation in cardiac cells. The data shown in Figure 3 compare the patterns of protein S-thiolation at 0.2 mM t-BuOOH, which depleted glutathione by 22% and produced GSH/GSSG = 2, with several diamide concentrations. With 0.2 mM diamide, glutathione was 54% depleted and GSH/GSSG = 1.8. Thus, with equivalent oxidation/reduction status there was much more extensive modification of proteins by diamide. With 0.1 mM diamide, glutathione was depleted by 29% and GSH/GSSG = 1.8. With equivalent depletion of the total glutathione pool, there was a very large difference in the S-thiolation state of creatine kinase (42 kDa) and phosphorylase (97 kDa), two of the more abundant proteins in cardiac cells. On the other hand, the S-thiolation of the 23 kDa protein (a much less abundant protein) was approximately equal. This experiment showed that the specificity of protein modification was dependent on the mechanism of protein oxidation. With t-BuOOH, S-thiolation of

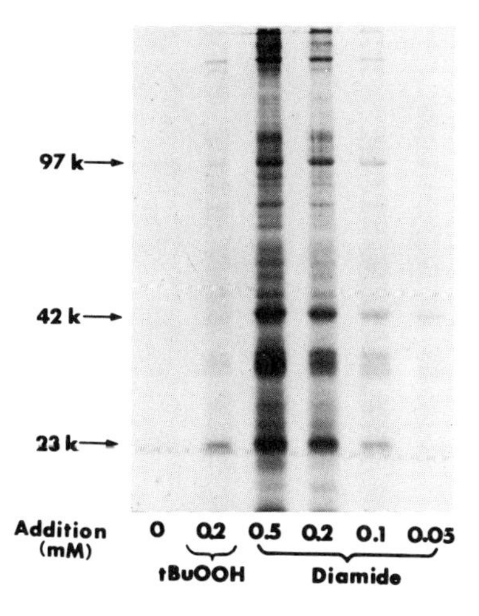

Autoradiogram

Figure 3. S-thiolation of soluble proteins in intact cardiac cells. Neonate rat heart cells were labeled with 35S-cystine as described in Methods and then treated with diamide or t-butylhydroperoxide for 3 min before isolation of proteins for analysis by SDS-page and autoradiography. Equal amounts of protein were applied to in each lane.

proteins was not impressive when compared to the effects of this agent on the glutathione pool.

Hepatocytes

Experiments with cultured hepatocytes showed the effect of t-BuOOH on these cells. Figure 4 shows the time course of S-thiolation in these cells when treated with 0.5 mM t-BuOOH. A rapid modification of many proteins occurred and the modification of the 30 kDa protein was especially prominant. The 30 kDa protein is the most abundant soluble protein in the soluble extract from these cells. It is a single subunit protein with one reactive sulfhydryl and a second sulfhydryl that reacts somewhat less readily. The glutathione pool was depleted by 20% at 3 min and GSH/GSSG = 0.64. The S-thiolation of all proteins was a transient response in these cells, with no discernable difference in the time sequence of events for individual proteins observed. This is typical of all cells studied. The medium from these cells at seven minutes still contained enough t-BuOOH to stimulate S-thiolation in other cells to a high level. Therefore, the transient nature of this response suggests the induction of dethiolation mechanisms during the stress.

Murine Macrophages

Macrophages were obtained from the peritoneal cavity of normal mice (resident macrophages) and mice that had been treated with lipopolysaccharide (LPS) four days prior to harvesting (activated macrophages). Activated macrophages can produce superoxide much more rapidly and in greater amounts than resident macrophages (Kitagawa and Johnston, 1985). Cells were treated with either opsonized zymosan (OZ) or phorbol myristate acetate (PMA) to stimulate the NADPH oxidase, producing large amounts of superoxide anion. Figure 5 shows the S-thiolation of soluble proteins from these cells in response to these stimuli. Part A compares the effect of zymosan and PMA on resident macrophages and Part B compares the effect of PMA on proteins from resident and LPS-activated macrophages. In no case was there any appreciable change in the redox status of the glutathione pool of these cells and the observed

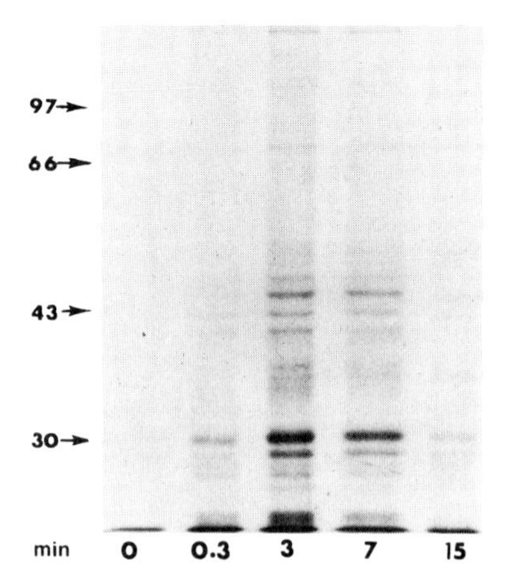

Figure 4. S-thiolation of soluble proteins in cultured rat hepatocytes. The glutathione pool in the hepatocytes was labeled with Tran35S-label (ICN, Inc.) as described in Methods. The cells were treated with 0.5 mM t-butylhydroperoxide and the individual cultures were harvested for protein analysis by SDS-page and autoradiography at the times indicated. Equal amounts of protein were applied in each lane.

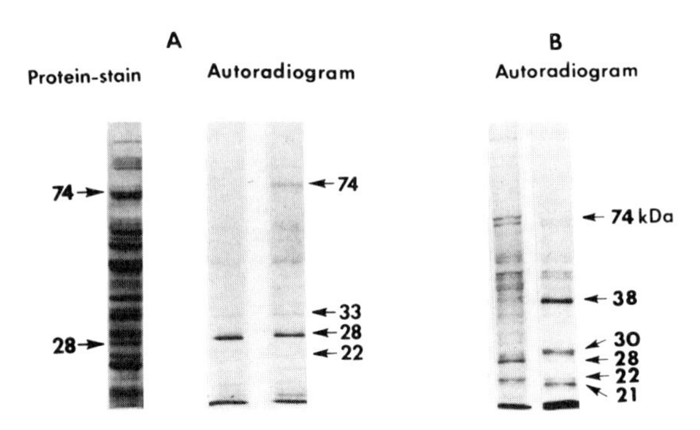

Figure 5. S-thiolation of soluble proteins in macrophages. Macrophages were obtained from the peritoneal cavity of normal mice and from mice that had been treated with 30 ug of lipopolysaccharide (LPS) for 4 days. The macrophages were attached to culture plates for 1 hour and labeled with 35S-cystine for four hours before addition of either 1 mg/ml opsonized zymosan (OZ) or 500 ng/ml phorbol myristyl acetate (PMA). Soluble proteins were extracted for SDS-page and autoradiographic analysis as described in Methods. (A) Normal resident macrophages. Left lane = OZ stimulated. Right lane = PMA stimulated. (B) Both lanes PMA stimulated. Left = resident macrophages. Right = LPS-activated macrophages.

protein S-thiolation cannot result from a thiol/disulfide exchange mechanism. This suggests that S-thiolation resulted from oxyradical-initiated protein S-thiolation as described in Fig. 1A. Each NADPH oxidase-stimulating agent produced a different S-thiolation pattern (Fig. 5A), and each cell type responded to PMA in a specific manner (Fig. 5B). These results suggest that individual protein modifications are related to the specific metabolic requirements of the affected cells. When macrophages were

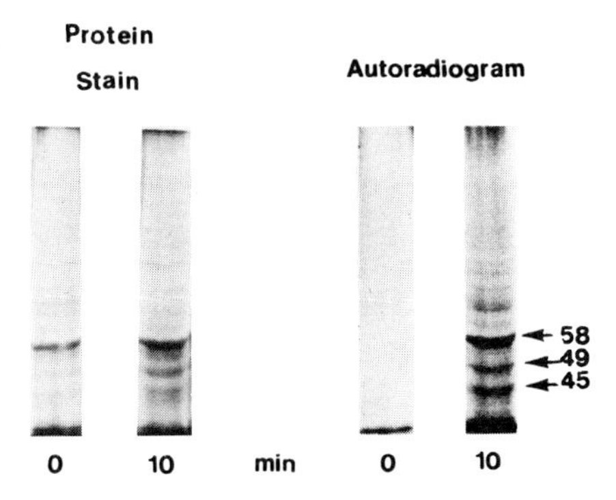

Figure 6. S-thiolation of membrane proteins in 3T3-L1 cells. Cells were labeled with 35S-cystine and membranes were isolated as described in Methods. S-thiolation was stimulated by addition of 0.5 mM diamide for 10 min.

treated with diamide the number of S-thiolated proteins was increased dramatically (data not shown).

3T3-L1 Cells

Since many membrane proteins are known to have reactive sulfhydryl groups it seemed possible that these proteins might also become S-thiolated under appropriate metabolic conditions. Thus, specific S-thiolation of membrane proteins from cultured 3T3-L1 cells was examined. Many solubleproteins were extensively S-thiolated within a few minutes in these cells and the glutathione pool was also rapidly and extensively depleted, while significant concentrations of glutathione disulfide were produced. Figure 6 shows the S-thiolation of specific membrane-associated proteins in these cells. S-thiolation of the membrane proteins was both rapid and extensive. This experiment shows that soluble thiols such as glutathione can form adducts with reactive membrane sulfhydryls.

DISCUSSION

Experiments described here clearly show that many different proteins in a variety of different cells are subject to S-thiolation/dethiolation. It is noteworthy that (1) the response of macrophages to NADPH-oxidase directed stimuli (zymosan and PMA) inclues S-thiolation of specific proteins, and that (2) both soluble and membrane-bound proteins are potential targets for S-thiolation. Table I shows a summary of major S-thiolated proteins that have already been identified in this report. The list of modified proteins is extensive. Only two of these proteins have been identified at present, i.e., cardiac creatine kinase and phosphorylase, but identification of other proteins is in progress. It should be emphasized that the studies summarized here have necessarily focused on the more abundant proteins of the cell fractions studied and many of the most abundant proteins in cells are included in the list of proteins modified by S-thiolation. These abundant proteins by themselves represent very substantial pools of reactive sulfhydryls that can act to sequester reactive oxygen intermediates as they are produced is specific reactions inside and outside cells. Thus, one can suggest that these major proteins by themselves may provide a significant reservoir of antioxidants proteins with specificity for certain reactive oxygen intermediates. These antioxidant proteins may be located in very sensitive locations within cells as a means of providing effective antioxidant capacity. The two-fold role of protein S-thiolation in cells may include important regulation of metabolic processes (Gilbert, H.F., 1984, and Brigelius, R., 1985) and oxyradical detoxification.

Table 1. Proteins S-Thiolated in Intact Cells

CELLS	MAJOR S-THIOLATED PROTEINS	STIMULUS	REFERENCE
SOLUBLE PROTEINS			
HEART	23, 97 kDa	t-BuOOH	Collison, et.al. 1986
	23, 36, 38, 42, 97 kDa	diamide	
LIVER	30 kDa	diamide, t-BuOOH menadione	Rokutan, et.al., 1989
	28, 30, 40, 43, 46, 52, 72 kDa	t-BuOOH	this report
MOUSE MACROPHAGES			
resident	22, 28 kDa	Zymosan	this report
	33, 74 kDa	PMA	
LPS-elicited	21, 22, 28, 30 33, 38, 74 kDa	PMA	this report
MEMBRANE PROTEINS			
3T3-L1	37, 45, 49, 58 kDa	diamide	this report.

These experiments also suggest that the role of glutathione in S-thiolation of individual proteins needs to be further studied. In some cells, protein S-thiolation was easily observed without significant alterations in the glutathione pool. With others, the extent of modification of individual proteins may have been related to either the total glutathione pool size or to the oxidation/reduction status of the glutathione pool. Thus, specific effects of certain chemical interventions may express one or the other effect more efficiently. Glutathione can participate in S-thiolation/dethiolation 1) as a co-reactant when proteins are activated by direct action of ROS (Figure 1A, 2) as a substrate (in the form of glutathione disulfide) that reacts directly with protein sulfhydryls to cause protein S-thiolation (Figure 1B, 3) as a substrate (in the form of reduced glutathione) that participates in protein dethiolation either by an enzymatic mechanism (Figure 2B) or by a non-enzymatic mechanism (Figure 2C). The oxidation/reduction status of glutathione may be more related to S-thiolation through items 1 and 2 cited above. On the other hand, the total glutathione pool may have a relationship to item 3, i.e., dethiolation. As an example, the enzymatic dethiolation of both phosphorylase b and the rat liver 30 kDa protein has a K_m of approximately 10 uM (Park and Thomas, 1989). Thus, even when glutathione pools are depleted in cells, the enzymatic dethiolation of these proteins should be unaffected. However, the non-enzymatic dethiolation of creatine kinase required millimolar glutathione, suggesting that the process would be slowed considerable when the glutathione pool was depleted. This observation is in keeping with the observed fact that S-thiolation of creatine kinase was increased in cardiac cells with a depleted pool of glutathione (Collison, et.al., 1986), e.g., a decreased rate of dethiolation could lead to more extensive S-thiolation. More definitive studies will be needed to clarify the potential regulatory roles of glutathione in protein S-thiolation/dethiolation.

ACKNOWLEDGEMENT

This research has been supported by grants from the American Heart Association and the American Diabetes Association to J.A.T., and by a grant from the U.S.P.H.S. to R.B.J, Jr.

REFERENCES

Brigelius, R., (1985) Mixed Disulfides: Biological Functions and Increase in Oxidative Stress, in *Oxidative Stress*, (H. Sies, ed.) pp. 243-272, Academic Press, London.
Collison, M.W., Beidler, D., Grimm, L.M., and Thomas, J.A., (1986). A Comparison of

Protein S-thiolation in Heart Cells Treated with t-Butylhydroperoxide or Diamide. *Biochim. Biophys. Acta*, **885**, 58-67.

Cuatrecasas, P., and Parikh, I., (1974). Insulin Receptors. *Meth. Enzymol.* **34**, 653-670.

Dotan, I., and Shechter, I., (1987). Isolation and Purification of a Rat Liver 3-Hydroxy-3-Methylglutaryl Coenzyme A Reductase Activating Protein. *J. Biol. Chem.*, **262**, 17058-17064.

Freedman, R.B., (1989). Protein Disulfide Isomerase: Multiple Roles in the Modification of Nascent Secretory Proteins. *Cell* **57**, 1069 1072.

Gilbert, H.F., (1984). Redox Control of Enzyme Activities by Thiol/Disulfide Exchange. *Meth. Enzymol.*, **107**, 330-351.

Grimm, L.M., Collison, M.W., Fisher, R.A., and Thomas, J.A., (1985). Protein Mixed Disulfides in Cardiac Cells. *Biochim. Biophys. Acta*, **844**, 50-54.

Kitagawa, S., and Johnston, R.B., Jr., (1985). Relationship Between Membrane Potential Changes and Superoxide-Releasing Capacity in Resident and Activated Mouse Peritoneal Macrophages. *J. Immunol.*, **135**, 3417-3423.

Mannervik, B., Axelsson, K., Sundewall, A-C., and Holmgren, A., (1983). Relative Contributions of Thioltransferase and Thioredoxin-Dependent Enzymes in Reduction of Low Molecular Mass and Protein Disulfides. *Biochem. J.*, **213**, 519-523.

Miller, R.M., Sies, H., Park, E-M., and Thomas, J.A., (1990). Phosphorylase and Creatine Kinase Modification by Thiol/disulfide Exchange and by Xanthin Oxidase-Initiated S-thiolation. *Arch. Biochem. Biophys.*, **276**, (in press).

Ornerod, M.G., and Singh, B.B., (1966). The Formation of Unpaired Electrons on Sulphur Atoms in Irradiated Dry Proteins as Studied by Electron Spin Resonance. *Biochim. Biophys. Acta*, **120**, 413-426.

Park, E-M., and Thomas, J.A., (1988). S-thiolation of Creatine Kinase and Glycogen Phosphorylase b Initiated by Partially Reduced Oxygen Species. *Biochim. Biophys. Acta*, **964**, 151-160.

Park, E-M., and Thomas, J.A., (1989). The Mechanisms of Reduction of Protein Mixed-Disulfides (Dethiolation) in Cardiac Tissue. *Arch. Biochem. Biophys.*, **274**, 47-54.

Rokutan, K., Thomas, J.A., and Sies, H., (1989). Specific S-thiolation of a 30 kDa Cytosolic Protein from Rat Liver under Oxidative Stress. *Eur. J. Biochem.*, **179**, 233-239.

Ziegler, D.M., (1985). Role of Reversible Oxidation-Reduction of Enzyme Thiols-Disulfides in Metabolic Regulation. *Ann. Rev. Biochem.*, **54**, 305-329.

CYTOCHROME P-450 AS A TARGET OF BIOLOGICAL REACTIVE INTERMEDIATES

James R. Halpert and Jeffrey C. Stevens

Department of Pharmacology and Toxicology, College of Pharmacy, University of Arizona, Tucson, AZ 85721

INTRODUCTION

In addition to being a primary source for the generation of biological reactive intermediates (BRI), cytochromes P-450 often serve as target molecules for these intermediates. Specific binding of BRI to individual forms of cytochrome P-450 may be exploited in the design of irreversible inhibitors known as mechanism-based inactivators. The high degree of selectivity often attainable with such compounds makes them particularly attractive for *in vivo* applications (Rando, 1984; Ortiz de Montellano et al., 1986).

The selectivity of a mechanism-based inactivator results from both the binding and catalytic specificity of the target enzyme and can be best understood by reference to the kinetic scheme shown in Figure 1. The inactivator is a substrate for the enzyme and first binds in the form of a reversible enzyme-substrate complex (ES). In the case of cytochromes P-450, NADPH and NADPH-cytochrome P-450 reductase are required for the substrate to be converted to a reactive form, still reversibly bound to the enzyme (EI). Reaction of this intermediate with an active-site residue or the heme prosthetic group of cytochrome P-450 results in inactive enzyme (E*-I). However, if the intermediate diffuses away from the active-site before reacting, it may covalently bind to other macromolecules or may rearrange to form a stable product (P). The inactivation generally obeys Michaelis-Menten kinetics (Rando, 1984). Analogous to the Vmax and Km for a normal enzyme catalyzed reaction, the inactivation process can be characterized by a maximal rate constant for inactivation (k inactivation) and an inhibitor constant (K_I). A marked difference in either constant between two cytochrome P-450 forms can result in selective enzyme inactivation. Another critical parameter is the efficiency of the inactivation process. The fewer turnovers required per inactivation event, the lesser the likelihood that the BRI generated will diffuse away from the cytochrome P-450 and react with other macromolecules.

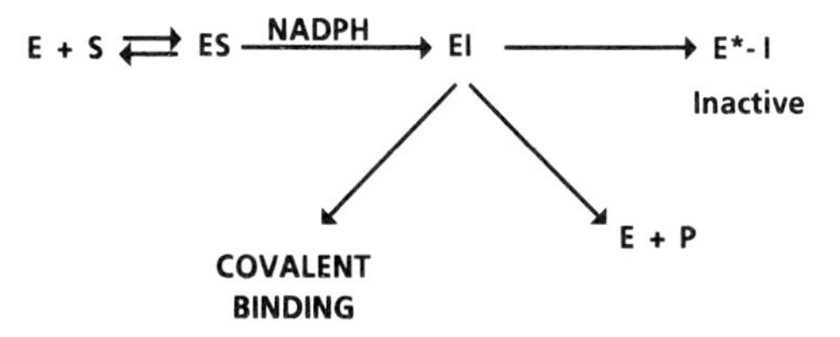

Figure 1. Kinetic scheme for mechanism-based inactivation of cytochrome P-450.

Biological Reactive Intermediates IV, Edited by C.M. Witmer *et al.*
Plenum Press, New York, 1990

Even in the case of a highly selective and efficient mechanism-based inactivator, the biological consequences of cytochrome P-450 modification may not be limited to enzyme inactivation (Figure 2). The reasons for this are multiple and can only be understood by examining the precise interactions between the inactivator and the enzyme. To date, three modes of cytochrome P-450 modification by BRI have been described (Osawa et al., 1989). The most thoroughly studied mechanism involves alkylation of a pyrrole nitrogen in the prosthetic heme group by BRI generated from such compounds as olefins, acetylenes, and dihydropyridines (Ortiz de Montellano et al., 1986). Cytochrome P-450 is the major consumer of heme in hepatocytes, and alkylation of P-450 heme often results in perturbation of hepatic heme metabolism and induction of porphyria (Marks et al., 1988). A second mechanism involves the generation of BRI that activate the heme moiety into a reactive species that binds to the protein moiety of cytochrome P-450. This may target the enzyme for proteolytic degradation, resulting in a decrease in P-450 protein in the microsomes (Osawa et al., 1989). The third and apparently least common mechanism involves the binding of BRI to amino acid residues in cytochrome P-450. The best studied examples of this are chloramphenicol (Halpert, 1981) and certain acetylenic fatty acids (CaJacob et al., 1988). It is not yet known whether protein or heme modification is the mechanism whereby certain compounds render cytochromes P-450 immunogenic, leading to the generation of anti-P-450 antibodies (Beaune et al., 1987).

DETERMINANTS OF THE MECHANISM OF CYTOCHROME P-450 MODIFICATION

To date, the factors that determine how a particular BRI will modify a given form of cytochrome P-450 are poorly understood in most cases. On the assumption that protein modification may be the mechanism of cytochrome P-450 inactivation least likely to cause other biological effects, our laboratory has focused on the chemical and enzymatic determinants of cytochrome P-450 protein modification. Extensive studies of a series of chloramphenicol analogs (Figure 3) indicate a requirement for a suitably substituted dichloroacetamido group for protein but no heme modification to occur (Halpert et al., 1986; Halpert et al., 1990). For example, replacement of the dichloroacetamido (-NH-CO-$CHCl_2$) group with a chlorofluoroacetamido (-NH-CO-CHClF) or dichloromethylketo (-CO-$CHCl_2$) group results in heme destruction. Interestingly, however, the chlorofluoroacetamides are more selective than their dichloroacetamide counterparts (Halpert et al., 1990).

1. **Enzyme Inactivation**
2. **Loss of apo-protein from microsomes**
   a. **Rapid (carbon tetrachloride)**
   b. **Delayed (chloramphenicol)**
3. **Perturbation of heme metabolism**
   a. **Ferrochelatase inhibition (dihydropyridines)**
   b. **Heme depletion (allylisopropylacetamide)**
4. **Anti-P-450 antibody formation (tienilic acid)**

Figure 2. Consequences of cytochrome P-450 modification.

R_3–C_6H_4–CH(R_2)–CH(R_1)–NH–C(=O)–CHClX

X = Cl, F, H

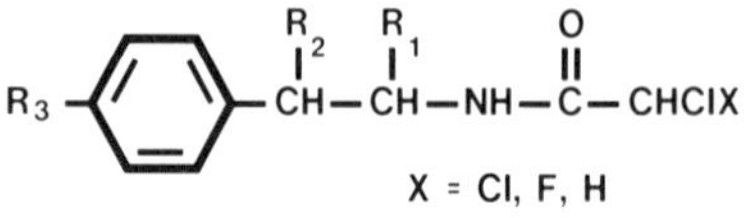

Figure 3. Generic structure of chloramphenicol analogs. In chloramphenicol, X = Cl, R_1 = CH_2OH, R_2 = OH, and R_3 = NO_3.

More recently we have turned our attention to inactivators of two adrenal microsomal cytochromes P-450 involved in steroid hormone biosynthesis. P-450 C-21 and P-450 17α catalyze the hydroxylation of progesterone at the 21- and 17- positions, respectively. The metabolic products of these reactions are important precursors in the production of mineralocorticoids, glucocorticoids, and sex hormones. In fact, an aberrant form of P-450 C-21 is one cause of the disease congenital adrenal hyperplasia, which can produce a virilization of the female or an inability to retain salt (Miller, 1988). By the incorporation of appropriate functional groups at the specific sites of P-450 metabolism of progesterone we have developed inactivators of varying specificity. In addition to studying the selectivity of different compounds as inactivators of P-450 17α and P-450 C-21, we have also investigated the mechanism of the inactivation process - the ability of each compound to 1) destroy the P-450 heme, 2) modify the P-450 protein, and 3) produce free, non-inactivating metabolites.

The substitutions made can be illustrated by reference to the structure of progesterone shown in Figure 4. Replacement of the 17-beta methylketo group of progesterone with a difluoromethylketo or dichloromethylketo function results in the two analogs 21,21-difluoro- and 21,21-dichloroprogesterone, respectively.

Our hypothesis was that these compounds might be selective and efficient inactivators of bovine adrenal P-450 C-21 by the scheme shown in Figure 5. With X representing a halogen atom, the P-450 would first hydroxylate at the 21-position. In contrast to the 21-hydroxyprogesterone metabolite (deoxycorticosterone) which is chemically stable, the putative metabolites of the halogenated analogs are likely to lose HF or HCl to form the acyl fluoride or acyl chloride reactive intermediates. The inherent chemical reactivity of these intermediates and the proximity of potentially modifiable sites on the enzyme probably determine whether the acyl halides bind irreversibly to the P-450 or undergo hydrolysis to the free acid metabolite (21-pregnenoic acid).

To first determine the mechanism of bovine adrenal P-450 C-21 inactivation by 21,21-dihalosteroids, the purified enzyme was incubated in a reconstituted system with the inhibitor and NADPH and then dialyzed extensively to remove any free compound. The residual P-450 content and progesterone 21-hydroxylase activity were then determined. For 21,21-dichloroprogesterone, approximately 14% of the enzyme activity and 39% of the spectrally detectable P-450 (as the reduced-CO complex) remained after a 10 minute incubation. This indicates that, contrary to the inactivation of the major phenobarbital-inducible rat liver cytochrome P-450 by chloramphenicol, this steroid inactivates P- 450 C-21 mainly by destroying the heme moiety. Inactivation by 21,21-difluoroprogesterone is less rapid (39% residual activity) and destroys less heme (74% P-450 left). Interestingly, in separate experiments using purified bovine adrenal P-450 17α and P-450 C-21, the difluoro but not the dichloro compound was found to be a selective inactivator of P-450 C-21. These results indicate that even a rational approach to the design of selective cytochrome P-450 inactivators must encompass a considerable amount of empiricism.

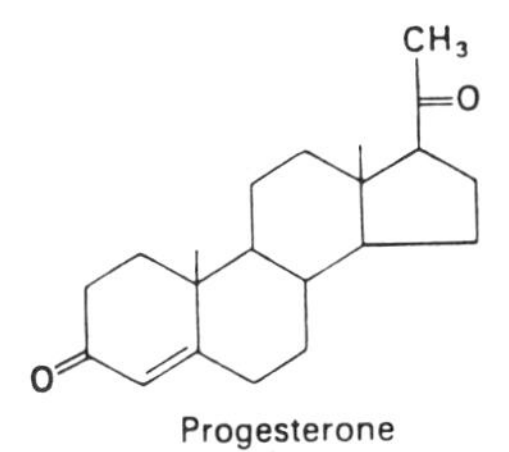

Figure 4. Structure of progesterone.

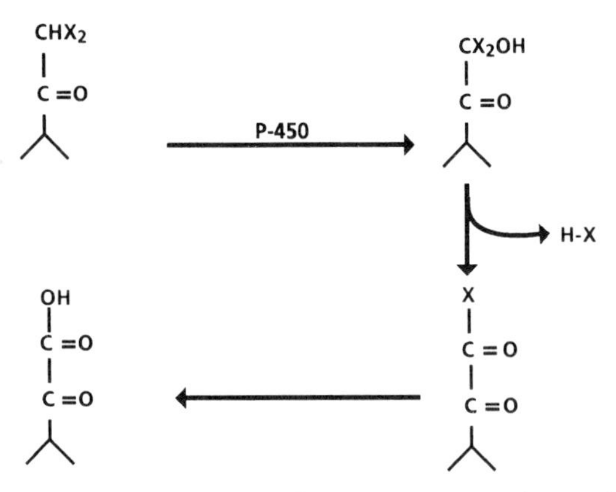

Figure 5. Proposed scheme for the metabolism of 21,21-dihalosteroids by bovine adrenal P-450 C-21.

An additional important question regarding the inactivation of bovine adrenal P-450 C-21 by our 21,21-dihalo steroids is the ratio of free metabolite produced to enzyme inactivated, the partition ratio. The ratio was determined by incubating purified P-450 C-21 with 21,21-dichloro- and 21,21-difluoroprogesterone, followed by high performance liquid chromatographic analysis of the organic extract. Initial experiments showed that the major metabolite formed from both steroids was 21-pregnenoic acid (see Figure 5). A representative chromatogram from an incubation with 21,21- dichloroprogesterone is shown in Figure 6. From these experiments we determined that 21-pregnenoic acid is formed from both steroids at a level of approximately 5 nmol of acid per nmol P-450 C-21 inactivated. The methyl ester derivative of the acid (formed by reaction with diazomethane) has a retention time of 14.8 minutes. The two peaks at 12.08 and 18.41 minutes did not appear in the minus NADPH samples and are unidentified. For 21,21-difluoroprogesterone, the more polar metabolite at 12.08 minutes appears to be formed at a much lower rate than for the dichloro compound.

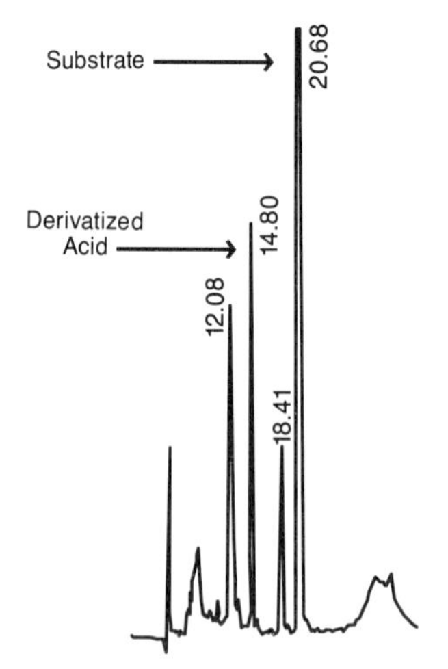

Figure 6. Chromatogram of the organic extract from the incubation of 21,21-dichloroprogesterone with bovine adrenal P-450 C-21 in a reconstituted system. The x-axis represents retention time in minutes and the y-axis absorbance at 254 nm.

CONCLUSIONS

By definition, mechanism-based inactivators require metabolism and must proceed through a BRI to produce enzyme inactivation. We have designed selective inactivators of rat liver and bovine adrenal cytochromes P-450 by synthesizing appropriate dihalomethyl compounds. These are converted by cytochromes P-450 to reactive acyl halides that may covalently bind to the protein moiety of the enzyme. In addition certain of the compounds appear to destroy the heme moiety of cytochromes P-450 by an unknown mechanism. The selectivity of the compounds and the mechanism of P-450 inactivation (heme vs. protein modification) are a function of the particular halogen substituents. Although these results show promise in better predicting inactivators of cytochromes P-450, the multiple variables in the inactivation process and the reactivity of generated BRI leave many questions to be investigated.

REFERENCES

Beaune, P., Dansette, R.M., Mansuy, D., Kiffel, L., Finck, M., Amar, C., Leroux, J.P., and Homberg, J.C. (1987). Human anti-endoplasmic reticulum autoantibodies appearing in a drug-induced hepatitis are directed against a human liver cytochrome P-450 that hydroxylates the drug. *Proc. Natl. Acad. Sci. USA* **84,** 551-555.

CaJacob, C.A., Chan, W.K., Shephard, E., and Ortiz de Montellano, P.R. (1988). The catalytic site of rat hepatic lauric acid ω-hydroxylase. Protein versus prosthetic heme alkylation in the ω-hydroxylation of acetylenic fatty acids. *J. Biol. Chem.* **263,** 18640-18649.

Halpert, J. (1981). Covalent modification of lysine during the suicide inactivation of rat liver cytochrome P-450 by chloramphenicol. *Biochem. Pharmacol* **30,** 875-881.

Halpert, J., Balfour, C., Miller, N.E., and Kaminsky, L.S. (1986). Dichloromethyl compounds as mechanism-based inactivators of rat liver cytochromes P-450 *in vitro*. *Mol. Pharmacol.* 30, 19-24.

Halpert, J., Jaw, J.-Y., Balfour, C., and Kaminsky, L.S. (1990). Selective inactivation by chlorofluoroacetamides of the major phenobarbital-inducible form(s) of rat liver cytochrome P-450. *Drug Metab. Disp.*, in press.

Marks, G.S., McCluskey, S.A., Mackie, J.E., Riddick, D.S., and James, C.A. (1988). Disruption of hepatic heme biosynthesis after interaction of xenobiotics with cytochrome P-450. *FASEB J.* **2,** 2774- 2783.

Miller, W. (1988). Molecular biology of steroid hormone synthesis. *Endocrine Rev.* **9,** 295-318.

Ortiz de Montellano, P.R., and Reich, N.O. (1986). Inhibition of cytochrome P-450 enzymes. In *Cytochrome P-450-Structure, Mechanism, and Biochemistry* (P.R. Ortiz de Montellano, Ed.) pp. 273-314. Plenum Press, New York.

Osawa, R., and Pohl, L.R. (1989). Covalent bonding of the prosthetic heme to protein: a potential mechanism for the suicide inactivation or activation of hemoproteins. *Chem. Res. Toxicol.* **2,** 131-141.

Rando, R.R. (1984). Mechanism-based enzyme inactivators. *Pharmacol. Rev.* **36,** 111-142.

HAPTEN CARRIER CONJUGATES ASSOCIATED WITH HALOTHANE HEPATITIS

Lance R. Pohl, David Thomassen, Neil R. Pumford, Lynn E. Butler, Hiroko Satoh, Victor J. Ferrans[+], Andrea Perrone, Brian M. Martin[*] and Jackie L. Martin

Laboratory of Chemical Pharmacology, National Heart, Lung, and Blood Institute, [+]Pathology Branch, Ultrastructure Section, National Heart, Lung, and Blood Institute, and [*]Clinical Neurosciences Branch
National Institute of Mental Health
National Institutes of Health
Bethesda, Maryland 20892

INTRODUCTION

The elucidation of the mechanism of hepatitis caused by the inhalation anesthetic agent halothane ($CF_3CHClBr$) (Satoh et al., 1987; Neuberger and Kenna, 1987; Weis and Engelhardt, 1989) is important since halothane is still widely used in adults in the United Kingdom and Europe (Neuberger and Kenna, 1987; Weis and Engelhardt, 1989) and is often an agent of choice for children (Hals et al., 1986; Whitburn and Sumner, 1986; Kenna et al., 1987a). Also, recent reports suggest that the structurally related inhalation anesthetic enflurane (CHF_2OCF_2CHFCl), which is widely used in adults in the United States, may cause liver damage by a similar process (Christ et al., 1988a; Christ et al., 1988b).

Several studies have suggested that halothane hepatitis is due to an immune response directed against novel hapten carrier protein conjugates (neoantigens) of the liver (for reviews see Satoh et al., 1987; Neuberger and Kenna, 1987a; Pohl et al., 1989a). The most informative findings have come from immunoblotting studies with the patients' sera. In these investigations, serum antibodies were found to recognize polypeptide fractions in liver microsomes of humans (Kenna et al., 1988b) and animals (Kenna et al., 1987b; Kenna et al., 1988a; Pohl et al., 1988) of 100 kDa, 76 kDa, 59 kDa, 57 kDa, and 54 kDa that had been covalently modified by the reactive trifluoroacetyl halide (CF_3COX, TFA-X) metabolite of halothane (Kenna et al., 1988a; Roth et al., 1988; Pohl et al., 1989a). The epitopes of the neoantigens recognized by the patients' antibodies appeared to consist of antigen-specific domains that had been modified by the TFA moiety (Kenna et al., 1988a; Pohl et al., 1989a). These findings and the observation that antibodies in the sera of different halothane hepatitis patients did not always recognize the same hapten carrier conjugates indicated that specific TFA altered carrier proteins are immunogenic in a given patient and possibly have an immunopathological role in the development of the patient's hepatitis.

In this report, we will review the results of the most recent studies dealing with the characterization of the TFA neoantigens. Several proteins have been identified, and appear to have important physiological functions.

Biological Reactive Intermediates IV, Edited by C.M. Witmer *et al.*
Plenum Press, New York, 1990

RESULTS

Characterization of the TFA 59 kDa hapten carrier conjugate

The basic approach used to characterize the TFA proteins has been to purify them from liver microsomes of rats treated with halothane and to determine whether the amino acid sequences of their N- terminals or internal peptides correspond to the sequences of known proteins.

Using this approach, we have purified the TFA 59 kDa protein from liver microsomes of halothane treated rats by first isolating the TFA proteins from liver microsomes by chromatography on an immunoaffinity column of anti-TFA IgG and then purifying the TFA 59 kDa protein from this mixture by anion exchange hplc. Based upon its apparent monomeric molecular mass, NH_2-terminal amino acid sequence, catalytic activity, and other physical properties, the protein was identified as a microsomal carboxylesterase (Satoh et al., 1989), which appeared to be identical with form E1 (Harano et al., 1988). This isozyme apparently corresponds to the form ES- 8/ES-10 (Mentlein et al., 1987).

Since at the time of the purification of the TFA 59 protein no complete primary sequence was known for any mammalian liver carboxylesterase, we cloned a liver carboxylesterase cDNA from a rat liver lambda gt11 expression library using a polyclonal antibody raised against the TFA 59 kDa protein as a screening reagent (Long et al., 1988). The clone encoded for a carboxylesterase that showed overall identity of 84 % with several peptides derived from the 59 kDa protein. It therefore appeared that an isoform of the 59 kDa protein had been cloned. Other investigators apparently have cloned the same rat liver carboxylesterase (Takagi et al., 1988). It is not yet known if any of the liver microsomal carboxylesterases isozymes biochemically characterized to date correspond to the cloned cDNA.

The carboxylesterases are a family of enzymes based upon their substrate specificity and electrophoretic and immunochemical properties (Mentlein et al., 1980; Robbi and Beaufay, 1983; Mentlein et al., 1984a). Although the genes of these enzymes have not yet been cloned, genetic studies in animals (Mentlein et al., 1987) as well as Southern blot analysis of genomic DNA (Long et al., 1988) suggest that the enzymes comprise a multigene family. These enzymes may have several functions in the body. They have been shown to metabolize not only ester, thioester, phosphoester, and amide bonds of xenobiotics, but also endogenous lipids, such as fatty acid esters of carnitine and CoA, acylglycerols, lysophospholipids, and phospholipids (Mentlein and Heymann, 1984; Mentlein et al., 1984b; Mentlein et al., 1985a; Mentlein et al., 1985b; Satoh, 1987). One of the carboxylesterases, also known as egasyn, forms a complex with β-glucuronidase. This interaction appears to anchor β-glucuronidase within the lumen of the endoplasmic reticulum (ER) (Medda et al., 1989). The 59 kDa carboxylesterase that we have purified, however, does not have this activity (Medda et al., 1989). Whether any of the carboxylesterases have a function of anchoring other proteins in the lumen of the ER is not known.

The potential biological importance of the carboxylesterases is also suggested by the finding that nearly all of the tissues studied to date contain carboxylesterase activities (Mentlein et al., 1987). However, there have not been any reports of comparative tissue concentrations or cellular localization of the carboxylesterases by immunochemical procedures. In this regard, we have found by quantitative immunoblotting of tissue homogenates with anti-59 kDa polyclonal antibodies that this protein is present at relatively high levels not only in the liver, but also in the testis, lung, fat, and adrenal gland (Fig. 1). The protein could not be detected in the kidney. Immunohistochemical studies showed that the protein was concentrated in the perivenular (zone 3) regions of the liver acinus (Fig. 2A) and in the interstitial spaces (possibly the Leydig cells) between the seminiferous tubules of the testis (Fig. 2B). This finding indicates that in the liver and testis, the cellular concentration of the 59 kDa carboxylesterase is higher than is suggested by the results of the immunoblotting studies with total tissue homogenates. Although the physiological significance of these results is not known, immunohistochemical studies have shown that the TFA hapten was also concentrated in the perivenular regions of the liver after the administration of halothane to rats (Satoh et al., 1985b). It is concluded from these results that the cytochromes P-450 that convert halothane into the reactive trifluoroacetyl halide metabolite and at least one of the proteins that traps this species are concentrated within the same cells. This circumstance would permit the formation of a relatively high level of hapten carrier conjugate, which is one of the criteria that probably determines, at least in part, the immunogenic properties of a neoantigen (Pohl, 1989b). It is noteworthy that recent studies indicate that the carboxylesterases in the liver, if not in other organs, are localized in the lumen of the endoplasmic reticulum (Robbi and Beaufay, 1987; Harano et al., 1988; Mentlein et al., 1988). As will be described shortly, this subcellular localization may have importance in halothane hepatitis, since many of the proteins that become labeled with the TFA moiety and are recognized by the patients' antibodies appear to be localized in this subcellular compartment.

Characterization of the other TFA hapten carrier conjugates

It was recently found that the 100 kDa, 76 kDa, 59 kDa, and 57 kDa TFA neoantigens could be selectively extracted from liver microsomes by treatment with 0.1 % deoxycholate detergent. The individual proteins were purified from this mixture by a combination of DEAE Sepharose chromatography and hplc (Fig. 3A). Purification of the TFA 59 kDa protein by this approach has been described (Medda et al., 1989) and the detailed methods for the purification and characterization of the other TFA proteins will be the subject of future papers. During the purification of the proteins, two additional TFA proteins were discovered. One had an apparent monomeric molecular mass of 58 kDa whereas the other was 63 kDa, as determined by SDS/PAGE. These proteins were not detected previously in liver microsomes by immunoblotting,

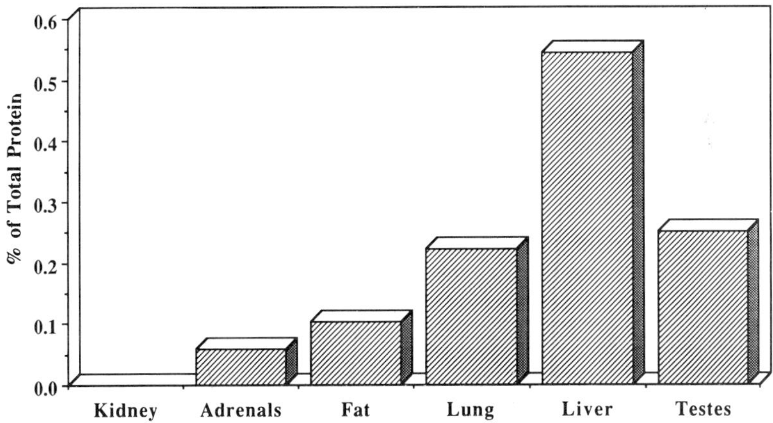

Figure 1. Distribution of the 59 kDa carboxylesterase in various tissues of rat as determined by quantitative immunoblotting with anti-59 kDa serum. SDS/PAGE and immunoblotting was conducted as previously described except that the goat second antibody was conjugated with alkaline phosphatase instead of horseradish peroxidase (Satoh, et al., 1989). Each lane contained 100 µg of tissue homogenate from untreated male Sprague-Dawley rats (200- 250 g). The amount of 59 kDa carboxylesterase in the homogenates was determined by comparing the intensity of the densitometric scans of the immunoblots of the tissue homogenates to that of a standard curve of the purified TFA 59 kDa protein (1 µg, 0.5 µg, 0.25 µg and 0.1 µg).

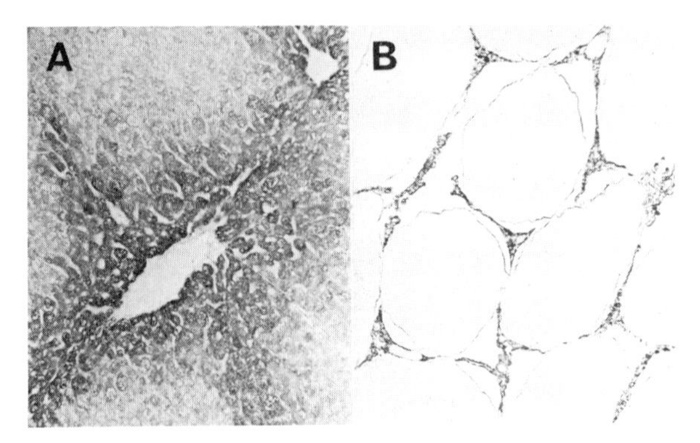

Figure 2. Immunohistochemical localization of the 59 kDa carboxylesterase in the liver (A) and testis (B) of rat. Liver and testis from untreated male Sprague-Dawley rats (200-250 g) were fixed in phosphate buffered formalin for 24 hr. Paraffin sections of the tissues (5-6 μm) were mounted onto glass slides. The paraffin was removed in the usual manner, endogenous peroxidase activity was abolished by placing the slides in 3 % hydrogen peroxide in methanol for 30 min, and non-specific binding sites were blocked with normal goat serum. Anti-59 kDa serum (1/1000 dilution) was placed onto the slides for 1 to 2 hrs. Immunoperoxidase detection was accomplished by an avidin-biotin peroxidase complex (ABC) procedure using an ABC kit from Vector Laboratory. No staining was observed when anti-TFA serum was used as a control in place of anti-59 kDa serum. In addition, the staining was blocked when the anti-59 kDa antibody was preincubated with purified 59 kDa carboxylesterase before it was added to the tissue sections.

because they were not resolved from the 59 kDa protein (Kenna et al., 1988a; Satoh et al., 1989; Pohl et al., 1989a). Moreover, the TFA protein in liver microsomes of halothane treated rats that has been designated as the TFA 76 kDa neoantigen (Kenna et al., 1988a; Satoh et al., 1989; Pohl et al., 1989a), was calculated by SDS/PAGE to have an apparent monomeric molecular mass of 80 kDa after it was purified. For this reason, the 76 kDa neoantigen will be designated in the future as the 80 kDa neoantigen.

Immunoblotting of the purified TFA neoantigens with hapten specific anti-TFA antiserum indicated that the 59 kDa protein had covalently bound to it the highest concentration of TFA haptens, while the lowest level of hapten was bound to the 63 kDa protein, which could only be faintly detected under the conditions of the immunoblotting procedure (Fig. 3B). The 57 kDa protein seemed to have a higher level of TFA hapten bound to it than the 58 kDa, 80 kDa, and 100 kDa proteins, which stained at approximately the same level of intensity. Determination of the reactivity of the TFA neoantigens with sera from ten patients with a clinical diagnosis of halothane hepatitis (Table 1) indicated that the relative level of covalent binding of the TFA hapten to the proteins (Fig. 3B) was not the major factor that determined the order of antigenicity of the proteins. This fact is illustrated by the low incidence of immunoreactivity of the patients' sera with the TFA 59 kDa neoantigen, even though this protein appeared to contain the highest level of TFA haptens. If it is assumed that the TFA proteins present in human liver (Pohl et al., 1989a) have the same relative level of TFA hapten groups bound to them as that of the rat proteins, this would suggest that the immunogenicity of the TFA proteins is more dependent upon the structural features of the specific proteins than on the level of TFA haptens bound to them. This idea is consistent with the results of immunochemical studies on the importance of the role of the carrier protein on the immunogenicity of synthetic hapten carrier conjugates (Pohl et al., 1988).

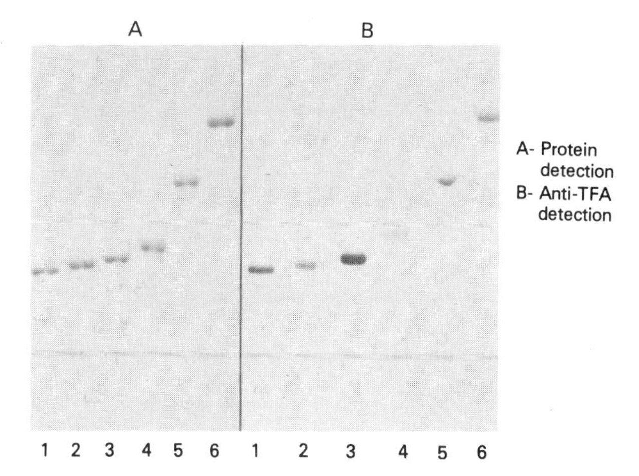

Figure 3. Protein staining (A) and anti-TFA reactivity (B) of TFA neoantigens purified from rat liver microsomes. Male Sprague-Dawley rats (200-250 g) were treated with halothane (20 mmole/kg dissolved in sesame oil). After 16 hr, microsomes were prepared and extracted with 0.1 % deoxycholate as previously described (Medda et al.1989). The TFA neoantigens were purified from the extract following the general method that was used to purify the TFA 59 kDa carboxylesterase (Medda et al.1989). SDS/PAGE and immunoblotting with anti-TFA antiserum was conducted as previously described except that the goat second antibody was conjugated with alkaline phosphatase instead of horseradish peroxidase (Kenna et al.1988a). Each lane contained 2 μg of the purified TFA proteins: 1-57 kDa, 2-58 kDa, 3-59kDa, 4-63 kDa, 5-80 kDa, and 6-100 kDa.

Recent studies, comparing the amino acid sequences of the N-terminals and internal peptides of the other purified proteins with the sequences of known proteins, have led to the identification of the 57 kDa protein and the tentative identifications of the 100 kDa and 63 kDa proteins. The TFA 57 kDa protein corresponds to protein disulfide isomerase (Martin et al., 1989). This protein appears to have multiple physiological functions. Not only does it catalyze the isomerization of both the intramolecular and intermolecular disulfide bonds of proteins, but it also appears to be the β-subunit of proline-4-hydroxylase, the major non-nuclear binding protein of 3,3',5-triodo-L-thyronine thyroid hormone, and the glycosylation site binding protein component of oligosaccharide transferase (Freedman, 1989). It is also concentrated within the lumen of the endoplasmic reticulum (Pelham, 1989).

The 100 kDa protein has shown high sequence homology with ERp99, which is believed to be identical with a 94 kDa glucose-regulated protein (grp94), and endoplasmin (Mazzarella and Green, 1987; Macer and Koch, 1988; Booth, and Koch, 1989; Thomassen et al., 1990). Although the physiological function of this protein is not known, it appears to have at least two activities. It is a calcium binding protein and may have a role in cellular calcium homeostasis (Macer, and Koch, 1988; Booth, and Koch, 1989). The protein is also thought to bind malfolded proteins. It has been speculated that this activity may have a role in preventing the secretion of malfolded proteins, in the disposal of malfolded protein aggregates or in the renaturation of malfolded proteins (Pelham, 1989; Deshaies et al., 1988). Like the 59 kDa and 57 kDa proteins, the 100 kDa protein is also concentrated within the lumen of the endoplasmic reticulum. Several studies have indicated that the reason the 100 kDa and 57 kDa proteins are localized in the lumen of the endoplasmic reticulum is due to the presence of a KDEL sequence at their C-terminals (Pelham, 1989). This structural motif, however, does not appear to be the only signal that retains proteins in this subcellular compartment (Pelham, 1989); for example, the C-terminal of the carboxylesterase encoded by a rat liver cDNA has the sequence TEHT (Long et al., 1988; Takagi et al., 1988).

Sequence analysis of the N-terminal of the TFA 63 kDa neoantigen has led to the tentative conclusion that this protein is calregulin (Van et al., 1989). The protein is localized in the endoplasmic reticulum and is thought to have a role in calcium homeostasis, since it appears to be a major site of calcium binding in this organelle (Van et al., 1989).

DISCUSSION

The understanding of how drugs cause allergic or hypersensitivity reactions has been limited mainly by the lack of knowledge about the immunogens responsible for these toxicities (Pohl et al., 1988; Pohl, 1989b). It is generally believed that in order for a drug to cause an allergic reaction, it or one its metabolites must become covalently bound to an endogenous carrier macromolecule to form a drug carrier conjugate (Park et al., 1987; Pohl et al.1988). As described in this study, the identity of several hapten carrier conjugates associated with halothane hepatitis have been defined. This permits several important studies to be done. For example, the TFA proteins can be used as antigens in a sensitive and specific ELISA procedure for detecting patients' sensitized to halothane (Table 1) (Martin et al., 1990). This would be important for several reasons. First, the presence of antibodies in the sera of patients previously exposed to halothane would indicate prior sensitization and thus identifies patients at increased risk for developing a hypersensitivity reaction upon further exposure to halothane. Second, these same antibodies have been shown to react with liver microsomal neoantigens produced by the structurally related acyl halide metabolite (CHF_2OCF_2COX) of the inhalation anesthetic enflurane. This finding suggested that a cross-sensitization reaction might occur in a patient administered enflurane if the patient had been previously sensitized to halothane (Christ et al., 1988a; Christ et al., 1988b). Third, the assay can be used to monitor sensitization to halothane in animal model studies of the mechanism of halothane hepatitis (Neuberger et al., 1987b; Callis et al., 1987; Roth et al., 1988).

It is now possible to examine the properties of the TFA carrier proteins that make them antigenic and presumably immunogenic. The results of this study indicate that the number of TFA haptens groups bound to the carrier proteins is not the major factor that determines their antigenicity (Fig. 3B and Table 1). Those structural features that make them antigenic can be determined by mapping their epitopes with serum antibodies from halothane hepatitis patients. The positions of TFA labeling of the proteins can be determined with hapten specific anti-TFA antibodies. Another factor that should contribute to the immunogenic potential of a TFA protein, if it can accumulate in high enough concentration (Pohl, 1989b; Kenna et al., 1990) and be recognized by the immune system, will be its efficiency of reaching the immune system. This should be important not only for the initial induction of an immune response, but also for the subsequent development of an immune-mediated toxicity. Although there are several possible pathways that could account for this interaction (Satoh et al., 1987), we believe that their formation in the ER and subsequent transfer out of this compartment is the predominant way the TFA neoantigens contact the immune system. As discussed earlier, current evidence indicates that the 100 kDa, 59 kDa, and 57 kDa proteins are concentrated within the lumen of the endoplasmic reticulum. Most proteins that are destined to be secreted or are to become components of the plasma membrane traffic through this region of the endoplasmic reticulum after they are synthesized in the rough endoplasmic reticulum (Pelham, 1989). A recent study indicates that proteins normally residing in the lumen of the endoplasmic reticulum can be secreted when a cell is perturbed, such as by an increase in intracellular calcium (Booth and Koch, 1989), or possibly if they are chemically altered by a reactive metabolite. It also may be possible for luminal proteins to be transferred to the plasma membrane of the cell by major histocompatibility complex (MHC) class I or class II molecules (Long, 1989). Although hepatocytes normally express only low levels of these immunologically important molecules, under certain conditions their levels may become elevated (Pohl, 1989b). In contrast, integral membrane proteins of the ER, such as cytochromes P-450, would not be expected to be exported from the ER to the plasma membrane or to the exterior of the cell as rapidly as proteins that reside in the lumen of the ER (Pelham, 1989). This may be the reason, at least in part, why the TFA 54 kDa neoantigen, which is likely a cytochrome P-450 (Satoh et al., 1985a), appears to be a minor immunogen (Kenna et al., 1988a).

Although most evidence indicates that halothane hepatitis has an immune basis, this idea has not been proven because, to the best of our knowledge, no one has been successful in developing an model of this toxicity (Farrell, 1988). One approach that has been used is to immunize animals with TFA labeled serum proteins to sensitize animals to the TFA hapten (Farrell, 1988). This resulted in the production of anti-TFA hapten antibodies, but it did not potentiate the hepatotoxicity of halothane. It is believed that one of the reasons why this approach has not worked, is because patients with halothane hepatitis are not sensitized to the TFA hapten, but instead to specific TFA carrier proteins. A more rational approach for sensitizing the animals to halothane would be to immunize them with the purified TFA carrier proteins before challenging them with halothane. This is the same basic approach that has proven to be successful in developing animal models of autoimmune diseases (Calder et al., 1989). If this strategy were successful, then it should be possible to determine whether the hepatocellular damage caused by halothane can be mediated by specific antibodies or specific sensitized T cells.

Table 1. Reactivity of antibodies in the sera of 10 halothane hepatitis with the purified TFA carrier proteins as determined by an enzyme-linked immunosorbent assay[a]

Patient	Antigen (kDa)					
	100	80	63	59	58	57
1		+				+
2	+				+	
3	+					
4	+	+	+			
5	+					
6	+			+		
7	+	+			+	
8	+				+	
9	+	+		+	+	
10	+	+			+	+

[a]The assay was conducted in microtiter plates using 1μg/well of TFA carrier protein as previously described (Martin et al.1990). A positive reaction is defined as a value that is two standard deviations above the range of results obtained from the sera of 10 control patients.

It will be important in the future to determine whether the proteins covalently modified by the CF_3COX metabolite of halothane are also targets of attack by reactive metabolites of other compounds. If they are, then they might be associated with other drug-induced allergic reactions. Also, if the physiological activities of the proteins described in this report can be altered by covalent modification by reactive metabolites, this might be a factor that contributes to the development not only of allergic reactions, but also of toxicities that may have a nonimmunological basis.

ACKNOWLEDGEMENTS

David Thomassen was supported by a Fellowship from Anaquest and Jackie L. Martin was supported in part by a Pharmacology Research Associated Fellowship from the National Institute of General Medical Sciences.

REFERENCES

Booth, C., and Koch, G. (1989). Perturbation of cellular calcium induces secretion of luminal ER proteins. *Cell* **59**, 729-737.

Calder, V., Owen, S., Watson, C., Feldmann, M., and Davison, A. (1989). MS: a localized immune disease of the central nervous system. Immunol. *Today* **10**, 99-103.

Callis, A.H., Brooks, S.D., Roth, T.P., Gandolfi, A.J., and Brown, B.R. (1987). Characterization of a halothane-induced immune response in rabbits. *Clin. Exp. Immunol.* **67**, 343-351.

Christ, D.D., Kenna, J.G., Kammerer, W., Satoh, H., and Pohl, L.R. (1988a). Enflurane metabolism produces covalently-bound liver adducts recognized by antibodies from patients with halothane hepatitis. *Anesthesiology* **69**, 833-838.

Christ, D.D., Satoh, H., Kenna, J.G., and Pohl, L.R. (1988b). Potential metabolic basis for enflurane hepatitis and the apparent cross-sensitization between enflurane and halothane. *Drug Metab. Disp.* **16**, 135-140.

Deshaies, R.J., Koch, B.D., and Schekman, R. (1988). The role of stress proteins in membrane biogenesis. Trends. *Biochem. Sci.* **13**, 384-388.

Farrell, G. (1988). Mechanism of halothane-induced liver injury:Is it immune or metabolic idiosyncrasy. *J. Gastroenterol. Hepatol.* **3**, 465-482.

Freedman, R.B. (1989). Protein disulfide isomerase: multiple roles in the modification of nascent secretory proteins. *Cell* **57**, 1069-1072.

Hals, J., Dodgson, M.S., Skulberg, A., and Kenna, J.G. (1986). Halothane-associated liver damage and renal failure in a young child. Acta Anaesthesiol. *Scand.* **30**, 651-655.

Harano, T., Miyata, T., Lee, S., Aoyagi, H., and Omura, T. (1988). Biosynthesis and localization of rat liver microsomal carboxyesterase E1. *J. Biochem.* **103**, 149-155.

Kenna, J.G., Neuberger, J., Mieli-Vergani, G., Mowat, A.P., and Williams, R. (1987a). Halothane hepatitis in children. *Br. Med. J.* **294**, 1209-1211.

Kenna, J.G., Neuberger, J., and Williams, R. (1987b). Identification by immunoblotting of three halothane-induced liver microsomal polypeptide antigens recognized by antibodies in sera from patients with halothane-associated hepatitis. *J. Pharmacol. Exp. Ther.* **242**, 733-740.

Kenna, J.G., Satoh, H., Christ, D.D., and Pohl, L.R. (1988a). Metabolic basis for a drug hypersensitivity:antibodies in sera from patients with halothane hepatitis recognize liver neoantigens that contain the trifluoroacetyl group derived from halothane. J. Pharmacol. *Exp. Ther.* **245**, 1103-1109.

Kenna, J.G., Neuberger, J.M., and Williams, R. (1988b). Evidence for expression in human liver of halothane induced neoantigens recognized by antibodies in sera from patients with halothane hepatitis. *Hepatology* **8**, 1635-1641.

Kenna, J.G., Martin, J.L., Satoh, H., and Pohl, L.R. (1990). Factors affecting the expression of trifluoroacetylated liver microsomal protein neoantigens in rats treated with halothane. *Drug Metab. Disp.* (in press)

Long, E.O. (1989). Intracellular traffic and antigen processing. *Immunol. Today* **10**, 232-234.

Long, R.M., Satoh, H., Martin, B.M., Kimura, S., Gonzalez, F.J., and Pohl, L.R. (1988). Rat liver carboxylesterase: cDNA cloning, sequencing, and evidence for a multigene family. *Biochem. Biophys. Res. Commun.* **156**, 866-873.

Macer, D.R., and Koch, G.L. (1988). Identification of a set of calcium-binding proteins in reticuloplasm, the luminal content of the endoplasmic reticulum. *J. Cell Sci.* **91**, 61-70.

Martin, J.L., Kenna, J.G., Martin B.M., and Pohl, L.R. (1989). Trifluoroacetylated protein disulfide isomerase is a halothane- induced neoantigen. *Toxicologist* **9**, 5.

Martin, J.L., Kenna, J.G., and Pohl, L.R. (1990). Antibody assays for the detection of patients sensitized to halothane. Anesth. Analg. (in press)

Mazzarella, R.A., and Green, M. (1987). ERp99, an abundant, conserved glycoprotein of the endoplasmic reticulum, is homologous to the 90-kDa heat shock protein (hsp90) and the 94-kDa glucose regulated protein (GRP94). *J. Biol. Chem.* **262**, 8875-8883.

Medda, S., Chemelli, R.M., Martin, J.L., Pohl, L.R., and Swank, R.T. (1989). Involvement of the carboxyl-terminal propeptide of beta-glucuronidase in its compartmentalization within the endoplasmic reticulum as determined by a synthetic peptide approach. *J. Biol. Chem.* **264**, 15824-15828.

Mentlein, R., Heiland, S., and Heymann, E. (1980). Simultaneous purification and comparative characterization of six serine hydrolases from rat liver microsomes. *Arch. Biochem. Biophys.* **200**, 547-559.

Mentlein, R., and Heymann, E. (1984). Hydrolysis of ester- and amide-type drugs by the purified isoenzymes of nonspecific carboxylesterase from rat liver. *Biochem. Pharmacol.* **33**, 1243-1248.

Mentlein, R., Schumann, M., and Heymann, E. (1984a). Comparative chemical and immunological characterization of five lipolytic enzymes (carboxylesterases) from rat liver microsomes. *Arch. Biochem. Biophys.* **234**, 612-621.

Mentlein, R., Suttorp, M., and Heymann, E. (1984b). Specificity of purified monoacylglycerol lipase, palmitoyl- CoA hydrolase, palmitoyl-carnitine hydrolase, and nonspecific carboxylesterase from rat liver microsomes. *Arch. Biochem. Biophys.* **228**, 230-246.

Mentlein, R., Berge, R.K., and Heymann, E. (1985a). Identity of purified monoacylglycerol lipase, palmitoyl- CoA hydrolase and aspirin-metabolizing carboxylesterase from rat liver microsomal fractions. A comparative study with enzymes purified in different laboratories. *Biochem. J.* **232**, 479-483.

Mentlein, R., Reuter, G., and Heymann, E. (1985b). Specificity of two different purified acylcarnitine hydrolases from rat liver, their identity with other carboxylesterases, and their possible function. *Arch. Biochem. Biophys.* **240**, 801-810.

Mentlein, R., Ronai, A., Robbi, M., Heymann, E., and von Deimling, O. (1987). Genetic identification of rat liver carboxylesterases isolated in different laboratories. *Biochim. Biophys. Acta* **913**, 27-38.

Mentlein, R., Rix-Matzen, H., and Heymann, E. (1988). Subcellular localization of non-specific carboxylesterases, acylcarnitine hydrolase, monoacylglycerol lipase and palmitoyl-CoA hydrolase in rat liver. *Biochim. Biophys. Acta* **964**, 319-328.

Neuberger, J., and Kenna, J.G. (1987a). Halothane hepatitis: A model of immune mediated drug hepatotoxicity. *Clin. Sci.* **72**, 263-270.

Neuberger, J., Kenna, J.G., and Williams, R. (1987b). Halothane hepatitis:Attempt to develop an animal model. *Int. J. Immunopharmacol.* **9**, 123-131.

Park, B.K., Coleman, J.W., and Kitteringham, N.R. (1987). Drug disposition and drug hypersensitivity. *Biochem. Pharmacol.* **36**, 581-590.

Pelham, H. (1989). Control of protein exit from the endoplasmic reticulum. *Annu. Rev. Cell Biol.* **5**, 1-23.

Pohl, L.R., Satoh, H., Christ, D.D., and Kenna, J.G. (1988). The immunologic and metabolic basis of drug hypersensitivities. *Ann. Rev. Pharmacol. Toxicol.* **28**, 367-87.

Pohl, L.R., Kenna, J.G., Satoh, H., Christ, D.D., and Martin, J.L. (1989a). Neoantigens associated with halothane hepatitis. *Drug Metab. Rev.* **20**, 203-217.

Pohl, L.R. (1989b). Drug-induced allergic hepatitis: does isaxonine fall into this category. *Hepatology* **9**, 785-788.

Robbi, M., and Beaufay, H. (1983). Purification and characterization of various esterases from rat liver. *Eur. J. Biochem.* **137**, 293-301.

Robbi, M., and Beaufay, H. (1987). Biosynthesis of rat liver pI-6.1 esterase, a carboxylesterase of the cisternal space of the endoplasmic reticulum. *Biochem. J.* **248**, 545-550.

Roth, T.P., Hubbard, A.K., Gandolfi, A.J., and Brown, B.R. (1988). Chronology of halothane-induced antigen expression in halothane-exposed rabbits. *Clin. Exp. Immunol.* **72**, 330-336.

Satoh, H., Gillette, J.R., Davies, H.W., Schulick, R.D., and Pohl, L.R. (1985a). Immunochemical evidence of trifluoroacetylated cytochrome P-450 in liver of halothane-treated rats. *Mol. Pharmacol.* **28**, 468-474.

Satoh, H., Fukuda, Y., Anderson, D.K., Ferrans, V.J., Gillette, J.R., and Pohl, L.R. (1985b). Immunological studies on the mechanism of halothane-induced hepatotoxicity:Immunohistochemical evidence of trifluoroacetylated hepatocytes. *J. Pharmacol. Exp. Ther.* **233**, 857-862.

Satoh, H., Davies, H.W., Takemura, T., Gillette, J.R., Maeda, K., and Pohl, L.R. (1987). An immunochemical approach to investigating the mechanism of halothane-induced hepatotoxicity. In *Progress in Drug Metabolism* (J.W. Bridges, L.F. Chasseaud, and G.G. Gibson, Eds.), pp. 187-206. Taylor & Francis, Philadelphia.

Satoh, H., Martin, B.M., Schulick, A.H., Christ, D.D., Kenna, J.G., and Pohl, L.R. (1989). Human anti-endoplasmic reticulum antibodies in sera of halothane hepatitis patients are directed against a trifluoroacetylated carboxylesterase. *Proc. Natl. Acad. Sci.* (USA) **86**, 322-326.

Satoh, T. (1987). Role of carboxylesterases in xenobiotic metabolism. Reviews in *Biochem. Toxicol.* **8**, 155-181.

Takagi, Y., Morohashi, K., Kawabata, S., Go, M., and Omura, T. (1988). Molecular cloning and nucleotide sequence of cDNA of microsomal carboxyesterase E1 of rat liver. *J. Biochem.* (Tokyo.) **104**, 801-806.

Thomassen, D., Martin, B.M., Martin, J.L., Pumford, N.R., and Pohl, L.R. (1990). Characterization of a halothane induced trifluoroacetylated 100 kDa neoantigen that is related to a glucose-regulated protein. *FASEB Journal* (in press)

Van, P.N., Peter, F., and Soling, H.D. (1989). Four intracisternal calcium-binding glycoproteins from rat liver microsomes with high affinity for calcium. No indication for calsequestrin-like proteins in inositol 1,4,5-trisphosphate-sensitive calcium sequestering rat liver vesicles. *J. Biol. Chem.* **264**, 17494-17501.

Weis, K.H., and Engelhardt, W. (1989). Is halothane obsolete? Two standards of judgement. *Anaesthesia* **44**, 97-100.

Whitburn, R.H., and Sumner, E. (1986). Halothane hepatitis in an 11-month-old child. *Anaesthesia.* **41**, 611-613.

METABOLISM OF DRUGS BY ACTIVATED LEUKOCYTES: IMPLICATIONS FOR DRUG-INDUCED LUPUS AND OTHER DRUG HYPERSENSITIVITY REACTIONS

Jack Uetrecht

Faculties of Pharmacy and Medicine
University of Toronto and Sunnybrook Medical Centre
Centre for Drug Safety Research
University of Toronto
Toronto, Canada M5S 2S2

INTRODUCTION

Adverse drug reactions represent a serious medical problem. Idiosyncratic reactions are especially disconcerting because of their unpredictable nature which makes them difficult to prevent, and because they are associated with a relatively high mortality rate. General agreement on the definition of an idiosyncratic reaction is lacking, but the term will be used to mean a reaction that does not occur in most patients given the drug even at a high dose, and one that does not represent a manifestation of the known pharmacological effects of a drug. The mechanism of idiosyncratic reactions is unknown, and several different mechanisms may be involved. The study of idiosyncratic reactions is very difficult because they do not occur in most patients, and by extension, they do not occur in most animals even at a very high dose. Thus, there are very few animal models of idiosyncratic reactions which would permit careful investigation of their mechanism. It is probable that if enough different types of animals were tried, an animal model could be developed, but it might take many thousands of different species before one was found that would have a reaction similar to the rare individual human that has a reaction. It should be noted that, although a specific adverse reaction to a specific drug may be rare, there are enough different drugs being given to enough different individuals that idiosyncratic reactions are very common.

Although the mechanism of idiosyncratic drug reactions is unknown, the characteristics of many idiosyncratic reactions suggest involvement of the immune system (Park, et al., 1987; Parker, 1980; Pohl, et al., 1988). In several cases antibodies have been detected in association with the toxicity that are specific for the drug or a metabolite of the drug bound to protein, but it is often difficult to prove such antibodies are pathogenic and not just part of the clean up of damaged cells. Other characteristic of these reactions that suggests involvement of the immune system is a delay between starting therapy and the development of toxicity which is usually more than a week and in some cases the delay is more than a year. Yet, if the patient is exposed to the same drug again, toxicity may be immediately manifest. The lack of toxicity in most patients and in animals even at high dose suggests that the mechanism does not involve direct cytotoxicity. It is likely that, with the exception of cancer chemotherapy, most adverse drug reactions do not involve direct cytotoxicity (Uetrecht, 1989).

Biological Reactive Intermediates IV, Edited by C.M. Witmer *et al.*
Plenum Press, New York, 1990

One type of idiosyncratic drug reaction that clearly involves the immune system is drug-induced lupus (Uetrecht, 1988; Uetrecht, 1990, in press). Lupus is an autoimmune disease which means that antibodies are produced that are against a patient's own tissue. The cause of most lupus is unknown and is referred to as idiopathic lupus; however, about one third of the cases are associated with the use of certain drugs. The characteristic autoantibody that is found in almost all lupus patients is the antinuclear antibody. However, many other types of autoantibodies are produced and even the antinuclear antibodies can have many different specificities, e.g. anti-double stranded DNA, anti-histone, etc.

Idiopathic Lupus

Idiopathic lupus is a serious disease which usually affects young women (Wallace and Dubois, 1987). Lupus can affect almost any tissue but the most serious manifestations involve the kidneys and the brain. The renal disease appears to be due to deposition of antigen antibody complexes in which the antigen is usually native DNA, but the pathogenic mechanisms of lupus are not completely understood.

It has been suggested that idiopathic lupus could be due to some environmental chemical (Reidenberg, 1983); however, there is little evidence to support this hypothesis. Another hypothesis is based on the observation that many of the ANA in idiopathic lupus also bind with high affinity to certain bacterial cell wall antigens (Schwartz and Stollar, 1985). This suggests that lupus could be due to an antibody synthesized to deal with an infection which also happens to cross-react with a "self" antigen. Such cross-reactivity is not rare and anti-DNA antibodies often cross-react with cardiolipin. At first glance DNA and cardiolipin have very different structures and such cross-reactivity would seem curious in view of the specific nature of antibodies; however, both molecules have exposed phosphate groups, and this is the likely site of recognition by the antibody. The cross-reactivity between the autoantibodies and bacterial antigens may not be accidental, but rather it may be due to an attempt by the bacteria to "hide" from the immune system by mimicking a self antigen. In some patients, the immune response generated against the bacteria could lead to an autoimmune syndrome.

Comparison Between Idiopathic and Drug-Induced Lupus

It is not clear to what degree the pathogenesis of idiopathic lupus and drug-induced lupus are similar. The clinical manifestations of drug-induced lupus tend to be less severe than those of idiopathic lupus, and involvement of the kidneys and brain are much less common in drug-induced lupus (Uetrecht and Woosley, 1981). However, in a specific patient it would be impossible to differentiate drug-induced and idiopathic lupus on purely clinical grounds. The specificity of the antinuclear antibodies in idiopathic and drug-induced lupus are said to be different. The antinuclear antibodies in idiopathic lupus tend to be very heterogeneous while those in drug-induced lupus are more homogeneous, and many bind to histone protein (Fritzler and Tan, 1978). It has been claimed that the antinuclear antibodies in drug-induced lupus are never against native DNA but this is controversial.

GENERAL MECHANISMS OF ANTIBODY INDUCTION

Although the exact role of antibodies in the pathogenesis of lupus is unclear, autoantibodies are the characteristic finding. In general, the induction of antibody synthesis involves the cooperation of several types of cells. The following steps are thought to occur. The first step appears to be the uptake and processing of an immunogen by a macrophage (Unanue and Allen, 1987). (An immunogen is the molecule that induces an antibody response. An antigen is any molecule for which an antibody has a high affinity, and it may not be immunogenic.) The processing of an immunogen appears to involve hydrolysis. Fragments of the processed immunogen are then carried to the surface of the macrophage in association with the class II major compatibility antigen (MHC II). The combination of the processed immunogen and MHC II is then recognized by specific clones of helper-T lymphocytes resulting in the release of cytokines that stimulate specific clones of B lymphocytes. B lymphocytes that also recognize the immunogen are stimulated of proliferate and differentiate into mature antibody-secreting plasma cells. In the absence of macrophages or T lymphocytes, the response of B lymphocytes to a potential immunogen is greatly attenuated (Park, et al., 1987).

Induction of Antibody Synthesis by Drugs

With a few possible exceptions, molecules of molecular weight less than 1,000 are not directly immunogenic (Pohl, et al., 1988). For small molecules to induce an immune response, they must bind to a macromolecule. This combination can be immunogenic even though neither the macromolecule nor the drug were immunogenic alone. The interaction between drug and macromolecule must be essentially irreversible, and in virtually all cases this requires a covalent bond. The requirement of a covalent bond is presumably due to the processing and presentation steps which would require an interaction that lasts long enough to survive these steps. Most drugs are not sufficiently reactive to form a covalent bond to macromolecules such as protein. However many, if not most, drugs are metabolized to reactive metabolites that could form a covalent bond to protein.

STRUCTURE OF DRUGS ASSOCIATED WITH DRUG-INDUCED LUPUS

It has long been recognized that certain structures are associated with a high incidence of idiosyncratic drug reactions. Arylamines and sulfhydryl containing drugs are common offenders. In the case of drug-induced lupus, most of the drugs associated with the highest incidence are arylamines and hydrazines (Figure 1). Reactive metabolites are central to many mechanistic hypotheses of toxicity as evidenced by the title of this symposium. The association between certain functional groups and a high incidence of idiosyncratic reactions is likely due to the ease with which these groups are metabolized to reactive metabolites. Arylamines are known to be oxidized to hydroxylamines and nitroso metabolites, and such metabolites are thought to be responsible for the association of polycyclic arylamines with cancer (Radomski, 1979). Hydrazines are known to be oxidized to reactive diazenes (Moloney and Prough, 1983).

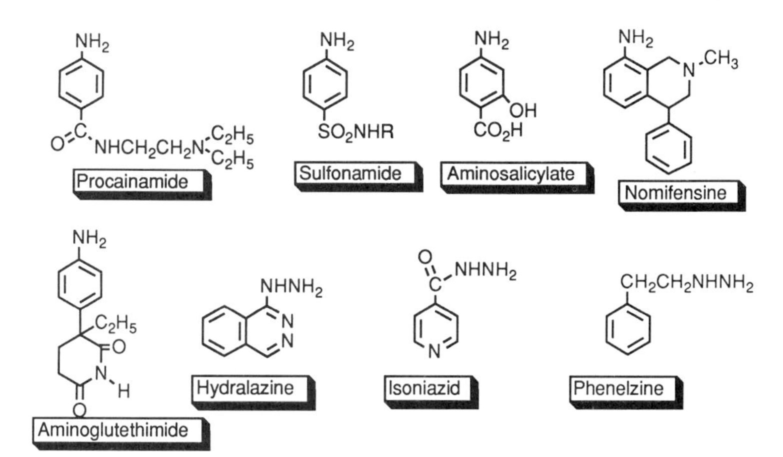

Figure 1. Structures of drugs that are arylamines or hydrazines and are associated with the induction of a lupus-like syndrome. These are not the only drugs that have been associated with lupus but this figure contains the structures of several associated with the highest incidence of this syndrome.

The drug associated with the highest incidence of drug-induced lupus is procainamide. It is an arylamine, and we have demonstrated that it is oxidized by human and rat hepatic microsomes to a hydroxylamine (Uetrecht, et al., 1981). The hydroxylamine is further oxidized to a nitroso metabolite which is chemically reactive (Uetrecht, 1985). The nitroso metabolite reacts with histone protein *in vitro*. Since the most common antinuclear antibodies in procainamide-induced lupus are reported to be against histone protein, the nitroso metabolite could bind to histone protein, and by acting as a hapten, induce antibodies against this adduct. In some patients, antibodies might be induced that cross-react with native histone protein and produce a true autoimmune disease. However, there are two facts that make this hypothesis unattractive. One is that we have been unable to detect the hydroxylamine or nitroso metabolites outside of the liver. A small amount of metabolite might escape hepatocytes, or the autoimmune syndrome might be induced when the hepatocytes die and are phagocytosed; however, it seems more likely that the reactive metabolite would be scavenged by glutathione and cause no toxicity. In addition, this hypothesis does not explain why antibodies against histone protein should predominate. It appears that the nitroso metabolite reacts with most proteins and being in the nucleus, histone protein is unlikely to be exposed to as high a concentration of reactive metabolite as endoplasmic reticulum or cytosolic proteins.

METABOLISM OF DRUGS BY ACTIVATED LEUKOCYTES

Since drug-induced lupus is an autoimmune disease, it would seem that formation of a reactive metabolite by the cells of the immune system would be more likely to lead to an autoimmune syndrome than if the metabolite is generated in the liver. Lymphocytes contain a small amount of cytochrome P-450; however, our initial experiments with human peripheral mononuclear leukocytes (mostly lymphocytes but with about 20% monocytes) failed to find any evidence of metabolism of procainamide (unpublished observation). Microsomes isolated from these cells also failed to produce detectable procainamide metabolism.

Another leukocyte system that might oxidize a drug is the myeloperoxidase system. Neutrophils and monocytes use the combination of myeloperoxidase, hydrogen peroxide and chloride ($MPO/H_2O_2/Cl^-$) to oxidize and kill bacterial and other pathogens (Lehrer, et al., 1988). This system generates hypochlorite or its enzymatic equivalent. It is not active in a resting leukocyte; however, when these cells are activated by the phagocytosis of bacteria or by most inflammatory conditions, NADPH oxidase is activated. NADPH oxidase reduces oxygen to superoxide which is in turn converted to hydrogen peroxide. Simultaneously, myeloperoxidase is released from granules.

Procainamide

When procainamide was incubated with neutrophils that had been activated by a phorbol ester or opsonized zymosan, it was oxidized to the hydroxylamine (Uetrecht, et al., 1988a). The hydroxylamine was further oxidized to the nitro analog with the nitroso analog being the presumed intermediate. If the cells were not activated, no significant degree of oxidation occurred. The oxidation was inhibited by azide (inhibitor of peroxidase) and catalase, but it was not inhibited by aspirin and indomethacin (inhibitors of prostaglandin synthetase pathway). Peripheral blood mononuclear cell which had failed to produce metabolites in the experiments described earlier, did metabolize procainamide to the hydroxylamine when they were activated with phorbol ester. When the peripheral blood mononuclear cells were separated into lymphocytes and monocytes, only the monocytes produced detectable quantities of the hydroxylamine (unpublished observation). This fits the myeloperoxidase hypothesis because monocytes contain myeloperoxidase and NADPH oxidase but lymphocytes do not. The combination of purified myeloperoxidase and hydrogen peroxide also converted procainamide to the hydroxylamine. In addition, in the presence of chloride, the arylamine group of procainamide was chlorinated (Uetrecht and Zahid, 1988b). This reactive chloramine rearranged spontaneously to o-chloroprocainamide. Neither N-chloroprocainamide nor o-chloroprocainamide were observed when procainamide

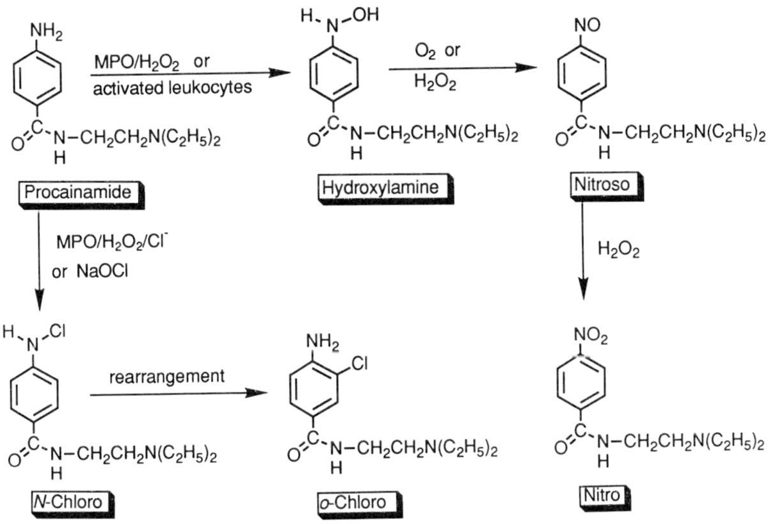

Figure 2. Metabolism of procainamide by activated neutrophils or mononuclear cells or catalyzed by myeloperoxidase.

was incubated with activated neutrophils, but *N*-chloroprocainamide appears to react rapidly with cells, and it is unlikely to be observed in a cellular system even if it is formed. In contrast to myeloperoxidase, horseradish peroxidase did not catalyze a significant degree of oxidation of procainamide in the presence of hydrogen peroxide. These reactions are summarized in Figure 2.

Incubation of radiolabeled procainamide with activated neutrophils or monocytes led to covalent binding of the drug, but covalent binding was not observed when it was incubated with lymphocytes treated in the same manner (unpublished observation). Thus, it is likely that covalent binding of procainamide to neutrophils and monocytes occurs in humans when the cells are activated by an infection or some inflammatory condition.

Other Drugs

Other arylamnes that we studied underwent similar oxidation although there were some differences. Dapsone underwent oxidation to the hydroxylamine metabolite with activated neutrophils or catalyzed by myeloperoxidase, but no *N*- or *o*-chlorodapsone was detected when dapson was incubated with $MPO/H_2O_2/Cl^-$ (Uetrecht, et al., 1988b). Sulfadiazine was also oxidized by $MPO/H_2O_2/Cl^-$ to the hydroxylamine (Uetrecht, et al., 1986), but the major metabolite of sulfamethoxazole appears to involve *N*-chlorination (unpublished observation).

Hydralazine, the drug associated with the second highest incidence of drug-induced lupus, was oxidized by activated neutrophils or $MPO/H_2O_2/Cl^-$ to phthalazine and phthalazinone (Hofstra and Uetrecht, 1988). The production of phthalazine had characteristics of a nonenzymatic free radical reaction, but the production of phthalazinone was catalyzed by myeloperoxidase. Phthalazinone and phthalazine are not reactive metabolites but the production of phthalazinone has been hypothesized to be associated with hydralazine-induced lupus, presumably because it is downstream of a reactive metabolite (Timbrell, et al., 1984). If the reactive metabolite is a diazene it is not surprising that it would not be observed directly. Unlike the other drugs associated with the induction of lupus, hydralazine has sufficient chemical reactivity as a nucleophile that it may not require metabolism in order to produce a significant degree of covalent binding. Isoniazid is also associated with drug-induced lupus and is also metabolized by $MPO/H_2O_2/Cl^-$ (Li and Uetrecht, 1988).

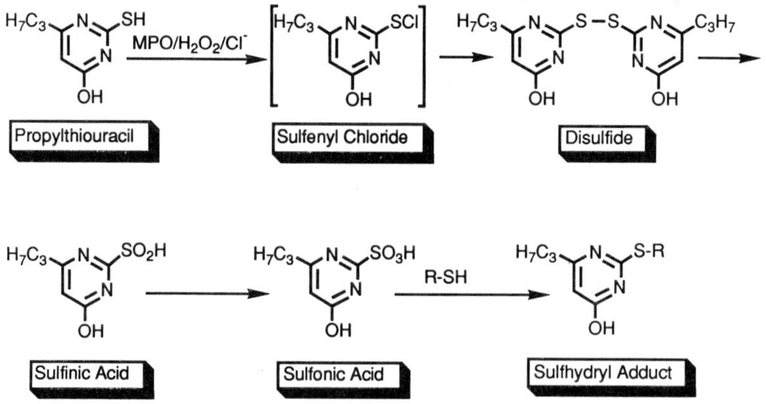

Figure 3. Metabolism of propylthiouracil by activated leukocytes and reaction of the sulfonic acid metabolite with sulfhydryl groups.

Propylthiouracil is associated with the induction of lupus (Amrhein, et al., 1970; Takuwa, et al., 1981), and it is the only drug known to induce a significant incidence of a lupus-like syndrome in an animal model. Alcoin has found that about 50% of cats treated with propylthiouracil develop antinuclear antibodies as well as lymphadenopathy, anemia, fever, and inflammation of joints (Aucoin, et al., 1985; Aucoin, et al., 1988). We and others have found that propylthiouracil is oxidized to several metabolites by activated neutrophils and $MPO/H_2O_2/Cl^-$ (Lee, et al., 1988b; Waldhauser and Uetrecht, 1988). The first observed metabolite is the disulfide, but it is probable that a reactive sulfenyl chloride is an intermediate. The disulfide is further oxidized to what appears to be a sulfinic acid and finally to a sulfonic acid. The sulfonic acid is also chemically reactive and forms an adduct with sulfhydryl reagents such as glutathione and *N*-acetylcysteine. These reactions are summarized in Figure 3.

Phenytoin is associated with the induction of lupus and we have found that it is chlorinated to N,N'-dichlorophenytoin (Uetrecht and Zahid, 1988a). Covalent binding of radiolabeled phenytoin to activated neutrophils also occurred, but to a lesser degree than the covalent binding of procainamide to neutrophils.

Therapy with the anticonvulsant, carbamazepine, is associated with a high incidence of antinuclear antibodies although clinically manifest lupus is less common (Alarcon-Segovia, et al., 1972). The oxidation of carbamazepine by activated neutrophils and $MPO/H_2O_2/Cl^-$ is quite interesting although we have not fully characterized it. Major metabolites are acridone and chloroacridone (unpublished observation). This represents a contraction of the seven-membered ring with loss of a carbon atom. It is possible that iminostilbene is an intermediate in this reaction because a small amount of it is detected and oxidation of iminostilbene gives similar metabolites. It does not appear that formation of iminostilbene represents a simple hydrolysis because the same products are formed by the reaction of carbamazepine with NaOCl. A more likely mechanism involves chlorination of the terminal $-NH_2$ of the urea group followed by elimination of isocyanic acid. The remaining nitrene would be the same as that which could be formed after direct chlorination of iminostilbene. Although this would be a reactive intermediate, it is not clear how it would lead to acridone and chloroacridone.

In summary, it is clear that many drugs that are associated with a high incidence of drug-induced lupus are oxidized to reactive metabolites by activated neutrophils and monocytes. We have also demonstrated that these reactive metabolites can result in covalent binding of the metabolite to leukocytes. How could such reactive metabolites lead to drug-induced lupus?

POSSIBLE MECHANISMS OF DRUG-INDUCED LUPUS

Drug-induced lupus involves the induction of autoantibodies. There are several mechanisms by which a leukocyte-generated reactive metabolite could lead to such an effect. Reactive metabolites generated by leukocytes would seem more likely to have specific effects on the immune system than reactive metabolites generated by other cells such as hepatocytes.

Direct Effect of a Reactive Metabolite on the Function of Cells That Control the Balance of Immune Response

Normal individuals produce a small amount of autoantibody. Therefore lupus may represent a loss in the balance of helper and suppressor influences leading to over activation of the immune system and an increase in the production of autoantibodies. We have found that the hydroxylamine of procainamide (probably due to the nonenzymatic generation of the more reactive nitroso analog) has a toxic effect on lymphocytes as measured by Trypan Blue exclusion at concentrations in the μM range while procainamide and other known hepatic metabolites are nontoxic (Rubin, et al., 1987). The hydroxylamine or some other metabolite such as *N*-chloroprocainamide could have more subtle functional effects on lymphocytes or other leukocytes at even lower concentrations. Because of the complexity of the immune system, such effects are unlikely to be the same *in vitro*, and it will be difficult to sort out whether an observed effect is a primary effect or secondary to some other effect.

Reactive Metabolite Acting as a Hapten To Induce An Immune Response

As mentioned earlier, the first step in the induction of an antibody is believed to be the uptake of an immunogen followed by processing and presentation of the immunogen by a macrophage. Since drugs are too small to act as immunogens they must first covalently bind to a macromolecule. It appears that the density of hapten molecules on the macromolecule is important in determining the immunogenicity of the adduct. A relatively high density is necessary for significant immunogenicity. One molecule of hapten per macromolecule is not sufficient for immunogenicity (Park, et al., 1987). In addition, the hapten density also appears to have an effect on the specificity of the antibodies produced. A very high hapten density tends to induce antibodies that recognize hapten alone, while a relatively low hapten density tends to induce antibodies in which the macromolecule is the major part of the epitope (the epitope is the part of an antigen that is recognized by an antibody). It appears that most reactions of drug or metabolite with macromolecules that occurs *in vivo* leads to a hapten density that is far too low to induce an immunological response. In contrast, when an attempting to induce specific anti-drug antibodies, an immunogen is used in which a very high hapten density has been produced by a chemical reaction. In cases where haptenization occurs *in vivo* it is probably much more likely that, when a sufficient hapten density is produced to induce an antibody response, the macromolecule will make up a large portion of the epitope, and in some cases true autoantibodies may be induced.

As mentioned earlier, monocytes contain MPO, can generate hydrogen peroxide when activated, and we have shown them to be capable of metabolizing drugs to reactive metabolites. When MPO is released from monocytes, it sticks to the surface of the cell. HADPH oxidase, which is indirectly the source of hydrogen peroxide, is also located on the outside surface of the cell. Therefore, reactive metabolites will be generated in relatively high concentration on the surface of activated monocytes. Since monocytes are also macrophages, and as mentioned earlier, macrophages play a critical role in the processing and presentation of immunogen to induce an immune response, the formation of a high concentration of reactive metabolite on the surface of the monocytes would seem to be optimal for the induction of an immunological reaction.

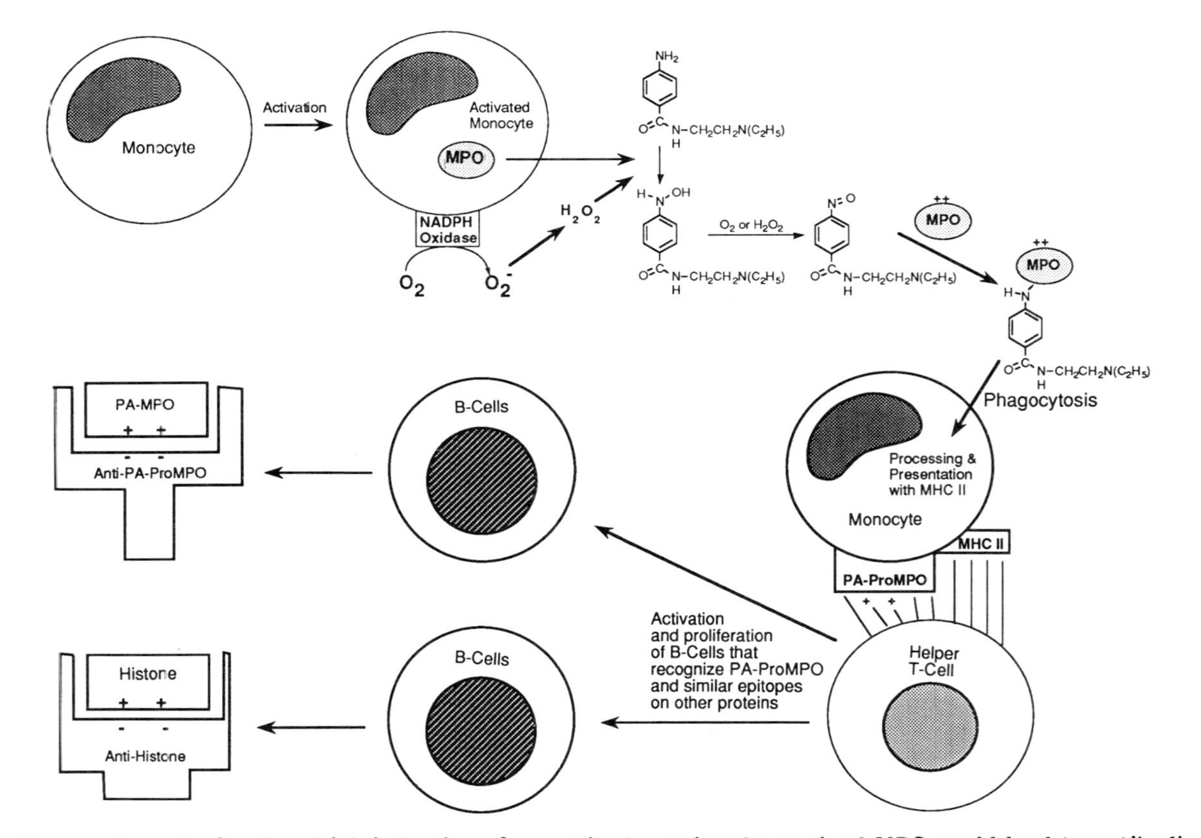

Figure 4. Proposed mechanism by which induction of an antibody against haptenized MPO could lead to antibodies that cross react with histone protein because of a similar positively charged epitope (MPO = myeloperoxidase, PA-ProMPO = procainamide adduct of processed MPO, PA-MPO = procainamide-MPO adduct, MHC II = class II major histocompatability antigen).

Since anti-histone antibodies are reported to be a major autoantibody associated with drug-induced lupus, one possibility is that the reactive metabolite covalently binds to histone protein, and the resulting adduct induces antibodies that can cross-react with native histone protein. As mentioned earlier, the hydroxylamine of procainamide (probably after nonenzymatic oxidation to nitroso procainamide) reacts with histone protein. However, the hydroxylamine appears to react with most proteins, and because of the location of histone proteins in the nucleus, it is unlikely that a high enough concentration of reactive metabolite would reach histone protein to produce a sufficient hapten density to induce an antibody response. N-chloroprocainamide is even less likely to reach the nucleus. Alternatively, the reactive metabolite could react with some protein which is on the monocyte cell membrane, and this immunogen could induce antibodies that cross-react with histone protein. The protein exposed to the highest concentration of reactive metabolite should be myeloperoxidase. Myeloperoxidase is a highly basic protein with a pI > 11. The major characteristic of histone protein is also its basicity; its basic amino acids neutralizing the negative charges of the DNA which it surrounds. Therefore, antibodies induced against a myeloperoxidase-procainamide adduct might, in some patients, be true autoantibodies and cross-react with native histone protein (Uetrecht, 1990, in press). This mechanism is analogous to that proposed for the mechanism of idiopathic lupus in which antibodies induced against bacterial antigens cross-react with native antigens. consistent with this hypothesis is the observation that the nucleus of neutrophils and monocytes contain myeloperoxidase (Murao, et al., 1988) and the antibodies that are reported to correlate with clinical evidence of drug-induced lupus only bind to the nuclei of leukocytes (Gorsulowshy, et al., 1985). This hypothesis is summarized in Figure 4. There are also other basic proteins in the granules of leukocytes that have surfactant properties and would be exposed to a relatively high concentration of reactive metabolite (Klebanoff and Clark, 1978).

Another protein that is likely to be haptenized to a high degree by leukocyte-generated reactive metabolites is MHC II because of its location on the surface of monocytes. Because an immunogen is much more effective in inducing an immunological response if it is presented along with MHC II, it has been suggested that direct binding of a reactive metabolite to MHC II would be a powerful stimulus to the immune system (Park, et al., 1987). The proposed generation of reactive metabolites by activated leukocytes would be the optimal circumstance to effect such a reaction. This mechanism does not explain why the antibodies produced should be selective for histone protein but the strength of this association is not clear.

OTHER IMPLICATIONS OF REACTIVE METABOLITES GENERATED BY ACTIVATED LEUKOCYTES

Other Idiosyncratic Drug Reactions

Most of the same drugs that are associated with the induction of lupus are also associated with a high incidence of other idiosyncratic drug reactions, especially agranulocytosis (a life-threatening decrease in the number of circulating neutrophils) (Uetrecht, 1989; Uetrecht, 1990 in press). the link between drug-induced lupus and drug-induced agranulocytosis is natural if drug-induced lupus involves the formation of reactive metabolites by the myeloperoxidase system in monocytes and drug-induced agranulocytosis involves the formation of the same reactive metabolites by the analogous system in neutrophils (or bone marrow cells that are precursors to neutrophils). Other idiosyncratic drug reactions often involve several different organs, especially the skin, liver, lungs, and kidneys, and these reactions could also be due to the formation of reactive metabolites by activated leukocytes such as monocytes.

Unlike cytochromes P-450, myeloperoxidase can only oxidize a relatively limited number of functional groups, e.g. arylamines, sulfhydryl groups and hydrazines. Thus the high incidence of idiosyncratic drug reactions associated with certain functional groups may be due to their ease of oxidation to reactive metabolites by the myeloperoxidase system.

Thyroid Function

These same functional groups often have an effect on the thyroid gland. Clinical use is made of this effect with propylthiouracil and methimazole which are used to treat hyperthyroidism. The mechanism of action of propylthiouracil appears to involve the formation of a reactive metabolite catalyzed by thyroid peroxidase that inhibits thyroid peroxidase and the formation of thyroxine (Engler, et al., 1982). Arylamines also inhibit thyroid function. Animals such as rats are more sensitive to this inhibition than humans, and drugs such as sulfonamides do not appear to case hypothyroidism at clinically relevant doses in man (Takayama, et al., 1986). Dapsone appears to cause thyroid cancer in rats (Griciute and Tomatis, 1980).

Recently we have seen three pediatric patients who became clinically hypothyroid shortly after recovery from a severe idiosyncratic drug reaction (Gupta, et al., 1988). These patients had anti-thyroid microsome antibodies (thyroid microsomes are chiefly thyroid peroxidase). Antimicrosomal antibodies are common in idiosyncratic autoimmune thyroid disease; however, this is an unusual condition in prepubertal children. In addition, the hypothyroidism resolved after a period of about a year which is very unusual. We therefore propose that these cases of thyroid disease were due to reactive metabolites of the drugs produced by thyroid peroxidase that haptenized thyroid microsomes and induced an autoimmune response.

Antiinflammatory Effects

Several of the drugs associated with a high incidence of idiosyncratic drug reactions have antiinflammatory effects, especially if the inflammation is mediated by neutrophils. The best example is dapsone. Dapsone is the drug of choice for conditions such as dermatitis herpetiformis in which there is an infiltration of neutrophils into the skin and the production of vesicles and itching. Dapsone has a dramatic effect on this condition, and evidence suggest that this effect is due to the inhibition of myeloperoxidase (Stendahl, et al., 1978). We propose that this inhibition involves the formation of reactive metabolites analogous to the formation of reactive metabolites of propylthiouracil by thyroid peroxidase that inhibit thyroxine synthesis. Conversely propylthiouracil inhibits neutrophil function, probably by a similar mechanism (Lee, et al., 1988a).

SUMMARY

Despite their importance, little is known about the mechanism of idiosyncratic reactions. many such reactions have characteristics that suggest an immune-mediated mechanism. This is particularly true of drug-induced lupus which is an autoimmune syndrome. Certain functional groups are associated with a high incidence of idiosyncratic reactions, probably reflecting the ease with which they are metabolized to reactive metabolites. Although the liver is the principal organ of drug metabolism, most reactive metabolites generated in the liver would not reach other organs in significant concentrations. Because of the function of leukocytes, especially monocytes, in the induction of an immune response, the generation of reactive metabolites by monocytes would seem likely to lead to an immune-mediated adverse reaction. We have found that drugs that are associated with drug-induced lupus are oxidized to reactive metabolites by the myeloperoxidase system of monocytes. The initial step in drug-induced lupus could be haptenization of a protein on the surface of monocytes by these reactive metabolites.

Other types of idiosyncratic drug reactions may involve a similar mechanism and the same drugs that induce lupus are usually associated with a high incidence of other types of idiosyncratic reactions. for example, procainamide, which causes the highest incidence of drug-induced lupus, also causes a relatively high incidence of agranulocytosis. Even some of the therapeutic effects of drugs may involve the production of reactive metabolites by myeloperoxidase or thyroid peroxidase.

ACKNOWLEDGEMENTS

This research was supported by grants from the Medical Research Council of Canada (MA-9336 and MA-10036.

REFERENCES

Alarcon-Segovia, D., Fishbein, E., Reyes, P.A., Dies, H. and Shwadsky, S. (1972). Antinuclear antibodies in patients on anticonvulsant therapy. *Clin. Exp. Immunol.* **12**, 39-47.

Amrhein, J.A., Kenny, F.M. and Ross, D. (1970). Granulocytopenia, lupus-like syndrome, and other complications of propylthiouracil therapy. *J. Pediat.* **76**, 54-63.

Aucoin, D.P., Peterson, M.E., and Hurvitz, A.I., (1985). Propylthiouracil-induced immune-mediated disease in the cat. *J. Pharmacol. Exp. Ther.* **234**, 13-18.

Aucoin, D.P., Rubin, R.L., and Peterson, M.E. (1988). Dose-dependent induction of anti-native DNA antibodies in cats by propylthiouracil. *Arthritis Rheum.* **31**, 688-692.

Engler, H., Taurog, A. and Nakashima, T. (1982). Mechanism of inactivation of thyroid peroxidase by thioureylene drugs. *Biochem. Pharmacol.* **31**, 3801-3806.

Fritzler, M.J. and Tan, E.M. (1978). Antibodies to histones in drug-induced and idiopathic lupus erythematosus. *J. Clin. Invest.* **62**, 560-567.

Gorsulowsky, D.C., Bank, P.W., Goldberg, A.D., Lee, T.G., Heinzerling, R.H. and Burnham, T.K. (1985). Antinuclear antibodies as indicators for the procainamide-induced systemic lupus erythematosus-like syndrome and its clinical presentations. *J. Am. Acad. Dermatol.* **12**, 245-253.

Gruciute,L. and Tomatis, L. (1980). Carcinogenicity of dapsone in mice and rats. *Int. J. Cancer* **25**, 123-129.

Gupta, A., Waldhauser, L., Reider, M. (1988). Drug-induced hypothyroidism: the thyroid as a target organ in hypersensitivity reactions. *Soc. Ped. Res.* **23**, 277A.

Hofstra, A. and Uetrecht, J. (1988). Oxidation of hydralazine by monocytes as a mechanism of hydralazine-induced lupus. *Pharmacologist* **30**. A99.

Klebanoff, S.J. and Clark, r.a. (1978). *The Neutrophil: Function and Clinical Disorders*, Elsevier/North-Holland Inc. Amsterdam.

Lee, E., Hirouchi, M., Hosokawa, M., Sayo, H., Kohno, M. and Kariya, K. (1988a). Inactivation of peroxidases of rat bone marrow by repeated administration of propylthiouracil is accompanied by a change in the heme structure. *Biochem. Pharmacol.* **37**, 2151-2153.

Lee, E., Miki, Y., Hosokawa, M., Sayo, H. and Kariya, K. (1988b). Oxidative metabolism of propylthiouracil by peroxidases from rat bone marrow. *Xenobiotica* **18**, 1135-1142.

Lehrer, R.I., Ganz, T., Selsted, M.E., Babior, B.M. and Curnutte, J.T. (1988). Neutrophils and host defense. *Ann. Intern. Med.* **109**, 127-142.

Li, A. and Uetrecht, J. (1988). Metabolism of isoniazid by activated neutrophils. *Pharmacologist* **30**, A99.

Moloney, S.J. and Prough, R.A. (1983). Biochemical toxicology of hydrazines. In *Reviews in Biochemical Toxicology*. (Hodgson, E., Bend, J.R., Philpot, R.M., ed), 313-348. Elsevier, New York.

Murao, S., Stevens, F.J., Ito, A. and Huberman, E. (1988). Myeloperoxidase: a myeloid cell nuclear antigen with DNA-binding properties. *Proc. Natl. Acad. Sci. USA* **85**, 1232-1236.

Park, B.K., Coleman, J.W. and Kitteringham, N.R. (1987). Drug disposition and drug hypersensitivity. *Biochem. Pharmacol.* **36**, 581-590.

Parker, C.W. (1980). Drug allergy. In *Clinical Immunology* (Parker C.W. ed), 1219-1260. W.B. Saunders, Philadelphia.

Pohl, L.R., Satoh, H., Christ, D.D. and Kenna, J.G. (1988). the immunologic and metabolic basis of drug hypersensitivities. *Ann. Rev. Pharmacol.* **28**, 367-387.

Radomski, J.L. (1979). the primary aromatic amines: their biological properties and structur-activity relationships. *Ann. Rev. Pharmacol. Toxicol.* **19**, 129-157.

Reidenberg, M.M. (1983). Aromatic amines and the pathogenesis of lupus erythematosus. *Am. J. Med.* **75**, 1037-1042.

Rubin, R.L., Uetrecht, J.P. and Jones, J.E. (1987). Cytotoxicity of oxidative metabolites of procainamide. *J. Pharmacol. Exp. Ther.* **242**, 833-841.

Schwartz, R.S. and Stollar, B>D. (1985). Origins of anti-DNA autoantibodies. *J. Clin. Invest.* **75**, 321-327.

Stendahl, O., Molin, L. and Dahlgren, C. (1978). the inhibition of polymorphonuclear keukocyte cytotoxicity by dapsone: a possible mechanism in the treatment of dermatitis herpetiformis. *J. Clin. Invest.* **62**, 214-220.

Takayama, S., Aihara, K., Onodera, T. and Akimoto, T. (1986). Antithyroid effects of propylthiouracil and sulfamonomethoxine in rats and monkeys. *Toxicol. Appl. Pharmacol.* **82**, 191-199.

Takuwa, N., Kojima, I. and Ogata, E. (1981). Lupus-like syndrome - a rare complication in thionamide treatment for Graves' disease. *Endocrinol. Japan.* **28**, 663-667.

Timbrell, J.A., Facchini, V., Harland, S.J. and Mansilla-Tinoco, R. (1984). Hydralazine-induced lupus: is there a toxic pathway? *Eur. J. Clin. Pharmacol.* **27**, 555-559.

Uetrecht, J. (1989). Mechanism of hypersensitivity reactions: proposed involvement of reactive metabolites generated by activated leukocytes. *Trends Pharmacol. Sci.* **10**, 463-467.

Uetrecht, J. and Zahid, N. (1988a). N-Chlorination of phenytoin by myeloperoxidase to a reactive metabolite. *Chem. Res. Toxicol.* **1**, 148-151.

Uetrecht, J. and Zahid, N. (1988b). Procainamide (PA) is N-chlorinated by myeloperoxidase (MPO)-implications for toxicity. *Pharmacologist* **30**, A98.

Uetrecht, J., Zahid, N. and Rubin, R. (1988a). Metabolism of procainamide to a hydroxylamine by human neutrophils and mononuclear leukocytes. *Chem. Res. Toxicol.* **1**, 74-78.

Uetrecht, J., Zahid, N. Shear, N.H. and Biggar, W.D. (1988b). Metabolism of dapsone to a hydroxylamine by human neutophils and mononuclear cells. *J. Pharmacol. Exp. Ther.* **245**, 274-279.

Uetrecht, J.P. (1985). reactivity and possible significance of hydroxylamine and nitroso metabolites of procainamide. *J. Pharmacol. Exp. Ther.* **232**, 420-425.

Uetrecht, J.P. (1988). Mechanism of drug-induced lupus. *Chem. Res. Toxicol.* **1**, 133-143.

Uetrecht, J.P. (1990). Drug metabolism by leukocytes, its role in drug-induced lupus and other idiosyncratic drug reactions. *CRC Crit. Rev. Tox*, in press.

Uetrecht, J.P., Shear, N. and Biggar, W. (1986). Dapsone is metabolised by human neutrophils to a hydroxylamine. *Pharmacologist* **28**, 239.

Uetrecht, J.P. and Woosley, R.L. (1981). Acetylator phenotype and lupus erythematosus. *Clin. Pharmacokin.* **6**, 118-134.

Uetrecht, J.P., Woosley, r.L., Freeman, r.W., Sweetman, B.J. and Oates, J.A. (1981). Metabolism of procainamide in the perfused rat liver. *Drug Metab. Disp.* **9**, 183-187.

Unanue, E.R. and allen, P.M. (1987). the basis for the immunoregulatory role of macrophages and other accessory cells. *Science* **236**, 551-557.

Waldhauser, L. and Uetrecht, J. (1988). Propylthiouracil (PTU) is metabolized by activated neutrophils-implications for agranulocytosis. *FASEB J.* **2**, A1134.

Wallace, D.J. and Dubois, E.L. (1987). *Dubois' Lupus Erythematosus* (3rd ed.), Lea and Febiger, Philadelphia.

FORMATION OF REACTIVE METABOLITES AND APPEARANCE OF ANTI-ORGANELLE ANTIBODIES IN MAN

D. Mansuy

Laboratoire de Chimie et Biochimie Pharmacologiques et Toxicologiques
Université René Descartes
URA 400 CNRS 45 rue des Saints-Péres, 75270
PARIS Cedex 06, France

INTRODUCTION

The toxic effects of many xenobiotics are very often not due to the parent compound itself but to reactive intermediates or metabolites formed *in situ* inside the cell. These reactive species may bind covalently to cell proteins leading to various cell disorders and eventually to toxic effects.

In some cases (*direct toxicity*), the final toxic effects are clearly related to the dose, occur with a high incidence, and can be easily reproduced in laboratory animals. Therefore, compounds, and, particularly drugs, which lead to such direct toxic effects, should be easily detected by usual toxicological studies in laboratory animals.

In some rarer cases (*idiosyncratic immunoallergic toxicity*), the toxic effects, of the immunoallergic type, occur with a very low incidence (generally one case for 10,000 patients), are not clearly related to the administered dose, are very dependent on the individual, and are not easily reproduced in laboratory animals. Thus, the compounds leading to such toxic effects cannot be easily detected by usual toxicological studies made on laboratory animals. In order to find out new tests for detecting such compounds, it is first necessary to understand the detailed mechanisms of the appearance of their toxic effects. In the case of some compounds leading to such immunoallergic effects, anti-organelle autoantibodies have been reported to appear in the sera of patients. They should be important tools for mechanistic studies. For instance, a report on 157 cases of drug-induced hepatitis showed the appearance of various anti-organelle autoantibodies in patient sera. Some of these autoantibodies appeared to be specifically related to the secondary immunoallergic toxic effects of a given drug. Thus, anti-M_6 (M for mitochondria) antibodies have been specifically associated to the secondary effects of iproniazid, whereas anti-LKM_2 (LKM for liver kidney microsomes) antibodies have been related to the secondary effects of tienilic acid (Homberg et al., 1985). In the following, recent results showing the nature of the human liver antigens which are responsible for the appearance of the anti-M6 and anti-LKM_2 antibodies will be described, and possible mechanisms for the appearance of these antibodies will be discussed.

Biological Reactive Intermediates IV, Edited by C.M. Witmer *et al.*
Plenum Press, New York, 1990

RESULTS AND DISCUSSION

Tienilic acid and anti-LKM_2 antibodies

Tienilic acid is an uricosuric diuretic drug which was found devoid of any direct hepatotoxic effect but which has led to rare cases (1 for 10,000) of immunoallergic hepatotoxic effects. The anti-LKM_2 antibodies detected in some patients suffering from tienilic acid-induced hepatitis appear directed against components of liver and kidney microsomes (Homberg et al., 1985). Recently, it has been found, by using immunoblot techniques, that anti-LKM_2 antibodies recognized almost only one protein from human liver microsomes (Beaune et al., 1987). This protein is a cytochrome P-450 of the P-450IIC subfamily which was called cytochrome P-450-8 and which belongs to the same subfamily than that which metabolizes mephenytoin (Beaune et al., 1987). Immunoblot analysis of human liver microsomes were done with 40 sera from patients or volunteers. From the 20 tested sera from patients suffering from hepatitis and containing anti-LKM_2 antibodies, 12 sera recognized almost exclusively P-450-8. On the contrary, the 20 control sera not containing anti-LKM_2 antibodies failed to recognize cytochrome P-450-8.

Tienilic acid (TA) is metabolized in man and rat with the formation of a major metabolite derived from the hydroxylation of its thiophene ring in position 5 (Mansuy et al., 1984). Metabolism of TA by human and rat liver microsomes led to the same major metabolite (5-OHTA) and also to reactive metabolites which bind covalently to microsomal proteins (Dansette et al., 1990) (Figure 1). Both 5-hydroxylation of TA by human liver microsomes and covalent binding of its reactive metabolite(s) to liver proteins were found to be almost completely inhibited by anti-cytochrome P-450-8 antibodies and by human sera containing anti-LKM_2 antibodies (Dansette et al., 1989). *These results provided the first evidence for the existence of circulating antibodies raised against a cytochrome P-450 in humans. They also showed for the first time that antibodies appearing in the sera of patients suffering from drug-induced hepatitis were directed against the liver enzyme which was mainly responsible for the metabolic activation of this drug.*

It was also shown that reactive metabolites formed by oxidation of ^{14}C-tienilic acid by human liver microsomes irreversibly bound to several microsomal proteins and particularly to cytochrome P-450-8 (H. Hoellinger, M. Sonnier, P. Dansette, D. Mansuy, unpublished results).

From these data, it was tempting to speculate that, after TA activation and P-450-8 akylation by TA metabolites, cytochrome P-450-8 (or alkylated cytochrome P-450-8) could migrate from the membranes of the endoplasmic reticulum to the outer membrane of the hepatocyte, and could be recognized there by the immune system. This would lead to the appearance of anti-LKM_2 (in fact anti-P-450-8) antibodies. In order to test this hypothesis, hepatocytes from rats treated by tienilic acid were incubated with anti-LKM_2 antibodies from three human sera, and the possible existence of new antigens recognized by the antibodies at the outer surface of the hepatocytes was studied by immunofluorescence after addition of anti-human IgG labelled with fluorescein. It was previously shown that human anti-LKM_2 antibodies were able to recognize one of the major P-450 isozyme of liver microsomes from control rats, called cytochrome P-450-UTA (P-450 II C11), which exhibits a 76% sequence homology with human liver cytochrome P-450-8 (P-450 II C8 for instance) (C. Pons, P. Dansette, D. Mansuy, in preparation).

Whereas incubation of control rat hepatocytes with three sera containing anti-LKM_2 antibodies failed to give appearance of any fluorescence at the surface of the hepatocytes, the same experiments performed with hepatocytes from rats treated by TA led to a weak fluorescence at the hepatocyte outer membrane. Pretreatment of rats by various inducers like phenobarbital or 3-methyl-cholanthrene, before treatment by TA, gave similar results, while pretreatment by clofibrate before TA led to the appearance of a very strong fluorescence clearly located at the hepatocyte surface (C. Pons, M. Sonnier, H. Hoellinger, P. Dansette, D. Mansuy, in preparation). It is noteworthy that identical experiments performed with hepatocytes from rats treated

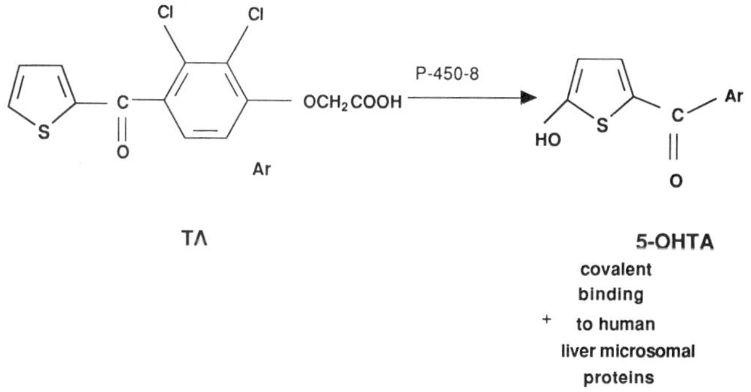

Cl Cl

TA —P-450-8→ 5-OHTA

covalent binding + to human liver microsomal proteins

Figure 1. Metabolic activation of tienilic acid by human liver microsomes.

by TA (and various inducing agents) incubated with control human sera not containing anti-LKM_2 did not give any appearance of fluorescence at the hepatocyte surface.

Although much work is still necessary to determine a detailed mechanism for the secondary hepatotoxic effects of TA, these data suggest the following steps for the appearance of anti-LKM_2 antibodies in tienilic acid-induced hepatitis (Figure 2).

Several steps of this mechanism have been proposed previously in the case of the appearance of anti-LKM1 antibodies in halothane-induced hepatitis (Pohl et al., 1988). However, the present results clearly establish the importance of a drug (TA) and the enzyme mainly involved in its metabolic activation (cytochrome P-450-8) in human liver, for the appearance of autoantibodies and of immunoallergic effects in humans.

Iproniazid and anti-M6 antibodies

Iproniazid is a monoamine oxidase inhibitor which was widely used as an antidepressive drug. It caused liver damage in 1% of recipients and it was suggested that the mechanism of injury was production of reactive hepatotoxic metabolites. The anti-M6 antibodies detected in the sera of patients suffering from iproniazid-induced hepatitis appear directed against components of liver mitochondria (Homberg et al., 1985). Since iproniazid is an irreversible inhibitor of monoamine oxidases (MAO), it was tempting to speculate that mitochondrial MAO could be involved in the appearance of anti-M6 antibodies. In order to study this hypothesis, mitochondrial MAO-A and MAO-B were labelled upon incubation of solubilized mitochondria from human liver with radioactive pargyline, a well-known inhibitor of MAOs which covalently binds to MAO-A and MAO-B after having been oxidized by these enzymes (7). Upon incubation of these labelled mitochondria with two sera from patients containing anti-M6 antibodies, in the presence of protein A sepharose, one observed a selective immuno-precipitation of MAOs. This was shown both by the appearance of labelled MAO in the pellet and by the disappearance of a MAO-dependent activity (tyramine oxidation) in the mitochondria-containing supernatant (C. Pons, P. Dansette, J.C. Hombert, D. Mansuy, in preparation).

On the contrary, control sera from humans not containing anti-M6 antibodies failed to give any immunoprecipitation of MAOs under identical conditions. In order to know whether MAO-A or MAO-B or both MAO-A and MAO-B were immunoprecipitated by anti-M6 antibodies, two sets of experiments were performed. In the first one, MAO-A was selectively labelled by pargyline upon incubation of this compound with human liver mitochondria which were first pretreated by deprenyl, a selective inhibitor of MAO-B. A similar experiment was performed to selectively label MAO-B by using a selective inhibitor of MAO-A, clorgyline. Immunoprecipitation of mitochondria, where either MAO-A or MAO-B has been selectively labelled as indicated before, by anti-M6 antibodies clearly showed that mainly MAO-B was

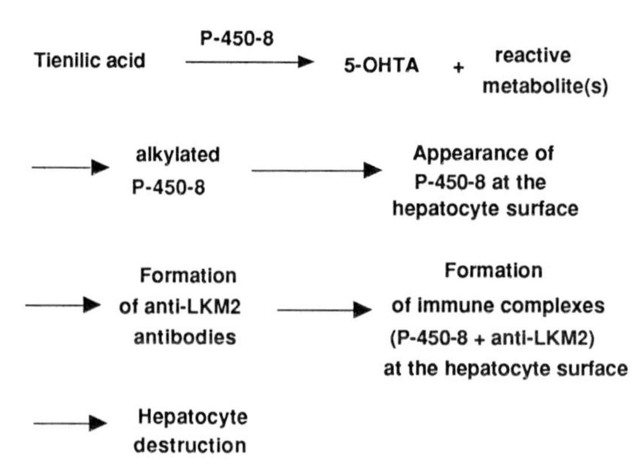

Figure 2. Possible schematic mechanism for the appearance of anti-LKM_2 antibodies in tienilic acid-induced hepatitis.

recognized by these antibodies. The second set of experiments used human placenta, which contains almost only MAO-A and confirmed that sera containing anti-M6 antibodies were unable to precipitate MAO-A.

These results provide *a first evidence that anti-M6 antibodies selectively recognize MAO-B of human liver mitochondria.* MAO-B is known to be irreversibly inactived by iproniazid and to oxidize various hydrazines to reactive metabolites which irreversibly bind to its flavine residue (Singer et al. , 1985). Thus, its seems likely that iproniazid itself or isopropylhydrazine, one of its metabolites in humans, is oxidized by MAO-B with formation of reactive metabolites able to covalently bind to MAO-B. This would lead to a very strong analogy between tienilic acid and cytochrome P-450-8 on one side and iproniazid and MAO-B on the other side. In both cases, the formation of an alkylated enzyme after oxidative activation of the drug would be the first step causing eventually the appearance of antibodies directed against the enzyme itself (anti-P-450-8 = anti-LKM_2 or anti-MAO-B = anti-M6 respectively). Therefore, it is tempting to generalize the mechanism proposed for tienilic acid (Figure 2) to the iproniazid case (Figure 3). In such a mechanism, the formation of anti-M6 antibodies would occur after oxidative activation of iproniazid (or of its isopropyl hydrazine metabolite) by MAO-B and alkylation of MAO-B by reactive metabolites, and recognition of alkylated MAO-B by the immune system which would result in anti-M6 antibodies formation. As far as the secondary hepatotoxic effects of iproniazid are concerned, one could also speculate the formation of an antigen-antibody (MAO-B-anti-M6) complex at the hepatocyte surface, and destruction of these activated hepatocytes by the immune system.

Very few is known so far on the mechanisms of the apperance of anti-organelle antibodies in drug-induced immunoallergic hepatitis.

However, the aforementioned results on tienilic acid and iproniazid show that the two proteins which have been found so far to selectively recognize anti-organelle antibodies appearing in drug-induced hepatitis and which could act as antigens for the appearance of these antibodies (P-450-8 and MAO-B) are able to oxidize these drugs into reactive alkylating intermediates.

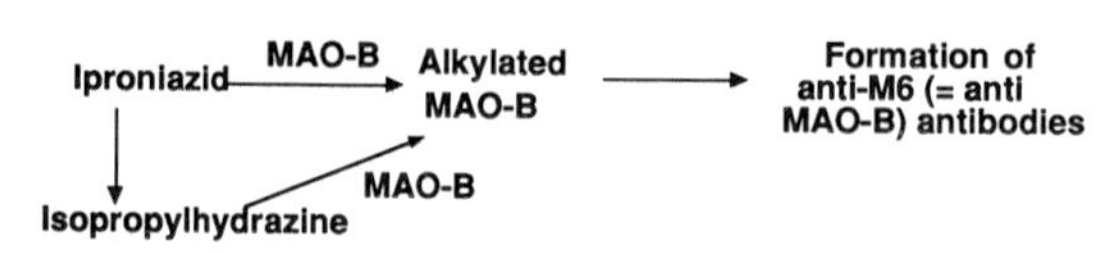

Figure 3. A possible mechanism for the appearance of anti-M6 antibodies in iproniazid-induced hepatitis.

Similar results have been obtained in the case of halothane and anti-LKM1 antibodies (Pohl et al., 1988). Therefore, it appears that a possible consequence of the covalent binding of reactive metabolites to cell proteins inside the cell could be the formation of antibodies against some of these proteins, and the eventual destruction of cells and tissues containing them by the immune system.

REFERENCES

Beaune, P., Dansette, P.M., Mansuy, D., Kiffel, L., Finck, M., Amar, C., Leroux, J.P. and Homberg, J.C. (1987). Human anti-endoplasmic reticulum autoantibodies appearing in a drug-induced hepatitis are directed against a human liver cytochrome P-450 that hydroxylates the drug. *Proc. of the National Academy of Sciences, USA*, **84**, 551-555.

Dansette, P.M., Amar, C., Smith, C.B., Neau, E., Pons, C. and Mansuy, D. (1989). *In vitro*, the 5-hydroxylation of tienilic acid is catalyzed by cytochrome P-450 in rat and human microsomes, in *Cytochrome P-450, Biochemistry and Biophysics*, Schuster I. Edit., Taylor and Francis, pp. 751-754.

Dansette, P.M., Amar, C., Smith, C.B., Pons, C. and Mansuy, D. (1990). Oxidative activation of the thiophene ring by hepatic enzymes: hydroxylation and formation of electrophilic metabolites during metabolism of tienilic acid and its isomer by rat liver microsomes. *Biochemical Pharmacology*, in press.

Homberg, J.C., Abuaf, N., Helmy-Khalil, S., Biour, M. et al. (1985). Drug-induced hepatitis with anti-intracytoplasmic organelle autoantibodies. *Hepatology*, **5**, 722-727.

Mansuy, D., Dansette, P.M., Foures, C., Jaouen, M., Moinet, G. and Bayer, N. (1984). Metabolic hydroxylation of the thiophene ring: isolation of 5-hydroxytienilic acid as major urinary metabolite in man and rat. *Biochemical Pharmacology*, **33**, 1429-1439.

Pohl, L.R., Satoh, H., Christ, D.D. and Kenna, J.G. (1988). The immunologicand metabolic basis of drug hypersensitivities. *Annual Review of Pharmacology*, **28**, 367-387.

Singer, T.P. (1985). Mitochondrial monoamine oxidase, in *Biochemical Pharmacology and Toxicology*, Zakim, D. and Vessey, D.A. Edit, Wiley J. and Sons, New-York, vol. 1, pp. 229-264.

FORMATION OF A PROTEIN-ACETALDEHYDE ADDUCT IN LIVER DURING CHRONIC ALCOHOL EXPOSURE

Renee C. Lin and Lawrence Lumeng

Departments of Medicine and Biochemistry
Indiana University School of Medicine, and the Veterans Affairs Medical Center, Indianapolis, IN. USA 46202

INTRODUCTION

Acetaldehyde is an intermediary metabolite of ethanol oxidation that forms in the liver during alcohol ingestion. Formation of acetaldehyde is mediated by two alcohol metabolizing systems: alcohol dehydrogenase (ADH) (Li, 1977) and the microsomal ethanol oxidizing system (MEOS or cytP450IIE1) (Lieber et al., 1970). In individuals who do not abuse alcohol, ADH is the most important enzyme responsible for the conversion of ethanol to acetaldehyde. However, MEOS is inducible by chronic alcohol consumption (Lieber, et al., 1972), therefore its contribution to the elimination of alcohol becomes more significant in chronic alcoholics.

Little acetaldehyde accumulates during alcohol oxidation because this intermediate is rapidly converted to acetate by aldehyde dehydrogenase (ALDH). Nevertheless, acetaldehyde is highly reactive. Various pathophysiologic abnormalities induced by alcohol drinking have been attributed to the accumulation of acetaldehyde (Salaspuro, et al., 1985). Biochemical studies in both humans and experimental animals have provided evidences that indicate acetaldehyde might be directly or indirectly hepatotoxic. For example, acetaldehyde may inhibit protein secretion by the liver resulting in accumulation of proteins and water in hepatocytes and leading to balloon degeneration (Baroana et al., 1975). A number of studies *in vitro* have shown that acetaldehyde can bind to plasma proteins (Lumeng et al., 1982), albumin (Lumeng et al., 1982; Donahue at al., 1983), erythrocyte membrane proteins (Gaines et al., 1977), tubulin (Jennett et al., 1980; Jennett et al., 1985), hepatic proteins (Nomura et al., 1981; Medina et al., 1985), a number of enzymes with critical lysine residues (Mauch et al., 1985), and to hemoglobin (Lumeng et al., 1982; Lumeng et al., 1985; Stevens et al., 1981). However, demonstration of the formation of protein-acetaldehyde adducts (protein-AAs) *in vivo* has been more difficult. Using a high performance liquid chromatography method, Lumeng, et al., (1985) were able to separate a fast-moving hemoglobin-AA peak from unmodified hemoglobin in mice fed alcohol chronically. The HPLC method, however, lacked sensitivity to detect the presence of hemoglobin-AA in human alcoholics. Barry et al., (1987) reported detection of a seemingly unstable liver plasma membrane protein-AA from alcohol-fed rats when the membranes were prepared by a rapid Percoll method and assessed by reversed phase liquid chromatography. The membrane protein involved was not characterized, and its molecular weight was not determined.

In 1986, Israel and his coworkers (1986) succeeded in raising antibodies that specifically recognized acetaldehyde-containing epitopes in protein-AAs. These investigators prepared the antigen by incubating keyhole limpet hemocyanin with

Biological Reactive Intermediates IV, Edited by C.M. Witmer *et al.*
Plenum Press, New York, 1990

acetaldehyde for 1h and then incubating the mixture with the addition of NaCNBH3 for an additional 3h. When the antigen was repeatedly injected into rabbits, the animals produced antibodies against hemocyanin-AA. These anti-hemocyanin-AA antibodies cross-react with both plasma protein-AAs and erythrocyte protein-AAs but did not cross-react with the respective unmodified carrier proteins. Using antibodies raised in this manner, we were able to detect for the first time the presence of a protein AA in the liver of rats fed alcohol-containing liquid diet chronically (Lin et al., 1988a). The protein-AA has an apparent molecular weight of 37,000. The adduct is stable and does not require borohydride reduction to stabilize it. Repeated acute administrations of ethanol by i.p. injections over 24h did not produce the 37KD protein-AA. Chronic administration of alcohol by forced feeding of ethanol in a liquid diet required approximately one week to produce this protein-AA *in vivo*. The presence of the 37KD protein-AA in the liver of the alcohol-fed rat has been recently confirmed by Israel and his coworkers (personal communication). Several protein-AAs that can form *in vivo* either in the rat or in man have been reported subsequently. These include: cytP450IIE1-AA (Behrens et al., 1988), hemoglobin-AA (Niemela et al., 1987a) and two serum protein-AAs (Lin et al., 1988b). In this report, studies that further characterize the 37KD protein-AA and its formation in cultured hepatocytes will be described.

METHODS

Detection of Liver Protein-AAs

Anti-protein-AA antibodies were raised in rabbits using hemocyanin-AA as the antigen as described previously (Lin et al., 1988a). Soluble proteins were extracted from rat livers or cultured rat hepatocytes and subjected to sodium dodecylsulfate polyacrylamide gel electrophoresis (SDS-PAGE). Protein bands were electrotransferred to a piece of nitrocellulose paper and then immunoblotted with rabbit anti-hemocyanin-AA IgG to locate protein-AAs (Lin et al., 1988a).

Animal Procedures

Male Wistar rats were pair-fed alcohol-containing and alcohol-free AIN"76 liquid diets for up to 7 weeks. The Lieber-DeCarli liquid diet was also used in some experiments as indicated. Ethanol constituted 35% of total calories in the alcohol-containing liquid diet and it was isocalorically substituted with maltose-dextrin in the control diet.

Isolation and Culture of Hepatocytes

Hepatocytes were isolated by perfusing a rat liver with collagenase (Type CLS II, Cooper Biomedical, Malvern, PA) and were cultured as monolayers in Waymouth's medium (Lin et al., 1975). The Waymouth's medium contained added trace minerals (Crabb et al., 1985), growth factors and hormones (Enat et al., 1984). It also contained 28 mM HEPES (pH 7.4) and 26 mM $NaHCO_3$ as buffers. The incubator was equilibrated with 95% O_2-5% CO_2 and was maintained at 37°C. When cells were treated with alcohol, Petri dishes with cultured hepatocytes were maintained in an incubator humidified with a pan containing 2 L of an ethanol solution that had the same ethanol concentration as that in the culture medium. The culture medium was changed daily and a dose of ethanol was added each day to the fresh medium to maintain the ethanol concentration at a near-constant level. At indicated time points, cell monolayers were washed and harvested with phosphate-buffered saline.

RESULTS

Formation of the 37KD Protein-AA In Vivo

When rats were pair-fed the alcohol-containing and alcohol-free liquid diets, a protein-AA (MW of 37,000) was detected in the liver of alcohol-fed rats but not their pair-fed controls. The 37KD protein-AA could be detected as early as 1 week of chronic alcohol consumption. While the intensity of the 37KD protein-AA band increased with prolonged alcohol feeding, no additional protein-AA bands could be found with our anti-hemocyanin-AA antibodies in the liver when rats were fed alcohol for up to 7 weeks. Feeding rats with the alcohol-containing AIN'76 or the alcohol-containing Lieber-DeCarli diet produced the same 37KD protein-AA of approximately the same intensities (Figure 1). The 37KD protein-AA by alcohol-feeding degraded *in vivo* with a half-life of approximately 4 days. When subcellular fractions, including mitochondria, microsomes and cytosol, were prepared from fresh liver samples of alcohol-fed rats, the 37KD protein-AA could only be detected in the crude homogenate and the cytosolic fraction (Figure 2).

Formation of the 37KD Protein-AA In Vitro

Cellular soluble proteins and microsomes were prepared daily from cultured rat hepatocytes to determine their ADH and MEOS activities. It was found that nearly 70% and 40%, respectively, of the initial ADH and MEOS activities remained after 4 days of culture. Thus cultured hepatocytes maintained their ability to metabolize ethanol reasonably well and they provided a useful model to study the formation of the 37KD protein-AA *in vitro*. As shown in Figure 3, hepatocytes treated with 40 mM ethanol produced an easily detectable 37KD protein-AA band as early as in 3 days. The intensities of the protein-AA band increased as a function of time with prolonged alcohol treatment when measured by a densitometer. No protein-AA formed in control cells which had not been exposed to alcohol (Figure 4, lane 1). When cells were cultured in the presence of various concentrations of ethanol, the 37KD protein-AA could be seen in cells treated with ethanol at concentrations as low as 5 mM (Figure 4). When the concentration of acetaldehyde in the culture medium was determined, it was evident that the intensities of the protein-AA rose in parallel with the acetaldehyde concentrations in the medium, and presumably also in cells (Figure 5).

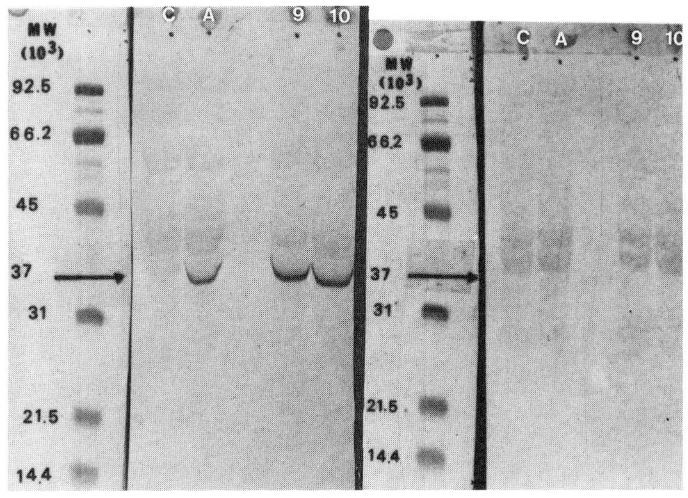

Figure 1. Formation of the 37KD liver protein-AA in rats fed alcohol-containing liquid diet for 7 weeks. Lane C, rats fed the control Lieber-DeCarli diet; Lane A, rat fed the alcohol-containing Lieber-DeCarli diet; Lane 9 and 10, rts fed the alcohol-containing AIN'76 diet. The left panel was immunoblotted with anti-hemocyanin-AA IgG while the right panel was immunoblotted with unimmunized control rabbit IgG.

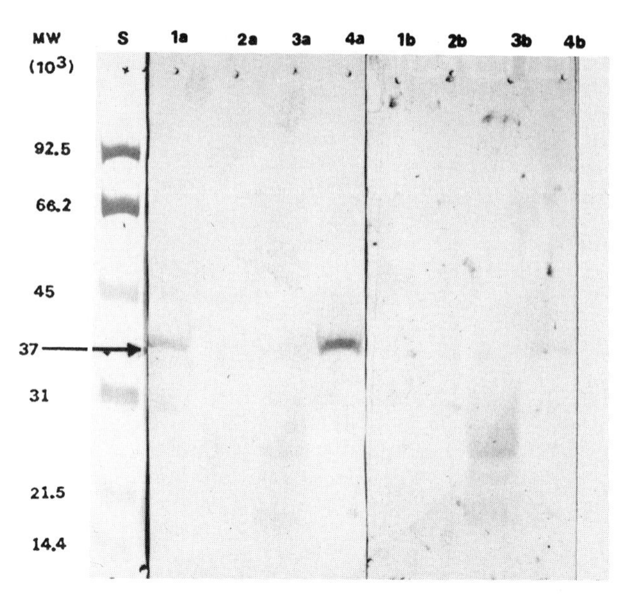

Figure 2. Subcellular localization of the 37KD liver protein-AA.Lanes denoted by a were immunoblotted with anti-hemocyanin-AA IgG; lanes denoted by b were immunoblotted with control rabbit IgG. S, molecular weight standards; 1, crude homogenate; 2, mitochondria; 3, microsomes; 4, cytosol.

Effects of pyrazole and cyanamide

Cyanamide is a well recognized drug to deter alcoholics from drinking. It selectively inhibits hepatic aldehyde dehydrogenase activity *in vivo* (Ritchie, 1985; Peachey et al., 1981). The mechanism of inhibition is thought to be mediated by one of its active metabolites, nitroxyl (DeMaster et al., 1989). Pyrazole and its derivatives, e.g. 4-methylpyrazole, are potent inhibitors of ADH (Li et al., 1969; Reynier, 1969; Goldberg et al., 1969). On the other hand, pyrazole has been found to induce microsomal oxidation of alcohols by increasing several cytochrome P-450 isozymes including cytP450IIE1 (MEOS) (Krikun et al., 1986). The effects of pyrazole and cyanamide on the formation of the 37KD liver protein-AA were studied both *in vivo* and *in vitro* (Lin et al., 1989; Lin et al., 1990). Six pairs of Wistar male rats were pair-fed the AIN'76 liquid diets with and without alcohol for 3 weeks and were further divided into 3 groups. Rats in Group I received no supplements. Animals in Group II were supplemented with pyrazole, 2 mM. Rats in Group III were supplemented with cyanamide, 100 mg/l. As shown in Figure 6, none of the rats fed the alcohol-free diet, with or without supplement, showed any protein-AA band in the liver extract. Both alcohol-fed rats in Group I developed the 37KD protein-AA. Cyamaide supplementation in the aclohol-containing diet intensified the 37KD protein-AA band (Group III, lane 9 and 11) . No protein-AA band was visible in rats in Group II (lane 5, 6, 7 and 8) which were supplemented with pyrazole. Microsomal MEOS activities in these three groups were measured by immunoblotting using anti-MEOS antibodies and quantitated by a densitometer (Figure 7). Results showed that alcohol-feeding increased cytP450IIE1 content in the liver. While pyrazole or cyanamide alone exhibited no effect, combining pyrazole and alcohol significantly increased this microsomal protein when compared to feeding alcohol only. Cyanamide did not interfere with the induction of cytP450IIE1 by alcohol-feeding. When plasma acetaldehyde concentrations of these rats were measured by gas-chromatography, it was found that cyanamide supplementation in the alcohol diet increased the acetal-

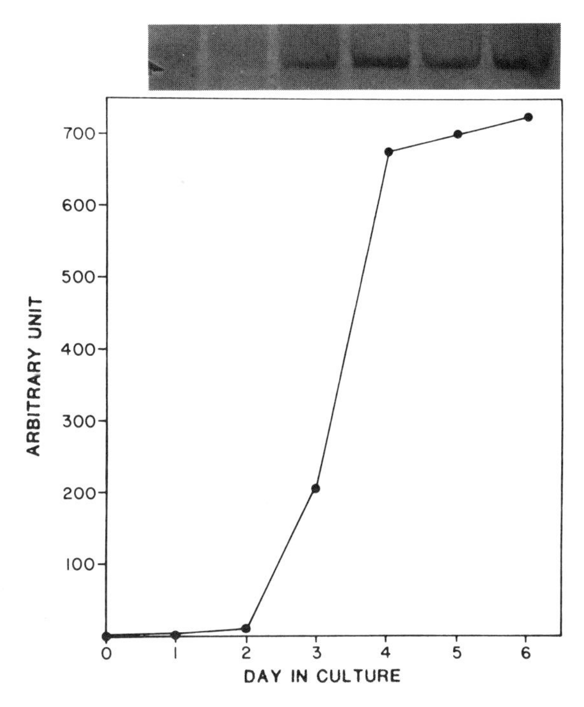

Figure 3. Time course for the formation of the 37KD protein-AA in hepatocyte culture. Half-tone photograph is the immunoblot of the 37KD protein-AA band formed as a function of days in culture. The intensities of the bands were determined by using a densitometer and were plotted as arbitrary units.

dehyde levels about 6 folds over feeding alcohol alone. The plasma acetaldehyde levels in rats fed alcohol and pyrazole were too low and below the detection limit. Inhibition of the formation of the 37KD protein-AA by 4-methylpyrazole was also demonstrated *in vitro*. When 4-methylpyrazole (10 μM) was added to the culture medium of hepatocytes, the 37KD protein-AA band disappeared (not shown). On the other hand, adding cyanamide to the culture medium at a concentration of 50 μM increased the intensity of the 37KD protein-AA band by 2 folds (Figure 8).

DISCUSSION

A 37KD protein in liver was found to form adducts with acetaldehyde when rats were fed alcohol chronically. This 37KD hepatic protein-AA could be detected within one week of alcohol ingestion. Both the Lieber-DeCarli and the AIN'76 liquid diets contain 35% of their calories as ethanol; however, the former diet contains 35% fat and 18% protein, while the latter contains 12% fat and 21% protein (Lieber et al., 1982). Because of these differences in fat and protein contents, rats fed the alcohol-containing Lieber-DeCarli formula are known to develop protein accumulation, fatty liver and hepatomegaly (Baroana et al., 1975). By contrast, rats fed the alcohol-containing AIN'76 diet are known to show only minimal hepatic changes (Lieber et al., 1989). In our studies, rats fed either the alcohol-containing Lieber-DeCarli or AIN'76 diet exhibited the same 37KD protein-AA band on immunotransblot. We could not detect other additional hepatic protein-AA bands with either diet. The results indicate that a 37KD protein in liver is unusually susceptible to acetaldehyde modification and this process proceeds with either the Lieber-DeCarli or the AIN"76 alcohol diet.

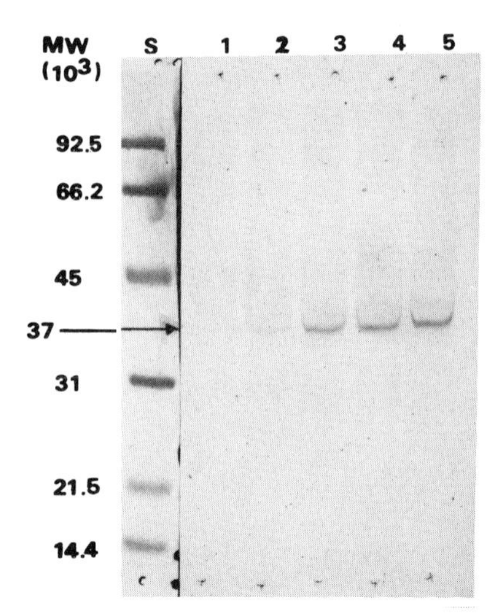

Figure 4. Effects of varying ethanol concentrations on the formation of the 37KD protein-AA. Hepatocytes were cultured for 4 days. Varying concentrations of ethanol were added to the culture medium daily. Lanes 1 to 5 correspond to ethanol concentrations of 0, 5, 10, 20, and 40 mM.

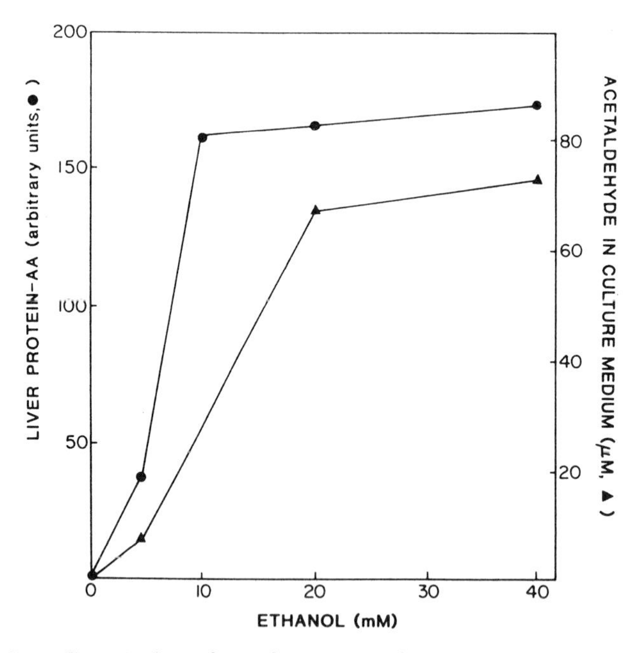

Figure 5. Correlation of the formation of the 37KD protein-AA and acetaldehyde concentrations in the culture medium. Intensities of the 37KD protein-AA bands formed with various ethanol concentrations were determined by densitometry. The acetaldehyde concentrations in the medium were determined with gas chromatography.

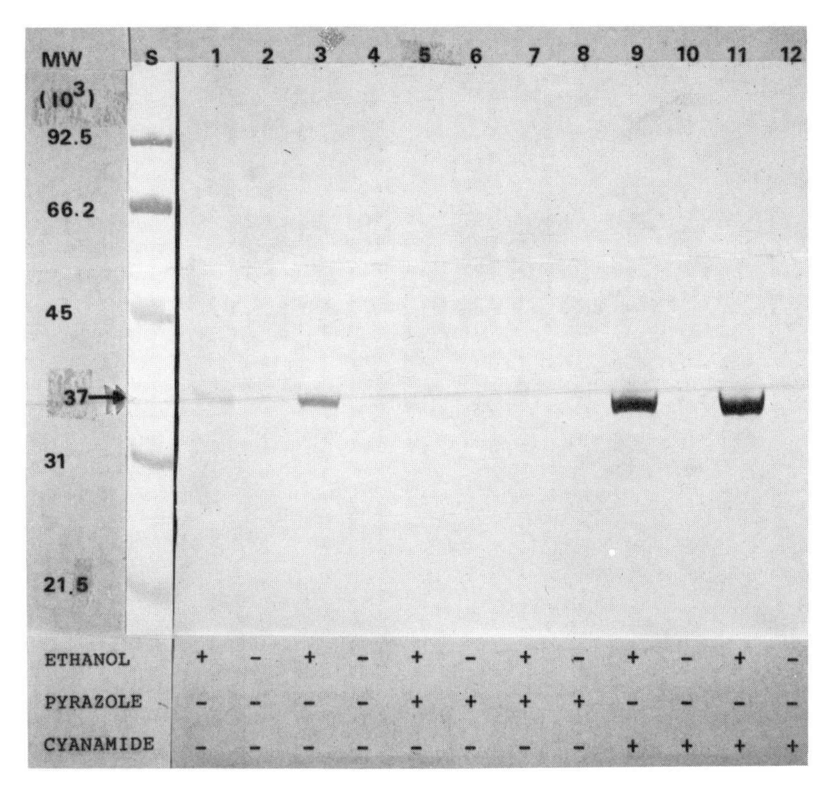

Figure 6. Effects of pyrazole and cyanamide on the formation of the 37KD liver protein *in vivo*. Rats were pair-fed the alcohol-containing (odd numbers) and the control (even numbers) diets. Group I (1, 2, 3, and 4) recieved no supplement. Group II (5, 6,7, and 8) was supplemented with pyrazole (2 mM). Group III (9, 10, 11, and 12) was supplemented with cyanamide (100 mg/L).

The 37KD protein-AA, once formed, was degradable *in vivo* when alcohol was removed from the diet. The half-life of this adduct *in vivo* was estimated to be four days. The protein formed only in the cytosol. We have shown previously that the 37KD protein is neither alcohol dehydrogenase nor a degradation product of aldehyde dehydrogenase. Thus the 37KD protein is not directly involved in the metabolism of alcohol or acetaldehyde. Its identity remains to be determined.

Using cultured hepatocytes as the experimental model, we were able to demonstrate that the 37KD protein-AA is most likely a product that results from a direct reaction between the parent protein and acetaldehyde. The 37KD protein-AA formed when the ethanol concentraion was as low as 5 mM, and its intensity further increased when cells were treated with higher concentrations of alcohol. These alcohol concentrations are achievable *in vivo* during alcohol drinking. Furthermore, the increase of the amounts of the 37KD protein-AA in hepatocytes paralleled the increase of acetaldehyde concentrations in the culture media (Figure 5). This strongly implicates a substrate-product relationship. However, there was one anomaly, i.e. a time delay or "window period" exists in the formation of the 37KD liver protein-AA. The 37KD protein-AA was detected in the liver of rats only after they had been fed alcohol for 1 week, and it was not detected in hepatocytes until the cells had been exposed to ethanol for 3 days. We have shown previously that the acetaldehyde

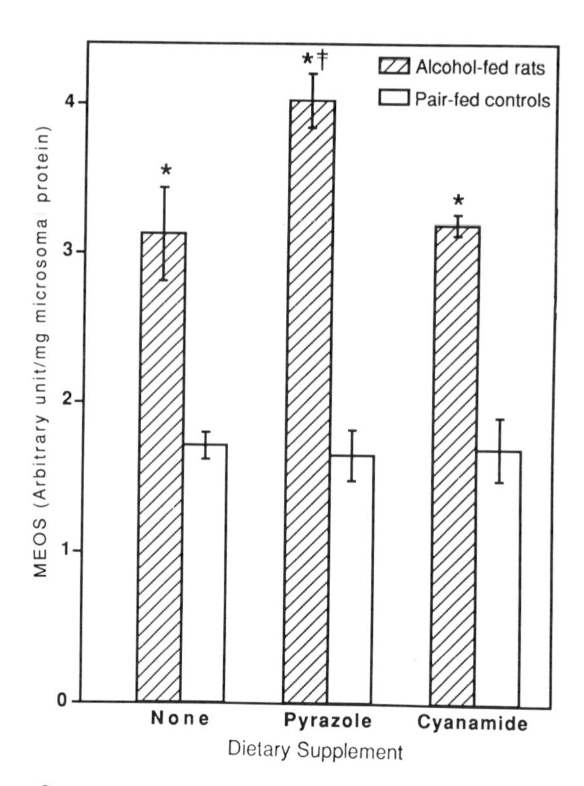

Figure 7. Cy t P450IIE1 content in rat livers. Rats were pair-fed the alcohol-containing and the control diets. One group was supplemented with pyrazole (2 mM) and the other was supplemented with cyanamide (100 mg/L). There were 4 pairs of rats in each group.

concentration in the culture medium of hepatocytes treated with alcohol for 1 day was the same as that of cells treated with alcohol for 4 days (Lin et al., 1990). Yet no protein-AA could be detected in 1-day-old cells. Thus the formation of the protein-AA must involve more complex mechanism but not the simple result of mixing acetaldehyde and the 37KD protein *in vivo* or *in vitro*. Several possible explanations for this "window period" required for the formation of the 37KD protein-AA are currently under investigation.

Feeding cyanamide to rats enhanced while feeding pyrazole blocked the formation of the 37KD protein-AA. The effects of cyanamide and pyrazole were also observed in cultured hepatocytes. These observations *in vivo* and *in vitro* further provide evidence that the formation of the protein-AA is dependent on acetaldehyde, and that the protein band detected by immunoblotting with anti-hemocyanin-AA IgG is indeed an acetaldehyde adduct.

The effect of pyrazole is most interesting. Rats fed alcohol supplemented with pyrazole (2 mM) exhibited elevated amounts of cytP450IIE1 in liver microsomes yet their plasma acetaldehyde levels were not detectable. Since pyrazole also abrogated the formation of the 37KD protein-AA *in vivo*, the aggregate of these findings indicates that the source of acetaldehyde for the formation of the 37KD protein-AA is alcohol dehydrogenase and not cytP450IIE1. Behrens et al., (1988) recently reported that cytP450IIE1 formed adducts with acetaldehyde *in vivo* when rats were fed alcohol chronically and they postulated that the source of acetaldehyde is most likely from

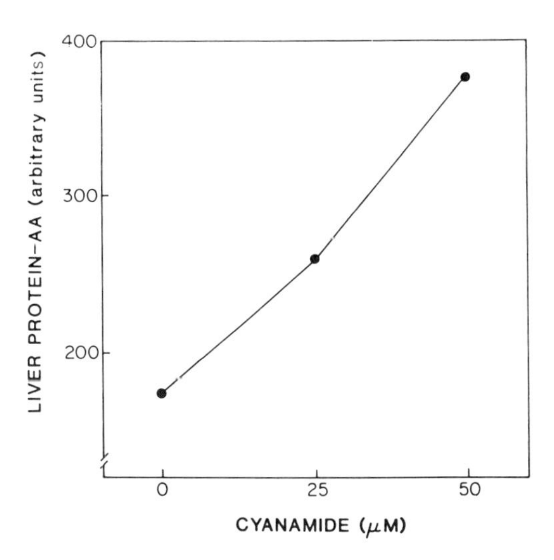

Figure 8. Effect of cyanamide on the formation of the 37KD protein-AA in hepatocytes. Cells were cultured for 4 days. Various concentrations of cyanamide were added to the culture medium when the latter was changed daily.

microsomal ethanol oxidation, i.e. via cytP450IIE1 per se. The divergent results reported by Behrens et al., and by us suggest that acetaldehyde available for protein-AA formation in liver might be compartmentalized, i.e., alcohol dehydrogenase generates acetaldehyde for the formation of the 37KD protein-AA in the cystolic compartment while cytP450IIE1 provides acetaldehyde for forming cytP450IIE1-AA in the microsomal compartment.

Using our antibodies, we were not able to detect any cytP450IIE1-AA. Likewise, Behrens et al., (1988) failed to demonstrate the presence of any cytosol protein-AA, although the existence of the 37KD protein-AA has been confirmed recently by Israel et al., (personal communication). It is important to point out that the procedures employed to prepare hemocyanin-AA used in raising polyclonal antibodies were different between Behrens' and our laboratories. Thus, it is possible that antibodies prepared by different methods may recognize different protein-AA epitopes. This leaves open the possiblity that other protein-AAs of different epitopes may exist in the liver of alcohol-fed rats. Furthermore, one can expect more protein-AAs to be formed with prolonged alcohol-feeding or with more advanced liver disease, e.g. alcoholic hepatitis and cirrhosis. Nonetheless, our studies seem to show that the 37KD liver protein is unusually susceptible to chemical modification by acetaldehyde *in vivo*.

The physiological consequence of *in vivo* formation of liver protein-AA is yet to be elucidated. Hypothesis that immune response plays a etiologic role in the development of alcohol liver diseases have been proposed (Neuberger et al., 1984; McFarlane, 1984). In 1986, Israel et al., (1986) reported that mice given alcohol in drinking water chronically developed serum antibodies that reacted against protein-AAs prepared *in vitro*. Furthermore, Hoerner et al., (1986) and Niemela et al., (1987b) reported that the titers of anti-protein-AA immunoglobulins were considerably higher in serum samples obtained from alcoholic patients than those from healthy, non-drinking subjects. We have recently obtained evidence showing that the 37KD protein-AA can accumulate in plasma membrane of the liver and it therefore has the potentials of being a surface-bound "neoantigen".

REFERENCES

Baroana, E., Leo, M.A., Borowsky, S.A. and Lieber, C.S. (1975). Alcoholic hepatomegaly: accumulation of protein in the liver. *Science* **190**, 794-795.

Barry, R.E., Williams, A.J.K. and McGivan, J.D. (1987). The detection of acetaldehyde/liver plasma membrane protein adduct formed *in vivo* by alcohol feeding. *Liver* **7**, 364-368.

Behrens, U.J., Hoerner, M., Lasker, J.M. and Lieber, C.S. (1988). Formation of acetaldehyde adducts with ethanol-inducible P450IIE1 *in vivo*. *Biochem. Biophys. Res. Commun.* **154**, 584-590.

Crabb, D.W. and Li, T.-K. (1985). Expression of alcohol dehydrogenase in primary monolayer cultures of rat hepatocytes. *Biochem. Biophys. Res. Commun.* **128**, 12-17.

DeMaster, E.G., Shirota, F.N., Redfern, B., Nagasawa, H.T. (1989). Identification of nitroxyl (HN=O) as a product of cyanamide-oxidation catalyzed by bovine liver catalase *in vitro*. *FASEB* **3**, A432.

Donohue, T.M., Tuma, D.J. and Sorrell, M.F. (1983). Acetaldehyde adducts with proteins: binding of 14C-acetaldehyde to serum albumin. *Arch. Biochem. Biophys.* **220**, 239-246.

Enat, R., Jefferson, D.M., and Ruiz-Opazo, N. (1984). Hepatocytes proliferation *in vitro*: its dependence on the use of serum-free hormonally defined medium and substrata of extracellular matrix. *PNAS, USA* **81**, 1411-1415.

Gaines, K.C., Salhany, J.M., Tuma, D.J. and Sorrell, M.F. (1977). Reactions of acetaldehyde with human erythrocyte membrane proteins. *FEBS* **75**, 115-119.

Goldberg, L. and Rydberg, U. (1969). Inhibition of ethanol metabolism *in vivo* by administration of pyrazole. *Biochem. Pharmacol.* **18**, 1749-1762.

Hoerner, M., Behrens, U.J., Worner, T. and Lieber, C.S. (1986). Humoral immune response to acetaldehyde adducts in alcoholic patients. *Res. Commun. Chem. Pathol. Pharm.* **54**, 3-12.

Israel, Y., Hurwitz, E., Niemela, O. and Ornon, R. (1986). Monoclonal and polyclonal antibodies against acetaldehyde-containing epitopes in acetaldehyde-protein adducts. *Proc. Natl. Acad. Sci. USA* **83**, 7923-7927.

Jennett, R.B., Tuma, D.J. and Sorrell, M.F. (1980). Effect of ethanol and its metabolites. *Pharmacology* **21**, 363-368.

Jennett, R.B., Johnson, E.L. and Sorrell, M.F. (1985). Covalent binding of acetaldehyde to tubulin is associated with impaired polymerization. *Hepatology* **5**, 1055.

Krikun, G., Feierman, D.E. and Cederbaum, A.I. (1986). Rat liver microsomal induction of the oxidation of drug and alcohols, and sodium dodecyl sulfate-gel profiles after *in vivo* treatment with pyrazole or 4-methylpyrazole. *J. Pharmacol. Exp. Ther.* **237**, 1012-1019.

Li, T.-K. and Theorell, H. (1969). Human liver alcohol dehydrogenase: Inhibition by pyrazole and pyrazole analogs. *Acta Chem. Scand.* **23**, 892-902.

Li, T.-K. (1977). Enzymology of human alcohol metabolism. In: *Advances in Enzymology* (A. Meister, Ed.) Vol. 45, pp. 427-483, John Wiley & Sons, Inc., New York.

Lieber, C.S. and DeCarli, L.M. (1970). Hepatic microsomal ethanol-oxidizing system. *in vitro* characteristics and adaptive properties *in vivo*. *J. Biol. Chem.* **245**, 2505-2512.

Lieber, C.S. and DeCarli, L.M. (1972). The role of the hepatic microsomal ethanol oxidizing sytem (MEOS) for ethanol metabolism *in vivo*. *J. Pharmacol. Exp. Ther.* **181**, 279-287.

Lieber, C.S. and DeCarli, L.M. (1982). The feeding of alcohol in liquid diets: two decades of applications and 1982 update. *Alcoholism: Clin. Exp. Res.* **6**, 523-531.

Lieber, C.S., DeCarli, L.M. and Sorrell, M.F. (1989). Experimental methods of ethanol administration. *Hepatology* **10**, 501-510.

Lin, R.C. and Snodgrass, P.J. (1975). Primary culture of normal adult rat liver cells which maintain stable urea cycle enzymes. *Biochem. Biophys. Res. Commun.* **64**, 725-734.

Lin, R.C., Smith, R.S. and Lumeng L. (1988a). Detection of a protein-acetaldehyde adduct in the liver of rats fed alcohol chronically. *J. Clin. Invest.* **81**, 615-619.

Lin, R.C., Lumeng, L, Kelly, T. and Pound, D. (1988b). Protein-acetaldehyde adducts

in serum of alcoholic patients. In: *Biomedical and Social Aspects of Alcohol and Alcoholism* (Kuriyama, K., Takada, A. and Ishii, H., Eds.), pp. 541-544, Excerpta Medica International Congress Series, Amsterdam.
Lin, R.C. and Lumeng, L. (1989). Further studies on the 37KD liver protein-acetaldehyde adduct that forms *in vivo* during chronic alcohol ingestion. *Hepatology* **10**, 807-814.
Lin, R.C., Fillenwarth, M.J., Minter, R. and Lumeng, L. (1990). Formation of the 37KDa protein-acetaldehyde adduct in primary cultured rat hepatocytes exposed to alcohol. *Hepatology* (in press).
Lumeng, L., Minter, R.G. and Li, T.-K. (1982). Distribution of stable acetaldehyde adducts in blood under physiological conditions. *Fed. Proc.* **41**, 765.
Lumeng, L. and Minter, R.G. (1985). Formation of acetaldehyde-hemoglobin adducts *in vitro* and *in vivo* demonstrated by high performance liquid chromatography. *Alcoholism: Clin. Exp. Res.* **9**, 209.
Mauch, T.J., Donohue, T.M., Zetterman, R.K. and Sorrell, M.F. (1985). Covalent binding of acetaldehyde to lysine-dependent enzymes can inhibit catalytic activity. *Hepatology* **5**, 1056.
McFarlane, I.G. (1984). Autoimmunity in liver disease. *Clin. Sci.* **67**, 569-578.
Medina, V.A., Donohue, T.M., Sorrell, M.F. and Tuma, D.J. (1985). Covalent binding of acetaldehyde to hepatic proteins during ethanol oxidation. *J. Lab. Clin. Med.* **105**, 5-10.
Neuberger, J., Crossley, I.R., Saunders, J., Davis, M., Portmann, B., Eddleston, A.L.W.F. and Williams, R. (1984). Antibodies to alcohol-altered liver cell determinants in patients with alcohol liver disease. *Gut* **25**, 300-304.
Niemela, O. and Israel, Y. (1987a). Immunological detection of acetaldehyde containing epitopes (ACES) in human hemoglobin: a new test in alcoholism (abstract). *Alcoholism: Clin. Exp. Res.* **11**, 203.
Niemela, O., Klajner, F., Orrego, H., Vidins, E., Blendis, L. and Israel, Y. (1987b). Antibodies against acetaldehyde-modified protein epitopes in human alcoholics. *Hepatology* **7**, 1210-1214.
Nomura, F. and Lieber, C.S. (1981). Binding of acetaldehyde to rat liver microsomes: enhancement after chronic alcohol consumption. *Biochem. Biophys. Res. Commun.* **100**, 131-137.
Peachey, J.E. and Sellers, E.M. (1981). The disulfiram and calcium carbimide acetaldehyde-mediated ethanol reactions. *Pharmacol. Ther.* **15**, 89-97.
Reynier, M. (1969). Pyrazole inhibition and kinetic studies of ethanol and retinol oxidation catalyzed by rat liver alcohol dehydrogenase. *Acta Chem. Scand.* **23**, 1119-1129.
Ritchie, J.M. (1985). The aliphatic alcohols. In: *The Pharmacological Basis of Therapeutics* (Gilman, A.G., Goodman, L.S, Gilman, A., Eds.) 7th Ed., pp. 372-386, McMillan, New York.
Salaspuro, M. and Lindros, K. (1985). Metabolism and toxicity of acetaldehyde. In: *Alcohol Related Diseases in Gastroenterology* (Seitz, H.K., Kommerell, B., Eds), pp. 105-123, Springer-Verlag, Berlin.
Stevens, V.J., Fantl, W.J., Newman, C.B., Sims, R.V., Cerami, A. and Peterson, C.M. (1981). Acetaldehyde adducts with hemoglobin. *J. Clin. Invest.* **67**, 361-369.

DOSE-RESPONSE RELATIONSHIPS IN CHEMICAL CARCINOGENESIS: FROM DNA ADDUCTS TO TUMOR INCIDENCE

Werner K. Lutz

Institute of Toxicology
Swiss Federal Institute of Technology and
University of Zurich
CH-8603 Schwerzenbach, Switzerland

Index Words: aflatoxin B1, 2-acetylaminofluorene, DNA, adduct, covalent, binding, carcinogen, dose, extrapolation, individual, susceptibility, heterogeneous population, risk, tumor.

SUMMARY

Mechanistic possibilities responsible for nonlinear shapes of the dose-response relationship in chemical carcinogenesis are discussed. (i) Induction and saturation of enzymatic activation and detoxification processes and of DNA repair affect the relationship between dose and steady-state DNA adduct level; (ii) The fixation of DNA adducts in the form of mutations is accelerated by stimulation of the cell division, for instance due to regenerative hyperplasia at cytotoxic dose levels; (iii) The rate of tumor formation results from a superposition of the rates of the individual steps. It can become exponential with dose if more than one step is accelerated by the DNA damage exerted by the genotoxic carcinogen. The strongly sigmoidal shapes often observed for dose-tumor incidence relationships in animal bioassays supports this analysis. A power of four for the dose in the sublinear part of the curve is the maximum observed (formaldehyde). In contrast to animal experiments, epidemiological data in humans rarely show a significant deviation from linearity. The discrepancy might be explained by the fact that a large number of genes contribute to the overall sensitivity of an individual and to the respective heterogeneity within the human population. Mechanistic nonlinearities are flattened out in the presence of genetic and life-style factors which affect the sensitivity for the development of cancer. For a risk assessment, linear extrapolation from the high-dose incidence to the spontaneous rate can therefore be appropriate in a heterogeneous population even if the mechanism of action would result in a nonlinear shape of the dose-response curve in a homogeneous population.

INTRODUCTION

The reaction of biological reactive intermediates with DNA can be investigated at low dose levels which would not give rise to a significant increase in tumor incidence in a standard animal bioassay on carcinogenicity. The shape of the dose-response curve for DNA adduct formation is therefore considered valuable information for a low-dose risk evaluation. In this presentation, general principles are derived to

Biological Reactive Intermediates IV, Edited by C.M. Witmer *et al.*
Plenum Press, New York, 1990

explain the available data. In addition, the value of these data for human risk evaluation is critically assessed.

From Carcinogen Dose TO DNA Adducts

Single-Dose Data: The processes and reactions leading to DNA adducts include diffusion as well as enzymatic and chemical reactions. At lowest dose levels, when the concentrations of the carcinogen and of its proximate and ultimate metabolite(s) are well below the Michaelis concentration of the enzymatic reactions, the formation of DNA adducts is expected to be proportional to dose. This hypothesis was first supported by data published by Neumann (1980) with trans-4-dimethylaminostilbene. Additional experiments which showed a linear dose-response relationship were reviewed by Lutz (1987). The only nonlinearity discussed in this review was the example of O6-methylguanine determined after administration of a methylating agent. It shows that the instantaneous repair by methyl transfer to an acceptor protein is exhausted above a certain level of DNA methylation.

Additional evidence of nonlinearities in an intermediate dose range of the dose-DNA adduct curve was reviewed by Swenberg et al. (1987). A sigmoidal shape of the dose-response curve was found with formaldehyde-induced DNA-protein crosslinks (Heck and Casanova, 1987), probably due to a saturation of metabolic inactivation. A superlinear shape was reported with the tobacco-specific nitrosamine 4-(N-methyl-N-nitrosamino)-1-(3-pyridyl)-1-butanone (NNK) and was explained on the basis of a low KM pathway for the metabolic activation (Belinsky et al., 1987).

A superlinear shape for DNA adduct formation is often seen at highest dose levels due to a saturation of the metabolic activation of the carcinogen.
In summary, while proportionality between dose and DNA adduct level is expected to hold at low doses, a number of saturation phenomena have been shown to generate nonlinear dose-response curves in an intermediate and high-dose range (Figure 1 top).

Multiple-Dose Data: The data referred to above are derived from single-dose experiments. For a correlation of DNA adduct formation with tumor induction, however, it is the level of dangerous adducts arising from chronic exposure which is the important variable. Since DNA repair is also an enzymatic reaction, the rate of DNA repair is expected to be proportional to the level of DNA adducts. Consequently, the DNA damage levels off to a steady-state when the number of newly-formed adducts equals the number of adducts repaired or lost during the same period of time. Proportionality between daily dose and steady-state adduct level has to be postulated under these conditions, unless DNA repair is induced. In this laboratory, DNA adduct levels were measured in rat liver after exposure for 4, 6, and 8 weeks to aflatoxin B1 and 2-acetylaminofluorene administered in the drinking water. Steady-state level of DNA adducts was reached during this period of treatment and was proportional to dose down to the limit of detection at one thousandth of the TD50 dose level (Buss and Lutz, 1988).

From DNA Adducts To Mutations

DNA adducts alone are not sufficient to generate a heritable, initiated phenotype of a cell because adducts cannot be copied onto the progeny DNA strand. Only upon DNA replication can mutations arise in the new strand opposite an adduct. Since DNA adducts can be repaired, the probability of a fixation of the DNA adduct in the form of a mutation is dependent on the relative rates of DNA repair and cell division. All acceleration of cell division will therefore have a synergistic effect on the formation of mutations.

DNA adduct-forming carcinogens do not bind exclusively to DNA, but always also to RNA and protein. Above a certain level of macromolecular binding, the cell dies. This can induce regenerative hyperplasia in the surrounding tissue. Genotoxic agents at high dose, therefore, can stimulate cell division and accelerate the process of mutagenesis and carcinogenesis in the surviving cells. High-dose exposure to a genotoxic carcinogen (which is the usual situation in a bioassay; Hoel et al., 1988) therefore results in a sublinear shape of the dose response curve for mutations (and tumors) in division-competent cells (Figure 1 middle).

In the low dose range which does not produce cytotoxicity, a linear relationship between level of DNA adducts and mutations can be expected. A nice example for such a linear -> sublinear shape can be given for the induction of liver tumors in female rats by nitrosodiethylamine (NDEA). In this study, NDEA was administered in the drinking water to groups of 60 rats at 15 dose levels (0.033 to 16.9 ppm; Peto et al., 1984). When the ratio observed/expected liver cancer incidence is plotted as a function of dose, a quadratic to cubic shape is seen. However, if only the low dose data are taken (up to 0.53 ppm), the increase in tumor response appears to be linear with dose and less rapid than the curve fitted to all data points (Zeise et al., 1987). This large experiment, using more than a thousand animals, afforded a unique opportunity to detect linearity at low dose levels and a power of the dose in the high-dose range. It is a good example to stress the idea that mode of action and potency of a carcinogen can change with the dose level.

From Mutations To Tumors

As a consequence of the multi-stage nature of carcinogenesis, the probability of cancer induction can be approximated by the product of the probabilities for the various steps (Armitage, 1985). Since all steps have a genetic basis to become heritable, DNA-damaging compounds could accelerate more than one step. The dose-response relationship then appears as a superposition of the curves for the individual steps. If, for instance, two events are affected, both proportionally with the level of DNA adducts, the product of the lines will be exponential (Figure 1 bottom).

Dose Response for Tumor Induction

The mechanistic analysis given above indicated a large number of possibilities for nonlinear dose-response relationships, especially at high exposure levels. In the high-dose range, data on tumor formation in animals and humans become available, and the following paragraphs will show whether the postulated steep slopes are indeed found.

Bioassays on Carcinogenicity

A comprehensive and recent review on this topic is available (Zeise et al., 1987). In the observable (high) dose range used in animal studies, several examples from well-documented studies suggest that nonlinearities are common. For instance, a sublinear curve is seen for bladder cancer caused by sodium saccharin and 2ãcetylaminofluorene, for liver cancer induced by diethylnitrosamine and dimethylnitrosamine, and for oesophageal cancer from diethylnitrosamine. In an extreme case, formaldehyde induced nasal tumors in rats with a frequency of 1 and 44 % after inhalation exposures of 5.6 and 14.3 ppm, respectively; Swenberg et al., 1983). Here, the tumor incidence increased with the fourth power of the dose. In all other examples, the respective exponent was smaller.

Epidemiological Evidence in Humans

In human studies, significant deviations from linearity are much more difficult to find. Only in two reports, namely lung cancer in smoking British physicians and bladder cancer from β-naphthylamine exposure is a sublinear fit clearly better than a linear one. The relationship in these two cases is approximately quadratic; higher exponents have not been observed (Zeise et al., 1987).

The fourth power of the dose seen in an animal experiment is clearly compatible with the multi-stage nature of carcinogenesis and with the mechanistic analysis discussed above. Astonishing at first glance is the fact that epidemiological investigations with humans in general showed a linear dose-response relationship, or, at most, a second order for the dose. The reasons for this difference between animals and humans might be found with the heterogeneity of the human population.

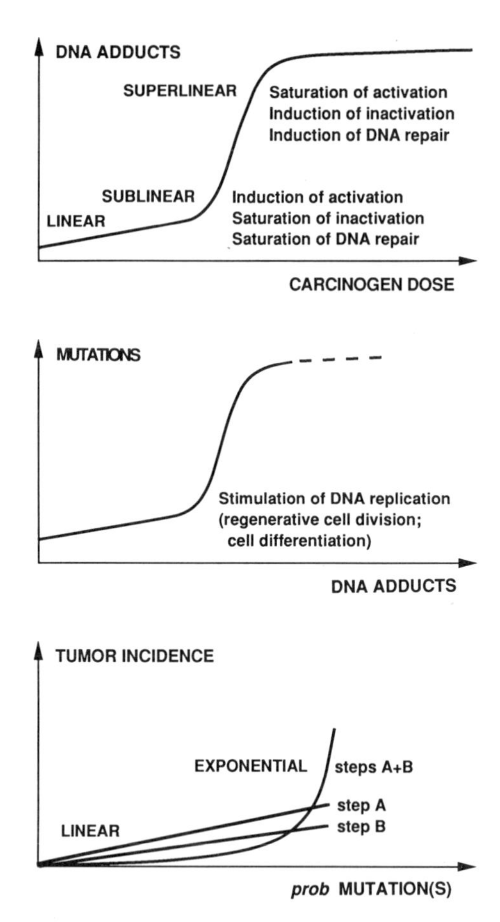

Figure 1. Schematic representation of the dose-effect relationship for three aspects of tumor induction by genotoxic agents.

Homogeneous vs. Heterogeneous Population

So far, carcinogenesis has been discussed mechanistically, as a process in one individual or in an animal population which is relatively homogeneous both with respect to the genetic predisposition and the life style (Fig. 2 top). The epidemiological evidence with humans, on the other hand, is based on a highly heterogeneous population. For instance, interindividual differences in the metabolism of chemical carcinogens and repair rates of DNA adducts are large and reflect acquired and inherited host factors that may influence an individual's risk for development of cancer (Harris, 1989). For instance, genetically controlled metabolic capacities contribute to a predisposition of humans to bladder cancer (Kaisary et al., 1987).

If one single gene was to govern the sensitivity to a carcinogen, the human population would be separated into a population A where a given dose leads to a tumor in almost all exposed individuals whereas the remaining people (population B) develop a tumor only at higher dose levels (Fig. 2 middle). The mechanistic aspects discussed above would then only become important for an evaluation of the dose-response curve within each subpopulation.

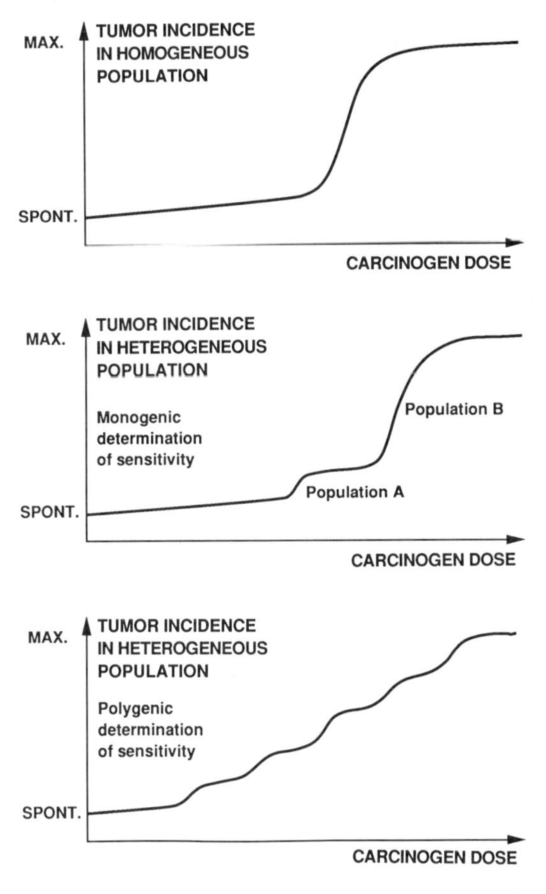

Figure 2. Schematic representation of the effect of population heterogeneity on the dose-effect relationship in chemical carcinogenesis. Top, homogeneous population; middle, heterogeneous populations with monogenic determination of the sensitivity; bottom, polygenic determination of the sensitivity.

Much more realistic is the situation where the sensitivity is governed by a large number of genes. As a result, the dose-response curve will become flatter with each modulatory factor (Fig. 2 bottom). If, in addition, the sensitivity is modulated by the lifestyle, near-linearity can be produced between the 'spontaneous' tumor incidence (Lutz, 1990) and the high-dose data. This polygenic-multifactorial situation might explain the finding that dose-response curves in human epidemiological studies are linear in most cases, and never show the steep slopes observed in animal studies.

Risk Extrapolation

This analysis has revealed a large number of mechanistic possibilities to generate nonlinear dose-response relationships in chemical carcinogenesis. On the other hand, the dose-response curve is linearized in a heterogeneous population, if the sensitivity of an individual is determined by its specific set of genetic factors and is further modulated by the lifestyle. The larger the number of these factors, the more will the dose-response curve become linear. Extrapolation of tumor incidence from the high dose range to low dose levels can therefore be linear in a heterogeneous population even if the mechanism of carcinogenesis in an individual produces a nonlinearity.

REFERENCES

Armitage, P. (1985). Multistage models of carcinogenesis. *Environ. Health Perspect.* **63**, 195-201.

Belinsky, S.A., White, C.M., Devereux, T.R., Swenberg, J.A., Anderson, M.W. (1987). Cell selective alkylation of DNA in rat lung following low dose exposure to the tobacco specific carcinogen 4-(N-methyl-N-nitrosamino)-1-(3-pyridyl)-1-butanone. *Cancer Res.* **47**, 1143-1148.

Buss, P., Lutz, W.K, (1988). Steady-state DNA adduct level in rat liver after chronic exposure to low doses of aflatoxin B1 and 2-acetylaminofluorene. *Proc. Amer. Assoc. Cancer Res.* **29**, 96.

Harris, C.C. (1989). Interindividual variation among humans in carcinogen metabolism, DNA adduct formation and DNA repair. *Carcinogenesis* **10**, 1563-1566.

Heck, H.A., Casanova, M. (1987). Isotope effects and their implications for the covalent binding of inhaled (3H) and (14C) formaldehyde in the rat nasal mucosa. *Toxicol. Appl. Pharmacol.* **89**, 122-134.

Hoel, D.G., Haseman, J.K., Hogan, M.D., Huff, J., McConnell, E.E. (1988). The impact of toxicity on carcinogenicity studies: implications for risk assessment. *Carcinogenesis* **9**, 2045-2052.

Kaisary, A., Smith, P., Jaczq, E., McAllister, C.B., Wilkinson, G.R., Ray, W.A., Branch, R.A. (1987). Genetic predisposition to bladder cancer: ability to hydroxylate debrisoquine and mephenytoin as risk factors. *Cancer Res.* **47**, 5488-5493.

Lutz, W.K. (1987). Quantitative Evaluation of DNA-Binding Data In Vivo for Low-Dose Extrapolations. *Arch. Toxicol.* (Suppl) **11**, 66-74.

Lutz, W.K. (1990). Endogenous genotoxic agents and processes as a basis of spontaneous carcinogenesis. *Mutat. Res.*, in press.

Neumann, H.G. (1980). Dose-response relationship in the primary lesion of strong electrophilic carcinogens. *Arch. Toxicol.* (Suppl) **3**, 69-77.

Peto, R., Gray, R., Brantom, P., Grasso, P. (1984). Nitrosamine carcinogenesis in 5120 rodents: chronic administration of sixteen different concentrations of NDEA, NDMA, NPYR and NPIP in the water of 4440 inbred rats, with parallel studies on NDEA alone of the effect of age of starting and of species. *IARC Sci. Publ.* **57**, 627-665.

Swenberg, J.A., Barrow, C.S., Boreiko, C.J., Heck, H.A., Levine, R.J., Morgan, K.T., Starr, T.B. (1983). Non-linear biological responses to formaldehyde and their implications for carcinogenic risk assessment. *Carcinogenesis* **4**, 945-952.

Swenberg, J.A., Richardson, F.C., Boucheron, J.A., Deal, F.H., Belinsky, S.A., Charbonneau, M., Short, B.G. (1987). High to low dose extrapolation: critical determinants involved in the dose response of carcinogenic substances. *Environ. Health Perspect.* **76**, 57-63.

Zeise, L., Wilson, R., Crouch, E.A.C. (1987). Dose-response relationships for carcinogens: a review. *Environ. Health Perspect.* **73**, 259-308.

THE SINGLE CELL GEL (SCG) ASSAY: AN ELECTROPHORETIC TECHNIQUE FOR THE DETECTION OF DNA DAMAGE IN INDIVIDUAL CELLS

R.R. Tice[1], P.W. Andrews[1], O. Hirai[1,2] and N.P. Singh[3]

Integrated Laboratory Systems[1]
P.O. Box 13501
Research Triangle Park, NC 27709,
Fujisawa Pharmaceutical Company[2]
Osaka, Japan, and
Eastern Washington University[3]
Cheney, WA 99004

INTRODUCTION

Techniques which permit the sensitive detection of DNA damage are needed to evaluate the genotoxic potential of biologically reactive intermediates. Since the effects of many reactive intermediates are tissue and cell-type specific, it is important to utilize techniques which can directly detect DNA damage in individual cells. Cytogenetic techniques, while providing information at the level of the individual cell, are largely limited to proliferating cell populations. Furthermore, these techniques require the processing of DNA damage into microscopically visible lesions. Biochemical techniques, such as alkaline elution and nucleoid sedimentation, circumvent these difficulties in that DNA damage can be evaluated directly in almost any cell population. However, the resulting data do not provide any information about the distribution of damage or repair among individual cells. Recently, an electrophoretic technique capable of detecting DNA single-strand breaks and alkali labile sites in individual cells was developed (Singh et al., 1988). Eukaryote cells are embedded in an agarose gel on a microscope slide, lysed by detergents and high salt at pH 10, and then electrophoresed for a short time under alkaline conditions. Cells with increased DNA damage display increased migration of the DNA from the nucleus towards the anode. The importance of this technique lies in its ability to detect intercellular differences in DNA damage in virtually any eukaryote cell population, and in its requirement for extremely small cell samples (from 1 to 10,000 cells). Data from three recent pilot studies, involving (i) the exposure of human lymphocytes to hydrogen peroxide, (ii) the incubation of isolated mouse hepatocytes with cyclophosphamide, and (iii) the treatment of mice with acrylamide, will be used to demonstrate the sensitivity and utility of this technique for evaluating intercellular differences in DNA damage.

The Single Cell Gel (SCG) Technique

The basic technique is described in detail in Singh et al. (1988). Briefly, up to 50,000 cells were mixed with 0.5% low- melting agarose at 37°C and then placed on a fully frosted microscope slide coated previously with 0.5% agarose. After the agarose had solidified, an additional layer of agarose was added. The slides were then immersed in a cold lysing solution (1% sodium sarcosinate, 2.5 M NaCl, 100 mM Na_2EDTA, 10 mM Tris, pH 10, 10% dimethylsulfoxide and 1% Triton X-100, added

Biological Reactive Intermediates IV, Edited by C.M. Witmer *et al.*
Plenum Press, New York, 1990

fresh) for at least 1 hour to lyse the cells and to permit DNA unfolding. The slides were then removed from the lysing solution and placed on a horizontal gel electrophoresis unit. The unit was filled with fresh electrophoretic buffer (1 mM Na_2EDTA and 300 mM NaOH; pH ~13) and the slides were allowed to sit in this alkaline buffer for 20 minutes to allow the DNA to unwind before electrophoresis. Depending on the basal level of DNA damage in the target cell population, electrophoresis was conducted for from 15 to 20 minutes at 25 volts. These steps were conducted under yellow light to prevent additional DNA damage. After electrophoresis, the slides were washed with 0.4 M Tris, pH 7.5, to remove alkali and detergents. The slides were then stained with ethidium bromide (10 ug/ml) in distilled water. Observations were made at 250x magnification using a fluorescent microscope equipped with an excitation filter of 515-560 nm and a barrier filter of 590 nm. Images of 25 to 50 randomly selected cells per sample were analyzed for migration length using a Cambridge Instruments Quantimet 520 image analyzer. The effect of dose on the mean length of the DNA migration was analyzed using a one-tailed trend test, with the alpha level set at 0.05. To determine the lowest dose at which a significant increase in the length of migration occurred, multiple pairwise comparisons were conducted between the control data and each dose using Student's t test. To evaluate the intercellular distribution of DNA migration patterns among cells within cultures and tissues, the dispersion coefficient H (Snedecore and Cochran, 1967), the ratio of the sample variance to the mean, was determined and evaluated as described for mean migration length.

Hydrogen Peroxide Induced DNA Damage in Human Peripheral Blood Lymphocytes

The SCG assay is an extremely sensitive technique for evaluating free-radical induced DNA damage, such as that caused by hydrogen peroxide (Singh et al, 1988). In this experiment, 5 uL samples of whole blood were obtained from a single male donor, and added to 1 ml of RPMI-1640 containing 10% fetal calf serum. After centrifugation, the pellet was suspended in Hanks calcium-magnesium free, balanced salt solution containing hydrogen peroxide (44 and 176 uM) for 1 hr at 4°C. The cells were collected by centrifugation, and processed for SCG analysis as described. The slides were electrophoresed for 20 minutes, and 50 cells were scored per treatment.

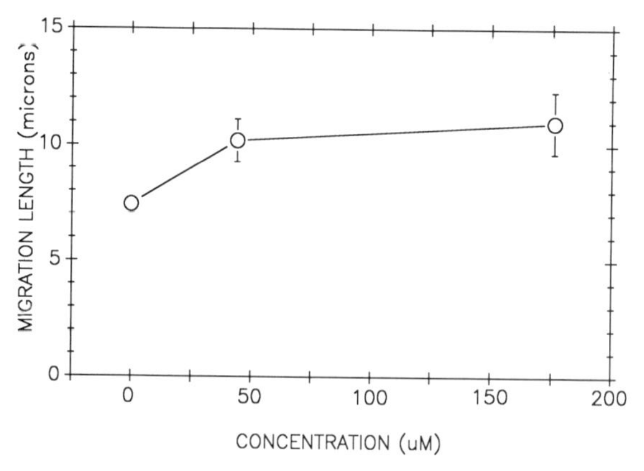

Figure 1. Induction of DNA damage by hydrogen peroxide in human peripheral blood leukocytes. Data are presented as the mean and standard error for 50 cells. Some error bars are smaller than the symbol used to designate the mean.

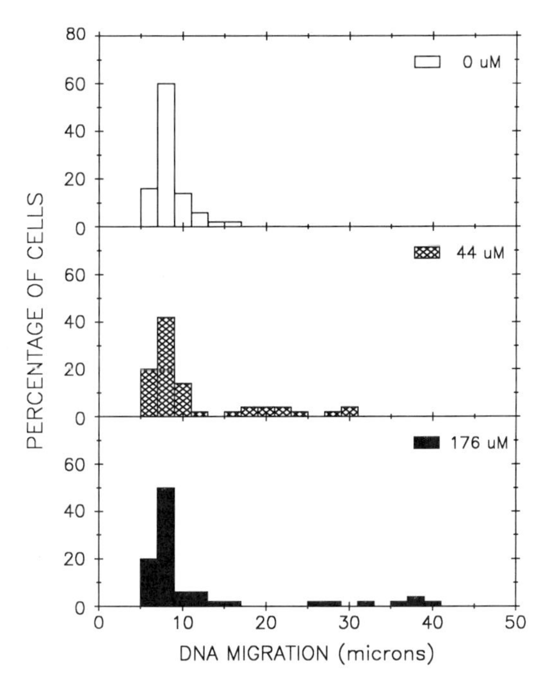

Figure 2. Distribution of DNA migration patterns among individual human peripheral blood lymphocytes exposed to hydrogen peroxide. 50 cells analyzed per sample. The width of each bar represents 2 microns.

At these concentrations, hydrogen peroxide induced a significant increase in the migration of DNA (Figure 1). The mean extent of DNA migration was not significantly different between the two concentrations of hydrogen peroxide. In contrast to the relatively homogeneous DNA migration patterns observed for lymphocytes irradiated with x-rays (Singh et al., 1988), exposure to hydrogen peroxide results in an extremely heterogenous intercellular response, with the majority of the cells exhibiting no damage (Figure 2). Several explanations are possible for this extensive heterogeneity in DNA damage: individual cells may vary in their permeability to hydrogen peroxide, their radical scavenging capabilities, the accessibility of DNA to the damaging species and other mechanisms which either enhance or diminish the effects of hydrogen peroxide.

Induction of DNA Damage In Mouse Hepatocytes In Vitro by Cyclophosphamide

Currently, one of the primary *in vitro* methods for ascertaining the ability of chemicals to induce DNA damage involves an evaluation of DNA repair synthesis (so-called unscheduled DNA synthesis or UDS) in rodent hepatocytes (Butterworth et al., 1987). This technique is based on an autoradiographic determination of the incorporation of tritiated thymidine into DNA repair sites. Rodent hepatocytes are used because the target cells themselves are metabolically competent, eliminating the need for an exogenous source of metabolic activity. The primary rat hepatocyte UDS assay is considered to be an excellent method for detecting the genotoxic activity and potential carcinogenicity of chemicals (Probst et al., 1980; Williams et al., 1982; Mitchell et al., 1983). However, the very nature of the technique limits its detection generally to those lesions requiring long patch repair, involves the use of radioactive precursors to DNA and requires an extended processing time. Furthermore, based on statistical grounds, the assay is reported as insensitive to weak mutagens (Margolin and Risko, 1988). Based on these considerations, we have begun studies to evaluate

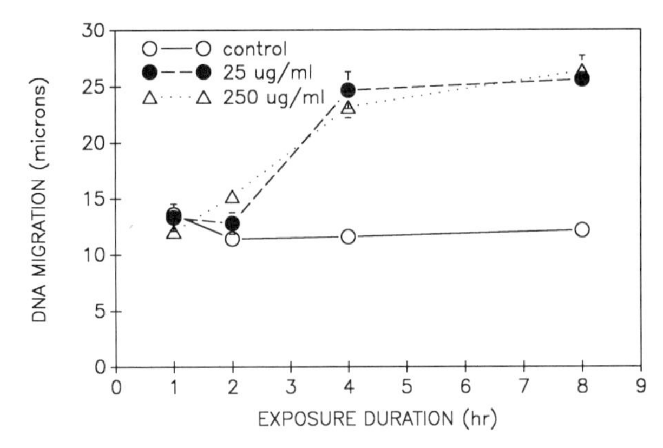

Figure 3. Time course for the induction of DNA damage in primary mouse hepatocytes by cyclophosphamide. Data are presented as the mean of triplicate cultures (25 cells were scored in each culture). Some error bars are smaller than the symbol used to designate the mean.

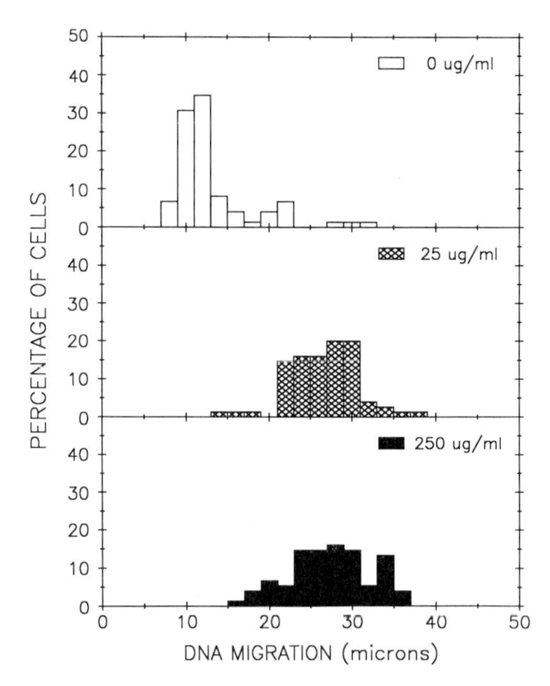

Figure 4. Distribution of DNA migration patterns among individual primary mouse hepatocytes exposed with cyclophosphamide for 8 hours. Data for each sample based on 75 cells. The width of each bar represents 2 microns.

the applicability of the SCG assay to a detection of DNA damaging agents using the *in vitro* rodent hepatocyte cell system. The data presented here comes from an initial experiment involving cyclophosphamide (CP), a well-known alkylating agent requiring metabolic activation.

Mouse primary hepatocytes were obtained as described in Butterworth et al. (1987), as modified by Ashby et al. (1985). Cell density and viability were determined using a hemocytometer and the trypan blue dye exclusion technique. About 1.5 x 10^5 viable cells in complete Williams media E were added to 3 cm petri dishes, and placed in a 37°C, 5% CO_2 incubator for 1.5-2 hours to allow cell attachment. The non-adhering cells were removed and the attached cells were exposed to CP (25 and 250 ug/ml) for from 1 to 8 hours. Following treatment, the cells were removed using a teflon scrapper and processed for evaluation in the SCG assay. An electrophoresis time of 15 minutes was used in these studies.

Exposure to CP induced a dose-dependent increase in DNA damage in mouse hepatocytes at the two hour sample time (Figure 3). After 4 hours of exposure, the level of DNA damage was not significantly different between the two doses and had appeared to have saturated. An analysis of the distribution of migration patterns among individual cells indicated a homogenous response among all hepatocytes in the 8 hour samples (Figure 4). Experiments to evaluate DNA damage using lower doses of CP are in progress. This preliminary experiment suggests that the SCG assay may be easily applied to the evaluation of genotoxic activity in the *in vitro* hepatocyte cell system.

Chemically-Induced DNA Damage In Vivo In Mice

One of the main advantages of the SCG assay is its requirement for extremely small cell samples (i.e., from a few to a few thousand cells). This advantage makes the assay extremely useful in studies where only small numbers of cells are available. One such area of research involves the evaluation of organ-specific levels of DNA damage induced by biologically reactive intermediates in animal models of disease. We have initiated studies to assess the feasibility of utilizing the SCG assay to detect DNA damage induced in various organs of the mouse. Preliminary data collected on mice treated with acrylamide (ACR) will be presented. Briefly, male B6C3F1 mice (10 - 13 weeks of age, 25 to 32 gm in body weight, 4 mice per group) were gavaged with 10 and 100 mg/kg ACR in PBS. Six hours after treatment, groups of mice were killed by CO_2 asphyxiation, and samples of liver, spleen and testis were removed from each mouse and stored on ice in Hank's buffer containing 20 mM EDTA. Blood samples were collected by adding 5uL to RPMI-1640 containing 10% fetal calf serum. The samples of liver, spleen and testis were minced, while the blood sample was centrifuged to recover the leukocytes. An aliquot of cells was mixed with low melting agarose and placed on microscope slides as described earlier. The cells were electrophoresed for 20 minutes, stained with ethidium bromide and 25 cells per cell type per sample were evaluated for DNA migration using the Quantimet 520 image analyzer. Total leukocytes were scored from blood, splenocytes from spleen, parenchymal and nonparenchymal cells from liver, while the analysis of damage in testis was limited to diploid cells.

Based on an analysis of mean DNA migration patterns, treatment with 100 mg/kg ACR induced a significant increase in DNA damage in all organs 6 hours after treatment (Figure 5), while only spleen exhibited a significant increase at 10 mg/kg. However, when the dispersion of DNA migration patterns within individual samples was used as the basis for the statistical analysis, a significant increase in damage was also detected in blood (Figure 6). This was due to the presence of a small population of damaged cells among a much larger population of unaffected cells. In comparing the response among various organs, leukocytes in the blood exhibited the greatest response, while diploid cells in the testis exhibited the lowest level of damage. In liver, both parenchymal and nonparenchymal cells exhibited a similar increase in DNA damage.

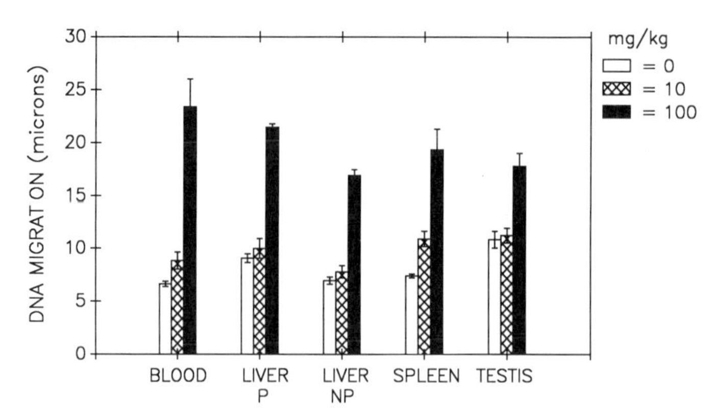

Figure 5. Organ-dependent induction of DNA damage by ACR (10 and 100 mg/kg) in male B6C3F1 mice sampled 6 hours after treatment by gavage. Data presented as the mean DNA migration and standard error of the mean for 4 mice per group. P = parenchymal; NP = nonparenchymal.

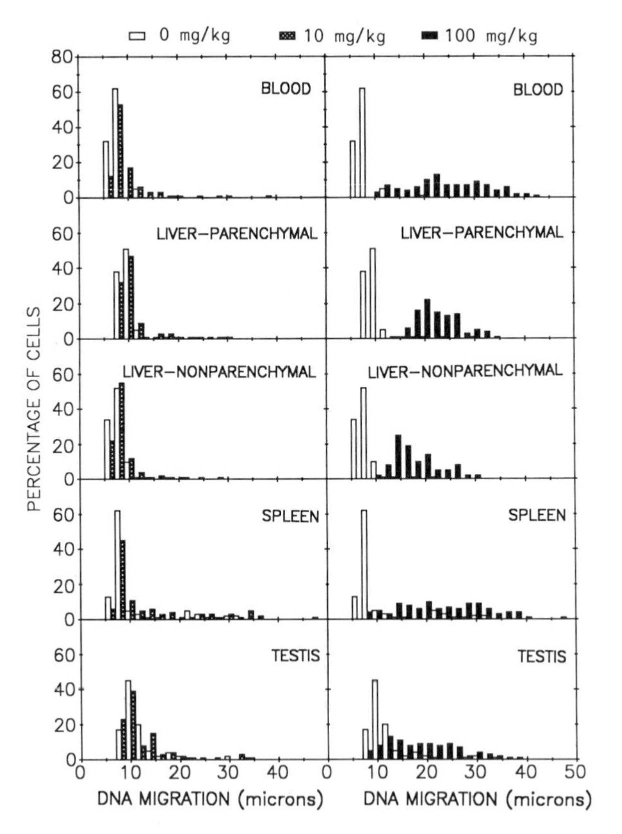

Figure 6. Distribution of DNA migration patterns among individual cells from various organs of male mice treated with ACR (10 and 100 mg/kg) by gavage and sampled 6 hours later. Data for each sample based on 100 cells (25 cells per cell type per mouse). The width of each bar represents 2 microns.

CONCLUSIONS

These studies demonstrate the sensitivity of the SCG technique for identifying the presence of DNA damage among a variety of cell types and the importance of data collected on a cell by cell basis. The technique is also relatively quick, i.e., data can be obtained at a rate of about 1 cell every 30 seconds within 3 hours after collection. *In vitro*, lymphocytes exposed to hydrogen peroxide exhibited considerable heterogeneity in damage, the extent of which is clearly not appreciated with techniques based on pooled cell samples. The data obtained from primary mouse hepatocytes exposed to cyclophosphamide suggests a sensitive approach for detecting genotoxic agents that require metabolic activation before becoming biologically reactive. The *in vivo* experiment indicates that acrylamide is an inducer of DNA damage in all tissues evaluated, including testis. Not unexpectedly, the induction of DNA damage *in vivo* is organ-dependent. Although not examined in this particular study, other organs such as bone marrow, lung and brain are equally amenable to analysis.

The advantages of the SCG technique include: (i) its sensitivity for detecting DNA damage, (ii) the collection of data at the level of the individual cell, allowing for more robust types of statistical analyses, (iii) the use of extremely small cell samples, and (iv) that virtually any eukaryote cell population is amenable to analysis. Unfortunately, mature sperm, due to the extreme condensation of the DNA, contain 10^5 to 10^6 alkali-labile sites (Singh et al., 1989), preventing an effective analysis of DNA damage in this cell type. The disadvantages of the assay include the necessity for single cell suspensions, and the fact that small cell samples may not be representative of the total cell population. We anticipate that this technique will play an increasingly important role in future studies to evaluate DNA damage in humans, in detecting organ-specific levels of damage in animals exposed to potential mutagens and carcinogens, in reproductive and teratological studies, and in studies to examine the basic mechanisms of DNA damage and repair.

ACKNOWLEDGEMENTS

Although portions of the research described in this article have been supported by the U.S EPA under contract 68-68-0069, it has not been subjected to Agency review and therefore does not necessarily reflect the views of the Agency and no official endorsement should be inferred.

REFERENCES

Ashby, J., Lefevre, P.A., Burlinson, B., and Penman, M.G. (1985). An assessment of the *in vivo* rat hepatocyte DNA-repair assay. *Mutat. Res.* **156**, 1.

Butterworth, B.E., Ashby, J., Bermudez, E., Casciano, D., Mirsalis, J., Probst, G., and Williams, G. (1987). A protocol and guide for the *in vitro* rat hepatocyte DNA-repair assay. *Mutat. Res.* **189**, 113.

Margolin, B.H. and Risko, K.J. (1987). The statistical analysis of *in vivo* genotoxicity data. Case studies of the rat hepatocyte UDS and mouse bone marrow micronucleus assays, In: "Evaluation of Short-Term Tests for Carcinogens. Report of the International Programme on Chemical Safety's Collaborative Study on *In vivo* Assays", J. Ashby, F.J. deSerres, M.D. Shelby, B.H. Margolin, M. Ishidate, Jr., and G.C. Becking, eds., Cambridge Univ. Press, Cambridge, U.K., Vol I, pp. 29.

Mitchell, A.D., Casciano, D.A., Meltz, M.L., Robinson, D.E., San, R.H.C., Williams, G.M., and von Halle, E.S. (1983). Unscheduled DNA synthesis test: A report of the Gene-Tox Program. *Mutat. Res.* **123**, 363.

Probst, G.S., Hill, L.E., and Brewsey, B.J. (1980). Comparison of three *in vitro* assays for carcinogen-induced DNA damage. *J. Toxicol. Environ. Hlth.* **6**, 333.

Singh, N.P., McCoy, M.T., Tice, R.R. and Schneider, E.L. (1988). A simple technique for quantitation of low levels of DNA damage in individual cells. *Exp. Cell Res.* **175**, 184.

Singh, N.P., Danner, D.B., Tice, R.R., McCoy, M.T., Collins, G.D.,and Schneider, E.L. (1989). Abundant alkali-labile sites in human and mouse sperm DNA. *Exp. Cell Res.*, **184**, 461.

Williams, G.M., Laspia, M.F., and Dunkel, V.C. (1982). Reliability of the hepatocyte primary culture/DNA repair test in testing of coded carcinogens and non/carcinogens. *Mutat. Res.* **97**, 359-370.

MONITORING HUMAN EXPOSURE TO ENVIRONMENTAL CARCINOGENS

Regina M. Santella, Yu Jing Zhang, Tie Lan Young, Byung Mu Lee, Xiao Qing Lu

Comprehensive Cancer Center and Division of Environmental Sciences,
School of Public Health
Columbia University
New York, NY 10032

INTRODUCTION

One of the long range goal of research in chemical carcinogenesis is the identification of individuals at increased risk of cancer development. Cancer is a multistep, multistage process in which many factors effect ultimate risk. The initiating event in the process of chemical carcinogenesis is the binding of the reactive electrophilic species of the carcinogen to nucleophilic sites in DNA. The extent of this reaction is influenced by a number of factors including metabolism to the active species or less toxic metabolites, detoxification of reactive intermediates and repair of adducts once formed. Thus, individuals with the same exposure may be at very different risk for cancer development because of differences in these processes due to genetic susceptibility. Suggestive evidence for this genetic susceptibility has come from epidemiologic studies demonstrating a higher proportion of individuals with a specific phenotype (e.g. poor metabolizers of debrisoquin or lacking specific glutathione transferase activity) among cancer cases than controls (Ayesh et al. 1984, Seidegard et al. 1986).

By measuring levels of carcinogen-DNA adducts in humans we may be able to obtain information directly related to that individuals risk for cancer development. Measurement of carcinogen-DNA adducts in the target tissue, termed biologically effective dose, is believed to be a more relevant arker of exposure than measurement of the chemical itself either in the environment or in body fluids (Perera and Weinstein 1982). Such assays take into account individual differences in absorption and metabolism of carcinogens as well as
repair of adducts.

Methods are now available for the sensitive detection and quantitation of carcinogen-DNA adducts which do not depend upon radiolabeled carcinogens. These methods are essential for the measurement of adducts in humans with exposure to environmental and occupational carcinogens. Immunologic methods for measurement of DNA adducts have utilized monoclonal and polyclonal antibodies recognizing a number of specific carcinogen-DNA adducts (Poirier 1984, Santella 1988). Antibodies can be developed against either the carcinogen-nucleoside adduct covalently coupled to carrier protein or the modified DNA electrostatically complexed to methylated bovine serum albumin. These antibodies can be used in highly sensitive competitive enzyme-linked immunosorbent assays (ELISA) with color- or fluorescence-endpoint detection. Since femtomole (10^{-15}) sensitivities are readily attainable, DNA adduct

Biological Reactive Intermediates IV, Edited by C.M. Witmer *et al.*
Plenum Press, New York, 1990

levels in the range of $1/10^8$ nucleotides can be measured. With monoadduct-specific antibodies, higher sensitivities may be obtainable if large amounts of DNA are available and the adduct is isolated by various chromatographic procedures before quantitation in the ELISA. A major advantage of immunologic methods for adduct detection in humans is that once a sensitive and specific method has been developed it can easily be applied to the large number of samples that are collected in epidemiologic studies. However, before an immunoassay can be developed the structure of the adduct of interest must be known and it must be possible to systhesize the adduct for development of the antibody. Table 1 lists some of the available polyclonal and monoclonal antisera recognizing carcinogen-DNA adducts. In addition, we have recently developed monoclonal antibodies to 8-oxoguanosine, 4-aminobiphenyl-guanosine and 7-hydroxyethylguanosine (unpublished studies). A number of the antibodies listed in Table 1, including those recognizing alkylation, aflatoxin, benzo(a)pyrene diol epoxide, cisplatinum and 8-methoxypsoralen-DNA adducts, have been applied to adduct detection in humans (reviewed in Farmer et al. 1987, Santella 1988).

An alternate method for adduct detection utilizes [^{32}P] postlabeling of adducts after enzymatic digestion of the DNA to 3'monophosphates (Randerath et al. 1981). Very small amounts of DNA are required (1-50ug) and prior knowledge of the identity of the adducts, essential for the immunologic approach, is not necessary. Four dimensional thin layer chromatography of the labeled nucleoside bisphosphates, followed by autoradiography allows separation of the normal nucleotides and visualization of adduct spots. Quantitative data is obtained by counting areas of the chromatogram containing adducts. Adducts cannot be identified but because of the limitations of the thin layer chromatography system, must result from the binding of bulky hydrophobic carcinogens. While the alkylated adducts would be lost with the normal nucleotides in the standard assay, HPLC methods have been developed for their quantitation (Reddy et al. 1984, Wilson et al. 1988). Several methods have also been developed to give high sensitivity (up to a reported 1 adduct/10^{9-10}nucleotides) including butanol extraction of adducts (Gupta and Earley 1988) and nuclease P1 digestion of normal nucleotides (Reddy and Randerath 1986) before labeling to enrich the sample in adducts. The advantages of the method, including high sensitivity, small sample size and ability to detect a broad range of adducts, make it ideal for studying individuals with exposure to complex mixtures. The major disadvantages are the utilization of [^{32}P], the complexity of the assay and the inability to measure some adducts by standard procedures. The method has been applied to the detection of adducts in placental (Everson et al. 1986, Everson et al. 1988) and lung (Phillips et al. 1988a) DNA of smokers and nonsmokers and white blood cell DNA of roofers (Herberts et al. in press) and foundry workers (Phillips et al. 1988b).

Several other methods have been developed for quantiation of DNA adducts but have not been as extensively applied to human adduct detection as have immunoassays and postlabeling. They include synchronous fluorescence spectroscopy (Harris et al. 1985) and gas chromatography/ mass spectroscopy (Weston et al. 1989).

While DNA is believed to be the critical target, chemical carcinogens also bind to RNA and proteins. Quantitation of protein adducts on either hemoglobin or albumin, has been used as an alternate marker of exposure to environmental carcinogens. Large amounts of protein can be obtained from blood samples (6mg albumin and 140mg hemoglobin/ml of whole blood). This can be contrasted with the 500-700ug of DNA normally obtained from 30-35ml of blood. Thus, much smaller amounts of blood are required for protein adduct measurement making these methods more generally applicable to routine occupational monitoring. In addition, a number of studies have demonstrated a correlation between DNA and protein adduct levels suggesting that it is an appropriate surrogate for DNA adduct measurement (Neumann 1984). No repair occurs on protein thus, chronic low levels of exposure may be measurable. Red blood cells have an average lifespan of 4 months while albumin has a half life of 21 days indicating that only recent exposure will be detectable.

Table 1. Antisera Recognizing Carcinogen-DNA Adducts

Adduct	References
acetylaminodimethyldipyridoimidazole (gluP3)-guanosine	(Hebert et al, 1985)
Acetylaminofluorene-DNA	(Ball et al. 1987, Leng et al. 1978, Sage et al. 1979)
Acetylaminofluoreneguanosine	(Baan et al. 1985, Guigues and Leng 1979, Poirier et al. 1977, Van der Laken et al. 1982)
Aflatoxin B_1-DNA	(Haugen et al. 1981, Hertzog et al. 1982, Hsieh et al 1988)
4-Aminobiphenylguanosine	(Roberts et al. 1988)
Aminofluoreneguanosine	(Poirier et al. 1983, Rio and Leng 1980)
Aminopyrene-DNA	(Hsieh et al. 1985)
Benzo(a)pyrene diol epoxide-DNA	(Poirier et al. 1980, Santella et al. 1984, Slor et al. 1981, van Schooten et al. 1987)
O^6-Butylguanosine	(Muller and Rajewsky 1981, Rajewsky et al. 1980)
Cisdiammine dichloroplatinum	(Fichtinger-Schepman et al. 1985, Malfoy et al. 1981, Mustonen et al. 1987, Poirier et al. 1982, Sundquist et al. 1987)
cyclic 1,N^2-propanoguanosine	(Foiles et al. 1986)
Ethenoadenine	(Eberle et al. 1989, Young and Santella 1988)
Ethenocytidine	(Eberle et al. 1989, Young and Santella 1988)
O^6Ethylguanosine	(Muller and Rajewsky 1980, Rajewsky et al. 1980, Van der Laken et al. 1982)
O^4Ethylthymidine	(Muller and Rajewsky 1981, Rajewsky et al. 1980)
Melphalan-DNA	(Tilby et al. 1987)
8-Methoxypsoralen-DNA	(Santella et al. 1985, Zarebska et al. 1984)
O^6Methylguanosine	(Muller and Rajewsky 1981, Wild et al. 1983)
N^6-Methyladenosine	(Munns et al. 1977)
7-Methylguanosine	(Degan et al. 1988, Meridith and Erlanger 1979, Munns et al. 1977)
iro-7-Methylguanosine*	(Stein et al. 1989)
O^2Methylthymidine	(Strickland and Boyle 1984)
4-Nitroquinoline-N-oxide-DNA	(Morita et al. 1988)
Trimethylangelicine-DNA	(Miolo et al. 1989)
Thymine dimer	(Ley 1983, Strickland and Boyle 1981, Wani et al. 1984)
Thymine glycol	(Leadon and Hanawalt 1983)

*iro, imidazole ring opened

Most data collected to date on protein adduct levels in humans has utilized GC/MS methods. Several different approaches have been used to increase the sensitivity of these measurements. For several aromatic amines including 4-aminobiphenyl, globin adducts are measured after cleavage of the amine from the protein followed by isolation and derivatizion (Bryant et al. 1987). Similarly, for benzo(a)pyrene adducts, acid treatment cleaves the adduct releasing BP tetrols, which can then be quantitated (Weston et al. 1989). In contrast, ethylene oxide globin adducts are measured by cleavage of the N terminal modified valine by an Edman degradation reaction (Tornqvist et al. 1986). All of these methods allow enrichment of the adduct to be measured by separation from the bulk of the nonmodified protein. Elevated levels of 4-aminobiphenyl-globin adducts (Bryant et al. 1987) and hydroxyethylvaline (Tornqvist et al. 1986) have been measured in smokers compared to nonsmokers.

A limited number of antibodies have been developed against protein adducts (Table 2). Only the antisera recognizing aflatoxin and BP adducts have been applied to human adduct detection. Data on aflatoxin have been published (Gan et al. 1988, Wild et al. 1990) and our initial studies in humans for BP adducts are summarized below.

Table 2. Antisera Recognizing Carcinogen-Protein Adducts

Acetaldehyde-protein	(Israel et al. 1986)
Acetaminophen-protein	(Roberts et al. 1987)
Aflatoxin B_1-albumin	(Gan et al. 1988, Wild et al. 1990)
Benzo(a)pyrene diol epoxide-protein	(Santella et al. 1986)
Ethylene oxide-hemoglobin	(Wraith et al. 1988)

MEASUREMENT OF EXPOSURE TO POLYCYCLIC AROMATIC HYDROCARBONS

Antibodies to Benzo(a)pyrene Diol Epoxide-DNA

Benzo(a)pyrene (BP), a polycyclic aromatic hydrocarbon (PAH), is a ubiquitous environmental pollutant found in cigarette smoke, various foods and in all products of combustion. BP is metabolized *in vivo* to benzo(a)pyrene diol epoxide (BPDE-I) which reacts with the N2 position of guanine to form a covalent DNA adduct (Jeffrey et al. 1977). We have developed both polyclonal and monoclonal antisera recognizing BPDE-I-DNA. These antisera have been used to quantitate adducts in biological samples obtained from humans with various occupational or environmental exposure to PAHs. When first developed, these antibodies were found to be highly specific for the modified DNA not recognizing BP itself nor nonmodified DNA. In addition, there was no crossreactivity with several other carcinogen modified DNAs, including acetylaminofluorene and 8-methoxypsoralen-DNA. More recently, the reactive diol epoxides of several other PAHs were synthesized. These were used to modify calf thymus DNA *in vitro* for testing antibody crossreactivity. Both the monoclonal and polyclonal antisera were found to crossreact with structurally related diol epoxide adducts of several other PAHs including chrysene and benz(a)anthracene (Santella et al. 1987). Polyclonal antisera #29, obtained from animals immunized with BPDE-I-DNA, recognizes DNA modified by chrysene-1,2-diol-3,4-epoxide more efficiently (50% inhibition at 18 fmol) than it recognizes BPDE-I-DNA (50% inhibition at 30 fmol). This antibody also reacts with DNA modified by benz(a)anthracene-8,9-diol-10,11-epoxide (50% inhibition at 42 fmol) and by 3,4-diol-1,2-epoxide (50% inhibition at 114 fmol). Humans are exposed to BP in complex mixtures containing a number of other PAHs. Multiple PAH adducts may be present but at such low levels they cannot be identified. Thus, the appropriate standard to use cannot be determined and absolute quantitation of adducts is not possible. BPDE-I-DNA is used as the standard since it was the antigen originally used for antibody development and measured values are expressed as femtomole equivalents of BP adducts which would cause a similar inhibition in the assay. Since a number of PAHs in addition to BP are carcinogenic, the ELISA provides a biologically relevant general index of DNA binding by this class of compounds.

We have also recently determined that polyclonal antisera #29 detects adducts more efficiently in highly modified DNA (1.2 adducts/100 nucleotides) than in low modified DNA ($1.5/10^5$) (Santella et al. 1988). This efficiency also varied with the type of ELISA used. With the color endpoint ELISA there was a 2.5 fold difference between the high and low modified DNA samples but with the fluorescence endpoint ELISA the difference was 10 fold (Figure 1). This antisera was obtained from animals immunized with highly modified DNA. Clustering of adducts or some unique determinents present on highly modified DNA may be responsible for the higher sensitivity with these samples. In our original studies, we utilized highly modified DNA in the standard curve which resulted in an underestimation of adduct levels. Similar preferential reactivity with highly modified DNA has been seen with other antisera against BPDE-I-DNA (van Schooten et al. 1987) and melphalan-DNA (Tilby et al. 1987). In contrast, antibodies recognizing 8-MOP-DNA have similar crossreactivity with adducts in high and low modified samples (Yang et al. 1987). These results demonstrate the importance of thorough characterization of antisera before application to human samples and the utilization of appropriate standards for analyzing biological samples. For adduct detection in humans with antibody #29, we currently use a low modified standard with fluorescence endpoint detection.

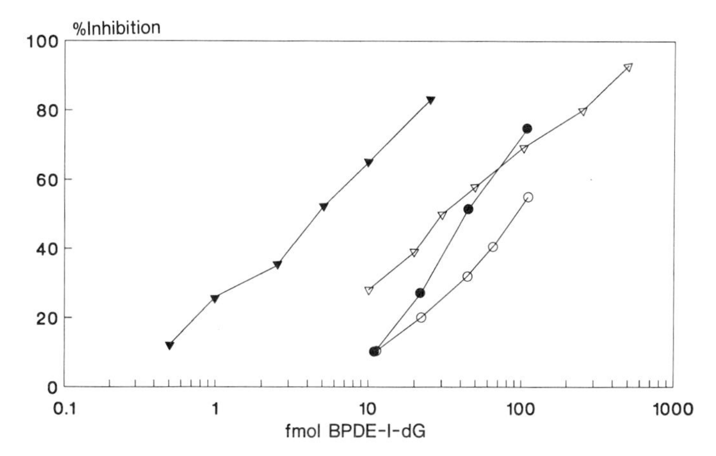

Figure 1. Competitive inhibition of antisera #29 binding to BPDE-I-DNA by high modified DNA (1.2adducts/100 nucleotides) and low modified DNA (1.5/105 nucleotides). Plates were coated with highly modified DNA and antisera diluted (1:30,000 for the color assay and 1:1,600,000 for the fluorescence assay) and mixed with the competitor before addition to the well. Incubation with the secondary antibody, goat anti-rabbit IgG-alkaline phosphatase was followed by the substrate, p-nitrophenylphosphate in the color assay and 4-methylumbelliferyl phosphate in the fluorescence assay. Details of the procedure have been published (Santella et al. 1988). High (▽) and low (○) modified DNA in the color assay and high (▼) and low (●) modified DNA in the fluorescence assay.

Antibodies recognizing PAH-DNA adducts have been used in several studies to quantitate adduct levels in humans. Since target tissue (eg, lung, liver) is not available on a routine basis from healthy individuals, white blood cells have been utilized in a number of biomonitoring studies as a surrogate. In collaboration with K. Hemminki, Institute of Occupational Health, Finland, adduct levels were measured in white blood cell DNA from foundry workers and controls (Perera et al. 1988). Workers were classified into high, medium exposure to low exposure to BP based on air monitoring data and an industrial hygenist's evaluation of the job description (Table 3). A dose response relationship was found between adduct levels and exposure. Adducts in these samples were also analyzed by two other laboratories by [^{32}P] postlabeling (Phillips et al. 1988). While adduct levels were lower in the postlabeling assay, there was a good correlation between the immunoassay and postlabeling data. A set of nine foundry workers were sampled after just returning from vacation and again three months later. Adduct levels after vacation were slightly elevated compared to controls (0.39adducts/10^7nucleotide for workers and 0.22/10^7 for controls) but increased dramatically after returning to thework environment. These results suggest a rather rapid repair of adducts over the one month vacation followed by a rise after reexposure.

Adduct levels were also quantitated in white blood cells of Polish coke oven workers, a local control group living near the plant and a rural control group from an unpolluted region of Poland (Table 3). Adducts in local controls were only slightly lower (1.3/10^7 nucleotides) than in coke oven workers (1.5/10^7) but rural controls were comparable to the Finish foundry worker controls. Similar data was obtained by the postlabeling assay (Hemminki et al. in press). Although the levels of BP are about 10-fold higher in the cokery than in the neighboring residential areas, there is only a modest difference in adduct levels between workers and local controls. It is likely that inhalation is only one component of the exposure; ingestion of PAHs in food and skin absorption may also be important.

Table 3. PAH-DNA Adducts In White Blood Cells of Foundry and Coke Oven Workers and Cigarette Smokers and Nonsmokers

	N	mean aducts/ 10^7 nucleotides
Foundry		
High exposure (>0.2ug BP/m^3)[a]	4	5.0
Medium exposure (0.05-0.2ug BP/m^3)	13	2.2
Low exposure (<0.05ug BP/m^3)	18	0.80
Control (<0.01ug BP/m^3)	10	0.22
At work	9	3.08
After vacation	9	0.39
Coke Oven (0.1-50ug BP/m^3)	41	1.5
Local Control (0.015-0.057ug BP/m^3)	15	1.3
Rural Control (<0.01ug BP/m^3)	13	0.23
Smokers		0.15
Nonsmokers		0.12

Data from (Perera et al. 1988, Perera et al. 1987) and unpublished data.
[a] Estimated air levels of BP.

In contrast to these results, several studies on adducts in smokers and nonsmokers have not detected significant differences between these populations. White blood cell DNA adducts were not significantly different in smokers compared to nonsmokers (Table 3) (Perera et al. 1987). Mean adduct levels in placental DNA were higher than in white blood cells but again no difference was seen between smokers ($1.9/10^7$) and nonsmokers ($1.2/10^7$) (Everson et al. 1986, Everson et al. 1988). These results probably reflect the ubiquitous exposure of the general population to PAHs from a number of sources including air pollution and, more importantly, diet. Another factor effecting adduct levels appears to be season of blood collection. We recently found higher adduct levels in blood samples collected during the late summer and early fall than in those collected during the other two seasons (Perera et al. 1989). This seasonal effect is consistent with observations of a peak in aryl hydrocabon hydroxylase (AHH) inducibility during this period (Paigen et al. 1981). While this initial data are from seasonal analysis of samples collected throughout the year, current studies involving the collection of repeat samples from the same individual should provide more conclusive evidence of a seasonal effect.

PAH Exposure in Coal Tar Treated Psoriasis Patients

Topical application of 2-5% crude coal tar followed by skin irradiation with UVB (Goeckerman theraphy) is used as an effective therapy for psoraisis, a hyperproliferative disease of the skin. Coal tars, including pharmaceutical-grade preparations, are complex mixtures of PAHs and well established mutagens and animal carcinogens (IARC 1985). A number of studies have confirmed that occupational exposure to coal tar results in elevated risk of cancer (IARC 1985). Isolated case reports have also suggested that therapeutic coal tar treatment can produce squamous and basal cell carcinomas (reviewed in Bickers 1981).

PAHs are known to be absorbed through the skin after topical application as demonstrated by elevated levels of several PAHs in blood (Storer et al. 1984). Urinary excretion of 1-hydroxypyrene, a fluorescent metabolite of pyrene, has also been used as a marker of internal dose of coal tar in treated psoriasis patients (Clonfero et al. 1989, Jongeneelen et al. 1985). Elevated levels of urinary mutagens in coal tar treated patients have also been found with the Salmonella mutagenesis assay (Clonfero et al. 1989, Wheeler et al. 1981). Recently, elevated levels of mutagens and PAHs in urine as well as sister chromatid exchange (SCE) and chromosomal aberrations in peripheral blood lymphocytes related to exposure level were seen in treated patients (Sato et al. 1989).

We have carried out a small pilot study on blood samples from coal tar treated psoriasis patients and controls with a panel of biomarkers for exposure to genotoxic agents as a model system for skin exposure in the occupational setting. The markers utilized include SCE and micronuclei (MN) in pheripheral lymphocytes, white blood cell PAH-DNA adducts measured by ELISA and [^{32}P] postlabeling and serum antibodies to BPDE-I-DNA adducts. Blood samples were obtained from 22 coal tar treated psoriasis patients and 5 controls. DNA was isolated from white blood cells and PAH-diol epoxide DNA adducts were determined by competitive ELISA. Only one of five controls had a detectable value while 13/22 patients were positive. The mean value (Table 4) for patients was $1.70/10^7$ compared to $0.79/10^7$ for controls. The control value is higher than seen in previous studies on nonoccupationally exposed smokers and nonsmokers (0.15-$0.23/10^7$). This was due to the very high adduct level, comparable to that of foundry workers, seen in the one positive control ($3.3/10^7$), a nurse on the dermatology floor. However, her occupational exposure to coal tar should be minimal and it is not clear what exposures resulted in the high adduct levels. Adducts in this sample were also high by postlabeling. Mean values by [^{32}P] postlabeling are also given in Table 4 and are slightly lower than the ELISA values.

Antibodies recognizing PAH-DNA adducts have been reported in the sera of humans with occupational or environmental exposures (Harris et al. 1985, Weston et al. 1987). To determine if the presence of serum antibodies to carcinogen adducts could serve as a marker of exposure, we determined the number of individuals with detectable levels of serum antibodies against BPDE-I-DNA as described (Harris et al.

1985). No significant difference in positive titer was seen between patients and controls suggesting this assay may not be a marker of exposure. SCE and MN were analyzed in a subset of 17 of the patients and in the 5 controls. Mean SCE levels were higher for patients (9.43) than controls (7.74). No significant difference was seen in MN between patients (13.59) and controls (13.75). We are currently carrying out a larger scale study of patients and controls.

Table 4. Mean Assay Results For Coal Tar Treated Psoriasis Patients and Controls

Assay	Patients	Controls
ELISA		
Mean adducts/10^7nucleotide	1.70+/-3.98	0.79+/-1.43
Range	ND-19	ND-3.3
Number detectables/assayed	13/22	1/5
^{32}P postlabeling		
Mean adducts/10^8nucleotide	0.54+/-1.14	0.35+/-0.50
Range	<0.01-5.5	0.050-1.2
Number detectable/assayed	22	5
Serum Antibodies to BPDE-I-DNA		
Number with positive titer/ assayed	13/22	4/5
Sister chromatid exchange		
Average number/metaphase	9.43+/-1.16	7.74+/-0.87
Range	8.30-11.46	6.74-8.72
Number assayed	17	5
Micronuclei		
Average number/1000cells	13.59+/-3.87	13.75+/-5.06
Range	8-21	8-20
Number assayed	17	5

Immunohistochemical Studies on Skin Biopsies

Antibodies to particular carcinogen-DNA adducts can also be utilized to investigate localization of adduct formation in specific cell or tissue types. Indirect immunofluorescence staining with primary antibodies followed by fluorescein isothiocyanate labeled secondary antibodies allows visualization of adducts. We obtained 4mm punch biopsies from several coal tar treated patients as well as from untreated volunteers. Sections were cut and stained with the polyclonal antisera used for quantitating PAH-DNA adducts in white blood cells. To enhance antibody binding, slides were first treated with RNase A, proteinase K and HCl. Staining of skin biopsy sections from coal tar treated psoriasis patients indicated specific nuclear staining in the stratum spinosum and granulosum of the epidermis (Figure 2). Staining was also scattered variably throughout the dermis with some localization to fibroblasts and vessels. No staining was visible in sections from an untreated control.

With the conventional immunofluorescence methods utilized here, quantitative adduct levels cannot be determined. However, they can be estimated by comparison to previous studies carried out in keratinocytes treated in culture with BP. In these studies, adduct levels in the range of $1/10^6$ gave detectable immunofluorescence (Poirier et al. 1982b). This is in the same range as found to result in detectable staining in immunofluorescence studies utilizing antibodies to 8-methoxypsoralen-DNA adducts (Yang et al. 1987).

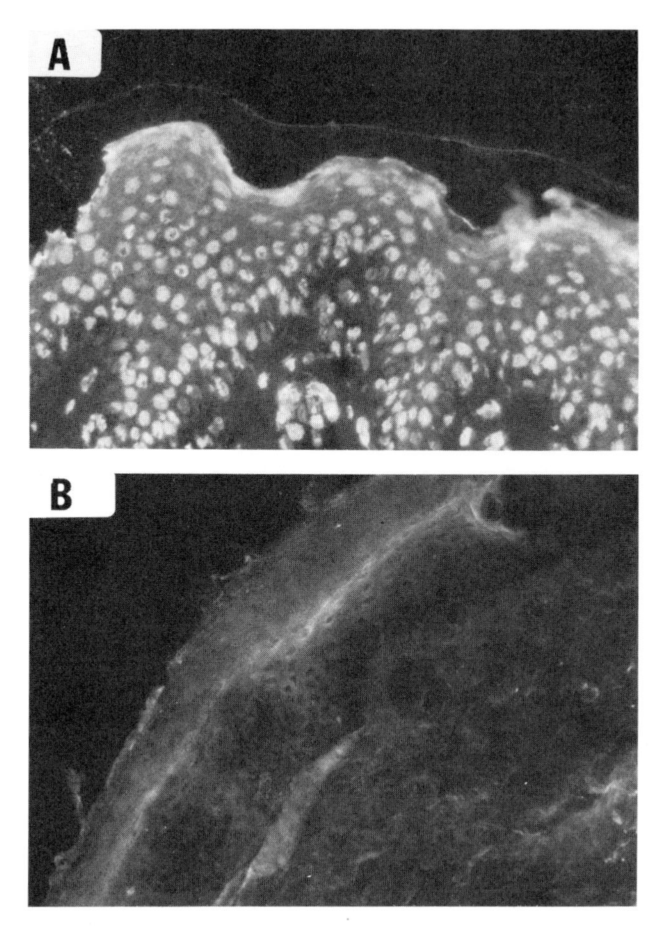

Figure 2. Indirect immunofluorescence staining of human skin biopsies from a patient treated topically with coal tar and a control. Slides were treated with RNase, proteinase K and 4N HCl before staining with antisera #29 (1:50 dilution) and goat anti-rabbit IgG antibody conjugated with fluorescein (1:30). Magnification was 220 fold. a) section from patient; b) section from control.

Computer-assisted video microscopy systems or the use of biotin-streptavidin staining should further increase sensitivity. It may then be possible to utilize these methods for adduct detection in human samples from occupational or environmental exposures. Since adducts can in theory be visualized in single cells, the small amount of material obtained at biopsy could be utilized.

Detection of PAH-Protein Adducts

Measurement of carcinogen-protein adducts has been used as an alternate marker of exposure for several carcinogens. To determine PAH-protein adduct levels, we have utilized a monoclonal antibody, 8E11, developed from animals immunized with BPDE-I-G coupled to bovine serum albumin (Santella et al. 1984). This antibody recognizes BPDE-I modified dG, DNA, and protein as well as BPDE-I-tetrols. More recently, we have further characterized this antibody in terms of crossreactivity with a number of other BP metabolites, other PAHs and other PAH diol epoxide modified DNAs. 8E11 crossreacts with a number of BP metabolites (Table 5), with higher sensitivity for the diols and weak recognition of phenols. There is some weak crossreactivity with other PAHs including pyrene, aminopyrene and nitropyrene. It also recognizes the diol epoxide adducts of several other PAHs. This antibody will thus recognize a number of PAHs and their metabolites with variable sensitivities.

Table 5. Competitive Inhibition of Antibody 8E11 Binding to BPDE-I-BSA

Competitors	femtomole causing 50% inhibition
BPDE-I-tetrol	350
BPDE-I-BSA digested	400
BPDE-I-BSA nondigested	1450
BP-7,8-diol	250
BP-9,10-diol	150
4-OH-BP	42700
5-OH-BP	$>1x10^5$
BP	6000
1-aminopyrene	70000
1-nitropyrene	16000
Dimetylbenz(a)anthracene	$>1x10^6$
1-OH-pyrene	3400
Pyrene	16200
BPDE-I-DNA	350
Chrysene diol epoxide-DNA	160
Benz(a)anthracene diol epoxide-DNA	1350

Plates were coated with 25ng BPDE-I-bovine serum albumin. Antibody 8E11 was used at a 1:30,000 dilution and mixed with competitor before addition to the plate. Goat anti-mouse IgG-alkaline phosphatase at a 1:500 dilution and p-nitrophenyl phosphate were used for antibody detection. Details of the assay have been published (Lee and Santella 1988).

Direct quantitation of adducts on intact protein cannot be carried out sensitively due to the low affinity of the antibody for the adduct in intact protein. This low sensitivity is probably due to burying of the adduct in hydrophobic regions of the protein. Others have suggested that release of tetrols with acid treatment is a sensitive method for determination of protein adducts (Shugart 1986, Weston et al. 1989). However, our initial studies in mice treated with radiolabeled BP indicated that only low levels of radioactivity could be released from globin by acid treatment (Wallin et al. 1987). For this reason, we used an alternate approach for measurement of protein adducts. Globin was enzymatically digested to peptides and amino acids before ELISA. When tested on protein modified *in vitro* with BPDE-I, a 3-4 fold increase in sensitivity resulted (Table 5). This assay was validated using globin isolated from animals treated with radiolabeled BP. The ELISA was able to detect 90-100% of the adducts measured by radioactivity (Lee and Santella 1988). These animal studies also indicated that adduct levels were about 10 fold higher in albumin than in globin. For this reason, our initial work on human samples has been with albumin isolated from workers occupationally exposed to PAHs. Albumin was isolated by Reactive blue 2-Sepharose CL-4B affinity chromatography and enzymatically digested with insoluble protease coupled to carboxymethyl cellulose which could be easily removed by centrifugation. Samples were then analyzed by competitive ELISA with antibody 8E11. Initial studies have been carried out on a small number of roofers occupationally exposed during the removal of an old pitch roof and application of new hot asphalt. These studies were carried out in collaboration with Dr. Robin Herbert, Mt. Sinai Medical Center, NY. Seventy percent of the roofers samples were positive with a mean level of 5.4fmol/ug while 62% of the controls had detectible adduct levels (mean of 4.0fmol/ug). In this small number of subjects there was a trend but no significant difference between roofers and controls. However, we are continuing studies of PAH-albumin adducts in a larger sample.

Quantitation of 8-methoxypsoralen and Aflatoxin-DNA Adducts

We have utilized two other antisera recognizing DNA adducts to monitor adducts in humans. 8-methoxypsoralen (8-MOP), a photoactivated drug, is used clinically for the treatment of psoriasis, a hyperproliferative disease of the skin. Patients take the drug orally followed by skin irradiation with UVA. The advantage of working with a clinically used drug, as opposed to environmental carcinogens such as BP where there is ubiquitous exposure, is that individuals with high, well defined exposures as well as control, unexposed individuals can be readily identified. Monoclonal antibodies recognizing both the thymine monoadducts and interstrand cross-linked adducts have been developed (Santella et al. 1985). The antisera has been used in indirect immunofluorescence studies to demonstrate adduct formation in the stratified squamous epithelium of the epidermis of patients immediately after treatment (Yang et al. 1989). No adducts were detectable by ELISA on DNA isolated from the white blood cells of these same patients. This was not surprising since 8-MOP must be photoactivated and the UV dose to the circulating cells is minimal with skin irradiation.

Antibodies have also been developed which recognize the stable imidazole ring opened form of the N7 aflatoxin-DNA adduct. This antibody was used in a pilot study to monitor adduct levels in liver samples obtained from hepatocellular cancer patients in Taiwan (Hsieh et al. 1988). Detectable adducts, some as high as $3/10^6$ nucleotides were found in a number of patients. We are currently expanding these studies to a larger sample size of tumor and nontumor liver tissue from Taiwan and autopsy liver tissue from the US. We hope these studies will provide information on the relationship between aflatoxin exposure, hepatitis carrier status and liver cancer incidence.

Determination of Multiple Adducts

Humans are usually exposed to complex mixtures of environmental carcinogens and multiple adducts may be formed. Because of the limited availability of DNA from white blood cells, we would like to be able to determine multiple adducts in the same sample. Initial work has involved quantitation of adducts in a DNA sample modified in vitro with BPDE-I and 8-MOP. This sample was assayed with a mixture of antibodies recognizing the two different adducts each at the appropriate final dilution for the ELISA. The samples were sequentially incubated first on BPDE-I-DNA coated plates then on 8-MOP-DNA coated plates. Each plate was then incubated with the appropriate alkaline phosphatase conjugate as in the standard assay. When compared to the standard assay for each antibody (Figure 3) no significant difference is seen between detection of adduct on the DNA containing a single adduct or two adducts. Therefore, it may be possible to mix a panel of antibodies recognizing a number of different carcinogen adducts and by sequential incubation on plates with the appropriate antigen coating, quantitate multiple adducts. Finally, since the ELISA does not destroy the adduct, the DNA can be recovered from the microwell and repurified.

These studies demonstrate that immunologic methods have sufficient sensitivity to monitor human exposure to environmental carcinogens. Immunoassays also have the advantage of ease of application to a large number of samples making them ideal for epidemiologic studies. While methods for the determination of DNA adducts in humans provides information about the biologically effective dose of a carcinogen, and can therefore be used as a marker of exposure, information about the relationship of these measurements to risk is unknown. Future epidemiologic studies are needed to provide this information.

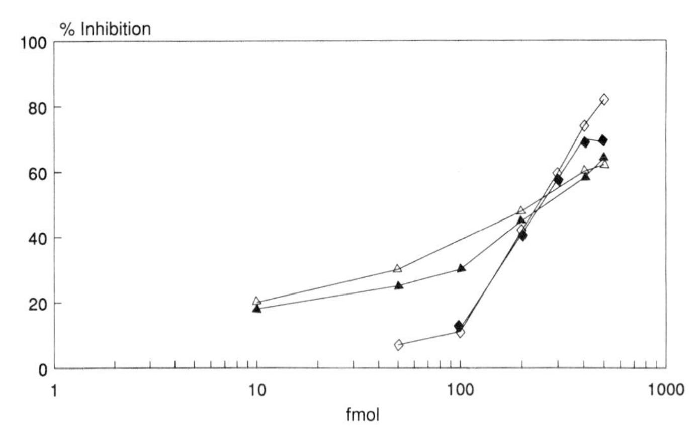

Figure 3. Multiple adduct analysis by competitive ELISA. The competitors were BPDE-I-DNA in the standard assay with antisera #29 (Δ) and 8-MOP-DNA in the standard assay with antibody 8G1 (◊). For the multiple adduct assay, DNA modified by BPDE-I and 8-MOP was mixed with antisera #29 and 8G1 and sequentially incubated on plates coated with BPDE-I-DNA (▲) and 8-MOP-DNA (♦).

ACKNOWLEDGEMENTS

Studies on DNA adducts in smokers and nonsmokers were carried out in collaboration with F. P. Perera and R. Everson. The work on adduct detection in Polish coke oven workers were carried out in collaboration with K. Hemminki, F.P. Perera, K. Randerath, D. Phillips, K. Twardowska-Saucha, J.W. Sroczynski, E. Grzybowka and M. Chorazy. Studies on coal tar treated psoriasis patients were carried out in collaboration with V. DeLeo, D. Warburton and M. Toor. This work was supported by NIH grants OH02622 and CA21111 and by a gift from the Lucille P. Markey Charitable Trust.

REFERENCES

Ayesh, R., Idle, J.R., Ritchie, J.C., Crothers, M.J., and Hetzel, M.R. (1984). Metabolic oxidation phenotypes as markers for susceptibility to lung cancer. *Nature*, **312**, 169-170.

Baan, R.A., Lansbergen, M.J., Bruin, P.A.F., Willems, M.I., and Lohman, P.H.M. (1985). The Organ-specific induction of DNA adducts in 2- acetylaminofluorence-treated rats, studied by means of a sensitive immunochemical method. *Mutation Res.*, **150**, 23-32.

Ball, S.S., Quaranta, V., Shadravan, F., and Walford, R.L. (1987). An ELISA for for detection of DNA-bound carcinogen using a monoclonal antibody to to N-acetoxy-2-acetylaminofluorene-modified DNA. *J. Immunol. Methods*, **98**, 195-200.

Bickers, D.R. (1981). The carcinogenicity and mutagenicity of therapeutic coal tar - A perspective. *J. Invest. Derm.*, **77**, 173-174.

Bryant, M.S., Skipper, P.L., and Tannenbaum, S.R. (1987). Hemoglobin adducts of 4-aminobiphenyl in smokers and nonsmokers. *Cancer Res.*, **47**, 602-608.

Clonfero, E., Zordan, M., Venier, P., Paleologo, M., Levis, A.G., Cottica, D., Pozzoli, L., Jongeneelen, F.J., Bos, R.P., and Ansion, R.B.B. (1989). Biological monitoring of human exposure to coal tar. *Int. Arch. Occup. Environ. Hlth.*, **61**, 363-368.

Degan, P., Montesano, R., and Wild, C.P. (1988). Antibodies against 7-methyl-

deoxyguanosine: Its detection in rat peripheral blood lymphocyte DNA and potential applications to molecular epidemiology. *Cancer Res.*, **48**, 5065-5070.
Eberle, G., Barbin, A., Laib, R.J., Ciroussel, F., Thomale, J., Bartsch, H. and Rajewsky, M.F. (1989). 1,N6-etheno-2'-deoxyadenosine and 3, N4-etheno-2'-deoxycytidine detected by monoclonal antibodies in lung and liver DNA of of rats exposed to vinyl chloride. *Carcinogenesis*, **10**, 209-212.
Everson, R.B., Randerath, E., Santella, R.M., Cefalo, R.C., Avitts, T.A., and Randerath, K. (1986). Detection of smoking-related covalent DNA adducts in human placenta. *Science*, **231**, 54-57.
Everson, R.B., Randerath, E., Santella, R.M., Avitts, T.A., Weinstein, I. B., and Randerath, K. (1988). Quantitative associations between DNA damage in human placenta and maternal smoking and birth weight. *J. Natl. Cancer. Inst.*, **80**, 567-576.
Farmer, P.B., Neumann, H.G., and Henschler, D. (1987). Estimation of exposure of man to substances reacting covalently with macromolecules. *Arch. Toxicol.*, **60**, 251-260.
Fichtinger-Schepman, A., Baan, R., Luiten-Schuite, A., VanDijk, M., and Lohman, P.H.M. (1985). Immunochemical quantitation of adducts induced in DNA by cis-diamminedichloroplatinum(II) and analysis of adduct related DNA-unwinding. *Chem. Biol. Inter.*, **55**, 275-288.
Foiles, P.G., Chung, F.L., and Hecht, S.S. (1986). Development of a monoclonal antibody based immunoassay for cyclic DNA adducts resulting from exposure to crotonaldehyde. *Cancer Res.*, **47**, 360-363.
Gan, L.S., Skipper, P.L., Peng, X., Groopman, J.D., Chen, J., Wogan, G.N., and Tannenbaum, S.R. (1988). Serum albumim adducts in the molecular epidemiolgy of aflatoxin carcinogenesis: correlation with aflatoxin B1intake and urinary excretion of aflatoxin M1. *Carcinogenesis*, **9**,1323-1325.
Guigues, M., and Leng, M. (1979). Reactivity of antibodies to guanosine modified by the carcinogen N-acetoxy-N-2-acetylaminofluorene. *Nucleic Acid Res.*, **6**, 733-744.
Gupta, R.C., and Earley, K. (1988). ^{32}P-adduct assay: comparative recoveries of structurally diverse DNA adducts in the various enhancement procedures. *Carcinognesis*, **9**, 1687-1693.
Harris, C.C., Vahakangas, K., Newman, J.M., Trivers, G.E., Shamsuddin, A., Sinopoli, N., Mann, D.L., and Wright, W.E. (1985). Detection of benzo[a]pyrene diol epoxide-DNA adducts in peripheral blood lymphocytes and antibodies to the adducts in serum from coke oven workers. *Proc. Natl. Acad. Sci. USA*, **82**, 6672-6676.
Haugen, A., Groopman, J.D., Hau, I.C., Goodrich, G.R., Wogan, G.W., and Harris, C.C. (1981). Monoclonal antibody to aflatoxin B1-modified DNA detected by enzyme immunoassay. *Proc. Natl. Acad. Sci. USA*, **78**, 4124-4127.
Hebert, E., Saint-Ruf, G., and Leng, M. (1985). Immunological tritration of 3-N-acetyl-hydroxyamino-4, 6-dimethyldipyrido(1,2-a:3',2'- d)imidazole-rat liver DNA adducts. *Carcinogenesis*, **6**, 937-939.
Hemminki, K., Twardowska-Saucha, K., Sroczynski, J.W., Grzybowka, E., Chorazy, M.,Putnam, K.L., Randerath, K., Phillips, D.H., Hewer, A., Santella, R.M., Young, T.L., and Perera, F.P. DNA adducts in humans environmentally exposed to aromatic compounds in an industrial area of Poland. *Carcinogenesis* (in press).
Herbert, R., Marcus, M., Wolff, M.S., Perera, F.P., Andrews, L., Godbold, J.H., Stefanidis, M., Rivera, M., Lu, X.Q., Landrigan, P.J., and Santella, R.M.Detection of DNA adducts in white blood cells of roofers by ^{32}P postlabeling. *Scand. J. Work. Environ. Hlth.* (in press).
Hertzog, P.J., Smith, J.R.L., and Garner, R.C. (1982). Production of monoclonal antibodies to guanine imidazole ring opened aflatoxin B1 DNA, the persistent DNA adduct *in vivo*. *Carcinogenesis*, **3**, 825-828.
Hsieh, L.L., Jeffrey, A.M., and Santella, R.M. (1985). Monoclonal antibodies to 1-aminopyrene-DNA. *Carcinogenesis*, **6**, 1289-1293.
Hsieh, L.L., Hsu, S.W., Chen, D.S., and Santella, R.M. (1988). Immunoligical detection of aflatoxin B1-DNA adducts formed *in vivo*. *Cancer Res.*, **48**,6328-6331.
International Agency for Research on Cancer (1985). IARC monographs on the evaluation of the carcinogenic risks of chemicals to humans: polynuclear aromatic compounds, Part 4. Lyon, IARC.

Israel, Y., Hurwitz, E., Niemela, O., and Arnon, R. (1986). Monoclonal and polyclonal antibodies against acetaldehydecontaining epitopes in acetaldehyde-protein adducts. *Proc. Natl. Acad. Sci. USA*, **83**, 7923-7927.

Jeffrey, A.M., Weinstein, I.B., Jennette, K.W., Grzeskowiak, K., Nakanishi, K., Harvey, R.G., Autrup, H., and Harris,C. (1977). Structures of benzo[a]pyrene-nucleic acid adducts formed in human and bovine bronchial explants. *Nature*, **269**, 348-350.

Jongeneelen, F.J., Anzion, R.B.M., Leijdekkers, C.M., Bos, R.P., and Henderson, P.T.(1985). 1-Hydroxypyrene in human urine after exposure to coal tar and a coal tar derived product. *Int. Arch. Occup. Environ. Hlth.*, **57**, 47-55.

Leadon ,S.A., and Hanawalt, P.C. (1983). Monoclonal antibody to DNA containing thymine glycol. *Mutation Res.*, **112**, 191-200.

Lee, B.M., and Santella, R.M. (1988). Quantitation of protein adducts as a marker of genotoxic exposure: immunologic detection of benzo(a)pyrene-globin adducts in mice. *Carcinogenesis*, **9**, 1773-1777.

Leng, M., Sage, E., Fuchs, R.P.P., and Daune, M.P. (1978). Antibodies to DNA Modified by the carcinogen N-acetoxy-N-2-acetylaminofluorene. *FEBS Letter*, **92**, 207.

Ley, R.D. (1983). Immunological detection of two types of cyclobutane pyrimidine dimers in DNA. *Cancer Res.*, **43**, 41-45.

Malfoy, B., Hartmann, B., Macquet, J.P., and Leng, M. (1981). Immunochemical studies of DNA modified by cis dichlorodiammine platinum (II). *Cancer Res.*, **41**, 4127-4131.

Meridith, R.D., and Erlanger, B.F. (1979). Isolation and characterization of rabbit anti-m7G-5'-P antibodies of high apparent affinity. *Nucleic Acids Res.*, **6**, 2179-2191.

Miolo, G., Stefanidis, M., Santella, R.M., Acqua, F., and Gasparro, F. (1989). 6,4,4'-Trimethylangelecin photoadduct formation in DNA: production and characterization of a specific monoclonal antibody. *Photochem. Photobiol.*, **3**, 101-112.

Morita, T., Ikeda, S., Minoura, Y., Kojima, M., and Tada, M. (1988). Polyclonal antibodies to DNA modified with 4-nitroquinoline 1-oxide Application for the detection of 4-nitroquinoline 1-oxide-DNA adducts *in vivo*. *Jpn. J. Cancer Res.*,(Gann) **79**, 195-203.

Muller, R., and Rajewsky, M.F. (1980). Immunological quantification by high-affinity antibodies of O6- ethyldeoxyguanosine in DNA exposed to N-ethyl-N-nitrosourea. *Cancer Res.*, **40**, 887-896.

Muller, R., and Rajewsky, M.F. (1981). Antibodies specific for DNA components structurally modified by chemical carcinogens. *J. Cancer Res. Clin. Oncol.*, **102**, 99-113.

Munns, T.W., Liszewski, M.K., and Sims, H.F. (1977). Characterization of antibodies specific for N6-methyladenosine and for 7-methylguanosine. *Biochem.*, **16-10**, 2163-2168.

Mustonen, R., Hemminki, K., Alhonen, A., Hietanen, P., and Kiilunen, M. (1987). Determination cis-diamminedichlorodiplatinum (II) in blood compartments of cancer patients. In: Detection methods for DNA damaging agents in humans: Applications in cancer epidemiology and prevention. IARC, Lyon, p. (eds) Hemminki,K., Bartsch, H., and O'Neill, I.K.

Neumann, H.G. (1984). Analysis of hemoglobin as a dose monitor for alkylating and arylating agents. *Arch. Toxicol.*, **56**, 1-6.

Paigen, B., Ward, E., Reilly, A., Houten, L., Gurtoo, H.L., Minowada, J., Steenland, K., Havens, M.B., and Satori, P. (1981). Season variation of aryl hydrocarbon hydroxylase activity in human lymphocytes. *Cancer Res.*, **41**, 2757-2761.

Perera, F.P., and Weinstein, I.B. (1982). Molecular epidemiology and carcinogen-DNA adduct detection: New approaches to studies of human cancer causation. *J. Chron. Dis.*, **35**, 581-600.

Perera, F.P., Santella, R.M., Brenner, D., Poirier, M.C., Munshi, A.A., Fischman, H.K., and VanRyzin, J. (1987). DNA adducts, protein adducts and SCE in cigarette smokers and nonsmokers. *J. Natl. Cancer Inst.*, **79**, 449-456.

Perera, F.P., Hemminiki, K., Young, T.L., Santella, R.M., Brenner, D., and Kelly, G. (1988). Detection of polycyclic aromatic hydrocarbon-DNA adducts in white blood cells of foundry workers. *Cancer Res.*, **48**, 2288-2291.

Perera, F., Mayer, J., Jaretzki, A., Hearne, S., Brenner, D., Young, T.L., Fischman, H.K., Grimes,M., Grantham,S.,Tang,M.X., Tsai,W-Y., Santella, R.M., (1989). Comparison of DNA adducts and sister chromatid exchange in lung cancer cases and controls. *Cancer Res.*, **49**, 4446-4451.

Phillips, D.H., Hewer, A., Martin, C.N., Garner, R.C., and King, M.M. (1988a). Correlation of DNA adduct levels in human lung with cigarette smoking. Nature (London), 336, 790-792.

Phillips, D.H., Hemminki, K., Alhonen, A., Hewer, A., and Grover, P.L. (1988b). Monitoring occupational exposure to carcinogens:detection by ^{32}P-postlabelling of aromatic DNA adducts in white blood cells from iron foundry workers. *Mutation Res.*, **204**, 531-541.

Poirier, M.C., Yuspa, S.H., Weinstein, I.B., and Blobstein, S. (1977). Detection of carcinogen-DNA adducts by radioimmunoassay. *Nature*, **270**, 186-188.

Poirier, M.C., Santella, R., Weinstein, I.B., Grunberger, D., and Yuspa, S.H. (1980). Quantitation of benzo[a]pyrene-deoxyguanosine adducts by radioimmunoassay. *Cancer Res.*, **40**, 412-416.

Poirier, M.C., Lippard, S., Zwelling, L.A., Ushay, M., Kerrigan, D., Santella, R.M., Grunberger, D., and Yuspa, S.H. (1982a). Antibodies elicited against cis-diamminedichloroplatinum(II)-modified DNA are specific for cis-diamminedichloroplatinum(II)-DNA adducts formed *in vivo* and *in vitro*. *Proc. Natl. Acad. Sci. USA*, **79**, 6443-6447.

Poirier, M.C., Stanley, J.R., Beckwith, J.B., Weinstein, I.B., and Yuspa, S.H., (1982b). Indirect immunofluorescent localization of benzo(a)pyrene adducted to nucleic acids in cultures mouse keratinocyte nuclei. *Carcinogenesis*, **3**, 345-348.

Poirier, M.C., Nakayama, J., Perera, F.P., Weinstein, I.B., and Yuspa, S.H. (1983). Identification of carcinogen-DNA adducts by immunoassays. In Milman, H.A. and Sell, S. (eds.), Application of biological markers to carcinogen testing. Plenum Publ. Corp., New York, pp. 427-440.

Poirier, M.C. (1984). The use of carcinogen-DNA adduct antisera for quantitation and localization of genomic damage in animal models and the human population. *Environ. Mutag.*, **6**, 879-887.

Rajewsky, M.F., Muller, R., Adamkiewicz, J., and Drosdziok, W. (1980). Carcinogenesis: fundamental mechanisms and environmental effects. Dordrecht, Holland, Reidel Press, pp. 207-218.

Randerath,D., Reddy, M.V., and Gupta, R.A.C. (1981). [^{32}P]-labelling test for DNA damage. *Proc. Natl. Acad. Sci. USA*, **78**, 6126-6129.

Reddy, M.V., Gupta, R.C., Randerath, E., and Randerath, K. (1984). ^{32}P-Postlabeling test for covalent DNA binding of chemicals in vivo: application to a variety of aromatic carcinogens and methylating agents. *Carcinogenesis*, **5**, 231-243.

Reddy, M.V., and Randerath, K. (1986). Nuclease P1-mediated enhancement of sensitivity of [^{32}P]- postlabeling test for structurally diverse DNA adducts. *Carcinogenesis*, **7**, 1543-1551.

Rio, P., and Leng, M. (1980). Antibodies to N-(guanosine-8-yl)-2-aminofluorene. *Biochem.*, **62**, 487-490.

Roberts, D.W., Pumford, N.R., Potter, D.W., Benson, R.W., and Hinson, J.A. (1987). A sensitive immunochemical assay for acetaminophen-protein adducts. *J. Pharmaco. Experim. Therapeu.*, **241**, 527-533.

Roberts, D.W., Benson, R.W., Groopman, J.D., Flammang, T.J., Nagle, W.A., Moss, A.J., and Kadlubar, F.F. (1988). Immunochemical quantitation of DNA adducts derived from the human bladder carcinogen, 4-aminobiphenyl. *Cancer Res.*, **48**, 6336-6342.

Sage, E., Fuchs, R.P., and Leng, M. (1979). Reactivity of the antibodies to DNA modified by the carcinogen N-acetoxy-N-acetyl-2-aminofluorene. *Biochem.*, **18**, 1328-1334.

Santella, R.M. (1988). Application of new techniques for the detection of carcinogen adducts to human population monitoring. *Mutation Res.*, **205**, 271-282.

Santella, R.M., Lin, C.D., Cleveland, W.L., and Weinstein, I.B. (1984). Monoclonal antibodies to DNA modified by a benzo[a]pyrene diol epoxide. *Carcinogenesis*, **5**, 373-377.

Santella, R.M., Dharmaraja, N., Gasparro, F.P., and Edelson, R.L. (1985). Monoclonal Antibodies to DNA modified by 8-methoxypsoralen and ultraviolet A light. *Nucleic Acids Res.*, **13**, 2533-2544.

Santella, R.M., Lin, C.D., and Dharmaraja, N. (1986). Monoclonal antibodies to a benzo[a]pyrene diolepoxide modified protein. *Carcinogenesis*, **7**, 441-444.

Santella, R.M., Gasparo, F.P., and Hsieh, L.L. (1987). Quantitation of carcinogen-DNA adducts with monoclonal antibodies. Progr. Experim. *Tumor Res.*, **31**, 63-75.

Santella, R.M., Weston, A., Perera, F.P., Trivers, G.T., Harris, C.C., Young, T.L.,Nguyen, D., Lee, B.M., Poirier, M.C. (1988). Interlaboratory comparison of antisera and immunoassays for benzo(a)pyrene-diol-epoxide-I-modified DNA. *Carcinogenesis*, **9**, 1265-1269.

Sato, F., Zordan, M., Tomanin, R., Mazzotti, D., Canova, A., Cardin, E.L., Bezze, G., and Levis, A.G. (1989). Chromosomal alterations in peripheral blood lymphocytes, urinary mutagenicity and excretion of polycyclic aromatic hydrocarbons in six psoriatic patients undergoing coal tar therapy. *Carcinogenesis*, **10**, 329-334.

Seidegard, J., Pero, R.W., Miller, D.G., and Beattie, E.J. (1986). A glutathione transferase in human leukocytes as a marker for the susceptibility to lung cancer. *Carcinogenesis*, **7**, 751-753.

Shugart, L. (1986). Quantifying adductive modification of hemoglobin from mice exposed to benzo[a]pyrene. *Anal. Biochem.*, **152**, 365-369.

Slor, H., Mizusawa, N., Nechart, T., Kakefuda, R., Day, R.S., and Bustin, M. (1981). Immunochemical visualization of binding of the chemical carcinogen benzo[a]pyrene diol epoxide to the genome. *Cancer Res.*, **41**, 3111-3117.

Stein, A.M., Gratzner, H.G., Stein, J.H., and McCabe, M.M. (1989). High avidity monoclonal antibody to imidazole ring-opened ethylguanine. *Carcinogenesis*, **10**, 927-973.

Storer, J.S., DeLeon, I., Millikan, L.E., Laseter, J.L., and Griffing, C. (1984). Human Absorption of crude coal tar products. *Arch. Derm.* **120** 874-877.

Strickland,P.T., and Boyle, J.M. (1981). Characterisation of two monoclonal antibodies specific for dimerised and non-dimerised adjacent thymidines in single stranded DNA. *Photochem. Photobiol.*, **34**, 595-601.

Strickland, P.T., and Boyle, J.M. (1984). Immunoassay of carcinogen-modified DNA. In Cohn, W.E. (ed.), Progress in Nucleic Acid Research & Molecular Biology. Academic Press, New York, pp. 1-58.

Sundquist, W.I., Lippard, S.J., and Stollar, B.D. (1987). Monoclonal antibodies to DNA modified with cis- or trans-diamminedichloroplatinum (II). *Biochem*, **84**, 8225-8229.

Tilby, M.J., Styles, J.M., and Dean, C.J. (1987). Immunological dectection of DNA damage caused by melphalan using monoclonal antibodies. *Cancer Res.*, **47**, 1542-1546.

Tornqvist, M., Osterman-Golkar, S., Kautiainen, S., Jensen, S., Farmer, P.B., and Ehrenberg, L. (1986). Tissue doses of ethylene oxide in cigarette smokers determined from adduct levels in hemoglobin. *Carcinogenesis*, **7**, 1519-1521.

van Schooten, F.J., Kriek, E., Steenwinkel, M.S.T., Noteborn, H.P.J.M.,Hillebrand, M.J.X., and vanLeeuwen, F.E. (1987). The binding efficiency of polyclonal and monoclonal antibodies to DNA modified with benzo[a]pyrene diol epoxide is dependent on the level of modification. Implications for quantitation of benzo[a]pyrene-DNA adducts *in vivo*. *Carcinogenesis*, **8**, 1263-1269.

Van der Laken, C.J., Hagenaars, A.M., Hermsen, G., Kriek, E., Kuipers, A.J., Nagel, J., Scherer, E., and Welling, M. (1982). Measurement of O6-ethyl-deoxyguanosine and N-(deoxyguanosine-8-yl)-N-acetyl-2-aminofluorene in DNA by high-sensitive enzyme immunoassays. *Carcinogenesis*, **3**, 569-572.

Wallin, H., Jeffrey, A.M., and Santella, R.M. (1987). Investigation of benzo[a]pyrene-globin adducts. *Cancer Let.*, **35**, 139-146.

Wani, A.A., Gibson-D'Ambrosio, R.E., and D'Ambrosio, M.D. (1984). Antibodies to UV irradiated DNA: the monitoring of DNA damage by Elisa and indirect immunofluorescence. *Photochem. Photobiol.*, **40**, 465-471.

Weston, A., Trivers, G., Vahakangas, K., Newman, M., and Rowe, M. (1987). Detection of carcinogen-DNA adducts in human cells and antibodies to these adducts in human sera. *Prog. Experim. Tumor Res.*, **31**, 76-85.

Weston, A., Rowe, M.I., Manchester, D.K., Farmer, P.B., Mann, D.L., and Harris, C.C. (1989). Fluorescence and mass spectral evidence for the formation of benzo(a)pyrene anti-diol-epoxide-DNA and -hemoglobin adducts in humans. *Carcinogenesis*, **10**, 251-257.

Wheeler, L.A., Saperstein, M.D., and Lowe, N.J. (1981). Mutagenicity of urine from psoriatic patients undergoing treatment with coal tar and ultraviolet light. *J. Invest. Derm.*, **77**, 181-185.

Wild, C.P., Jiang, Y.Z., Sabbioni, G., Chapot, B., and Montesano, R. (1990). Evaluation of methods for quantitation of aflatoxin-albumim and their application to human exposure assessment. *Cancer Res.*, **50**, 245-251.

Wild, C.P., Smart, G., Saffhill, R., and Boyle, J.M. (1983). Radioimmunoassay of O6 methyldeoxyguanosine in DNA of cells alkylated *in vitro* and *in vivo*. *Carcinogenesis*, **4**, 1605-1609.

Wilson, V.L., Basu, A.K., Essigmann, J.M., Smith, R.A., and Harris, C.C. (1988). O6-Alkyldeoxyguanosine detection by ^{32}P-postlabeling and nucleotide chromatographic analysis. *Cancer Res.*, **48**, 2156-2161.

Wraith, M.J., Watson, W.P., Eadsforth, C.V., van Sittert, N.J., and Wright, A.S. (1988). An immunoassay for monitoring human exposure to ethylene oxide. In Bartsch, H., Hemminki, K., and O' Neill,I.K. (eds.), Methods for detecting DNA damaging agents in humans: applications in cancer epidemiology and prevention. IARC Publications, Lyon, pp. 271-274.

Yang, X.Y., DeLeo, V., and Santella, R.M. (1987). Immunological detection and visualization of 8-methoxypsoralen-DNA photoadducts. *Cancer Res.*, **47**, 2451-2455.

Yang, X.Y., Gasparro, F.P., DeLeo, V.A., and Santella, R.M. (1989). 8-Methoxypsoralen-DNA adducts in patients treated with 8-methoxypsoralen and ultraviolet a light. *J. Invest. Derm.*, **92**, 59-63.

Young, T.L., and Santella, R.M. (1988). Development of techniques to monitor for exposure to vinyl chloride: monoclonal antibodies to ethenoadenosine and ethenocytine. *Carcinogenesis*, **9**, 589-592.

Zarebska, Z., Jarbabek-Chorzelska, M., Chorzelski, T., and Zablonska, S. (1984). Immune serum against anti DNA-8-methoxypsoralen photoadduct. *Z. Naturforsch*, **39**, 136-140.

COMPARING THE FREQUENCY AND SPECTRA OF MUTATIONS INDUCED WHEN AN SV40-BASED SHUTTLE VECTOR CONTAINING COVALENTLY BOUND RESIDUES OF STRUCTURALLY-RELATED CARCINOGENS REPLICATES IN HUMAN CELLS

Veronica M. Maher, Jia-Ling Yang, M. Chia-Miao Mah, Yi-Ching Wang, Janet Boldt, and J. Justin McCormick

Carcinogenesis Laboratory-Fee Hall
Department of Microbiology and Department of Biochemistry
Michigan State University
East Lansing, MI 48824-1316

INTRODUCTION

It is widely recognized that the transformation of normal cells into malignant cells is a multistep process, and there is substantial evidence that mutations and gene rearrangements are involved in causing one or more of the required changes. However, the molecular mechanisms by which carcinogens induce such changes in mammalian cells are not well understood. One reason for this has been the difficulty involved in sequencing the newly mutated genes to determine the kinds of changes that have taken place. The recent development of shuttle vectors, i.e., small circular plasmid DNA molecules that are capable of replicating in mammalian cells and also in bacteria, has greatly reduced the level of difficulty (Calos et al.,1983; Sarkar et al., 1984; Seidman et al., 1985). It is now feasible to investigate at the DNA sequence level the nature of mutations formed when carcinogen-treated plasmids replicate in human cells.

We have been making use of a shuttle vector, pZ189, containing a small target gene, *supF*, to examine the frequency and kinds of mutations induced by a series of N-substituted aryl compounds and structurally-related carcinogens, as well as their specific location in the target gene (spectrum of mutations) (Yang et al., 1987a, 1987b, 1988; Mah et al., 1989). The target gene, *supF*, codes for a tRNA. Because the structure of the tRNA is essential for its purpose, a single base pair (bp) change at almost any site of the 85 bp structure of the tRNA results in a mutant phenotype, making this gene exceptionally responsive to mutagens because of the low percentage of silent mutations (Kraemer and Seidman 1989). The small size of the target gene greatly facilitates sequence analysis and determination of "hot spots" and "cold spots" for mutation induction by particular carcinogens.

The plasmids are treated with radiolabeled carcinogens in vitro. The DNA is then transferred into human cells where the plasmids replicate using the DNA replication apparatus of the human cells. Treating the plasmid in vitro allows one to determine the number of carcinogen residues covalently bound to the DNA (adducts) which makes it possible to compare the various agents for their ability to produce mutations as a function of the number of adducts per plasmid. It also facilitates determining the nature of the adducts using high performance liquid chromatography. After the plasmids have replicated in the host cells, they are recovered and the progeny plasmids are analyzed for mutations in the target supF gene. The kinds and

Biological Reactive Intermediates IV, Edited by C.M. Witmer *et al.*
Plenum Press, New York, 1990

spectra of mutations are compared in order to identify common features in the modes of action of the various agents and determine how they differ from each other.

The compounds studied are (±) -7β, 8α-dihydroxy-9α,10α-epoxy-7,8,9,10-tetrahydrobenzo[a]pyrene (BPDE), 1-nitrosopyrene (1-NOP), 1-nitro-6-nitrosopyrene (1-N-6-NOP), N-acetoxy-2-acetylaminofluorene (N-AcO-AAF) and its trifluoro acetyl derivative (N-AcO-TFA-AF). BPDE forms its predominant DNA adduct at the N2 position of guanine, (Weinstein et al., 1976); each of the other four carcinogens forms adducts predominantly, or exclusively, at the C8 position of guanine (Poirier et al., 1980; Heflich et al., 1985, 1988; Beland et al., 1986; Djuric et al., 1988; Mah et al., 1989). The results we have obtained to date indicate that the frequency of *supF* mutants induced in plasmids carrying BPDE adducts is approximately four times higher than in those with adducts formed by the four carcinogens, but the *kinds* of mutations are very similar, i.e., predominantly base pair substitutions, mainly G·C to T·A transversions. Our data also show that each carcinogen produces its own unique spectrum of mutational hot spots.

MATERIALS AND METHODS

Cells and Plasmid

The materials and methods used for these studies have been described (Yang et al., 1988; Mah et al., 1989). Briefly, the host cells are the human kidney cell line 293 (Graham et al., 1977). The bacterial host cell, *E. coli* SY204, is ampicillin sensitive and carries an amber mutation in the β-galactosidase gene (Sarkar et al., 1984). The 5.5 kbp shuttle vector, pZ189, constructed by Seidman et al. (1985), contains the *supF* gene flanked by two genes essential for recovery in *E. coli*, the ampicillin gene and the bacterial origin of replication. It also carries the origin of replication and large T-antigen gene from simian virus-40.

Treatment of Plasmid with Carcinogens

BPDE, obtained from the Carcinogenesis Program of the National Cancer Institute, was dissolved in dimethylsulfoxide and added to an 0.5 mg/ml solution of DNA in 10 mM Tris-HCl/1 mM EDTA buffer, pH 8.0. The solution was incubated for 2 hours at room temperature. For exposure to 1-NOP and 1-N-6-NOP, supplied by Dr. F. A. Beland of the National Center for Toxicological Research, the plasmid was suspended in helium-purged Na citrate (10mM) buffer, pH 5.0 at 0.3 mg/ml. A freshly prepared solution of ascorbic acid (20 mM in H_2O) was added to provide needed reduction (Heflich et al., 1985), followed by radiolabeled carcinogen dissolved in dimethylsulfoxide. The solutions were incubated for 2 hours at room temperature. N-AcO-AAF and N-AcO-TFA-AF, supplied by Dr. C. M. King of Michigan Cancer Foundation, was dissolved in anhydrous ethanol and added to DNA dissolved in 2mM Na citrate buffer, pH 7.0 at 1 mg/ml. The solution was incubated for 30 minutes at 37°C. In every case, unbound compound was removed by five successive ether extractions, followed by extensive purification using phenol and one or more ethanol precipitations. The number of residues bound per mole of plasmid was calculated from the A_{260} absorption profile of the DNA and the specific activity of the labeled carcinogen.

Transfection of Host, Rescue of Replicated Plasmid, and Assaying for supF Mutants

A number of independent cultures of human 293 cells were transfected with plasmid, and after 44 hours the progeny plasmids were extracted and separated from cellular DNA as described (Hirt, 1967). The DNA from each set of cells was purified and assayed for *supF* mutants separately in order to be able to distinguish independent mutants with identical mutations from putative siblings derived from the same set of harvested cells. The plasmids were assayed for mutant *supF* genes by transforming *E. coli* and selecting the transformants on agar plates containing ampicillin, X-Gal, and an inducer of the β-galactosidase gene. On these plates transformants containing plasmids lacking a functioning *supF* gene form white or light blue colonies, rather than dark blue colonies.

Characterization of Mutants and Determination on Sites of Adducts

Plasmids from white or light blue colonies were analyzed for altered DNA mobility (gross alteration) by electrophoresis on 0.8% agarose gels. The *supF* genes in plasmids lacking gross alterations were sequenced using the dideoxyribonucleotide method, using ^{35}S α-dATP and buffer gradient polyacrylamide gels for greater resolution. The positions of carcinogen adducts in the *supF* gene were determined by the *in vitro* DNA polymerase)stop assay of Moore et al. (1982). Plasmids were denatured and annealed with primer, and polymerization was carried out as for the sequencing reaction, except that the dideoxy)nucleotides were omitted. DNA from the four dideoxy sequencing reactions, carried out on an untreated *supF* template, was electrophoresed on the same gel to serve as DNA size markers. The relative intensities of the band on the autoradiograph of the gel were determined by a laser densitometer.

RESULTS

Comparing the Biologic Effectiveness of the Adducts Induced by These Agents

The ability of these carcinogens to react with DNA to form covalently-bound residues is shown in Figures 1A, 2A, and 3A. 1-NOP appears to be twice as reactive as BPDE or N-AcO-TFA-AF, but this merely reflects the use of ascorbic acid to activate 1-NOP. N-AcO-AAF was approximately equal to N-AcO-TFA-AF in reactivity and 1-N-6-NOP was similar to 1-NOP (data not shown).

Once formed, BPDE and 1-NOP were virtually equal in their ability to interfere with bacterial transformation (Figures 1B and 2B), and were approximately four times more effective than the adducts formed by N-AcO-TFA-AF (Figure 3B). These differences may reflect the structure of the adducts, but could also reflect differences in the rate of removal of specific adducts by the bacterial cells before transformation to ampicillin resistance takes place. HPLC analysis showed that plasmids treated with N-AcO-TFA-AF contain only one adduct, N-(deoxyguanosin-8-yl)-2-aminofluorene (dG-C8-AF), and plasmids reacted with 1-NOP contain only N-(deoxyguanosin-8-yl)-1-aminopyrene (dG-C8-AP), a closely related adduct. As expected, N-AcO-AAF and 1-N-6-NOP were also shown by HPLC to bind to DNA at the C8 position of guanine (data not shown).

Plasmids containing these adducts were transfected into human 293 cells and allowed to replicate, and the progeny plasmids were assayed for mutants. As shown in Figures 1C, 2C, and 3C, each carcinogen induced a linear increase in the frequency of *supF* mutants as a function of the number of adducts per plasmid. Significantly, the frequency induced per adduct was 3.7 fold higher for BPDE than for the other four agents. For example, interpolation of the data in Figure 1C indicate that 20 BPDE adducts per plasmid increased the mutant frequency from 1.4×10^{-4} to 40×10^{-4} whereas in Figures 2C and 3C, 20 1-NOP or AF adducts per plasmid increased it only to 10 to 12×10^{-4}. 20 AAF adducts or 1-N-6-NOP adducts per plasmid also increased the frequency to 10 to 14×10^{-4} (data not shown).

The fact that all four of the agents that yield C8 guanine adducts in DNA gave virtually identical induced mutation frequencies, whereas BPDE, the only agent tested that binds to the N2 position, gave much higher frequencies strongly suggests that location of the adduct in the DNA plays a more critical role than does the specific structure of the adduct. However, this difference in mutant frequency might reflect a difference in the rate of excision of adducts from the plasmid by the host cells. This is because the number of adducts bound refers to the number initially formed, whereas the introduction of mutations occurs at a much later time, after the plasmids have entered the nucleus of the 293 cells (Calos et al., 1983). Although it was not possible to determine the rate of removal of these adducts from the plasmids, Yang et al. (1988) compared the rate of removal of BPDE and 1-NOP adducts from the genomic DNA of the host cells. The results showed that 1-NOP residues were removed at a rate 3.7 fold faster than were BPDE-residues which supports the hypothesis that the difference in mutation frequency reflects a difference in rate of repair.

Kinds of Mutations Induced by BPDE, 1-NOP, and N-AcO-TFA-AF

A total of 40 spontaneous, 114 BPDE-induced, 88 1-NOP induced, and 118 N-AcO-TFA-AF-induced mutants were assayed by gel electrophoresis for gross rearrangements. Those that did not show such changes, i.e., 26 spontaneous, 93 BPDE-induced, 64 1-NOP-induced, and 50 N-AcO-TFA-AF-induced mutants, were sequenced (Yang et al., 1987b, 1988; Mah et al., 1989). The results showed that only 30% of the spontaneous mutants contained point mutations (substitution, deletion, or insertion of 1 to 3 base pairs), compared to 83% of these from BPDE-treated plasmids, 88% of those from 1-NOP-treated plasmids and 90% from N-AcO-TFA-AF-treated plasmids. Examination of 86 unequivocally-independent mutants induced by BPDE, 60 by 1-NOP, and 50 by N-AcO-TFA-AF showed that the majority (70% for BPDE; 85% for 1-NOP and 90% for N-AcO-TFA-AF) consisted of base substitutions (Table 1), with 61% to 65% of these being G·C → T·A transversions (Table 2). 87% to 90% of the base substitutions found with BPDE and 1-NOP and 100% of those from N-AcO-TFA-AF involved G·C base pairs. These results are consistent with the DNA binding pattern of these carcinogens, since unlike N-AcO-TFA-AF, BPDE and 1-NOP bind to adenine at a low frequency.

Spectra of Point Mutations Induced by BPDE, 1-NOP, and N-AcO-TFA-AF

As shown in Figure 4, N-AcO-TFA-AF, 1-NOP, and BPDE exhibited a unique spectrum of base changes in the *supF* gene. There were four major hot spots with N-AcO-TFA-AF, five with 1-NOP, and three with BPDE. Hot spots (defined as places where 6% or more of the observed mutations were located) produced by one agent were often cold spots for another agent. The majority of the bases that were not changed in plasmids treated with any of the three carcinogens cannot be considered as "silent" since they have been demonstrated by other investigators to be positions that will show a mutant phenotype, if altered (Kraemer and Seidman, 1989).

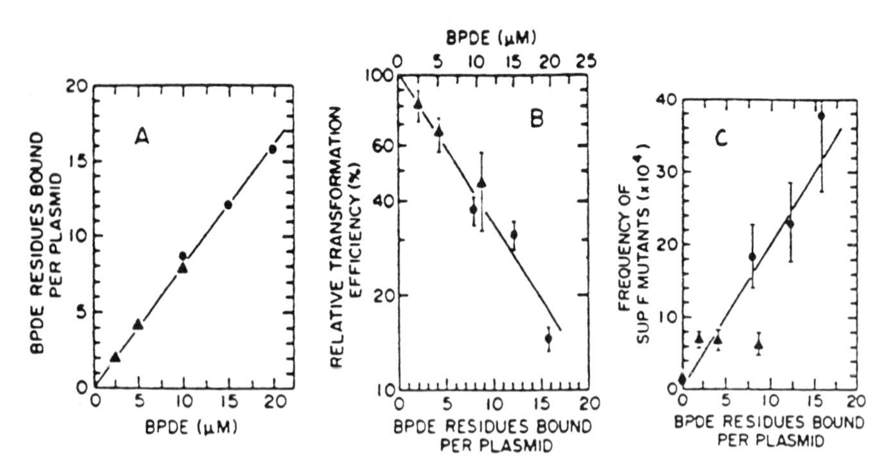

Figure 1. Adduct formation as a function of applied concentration of BPDE (A); degree of interference with transformation (B); and frequency of *supF* mutants (C), as a function of the number of BPDE adducts per plasmid. (From Yang et al., 1987b with permission.)

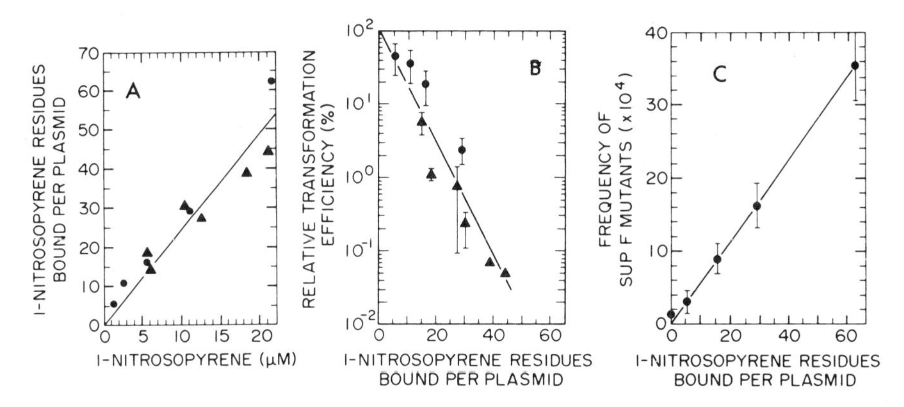

Figure 2. Adduct formation as a function of applied concentration of 1-NOP (A); degree of interference with transformation (B); and frequency of *supF* mutants (C), as a function of the number of 1-NOP adducts per plasmid. (From Yang et al., 1988 with permission.)

To see if the mutation hot spots resulted from hot spots for carcinogen binding in the target gene, we carried out the DNA synthesis)stop assay (Moore et al. 1982) in which bulky adducts interfere with DNA replication. The gel pattern of the bands obtained with plasmid DNA containing BPDE or 1-NOP, residues corresponded to positions one nucleotide 5' to virtually every cytosine in the DNA sequencing standard lane, indicating that DNA synthesis was terminated one base prior to each guanine in the template (Yang et al., 1988). The gel pattern of bands obtained with N-AcO-TFA-AF-treated plasmids corresponded to positions one nucleotide 5' to every cytosine or opposite every cytosine except one (Mah et al., 1989). No other bands were seen, and there was no evidence of any interference with polymerization using untreated template. If one assumes that these bands were generated by the Klenow fragment of DNA polymerase I falling off the template at the sites of bulky adducts, the results indicate that in some cases there was good correlation between the extent of binding and the frequency of mutation induction. However, in some cases this was not true. Some of the lack of correlation can be accounted for by "silent mutations," but not all of it.

DISCUSSION

Several lines of evidence suggest that the mutations observed in our studies were targeted to the adducts formed. For example, the increase in frequency of mutants was linearly correlated with the number of adducts per plasmid. Among the mutants obtained from carcinogen-treated plasmids, the frequency of those containing gross rearrangements, or deletions or insertions greater than 1 to 3 bp was approximately equal to the background frequency. The rest carried point mutations exclusively or predominantly involving G·C base pairs. A minor fraction of the base substitutions observed with BPDE- or 1-NOP-treated plasmids involved AT base pairs, and these agents bind to adenine at a low, but detectable frequency. None of the mutations in the 5' to 3' strand were located at sites where termination of polymerization on a 3' to 5' template by the Klenow fragment was not detected, (cf. Yang, et al., 1988 and Mah et al., 1989), and the two mutational hot spots for which the dG-C8-AF or dG-C8-AP adduct would be located in the 3' to 5' strand and the most prominent hot spot for BPDE were also prominent positions for termination of polymerization in the stop assay.

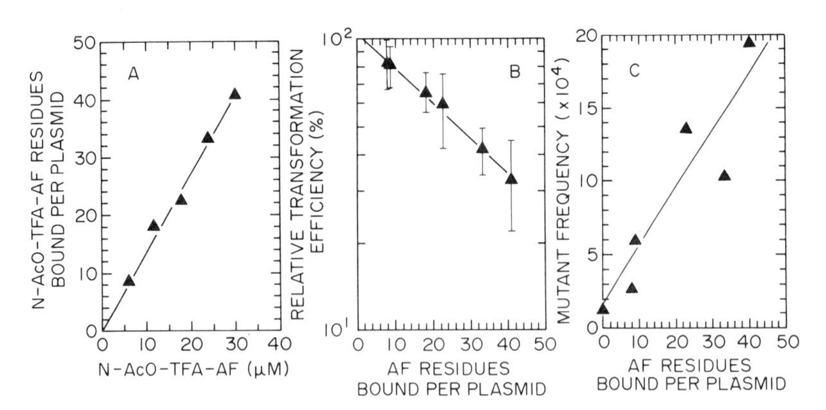

Figure 3. Adduct formation as a function of applied concentration of N-AcO-TFA-AF (A); degree of interference with transformation (B); and frequency of *supF* mutants (C), as a function of the number of AF adducts per plasmid. (From Mah et al., 1989 with permission.)

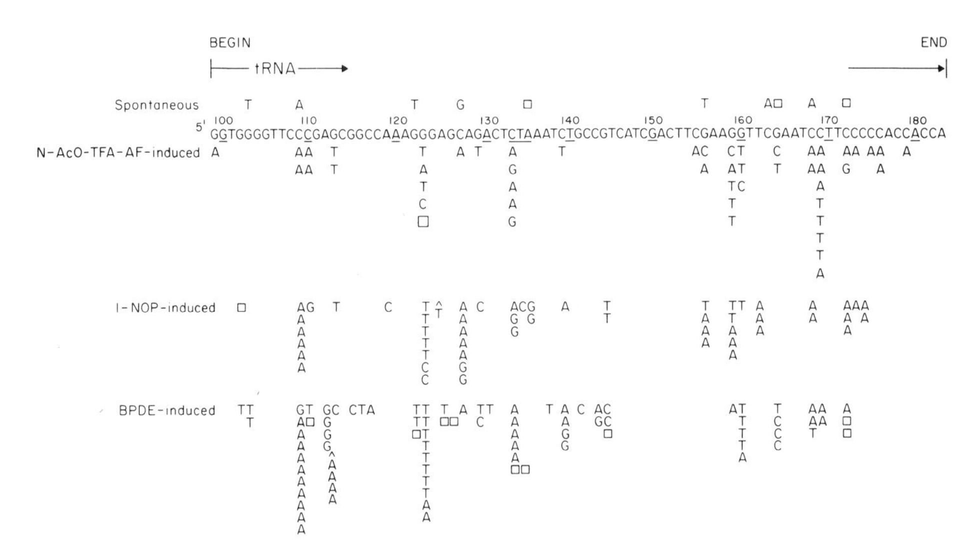

Figure 4. Location of N-AcO-TFA-AF-, BPDE-, and 1-NOP-induced point mutations in *supF*. The DNA strand shown is the counterpart of the structural region of the tRNA. The open squares represent base deletions; the crets indicate where a base has been inserted. Every 10th base and the anticodon triplet have been underlined. (Data taken from Yang et al., 1987b and 1988, and Mah et al., 1989 with permission.)

One possible explanation for why the majority of the base substitutions were G·C → T·A transversions is that a DNA polymerase preferentially inserts adenosine triphosphate opposite the adduct. However, recent studies by Norman et al., (1989) suggest that adenosine triphosphate can be accommodated opposite dG-C8-AF. If this occurred during replication of the plasmid, it would result in the type of transversion observed most frequently with N-AcO-TFA-AF-treated plasmids. It may be that stable base pairing of adenosine triphosphate opposite dG-C8-AP adducts also occurs since the two C8 guanine adducts are so similar. If BPDE N2 guanine adducts can also base pair stably with adenosine triphosphate, this would explain why BPDE also induces predominantly G·C → T·A transversions.

The correlation between hot spots for initial adduct formation (as judged by the stop assay) and hot spots for base substitution was not perfect. This may reflect differential DNA repair of the adducts from the plasmids in the 293 cells. That is, nucleotide excision repair might preferentially remove adducts from particular locations in the target gene. Another possible explanation is that the chance of a mutation depends not only on the presence of an adduct, but also on the neighboring bases.

The data in Figures 1C, 2C, and 3C suggest that BPDE adducts are 3.7 times more mutagenic than adducts formed by the other two compounds. They are also more mutagenic than AAF or 1-N-6-NOP adducts (data not shown). However, Yang et al. (1988) showed that 1-NOP adducts were lost from genomic DNA of 293 cells at a rate three to four times faster than BPDE adducts. This suggests that following transfection, 1-NOP adducts (and perhaps 1-N-6-NOP, AAF and AF adducts) are removed from plasmid DNA faster than BPDE adducts. If the frequency of adducts remaining in the plasmids at the time replication occurs were three to four times lower than the initial frequency shown in Figures 2C and 3C, such adducts might actually be equal to BPDE adducts in their effectiveness in inducing *supF* mutants during replication in human cells.

Table 1. Comparison of Sequence Alterations Generated in *supF* by Replication of Carcinogen-treated Plasmid in Human Cells

Sequence Alterations	Number of Independent Mutants Sequenced							
	Control		BPDE		1-NOP		N-AcO-TFA-AF	
<u>Single Base Substitution</u>:	3		51		48		42	
<u>Two Base Substitutions</u>:								
Tandem	0		3		2		1	
≤ 20 bases apart	2	24%	3	70%	0	85%	0	90%
> 20 bases apart	0		3		1		1	
<u>Three Base Substitutions</u>:								
≥ 20 bases apart	0		0		0		1	
<u>Deletions</u>:								
Single G·C pair	2		5		1		1	
Single A·T pair	1		0		0		0	
Tandem bases	0	66%	2	24%	0	10%	0	8%
4-20 bases	4		3		3		2	
> 20 bases	7		11		2		1	
<u>Insertions</u>:								
Single A·T pair	0	10%	4	6%	1	5%	0	2%
≤ 20 bases	2		1		2		1	
Total	21		86		60		50	

Table 2. Comparison of the Kinds of Base Substitutions Generated in *supF* during Replication of Carcinogen-treated Plasmid in Human Cells

	Number of Mutants Observed			
Base Change	Control	BPDE	1-NOP	N-AcO-TFA-AF
Transversions:				
G·C ---> T·A	6 (86%)	45 (63%)	33 (61%)	32 (65%)
G·C ---> C·G	1 (14%)	13 (18%)	8 (15%)	8 (17%)
A·T ---> T·A	0	3 (4%)	3 (5.5%)	0
A·T ---> C·G	0	0	1 (2%)	0
Transitions:				
G·C ---> A·T	0	6 (9%)	6 (11%)	9 (18%)
A·T ---> G·C	0	4 (6%)	3 (5.5%)	0
Total	7	71	54	49

ACKNOWLEDGEMENTS

We thank our colleagues Dr. Frederick A. Beland of the National Cancer for Toxicological Research for providing us with the tritiated 1-NOP and 1-N-6-NOP and for HPLC analysis of the adducts formed on the plasmid by these agents. We thank Drs. Charles M. King and Thomas M. Reid of the Michigan Cancer Foundation for reacting the plasmid with N-AcO-TFA-AF and for HPLC analysis of the adducts formed. The research was supported by DHHS grant CA21253 from the National Cancer Institute and by Contract 87-2 from the Health Effects Institute, an organization jointly funded by the U.S. EPA and automotive manufacturers. It is currently under review by the Institute. The contents of this article do not necessarily reflect the policies of EPA, or automotive manufacturers.

REFERENCES

Beland, F. A., Ribovich, M., Howard, P. C., Heflich, R. H., Kurian, P., and Milo G. E. (1986). Cytotoxicity, cellular transformation and DNA adducts in normal human diploid fibroblasts exposed to 1-nitrosopyrene, a reduced derivative of the environmental contaminant, 1-nitropyrene. *Carcinogenesis* **7**, 1279-1283.

Calos, M. P., Lebkowski, J. S., and Botchan M. R. (1983). High mutationfrequency in DNA transfected into mammalian cells. *Proc. Natl. Acad. Sci. U.S.A.* **80**, 3015-3019.

Djuric, Z., Fifer, E. K., Yamazoe, Y., and Beland F. A. (1988). DNA binding by 1-nitropyrene and 1,6-dinitropyrene *in vitro* and *in vivo*: effects of nitroreductase induction. *Carcinogenesis* **9**, No. 3, 357-364.

Graham, F. L., Smiley, J., Russell, W. C., and Nairn, R. (1977). Characteristics of a human cell line transformed by DNA from human adenovirus type 5. *J. Gen. Virol.* **36**, 59-74.

Heflich, R. H., Howard, P. C., and Beland, F. A. (1985). 1-Nitrosopyrene: An intermediate in the metabolic activation of 1-nitropyrene to a mutagen in *Salmonella typhimurium* TA1538. *Mutation Res.* **149**, 25-32.

Heflich, R. H., Djuric, Z., Zhuo, Z., Fullerton, N. F., Casciano, D. A., and Beland, F. A. (1988). Metabolism of 2-acetylaminofluorene in the Chinese hamster ovary cell mutation assay. *Environ. Mol. Mutag.* **11**, 167-181.

Hirt, B. (1967). Selective extraction of polyoma DNA from infected mouse cell cultures. *J. Mol. Biol.* **26**, 365-369.

Kraemer, K. H., and Seidman, M. M. (1989). Use of *supF*, an *Escherichia coli* tyrosine suppressor tRNA gene, as a mutagenic target in shuttle-vector plasmids. *Mutat. Res.* **220**, 61-72.

Mah, M. C.-M., Maher, V. M., Thomas, H., Reid, T. M., King, C. M., and McCormick,J. J. (1989). Mutations induced by aminofluorene-DNA adducts during replication in human cells. *Carcinogenesis* **10**, 2321-2328.

Moore P., Rabkin, S. D., Osborn, A. L., King C. M., and Strauss, B. S. (1982). Effect of acetylated and deacetylated 2-aminofluorene adducts on *in vitro* DNA synthesis. *Proc. Natl. Acad. Sci. U.S.A.* **79**, 7166-7170.

Norman, D., Abuaf, P., Hingerty, B. E., Live, D., Grunberger, D., Broyde, S., and Patel, D. J. (1989). NMR and computational characterization of the N-(deoxyguanosin- 8-yl)aminofluorene adduct [(AF)G] opposite adenosine in DNA: (AF)G[syn]·A[anti] pair formation and its pH dependence. *Biochemistry* **28**, 7462-7476.

Poirier, M. C., Williams, G. M., and Yuspa, S. H. (1980). Effect of culture conditions, cell type, and species of origin on the distribution of acetylated and deacetylated deoxyguanosine C)8 adducts of CCNDD-acetoxy-2-acetylaminofluorene. *Mol. Pharmacol.* **18**, 581-587.

Sarkar, S., Dasgupta, U. B., and Summers, W. C. (1984). Error-prone mutagenesis detected in mammalian cells by a shuttle vector containing the *supF* gene of *Escherichia coli. Mol. Cell. Biol.* **4**, 2227-2230.

Seidman, M. M., Dixon, K., Rassaque, A., Zagursky, R. J., and Berman M. L. (1985). A shuttle vector plasmid for studying carcinogen-induced point mutations in mammalian cells. *Gene.* **389**, 233-237.

Weinstein, I, B., Jeffrey, A. M., Jennette, K. W., Blobstein, S. H., Harvey, R. G., Harris, C., Autrup, H., Kasai, H., and Nakanishi K. (1976). Benzo[a]pyrene diol epoxides as intermediates in nucleic acid binding *in vitro* and *in vivo. Sci.* **193**, 592-595.

Yang, J.-L., Maher, V. M., and McCormick, J. J. (1987a). Kinds of mutations formed when a shuttle vector containing adducts of benzo[a]pyrene-7,8-diol-9,10-epoxide replicates in COS7 cells. *Mol. Cell. Biol.* **7**, 1267-1270.

Yang, J.-L., Maher, V. M., and McCormick, J. J. (1987b). Kinds of mutations formed when a shuttle vector containing adducts of (±)-7β,8α-dihydroxy-9α,10α-epoxy-7,8,9,10-tetrahydrobenzo[a]pyrene replicates in human cells. *Proc. Natl. Acad. Sci. U.S.A.* **84**, 3787-3791.

Yang, J.-L., Maher, V. M., and McCormick, J. J. (1988). Kinds and spectrum of mutations induced by 1-nitrosopyrene adducts during plasmid replication in human cells. *Mol. Cell. Biol.* **8**, 3364-3372.

MOLECULAR TARGETS OF CHEMICAL MUTAGENS

Bradley D. Preston[1] and Rupa Doshi

Department of Chemical Biology
Rutgers University
College of Pharmacy
Piscataway, NJ 08855-0789

Most mutagens are biologically active by virtue of their chemical reactivity. The pioneering work of the Millers (reviewed in Miller and Miller, 1974, 1981) and others (Searle, 1984 and references therein) clearly established that most chemical mutagens (or their metabolites) are electrophilic and covalently bind to cellular macromolecules. Proteins were first identified as cellular targets (Miller, 1951), and it was soon recognized that RNA and DNA are also extensively damaged by mutagens (for reviews see: Searle, 1984; Miller and Miller, 1974). Theoretically, every cellular nucleophile is a potential target for damage by electrophilic mutagens.

If one defines a mutation as any heritable change in the nucleotide sequence of DNA, then mutations must ultimately arise either as a consequence of errors during DNA synthesis or via anomalous genetic recombination. Thus, in order to induce a mutation, a mutagen must somehow disrupt the function of molecules that preserve genomic integrity. It is estimated that the spontaneous mutation frequency of a cell is extremely low, ranging from about 10^{-9} to 10^{-11} mutations per nucleotide per round of genomic replication (Drake, 1969; Wabl et al., 1985). To achieve this extraordinary fidelity, cells have evolved elaborate biochemical systems that prevent mutation (Figure 1). There are four general mechanisms that help protect the cell from genetic change: 1) prevention of molecular damage by removal of bioreactive agents; 2) repair of promutagenic DNA lesions; 3) faithful template-directed DNA synthesis from deoxyribonucleoside triphosphate (dNTP) precursors, and; 4) postreplication repair of mismatched nucleotides. The cumulative effects of these protective systems ensures that genetic information is faithfully passed from parental to daughter cells. It follows, therefore, that disruption of any these systems will put a cell at risk for mutations.

The critical role of the replication fork in determining the fidelity of DNA synthesis suggests that components of the replication and repair machinery are likely targets of mutagens. In this review we will consider the potential mutagenic consequences of damage to four classes of molecules that are necessary for faithful DNA synthesis: 1) the DNA template; 2) dNTPs and their precursors; 3) DNA replication proteins, and; 4) DNA repair proteins. Our discussion will emphasize non-DNA molecules, since damage to DNA and its role in mutagenesis has been extensively reviewed elsewhere (Singer and Grunberger, 1983; Searle, 1984).

1. Address Correspondence to: Dr. Bradley D. Preston, College of Pharmacy, Rutgers University, P.O. Box 789, Piscataway, NJ 08855-0789

Biological Reactive Intermediates IV, Edited by C.M. Witmer *et al.*
Plenum Press, New York, 1990

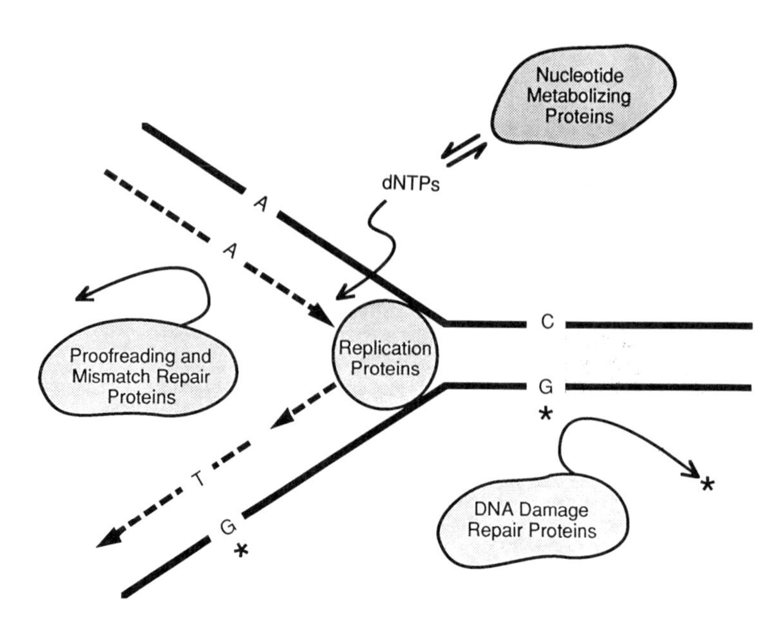

Figure 1. Schematic of DNA replication fork emphasizing pathways and substrates critical to faithful DNA synthesis. Template and nascent DNA strands are indicated by solid and dashed lines, repectively. The replication fork is moving from left to right with leading strand DNA synthesis occuring on the upper template strand and lagging strand synthesis occurring on the lower template strand. * = DNA damage.

DNA AS A CRITICAL TARGET

The DNA template is clearly an important target for mutagens (Singer and Grunberger, 1983). Although the potential genetic consequences of DNA damage have long been recognized (Loveless, 1969 and references therein), proof that DNA damage *per se* is sufficient to cause mutations was not forthcoming until it was possible to transform cells in culture with biologically active DNA. This permitted experiments whereby naked plasmid or virus DNA could be damaged by mutagens *in vitro* and subsequently transfected into and replicated in appropriate host cells. By designing the target DNAs to contain phenotypically scorable genes, it is possible to measure mutations resulting from direct damage to DNA without damage to other cellular molecules.

Initial studies using this approach focused primarily on prokaryotic replication vectors and phage DNAs as targets for damage (Fuchs et al., 1981; LeClerc and Istock, 1982; Loeb et al., 1988). More recent studies on shuttle vectors (Lebkowski et al., 1985; Seidman et al., 1985; Kraemer and Seidman, 1989; Yang et al., 1987; Yang et al., 1988) and eukaryotic viral DNAs (Gentil et al., 1986; Gentil et al., 1984) have permitted an assessment of the role of DNA damage in mammalian cell mutagenesis. An important observation from these studies in both prokaryotic and eukaryotic systems is that different mutagens tend to yield different distributions and types of mutations throughout the target gene sequence. For example, DNA damaged by ultraviolet (UV) irradiation yields predominantly transition mutations at thymine residues (LeClerc and Istock, 1982; Protic-Sabljic et al., 1986), while acetyl aminofluorene (AAF)-damaged DNA yields small frameshift mutations (Fuchs et al., 1981).

Recent studies using site-specifically damaged replication vectors have provided detailed insights into the mechanisms of mutagenesis by DNA damage (Basu and Essigmann, 1988). In order for a damaged base to induce a mutation it must affect the fidelity of DNA synthesis. Mutations occurring at the site of damage presumably arise as a result of either ambiguous base-pairing (Singer and Grunberger, 1983) or error-prone repair of the damaged base (Walker, 1984; Witkin, 1976). Probably the best characterized promutagenic DNA adducts are those formed by simple alkylating mutagens such as the dialkyl sulfates, sulfonates and nitrosamines. These mutagens are capable of forming alkyl adducts at virtually every nitrogen and oxygen atom on DNA (Singer and Grunberger, 1983). In order to identify those adducts which are critical to mutagenesis, several laboratories have constructed genes *in vitro* that contain single alkyl adducts at predetermined sites (Preston et al., 1986; Preston et al., 1987; Loechler et al., 1984; Bhanot and Ray, 1986; Chambers et al., 1985; Basu and Essigmann, 1990). By transfecting these site-specifically adducted genes into host cells and allowing DNA synthesis to proceed, it is possible to determine the mutagenic potency and base-pairing preferences of each adduct. In this way, it has been shown that O^6-alkylguanine and O^4-alkylthymine adducts exhibit ambiguous base-pairing properties and characteristically induce G·C → A·T and A·T → G·C transition mutations, respectively (Loechler et al., 1984; Preston et al., 1986). Recent studies on several other site-specific DNA lesions further support the notion that individual adducts exhibit specific base-pairing preferences (Basu and Essigmann, 1988; Basu and Essigmann, 1990). By comparing the types of mutations induced by site-specific DNA adducts with the mutation spectrum resulting from treatment of whole cells with a given mutagen, it should be possible to identify those adducts which contribute to mutagenesis.

EFFECTS OF MUTAGENS ON NUCLEOTIDE POOLS

Although the simplest mechanism for mutagenesis is via direct damage to the DNA template, mutagens may also act indirectly by damage to other components of the replication fork that are essential to faithful daughter-strand synthesis (Figure 1). Deoxyribonuleoside triphosphates (dNTPs) are the building blocks of the nascent DNA. It follows, therefore, that damage to the cellular dNTP pools might affect mutagenesis. At least two mechanisms could contribute to mutagenesis via the disruption of dNTP pools. First, a mutagen might directly form promutagenic adducts with dNTPs or nucleotide precursors. The resultant dNTP adducts could then be incorporated into the growing DNA strand and yield mutations after repair of ambiguously incorporated adducts or after subsequent rounds of DNA replication (Snow and Mitra, 1987; Snow and Mitra, 1988). Alternatively, a mutagen might disrupt the regulation of dNTP synthesis in such a way as to cause an imbalance in the ratios of the four normal nucleotides (Kunz, 1988). Since cellular dNTP levels appear to be present in limited amounts and are tightly regulated with the cell cycle (Reichard, 1988), a nucleotide pool imbalance will force the replicative polymerases to occasionally misincorporate those dNTPs present in relative excess (Meuth, 1989).

Direct Damage to dNTPs or Nucleotide Precursors

The absence of hydrogen bonding in nonpolymerized DNA precursors exposes several nucleophilic sites that are sterically hindered and unavailable for reaction in double-stranded DNA polymers (Singer and Grunberger, 1983). Thus, unincorporated nucleotides, nucleosides and bases are excellent targets for damage by mutagens *in vitro* and similarly are likely targets *in vivo*. The potential importance of DNA precursor pools as targets of alkylating agents is supported by the observation that N-methyl-N-nitrosourea (MNU) produces quantitatively more deoxynucleotide adducts in the unincorporated nucleotide pools than in DNA (Topal and Baker, 1982). These data suggest that free nucleotides in mammalian cells are more susceptible to alkylation than deoxyribonucleotide residues in DNA. To assess the role of free nucleotide adducts in mutagenesis, Arecco et al. (Arecco et al., 1988) conducted a series of experiments in which individual nucleotide pool sizes were transiently expanded immediately prior to treatment with MNU. It was reasoned that if a nucleotide pool is a mutagenic target, then expansion of that pool should increase the target size within the cell and enhance mutagenesis. Conditions that expanded pools of dATP, dTTP or dGTP were found to stimulate MNU mutagenesis by 2- to 6-fold, suggesting that nucleotide pools may serve as critical targets for alkylating mutagens. The levels of dNTP pools also affect the mutagenicities of ultraviolet light (UV) (Prakash et al., 1978; McKenna and Hickey, 1981; Baranowska et al., 1987), 2-aminopurine (Hopkins and Goodman, 1985; Caras et al., 1982), 6-amino-2-hydroxypurine, 5-bromodeoxyuridine (Meuth and Green, 1974; Kaufman, 1986), mitomycin C, and several other monofunctional alkylating agents (Meuth, 1983; Önfelt and Jenssen, 1982) including ethyl methanesulfonate (EMS), methyl methanesulfonate (MMS), ethyl nitrosourea (ENU) and N-methyl-N'-nitro-N-nitrosoguanidine (MNNG) (Roguska and Gudas, 1984; McKenna, et al., 1985; Peterson, et al., 1985; Ayusawa, et al., 1983; Randazzo, et al., 1987; Kunz, 1982; McKenna and Yasseen, 1982; Kunz, 1988). However, the mechanisms underlying many of these observations are difficult to determine, since the effects of DNA precursor levels on induced mutagenesis appear to be dependent on the particular dNTP pool, the organism and the mutational system under study. A further complication is that dNTP pools also affect the fidelity of DNA repair, and many mutagens themselves induce dNTP pool imbalances (see below).

One mechanism whereby modified dNTPs might induce mutations is via ambiguous base-pairing during or after incorporation into DNA. Early studies on the prototypic mutagens 2-aminopurine (Mhaskar and Goodman, 1984) and 5-bromouracil (Lasken and Goodman, 1984) clearly show that mutations can arise via incorporation of modified bases into newly synthesized DNA. In order for this to occur, the dNTP adducts must serve as substrates for DNA polymerases. Studies using purified DNA polymerases show that many dNTP analogues are efficiently incorporated into DNA *in vitro* (Preston and Loeb, 1988). The efficiency of incorporation varies among the different adducts studied and appears to be dependent on both the size of the adducted side group as well as the position of modification on the dNTP molecule (Preston et al., 1987; Singer et al., 1987). The efficiency of incorporation of dNTP adducts is presumably determined by the tolerance of the polymerase's dNTP binding site for varied dNTP structures. This is consistent with the observation that certain alkylated dNTPs (e.g., O^2-methyl- and O^4-isopropyl-dTTPs) are more efficiently incorporated by the error-prone AMV reverse transcriptase compared to the highly accurate *E. coli* DNA polymerase I (pol I) (Preston et al., 1987). Thus, in eukaryotic cells dNTP adducts might be most readily incorporated during DNA repair synthesis by DNA polymerase β (pol β), since pol β is highly error-prone (Kunkel, 1985) and presumably has a relatively high tolerance for unusual dNTP structures. The size limits of dNTP adducts as polymerase substrates are not known. However, the ability of polymerases to incorporate biotinylated dNTPs with long side chains (Brigati et al., 1983; Langer et al., 1981) suggests that at least some "bulky" dNTP adducts might be incorporated by this mechanism. It is likely that bulky side-groups positioned at hydrogen bonding atoms will block incorporation. Thus, the best bulky dNTP substrates should be those with side-groups at positions 5 or 6 of pyrimidines or on the imidazole ring of purines. Preliminary data from our laboratory suggest that C8-aminofluorene-dGTP is incorporated by some DNA polymerases *in vitro* (Preston et al., 1988b). There are no reported studies on other bulky dNTP adducts.

Induction of dNTP Pool Imbalances

It is well established that perturbations of dNTP pools elicit a broad range of genetic changes, including mutations, recombination, chromosome loss, chromosome aberrations and breakage, sister chromatid exchange, and DNA strand breaks (Haynes and Kunz, 1988). Mutations by dNTP pool imbalances are thought to occur via a decrease in the fidelity of DNA synthesis (Loeb and Kunkel, 1982). Studies *in vitro* show that the ratios of dNTPs during DNA polymerization strongly influence the fidelity of DNA polymerases to favor misincorporation of the nucleotides in excess (Fersht, 1979; Kunkel and Loeb, 1979; Hibner and Alberts, 1980; Kunkel et al., 1982; Zakour and Loeb, 1982). DNA sequence analyses of mutations induced by nucleotide pool imbalances in cells in culture strongly support this kinetic mechanism of mutagenesis (Meuth, 1989). Disruptions of nucleotide pool ratios also produce a "next nucleotide effect" on mutagenesis that appears to be due to a decrease in the efficiency of mispair excision by proofreading 3' → 5' exonucleases that are associated with some DNA polymerases (Fersht, 1979; Kunkel et al., 1981; Phear et al., 1987). Thus, dNTP imbalances can affect polymerase fidelity in two ways: by increasing the rate of misincorporation and by decreasing proofreading efficiency.

Several studies have shown that treatment with physical and chemical mutagens causes changes in the intracellular levels of DNA precursors (Kunz, 1988). UV irradiation of *E. coli* induces a rapid elevation of dATP and dTTP pools with little effect on the levels of dCTP and dGTP (Das and Loeb, 1984; Suzuki et al., 1983). A variety of other mutagens also disrupt intracellular dNTP pools in eukaryotic cells. However, the nucleotide specificity and extent of imbalance appears dependent on the mutagen as well as the cell line under study. Thus, exposure of Chinese hamster V79 cells to UV, MNNG, mitomycin C or cytosine arabinoside results in dATP and dTTP pool expansions with little effect on the levels of dCTP and dGTP (Das et al., 1983). In Chinese hamster ovary cells, UV also induces a rapid increase in dTTP, but there is a concurrent rapid decrease in dCTP with a more gradual reduction of dATP and no change in dGTP (Newman and Miller, 1983a; Newman and Miller, 1983b). Mouse FM3A cells, on the other hand, appear refractory to significant pool imbalances induced by UV, MNNG or mitomycin C (Hyodo et al., 1984), and treatment of mammalian cells with the classic mutagen bromouridine results in a rapid and specific decrease in only the dCTP pool (Ashman and Davidson, 1981; Meuth and Green, 1974).

The mechanisms of mutagen-induced dNTP pool imbalances are not entirely clear. In order to maintain normal dNTP levels, cells have evolved very intricate and exquisite controls of nucleotide synthesis and turnover (Reichard, 1988; Adams et al., 1986). It is reasonable, therefore, that mutagens might affect the balance of dNTPs by disrupting the regulation of nucleotide metabolism. In principle this could occur at several levels. First, mutagens could directly degrade dNTPs thereby decreasing the intracellular concentration of the damaged dNTP and/or shunting the degradation products into the synthesis of other dNTPs (Newman and Miller, 1985). This mechanism might account for the effects of high doses of mutagens. However, it is unlikely to contribute to mutagenesis under most circumstances, since a large proportion of dNTPs would have to be damaged in order to affect a significant reduction in dNTPs available for DNA synthesis. Although dNTPs are a significant target for mutagen damage, the modified dNTPs represent less than 1% of the total dNTPs in the cell (Snow and Mitra, 1988). A second mechanism for induced pool imbalances could be through the altered expression of genes coding for enzymes or regulatory proteins involved in nucleotide metabolism. This mechanism is supported by the finding that the expression of either dTMP kinase or ribonucleotide reductase is increased in yeast following DNA damage (Elledge and Davis, 1987). A third and potentially very important mechanism for induced pool imbalances is via the direct inhibition of nucleotide metabolizing proteins by mutagenic agents.

There is strong evidence showing that the disruption of nucleotide metabolism induces dNTP pool imbalances and causes mutations. Genetic studies on genes coding for CTP synthase, dCMP deaminase, dCMP hydroxymethylase, deoxyuridine triphosphatase, ribonucleotide reductase and thymidylate synthase show that disruption of the activity of any one of these enzymes leads to very specific and predictable dNTP imbalances and confers a mutator phenotype on cells in culture (Arpaia et al., 1983; Ayusawa et al., 1983; Hochauser and Weiss, 1978; Maus, et al., 1984; Roguska and Gudas, 1984; Sargent and Mathews, 1987; Sedwick, et al., 1986; Trudel, et al., 1984; Weinberg, et al., 1981; Weinberg, et al., 1985; Williams and Drake, 1977). Drugs that block nucleotide synthesis also induce mutations as well as mitotic recombination. Thus, inhibition of thymidylate synthase by fluorodeoxyuridine or inhibition of dihydrofolate reductase by the antifolates aminopterin and methotrexate results in the blockage of dTMP biosynthesis and induction of mutations (Abersold, 1983; Hoar and Dimnik, 1985; Kunz et al., 1986; Peterson et al., 1983; Wurtz et al., 1979). Other agents known to inhibit nucleotide metabolizing enzymes and to affect dNTP pools include hydroxyurea which inhibits ribonucleotide reductase (Arecco et al., 1988), 8-aminoguanosine (Kazmers et al., 1981) and 8-amino-9-benzylguanine (Shewach et al., 1986) which inhibit purine nucleoside phosphorylase, and azaserine and 6-diazo-5-oxonorleucine which inhibit phosphoribosyl pyrophosphate amidotransferase (Adams et al., 1986).

These studies suggest that some mutagens might act indirectly by damaging proteins critical to the maintenance of faithful DNA synthesis. Intracellular dNTP pools are synthesized by a network of nucleotide interconversions catalyzed by multiple enzymes that are tightly regulated. Some of the key enzymes involved in de novo nucleotide synthesis are: ribonucleotide reductase, dCDP phosphatase, dCMP deaminase, dUTPase, thymidylate synthase, dihydrofolate reductase, serine hydroxymethyltransferase, dTMP kinase, and nucleoside diphosphate kinases. Enzymes catalyzing salvage pathways of synthesis include: phosphoribosyl transferases, nucleoside phosphorylases, nucleoside kinases, nucleoside transglycosylases, deaminases, nucleoside monophosphate kinases, and nucleoside diphosphate kinases (for reviews see: Adams, et al., 1986; Reichard, 1988; Kornberg, 1980). Mutagen damage to any one of these enzymes could bring about a change in the ratios of dNTPs at the replication fork and thus force a polymerase error.

EFFECTS OF MUTAGENS ON DNA REPLICATION PROTEINS

Although the template DNA and the dNTPs are essential building blocks for daughter-strand DNA synthesis, the accuracy of synthesis is ultimately determined by the proteins at the replication fork. Central to this process is the DNA polymerase itself which plays a critical role in "reading" the template DNA and selecting correct dNTPs for incorporation (Loeb and Kunkel, 1982; Goodman, 1988; Kunkel and Bebenek, 1988; Preston et al., 1988c). Thermodynamic measurements indicate that the hydrogen bonding energies of G·C and A·T base pairs are only sufficient to direct DNA synthesis with an accuracy of about 1 error per 100 nucleotides polymerized (Mildvan and Loeb, 1979; Loeb and Kunkel, 1982). Yet DNA synthesis *in vitro* with purified DNA polymerases is highly accurate showing error rates as low as 1 in 10^4 to 1 in 10^5 nucleotides copied in the absence of exonucleolytic proofreading (Preston et al., 1988c; Loeb and Kunkel, 1982; Loeb and Reyland, 1987). Thus, DNA polymerases contribute at least 100-fold to the fidelity of DNA synthesis. In the presence of proofreading $3' \rightarrow 5'$ exonucleases, this accuracy is increased another 10- to 100-fold to achieve an error rate of 10^{-6} to 10^{-7} for certain polymerases (Preston et al., 1988c). Single-strand DNA binding proteins also increase the fidelity of DNA synthesis (Kunkel et al., 1979), and processivity factors such as proliferating cell nuclear antigen (PCNA) are likely to affect the frequency of certain types of base substitution mutations (Bebenek et al., 1989). Other proteins essential to DNA replication include primase, RNase H, ligase, helicase, topoisomerase, gyrase, and initiation factors (Kornberg, 1980; Adams et al., 1986). Since these enzymes are closely associated with the processes of DNA replication and repair, damage to any of these could cause significant aberrations in DNA structure and result in mutations.

A variety of electrophilic mutagens react with and decrease the catalytic activity of DNA polymerases (Busbee et al., 1984; Chan and Becker, 1979; Chan and Becker, 1981; Jimenez-Sanchez, 1976; Miyaki et al., 1983; Norman et al., 1986; Saffhill, 1974; Salazar et al., 1985). However, inactivation of polymerases alone is insufficient to induce mutations and would predominantly result in blocked DNA synthesis. In order for polymerase damage to contribute to mutagenesis, the activity of the enzyme itself must be preserved and the damage must somehow diminish the ability of the enzyme to distinguish correct and incorrect basepairs. Studies on several mutagens suggest that the effects on polymerase fidelity are dependent both on the nature of the mutagen and the source of the polymerase. Thus, the fidelity of *E. coli* pol I is decreased by treatment with γ-rays, MNU, and MNNG but not by MMS or dimethyl sulfate (DMS) (Miyaki et al., 1983; Saffhill, 1974). Although the effects of MNNG are partially due to a differential inhibition of $3' \rightarrow 5'$ exonucleolytic proofreading (see below), γ-rays and MNU decrease fidelity without changing proofreading efficiency. In contrast to pol I, the fidelities of purified mammalian DNA polymerases α and β do not appear to be affected by damage *in vitro* with either MNU or N-acetoxy-AAF (Chan and Becker, 1981). The fidelity of a cytoplasmic form of pol α is, however, dramatically decreased following treatment of rats with AAF *in vivo* (Chan and Becker, 1979). The differential effects of some mutagens on some polymerases presumably reflects the varied primary and secondary structures of different polymerases and the chemical reactivity of polypeptide domains critical to base selection and fidelity. Also, many of the fidelity studies on mutagen-damaged polymerases have been conducted using relatively insensitive assays of fidelity with homopolymeric templates. It is possible that the apparently inactive mutagens will prove to affect fidelity when retested in more sensitive fidelity assays on heteropolymeric biological templates such as ϕX174 or M13 DNA (Loeb and Reyland, 1987).

Metals are an interesting class of mutagens that act, at least in part, by direct interaction with dNTPs at the polymerase active site. Polymerases require divalent metal cations for catalysis, and the natural cation *in vivo* is Mg^{2+} (Kornberg, 1980). Studies on DNA polymerization *in vitro* show that Mn^{2+}, Co^{2+}, Ni^{2+} or Zn^{2+} can substitute for Mg^{2+} as metal activators for DNA polymerases from divergent organisms (Sirover and Loeb, 1976b; Mildvan and Loeb, 1979). Polymerization in the presence of Mn^{2+} or Co^{2+} proceeds with a relatively high error rate when compared to that with Mg^{2+} (Sirover et al., 1979; Kunkel and Loeb, 1979). When a variety of metals were screened for their effects on polymerase fidelity, there was a good correlation between their mutagenicity *in vivo* and their ability to perturb fidelity during DNA synthesis *in vitro* (Sirover and Loeb, 1976a). The mechanism for metal-induced infidelity appears to depend on the type of metal salt as well as its concentration in the reaction mixture. At low concentrations, Mn^{2+} causes errors via its interaction with either the DNA template or the dNTPs, but at high concentrations errors appear to result from binding to single-stranded regions of the DNA or to accessory sites on the polymerase molecule (Beckman et al., 1985). In contrast, the induction of error-prone synthesis by beryllium occurs by weak and completely reversible binding of the metal cation to the DNA polymerase (Sirover and Loeb, 1976c). The mechanism of action of other metal mutagens that affect fidelity (Ag, Be, Cd, Co, Cr, Ni, Pb, and Cu) are not known. However, the fact that some of these can serve as activators of polymerase catalysis suggests that they may perturb base selection by direct interactions at the polymerase active site.

EFFECTS OF MUTAGENS ON DNA REPAIR PROTEINS

Cells have evolved a multitude of repair systems that protect the integrity of genomic sequences (Friedberg, 1985). The major pathways of DNA repair fall into two general categories (Figure 1). The first involves the repair of DNA damage and can act either prior to or after DNA synthesis. The second set of pathways act postreplicationally to repair nucleotide mismatch errors made by DNA polymerases. The importance of these repair systems is clearly demonstrated by genetic studies which show that removal of functional repair proteins results in a mutator phenotype (Friedberg, 1985). It is reasonable, therefore, that a mutagen could similarly affect mutagenesis by damaging one or more of these repair proteins.

Studies on the repair of DNA adducts in cells exposed to either UV light or MMS or both show that MMS inhibits the repair of UV lesions, whereas UV has no effect on alkyl adduct repair (Park et al., 1981). The inhibition of UV-damage repair is also mediated by other alkylating agents (e.g. DMS and MNNG) and can occur by exposure to the alkylating agent either before or after UV irradiation (Cleaver, 1982; Correia and Tyrrell, 1979; Gruenert and Cleaver, 1981). Because UV lesions and alkyl adducts are repaired by different repair pathways (Friedberg, 1985), these data suggest that repair inhibition is due to alkylation of excision repair enzymes (Park et al., 1981). Other DNA damage repair systems might also be inhibited directly or indirectly by mutagenic agents. The repair protein O^6-alkylguanine-DNA alkyltransferase should be an excellent target for chemical inactivation, since this protein catalyzes the irreversible transfer of alkyl groups from promutagenic O^6-alkylguanine-DNA adducts to a specific cysteine -SH group on the transferase molecule (Lindahl et al., 1982; Demple et al., 1985). This irreversible alkylation of cysteine results in "suicide" inactivation of the repair protein (Lindahl et al., 1982). Thus, cells treated with low levels of MNNG or substrate analogs (e.g. O^6-methylguanine as free base) show a significant decrease in alkyltransferase activity with a concomitant increase in sensitivity to subsequent doses of mutagenic alkylating agents (Loechler et al., 1984; Dolan et al., 1985; Yarosh et al., 1986; Karran, 1985). The repair of adducts could also be inhibited by the formation of closely positioned adducts on a DNA molecule, where one adduct prevents the repair of the other. It is well established that mutagen-damaged DNAs are poor substrates for a variety of DNA-specific enzymes including DNase I (Maniatis et al., 1982), DNA polymerases (Larson et al., 1985; Moore et al., 1982), restriction endonucleases and methyltransferases (Wojciechowski and Meehan, 1984). Such an indirect mechanism of repair inhibition would be significant to mutagenesis if an otherwise non-mutagenic adduct (e.g. N^6-methyladenine) blocks the repair of a mutagenic lesion (e.g., an abasic site or O^6-methylguanine). Taken together, these studies suggest that bioreactive agents can indirectly increase the levels of promutagenic DNA adducts by: 1) direct inactivation of DNA repair proteins, and/or; 2) formation of non-mutagenic DNA adducts that block the repair of proximal mutagenic adducts. Recent progress in the purification of DNA damage repair proteins should facilitate the direct testing of these mechanisms.

Although DNA damage is crucial to induced mutagenesis, the greatest number of spontaneous nucleotide sequence changes in a cell are likely to result from errors made by DNA polymerases during normal DNA replication. It is estimated that non-proofreading DNA polymerases make errors about once in every 10^4 nucleotides polymerized (Preston et al., 1988c; Loeb and Reyland, 1987; Kunkel and Bebenek, 1988). Thus, during the replication of a normal human diploid cell, about 600,000 nucleotide mispairs are made as a result of polymerase errors. The accumulation of these errors in the daughter DNA is prevented by two major types of repair systems that recognize and correct nucleotide mismatches: 1) 3' → 5' exonucleolytic proofreading (Kunkel, 1988) and 2) postreplicational mismatch repair (Radman and Wagner, 1986; Brown and Jiricny, 1987; Lahue et al., 1989; Kramer et al., 1989; Bishop et al., 1989). These repair systems decrease the spontaneous mutation frequency of a cell by some 10^5-fold (Cox, 1976; Radman and Wagner, 1986). Thus, any bioreactive mutagen that interferes with these repair pathways would significantly affect mutagenesis.

Studies on purified DNA polymerases containing integral 3' → 5' exonucleolytic proofreading activity show that certain mutagens inhibit 3' → 5' exonuclease activity and decrease the fidelity of DNA synthesis. Reaction of *E. coli* pol I with MNNG results in a differential inhibition of the pol I-associated 3' → 5' exonuclease compared to the polymerase activity (Miyaki et al., 1983). Thus, MNNG changes the exonuclease:polymerase ratio and alters the error rate of the enzyme. Interestingly, the effect on error rate is different for different nucleotide mispairs. The frequencies of misincorporation of dA and dT into template poly (dG·dC) are increased while misincorporation of dG and dC into template poly (dA·dT) is decreased. The fidelity of pol I is not affected by MMS or UV damage (Miyaki et al., 1983), but treatment with the acylating agent *N*-carboxymethylisatoic acid anhydride in the presence of dNTP and Mg^{++} results in a selective inhibition of 3' → 5' exonuclease with little effect on polymerase activity (Klenow and Overgaard-Hansen, 1981). Although the effects of this selective inhibition on fidelity were not measured, one would predict that the error rate of the damaged pol I would be inversely proportional to its 3' → 5' exonuclease activity. A similar inverse relationship of error rate and exonuclease activity has been observed with a variety of proofreading polymerases in which the 3' → 5' exonuclease activity is selectively inhibited either by mutation of the exonuclease domain (Tabor and Richardson, 1989) or by incubation with deoxynucleoside 5'-monophosphates (dNMP) (Loeb and Kunkel, 1982; Kunkel, 1988) or mutagenic dNMP analogs (Byrnes et al., 1977). It should be noted that in order for the fidelity of polymerization to be affected, exonucleolytic proofreading must be preferentially inhibited with minimal inhibition of polymerase activity. This will decrease the 3' → 5' exonuclease:polymerase ratio and permit continued, uncorrected synthesis of DNA following misincorporation of nucleotides by the polymerase. Mutagens inhibiting both activities equally will not affect the fidelity of DNA synthesis by this mechanism.

Perhaps the most dramatic effect of a bioreactive agent on nucleotide mismatch repair is that observed following damage of T7 DNA polymerase (T7 pol) with reactive oxygen species. Reaction of T7 DNA polymerase with molecular oxygen in the presence of Fe^{2+} and a reducing agent results in a very selective oxidation of the exonuclease domain of the polymerase (Tabor and Richardson, 1987). This selective oxidation occurs only at Fe^{2+} concentrations less than or equimolar to that of the polymerase protein and results in inactivation of the proofreading exonuclease activity at a rate >100 times faster than inactivation of polymerase activity. This powerful inhibition of exonucleolytic proofreading appears to be mediated by Fe^{2+}-catalyzed activation of molecular oxygen to generate reactive oxygen species proximal to the metal binding site of the exonuclease polypeptide domain (Tabor and Richardson, 1989). The site-specifically generated reactive oxygen species presumably modify nearby amino acid residues essential to exonuclease activity. This *in situ* generation of oxygen radicals provides a highly efficient mechanism for enzyme inactivation, since damage is targeted to the metal binding site which is essential for enzyme activity and presumably constitutes part of the enzyme's active site (normally occupied by Mg^{++} as a divalent metal co-substrate). However, not all metal-binding sites are susceptible to this mechanism of "internal" oxidation, as evidenced by minimal inactivation of the T7 polymerase domain (a Mg^{++}-dNTP binding site) and the inability of Fe^{2+}/O_2 treatment to affect 3' → 5' exonuclease activity of the homologous polI protein (Tabor and Richardson, 1987).

To determine the potential mutagenic consequences of exonuclease damage by oxygen radicals, we have recently measured the fidelity of oxygen-damaged T7 pol during DNA polymerization *in vitro* (Doshi and Preston, 1990). Incubation of oxygen-damaged T7 pol (Sequenase) with oligonucleotide-primed, single-stranded ϕX174 DNA in the presence of only three dNTPs (Preston et al., 1988a) gave high yields of extended primers (>90%) with presumed mismatches at template positions complementary to the missing dNTPs. In contrast, undamaged T7 pol was highly accurate as indicated by the almost quantitative cessation of primer extension immediately prior to the target template positions. A comparison of the relative yields of mispairs showed that oxygen-damaged T7 pol preferentially forms mispairs opposite template G and A residues with an error rate at least 50-fold greater than that of the undamaged enzyme and comparable to that of a mutant T7 pol completely defective in exonucleolytic proofreading activity.

These data show that oxygen radicals can greatly influence the fidelity of a DNA polymerase and suggest an epigenetic mechanism for oxygen radical mutagenesis based on selective damage to a proofreading repair enzyme. The effects of oxygen radicals on the fidelities of eukaryotic proofreading DNA polymerases (e.g. pol δ and pol γ) are not known. It is likely that selective inhibition of proofreading will be governed by the structural and chemical properties of each enzyme's exonuclease domain and by the accessability of this domain to molecular oxygen and catalytic metals. The diversity of polymerase and exonuclease primary structures (Bernad et al., 1989) suggests that different enzymes will respond differently to different bioreactive agents. The effects of mutagens on endonucleolytic postreplication mismatch repair is largely unexplored, but studies in this area should be greatly facilitated by recent efforts to purify and reconstitute these repair systems *in vitro* (Lahue et al., 1989).

SUMMARY AND PERSPECTIVES

In this review, we have briefly discussed known and potential molecular targets of mutagens. In order for a target to be critical to mutagenesis, damage to this target must somehow lead to a heritable change in the nucleotide sequence of the cellular genome. EfEThus, by definition a mutation can only arise as a consequence of DNA replication, and mutagens can be thought of as agents that perturb the fidelity of DNA polymerization and/or recombination. The fidelity of DNA synthesis is stringently maintained by: 1) the base-pairing properties of the DNA template; 2) the structure and ratios of dNTP pools; 3) the accuracy of DNA polymerases and their accessory proteins, and; 4) the efficiency of DNA repair proteins. The disruption of any of these processes could result in mutation, and all molecules involved in the maintenance of genomic integrity must be considered potential targets of mutagens.

DNA is clearly a critical target for mutagens. However, several mutagens also damage nucleotide pools as well as proteins involved in DNA replication and repair, and in many cases damage to these non-DNA molecules results in significant increases in errors during DNA polymerization *in vitro*. Thus, damage to DNA, nucleotide pools and proteins may all contribute to mutagenesis. The relative contribution of damage to each of these targets will be determined by the chemical reactivity and cellular distribution of the mutagen as well as the extent of damage and the functional competence of the damaged molecules.

The level of damage is particularly significant in mechanisms of mutagenesis involving damage to non-DNA targets. Unlike DNA, which is present nominally at a fixed level of 2 copies per somatic cell, non-DNA targets (e.g. proteins and dNTPs) are usually present in multiple copies per cell, and cells have the capacity to make many more copies when needed. Thus, certain conditions must be met in order for non-DNA molecules to serve as primary targets during mutagenesis. The relative importance of damage to a particular protein target will be determined by the function of the protein in the cell, its relative abundance (i.e., number of copies per cell) and its rate of synthesis and degradation (i.e., turnover). Proteins functioning to maintain genomic integrity are obvious potential targets for mutagens. But the abundance and turnover rates of these proteins in the cell are also critical in determining their importance as mutagen targets. Proteins most at risk are those that: 1) are essential to accurate DNA replication; 2) turnover very slowly, and; 3) are present in limiting concentrations in the cell. Replication proteins such as DNA polymerases, proofreading exonucleases and some nucleotide metabolizing enzymes may be important mutagen targets, since their levels in the cell appear to be rate limiting even during S phase and their turnover is tightly regulated by the cell cycle (Alberts et al., 1989; Kornberg, 1980; Reichard, 1988; Wahl et al., 1988; Wong et al., 1988). Some repair proteins are also present at very low levels in the cell but are highly inducible (e.g., *E. coli* O^6-methylguanine-DNA alkyltransferase; ~10 copies per cell) (Lindahl, 1982). Thus, damage to inducible repair proteins must occur at or very near the time of DNA replication in order for the protein damage to contribute to mutagenesis. Otherwise the induced cell will quickly make new undamaged proteins that will dilute out the damaged molecules and thus maintain a repair-competent phenotype. Clearly, any model for mutagenesis involving damage to proteins must consider the biochemistry, molecular biology and cellular dynamics of the target protein.

Studies during the past 20 years have focussed heavily on DNA adducts as promutagenic lesions. However, the data presented in this review suggest that mutagens might act via multiple pathways involving damage to several molecular targets in the cell. These pathways may contribute singly or concurrently to mutagenesis, and the relative contribution of each pathway is presumably a function of mutagen chemistry as well as cellular physiology. The evidence for multiple molecular targets is based largely on the observed reactivity of mutagens toward purified molecules and the effects of this damage on biochemical function measured *in vitro*. However, the relative importance of damage to different molecular targets needs to be determined *in vivo*. Advances in techniques of molecular biology and genetic engineering suggest several experimental approaches to this central question. Recent studies characterizing the types of nucleotide sequence changes induced by mutagens *in vivo* show that different classes of mutagens induce unique base changes with unique distribution along the sequence of a given gene (Carothers et al., 1989; Miller, 1983; Mazur and Glickman, 1988; Vrieling et al., 1989). Thus, *in vivo* each mutagen (or class of mutagens) induces a distinct spectrum of mutations that reflects the various pathways contributing to mutagenesis by that particular agent. Comparisons of these mutational spectra or "fingerprints" with established pathways of mutagenesis allows one to indirectly deduce the likely molecular event leading to a particular nucleotide sequence change. In this way, it has been possible to identify O^6-alkyl-G (Loechler, et al., 1984) and O^4-alkyl-T (Preston, et al., 1986) adducts as probable promutagenic lesions induced by alkylating agents *in vivo*. Similar comparisons of *in vivo* mutation spectra with the unique and predictable base changes induced by nucleotide pool imbalances (Meuth, 1989), error-prone DNA polymerases (Roberts and Kunkel, 1986) and deficiencies in DNA repair (Schaaper and Dunn, 1987; Schaaper, 1988) should help to identify the contributions of these pathways to mutagenesis by bioreactive agents.

ACKNOWLEDGMENTS

This work was supported by Grant CA48174 from the National Institutes of Health, an American Cancer Society Junior Faculty Research Award (#JFRA-245), and a Henry Rutgers Research Fellowship.

REFERENCES

Abersold, P. M. (1983). Mutation induction by 5-fluorodeoxyuridine in synchronous Chinese hamster cells. *Cancer Res.* **39**, 808-810.

Adams, R. L. P., Knowler, J. T., and Leader, D. P. (1986). *The Biochemistry of the Nucleic Acids.* Chapman and Hall, New York.

Alberts, B., Bray, D., Lewis, J., Raff, M., Roberts, K., and Watson, J. D. (1989). *Molecular Biology of the Cell.* Garland Publishing, Inc., New York.

Arecco, A., Mun, B.-J., and Mathews, C. K. (1988). Deoxyribonucleotide pools as targets for mutagenesis by N-methyl-N-nitrosourea. *Mutation Res.* **200**, 165-175.

Arpaia, E., Ray, P. N., and Siminovitch, L. (1983). Isolations of mutants of CHO cells resistant to 6-(*p*-hydrophenylazo)-uracil, II. Mutants auxotrophic for deoxypyrimidines. *Somat. Cell. Genet.* **9**, 287-297.

Ashman, C. R., and Davidson, R. L. (1981). Bromodeoxyuridine mutagenesis in mammalian cells is related to deoxyribonucleotide pool imbalance. *Mol. Cell. Biol.* **1**, 254-260.

Ayusawa, D., Iwata, K., and Seno, T. (1983). Unusual sensitivity to bleomycin and joint resistance to 9-β-D-arabinofuranosyladenine and 1-β-D-arabinofuranosylcytosine of mouse FM3A cell mutants with altered ribonucleotide reductase and thymidylate synthase. *Cancer Res.* **43**, 814-818.

Baranowska, H., Zaborowska, D., and Zuk, J. (1987). Decreased u.v. mutagenesis in an excision-deficient mutant of yeast. *Mutagenesis* **2**, 1-6.

Basu, A. K., and Essigmann, J. M. (1988). Site-specifically modified oligodeoxynucleotides as probes for the structural and biological effects of DNA damaging agents. *Chemical Research in Toxicology* **1**, 1-18.

Basu, A. K., and Essigmann, J. M. (1990). Site-specifically alkylated oligodeoxynucleotides: Probes for mutagenesis, DNA repair and the structural effects of DNA damage. Mutation Res., in press.

Bebenek, K., Abbotts, J., Roberts, J. D., Wilson, S. H., and Kunkel, T. A. (1989). Specificity and mechanism of error-prone replication by human immunodeficiency virus-1 reverse transcriptase. *J. Biol. Chem.* **264**, 16948-16956.

Beckman, R. A., Mildvan, A. S., and L.A., L. (1985). On the fidelity of DNA replication: Manganese mutagenesis *in vitro*. *Biochemistry* **24**, 5810-5817.

Bernad, A., Blanco, L., L@zaro, J. M., Martin, G., and Salas, M. (1989). A conserved 3' → 5' exonuclease active site in prokaryotic and eukaryotic DNA polymerases. Cell 59,219-228.

Bhanot, O. S., and Ray, A. (1986). The *in vivo* mutagenic frequency and specificity of O^6-methylguanine in ϕX174 replicative form DNA. *Proc. Natl. Acad. Sci. USA* **83**, 7348-7352.

Bishop, D. K., Andersen, J., and Kolodner, R. D. (1989). Specificity of mismatch repair following transformation of *Saccharomyces cerevisiae* with heteroduplex plasmid DNA. *Proc. Natl. Acad. Sci. USA* **86**, 3713-3717.

Brigati, D. J., Myerson, D., Leary, J. J., Spalholz, B., Travis, S. Z., Fong, C. K. Y., Hsiung, G. D., and Ward, D. C. (1983). Detection of viral genomes in cultured cells and paraffin-embedded tissue sections using biotin-labeled hybridization probes. *Virology* **126**, 32-50.

Brown, T. C., and Jiricny, J. (1987). A specific mismatch repair event protects mammalian cells from loss of 5-methylcytosine. *Cell* **50**, 945-950.

Busbee, D. L., Joe, C. O., Norman, J. O., and Rankin, P. W. (1984). Inhibition of DNA synthesis by an electrophilic metabolite of benzo[a]pyrene. *Proc. Natl. Acad. Sci. USA* **81**, 5300-5304.

Byrnes, J. J., Downey, K. M., Que, B. G., Lee, M. Y. W., Black, V. L., and So, A. G.

(1977). Selective inhibition of the 3' to 5' exonuclease activity associated with DNA polymerases: A mechanism of mutagenesis. *Biochemistry* **16**, 3740-3746.
Caras, I. W., MacInnes, M. A., Persing, D. H., Coffino, P., and Martin Jr., D. W. (1982). Mechanism of 2-aminopurine mutagenesis in mouse T-lymphosarcoma cells. *Mol. Cell. Biol.* **2**, 1096-1103.
Carothers, A. M., Steigerwalt, R. W., Urlaub, G., Chasin, L. A., and Grunberger, D. (1989). DNA base changes and RNA levels in *N*-acetoxy-2-acetylaminofluorene-induced dihydrofolate reductase mutants of Chinese hamster ovary cells. *J. Mol. Biol.* **208**, 417-428.
Chambers, R. W., Sledziewska-Gojska, W., Hirani-Hojatti, S., and Borowy-Borowski, H. (1985). uvrA and recA mutations inhibit a site-specific transition produced by a single O6-methylguanine in gene G of bacteriophage ϕX174. *Proc. Natl. Acad. Sci. USA* **82**, 7173-7177.
Chan, J. Y. H., and Becker, F. F. (1979). Decreased fidelity of DNA polymerase activity during N-2-fluorenylacetamide hepatocarcinogenesis. *Proc. Natl. Acad. Sci. USA* **76**, 814-818.
Chan, J. Y. H., and Becker, F. F. (1981). Fidelity of DNA synthesis *in vitro* by carcinogen-reacted DNA polymerases and carcinogen modified templates. *J. Supramol. Struc. Suppl.* **5**, 209.
Cleaver, J. E. (1982). Inactivation of ultraviolet repair in normal and xeroderma pigmentosum cells by methyl methansulfonate. *Cancer Res.* **42**, 860-863.
Correia, I. S., and Tyrrell, R. M. (1979). Lethal interaction between ultraviolet radiations and methyl methane sulfonate in repair proficient and repair deficient strains of Escherichia coliDD. *Photochem. Photobiol.* **29**, 521-526.
Cox, E. C. (1976). Bacterial mutator genes and the control of spontaneous mutation. *Ann. Rev. Genet.* **10**, 135-156.
Das, S. K., Benditt, E. P., and Loeb, L. A. (1983). Rapid changes in deoxynucleoside triphosphate pools in mammalian cells treated with mutagens. *Biochem. Biophys. Res. Comm.* **114**, 458-464.
Das, S. K., and Loeb, L. A. (1984). UV irradiation alters deoxynucleoside triphosphate pools in *Escherichia coli*. *Mutation Res.* **131**, 97-100.
Demple, B., Sedgwick, B., Robins, P., Totty, N., Waterfield, M. D., and Lindahl, T. (1985). Active site and complete sequence of the suicidal methyltransferase that counters alkylation mutagenesis. *Proc. Natl. Acad. Sci. USA* **82**, 2688-2692.
Dolan, M. E., Morimoto, K., and Pegg, A. E. (1985). Reduction of O6-alkylguanine-DNA alkyltransferase activity in HeLa cells treated with O6-alkylguanines. *Cancer Res.* **45**, 6413-6417.
Doshi, R., and Preston, B. D. (1990). Effect of oxidative exonuclease damage on the fidelity of T7 DNA polymerase. *Proc. Amer. Assoc. Cancer Res.* **31**, 100.
Drake, J. W. (1969). Comparative rates of spontaneous mutation. *Nature* **221**, 1132.
Elledge, S. J., and Davis, R. W. (1987). Indentification and isolation of the gene encoding the small subunit of ribonucleotide reductase from *Saccharomyces cerevisiae*: DNA damage-inducible gene required for mitotic viability. *Mol. Cell. Biol.* **7**, 2783-2793.
Fersht, A. R. (1979). Fidelity of replication of phage ϕX174 DNA by DNA polymerase III holoenzyme: spontaneous mutation by misincorporation. *Proc. Natl. Acad. Sci. USA* **76**, 4946-4950.
Friedberg, E. C. (1985). DNA Repair. W.H. Freeman and Company, New York.
Fuchs, R. P. P., Schwartz, N., and Duane, M. P. (1981). Hot spots of frameshift mutations induced by the ultimate carcinogen N-acetoxy-N-2-acetylaminofluorene. *Nature* **294**, 657-659.
Gentil, A., Margot, A., and Sarasin, A. (1984). Apurinic sites cause mutations in simian virus 40. *Mutation Res.* **129**, 141-147.
Gentil, A., Margot, A., and Sarasin, A. (1986). 2-(N-acetoxy-N-acethylamino)fluorene mutagenesis in mammalian cells: Sequence-specific hot spot. *Proc. Natl. Acad. Sci. USA* **83**, 9556-9560.
Goodman, M. F. (1988). DNA replication fidelity: kinetics and thermodynamics. *Mutation Res.* **200**, 11-20.
Gruenert, D. C., and Cleaver, J. E. (1981). Repair of ultraviolet damage in human cells also exposed to agents that cause strand breaks, crosslinks, monoadducts and alkylations. *Chem.-Biol. Interact.* **33**, 163-177.
Haynes, R. H., and Kunz, B. A. (1988). Metaphysics of regulated deoxyribonucleotide biosynthesis. *Mutation Res.* **200**, 5-10.

Hibner, U., and Alberts, B. M. (1980). Fidelity of DNA replication catalysed *in vitro* on a natural DNA template by the T4 bacteriophage multi)enzyme complex. *Nature* **285**, 300-305.

Hoar, D. I., and Dimnik, L. S. (1985). Induction of mitochondrial mutations in human cells by methotrexate. In *Genetic Consequences of Nucleotide Pool Imbalance* (F. J. de Serres, Ed.), pp. 265-282. Plenum Press, New York.

Hochauser, S. J., and Weiss, B. (1978). *Escherichia coli* mutants defective in deoxyuridine triphosphatase. *J. Bacteriol.* **134**, 157-166.

Hopkins, R. L., and Goodman, M. F. (1985). Ribonucleoside and deoxyribonucleoside triphosphate pools during 2-aminopurine mutagenesis in T4 mutator-, wild type-, and antimutator-infected *Escherichia coli*. *J. Biol. Chem.* **260**, 6618-6622.

Hyodo, M., Ito, N., and Suzuki, K. (1984). Deoxynucleoside triphosphate pool of mouse FM3A cell lines unaffected by mutagen treatment. *Biochem. Biophys. Res. Commun.* **122**, 1160-1165.

Jimenez-Sanchez, A. (1976). The effect of nitrosoguanidine upon DNA synthesis *in vitro*. *Molec. Gen. Genet.* **145**, 113-117.

Karran, P. (1985). Possible depletion of a DNA repair enzyme in human lymphoma cells by subversive repair. *Proc. Natl. Acad. Sci.* **82**, 5285-5289.

Kaufman, E. R. (1986). Altered CTP synthetase activity confers resistance to 5-bromodeoxyuridine toxicity and mutagenesis. *Mutation Res.* 161, 19-27.

Kazmers, I. S., Mitchell, B. S., Dadonna, P. E., Wotring, L. L., Townsend, L. B., and Kelley, W. N. (1981). Inhibition of purine nucleoside phosphorylase by 8-aminoguanosine: Selective toxicity for T lymphoblasts. *Science* **214**, 1137-1139.

Klenow, H., and Overgaard-Hansen, K. (1981). Differential effects of N-carboxymethylisatoylation on the DNA polymerase activity, the 5' → 3'-exonuclease activity and the 3'→5'-exonuclease activity of DNA polymerase I of *Escherichia coli*. *Biochim. Biophys. Acta* **654**, 187-193.

Kornberg, A. (1980). *DNA Replication*. W.H. Freeman & Co., San Francisco.

Kraemer, K. H., and Seidman, M. M. (1989). Use of supF, an *Escherichia coli* tyrosine suppressor tRNA gene, as a mutagenic target in shuttle-vector plasmids. *Mutation Res.* **220**, 61-72.

Kramer, B., Kramer, W., Williamson, M. S., and Fogel, S. (1989). Heteroduplex DNA correction in *Saccharomyces cerevisiae* is mismatch specific and requires functional PMS genes. *Molec. Cell. Biol.* **9**, 4432-4440.

Kunkel, T. A. (1985). The mutational specificity of DNA polymerase-β during *in vitro* DNA synthesis. Production of frameshift base substitution, and deletion mutations. *J. Biol. Chem.* **260**, 5787-5796.

Kunkel, T. A. (1988). Exonucleolytic proofreading. *Cell* **53**, 837-840.

Kunkel, T. A., and Bebenek, K. (1988). Recent studies of the fidelity of DNA synthesis. *Biochem. Biophys. Acta* **951**, 1-15.

Kunkel, T. A., and Loeb, L. A. (1979). On the fidelity of DNA replication. Effect of divalent metal ion activators and deoxyribonucleoside triphosphate pools on *in vitro* mutagenesis. *J. Biol. Chem.* **254**, 5718-5725.

Kunkel, T. A., Meyer, R. R., and Loeb, L. A. (1979). Single-strand binding protein enhances fidelity of DNA synthesis *in vitro*. *Proc. Natl. Acad. Sci. USA* **76**, 6331-6335.

Kunkel, T. A., Schaaper, R. M., Beckman, R. A., and Loeb, L. A. (1981). On the fidelity of DNA replication. Effect of the next nucleotide on proofreading. *J. Biol. Chem.* **256**, 9883-9889.

Kunkel, T. A., Silber, J. R., and Loeb, L. A. (1982). The mutagenic effect of deoxynucleotide substrate imbalances during DNA synthesis with mammalian DNA polymerases. *Mutation Res.* **94**, 413-419.

Kunz, B. A. (1982). Genetic effects of deoxyribonucleotide pool imbalances. *Environ. Mutagenesis* **4**, 695-725.

Kunz, B. A. (1988). Mutagenesis and deoxyribonucleotide pool imbalance. *Mutation Res.* **200**, 133-147.

Kunz, B. A., Taylor, G. R., and Haynes, R. H. (1986). Intrachromosomal recombination is induced in yeast by inhibition of thymidylate biosynthesis. *Genetics* **114**, 375-392.

Lahue, R. S., Au, K. G., and Modrich, P. (1989). DNA mismatch correction in a defined system. *Science* **245**, 160-164.

Langer, P. R., Waldrop, A. A., and Ward, D. C. (1981). Enzymatic synthesis of biotin-labeled polynucleotides: Novel nucleic acid affinity probes. *Proc. Natl. Acad. Sci. USA* **78**, 6633-6637.

Larson, K., Sahm, J., Shendar, R., and Strauss, B. (1985). Methylation-induced blocks to *in vitro* DNA replication. *Mutation Res.* **150**, 77-84.

Lasken, R. S., and Goodman, M. F. (1984). The biochemical basis of 5-bromouracil-induced mutagenesis. *J. Biol. Chem.* **259**, 11491-11495.

Lebkowski, J. S., Clancy, S., Miller, J. H., and Calos, M. P. (1985). The lacI shuttle: Rapid analysis of the mutagenic specificity of ultraviolet light in human cells. *Proc. Natl. Acad. Sci. USA* **82**, 8606-8610.

LeClerc, J. E., and Istock, N. L. (1982). Specificity of UV mutagenesis in the lac promoter of M13lac hybrid phage DNA. *Nature* **297**, 596-598.

Lindahl, T. (1982). DNA repair enzymes. *Ann. Rev. Biochem.* **51**, 61-87.

Lindahl, T., Demple, B., and Robins, P. (1982). Suicide inactivation of the *E. coli* O^6-methylguanine-DNA methyl transferase. *EMBO J.* **1**, 1359-1363.

Loeb, L. A., James, E. A., Waltersdorph, A. M., and Klebanoff, S. J. (1988). Mutagenesis by the autoxidation of iron with isolated DNA. *Proc. Natl. Acad. Sci. USA* **85**, 3918-3922.

Loeb, L. A., and Kunkel, T. A. (1982). Fidelity of DNA synthesis. *Ann. Rev. Biochem.* **52**, 429-457.

Loeb, L. A., and Reyland, M. E. (1987). Fidelity of DNA synthesis. In *Nucleic Acids and Molecular Biology* (F. Eckstein and D. M. J. Lilley, Ed.), pp. 157-173. Springer-Verlag, Heidelberg.

Loechler, E. L., Green, C. L., and Essigmann, J. M. (1984). *In vivo* mutagenesis by O^6-methylguanine built into a unique site in a viral genome. *Proc. Natl. Acad. Sci. USA* **81**, 6271-6275.

Loveless, A. (1969). Possible relevance of O-6 alkylation of deoxyguanosine to the mutagenicity and carcinogenicitity of nitrosoamines and nitrosamides. *Nature* **223**, 206-207.

Maniatis, T., Fritsch, E. F., and Sambrook, J. (1982). *Molecular Cloning. A Laboratory Manual.* Cold Spring Harbor Laboratories, Cold Spring Harbor.

Maus, K. L., McIntosh, E. M., and Haynes, R. H. (1984). Defective dCNP deaminase confers a mutator phenotype on *Saccharomyces cerevisiae. Environ. Mutag.* **6**, 415.

Mazur, M., and Glickman, B. W. (1988). Sequence specificity of mutations induced by benzo[a]pyrene-7,8-diol-9,10-epoxide at endogenous aprt gene in CHO cells. *Somatic Cell Molec. Gen.* **14**, 393-400.

McKenna, P. G., and Hickey, I. (1981). UV sensitivity in thymidine kinase deficient mouse erythroleukemia cells. *Cell Biol. Int. Rep.* **5**, 555-561.

McKenna, P. G., and Yasseen, A. A. (1982). Increased sensitivity to cell killing and mutagenesis in thymidine kinase-deficient subclones of a Friend murine leukemia cell line. *Genet. Res.* **40**, 207-212.

McKenna, P. G., Yasseen, A. A., and McKelvey, V. J. (1985). Evidence for direct involvement of thymidine kinase in excision repair processes in mouse cell lines. *Somat. Cell Mol. Genet.* **11**, 239-246.

Meuth, M. (1983). Deoxycytidine kinase-deficient mutants of Chinese hamster ovary cells are hypersensitive to DNA alkylating agents. *Mutation Res.* **110**, 383-391.

Meuth, M. (1989). The molecular basis of mutations induced by deoxyribonucleoside triphosphate pool imbalances in mammalian cells. *Exp. Cell. Res.* **181**, 305-316.

Meuth, M., and Green, H. (1974). Induction of a deoxycytidineless state in cultured mammalian cells by bromodeoxyuridine. *Cell* **2**, 109-112.

Mhaskar, D. N., and Goodman, M. F. (1984). On the molecular basis of transition mutations. Frequency of forming 2-aminopurine-cytosine base mispairs in the G·C → A·T mutational pathway by T4 DNA polymerase *in vitro. J. Biol. Chem.* **259**, 11713-11717.

Mildvan, A. S., and Loeb, L. A. (1979). Role of metal ions in the mechanisms of DNA and RNA polymerases. *Crit. Rev. Biochem.* **6**, 219-244.

Miller, E. C. (1951). Studies on the formation of protein-bound derivatives of 3,4-benzpyrene in the epidermal fraction of mouse skin. *Cancer Res.* **11**, 100-108.

Miller, E. C., and Miller, J. A. (1974). Biochemical mechanisms of chemical carcinogenesis. In *The Molecular Biology of Cancer* (H. Busch, Ed.), pp. 377-402. Academic Press, New York.

Miller, E. C., and Miller, J. A. (1981). Searches for ultimate chemical carcinogens and their reactions with cellular macromolecules. *Cancer* **47**, 2327-2345.

Miller, J. H. (1983). Mutational specificity in bacteria. *Ann. Rev. Genet.* **17**, 215-238.

Miyaki, M., Suziki, K., Aihara, M., and Ono, T. (1983). Misincorporation in DNA synthesis after modification of template or polymerase by MNNG, MMS and UV radiation. *Mutation Res.* **107**, 203-218.

Moore, P. D., Rabkin, S. D., Osborn, A. L., King, C. M., and Strauss, B. S. (1982). Effect of acetylated and deacetylated 2-aminofluorene adducts on *in vitro* DNA synthesis. *Proc. Natl. Acad. Sci. USA* **79**, 7166-7170.

Newman, C. N., and Miller, J. H. (1983a). Kinetics of UV-induced changes in deoxynucleoside triphosphate pools in Chinese hamster ovary cells and their effect on measurements of DNA synthesis. *Biochem. Biophys. Res. Commun.* **116**, 1064-1069.

Newman, C. N., and Miller, J. H. (1983b). Mutagen-induced changes in cellular deoxycytidine triphosphate and thymidine triphosphate in Chinese hamster cells. *Biochem. Biophys. Res. Commun.* **114**, 34-40.

Newman, C. N., and Miller, J. H. (1985). Mechanism of UV-induced deoxynucleoside triphosphate pool imbalance in CHO-Kl cells. *Mutation Res.* **145**, 95-101.

Norman, J. O., Joe, C. O., and Busbee, D. L. (1986). Inhibition of DNA polymerase activity by methyl methanesulfonate. *Mutation Res.* **165**, 71-79.

Önfelt, A., and Jenssen, D. (1982). Enhanced mutagenic response of MNU by post-treatment with methylmercury, caffeine or thymidine in V79 Chinese hamster cells. *Mutation Res.* **106**, 297-303.

Park, S. D., Choi, K. H., Hong, S. W., and Cleaver, J. E. (1981). Inhibition of excision-repair of ultraviolet damage in human cells by exposure to methyl methanesulfonate. *Mutation Res.* **82**, 365-371.

Peterson, A. R., Dananberg, P. V., Ibric, L. L. V., and Peterson, H. (1985). Deoxyribonucleoside-induced selective modulation of cytotoxicity and mutagenesis. In *Genetic Consequences of Nucleotide Pool Imbalance* (F. J. de Serres, Ed.), pp. 313-334. Plenum Press, New York.

Peterson, A. R., Peterson, H., and Danenberg, P. V. (1983). Induction of mutations by 5-fluorodeoxyuridine: a mechanism of self-potentiated drug resistance? *Biochem. Biophys. Res. Commun.* **110**, 573-577.

Phear, G., Nalbantoglu, J., and Meuth, M. (1987). Next-nucleotide effects in mutations driven by DNA precursor pool imbalances at the aprt locus of Chinese hamster ovary cells. *Proc. Natl. Acad. Sci. USA* **84**, 4450-4454.

Prakash, L., Hinkle, D., and Prakash, S. (1978). Decreased UV mutagenesis in cdc8, a DNA replication mutant of *Saccharomyces cerevisiae*. *Mol. Gen. Genet.* **172**, 249-258.

Preston, B. D., and Loeb, L. A. (1988). Enzymatic synthesis of site-specifically modified DNA. *Mutation Res.* **200**, 21-35.

Preston, B. D., Poiesz, B. J., and Loeb, L. A. (1988a). Fidelity of HIV-1 reverse transcriptase. *Science* **242**, 1168-1171.

Preston, B. D., Singer, B., and Loeb, L. A. (1986). Mutagenic potential of O^4-methylthymine *in vivo* determined by an enzymatic approach to site-specific mutagenesis. *Proc. Natl. Acad. Sci. USA* **83**, 8501-8505.

Preston, B. D., Singer, B., and Loeb, L. A. (1987). Comparison of the relative mutagenecities of O-alkylthymines site-specifically incorporated into ϕX174 DNA. *J. Biol. Chem.* **262**, 13821-13827.

Preston, B. D., Wu, D., Reid, T. M., King, C. M., and Loeb, L. A. (1988b). Site-specific incorporation of 2-aminofluorene (AF)- and N-acetyl-2-aminofluorene (AAF)-deoxyguanosine triphosphate adducts by DNA polymerases. *J. Cell. Biochem.* **12A**, 348.

Preston, B. D., Zakour, R. A., Singer, B., and Loeb, L. A. (1988c). Fidelity of base selection by DNA polymerases: Studies on site-specific incorporation of base analogues. In *DNA Replication and Mutagenesis* (R. E. Moses, and W. C. Summers, Eds.), pp. 196-207. American Society of Microbiology, Washington, D.C.

Protic-Sabljic, M., Tuteja, N., Munson, P. J., Hauser, J., Kraimer, K. H., and Dixon, K.

(1986). UV light-induced cyclobutane pyrimidine dimers are mutagenic in mammalian cells. *Molec. Cell. Biol.* **6**, 3349-3356.
Radman, M., and Wagner, R. (1986). Mismatch repair in *Escherichia coli*. *Ann. Rev. Genet.* **20**, 523-538.
Randazzo, R., Di Leonardo, A., Bonatti, S., and Abbondandolo, A. (1987). Modulation of induced reversion frequency by nucleotide pool imbalance as a tool for mutant characterization. *Environ. Mol. Mutagen.* **10**, 17-26.
Reichard, P. (1988). Interactions between deoxyribonucleotide and DNA synthesis. *Ann. Rev. Biochem.* **57**, 349-374.
Roberts, J. D., and Kunkel, T. A. (1986). Mutational specificity of animal cell DNA polymerases. *Environ. Mutagen.* **8**, 769-789.
Roguska, M. A., and Gudas, L. J. (1984). Mutator phenotype in a mutant of S49 mouse T-lymphoma cells with abnormal sensitivity to thymidine. *J. Biol. Chem.* **259**, 3782-3790.
Saffhill, R. (1974). The effect of ionising radiation and chemical methylation upon the activity and accuracy of *E. coli* DNA polymerase I. *Biochem. Biophys. Res. Commun.* **61**, 802-808.
Salazar, I., Tarrago-Litvak, L., Litvak, S., and Gil, L. (1985). Effect of benzo(a)pyrene on DNA synthesis and DNA polymerase activity of rat liver nuclei. *Biochem. Pharmacol.* **34**, 755-762.
Sargent, R. G., and Mathews, C. K. (1987). Imbalanced deoxyribonucleoside triphosphate pools and spontaneous mutation rates determined during dCMP deaminase-defective bacteriophage T4 infections. *J. Biol. Chem.* **262**, 5546-5553.
Schaaper, R. M. (1988). Mechanisms of mutagenesis in the *Escherichia coli* mutator *mutD5*: Role of DNA mismatch repair. *Proc. Natl. Acad. Sci. USA* **85**, 8126-8130.
Schaaper, R. M., and Dunn, R. L. (1987). Spectra of spontaneous mutations in *Escheichia coli* strains defective in mismatch correction: The nature of *in vivo* DNA replication errors. *Proc. Natl. Acad. Sci. USA* **84**, 6220-6224.
Sedwick, W. D., Brown, O. E., and Glickman, B. W. (1986). Deoxyuridine misincorporation causes site-specific mutational lesions in the *lacI* gene of *Escherichia coli*. *Mutation Res.* **162**, 7-20.
Seidman, M. M., Dixon, D., Razzaque, A., Zagursky, R. J., and Berman, M. L. (1985). A shuttle vector plasmid for studying carcinogen-induced point mutations in mammalian cells. *Gene* **38**, 233-237.
Shewach, D. S., Chern, J.-W., Pillote, K. E., Townsend, L. B., and Dadonna, P. E. (1986). Potentiation of 2'-deoxyguanosine cytotoxicity by a novel inhibitor of purine nucleoside phosphorylase, 8-amino-9-benzylguanine. *Cancer Res.* **46**, 519-523.
Singer, B., and Grunberger, D. (1983). *Molecular Biology of Mutagens and Carcinogens*. Plenum Press, New York.
Singer, B., Spengler, S. J., Chavez, F., Sagi, J., Ku'smierek, J. T., Preston, B. D., and Loeb, L. A. (1987). O-Alkyl deoxythymidines are recognized by DNA polymerase I as deoxythymidine or deoxycytidine. In *N-NItroso Compounds: Occurrence, Biological Effects and Relevance to Human Cancer* (J. K. O'Neill et al., Eds.), pp. 37-41. Oxford University Press, Oxford.
Sirover, M. A., Dube, D. K., and Loeb, L. A. (1979). On the fidelity of DNA replication. Metal activation of *Escherichia coli* DNA polymerase I. *J. Biol. Chem.* **254**, 107-111.
Sirover, M. A., and Loeb, L. A. (1976a). Infidelity of DNA synthesis *in vitro*: screening for potential metal mutagens or carcinogens. *Science* **194**, 1434-1436.
Sirover, M. A., and Loeb, L. A. (1976b). Metal activation of DNA synthesis. *Biochem. Biophys. Res. Commun.* **70**, 812-817.
Sirover, M. A., and Loeb, L. A. (1976c). Metal-induced infidelity during DNA synthesis. *Proc. Natl. Acad. Sci. USA* **73**, 2331-2335.
Snow, E. T., and Mitra, S. (1987). Do carcinogen-modified deoxynucleotide precursors contribute to cellular mutagenesis? *Cancer Investigation* **5**, 119-125.
Snow, E. T., and Mitra, S. (1988). Role of carcinogen-modified deoxynucleotide precursors in mutagenesis. *Mutation Res.* **200**, 157-164.
Suzuki, K., Miyaki, M., Ono, T., Mori, H., Moriya, H., and Kato, T. (1983). UV-induced

imbalance of the deoxynucleoside triphosphate pool in *E. coli*. *Mutation Res.* **122**, 293-298.
Tabor, S., and Richardson, C. C. (1987). Selective oxidation of the exonuclease domain of bacteriophage T7 DNA polymerase. *J. Biol. Chem.* **262**, 15330-15333.
Tabor, S., and Richardson, C. C. (1989). Selective inactivation of the exonuclease activity of bacteriophage T7 DNA polymerase by *in vitro* mutagenesis. *J. Biol. Chem.* **264**, 6447-6458.
Topal, M. D., and Baker, M. S. (1982). DNA precursor pool: a significant target for N-methyl-N-nitrosourea in C3H/10T1/2 clone 8 cells. *Proc. Natl. Acad. Sci. USA* **79**, 2211-2215.
Trudel, M., Van Genechten, T., and Meuth, M. (1984). Biochemical characterization of the hamster thy mutator gene and its revertants. *J. Biol. Chem.* **259**, 2355-2359.
Vrieling, H., Van Rooijen, M. L., Groen, N. A., Zdzienicka, M. Z., Simons, J. W. I. M., Lohman, P. H. M., and van Zeeland, A. A. (1989). DNA strand specificity for UV-induced mutations in mammalian cells. *Mol. Cell. Biol.* **9**, 1277-1283.
Wabl, M., Burrows, P. D., von Gabain, A., and Steinberg, C. (1985). Hypermutation at the immunoglobulin heavy chain locus in a pre-B-cell line. *Proc. Natl. Acad. Sci. USA* **82**, 479-482.
Wahl, A. F., Geis, A. M., Spain, B. H., Wong, S. W., Korn, D., and Wang, T. S.-F. (1988). Gene expression of human DNA polymerase α during cell proliferation and the cell cycle. *Mol. Cell. Biol.* **8**, 5016-5025.
Walker, G. C. (1984). Mutagenesis and inducible responses to deoxyribonucleic acid damage in *Escherichia coli*. *Microbiol. Rev.* **48**, 60-93.
Weinberg, G., Ullman, B., and Martin Jr., D. W. (1981). Mutator phenotypes in mammalian cell mutants with distinct biochemical defects and abnormal deoxyribonucleoside triphosphate pools. *Proc. Natl. Acad. Sci. USA* **78**, 2447-2451.
Weinberg, G. L., Ullman, B., Wright, C. M., and Martin Jr., D. W. (1985). The effects of exogenous thymidine on endogenous deoxynucleotides and mutagenesis in mammalian cells. *Somat. Cell. Mol. Genet.* **11**, 413-419.
Williams, W. E., and Drake, J. W. (1977). Mutator mutations in bacteriophage T4 gene 42 (dHMC hydroxymethylase). *Genetics* **86**, 501-511.
Witkin, E. M. (1976). Ultraviolet mutagenesis and inducible DNA repair in *Escherichia coli*. *Bacteriol. Rev.* **40**, 869-907.
Wojciechowski, M. F., and Meehan, T. (1984). Inhibition of DNA methyltransferases *in vitro* by benzo(a)pyrene diol epoxide-modified substrates. *J. Biol. Chem.* **259**, 9711-9716.
Wong, S. W., Wahl, A. F., Yuan, P.-M., Arai, N., Pearson, B. E., Arai, K.-I., Korn, D., Hunkapiller, M. W., and Wang, T. S.-F. (1988). Human DNA polymerase α gene expression is cell proliferation dependent and its primary structure is similar to both prokaryotic and eukaryotic replicative DNA polymerases. *EMBO J.* **7**, 37-47.
Wurtz, E. A., Sears, B. B., Rabert, D. K., Shepard, H. S., Gillham, N. W., and Boynton, J. E. (1979). A specific increase in chloroplast gene mutations following growth of Chlamydomonas in 5-flurodeoxyuridine. *Mol. Gen. Genet.* **170**, 235-242.
Yang, J.-L., Maher, V. M., and McCormick, J. J. (1987). Kinds of mutations formed when a shuttle vector containing adducts of (±)-7β,8α-dihydroxy-9α,10α-epoxy-7,8,9,10-tetrahydrobenz[a]pyrene replicates in human cells. *Proc. natl. Acad. Sci. USA* **84**, 3787-3791.
Yang, J. L., Maher, V. M., and McClormick, J. J. (1988). Kinds and spectrum of mutations induced by 1-nitrosopyrene adducts during plasmid replication in human cells. *Molec. Cell. Biol.* **8**, 3364-3372.
Yarosh, D. B., Hurst-Calderone, S., Babich, M. A., and Day, R. S. I. (1986). Inactivation of O^6-methylguanine-DNA methyltransferase and sensitization of human tumor cells to killing by chloroethylnitrosourea by O^6-methylguanine as a free base. *Cancer Res.* **46**, 1663-1668.
Zakour, R. A., and Loeb, L. A. (1982). Site-specific mutagenesis by error-directed DNA synthesis. *Nature* **295**, 708-710.

MUTAGENIC CONSEQUENCES OF THE ALTERATION OF DNA BY CHEMICALS AND RADIATION

Bernard Strauss, Edith Turkington, Jhy Wang, and Daphna Sagher

Department of Molecular Genetics and Cell Biology
The University of Chicago
Chicago, IL 60637, USA.

INTRODUCTION

The induction of mutations *in vivo* is a process that involves the interaction of exogenous agents, the Biologically Reactive Intermediates (BRIs), with particular nucleotides in the DNA. The finding of "hot spots" (see Hsia et al., 1989) and of "mutator" strains of organisms (Modrich, 1987) indicates that superimposed on the primary interaction of BRIs and nucleotides is an effect of DNA sequence (Burns et al., 1987) and of the proteins involved in replication and in the monitoring of the DNA. Many of the primary interactions of BRIs with DNA result in alterations which block DNA synthesis, at least *in vitro*. It appears to be a truism that mutation, at least point mutation as a result of damage induced by an agent which inhibits DNA synthesis, requires that the DNA synthetic machinery should bypass the damage by some mechanism. An understanding of the phenomena of mutation therefore requires knowledge of the relationships between the altered DNA bases, the arrest of DNA synthesis, and the location of the damage within the DNA sequence. For example, one might assume that the sites most subject to modification by BRIs are those at which mutation occurs most readily. In fact, it appears that this simplest of hypotheses is not inevitably so, and that other factors may intervene (Brash et al., 1987).

In order to study the mechanistic details of mutation we employ a model system in which single stranded bacteriophage DNA containing a bacterial reporter gene (the a complementing peptide of β galactosidase) is replicated *in vitro* (Sahm et al., 1989). This system permits us to replicate a DNA sequence previously treated with different mutagens by different polymerases and enzyme systems. The reporter gene is subject to mutation at a number of places: at least 173 different changes can be detected (Kunkel & Alexander, 1986) permitting the role of neighboring sequences in mutation to be observed. It is possible to study the termination of DNA synthesis and mutation on the same template. Our experiments have led us to ask whether there is a relationship between the sites of spontaneous termination of DNA synthesis and of *induced* mutation. In this paper we would like to discuss: a) the modification of our system for the study of mutation induced by base damage caused by ionizing radiation; b) the alteration of a DNA lesion *in situ* which permits us to define the reaction sites; and c) some preliminary data which lead us to suspect that there is a relation of the sites of induced termination of DNA synthesis, and probably of mutation, to the spontaneous sites of termination.

In our previous studies (Sahm et al., 1989) we used single stranded DNA of bacteriophage M13mp2 grown on an *ung* $^-$ *dut* $^-$ bacterial host so that about 5 percent of the T's were substituted by U. Such DNA will not survive in an *ung* $^+$ *dut* $^+$ host.

Biological Reactive Intermediates IV, Edited by C.M. Witmer *et al.*
Plenum Press, New York, 1990

This single stranded DNA was reacted with mutagen and then hybridized to a non-uracil DNA fragment prepared from RF which had been restricted and purified to remove the complementing a peptide insert. The product of this hybridization was a double stranded circular molecule with a single stranded U-containing gap. This molecule transfects *ung* + *dut* + strains with low efficiency. DNA synthesis across the gap increases transfection efficiency in *ung* + *dut* + strains by producing a complete (-) strand without uracil. The presence of DNA synthesis-arresting lesions in the gap inhibits this increase in infectivity and selects for molecules in which synthesis across the lesion (bypass) has occurred. If such bypass is mutagenic, it will be observed as an increased frequency of mutants, recognized by the light blue or colorless nature of the plaques on suitable indicator medium. This system has previously been used to study the mutations induced by aminofluorene adducts in DNA using an altered T7 DNA polymerase, Sequenase, for the *in vitro* synthesis (Sahm et al., 1989).

RESULTS

The system used previously for the study of mutation *in vitro* is also applicable to the study of base damage induced by ionizing radiation because the protocol filters out damage induced by breaks in the DNA chain within the reporter gene sequence (Figure 1). Breaks produced outside of the reported gene sequence will not inhibit circularization and formation of a gapped template on hybridization with RF from which the reporter sequence has been removed by restriction enzyme treatment and purification. However, because of the absence of a complementary DNA chain in this region, breaks within the reporter sequence lead only to linear molecules on hybridization and these transfect poorly, even on synthesis (Figure 1).

This theoretical argument has been validated by the following experiments. As is well established, single stranded circular DNA is more susceptible to inactivation by ionizing radiation than is the double stranded replicative form, RF , because of the preponderance of single over double strand breaks and because of the possibility of DNA excision repair in double stranded molecules (van Touw et al., 1985). Hybridization with a DNA fragment, not containing the a complementing region, increases survival (Figure 2). Transfection of the fragment by itself (not shown in the figure) results in a very low transfection efficiency, about 0.22 pfu/ng DNA as compared with 860 pfu/ng of single stranded DNA or 2270 pfu/ng of RF DNA for unirradiated DNA (Figure 2). Mixing fragments with irradiated DNA increased survival and an additional increase in infectivity was obtained after annealing at 60° for 30 min. We interpret these increases in transfection efficiency as due to the assembly of circular molecules from the linear molecules formed as a result of radiation induced breaks.

These results were obtained with DNA that contained no uracil. When a similar experiment was carried out with uracil-containing DNA assayed in *ung* + *dut* + strain, the transfection efficiency remained low (Table 1). Use of the hybridized, gapped molecule as a substrate for DNA synthesis in a reaction mixture containing the normal dNTPs and catalyzed by the altered T7 DNA polymerase, Sequenase 2.0, resulted in a tenfold increase in transfection activity for unirradiated DNA but only a 3-5 fold increase for irradiated DNA as would be expected if the irradiated DNA contained base damage blocking DNA synthesis (Table 1). In these experiments the reaction mixture was transfected into a *rec* ⁻ recipient to decrease the probability of repair or bypass reactions in the host cell.

Plating of M13 mp2 on indicator plates yields dark blue plaques. Mutants can be identified as producing light blue or colorless plaques. Our experience and that of others is that the light blue plaques are almost exclusively due to base substitution whereas the colorless plaques include both base substitution and deletion mutations. Synthesis with Sequenase 2.0 on irradiated templates increased the number and frequency of mutations isolated (Table 2). The large increase in the frequency of light blues indicates that base substituiton mutations were induced. We are now sequencing these mutants to see whether the distribution is different from those induced on non-irradiated templates.

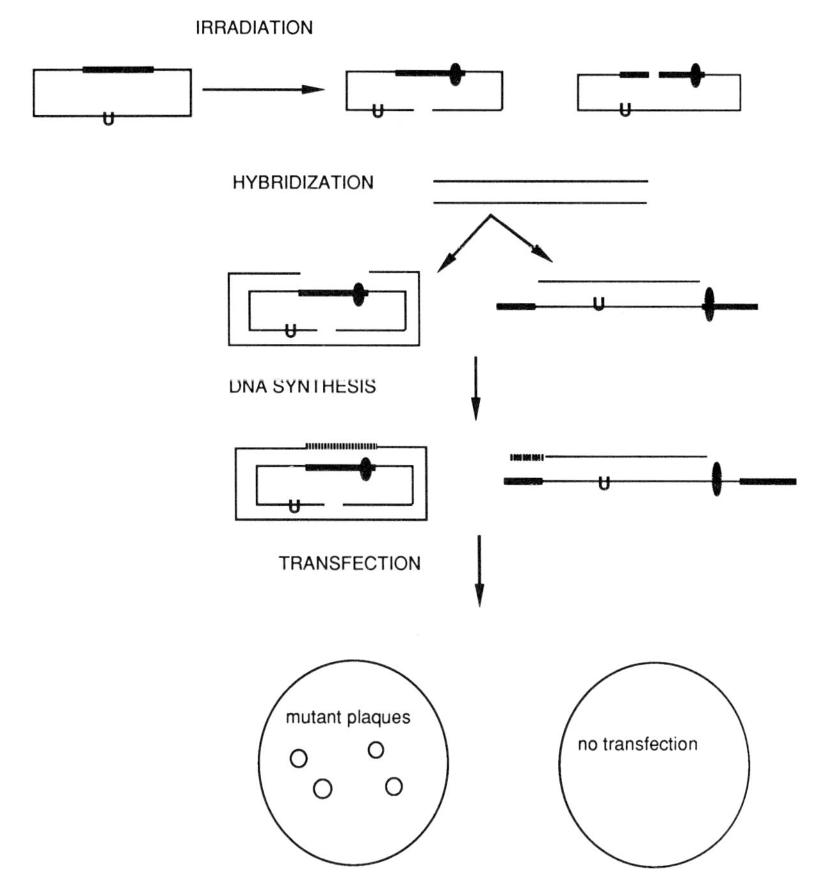

Figure 1. Scheme for the detection of ionizing radiation-induced mutations resulting from base damage in DNA.

Two of the most studied modifications are DNA with either acetyl aminofluorene or aminofluorene adducts. The presence of the acetyl group on an aminofluorene residue bound to the C-8 position of guanine makes a major difference in the structure, since the acetyl aminofluorene (AAF) derivative occurs mostly in the *syn* configuration whereas the aminofluorene (AF) derivative remains *anti* (Norman et al., 1989). AAF adducts in DNA are absolute blocks to *in vitro* DNA synthesis (Moore et al., 1980) whereas the AF residues are less lethal (Lutgerink et al., 1985) and are more likely to be bypassed *in vitro* (Sahm et al., 1989). M13mp2 DNA reacted with N-hydroxy-2-aminofluorene gave fewer termination bands with both *E. coli* DNA polymerase I (large fragment) [pol I (Kf)] and with Sequenase than did DNA reacted with acetoxy acetyl aminofluorene (Sahm et al., 1989). A difficulty with the experiments was that it was not possible to exclude the possibility that there was differential reactivity at the relatively low levels used (6 adducts per M13mp2 molecule), so that the termination sites and the mutations observed might have reflected the reactivity of special sites in DNA. We therefore used alkali to convert the acetylaminofluorene derivative of guanine to aminofluorene by alkaline hydrolysis *in situ* (Figure 3, Kriek et al, 1967; Norman et al., 1989). Reaction at a site is indicated by the demonstration that termination bands occur after reaction with acetoxy acetylaminofluorene (AAAF). If after *in situ* conversion of AAF-DNA to AF-

DNA the bands disappear bypass of a reacted site has presumably occurred. If two sites give equal termination bands with AAF-DNA but differential termination after *in situ* conversion to AF-DNA, we can conclude that an AF adduct present at the one site is more likely to produce termination of DNA synthesis than at the other.

We reacted M13mp2 DNA with [^{3}H] labeled N-acetoxy-2-acetylaminofluorene to give approximately 11 adducts per molecule. The transfection efficiency of this DNA was reduced from the control value of 721 pfu/ng DNA to 0.11 pfu/ng. We reacted this DNA with 1.0 M NaOH in the presence of §-mercaptoethanol (Norman et al., 1989) and monitored the conversion of G-8 AAF-DNA to AF-DNA by both HPLC (Figures 4,5) and by transfection efficiency (Figure 5). Maximum transfection efficiency was reached after 45 minutes incubation at room temperature with 1M NaOH. The approximately 900 fold increase in transfection efficiency of the AAF-containing DNA was accompanied by a decrease in the transfection efficiency of control DNA (no AAF residues) on treatment with alkali from 721 to 338 pfu/ng DNA. Based on the Poisson distribution, approximately 75 percent of the control DNA has suffered one or more lethal hits, as a result of a treatment which has converted 87 percent of 11 AAF lesions to AF.

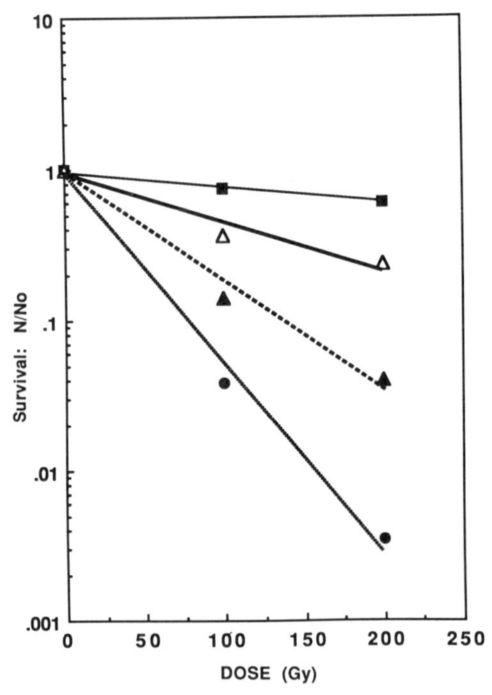

Figure 2. M13mp2 single stranded DNA in HE buffer was irradiated with a Co-60 g source. The DNA was then hybridized with the large PvuI/AvaII fragment of M13mp2 RF DNA as previously described. Closed squares, RF DNA; closed circles; single strand M13mp2; closed triangles, single strand M13mp2 mixed with double stranded fragment but not annealed; open triangles, single strand M13mp2 hybridized with double stranded fragment by heating.

Table 1. Effect of synthesis with an altered T7 DNA polymerase (Sequenase 2.0) on the transfection efficiency of irradiated, hybridized, uracil-containing M13mp2 DNA.

	Radiation Dose - ^{60}Co γ rays(Gy)		
	0	300	600
DNA			
ss DNA - no uracil	97.		
RF DNA - no uracil	114.		
ss DNA - containing uracil*	0.01		
uracil ssDNA- hybridized	1.5	0.42	0.36
above after DNA synthesis with "Sequenase" II	15.2	1.40	1.87

recorded plaque forming units / ng DNA
* transfection efficiency determined in a separate experiment

Table 2. Mutations produced by synthesis with an altered T7 DNA polymerase. (Sequenase 2.0) on a template irradiated with ^{60}Co γ rays. "Blue" mutants give light blue plaques.

DNA	Synthesis	Total Plaques	Blue	Colorless	Total%	Blue%
no irradiation	0	2928	1	7	0.27	0.03
	+	5432	13	16	0.53	0.24
400 Gy	0	1865	0	6	0.32	0
	+	4522	36	24	1.3	0.80
800 Gy	0	1826	1	17	0.99	0.05
	+	5628	36	36	1.28	0.64

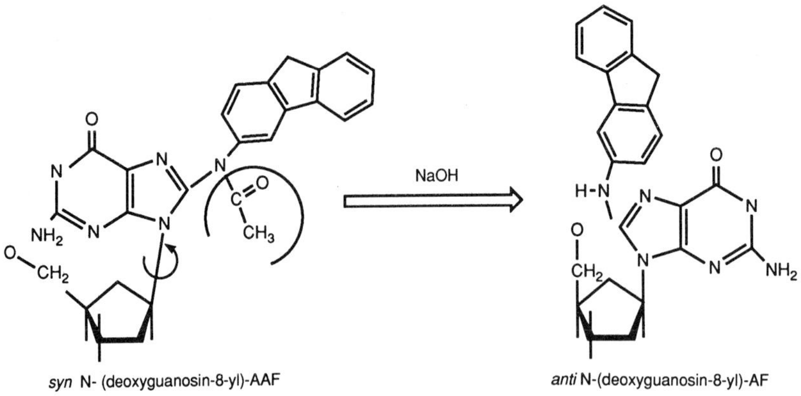

NaOH

syn N- (deoxyguanosin-8-yl)-AAF

anti N-(deoxyguanosin-8-yl)-AF

Figure 3. Conversion of C-8 guanyl acetyl aminofluorene in DNA to C-8 guanyl aminofluorene by alkali.

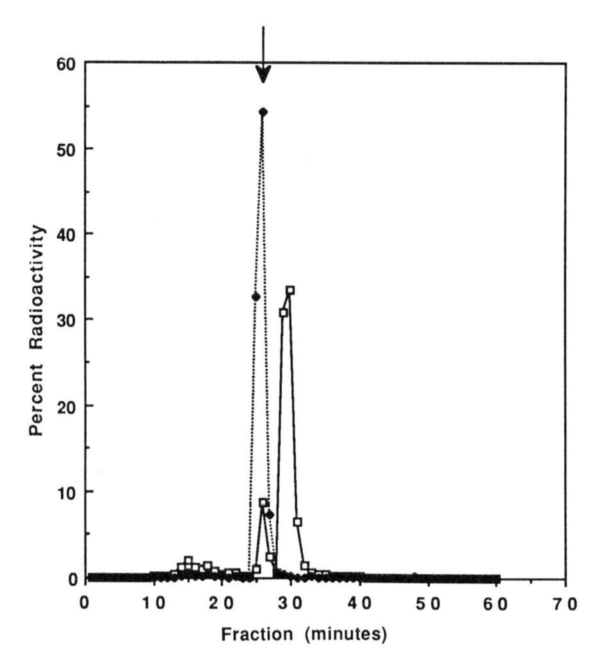

Figure 4. Analysis of DNA reacted 60 min at 37°C with 10 μM N-acetoxy-2-acetylamino [ring-G-^{3}H] fluorene (9.3 Ci/mmol). M13mp2 DNA was reacted to give about 11 residues per molecule. A portion was precipitated, dried, taken up in water and incubated at approx. 2 mg/ml with 1M NaOH + 0.04 M β-mercaptoethanol (Norman et al., 1989) for 45 min. at room temperature. The alkali treated DNA had a radioactivity of 106.2 dpm/ng DNA as compared to 94.3 dpm/ng for the unhydrolyzed control. Both DNAs were then hydrolyzed with trifluoroacetic acid and analyzed by the method of Tang and Lieberman (1983). The arrow indicates the position of a sample of authentic N-(guanin-8-yl)-2-acetylaminofluorene cochromatographed with our sample. We thank Dr. J. Miller for providing this material.

The inactivation data indicate the acetoxy acetylaminofluorene to have produced 8.8 lethal hits in the phage. After alkali-treatment, there remain 1.23 lethal hits in the DNA (using alkali-treated DNA as the control). This residual lethality is probably due to both the residual AAF-DNA, and to the AF-DNA lesions. The HPLC data indicate there to be an average of 1.4 AAF residues per molecule remaining in the DNA. Lutgerink et al. (1985) report that 7 AF-DNA lesions are required to produce a single lethal hit in single stranded phage ∅X174 DNA. The data do show that one can convert AAF-DNA to AF-DNA *in situ* with an accompanying large increase in biological activity.

We used the AAF-DNA and its alkali-treated product as substrates in DNA synthesis-termination experiments. Our protocol for these experiments involves synthesis with 150 μM each of dCTP, dGTP, and dTTP and 0.5 μM deoxyadenosine 5' - α- [^{35}S] thiotriphosphate for 15 minutes followed by an additional 15 min chase with 150 μM of dATP. It is possible that the use of the S-labeled dATP in low concentration has led to some enzyme specific artefact, and these experiments are now being repeated and quantitated using [^{32}P] end-labeled primer. Notwithstanding this possibility, the experiments reveal interesting differences between the behavior of DNA polymerases on our templates.

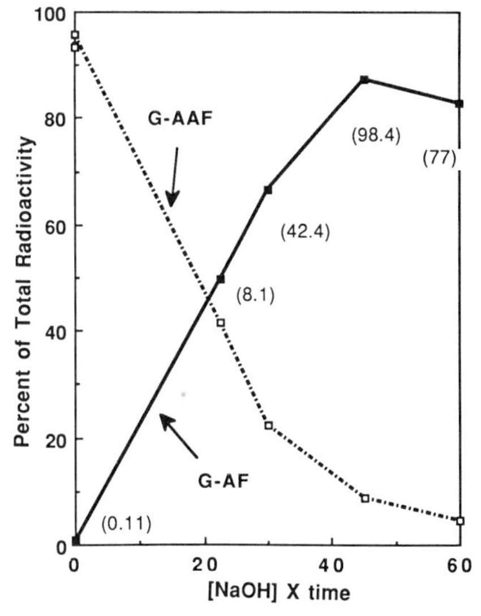

Figure 5. Effect of time of alkali treatment on the conversion of AAF- to AF-DNA. DNA was reacted with [^{3}H] N-acetoxy-2-acetylaminofluorene as described above and was then precipitated, dried, taken up in water and incubated with 0.5 M NaOH or 1M NaOH plus 0.04M β-mercaptoethanol for increasing times. At the conclusion of incubation the NaOH was neutralized with 1M ammonium acetate, pH 4.6, ethanol precipitated, dried and taken in water. A portion was analyzed as described in the legend to Fig. 4 and a second portion was analyzed for transfection efficiency. The values in parenthesis give the transfection efficiency in pfu/ng DNA. [NaOH] X time is the product of the concentration of NaOH (M) multiplied by the time of incubation in minutes.

The enzymes we have used (*E. coli* polI (Kf); Sequenase 2.0, AMV reverse transcriptase, HIV reverse transcriptase) have different and unique patterns of spontaneous termination on control DNA under our experimental conditions (Figure 6-9). With Sequenase 2.0 there does not seem to be a difference in the pattern of termination on alkali-treated control DNA as compared to untreated DNA (Figure 7). The termination pattern observed with *E. coli* pol I (Kf) (Figure 6) is in accord with an earlier study from this laboratory which showed that this polymerase stopped prior to AAF-DNA lesions but opposite AF-DNA lesions (Moore et al., 1982). [We have evidence based on sequence reactions using AAF-DNA as a template (Figure 9) that DNA in our current termination reactions is retarded by about one nucleotide in its migration as compared to the sequencing standards prepared using Sequenase 2.0 in a kit supplied by U.S. Biochemical Corporation reaction.]. There are also AF-DNA sites at which *E. coli* pol I (Kf) does not appear to terminate (e.g. position 126, Figure 6).

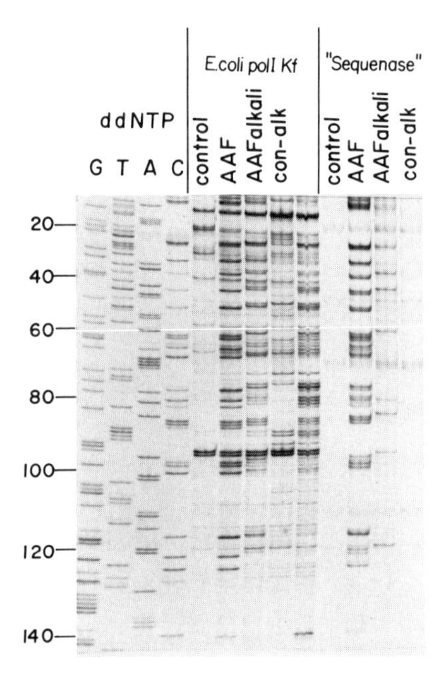

Figure 6. Pause sites for DNA synthesis catalyzed by *E. coli* pol I (Kf) (3 units)and Sequenase 2.0 (2 units) on AAF-DNA templates (11 residues per molecule). Synthesis starts at position 183 (Kunkel and Alexander, 1986). Reaction conditions were as previously described (Sahm et al., 1989) except that we used final concentrations of dGTP, dCTP and dTTP of 150µM. 35S-labeled dATP was added at 0.8 µM and then after 15 minutes dATP was added to a final concnetration of 150µM and incubation was continued for an additional 15 min. *E. coli* pol I (Kf) synthesis was at 10°C, Sequenase 2.0 was at 37°. Approximately 7000 cpm of pol I product and 50,000 cpm of Sequenase product was loaded in each lane. AAF-alkali and con-alk samples were treated with 1.0 M NaOH for 45 minutes as described in the legend to Fig. 5.

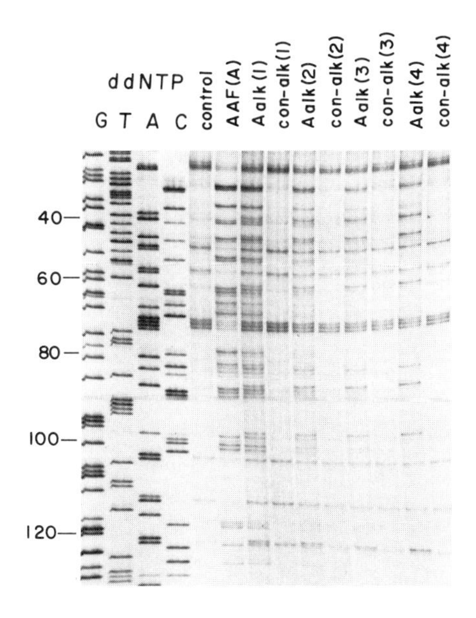

Figure 7. Effect of alkali on the pattern of pause sites for DNA synthesis catalyzed by Sequenase 2.0. Reactions with Sequenase 2.0 (2 units) were performed as indicated in the legend to Fig. 6. Approximately 28,000 cpm of product was loaded in each lane. AAF, AAF-DNA, 11 adducts per molecule; alk(1), 0.5 M NaOH, 45 min at r.t.; alk(2) 1.0M NaOH, 30 min at r.t.; alk (3), 1.0 M NaOH, 45 min at r.t.; alk (4) 1.0M NaOH, 60 min at r.t.

Sequenase 2.0 appears to bypass AF residues more readily than does *E. coli* pol I (Kf) as has been noted previously. Comparison of the bands produced with AAF-DNA substrates after treatment with increasing amounts of alkali demonstrates an interesting relationship between induced termination, spontaneous termination and the position of T's in the template. First, it can be seen (Figures 7, 9) that spontaneous Sequenase 2.0 termination bands occur before runs of T's in the template. Although termination may be related to the particular conditions under which the reaction is run (see above), this pattern of termination or pause is seen only with Sequenase 2.0. Many (but not all) of the residual terminations by Sequenase 2.0 on AF-DNA occur just before T's as the next base in the reacted template (e.g. positions 123, 101, 88, 62). In the case of AMV reverse transcriptase it also appears that strong termination bands on AF-DNA templates occur at a number of sites at positions which show weak spontaneous termination sites (Figure 7). We carry out our experiments by loading equal amounts of radioactivity to all of the lanes which means that for the non-reacted templates, most of the radioactivity is found at higher molecular weight products at the top of the gel.

DISCUSSION

As has been pointed out before (Sahm et al., 1989), the experiments show that the response to an altered site in the DNA is polymerase specific. In the case of the G-8 acetyl aminofluorene derivative, the lesion is apparently so deforming that it acts as a block for all polymerases and the differences in response are matters of detail. The AF-DNA lesion, at the same sites, is more permissive and the different properties of the polymerases become more important.

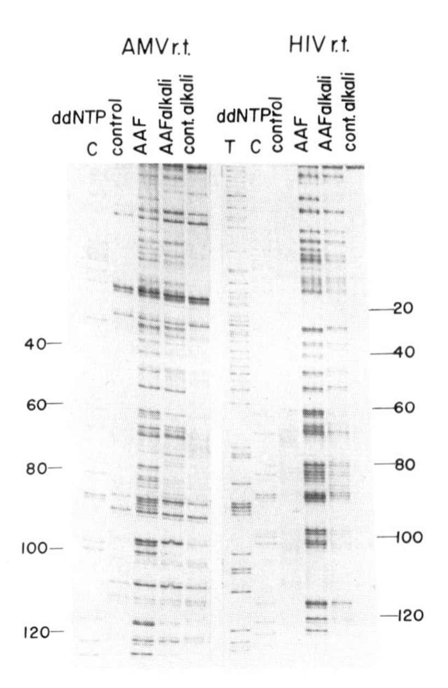

Figure 8. Pause sites for AMV reverse transcriptase and HIV reverse transcriptase catalyzed DNA synthesis on an AAF-DNA template treated with alkali. Conditions were as described in the legend to Fig. 6 above. Twenty units of AMV reverse transcriptase were used. Approximately 13,500 cpm of the AMV product and 24,000 cpm of the HIV product were loaded. Substrates were as described in the legend to Fig. 6. We thank Dr. Samuel Wilson for his gift of HIV reverse transcriptase.

The observation that Sequenase 2.0 tends to stop at AF-DNA lesions when these occur just before T's in the templates makes it profitable to reexamine some data on *in vitro* mutagenesis obtained with AF-DNA templates. We observed Sequenase 2.0 induced base substitutions and deletions at a number of positions. It is interesting (Figure 10) to note that the hotspots for single deletion (position 90-93) or double deletion (position141-140) occur just before (3') T's or runs of T's in the template. Most, but not all, of the other deletions reported also occur before T's in the template as might occur, for example, if a processive polymerase paused and then "fell off" at an AF-G prior to a natural pause site (T) and then, as a result of folding of the template, reinitiated at some place on the template 5' to the last inserted nucleotide. While this is speculation based on what must be considered preliminary results, it raises the possibility that the sites of induced mutation may be related to the sites of spontaneous termination and of spontaneous mutation for *in vitro* systems. If this should be so, then one might expect to find enzyme specific patterns of induced *in vitro* mutation for a variety of lesions, including those induced by ionizing radiation. Sequencing the mutants we have obtained (Table 2) should provide useful information. Whether this has any *in vivo* significance, e.g. whether there are *in vivo* pause sites for the replication apparatus and whether these are related to spontaneous and induced mutation remains to be seen.

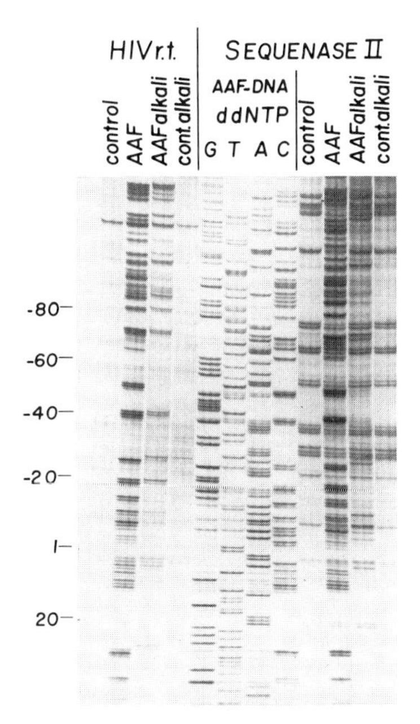

Figure 9. Pause sites for Sequenase 2.0 and HIV reverse transcriptase on an AAF-DNA template treated with alkali. Approximately 25,000 cpm of product were loaded in each lane. Synthesis starts at position 70. The sequence ladder was prepared with Sequenase 2.0 using a template containing AAF lesions.

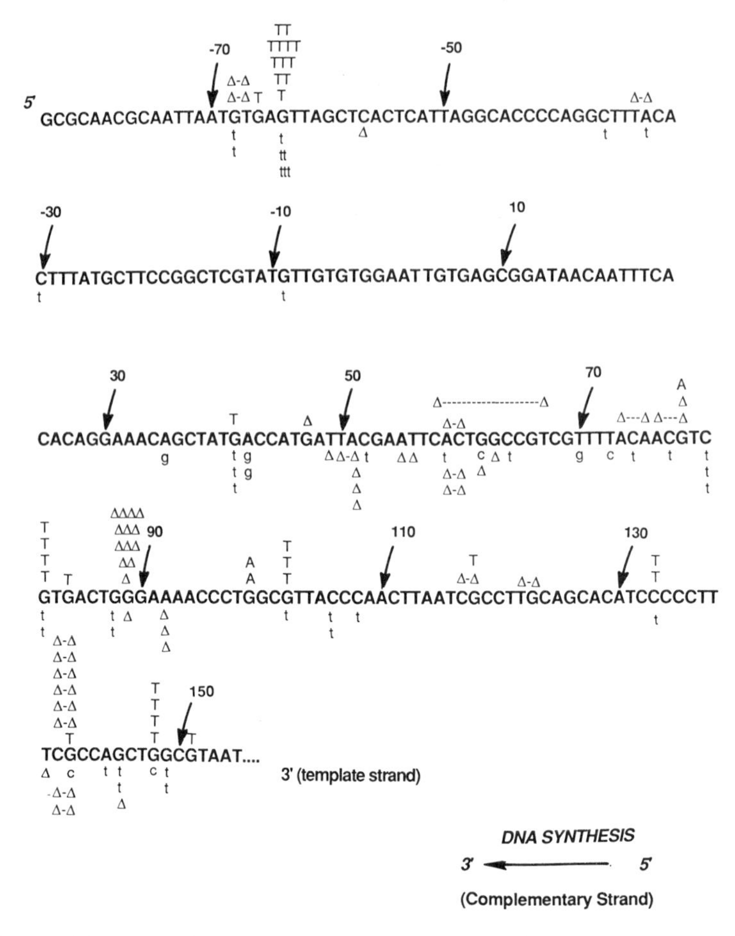

Figure 10. Location and nature of mutants obtained *in vitro* by synthesis with Sequenase (original version) on an AF-containing template. Lower-case letters or symbols below the sequences indicate "spontaneous" mutants, i.e. mutants obtained by synthesis on a nontreated, uracil-containing, gapped template. Upper-case letters and symbols above the line indicate mutants obtained after synthesis on an AF-containing, uracil, gapped template. (Taken from Sahm et al., 1989).

ACKNOWLEDGEMENT

The work reported in this paper was supported in part by grants from the National Institute of General Medical Sciences (GM 07816), the National Cancer Institute (GM 07816) and the Department of Energy (DE-FG02-88ER60678).

REFERENCES

Brash, D., Seetharam, S., Kraemer, K., Seidman, M. and Bredberg, A. (1987). Photoproduct frequency is not the major determinant of UV base substitution hot spots or cold spots in human cells. *Proc. Natl. Acad. Sci. USA.* **84**, 3782-3786.

Burns, P., Gordon, A. and Glickman, B. (1987). Influence of neighbouring base seqeunce on N-methyl-N'-nitro-N-nitrosoguanidine mutagenesis in the lacI gene of Escherichia coli. *J. Mol. Biol.* **194**, 385-390.

Hsia, H., Lebkowski, J., Leong, P., Calos, M. and Miller, J. (1989). Comparison of ultraviolet-induced mutagenesis of the lacI gene in Escherichia coli and in human 293 cells. *J. Mol. Biol.* **205**, 103-113.

Kriek, E., Miller, J., Juhl, U. and Miller, E.C. (1967). 8-(N-2-Fluorenylacetamido) guanosine, an arylamidation reaction product of guanosine and the carcinogen N-acetoxy-N-2-fluorenylacetamide in neutral solution. *Carcinogenesis* **6**, 177-182.

Kunkel, T. and Alexander, P. (1986). The base substitution fidelity of eucaryotic DNA polymerases. Mispairing frequencies, site-preferences, insertion preferences and base substitution by dislocation. *J. Biol. Chem.* **261**, 160-166.

Lutgerink, J., Retel, J., Westra, J., Welling, M., Loman, H. and Kriek, E. (1985). By-pass of the major aminofluorene-DNA adduct during *in vivo* replication of single- and double-stranded ¿X174 DNA treated with N-hydroxy-2-aminofluorene. *Carcinogenesis* **6**, 1501-1506.

Modrich, P. (1987). DNA mismatch correction. *Ann. Rev. Biochem.* **56**, 435-466.

Moore, P., Rabkin, S. and Strauss, B.S. (1980). Termination of *in vitro* DNA synthesis at AAF adducts in the DNA. *Nucleic Acids Res.* **8**, 4473-4484.

Moore, P., Rabkin, S., Osborn, A., King, C. and Strauss, B.S. (1982). Effect of acetylated and deacetylated 2-aminofluorene adducts on *in vitro* DNA synthesis. *Proc. Natl. Acad. Sci. USA.* **79**, 7166-7170.

Norman, D., Abuaf, P., Hingerty, B., Live, D., Grunberger, D., Broyde, S. and Patel, D. (1989). NMR and computational characterization of the N-(Deoxyguanosin-8-yl) aminofluorene adduct [(AF)G] opposite adenosine in DNA: (AF)G[syn¥ A[anti] pair formation and its pH dependence. *Biochemistry* **28**, 7462-7476.

Sahm, J., Turkington, E., LaPointe, D. and Strauss, B. (1989). Mutation induced *in vitro* on a C-8 Guanine Aminofluorene containing template by a modified T7 DNA polymerase. *Biochemistry* **28**, 2836-2843.

Tang, M., and Lieberman, M. (1983). Quantification of adducts formed in DNA treated with N-acetoxy-2-acetylaminofluorene or N-hydroxy-2-aminofluorene: comparison of trifluoroacetic acid and enzymatic degradation. *Carcinogenesis* **4**, 1001-1006.

van Touw, J., Verberne, J., Retel, J. and Loman, H. (1985). Radiation-induced strand breaks in ¿X174 replicative form DNA: an improved experimental and theoretical approach. *Int. J. Radiat. Biol.* **48**, 567-578.

GENE SPECIFIC DAMAGE AND REPAIR AFTER TREATMENT OF CELLS WITH UV AND CHEMOTHERAPEUTICAL AGENTS

Vilhelm A. Bohr

Laboratory of Molecular Pharmacology
Division of Cancer Treatment
National Cancer Institute, NIH
Bethesda, Maryland 20892

ABSTRACT

We have previously demonstrated preferential DNA repair of active genes in mammalian cells. The methodology involves the use of a specific endonuclease or other more direct approaches to create nicks at sites of damage followed by quantitative Southern analysis and probing for specific genes. Initially, we used pyrimidine dimer specific endonuclease to detect pyrimidine dimers after UV irradiation. We now also use the bacterial enzyme ABC excinuclease to examine the DNA damage and repair of a number of adducts other than pyrimidine dimers in specific genes. We can detect gene specific alkylation damage by creating nicks via depurination and alkaline hydrolysis. In our assay for preferential repair, we compare the efficiency of repair in the DHFR gene to that in the 3' flanking, non-coding region to the gene. In CHO cells, UV induced *pyrimidine dimers* are efficiently repaired from the active DHFR gene, but not from the inactive region. We have demonstrated that the *6-4 photoproducts* are also preferentially repaired and that they are removed faster from the regions studied than pyrimidine dimers. Using similar approaches, we find that DNA adducts and crosslinks caused by *cisplatinum* are preferentially repaired in the active gene compared to the inactive regions and to the inactive c-fos oncogene. Also, *nitrogen mustard* and *methylnitrosurea* damage is preferentially repaired whereas *dimethylsulphate* damage is not. *NAAAF* adducts do not appear to be preferentially repaired in this sytem.

In an attempt to correlate the repair events in specific regions with genomic translocations we have studied DNA damage and repair in the murine c-myc oncogene after UV damage and in the human c-myc after alkylation damage; these results will be discussed.

METHODOLOGICAL APPROACH

Techniques have now been developed to study damage and repair in genes and other specific sequences (Bohr, V.A. et al., 1987, 1988; Thomas, D.C., et al. 1988). In general this approach can be used to determine the frequency of strand breaks in any restriction fragment of interest. Some agents cause strand breaks directly (e.g. ionizing irradiation and bleomycin) and in other cases the DNA lesions are detected with specific endonucleases or after depurination followed by alkaline hydrolysis. The frequency of strandbreaks in specific restriction fragments is determined through

Biological Reactive Intermediates IV, Edited by C.M. Witmer *et al.*
Plenum Press, New York, 1990

quantitative Southern analysis and probing of denaturing gels. The technique is outlined in Fig 1, and will be briefly discussed. Cells are uniformly prelabeled with ^{3}H-thymidine to tag and quantify the DNA. After damage the cells are incubated for repair in the presence of the heavy thymidine analog bromodeoxyuridine (BrdUrd) to allow us to later separate the (semiconservatively) replicated DNA from the parental. This step is required in most repair experiments since the replicated (lesion free) DNA can mistakenly be assayed as repaired. The DNA is then isolated, restricted and the parental DNA is separated on CsCl gradients. After quantitation, strand breaks are generated by endonuclease treatment or depurination followed by alkaline hydrolysis, and the DNA is electrophoresed on alkaline gels. After transfer, the membranes are hybridized with appropriate DNA probes and subjected to autoradiography. Bands are usually quantitated by densitometry. The number of lesions per fragment is calculated on the basis of the zero class, i.e. the fraction of fragments free of damage using the Poisson distribution. Initially, we used a specific endonuclease (T4 endonuclease V) to detect pyrimidine dimers after UV irradiation. We now also use the bacterial enzyme ABC excinuclease to examine the DNA damage and repair of bulky adducts other than pyrimidine dimers in specific genes (Thomas D.C., et al. 1988). In another approach, strand breaks are generated after depurination followed by alkaline hydrolysis. In our standard assay for preferential repair, the efficiency of DNA repair in the DHFR gene is compared to that in the 3' flanking, non-coding region to the gene, Figure 2.

Use of the ABC Excinuclease

The *E. coli* uvr ABC gene products together form the ABC excinuclease which has proven to be one of the most versatile DNA repair enzymes known. It recognizes a

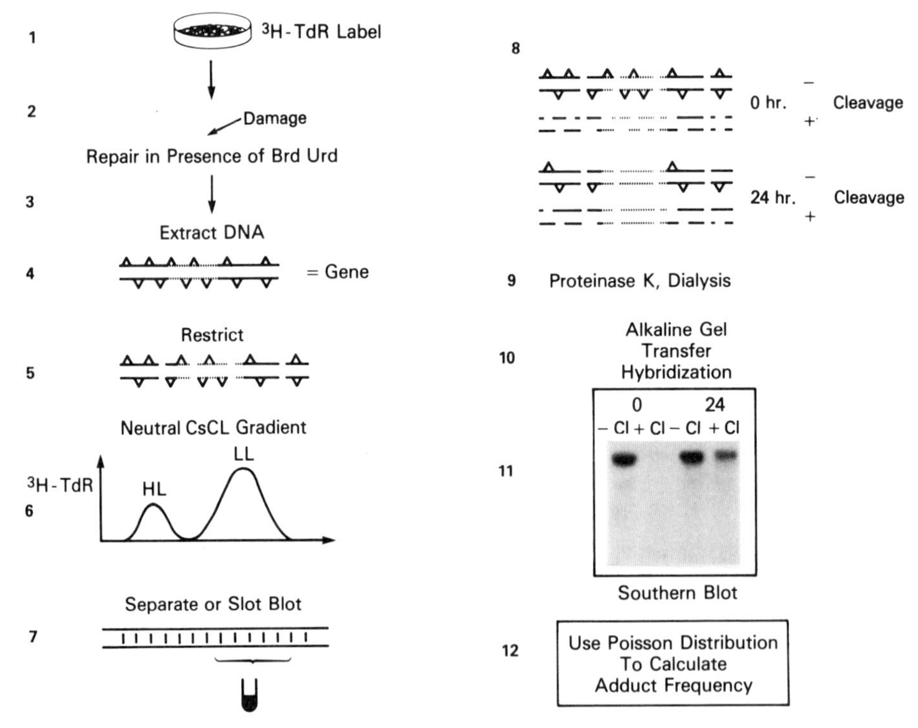

Figure 1. Flow chart illustrating the method for gene specific DNA damage and repair. Described in detail in text.

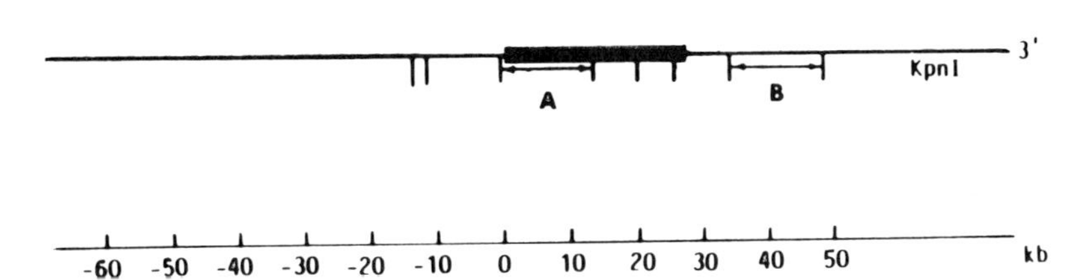

Figure 2. CHO DHFR gene. Vertical lines are restriction sites for KpnI. A is the 14 kb intragenic fragment generally used for our repair assay in the CHO DHFR gene. B is the downstream, non-transcribed fragment used for measurements of DNA repair in non-transcribed DNA. The support membrane is often probed for fragment A, then stripped and reprobed for fragment B. Repair can thus be compared in transcribing and nontranscribing regions in the same experiment using the same DNA.

wide variety of DNA damaging agents and cleaves the DNA at the site of the damage. This enzyme complex has been of considerable interest in the DNA repair field and the enzymology has recently been reviewed (Sancar, A. and Sancar, G.B., 1988). In collaboration with Dr. A. Sancar, Univ. North Carolina, who originally cloned these enzymes, we have developed methodology to study damage and repair in specific genes after cisplatinum, psoralens and 4NQO (Thomas, D.C., et al. 1988). These agents are recognized by the ABC excinuclease which generates strand breaks at sites of damage. The difficulty which we had to overcome in order to quantitatively use the ABC excinuclease was to find a way to get the enzyme out of the reaction mixture after reacting. We have achieved that by use of proteinase K and dialysis.

Recently, we have been able to detect **6-4 photoproducts** in specific genes with the use of the ABC excinuclease (Thomas, D.C., et al. 1989). There has previously not been available methodology for this. The approach involves the use of the photolyase enzyme which reverts pyrimidine dimers. When UV irradiated DNA is reacted with this enzyme, the cyclobutane pyrimidine dimers are removed leaving all other photoproducts. However, these other photoproducts are almost exclusively 6-4 photoproducts (Patrick, M.H., and Rahn, R.D., 1976). And these adducts are then detected with the ABC excinuclease.

In collaboration with Dr. M-S. Tang (Univ. Texas, MD Anderson), we have established methodology to detect the formation and repair of adducts formed after treatment of cells with **N-acetoxy-acetaminofluorene (NAAAF)** in specific genes using ABC excision nuclease. One of the most mutagenic compounds known is **IQ (2-amino-3-methyl-imidazo (4,5 F) quinoline**, a highly mutagenic, heterocyclic compound) which we can detect.

We have been able to detect **cisplatinum adducts** in specific regions *in vivo* by use of the ABC excinuclease complex (not yet published). In addition to the assessment of the total cisplatinum adducts using ABC excinuclease, we can now also separately visualize the **crosslinks** in specific genes. Whereas we usually run alkaline gels to study single stranded DNA, we run neutral gels in the assay for crosslinks. The drug treated DNA is denatured (30 mM NaOH for 15 min), allowed to reanneal briefly (10 min on ice) and then electrophoresed. At present we have preliminary results showing formation and repair of cisplatinum crosslinks in the CHO DHFR gene. The crosslinks, as would be expected, are resistant to single stranded S_1 nuclease. We plan to use this assay widely after cisplatinum treatment, alkylation treatment and other drug treatments which cause crosslink formation.

DNA REPAIR STUDIES

DNA Repair Mechanism

Our genetic material is constantly exposed to endogenous and exogenous damage caused by various agents including chemotherapeutic agents, irradiation and carcinogens. This damage can give rise to malignant transformation, in part via mutations at specific sites in certain oncogenes. Most of this damage is removed by the DNA repair mechanisms, and were it not for these enzymes, we would not survive this exposure. There is good evidence that DNA repair plays an important role in the prevention of cancer. For example, certain human disorders where the patients have lowered DNA repair capacity are associated with higher incidence of cancer. Although the DNA repair mechanism is as important as the related processes replication and transcription, it is not as well understood. However, lately there have been some important improvements in our understanding. One of these is the increased understanding of the fine structure of the repair process through new methodology which allows us to study these processes at more refined levels.

The enzymatic pathways involved in DNA repair appear to vary considerably with the type of damage introduced, but the most general and well characterized mechanism is the nucleotide excision repair process which is responsible for the removal of UV-induced damage and of bulky chemical adducts. The sequential steps in nucleotide excision repair include: 1) pre-incision recognition of damage, 2) incision of the damaged DNA strand near the site of the defect, 3) excision of the defective site and localized degradation of the affected strand, 4) repair replication to replace the excised region with a corresponding stretch of normal nucleotides, and finally 5) ligation to join the repair patch at its 3' end to the contiguous parental DNA.

It is well recognized that DNA in the nucleus of mammalian cells exists in the form of chromatin which is organized and packaged into higher order structures. This highly complex structural organization must affect all nuclear reactions including replication, transcription, recombination and DNA repair. The mammalian genome contains on the order of 10^5 genes which constitute on the order of one percent of the total DNA, the remainder being non-coding sequences. For many years determinations of DNA repair have been based on measurements over the total cellular DNA which provide an average measurement for the entire genome. However, it is now possible to measure DNA repair in parts of the genome such as individual genes.

DNA Repair At the Level of the Gene, Preferential DNA Repair

Most available results on repair in specific sequences have been obtained after UV damage to the cells. It was initially demonstrated that repair of UV damage in the essential gene for dihydrofolate reductase (DHFR) in Chinese hamster ovary (CHO) cells was much more efficient than the repair in the overall genome (Bohr, V.A., et al. 1985). These results may explain the paradox that rodent cells are deficient in overall genome DNA repair but are as resistant to UV damage as normal, repair proficient human cells: They survive by selectively repairing the vital sequences. As with CHO cells, cell lines from patients with the human disorder xeroderma pigmentosum (XP) repair only a small fraction of the induced pyrimidine dimers in the overall genome. Based on experiments measuring DNA repair after UV damage in the DHFR gene in CHO cells, XP cells and normal human cells, it has been suggested that the determination of repair in essential genomic regions is an important feature for comparisons between DNA repair and cellular survival (Bohr, V.A., et al. 1986).

In normal, repair proficient human cells, the whole genome is repaired during 24 hours after UV damage. However, essential genes are repaired faster than non-coding sequences in the genome (Mellon, I.M., et al. 1986). Some general conclusions from the gene-repair studies in hamster and human cells are summarized in Table 1.

Table 1. Preferential DNA Repair

	Percent Repair After 8 hours	After 24 hours
CHO cells		
Bulk DNA		15
Non-coding regions		15
An active gene	50	75
Human cells		
Bulk DNA	35	80
An active gene	75	80
Mouse cells		
c-abl proto oncogene		80
c-mos proto oncogene		20

In **rodent** cells, the active genes are efficiently repaired whereas the non-coding regions or inactive genes have minimal repair.

In **human** cells, the active genes are repaired much faster than the bulk of the genome or inactive genes.

Repair has also been examined in a number of different genes, and in cells of a variety of different species. The data suggest that the preferential DNA repair of pyrimidine dimers in active genes after UV damage is not only a general phenomenon in mammalian cells, but is also seen in species as diversified as gold fish, yeast (Terleth, C., et al. 1989), Drosophila and *E. coli* (Mellon, I. and Hanawalt, P.C., 1989). A DNA repair domain has been described in the DHFR locus: The preferential repair of the CHO DHFR gene is confined to a 60-80 kb region centered around the 5'end of the gene (Bohr, V.A., et al. 1986). The initial frequency of pyrimidine dimers is similar in all fragments in and around the DHFR gene whereas the repair differs considerably and is maximal at the 5' end of the gene. The size of this DNA repair domain is similar to a loop or higher order structure in chromatin and suggests a relation between DNA repair processes and chromatin structure.

Different genes within the same cell can be repaired with different efficiencies. In a study on the repair of a number of proto-oncogenes in mouse cells it was found that the *c-abl* gene is repaired much more efficiently than the *c-mos* gene (Madhani, H.D., et al. 1986). There are many important differences between these two oncogenes, and one is that the *c-abl* gene is actively transcribed in the cells whereas the *c-mos* is not. These experiments indicate that a correlation exists between the level of transcription and the efficiency of repair for a gene. This was further supported by studies on the repair of UV damage in the metallothionein gene in CHO cells (Okumoto, D.S., and Bohr, V.A., 1987) and in human cells (Leadon, S.A., and Snowden, M.M., 1988). The repair in this gene is markedly more efficient when the gene is transcriptionally active than when it is not. These findings suggest that there is some cellular association between the repair machinery and the transcription machinery.

Human Repair Gene, ERCC-1

Several studies over the last decade have suggested that DNA damage and repair in the mammalian genome is governed by the "openness" or accessibility of chromatin (Wilkins, R.J., and Hart, R.W., 1974; van Zeeland, A.A., et al. 1981). We recently studied the preferential DNA repair after UV damage of the DHFR gene in wild type CHO cells and UV sensitive CHO cells, some of which were transfected with repair genes (Bohr, V.A., and Hanawalt, P.C., 1987; Bohr, V.A., et al., 1988). The transfectants contained either the bacteriophage gene *denV* coding for the pyrimidine dimer specific repair enzyme T4 endonuclease V or the well characterized human repair gene, ERCC-1 (van Diun, M., et al. 1986; ,Zdzienicka, M.Z., et al., 1987). The UV sensitive mutants did not repair UV damage, but the repair was restored in the transfectants. Cells containing the *denV* gene repaired all genomic sequences equally efficiently whereas cells containing the human ERCC-1 gene repaired the active DHFR gene much more efficiently than the non-coding sequences. Evidently, the ERCC-1 gene product reacts more like the endogenous enzyme involved in normal repair in mammalian cells. Since the T4 endonuclease V is a small 16 kD enzyme (half the estimated size of the ERCC-1 gene product), these results support that the preferential repair of active genes is related to chromatin accessibility. The ERCC-1 gene product may also be a part of a large enzyme complex responsible for excision. Incidentally, these transfected cell lines constitute an example of gene therapy for DNA repair. In the future we may be able to insert repair genes as a therapy in some cases of defective DNA repair.

6-4 Photoproducts

UV irradiation introduces different photoproducts in the DNA of which the major one (60-80% of total) is the (cyclobutane) pyrimidine dimer. However, other photoproducts are also formed, notably the 6-4 photoproduct. Almost all the work done on UV damage and repair is based on analysis of pyrimidine dimers, and although recent data suggest that the 6-4 photoproduct plays a major role in mutagenesis, little is known about the fate of this photoproduct (Mitchell, D.L., 1988, 1988, 1989). As mentioned, we now have an assay to determine the formation and repair of 6-4 photoproducts in individual genes. We have found that the frequency of 6-4 photoproducts is about 40% of that of pyrimidine dimers in the DHFR gene; this frequency is somewhat higher than previous estimations of frequencies in the overall genome. We have further demonstrated that the 6-4 photoproducts are preferentially repaired in the DHFR gene as compared to non- transcribed genomic regions (Thomas, D.C., et al. 1989).

NAAAF

After treatment of CHO cells with N-acetoxy-acetamino-fluorene (NAAAF), we found (by HPLC) that the main adduct formed in the DNA was the dG-C-8-AF (Tang, M-S., et al. 1989). We quantitated the NAAAF adduct formation and repair in defined DNA sequences in mammalian cells using the UVR ABC excinuclease and in the overall genome using HPLC. It appeared that the adduct was formed with similar frequency and repaired with similar efficiency in the average, overall genome, in the DHFR gene and in its 3' flanking and in the non-coding sequences (Tang, M-S., et al. 1989). Thus we found no preferential damage or repair of this compound in expressed gene regions where the chromatin structure supposedly is less compact.

Chemotherapeutics

We have recently found that cisplatin adducts are preferentially repaired in the CHO DHFR gene. Also, nitrogen mustard adducts are preferentially repaired. In contrast, damage caused by dimethyl sulphate is not preferentially repaired. This latter finding is in acordance with other observations (Scicchitano, D.A. and Hanawalt, P.C., 1989). It is surprising that for two related alkylating agents, one is repaired preferentially and the other not.

Preferential repair of genes might be ascribed to the more "open" chromatin structure in actively transcribed genomic regions. The recent demonstration that DNA repair shows strand specificity towards the transcribing strand (Mellon, I.M., et al. 1986; Mellon, I.M., et al. 1987), however, suggests that repair is directed towards certain genomic regions rather than just dependent upon chromatin accessibility. Repair enzymes may be linked with the transcription complex. Further studies are needed to examine the relative importance of local chromatin structure, primary DNA sequence and function of the DNA for the determination of efficiency and organization of DNA repair.

Certain protein molecules may play important roles in locating vital sequences to be dealt with by the DNA repair machinery. Also the importance of local levels of methylation may be of importance. We have recently reported (Ho, L., et al. 1989) that demethylation of CHO cells by growth of the cells for many generations in azacytidine enhanced the overall genome DNA repair and changed the fine structure organization of the DNA repair domain in the DHFR gene.

Furthermore, different results suggest or at least provide a working hypothesis that there may be enzymatic differences between the "average" repair pathway in the cell or bulk of the DNA and that responsible for the preferential repair seen in active genes.

Gene Specific Repair and Cancer Risk

It is thought that protooncogenes are among the cellular targets of physical and chemical carcinogens. A number of protooncogenes have been found to be mutated, rearranged, translocated, and/or amplified in a number of tumors. It is evident that the efficient removal of DNA lesions from protooncogenes at risk could represent a critical step in tumorigenesis.

Plasma cell tumors as well as several other tumors of B-cell lineage are characterized by chromosomal rearrangements that activate the c-myc proto-oncogene. The preponderance of these rearrangements in B-cell tumors demonstrates the crucial role that c-myc plays in B-cell growth and pathophysiology (Potter, M.1987; Mushinski, J.F., et al. 1987). At present, it is not clear what molecular mechanisms are involved in creating the observed c-myc rearrangements nor is it clear why, in experimental mice some inbred strains are susceptible while other strains appear to be resistant to the formation of B-cell tumors. Cytogenetic studies performed on cells from several inbred strains of mice have shown quantitative differences between strains in the number of X-ray induced chromosomal abnormalities. These differences may reflect intrinsic variations between strains in the ability to efficiently repair DNA. Cells from strains DBA/2N, C3H/HeN, STS/A, C57BL/6N, Balb/cj and AKR/N show an efficient repair phenotype while cells from Balb/cAn and NZB/NJ show an inefficient repair phenotype (Potter, M., 1988). Balb/cAn and NZB/NJ are unique among these strains in that both show susceptibilities to the induction of plasmacytomas.

A correlation between the inability to repair DNA and the formation of specific chromosomal translocation has not been established and these translocations are integral parts of the development of B-cell tumors. Regardless of the mechanisms involved, the formation of non-random chromosomal rearrangements requires strand cleavage events and subsequent error prone ligation in defined genomic regions. We wish to address the following questions: Is the c-myc gene preferentially repaired in murine B-cells? Is there a difference in the repair of c-myc in resistant versus susceptible strains of mice both at the level of the overall genome and in the gene? Is the repair of the c-myc gene dependent upon the differentiational stage of the cell? Is the repair of the c-myc gene linked to its transcriptional stage? Is there a correlation between observed breakpoint clusters and inefficient repair in those genomic regions?

At this point we already have preliminary data on the repair of UV damage in the myc gene of some of the mouse strains. Interestingly, results indicate that there is more efficient repair in the 5' end of the c-myc gene in the resistent cell line DBA/2N than in the sensitive line Balb/cAN. The repair of the 3' end of c-myc and of the DHFR gene are the same in the two strains.

This finding is presently being further explored. It could suggest an important role of DNA repair in relation to tumor susceptibility or cancer risk.

REFERENCES

Bohr, V.A., Smith, C.A., Okumoto, D.S., Hanawalt, P.C. (1985). DNA repair in an active gene: Removal of pyrimidine dimers from the DHFR gene of CHO cells is much more efficient than the genome overall. *Cell* **40,** 359-369.

Bohr, V.A., Okumoto, D.S., Hanawalt, P.C. (1986). Survival of UV-irradiated mammalian cells correlates with efficient DNA repair in an essential gene. *Proc Natl Acad Sci* USA **83** ,3830-3837.

Bohr, V.A., Okumoto, D.S., Ho, L., Hanawalt, P.C. (1986). Characterization of a DNA repair domain containing the dihydrofolate reductase gene in CHO cells. *J. Biol. Chem* **261,** 16666.

Bohr, V.A., and Hanawalt, P.C. (1987). Enhanced repair of pyrimidine dimers in coding and non-coding genomic sequences in CHO cells expressing a prokaryotic DNA repair gene. *Carcinogenesis* **8,** 1333-1337.

Bohr, V.A., Phillips, D.H., Hanawalt, P.C. (1987). Heterogeneous DNA damage and repair in the mammalian genome. *Cancer Res* **47,** 6426-36.

Bohr, V.A. and Wassermann, K. (1988). DNA repair at the level of the gene. *Trends in Biochemical Sciences* **13,** 429-435.

Bohr, V.A., Okumoto, D.S. (1988). Analysis of frequency of pyrimidine dimers in specific genomic sequences. In DNA repair: A laboratory manual. Edited by Friedberg EC, Hanawalt PC. Vol. 3, p 347. Marcel Dekker, New York.

Bohr, V.A., Chu, E., van Duin, M., Hanawalt, P.C. and Okumoto, D. (1988). Human DNA repair gene restores pattern of preferential DNA repair. *Nucl Acids Res* **16,** 7397-7403.

Ho, L., Bohr, V.A. and Hanawalt, P.C. (1989). Demethylation enhances removal of pyrimidine dimers from the overall genome and from specific DNA sequences in CHO cells. *Mol. Cell. Biol.* **9,** 1594-1603.

Leadon, S.A., and Snowden, M.M. (1988). Differential repair of DNA damage in the human metallothionein gene family. *Molec Cell Biol* **8,** 5331-5339.

Madhani, H.D., Bohr, V.A., and Hanawalt, P.C. (1986). Differential DNA repair in transcriptionally active and inactive proto-oncogenes: c-abl and c-mos. *Cell* **45,** 417-424.

Mellon, I.M., Bohr, V.A., Smith, C.A., Hanawalt, P.C. (1986). Preferential DNA repair of an active gene in human cells. *Proc Natl Acad Sci* USA **83,** 8878-8888.

Mellon, I.M., Spivak, G., and Hanawalt, P.C. (1987). Selective removal of transcription-blocking DNA damage from the transcribed strand of the mammalian DHFR gene. *Cell* **51,** 241-247.

Mellon, I. and Hanawalt, P.C. (1989). Induction of the E. coli lactose operon selectively increases repair of its transcribed DNA strand. *Nature* **342,** 95-98.

Mitchell, D.L. (1988). The relative cytotoxicity of (6-4) photoproducts and cyclobutane dimers in mammalian cells. *Photochem. Photobiol.* **48,** 51-57.

Mitchell, D.L. (1988). The induction and repair of lesions produced by the photolysis of (6-4) photoproducts in normal and UV-hypersensitive human cells. *Mutat Res* **194,** 227-37.

Mitchell, D.L. and Nairn, R.S. (1989). The biology of the (6-4) photoproduct. *Photochem. and Photobiol.* **49,** 805-819.

Mushinski, J.F., Davidson, W.F. and Morse, H.C. (1987). Activation of Cellular oncogenes in human and mouse leukemia-lymphomas, Spontaneous and induced oncogene expression in murine B lymphocytic neoplasms. *Cancer Investigation* **5**(4), 345-368.

Okumoto, D.S., and Bohr, V.A. (1987). DNA repair in the metallothionein gene increases with transcriptional activation. *Nucl Acids Res* **15,** 10021-10031.

Patrick, M.H., and Rahn, R.D. (1976). In Photochemistry and Photobiology of Nucleic

Acids (S.Y. Wang, ed.) Vol II, pp 35-145. Acad. Press, N.Y..
Potter, M. (1987). Mechanisms of B-cell neoplasia. Eds. Melchers, F. and Potter, M. Hoffman-La Roche and Co. Basle, Switzerland.
Potter, M., Sanford, K.K., Parshad, R., Tarone, R.E., Price, F.M., Mock, B. and Huppi, K. (1988). Genes on chromosomes 1 and 4 in the mouse are associated with repair of radiation induced chromatin damage. *Genomics* **2**, 1-6.
Sancar, A., and Sancar, G.B. (1988). DNA repair enzymes. *Ann. Rev. Biochem.* **57**, 29-67.
Scicchitano, D.A. and Hanawalt, P.C. (1989). Repair of N-methylpurines in specific DNA sequences in Chinese hamster ovary cells: Absence of strand specificity in the DHFR gene. *Proc. Nat. Acad. Sci.* USA **86**, 3050-3054.
Tang, M-S., Bohr, V.A., Xang, X-S., Pierce, J. and Hanawalt, P.C. (1989). Quantitation of aminofluorene adduct formation and repair in defined DNA sequences in mammalian cells using UVRABC nuclease. *Journ. Biol. Chem.* **264**, 14455-14462.
Terleth, C., van Sluis, C.A. and van de Putte, P. (1989). Differential repair of UV damage in Saccharomyces cerevisciae. *Nucleic Acids Res.* **17**, 4433-4439.
Thomas, D.C., Morton, A.G., Bohr, V.A., Sancar, A. (1988. A general Method for Quantitating bulky adducts in mammalian genes. *Proc Nat Acad Sci* USA **85**, 3723-3727).
Thomas, D.C., Okumoto, D.S., Sancar, A. and Bohr, V.A. (1989). Preferential DNA repair of (6-4) photoproducts from the CHO DHFR gene. *J. Biol. Chem* **264**, 18005-18010.
van Zeeland, A.A., Smith, C.A., and Hanawalt, P.C. (1981). Introduction of T4 endonuclease V into frozen and thawed cells. *Mutat Res* **82**, 173-179.
van Diun, M., De Wit, J., Odijik, H., and Westerveld, A. (1986). Molecular characterization of the human DNA repair gene ERCC-1: cDNA cloning and amino acid homology with the yeast DNA repair gene RAD 10. *Cell* **44**, 913-920.
Wilkins, R.J., and Hart, R.W. (1974). Preferential DNA repair in human cells. *Nature* **247**, 35-45.
Zdzienicka, M.Z., Roza, L., Westerveld, A., Bootsma, D. and Simons, J.W.I.M. (1987). Biological and biochemical consequences of the human repair gene ERCC-1 after transfection into a repair deficient CHO cell line. *Mutat. Res.*, DNA Repair Reports **183**, 69-75.

PROTO-ONCOGENE ACTIVATION IN RODENT AND HUMAN TUMORS

Marshall W. Anderson, Ming You, and Steven H. Reynolds

Department of Health and Human Services
National Institutes of Health
Research Triangle Park, NC 27709

INTRODUCTION

The process of cell transformation is a multistep phenomenon. Increasing evidence suggests that a small set of cellular genes appear to be targets for genetic alterations that contribute to the neoplastic transformation of cells. The development of neoplastia may, in many cases, require changes in at least two classes of cellular genes: proto-oncogenes (Bishop, 1987; Anderson and Reynolds, 1989) and tumor suppressor genes (Weinberg, 1989; Hansen and Cavenee, 1988; Barrett, 1987). For example, both the activation of ras oncogenes and the inactivation of several suppressor genes have been observed in the development of human colon tumors (Stanbridge, 1990) and human lung tumors (Tabahashi et al., 1989; Weston et al., 1989; Reynolds et al., 1989; Minna et al., 1989; Rodenhuis et al., 1988). These two examples illustrate that a cell accumulates several types of genetic alterations in its evolution to a malignant phenotype. The focus of this chapter is to discuss the activation of proto-oncogens in human and rodent tumors and the pattern of mutations in *ras* oncogenes detected in human and rodent tumors.

The Activation of Proto-oncogenes in Human and Rodent Tumors

Proto-oncogenes were initially discovered as the transduced oncogenes of acute transforming retroviruses (Bishop, 1985). Viral oncogenes arose by recombination between cellular proto-oncogenes and the genome of nontransforming retroviruses. It has now been established that proto-oncogenes can also be activated by mechanisms independent of retroviral involvement (Bishop, 1989; Westin, 1989; Anderson and Reynolds, 1989; Varmus, 1989). These mechanisms include point mutations as well as gross DNA rearrangements such as translocation and gene amplification.

Proto-oncogenes are expressed during 'regulated growth' such as embryogenesis, regeneration of damaged liver and stimulation of cell mitosis by growth factors. Proto-oncogenes are highly conserved, being detected in species as divergent as yeast, *Drosophila* and humans. Included are genes which encode for growth factors, growth factor receptors, regulatory proteins, and tyrosine and serine/threonine kinases (Table 1). Thus, the encoded proteins appear to play a crucial role in normal cellular growth and differentiation.

The activation of proto-oncogenes in human and rodent tumors has been studied in great detail over the past several years. The induction of aberrant expression of proto-oncogenes by gene amplification or chromosomal translocation has been observed in a variety of tumors (Anderson and Reynolds, 1989; Westin, 1989; Varmus,

Biological Reactive Intermediates IV, Edited by C.M. Witmer *et al.*
Plenum Press, New York, 1990

Table 1. Proto-Oncogene Families

Function	Proto-oncogene
Growth Factor	sis (PDGF), int-2, hst
Growth Factor Receptor	erb B (EGF receptor), fms (CSF receptor), neu, ros
Nuclear Regulatory Protein	myc, myb, fos, **jun**
Regulatory Protein in	Ha-*ras*, K-*ras*, N-*ras*
Tyrosine Kinase	src, abl, met
Serine/threonine Kinase	mos, raf

1989; Bishop, 1989). Proto-oncogene amplification is observed as both a low-frequency event in diverse tumor types and as a high-frequency event in specific tumor types. Several proto-oncogenes have been observed at relatively high frequency as components of amplified DNA in specific human tumor types: 1) *erbB1* in squamous cell carcinomas; 2) c-*myc* in breast carcinomas, cervix carcinomas, small cell lung cancer and morphological variant cell lines from small cell lung cancer, and rat skin tumors; 3) N-*myc* in small cell lung cancer, cell lines from small cell lung cancer, and neuroblastoma; 4) L-*myc* in small cell lung cancer and cell lines from small cell lung cancer; 5) *HER-2/neu* in breast carcinomas and salivary gland carcinomas; 6) *int*-1 in carcinoma of the breast. Refer to Table 1 in Anderson and Reynolds, 1989, and Table 2 in Bishop, 1989, for references. It appears that proto-oncogene amplification is usually associated with neoplastic progression rather than with the initiation of tumorigenesis. For example, the degree of *HER-2/neu* amplification is inversely related to survival and time to relapse in women with breast cancer (Slamon et al., 1987). Gene amplification has been observed more frequently in human tumors than in rodent tumors.

Genomic rearrangements are commonly encountered in the karotypes of cancer cells. The significance of these rearrangements has been elucidated in several cases by identifying the altered gene to be a proto-oncogene. For example, c-*myc* is joined to various immunoglobulin genes in cells of Burkitt lymphoma and c-murine plasmacytomas. The translocations probably perturb transcriptional control of c-*myc* and potentiates oncogenic growth of B cells (Leder et al., 1984). The Philadelphia chromosome, found in over 90% of chronic myelogenous leukemias, creates a fusion protein where the tyrosine kinase domain of the c-*abl* proto-oncogene located on 9q is moved to chromosome 22 and joined to an undefined genetic locus termed *bcr* to create a hybrid *bcr-abl* protein with oncogenic potential (Konopka et al., 1984). Chromosome translocation may help elucidate the presence and chromosomal location of unknown putative oncogenes; for example, *bcl*-2 has been cloned, its protein products identified, and is frequently activated by translocation in follicular lymphomas (Tsujimoto and Croci, 1986; and Tsujimoto et al., 1988); also, refer to Table 1 in Bishop, 1989, for other examples.

A number of oncogenes in both human and animal tumors have been detected by gene transfer methods. This test involves the ability of a recipient cell, such as the NIH/3T3 mouse fibroblast, to accept and express genes from donor tumor DNA, resulting in the formation of transformed cells. Shih et al., 1979, were the first to show that DNA from carcinogen-transformed cell lines could cause morphologic transformation of NIH/3T3 cells after transfection. The transformed cells also have an increase in refractility and exhibit anchorage-independent growth. An extension of the NIH/3T3 transfection assay that affords greater sensitivity is the nude mouse tumorigenicity assay which involves the cotransfection of NIH/3T3 cells with tumor DNA and a selectable marker gene (Fasano et al., 1984). The selected cell populations are then subcutaneously injected into immunocompromised mice, and tumor formation observed. The tumors that develops in the nude mouse are then analyzed to characterize the transfected oncogenes.

Since the initial experiment with the gene transfer assay, many investigations have utilized this methodology to analyze tumors for the presence of dominant transforming genes. Examples of oncogenes detected in human tumors are presented in Table 2.

Member of the *ras* gene family were the first activated proto-oncogenes detected in the NIH/3T3 assay and have been detected in more tumor types and at higher frequency than any other oncogene. Activated *ras* proto-oncogenes have been detected in a high percentage of human colon carcinomas (50%), lung adenocarcinomas (40%), pancreatic carcinomas (75%), and cholangiocarcinomas (80%), and to a lesser

Table 2. Activated Oncogenes Detected in Human Tumors By DNA Transfection Assays[a]

Tumor	Number Positive/ Number Tested	Oncogene	Ref.
Colon			
Adenoma(FAP)[b]	10/115	K-ras(10)[c]	[1,2][d]
Adenomas	36/84	K-ras(35), N-ras(1)	[2,3,4]
Carcinomas	126/268	K-ras(121), N-ras(5)	[2,3,4,5]
Metastases	11/171	K-ras(11)	[3]
Pancreatic Carcinomas	146/181	K-ras(146)	[3,6-8]
Lung Adenocarcinomas	31/72	K-ras(26), H-ras(1) N-ras(1), raf(1), NC(2)[b]	[9,10]
AML[b]	12/45	K-ras(10), N-ras(2)	[11]
Pre-AML Cells, MDS[b]	3/8	N-ras(3)	[12]
Pre-AML Cells, MDS	2/4	K-ras(2)	[13]
CML[b]-Chronic Phase	2/6	H-ras(1), NC(1)	[14]
CML-Blast Phase	4/6	H-ras(2), N-ras(1), NC(1)	[14]
Squamous Cell Carcinomas	4/6	H-ras(4)	[15]
Cutaneous Melanoma	7/37	N-ras(7)	[16]
Thyroid Papillary Carcinomas	5/20	NC(5)	[17]
Stomach Carcinomas	3/58	hst(3)	[18]
Hepatocellular Carcinomas	2/11	lca(2)	[19]
Bladder Carcinomas	7/62	H-ras(7)	[20,21]
Cholangocarcinomas	6/9	K-ras(7)	[22]

[a] The basis for detection of the oncogenes listed in this table is the DNA transfection assay. Biochemical methods have enhanced the sensitivity to detect mutation in the ras genes. In some cases, the biochemical methods was the only assay used that thus, non-ras oncogenes would not be dtected.

[b] FAP: Familial polyposis coli patients; NC: refers to putative oncogene detected by transfection assay but has not been characterized to data; AML: Acute myeloid leukemia; MDS: Myelodisplastic syndrome; CML: Chronic myeloid leukemia

[c] Number in parenthesis indicate the number of positive samples with that oncogene.

[d] [1] Farr et al., 1988; [2] Vogelstein et al., 1988; [3] Perucho et al., 1989; [4] Brumer and Loeb, 1989; [5] Bos et al., 1987; [6] Shibata et al., 1990; [7] Smit et al., 1987; [8] Krinewald et al., 1989; [9] Rodenhuis et al., 1988; [10] Reynolds and Anderson, unpublished data; [11] Bos, et al., 1987; [12] Hirari et al., 1987; [13] Liu et al., 1987; [14] Liu et al., 1988; [15] Ananthaswany et al., 1987; [16] Veer et al., 1989; [17] Fusco et al., 1987; [18] Sakamoto et al., 1986; [19] Ochija et al., 1986; [20] Visvanathan et al., 1988; [21] Fujita et al., 1985; [22] Tada et al., 1990.

extent in several other human tumor types (Table 2). Human mammary tumor is a major human tumor type for which activated *ras* proto-oncogenes have not been detected at any appreciable frequency. Activated *ras* genes have also been detected in tumors generated from numerous rodent model systems (Anderson and Reynolds, 1989; Barbacid, 1987; Brown, et al., 1990; Balmain and Brown, 1988; Guerrero and Pellicer, 1987; Sukumar, 1988; Bailleul et al., 1989; Newcomb et al., 1989). The mutation spectrum in *ras* genes detected in spontaneous and chemical-induced mouse tumors will be discussed in the next section.

H-*ras*, K-*ras*, and N-*ras* can acquire transforming activity by a point mutation in their coding sequence. *In vivo*, activating point mutations have been observed in codons 12, 13, 61, 117, and 146 (Sloan et al., 1990; Barbacid, 1987; Reynolds et al. 1987). Also we have recently detected a 30 base repeat around codon 61 in the K-*ras* gene in several mouse tumors. This repeat appears to activate the K-*ras* gene. The *ras* gene products are 21,000-dalton proteins (p21) which bind guanine nucleotides with high affinity and are thought to be involved in various signal transduction pathways in many cell types (McCormick, 1989). The crystal structure of p21 has recently been determined by Pai et al. (1989) and Milbrum et al. (1990).

Activated proto-oncogenes other than *ras* have also been detected by DNA transfection methods in both human and rodent tumors (refer to Tables 3, 4 and 5 in Anderson and Reynolds, 1989; Bishop, 1989). Two proto-oncogenes which display tyrosine kinase activity, *ret*, and *trk*, have been found to be activated in human thyroid carcinomas (Bongarzone et al., 1989). The *neu* oncogene was originally detected in ENU-induced neuroblastoma of rats (Scherhter et al., 1984) and was activated by a point mutation in the transmembrane region of the putative growth factor receptor encoded by *neu*. The *hst* oncogene was isolated from a human gastric carcinoma (Sakamoto et al., 1986) and, perhaps, has been found in other human tumor types (Yoshida et al., 1987). A transforming gene called *lca* (Ochija et al., 1986) was detected in two human hepatocellular carcinomas. Several other potentially new types of activated oncogenes which await characterization have been detected in human chromic myelogenous leukemias (Liu et al., 1988), human lung tumors (Anderson and Reynolds, unpublished data) and several types of rodent tumors (refer to Table 5 in Anderson and Reynolds, 1989). The prevalence and mode of activation of these non-*ras* genes is not clear at present.

Mutation Spectrum in K-ras Oncogenes Detected in Mouse lung Tumors

The reproducible activation of *ras* oncogenes in chemical induced tumors in rodents has made it possible to correlate the activating mutations with the promutagenic adducts formed from direct or metabolic activation of the carcinogen. Mutation spectrums have been determined in *ras* oncogenes detected in tumors generated in mouse lymphomas (Newcomb et. al., 1989), liver (Wiseman et. al., 1986; Reynolds et.al, 1987), mouse skin (Brown, et. al., 1990), mouse lung (You et. al., 1988; Belinsky et. al., 1989), and rat mammary (Barbacid, 1987). The data in Table 3 illustrates the spectrums observed in the K-*ras* oncogenes detected in spontaneous and chemical-induced lung tumors of the A/J mouse. The selectivity of adduct formation is very striking for the various chemicals examined and the mutation pattern of the K-*ras* oncogene detected in each of the chemical induced tumors appear to be distinctly different from the spontaneous spectrum. A detailed discussion of the relationship between the mutation spectrums in Table 3 and the known DNA binding properties are presented in You et. al. (1988) and Belinsky et. al. (1989) as well as relevant references. The following is a brief summary of their discussions:

1. BP and DMBA are polycyclic aromatic hydrocarbons. The bay-region diol epoxide metabolite of BP, r-7, t-8- dihydroxy-t-9, 10-oxy-1,8,9,10-tetrahydrobenzo(a)pyrene (anti-BPDE), binds to guanosine at the N^2-position in native DNA and also forms lesser amounts of adducts with deoxyadenosine. the anti-BPDE induced predominantly GC to TA transversions in the lac I gene of E. coli and the APRT gene in Chinese hamster ovary cells. Consistent with these mutagenicity studies, a high prevalence for the GC to TA mutations (9/14) were observed in codon 12 of the K-*ras* gene of the BP-induced lung tumors (Table 3). These same GC to TA mutations were also observed in codon 12

of the K-*ras* gene of BP-induced mouse forestomach tumors (unpublished data). The mutation observed in DMBA-induced tumors suggest that DMBA metabolites activate the K-*ras* gene via adenosine adducts. DBMA metabolites are known to bind to deoxyadenosine at a much greater rate than BP metabolites.

2. Specific activation of H-*ras* and K-*ras* oncogenes in the second base of codon 61 by EC and vinyl carbamate (a potential metabolite of EC) has been observed in mouse skin, mouse liver, and mouse lung tumors. However, the only reported adduct to date is a deoxyguanosine adduct (Miller and Miller, 1986) which is not consistent with the observed *ras* mutations.

3. MNU is a direct acting methylating agent whereas NNK and DMN can be activated via α-hydroxylation to methylating agents. The GC to AT transition in the second base of codon 12 of the K-*ras* gene detected in MNU and NNK-induced lung tumors (Table 3) has also been observed in several other types of rodent tumors induced by methylating agents (refer to Table 3, Wang et. al. 1990). The GC to AT mutation results from the O^6-methylguanine adduct. The AT to GC transition in codon 61 of K-*ras* detected in NNK and DMN induced lung tumors are consistent with the formation of promutagenic O^4-methylthymidine adducts by these methylating agents.

The data presented in Table 2 suggest that the activating point mutations in K-*ras* genes detected in these chemically induced lung tumors result from genotoxic effects of the chemicals. Activated *ras* genes, primarily K-*ras*, have also been detected in approximately 40% of human pulmonary adenocarcinomas (Table 2). Cigarette smoking is widely accepted as the major cause of human lung cancer and linear relationships have been established between the number of cigarettes smoked and lung cancer risk (Wynder, 1972). Although many mutagenic and carcinogenic compounds such as polycyclic aromatic hydrocarbons and nitrosamine have been identified in cigarette smoke, it is not clear which of these compounds contribute to the development of lung tumors in smokers. Philips et. al. (1988) have shown a linear relationship between numbers of cigarettes smoked and DNA adduct levels in the lung. Thus the activated *ras* genes detected in human pulmonary adenocarcinomas from

Table 3. Mutation Spectrum in K-*ras* Oncogenes Detected in Spontaneous and Chemical-Induced Lung Tumors of the Strain A Mouse[a]

		Codon 12 Mutations				Codon 61 Mutations			
		GGT → Gly	TGT Cys	GTT Val	GAT Asp	CGT Arg	CAA → Gin	CTA Leu	CGA Arg	CAT His	
Treatment	# of Tumors with K-*ras*									NC
Control	19	0	4	4	1	0	7	2	1
BP[b]	14	8	1	4	0	0	0	0	1
DMBA	10	0	0	0	0	10	0	0	0
EC	10	0	1	0	0	7	2	0	0
MNU	15	0	0	15	0	0	0	0	0
NMK	11	0	1	8	0	0	2	0	0
DMN	10	0	0	7	0	0	3	0	0

[a]Data from You et. al. (1989), Belinsky et.al. (1989), and unpublished observations.

[b]Abbreviations: BP- benzo(a)pyrene; DMBA- dimethylbenz(a)anthracene; EC- ethyl carbamate; MNU- methylnitrosourea; NMK- 4-(N-methyl-N-nitrosamino)-1-(3-pyridyl)-1-butanone; DMN- dimethylnitrosamine

smokers probably result from a direct genotoxic effect of carcinogens present in tobacco smoke. Comparison of the frequency of detection of *ras* genes between smokers and non-smokers strongly supports this hypothesis. Activated *ras* proto-oncogenes were not detected in six of six non-smokers and three of four patients who had stopped smoking for at least five years before diagnosis whereas 44% (27/62) of adenocarcinomas from smokers had activated *ras* genes (Table 2 and Rodenhuis et. al., 1988).

SUMMARY

The transformation of a normal cell into a tumorigenic cell involves both the activation and concerted expression of proto-oncogenes and inactivation of suppressor genes. The activation of *ras* proto-oncogenes represents one step in the multistep process of carcinogenesis for a variety of rodent and human tumors. This activation is probably an early event in tumorigenesis in many cases and may be the `initiation' event in some cases. Thus, a chemical that induces rodent tumors by activation of *ras* proto-oncogenes can potentially invoke one step of the neoplastic process in humans exposed to the chemical. Moreover, dominant transforming oncogenes other than *ras* have been detected in human tumors as well as rodent tumors. The involvement of these putative proto-oncogenes in the development of neoplasia is unclear at present.

REFERENCES

Anathaswamy, H.N., Price, J.E., Goldberg, L.H., and Straka, C. (1987). Simultaneous transfer of tumorigenic and metastatic phenotypes by transfection with genomic DNA from a human cutaneous squamous cell carcinoma. *Proc. Am. Assoc. Cancer Res.* **28**, 69.

Anderson, M.W., and Reynolds, S.H. (1989). Activation of oncogenes by chemical carcinogens. In *The Pathobiology of Neoplasia* (A. Sirica, Ed.), pp.291-304. Plenum Press, New York, NY.

Bailleul, B., Brown, K., Ramsden, M., Akhurst, R.J., Fee, F., and Balmain, A. (1989). Chemical induction of oncogene mutations and growth factor activity in mouse skin carcinogenesis. *Environmental Health Perspectives* **81**, 23-27.

Balmain, A., and Brown, K. (1988). Oncogene activation in chemical carcinogenesis. *Advances in Cancer Res.* **51**, 147-182.

Barrett, J.C., Oshimura, M., and Koi, M. (1987). Role of oncogenes and tumor suppressant genes in a multistep model of carcinogenesis. In *Symposium on Fundamental Cancer Research* (F. Becker, Ed.), Vol. 38, pp.45-56.

Barbacid, M. (1987). *Ras* genes. *Ann. Rev. Biochem.* **56**, 780-813.

Belinsky, S., Devereux, T., Maronpot, R., Stoner, G., and Anderson, M. (1989). Relationship between the formation of promutagenic adducts and the activation of the K-*ras* protooncogene in lung tumors from A/J mice treated with nitrosamines. *Cancer Res.* **49**, 5305-5311.

Bishop, J.M. (1989). Oncogenes and clinical cancer. In *Oncogenes and Clinical Cancer*, pp.327-358.

Bishop, J.M. (1987). The molecular genetics of cancer. *Science (Wash. DC)* **235**, 303-311.

Bishop, J.M. (1985). Viral oncogenes. *Cell* **42**, 23-38.

Bongarzone, I., Pierotti, M.A., Monzini, N., Mondellini, P., Manenti, G., Donghi, R., Pilotti, S., Grieco, M., Santoro, M., Fusco, A., Vecchio, G., and Della Porta, G. (1989). High frequency of tyrosine kinase oncogenes in human papillary thyroid carcinoma. *Oncogene* **4**, 1457-1462.

Bos, J.L., Fearon, E.R., Hamilton, S.R., Verlaan-de Vries, M., van Boom, J.H., van der Eb, A.J., and Vogelstein, B. (1987). Prevalence of *ras* gene mutations in human colorectal cancers. *Nature* **327**, 293-299.

Bos, J.L., Verlaan-de Vries, M., van der EB, A.J., Janssen, J.W.G., Delwel, R., Lowen berg, B., and Colly, L.P. (1987). Mutations in N-*ras* predominate in acute myeloid leukemia. *Blood* **69**, 1237-1241.

Brown, K., Buchmann, A., and Balmain, A. (1989). Carcinogen-induced mutations in the mouse c-Ha-*ras* gene provide evidence of multiple pathways for tumor progression. *Proc. Natl. Acad. Sci. USA* **87**, 538-542.

Burmer, G.C., and Loeb, L.A. (1989). Mutations in the KRAS2 oncogene during progressive stages of human colon carcinoma. *Proc. Natl Acad. Sci. USA* **86**, 2403-2407.

Farr, C.J., Marshall, C.J., Easty, D.J., Wright, N.A., Powell, S.C., and Paraskeva, C. (1988). A study of *ras* gene mutations in clonic adenomas from familial polyposis coli patients. *Oncogene* **3**, 673-678.

Fasano, O., Birnbaum, D., Edlund, L., Fogh, J., and Wigler, M. (1984). New human genes detected by tumorigenicity assay. *Mol. Cell. Biol.* **4**, 1695-1705.

Fujita, J., Srivastava, S.K., Kraus, M.H., Rhim, J.S., Tronick, S.R., and Aaronson, S.A. (1985). Frequency of molecular alterations affecting *ras* protooncogenes in human urinary tract tumors. *Proc. Natl. Acad. Sci. USA* **82**, 3849-3853.

Grunewald, K., Lyons, J., Frohlich, A., Feichtinger, H., Weger, R.A., Schwab, G., Janssen, J.W.G., and Bartram, C.R. (1989). High frequency of Ki-*ras* codon 12 mutations in pancreatic adenocarcinomas. *Int. J. Cancer* **43**, 1037-1041.

Guerrero, I., and Pellicer, A. (1987). Mutational activation of oncogenes in animal model systems of carcinogenesis. *Mutation Research* **185**, 293-308.

Hansen, M.F., and Cavenee, W.K. (1988). Tumor suppressors: recessive mutations that lead to cancer. *Cell* **53**, 172-173.

Hirari, H., Kobayashi, Y., Mano, H., Hagiwara, K., Maru, Y., Omine, M., Mizoguxhi, H., Nishida, J., and Takaku, F. (1987). A point mutation at codon 13 of the N-*ras* oncogene in myelodysplastic syndrome. *Nature (Lond.).* **327**, 430-432.

Konopka, J.B., Watanabe, S.M., and Witte, O.N. (1984). An alteration of the human c-abl protein in K562 leukemia cells unmasks associated tyrosine kinase activity. *Cell* **37**, 1035-1042.

Leder, P., Battery, J., Lenior, G., Moulding, C., Murphy, W., Potter, H., Stewart, T., and Taub, R. (1984). Translocations among antibody genes in human cancer. *Science* **22**, 765-771.

Liu, E., Hjelle, B., and Bishop, M. (1988). Transforming genes in chronic myelogenous leukemia. *Proc. Natl. Acad. Sci. USA* **85**, 1952-1956.

Liu, E., Hjelle, B., Morgan, R., Hecht, F., and Bishop, M. (1987). Mutations of the kirsten-*ras* proto-oncogene in human preleukaemia. *Nature* **330**, 186-188.

McCormick, F. (1989). *Ras* oncogenes. In *Oncogenes and the Molecular Origin of Cancer*. (R.A. Weinberg, Ed.), pp.125-145, Cold Spring Harbor, NY.

Milburn, M.V., Tong, L., DeVos, A.M., Brunger, A., Yamaizumi, Z., Nishimura, S., and Kim, S. (1990). Molecular switch for signal transduction: Structural differences between active and inactive forms of protooncogenic *ras* proteins. *Science* **247**, 939-945.

Minna, J., Schutte, J., Viallet, J., Thomas, F., Kaye, F., Takahashi, T., Nau, M., Whang-Peng, J., Birrer, M., and Gazdar, A.F. (1989). Transcription factors and recessive oncogenes in the pathogenesis of human lung cancer. *Int. J. Cancer* **4**, 32-34.

Miller, J., and Miller, E. (1983). The metabolic activation and nucleic acid adducts of naturally-occurring carcinogens: recent results with ethyl carbamate and the spice flavors safrole and estragole. *Br. J. Cancer* **48**, 1-15.

Newcomb, E.W., Diamond, L.E., Sloan, S.R., Corominas, M., Gurrerro, I., and Pellicer, A. (1989). Radiation and chemical activation of *ras* oncogenes in different mouse strains. *Environmental Health Perspectives* **81**, 33-37.

Ochiya, T., Fujiyama, A., Fukushige, S., Hatada, I., and Matsubara, K. (1986). Molecular cloning of an oncogene from a human hepatocellular carcinoma. *Proc. Natl. Acad. Sci. USA* **83**, 4993-4997.

Pai, E., Kabsch, W., Krengel, U., Holmes, K., John, J., and Wittinghofer, A. (1989). Structure of the guanine- nucleotide-binding domain of the Ha-*ras* oncogene product p21 in the triphosphate conformation. *Nature* **341**, 209- 214.

Perucho, M., Forrester, K., Almoguera, C., Kahn, S., Lama, C., Shibata, D., Arnheim, N., and Grizzle, W.E. (1989). Expression and mutational activation of the c-Ki-*ras* gene in human carcinomas. *Cancer Cells* **7**, 137-141.

Philips, D.H., Hewer, A., Martin, C.N., Garner, R.C., and King, M.M. (1988). Correlation of DNA adduct levels in human lung with cigarette smoking. *Nature* **336**, 790-792.

Reynolds, S.H., Hunnicutt, C.K., Brown, K.C., Beattie, T., Pero, R., and Anderson,

M.W. (1989). *Ras* oncogenes in human lung tumors associated with exposure to cigarette smoke. *J. Cell. Biochemistry*.

Reynolds, S.H., Stowers, S.J., Maronpot, R.R., Aaronson, S.A., and Anderson, M.W. (1987). Activated oncogenes in B6C3F1 mouse liver tumors: implications for risk assessment. *Science* **237**, 1309-1317.

Rodenhuis, S., Slebos, R.J.C., Boot, A.J.M., Evers, S.G., Mooi, W.J., Wagenaar, S.S., Bodegom, P.C., and Bos, J.L. (1988). Incidence and possible clinical significance of K-*ras* oncogene activation in adenocarcinoma of the human lung. *Cancer Research* **48**, 5738-5741.

Sakamoto, H., Mori, M., Tara, M., Yoshida, T., Matsukawa, S., Shimizu, K., Sekiguchi, M., Terada, M., and Sugimura, T. (1986). Transforming gene from human stomach cancers and a noncancerous portion of stomach mucosa. *Proc. Natl. Acad. Sci. USA* **83**, 3997-4001.

Schechter, A.L., Stern, D.F., Vaidyanathan, L., Decker, S.J., Drebin, J.A., Greene, M.I., and Weinberg, R.A. (1984). The neu oncogene: An erb-B-related gene encoding a 185,000-M, tumor antigen. *Nature (Lond.)*. **312**, 513- 516.

Shibata. D., Almoguera, C., Forrester, K., Dunitz, J., Martin, S.E., Cosgrove, M.M., Perucho, M., and Arnheim, N. (1990). Detection of c-K-*ras* mutations in fine needle aspirates from human pancreatic adenocarcinomas. *Cancer Res.* **50**, 1279-1283.

Shih, C.S., Shilo, B., Goldfarb, M.P., Dannenberg, A., and Weinberg, R.A. (1979). Passage of phenotypes of chemically transformed cells via transfection DNA and chromatin. *Proc. Natl. Acad. Sci. USA* **76**, 5714-5718.

Slamon, D.J., Clark, G.M., Wong, S.G., Levin, W.J., Ullrich, A., and McGuire, W. (1987). Human breast cancer: correlation of relapse and survival with amplification of the HER-2/neu oncogene. *Science* **235**, 177-182.

Sloan, S.R., Newcomb, E.W., and Pellicer, A. (1990). Neutron radiation can activate K-*ras* via a point mutation in codon 146 and induces a different spectrum of *ras* mutations than does gamma radiation. *Mol. and Cell. Biol.* **10**, 405-408.

Smit, V.T.H.B.M., Boot, A.J.M., Smits, A.M.M., Fleuren, G, Cornelisse, C.J., and Bos, J.L. (1988). KRAS codon 12 mutations occur very frequently in pancreatic adenocarcinomas. *Nucleic Acids Research* **16**, 7773-7782.

Stanbridge, E.J. (1990). Identifying tumor suppressor genes in human colorectal cancer. *Science* **247**, 12-13.

Sukumar, A. (1988). Involvement of oncogenes in carcinogenesis. In *Cellular and Molecular Biology of Mammary Cancer* (Medina, D., Kidwell, W., Heppner, G., and Anderson, E., Eds.), pp.381-398. Plenum Press, New York, London.

Tada, M., Omata, M., and Ohto, M. (1990). Analysis of *ras* gene mutations in human hepatic malignant tumors by polymerase chain reaction and direct sequencing. *Cancer Res.* **50**, 1121-1124.

Takahashi, T., Nau, M.M., Chiba, I., Birrer, M.J., Rosenberg, R.K., Vincour, M., Levitt, M., Pass, H., Gazdar, A.F., Minna, J.D. (1989). p53: A frequent target for genetic abnormalities in lung cancer. *Science* **240**, 491- 494.

Tsujimoto, Y., Ikegaki, Y.N., and Croce, C.M. (1988). Characterization of the protein product of bcl-2, the gene involved in human follicular lymphoma. *Oncogene* **2**, 3-9.

Tsujimoto, Y., and Croce, C.M. (9186). Analysis of the structure, transcripts, and protein products of the bcl- 2, the gene involved in human follicular lymphoma. *Proc. Natl. Acad. Sci..* **83**, 5214-5218.

Varmus, H. (1989). An historical overview of oncogenes. In *Oncogenes and the Molecular Origin of Cancer* (R.A. Weinberg, Ed.), pp.3-44. Cold Spring Harbor Lab. Press, Cold Spring Harbor, NY.

Van 'T Veer, L.J., Burgering, B.M.T., Versteeg, R., Boot, A.J.M., Ruiter, D.J., Osanto, S., Schrier, P.I., and Bos, J.L. (1989). N-*ras* mutations in human cutaneous melanoma from sun-exposed body sites. *Mol. Cell. Biol.* **9**, 3114-3116.

Visvanathan, K., Pocock, R.D., Summerhayes, I.C. (1988). Preferential and novel activation of H-*ras* in human bladder carcinomas. *Oncogene Res.* **3**, 77-86.

Vogelstein, B., Fearon, E.R., Hamilton, S.R., Kern, S.E., Presinger, A.C., Leppert, M., Nakamura, Y., White, R., Smits, A.M.M., and Bos, J.L. (1988). Genetic alterations during colorectal-tumor development. *The New England J of Med.* **319**, 525-532.

Wang, Y., You, M., Reynolds, S., Stoner, G., and Anderson, M. (1990). Mutational

activation of the cellular Harvey *ras* oncogene in rat esophageal papillomas induced by methylbenzylnitrosamine. *Cancer Res.* **50**, 1591-1595.
Weinberg, R.A. (1985). Oncogenes, antioncogenes, and the molecular basis of multistep carcinogenesis. *Cancer Res.* **49**, 3713-3721.
Westin, E.H. (1989). Oncogenes. In *The Pathobiology of Neoplasia* (A. Sirica, Ed.), pp.275-290. Plenum Press, New York, NY.
Weston, A., Willey J.C., Modali, R., Sugimura, H., McDowell, E.M., Resau, J., Light, B., Haugen, A., Mann, D.L., Trump, B.F., and Harris, C.C. (1989). Differential DNA sequence deletions from chromosomes 3, 11, 13, and 17 in squamous-cell carcinoma, large-cell carcinoma, and adenocarcinoma of the human lung. *Proc. Natl. Acad. Sci. USA.* **86**, 5099-5103.
Wiseman, R., Stowers, S., Miller, E., Anderson, M., and Miller, J. (1986). Activating mutations of c-Ha-*ras* protooncogenes in chemically induced hepatomas of the male B6C3F1 mouse. *Proc. Natl. Acad. Sci. USA.* **83**, 5285-5289.
Wynder, E.L. (1972). Etiology of lung cancer. Reflections on two decades of research. *Cancer* **30**, 1332-1337.
Yoshida, T., Miyagawa, K., Odagiri, H., Sakamoto, H., Little, P.F.R., Terada, M., and Sugimura, T. (1987). Genomic sequence of hst, a transforming gene encoding a protein homologous to fibroblast growth factors and the int-2-encoded protein. *Proc. Natl. Acad. Sci. USA.* **84**, 7305-7310.
You, M., Candrian, U., Maronpot, R., Stoner, G., and Anderson, M. (1989). Activation of the K-*ras* protooncogene in spontaneously occurring and chemically-induced lung tumors of the strain A mouse. *Proc. Natl. Acad. Sci. USA.* **86**, 3070-3074.

TOXICITY OF 3-METHYLENEOXINDOLE, A PROPOSED REACTIVE INTERMEDIATE IN THE METABOLISM OF 3-METHYLINDOLE

Martin L. Appleton, Douglas N. Larson, Gary L. Skiles, William K. Nichols, and Garold S. Yost

Department of Pharmacology and Toxicology
University of Utah
Salt Lake City, Utah 84112

INTRODUCTION

3-Methylindole (3MI) is constitutively produced by anaerobic bacterial fermentation of tryptophan in mammalian intestines. In the case of ruminants excessive 3MI production causes selective pulmonary toxicity which can be fatal (Carlson and Yost, 1989; Yost, 1989). Previous work with ruminants has demonstrated that type I alveolar epithelial and nonciliated bronchiolar epithelial (Clara) cells are the most susceptible pulmonary cells. It has been suggested (Yost, 1989) that this pneumotoxicity results from cytochrome P-450 activation of 3MI to reactive intermediates (Figure 1) which react with macromolecules to produce the toxicity. One of the proposed intermediates is 3-methyleneoxindole (MEOI) which is known to react covalently with sulfhydryl groups (Still et al., 1965). This paper presents *in vivo* and *in vitro* findings about the toxicity of MEOI.

MATERIALS AND METHODS

All chemical and biochemical reagents were at least reagent grade and were commercially available.

Synthesis of 3-Methyleneoxindole (MEOI)

3-Methyleneoxindole (MEOI) was synthesized from indole-3-acetic acid via 3-bromooxindole-3-acetic acid. The 3-bromooxindole-3-acetic acid was synthesized using the procedure of Hinman and Bauman (1964). Conversion of the 3-bromo compound to MEOI was carried out according to the method of Andersen et al. (1972). Direct-probe electron impact mass spectrometry produced the following ions: m/z (relative abundance) 145 (100.0), 117 (99.5), and 90 (96.1) all consistent with the structure of MEOI. The ^1H NMR spectrum (200 MHz) of the product in acetone-d_6 (24°, referenced to acetone δ 2.04 ppm) showed: δ 6.20 [s, 2 H], 6.90 [d, J = 7.79 Hz, 1 H], 6.97 [t, J = 7.54 Hz, 1 H], 7.25 [t, J = 7.69 Hz, 1 H], 7.58 [d, J = 7.51 Hz, 1 H] again consistent with the structure of MEOI.

Animal Dosing

Male Swiss-Webster mice (approximately 30 g) were dosed by ip injection (10 µl/g) with the individual indole derivatives dissolved in corn oil (propylene glycol for MEOI) and observed for 24 hours. Approximate lethal doses (Table 1) were determined

Biological Reactive Intermediates IV, Edited by C.M. Witmer *et al.*
Plenum Press, New York, 1990

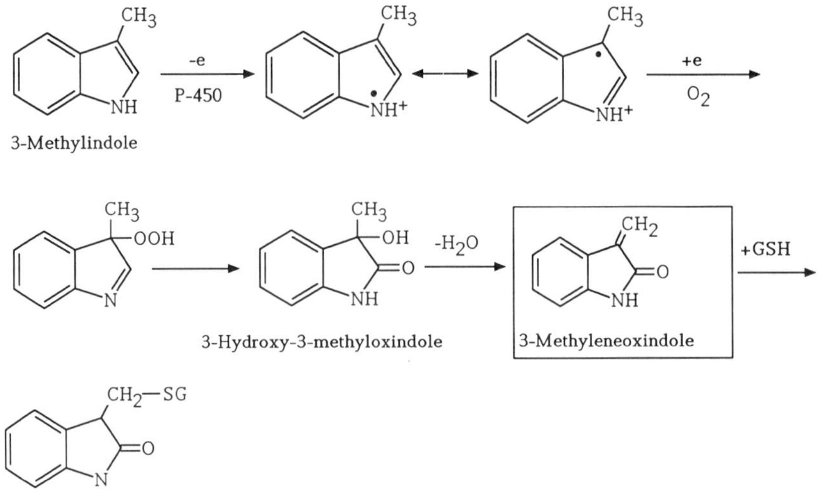

Figure 1. A possible metabolic pathway for 3-methylindole (3MI), showing 3-methyleneoxindole (MEOI) and conjugation with glutathione (GSH).

using 3 to 6 mice. Doses were increased in approximately 50 (MEOI) or 100 (others) mg/kg increments until animals expired.

Cell Isolation and Toxicity

Rabbit lung cells were isolated using centrifugal elutriation of protease-treated lung tissue as described by Devereux and Fouts (1981) and modified by Domin et al. (1986). Final purification of Clara and type II cells showed enrichment to 40-70% and 50-80%, respectively. Macrophage contamination in Clara cell fractions was reduced to <100% by panning cell suspensions for one hour on IgG-coated petri dishes as described by Dobbs et al. (1986). Macrophages were highly enriched in alveolar lavage fluids (>95%). Isolated lung cell populations were incubated in Dulbecco's modified Eagle medium (DMEM) containing 15 mM HEPES in the presence and absence of test compounds for 3 hours. Cell viability was evaluated using trypan blue exclusion.

RESULTS

Table 1 shows the approximate lethal doses of 3MI, MEOI and several related indole derivatives, which are also possible metabolites, administered to mice. MEOI was shown to be approximately 4-7 times more toxic than the other indole derivatives. Urine from the MEOI-treated mice was extracted with ethyl acetate and the non-polar metabolites were analyzed by reverse-phase HPLC. Comparisons of peak retention times with available standards (compounds in Table 1) showed the presence of the reduction product 3-methyloxindole as a significant metabolite, but no other metabolites could be identified.

The cytotoxicities of 3MI and MEOI to isolated rabbit pulmonary cells are shown in Table 2. 3MI (1 mM) showed the following order of cell susceptibilities: Clara > type II > macrophages. This is in contrast to MEOI (0.05 mM) which showed nonselective cytotoxicity. Pulmonary cell viabilities were also determined in the presence of 3MI or MEOI with 1-aminobenzotriazole (ABT). ABT is a potent cytochrome P-450 suicide substrate inhibitor which inactivates the enzyme by irreversible alkylation of the heme prosthetic group (Ortiz de Montellano and Mathews, 1981). As the data indicate, 3MI required activation by cytochrome P-450, whereas MEOI did not.

Table 1. Approximate Lethal Doses of 3-Methylindole and Derivatives Administered to Mice

Compound	Appr. Lethal Dose (mg/kg)
3-Methylindole	700
3-Hydroxy-3-methyloxindole	>650
3-Methyloxindole	400
Indole-3-carbinol	550
3-Methyleneoxindole	100

Table 2. Role of Bioactivation in 3MI and MEOI Induced Cytotoxicity

	Cell Viability (% of control)		
Compound(s)	Clara	Type II	Macrophage
3MI (1 mM)	54 ± 6	72 ± 7	85 ± 1
+ABT (5 mM)	87 ± 8	93 ± 1	96 ± 3
MEOI (0.05 mM)	23 ± 6	35 ± 17	17 ± 4
+ABT (5 mM)	35 ± 14	40 ± 7	23 ± 12

Cells (10^6) were incubated for 3 hours with compounds. ABT alone had no effect on cell viability. Values are averages ± one standard deviation (N=4 for 3MI; 3 for MEOI).

MEOI was approximately 7 times more toxic to mice than 3MI and was greater than 20 times more toxic to isolated rabbit pulmonary cells. This pulmonary cell toxicity was non-selective and was not inhibited by ABT. The results indicate that MEOI does not require bioactivation by cytochrome P-450 in the same way that 3MI does. Based on these results MEOI can be considered a good candidate for a reactive intermediate of 3MI metabolism and may contribute significantly to the pneumotoxicity of 3MI.

ACKNOWLEDGEMENTS

This work was supported by USPHS Grant HL13645. GSY is a USPHS Research Career Development Awardee (HL02119).

REFERENCES

Andersen, A.S., Moller, I.B. and Hansen, J. (1972). 3-Methyleneoxindole and plant growth regulation. *Physiol. Plant.* **27**, 105-108.

Carlson, J.R. and Yost, G.S. (1989). 3-Methylindole-induced acute lung injury resulting from ruminal fermentation of tryptophan. In *Toxicants of Plant Origin* (P.R. Cheeks, Ed.) pp. 108-123. CRC press, Boca Raton.

Deuereux, T.R. and Fouts, J.R. (1981). Isolation of pulmonary cells and use in studies of xenobiotic metabolism. *Methods Enzymol.* **77**, 147-154.

Dobbs, L.G., Gonzalez, R., and Williams, M.C. (1986). An improved method for isolating type II cells in high yield and purity. *Am. Rev. Respir. Dis.* **134**, 141-145.

Domin, B.A., Devereux, T.R., and Philpot, R.M. (1986). The cytochrome P-450 monooxygenase system of rabbit lung: Enzyme components, activities, and induction in the nonciliated bronchiolar epithelial (Clara) cell, alveolar type II cell, and alveolar macrophage. *Mol. Pharmacol.* **30**, 296-303.

Hinman, R.L. and Bauman, C.P. (1964). Reactions of N-bromosuccinimide and indoles. A simple synthesis of 3-bromooxindoles. *J. Org. Chem.* **29**, 1206-1215.
Ortiz De Montellano, P.R. and Mathews, J.M. (1981). Autocatalytic alkylation of the cytochrome P-450 prosthetic haem group by 1-aminobenzotriazole. *Biochem. J.* **195**, 761-764.
Still, C.C., Fukuyama, T.T., and Moyed, H.S. (1965). Inhibitory oxidation products of indole-3-acetic acid. *J. Biol. Chem.* **240**, 2612-2618.
Yost, G.S. (1989). Mechanisms of 3-methylindole pneumotoxicity. *Chem. Res. Toxicol.* **3**, 273-279.

THE ROLE OF CYTOCHROME P450IIE1 IN BIOACTIVATION OF ACETAMINOPHEN IN DIABETIC AND ACETONE-TREATED MICE

E.H. Jeffery[1], K. Arndt and W.M. Haschek,

Division of Nutritional Sciences
College of Veterinary Medicine
Institute for Environmental Studies
University of Illinois at Urbana-Champaign
Urbana, IL 61801

Of a number of purified rabbit hepatic isozymes of P450, two isozymes exhibit appreciable activity in the bioactivation of acetaminophen (AP), isozymes P450IIE1 and P450IA2 (originally known as 3a and 4, respectively) [Morgan et al., 1983]. Using monoclonal antibodies to these isozymes Raucy and coworkers have determined the relative importance of these isozymes in human hepatic microsomal AP bioactivation. They found that P450IIE1 and P450IA2 were approximately equally responsible for AP bioactivation when human hepatic microsomes were incubated with 10 mM AP [Raucy et al., 1989]. In this study we have attempted to relate the role of P450IIE1 in AP bioactivation in murine hepatic microsomes to AP hepatotoxicity in the whole mouse. We have evaluated microsomal bioactivation in the presence and absence of an inhibitor of P450IIE1, dimethylsulfoxide (DMSO), and compared the effect of induction of P450IIE1 on hepatic microsomal AP bioactivation and on AP toxicity in the whole mouse.

MATERIALS AND METHODS

Female Swiss outbred mice (20 - 25g) were given either: acetone (1% in the drinking water) for 10 days, streptozotocin (180 mg/kg ip in citrate) once 10 days before they were killed, phenobarbital (50 mg/kg ip in saline) daily for two days, or 3-methylcholanthrene (10 mg/kg ip in corn oil) daily for two days.

Mice were killed by cervical dislocation, and hepatic microsomes prepared by differential centrifugation. Dimethylnitrosamine (DMNA) demethylation was estimated as the rate of formaldehyde formation [Jeffery et al., 1988] and acetaminophen bioactivation by HPLC separation and UV detection of the acetaminophen-glutathione conjugate (a standard was kindly given us by Dr. J. Sinclair; VA Hospital, White River Junction, VT) [Howie et al., 1977].

Mice, 4 per group, were given AP (600 mg/kg ip in hot basic saline) or vehicle 6 hours before death. Livers were examined by light microscopy. The severity of lesions was qualitatively graded on a scale of 0 - 4, as follows: 0 = no lesions; 1 = mild centrilobular congestion and degeneration; 2 = 1/3 - 1/2 lobule affected by congestion, degeneration, necrosis and hemorrhage; 3 = 2/3 lobule affected by necrosis and hemorrhage; 4 = most of lobule affected.

[1] Author for correspondence: Dr. E.H. Jeffery, University of Illinois, 1005 W. Western Ave., Urbana, IL 61801

Biological Reactive Intermediates IV, Edited by C.M. Witmer *et al.*
Plenum Press, New York, 1990

RESULTS AND DISCUSSION

At low concentrations of DMNA, demethylation was induced by acetone and competitively inhibited by DMSO, K_i 0.93 mM. When high concentrations of DMNA were used, and the low K_m activity subtracted, no induction was seen by acetone and no inhibition by DMSO was observed [Jeffery et al., 1988].

Acetone produced a 3-fold increase in DMSO binding to hepatic microsomal P450 (reverse type I), with no change in K_s (untreated 13.04 mM, acetone-treated 13.21 mM). Morgan, Koop and Coon had previously shown that DMSO binds to P450IIE1 purified from rabbits, with a K_s of 73 mM, and that DMSO competitively inhibits purified P450IIE1- dependent ethanol oxidation [Morgan et al., 1983]. Also, Yang and co-workers have shown that the low-K_m DMNA demethylase in rat hepatic microsomes is P450IIE1-dependent [Peng et al., 1983]. We conclude that in mice DMSO preferentially inhibits the acetone-inducible isozyme, P450IIE1.

Diabetes, like acetone, causes the induction of P450IIE1 [Peng et al., 1983]. We have estimated the bioactivation of AP in hepatic microsomes from mice treated with acetone or streptozotocin, as well as mice treated with xenobiotics known to induce isozymes of P450 other than P450IIE1, phenobarbital (PB) and 3-methylcholanthrene (3-MC), Table 1. Acetone treatment and diabetes cause a considerable increase in AP bioactivation. No significant increase was seen after PB or 3-MC treatment. The inhibition of AP bioactivation by DMSO was of the order of 70% (Table 1), suggesting that P450IIE1 plays a major role in murine hepatic microsomal bioactivation of AP.

Table 1. Effect of Inducers on Mouse Hepatic Microsomal Acetaminophen Bioactivation

Treatment	Glutathione Conjugate (nmol/min/mg) - DMSO	+DMSO	% Inhibition
Control	0.58 ± 0.03	0.16 ± 0.01	73
Diabetic	1.15 ± 0.10	0.35 ± 0.06	70
Acetone	1.40 ± 0.04	0.12 ± 0.01	91
3-MC	0.51 ± 0.03	0.15 ± 0.01	70
PB	0.60 ± 0.03	0.24 ± 0.01	60

When toxicity of AP was compared across untreated, acetone-treated and streptozotocin-treated mice, the development of diabetes was found to slightly protect mice from AP toxicity, rather than to potentiate toxicity, as would be expected from the microsomal data, Table 2. We conclude that while P450IIE1 appears to play a major role in murine hepatic microsomal bioactivation of AP, a direct extrapolation from microsomal bioactivation to toxicity in the whole animal is not possible.

Table 2. Effect of Diabetes and Acetone-Treatment on Acetaminophen Hepatotoxicity[a]

Treatment	Control	Diabetic	Acetone
Saline	0.0 ± 0.0	0.0 ± 0.0	0.0 ± 0.0
Acetaminophen	2.0 ± 0.4	1.0 ± 0.3	3.2 ± 0.3

[a] See methods for toxicity scoring, scale 0-4.

REFERENCES

Howie, D, Adriaenssens, P.I. and Prescott, L.F. (1977). Paracetamol metabolism following overdosage: application of high performance liquid chromatography. *J. Pharm. Pharmacol.* **29**, 235-237.

Jeffery, E.H., Arndt, K. and Haschek, W.M. (1988). Mechanism of inhibition of hepatic bioactivation of paracetamol by dimethyl sulfoxide. *Drug Metab. Drug Interact.* **6**, 413-424.

Morgan, E.T., Koop, D.R. and Coon, M.J.(1982). Catalytic activity of cytochrome P-450 isozyme 3a isolated from liver microsomes of ethanol-treated rabbits. *J. Biol. Chem.* **257**, 13951-13957.

Morgan, E.T., Koop, D.R. and Coon, M.J. (1983). Comparison of six rabbit liver cytochrome P-450 isozymes in formation of a reactive metabolite of acetaminophen. Biochem. *Biophys. Res. Commun.* **112**, 8-13.

Raucy, J.L., Lasker, J.M., Lieber, C.S. and Black, M. (1989). Acetaminophen activationby human liver cytochrome P450IIE1 and P450IA2. *Arch. Biochem. Biophys.* **271**, 270-283.

Peng, R.X., Tennant, P., Lorr, N.A. and Yang, C.S. (1983). Alterations of microsomal monooxygenase system and carcinogen metabolism by streptozotocin-induced diabetes in rats. *Carcinogenesis*, **4**, 703-708.

ARYLAMINE-INDUCED HEMOLYTIC ANEMIA: ELECTRON SPIN RESONANCE SPECTROMETRY STUDIES

Timothy P. Bradshaw, D.C. McMillan, R.K. Crouch and D.J. Jollow

Departments of Pharmacology and Ophthamology
Medical University of South Carolina
Charleston, SC 29425

A variety of arylamine derivatives (e.g. dapsone, primaquine) are known to produce hemolytic anemia in man and experimental animals. Extensive studies in the 1950's and 60's established that hemolytic drugs induced methemoglobinemia and loss of erythrocytic reduced glutathione (GSH), and that individuals deficient in erthrocytic glucose-6-phosphate dehydrog-enase displayed enhanced susceptibility to drug-induced hemolytic anemia (for review, see E. Beutler, 1969; E. Beutler, 1972). Since drugs such as primaquine were active *in vivo* but not *in vitro*, the concept arose that the drugs were metabolized in the liver to active/reactive metabolites which, on entry to the red cell, produced a state of "oxidative stress". The resulting oxidative damage to critical sites within the red cell has been considered to lead to their "premature aging" and premature removal from the circulation by the spleen (A.R. Tarlov, et al., 1962; F.C. Gooden-Smith, et al., 1974; G. Cohen et al., 1964; A. Miller et al., 1970; R.W. Carrell, et al., 1975). The nature of the oxidant stress and the identity of the critical sites, however, are still unclear. Since the oxidation of hemoglobin to methemoglobin is known to be associated with the reduction of oxygen, much work has centered around the role of active oxygen species and free radicals as molecules capable of attacking cellular components (G. Cohen, et al., 1964; R.W. Carrell, et al., 1975; H.P. Misra et al., 1976; H.A. Itano, et al., 1977; B. Goldberg, et al., 1977). We have recently shown that phenylhydroxylamine (PHA) is a direct acting hemotoxin, capable of damaging the red blood cell during *in vitro* incubation such that, when readministered to isologous rats, the cells are rapidly sequestered by the spleen (J.H. Harrison, et al., 1986). This communication describes spin trap studies aimed at determination of whether or not free radical specie(s) are formed in the red cell in response to PHA, and if so, the identification of these specie(s).

Phenylhydroxylamine was synthesized as described previously (J.H. Harrison, et al., 1983). The spin-trap, 5,5'-dimethyl-1-pyrroline-N-oxide (DMPO), was purchased from Aldrich Chemical Co. (Milwaukee, WI); diethylenetriaminepentaacetic acid (DTPA) was obtained from Sigma Chemicals (St. Louis, MO). All other chemicals were reagent grade and used without further purification. The ESR instrument was a Varian E-4 EPR spectrometer.

Erythrocytes were collected from the descending aorta of male Sprague-Dawley rats (200-300 g, Camm Research Inc., Wayne, NJ) and washed with three volumes of phosphate-buffered saline, pH 7.4, containing 10 mM glucose (PBSG 7.4). Reaction mixtures consisted of 100 mM DMPO, 0.1 mM DTPA, and either lysed or intact red cells in PBSG 7.4 at 4°C. After a 5 min. incubation, phenylhydroxylamine (PHA) was added and the sample placed in a quartz ESR flat-cell. The ESR spectrum was then recorded at 25°C over an 8 min. scan time. Photolytic generation of the DMPO-

Biological Reactive Intermediates IV, Edited by C.M. Witmer *et al.*
Plenum Press, New York, 1990

hydroxyl radical adduct (DMPO-OH) was accomplished with H_2O_2 and UV light from a UVS-11 mineralight (Ultra-Violet Products Inc., San Gabriel, CA). The DMPO-glutathione adduct (DMPO-SG) was produced with glutathione disulfide (GSSG) by focusing an unfiltered 85W mercury lamp through a 50% transmission grating and onto the cell inserted into the cavity of the spectrometer.

As shown in Figure 1A, when intact red cells were incubated with the spin-trap and PHA under aerobic conditions, a four-line 1:2:2:1 ESR signal was produced. No signal was produced if any of the components (PHA, DMPO, or red cells) were omitted from the reaction mixture. Of interest, the intensity of the ESR signal increased as the concentration of PHA was increased from 5 to 200 μM, indicating concentration dependence towards PHA. In parallel experiments using lysed cell preparations, inclusion of either superoxide dismutase or catalase suppressed the ESR signal, suggesting a role for active oxygen in the generation of the spin trap signal.

The use of DMPO as a spin-trap is well documented in the literature, and although primarily used in the trapping of hydroxyl radicals and superoxide anions (B.E. Britigan, et al., 1987; P.J. Thornalley, et al., 1989; G.R. Buettner, 1989), it has recently found use for the detection of glutathione thiyl radicals (G.R. Buettner, 1987; D. Ross, et al., 1985; M.J. Davies, et al., 1987). The ESR spectra of the hydroxyl and thiyl adducts of DMPO are very similar. As illustrated in Figure 1B and C, DMPO-OH and DMPO-SG generated photolytically from H_2O_2 and GSSG, respectively, yield 1:2:2:1 spectra of similar structure. In agreement with the literature, the photolytically-generated DMPO-OH adduct had a nitrogen coupling constant (a^N) and a b-hydrogen splitting (a^H) of 14.9 G (M.J. Davies, et al., 1987). The DMPO-SG adduct had coupling constants of 15.4 (a^N) and 15.8 (a^H); both values are well within reported ranges (G.R. Buettner, 1987; D. Ross, et al., 1985; M.J. Davies, et al., 1987). The DMPO adduct formed experimentally by exposing red cells to PHA (Figure 1A) had hyperfine splittings of 15.3 (a^N) and 15.8 (a^H) which is within the reported range for the glutathione derived radical. However, the two adducts differed markedly in regard to stability. The DMPO-OH adduct is quite stable at room temperature, with a lifetime approaching hours, whereas the DMPO-SG adduct is a transient species which can not be detected by ESR one minute after removal of the light source. Thus the signal generated by the addition of PHA to DMPO-treated red cells and red cell lysate shows characteristics of both the DMPO-OH and DMPO-SG adducts; the coupling

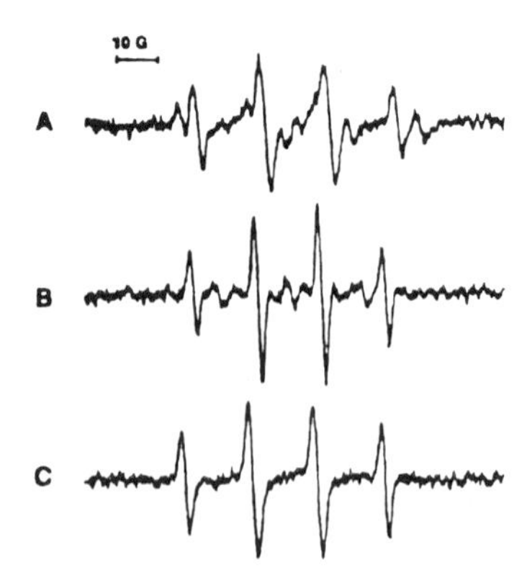

Figure 1. EPR spectra of DMPO trapped free radicals: (A) 100 mM DMPO, 0.1 mM DTPA, 15% red cells and 10 μM PHA in PBSG 7.4; (B) 20 mM DMPO, 25 mM H_2O_2 in PGSG 7.4, expose to uv light for 60 s; (C) 10 mM DMPO, 50 mM GSSG in PBSG 7.4, continuous exposure to uv light. Instrument settings: field set 3380 G. Time constant 1.0. Receiver gain 12,500. Modulation amplitude 2.5. Microwave power 10 mW. Frequency 9..45 GHz.

constants are similar to those of the glutathionyl adduct whereas the stability of the complex once formed, is akin to that of the hydroxyl adduct. Of particular importance, addition of red cell lysate to a photolytically generated DMPO-SG adduct maintained in the spectrometer with constant exposure to uv light, led to rapid quenching of the DMPO-SG signal. This effect was not seen with the DMPO-OH adduct.

In summary, exposure of rat red blood cells to phenylhydroxylamine resulted in the production of free radical specie(s) which may be trapped by DMPO. The intensity of the signal was directly dependent on the concentration of PHA added to the incubate (5 to 200 μM) and was most marked at concentrations of PHA which elicit a hemolytic response when the cells are readministered to isologous rats (> 50 μM). The identity of the radical species is not yet certain since the observed response shows characteristics of both a glutathionyl- and hydroxyl-DMPO adduct. In view of the ability of superoxide dismutase and catalase to suppress signal formation, it is likely that both hydroxyl and thiyl radicals are generated in the cell by PHA. This possibility is compatable with our overall hypothesis that active oxygen species formed by the interaction of PHA with oxyhemoglobin act to mediate glutathionyl and protein-cysteinyl radical formation which in turn are causal in the hemolytic sequalae.

ACKNOWLEDGEMENTS

This work was supported by Grants HL30038 and EY05757 from the U.S. Public Health Service.

REFERENCES

Beutler,E. (1969). *Pharmacol. Reviews* **21**, 73.
Beutler, E. (1972). In: *The Metabolic Basis of Inherited Disease* (J.B. Stanbury, J.B. Wyngaarden, and D.S. Fredrickson, eds.) McGraw Hill 3rd Edition, 1358-1388.
Britigan, B.E., Cohen, M.S. and Rosen, G.M. (1987). *J. Leuk. Biol.* **41**, 349.
Buettner, G.R. (1987). *Free Rad. Biol. Med.* **3**, 259.
Buettner, G.R. (1989). In: *Handbook of Methods for Oxygen Radical Research* (R.A. Greenwald, ed.) CRC Press, Cleveland, Ohio.
Carrell, R.W., Winterbourn, C.C., Rachmilewitz, E.A. (1975). *Brit. J. Haematol.* **30**, 259.
Cohen, G. and Hochstein, P. (1964). *Biochemistry* **3**, 895.
Davies, M.J., forni, L.G. and Shuter, S.L. (1987). *Chem.-Biol. Interactions* **61**, 177.
Goldberg, B. and Stern, A. (1977). *Mol. Pharmacol.* **13**, 832.
Gooden-Smith, F.C. and White, J.M. (1974). *Brit. J. Haematol.* **26**, 573.
Harrison, J.H. and Jollow, D.J.(1986). Pharmacol. Exp. Ther. **238**, 1045.
Harrison, J.H. and Jollow, D.J. (1986). *Chromatogr.* **277**, 173.
Itano, H.A., Hirota, K., and Vedvick, T.S. (1977). *Proc. Nat. Acad. Sci.* **74**, 2556.
Miller, A. and Smith, H.C. (1970). *Brit. J. Haematol.* **19**, 417.
Misra, H.P. and Fridovich, I. (1976). *Biochemistry* **15**, 681.
Ross, D., Norbeck, K. and Moldeus, P.J. (1985). *Biol. Chem.* **260**, 15028.
Tarlov, A.R., Brewer, G.J., Carson, P.E., and Alving, A.S. (1962). *Arch. Internal Med* **109**, 137.
Thornalley, P.J. and Bannister, J.V. (1989). In: *Handbook of Methods for Oxygen Radical Research* (R.A. Greenwald, ed.) CRC Press, Cleveland, Ohio.

SELECTIVE ALTERATIONS IN THE PROFILES OF NEWLY SYNTHESIZED PROTEINS BY ACETAMINOPHEN (APAP) AND ITS DIMETHYLATED ANALOGUES: RELATIONSHIP TO OXIDATIVE STRESS

Mary K. Bruno*, Steven D. Cohen** and Edward A. Khairallah*

Departments of *Molecular and Cell Biology and **Pharmacology and Toxicology,
University of Connecticut
Storrs, CT 06269

Acetaminophen (N-acetyl-p-aminophenol, APAP), one of the most widely used analgesic, antipyretic drugs currently available, when taken in excess of therapeutic doses can be activated by cytochrome P-450 to a highly reactive metabolite, N-acetylbenzoquinoneimine (NAPQI) (Dahlin, et al., 1984). NAPQI has been characterized as a strong electrophile and a potent oxidizing agent (Blair, et al., 1980) and both properties can lead to adverse effects on cellular metabolism (Albano, et al., 1985; Porubek, et al., 1987; Birge, et al., 1988). In order to more effectively evaluate the mechanisms of action of APAP and their physiological consequences, it becomes crucial not only to identify the early metabolic events that are altered, but also whether the functional impairments can be restored or have become irreversible.

Decreases in protein synthesis have been reported following administration of a hepatotoxic dose of APAP in vivo (Thorgeirsson, et al., 1976) and following exposure of hepatocytes in vitro (Gwynn, et al., 1979; Beales, et al., 1985). The objective of this study was to ascertain (1) whether the inhibition of protein synthesis during treatment and recovery from APAP are associated with a selective induction and/or repression of the synthesis of individual proteins as distinct from an overall general effect on total cellular protein synthesis, and (2) whether APAP's inhibition of protein synthesis is a related to its oxidative or arylative properties.

METHODS

Hepatocytes were isolated by collagenase perfusion from 3 month old C57Bl/6J male mice (Jackson Laboratories, Bar Harbor, ME) fed ad libitum and cultured as described (Bruno et al., 1988). The cells were maintained in culture for at least 18 hr following isolation prior to the initiation of experiments in fresh media lacking nicotinamide. Hepatocytes were exposed to either 10 mM APAP or 2,6-dimethyl acetaminophen (2,6-DMA), or 5 mM 3,5-dimethyl acetaminophen (3,5-DMA) for up to 12 hr (treatment phase). In some experiments the media was removed at the end of the exposure period and the cells were permitted to recover in drug-free media for an additional 4 hr (recovery phase). Control cells were run in parallel for each exposure and recovery period. Paired control and treated cells were pulse-labelled for 30 min with 30 µCi ^{35}S-methionine (sp. activity approximately 1 Ci / µmole) in methionine-free MEM containing 10% FBS. The pulse was terminated by rapidly washing the cells twice with ice cold phosphate buffered saline containing 0.1 mM phenylmethsulfonylfluoride and 10 mM unlabelled methionine followed by direct solubilization in 300 µl of electrophoresis sample gel buffer (0.0625 M Tris (pH 7.0), 1% SDS, and 10% glycerol). Aliquots were sampled for the determination of protein content and acid precipitable radioactivity prior to adjusting the remaining volume of solubilized cells to a final concentration of 0.1% 2-mercaptoethanol and 0.0025% phenol red as the tracking dye. Cell extracts containing equivalent amounts of radioactivity were analyzed on 10% polyacrylamide gels (Laemmli, 1970). Estimation of molecular weights was achieved by including ^{14}C-labelled protein molecular weight markers. Following electrophoresis, gels were fixed, impregnated with EN^{3}HANCE, and dried. Fluorography was performed at -70°C using preflashed Kodak XAR-5 X-ray film.

RESULTS AND DISCUSSION

Reduced rates of protein synthesis coincident with APAP binding and a 40 % decrease in cellular glutathione were exhibited prior to overt toxicity by cultured mouse hepatocytes exposed to 10 mM APAP for 4 hr (Bruno, et al., 1985; 1986). Upon removal of APAP and incubation of cultures in drug free media for an additional 4 hr, this inhibition of protein synthesis was fully reversible. After 12 hr of treatment, protein synthesis was diminished to 50% of control with minimal evidence of toxicity, however the inhibition was no longer fully reversible after the removal of APAP.

To evaluate whether the inhibition of protein synthesis was the result of a general diminution in the biosynthesis of all cellular proteins or reflective of a specific decrease in the synthesis of a few proteins, cultures were exposed to APAP for up to 12 hr and pulse-labelled with ^{35}S-methionine for subsequent electrophoretic analysis. If the newly synthesized proteins were analyzed on the basis of equivalent amounts of protein per lane, the gradual decrease in protein synthesis was reflective of diminished labelling of nearly all proteins resolvable by single dimension SDS-PAGE (data not shown). However, if comparisons between control and APAP-treated cultures were made on the basis of applying equivalent amounts of radioactivity per lane, a selective alteration in the relative rate of synthesis of some proteins was observed as early as 8 hr after treatment (Figure 1, compare lanes 1 and 3 of panel A). Specifically, the synthesis of a protein(s) migrating at an apparent molecular weight of 56-58 kDa (P_{56-58}) was gradually diminished relative to control, such that by 12 hrs the synthesis of this protein was virtually undetectable. Co-incident with the decline in the synthesis of P_{56-58} was an increase in the biosynthesis of another protein of 32 kDa (P_{32}).

If the diminished synthesis of P_{56-58} were functionally related to APAP toxicity, then one would predict that cellular recovery, upon APAP removal, would be accompanied by resynthesis of P_{56-58}. By contrast, cells that do not recover from the APAP insult should continue to exhibit a diminished capacity to synthesize this protein. Lane 2 of panel A revealed that in cultures recovering from an 8 hr APAP exposure, the synthesis of P_{56-58} was partially restored while the synthesis of P_{32} continued to remain elevated relative to control. By contrast, cells recovering from a 12 hr exposure did not demonstrate a recovery in the synthesis of P_{56-58} even though the enhanced synthesis of P32 persisted (data not shown).

In an effort to determine if the alterations in the synthesis of P_{56-58} and P_{32} were related to the arylative or oxidative properties of NAPQI, comparisons were made between APAP and its dimethylated analogues. 2,6-DMA, which has been reported to bind covalently to proteins without producing oxidative stress (Birge, et al., 1988; 1989), did not significantly alter protein synthesis (data not shown). By contrast, the greater cytotoxicity observed following 3,5-DMA, relative to APAP, has been attributed solely to the oxidative properties of the analogue (Fernando, et al., 1980; Porubek, et al., 1987; Birge, et al., 1988). Consistent with these observations, 3,5-DMA was a more potent inhibitor of protein synthesis than equimolar concentrations of APAP; within 8 hr, 5 mM 3,5-DMA was as effective as 10 mM APAP in inhibiting protein synthesis (data not shown).

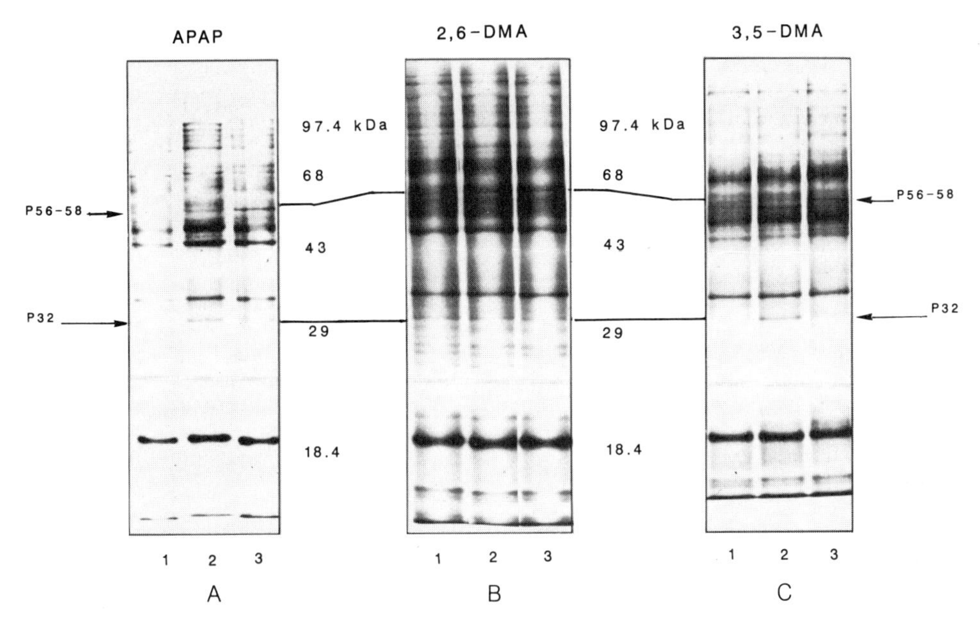

Figure 1. Mouse hepatocyte cultures were incubated with either 10 mM APAP (panel A), 10 mM 2,6-DMA (panel B), or 5 mM 3,5-DMA (panel C) directly solubilized in culture media for 8 hr. Cultures were pulse-labelled with ^{35}S-methionine in the absence of drug either immediately after treatment (lanes 1 of all panels) or following a 4 hr recovery in drug-free media (lanes 2 of all panels). Control cultures are represented in lanes 3 of all panels. Equivalent amounts of radioactivity were applied to each lane. Molecular weight markers are noted and newly synthesized P_{56-58} and P_{32} are indicated by the arrows.

Electropherograms obtained from cultures treated with or recovering from an 8 hr exposure to 10 mM 2,6-DMA or 5 mM 3,5-DMA are presented in panels B and C, respectively. Inspection of the profiles of newly synthesized proteins obtained from cultures incubated with 2,6-DMA for 8 hrs and allowed to recover revealed neither a significant decrease in P_{56-58}, nor was P_{32} induced above basal levels. By contrast, 3,5-DMA treated cultures exhibited diminished synthesis of P_{56-58} during the first 8 hrs with minimal regain of the protein during the 4 hr recovery period. However, the most prominent change observed in hepatocytes recovering from 3,5-DMA was an induction in P_{32} that exceeded levels detected in cultures recovering from APAP exposure.

In summary, these data indicate that the ability of hepatocytes to resynthesize P_{56-58} during and after APAP insult was reflective of the extent and reversibility of hepatocelluar damage induced by APAP. Furthermore, since both APAP and 3,5-DMA, by contrast to 2,6-DMA are metabolized to oxidatively reactive quinones (Blair, et al., 1980; Dahlin, et al., 1984; Fernando, et al., 1980) the selective induction of P32 synthesis both during drug exposure and recovery suggests that this effect may be in response to oxidative stress.

ACKNOWLEDGEMENTS

Supported in part by the Center for Biochemical Toxicology, The University of Connecticut Research Foundation and PHS Grant GM 31460.

REFERENCES

Albano, E., Rundgren, M., Harvison, P.J., Nelson, S.D. and Moldeus, P. (1985). Mechanism of N-acetyl-p-benzoquinone imine cytotoxicity. *Mol. Pharmacol.* **28**, 306-311.

Beales, D., Hue, D.P. and McLean, A.E.M. (1985). Lipid peroxidation, protein synthesis, and protection by calcium EDTA in paracetamol injury to isolated hepatocytes. *Biochem. Pharmacol.* **34**, 19-23.

Birge, R.B., Bartolone, J.B., Nishanian, E.V., Bruno, M.K., Mangold, J.B., Cohen, S.D. and Khairallah, E.A. (1988). Dissociation of covalent binding from the oxidative effects of acetaminophen: studies using dimethylated acetaminophen derivatives. Biochem. *Pharmacol.* **37**, 3383-3393.

Birge, R.B., Bartolone, J.B., McCann, D.J., Mangold, J.B., Cohen, S.D. and Khairallah, E.A. (1989). Selective protein arylation by acetaminophen and 2,6-dimethyl-acetaminophen in cultured hepatocytes from phenobarbital-induced and uninduced mice. *Biochem. Pharmacol.* **38**, 4429-4438.

Blair, I.A., Boobis, A.R., Davies, D.S., and Cresp, T.M. (1980). Paracetamol oxidation: Synthesis and reacivity of N-acetyl-p-benzoquinoneimine. *Tetrahedron Lett.* **21**, 4947-4950.

Bruno, M.K., Bartolone, J.B., Cohen, S.D. and Khairallah, E.A. (1985). Cultured mouse hepatocytes as a valid model for acetaminophen hepatotoxicity. *Pharmacologist* **27**, 482.

Bruno, M.K., Cohen, S.D. and Khairallah, E.A. (1986). Perturbations in protein metabolism induced by acetaminophen are not the result of GSH depletion. *Fedn. Proc.* **45**, 1932.

Bruno, M.K., Cohen, S.D. and Khairallah, E.A. (1988). Antidotal effectiveness of N-acetylcysteine in reversing acetaminophen-induced hepatotoxicity. Enhancement of the proteolysis of arylated proteins. *Biochem. Pharamcol.* **37**, 4319-4325.

Dahlin, D.C., Miwa, G.T., Lu, A.Y. and Nelson, S.D. (1984). N-Acetyl-p-benzoquinone imine: a cytochrome P-450-mediated oxidation product of acetaminophen. *Proc. Natl. Acad. Sci. USA* **81**, 1327-1331.

Fernando, C.R., Calder, I.C. and Ham, K.N. (1980). Studies on the mechanism of toxicity of acetaminophen. Synthesis and reactions of N-acetyl-2,6-dimethyl- and N-acetyl-3,5-dimethyl-p-benzoquinone imines. *J. Med. Chem.* **23**, 1153-1158.

Gwynn, J., Fry, J.R. and Bridges, J.W. (1979). The effect of paracetamol and other foreign compounds on protein synthesis in isolated adult rat hepatocytes. *Biochem. Soc. Trans.* **7**, 117-119.

Laemmli, U.K. (1970). Cleavage of structural proteins during the assembly of the head bacteriophage T4. *Nature* **227**, 680-685.

Porubek, D.J., Rundgren, M., Harvison, P.J., Nelson, S.D. and Moldeus, P. (1987). Investigation of the mechanism of acetaminophen toxicity in isolated hepatocytes with the acetaminophen analogues 3,5-dimethylacetaminophen and 2,6-dimethylacetaminophen. *Mol. Pharmacol.* **31**, 647-653.

Thorgeirsson, S.S., Sasame, H.A., Mitchell, J.R., Jollow, D.J. and Potter, W.Z. (1976). Biochemical changes after hepatic injury from toxic doses of acetaminophen or furosemide. *Pharmacology* **14**, 205-217.

BENZENE METABOLISM BY TWO PURIFIED, RECONSTITUTED RAT HEPATIC MIXED FUNCTION OXIDASE SYSTEMS

Thomas A. Chepiga[1], Chung S. Yang[2] and Robert Snyder[1]

[1]Joint Graduate Program in Toxicology and [2]Department of Chemical Biology and Pharmacognosy, College of Pharmacy
Rutgers University/Robert Wood Johnson Medical School
Piscataway, NJ 08855-0789

Benzene, a bone marrow depressant, requires metabolism in order to exert its hematopoetic toxicity (Snyder et al., 1980). In mammals, this metabolism ccurs primarily in the liver and is catalyzed by the cytochrome P-450 containing mixed function oxidase (MFO) system (Gonasun et al., 1973). Post and Snyder (1983) demonstrated that rat liver microsomes contain at least two, distinct MFO activities which can metabolize benzene. One is induced by phenobarbital (PB) pretreatment and displays a Km value greater than 10mM; the other is induced by benzene pretreatment and displays a Km value equal to approximately 0.1mM. PB pretreatment results in the induction of cytochrome P450IIB1 (Ryan et al., 1979), while benzene pretreatment appears to induce cytochrome P450IIE1 (Ingelman-Sundberg and Johansson, 1984). Koop et al. (1989) demonstrated that P450IIE1, purified from rabbit liver microsomes, is an effective benzene hydroxylase. In this study, we have examined benzene metabolism by two purified, reconstituted rat hepatic MFO systems containing either cytochrome P450IIB1 or P450IIE1.

MATERIALS AND METHODS

All proteins were purified using rat hepatic microsomes (Lu and Levin, 1972). Cytochrome P450IIB1 was isolated from PB-treated rats according to the procedure of Ryan et al. (1982). The final preparation had a specific content of 16.2 nmol cytochrome P-450/mg protein. Cytochrome P450IIE1 was purified from isoniazid-treated rats according to the method of Ryan et al. (1985). Two preparations with specific contents of 9.6 and 11.3 nmol cytochrome P-450/mg protein, respectively, were isolated for these experiments.

NADPH-cytochrome c (P-450) reductase was purified according to a combination of the methods of Dignam and Stroebel (1975) and Yasukochi and Masters (1976). The final preparation had a specific content of 34,000 units/mg protein (one unit of reductase is that amount capable of catalyzing the reduction of 1 nmole of cytochrome c/minute at 22°C). Cytochrome b_5 was purified according to the procedure of Tamburini et al. (1985) to a final specific content of 45 nmol/mg protein.

Incubation volumes contained 0.1 nmole cytochrome P-450 (either P450IIE1 or P450IIB1), 1200 units NADPH-cytochrome c (P-450) reductase and 15 ug lipid (dilauroylphosphatidylcholine). The lipid was sonicated in water immediately prior to use, combined with the cytochrome P-450 and reductase, and allowed to stand at room temperature for 5 minutes before addition to reaction vials. Cytochrome b_5, when

Biological Reactive Intermediates IV, Edited by C.M. Witmer *et al.*
Plenum Press, New York, 1990

present, was added to the cytochrome P-450/reductase mixture, prior to the addition of the lipid, in equimolar concentration to cytochrome P-450.

Enzyme mixtures were added to reaction vessels containing potassium phosphate buffer (pH 7.4) and kept on ice. Radiolabelled benzene, diluted in buffer, was then added and the vials preincubated at 37°C for three minutes. Reactions were initiated by the addition of NADPH (final concentration, 1.0 mM) to a final volume of 0.25 ml. Reactions were stopped after 10 (P450IIE1) or 20 (P450IIB1) minutes by the addition of 25 ul of 15% formic acid (reaction vial pH = 3.0). Twenty-five microliters of an L-ascorbic acid solution (final concentration, 10 mM) were immediately added and the vials kept frozen at -70°C until analyzed (<30 hrs).

A thawed aliquot of the reaction vial contents was filtered through a 0.45 um nylon filter and spiked with a solution containing non-radioactive standards (i.e. hydroquinone, catechol, phenol). A portion of this filtered, spiked sample was then injected onto a standard analytical ODS C-18 reverse phase HPLC column in-line with a flow-through UV spectrophotometer. Metabolite separation was accomplished utilizing a water/methanol gradient and the column eluate was monitored at 280 nm. Fractions were collected at one-half minute intervals. Products were identified by comparing the retention times of the non-radioactive standards, determined spectrophotometrically, with the retention times of the radioactive peaks and quantitated by liquid scintillation counting.

All results represent the mean of at least two experiments run in duplicate or triplicate.

RESULTS AND DISCUSSION

Benzene metabolism by cytochromes P450IIE1 and P450IIB1 was examined both in the absence and presence of cytochrome b5. Cytochrome b5 has been demonstrated to enhance the cytochrome P-450-catalyzed metabolism of a number of substrates (Tamburini et al., 1985). Its influence seems to depend both upon the particular P-450 isozyme examined and the substrate utilized.

The results of the experiments examining benzene metabolism by the MFO system containing P450IIE1 are presented in Figure 1 and the upper portion of Table 1. Metabolism was examined over a substrate concentration range of 20 uM to 4.0 mM benzene. At all substrate concentrations, hydroquinone and phenol were the only detectable metabolites formed.

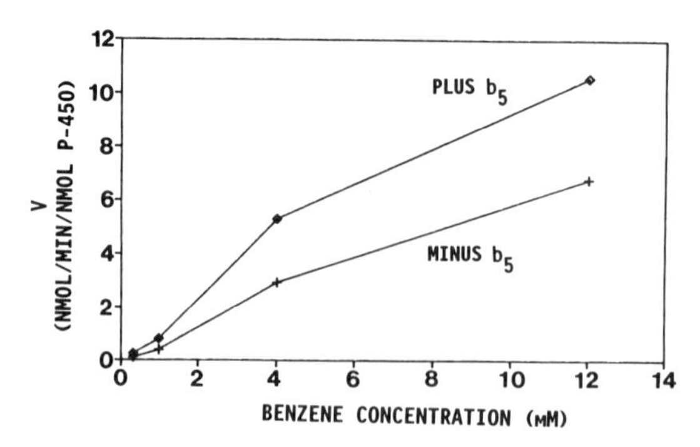

Figure 1 Benzene metabolism by cytochrome P450IIE1 plus/minus cytochrome b5

Table 1. Effect of Cytochrome b_5 on Phenol and Hydroquinone Formation

	BENZENE CONCENTRATION	MINUS CYTOCHROME b_5		PLUS CYTOCHROME b_5	
		PHENOL	HYDROQUINONE	PHENOL	HYDROQUINONE
CYTOCHROME P450IIE1	20 uM	0.16	0.03	0.70	0.46
	80 uM	0.46	0.04	2.04	0.77
	320 uM	1.18	0.08	4.68	0.76
	800 uM	2.40	0.15	7.45	0.76
	2.0 mM	3.90	0.21	8.87	0.63
	4.0 mM	5.28	0.29	9.84	0.50
CYTOCHROME P450IIB1	320 uM	0.11	ND	0.25	ND
	1.0 mM	0.35	0.04	0.75	0.05
	4.0 mM	2.78	0.16	5.02	0.28
	12.0 mM	6.48	0.32	10.14	0.44

UNITS = nmol product formed/minute/nmol P-450

ND = not detected

Cytochrome b_5 was found to increase overall benzene metabolism 6-fold at the 20 and 80 uM substrate concentrations but only 2 to 4-fold at higher benzene concentrations when cytochrome P450IIE1 was used. The increase in phenol formation paralleled this overall increase, while cytochrome b_5 stimulated hydroquinone formation approximately 15-fold at the three lowest substrate concentrations (20, 80, 320 uM), but only about 3.5-fold at higher benzene concentrations.

In studies examining cytochrome P450IIB1-catalyzed benzene metabolism, a greater substrate concentration range was selected since the Km for benzene metabolism in PB-induced microsomes is >10 mM. The results of these experiments are presented in Fig. 2 and the lower portion of Table 1. Hydroquinone and phenol were the only metabolites detected. Cytochrome b_5 increased cytochrome P450IIB1-catalyzed metabolism 1.5 to 2-fold at each substrate concentration examined. At the lower substrate concentrations of 320 uM and 1.0 mM benzene, cytochrome P450IIB1 did not metabolize benzene efficiently, even in the presence of cytochrome b_5. However, this rate of metabolism increases substantially at higher benzene concentrations. Cytochrome b_5 enhanced the rate of formation of both phenol and hydroquinone approximately 1.5 to 2-fold. This is in contrast to the differential stimulation observed in the MFO system containing cytochrome P450IIE1.

Based on these results, it is evident that cytochrome P450IIE1 and cytochrome P450IIB1 possess different benzene monooxygenase activities in purified, reconstituted rat hepatic MFO enzyme systems. P450IIB1 represents a relatively low affinity form of cytochrome of P-450 with respect to benzene metabolism, while P450IIE1 is metabollically more efficient at lower substrate concentrations. At the common substrate concentration of 320 uM benzene, P450IIB1 exhibited turnover numbers of

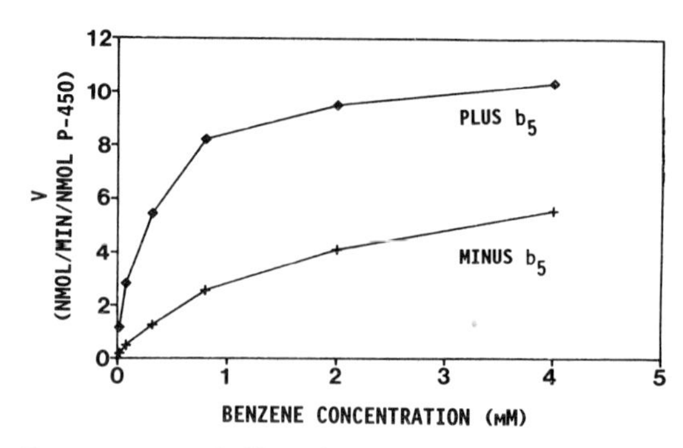

Figure 2. Benzene metabolism by cytochrome P450IIB1 plus/minus cytochrome b_5

0.11 (minus b_5) and 0.25 (plus b_5), while P450IIE1 metabolized benzene at the rates of 1.26 (minus b_5) and 5.44 (plus b_5) [units = nmol/min/nmol P-450]. This corresponds to 10 and 20-fold metabolic rate differences, respectively, between the two enzyme systems. It is probable that the different MFO activities reported by Post and Snyder (1983) in hepatic microsomes from rats treated with either phenobarbital or benzene are a result of the respective induction of P450IIB1 and P450IIE1 in these animals.

Cytochrome b_5 stimulated the cytochrome P-450-catalyzed metabolism of benzene in both of the purified MFO systems examined. However, this enhancement by b_5 was found to be differential, stimulating the metabolic rate of P450IIB1 approximately 2-fold, while enhancing P450IIE1 benzene metabolism up to 6-fold. Additionally, cytochrome b_5-enhanced P450IIE1 was the system most effective in forming hydroquinone, a metabolite which is thought to contribute to the production of benzene toxicity.

REFERENCES

Dignam, J.D. and Stroebel, H.W. (1975). Preparation of homogenous NADPH-cytochrome P-450 reductase from rat liver. *Biochem. Biophys. Res. Commun.* **63**, 845-852.

Gonasun, L., Witmer, C., Kocsis, J.J. and Snyder, R. (1973). Benzene metabolism in mouse liover microsomes. *Toxicol. Appl. Pharmacol.* **26**,398-406.

Ingelman-Sundberg, M., and Johansson, I. (1984). Mechanisms of hydroxyl radical formation and ethanol oxidation by ethanol-inducible and other forms of rabbit liver microsomal cytochromes P-450. *J. Biol. Chem.* **259**, 6447-6458.

Koop, D.R., Laethem, C.L. and Schnier, G.G. (1989). Identification of ethanol-inducible P450 isozyme 3a (P450IIE1) as a benzene and phenol hydroxylase. *Toxicol. Appl. Pharmacol.* **98**, 278-288.

Lu, A.Y.H. and Levin, W. (1972). Partial purification of cytochromes P-450 and P-448 from rat liver microsomes. *Biochem. Biophys. Res. Commun.* **46**, 1334-1339.

Post, G.B. and Snyder, R. (1983). Effects of enzyme induction on microsomal benzene metabolism. *J. Toxicol. Environ. Health.* **11**, 811-825.

Ryan, D.E., Ramanathan, L., Iida, S., Thomas, P.E., Haniu, M., Shively, J.E., Lieber, C.S. and Levin, W. (1985). Characterization of a major form of rat hepatic microsomal cytochrome P-450 induced by isoniazid. *J. Biol. Chem.* **260**, 6385-6393.

Ryan, D.E., Thomas, P.E., Korzeniowski, D. and Levin, W. (1979). Separation and characterization of highly purified forms of liver microsomal cytochrome P-450 from rats treated with polychlorinated biphenyls, phenobarbital, and 3-methylcholanthrene. *J. Biol. Chem.* **254**, 1365-1374.

Ryan, D.E., Thomas, P.E. and Levin, W. (1982). Purification and characterization of a minor form of hepatic microsomal cytochrome P-450 from rats treated with polychlorinated biphenyls. *Arch. Biochem. Biophys.* **216**, 272-288.

Snyder, R., Sammett, D., Witmer, C., Kocsis, J.J. and Snyder, R. (1980). An overview of the problem of benzene toxicity and some recent data on the relationship of benzene metabolism to benzene toxicity. In: Genotoxic Effects of Airborne Agents. Tice, R.R., Costa, D.L. and Schaich, K.M. (eds.) Plenum Press, New York.

Tamburini, P.P., White, R.E. and Schenkman, J.B. (1985). Chemical characterization of protein-protein interactions between cytochrome P-450 and cytochrome b_5. *J. Biol. Chem.* **260**, 4007-4015.

Yasukochi, Y. and Masters, B.S.S. (1976). Some properties of a detergent-solubilized NADPH-cytochrome c (cytochrome P-450) reductase purified by biospecific affinity chromatography. *J. Biol. Chem.* **251**, 5337-5344.

STEREOCHEMICAL INDUCTION OF CYTOCHROME P450IVA1 (P452) AND PEROXISOME PROLIFERATION IN MALE RAT

Edwin C. Chinje and G. Gordon Gibson

Molecular Toxicology Group and Department of Biochemistry
University of Surrey
Guildford, SURREY. GU2 5XH , England, U.K.

Peroxisome proliferator xenobiotics (including clofibrate, a hypolipidaemic drug) are a class of non-genotoxic hepatocarcinogen in sensitive species such as the rat and mouse and result in characteristic liver responses including hepatomegaly, peroxisome proliferation and induction of microsomal cytochrome P450IVA1 (P452)-dependent fatty acid hydroxylase.

It has recently been proposed that catalytically-competent cytochrome P452 is a necessary prerequisite for peroxisome proliferation by these xenobiotics (Lake, B.G. et al., 1984). In a scheme presented in the paper, the initial liver response to peroxisome proliferator increase in long chain omega-hydroxy fatty acids that are subsequently oxidized to the corresponding long chain dicarboxylic acids. As these latter metabolites are preferentially degraded by beta-oxidation in the peroxisome, it was proposed that dicarboxylic acids then form the proximal stimulus for peroxisome proliferation in an attempt by the cell to maintain lipid homeostasis (Sharma et al., 1988).

It is the purpose of this short communication to examine the above hypothesis further by comparing the relative abilities of optically-active enantiomers of a clofibrate analogue to induce both microsomal cytochrome P450IVA1 and peroxisome proliferation.

MATERIALS AND METHODS

Enantiomers

Optically pure enantiomers and the corresponding racemic mixture of the clofibrate analogue [2-(4-p-chlorophenyl-oxy)-2-phenyl ethanoic acid] were kindly provided by ICI Pharmaceuticals, Cheshire, England.

Animals and Chemical Pretreatment

Male Long Evans hooded rats (125-150g body weight) were pretreated by gavage once daily for 3 consecutive days with the enantiomers or racemate at a dose level of 80mg/Kg. Compounds were administered in gum tragacanth as vehicle and control groups received the vehicle only. All animals were killed at the start of the fourth day, i.e 24 hours after the last dose and liver homogenates and microsomal fractions were prepared as previously described (Sharma et al., 1988).

Biological Reactive Intermediates IV, Edited by C.M. Witmer *et al.*
Plenum Press, New York, 1990

Enzyme Assays

Total carbon monoxide-discernible cytochrome P450, specific ELISA-based cytochrome P450IVA1 determination, 11- and 12-hydroxylation of lauric acid, KCN-insensitive palmitoyl CoA oxidation and carnitine acetyl transferase were determined as previously described (Sharma, et al., 1988). ELISA quatitation of bifunctional protein of peroxisomal beta-oxidation spiral was as described elsewhere (Milton, M.N., 1989).

Statistical Analysis

Statistical data evaluation was performed using the Students t-test.

RESULTS AND DISCUSSION

The effects of administration of the pure enantiomers and racemate of the clofibrate analogue on liver / body weight ratio (a measure of hepatomegaly), total carbon monoxide-descernible cytochrome P450 and specific cytochrome P450IVA1 isoenzyme levels are presented in Table 1.

The data presented for hepatomegaly studies show the S(+)-isomer causing the highest increase in liver/body weight ratio. This increase was found to be significantly different (P <0.01) from control, whereas the R(-) antipode and racemate were not at this level of test. Hepatomegaly occurs within a few days of administration of test compound and has been shown to be dose-dependent (Lake, B.G. et al., 1984). The present finding is in agreement with previous reports since the most potent of the enantiomeric forms (i.e the eutomer) produced the greatest extent of hepatomegaly at the dose and duration of these studies. Similar results were obtained for the CO-descernible cytochrome P450 which were not indicative of the changes in specific isoenzyme composition.

Induction of specific cytochrome P450IVA1 is more accurately reflected in the enzyme-linked immunosorbent assay (ELISA). The data in Table 1 shows the isoenzyme level in control group to be about 5% of total cytochrome P450 population, and induction by the isomers and racemate led to 1.5 to 4 fold induction over constitutive levels at the dose level studied. Again the highest fold increase was with the S(+)-isomer having an eudismic ratio (S/R activity ratio) of about 3. These results indicate that there is a high degree of enantioselectivity in this isoenzyme induction.

Table 1. Stereoselective Induction of Hepatomegaly and Cytochrome P450/P452 Content in Male Long Evans Hooded Rat

Treatment	Liver/body weight ratio(%)	Total Cyt.P450 specific content (nmol/mg)	Cyt.P450IVA1 quantitation: Specific Cyt.P452 (nmol/mg)	Cyt.P450IVA1 quantitation: % Total Cyt.P450
CONTROL	5.64 ± 0.31	0.48 ± 0.06	0.028 ± 0.003	5.95 ± 0.88
S(+)ISOMER	6.67 ± 0.41 *	0.57 ± 0.06 *	0.107 ± 0.012 ***	19.40 ± 3.27 ***
R(-)ISOMER	5.83 ± 0.19	0.49 ± 0.07	0.037 ± 0.007 **	7.44 ± 0.41 **
RACEMIC MIXTURE	5.97 ± 0.26	0.55 ± 0.09 *	0.059 ± 0.006 ***	10.88 ± 0.75 ***

Values are mean ± standard deviation of six animals (control) or three animals in the test group. *P < 0.01, **P < 0.005, ***P < 0.001 .

Reverse phase HPLC analysis on the influence of the different treatments on the 11- and 12-hydroxylation of lauric acid is shown on Table 2. Total 11- and 12-hydroxylation of lauric acid has been shown to be induced by other workers. Also preferential induction of laurate 12-hydroxylase activity was demonstrated in our laboratory (Lake, B.G. et al., 1984) and the present findings agree with previous reports. Here again we observe an eudismic ratio of about 3 for this activity being further evidence of enantioselectivity in the induction process. T he significance of this preferential induction in endogenous metabolism is not clear.

The effect of the various treatments on the capacity of some peroxisomal beta-oxidation enzymes is also included in Table 2. Data presented for both cyanide-insensitive palmitoyl CoA (pCoA) and bifunctional protein of peroxisomal beta-oxidation spiral show clear consistency in the stereoselective induction of this organelle with the S(+)-isomer causing 3 fold and 6 fold increases respectively. These corresponded to eudismic ratios of 2.3 and 5.5. Increases in the activity of hepatic carnitine acetyl transferase (peroxisomal and mitochondrial) was also demonstrated to be highly stereoselective (see Table 2). Again consistent with the other results, the S(+)-isomer caused about 9 fold and about 5 and 2 for the racemate and R(-)-isomer respectively. It therefore becomes increasingly evident that as with the microsomal system, there is also a high degree of stereoselectivity in the capacity of the peroxisomal beta-oxidizing machinery following treatment by these compounds.

In summary the results presented in this study show the S(+)-isomer to be a more potent inducer i.e the eutomer, of microsomal cytochrome P450IVA1 and its associated lauric acid hydrxylase activity than its corresponding R(-) antipode, i.e the distomer, with the racemate exhibiting an intermediary potency. Also an identical enantiomeric selectivity was observed for the phenomenon of peroxisome proliferation by these compounds. Thus taken collectively, the above data is not inconsistent with the previously stated hypothesis that cytochrome P450IVA1 induction and peroxisome proliferation are intimately linked. Whether the observed stereochemical induction is in recognition or disposition still remains to be elucidated.

Table 2. Differential Induction of Rat Hepatic Microsomal and Peroxisomal Enzyme Activities

Treatment	Laurate 12-hydroxylase activity (nmol/min/mg)	PCoA oxidase (nmolNADH/ min/mg)	Peroxisomal bifunctional protein(a)	Carnitine acetyl transferase (nmolCoA/min/mg)
CONTROL	3.52 ± 0.62	4.42 ± 0.91	1.24 ± 0.16	1.42 ± 0.44
S(+)ISOMER	11.80 ± 1.53 ***	11.67 ± 2.66 ***	7.58 ± 1.50 ***	14.68 ± 4.90 ***
R(-)ISOMER	4.49 ± 0.59 **	5.20 ± 0.62 *	1.39 ± 0.29 *	2.71 ± 0.79 **
RACEMIC MIXTURE	6.85 ± 0.64 ***	8.71 ± 1.42 ***	4.91 ± 0.81 ***	6.01 ± 0.61 ***

Values are mean ± standard deviation of either six animals (control) or three animals in the test group. *P < 0.1, **P < 0.01, ***P < 0.001 .
(a)= Units corresponding to mg control protein equivalents/mg protein loaded.

ACKNOWLEDGEMENTS

This work was supported by a pre-doctoral studentship (E.C.C.) of the Ministry of Higher Education and Scientific Research, Republic of Cameroon and project grants from the MRC and Wellcome Trust (G.G.G.).

REFERENCES

Lake, B.G., Gray, T.J.B., Pels Rijcken, W.R., Beamand, J.A. and Gangolli, S.D. (1984). *Xenobiotica* **14**, 269-276.

Milton, M.N. (1989). PhD. Thesis, University of Surrey, England.

Sharma, R., Lake, B.G. and Gibson, G.G. (1988). *Biochem. Pharmacol.* **37**, 1203-1206 .

Sharma, R., Lake,B.G., Foster, J. and Gibson, G.G. (1988). *Biochem Pharmacol.* **37**, 1193-1201.

CYANIDE LIBERATION AND OXIDATIVE STRESS BY ORGANOTHIOCYANATES, ORGANONITRILES AND NITROPRUSSIDE IN ISOLATED HEPATOCYTES

Sahar Elguindi and Peter J. O'Brien[1]

Faculty of Pharmacy
University of Toronto
Toronto, Ontario CANADA M5S 1A1

Organothiocyanates and organonitriles are effective insecticide and fungicide chemicals. Glutathione-S-transferases may play a role in the liberation of HCN from organothiocyanates whereas mixed-function oxidases may play a role in HCN liberation from various organonitriles [Ohkawa et al., 1972].

Nitroprusside is a widely used potent hypotensive agent to stop bleeding during surgery. The most widely held mechanism for cyanide liberation from nitroprusside *in vivo* involves a reaction with oxyhemoglobin possibly via hemoglobin sulfydryl groups in the erythrocyte which results in the production of cyanomethemoglobin [Smith et al., 1974]. Nitroprusside has also been shown to release cyanide with chemical reductants and sulphur nucleophiles *in vitro* [Butler et al., 1974]. However, the hypotensive effects of nitroprusside infused before the hind-leg, the liver and the head were considerably reduced in comparison to the responses of i.v. infused nitroprusside [Kreye et al., 1982]. Furthermore, the biological half-life is also two orders of magnitude shorter than its half-life in blood *in vitro*. Erythrocytes, therefore, may not play such an important role in nitroprusside inactivation *in vivo* [Kreye et al., 1982]. Recently, perfusion of isolated liver with nitroprusside was also found to release cyanide [Devlin et al., 1989].

The following shows that cyanide was released from the above compounds, including nitroprusside, in isolated hepatocytes and this cyanide release was associated with the cytotoxicity of these compounds. Furthermore, cytotoxicity could be prevented with sodium thiosulphate.

MATERIALS AND METHODS

All organothiocyanates and organonitriles were obtained from Aldrich Chemical Co. (Milwaukee, U.S.A). Nitroprusside, sodium thiocyanate and potassium cyanide were obtained from Fischer Scientific Co. (Fairlawn, NJ).

Male Sprague-Dawley rats (200-220 g) were obtained from Charles River (St Constant, Quebec) and fed a standard chow diet. Hepatocytes were prepared by collagenase perfusion of rat liver as previously described [Moldeus et al., 1978]. Isolated cells were suspended in Krebs-Henseleit buffer, pH 7.4 containing 12.5 mM Hepes at a concentration of 1 x 10^6 cells per ml and incubated in rotating round bottom flasks in a water bath (37°C) under an atmosphere of 95% O_2/5% CO_2. Cell

[1]. Address all correspondence to: P.J. O'Brien, Faculty of Pharmacy, 19 Russell St., Toronto, Ontario, CANADA M5S 1A1

Biological Reactive Intermediates IV, Edited by C.M. Witmer *et al.*
Plenum Press, New York, 1990

viability (normally 80-90%) was determined by trypan blue exclusion (final concentration 0.16%). Cells were preincubated for 30minutes prior to the addition of any chemicals.

Oxygen uptakes was determined using a Clark type oxygen electrode (Yellow Springs Instrument Co., Model 5300) using a thermostated 2 ml chamber at 37°C.

RESULTS AND DISCUSSION

Experiments incubating isolated hepatocytes with various concentrations of organothiocyanates and organonitriles indicated that the order of effectiveness at inducing hepatocyte death was: benzylthiocyanate >octylthiocyanate > potassium cyanide > lethane 384 >lactonitrile > benzyl cyanide > ethylthiocyanate > >nitroprusside > butyronitrile. Benzylthiocyanate (1 mM) caused 50% toxicity within 2 hours whereas sodium thiocyanate was not toxic even at 10 mM. (data not shown). Since the physiological mechanism for the irreversible conversion of the toxic cyanide to the non-toxic sodium thiocyanate involves sodium thiosulphate and rhodanese, the protective effect of sodium thiosulphate, if any, was investigated so as to ascertain whether these compounds exert their cytotoxic effect to isolated hepatocytes by releasing cyanide. Among mammalian tissues, liver rhodanese activity was highest in the liver and was located in the mitochondria [Tanka et al., 1983]. Sodium thiosulphate (10mM) was added to cells 45 minutes prior to addition of the investigated compounds and protection was observed (Table 1).

Table 1. Pretreatment of Isolated Rat Hepatocytes with 10mM Sodium Thiosulphate

	Concentration mM	Per Cent Trypan Blue Uptake at 4 hours	
		With thiosulphate	Without thiosulphate
Control	--	19 ± 4	20 ± 4
Potassium Cyanide	1	29 ± 7	74 ± 5
Benzyl Cyanide	4	28 ± 6	100
Benzylthiocyanate	2	45 ± 5	100
Nitroprusside	10	78 ± 6	73 ± 7
Ethylthiocyanate	10	65 ± 4	100
Nitroprusside + Dithiothreitol	5 2	38 ± 5	100
Nitroprusside + Sodium borohydride	5 2	40 ± 7	100

Trypan blue uptake is an indictor of cell viability. Values are reported as the mean ± SD of at least 3 experiments. Isolated rat hepatocytes were pretreated with 10 mM sodium thiosulphate 45 minutes prior to addition of compound.

Nitroprusside readily released cyanide on additon of thiols, glutathione, dithiothreitol or reductants such assodium borohydride *in vitro* (results not shown). Furthermore, as shown in Figure 1, glutathione, dithiothreitol or sodium borohydride enhanced the toxicity of nitroprusside (5 mM) in isolated hepatocytes. However, this enhancement was eliminated by pretreatment with 10 mM sodium thiosulphate (Table 1). Because sodium thiosulphate did not protect hepatocytes from nitroprusside alone,

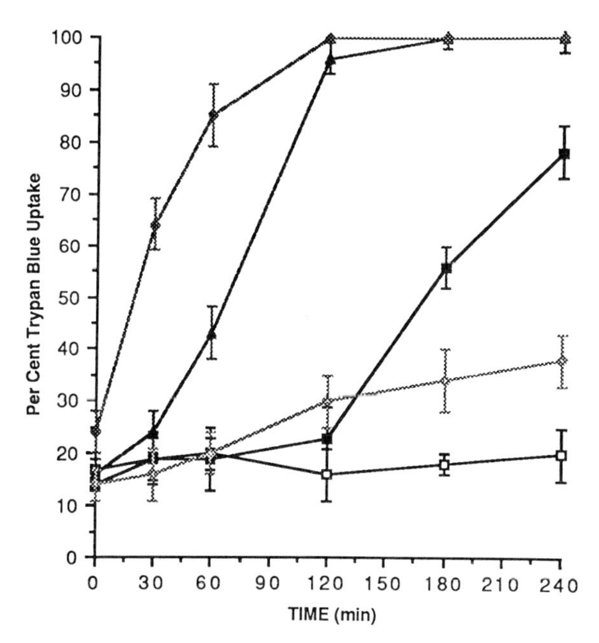

Figure 1. Isolated hepatocytes were incubated with 1 mM nitroprusside (◊), 5 mM nitroprusside (■), 1 mM nitroprusside + 50 μM BCNU (♦), 5 mM nitroprusside + 4 mM Azide (▲) and no additions or 4 mM azide or 50 μM BCNU alone (□) for four hours. Cell viability was determined by the per cent uptake of trypan blue after 0, 30, 60, 120, 180 and 240 minutes of incubation. Experiments were repeated 3 times with 3 different batches of cells.

other mechanisms of cytotoxicity were investigated for nitroprusside. The effect on hepatocyte susceptibility to nitroprusside when the enzymes involved in hydrogen peroxide metabolism (e.g. glutathione reductase and catalase) were inactivated was therefore investigated. As shown in Figure 1, glutathione reductase inactivation by 50 μM 1,3 bis (2-chloroethyl)-1-nitrosourea (BCNU) or catalase inactivation by 4mM azide, markedly enhanced nitroprusside toxicity. This suggested that nitroprusside cytotoxicity in isolated hepatocytes may involve oxidative stress. Further evidence for this was the marked increase in cyanide resistant respiration (results not shown).

In summary, cyanide release plays a major role in the toxicity of organothiocyanates, organonitriles and nitroprusside. These compounds are capable of releasing cyanide by reduction pathways as well as reaction with glutathione. Pretreatment with sodium thiosulphate, a substrate for the physiological mechanism for the removal of cyanide, prevented hepatocyte cytotoxicity. It is therefore concluded that cyanide antidote regimens should be considered in the therapy of humans who may have been poisoned with such compounds.

REFERENCES

Butler, A.R. and Glidewell, C. (1987). Recent chemical studies of sodium nitroprusside relevent to its hypotensive action. *Chem. Soc. Rev.* **16**, 361-380.

Devlin, D.J., Smith, R.P., and Thron, C.D. (1989). Cyanide release from nitroprusside in the isolated, perfused, bloodless liver and hindlimbs of the rat. *Toxicol. Appl. Pharmacol.* **99**,354-356.

Kiese, M. (1974). In *Methemoglobinemia: A Comprehensive Treatise*, pp. 3-7. CRC Press, Cleveland, Ohio.

Kreye, V.A.W., and Reske, S.N. (1982). Possible site of the *in vivo* disposition of sodium nitroprusside in the rat. *Naunyn-Schmiedeberg's Arch. Pharmacol.* **320**, 260-265.

Moldeus, P., Hogberg, J. and Orrenius, S. (1978). In *Methods in Enzymology* (Eds. P.A. Hoffee and M.E.Jones), Vol. 51, pp. 60-71. Academic Press, New York.

Ohkawa, H., Ohkawa, R., Yamamoto, I., and Casida, J.E. (1972). Enzymatic mechanisms and toxicological significance of hydrogen cyanide liberation from various organothiocyanates and organonitriles in mice and houseflies. *Pest. Biochem. and Physiol.* **2**, 95-112.

Smith, R.P., and Kruszyna, H. (1974). Nitroprusside produces cyanide poisoning via a reaction with hemoglobin. *J. Pharmacol. Exp. Therap.* **191**, 557-563.

Tanka, D. and Gatal, K. (1983). Histochemical detection of thiosulphate sulphurtransferase (rhodanese) activity. *Histochemistry* **77**, 285-288.

MIXED FUNCTION OXIDASE ENZYME RESPONSES TO *IN VIVO* AND *IN VITRO* CHROMATE TREATMENT

Ellen C. Faria and Charlotte M. Witmer

Department of Pharmacology and Toxicology
Rutgers, The State University of New Jersey
Piscataway, N.J. 08855-0789

Chromium (Cr) is an essential nutrient in humans which aids in the metabolism of cholesterol, glucose and fats. The trivalent state (Cr(III)) is essential in trace doses, while Cr(VI) is toxic to mammals in acute and subchronic doses [Langard and Norseth, 1986] and long term exposure has been associated with respiratory cancer [Chiazze and Wolf, 1980; Langard and Norseth, 1975; Mancuso, 1975]. Unusually high rates of lung cancer have been reported in workers in chrome plating, leather tanning, and other Cr related industries [Chiazze and Wolf, 1980; Langard and Norseth, 1975; Mancuso, 1975]. The mechanisms for the toxic and genotoxic effects of Cr are only partially understood. The differences in toxicity of the two most common oxidation states, Cr(VI) and Cr(III), are due to the relative lack of ability of cationic Cr(III) compounds to cross cell membranes, while Cr(VI) as the chromate anion, crosses biological membranes freely [Aaseth et al., 1982; Wiegand et al., 1985]. The intracellular reduction of Cr(VI) has recently been shown to result in both Cr(V) and Cr(III) production, both of which are putative DNA damaging agents [Goodgame et al., 1982; Jennette, 1982]. Compounds such as glutathione (GSH) [Aaseth et al., 1982; Connett and Wetterhahn, 1983], ascorbate [Connett and Wetterhahn, 1983], and hydrogen peroxide [Cupo and Wetterhahn, 1985] participate in the intracellular reduction of Cr(VI). Reduction takes place both in mitochondria [Alexander et al., 1982], and the endoplasmic reticulum [Gruber and Jennette, 1978]. When GSH is the reductant a toxic glutathionyl radical (GS) may be formed [Wetterhahn, 1990, in press]. Cr(III) binds to nucleophiles including some sulfhydryl (SH) containing enzymes, with some resulting enzyme inhibition. Thus the metabolism of Cr(VI) is important for the interaction of Cr with DNA [Tsapakos and Wetterhahn, 1983], with GSH [Wiegand et al., 1985] and with SH groups on other cellular macromolecules. Previous studies with microsomes have indicated that cytochrome P-450 (P-450), an SH containing enzyme, acts as an electron donor in the microsomal reduction of Cr(VI) [Gruber and Jennette, 1978].

In the present study we investigate the *in vivo* and *in vitro* effects of the interaction of Cr(VI) with the hepatic microsomal mixed function oxidase system. This system includes P-450, cytochrome b_5 (b_5) and NADPH-Cytochrome c reductase (reductase). We report that Cr(VI) decreases P-450 content both *in vivo* and *in vitro*, while b_5 content is not affected. The reductase activity is inhibited *in vitro* but not *in vivo*.

MATERIALS AND METHODS

In vivo experiments were carried out using male and female Sprague Dawley (SD) rats and female Wistar rats, 100-120 g body weight. The rats were housed in plastic cages, with food and water supplied *ad libitum*. The dichromate ($K_2Cr_2O_7$) was given every 48 hours by i.p. injection for a total of 6 injections, in neutral 0.9% saline or 0.1M phosphate buffer containing 0.15M KCl (PO_4 buffer), pH adjusted to 7.4. The doses administered were 68, 130, 160, and 192 μmoles Cr/kg body weight/single injection. Control animals underwent the same procedure but were administered the appropriate vehicle, pH 7.4. The rats were fasted and sacrificed 24 hours after the last injection, by ether, and the livers excised and perfused with ice cold 0.9% saline. Hepatic microsomes were prepared by differential centrifugation [Gonasun et al., 1973]. The microsomal protein content was determined by the biuret method [Layne, 1957]. Hepatic microsomal P-450 and b_5 content were determined by the method of Omura and Sato [1964] and the reductase activity was determined by the method of Phillips and Langdon [1962].

The *in vitro* effects of chromate on P-450 and b_5 were examined by incubating hepatic microsomes with 25.0, 50.0 and 100.0 μmoles $K_2Cr_2O_7$/nmole P-450 at 37° for 15 min. The microsomal samples were then washed with ice cold PO_4 buffer, and the microsomes repelleted by centrifugation at 105,000g, were resuspended in PO_4 buffer and the P-450 and b_5 content determined. The protein content was again determined by the method of Lowry [Lowry et al., 1951]. The *in vitro* effects of chromate on the reductase activity were examined by the addition to hepatic microsomes, without incubation, of 1.0, 2.0, and 4.0 μmoles K_2CrO_4/3.2 ml total assay volume. Each 3.2 ml

Table 1. Hepatic Microsomal Cytochrome P-450 Content and NADPH-Cytochrome c Reductase Activity after $K_2Cr_2O_7$ Treatment of Rats.

Animal Treatment	%Lethality	nmoles cyt P-450 / mg protein	nmoles red. cyt c / mg protein
68 umoles Cr/kg:A			
♀ Wistar - Control (5)	0	0.57 ± 0.09	134.7 ± 12.2
♀ Wistar - Cr (6)	0	0.64 ± 0.12	123.5 ± 15.4
♂ SD - Control (5)	0	0.77 ± 0.12	125.1 + 9.2
♂ SD - Cr (6)	0	0.79 ± 0.07	152.7 ± 19.3
130 umoles Cr/kg:B			
♂ SD - Control (9)	0	0.84 ± 0.09	242.1 ± 18.5
♂ SD - Cr (12)	0	0.59 ± 0.09 *	224.6 ± 37.6
♀ SD - Control (5)	0	0.61 ± 0.08	247.3 ± 51.0
♀ SD - Cr (7)	0	0.63 ± 0.06	289.3 ± 43.6
160 umoles Cr/kg:B			
♂ SD - Control (5)	0	0.84 ± 0.09	242.1 ± 18.5
♂ SD - Cr (7)	7/7 (100%)	———	———
192 umoles Cr/kg:A			
♀ Wistar - Control (5)	0	0.57 ± 0.09	134.7 ± 12.2
♀ Wistar - Cr (6)	7/7 (100%)	———	———

Animals dosed as described in Materials and methods. Numbers in parentheses equal the number of animals per group. Values are expressed as mean ± S.D.
A = $K_2Cr_2O_7$ (pH 7.4) in 0.1M PO_4 + 0.15M KCl (pH 7.4) buffer.
B = $K_2Cr_2O_7$ (pH 7.4) in 0.9% saline.
* values significantly different from control ($p < 0.001$).

contained the same amount of reductase activity (50 nmoles Cyt *c* reduced/min). The microsomes used in the *in vitro* studies were obtained from male SD or female Wistar rats which were untreated or pretreated with phenobarbital (PB 60 mg/kg) for 3 days; PB (60 mg/kg) for 3 days followed by beta naphthoflavone (BNF 80 mg/kg) 48 hours prior to sacrifice; or $K_2Cr_2O_7$ (10 mg/kg/48 hrs 6 times), as designated in each figure.

RESULTS AND DISCUSSION

Administration of 68 μmoles Cr/kg body weight/single i.p. injection as $K_2Cr_2O_7$ (pH 7.4), following the dosage regime outlined in *Materials and Methods*, to either female Wistar rats or to male SD rats did not reduce hepatic microsomal P-450 content nor reductase activity as compared to control levels (Table 1). Nor was b_5 content reduced in the treated animals (data not shown). Administration of 130 μmoles Cr/kg body weight to male SD rats did decrease ($p < 0.001$) the hepatic microsomal P-450 content (Table 1), to 70% of control values. Neither the reductase activity (Table 1) nor the b_5 content (data not shown) were significantly reduced with this dose, however, administration of 160 and 192 μmoles Cr/kg body weight doses to male SD and female Wistar rats, respectively, resulted in 100% mortality after 3 injections (Table 1).

The *in vitro* incubation of 25.0, 50.0, and 100.0 μmoles $K_2Cr_2O_7$/nmole P-450 with hepatic microsomes from $K_2Cr_2O_7$ pretreated SD rats caused decreases in P-450 content to an average of 58, 55, and 32% of controls, respectively, (Figure 1). The same incubation conditions with hepatic microsomes from $K_2Cr_2O_7$ pretreated female Wistar rats yielded decreases in P-450 content to an average of 65, 49, and 30% of control, respectively (Figure 1). Similar reductions in P-450 content (average difference = 5.5%) resulted using hepatic microsomes from SD and Wistar rats which were untreated and PB pretreated (data not shown), and with PB-BNF pretreated rats (Figure 2).

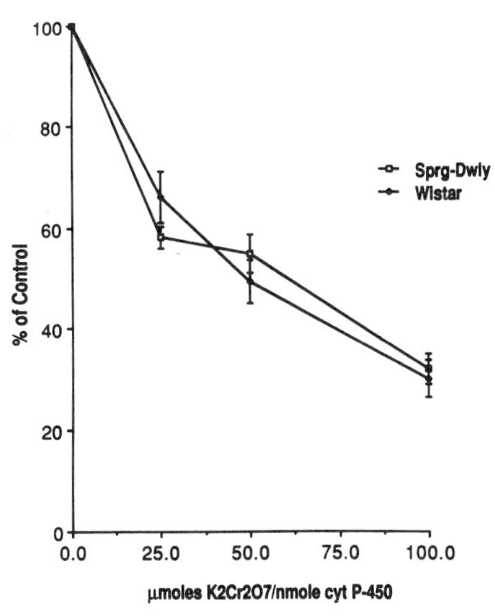

Figure 1. The effects of $K_2Cr_2O_7$ incubation on hepatic microsomal cytochrome P-450 content from $K_2Cr_2O_7$ pretreated rats. Experimental procedure is described in text. SD rats average control value = 0.70 ± 0.11 nmoles P-450/mg microsomal protein. Wistar rats average control value = 0.70 ± 0.07 nmoles P-450/mg microsomal protein. (n = 6 for each data point).

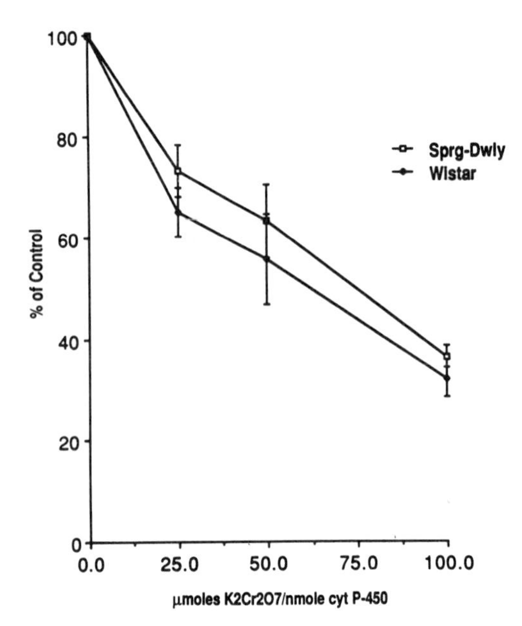

Figure 2. The effects of $K_2Cr_2O_7$ incubation on hepatic microsomal cytochrome P-450 content from PB-BNF pretreated rats. Experimental procedure is described in text. SD rats average control value = 1.66 ± 0.11 nmoles P-450/mg microsomal protein. Wistar rats average control value = 1.97 ± 0.45 nmoles P-450/mg microsomal protein. (n = 6 for each data point).

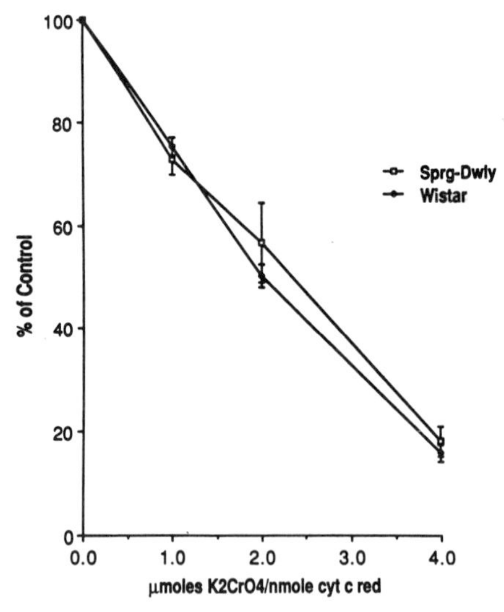

Figure 3. The effects of K_2CrO_4 addition on hepatic microsomal NADPH-Cytochrome c reductase activity from untreated rats. Experimental procedure is described in text. SD rats average control value = 147.2 ± 37.7 nmoles reduced Cyt c/min/mg microsomal protein. Wistar rats average control value = 64.1 ± 5.3 nmoles reduced Cyt c/min/mg microsomal protien. (n=6 for each data point).

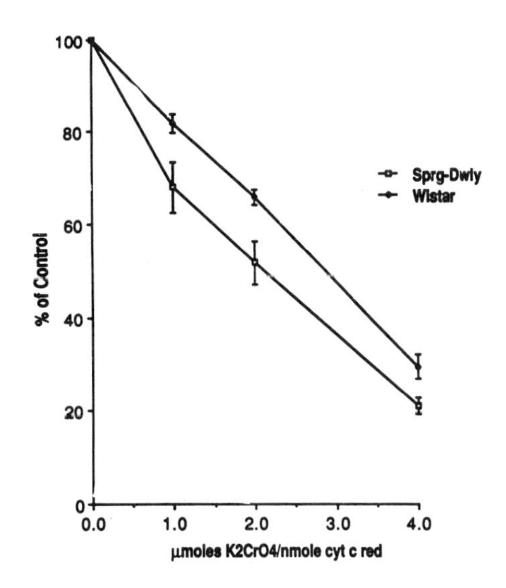

Figure 4. The effects of K_2CrO_4 addition on hepatic microsomal NADPH-Cytochrome c reductase activity from PB-BNF pretreated rats. Experimental procedure is described in text. SD rats average control value = 264.3 ± 31.1 nmoles reduced Cyt c/min/mg microsomal protein. Wistar rats average control value = 250.7 ± 24.5 nmoles reduced Cyt c/min/mg microsomal protein. (n=6 for each data point).

Addition, without incubation, of 1.0, 2.0, and 4.0 µmoles K_2CrO_4 to samples of hepatic microsomes from control SD rats containing equivalent amounts of reductase activity, resulted in decreases in reductase activity to an average of 68, 52, and 21% of controls, respectively (Figure 3). The same treatment of hepatic microsomes from untreated female Wistar rats resulted in decreases in reductase activity to an average of 82, 66, and 29% of controls, respectively (Figure 3). Similar changes in reductase activity (average difference = 6.3%) resulted using hepatic microsomes from SD and Wistar rats which were $K_2Cr_2O_7$ and PB pretreated (data not shown), and with PB-BNF pretreated rats (Figure 4).

The two enzymes studied, P-450 and the reductase, are both SH containing hemoproteins important in the metabolism of xenobiotics. Previous studies have demonstrated that Cr(VI) is a substrate of P-450 therefore reduces Cr(VI) to Cr(V) or Cr(III) (Gruber and Jennette, 1978). The treatment of male SD rats with 130 µmoles Cr/kg body weight as Cr(VI) induced a significant decrease in P-450 levels, while the same treatment of female SD rats caused no similar reduction. These different susceptibilities may reflect some structural differences in the P-450s. The Cr(III) produced from the reduction of Cr(VI) is known to form relatively inert complexes with several different ligands, thus the Cr(III) formed might bind to and inactivate specific P-450s.

The *in vitro* incubation of Cr(VI) with hepatic microsomes resulted in dose response decreases in P-450 levels. These reductions were similar for hepatic microsomes from control, PB, PB-BNF, and $K_2Cr_2O_7$ pretreated animals. This suggests that no specific isoenzyme of P-450 is preferentially attacked. The *in vitro* addition of Cr(VI) to hepatic microsomes also resulted in dose response decreases in reductase activity. These reductions are in contrast to the effect found with *in vivo* treatment of animals with Cr(VI). Treatment of male and female SD rats with 130 µmoles Cr/kg body weight as Cr(VI) did not reduce any reductase activity, but did decrease the P-450 content. This may be the result of a low availability of the reductase to the Cr(VI) *in vivo*. The studies resulted in a high degree of lethality with higher Cr(VI) doses (160 and 192 µmoles/kg). The cause(s) of lethality have not been well defined. The decrease in P-450 observed in Cr(VI) male SD treated animals suggests that the metabolism of other xenobiotics would be affected. Exposure of humans to Cr(VI) normally occurs with a co-exposure to a mixture of xenobiotics which are substrates for P-450. Whether Cr(VI) exposure decreases humans P-450 is not known but such a decrease would perhaps have important consequences.

REFERENCES

Aaseth, J., Alexander, J., and Norseth, T. (1982). Uptake of ^{51}Cr-chromate by human erythrocytes - a role of glutathione. *Acta. Pharmacol. Toxicol.* **50**, 310-315.

Alexander, J., Aaseth, J., and Norseth, T. (1982). Uptake of chromium by rat liver mitochondria. *Toxicology* **24**, 115-122.

Chiazze, L., Jr., and Wolf, P.H. (1980). Epidemiology of respiratory cancer and other health effects among workers exposed to chromium. In *Proc. Chromates Symposium 80: Focus of a Standard*, pp. 110. Industrial Health Foundation, Pittsburgh, Pa.

Connett, P. H., and Wetterhahn, K. E. (1983). Metabolism of the carcinogen chromate by cellular constituents. *Structure and Bonding* **53**, 93-125.

Cupo, D. Y., and Wetterhahn, K. E. (1985). Modification of chromium (VI)-induced DNA damage by glutathione and cytochrome P-450 in chicken embryo hepatocytes. *Proc. Natl. Acad. Sci. U.S.A.* **82**, 6755-6759.

Gonasun, L. M., Witmer, C. M., Kocsis, J. J., and Snyder, R. (1973). Benzene metabolism in mouse liver microsomes. *Toxicol. Appl. Pharmacol.* **26**, 398-406.

Goodgame, D. M. L., Hayman, P. B., and Hathaway, D. E. (1982). Carcinogenic chromium(VI) forms chromium(V) with ribonucleotides but not with deoxyribonucleotides. *Polyhedron.* **1**, 497-499.

Gruber, J.E., and Jennette, K. W. (1978). Metabolism of the carcinogen chromate by rat liver microsomes. *Biochem. Biophys. Res. Commun.* **82**, 700-706.

Jennette, K.W. (1982). Microsomal reduction of the carcinogen chromate produces Chromium (V). *J. Am. Chem. Soc.* **104**, 874-875.

Langard, S., and Norseth, T. (1975). A cohort study of bronchial carcinomas in workers producing chrome pigments. *Br. J. Ind. Med.* **32**, 62-65.

Langard, S., and Norseth, T. (1986). Chromium. In *Handbook on the Toxicology of Metals*, Vol. II (L. Friberg, G. F. Nordberg, and V. B. Vouk, Eds.), pp. 185-210. Elsevier Science Publishers, Amsterdam.

Layne, E. (1957). Spectrophotometric and turbidimetric methods for measuring proteins. In *Methods In Enzymology*, Vol. III (S. P. Colowick and N. O. Kaplan, Eds.), pp. 447-454. Academic Press, New York.

Lowry, O. H., Rosebrough, J. N., Farr, A. L., and Randall, R. J. (1951). Protein measurement with the Folin phenol reagent. *J. Biol. Chem.* **193**, 265-275.

Mancuso, T.F. (1975). Consideration of chromium as an industrial carcinogen. In *Proceedings of the International Conference on Heavy Metals in the Environment* (T.C. Hutchinson, Ed.), pp. 343-356. Toronto Institute for Environmental Studies, Toronto, Canada.

Phillips, A. H., and Langdon, R. G. (1962). Hepatic Triphosphopyridine nucleotide-cytochrome c reductase: isolation, characterization and kinetic studies. *J. Biol. Chem.*, **237**, 2652-2660.

Omura, T., and Sato, R. (1964). The carbon monoxide binding pigment of liver microsomes. I. Evidence for its hemoprotein nature. *J. Biol. Chem.*, **239**, 2370-2378.

Tsapakos, M. J., and Wetterhahn, K. E. (1983). The interaction of chromium with nucleic acids. *Chem.-Biol. Interact.* **46**, 265-277.

Wetterhahn, K. (1990). In press.

Wiegand, H. J., Ottenwaelder, H., and Bolt, H. J. (1985). Fast uptake kinetics *in vitro* of ^{51}Cr(VI) by red blood cells of man and rat. *Arch. Toxicol.* **57**, 31-34.

FATTY ACID β-OXIDATION-DEPENDENT BIOACTIVATION OF 5,6-DICHLORO-4-THIA-5-HEXENOATE AND ANALOGS IN ISOLATED RAT HEPATOCYTES

M. E. Fitzsimmons and M. W. Anders

Department of Pharmacology
University of Rochester
Rochester, NY 14642, U.S.A.

INTRODUCTION

S-(1,2-Dichlorovinyl)-L-cysteine (DCVC) is a nephrotoxic agent found in trichloroethene-extracted soybean meal (McKinney et al., 1957). Structure-toxicity studies with DCVC showed that the decarboxylated analog *S*-(1,2-dichlorovinyl)cysteamine was nontoxic whereas the desamino analog 5,6-dichloro-4-thia-5-hexenoate (DCTH) was a potent inhibitor of mitochondrial respiration (Parker et al., 1965; Stonard et al., 1977). The bioactivation mechanism of DCVC has been elucidated (Anders et al., 1988), but the bioactivation mechanism of DCTH has, however, not been defined. The objective of the present study was to elucidate the bioactivation mechanism of DCTH. This research tested the hypothesis that DCTH is bioactivated by enzymes of the fatty acid β-oxidation pathway to cytotoxic intermediates (Fig. 1). Specifically, DCTH may be metabolized by fatty acid acyl-CoA synthetase to the corresponding CoA thioester 1, which may be metabolized by fatty acid acyl-CoA dehydrogenase to 5,6-dichloro-4-thia-2,5-hexadienoate 2. Dienoate 2 may be metabolized by enoyl-CoA hydratase to the hemimercaptal 5,6-dichloro-4-thia-3-hydroxy-5-hexenoate 3, which may eliminate the unstable enethiol 1,2-dichloroethenethiol 4 whose formation is associated with the toxicity of DCVC (Anders et al., 1988).

MATERIALS AND METHODS

Syntheses

5,6-Dichloro-4-thia-5-hexenoate was synthesized according to the method of McKinney et al. (1957). 6,7-Dichloro-5-thia-6-heptenoate, 7,8-dichloro-6-thia-7-octenoate, and 8,9-dichloro-7-thia-8-nonenate were prepared in an analogous manner. 5,6,7,8,8-Pentachloro-4-thia-5,7-octadienoate and 6-chloro-5,5,6-trifluoro-4-thiahexenoate were synthesized by treating the dilithium salt of 3-mercaptoprionate with excess hexachloro-1,3-butadiene or chlorotrifluoroethene in THF.

Hepatocyte preparation

Isolated rat hepatocytes were prepared from male Long-Evans rats (Charles River, Wilmington, MA; 200-250 g) by the method of Moldéus et al. (1978). Hanks and Krebs-Henseleit buffers were supplemented with 25 mM HEPES and 2% BSA. Cell viability was measured by trypan blue exclusion. Hepatocyte suspensions were diluted to 1 to 2 x 10^6 cells/ml and incubated in 25 ml flasks at 37°C under 95% O_2/5% CO_2.

Biological Reactive Intermediates IV, Edited by C.M. Witmer *et al.*
Plenum Press, New York, 1990

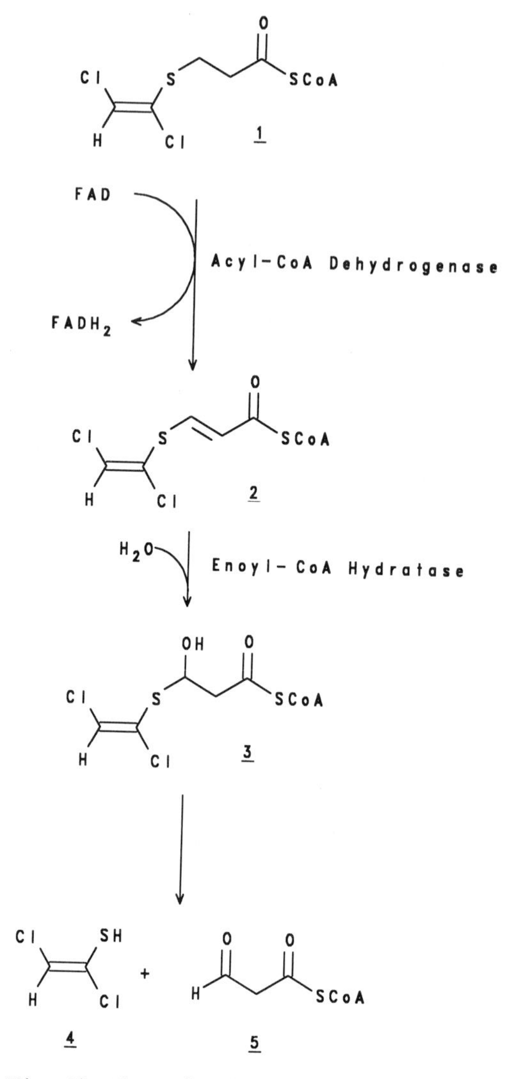

Figure 1. Bioactivation of 5,6-dichloro-4-thia-5-hexenoate 1; 2, 5,6-dichloro-4-thia-2,5-hexadienoate; 3, 5,6-dichloro-4-thia-3-hydroxy-5-hexenoate; 4, 1,2-dichloroethenethiol; 5, malonyl-CoA semialdehyde.

Oxygen consumption

Oxygen consumption by hepatocytes was measured with a Clark-type electrode at 37°C. The oxygraph was calibrated with solutions of known oxygen concentration.

RESULTS

Isolated hepatocytes incubated with 50 to 200 μM DCTH showed time- and concentration-dependent cytotoxicity (Figure 2). DCTH (50 μM) decreased hepatocyte viability to 60% by 3 hr. Similarly, 7,8-dichloro-6-thia-7-octenoate (50 μM) reduced cell viability to 50% by 3 hr. In contrast, the odd chain-length compounds 6,7-dichloro-5-thia-6-heptenoate (1 mM) and 8,9-dichloro-7-thia-8-nonenate (1 mM) were not cytotoxic. The cytotoxicity of DCTH was reduced by incubating the hepatocytes for 10 min with either sodium benzoate (1 mM) or octanoic acid (1 mM) before adding the DCTH. The viability of hepatocytes incubated with sodium benzoate and subsequently exposed to DCTH (100 μM) was reduced by 35% compared with the 90% reduction in viability of hepatocytes exposed to DCTH (100 μM) alone. Octanoic acid was not as effective as sodium benzoate in blocking the cytotoxicity of DCTH, but decreased DCTH cytotoxicity by 30%. Respiration of hepatocytes after DCTH exposure was also investigated. At concentrations of 50 and 100 μM, oxygen consumption was reduced by 55 and 85% by 3 h, respectively, when compared with control hepatocytes.

The cytotoxicity of two other 4-thiaalkanoates was also investigated. Isolated rat hepatocytes were incubated with 5,6,7,8,8-pentachloro-4-thia-5,7-octadienoate (100 μM) and 6-chloro-5,6,6-trifluoro-4-thiahexenoate (100 μM), and hepatocyte viability was determined. Both compounds decreased hepatocyte viability by to 40 to 50%.

DISCUSSION

The results of the present investigation show that the even chain-length 4- and 6-thiaalkenoates are cytotoxic in isolated rat hepatocytes, but that the odd chain-length 5- and 7-thiaalkenoates are nontoxic. Both DCTH and 7,8-dichloro-6-thia-7-octenoate as well as 6-chloro-5,6,6-trifluoro-4-thiahexenoate and 5,6,7,8,8-pentachloro-4-thia-5,7-octadienoate produce time- and concentration-dependent cytotoxicity in isolated hepatocytes. These data are consistent with the hypothesis that these thiaalkanoates are bioactivated by the enzymes of fatty acid β-oxidation (Figure 1). 7,8-Dichloro-6-thia-7-octenoate may be metabolized by fatty acid β-oxidation to DCTH, which may be metabolized to 5,6-dichloro-4-thia-3-hydroxy-5-hexenoate 3. Hemimercaptal 3 may nonenzymatically eliminate the unstable intermediate 1,2-dichloroethenethiol 4, which is the cytotoxic metabolite of DCVC (Anders et al., 1988). 6,7-Dichloro-5-thia-6-heptenoate and 8,9-dichloro-7-thia-8-nonenate, which cannot be metabolized to hemimercaptal 3, and hence are unable to eliminate the ethenethiol 4, were not cytotoxic. Furthermore, 6-chloro-5,5,6-trifluoro-4-thiahexanoate and 5,6,7,8,8-pentachloro-4-thia-5,7-octadienoate may also be metabolized to the corresponding hemimercaptals, which may eliminate 2-chloro-1,2,2- trifluoroethanethiol and 1-mercapto-1,2,3,4,4-pentachlorobuta-1,3-diene, which are known cytotoxic metabolites. To investigate further the role of the fatty acid β-oxidation pathway, hepatocytes were treated with sodium benzoate prior to incubation with DCTH. Sodium benzoate, which depletes coenzyme A, reduced the cytotoxicity of DCTH in isolated hepatocytes, indicating a role of coenzyme A in the bioactivation of DCTH. In addition, octanoic acid partially blocked the cytotoxicity of DCTH in hepatocytes, perhaps by competing for CoA thioester formation or for the fatty acid β-oxidation enzymes. These results indicate that the metabolic scheme (Figure 1) is involved in the bioactivation of DCTH and analogues.

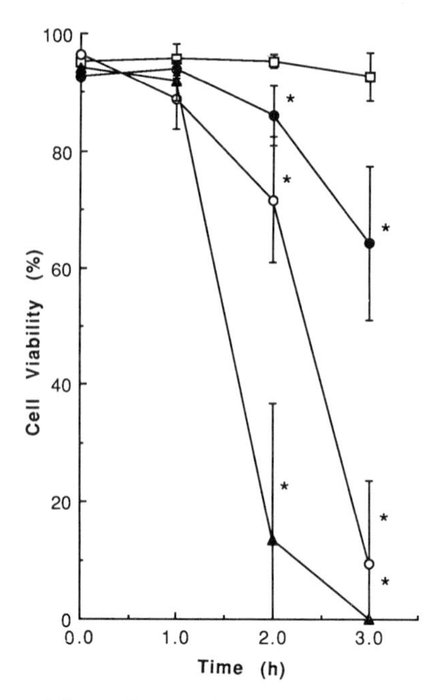

Figure 2. Cytotoxicity of 5,6-dichloro-4-thia-5-hexenoate. Rat hepatocytes were incubated 0 (□), 50 (•), 100 (°), and 200 μM (▲) 5,6-dichloro-4-thia-5-hexenoate. Cell viability was measured by trypan blue exclusion. Values are shown as mean ± SD. *Significantly different from control hepatocytes.

Finally, DCTH reduced oxygen consumption in isolated hepatocytes in a time- and concentration-dependent manner. These data indicate that the mitochondria are an important cellular target of DCTH, which is consistent with the generation of reactive intermediates within the mitochondria by the fatty acid β-oxidation pathway.

ACKNOWLEDGEMENT

This research was supported by NIEHS grant ES03127 to M.W.A.

REFERENCES

Anders, M. W., Lash, L. H., Dekant, W., Elfarra, A. A., and Dohn, D. R. (1988). Biosynthesis and biotransformation of glutathione *S*-conjugates to toxic metabolites. *CRC Crit. Rev. Toxicol.* **18**, 311-341.

McKinney, L. L., Weakley, F. B., Eldridge, A. C., Campbell, R. E., Cowan, J. C., Picken, J. C., and Biester, H. E. (1978). *S*-(Dichlorovinyl)-L-cysteine: An agent causing fatal aplastic anemia in calves. *J. Am. Chem. Soc.* **79**, 3932-3933.

Moldéus, P., Högberg, J., and Orrenius, S. (1978). Isolation and use of liver cells. *Methods Enzymol.* **52**, 60-70.

Parker, V. H. (1965). A biochemical study of the toxicity of S- dichlorovinyl-L-cysteine. *Fd. Cosmet. Toxicol.* **3**, 75-84.

Stonard, M. D., and Parker, V. H. (1971). 2-Oxoacid dehydrogenases of rat liver mitochondria as the site of action of S-(1,2-dichlorovinyl)-L-cysteine and S-(1,2-dichlorovinyl)-3-mercaptopropionic acid. *Biochem. Pharmacol.* **20**, 2417-2427.

ROLE OF THE ACETONE INDUCIBLE P-450IIE1 IN THE DEETHYLATION OF DIETHYLNITROSAMINE IN HAMSTER

P.G. Gervasi, P. Puccini, V. Longo and A. Lippi

Istituto di Mutagenesi e Differenziamento, CNR
Via Svezia 10, 56100 - Pisa, Italy

Nitrosamines are well known carcinogenic compounds that require a metabolic activation to generate electrophilic species able to elicit toxic, mutagenic or carcinogenic damage in cells. This activation process, generally involving an α-C-hydroxylation has been shown to be a cytochrome P-450 dependent reaction.

It has recently been shown that P-450IIE1 (inducible by ethanol, acetone, imidazole, isopropanol, pyrazole or fasting and diabetes), at least in rats and rabbits is capable of selectively metabolizing dimethylnitrosamine (DMN) when it is present at low concentration (0.2-2mM). The enzyme has a low K_m (40-70 μM) for DMN demethylation (Yang, et al., 1985; Patten, C.J. et al., 1986). Various P-450 isozymes are involved in the metabolism of diethylnitrosamine (DEN) in different species (Chau, I.Y. et al., 1978; Longo V. et al., 1986; Ding X. et al., 1989; Lee, M. et al, 1989), however the role of P-450IIE1 in DEN deethylation is unclear. Recently we have found that microsomes from acetone-treated rats (Puccini, P. et al., 1989) and acetone-treated mice (unpublished results) show an high affinity form of DEN deethylase activity probably due to P-450IIE1.

We report herein that the P-450IIE1 orthologue in hamster liver microsomes metabolizes DEN in a different way: the enzyme exibits a low affinity for DEN.

MATERIALS AND METHODS

Male Syrian Golden hamsters (4-5 weeks old, from Charles River FRG) were used. Hamsters were treated with acetone 3% in the drinking water for 9-10 days before they were killed by CO_2 asphyxia. Microsomes were prepared as previously described (Longo, V., et al., 1986) and the washed microsomal pellets were resuspended in Tris-buffer (100mM, pH 7.4) 1mM EDTA and stored at -80°C. The extent of oxidative deethylation of DEN was determined by measuring acetaldehyde formation by HPLC according to Farrelly (1951) as previously reported (Longo, V., et al., 1986). Microsomal proteins were assayed by the method of Lowry et al. (1951).

The acetone-inducible P-450IIE1 orthologue (ham-P-450IIE1) has been purified from acetone treated hamster liver following a procedure similar to that used for the purification of P-450j from pyrazole-treated rats (Palakodety, R.B. et al., 1988). NADPH-cytochrome P-450 reductase was purified according to Yasukochi and Masters (1976). The reconstituted system consisted of 0.2 μM P-450IIE1, 0.6 μM NADPH-cytochrome P-450 reductase and 30μg/ml dilauroyl phosphatidylcholine added as sonicated micelles. The incubation mixtures, containing 100mM phosphate buffer (pH 7.4) various concentration of DEN as substrate and the reconstituted monooxygenase

Biological Reactive Intermediates IV, Edited by C.M. Witmer *et al.*
Plenum Press, New York, 1990

system, were initiated with 0.5mM NADPH and incubated for 15 min at 37°C. The apparent kinetic costants (V_{max}, K_m) were calculated from the Lineweaver-Burk plots of the acetaldehyde formation data.

RESULTS AND DISCUSSION

Pretreatment of hamsters with acetone caused an increase of P-450 content and of some microsomal P-450IIE1-linked monooxygenase activities such as aniline hydroxylase, p-nitrophenol hydroxylase and acetone hydroxylase (data not shown). Similar results have also been found with acetone-treated rats (Puccini, P. et al., 1989) or mice. Hepatic microsomes from acetone-treated hamsters metabolized DEN in a similar manner to microsomes from control hamsters, both following a biphasic Michaelis-Menten kinetic and exhibiting the same values for K_{m1} and K_{m2} (Table 1). Acetone-treated microsomes as compared to controls only increased (about 6-fold) the activity of the lower affinity component (K_{m2}) of DEN deethylase. The present results differ from those obtained using microsomes from rats or mice where acetone pretreatment caused the appearance of a new high affinity form of P-450, not detectable in control microsomes, capable of deethylating DEN when it is present at low concentration (<1mM) (Puccini, P. et al., 1989). In rats and mice this high affinity form of DEN-deethylase was associated to P-450IIE1.

To clarify whether this low affinity DEN deethylase of acetone-treated hamster microsomes was due to the catalitic activity of the orthologue P-450IIE1, we have purified the ham-P-450IIE1. The ham-P-450IIE1, when reconstituted with NADPH cytochrome P-450 reductase and phosphatidylcoline, was catalytically very active towards DEN. Deethylation of DEN was monophasic and the enzyme shows a high turnover number (8-fold higher than that observed with acetone-treated microsomes), but a high K_m very similar to the K_{m2} shown by acetone treated-microsomes (Table 1). Thus, the low affinity of purified ham-P-450IIE1 toward the DEN deethylase activity is in agreement with the data observed with hepatic microsomes from acetone-treated hamsters. Purified enzyme shows a single K_m and that is similar to the K_{m2} of microsomes.

Table 1. Values of Apparent Kinetic Constants for DEN-deethylase in Various P-450 Containing Hamster System

	DEN DEETHYLASE				
Assay system	K_m1 (mM)	K_m2 (mM)	$V_{max}1$ nmol / mg prot. min	$V_{max}2$ nmol / mg prot. min	nmol
Control microsomes (a)	0.19 ± 0.07	3.09 ± 1.0	1.64 ± 0.3	2.7 ± 0.5	2.2 ± 0.5
Acetone microsomes (a)	0.13 ± 0.08	3.33 ± 0.9	1.33 ± 0.4	16.0 ± 1.1(c)	8.0 ± 0.6(c)
Reconstituted system with ham-P-450IIE1 (b)	-	2.5 (d)	-	-	61 (d)

(a) Incubations were performed for 15 min at 37°C with a protein cocentration of 0.8 mg/ml. The values are presented as the means ± SD of 3 separate experiments.
(b) The reconstituted system contained in a 2.5 ml volume: 0.5 nmol ham-P-450IIE1, 1.5 nmol NADPH cyt. P-450 reductase, 75 μg dilauroylphosphatidilcholine and cofactors.
(c) Significantly different from control microsomes by Student's t-test, $p<0.01$.
(d) The values of K_m and V_{max} are the mean of two experiments.

In conclusion the P-450IIE1 in hamster, present costitutively and easily inducible by acetone, has a low affinity for DEN and thereby ham-P450IIE1 is someway catalitically different from the orthologue P-450IIE1 of rat and mouse. Probably this result reflects a subtle structural difference in the active site of the ham-P-450IIE1 compared to the P-450IIE1 of the other species.

REFERENCES

Chau, I.Y., Degani, D. and Archer, M.C. (1978). Kinetic studies on the hepatic microsomal metabolism of dimethylnitrosamine, diethylnitrosamine and methylethylnitrosamine in the rat. *J. Natl. Cancer Inst.* **61**, 517-520.

Ding, X. and Coon, M.J. (1989). Cytochrome P-450-dependent formation of ethylene from N-nitrosoethylamines. *Drug Metab. Disp.* **16**, 265-269.

Farrelly, J.G. (1980). A new assay for the microsomal metabolism of nitrosamines. *Cancer Res.* **40**, 3241-3244.

Lee, M., Ishizaki, H., Brady, J.F. and Yang, C.S. (1989). Substrate specificity and alkyl group selectivity in the metabolism of N-nitrosodialkylamines. *Cancer Res.* **49**, 1470-1474.

Longo, V., Citti, L., and Gervasi, P.G. (1986). Metabolism of diethylnitrosamine by nasal mucosa and hepatic microsomes from hamster and rat: species specificity of nasal mucosa. *Carcinogenesis* **7**, 1323-1328.

Lowry, O.H., Rosebrough, N.J., Farr, A.L. and Randall, R.J. (1951). Protein measurement with the Folin phenol reagent. *J. Biol. Chem.* **193**, 265-275.

Palakodety, R.B., Clejan, L.A., Krikun, G., Feierman, D.E. and Cederbaum, A.I. (1988). Characterization and identification of a pyrazole-inducible form of P-450. *J. Biol. Chem.* **263**, 878-884.

Patten, C.J., Ning, S.M., Lu, A.Y.H. and Yang, C.S. (1986). Acetone-inducible cytochrome P-450: purification, catalytic activity and interaction with cytochrome b_5. *Arch. Biochem. Biophys.* **251**, 629-638.

Puccini, P., Fiorio, R., Longo, V. and Gervasi, P.G. (1989). High affinity diethylnitrosamine-deethylase in liver microsomes from acetone-induced rats. *Carcinogenesis* **10**, 629-1634.

Yang, C.S., Tu, Y.Y., Koop, D.R. and, Coon M.J. (1985). Metabolism of nitrosamines by purified rabbit liver cytochrome P-450 isozymes. *Cancer Res.* **45**, 1140-1145.

Yasukochi,Y. and Masters, B.S.S. (1976). Some properties of a detergent-solubilized NADPH-cytochrome c (cytochrome P-450) reductase purified by biospecific affinity chromatography. *J. Biol. Chem.* **251**, 5337-5344.

STEREOCHEMISTRY OF THE MICROSOMAL GLUTATHIONE *S*-TRANSFERASE-CATALYZED ADDITION OF GLUTATHIONE TO CHLOROTRIFLUOROETHENE IN ISOLATED RAT HEPATOCYTES

Sally J. Hargus, Michael E. Fitzsimmons and M.W. Anders

Department of Pharmacology
University of Rochester
Rochester, NY 14642

INTRODUCTION

Glutathione *S*-transferases catalyze the nucleophilic addition of glutathione to electrophilic substrates. Glutathione *S*-conjugate formation, followed by conversion to mercapturates, serves to detoxify potentially harmful xenobiotics. Alternatively, glutathione *S*-conjugate formation is an important bioactivation mechanism for several classes of compounds (Anders et al., 1988). The nephrotoxicity of several haloalkenes, including chlorotrifluoroethene (CTFE), is attributable to glutathione *S*- conjugate formation, metabolism of the glutathione *S*-conjugate to the corresponding cysteine *S*-conjugate, translocation to the kidney, and bioactivation by cysteine conjugate β-lyase (Figure 1).

Microsomal glutathione *S*-transferase (GST_m)-catalyzed *S*-(2-chloro-1,1,2-trifluoroethyl)glutathione (CTFG) production is under stereochemical control and generates one diastereomer preferentially, whereas the cytosolic glutathione *S*-transferase (GST_c)-catalyzed reaction yields a 1:1 mixture of diastereomers (Dohn et al., 1985).

The objectives of the present study were to determine the absolute configuration of the new chiral center in CTFG, to develop a method for quantification of CTFG diastereomers, to study the stereochemical course of the GST_m-catalyzed addition of glutathione to CTFE in hepatocytes, and to determine the relative contributions of GST_c and GST_m to CTFG formation in hepatocytes.

MATERIALS AND METHODS

Male Long-Evans rats were obtained from Charles River (Charleston, DC). Ethyl chlorofluoroacetate was obtained from PCR Incorporated (Gainesville, FL). Chlorotrifluoroethene was purchased from Mathison Gas Products (Buffalo, NY). All other reagents were obtained from Aldrich (Milwaukee, WI). γ-Glutamyltranspeptidase (type I) was obtained from Sigma (St. Louis, MO). Biosynthetic CTFG from liver subcellular fractions was prepared as previously described (Dohn et al., 1985). *N*-Dodecylpyridoxal bromide was prepared by the method of Kondo et al. (1985). A Bruker 270 MHz NMR was used to record ^{19}F and ^{1}H NMR spectra. A Hewlett-Packard 5970 mass selective detector coupled to a 5880A and ^{1}H gas chromatograph was used for GC/MS analyses.

Biological Reactive Intermediates IV, Edited by C.M. Witmer *et al.*
Plenum Press, New York, 1990

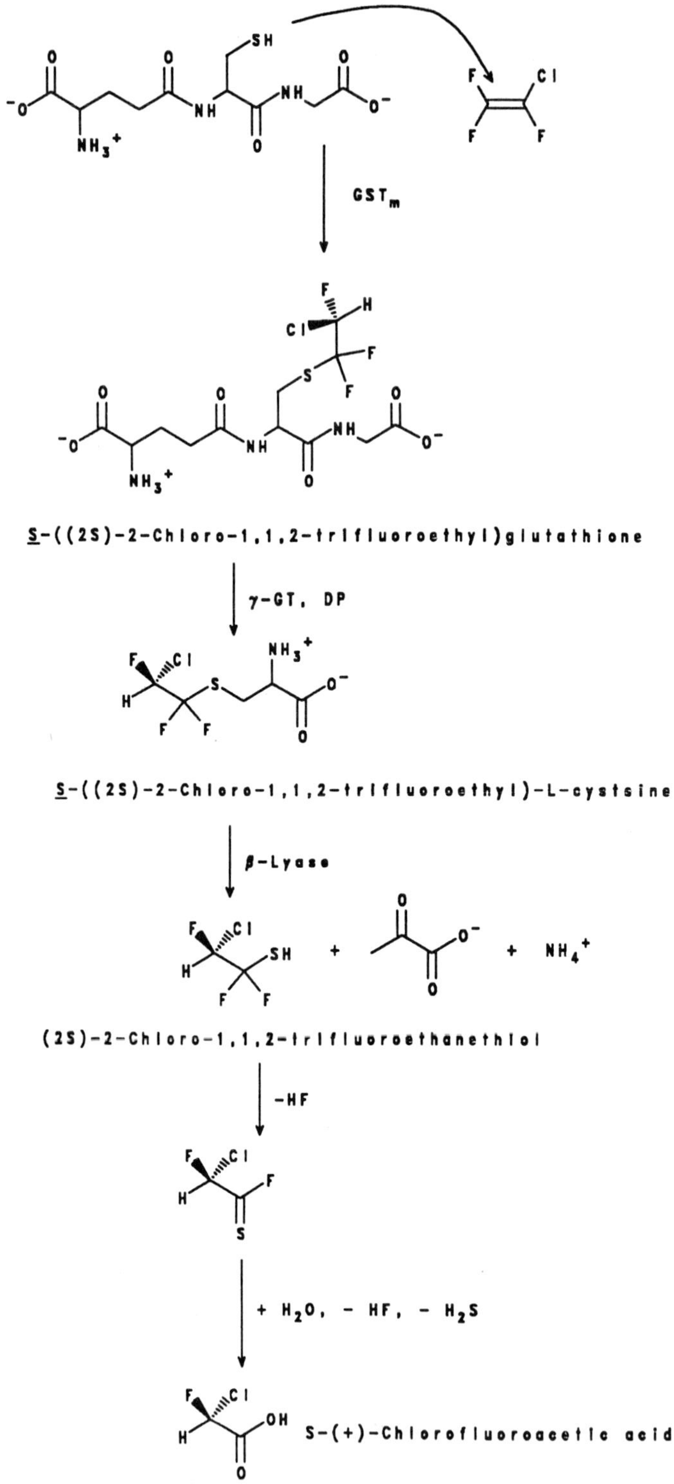

SH O O NH O NH O⁻ ⁻O NH$_3^+$ O F Cl F F

GST_m

F H Cl F S F O O O ⁻O NH NH O⁻ NH$_3^+$ O

S-((2S)-2-Chloro-1,1,2-trifluoroethyl)glutathione

γ-GT, DP

F Cl NH$_3^+$ S O⁻ H F F O

S-((2S)-2-Chloro-1,1,2-trifluoroethyl)-L-cystsine

β-Lyase

F Cl SH + O O⁻ + NH_4^+ H F F O

(2S)-2-Chloro-1,1,2-trifluoroethanethiol

-HF

F Cl F H S

+ H_2O, - HF, - H_2S

F Cl OH H O S-(+)-Chlorofluoroacetic acid

Figure 1. Bioactivation of CTFE

Resolution of Chlorofluoroacetic Acid

(RS)-chlorofluoroacetic acid (CFAA) was resolved according to the method of Bellucci et al. (1969). Melting points and optical rotations were in agreement with literature values. The optical purity of the resolved acids was determined by derivatization with S-(-)-α-methylbenzylamine and analysis by GC/MS.

Synthesis of N-[(S)-Ω-methylbenzyl](R,S)-chlorofluoroacetamide

N-[(S)-α-Methylbenzyl](R/S)-chlorofluoroacetamide was prepared by reacting 1 equivalent of S-(-)-α-methylbenzylamine, 0.5 equivalent of (R,S)- CFAA, 1 equivalent of 1,3-dicyclohexylcarbodiimide, and 1.3 equivalents of 1-hydroxybenzotriazole hydrate in 5 ml dry tetrahydrofuran at 0°C for 12 hr. The solvent was removed, and the residue was suspended in water. The suspension was washed successively twice with diethyl ether; the organic layer was washed with 1N HCl, with saturated $NaHCO_3$, and with saturated NaCl and then dried over anhydrous Na_2SO_4 and filtered. The white, crystalline product melted at 55-56°C, 1H NMR ($CDCl_3$) reported in ppm downfield from tetramethylsilane: δ 7.4 (m, 5H), δ 6.5 (s, 1H), δ 6.3 (d, J_{H-F}= 50 Hz 1H), diastereomer is shifted upfield 0.03 ppm, δ 5.2 (m, 1H), δ 1.5 (m, 3H). GC analysis gave two peaks which showed identical mass spectra, with retention times of 21.5 min and 22.3 min, for the two diastereomers; MS: m/z (relative abundance) 215 (28%, M^+), 180 (100%, M-Cl), 105 (79%, $M-CHCH_3C_6H_5$).

Degradation of CTFG to CFAA

CTFG was suspended in 2.5 ml phosphate buffer (100 mM, pH 8.5). γ-Glutamyltranspeptidase (10 u, type I) and 30 mg glycylglycine were added and allowed to react 12 hr at room temperature, and the mixture was then lyophilized. The degradation product *S*-(2-chloro-1,1,2-trifluoroethyl)-L-cysteine (CTFC) was purified by preparative HPLC on a Whatman ODS column with 10% acetonitrile/85% water + 0.01 mM trifluoroacetic acid as the mobile phase.

Degradation of CTFC to CFAA was accomplished by suspending lyophilized CTFC in 5 ml phosphate buffer (100 mM, pH 8.5); 20 μl cetyltrimethylammonium chloride and 1 mg N-dodecylpyridoxal bromide were added and allowed to react for 2 hr at 37°C.

In Situ Derivatization of CFAA

CFAA was derivatized after degradation of CTFC by the pyridoxal model. The reaction mixture was acidified to pH 2 with concentrated HCl, and solid KCl was added to make a 2% (w/v) solution. A solution of 0.1 mmol S-(-)-α-methylbenzylamine and 0.25 mmol 1,3-dicyclohexylcarbodiimide in 1 ml ethyl acetate was mixed with the CTFC-containing solution, and 0.75 mmol 1-hydroxybenzotriazole and additional 3 ml ethyl acetate were added. The mixture was allowed to react for 12 hr at room temperature. Solid KCl was added to a final concentration of 10% (w/v), and the solution was extracted twice with ethyl acetate. The organic layer was washed successively with 1 N HCl, with saturated $NaHCO_3$ and with saturated NaCl and then dried over anhydrous Na_2SO_4. The resulting brown liquid was eluted through a silica gel cartridge (500 mg, Alltech) and analyzed by GC/MS (isothermal, 120°C).

Hepatocyte Isolation and Incubation with CTFE Isolation

Hepatocytes were isolated by the method of Moldeus et al. (1978) and incubated at 37°C in 1-liter filter flasks fitted with septa and a large balloon. After flushing the flask with 95% O_2/5%CO_2, approximately 200 ml of CTFE (at room temp.) was added and the mixture was incubated for 75 min. Initial viability of all preparations was > 95%. At the end of incubation, the viability of treated hepatocytes was typically about 75%. Untreated hepatocytes were > 85% viable at 75 min. Ethanol (50% final concentration) was added to precipitate cellular proteins. Samples were lyophilized and then purified as previously described (Dohn et al., 1985).

Statistical analyses

Statistical significance was assessed by application of Student's t-test and analysis of variance.

RESULTS AND DISCUSSION

Results are summarized in Table 1. The first experimental objective was to determine the absolute configuration of the new chiral center in S-(2-chloro-1,1,2-trifluoroethyl)glutathione catalyzed by GST_m, and to develop a method of quantification of CTFG diastereomers. Attempts to resolve the methyl ester of CTFC (the partial degradation product of CTFG) by GC/MS failed. Furthermore, ^{19}F NMR analysis of CTFG did not provide an accurate determination of the relative quantities of diastereomers of the S-conjugate.

The objective was accomplished by degrading CTFG to CFAA, which retains the configuration of the new chiral center from CTFG. The absolute configuration of chlorofluoroacetic acid (CFAA) is known (Bellucci et al., 1969). Peaks on the GC chromatogram were identified as (S,R)-N-(α-methylbenzyl)chlorofluoroacetamide (21.5 min) and (S,S)-N-(α-methylbenzyl)chlorofluoroacetamide by resolution of CFAA, followed by derivatization of optically pure (R)- or (S)-CFAA with the amine. Degradation of GST_m-catalyzed CTFG to CFAA, followed by derivatization and GC/MS analysis showed that GSTm catalyzed the formation of 80% (2S)-CTFG, whereas GST_c catalysis yielded (2R)- and (2S)-CTFG in a 1:1 ratio. To investigate the possibility that another enzyme activity in the microsomal preparation might selectively degrade (2R)-CTFG, microsomes were incubated with chemically synthesized CTFG, which has a one-to-one ratio of diastereomers, then degraded and derivatized. No selective loss of (2R)-CTFG was observed.

Isolated hepatocytes were incubated with CTFE to investigate the stereochemical course of CTFG formation. GC/MS of resulting products showed that 75% (2S)-CTFG was formed in hepatocytes, indicating that GSTm preferentially catalyzes CTFE addition to glutathione in isolated hepatocytes. To determine whether nonviable cells could contribute significantly to CTFG catalysis, cells (106/ml) were freeze-thawed three times to cause lysis and incubated with CTFE. CTFG formation was not detectable by HPLC, presumably because the glutathione concentration became diluted after cell lysis and little CTFG formation occurred.

Rat 9000 x g supernatant was incubated with CTFE to determine whether this system could be useful as a model for studying the stereochemistry of CTFG formation in cells. The ratio of diastereomers formed by rat 9000 x g supernatant was not different from that found in isolated hepatocytes (Table 1).

Table 1. Percent (2S)-CTFG formed in hepatocytes and subcellular fractions.

synthetic	51 ± 1^{a}
cytosol	45 ± 1^{b}
microsomes	$81 \pm 1^{a,b}$
hepatocytes	$75 \pm 1^{a,b}$
9000 x g supernatant	$79 \pm 4^{a,b}$

n=3; reported as mean ± SEM
a,b = <S Values with same superscript are significantly different (P < 05).

CONCLUSIONS

Based on the results of this study, we conclude that GST_m makes the largest contribution to glutathione conjugation with CTFE in isolated hepatocytes. GST_m is known to show high specific activity for halogenated olefins such as CTFE (Dohn and Anders, 1984; Wolf et al., 1984), but previous studies have not addressed the stereochemistry of CTFG formation in isolated cells or *in vivo*, but have focused on the stereochemistry of CTFG formation in subcellular fractions (Dohn et al., 1985). The relatively small contribution of GST_c to CTFG formation in isolated cells was unexpected because, in theory, CTFE would be exposed to cytosolic enzymes prior to microsomal enzymes. It is possible, however, that CTFE is bound to carrier proteins in the cytosol, and isn't available for binding to GST until it contacts the lipophilic endoplasmic reticulum, where it can dissolve readily in the membrane.

We also conclude that rat 9000 x g supernatant is a convenient model for future stereochemistry studies with compounds structurally similar to CTFE. These studies will enhance our understanding of the role of GST_m in *S*-conjugate formation in cells and eventually, *in vivo*.

ACKNOWLEDGEMENT

This research was supported by NIEHS grant ES03127 to M.W.A.

REFERENCES

Anders, M. W., Lash, L., Dekant, W., Elfarra, A. A., and Dohn, D. R. (1988). Biosynthesis and biotransformation of glutathione *S*-conjugates to toxic metabolites. *CRC Crit. Rev. Toxicol.* **18**, 311-341.

Bellucci, G., Berti, G., Borraccini, A., and Macchia, F. (1969). The preparation of optically active chlorofluoroacetic acid and chlorofluoroethanol. *Tetrahedron* **25**, 2979-2985.

Dohn, D. R., Quebbemann, A. J., Borch, R. F., and Anders, M. W. (1985). Enzymatic reaction of chlorotrifluoroethene with glutathione: 19F NMR evidence for stereochemical control of the reaction. *Biochemistry* **24**, 5137-5143.

Dohn, D. R., and Anders, M. W. (1982). The enzymatic reaction of chlorotrifluoroethylene with glutathione. *Biochem. Biophys. Res. Commun.* **109**, 1339-1345.

Moldéus, P., Högberg, J., and Orrenius, S. (1978). Isolation and use of liver cells. *Methods Enzymol.* **52**, 60-70.

Wolf, C. R., Berry, P. N., Nash, J. A., Green, T., and Lock, E. A. (1984). Role of microsomal and cytosolic glutathione *S*-transferases in the conjugation of hexachloro-1,3-butadiene and its possible relevance to toxicity. *J. Pharmacol. Exp. Ther.* **228**, 202-209.

THE PATHOPHYSIOLOGICAL SIGNIFICANCE OF REACTIVE OXYGEN FORMATION IN RAT LIVER

Hartmut Jaeschke*, Arthur E. Benzick*, Charles V. Smith*† and Jerry R. Mitchell*‡

Center for Experimental Therapeutics*, Department of Medicine and Department of Pediatrics†
Baylor College of Medicine
Houston, Texas and
The Upjohn Company‡
Kalamazoo, Michigan

INTRODUCTION

Reactive oxygen species (ROS) are thought to be involved in the pathogenesis of various important human diseases and their therapeutic interventions, e.g. ischemia/reperfusion injuries such as myocardial infarction, stroke, shock and organ transplantation (McCord, 1985). Evidence for the involvement of ROS is mainly based on pharmacological interventions. Since in most cases several potential damaging mechanisms could contribute to cell and organ injury, the exact role of ROS remains unknown. We recently have been studying the quantification of intracellular ROS formation in various models of no-flow ischemia (Jaeschke et al., 1988) and hypoxia (Jaeschke et al., 1988; Jaeschke and Mitchell, 1989; Jaeschke, 1990a) in the liver by measurement of the biliary and sinusoidal release of glutathione disulfide (GSSG). It was found that during hypoxia even the physiological formation of ROS was suppressed by more than 80% (Jaeschke, 1990a). During reoxygenation ROS formation was estimated as 0.15 µmol O_2^-/min/g liver wt., a 3-4 fold increase above physiological values (Jaeschke et al., 1988; Jaeschke and Mitchell, 1989). Formation of ROS caused by reperfusion after hepatic no-flow ischemia was estimated by these methods to be less than 0.05 µmol O_2^-/min/g (Jaeschke et al., 1988).

In order to assess the pathophysiological importance of enhanced ROS formation, a model of selective oxygen toxicity (diquat) was studied in the isolated perfused rat liver. The redox-cycler diquat is thought to cause liver injury through ROS formation and lipid peroxidation (Smith, 1987); no evidence was found for the significant binding of a reactive metabolite to tissue proteins or lipids (Spalding et al., 1989). Thus, we used the model of diquat-induced generation of ROS in the isolated perfused liver to address the questions: 1. How much ROS can be generated intracellularly without cell and organ damage? 2. What is the relative importance of various defense mechanisms against intracellular reactive oxygen?

MATERIALS AND METHODS

The livers from male Fischer-344 rats (Harlan Sprague Dawley Inc., Houston, Texas) were isolated and perfused in a single pass mode with hemoglobin-free Krebs-Henseleit bicarbonate buffer gassed with carbogen (pH 7.4; 37°C). The model used has

Biological Reactive Intermediates IV, Edited by C.M. Witmer *et al.*
Plenum Press, New York, 1990

been described and discussed in detail (Sies, 1978, Jaeschke et al., 1983). The livers were preperfused for 30 min followed by perfusion with or without 200 µM diquat (a gift from Dr. Ian Wyatt, Imperial Chem. Industries, Macclesfield, England) for 60 min. Some of the animals were pretreated *in vivo*:

1. **Phorone:** 200 mg phorone/kg b.wt. was dissolved in corn oil and injected i.p. 1.5 h prior to the perfusion.
2. **Iron:** 100 mg FeSO4/kg b.wt. was dissolved in saline and injected i.p. 2.5 h prior to the perfusion.
3. **Iron and phorone:** combined treatment as decribed in 1 and 2.

All analytical procedures were used as described in detail: 1. Lactate dehydrogenase (Bergmeyer, 1974); 2. GSH and GSSG (Jaeschke and Mitchell, 1990).

RESULTS AND DISCUSSION

Infusion of 200 µM diquat into the isolated perfused liver resulted in a steady state secretion rate of GSSG of 22 nmol GSH-eq./min/g liver wt. for 1 hr (Table 1). About 80% of the GSSG released was exported into bile and 20% into the perfusate. Based on experiments with the uptake and metabolism of a selective glutathione peroxidase substrate (t-butyl hydroperoxide), it can be estimated that about 3-4% of the GSSG formed intracellularly is released from the liver (Sies and Summer, 1975; Jaeschke et al., 1988). Since the catalase inhibitor aminotriazole increased the GSSG secretion in the diquat experiments by 20-30% (data not shown), we estimate that 70-80% of the H_2O_2 generated is detoxified by glutathione peroxidase. Based on these quantitative relationships, the rate of superoxide formation during diquat infusion can be estimated as 1 µmol O_2^-/min/g liver wt., or about 60 µmol O_2^-/g in the course of 1 hr. These values reflect an oxidant stress which is orders of magnitude higher than seen during reperfusion after hepatic ischemia (about 0.036 µmol O_2^-/g liver) (Jaeschke et al., 1988). Despite the higher steady state intracellular oxidant stress for a much longer duration in the diquat perfused livers, only a minor release of lactate dehydrogenase (cell damage) was observed through 1 hr of diquat perfusion (Table 1).

Table 1. Hepatic Efflux of LDH and GSSG during Diquat Perfusion

	LDH mU/min/g	Bile GSSG nmol/min/g	Perfusate GSSG nmol/min/g	Bile Flow µl/min/g
A. Control	12 ± 4	1.3 ± 0.1	0.2 ± 0.1	1.23 ± 0.05
B. DQ	52 ± 9*	17.0 ± 2.2*	4.8 ± 0.4*	1.47 ± 0.06*
A. Fe	53 ± 16	1.4 ± 0.1	0.8 ± 0.2	1.00 ± 0.04
B. Fe+DQ	107 ± 60	14.5 ± 1.9*	3.0 ± 0.3*	1.28 ± 0.03*
A. PH	50 ± 23	0.1 ± 0.1	0.4 ± 0.1	0.79 ± 0.05
B. PH+DQ	105 ± 27	2.0 ± 0.3*	0.8 ± 0.2*	0.69 ± 0.05
A. Fe+PH	63 ± 18	0.1 ± 0.1	0.2 ± 0.1	0.71 ± 0.02
B. Fe+PH+DQ	260 ± 30*	2.8 ± 0.5*	0.5 ± 0.1*	0.89 ± 0.04*

Animals were used untreated or pretreated either with iron (Fe), phorone (PH) or a combination of both. Livers were isolated and perfused for 30 min followed by a 60 min perfusion with 200 µM diquat (B) or with plain perfusate (A). The hepatic release of lactate dehydrogenase (LDH) was determined in the perfusate at the end of the experiment (90 min). The average secretion rate of GSSG into bile and perfusate as well as the average bile flow is given as mean ± SEM (n=3-5) for the 60 min period.
* $p < 0.05$ (A versus B)

In order to test whether impairment of the detoxification system for ROS aggravates hepatic injury during the massive diquat-induced oxidant stress, animals were pretreated *in vivo* with iron (Smith, 1987), phorone (Jaeschke et al., 1987) or a combination of both. As shown in Table 1, only the depletion of hepatic glutathione by 90% (Control: 7.4 ± 0.5 μmol/g; Phorone: 0.8 ± 0.1 μmol/g) and iron-loading significantly enhanced cell damage. However, even under these conditions the damage is only moderate compared with the LDH efflux rates of 5000 - 12000 mU/min/g liver wt. during hypoxic injury (Jaeschke et al., 1988; Jaeschke and Mitchell, 1989). Our results demonstrate a high tolerance of the intact liver to intracellular ROS formation even in this highly susceptible strain (Smith et al., 1985).

These data can be used as a reference to assess the direct contribution of intracellular ROS formation to cell injury in the pathophysiology of various liver diseases. For example, direct damage by ROS is unlikely to cause ischemia/reperfusion injury in the liver with an estimated additional ROS formation of 0.036 μmol O_2^-/g during reperfusion (Jaeschke et al., 1988). On the other hand, a significant extracellular oxidant stress was identified during the early reperfusion phase after hepatic ischemia *in vivo* (Jaeschke, 1990b). The main source of extracellular ROS formation in the liver under these conditions seems to be Kupffer cells rather than accumulating neutrophils (Jaeschke and Farhood, 1990). The exact role of the extracellular oxidant stress in hepatic ischemia and reperfusion injury needs to be elucidated; our present results indicate that it is surprisingly difficult, even with the impairment of detoxification mechanisms, to cause liver injury selectively through the intracellular formation of reactive oxygen.

REFERENCES

Bergmeyer, H.U. (editor), (1974). *Methods in Enzymatic Analysis.* Academic Press, New York.

Jaeschke, H. (1990a). Glutathione disulfide as index of oxidant stress in rat liver during hypoxia. *Am. J. Physiol.*, in press.

Jaeschke, H. (1990b). Vascular oxidant stress and hepatic ischemia/reperfusion injury. *Free Rad. Res. Commun.*, in press

Jaeschke, H., and Mitchell, J.R. (1989). Mitochondria and xanthine oxidase both generate reactive oxygen species in isolated perfused rat liver after hypoxic injury. *Biochem. Biophys. Res. Commun.* **160**, 140-147.

Jaeschke, H., and Mitchell, J.R. (1990). The use of isolated perfused organs in hypoxia and ischemia/reperfusion oxidant stress. *Methods Enzymol.* Vol. 186, in press.

Jaeschke, H., and Farhood, A. (1990). Macrophages and neutrophils as potential sources of reactive oxygen in hepatic ischemia/ reperfusion injury. *N.S. Arch. Pharmacol.* (abstr.) in press

Jaeschke, H., Krell, H., and Pfaff, E. (1983). No increase of biliary permeability in ethinylestradiol-treated rats. *Gastroenterology* **85**, 808-814.

Jaeschke, H., Kleinwaechter, C., and Wendel, A. (1987). The role of allyl alcohol-induced lipid peroxidation and liver cell damage in mice. *Biochem. Pharmacol.* **36**, 51-57.

Jaeschke, H., Smith, C.V., and Mitchell, J.R. (1988). Reactive oxygen species during ischemia-reflow injury in isolated perfused rat liver. *J. Clin. Invest.* **81**, 1240-1246.

Jaeschke, H., Smith, C.V., and Mitchell, J.R. (1988). Hypoxic damage generates reactive oxygen species in isolated perfused rat liver. *Biochem. Biophys. Res. Commun.* **150**, 568-574.

McCord, J.M. (1985). Oxygen-derived free radicals in postischemic tissue injury. *N. Engl. J. Med.* **312**, 159-163.

Sies, H. (1978). The use of perfusion of liver and other organs for the study of microsomal electron-transport and cytochrome P-450 systems. *Methods Enzymol.* **52**, 48-59.

Sies, H., and Summer, K.H. (1975). Hydroperoxide-metabolizing systems in rat liver. *Eur. J. Biochem.* **57**, 503-512.

Smith, C.V. (1987). Evidence for the participation of lipid peroxidation and iron in diquat-induced hepatic necrosis *in vivo*. *Mol. Pharmacol.* **32**, 417-422.

Smith, C.V., Hughes, H., Lauterburg, B.H., and Mitchell, J.R. (1985). Oxidant stress and hepatic necrosis in rats treated with diquat. *J. Pharmacol. Exp. Ther.* **235**, 172-177.

Spalding, D.J.M., Mitchell, J.R., Jaeschke, H., and Smith, C.V. (1989). Diquat hepatotoxicity in the Fischer-344 rat: the role of covalent binding to tissue proteins and lipids. *Toxicol. Appl. Pharmacol.* **101**, 319-327.

OXIDATIVE STRESS DURING HYPOXIA IN ISOLATED-PERFUSED RAT HEART

James P. Kehrer and Youngja Park

Division of Pharmacology and Toxicology
College of Pharmacy
The University of Texas at Austin
Austin, TX 78712

INTRODUCTION

Reoxygenation injury in the heart, also known as the "oxygen paradox", is characterized by the sudden release of cytoplasmic constituents upon reoxygenation of the myocardium following a sustained period of hypoxia. In recent years, reactive oxygen species have been indirectly implicated in producing cardiac reoxygenation injury in some model systems through the protection provided by various antioxidants (Ambrosio et al., 1987; Granger et al., 1986; Hess and Manson, 1984). However, other studies have failed to find such protection (Gallagher et al., 1986; Klein et al., 1988; Nejima et al., 1989; Uraizee et al., 1987) and our results (Kehrer et al., 1987), and those of other investigators (Bindoli et al., 1988; Vander Heide et al., 1987), indicated extracellular oxygen radicals are not involved in the oxygen paradox. Nevertheless, a variety of electron spin resonance studies have indicated that oxygen radicals are produced at reoxygenation (Arroyo et al., 1987; Baker et al., 1988; Garlick et al., 1987; Zweier, 1988). The production of free radicals during the ischemic period has also been reported (Baker and Kalyanaraman, 1989; Maupoil and Rochette, 1988; Rao et al., 1983). The unanswered question is whether any of these radicals, which may be generated intracellularly, are actually interacting with various tissue components to produce damage.

Maximum enzyme release from rat heart tissue subjected to prolonged hypoxia occurs 4 min after reoxygenation (Kehrer et al., 1988). The data presented here demonstrate that changes are evident in a variety of indices of oxidative stress in isolated perfused rat heart tissue subjected to as little as 10 min of hypoxia and that these changes are not exacerbated by reoxygenation. In addition, it appears that these changes are concentrated in the mitochondria suggesting this organelle produces an excess of free radicals under conditions of low oxygen tension.

MATERIALS AND METHODS

Animals and perfusion apparatus

Male rats (Sprague-Dawley, 160 to 230 g) were obtained from Harlan Sprague Dawley (Indianapolis, IN). The animals (unfasted) were anesthetized with pentobarbital and dosed with 150 IU heparin through the inferior vena cava. Hearts were removed and placed into ice-cold Krebs-Henseleit bicarbonate solution containing 2.5 mM $CaCl_2$ and 10 mM glucose. Hearts were retrogradely perfused with

Biological Reactive Intermediates IV, Edited by C.M. Witmer *et al.*
Plenum Press, New York, 1990

the same solution in a non-recirculating Langendorff (1895) apparatus with a constant pressure of 80 cm water at 36°C. The standard perfusion medium was equilibrated with 95% O_2, 5% CO_2, and all hearts were allowed to stabilize for 30 min before hypoxic perfusions (without any ischemia) were performed by equilibrating the same perfusion medium with 95% N_2, 5% CO_2. Mitochondria were isolated from heart ventricles by differential centrifugation according to the method of Mela and Seitz (1979).

Samples of the coronary effluent were collected at selected intervals for determination of coronary flow rates and lactate dehydrogenase (LDH) release. LDH activity, at 30°C, was determined spectrophotometrically at 340 nm in 100 mM triethanolamine HCl buffer, pH 7.6, containing 0.15 mM NADH, 1 mM EDTA, and 1.5 mM pyruvate.

Chemicals

NADH, $NADP^+$, NADPH, pyruvate, 4,4'-bis(dimethylamino)benzhydrol, d-α-tocopherol, 5,5'-dithiobis-(2-nitrobenzoic acid) (DTNB), N-ethylmaleimide (NEM), glucose, glucose 6-phosphate, glucose 6-phosphate dehydrogenase, glutathione (GSH), glutathione disulfide (GSSG), 2-thiobarbituric acid, glutathione reductase, and guanidine hydrochloride were obtained from Sigma Chemical Co. (St. Louis, MO). 2,4-Dinitrophenylhydrazine (DNP), 2,2'-dipyridyl and dithiothreitol were obtained from Aldrich Chemical Co. (Milwaukee, WI). Bio-beads™ were obtained from Bio-Rad Laboratories (Richmond, CA). All other chemicals were reagent or spectrophotometric grade.

Biochemical analyses

Vitamin E was measured in freeze-clamped heart tissue. 200 mg of the frozen tissue powder was extracted for lipids, including tocopherols, and high-performance liquid chromatography with electrochemical detection was used for analysis of tocopherol compounds as described in detail elsewhere (Murphy and Kehrer, 1987). Protein carbonyl groups (the result of oxidative modification of proteins) were estimated by the technique of Oliver et al. (1987) as previously described (Murphy and Kehrer, 1989) by measuring the A_{370} of the DNP-derivatives and subtracting the absorption of the non-derivatized protein. GSH, GSSG, acid-soluble thiols, protein GSH mixed-disulfides, and total reduced protein thiols were measured by the general procedures recommended by Lou et al. (1987) as described previously (Murphy and Kehrer, 1989). Lipid peroxidation was assessed by determining the tissue content of thiobarbituric acid reactive substances (TBARS). 200 mg of heart tissue were mixed with 2 ml of 20% TCA containing 0.01% BHT. After centrifugation at 1000 g for 15 min, 1 ml of the supernatants were mixed with 1 ml 100mM thiobarbituric acid, capped and heated for 15 min in a boiling water bath. The absorbance of this solution at 532 nm was determined and TBARS were quantitated using a millimolar extinction coefficient of 146.

Statistical analyses

Comparisons among multiple groups were made using an ANOVA program with Student-Newman-Kuels post-hoc comparisons. All data are presented as mean ± S.E. A *p* value of less than 0.05 was considered significant.

RESULTS AND DISCUSSION

In agreement with previous results (Kehrer et al., 1987; Kehrer et al., 1988), an increased release of LDH, peaking after 4 min, was evident upon reoxygenation after 60 min of hypoxia in the presence of glucose. This demonstrated that these conditions were sufficient to produce damage consistent with the oxygen paradox.

Analyses of whole heart tissue after 10 min hypoxia revealed significant increases in the cardiac content of TBARS, GSSG and protein carbonyl groups, and a significant decrease in GSH (Figure 1). The increase in GSSG was not sufficient to explain the loss of GSH, although changes in mixed disulfides, which became significant after 60 min hypoxia (Figure 1) may account for this difference. Oxidative changes in whole heart tissue became significant (mixed disulfides) and were enhanced (TBARS) or unchanged (GSH, GSSG, protein carbonyls) after 60 min hypoxia, compared to oxygenated controls. After 4 min reoxygenation, no index of oxidative stress was enhanced above the level seen with hypoxia alone (Figure 1). There was, however, a decrease in cardiac TBARS, possibly due to the release of this water soluble compound along with other intracellular constituents during the massive cell lysis which occurs at this time. GSSG levels also returned to normal after reoxygenation, perhaps due to reduction by glutathione reductase as aerobic processes resumed. In general, these results are consistent with earlier work examining rat cardiac glutathione content where a 40-50% decline in myocardial GSH was found after 20-90 min hypoxia or ischemia (Arduini et al., 1988; Guarnieri et al., 1980; Julicher et al., 1984).

The site at which oxidative stress is occurring in hypoxic heart tissue is not known. It may occur diffusely throughout this tissue, or it could be localized at one or more subcellular sites. Mitochondria have been implicated as one of several sites of oxidative injury during during ischemia (Vaughn et al., 1988). A large increase in GSSG in mitochondria from ischemic dog hearts is also indicative of significant oxidative stress (Pauly et al., 1987). Analyses of mitochondria isolated from rat heart tissue after 60 min hypoxia revealed changes in various indices of oxidative stress which were similar, but more pronounced than those in whole heart tissue. Compared to 90

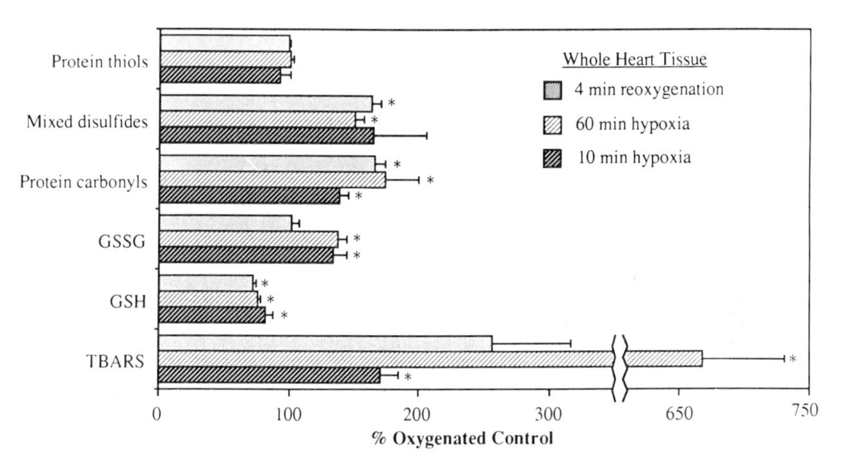

Figure 1. Indices of oxidative stress in whole heart tissue perfused with oxygenated medium for 30 min followed by: hypoxic medium for 10 or 60 min, or hypoxic medium for 60 min plus 4 min reoxygenation. Data are expressed as the mean ± S.E. of the percentage of the control values measured after perfusion with oxygenated medium for 30 or 90 min. (n = 3-4) 30 min control values were: TBARS (71 ± 3 nmol/g protein); GSH (1146 ± 34 nmol/g wet wt); GSSG (8.5 ± 0.6 nmol/g wet wt); protein carbonyl groups (2.9 ± 0.3 nmol/mg protein); mixed disulfides (0.14 ± 0.02 nmol GSH equivalents/mg protein); protein thiols (88 ± 5 nmol/mg protein). 90 min control values were: TBARS (23 ± 7 nmol/g protein); GSH (885 ± 12 nmol/g wet wt); GSSG (11.0 ± 0.1 nmol/g wet wt); protein carbonyl groups (2.3 ± 0.2 nmol/mg protein); mixed disulfides (0.30 ± 0.04 nmol GSH equivalents/mg protein); protein thiols (88 ± 2 nmol/mg protein). *Significantly different from the appropriate control ($p < 0.05$).

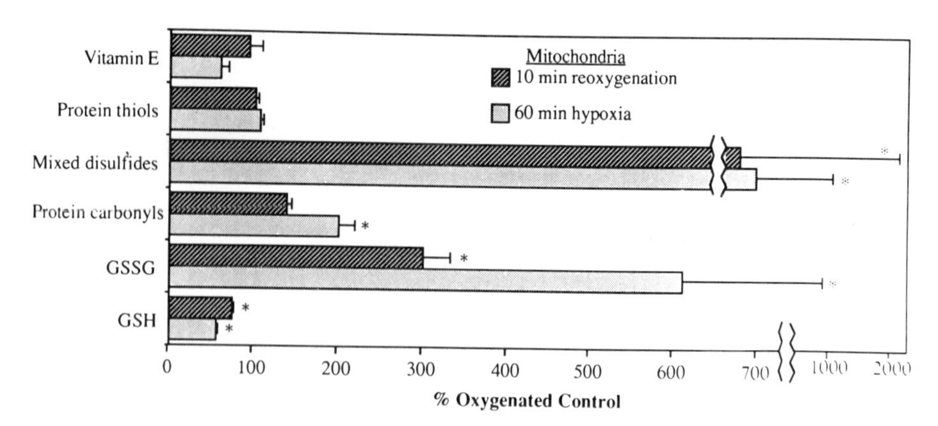

Figure 2. Indices of oxidative stress in mitochondria isolated from rat heart tissue perfused with oxygenated medium for 30 min followed by hypoxic medium for 60 min, or the same protocol plus 10 min reoxygenation. Data are expressed as mean ± S.E. of the percentage of control values measured in mitochondria isolated from hearts perfused with oxygenated medium for 90 min. (n = 3-4)
Control values were: GSH (3.2 ± 0.4 nmol/mg protein); GSSG (9.0 ± 7.0 pmol/mg protein); protein carbonyl groups (2.0 ± 0.2 nmol/mg protein); mixed disulfides (1.7 ± 1.7 nmol GSH equivalents/mg protein); protein thiols (62 ± 1 nmol/mg protein); vitamin E (66 ± 7 pmol/mg protein).
*Significantly different from control ($p < 0.05$).

min oxygenated controls, mitochondria isolated from hearts subjected to 60 min of hypoxia contained decreased GSH, and increased GSSG, protein carbonyls, and mixed disulfides (Figure 2). These changes were not enhanced in mitochondria isolated from hypoxic (60 min)/reoxygenated (10 min) heart tissue. However, it is possible that the "reoxygenation" inherent in the isolation of mitochondria from hypoxic heart tissue prevented detection of such an effect. Both protein thiols and vitamin E contents of mitochondria were unchanged by either hypoxia or reoxygenation (Figure 2). This was similar to the results in whole heart (Figure 1) and may be related to the relatively large quantities of these molecules present in cells which may have masked smaller changes evident in other indices with much small baseline values.

The source of reactive oxygen species during hypoxia remains speculative, but seems likely to be related to the site of oxidative modifications. Both hypoxic and ischemic tissues contain some oxygen, and it has been shown that free radical production remains significant even at a tissue pO_2 of 1 mm Hg (normal levels are 35 mm Hg) (Rao and Mueller, 1983). Our hypoxic perfusion medium contained less than 5 µM (about 3 mm Hg) oxygen which may have been sufficient to support some oxidative processes. Several studies have suggested that mitochondrial components are progressively reduced with the onset of oxygen deprivation. Thus, the auto-oxidation of one or more of these components may be a source of oxygen radicals.

SUMMARY

These data suggest that oxidative stress occurs at the low oxygen tensions which exist during perfusion of rat heart tissue with hypoxic medium. Importantly, no evidence was found for additional oxidative injury after 4 min reoxygenation when enzyme release is maximal in this system suggesting the oxygen paradox is unrelated to oxidative stress. However, the oxidative changes evident after 10-15 min of hypoxia do support the occurrence of free radical mediated injury at low oxygen tensions, and it is possible this injury is involved in the changes which lead to cell lysis at reoxygenation. The source of this oxidative stress is not known, but appears to be greater in mitochondria and may arise from an increased production of reactive oxygen species by this organelle. Whether the observed oxidative changes are directly injurious to a cell is not yet clear.

REFERENCES

Ambrosio, G., Weisfeldt, M.L., Jacobus, W.E. and Flaherty, J.T. (1987). Evidence for a reversible oxygen radical-mediated component of reperfusion injury: reduction by recombinant human superoxide dismutase administered at the time of reflow. *Circulation* **75**, 282-291.

Arduini, A., Mezzetti, A., Porreca, E., Lapenna, D., DeJulia, J., Marzio, L., Polidoro G., and Cuccurullo, F. (1988). Effect of ischemia and reperfusion on antioxidant enzymes and mitochondrial innner membrane proteins in perfused rat heart. *Biochim. Biophys. Acta.* **970**, 113-121.

Arroyo, C.M., Kramer, J.H., Dickens, B.F. and Weglicki, W.B. (1987). Identification of free radicals in myocardial ischemia/reperfusion by spin trapping with nitrone DMPO. *FEBS Lett.* **221**, 101-104.

Baker, J.E., Felix, C.C., Olinger, G.N. and Kalyanaraman, B. (1988). Myocardial ischemia and reperfusion: Direct evidence for free radical generation by electron spin resonance spectroscopy. *Proc. Natl. Acad. Sci. USA* **85**, 2786-2789.

Baker, J.E. and Kalyanaraman, B. (1989). Ischemia-induced changes in myocardial paramagnetic metabolites: implications for intracellular oxy-radical generation. *FEBS Lett.* **244**, 311-314.

Bindoli, A., Cavallini, L., Rigobello, M.P., Coassin, M., and Di Lisa, F. (1988). Modification of the xanthine-converting enzyme of perfused rat heart during ischemia and oxidative stress. *Free Rad. Biol. Med.* **4**, 163-167.

Gallagher, K.P., Buda, A.J., Pace, D., Gerran, R.A., and Shlafer, M. (1986). Failure of superoxide dismutase and catalase to alter size of infarction in conscious dogs after 3 hours of occlusion followed by reperfusion. *Circulation* **73**, 1065-1076.

Garlick, P.B., Davies, M.J., Hearse, D.J., and Slater, T.F. (1987). Direct detection of free radicals in the reperfused rat heart using electron spin resonance spectroscopy. *Circ. Res.* **61**, 757-760.

Granger, D.N., H?llwarth, M.E., and Parks, D.A. (1986). Ischemia-reperfusion injury: role of oxygen-derived free radicals. *Acta Physiol. Scand.* **548**, (Supplement) 47-63.

Guarnieri, C., Flamigni, F. and Caldarera, C.M. (1980). Role of oxygen in the cellular damage induced by re-oxygenation of hypoxic heart. *J. Mol. Cell. Cardiol.* **12**, 797-808.

Hess, M.L. and Manson, N.H. (1984). Molecular oxygen: friend or foe. The role of the oxygen free radical system in the calcium paradox, the oxygen paradox and ischemia/reperfusion injury. *J. Mol. Cell. Cardiol.* **16**, 969-985.

Julicher, R.H.M., Tijburg, L.B.M., Sterrenberg, L., Bast, A., Koomen, J.M. and Noordoek, J. (1984). Decreased defense against free radicals in rat heart during normal reperfusion after hypoxic, ischemic and calcium-free perfusion. *Life Sci.* **35**, 1281-1288.

Kehrer, J.P., Park, Y. and Sies, H. (1988). Energy-dependence of enzyme release from hypoxic isolated-perfused rat heart tissue. *J. Appl. Physiol.* **65**, 1855-1860.

Kehrer, J.P., Piper, H.M. and Sies, H. (1987). Xanthine oxidase is not responsible for reoxygenation injury in isolated-perfused rat heart. *Free Rad. Res. Commun.* **3**, 69-78.

Klein, H.H., Pich, S., Lindert, S., Buchwald, A., Nebendahl, K. and Kreuzer, H. (1988). Intracoronary superoxide dismutase for the treatment of "reperfusion injury". A blind randomized placebo-controlled trial in ischemic, reperfused porcine hearts. *Basic Res. Cardiol.* **83**, 141-148.

Langendorff, O. (1895). Untersuchungen am überlebenden Säugetierherzen. *Pflügers Arch.* **61**, 291-332.

Lou, M.F., Poulslen, L. and Ziegler, D.M. (1987). Cellular protein-mixed disulfides. *Methods Enzymol.* **143**, 124-129.

Maupoil, V. and Rochette, L. (1988). Evaluation of free radical and lipid peroxide

formation during global ischemia and reperfusion in isolated perfused rat heart. *Cardiovasc. Drugs Therap.* **2**, 615-621.
Mela, L. and Seitz, S. (1979). Isolation of mitochondria with emphasis on heart mitochondria from small amounts of tissue. *Methods Enzymol.* **55**, 39-45.
Murphy, M.E. and Kehrer, J.P. (1987). Simultaneous measurement of tocopherols and tocopheryl quinones in tissue fractions using high performance liquid chromatography with redox-cycling electrochemical detection. *J. Chromatog.* **421**, 71-82.
Murphy, M.E. and Kehrer, J.P. (1989). Oxidation state of tissue thiol groups and content of protein carbonyl groups in chickens with inherited muscular dystrophy. *Biochem. J.* **260**, 359-364.
Nejima, J., Knight, D.R., Fallon, J.T., Uemura, N., Manders, W.T., Canfield, D.R., Cohen, M.V. and Vatner, S.T. (1989). Superoxide dismutase reduces reperfusion arrhythmias but fails to salvage regional function or myocardium at risk in conscious dogs. *Circulation* **79**, 143-153.
Oliver, C.N., Ahn, B., Moerman, E.J., Goldstein, S. and Stadtman, E.R. (1987). Age-related changes in oxidized proteins. *J. Biol. Chem.* **262**, 5488-5491.
Pauly, D.F., Yoon, S.B. and McMillin, J.B. (1987). Carnitine-acylcarnitine translocase in ischemia: evidence for sulfhydryl modification. *Am. J. Physiol.* **253**, H1557-H1565.
Rao, P.S., Cohen, M.V. and Mueller, H.S. (1983). Production of free radicals and lipid peroxides in early experimental myocardial ischemia. *J. Mol. Cell. Cardiol.* **15**, 713-716.
Rao, P.S. and Mueller, H.S. (1983). Lipid peroxidation and acute myocardial ischemia. *Adv. Exp. Med. Biol.* **161**, 347-363.
Uraizee, A., Reimer, K.A., Murry, C.E. and Jennings, R.B. (1987). Failure of superoxide dismutase to limit size of myocardial infarction after 40 minutes of ischemia and 4 days of reperfusion in dogs. *Circulation* **75**, 1237-1248.
Vander Heide, R.S., Sobotka, P.A. and Ganote, C.E. (1987). Effects of the free radical scavenger DMTU and mannitol on the oxygen paradox in perfused rat hearts. *J. Mol. Cell. Cardiol.* **19**, 615-625.
Vaughan, D.M., Koke, J.R. and Bittar, N. (1988). Ultrastructure, peroxisomes and lipid peroxidation in reperfused myocardium. *Cytobios* **55**, 71-80.
Zweier, J.L. (1988). Measurement of superoxide-derived free radicals in the reperfused heart. Evidence for a free radical mechanism of reperfusion injury. *J. Biol. Chem.* **263**, 1353-1357.

ALTERATION OF GROWTH RATE AND FIBRONECTIN BY IMBALANCES IN SUPEROXIDE DISMUTASE AND GLUTATHIONE PEROXIDASE ACTIVITY

Michael J. Kelner[1], and Richard Bagnell

H-720-T
University of California San Diego
225 Dickinson Street
San Diego, California 92103

The free radical theory of aging, originally postulated by D. Harman (1956) assumes there is a single basic cause of aging which is modified by genetic and environmental factors (Harman, 1984). Oxidative free radicals, produced by normal metabolic processes, over time result in a progressive accumulation of damage that can not be fully repaired by cells. Recent work, however, suggests that cellular senescence is the result of programmed internal changes, and not accumulation of oxidative genetic damage. Results obtained by fusing different types of immortalized cells suggest that immortalization results from mutations in a small number of genes which other wise act to restrict the cell's proliferative potential. When different types of immortalized cells were fused, some hybrids became mortal again, senesced, and died (Pereira-Smith, 1988). Thus, the phenotype of cellular senescence is dominant and immortal cells arise due to recessive changes in growth control mechanisms. During the normal aging process then, cellular control of gene expression is gradually lost. One of the mRNAs of these "senescence" genes was recently cloned and coded for fibronection, a membrane associated protein (Pereira-Smith, 1988).

Oxygen radicals, however, could still be involved in these pre-programmed genetic changes as anti-oxidant enzymes may either be among the "senescent " genes, or alternatively, anti-oxidant enzymes may alter expression of senescent genes by controlling the redox state of the cell. If this hypothesis is correct, controlling ROS inside a cell may modulate the rate of growth and protein expression of "senescence genes" such as fibronectin. To investigate this possibility we genetically altered cells by stable transfection of a selectable expression vector for copper-zinc superoxide dismutase (CuZnSOD).

MATERIALS AND METHODS

Cell culture

NIH/3T3 cells were grown in DME media plus 10% bovine serum and 2mM glutamine (Kelner, et al., 1987), and were routinely screened for mycoplasm contamination. Alterations in glutathione peroxidase activity due to variations in calf serum selenium content were eliminated by supplementing media with selenium to a final concentration of 50 ng/ml (Speier, et al., 1985). For accurate quantification when

[1] To whom correspondence should be addressed.

Biological Reactive Intermediates IV, Edited by C.M. Witmer *et al.*
Plenum Press, New York, 1990

comparing growth rates between similar clones the rate of daily increase in thymidine incorporation is used.

Analysis

All water used was filtered to remove trace metal contaminates and increase resistance to greater than 18 megaohms/cm. Enzymes assays were performed with cells in log growth phase to eliminate cell cycle dependency. CuZnSOD and MnSOD (E.C. 1.15.1.1) were quantified using the NBT reduction method (Spitz and Oberley, 1989). SOD units for the clones, however, are expressed as units of SOD capable of inhibiting the reduction of cytochrome c by xanthine oxidase (McCord and Fridovich, 1969). The calibration curve for the NBT assay was constructed by plotting one over the rate of NBT reduction (1/rate) versus units of commercial bovine CuZnSOD (Flohe and Otting, 1984) which was previously calibrated by the xanthine/cytochrome c assay with 50 uM of potassium cyanide present to inhibit any cytochrome c oxidase present. The commercial SOD was also determined to be free of catalase and xanthine oxidase activity. SOD assays were ran immediately after tissue preparation as activity of either dismutase decreases with storage (Spitz and Oberley, 1989). Glutathione peroxidase (E.C. 1.11.1.0) was assayed using reduced glutathione and t-butyl hydroperoxide as substrates (Flohe and Gunzler, 1984). Cytosolic glutathione-transferase (E.C.2.5.1.18) activity was assayed with chlorodintrobenzene (Habig, et al., 1974). There was no detectable GSHTR (nonselenium) peroxidase activity using cumene hydroperoxide (Burk, et al., 1980). Catalase activity was assayed using hydrogen peroxide (Aebi, 1984) and the more sensitive Purpaldehyde method (Johansson and Borg, 1988). Fibronectin was determined by ELISA using a commercial antibody (Collaborative Research, Bedford, MA)(Chakrabarty, et al., 1987). Protein content was assayed using Coomasie blue with albumin as the standard. Results are expressed as mean ± SD, with N = 3 to 7 different determinations. Analysis of results was performed by using the Student t-test. Probability values (p value) were calculated from t-values and values of less than 0.05 were considered significant. Regression lines were fitted to data using a least squares method and correlation coefficients derived.

Superoxide dismutase expression vector and transfection

The SV-cDNA(CuZnSOD)-SVneo expression vector (Elroy-Stein, et al., 1986) has two regulatory units, and each unit contains its own early promoter, enhancer, termination and polyadenylation sites (Figure 1). One unit controls the cDNA of human copper-zinc superoxide dismutase and the other unit controls the gene for neomycin (G418) resistance. As each unit functions independently, the G418 resistance develops independent of expression of CuZnSOD. This allows for selection of clones with different amounts of dismutase activity. Transfection of the expression vector into mammalian cell lines was performed using calcium phosphate (Graham and Van der Eb, 1973), except cells were rinsed with a 20% DMSO solution to enhance DNA uptake. The cells were incubated undisturbed for 48 hours after transfection, split 1:6 at this time, then incubated in media supplemented with G418 (200 ug/ml).

RESULTS AND DISCUSSION

We noted that 3% of the clones were slow growing and these slow growing clones were not previously reported by other groups (Elroy-Stein, et al., 1986; Krall, et al., 1988). Normal NIH/3T3 fibroblasts can be easily removed with an EDTA/trypsin solution. The slow growth clones, however, were markedly adhesive to the culture flasks, resistant to trypsin treatment, required mechanical scraping for removal, and displayed a low saturation density. In contrast, fast growing clones attached minimally to the flask.

The activity of CuZnSOD was increased in all clones compared to the parent cells (Table 1). A wide variation in CuZnSOD activities between clones was observed. There was an increase in selenium-dependent GSHPX activity in fast growing clones compared to the parental cells. The GSHPX activity, however, was not increased in

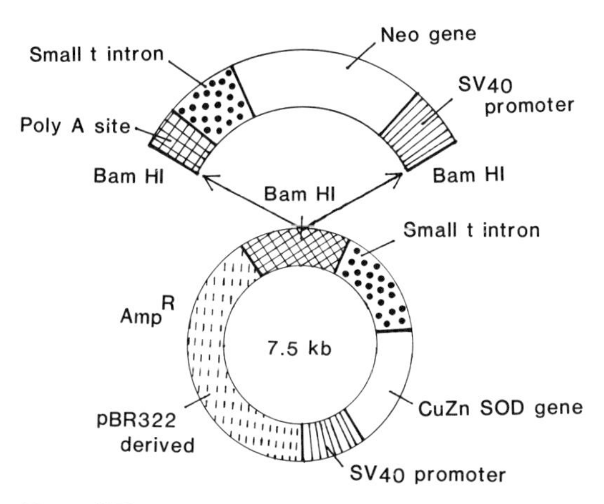

Figure 1. The pSV2-human copper-zinc superoxide dismutase-SVneo expression vector (Elroy-Stein, 1986) used to transfect the NIH/3T3 cells. The 2.7 kb SVneo portion inserted into the BamHI site is enlarged for detail. The open arcs represent either the neo gene or the copper-zinc superoxide dismutase genes; the hatched arcs represent the SV40 origin containing the early promoter and enhancer; the dotted arcs represent the SV40 small-t antigen intron; the cross-hatched arcs represent the site where the SV40 early transcript is polyadenylated; the dashed arc represents a portion of pBR322 containing the pBR322 origin of DNA replication and the ampicillin resistance gene.

Table 1. Comparison of superoxide dismutase activities and glutathione peroxidase activity to growth rate in G418 resistant clones (N = 3 to 7).

Clone	CuZnSOD[a]	GSHPX	MnSOD	doubling time (hrs)
Parent	9 ± 1	49 ± 3	0.85 ± 0.05	19 ± 3
FD5	13 ± 1	104 ± 11	0.51 ± 0.09	17 ± 1
FE5	13 ± 1	78 ± 6	0.50 ± 0.05	16 ± 1
FC6	14 ± 1	54 ± 2	0.56 ± 0.08	15 ± 1
FD4	16 ± 1	74 ± 4	0.54 ± 0.16	17 ± 1
FA1	21 ± 1	80 ± 6	0.39 ± 0.11	16 ± 1
FA4	19 ± 1	124 ± 22	0.30 ± 0.08	15 ± 2
FC1	29 ± 2	109 ± 7	0.40 ± 0.04	13 ± 1
FE4	26 ± 1	139 ± 20	0.39 ± 0.05	14 ± 1
FD1	21 ± 2	174 ± 24	0.35 ± 0.08	13 ± 1
<u>Slow Growing</u>				
S2F6	22 ± 2	45 ± 2	0.94 ± 0.10	21 ± 3
S1C4	27 ± 5	38 ± 4	1.11 ± 0.20	22 ± 3
S3A7	20 ± 3	48 ± 6	0.95 ± 0.06	24 ± 3
S5A7	29 ± 4	47 ± 6	1.13 ± 0.10	27 ± 5
S4C9	50 ± 7	47 ± 8	1.32 ± 0.11	32 ± 5

[a] Units for sueroxide dismutases are based on ability to inhibit cytochrome c reduction by xanthine-oxidase under previously defined conditions (McCord and Fridovich, 1969), and nanomoles per minute per milligram of protein for glutathione peroxidase.

slow growing clones. Surprisingly, the MnSOD activity was decreased in fast growing clones, but increased in slow growing clones. The change in MnSOD activity correlated with the ratio of CuZnSOD to GSHPX activity ($r = 0.888$, $p < 0.001$). The growth rate of the clones, however, correlated to MnSOD activity ($r = 0.941$, $p < 0.001$). The clones were screened for catalase and GSHTR. The activity of both enzymes was equivalent in both types of clones and the parent cell line (data not shown).

As fibronectin is reportedly a "senescent" gene (Pereira-Smith, 1988) and a marked difference was noted between fast and slow growth clones in their ability to adhere to the culture flasks, the fibronectin content was assayed in several clones. Fibronectin content of the parent cells was 3.1 ± 0.4 ug/mg of protein. Fibronectin was increased in slow growing clones S2F6 and S4C9 at 5.0 ± 0.8 and 8.1 ± 0.9, respectively. The fibronectin content in fast growing clone FC1 was decreased at 1.6 ± 0.3. We were aware decreases in fibronectin in the fast growing clones could be secondary to transformation (Chakrabarty, et al., 1987; Smith, et al., 1981; Hayman, et al., 1981). This was unlikely as the NIH/3T3 cells were extensively monitored for foci. Transformed cells have a distinct appearance, whereas the fast growing clones resemble the parent cell line in morphology. As a precaution, however, two fast clones (FD1 and FC1) and pEJ-(ras oncogene)-transfected transformed cells (as a positive control) were injected into athymic Balbc/c nude mice. The pEJ transformed cells produced tumors, but the two fast clones did not.

The exact mechanism by which an imbalance in the ratio of CuZnSOD and GSHPX (both cytosolic enzymes) alters the growth rate and fibronectin is unclear. It may involve a direct effect by the altered redox state. The correlation coefficients, however, suggest the effect is mediated through MnSOD expression. The alterations induced in MnSOD (by the imbalance between CuZnSOD and GSHPX) then result in alterations of growth rate.

In summary, we developed a cell culture model in which anti-oxidant enzymes can influence cell growth and phenotypic changes such as adhesion, saturation density, or fibronectin content. All of these traits are altered in fibroblasts upon aging or senescence (Macieira-Coelho, et al., 1966; Chandrasekhar and Millis, 1980; Sorrentino and Millis, 1984). Our results suggest an alternative hypothesis combining the free radical and the programmed (senesence) gene theories of aging. The alteration of anti-oxidant enzymes leads to alterations in ROS concentrations and intracellular redox states, which mediates expression of senescence genes such as fibronectin.

Abreviations: SOD, superoxide dismutases; CuZnSOD, copper-zinc superoxide dismutase; MnSOD, manganese superoxide dismutase; NBT, nitrobluetetrazolium; GSHPX, glutathione peroxidase (selenium-dependent); GSHTR, glutathione transferase; ROS, reactive oxygen species; DMSO, dimethylsulfoxide;

ACKNOWLEDGMENTS

Use of animals in this project was approved by the University of California, San Diego Animal Subjects Committee, protocol # 239-8.

REFERENCES

Aebi, H. (1984). Catalase *in vitro*. In: *Methods of Enzymology* Vol 105, (Packer, L., ed.). Academic Press, Inc. Orlando, Florida, pp. 121-126.

Burk, R.F., Lawrence, R.A., Lane, J.M. (1980). Liver necrosis and lipid peroxidation in the rat as a result of paraquat and diaquat administration: Effects of selenium deficiency. *J Clin Invest* **65**, 1024-1031.

Chakrabarty, S., Brattain, M.G., Ochs, R.L., and Varani, J. (1987). Modulation of fibronectin, laminin, and cellular adhesion in the transformation and differentiation of murine AKR fibroblasts. *J Cell Physiol.* **133**, 415-425.

Chandrasekhar, S., and Millis, A.J.T. (1980). Fibronectin from aged fibroblasts is

defective in promoting adhesion. *J Cell Physiol* **103**, 47-54.
Elroy-Stein, O., Bernstein, Y., and Groner, Y. (1986). Overproduction of human Cu/Zn superoxide dismutase in transfected cells. *EMBO J* **5**, 615-622.
Flohe, L., and Gunzler, W.A. (1984). Assays of glutathione peroxidase. In: *Methods of Enzymology* Vol 105, (Packer, L., ed.). Academic Press, Inc. Orlando, Florida, pp. 114-121.
Flohe, L., and Otting, F. (1984). Superoxide dismutase assays. In: *Methods of Enzymology* Vol 105, (Packer, L., ed.). Academic Press, Inc. Orlando, Florida,pp. 93-104.
Graham, F.L., and Van der Eb, A.J. (1973). A new technique for the assay of infectivity of human adenovirus 5 DNA. *Virology* **52**, 456-67.
Habig, W.H., Pabst, M.J., and Jakoby, W.B. (1974). Glutathione S-transferases. *J Biol Chem* **249**, 7130-7139.
Harman, D. (1956). Aging: A theory based on free radical and radiation chemistry. *J Gerontol* **11**, 298-300.
Harman, D. (1984). Free radical theory of aging. *Age* **7**, 111-131.
Hayman, E.G., Engvall, E., and Ruoslahti, E. (1981). Concomitant loss of cell surface laminin and fibronectin from transformed rat kidney cells. *J Cell Biol* **88**, 352-357.
Johansson, L.H., and Hakan Borg, L.A. (1988). A spectrophotometric method for determination of catalase activity in small tissue samples. *Anal Biochem* **174**, 331-336.
Kelner, M.J., McMorris, T.C., Beck, W.T., Zamora, J.M., and Taetle, R. (1987). Preclinical evaluation of illudins as anticancer agents. *Cancer Research* **47**, 3186-9.
Krall, J., Bagley, A.C., Mullenbach, G.T., Hallewell, R.A., and Lynch, R.E. (1988). Superoxide mediates the toxicity of paraquat for cultured mammalian cells. *J Biol Chem* **263**, 19190-1914.
Maceiera-Coelho, A., and Lima, L. (1973). Aging *in vitro*: Incorporation of RNA and protein precursors and acid phosphatase activity during the life span of chick embryo fibroblasts. *Mech Ageing Dev* **2**, 13-18.
Macieira-Coelho, A., Ponten, J., Philipson, L. (1966). Inhibition of the division cycle in confluent cultures of human fibroblasts *in vitro*. *Exp Cell Res* **43**, 20-29.
McCord, J., and Fridovich, I. (1969). Superoxide dismutase: An enzymatic function for erythrocuprein. *J Biol Chem* **244**, 6049-6055.
Pereira-Smith, O.M., and Smith, J.R. (1988). Genetic analysis of indefinite division in human cells: Identification of four complementation groups. *Proc Natl Acad Sci USA* **85**, 6042-6046. Are Aging and Death Programmed in our genes? *Science* **242**, 33.
Smith, H.S., Riggs, J.L., and Mosessan, M.W. (1981). Production of fibronectin by human epithelial cells in culture. *Cancer Res* **39**, 4138-4144.
Sorrentino, J.A., and Millis, A.J.T. (1984). Structural comparisons of fibronectins isolated from early and late passage cells. *Mech Ageing Dev* **28**, 83-97.
Speier, C., Baker, S.S., and Newburger, P.E. (1985). Relationships between *in vitro* selenium supply, glutathione peroxidase activity, and phagocytic function in the HL-60 human myeloid cell line. *J Biol Chem* **260**, 8951-8955.
Spitz, D.R., and Oberley, L.W. (1989). An assay for superoxide dismutase activity in mammalian tissue homogenates. *Anal Biochem* **179**, 8-18.

THE ANTIDOTAL ACTIVITY OF THE THIOL DRUG DIETHYLDITHIOCARBAMATE AGAINST N-ACETYL-p-BENZOQUINONE IMINE IN ISOLATED HEPATOCYTES

Veronique V. Lauriault and Peter J. O'Brien[1]

Faculty of Pharmacy
University of Toronto
Toronto, Ontario, Canada, M5S 2A2

Abbreviations used: DEDC: diethyldithiocarbamate; DS: disulfiram; DTT: dithiothreitol; NAPQI: N-Acetyl-p-Benzoquinone Imine; GSH: reduced glutathione; GSSG: oxidised glutathione; Hepes: 4-(2-hydroxyethyl)-1-piperazine-ethane-sulphonic acid; HPLC: high performance liquid chromatography; FAB: fast atom bombardment.

N-Acetyl-p-benzoquinone imine (NAPQI) is the putativereactive metabolite implicated in acetaminophenhepatotoxicity (Crocoran et al., Dahlin et al., Albano et al., and Rundgren et al.) which is a highly reactive electrophile as well as an oxidant (Rundgren et al.). Toxicity in isolated hepatocytes has been attibuted to the alkylation and/or oxidation of protein thiol groups (Albano et al. and Rundgren et al.).

Diethyldithiocarbamate (DEDC) is a thiol drug which has been reported to protect mice *in vivo* against such toxicants as carbon tetrachloride, allyl alcohol, bromobenzene and thioacetamide (Siegers et al.). Younes et al. (1980) later found that DEDC protected phenobarbital-treated rats *in vivo* against acetaminophen-induced hepatotoxicity. Furthermore, DEDC was also found to be a potent antidote against hepatic damage caused by acetaminophen in mice (Strubelt et al.). Other investigators also showed that DEDC was not only capable of suppressing acetaminophen-induced hepatotoxicity and nephrotoxicity but also inhibited the acetaminophen-induced glutathione depletion in the liver and kidney (Younes et al.). The oxidised metabolite of DEDC, disulfiram was also found to protect rats against acetaminophen induced hepatic damage (Jorgensen et al.). Earlier studies showed that a single intraperitoneal injection of DEDC or disulfiram caused a significant decrease in cytochrome P-450 but only after 6-24 hours in rats (Hunter et al. and Zemaitis et al.). From these studies it was suggested that DEDC or disulfiram (injected prior to acetaminophen) protects against acetaminophen-induced cytotoxicity by virtue of its capability to inhibit the mixed-function oxidase believed to catalyse the oxidation of acetaminophen to NAPQI (Siegers et al., Strubelt et al., Younes et al., and Jorgensen et al.).

However, suppression of acetaminophen bioactivation by inhibiting the mixed function oxidase may not explain the protective effect of DEDC *in vivo* when administered one hour after the acetaminophen. The present study elucidates another mechanism for the protection in order to better understand its antidotal activity *in vivo*. In the following, DEDC was found to form a non-toxic stable conjugate with NAPQI which prevented the complete alkylation of GSH, protein thiols and critical macromolecules and thus protected against cell death.

[1] Correspondence to: Dr. Peter J. O'Brien, Faculty of Pharmacy, 19 Russell Street, University of Toronto,Toronto, Ontario, Canada M5S 2A2. Tel: (416) 978-2716.

Biological Reactive Intermediates IV, Edited by C.M. Witmer *et al.*
Plenum Press, New York, 1990

MATERIAL AND METHODS

Chemicals

Diethyldithiocarbamate (DEDC), disulfiram (DS), acetaminophen, Trypan blue, fluoro-2,4-dinitrobenzene, iodoacetic acid, GSG, GSSG were obtained from Sigma Chemical Co., (St Louis, MO). Dithiothreitol (DTT) was obtained from Aldrich (Milwaukee,WI). N-Acetyl-p-benzoquinone imine (NAPQI) was purchased form Dalton Chemichal Co., (Mississauga, Canada). Collagenase (from *Clostridium histoliticium*), Hepes and Bovine Serum Albumin (BSA) were purchased from Boehringer-Mannheim (Montréal, Canada). HPLC solvents were purchased form Caledon laboratories Ltd (Georgetown, Canada).

Isolation of Rat Hepatocytes and Cytotoxicity Study

Adult male Sprague-Dawley rats (190-210g) fed *ad libitum* were used to prepare hepatocytes. Isolated hepatocytes were prepared by collagenase perfusion of the liver (Moldeus et al.). Cell viability was measured by trypan blue exclusion (final concentration: 0.16% w/v). Cells (15mls of $1X10^6$ cells/ml) were suspended in round bottomed flasks continuously rotating in a water bath at 37°C inKrebs-Henseleit buffer, pH 7.4 supplemented with 12.5mM Hepes under 95% O_2 and 5% CO_2. DEDC and DTT were dissolved in incubation buffer. NAPQI was dissolved in 1% DMSO. These solutions were prepared immediately prior to use and added to hepatocytes after 30 mins of preincubation. Diethyldithiocarbamate (DEDC) was added 60s after NAPQI was added to the hepatocyte incubate.

Glutathione Assay

The total amount of GSH and GSSG in isolated hepatocytes were measured on deproteinized samples (5% metaphosphoric acid) after derivatisation with iodoacetic acid and fluoro-2,4-dinitrobenzene (Reed et al.). Quantitation was carried out on a µBondapak NH2 column using a Water 6000A Solvent delivery system, equipped with a Model 660 solvent programmer, a WISP 710A automatic injector and a Data Module (Water Associates,Milford, MA). GSH and GSSG were used as external standards.

High Pressure Liquid Chromatography

Large scale isolation of the NAPQI:DEDC conjugate was carried out using HPLC (Beckman 421 controller, 110B solvent delivery module). NAPQI (200 µM) and DEDC (200 µM) were reacted in 0.1M Tris-HCl buffer pH 7.4 and analysed using a Water Millipore (Milford, MA) C18 µBondapak reverse phase column (0.3mm x 30cm) eluted at a flow rate of 1ml/min with a linear gradient system of methanol/water (10:90 to 100:0) over 25 minutes (U.V 260nm). The NAPQI:DEDC conjugate was eluted with retention time of 17.5 minutes. Methanol was removed from pooled samples with nitrogen compressed gas and the samples were lyophilised and stored at -20°C under nitrogen for mass spectrometry analysis.

Mass Spectrometry

Mass spectrum for the NAPQI:DEDC conjugate was recorded using a VG Analytical ZAB-1F instrument equipped for fast atom bombardment (FAB). For FAB analysis, the sample was dissolved in thioglycerol and placed directly on the probe.

RESULTS AND DISCUSSION

As shown in Figure 1A, the reactive metabolite of acetaminophen, NAPQI (300 µM) caused 60% cytotoxicity in isolated hepatocytes after 90 minutes. Addition of DEDC (1mM) 60s after NAPQI (300 µM) protected isolated hepatocytes against NAPQI-induced cytotoxicity. NAPQI reacted very rapidly to depleted cellular GSH with or without DEDC. However, in the presence of DEDC, approximately 50% of the cellular GSH levels recovered while GSSG levels remained low suggesting that DEDC

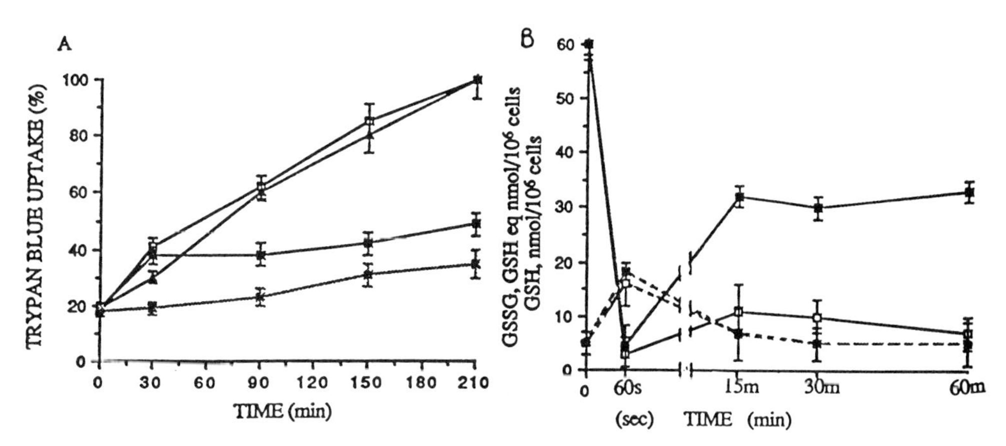

Figure 1. Hepatocytes (10^6 cells/ml) cytotoxicity induced by NAPQI (A). Cell viability was determined by Trypan blue exclusion. Effect of NAPQI (300 μM) (□) with DEDC (1mM) (■) added at 60s or with DTT (10 mM) added at 10 mins (▲) and control cells (X). Effect of NAPQI (300 μM) (□) with DEDC (1mM) (■) added at 60s on hepatocyte GSH and GSSG levels (B). GSSG is represented by dotted line. Values represent averages of three experiments with error bars representing the standard deviations.

prevented the complete alkylation of GSH and protein thiols as shown in Figure 1B. However, the reducing thiol, dithiothreitol (10mM) did not prevent subsequent NAPQI (300 μM)-induced cytotoxicity or restore total GSH levels (data not shown) when added 10 minutes after NAPQI suggesting that at this time the alkylation of GSH and protein thiols by NAPQI is irreversible. Furthermore, cytotoxicity induced by NAPQI may therefore be due mainly to the alkylation of GSH and protein thiols rather than oxidative stress.

Recently, Rundgren (1988) postulated that the primary cause of GSH depletion in hepatocytes by NAPQI is the result of GSH depletion in hepatocytes by NAPQI is the result of GSH conjugate formation. GSSG formation was attributed to GSH reacting with an ipso-conjugate. As shown in Figure 2, FAB mass spectral analysis of the isolated new peak from the reaction of DEDC with NAPQI with retention time of 17.5 minutes, yielded a base peak with m/e 299 corresponding to the (MH+) ion. This was consistent with the structure of N-acetyl-3-(diethyldithiocarbamate-S-yl)-4-hydroxyaniline. DEDC probably reacts at the C-3 position of the NAPQI aromatic ring to form a similar NAPQI conjugate as that described for GSH (Albano et al.).

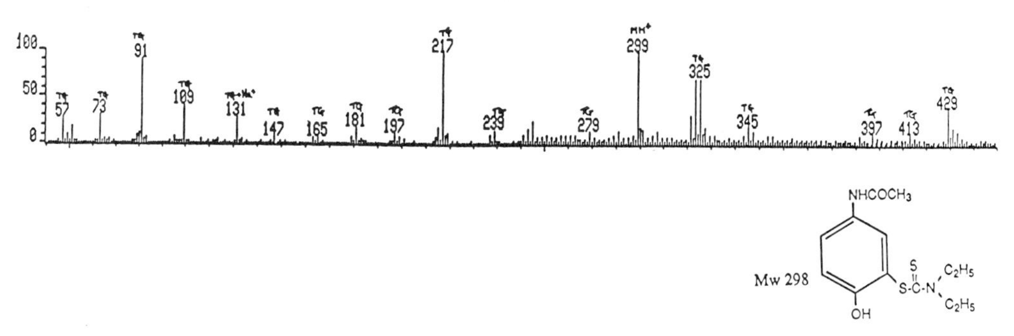

Figure 2. FAB mass spectra of DEDC:NAPQI conjugate identified as N-acetyl-3-(diethyldithiocarbamate-S-yl)-4-hydroxy-aniline.

The DEDC:NAPQI conjugate formed explains why GSH levels were not completely depleted since NAPQI was not available in the medium to alkylate GSH and protein thiols. Therefore, DEDC protected isolated hepatocytes from NAPQI-induced cytotoxicity by conjugating NAPQI and thereby preventing extensive GSH depletion and alkylation of those protein thiols which result in cytotoxicity.

In conclusion, DEDC acts as an antidote against NAPQI-induced cytotoxicity by forming a non-toxic DEDC:NAPQI conjugate. Such a mechanism could explain how DEDC prevents *in vivo* acetaminophen induced cytotoxicity.

ACKNOWLEDGEMENT

The authors would like to thank the Medical Research Council of Canada for supporting this work. The authors would also like to thank Dr. H. Pang of the Carbohydrate Institute, University of Toronto, for running the FAB mass spectra.

REFERENCE

Albano, E., Rundgren, M., Harvison, P.J., Nelson, S.D., Moldeus, P. (1985). Mechanisms of N-acetyl-p-benzoquinoneimine cytotoxicity. *Mol. Pharmacol.* **28,** 306-311.

Corcoran, G.B., Mitchell, J.M., Vaishnav, Y.N., Horning, E.C. (1980). Evidence that acetaminophen and n-hydroxyacetaminophen form a common arylating Intermediated; N-Acetyl-p-benzoquinone imine. *Mol.Pharmacol.* **18,** 536-542.

Dahlin, D.C., Miwa, G.T., Lu, A.Y.H., Nelson, S.D. (1984). N-Acetyl-p-benzoquinone imine: A cytochrome p-450-mediated oxidation product of acetaminophen. *Biochem.* **81,**1327-1331.

Jorgensen, L., Thomsen, P., Poulsen, H.E. (1988). Disulfiram prevents acetaminophen hepatotoxicity in rats. *Pharmacol.Toxicol.,* **62,** 267-271.

Hunter, A.E., Neal, R.A. (1975). Inhibition of hepatic mixed-funtion oxidase activity *in vitro* and *in vivo* by various thiono-sulfur-containing compounds. *Biochem. Pharmacol.,* **24,** 2199-2205.

Moldeus, P., Hogberg, J., Orrenius, S. (1979). Isolation and use of liver cells. *Methods Enzymol.* **52,** 60-71.

Reed, D.J., Babson, J.R., Beatty, B.W., Brodie, A.E, Ellis, W.W., Potter, D.N. (1978). High performance liquid chromatography analysis of nanomoles levels of glutathione, glutathione disulfide and related thiols and disulfides. *Anal. Biochem.* **106,** 55-62.

Rundgren, M., Porubek, D.J., Harvison, P.J., Cotgreave, I.A., Moldeus, P., Nelson, S.D. (1988). Comparitive cytotoxic effects of n-acetyl-p-benzoquinone imine and two dimethylated analogues. *Mol. Pharmacol.* **84,** 566-572.

Siegers, C.-P., Strubelt, O., Volpel, M. (1978). The antihepatotoxic activity of dithiocarb as compared to six other thiol compounds in mice. *Arch. Toxicol.* **41,** 79-88.

Strubelt, O., Siegers, C.-P., Schutt, A. (1974). The curative effects of cystamine, cysteine and dithiocarb in experimental paracetamol poisoning. *Arch Toxicol.* **33,** 55-64.

Younes, M, Siegers C.-P. (1980). Inhibition of the hepatotoxicity of paracetamol and its irreversible binding to rat liver microsomal protein. *Arch. Toxicol.* **45,** 61-65.

Younes, M., Sause, C., Siegers, C.-P. (1988). Effect of desferrioxamine and diethyldithiocarbamate on paracetamol-induced hepato- and nephrotoxicity. The role of lipid peroxidation. *J. Appl. Toxicol.* **8,** 261-265.

Zemaitis, M.A., Greene, F.E. (1979). *In vivo* and *in vitro* effect of thiuram disulfides and dithiocarbamates on hepatic microsomal drug metabolism in the rat. *Toxicol. Appl. Pharmacol.* **48,** 342-350.

TWO CLASSES OF AZO DYE REDUCTASE ACTIVITY ASSOCIATED WITH RAT LIVER MICROSOMAL CYTOCHROME P-450

Walter G. Levine and Shmuel Zbaida

Department of Molecular Pharmacology
Albert Einstein College of Medicine
Bronx, NY 10461, U.S.A.

Reduction of carcinogenic and other azo dyes by rat liver microsomes is catalyzed by NADPH-cytochrome P-450 reductase and cytochrome P-450 (P-450). Dimethylaminoazobenzene (DAB), a lipid soluble hepatocarcinogen, is reduced by a form of P-450 which is selectively induced by clofibrate but not other commonly used inducing agents; reduction is insensitive to O_2 and CO (Raza and Levine, 1986; Levine and Raza, 1988). In contrast, microsomal reduction of amaranth, a water soluble dye, is catalyzed by forms of P-450 which are induced by phenobarbital and methylcholanthrene; reduction is inhibited by O_2 and CO (Fujita and Peisach, 1978). Oxygen sensitivity of azoreduction is attributed to reoxidation of the 1-electron reduced free radical intermediate with formation of superoxide (Mason et al., 1978; Peterson et al., 1988). It follows that reduction of 2'-COOH-DAB (methyl red) by NAD(P)H:quinone reductase is oxygen insensitive since this enzyme contains two molecules of FAD and catalyzes a 2-electron reduction of the dye (Huang et al., 1979).

Substrate requirements for the clofibrate)inducible azoreductase have been defined. Electron-donating polar groups para to the azo linkage are required for binding to cytochrome P-450 (Zbaida et al., 1989). Azobenzene and p-isopropylazobenzene, lacking these requirements, are inactive. Attempts at 2-electron chemical reduction of azodye substrates with Zn^o yielded only fully reduced primary amines with the exception of DAB which yielded a spectrally definable hydrazo intermediate under rigorously anaerobic conditions. It was immediately reoxidized in air and, even under anaerobic conditions, disproportionated upon addition of water.

The present study extends the structure-activity considerations and indicates that there are at least two distinguishable classes of P-450 azoreductases.

A series of azo dyes related to DAB were purchased or synthesized and assessed as substrates for microsomal azoreduction. Microsomes were isolated from rat liver homogenates by differential centrifugation. They were incubated with dyes in the presence of a NADPH generating system. Primary amine products were extracted and assayed fluorometrically. All substrates required at least a para electron-donating substituent on one ring. If there were additional electron donating substituents on either ring, the substrates behaved as did DAB. Reduction was insensitive to O_2 and CO. These compounds were designated as I (insensitive) substrates (Figures 1 and 3A). If there were additional electron withdrawing sunstituents on the opposite ring, reduction was almost entirely inhibited by O_2 and CO. These compounds were designated as S (sensitive) substrates (Figures 2 and 3B). It was found that as little as 3% oxygen completely inhibited reduction of S substrates although I substrates were

Biological Reactive Intermediates IV, Edited by C.M. Witmer *et al.*
Plenum Press, New York, 1990

little affected at 20% (air). Reduction of both S (o-methyl red) and I (DAB) substrates was suppressed by standard inhibitors of P-450 activity, piperonyl butoxide, octylamine, alpha)naphthoflavone, SKF 525-A (Figure 4A).

Azoreduction of these substrates was also measured in a reconstituted P-450 system using highly purified NADPH-cytochrome P-450 reductase, P-450 and lipid. As shown previously for DAB (Levine and Raza, 1988), I substrates were weakly reduced by flavoprotein alone but activity was greatly enhanced upon addition of P-450 (Table 1). As seen in the microsomal system, reduction of representative I substrates was still mainly resistant to O_2 and CO (Figure 5A). Reduction of S substrates in the reconstituted system, unlike the microsomal system, appeared to involve primarily flavoprotein since addition of P-450 did not appreciably increase the rate of reduction (Table 1). Consequently, reduction was no longer sensitive to CO (Figure 5B). This was attributable to imperfect coupling between flavoprotein and P-450.

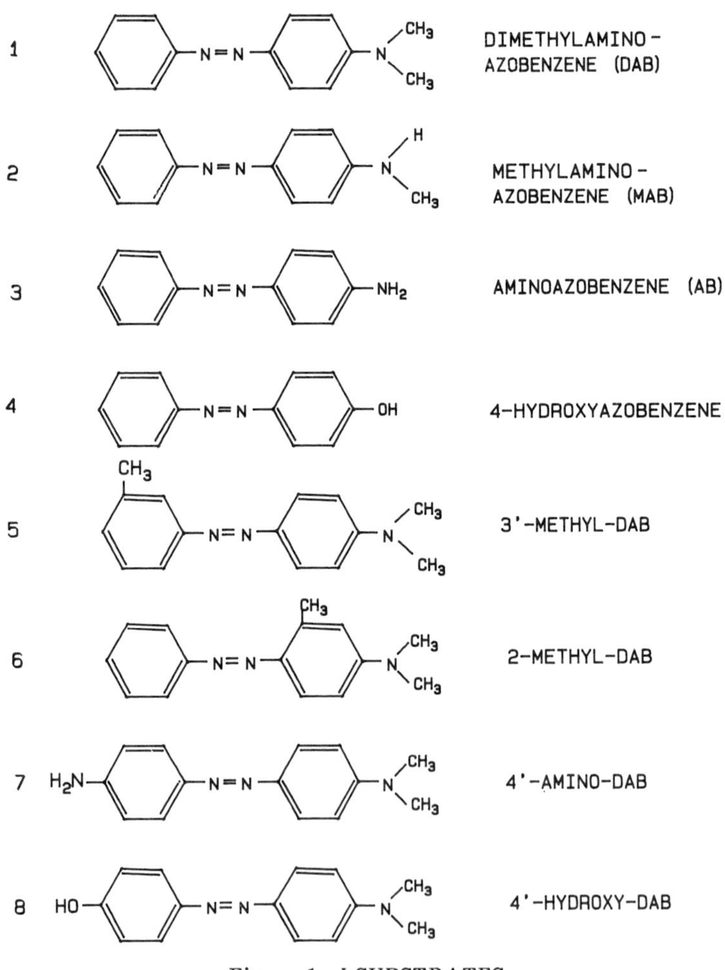

Figure 1. I SUBSTRATES
(insensitive substates)

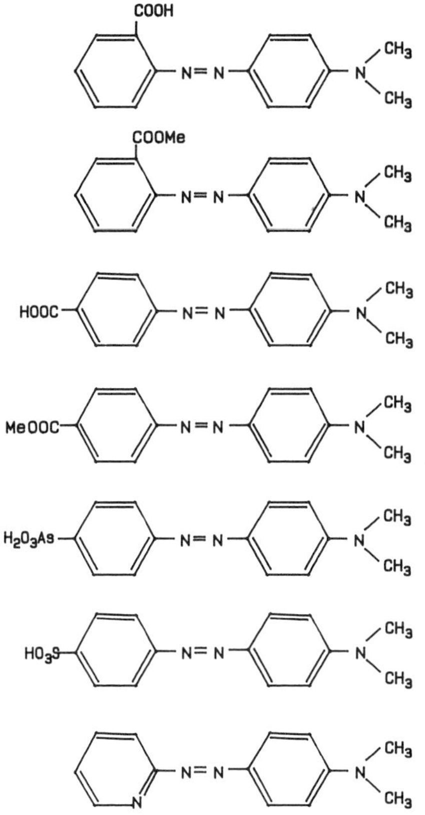

1 2'-CARBOXY-DAB (o-METHYL RED)

2 2'-CARBOXY-DAB METHYL ESTER

3 4'-CARBOXY DAB (p-METHYL RED)

4 4'-CARBOXY-DAB METHYL ESTER

5 DAB-ARSONATE

6 DAB-SULFONATE (METHYL ORANGE)

7 4-(2-PYRIDYLAZO)-N, N-DIMETHYLANILINE

Figure 2. S SUBSTRATES
(sensitive substrates)

Response to inducing agents also supported two classes of P-450 azoreductases. Reduction of I substrates (DAB, 4'-OH-DAB) was induced only by clofibrate, whereas reduction of S substrates (o-methyl red) responded to several inducing agents (Figure 4B). The O_2 insensitivity exhibited by I substrates is difficult to reconcile with the fact that chemically-reduced DAB is exquisitely sensitive to air (Zbaida et al. 1989). It is postulated that DAB and other lipophilic I substrates are bound at a hydrophobic site protected from O_2 and water. Upon accepting 2 electrons from P-450, the hydrazo intermediate would be released and spontaneously degrade to free amines, or disproportionate to form parent dye and reduced products. S substrates, being somewhat more polar, would not have access to the hydrophobic site and would therefore be exposed to O_2 in an aerobic environment. An alternate explanation for differences in oxygen sensitivity may lie in the chemical response of each of the substrates to oxygen independent of enzymic considerations. Preliminary evidence obtained through determination of oxidation)reduction potentials for each of the azo dyes suggests that reduced intermediates of S substrates persist long enough to interact with O_2 while those from I substrates do not. Further explorations in this direction are needed to elucidate the true mechanism.

Table 1. Reduction of S and I substrates by flavoprotein and by flavoprotein plus cytochrome P-450 (reconstituted system).*

S Substrates	pmol/min ± SD (1) Flavoprotein alone	(2) Flavoprotein + cytochrome P-450	Ratio (2)/(1)
1. 2'-carboxy-DAB (o-methyl red)	500 ± 180	631 ± 161	1.3
2. 2'-carboxy-DAB methyl ester	471 ± 102	633 ± 35	1.3
3. 4'-carboxy-DAB (p-methyl red)	69.3 ± 0	109 ± 44	1.6
4. 4'-carboxy-DAB methyl ester	6.1 ± 0.1	37.3 ± 5.7	6.1
5. DAB arsonate	68.9 ± 15.8	143 ± 21	2.1
6. DAB sulfonate (methyl orange)	212 ± 17	432 ± 47	2.0
I Substrates			
1. DAB	14.4 ± 3.6	202 ± 7.2	14.0
2. MAB	49.7 ± 21.0	227 ± 27	4.6
3. AB	48.0 ± 17.6	139 ± 22	2.9
4. p-hydroxyazobenzene	6.0 ± 1.4	44.8 ± 11.3	7.5
5. 3'-Me-DAB	6.0 ± 0.0	59.3 ± 10.3	9.9
6. 2-Me-DAB	18.8 ± 1.9	108 ± 41	5.7
7. 4'-NH_2-DAB	11.7 ± 1.7	71.6 ± 23.2	6.1
8. 4'-OH-DAB	19.0 ± 1.0	119 ± 10	6.3

*In reactions involving flavoprotein alone, all reactants of the reconstituted system were present except cytochrome P-450. All other conditions were identical for the two systems. Varying amounts of each of the enzymes were used, although the ratio of 2:1, reductase to cytochrome was maintained throughout. Because two enzyme systems were compared, rates are given as pmol/min.

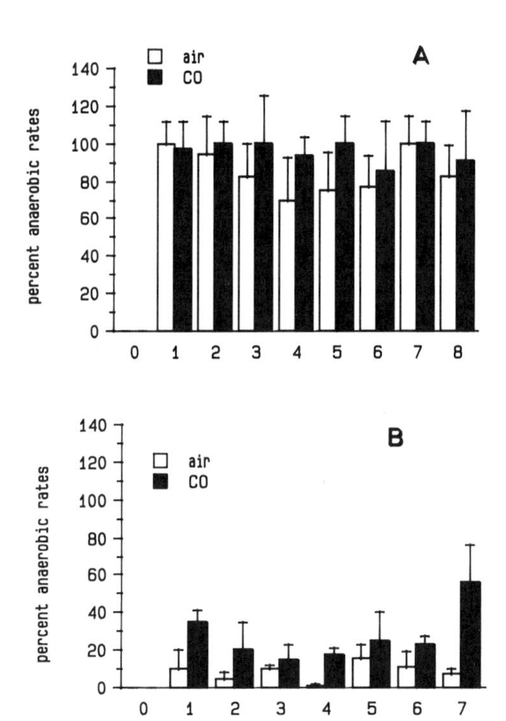

Figure 3. Effect of air and carbon monoxide on the reduction of azo dyes by rat liver microsomes. A: I substrates. B: S substrates. Enzyme rates are indicated as the percent of activity in a nitrogen atmosphere. The numbers refer to the structures in Fig. 1 and 2.

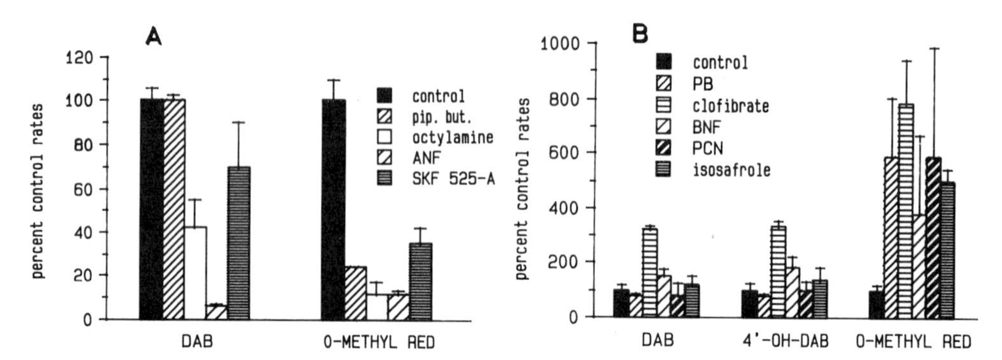

Figure 4. Effect of inhibitors (A) and induvcers (B) of cytochrome P-450 activity on the reduction of I and S azo dyes by rat liver microsomes.

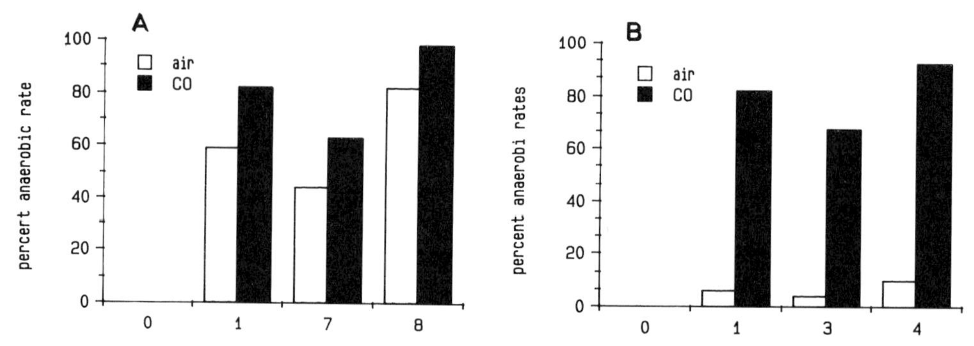

Figure 5. Effect of air and carbon monoxide on the reduction of azo dyes by a reconstituted system of purified NADPH)cytochrome P-450 reductase, partially purified cytochrome P-450 and lipid. A: I substrates. B: S substrates. Enzyme rates are indicated as the percent of activity in a nitrogen atmosphere. The numbers refer to the structures in Fig. 1 and 2.

Differential sensitivity of azoreduction to CO is not readily explainable. Binding of CO to the sixth ligand position of P-450 excludes O_2, which, however, does not enter into the reduction process. Possible mechanisms may relate to changes in oxidation)reduction potentials of the various P-450s which could alter the ability to transfer electrons to reducible substrates. Ultimately, it is hoped that measurement of oxidation-reduction potentials for this system will shed light on the mechanism of these differences.

In conclusion, at least two classes of azoreductase activity have been demonstrated for microsomal P-450. They differ in sensitivity to O_2 and CO, and selective responses to inducing agents can be demonstrated.

REFERENCES

Fujita, S. and Peisach, J. (1978). Liver microsomal cytochrome P-450 and azoreductase activity. *J. Biol. Chem.* **253**, 4512-4513.

Hernandez, P. H., Gillette, J. H., Mazel, P. (1967) . Studies on the mechanism of mammalian hepatic azoreductase. I. Azoreductase activity of reduced nicotinamide adenine dinucleotide phosphate cytochrome c reductase. *Biochem. Phramacol.* **16**, 1859-1875.

Huang, M)T., Miwa, G. T., Cornheim, N. and Lu, A. Y. H. (1979). Rat liver cytosolic azoreductase. Electron transport properties and the mechanism of dicumarol inhibition of the purified system. *J. Biol. Chem.* **254**, 11223-11227.

Levine, W. G. and Raza, H. (1988). Mechanism of azoreduction of dimethylamino azobenzene by rat liver NADPH-cytochrome P-450 reductase and partially purified cytochrome P-450. Oxygen and carbon monoxide sensitivities and stimulation by FAD and FMN. *Drug Metab. Dispos.* **16**, 441-448.

Mason, R. P., Peterson, F. J. and Holtzman, J. L. (1978). Inhibition of azoreductase by oxygen. The role of the azo anion free radical metabolite in the reduction of oxygen to superoxide. *Mol. Pharmacol.* **4**, 665-671.

Peterson, F. J., Holtzman, J. L., Crankshaw, D. and Mason, R. P. (1988). Two sites for azoreduction in the monoxygenase system. *Mol. Pharmacol.* **34**, 597-603.

Raza, H. and Levine, W. G. (1986). Azoreduction of N,N-dimethyl-4-aminoazobenzene (DAB) by rat hepatic microsomes. Selective induction by clofibrate. *Drug Metab. Dispos.* **14**, 19-24.

Zbaida, S., Stoddart, A.M. and Levine (1989). Studies on the mechanism of reduction of azo dye carcinogens by rat liver microsomal cytochrome P-450. *Chem. -Biol. Interact.* **69**, 61-71.

EXPRESSION OF A cDNA ENCODING RAT LIVER DT-DIAPHORASE IN *ESCHERICHIA COLI*

Qiang Ma*, Regina Wang+, Anthony Y.H. Lu+, and Chung S. Yang*

Joint Graduate Program in Toxicology and Department of Chemical Biology & Pharmacognosy*
College of Pharmacy
Rutgers University, Piscataway, NJ 08855
Department of Animal & Exploratory Drug Metabolism+
Merck Sharp & Dohme Research Laboratories
Rahway, NJ 07065

Quinones and their precursors are widely distributed in nature, both as natural compounds and as environmental pollutants (Smith,1985). The toxicity and mutagenicity of quinones are believed to be due to their enzymatic one electron reduction by flavoenzymes to form semiquinones which can alkylates critical nucleophiles or generate superoxide anion radical through a redox cycling process (Thor et al., 1985; Chesis et al., 1984). DT-diaphorase (NAD(P)H: quinone oxidoreductase, EC 1.6.99.2.) is a unique flavoprotein in that it catalyzes the obligatory two electron reduction of quinones to hydroquinones via a pathway bypassing the formation of semiquinone; thus, playing a protective role against quinone toxicity (Ernster et al., 1986; Atallah et al., 1988).

To study the structure and function of DT-diaphorase, it is desirable to obtain large quantities of pure enzyme. Recently, cDNA clones complementary to the rat liver cytosolic DT-diaphorase mRNA have been constructed (Williams et al., 1986; Robertson et al., 1986). One of these, pDTD55, is complementary to liver cytosol DT-diaphorase mRNA of 3-MC treated rats, although the 5'-end of pDTD55 cDNA coding strand is incomplete (Bayney and Pickett, 1988). In this study, polymerase chain reaction (PCR) was used to extend the 5'-end sequence of pDTD55 cDNA to include the N-terminal codon and the ATG initiation codon. The resulting full length cDNA was expressed in *Escherichia coli* using an expression vector pKK2.7, thus, permitting the study of structure and function of DT-diaphorase with the expressed enzyme.

MATERIALS AND METHODS

Construction of the expression plasmid. Two oligonucleotide primers were synthesized (Cyclone™ DNA synthesizer, Biosearch, Inc.). Primer A consists of an EcoRI site at 5'-end, a 23 nucleotide sequence at 3'-end homologous to 5'-end of pDTD55 cDNA coding strand, and a 36 nucleotide sequence in the middle which is missed in the 5'-end of pDTD55 cDNA. Primer B has a HindIII site at 5'-end and a 26 nucleotide sequence complementary to the 3'-end of pDTD55 cDNA coding strand. PCR was conducted with the two primers to extend the 5'-end sequence of pDTD55 cDNA. In addition, an EcoRI restriction site and a HindIII site were incorporated at 5'- and 3'-ends of the cDNA respectively. The resulting full length cDNA was digested with EcoRI and HindIII restriction enzyme (BRL), and was inserted into the EcoRI and

Biological Reactive Intermediates IV, Edited by C.M. Witmer *et al.*
Plenum Press, New York, 1990

HindIII sites of the dephosphorylated expression vector pKK2.7 (Wang et al., 1989). The constructed plasmid was purified. The insertion of the cDNA was confirmed by digestion with restriction enzymes and DNA sequence analysis.

Bacterial Trasformation and Cell Growth

E. coli strain AB1899 (the ion-1 protease-deficient strain, *E. coli* Genetic Stock Center, Yale University) was transformed with the constructed plasmid by the calcium cloride procedure (Mandel and Higa, 1970; Cohen et al., 1972). Colonies were selected by ampicillin resistance and screened by colony hybridization. The transformants were grown as described (Wang et al., 1989).

Purification of the Expressed DT-Diaphorase

E. coli cells (4.3 liters) transformed with pKK-DTD4 were centrifuged at 10,000g for 20 min. to harvest cells. The cells were disrupted by sonication in 10 mM $NaPO_4$,+ pH 7.4, containing 5 mM EDTA, and were centrifuged at 10,000g for 25 min. The cytosolic fraction was obtained by centrifugation of the 10,000g supernatant at 100,000g for 60 min. The cytosol was used for purification of the expressed DT-diaphorase using cibacron blue-agarose affinity chromatography (Prochaska, 1988).

SDS-PAGE and Western Blot Analysis

E. coli lysate or various purification fractions were separated on 12.5% SDS polyacrylamide gels by the method of Laemmli (Laemmli, 1970). Western blot was done using polyclonal antibodies against rat liver DT-diaphorase as described (Towbin et al., 1979). Alkaline phosphatase-conjugated goat anti-rabbit antibodies were used as 2nd antibody in the staining reaction (Calbiochem®).

DNA Sequence Analysis

The cDNA in plasmid pKK-DTD4 was sequenced by dideoxy method (Sanger et al., 1977).

Assays

Typical assays for DT-diaphorase is described in the legend for Table I. Protein concentration was determined by the method of Lowry (Lowry et al., 1951).

RESULTS AND DISCUSSONS

Table 1 shows the results for the assays of DT-diaphorase activity in the cytosol of *E. coli* strains AB1899 and JM105. The cytosolic fractions from both strains were able to reduce menadione with either NADPH or NADH as electron donor. However, the activity levels were much lower than that of rat liver cytosol, and were not sensitive to dicumarol inhibition at the concentration of 10^{-5} M with either NADPH or NADH as reductant. Western blot analysis using polyclonal antibodies against rat liver DT-diaphorase showed that DT-diaphorase was not detected in the cytosols of either strain. These results indicate that *E. coli* strains, AB1899 and JM105, do not have the dicumarol sensitive DT-diaphorase.

In the expression study, PCR was used to extend the 5'-end sequence of pDTD55 cDNA to include the N-terminal codon and the ATG initiation codon. In addition, an EcoRI site and a HindIII site were incorporated at the 5'- and 3'-ends of the cDNA respectively. The cDNA was inserted into an expression vector pKK2.7 under the control of the tac promotor. *E. coli* strain AB1899 was transformed with the constructed vector, and a positive clone (pKK-DTD4) was selected. Restriction enzyme analysis of this clone with EcoRI, HindIII and PstI showed that a single copy of the cDNA was inserted at the correct site. This was further proved by sequence analysis.

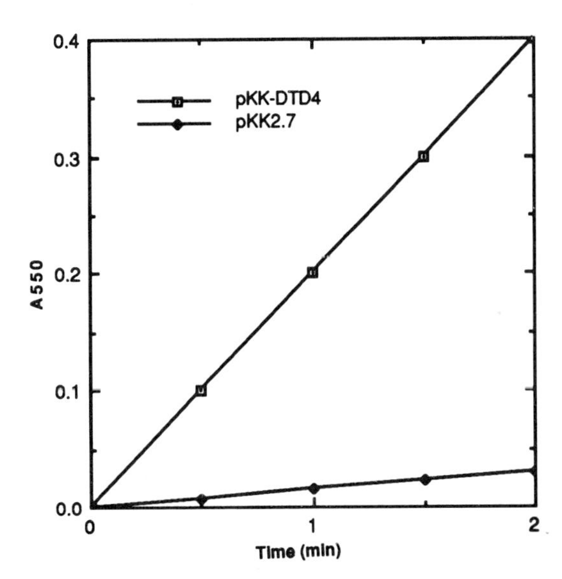

Figure 1. Quinone reductase activity in cell free extracts of *E. coli* strain AB1899. The *E. coli* cells containing vector pKK2.7 or vector pKK-DTD4 were grown overnight. The cells were disrupted by sonication and the cell free extracts were prepared and assayed for enzyme activity. The total amount of portein used in the assays were as follows: 0.464 µg for pKK-DTD4 containing cells and 0.832 µg for pKK2.7 containing cells.

Table 1. Effects of dicumarol on quinone reductase activities[a]

Conditions[b]		Rat liver	AB1899	JM105
NADPH				
-	dicumarol	0.30	0.32	0.19
+	dicumarol	0.01	0.30	0.20
+	NaOH	0.31	0.32	0.20
NADH				
-	dicumarol	0.26	0.23	0.12
+	dicumarol	0.02	0.22	0.12
+	NaOH	0.16	0.26	0.12

[a] Activities are expressed as A550.min. Enzyme sources are as follows: Rat liver: 1:100 dilution of 3-MC treated rat liver cytosol (S.A.=9.5 µmol/min/mg); AB1899: cytosol of *E. coli* strain AB1899 (S.A.=0.8 µmol/min/mg); JM105: cytosol of *E. coli* strain JM105 (S.A.=0.4 µmol/min/mg).
[b] 10 µl of cytosol was used in enzyme assay. Menadione of 0.01 mM was used as substrate. Dicumarol of 10^{-5} M was used in inhibition assay. NaOH of 2.4 x 10^{-5} M was used as control, since dicumarol was dissolved in NaOH.

Figure 1 shows the activities of DT-diaphorase in the cell free extracts of the transformed cells. High activity was observed in the cell free extracts of *E. coli* cells transformed with pKK-DTD4 (20.6 µnit/mg - µnit=1 µmol/min) and the activity was inhibited by dicumarol at the concentration of 10^{-5} M. In comparison, the activity of the cells transformed with pKK2.7 was 0.8 µnit/mg. Western blot analysis showed that a protein band with molecular weight equivalent to that of the rat liver cytosolic DT-diaphorase was recognized by the polyclonal antibodies. Two faint minor bands of lower molecular weight were also recognized by the antibodies. The expressed DT-diaphorase was purified and further characterization of the purified protein is being undertaken.

In summary, PCR was used to extend the 5'-end sequence of pDTD55 cDNA. The full length cDNA was expressed in *E. coli* using expression vector pKK2.7. The expressed DT-diaphorase showed high activity of menadione reduction and was inhibitable by dicumarol. The protein was recognized by antibodies specific for rat liver DT-diaphorase. Characterization of the purified expressed enzyme is in progress.

ACKNOWLEDGEMENTS

We thank Dr. Thomas H. Ruchmore and Dr. Su-Er Wu Huskey for valuable discussions, and Dr. Cecil B. Pickett for providing us pDTD55.

REFERENCES

Atallah, A.S., Landolph, J.R., Ernster, L. and Hochstein P. (1988). DT-diaphorase activity and the cytotoxicity of quinones in C3H/10T1/2 mouse embryo cells. *Biochem. Pharmacol.* **37**, 2451-2459.

Bayney, R.M. and Pickett, C.B. (1988). Rat liver NAD(P)H:quinone reductase: isolation of a quinone reductase structural gene and prediction of the NH2 terminal sequence of the protein by double-stranded sequencing of exons 1 and 2. *Archs. Biochem. Biophys* **260**, 847.

Chesis, P.L., Levin, D.E., Smith, M.T., Ernster, L. and Ames, B.N. (1984). Mutagenicity of quinones: pathways of metabolic activation and detoxification. *Proc. Natl. Acad. Sci., USA* **81**, 1696-1700.

Cohen, S.N., Chang, A.C.Y. and Hsu, L. (1972). Nonchromasomal antibiotic resistance in bacteria: Genetic transformation of *Eschetichia coli* by R-facter. *Proc. Natl. Acad. Sci, USA* **69**, 2110-2114.

Ernster, L., Estabrook, R.W., Hochstein, P. and Orrenius, S., eds. (1987). In, DT-diaphorase a quinone reductase with special functions in cell metabolism and detoxication. *Chem. Scripta* 27A.

Laemmli, U.K. (1970). Cleavage of struatural proteins during the assembly of the head of bacteriaphage T4. *Nature* (London) **227**, 680-685.

Mandel, M. and Higa, A. (1970). Calcium-depent bacteriophage DNA infection. *J. Mol. Biol.* **53**, 159-162.

Prochaska, H.J. (1988). Purification and crystallization of rat liver NAD)P)H:(quinone-acceptor) oxidoreductase by cibacron blue affinity chromatography: identification of a new and potent inhibitor. *Archs. Biochem. Biophys.* **267**, 529-538.

Robertson, J.A., Chen, H. and Nebert, D.W. (1986). NAD(P)H: Menadione oxidoreductase, novel purification of enzyme, cDNA and complete amino acid sequence, and gene regulation. *J. Biol. Chem.* **261**, 15794-15799.

Sanger, F., Nicklen, S. and Coulson, A.R. (1977). DNA sequencing with chain-terminating inhibitors. *Proc. Natl. Acad. Sci. USA* **74**, 5463-5467.

Smith, M.T. (1985). Quinones as mutagens, carcinogens and anticancer agents: introduction and overview. *J. Toxicol. Environ. Health* **16**, 665-672.

Thor, H., Smith, M.T., Hartzell, P., Bellomo, G., Jewell, S.A. and Orrenius, S. (1985). The metabolism of menadione (2-methyl-1,4-naphthoquinone) by isolated hepatocytes: a study of the implications of oxidative stress in intact cells. *J. Biol. Chem.* **257**, 12419-12424.

Towbin, H., Staehelin, T. and Gordon, J. (1979). Electrophoretic transfer of protein from polyacrylamide gels to nitrocellulose sheets: Procedure and some applications. *Proc. Natl. Acad. Sci. USA* **76**, 4350-4354.

Wang, R.W., Pickett, C.B. and Lu, A.Y.H. (1989). Expression of a cDNA encoding a rat liver glutathione S-transferase Ya subunit in *Escherichia coli*. *Archs. Biochem. Biophys.* **269**, 536-543.

Williams, J.B., Lu, A.Y.H., Cameron, R.G. and Pickitt, C.B. (1986). Rat liver NAD(P)H: quinone reductase, construction of a quinone reductase cDNA clone and regulation of quinone reductase mRNA by 3-methylcholanthrene and in persistent hepatocyte nodules induced by chemical carcinogens. *J. Biol. Chem.* **261**, 5524.

SUICIDAL INACTIVATION OF CYTOCHROME P-450 BY HALOTHANE AND CARBON TETRACHLORIDE

Maurizio Manno[1], Michela Rezzadore and Stefano Cazzaro

Institute of Occupational Medicine
University of Padua Medical School, Via J. Facciolati, 71
35127 Padova, Italy

Abbreviations used: H, halothane; -NF, -naphthoflavone PB, phenobarbitone.

INTRODUCTION

The reductive metabolism of carbon tetrachloride (CCl_4) and halothane (H) by liver microsomal cytochrome P-450 leads to the formation of reactive intermediates and results in the inactivation of the enzyme and loss of its prosthetic group haem (de Groot H. et al., 1981; Krieter P.A. et al., 1983). Previous work has shown that the haem moiety is the critical target of CCl_4 and H activation by cytochrome P-450, indicating that, with both substrates, a suicide type of reaction is probably involved (Manno M., et al., 1988; Manno M., et al., 1989). Indeed, haem alone, in the absence of the apoprotein, was able to undergo CCl_4- or H-dependent suicidal inactivation when incubated anaerobically with each substrate in presence of a reducing agent (Manno, M., et al., 1989; Manno M., 1989). The CCl_4-dependent loss of haem is due to attack by a single reactive metabolite molecule (Manno, M., et al., 1989). In the present study we compared the mechanisms of the suicidal, reductive inactivation of cytochrome P-450 by CCl_4 and H.

MATERIALS AND METHODS

Chemicals and Biochemicals

Halothane (2-bromo-2-chloro-1,1,1-trifluoro-ethane) and β-naphthoflavone (β-NF) were purchased from Aldrich Chimica (Milano, Italy). Phenobarbitone (PB) was from BDH Chemicals Ltd (Poole, Dorset, U.K.). CO and O_2-free nitrogen were purchased from SIO (Milano, Italy). $NADP^+$, NADPH, catalase (1.11.1.6), glucose oxidase (1.1.3.4), glucose-6-phosphate dehydrogenase (1.1.1.49), glucose-6-phosphate, haemin and protoporphyrin IX dimethyl ester were purchased from Sigma Chemical Co. (St. Louis, MO, U.S.A.).

---

1. Address for correspondence: Dr. Maurizio Manno, Istituto di Medicina del Lavoro, Universita' di Padova, Via J. Facciolati, 71,35127 Padova, Italy. Tel.: 39 49 821 6647; FAX: 39 49 821 6621

Treatment of Animals and Microsome Preparation

Male Wistar Albino rats (170-230 g) obtained from the Institute of General Pathology Animal House, University of Padua Medical School, were either untreated or pretreated i.p. with PB or β-NF (80 mg/kg b.w. daily for 3 days in saline or corn oil, respectively). Control animals were injected with saline or corn oil (5 ml/kg b.w. daily for 3 days). Livers were perfused *in situ* with saline and homogenized as described previously (Manno, M., et al., 1988). The microsomal fraction was prepared by a slight modification of the method of Kamath et al. (Kamath, S.A., et al., 1971) and finally resuspended in 0.1 M Na_2HPO_4 buffer, pH 7.4, containing 20% glycerol.

Incubations

Anaerobic incubations were performed in rubber-stoppered glass tubes containing 0.1 M Na_2HPO_4 buffer, pH 7.4, previously bubbled through with O_2-free nitrogen and supplemented with an oxygen scavenging system (Manno, M., et al., 1988). In some incubations the following NADPH-generating system was used: 0.2 IU/ml glucose-6-phosphate dehydrogenase, 4 mM glucose-6-phosphate, 0.2 mM $NADP^+$ and 6 mM $MgCl_2$.

Assays

Cytochrome P-450 was measured by the method of Omura & Sato (Omura T. and Sato R. 1964). Haem was assayed by two different methods: the pyridine/haemochromogen method (Paul K.G., et al., 1953) by recording the reduced absolute spectrum in pyridine/NaOH and using the $\delta mM_{557-541}$ 20.7 reported by Falk (1964), or by the porphyrin fluorescence technique (Morrison, G.R. 1965).

RESULTS AND DISCUSSION

A maximal loss of cytochrome P-450 of about 60 and 40% was observed in 20-30 minutes on addition of 1 mM CCl_4 and 5 mM H, respectively, to anaerobic incubations containing liver microsomes from PB-treated rats. With both compounds the reaction was saturable and showed biphasic, pseudo first-order kinetics of inactivation (data not shown), as one would expect for a suicide type of reaction (Waley, S.G. 1980). This suggested that at least two cytochrome P-450 isoenzymes were involved.

With both substrates the NADPH- or sodium dithionite-dependent loss of haem was accompanied by a significant loss of haem-derived protoporphyrin IX (Table 1), indicating that the loss of haem was due, at least in part, to a modification of its tetrapyrrolic ring.

Pretreatment of rats with PB or Aroclor 1254 increased, and that with β-NF decrease the CCl_4- or H-dependent loss of microsomal haem when compared to the loss observed with liver microsomes obtained from corn-oil- or saline-treated animals (data not shown). With both substrates a strong inhibition of the reaction was found in presence of saturating concentrations of CO (data not shown), suggesting that binding of the substrate to the haem iron is required for suicide activation to occur.

When metabolic turnover of the substrate and enzyme inactivation were compared using limiting concentrations of each substrate, mean partition ratios of about 121 and 27 were calculated for CCl_4 and H, respectively, indicating that CCl_4 is more effective than H as a suicide substrate of cytochrome P-450.

In conclusion, the present results seem to suggest a similar mechanism of reductive, suicidal inactivation of microsomal cytochrome P-450 by CCl_4 and H, where haem plays the double role of site of activation and critical target. Based on our values of partition ratio, CCl_4 appears to be approximately 5 times more effective, as a suicide substrate, than H. It is predicted that H, like CCl_4, inactivates cytochrome P-450 through covalent binding of a single reactive metabolite molecule to the haem moiety of the cytochrome.

Table 1. Substrate-dependent loss of haem and protoporphyrin IX from NADPH- or sodium dithionite-reduced rat liver microsomes.

Incub.	Reducing agent	Haem (nmol/ml)	(%)	Protoporphyrin IX (f.u.) #	(%)
Control	NADPH	2.51 ± 0.13	(100)	68.6 ± 1.0	(100)
	dithionite	2.04 ± 0.12	(100)	27.9 ± 1.3	(100)
Halothane	NADPH	1.63 ± 0.09 *	(65)	54.6 ± 1.9 *	(80)
	dithionite	1.22 ± 0.04 *	(60)	22.1 ± 1.0 *	(79)
CCl_4	NADPH	0.95 ± 0.06 *	(38)	37.4 ± 0.3 *	(55)
	dithionite	1.24 ± 0.07 *	(61)	21.2 ± 0.3 *	(76)

* p 0.001, when compared to the corresponding control value (Student's t test). # fluorescence units.
Anaerobic incubations were at 37°C for 15 minutes and contained 1.47 mg liver microsomal protein (2.26 nmol cytochrome P-450/mg) from PB-treated rats and a NADPH generating system or 1 mM sodium dithionite in 1.5 ml 0.1 M Na_2HPO_4 buffer. At time 0, CCl_4 or halothane (5 mM for both) was added to start the reaction. No substrate was added to control incubations. Values are mean ± SD from 4-5 incubations.

ACKNOWLEDGEMENTS

Supported in part by the CNR (grant No. 880055204) and the Italian Ministry of Education.

REFERENCES

de Groot, H. and Haas, W. (1981). Self-catalysed, 02-independent damage of cytochrome P-450 by carbon tetrachloride. *Biochem. Pharmac.* **30**, 2343-2347.

Falk, J.E. (1964). Absorption spectra. In *Porphyrins and metalloporphyrins* (Falk, J.E. Ed.), pp. 231-246, Elsevier, Amsterdam.

Kamath, S.A., Kummerow, F.A. and Narayan, K.A. (1971). A simple procedure for the isolation of rat liver microsomes. *FEBS Lett.* **17**, 90-92.

Krieter, P.A. and Van Dyke, R.A. (1983). Cytochrome P-450 and halothane metabolism. Decrease in rat liver microsomal P-450 *in vitro*. *Chem-Biol. Interactions* **44**, 219-235.

Manno, M., De Mateis, F. and King, L.J. (1988). The mechanism of the suicidal, reductive inactivation of microsomal cytochrome P-450 by carbon tetrachloride. *Biochem. Pharmac.* **37**, 1981-1990.

Manno, M., Cazzaro, S. and Rezzadore, M. (1989). Suicidal inactivation of cytochrome P-450 by halothane. In *V International Congress of Toxicology*, Brighton, 16-21 July, abstracts, p. 55.

Manno, M., King, L.J. and De Matteis, F. (1989). The degradation of haem by carbon tetrachloride: metabolic activation requires a free axial coordination site on the haem iron and electron donation. *Xenobiotica* **19**, 1023-1035.

Manno, M. (1989). Suicidal inactivation of haem by halothane. *Adv. Biosci.* **76**, 107-115.

Morrison, G.R. (1965). Fluorimetric microdetermination of heme protein. *Anal. Chem.* **37**, 1124-1126.

Omura, T. and Sato, R. (1964). The carbon monoxide-binding pigment of liver microsomes. I. Evidence for its hemoprotein nature. *J. Biol. Chem.* **239**, 2370-2378.

Paul, K.G., Theorell, H. and Akeson, A. (1953). The molar light absorption of pyridine ferroprotoporphyrin (pyridine haemochromogen). *Acta. Chem. Scand.* **7**, 1284-1287.

Waley, S.G. (1980). Kinetics of suicide substrates. *Biochem. J.* **185**, 771-773.

STRUCTURE-ACTIVITY RELATIONSHIPS OF ACRYLATE ESTERS: REACTIVITY TOWARDS GLUTATHIONE AND HYDROLYSIS BY CARBOXYLESTERASE *IN VITRO*

T.J. McCarthy and G. Witz

Joint Graduate Program in Toxicology
Rutgers University/UMDNJ-RW Johnson Medical School
Piscataway, NJ

INTRODUCTION

Acrylate esters are high volume, versatile chemicals in the plastics industry. Long)term industrial exposure to these esters has caused contact hypersensitivity and dermatitis, while laboratory studies have shown these esters to be strong irritants, sensitizers, clastogens, and some have been shown to be weak carcinogens (Andrews and Clary, 1986). Numerous members make up this class of chemicals sharing an α, β-unsaturated ester structure. The individual members differ from each other due to structural features added onto the α, β-unsaturated ester.

The goal of this research is to investigate the structure-activity relationships (SAR) of these esters in order to elucidate their mechanism(s) of toxicity. The structural features under investigation are: (i) α-methyl substitution, (ii) alcohol chain length, and (iii) mono- vs. bifunctionality. It is postulated that all acrylates react by the same mechanism, i.e. alkylation of cellular nucleophiles, and that their biological reactivities differ depending upon these modifying structural features.

The emphasis for occupational health is to find members of this family of chemicals less detrimental to worker's safety. Both a chemical and biological SAR are under investigation, and the goal of our studies is to determine whether a chemical SAR may be used as a predictive tool for the biological SAR. To address this goal, the endpoints under investigation are: (i) electrophilic reactivity with glutathione, both in a cell-free system and a rat red blood cell system to determine the second order rate constants for the spontaneous reaction of glutathione with the esters and to observe how a cellular system may modify this reactivity, (ii) the hydrolysis to an acid and alcohol or glycol by porcine liver carboxylesterase *in vitro*, as the acid does not alkylate cellular nucleophiles under physiological conditions and thus hydrolysis may be a detoxification step, and (iii) adduct formation with deoxyribonucleic acids and DNA to determine potential initiating activity of these esters.

METHODS

To determine the second order rate constant for the spontaneous reaction of glutathione and ester, glutathione in phosphate buffered saline (PBS) was mixed with an equal volume of ester in PBS containing 5% (final) propylene glycol at 37°C. At specific time points, aliquots were removed and added to Ellman's reagent [5,5'-dithio-bis(2-nitrobenzoic acid)] in 1% sodium citrate. Optical density was measured at 412 nm. From the loss of optical density over time (i.e. loss of free sulfhydryls), the second order rate constant for the spontaneous reaction of ester and gluthathione was calculated.

To determine the depletion of cellular glutathione by acrylate esters, red blood cells (5% HCT) suspended in PBS, pH 7.4, from female Sprague-Dawley rats were exposed to ester for 1 hr at 37°C. After removal of unreacted chemical, glutathione content was measured colormetrically using the method of Beutler (Beutler et al., 1963).

To calculate the kinetic constants for the enzymatic hydrolysis of ester by carboxylesterase, esters were initially diluted in ethanol followed by serial dilutions with TRIS, pH 8.0. Carboxylesterase (porcine liver, Sigma) was added, and the mixture was incubated for 20 min at 37°C. Reactions were acidified and terminated with conc. HC1, and organic acid was extracted with ethyl acetate. After drying the organic phase with $MgSO_4$ and concentrating under a stream of N_2, samples were injected onto a gas chromatograph equipped with a flame ionization detector. From the peak area of the acid produced, the K_m and V_{max} for the hydrolysis of ester was calculated.

RESULTS AND CONCLUSIONS

The reaction between glutathione and acrylate ester was second order, first order in each reactant. The second order rate constant for the spontaneous reaction of glutathione and methyl acrylate is 52.0 ± 5.0 liters/mole/min and the reaction with methyl methacrylate is 0.325 ± 0.059 liters/mole/min. These results indicate that methyl acrylate reacts 160 times faster with glutathione compared with the methacrylate analog. The rate constant for the reaction of glutathione and butyl acrylate is 38.7 ± 3.3 liters/mole/min. The difference in rate constants between the methyl and butyl acrylate may be due more to their difference in molecular weights rather than electrophilic reactivity, as the kinetic energy of a molecule is inversely proportional to the square root of its molecular weight. In this case MW_{BA}/MW_{MA} = 1.2, which is close to the inverse ratio of rate constants, K_{MA}/K_{BA} = 1.3. The relative electrophilic reactivity of acrylates in the cellular system was determined by comparing the ester concentration depleting 20% of rat red blood cell glutathione. The EC_{20} for methyl acrylate is 0.063 mM and for methyl methacrylate is 2.5 mM. The relative reactivity of acrylate vs. methacrylate ester towards gluthathione in the two systems, i.e. cell-free vs. cellular, is not the same. The cellular system differs from the cell-free system in two important features which may explain the discrepancy. The cell membrane has protein nucleophiles which could deplete acrylate concentration before the ester can react with intracellular glutathione. The depletion of a more reactive ester at the cell membrane is expected to be greater than that for a less reactive ester. The second difference is that both enzymatic and spontaneous conjugation may occur in the cell; the contribution of the former has not yet been characterized. In either system however, α-methyl substitution has a profound effect upon electrophilic reactivity.

The K_m for the carboxylesterase hydrolysis of ethyl acrylate (134 ± 16 μM) is not significantly different compared with the K_m for ethyl methacrylate. The V_{max} for ethyl acrylate (8.9 ± 2.0 nmole/min) is significantly higher compared with that for ethyl methacrylate (5.2 ± 2.5), although this difference is small compared to the differences in electrophilic reactivities towards glutathione. Unlike the results for the electrophilic reactivities, α-methyl substitution does not appear to affect greatly the enzymatic hydrolysis of these esters. The K_m and V_{max} values for the hydrolysis of butyl acrylate are 33.3 ± 8.5 μM and 1.49 ± 0.83 nmoles/min, respectively. The kinetic constants for the hydrolysis of butyl methacrylate are not significantly different compared with those of butyl acrylate, but both the K_m and V_{max} values for the butyl esters are significantly different compared with those of their methyl ester analogs. From this data, increased alcohol chain length increases substrate affinity, yet decreases turnover for the enzymatic hydrolysis of these esters.

Although all acrylates are direct)acting alkylating agents via Michael addition reactions, their chemical and biological activity differ depending upon structural features added to the α, β-unsaturated ester moiety. α-Methyl substitution has been shown to have a great effect upon electrophilic reactivity (i.e. toxicity) yet only a minor effect upon hydrolysis (detoxification). Alcohol chain length appears to have little effect upon electrophilic reactivity, yet it affects hydrolysis. Thus it may be possible to select acrylates for commercial use which pose the least health risk to workers.

REFERENCES

Andrews, L.S., and Clary, J.J. (1986). Review of the toxicity of multifunctional acrylates. *J. Toxicol. Environ. Health* **19**, 149-164.

Beutler, E., Duron, E., and Kelly, B. (1961). Improved method for determining blood glutathione. *J. Lab. Clin. Med.* **61**, 882-888.

Miller, R.R., Ayres, J.A., Rampy, L.W., and McKenna, M.J. (1981). Metabolism of acrylate esters in rat tissue homogenates. *Fundam. Appl. Toxicol.* **1**, 410-414.

GLUTATHIONE CONJUGATES OF HYDRALAZINE FORMED IN THEPEROXIDASE/ HYDROGEN PEROXIDE/GLUTATHIONE SYSTEM

Larry G. McGirr and Peter J. O'Brien

Faculty of Pharmacy
University of Toronto
Toronto,Ontario, Canada, M5S 1A

The antihypertensive drug hydralazine is associated withthe induction of symptoms similar to systemic lupuserythematosus (Perry, 1973); it is also associated withlung tumors in mice (Toth, 1978) , is mutagenic inseveral short term bacterial and mammalian cell assays (Williams et. al., 1980) and results in the irreversible inactivation of several amine oxidases (Lyles et. al., 1983, Numata et. al., 1981). Mechanisms responsible forthese conditions are unknown.

Human hydralazine metabolism results in the formation of several oxidised urinary compounds including phthalazone (Zak et. al., 1974). Similar metabolites have been found in *in vitro* rat liver microsomal or horseradish peroxidase (HRP) incubations (Streeter and Timbrell, 1985, LaCagnin et. al., 1986). In addition, covalent protein binding occurs in rat liver microsomal systems. Metabolism initially proceeds through the hydralazine radical ($RNHNH_2$) (Sinha, 1983). However, the pathway of metabolism and species responsible for covalent binding are still unknown.

In the following report, the nature of the reactive intermediates formed in the horseradish peroxidase system and the pathway of metabolism was investigated by trapping the reactive intermediates formed with glutathione (GSH).

MATERIALS AND METHODS

Hydralazine hydrochloride, glutathione, oxidised glutathione (GSSG), horseradish peroxidase (type VI), hydrogen peroxide (H_2O_2), fluoro-2,4-dinitrobenzene and iodoacetic acid were purchased from Sigma Chemical Co. (St. Louis, M.O.). Glutathione [glycine-2-^{3}H] (sp. act.1.0Ci/mmol) was obtained from Dupont (Boston, MA). All high pressure liquid chromatography solvents were purchased from Caledon Laboratories Ltd (Georgetown,Ontario). Phthalazine and phthalazone were gifts from Dr. J.Uetrecht, Faculty of Pharmacy, University of Toronto.

Metabolites were examined in standard reaction mixtures containing hydralazine (100 μM), horseradish peroxidase (1 μg/ml), hydrogen peroxide (100 μM) and glutathione (100 μM or 1mM) incubated for 10 minutes at room temperature in 1.0 ml of 0.1M Tris buffer, pH 7.4. Glutathione and oxidised glutathione were determined by the HPLC method described by Reed et.al., (1980). Metabolites were analysed by HPLC (Gilson model 302) using a 5μ LC-18 column (4.6 mm x 25 cm) column (Supelco) eluted with a methanol:water:acetic acid: triethyamine (70:30:2:0.12) sytem (1 ml/min). Metabolites were isolated from incubation mixtures containing 10 mM hydralazine (reaction mixtures were scaled up 100 fold in terms of reactant concentrations) by serial collections from several HPLC runs; methanol was removed under N_2 and the remaining water solution lyophilised prior to structural determination. FAB mass spectral analysis was conducted using a VG Analytical ZAB-1F instrument using a glycerol or thioglycerol matrix. NMR analysis of samples dissolved in D_2O was conducted using a Bruker 300MHz instrument.

RESULTS AND DISCUSSION

The peroxidase catalysed oxidation of hydralazine under the standard reaction conditions in the presence of GSH (100 μM) did not lead to the formation of GSSG. However, 50% of the GSH was used up in the reaction presumably as a result of

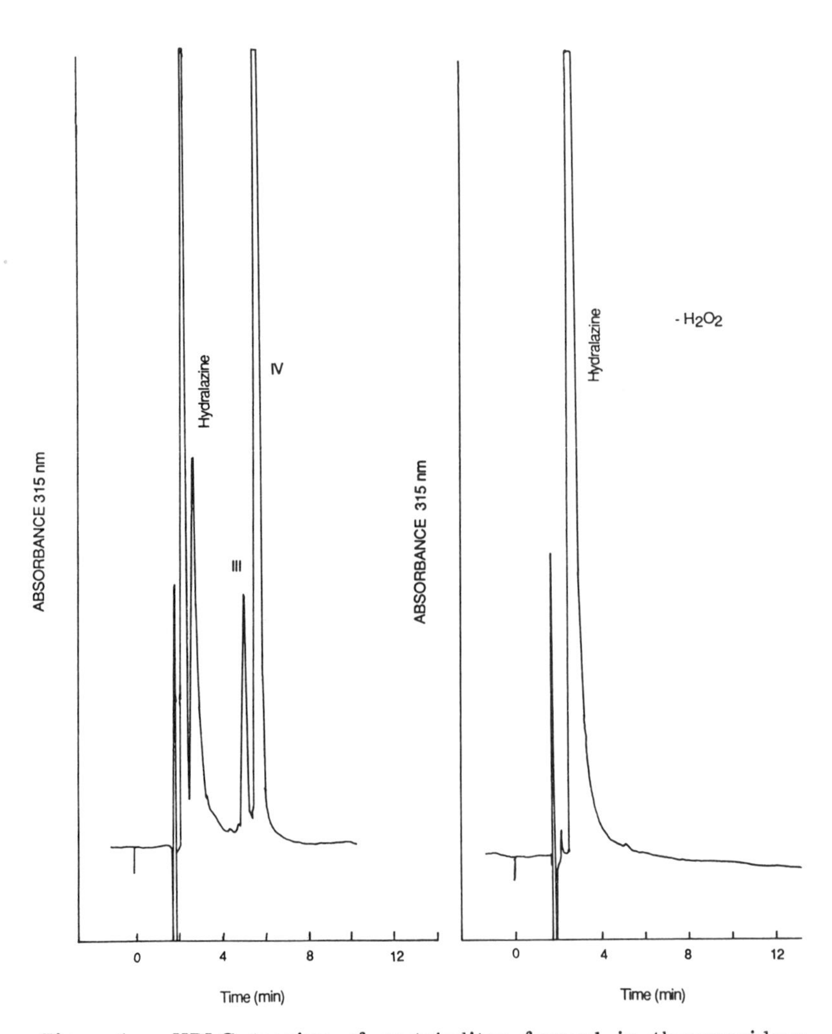

Figure 1. HPLC tracing of metabolites formed in theperoxidase catalysed oxidation of hydralazine in thepresence of glutathione. Hydralazine (1mM), HRP (10afag/ml), GSH (1mM) and H_2O_2 (1mM) incubated for 5 minutesin 1.0 ml of 0.1 M Tris buffer, pH 7.4.

conjugate formation. When the reaction mixture contained 1 mM GSH a small amount of GSSG (50 μM) was formed . Optical spectroscopy of the standard reaction mixture containing 100 μM GSH indicated the formation of a product with an absorption at 315 nm and a broad shoulder at 340 nm. The product remained in the water layer after ethyl acetateextraction. Similar reactions in the absence of GSH lead to the disappearance of the hydralazine absorbances (215, 265 and 315 nm) and the appearance of absorbances at 230, 280 and 380 nm. The 380 nm absorbing product(s) is probably the dimerization product of hydralazine (LaCagninet. al., 1986, Kanazawa et al., 1986).

HPLC analysis of the reaction mixture indicated the formation of at least three unknown peaks (Figure 1), one eluting prior to hydralazine and two eluting afterhydralazine. The addition of [^{3}H]-GSH to the reaction mixture and HPLC analysis of the reaction mixture indicated that all three unknown peaks contained radioactivity and hence were glutathione conjugates. Isolated peak IV has an optical spectru m at 240, 315 and 340 nm and was bright yellow in color accounting for the observed optical spectrum of the reaction mixture. Examination of urine from hydralazine treated patients had previously indicated the presence of sulphydryl hydralazine conjugates (Perry, 1953).

FAB/MS analysis (Figure 2) of peak IV in a glycerol matrix indicated a molecular ion of 436 consistent with a glutathione conjugate of phthalazine (MW-130). FAB/MS analysis of peak IV in a thioglycerol matrix led to the release of GSH (M + 1 308) and the formation of a thioglycerol/phthalazine conjugate (M + 1 237). This suggested that the phthalazine must be substituted at the N atom in order for it to undergo such an easy substitution reaction.

High field NMR analysis confirmed that the C_2 or C_9 protons (9.30 ppm) of phthalazine were still present however one was shifted upfield to 8.20 ppm consistent with substitution at the N atom. This was consistent with the structure 2-(N-glutathion-S-yl)-2,3-benzodiazene.

FAB/MS analysis of peak I resulted in the same molecular ions which appeared with peak IV indicating that peak I had undergone chemical breakdown during isolation. The structure of the two remaining conjugates is still under investigation.

The ratio of the three glutathione conjugates could be changed depending on the relative amount of glutathione in the reaction mixture. Low amounts of glutathione favored the formation of peak IV but high concentrations of glutathione favored the formation of peak I (Figure 3). Peak I was always formed early in the reaction indicating that this conjugate resulted from an early formed reactive intermediate. This suggested that each of the conjugates represented different reactive intermediates in the oxidation sequence. Regardless of this relative peak distribution, the incorporation of [^{3}H]-GSH into peak I even under conditions of low GSH concentration (100 μM) was twice that of peak IV indicating that peak I was the major conjugate formed under both conditions.

The standard reaction mixture supplemented with 1 mM GSH also resulted in increased oxygen uptake determined with a Clark type oxygen electrode. Oxygen uptake occurred after a lag period of approximately three minutes. Since GSH inhibited the rate of oxidation of hydralazine then GSH most likely does not reduce the initially formed hydralazine radical but must have interacted with another free radical in the pathway; possibly the diazenyl radical. This reduction reaction would lead to the formation of the thiyl radical with subsequent oxygen uptake (O'Brien, 1988).

The mechanism of formation of a N substituted glutathione conjugate of this nature is difficult to envision unless it results from the condensation reaction between a thiyl radical and the phthalazine radical. A postulated mechanism to explain the above results is presented in Figure 4.

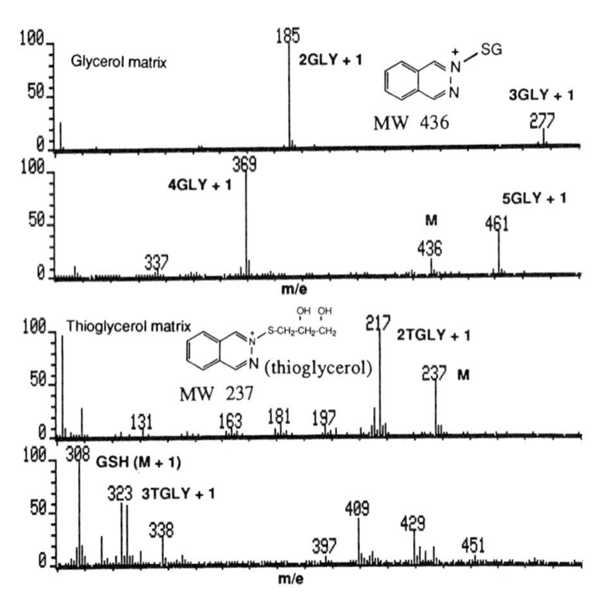

Figure 2. FAB mass spectral tracings of isolated metaboliteIV using a glycerol or thioglycerol matrix.

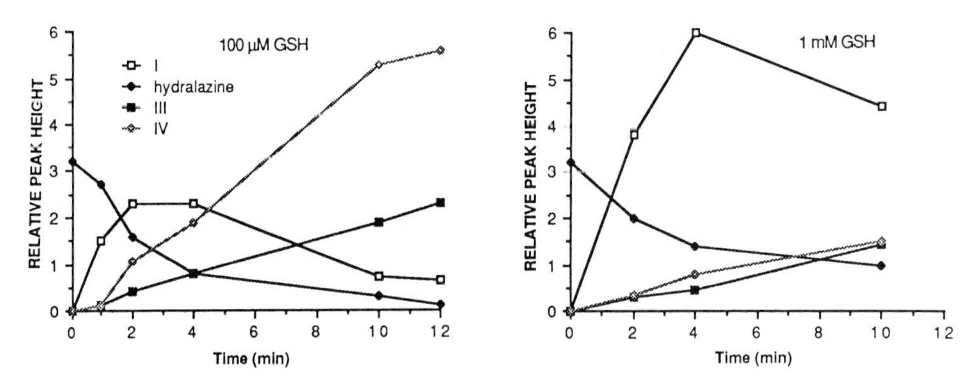

Figure 3. Time course of metabolites formed in the peroxidase catalysed oxidation of hydralazine in thepresence of glutathione. Incubation conditions: hydralazine (100 µM), HRP (1 µg/ml), GSH (100 µM or 1mM) and H_2O_2 (100 µM) incubated in 1.0 ml of 0.1M Tris buffer, pH 7.4. Metabolites were quantitated by HPLC.

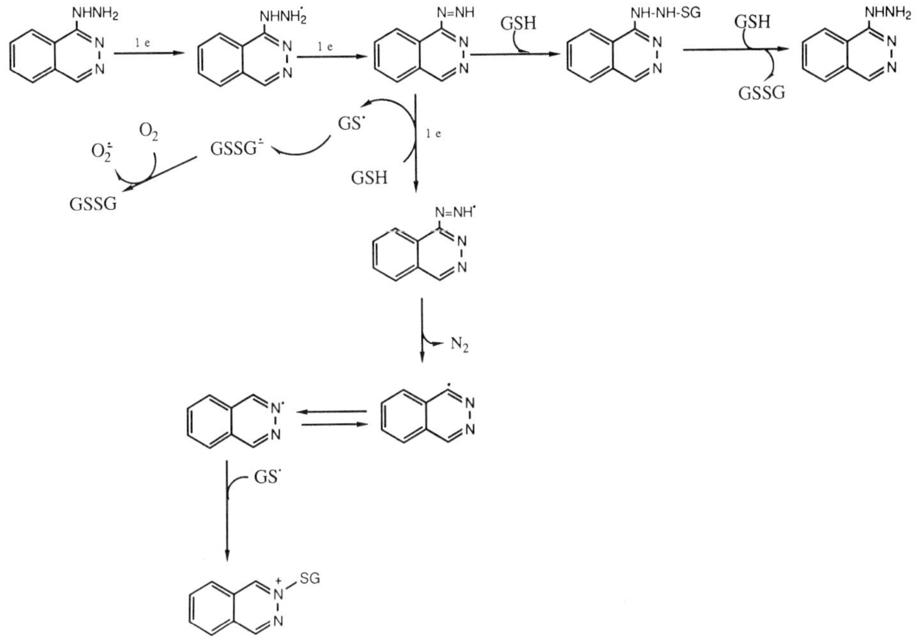

Figure 4. Postulated mechanisms of the peroxidase catalyzed oxidation of hydralazine in the presence of GSH.

The oxidation of hydralazine in the presence ofglutathione leads to the formation of several reactive intermediates which may be responsible for the protein binding observed in the microsomal systems and may be involved in the toxic mechanisms of this drug.

ACKNOWLEDGEMENTS

We would like to thank Dr. H. Pang and Dr. A. Gray of the Carbohydrate Institute, University of Toronto for the FAB mass spectral and NMR analyses respectively. This work was supported by the Medical Research Council and the National Science and Engineering Council of Canada. *Corresponding author: Dr. P.J. O'Brien, Faculty of Pharmacy, 19 Russell St., University of Toronto, Toronto ,Ontario, Canada, M5S 1A1.

REFERENCES

Kanazawa, H., Hara, T., Tsutsumino, K., Shibata, Y.,Tamura, Z., and Senga, K. (1986). New degradation products in an aqueous solution of hydralazine hydrochloride. *Chem. Pharm. Bull.* **34**, 1840-1842.

LaCagnin, L.B., Colby, H. D., and O'Donnell, J.P.(1986). The oxidative metabolism of hydralazine by rat liver microsomes. *Drug Metab. Dispos.* **14**, 549-554.

Lyles, G.A., Garcia-Rodriguez, J., and Callingham, B.A. (1983). Inhibitory actions of hydralazine upon monoamine oxidizing enzymes in the rat. *Biochem. Pharmacol.* **32**, 2515-2521.

Numata, Y., Takei, T., and Hayakawa,T. (1981). Hydralazine as an inhibitor of lysyl oxidase activity. *Biochem. Pharmacol.* **30**, 3125-3126.

O'Brien, P.J. (1988). Radical formation during the peroxidase catalyzed metabolism of carcinogens and xenobiotics: the reactivity of these radicals with GSH, DNA and unsaturated lipid. *Free Rad. Biol. Med.* **4**, 169-183.

Perry, Jr., H.M. (1953). A method of quantitating1-hydralazinophthalazine in body fluids. *J. Lab. Clin. Med.* **41**, 566-572.

Perry, H. M. Jr. (1973). Late toxicity to hydralazine resembling systemic lupus erythematosus or rheumatoid arthritis. *Am. J. Med.* **54**, 58-72.

Reed, D.J., Babson, J.R., Beatty, P.W., Brodie, A.E.,Ellis, W.W., and Potter, D.W. (1980). High performance liquid chromatographic analysis of nanomole levels of glutathione disulfide and related thiols. *Anal. Biochem.* **106**, 55-62.

Sinha, B.K. (1983). Enzymatic activation of hydrazinederivatives. A spin trapping study. *J. Biol. Chem.* **258**, 786-801.

Streeter, A.J., and Timbrell, J.A. (1985). Enzyme mediated covalent binding of hydralazine to rat liver microsomes. *Drug Metab. Dispos.* **13**, 255-259.

Toth, B. (1978). Tumorigenic effect of 1-hydrazinophthalazine hydrochloride in mice. *J. Natl.Cancer Inst.* **61**, 1363-1365.

Williams, G.M., Mazue, G., McQueen, C.A ., and Shimada, T. (1980). Genotoxicity of the antihypertensive drugs hydralazine and dihydralazine. *Science* **210**, 329-330.

Zak, S.B., Gilleran , T.G., Kavliner, J., and Lukas,G. (1974). Identification of two new metabolites of HP from human urine. *J. Med. Chem.* **17**, 381-382.

CONTRIBUTION OF 3,4-DICHLOROPHENYLHYDROXYLAMINE IN PROPANIL-INDUCED HEMOLYTIC ANEMIA

David C. McMillan[1], Timothy P. Bradshaw[1], JoEllyn M. McMillan[1], Jack A. Hinson[2], and David J. Jollow

Department of Pharmacology[1]
Medical University of South Carolina
Charleston, SC 29425 and
National Center for Toxicological Research[2]
Jefferson, AR

The methemoglobinemia and hemolytic anemia observed in experimental animals given aniline has been shown to be mediated by its N-oxidation metabolite, phenylhydroxylamine (Harrison, J.H. et al., 1987; Harrison, J.H. et al., 1986). The aniline derivative, propanil (3,4-dichloropropionanilide), is a widely used arylamide herbicide that has been shown to induce methemoglobinemia in experimental animals following conversion of the parent amide to one or more oxidized metabolites (Singleton, S>D. et al., 1973; Chow, A.Y.K. et al., 1975). Two methemoglobinemic metabolites of propanil have been identified, 3,4-dichlorophenylhydroxylamine (N-hydroxy-3,4-dichloroaniline) and 6-hydroxy-3,4-dichloroaniline (McMillan, D.C. et al., 1990). In view of the hemolytic activity of phenylhydroxylamine, we have examined the hemolytic potential of propanil and its metabolites in rats.

Propanil, N-hydroxy-3,4-dichloroaniline, and 6-hydroxy-3,4-dichloroaniline were synthesized as described previously (McMillan, D.C. et al., 1990; Lay, J.O. et al., 1986), and 3,4-dichloroaniline was purchased from Aldrich Chemical Co. (Milwaukee, WI). All other chemicals were reagent grade and were used without further purification.

Erythrocytes were collected from the abdomenal aorta of isologous male Sprague-Dawley rats (175-200 g, Camm Research Inc., Wayne, N.J.) and labelled with radioactive sodium chromate (New England Nuclear, Billerica, MA; 0.1 mCi/ml of packed cells) as described previously (Harrison, J.H. et al., 1986; Grossman, S.J. et al., 1987). Rats were administered i.v. ^{51}Cr-labelled erythrocytes (0.5 ml) 24 hr prior to i.p. administration of propanil dissolved in corn oil (5 ml/kg), and the time-dependent decrease in blood radioactivity was recorded as a measure of *in vivo* erythrocyte survival.

Administration of propanil induced a dose-dependent decrease in the half-life (T_{50}) of ^{51}Cr-labelled erythrocyts (Table 1). The TD_{50} for this effect in rats was approximately 1.80 mmol/kg. Administration of the deacylated propanil metabolite, 3,4-dichloroaniline, induced a similar effect on the survival of erythrocytes. N-Hydroxy-3,4-dichloroaniline induced a much greater effect (ca 10x) in erythrocyte survival than either propanil or 3,4-dichloroaniline, which suggested that this metabolite may mediate propanil-induced hemolytic anemia.

In experiments to determine if propanil or its metabolites were directly hemotoxic, erythrocyte suspensions (50% hematocrit) were incubated with the test compounds for 2 hr at 37°C in phosphate-buffered saline with glucose (110 mM NaCl, 200 mM Na_2HOP_4, 4 mM KH_2PO_4 and 10 mM glucose, pH 7.4), prior to the administration of the cells to isologous rats. N-Hydroxy-3,4-dichloroaniline was the only test compound that induced a concentration-dependent reduction in the T_{50} of labelled erythrocytes (EC_{50} ca. 150 μM), which is consistent with previous observations regarding the direct hemotoxicity of phenylhydroxylamine (Harrison, J.H. et al., 1986). In contrast, the other methemoglobinemic propanil metabolite, 6-hydroxy-3,4-dichloroaniline, had no effect on erythrocyte survival.

Although N-hydroxy-3,4-dichloroaniline and phenylhydroxylamine induced similar decreases in erythrocyte survival as well as similar changes in the electrophoretic pattern of erythrocyte membrane proteins (data not shown), there appear to be some differences between the two hydroxylamines regarding the nature of the interaction of these compounds with hemoglobin in the erythrocyte. For example, preliminary data suggest that N-hydroxy-3,4-dichloroaniline may be a slightly more potent hemolytic compound that phenylhydroxylamine, while N-hydroxy-3,4-dichloroaniline was much more potent than phenylhydroxylamine in its ability to induce methemoglobin formation *in vitro* (as determined by the area under the curve for the time-course of methemoglobin formation). An explanation for these differences may involve the relative stabilities of these compounds in erythrocyte suspensions, and/or their possible effects on NADH-dependent methemoglobin reductase.

Table 1. Effect of Dose of Propanil on the Survival of ^{51}Cr-labelled Erythrocytes in Rats

Propanil Dose (mmol/kg)	Erythrocyte T_{50}* (days)
0	9.81 ± 1.84
1.0	6.91 ± 0.85**
1.30	5.58 ± 0.58**
1.80	4.07 ± 0.37**

Isologous rats were administered i.v. ^{51}Cr-labelled erythrocytes 24 hr prior to i.p. administration of propanil dissolved in corn oil. Erythrocyte survival was assessed by measuring the decrease in the blood radioactivity over time. Values are means ± SD (n=7).
*T_{50}, time required for blood radioactivity to decrease by 50%.
** $p < 0.05$ by student's t-test.

In summary, these results support the hypothesis that propanil is converted to N-hydroxy-3,4-dichloroaniline in the liver. N-Hydroxy-3,4-dichloroaniline then enters the erythrocyte where it undergoes a coupled oxidation with oxyhemoglobin to form methemoglobin, 3,4-dichloronitrosobenzene, and partially reduced oxygen species (Kiese, M., 1974). This active oxygen species is then thought to induce changes in erythrocyte membrane proteins that results in premature splenic sequestration of erythrocytes.

ACKNOWLEDGEMENTS

This work was supported by Grant HL30038 from the U.S. Public Health Service.

REFERENCES

Chow, A.Y.K. and Murphy, S.D. (1990). *Toxicol. Appl. Pharmacol.* **33**, 14.
Grossman, S.J. and Jollow, D.J. (1987). *J. Pharmacol. Exp. Ther.* **244**, 118.

Harrison, J.H. and Jollow, D.J.(1987). *Molec. Pharmacol.* **32,** 423.
Harrison, J.H. and Jollow, D.J. (1986). *J. Pharmacol. Exp. Ther.* **238,** 1045.
Kiese, M. (1974). *Methemoglobinemia: A Comprehensive Treatise*, CRC Press, Cleveland, OH.
Lay, J.O. Evans, F.E. and Hinson, J.A. (1986). *Biomed. Environ. Mass Spectrom.* **13,** 495.
McMillan, D.C., Freeman, J.P. and Hinson, J.A. (1986). *Toxicol. Appl. Pharmacol.* **102,** in press.
Singleton, S.D. and Murphy, S.D. (1973). *Toxicol. Appl. Pharmacol.* **25,** 20.

A REDUCTION IN MIXED FUNCTION OXIDASES AND IN TUMOR PROMOTING EFFECTS OF ETHANOL IN A NDEA-INITIATED HEPATOCARCINOGENESIS MODEL

Siraj I. Mufti[1] and I. Glenn Sipes

Department of Pharmacology and Toxicology
The University of Arizona
College of Pharmacy
Tucson, Arizona 85721

Epidemiologic studies associate an increased cancer risk with chronic alcohol consumption [for example, Hakulinen et al., 1974; Breslow and Enstrom, 1974; Feldman et al., 1985; Williams and Horn, 1977; Schottenfeld, 1979; Kono and Ikeda, 1979; Tnyns, 1979; Doll and Peta, 1981]. However, the experimental evidence for this association is not clear. While a number of studies indicate that ethanol, the active ingredient of alcoholic beverages, increases the incidence of chemically induced tumors [for example, Gibel, 1967; Griciute et al., 1982; Gabriel et al., 1982; McCoy et al., 1986], other studies do not show such an effect [Schmahl et al., 1965; Schmahl, 1976; Habs and Schmahl, 1981; Teschke et al., 1983]. We recently reported studies that showed that ethanol increased tumor incidence only when administered as a tumor promoter after treatment with an esophagus specific carcinogen, N-nitrosomethylbenzlamine (NMB_ZA) was completed [Mufti et al., 1989]. In these studies we wanted to determine the effect of ethanol when given as a tumor promoter on liver carcinogenesis induced by N-nitrosodiethylamine (NDEA). The paper offers an explanation for the negative results obtained in these studies.

MATERIALS AND METHODS

Chemicals

NDEA was purchased from Eastman Kodak Company. Purity of the chemical was verified as 99% through mass spectrometric-gas chromatographic analysis by Dr. Karl Schram of our clinical chemistry department. Tightly sealed carcinogen was stored under refrigeration and only small aliquots were used for making fresh stock solution in 0.9% NaCl each time animals were treated. Guidelines of the Office of Risk Management University of Arizona were strictly abided by in handling carcinogens and chemicals.

Other chemicals and dietary items were purchased from ICN or Sigma chemical companies.

Animals and Their Diet

Male Sprague-Dawley rats weighing 100-120 g were purchased from Hilltop laboratories. Following a week of isolation after arrival, the animals were housed in

[1] To whom correspondence should be addressed.

Biological Reactive Intermediates IV, Edited by C.M. Witmer *et al.*
Plenum Press, New York, 1990

suspended cages in a room with regulated temperature (23°C) and humidity (30-40%) and a day and night cycle of 16 and 8 h respectively. Health of the animals was closely monitored by the veterinarians of the Division of Animal Resources and all animals were cared for and maintained according to the guidelines of the Committee on Care and Use of Laboratory Animals, University of Arizona.

Rats were fed regular Purina chow before and during treatment with NDEA, 2-acetylaminofluorene (2AAF) and CCl_4. A week after receiving the carcinogenic regimen, rats were fed an isocaloric liquid diet prepared in our laboratory according to the formulations of Lieber and DeCarli [1982] and American Institute of Nutrition [1977]. The liquid diet with or without ethanol provided 1 kcal/ml; ethanol substituted for part of dextrin-maltose of the control diet such that it provided 36% of total calories or 7% of the total diet in gram weight. The animals were pair fed; the control rats were restricted in their feeding so that they received no more than that consumed by ethanol-consuming rats. The rats received ethanol until dead or when terminated at 16 months of age. The blood ethanol content was measured randomly every 3 months employing a gas chromatographic procedure [Taylor et al., 1984] and showed no significant variation through the course of studies (values ranged 0.05 to 0.18%, mean = 0.08%).

Treatment of Animals

Rats received the following carcinogenic regimen. A single dose of NDEA in 0.9% NaCl solution was administered at a rate of 100 mg/kg bodyweight. Two weeks after the NDEA treatment, the rats were administered by gavage 2 acetylaminofluorene (2AAF) at a rate of 6 mg/rat/day for one week and on day 3 of this treatment received a single dose of CCl_4 at 2 ml/kg. This protocol is a modification of the procedure used by Lans et al. [1983] and the regimen subjects the NDEA initiated hepatocytes to appreciable selective pressures for hepatocarcinogenesis that are induced by 2AAF and CCl_4.

Processing of Hepatic Tissue/histochemical Characterization of Gamma Glutamyltranspeptidase (GGT) Altered Areas

The animals were killed by cervical dislocation, their abdomens were opened and livers excised. Cross sections approximately 1 cm in thickness were taken from the right anterior sublobe and quick frozen or stored in buffered formalin.

Liver slices were fixed in cold acetone for histochemical characterization of GGT. After paraffin embedding serial sections were cut, dried and deparaffinized and immediately incubated in a freshly prepared medium containing gamma-glutamyl-4-methoxy-2-naphthylamide as a substrate as described by Rutenberg et al. [1969]. Kidney sections were run alongside the liver as a positive control.

The liver sections were graded for GGT in a blind manner first visually under a dissecting microscope and then verified by weighing the colored areas. The most positive sections measured 50% as much as kidneys and the least positives measured 10% or less. Tumors were verified by histological staining for hematoxylin and eosin.

Isolation of Microsomes and Measurement of Cytochrome P-450 and NADPH Cytochrome c Reductase

About 2 g of liver tissues were homogenized in a glass teflon homogeniser and microsomes isolated by centrifugations in a ultracentrifuge. Cytochrome P-450 concentration was determined according to the procedure of Omura and Sato [1964] and cytochrome c reductase activity according to the procedure of Masters and Okits [1980]. Cytochrome P-450 is a hemoprotein that binds to carbon monoxide. In brief, it is estimated by adding sodium dithionate that reduces the cytochrome P-450 in the sample; carbon monoxide is then bubbled through the sample to produce the spectra that are characteristic of the cytochrome peaking at 450 nm. It is the terminal oxidase in the mono-oxygenase system that is involved in catalyzing the oxidation of a wide variety of compounds. NADPH cytochrome c reductase is responsible for the direct transfer of electrons in endoplasmic reticulum from NADP to cytochrome P-450. The activity of cytochrome c reductase present in the sample is measured by monitoring absorbance at 550 nm.

RESULTS AND DISCUSSION

These experiments were designed to determine the effect of ethanol administered as a tumor promoter on carcinogenesis induced by the well-known hepatocarcinogen, NDEA, in conjunction with 2AAF and CCl_4 (2AAF and CCl_4 generate conditions that are favorable for hepatocarcinogenesis by NDEA) [Lans et al., 1983]. Male Sprague-Dawley rats were used for these studies and the animals received the NDEA-2AAF-CCl_4 hepatocarcinogenesis protocol as described above in 'Materials and Methods'. Ethanol was given as part of isocaloric liquid diet a week after the hepatocarcinogenesis regimen and continued until termination of the experiment at 16 months of animal's age. The livers of terminated rats were excised and incidence of tumors and GGT-altered islands determined. The results of measurements of GGT-altered islands are presented in Table 1. It was observed that 2, 4 and 6 of a total of 31 ethanol-receiving rats exhibited 50%, 40% and 30% of GGT positive areas in their livers compared to 4, 3 and 6 of control rats (out of a total of 25) showing such areas respectively. The differences are not significant. The number of rats that showed 20% GGT positive areas were similar for both ethanol and control treatments. Apparently, the difference of 11 vs. 4 of ethanol-fed vs. control-fed rats showing 10% or less of GGT-positive foci is artefactual since ethanol in the blood could directly effect GGT [for a review Mufti, 1989]. The overall differences in the two treatments are also not significant. Additionally, there was one hepatic tumor in ethanol-fed rats and 2 in control-fed rats. Therefore, the results of these experiments do not indicate any tumor promoting effects that could be attributed to consumption of ethanol.

Table 1. Incidence of GGT-altered islands in a NDEA-initiated hepatocarcinogenesis model with ethanol used as a tumor promoter

GGT-positive %	Number of rats	
	Ethanol	Control
50%	2	4
40%	4	3
30%	6	6
20%	8	8
10% or less	11	4
Total	31	25

Young male Sprague Dawley rats were treated with NDEA, 2-AAF, CCl_4 regimen as described in the text. The animals received ethanol a week after completion of the regimen until their termination. Liver sections were stained for GGT. GGT% corresponds to the extent of GGT positive areas with a kidney run in parallel and graded as 100%. The differences between ethanol and control consuming rats are not significant.

The above experiments were prompted by our recent observations that ethanol caused an increase in incidence of tumors when used as a promoter i.e. administered after treatment with the esophagus specific carcinogen, methylbenzylnitrosamine was completed [Mufti et al., 1989]. Since, as discussed above, ethanol did not stimulate an increase in NDEA-initiated hepatocarcinogenesis, we wanted to explore the mechanism of ethanol-induced promotion that would explain the contradictory results observed with NDEA viz a viz NMB_zA related carcinogenetic process. In other related studies we have observed that whereas ethanol-induced increase in free radicals as measured by generation of lipid peroxidation products was present in NMB_zA-induced carcinogenesis, such an increase was absent in NDEA-induced carcinogenesis [Mufti, 1990]. Furthermore, with NDEA treatment there was significant increase in the contents of glutathione per gram of liver compared to NMB_ZA treatment [Mufti, 1990]. These observations suggest that there is a role for free radicals in tumor promotion associated with ethanol. Since microsomal ethanol oxidizing system is involved in producing free radicals [Lieber & Savolainen, 1984] therefore, we also wanted to determine the status of mixed function oxidase system in the NDEA-initiated hepatocarcinogenesis model. We were particularly interested to determine cytochrome P-450 since ethanol is known to induce a specific form of this system [Koop et al., 1982; Ryan et al., 1985, 1986]. The results of the effect of chronic ethanol consumption on induction of mixed function oxidases in hepatocarcinogenesis model are presented in Table 2. The data show that while there was no appreciable difference in the concentrations of cytochrome P-450 levels in ethanol-fed rats compared to their controls, the activity of cytochrome c reductase significantly increased with ethanol consumption. A particularly noteworthy finding was a remarkable decrease in concentrations of cytochrome P-450 with NDEA treatment and its drastic reduction with subsequent ethanol administration. Also, a significant decrease in activity of cytochrome c reductase was observed with NDEA treatment and ethanol consumption. NDEA leads to functional alterations in the mixed function oxidase system of the hepatocytic endoplasmic reticulum [Garcea et al., 1984] and therefore it appears that ethanol exacerbates changes that are induced by NDEA in this system.

Table 2. Effect of Chronic Ethanol Consumption on Induction of Mixed Function Oxidases in NDEA-initiated Hepatocarcinogenesis Model

Treatment	No. of Animals	Cytochrome P-450 nmol/mg protein	Cytochrome c reductase nmol/min/mg protein
Untreated Control	3	0.37 ± 0.025	32.4 ± 4.09
Untreated Ethanol	3	0.31 ± 0	49.6 ± 5.98
NDEA Control	3	0.11 ± 0.054	21.6 ± 1.29
NDEA Ethanol	3	0.06 ± 0	13.5 ± 4.76

Young Sprague-Dawley rats were treated with a NDEA, 2AAF, CCl_4 regimen as described. The animals received 7% ethanol diet a week after completion of the carcinogen regimen and continued on this diet until their termination. Livers of rats were excised, microsomes isolated and cytochrome P-450 and cytochrome c reductase determined. Cytochrome P-450 concentrations and activity of cytochrome c reductase are significantly different ($p < 0.05$) in the treatments indicated.

Based on the above observations, we hypothesize that an absence of tumor promoting effects of ethanol on hepatic enzyme-altered islands and tumors may be due to a lack of reactive oxygen intermediates. Such reactive species are normally generated with ethanol consumption in the endoplasmic reticulum by the enzymes associated with microsomal ethanol metabolizing system [Lieber & Savolainen, 1984].

ACKNOWLEDGMENTS

The research described here was supported by the U.S. National Institutes of Health grants no. ES03438 and CA 51088.

REFERENCES

American Institute of Nutrition (1977). Report of the American Institute of Nutrition Ad Hoc Committee on standards for nutritional studies. *J. Nutr.* **107**, 1340-1348.

Breslow, N.E. and Enstrom, J.E. (1974). Geographic correlations between cancer mortality rates and alcohol-tobacco consumption in the United States. *J. Natl. Cancer. Inst.* **53**, 631-639.
Doll, R. and Peto, R. (1981). The causes of cancer: quantitative estimates of avoidable risk of cancer in the United States today. *J. Natl. Cancer. Inst.* **66**, 1191-1308,.
Feldman, J.G., Haan, M., Nagarajen, M. and Kissin, B. (1975). A case-control investigation of alcohol, tobacco, and diet in head and neck cancer. *Prev. Med.* **4**, 444-463.
Gabrial, G.N., Schrager, T.F. and Newberne, P.M. (1982). Zinc deficiency, alcohol and a retinoid: association with esophageal cancer in rat. *J. Natl. Cancer Inst.* **68**, 785-789.
Garcea, R., Canuto, R.A., Biocca, M.E., Muzio, G., Rossi, M.A. and Dianzani, M.U. (1984). Functional alterations of the endoplasmic reticulum and the detoxification systems during diethyl-nitrosamine carcinogenesis in rat liver. *Cell Biochem. Funct.* **2**, 177-181.
Gibel, Von W. (1967). Experimental studies on syncarcinogenesis of oesophageal cancer. *Arch. Geschwulstforsch* **30**, 181-189.
Griciute, L., Castegnaro, M. and Bereziat, J-C. (1982). Influence of ethyl alcohol in the carcinogenic activity of N-nitrosodi-n-propylamine. In Bartsch H, Castegnaro M, O'Neill IK, Okada M and Davis W. (eds). N-Nitroso Compounds: Occurrence and Biological Effects. IARC Scientific Publication, Lyon, France, Vol. 41; pp. 643-648.
Habs, M. and Schmahl, D. (1981). Inhibition of the hepatocarcinogenic activity of di-ethylnitrosamine (DENA) by ethanol in rats. *Hepato-gastroenterol* **28**, 242-244.
Hakulinen, T., Lehtimaki, L., Lehtonen, M. and Teppo, L. (1974). Cancer morbidity among two male cohorts with increased alcohol consumption in Finland. *J. Natl. Cancer Inst.* **52**, 1711-1714.
Kono, S. and Ikeda, M. (1979). Correlation between cancer mortality and alcoholic beverage in Japan. *Br. J. Cancer* **40**, 449-455.
Koop, D.R., Morgan, E.T., Tarr, G.E. and Coon, M.J. (1982). Purification and characterization of a unique isozyme of cytochrome P-450 from liver microsomes of ethanol-treated rabbits. *J. Biol. Chem.* **257**, 8472-8480.
Lans, M., DeGerlache, J., Taper, H.S., Preat, V. and Roberfroid, M.B. (1983). Phenobarbital as a promoter in the initiation/selection process of experimental rat hepatocarcinogenesis. *Carcinogenesis* **4**, 141-144.
Lieber, C.S. and DeCarli, L.M. (1982). The feeding of alcohol in liquid diets: two decades of applications and 1982 update. *Alcohol Clin. Exp. Res.* **6**, 523-531.
Lieber, C.S. and Savolainen, M. (1984). Ethanol and lipids. *Alcohol Clin. Exp. Res* .**8**, 409-423.
Masters, B. and Okits, R.T.,, (1980). The history, properties and function of NADPH cytochrome P-450 reductase. *Pharmacol. Therapeut.* **9**, 227-244.
McCoy, C.D., Hecht, S.S. and Furuya, K. (1986). The effect of chronic ethanol consumption on the tumorigenicity of N-nitrosopyrrolidine in male Syrian golden hamsters. *Cancer Lett.* **33**, 151-159.
Mufti, S.I. (1989). An evaluation of currently used laboratory tests to detect alcoholism and alcohol abuse and suggestions for improvement. In: *Biochemistry and Physiology of Substance Abuse* (Ed. Watson RR), in press, CRC Press, Boca Raton, FL.
Mufti, S.I., Becker, G. and Sipes, I.G. (1989). Effect of chronic dietary ethanol consumption on the initiation and promotion of chemically-induced esophageal carcinogenesis in experimental rats. *Carcinogenesis* **10**, 303-309.
Mufti, S.I. (1990). Generation of free radicals in ethanol metabolism may be responsible for tumor promoting effects of ethanol. These proceedings.
Omura, T. and Sato, R. (1964). The carbon monoxide-binding pigment of liver microsomes. *J. Biol. Chem.* **239**, 2370-2385.
Rutenberg, A.M., Kim, H., Frishbein, J.W., Hanker, J.S., Wasserkrug, H.L. and Seligman, A. (1969). Histochemical and ultrastructural demonstration of γ-glutamyl transpeptidase activity. *J. Histochem. Cytochem.* **17**, 517-526.
Ryan, D.E., Koop, D.R., Thomas, P.E., Coon, M.J. and Levin, W. (1986). Evidence that

isoniazed and ethanol induce the same microsomal cytochrome P-450 in rat liver, an isozyme homologous to rabbit liver cytochrome P-450 3a. *Arch. Biochem. Biophys.* **246**, 633-644.

Rayan, D.E., Ramathan, L.. Lida, S., Thomas, P.E., Haniu, M., Shively, J.E., Lieber, C.S. and Levin, W. (1985). Characterization of a major form of rat hepatic cytochrome P-450 induced by isoniazid. *J. Biol. Chem.* **260**, 6385-6393.

Schmahl, D. (1976). Investigations on esophageal carcinogenicity by methyl-phenyl-nitrosamine and ethyl alcohol in rats. *Cancer Lett.* **1**, 215-218.

Schmahl, D., Thomas, C., Sattler, W. and Scheld, G.F. (1965). Experimental studies of syncarcinogenesis: III. Attempts to induce cancer in rats by administering di-ethylnitrosamine and CCl_4 (or ethyl alcohol) simultaneously. In addition, an experimental contribution regarding 'alcoholic cirrhosis'. *Z Krebsforsch* **66**, 526-532.

Schottenfeld, D. (1979). Alcohol is a co-factor in the etiology of cancer. *Cancer* **43**, 1961-1966.

Taylor, G.F., Turrill, G.H. and Carter, G. (1984). Blood alcohol analysis: a comparison of the gas chromatographic assay with an enzyme assay. *Pathology* **16**, 157-159.

Teschke, R., Minzlaff, M., Oldiges, H. and Frenzel, H. (1983). Effect of chronic alcohol consumption on tumor incidence due to dimethylnitrosamine administration. *J. Cancer Res. Clin. Oncol.* **106**, 58-64.

Tuyns, A.J. (1979). Epidemiology of alcohol and cancer. *Cancer Res.* **39**, 2840-2843.

Williams, R.R., and Horm, J.W. (1977). Association of cancer sites with tobacco and alcohol consumption and socioeconomic status of patients: interview study from the Third National Cancer Survey. *J. Natl. Cancer. Inst.* **58**, 525-547.

SELECTIVE INDUCERS OF THE Coh-LOCUS ENHANCE THE METABOLISMS OF COUMARIN- AND OF QUINOLINE-DERIVATIVES BUT NOT THAT OF NAPHTHALENES

K.J. Netter, B. Hahnemann, S.A. Mangoura, F. Feil, M. Tegtmeier, R.T. Mayer* and W. Legrum

Institute of Pharmacology and Toxicology
Philipps-University
Lahnberge, D-3550 Marburg, FRG and
U.S. Department of Agriculture*, Agricultural Research Service
U.S. Horticultural Research Laboratory
Orlando, FL 32803

The Coh-locus is a gene area on mouse chromosome 7 which was described first by Wood and Conney in 1974 (Wood, et al.) from breeding experiments. The name was coined because the gene locus was characterized by using coumarin as substrate for cytochrome P-450 (Coh = coumarin 7-hydroxylase). Later Wood and Taylor (1979) distinguished two forms of the locus, one coding for a cytochrome P-450 with low activity towards coumarin (Coh^l in B6-mice) and another coding for a high activity variety of the coumarin 7-hydroxylase (Coh^h in D2-mice).

The Coh-locus expresses a form of cytochrome P-450 that is furthermore characterized by its comparatively low molecular weight (48.5 kDa) (Juvonen, R.O. et al., 1985; Kling, L. et al., 1985) as well as its marked activity towards coumarin derivatives such as particularly 7-ethoxycoumarin (Juvonen, R.O. et al., 1985).

The cytochrome $P\text{-}450_{Coh}$ is selectively induced by a variety of unrelated substances such as heavy metals and N-containing heteroaromates. Particularly cobalt (Legrum, W. et al., 1979) and indium (Mangoura S.A. et al., 1989) proved to be inducers of this enzyme. On the other hand 3-amino-1,2,4-triazole (Legrum, W. et al., 1988), pyrazine (Hahnemann, B. et al., 1989) and pyrazole (Juvonen, R.O. et al., 1985) also induce this cytochrome.

So far there are no reports on substrates other than coumarins exhibiting a comparatively high increase in metabolism after Coh-locus inducers. In this situation it seemed logical to test also other related substrates in order to better describe the optimal structural requirement for substrate-enzyme interaction. Therefore, structure-specificity studies with derivatives of the two-membered ring system are in place to elucidate the possible role of heteroatoms in recognizing a particular induced form of cytochrome P-450. For this reason structural analogues of the coumarin ring system were investigated: quinolines (Mayer, R.T. et al., 1989) and naphthalenes (Netter, K.J., 1969). In detail, the ethyl ethers of 7-quinolinol and 2-naphthol served as substrates and were compared to the oxygen containing double ring system, namely umbelliferone (i.e. 7-hydroxycoumarin). The expected results might prove helpful in the elucidation of the action of standard and Coh-inducers.

Biological Reactive Intermediates IV, Edited by C.M. Witmer *et al.*
Plenum Press, New York, 1990

MATERIALS AND METHODS

Animals and Sample Preparation

Male mice of the strains NMRI, C57B1/6JHan (B6) and DBA/2JHan (D2) (7 weeks old, body weight 15-25 g) were used in this study. Mice were supplied by the Zentralinstitut für Versuchstierzucht (Hannover, FRG).

Groups of six animals each were pretreated with the standard inducers phenobarbital and 3-methylcholanthrene as follows: 80 mg PB/kg b. wt. (i.p.), daily for two days (PB) and 30 mg MC/kg b. wt. (i.p.), daily for two days (dissolved in arachis oil) (MC). The metals were applied as follows: 40 mg of $CoCl_2$/kg b. wt. (s.c.), daily for two days (Co) and 100 mg of $In_2(SO_4)_3$/kg b. wt. (s.c.), once for two days (In). The heteroaromates were all used in higher doses, namely 200 mg/kg, given intraperitoneally once daily for two days and labeled PL for pyrazole, PN for pyrazine and AT for aminotriazole. The controls received 0.1 ml of isotonic NaCl per 10 g body weight intraperitoneally. Hepatic microsomes were prepared individually from six mice per group according to Netter (1960). Cytochrome P-450 was measured spectrophotometrically as described by Omura and Sato (1964) using an extinction coefficient of 91 $mM^{-1} \cdot cm^{-1}$.

Substrates and Deethylation Assays

The ethyl ethers of umbelliferone (i.e. 7-ethoxycoumarin, 7EC) and 2-naphthol (i.e. 2-ethoxynaphthalene, 2EN) were synthesized according to Allen and Gates (1955) ethylating umbelliferone (Aldrich Europe, Düsseldorf, FRG) or 2-naphthol (E. Merck, Darmstadt, FRG) with iodoethane. 7-Ethoxyquinoline (7EQ) was prepared as described by Mayer et al. (1989). Figure 1 shows the structures of the compounds investigated and the free phenols measured fluorimetrically in alkaline solutions according to Aitio (1978). Furthermore the wavelengths of the maxima of excitation and emission of the phenolates in glycine-NaOH buffer, pH 10.5, are indicated. Products of the O-deethylations exert a severalfold higher fluorescence than the substrates which is expressed as EM_{rel} (see Figure 1). This demonstrates that phenolates are suitable for measurement at both, alkaline pH and physiological pH (Mayer, R.T. et al., 1989).

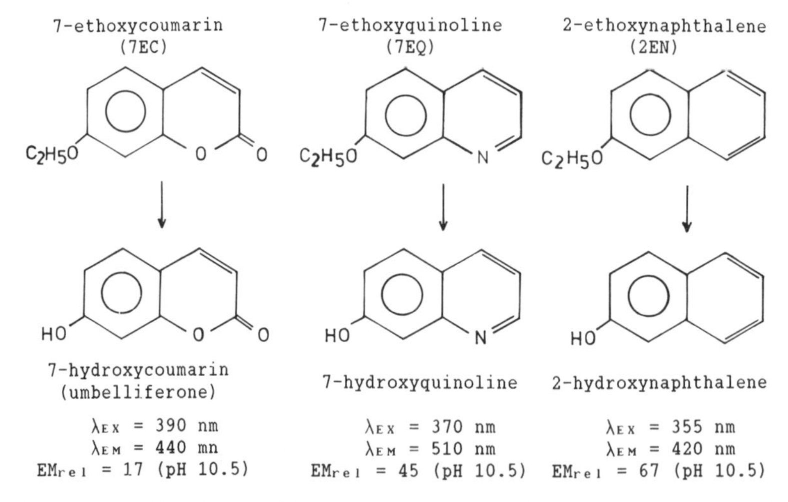

Figure 1. Structures of the investigated analogues. $Lambda_{EX}$ and $lambda_{EM}$ denote the wavelengths of excitation and emission, respectively. EM_{rel} is the intensity of emission of the phenolate in comparison to that of the ether.

During O-deethylation the concentration of the substrates 7EC and 7EQ was 10^{-4} M. Within a reaction time of 10 min the product formation was linear with time using a NADPH-regenerating system as described by Aitio (1978). In the case of 2EN as substrate the reaction time was reduced to 2 min while increasing the substrate concentration to $5 \cdot 10^{-4}$M in order to diminish the interaction of cytochrome P-450 and 2-naphthol which acts as an inhibitor (Feil F. et al., 1990). The O-deethylation rates are expressed as "molecular activities" (nmol product/nmol cytochrome P-450·min).

RESULTS AND DISCUSSION

In the mouse strains investigated the basic molecular activity of the O-deethylase prior to enzyme induction is higher for the oxygen and nitrogen containing substrates, namely 7-ethoxycoumarin and 7-ethoxyquinoline, than for the respectively substituted naphthol. The difference does, however, not exceed one order of magnitude (Table 1). Pretreatments with Coh-locus inducers produce complex effects. In B6 mice possessing the Coh^{l} activity (cf. Table 1) they increase the O-deethylation of 7EC and 7EQ, generally causing a 2-3 fold increase in molecular activity. This increase is more pronounced than that by standard inducers PB and MC (Table 2). In contrast there are no changes for the substrate 2EN; this shows that a pure hydrocarbon as substrate like 2EN obviously does not reflect an induction of the Coh-locus. D2 mice which possess the Coh^{h} gene exert high activities towards 7EC, 7EQ (cf. Table 1) and coumarin. Table 3 shows that in these mice Coh-locus inducers increase the O-deethylation of 7EC clearly more than that of 7EQ. Thus 7EQ seems to differentiate between the high and low activity form of the Coh expressed after pretreatment with Coh-locus inducers. Comparing the data of Table 1 with those of Table 2 and 3 it becomes evident that the extent of augmentation of the molecular activities is not correlated to the constitutive level of activity.

Table 1. Molecular Activities of the O-deethylation

strain:	NMRI	B6	D2
substrate			
7EC	11	17	57
7EQ	24	45	78
2EN	7	13	10

Values are expressed as nmol of product per nmol cytochrome P450·min. Number of experiments was six. Standard deviations are less than 10% and are therefore not indicated. For explanation of the abbreviations of the substrates see Figure 1.

Table 2. Induction of the O-deethylation in Male B6 Mice

inducer:	heteroaromates			metals		standard inducers	
	PL	PN	AT	Co	In	PB	MC
substrate							
7EC	3.5	3.0	2.5	3.0	2.4	1.4	1.6
7EQ	2.3	2.1	1.9	2.0	2.4	1.5	1.7
2EN	1.0	0.7	0.9	1.1	1.5	0.9	0.8

Data are multiples ("fold induction") of control molecular activities (control = 1.0). For absolute values cf. Table 1. Data are the ratios of the mean values of six experiments.

Finally Table 4 summarizes the affinities (K_m-values) of 7EC and 7EQ towards cytochrome P-450 after different pretreatments of D2 mice. As expected from the data of Table 3, with 7EC as substrate the K_m-values are clearly reduced after pretreatment with heteroaromates, while PB as standard inducer hardly affects the affinity. With 7EQ the increase in affinity is less pronounced after Coh-locus inducers as it is the representation of the Coh-locus inducers shown in Table 3.

Table 3. Induction of the O-deethylation in Male D2 Mice

inducer:	heteroaromates			metals		standard inducers	
	PL	PN	AT	Co	In	PB	MC
substrate							
7EC	5.2	7.7	3.9	4.7	1.8	1.6	1.7
7EQ	1.3	2.7	1.0	1.6	0.8	1.4	1.7

Data are multiples ("fold induction") of control molecular activities (control = 1.0). For absolute values cf. Table 1. Data are the ratios of the mean values of six experiments.

Table 4. Affinities Towards 7EC and 7EQ After Different Inducers (Male D2 Mice)

inducer:	heteroaromates			metals		std. ind.		control
	PL	PN	AT	Co	In	PB	MC	-
substrate								
7EC	5.0	5.0	5.9	nd	nd	11.2	nd	13.8
7EQ	1.9	5.3	2.0	3.2	3.9	3.8	6.7	7.1

Enzyme kinetic experiments were evaluated according to Hofstee (1952) leading to the tabulated K_M-values (µM). Concentrations of the substrates were in the range of $1 \cdot 10^{-5}$ M to $2 \cdot 10^{-4}$ M. Number of experiments was six. "Std.ind." denotes standard inducers and "nd" not determined.

In conclusion, the heteroatoms nitrogen and oxygen in the bimembered ring system increase the suitability of respective ethoxy substrates to recognize induced forms of cytochrome P-450, particularly cytochrome $P\text{-}450_{Coh}$.

REFERENCES

Aitio A. (1978). A simple and sensitive assay of 7-ethoxycoumarin deethylation. *Anal. Biochem.* **85**, 488-491.

Allen, C.F.H., and Gates, J.W. (1955). o-n-Butoxynitrobenzene (ether, butyl-o-nitrophenyl). *Org. Syntheses* Coll III, 140-141.

Feil, F., Tegtmeier, M., Netter, K.J. (1990). Ethyl ethers of 2-naphthol and 7-quinolinol in comparison to that of umbelliferone as substrates of the murine hepatic monooxygenase system. *Naunyn-Schmiedeberg's Arch. Pharmacol.* **341**, Suppl:A24.

Hahnemann, B., Kühn, B., Heubel, F. and Legrum, W. (1989). Selective induction of coumarin hydroxylase by N-containing heteroaromatic compounds. *Arch. Toxicol.* **13**, Suppl:297-301.

Hofstee, B.H.J. (1952). On the evaluation of the constants V_M and K_M in enzyme reactions. *Science* **116**, 329-331.

Juvonen, R.O., Kaipainen, P.K. and Lang, M.A. (1985). Selective induction of coumarin 7-hydroxylase by pyrazole in D2 mice. *Eur. J. Biochem.* **152**, 3-8.

Kling, L., Legrum, W. and Netter, K.J. (1985). Induction of liver cytochrome P-450 in mice by Warfarin. Comparison of Warfarin-, Phenobarbitone-, and Cobalt-induced hepatic microsomal protein patterns by PAGE after partial purification on Octyl-Sepharose CL-4B. *Biochem. Pharmacol.* **34**, 85-91.

Legrum, W., Stuehmeier, G. and Netter, K.J. (1979). Cobalt as a modifier of microsomal monooxygenase in mice. *Toxicol. Appl. Pharmacol.* **48**, 195-204.

Legrum, W., Hahnemann, B., and Mangoura, S.A. (1988). Selective inducers of hepatic coumarin metabolizing monooxygenases - A new type and principle of induction. *Naunyn-Schmiedeberg's Arch. Pharmacol.* **337**, Suppl:R14.

Mangoura, S.A., Strack, A., Legrum, W. and Netter, K.J. (1989). Indium selectively increases the cytochrome P-450 dependent O-dealkylation of coumarin derivatives in male mice. *Naunyn-Schmiedeberg's Arch. Pharmacol.* **339**, 596-602.

Mayer, R.T., Netter, K.J., Heubel, F., Buchheister, A. and Burke, M.B. (1989). Fluorimetric assay of hepatic microsomal monooxygenases by use of 7-methoxyquinoline. *Biochem. Pharmacol.* **38**, 1364-1368.

Netter, K.J. (1969). Untersuchungen zur mikrosomalen Naphthalinhydroxylierung. *Naunyn-Schmiedeberg's Arch. Pharmacol.* **262**, 375-387.

Netter, K.J. (1960). Eine Methode zur direkten Messung der O-Demethylierung in Lebermikrosomen und ihre Anwendung auf die Mikrosomenhemmwirkung von SKF 525-A. *Naunyn-Schmiedeberg's Arch. Pharmacol.* **238**, 292-300.

Omura, T., and Sato, R. (1964). The carbon monoxide-binding pigment of liver microsomes. I. Evidence for its hemoprotein nature. *J. Biol. Chem.* **239**, 2370-2378.

Wood, A.W., and Conney, A.H. (1974). Genetic variation in coumarin hydroxylase activity in the mouse (mus musculus). *Science* **185**, 612-614.

Wood, A.W., and Taylor, B.A. (1979). Genetic regulation of coumarin hydroxylase activity in mice. Evidence for a single locus control on chromosome 7. *J. Biol. Chem.* **254**, 5647-5651.

Wood, A.W. (1979). Genetic regulation of coumarin hydroxylase activity in mice. Biochemical characterization of the enzyme from two inbred strains and their F_1 hybrid. *J. Biol. Chem.* **254**, 5641-5646.

PRIMAQUINE-INDUCED OXIDATIVE STRESS IN ISOLATED HEPATOCYTES AS A RESULT OF REDUCTIVE ACTIVATION

José M. Silva and Peter J. O'Brien[1]

Faculty of Pharmacy
University of Toronto
Toronto,Ontario, Canada, M5S 1A1

Abbreviations used: GSH: reduced glutathione; GSSG:oxidised glutathione; Hepes:4-(2-hydroxyethyl)-1-piperazine-ethane-sulphonic acid; HPLC: high performance liquid chromatography; DETAPAC: Diethylenetriamine penta-acetic acid; H_2O_2: hydrogen peroxide.

Primaquine,6-methoxy-8-(4-amino-1-methylbutylamino)quinoline, is one of the most effective antimarial agents but because of its low therapeutic index its use is limited. Unlike the 4-aminoquinolines it is believed to act on the liver schizont stage and not the erythrocyte stage of the parasites life cycle. However the drug is reported to cause hemolytic anemia, especially in individuals whose erythrocytes are deficient in glucose-6-phosphate dehydrogenase (Beutler, 1969). In erythrocytes, primaquine or its liver metabolite 5-OH-primaquine is reported to cause intraerythrocytic superoxide and hydrogen peroxide formation, methemoglobinemia, the formation of Heinz bodies, and occasionally the induction of hemolysis (Cohen and Hochstein, 1964; Summerfield and Tudhope, 1978).

The mechanism for the toxicity of primaquine towards malarial parasites in the liver or erythrocytes is not known. However, it is generally believed that primaquine undergoes a cytochrome P450 dependent mixed function oxidase catalysed oxidative metabolism to a quinoneimine which causes erythrocyte oxidative stress by undergoing redox cycling with oxyhemoglogin (Strother, 1984). The following shows however that oxygen activation also occurs when primaquine is reduced by isolated rat hepatocytes.

MATERIALS AND METHODS

Chemicals

Primaquine, azide, ascorbate, KCN, trypan blue, GSH, GSSG, thiobarbituric acid and fluoro-2,4-dinitrobenzene were obtained from Sigma (St.Louis, MO). Collagenase (from *Clostridium histolicum*) and Hepes were purchased from Boehringer-Mannheim (Montreal,Canada). Other chemicals were of the highest grade available commercially.

1. Correspondence to: Dr. Peter J. O'Brien, Faculty of Pharmacy, 19 Russell Street, University of Toronto, Toronto, Ontario, Canada M5S 2A2. Tel: (416) 978-2716.

Biological Reactive Intermediates IV, Edited by C.M. Witmer *et al.*
Plenum Press, New York, 1990

Animals

Male Sprague-Dawley rats (body weight 180-250g) fed a standard chow diet and tap water *ad libitum* were used in the hepatocyte preparation.

Isolation and Incubation of Hepatocytes

Hepatocytes were obtained by collagenase perfusion of the liver as described by Moldeus et al., 1978. Approximately 90% ofthe freshly isolated hepatocytes excluded trypan blue. Cells were incubated at a concentration of $1x10^6$ cells/ml in rotating, round-bottom flasks at 37°C in Krebs-Henseleit buffer, pH 7.4, supplemented with 12.5 mM Hepes under an atmosphere of 95% O_2, 5% CO_2. The final incubation volume was 20 ml with a cell concentration of 10^6 /ml.

Glutathione Reductase Deficient Cells

Hepatocytes were pretreated with 50 μM 1,3-bis(2-chloroethyl)-1-nitrosourea (BCNU) for 45 min to inactivate glutathione reductase (>90%), along with 1 mM methionine to resynthesize depleted GSH. Under these conditions GSH concentration was found to be within 80% of untreated cells.

Glutathione Determination

The total GSH and GSSG content of hepatocytes was measured, on deproteinized samples (5% metaphosphoric acid), after derivatization with iodoacetic acid and fluoro-2,4-dinitrobenzene, by high-pressure liquid chromatography, using a μBondapak NH_2 column (Waters Associates, Milford, MA) (Reed et al., 1980). GSH and GSSG were used as external standards. A Waters 6000A solvent delivery system, equipped with a Model 660 solvent programmer, a Wisp 710A automatic injector, and a Data Module were used for analysis.

Oxygen Consumption

Oxygen uptake was measured by a Clark-type electrode (Model 5300; Yellow-Spring Instrument Co., Inc.) in a 2-ml chamber, maintained at 37°C. Before use, hepatocytes were kept at 37°C in Krebs-Henseleit buffer, plus Hepes (12.5mM), pH 7.4, under a stream of 95% air and 5% CO_2. KCN (2 mM, neutralized with HCl) was added to inhibit mitochondrial respiration.

RESULTS AND DISCUSSION

Ascorbate is known to act as a reductant of various quinone and nitro compounds causing extensive redox cycling of the compound under aerobic conditions (Gutierrez, 1888). Addition of ascorbate (10mM) to primaquine resulted in extensive consumption of oxygen (Table 1). This suggests that ascorbate can reduce primaquine to an autooxidizable species. Others have also reported that primaquine can form a charge transfer complex with NAD (P)H and transition metals in aerobic solutions which results in reductive oxygen activation as determined by *in vitro* spin trapping experiments (Thornalley et al., 1983; Augusto et al., 1986).

As shown in Table 1, primaquine was effective at stimulating respiration with intact hepatocytes in the presence of KCN (added to inhibit O_2 uptake due to mitochondrial respiration). If ascorbate was also included in the incubation mixture the cyanide resistant respiration increased markedly (Table 1). Addition of catalase near the end of the reaction released nearly 50% of the oxygen taken indicating that H_2O_2 was formed stoichiometrically.

Table 1. Primaquine-mediated oxygen activation withhepatocytes and ascorbate

ADDITION respiration	O_2 uptake without cells (nmol/min)	Cyanide-resistant (nmol/min/10^6cells)
None	0.03 ± 0.01	1.23 ± 0.4
Primaquine (500afaM)	0.03 ± 0.01	73.2 ± 18.4
Primaquine (500afaM) + ascorbate (10mM)	60.7 ± 12.7	124.3 ± 23.7
ascorbate (10mM)	0.04 ± 0.01	0.07 ± 0.02

Hepatocytes (10^6 cells/ml) were incubated in Krebs-Henseleit buffer, pH 7.4, containing Hepes (12.5 mM) and 1 mM KCN at 37°C, with the additions indicated above. Experiments were performed in 0.1 mM Tris-HCL buffer containing 2mM DETAPAC, pH 7.4. Oxygen uptake was measured using a Clarke electrode. Values are means ± S.D.

The effect of primaquine on the viability of isolated rat hepatocytes is shown in Table 2. Primaquine caused cell death in a time and dose dependent manner as shown by the trypan blue exclusion test. Since Table 1 suggest that H_2O_2 is produced from the addition of primaquine to hepatocytes we decided to investigate the role of H_2O_2 inprimaquine-induced cell toxicity. H_2O_2 is catabolized to H_2O by either catalase or glutathione peroxidase and azideis an effective inhibitor of catalase at concentrations which are nontoxic to hepatocytes (Rossi et al., 1988). Hepatocytes whose catalase had been inactivated with azide were considerably more susceptible to primaquinethan normal cells (Table 2). This potentiation in cytotoxicity could be substantially further increased by including ascorbate in the incubation mixture to cause extracellular redox cycling.. This again indicates that H_2O_2 formed as a result of redox cycling leads to cell death. Finally, as shown in Table 2, inhibition of hepatocyte glutathione reductase with BCNU also markedly increased the susceptibility of hepatocytes to primaquine. This indicates that inactivation of the other intracellular H_2O_2 defense system, GSH-peroxidase/GSH-reductase enables primaquine redox cycling to generate sufficient H_2O_2 to cause cytotoxicity.

Table 2. Primaquine-induced cytotoxicity in isolated hepatocytes

Treatment	Cytotoxicity (% trypan blue uptake) 30	60	120	180
	(mins)			
Control cells	13 ± 2	16 ± 3	18 ± 3	20 ± 3
+ Primaquine 200 ± M	13 ± 2	18 ± 3	23 ± 3	26 ± 3
+ Primaquine 500 ± M	22 ± 3	27 ± 3	45 ± 3	100
+ Primaquine 2500 ± M	25 ± 3	43 ± 4	100	
+ Primaquine 200 ± M + azide	14 ± 2	24 ± 3	37 ± 3	73 ± 4
+ Primaquine 200 ± M + azide + ascorbate	24 ± 2	46 ± 3	100	
BCNU treated cells	14 ± 2	17 ± 3	20 ± 3	22 ± 3
+ Primaquine 200afaM	14 ± 2	23 ± 3	36 ± 3	61 ± 4

Control hepatocytes (10^6 cells/ml) were incubated in Krebs-Henseleit buffer, pH 7.4, containing Hepes (12.5 mM)at 37°C. Primaquine was then added to cells and viability was assesed by trypan blue uptake. Where indicated azide (4mM) and ascorbate (5mM) were preincubated for 5 min. Values are means ± standard error.

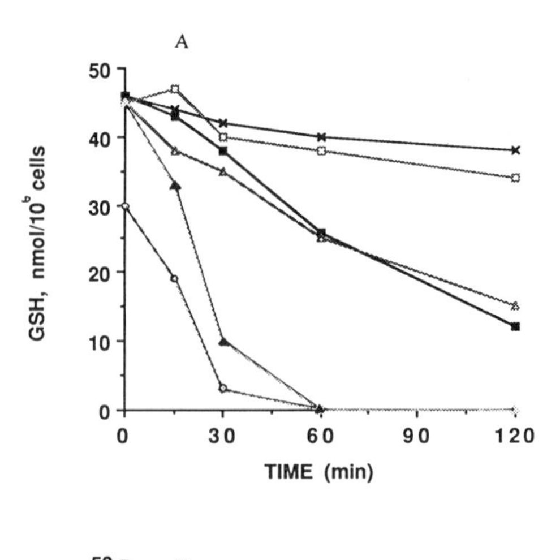

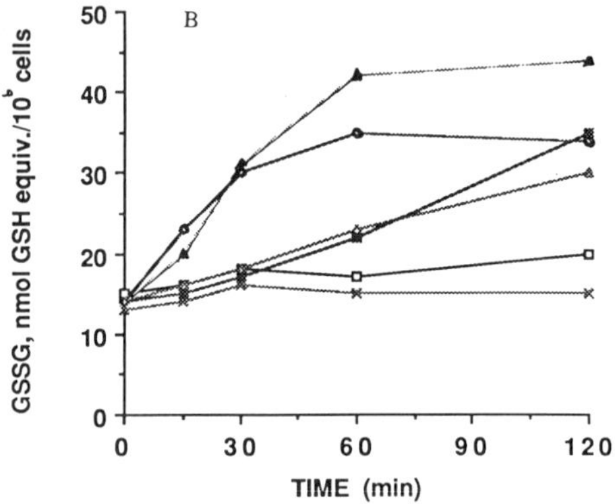

Figure 1. GSH depletion (A) and GSSG formation (B) induced by primaquine in isolated hepatocytes. Hepatocytes 10^6 cells/ml, were incubated with 500 μM primaquine (Δ), 200 μM primaquine (□), 200 μM primaquine + 4mM azide (■), 200 μM primaquine + 4mM azide + 5mM ascorbate (▲), 200 μM primaquine in BCNU treated cells (o). (x) represents control cells.

Xenobiotic induced liver toxicity is generally preceeded by the depletion of cellular glutathione. As shown in Figure 1 incubation of a toxic dose of primaquine also resulted in GSH depletion with the concurrent formation of oxidized GSH (GSSG). The appearance of GSSG coincided with onset of cytotoxicity except when ascorbate was also included or in BCNU treated cells in which case GSH oxidation preceeded cell death (Figure 1B). This suggested that GSH oxidation in primaquine treated hepatocytes in the absence of ascorbate was possibly due to membrane leakage of GSH and autooxidation to GSSG in the extracellular medium. To further examine this possibility the membrane impermeant cystine (200 μM) was included in the incubation medium. When cystine was present GSH depletion still occured but GSSG was not detected and mixed disulfides with cystine were formed instead (data not shown). However cystine did not prevent formation of GSSG in the presence of both azide and ascorbate in primaquine treated hepatocytes. This suggests that GSH-reductase is very efficient in maintaining reduced GSH in the presence of primaquine and is only over whelmed when the rate of redox cycling and oxygen activation is increased by adding azide and ascorbate together with primaquine.

In summary primaquine was shown to be toxic to isolated rat hepatocytes. The fact that cyanide resistant respiration was induced upon incubation of primaquine with hepatocytes and toxicity was markedly increased by compromizing the cells defense system against H_2O_2 indicates that H_2O_2 is being produced. Both of these effects were also further enhanced by extracellular ascorbate suggesting that H_2O_2 may be formed as a result of a reduced primaquine radical redox cycling with molecular oxygen resulting in the production of superoxide radicals which dismutate to H_2O_2. Recently a one-electron reduced product of primaquine has been reported using pulse radiolysis (Bisby, 1988) and oxygen activation by a NADPH:primaquine charge transfer complex observed by Thornelly et al., 1983 was attributed to NAD· and reduced primaquine radicals. This is the first report that primaquine reduction by ascorbate or hepatocyte reductases result by itself in oxygen activation. Others have also shown primaquine to be therapeutically more effective than its oxidative metabolites (Hopkins and Tudhope, 1974). The formation of H_2O_2 by primaquine in the liver cell could also explain its effectiveness in killing the malarial parasite.

ACKNOWLEDGEMENT

The authors would like to thank the Medical Research Council of Canada for supporting this work.

REFERENCES

Augusto, O., Weingrill, C. L., Schreier, S. and Amemiya,H. (1986). Hydroxyl radical formation as a result of the interaction between primaquine and reduced pyridine nucleotides. *Arch. Biochem. Biophys.* **244**, 147-155.

Beutler, E. (1969). Drug-induced hemolytic anemia. *Pharmacol. Rev.* **21**, 73-97.

Bisby, R. H. (1988). One-electron reduction of the antimalarial drug primaquine, studied by pulse radiolysis. *Free Rad. Res. Comms.* **5**, 117-124.

Cohen, G. and Hochstein, P. (1964). Glutathione peroxidase:The primary agent for the elimination of hydrogen peroxide in erythrovytes by hemolytic agents. *Biochem.* **3**, 895-903.

Gutierrez, P. L. (1988). The influence of ascorbic acid on the free-radical metabolism of xenobiotics: The example ofdiaziquone. *Drug Metab. Rev.* **19**, 319-343.

Hopkins, J. and Tudhope, G. R. (1978). Studies with primaquine *in vitro*: superoxide radical formation and oxidation of haemoglobin. *Br. J. Clin. Pharmacol.* **6**,319-323.

Moldeus, P., Hogberg, J. and Orrenius, S. (1978). Isolation and use of liver cells. *Methods. Enzymol.* **52**, 60-71.

Reed, D. J., Babson, J. R., Beatty, P. W., Brodie, A. F.,Ellis, W. W. and Potter, D. W. (1980). High-performance liquid chromatogrophy analysis of nanomole levels of glutathione, glutathione disulfide, and related thiols and disulfides. *Anal. Biochem.* **106**, 55-62.

Rossi, L., Silva, J. M., McGirr, L. and O'Brien, P. J.,(1988). Nitrofurantoin-mediated oxidative stress cytotoxicity in isolated hepatocytes. *Biochem. Pharmacol.* **37**, 3109-3117.

Strother, A., Allahyari, R., Buchholz, J., Fraser, I.M. and Tilton, B. E. (1984). *In vitro* metabolism of the antimalarial agent primaquine by mouse liver enzymes and identification of a methemoglobin-forming metabolite. *Drug Metab. Dispos.* **12**, 35-44.

Summerfield, M. and Tudhope, G. R. (1978). Primaquine and oxidation of hemoglobin. *Br. J. Pharmacol.* **6**, 319-323.

Thornalley, P.J.,Stern, A. and Bannister, J. V. (1983). A mechanism for primaquine mediated oxidation of NADPH in red blood cells. *Biochem. Pharmacol.* **32**, 3571-3575.

NITROPRUSSIDE: A POTPOURRI OF BIOLOGICALLY REACTIVE INTERMEDIATES

Roger P. Smith*[1], Dean E. Wilcox+, Harriet Kruszyna*, and Robert Kruszyna*

*Department of Pharmacology and Toxicology and
+Department of Chemistry
Dartmouth Medical School and Dartmouth College
Hanover, NH 03756

Sodium nitroprusside, $Na_2[Fe(CN)_5NO]$, (SNP) has been recognized as a potent, directly acting vasodilator for over a half-century. More recently, it has been found also to inhibit blood platelet aggregation and adhesion. These biologicial activities are ascribed to the nitrosyl ligand on SNP which is thought to activate guanylate cyclase by binding to its critical heme group. c-GMP in turn initiates a cascade of kinase reactions which result in the utimate biological effects. Thus, SNP is one of the compounds known as the nitric oxide (NO) vasodilators (Ignarro, 1989) which mimic the effects of the so-called endothelium-derived relaxing factor (Furchgott and Zawadzki, 1980), an endogenous mediator which many are convinced is identical to NO. The same or a similar factor is believed to play in role in excitatory amino acid transmission in the central nervous system, and to be the cytotoxic factor synthesized by neutrophils and macrophages (Collier and Vallance, 1989).

Considerable confusion exists in the literature about the chemistry and biotransformation pathways for SNP both *in vivo* and *in vitro*. In addition to the nitrosyl ligand, SNP releases free cyanide, and in acute overdoses fatal cyanide poisoning can result (Smith and Kruszyna, 1974). This complete decomposition of SNP can be initiated by a redox reaction with hemoglobin, but the reaction also occurs in cells and tissues in the absence of hemoglobin (Devlin and Smith, 1989). Surprisingly, there are those who believe that SNP spontaneously releases its nitrosyl ligand as NO (Brune and Lapetina, 1989), as well as those who believe that the iron-cyanide species resulting from the loss of the NO is a thermodynamically stable molecule with such high formation constants that the release of free cyanide under biological conditions is impossible (Butler and Glidewell, 1987). This study was undertaken to determine the mechanism of the reaction of hemoglobin with SNP as a possible model for the reaction that occurs in responsive cells and tissues and involves heme groups.

MATERIALS AND METHODS

The reaction between hemoglobin and SNP was studied by visible absorption spectrophotometry. Under aerobic or anaerobic conditons the product was cyanmethemoglobin, whereas under anaerobic conditons in the presence of a reducing agent, such as dithionite for solutions or methylene blue/glucose for intact cells the product was nitrosylhemoglobin. Mortality studies in mice established that (1) SNP

1. To whom correspondence should be addressed.

produced death by acute cyanide poisoning because it resulted in central respiratory arrest which was delayed relative to sodium cyanide, i. e., 30 vs. 5 min, (2) the blood levels of free cyanide were comparable to those found at death after exposure to sodium cyanide, and (3) there was protection against death by two established cyanide antagonists with different mechanisms of action. Comparison of the LD50 of SNP and sodium cyanide suggested that all five equivalents of cyanide are released *in vivo* (Smith and Kruszyna, 1974).

The release of NO by SNP in red blood cell suspensions and in intact mice was established by observation of the characteristic electron paramagnetic resonance (EPR) signals of nitrosylated forms of hemoglobin; these include the fully reduced, nitrosylated species and the two valency hybrid species in which either the α- or the β-subunits were oxidized and the other subunits were reduced and nitrosylated. In the cell suspension experiments this reaction must have occurred intracellularly; however, it is not clear whether it occured intracellularly *in vivo*, or at some other site from which the NO diffused into red cells where it was trapped by the hemoglobin (Kruszyna et al., 1987; 1988). There is no reaction between methemoglobin and SNP under these conditions. Incubation of hemoglobin with SNP in the presence of an excess of free ^{13}C-cyanide leads to an exchange of the trans-cyanide ligand on SNP with free cyanide in solution (Shafer et al., 1989). Electrolytic reduction of SNP under nitrogen gives two EPR signals, the dominant one being that of $[(CN)_5FeNO]^{-3}$. Subtraction of its signal from the total spectrum resulted in one very similar to that observed for $[(CN)_4FeNO]^{-2}$ resulting from the loss of the *trans*-cyanide (Wilcox et al., 1990).

Surprisingly, the relaxant activity of SNP on vascular smooth muscle was at least partially blocked or prevented by an excess of free cyanide ion and this effect was accompanied by a parallel influence on the tissue levels of c-GMP (Kruszyna et al., 1982). Although cyanide also blocks or reverses the biological effects of other NO-vasodilators under certain conditions, its effect on nitroprusside was the most consistent and extended to human blood platelets (Schwerin et al., 1983) and other smooth muscles such as guinea pig ileum, rabbit gall bladder and rat pulmonary artery (Kruszyna et al., 1985). This effect of cyanide had many of the aspects of competitive antagonism in that the antagonist (cyanide) had no action of its own in the concentrations tested, the effects of both the antagonist and the agonist (SNP) were reversible, the antagonist was selective in affecting some but not all of a group of similar agonists, the antagonist produced parallel shifts to the right in the log-dose response curves for the agonist and the data could be made to fit a Schild plot (Wilcox et al., 1990).

RESULTS AND DISCUSSION

The single electron reduction of SNP leads to a sequential decomposition which results in a variety of reactive intermediates and final products as summarized in Figure 1. As shown in step (1) oxy- or deoxy-hemoglobin is capable of reducing SNP to $[(CN)_5FeNO]^{-3}$, species II. Concomitantly, the heme iron is oxidized resulting in the formation of methemoglobin.

The trans-cyanide ligand on the one electron reduced SNP, species II, is labile, and it can exchange with free cyanide in solution. In the absence of free cyanide it may dissociate completely giving $[(CN)_4FeNO]^{-2}$, species III, and the free cyanide may bind to oxidized heme on methemoglobin forming cyanmethemoglobin. Dissociation of the *trans*-cyanide results in the labilization of the NO ligand of species III which allows species III to transfer the NO to an appropriate acceptor such as reduced heme on hemoglobin generating nitrosyl hemoglobin as observed by EPR. In the *in vitro* systems which we have studied, the reducing agents, dithionite or methylene blue, may reduce the oxidized heme back to the reduced form which can accept NO. Hemoglobin is a widely used inhibitor of the effects of EDRF and NO-vasodilator drugs apparently because of this ability to scavenge NO. In the presence of oxygen, however, nitrosyl hemoglobin is unstable to oxidation resulting in the formation of methemoglobin and nitrite. The nitrosylated valency hybrid species are notable exceptions (Kruszyna et

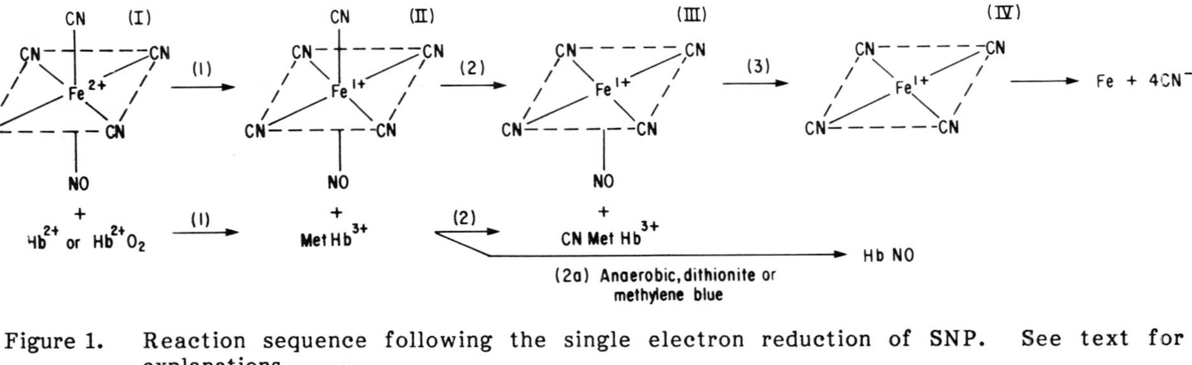

Figure 1. Reaction sequence following the single electron reduction of SNP. See text for explanations.

al., 1987a). The next intermediate, species IV, then dissociates all remaining cyanide ligands. It is species III, however, that is removed by excess of free cyanide in solution. Apparently the biological activity of SNP is inhibited or reversed through a decreased concentration of this species which is capable of transferring its NO ligand to the biological acceptor which may be the heme group on guanylate cyclase. Presumably this molecular interaction between cyanide and SNP accounts for the superficial resemblance of their biological interaction to competitive antagonism even though no drug "receptor" is known to be involved in its activity. Methylene blue has also been used to inhibit the effects of NO-vasodilators and EDRF. It is generally assumed that this redox dye inhibits guanylate cyclase, but no direct evidence appears to support that mechanism. Since methylene blue penetrates into cells, it has a wider spectrum of inhibitory effects than hemoglobin (Furchgott and Vanhoutte, 1989).

In summary, the reaction of SNP with hemoglobin may serve as a model for the actual biochemical mechanisms at work in its biological activity including the antagonistic effects of cyanide. An alternative mechanism for SNP activation, which has been less well characterized is the reaction between SNP and reduced glutathione (Wilcox et al., 1990)

ACKNOWLEDGEMENTS

This work was supported by Grant HL 14127 from the National Heart, Lung and Blood Institute (RPS). The Bruker ESP-300 spectrometer was purchased with funding from the NSF (Grant CHE-8701406). We thank Dr. E. Lucile Smith, Department of Biochemistry, Dartmouth Medical School for critical suggestions.

REFERENCES

Brune, B., and Lapetina, E. G. (1989). Activation of a cytosolic ADP-ribosyl--transferase by nitric oxide-generating agents. *J. Biol. Chem.* **264**, 8455-8458.

Butler, A. R., and Glidewell, C. (1987). Recent chemical studies of sodium nitroprusside relevant to its hypotensive action. *Chem. Soc. Rev.* **16**, 361-380.

Collier, J., and Vallance, P. (1989). Second messenger role for NO widens to nervous and immune systems. *TIPS* **10**, 427-431.

Devlin, D. J., Smith, R. P., and Thron, C. D. (1989). Cyanide release from nitroprusside in the isolated, perfused, bloodless liver and hindlimbs of the rat. *Toxicol. Appl. Pharmacol.* **99**, 354-356.

Furchgott, R. F., and Zawadski, J. V. (1980). The obligatory role of endothelial cells in the relaxation of arterial smooth muscle by acetylcholine. *Nature* **288**, 373-376.

Furchgott, R. F., and Vanhoutte, P. M. (1989). Endothelium-derived relaxing and contracting factors. *FASEB J.* **3**, 2007-2018.

Ignarro, L. J. (1989). Heme-dependent activation of soluble guanylate cyclase by nitric oxide: Regulation of enzyme activity by porphyrins and metalloporphyrins. *Semin. Hematol.* **26**, 63-76.

Kruszyna, H., Kruszyna, R., and Smith, R. P. (1982). Nitroprusside increases guanylate monophosphate concentrations during the relaxation of rabbit aortic strips and both effects are antagonized by cyanide. *Anesthesiology* **57**, 303-308.

Kruszyna, H., Kruszyna, R., and Smith, R. P. (1985). Cyanide and sulfide interact with nitrogenous compounds to influence the relaxation of various smooth muscles. *Proc. Soc. Exp. Biol. Med.* **179**, 44-49.

Kruszyna, H., Kruszyna, R., Smith, R. P., and Wilcox, D. E. (1987). Red blood cells generate nitric oxide from directly acting nitrogenous vasodilators. *Toxicol. Appl. Pharmacol.* **91**, 429-438.

Kruszyna, R., Kruszyna, H., Smith, R. P., Thron, C. D., and Wilcox, D. E. (1987a). Nitrite conversion to nitric oxide in red cells and its stabilization as a nitrosylated valency hybrid of hemoglobin. *J. Pharmacol. Exp. Ther.* **241**, 307-313.

Kruszyna, R., Kruszyna, H., Smith, R. P., and Wilcox, D. E. (1988). Generation of valency hybrid species of hemoglobin in mice by nitric oxide vasodilators. *Toxicol. Appl. Pharmacol.* **94**, 458-465.

Schwerin, F. T., Rosenstein, R., and Smith, R. P. (1983). Cyanide prevents the inhibition of platelet aggregation by nitroprusside, hydroxylamine and azide. *Thromb. Haemostas.* **50**, 780-783.
Shafer, P. R., Wilcox, D. E., Kruszyna, H., Kruszyna, R., and Smith, R. P. (1989). Decomposition and specific exchange of the trans-cyanide ligand on nitroprusside is facilitated by hemoglobin. *Toxicol. Appl. Pharmacol.* **99**, 1-10.
Smith, R. P., and Kruszyna, H. (1974). Nitroprusside produces cyanide poisoning via a reaction with hemoglobin. *J. Pharmacol. Exp. Ther.* **191**, 557-563.
Wilcox, D. E., Kruszyna, H., Kruszyna, R., and Smith, R. P. (1990). Effect of cyanide on the reaction of nitroprusside with hemoglobin: Relevance to cyanide interference with the biological activity of nitroprusside. *Chem. Res. Toxicol.* In press.

DENITROSATION OF *N*-NITROSODIMETHYLAMINE IN THE RAT *IN VIVO*

Anthony J. Streeter, Raymond W. Nims, Pamela R. Sheffels, and Larry K. Keefer

Chemistry Section
Laboratory of Comparative Carcinogenesis
National Cancer Institute
Frederick Cancer Research Facility
Frederick, MD 21701

The potent hepatocarcinogen N-nitrosodimethylamine (NDMA) has been shown to be metabolized by liver microsomes from acetone-induced rats via two pathways (Figure 1). While the denitrosation pathway has been studied fairly extensively *in vitro* (Keefer et al., 1987b; Wade et al., 1987; Amelizad et al., 1988), only a few tentative observations about its course in the intact animal have so far appeared in the literature (Heath and Dutton, 1958; Keefer et al., 1987a; Magee et al., 1988). The easiest and most direct assessment of denitrosation *in vivo* should come from measuring the extent of conversion of NDMA to MA, if the NDMA is labeled with carbon-14 so that the metabolically produced MA can be differentiated from the endogenous material that is known to be present (Davis and de Ropp, 1961).

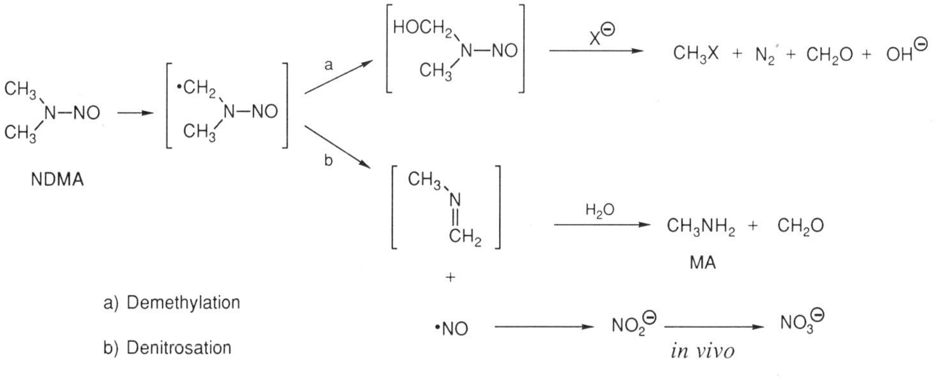

$(CH_3)_2N{-}NO$ (NDMA) $\rightarrow [\cdot CH_2(CH_3)N{-}NO]$

a: $\rightarrow [HOCH_2(CH_3)N{-}NO] \xrightarrow{X^{\ominus}} CH_3X + N_2 + CH_2O + OH^{\ominus}$

b: $\rightarrow [CH_3N{=}CH_2] \xrightarrow{H_2O} CH_3NH_2$ (MA) $+ CH_2O$

$+ \cdot NO \rightarrow NO_2^{\ominus} \xrightarrow{in\ vivo} NO_3^{\ominus}$

a) Demethylation

b) Denitrosation

Figure 1. Postulated mechanisms of NDMA metabolism. Path a represents the demethylation pathway proceeding via an ultimately genotoxic methylating species. Path b represents the course of enzymatic denitrosation, a potentially inactivating elimination route that occurs in parallel with demethylation.

Biological Reactive Intermediates IV, Edited by C.M. Witmer *et al.*
Plenum Press, New York, 1990

MATERIALS AND METHODS

N-Nitrosomethyl[^{14}C]methylamine ([^{14}C]NDMA, 40 mCi/mmol) and *N*-nitroso[$^{2}H_3$]methyl[^{14}C, $^{2}H_3$]methylamine ([^{14}C]NDMA-d_6, 40 mCi/mmol) were prepared by SRI International (Menlo Park, CA). The radiochemical purities were determined by HPLC to be 99.8 and 99.2%, respectively. [^{14}C]Methylamine ([^{14}C]MA, 40 mCi/mmol or 8.9 mCi/mmol) was bought from New England Nuclear (Boston, MA), methyl[^{14}C]methylamine ([^{14}C]DMA, 58 mCi/mmol) was purchased from Amersham (Arlington Heights, IL), and [^{14}C, $^{2}H_3$]methylamine ([^{14}C]MA-d_3, 20 mCi/mmol) was obtained from Dr. J. Hrabie (Program Resources, Inc., Frederick, MD).

Male Fischer 344 rats (8 weeks old) were given a single dose of either [^{14}C]NDMA, [^{14}C]MA, [^{14}C]NDMA-d_6, or [^{14}C]MA-d_3 by gavage (i.g.) or intravenous (i.v.) bolus injection into the tail vein (as indicated) between 8:40 a.m. and 1:44 p.m. Approximately 20 serial 50 µl retroorbital blood samples from each rat were collected into microcentrifuge tubes containing 20 µl of heparin (100 units/ml). The protein was then precipitated by the addition of 70 µl of saturated barium hydroxide solution. Aliquots of 70 µl of each supernatant fraction were then assayed by HPLC for NDMA and/or MA by quantifying the peaks of radioactive material that eluted at the retention times of authentic NDMA and MA and comparing those values to standard curves contructed by spiking blood samples with [^{14}C]NDMA, [^{14}C]MA, [^{14}C]NDMA-d_6, or [^{14}C]MA-d_3. The HPLC system used was essentially the same as that of Heur et al. (1989), except that it was eluted with 200 mM ammonium phosphate buffer (pH 3.0) at 1.0 ml/min.

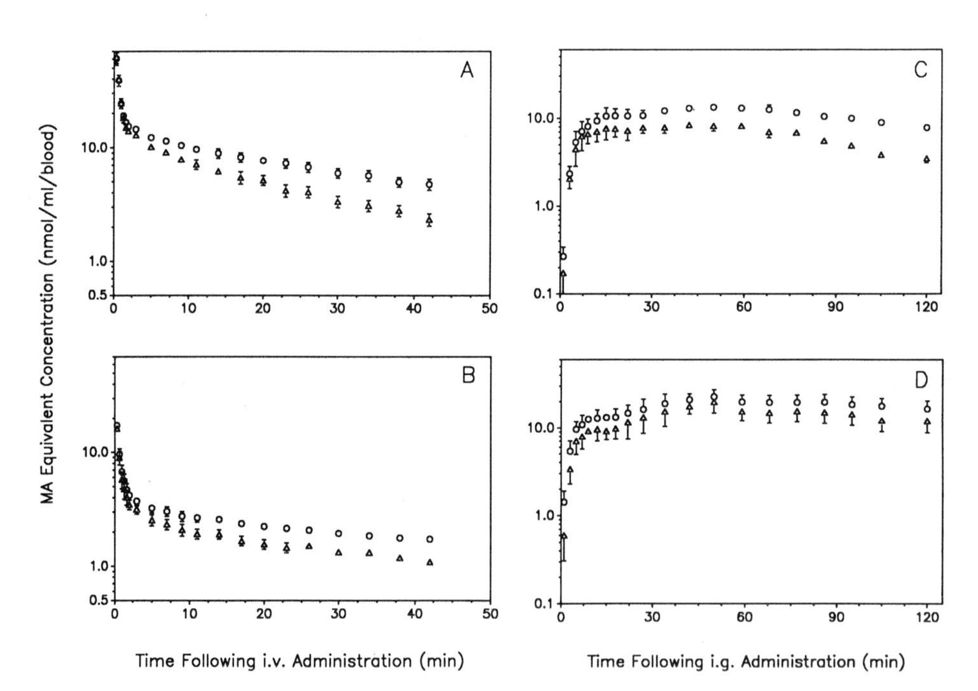

Figure 2. Blood concentrations (mean ± SE) of unchanged compound (Δ) or total radioactivity (o), as a function of time for groups of male F344 rats after i.v. bolus administration of (A) 18.9 ± 1.5 µmol/kg [^{14}C]MA (N=4) or (B) 3.5 ± 0.2 µmol/kg [^{14}C]MA-d_3 (N=4), or after i.g. administration of (C) 81.9 ± 1.3 µmol/kg [^{14}C]MA (N=4) or (D) 89.6 ± 1.9 µmol/kg [^{14}C]MA-d_3 (N=3).

Separate groups of rats were used to estimate the total amounts of NDMA and MA excreted renally. After i.v. administration, 4 rats/compound were placed into individual metabolism chambers and urine was collected at 24 hour intervals for 3 days. Aliquots of 100 μl of urine were assayed as above.

The basic toxicokinetic parameters and the extent of reversible protein binding in blood were obtained for each compound exactly as described previously (Streeter et al., 1989).

RESULTS AND DISCUSSION

In order to accurately estimate the extent of formation of MA from NDMA, it was necessary to define the elimination parameters of MA itself. The single-dose toxicokinetics of MA has been characterized for the first time in the rat (Figure 2). The plots of concentration of unchanged compound in whole blood versus time following an i.v. bolus dose revealed biphasic first-order elimination kinetics for both MA and MA-d_3. In the case of MA, values were calculated for $t_{1/2}(\beta)$[1], CL, V_{ss}, and F of 19.1 ± 1.3 min, 53.4 ± 3.5 ml/min/kg, 1216 ± 93 ml/kg, and 69 ± 3%, respectively. A comparison of MA-d_3 to its non-deuterated analogue revealed values for the same parameters of 37.0 ± 4.2 min, 24.4 ± 0.8 ml/min/kg, 1168 ± 122 ml/kg, and 121 ± 2%, respectively. Administration of i.v. bolus doses of MA (20.5 ± 1.4 μmol/kg) or MA-d_3 (3.96 ± 0.10 μmol/kg) resulted in the urinary excretion over 3 days of 10.7 ± 1.0% and 47.9 ± 0.7% of the dose as unchanged compound, respectively, accounting for almost all of the radioactivity that was excreted by that route.

Plots of concentration of unchanged compound against time following i.v. bolus administration of NDMA or NDMA-d_6 were also found to show biphasic first-order elimination kinetics (Figure 3). Values were obtained for $t_{1/2}(\beta)$, CL, V_{ss}, F of 10.7 ± 0.4 min, 50.2 ± 4.8 ml/min/kg, 630 ± 44 ml/kg, and 13 ± 1% for NDMA, and 14.9 ± 0.8 min, 28.4 ± 2.1 ml/min/kg, 563 ± 27 ml/kg, and 31 ± 2% for NDMA-d_6, respectively. Inspection of the terminal elimination (β) phases revealed that the unchanged compounds were being removed from the systemic circulation at a much greater rate than the total radioactivity, indicating that metabolites were being eliminated more slowly than they were being formed. One metabolite was unambiguously identified as MA. Dimethylamine was not detected. Following i.v. bolus doses of NDMA (14.1 ± 0.4 μmol/kg) or NDMA-d_6 (6.36 ± 0.54 μmol/kg), no detectable unchanged NDMA was excreted in the urine over 3 days while only 0.1% of the dose of NDMA-d_6 was excreted unchanged by that route. Of the 5.22 ± 0.175 and 11.71 ± 0.25% of the total radioactivity given as NDMA and NDMA-d_6 that was recovered in the urine, 5.63 ± 0.28% and 12.90 ± 1.96% was accounted for as metabolically produced MA, respectively (assuming that the MA was equally likely to be derived from the labeled and non-labeled carbon atoms). NDMA, MDNA-d_6, MA, and MA-d_3 all showed negligible reversible protein binding in rat plasma as determined by equilibrium dialysis.

The main reason for studying the deuterated compounds was to explain the observed reduction of hepatocarcinogenesis (Keefer et al., 1973) and DNA methylation (Swann et al., 1983) seen in rats when NDMA-d_6 was substituted for NDMA. These *in vivo* findings were apparently in contradiction to studies in hepatic microsomes from acetone-induced rats which showed that the proportions of nitrosamine metabolized via demethylation and denitrosation pathways remained the same for both substrates (Wade et al., 1987). However, such a situation does not appear to exist *in vivo* since, despite the urinary evidence indicating complete metabolism of both compounds, estimation of the fraction of the dose converted to MA demonstrated more denitrosation of NDMA-d_6 than NDMA (Table 1).

---

[1] $t_{1/2}(\beta)$, half-life of the terminal elimination phase following an i.v. dose; CL, systemic blood clearance; Vss, apparent steady-state volume of distribution; F, systemic bioavailability of an i.g. dose.

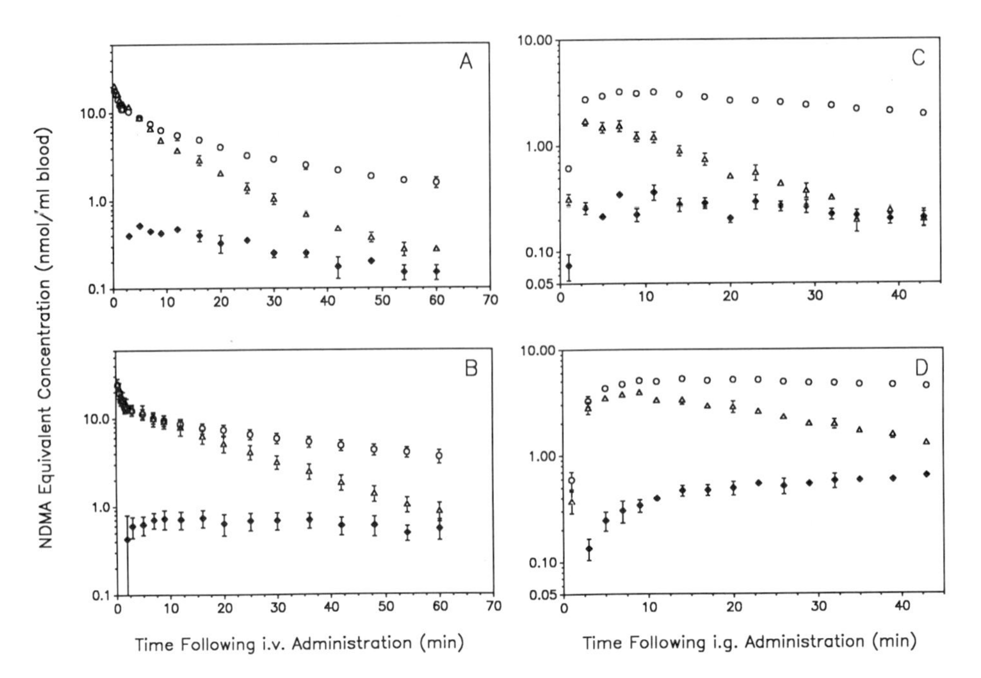

Figure 3. Blood concentrations (mean ± SE) of unchanged compound (Δ), MA (♦) or total radioactivity (o), as a function of time for groups of male F344 rats after i.v. administration of (A) 8.4 ± 1.1 μmol/kg [^{14}C]NDMA (N=4) or (B) 8.3 ± 1.0 μmol/kg [^{14}C]NDMA-d_6 (N=6), or after i.g. administration of (C) 13.5 ± 0.3 μmol/kg [^{14}C]NDMA (N=4) or (D) 14.6 ± 0.3 μmol/kg [^{14}C]NDMA-d_6 (N=4).

Table 1. Contribution of Metabolic Denitrosation to the Total Elimination of NDMA and NDMA-d_6 in Male F344 Rats

Rat no.	NDMA*		Rat no.	NDMA-d_6*	
	Method 1	Method 2		Method 1	Method 2
1	0.189	0.129	1'	0.532	0.645
2	0.211	0.144	2'	0.335	0.406
3	0.201	0.137	3'	0.194	0.235
4	0.251	0.171	4'	0.238	0.289
			5'	0.316	0.383
			6'	0.774	0.939
	----	----		----	----
Mean ± SE	0.213 ±0.013†	0.145 ±0.009‡	Mean ± SE	0.398 ±0.089†	0.483 ±0.108‡

*Values given for the individual rats are of fraction metabolized (f_m) to MA, calculated by one of two methods (Houston, 1986):

Method 1: $f_m = (AUC_{(MA),i.v.} \times Dose_{MA,i.g.}) / (AUC_{MA,i.g.} \times Dose_{NDMA,i.v.})$

Method 2: $f_m = (AUC_{(MA),i.v.} \times Dose_{MA,i.v.}) / (AUC_{MA,i.v.} \times Dose_{NDMA,i.v.})$

where $AUC_{(MA),i.v.}$ is the area under the curve for metabolic MA following an i.v. administration of NDMA ($Dose_{NDMA,i.v.}$), and $AUC_{(MA),i.v.}$ and $AUC_{MA,i.g.}$ are the areas under the curves for MA following i.v. ($Dose_{MA,i.v.}$) or i.g. ($Dose_{MA,i.g.}$) administration of MA, respectively.

† $P < 0.05$, significantly different (modified one-tailed Student's t test)

‡ $P < 0.02$, significantly different (modified one-tailed Student's t test)

In conclusion, the presence of MA as a circulating and urinary metabolite of NDMA has been confirmed and a significant increase, due to complete deuterium substitution, in the extent of *in vivo* denitrosation has been found which indicates that the microsomally-derived metabolic scheme with a common intermediate generated by cytochrome P450IIE1 is an oversimplification of the situation occurring in the intact animal.

REFERENCES

Amelizad, Z., Appel, K.E., Oesch, F., and Hildebrandt, A.G. (1988). Effect of antibodies against cytochrome P-450 on demethylation and denitrosation of *N*-nitrosodimethylamine and N-nitrosomethylaniline. *J. Cancer Res. Clin. Oncol.* **114**, 380-384.

Davis, E.J., and DeRopp, R.S. (1961). Metabolic origin of urinary methylamine in the rat. *Nature (London)* **190**, 636-637.

Heath, D.F., and Dutton, A. (1958). The detection of metabolic products from dimethylnitrosamine in rats and mice. *Biochem. J.* **70**, 619-626.

Heur, Y.-H., Streeter, A.J., Nims, R.W., and Keefer, L.K. (1989). The Fenton degradation as a nonenzymatic model for microsomal denitrosation of *N*-nitrosodimethylamine. *Chem. Res. Toxicol.* **2**, 247-253.

Houston, J.B. (1986). Drug metabolite kinetics. In *Pharmacokinetics: Theory and Methodology* (M. Rowland, and G. Tucker, Eds.), pp. 131-162, International Encyclopedia of Pharmacology and Therapeutics, Section 122, Pergamon Press, New York.

Keefer, L.K., Lijinsky, W., and Garcia, H. (1973). Deuterium isotope effect on the carcinogenicity of dimethylnitrosamine in rat lover. *J. Natl. Cancer Inst.* **51**, 299-302.

Keefer, L.K., Anjo, T., Heur, Y.-H., Yang, C.S., and Mico, b.A. (1987a). Potential for metabolic deactivation of carcinogenic *N*-nitrosodimethylamine *in vivo*. In *Relevance of N-Nitroso Compounds to Human Cancer: Exposure and Mechanisms* (H. Bartsch, I.K. O'Neill, and R. Schulte-Hermann, Eds.), pp. 113-116, IARC Scientific Publications No. 84, International Agency for Research on Cancer, Lyon, France.

Keefer, L.K., Anjo, T., Wade, D., Wang, T., and Yang, C.S. (1987b). Concurrent generation of methylamine and nitrite during denitrosation of *N*-nitrosodimethylamine by rat liver microsomes. *Cancer Res.* **47**, 447-452.

Magee, P.N., Harrington, G.W., Pylypiw, H., dollard, D., Kozeniauskas, R., Bevill, R.F., Thurmon, J., and Nelson, D. (1988). The metabolism of *N*-nitrosodimethylamine in the pig. *Proc. Am. Assoc. Cancer Res.* **29**, 124.

Streeter, a.J., Nims, R.W., Hrabie, J.A., Heur, Y.-H., and Keefer, L.K. (1989). Sex differences in the single-dose toxicokinetics of *N*-nitrosomethyl (2-hydroxyethyl)amine in the rat. *Cancer Res.* **49**, 1783-1789.

Swann, P.F., Mace, R., Angeles, R.M., and Keefer, L.K. (1983). Deuterium isotope effect on metabolism of *N*-nitrosodimethylamine *in vivo* in rat. *Carcinogenesis (London)* **4**, 821-825.

Wade, D., Yang, C.S., Metral, C.J., roman, J.M., Hrabie, J.A., Riggs, C.W., Anjo, T., Keefer, L.K., and Mico, B.A. (1987). Deuterium isotope effect on denitrosation and demethylation of *N*-nitrosodimethylamine by rat liver microsomes. *Cancer Res.* **47**, 3373-3377.

EFFECT OF PHENOL AND CATECHOL ON THE KINETICS OF HUMAN MYELOPEROXIDASE-DEPENDENT HYDROQUINONE METABOLISM

Vangala V. Subrahmanyam, Prema Kolachana and Martyn T. Smith[1]

Department of Biomedical and Environmental Health Sciences
School of Public Health
University of California at Berkeley
Berkeley, CA 94720

INTRODUCTION

Benzene, is hematotoxic and leukemogenic in humans (Aksoy, M., 1987). It is widely accepted that benzene requires metabolic activation to exert its toxic and carcinogenic effects (Cooper, K.R. and Snyder, R. 1987). The majority of benzene metabolism occurs in liver with phenol (PH), hydroquinone (HQ) and catechol (CAT) being the major metabolites formed. Benzene is also oxidized to a ring-opened product trans, trans-muconic acid.

Since the rate of benzene metabolism in bone marrow is very low, it is believed that metabolites generated in liver are transported to bone marrow. Bone marrow, the target organ for benzene toxicity contains high levels of myelo- and eosinophil peroxidases (Kariya, K., 1982). Phenolic compounds are known to be excellent reducing cosubstrates for peroxidases. We and others have recently shown that PH, HQ and CAT can be metabolized by horseradish peroxidase (HRP) and myeloperoxidase (MPO) to reactive intermediates (Eastmond, D.A. et al., 1986; Smith, M.T. et al., 1989; Sadler, A. et al., 1988; Subrahmanyam, V.V. et al., 1985). Interestingly, HQ metabolism by HRP, human MPO and by bone marrow cells has been shown to be stimulated by PH (Smith, M.T. et al., 1989; Subrahmanyam, V.V. et al., 1989). However, the exact mechanism of this stimulation is not clear.

In this communication, we report results describing the kinetics of myeloperoxidase-dependent PH, HQ and CAT metabolism, studied in an attempt to understand the mechanism of oxidation of these phenolics by MPO. The influence of PH and CAT on the kinetics of MPO-dependent HQ metabolism are also described. The potential toxicological significance of MPO-dependent phenolic compound metabolism in benzene-induced myelotoxicity and leukemogenicity is discussed.

MATERIALS AND METHODS

Phenol, catechol and hydroquinone were purchased from Sigma Chemical Company (St. Louis, MO). Hydrogen peroxide (30%) was obtained from Fisher

---

1. To whom correspondence should be addressed: Department of Biomedical and Environmental Health Sciences, School of Public Health, University of California at Berkeley, Berkeley, CA 94720

Biological Reactive Intermediates IV, Edited by C.M. Witmer *et al.*
Plenum Press, New York, 1990

Chemical Company (Los Angeles, CA). Human MPO was purchased from Calbiochem Corporation (La Jolla, CA).

Incubations were performed in 1 ml of 0.1 M sodium phosphate buffer pH 7.4 at room temperature. For kinetic studies the incubations consisted of either HQ (0-100 μM) or PH (0-100 μM) or CAT (0-100 μM), MPO (0 -0.2 U) and hydrogen peroxide (200 μM). The reactions were initiated with the addition of hydrogen peroxide and terminated with the addition of 100 μl of 70% perchloric acid. Apparent K_m and V_{max} values were derived from double-reciprocal plots of the substrate concentrations versus the initial velocities of substrate removal as measured by HPLC.

HPLC-Electrochemical Detection

A Beckman model 100A solvent delivery system equipped with C_{18} column (supelco; 4.6 mm x 25 cm) and an electrochemical detector was used for all HPLC analyses. For HQ and CAT analysis, aliquots (typically 10 μl) of reaction mixtures of HQ and CAT were analyzed by HPLC developed with 0.1 M ammonium acetate (pH 4) containing 1% acetonitrile with a flow-rate of 1.5 ml/min (Eastmond, D.A. et al., 1986). For the analysis of PH, the column was developed with 0.1 M ammonium acetate containing 10% acetonitrile with a flow rate of 2 ml/min (Eastmond, D.A. et al., 1986).

RESULTS AND DISCUSSION

Rickert et al. (1979) showed that significant levels of PH, HQ and CAT can be detected in bone marrow after administration of benzene to rats (Rickert, D.E. et al., 1979). However, administration of these phenolic metabolites alone does not result in myelotoxicity (Cooper, K.R. and Snyder, R. 1987). The administration of a combination of PH and HQ can decrease bone marrow cellularity in B6C3F1 mice, indicating that interactions of benzene metabolites in bone marrow could play an important role in benzene-in?duced myelotoxicity (Eastmond, D.A. et al., 1987). In support of this hypothesis, we found that HQ metabolism by human MPO or bone marrow cells is stimulated by PH (Smith, M.T. et al., 1989; Subrahmanyam, V.V. et al., 1989). Catechol, another metabolite of benzene, can also potentiate MPO-dependent HQ metabolism significantly (Steinmetz, K.L. et al., 1989). A comparison of the requirement of relative concentrations of PH and CAT for the stimulation of HQ metabolism showed that CAT even under equimolar concentrations to that of HQ can significantly stimulate the metabolism of HQ (data not shown). PH, however, must be in at least a 10-fold excess of HQ to significantly stimulate HQ metabolism. Since studies by Rickert and coworkers (Rickert, D.E. et al., 1979) show that CAT is also detected in bone marrow, in addition to PH and HQ, after administration of benzene to rats, it is possible CAT could play some role in the potentiation of HQ metabolism by MPO to 1,4-benzoquinone, and thus in benzene-induced myelotoxicity.

Table 1. Kinetic constants for hydroquinone, phenol and catechol metabolism by human myeloperoxidase

Substrate	K_m (μM)	V_{max} (μM/U/min)	V_{max}/K_m	E1/2(V)[T]
Hydroquinone	177 ± 35 (5)	1471 ± 153 (5)	8.36 ± 1.78 (5)	0.05
Phenol	67 ± 8 (4)	38 ± 14 (4)	0.53 ± 0.22 (4)	0.35
Catechol	189 ± 21 (3)	211 ± 42 (3)	1.68 ± 0.31 (3)	0.16

T = Taken from ref. 17.

Phenolic substrate (0 - 100 μM) was incubated with MPO (0.2 U) and hydrogen peroxide (200 μM) in 1 ml of 0.1 M sodium phosphate buffer, pH 7.4 for 15 s at room temperature. The metabolism of the phenolic substrates was measured by HPLC as described in Materials and Methods.

Studies by Dunford and coworkers indicated that the oxidation of phenolic compounds by HRP is dependent on the redox potentials of the phenolic substrates (Job, D. and Dunford, H.B., 1976). In this study, we determined the kinetic constants for MPO-dependent metabolism of PH, HQ and CAT (Table 1). The results indicate that the apparent K_m of HQ for MPO is 2 to 3-fold higher than that of PH and almost equal to that of CAT. This may indicate that PH has the higher affinity for MPO compared to HQ or CAT. However, the turnover rates (V_{max}) show that HQ is oxidized 7- and 39-fold faster than CAT or PH respectively. Calculation of the productive binding rates (V_{max}/K_m) demonstrates that HQ has 5- and 16-fold greater specificity for MPO than that of CAT and PH respectively (Table 1). This indicated that the binding affinity of these compounds for MPO may play little role in the rapidity with which they are oxidized. Thus, HQ which has a low redox potential is oxidized rapidly compared to CAT and PH which have comparatively higher redox potentials.

Stimulation of the oxidation of one substrate by other substrates or compounds in reactions catalyzed by HRP has been known for some time (Smith, M.T. et al., 1989; Subrahmanyam, V.V. et al., 1985; Eastmond, D.A. et al., 1987; Danner, D.J. et al., 1973; Che-Fu Koo et al., 1987; Van Steveninck, J. et al., 1987; Fox, R.L. et al., 1968). The exact nature of this effect is not known, but several investigators have suggested that radicals generated from the stimulant could react with the substrates whose reaction rates are enhanced (Smith, M.T. et al., 1989; Subrahmanyam, V.V. et al., 1985; Eastmond, D.A. et al., 1987; Fox, R.L. et al., 1968). Protein modification (Danner, D.J. et al., 1973; Che-Fu Koo et al., 1987) and protection of the enzyme from possible inactivation by oxygen radicals (Van Steveninck, J. et al., 1987) have also been suggested to be alternative mechanisms for the rate enhancement. We have previously postulated that phenoxy radicals generated during peroxidase-dependent metabolism could react with HQ and potentiate its metabolism (Smith, M.T. et al., 1989; Subrahmanyam, V.V. et al., 1989; Eastmond, D.A. et al., 1987). The results presented here, however, suggest that redox potentials are the major factor regulating the oxidation of HQ, CAT and PH metabolism by MPO, and that the HQ is a better substrate than either CAT or PH. Thus, the potentiation of HQ metabolism by PH and CAT may not be simply explained by a phenoxy radical dependent stimulation of HQ oxidation. Other mechanisms must therefore be involved. These are currently under investigation in our laboratory.

Table 2. Effect of phenol and catechol on kinetics of hydroquinone metabolism by human myeloperoxidase

Substrate(s)	K_m (μM)	V_{max} (μM/U/min)	V_{max}/K_m
Hydroquinone	177 ± 35 (5)	1471 ± 153 (5)	8.36 ± 1.78
+ Phenol (100 μM)	42 ± 8 (6)	620 ± 104 (6)	15.29 ± 3.75*
+ Catechol (20 μM)	23 ± 11 (4)	550 ± 89 (4)	23.89 ± 2.8*

*$p < 0.05$; p values were determined using two-tailed student's t test.
Hydroquinone removal (0 - 100 μM) incubated with MPO (0.2 U) and hydrogen peroxide (200 μM) in 1 ml of 0.1 M sodium phosphate buffer, pH 7.4, for 15 s at room temperature, was measured by HPLC as described in Materials Methods.

Interestingly, PH and CAT significantly decreased the K_m of HQ for MPO. While PH (100 μM) decreased the K_m of HQ for MPO by a factor of 4.2, CAT (20 μM) decreased the K_m of HQ for MPO by a factor of 7.7 (Table 2). The decrease in K_m was, however, also accompanied by a significant decrease in V_{max} by factors of 2.4 and 2.7 with PH and CAT respectively. The decrease in V_{max} was not as sharp as the decrease in K_m and thus the specificity (V_{max}/K_m) of MPO for HQ was raised in the presence of PH and CAT by a factor of 1.8 ($p < 0.05$) and 2.8 ($p < 0.05$) respectively (Table 2).

In summary, the results presented here indicate that HQ metabolism by purified human MPO is stimulated by PH at toxicologically relevant concentrations. In addition, CAT another metabolite of benzene, also stimulates the MPO-dependent metabolism of HQ. Determination of productive binding rates (V_{max}/K_m) of PH, HQ and CAT under identical conditions, indicated that HQ is metabolized by MPO at a rate of 16-fold and 5-fold faster than PH and CAT respectively. Thus, the oxidation of phenolics by MPO is primarily dependent upon their redox potentials but not on their affinities (K_m) for the enzyme. Furthermore, apparent K_m and V_{max} values obtained for PH metabolism by MPO seem to vary with the MPO and hydrogen peroxide concentrations (data not shown). The physiological significance of K_m values obtained for peroxidase-dependent metabolism of xenobiotics should therefore, be interpreted with caution. The results presented here also show that PH and CAT decrease the K_m and V_{max} values of HQ for MPO, but increase the specificity (V_{max}/K_m) of HQ for MPO. The increased specificity of HQ for MPO in the presence of PH and CAT could play some role in benzene-induced myelotoxicity and leukemogenicity.

ACKNOWLEDGEMENTS

This work was supported by NIH grants P42 ES04705 and P30 ESO1896.

REFERENCES

Aksoy, M. (1987). Benzene hematotoxicity; in *Benzene Carcinogenicity*. (Aksoy, M. ed.) pp. 59-112, CRC Press, New York.

Che-Fu Koo and Fridovich, I. (1987). Stimulation of the activity of horseradish peroxidase by nitrogenous compounds. *J. Biol. Chem.*, **263**, 3811-3817.

Cooper, K.R. and Snyder, R. (1987). Benzene metabolism. (Toxicokinetics and the molecular aspects of benzene toxicity); in *Benzene Carcinogenicity*. (Aksoy, M. ed.) pp. 33-58. CRC Press, New York.

Danner, D.J., Brignac, Jr. P.J. and Patel, Y. (1973). The oxidation of phenol and its reaction product by horseradish peroxidase and hydrogen peroxide. *Arch. Biochem. Biophys.* **156**, 759-763.

Eastmond, D.A., Smith, M.T., Ruzo, L.O. and Ross, D. (1986). Metabolic activation of phenol by human myeloperoxidase and horseradish peroxidase. *Molec. Pharmacol.*, **30**, 674-679.

Eastmond, D.A., Smith, M.T. and Irons, R.D. (1987). An interaction of benzene metabolites reproduces the myelotoxicity observed with benzene exposure. *Toxicol. Appl. Pharmacol.* **91**, 85-95.

Fox, R.L. and Purves, W.K. (1968). Mechanism of enhancement of IAA oxidation by 2,4-dichlorophenol. *Plant Physiol.* **43**, 454-456.

Job, D. and Dunford, H.B. (1976). Substituent effect on the oxidation of phenols and aromatic amines by horseradish peroxidase compound I. *Eur. J. Biochem.* **66**, 607-614.

Kariya, K., Lee, E., Hirouchi, M., Hosokawa, M. and Sayo, H. (1982). Purification and some properties of peroxidases of rat bone marrow. *Biochim. Biophys. Acta.* **99**, 95-101.

Pellack-Walker, P., Ken Walker, J., Evans, H.H. and Blumer, J.L. (1985). Relationship between the oxidation potential of benzene metabolites and their inhibitory effect on DNA synthesis in L5178YS cells. *Molec. Pharmacol.* **28**, 560-566.

Rickert, D.E., Baker, T.S., Bus, J.S., Barrow, C.S. and Irons, R.D. (1979). Benzene disposition in the rat after exposure by inhalation. *Toxicol. Appl. Pharmacol.* **49**, 417-423.

Sadler, A., Subrahmanyam, V.V. and Ross, D. (1988). Oxidation of catechol by horseradish peroxidase and human leukocyte peroxidase. *Toxicol. Appl. Pharmacol.* **93**, 62-71.

Smith, M.T., Yager, J.W., Steinmetz, K.L. and Eastmond, D.A. (1989). Peroxidase-dependent metabolism of benzene's phenolic metabolites and its poten?tial role in benzene toxicity and carcinogenicity. *Environ. Hlth. Perspect.* **82**, 23-39.

Steinmetz, K.L., Eastmond, D.A. and Smith, M.T. (1989). Hydroquinone metabolism by human myeloperoxidase is stimulated by phenol and several other compounds. *Toxicologist* **9**, abst # 1125.

Subrahmanyam, V.V. and O'Brien, P.J. (1985). Phenol oxidation products formed by a peroxidase-reaction that bind to DNA. *Xenobiotica* **15**, 873-885.

Subrahmanyam, V.V., Sadler, A., Suba, E. and Ross, D. (1989). Stimulation of *in vitro* bioactivation of hydroquinone by phenol in bone marrow cells. *Drug Metab. Disp.* **17**, 348-350.

Van Steveninck, J., Boegheim, J.P.J., Dubbelman, T.M.A.R. and Van der Zee, J. (1987). The mechanism of potentiation of horseradish peroxidase-catalyzed oxidation of NADPH by porphyrins. *Biochem. J.* **242**, 611-613.

ACTIVATION OF 1-HYDROXYMETHYLPYRENE TO AN ELECTROPHILIC AND MUTAGENIC METABOLITE BY RAT HEPATIC SULFOTRANSFERASE ACTIVITY

Young-Joon Surh, Judith C. Blomquist and James A. Miller

McArdle Laboratory for Cancer Research &
Environmental Toxicology Center
University of Wisconsin Medical School
Madison, WI 53706

INTRODUCTION

Hydroxylation of meso-methyl groups with subsequent formation of reactive benzylic esters bearing a good leaving group (e.g., sulfate, phosphate, acetate, etc.) has been proposed as a possible activation pathway in the DNA binding and carcinogenicity of some methyl-substituted polycyclic aromatic hydrocarbons (Flesher, et al., 1971; Flesher, et al., 1973). The metabolic formation of such reactive esters has recently been reported by Watabe et al. (1982, 1986, 1987; Okuda, H. et al., 1986). Their studies demonstrated the formation of electrophilic and mutagenic sulfuric acid ester metabolites from 7-hydroxymethyl-12-methylbenz[a]anthracene (HMBA) and related aromatic hydrocarbons by rat liver cytosolic sulfotransferase activity (reviewed in ref. 7). Non-enzymatic interaction of these reactive esters with the amino groups of guanine and adenine residues in calf thymus DNA produced the benzylic adducts. More recently we have noted the formation of such benzylic DNA adducts *in vivo* in the livers of infant rats and mice treated with HMBA, 6-hydroxymethylbenzo[a]pyrene (HMBP), or their electrophilic sulfuric acid esters (Surh, Y.-J. et al., 1987, 1989). The chemically synthesized sulfuric acid ester of 1-hydroxymethylpyrene (HMP) was also found to be highly mutagenic toward *Salmonella typhimurium* TA98 (Watabe, T. et al., 1982) and this intrinsic mutagenicity increased dramatically in the presence of chloride ion (Henschler, R., et al. 1989). However, the biological formation of such a labile and reactive sulfuric acid ester of HMP has not been reported.

In the present study, we examined the sulfotransferase activity for HMP in rat hepatic cytosols and the formation of benzylic DNA adducts from HMP and its sulfuric acid ester, 1-sulfooxymethylpyrene (SMP) in rat liver *in vivo* as well as *in vitro*. The possible role of chloride ion on the formation of these benzylic DNA adducts was also investigated.

MATERIALS AND METHODS

Chemicals

HMP was synthesized by reduction of 1-pyrenecarboxaldehyde with $LiAlH_4$ as reported previously (Rogan, E.G. et al., 1986) except that reaction was performed at room temperature for 20 min. [^{3}H]HMP (>98% radiochemical purity) was prepared from the aldehyde by treatment with [^{3}H]sodium borohydride (Amersham Co.,

Arlington Heights, IL). SMP was obtained by reaction of HMP with sulfuric acid using *N-N'*-dicyclohexylcarbodiimide (DCCI) as a condensing agent. Briefly, to a solution of DCCI (5mmol) in 10 ml DMF cooled in an ice bath was added HMP (1 mmol) in 10 ml DMF with vigorous stirring. Sulfuric acid (1.5 mmol) in 1.5 ml cold DMF was then slowly added to the above mixture and the reaction was continued at 0°C for 30 min. The solution was filtered and the filtrate was carefully neutralized with 2 M methanolic NaOH. The neutralized solution was further filtered and the organic solvent was removed by vacuum distillation. The residue was suspended in DMF (4 ml) and ethanol (4 ml) and insoluble salts were removed by centrifugation. To the supernatant was slowly added ten vol of dry ether with mild stirring. The white precipitate containing SMP (Na salt) was collected by centrifugation and rinsed repeatedly with dry ether. The structural identity of SMP was confirmed by UV, IR, and ^{1}H-NMR spectroscopy. The acetic acid ester of HMP was prepared according to the previously reported method (Rogan, E.G. et al., 1986). 1-Chloromethylpyrene (CMP) was obtained by reaction of HMP with thionyl chloride in dioxane at room temperature for 1 hr in the presence of zinc chloride as a catalyst (Squires, T.G. et al., 1975). 3'-Phosphoadenosine-5'-phosphosulfate (PAPS) and glutathione were products of Sigma Chemical Co. (St. Louis, MO) and Boeringer Mannheim biochemicals (Indianapolice, IN), respectively. Other chemicals and solvents were of reagent grade.

Sulfotransferase Assay

Determination of hepatic cytosolic sulfotransferase activity for HMP was performed as described previously (Surh, Y.-J. et al., 1987, 1989) using [^{3}H]guanosine as a nucleophilic acceptor. The resulting guanosine adduct was analyzed by radiomatic HPLC with a 25-min linear gradient of 60-100% methanol.

Bacterial Mutagenicity Assays

Sulfotransferase-mediated mutagenicity of HMP and the intrinsic mutagenicity of SMP were determined using *S. typhimurium* TA98 by a liquid preincubation method as performed by Watabe et al. (1982).

Effect of Chloride Ion and Other Anions on Chemical Reactivity of SMP

SMP (0.67 mM) was reacted with [^{3}H]guanosine (1.67 mM) at 37°C for 30 min in the presence or absence of an anion (154 mM) in a final volume of 150 µl H_2O. The resulting guanosine adduct was analyzed by reverse-phase HPLC under the same condition as described in the above sulfotransferase assay.

Benzylic DNA Adduct Formation from HMP and Its Sulfooxymethyl and Chloromethyl Derivatives In Vitro

For hepatic cytosolic sulfotransferase-mediated binding of HMP to DNA, the incubation mixture contained the hydrocarbon (0.1 µmol), calf thymus DNA (2 mg), liver cytosol (~ 2 mg protein), PAPS (0.1 µmol), $Mg(OAc)_2$ (10 µmol), and Bis-Tris buffer (100 µmol), pH 7.4 in a final volume of 2 ml. Incubations for non-enzymatic modification of DNA by SMP and CMP were carried out in the absence of cytosol and PAPS. After incubation at 37°C for 30 min, DNA was isolated and enzymatically hydrolyzed as described previously (Surh, Y.-J. et al., 1989). Benzylic DNA adducts extracted with chloroform: n-propanol (7:3, v/v) were analyzed by reverse-phase HPLC with a 30-min linear gradient of 60-100% MeOH. Adducts were monitored by following their fluorescence at 275 nm (emission, 389 nm) and their amounts were determined on the basis of fluorescence of standard adducts of known specific radioactivity.

Hepatic Benzylic DNA Adduct Formation from HMP, SMP, and CMP in Infant Male Rats

Male 12-day-old Sprague-Dawley rats were injected i.p. with 0.25 µmol of HMP or its derivatives/5 µmol DMSO/g body wt. Control animals received the same volume of the vehicle only. Animals were sacrificed 6 hr later and their livers were removed for isolation of nucleic acids. DNA isolation, digestion and analysis of benzylic DNA adducts were performed as described above for *in vitro* experiments.

Initiation of Mouse Skin Tumors

Groups of 30 female CD-1 mice received ten topical applications of the test compound in 0.2 ml of acetone: DMSO (85:15, v/v) on alternate days. Controls were treated with the vehicle only. One week after the last initiation dose, promotion was begun by applications twice weekly of 2.5 μg of 12-O-tetradecanoylphorbol-13-acetate (TPA) in 0.1 ml acetone for 22 weeks.

RESULTS AND DISCUSSION

In the present study we observed the existence of sulfotransferase activity for HMP in hepatic cytosols of infant male rats. The sulfotransferase activity for HMP, as determined by measuring the radioactivity (^3H) of the guanosine adduct of HMP, was completely dependent upon the presence of the cytosol and the sulfo-group donor, PAPS. Omission of either component from the incubation mixture resulted in no appreciable enzyme activity. As observed previously for other hydrocarbons such as HMBA (Surh, Y.-J. et al., 1987) and HMBP (Surh, Y.-J. et al., 1989), the hepatic cytosolic sulfotransferase activity for HMP was strongly inhibited by dehydroepiandrosterone (DHEA), a typical substrate for hydroxysteroid sulfotransferases (Table 1). On the contrary, pentachlorophenol which is a potent inhibitor of phenol sulfotransferases and hepatic sulfotransferase activities for several carcinogens including *N*-hydroxy-2-aminofluorene (Lai, C.-C. et al., 1985) and 1'-hydroxysafrole (Boberg, E.W. et al., 1983) did not show any significant inhibitory effect.

The activation of HMP by hepatic sulfotransferase activity was further confirmed by examining the PAPS plus S-105-dependent mutagenicity of HMP in bacteria. As shown in Table 2, HMP induced His^+ revertants in the TA98 strain of *S. typhimurium* in the presence of both hepatic cytosol and PAPS. This sulfotransferase-mediated bacterial mutagenicity of HMP was significantly inhibited by DHEA (Table 2), an inhibitor of hepatic cytosolic sulfotransferase activity for the hydrocarbon (Table 1). As previously reported by others (Watabe, T. et al., 1982; Henschler, R., et al., 1989), the chemically synthesized sulfuric acid ester of HMP was directly mutagenic toward *S. typhimurium* TA98. This intrinsic mutagenicity was significantly inhibited by glutathione (GSH) in the presence of hepatic cytosol (Figure 1); GSH alone was not effective in blocking the mutagenicity of SMP. The antimutagenic activity of GSH in the presence liver cytosol was markedly blocked by bromosulfophthaleine, a general inhibitor of GSH-*S*-transferase activity. Therefore, it is likely that the inactivation of SMP by GSH is catalyzed by GSH-*S*-transferase(s).

Table 1. Inhibition of Sulfotransferase Activity for HMP[a]

Inhibitor	Conc. (μM)	Sulfotransferase activity[b]	% Inhibition
None	-	200 ± 47	-
PCP	100	178 ± 14	13
DHEA	100	41 ± 7[c]	80

a The incubation was performed at 37°C for 30 min in a final volume of 150 μl containing $Mg(OAc)_2$ (1.0 μmol), Bis-Tris buffer (10 μmol), pH 7.4, [8-^3H]guanosine (0.25 μmol), 25 μl of liver cytosols from 12-day-old-male rats (20-30 mg protein/ml), PAPS (0.1 μmol) and HMP (0.025 μmol) in 5 μl DMSO in the presence or absence of a sulfotransferase inhibitor.

b Enzyme activity is expressed as pmol of [^3H]guanosine adduct produced/mg cytosolic protein/30 min.

c Statistically significant ($P<0.01$).

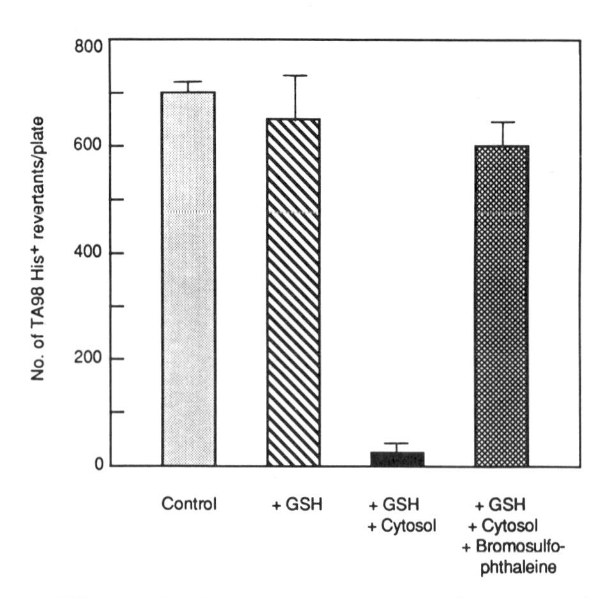

Figure 1. Effect of GSH and GSH-S-transferase activity for intrinsic mutagenicity of SMP in *S. typhimurium* TA98. The standard incubation mixture contained SMP (0.5 μM) and 2.6 x 10^9 bacteria in a final volume of 0.5 ml phosphate buffer (0.1 M), pH 7.4. Where needed, GSH (2 mM), cytosol from 12-day-old male rat liver (1.4 mg protein), or bromosulfophthaleine (1 mM) was added to the above incubation mixture. After 30 min incubation at 37°C, the whole content was mixed with soft agar and further incubated for 48 hours.

Table 2. Sulfotransferase-Mediated Mutagenicity of HMP in *S. typhimurium* TA98

Experimental conditions	Number of His $^+$ revertants/plate
Complete[a]	4296 ± 465
Boiled cytosol replacement	39 ± 3
- PAPS	28 ± 5
+ PCP (50 μM)	5304 ± 685
+ DHEA (50 μM)	2931 ± 211

[a] The complete incubation mixture contained HMP (2μM), $MgCl_2$ (3 mM), EDTA (0.1 mM), PAPS (0.2 mM), rat liver cytosol (1.4 mg protein) and 4 x 10^9 bacteria in a final volume of 0.5 ml phosphate buffer (0.1 M), pH 7.4.

Incubation of HMP with calf thymus DNA in the presence of PAPS and hepatic cytosols from 12-day-old male rats resulted in the formation of benzylic adducts with deoxyguanosine and deoxyadenosine residues in the DNA (Figure 2). The same adducts were produced in higher yield by non-enzymatic reactions of SMP and CMP with calf thymus DNA. Although CMP was found to be more mutagenic than SMP in *S. typhimurium* TA98 (Henschler, R. et al., 1989), both compounds showed the similar degree of chemical reactivity as determined by formation of above benzylic DNA adducts (Figure 2). The similar result was obtained when their reactivities were determined on the basis of [^{3}H]guanosine adduct formation (data not shown). In addition, the intrinsic chemical reactivity of SMP was not significantly affected by

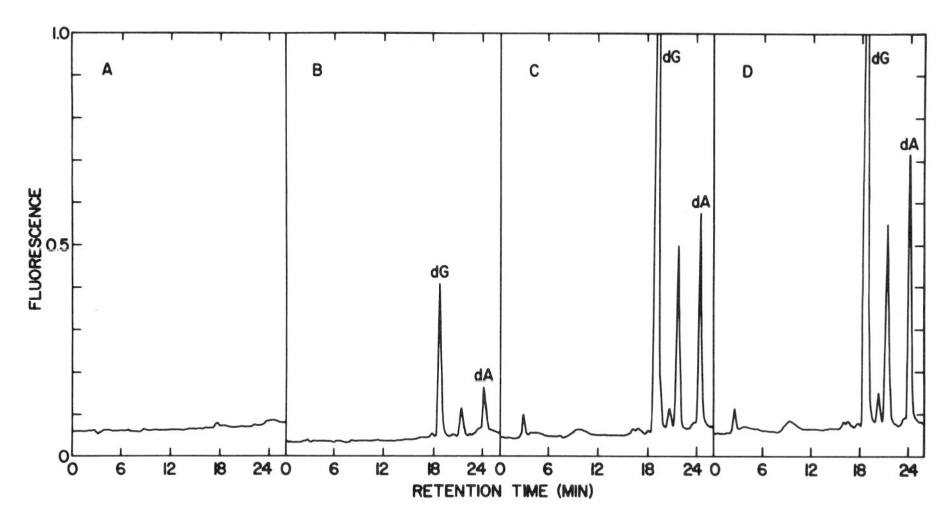

Figure 2. HPLC analysis of benzylic adducts from hydrolysates of 0.06 mg calf thymus DNA incubated with HMP and hepatic cytosol in the absence (A) or presence (B) of PAPS or reacted non-enzymatically with SMP (C) or CMP (D) only. dG and dA represent the deoxyguanosine and deoxyadenosine adducts, respectively. The amounts of the major benzylic DNA adduct (dG) in terms of pmol/mg DNA were 644 for (B), 3246 for (C) and 3781 for (D).

addition of chloride ion (Table 3). Based on these results, it is unlikely that the increased bacterial mutagenicity of SMP in the presence of chloride ion, previously reported by Henschler et al. (1989), is due to the increased chemical reactivity of CMP formed by reaction of SMP with Cl^-.

Table 3. Effect of Chloride Ion and Other Anions on the Reactivity of SMP

Anion	Reactivity as determined by formation of [^{3}H]guanosine adduct (pmol/30 min)[a]
None	1649 ± 165
Cl^-	1447 ± 276
Br^-	1184 ± 225[b]
I^-	267 ± 65[c]
CH_3COO^-	1539 ± 351

a The reactivity of SMP in the presence or absence of various anions was determined as described under Materials and Methods.

b.c Statistically significant ($P<0.05^b$; $P<0.01^c$).

I.p. injection of HMP (0.25 μmol/g body wt.) in preweanling male rats produced the benzylic adducts with deoxyguanosine and deoxyadenosine residues in the liver DNA (Figure 3). The formation of these benzylic adducts which account for about 60 to 70% of total HMP residues bound to the hepatic DNA was significantly inhibited by pretreatment of rats with DHEA (~ 45% inhibition). Both SMP and CMP produced higher levels of the hepatic benzylic DNA adducts than did the same dose of the parent hydroxymethyl hydrocarbon (Figure 3), but the chloromethyl derivative formed 6-7 times fewer adducts than the sulfuric acid ester (Table 4) in contrast to its significantly higher mutagenicity than SMP in bacteria. This difference in the levels

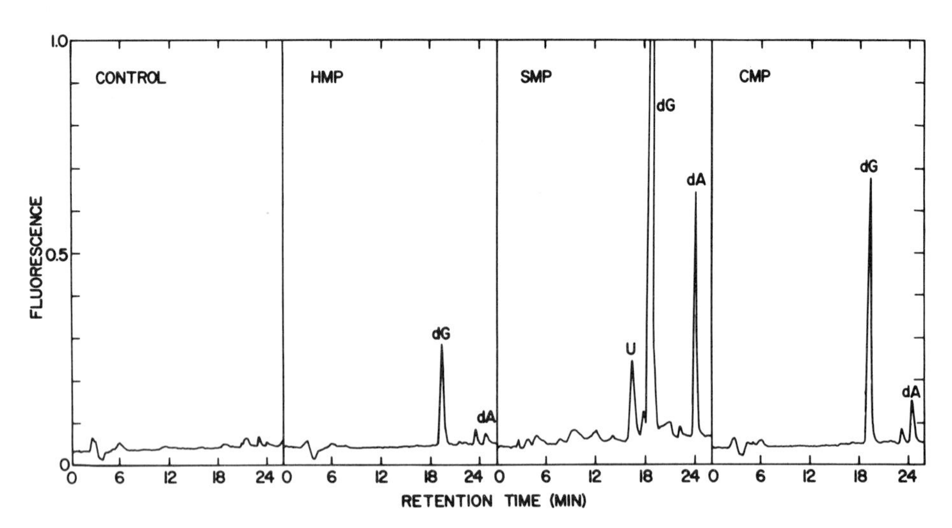

Figure 3. HPLC profiles of hydrolysates of the hepatic DNAs from 12-day-old male rats injected i.p. with only the vehicle, HMP, SMP, or CMP. Benzylic DNA adducts were extracted and aliquots corresponding to 0.4-0.5 mg hepatic DNA were analyzed as described in the text. dG and dA are abbreviated as described in the legend of Figure 2. U denotes an uncharacterized adduct.

of benzylic DNA adducts formed from SMP and CMP in rat liver *in vivo* appears to be associated with their different physico-chemical properties. Thus, as pointed out by J.A. Miller et al. elsewhere in this volume, it may be possible that the greater lipophilicity of the less polar chloromethyl derivative leads to the more solubilization in the fat depots in surrounding tissues, thereby decreasing the concentration of reactive species which ultimately interact with the hepatic DNA. In any case, a part of hepatic benzylic DNA adducts formed from HMP by metabolic activation *via* SMP might be derived from the chloromethyl hydrocarbon if such derivative could be generated from SMP as a result of interaction with chloride ion in the liver. This concept is represented in the Figure 4.

In addition to the sulfuric acid ester metabolite, other electrophilic benzylic esters of HMP might be formed *in vivo* to produce the benzylic DNA adducts. The acetic acid ester of HMP was found to be much less chemically reactive and

Table 4. Comparative Benzylic DNA Adduct Formation from HMP, SMP and CMP in Rat Liver *In Vivo*

Hydrocarbon	Dose (μmol/g body wt.)	Benzylic dGuo adduct (pmol/mg hepatic DNA)
HMP	0.25	94 ± 20
SMP	0.25	1458 ± 495
CMP	0.25	222 ± 82

Animals were killed 6 hr after treatment of hydrocarbons and hepatic DNA was isolated and digested. Benzylic DNA adducts were analyzed by HPLC as described under the Materials and Methods.

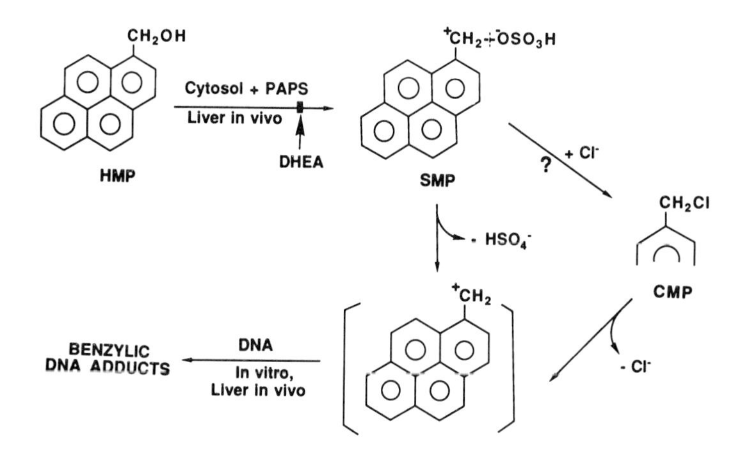

Figure 4. Proposed pathway for the formation of benzylic DNA adducts from HMP via an electrophilic sulfuric acid ester metabolite or possibly via a chloromethyl derivative in rat liver *in vivo* and in reactions *in vitro.*

mutagenic in bacteria than the sulfuric acid ester (Surh et al., unpublished observation). In addition, the carcinogenicity of 1-acetoxymethylpyrene (AMP), in terms of initiation of mouse skin tumors, was not significantly different from that of the solvent-treated controls (Table 5). the phosphoric acid ester of HMP might also be involved in the formation of benzylic DNA adducts *in vivo* if its electrophilicity is high enough to interact with cellular nucleophiles. The non-enzymatic formation of a benzylic phosphoric acid ester by trans-esterification of the corresponding sulfuric acid ester in the phosphate buffer has been reported very recently by Watabe and his co-workers (Okuda, H. et al., 1989). Their studies demonstrated that the nucleophilic attack of phosphate ion on the sulfuric acid ester of 9-hydroxymethyl-10-methylanthracene produced the corresponding phosphoric acid ester which was found to be as mutagenic in bacteria as 9-sulfooxymethyl-10-methylanthracene. Further studies are needed to elucidate if a similar phosphoric acid ester is formed from HMP or SMP.

Table 5. Comparative Activities of HMP and its Acetic Acid and Sulfuric Acid Esters for Initiation of Mouse Skin Tumors

Initiator	Dose[a] (μmol)	No. of mice at 22 wk	% tumor bearing mice	Average number of papillomas/mouse
HMP	0.5 x 10	30	10	0.2 ± 0.5
	0.25 x 10	30	3	0.03 ± 0.2
AMP	0.5 x 10	30	10	0.1 ± 0.3
	0.25 x 10	30	10	0.1 ± 0.4
SMP	0.5 x 10	30	30	0.5 ± 0.9[b]
	0.25 x 10	30	23	0.4 ± 0.8[b]
Solvent only		30	10	0.1 ± 0.3

a Applied topically as 10 subdoses on alternate days.
b Significantly different fromthe solvent control ($P<0.05$).

The ultimate goal of our present research is to demonstrate the role of the electrophilic sulfuric acid metabolites in carcinogenesis by parent hydrocarbons. However, it is difficult to determine the carcinogenicity of water-labile sulfuric acid esters formed metabolically in internal organs such as the liver. Alternatively, direct application of non-toxic doses of these short-lived esters to the skin could be considered. When topically applied on the mouse skin, the sulfuric acid ester of HMP revealed a weak tumor-initiating activity, but it was more active than the parent hydroxymethyl hydrocarbon in this regard (Table 5).

REFERENCES

Boberg, E.W., Miller, E.C., Miller, J.A., Poland, A. and Liem, A. (1983). Strong evidence from studies with brachymorphic mice and pentachlorophenol that 1'-sulfooxysafrole is the major ultimate electrophilic and carcinogenic metabolite of 1'-hydroxysafrole in mouse liver. *Cancer Res.* **43,** 5163-5173.

Flesher, J.W. and Sydnor, K.L. (1971). Carcinogenicity of derivatives of 7,12-dimethylbenz[a]anthracene. *Cancer Res.* **31,** 1951-1954.

Flesher, J.W. and Sydnor, K.L. (1973). Possible role of 6-hydroxymethylbenzo[a]-pyrene as a proximate carcinogen of benzo[a]pyrene and 6-methylbenzo[a]pyrene. *Int. J. Cancer* **11,** 433-437.

Henschler, R., Seidel, A. and Glatt, H.R. (1989). Phase-II-metabolite genotoxicity of polycyclic aromatic hydrocarbons: a chloromethyl derivative as a mutagenic intermediate from a benzylic sulfate ester. *Naunyn-Schmiedeberg's Arch. Pharmacol.* **339** (Suppl.), R26.

Henschler, R., Pauly, K., Godtel, U., Seidel, a., Oesch, F. and Glatt, H.R. (1989). The mutagenicity of sulfate esters in *Salmonella typhimurium* is strongly affected by the ions present in the exposure medium. *Mutagenesis* **4,** 309 (GUM abstracts).

Lai, C.-C., Miller, J.A., Miller, E.C. and Liem, A. (1985). N-Sulfooxy-2-amino-fluorene is the major ultimate electrophilic and carcinogenic metabolite of N-hydroxy-2-acetylaminofluorene in the livers of infant male C57BL/6J x C3H/HeJF_1 (B6C3F_1) mice. *Carcinogenesis* **6,** 1037-1045.

Okuda, H., Hiratsuka, A., Nojima, H. and Watabe, T (1986). A hydroxymethyl sulfate ester as an active metabolite of the carcinogen, 5-hydroxymethylchrycene. *Biochem. Pharmacol.* **35,** 535-538.

Okuda, H. and Watabe, T. (1989). Formation and metabolism of the active metabolite, 9-hydroxymethyl-10-methylanthracene phosphate, in rat liver. *Proc. Jpn. Cancer Assoc.* **44,** (abstract #158).

Okuda, H., Yoshioka, S. and Watabe, T. (1989). Sulfotransferase-mediated activation of 9-hydroxymethyl-10-methylanthracene, a major metabolite of the carcinogen 9,10-dimethylanthracene. *Mut. Res.* **216,** 372 (abstract #38).

Rogan, E.G., Cavalieri, E.L. Walker, B.A., Balasubramanian, R., Wislocki, P.G., Roth, R.W. and Saugier, R.K. (1986). Mutagenicity of benzylic acetates, sulfates and bromides of polycyclic aromatic hydrocarbons. *Chem. -Biol. Interact.* **58,** 253-275.

Squires, T.G., Schmidt, W.W. and McCandlish, C.S., Jr. (1975). Zinc chloride catalysis in the reaction of thionyl halides with aliphatic alcohols. *J. Org. Chem.* **40,** 134-136.

Surh, Y.-J., Lai, C.-C., Miller, J.A. and Miller, E.C. (1987). Hepatic DNA and RNA adduct formation from the carcinogen 7-hydroxymethyl-12-methylbenz[a]anthracene and its electrophilic sulfuric acid ester metabolite in preweanling rats and mice. *Biochem. Biophys. Res. Commun.* **144,** 576-582.

Surh, Y.-J., Liem, A., Miller, E.C. and Miller, J.A. (1989). Metabolic activation of the carcinogen 6-hydroxymethylbenzo[a]pyrene: formation of an electrophilic sulfuric acid ester and benzylic DNA adducts in rat liver *in vivo* and in reactions *in vitro*. *Carcinogenesis* **10,** 1519-1528.

Watabe, T., Ishizuka, T., Isobe, M. and Ozawa, N. (1982). A 7-hydroxymethyl sulfate ester as an active metabolite of 7,12-dimethylbenz[a]anthracene. *Science* **215,** 403-405.

Watabe, T., Hakamata, Y., Hiratsuka, A. and Ogura, K. (1986). A 7-hydroxymethyl sulfate ester as an active metabolite of the carcinogen, 7-hydroxymethylbenz[a]-anthracene. *Carcinogenesis* **7,** 207-214.

Watabe, T., Hiratsuka, A. and Ogura, K. (1987). Sulfotransferase-mediated covalent binding of the carcinogen, 7,12-dihydroxymethylbenz[a]anthracene to calf thymus DNA and its inhibition by glutathione transferase. *Carcinogenesis* **8,** 445-453.

Watabe, T., Ogura, K., Okuda, H. and Hiratsuka, A. (1989). Hydroxymethyl sulfate esters as reactive metabolites of the carcinogens, 7-methyl- and 7,12-dimethylbenz[a]anthracenes and 5-methylchrycene. In *Xenobiotic Metabolism and Disposition* (Kato, R., Estabrook, R.W. and Cayen, M.N. eds.), pp. 393-400, Taylor and Francis, London.

BIOACTIVATION OF 2,6-DI-TERT-BUTYL-4-METHYL PHENOL (BHT) AND HYDROXYLATED ANALOGUES TO TOXIC QUINOID METABOLITES

John A. Thompson, Judy L. Bolton, Kathleen M. Schullek and Hubert Sevestre

Molecular Toxicology and Environmental Health Sciences Program,
School of Pharmacy
University of Colorado
Boulder, Colorado 80309-0297

Phenolic antioxidants produce a variety of effects in animals. For example, butylated hydroxytoluene (BHT) has either anticarcinogenic or tumor promoting activity, depending on the dosing regimen (Malkinson, 1983). BHT also produces acute pulmonary toxicity in mice (Witschi et al., 1989) and hepatotoxicity in rats (Mizutani et al., 1987), and these effects are believed to be due to its oxidative biotransformation to reactive metabolites. As shown in Figure 1, mouse liver and lung cytochrome P-450's catalyze the metabolism of BHT along three pathways: (A) hydroxylation of the 4-methyl group (forming BHT-MeOH), (B) hydroxylation of a tert-butyl group (forming BHT-OH) and (C) oxidation of the pi-electron system to a phenoxy radical that leads to quinoid products (Thompson et al., 1987). Loss of a second electron from this radical generates the quinone methide (BHT-QM), and combination with molecular oxygen yields the peroxyquinol (BHT-PQ) that is rapidly degraded by P-450 to a variety of products (Wand and Thompson, 1986).

We have shown that the hydroxylated metabolite BHT-OH is considerably more toxic to mouse lung than either BHT or BHT-MeOH (Malkinson et al., 1989; Thompson et al., 1989). In previous work, it was shown also that BHT-PQ is substantially more cytotoxic to isolated rat hepatocytes than BHT (Thompson and Ross, 1988). The following report summarizes our investigation into the contributions of the quinone methide and peroxyquinol derivatives of BHT-OH (Figure 1) to the high cytotoxicity exhibited by this BHT metabolite.

MATERIALS AND METHODS

BHT, BHT-MeOH and DBQ were purchased from Aldrich. The synthesis of BHT-OH was described previously (Thompson et al., 1987), and *iso*BHT-OH was a side-product of this synthesis. The conversion of BHT to BHT-PQ has been described (Thompson and Wand, 1985), and the conversions of BHT-OH and *iso*BHT-OH to their respective peroxyquinols was accomplished by reaction with singlet oxygen (Adam and Lupon, 1988). All products were purified by semi-preparative HPLC on a 10 x 250 mm column containing Ultrasphere-ODS (Beckman) with acetonitrile-water gradients. Quinone methides were synthesized and assayed as their glutathione conjugates (Bolton et al., 1990). Hepatocytes were isolated from male Sprague Dawley rats (Moldeus et al, 1978) and assayed for viability by trypan blue exclusion. The viability prior to experiments was in the range 82-88%. Incubations were conducted by adding substrates as solutions in 10 ul of DMSO to each 1.0 ml of hepatocyte suspension (10^6

Biological Reactive Intermediates IV, Edited by C.M. Witmer *et al.*
Plenum Press, New York, 1990

cells/ml) and incubating under O_2:CO_2 (95:5) at 37°C for the indicated times. Control incubations were conducted by omitting the test compound, but including DMSO. Glutathione was measured by HPLC as described (Reed et al., 1980). Rat liver microsomes were prepared from phenobarbital-treated animals and incubations conducted as detailed earlier (Bolton et al., 1990).

RESULTS AND DISCUSSION

As shown in Figure 2, BHT caused very little toxicity to rat hepatocytes at a concentration of 1 mM. In contrast, 0.2 mM BHT-OH killed all of the cells within 60 min. The fact that BHT-MeOH exhibited a very low toxicity indicates that the toxic effect of BHT-OH is not simply due to its increased hydrophilicity relative to BHT. In order to probe the mechanism of cytotoxicity, two reactive metabolites of BHT-OH were investigated, the electrophilic quinone methide and the peroxyquinol which gives rise to free radicals (Wand and Thompson, 1986).

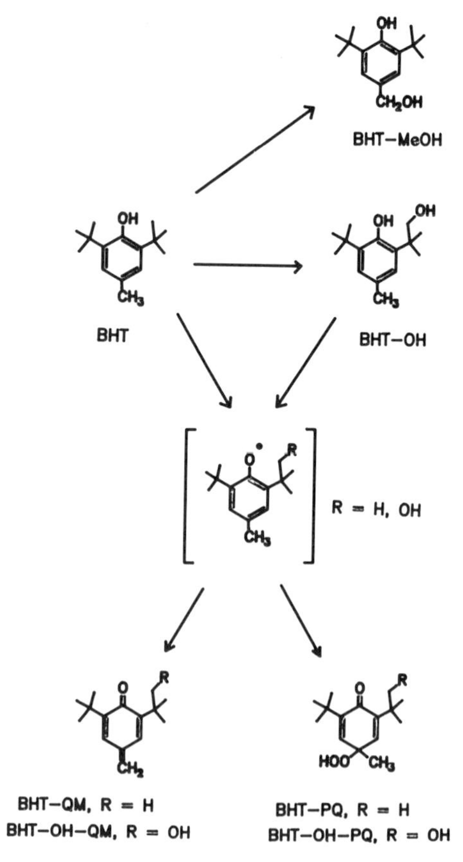

Figure 1. Partial scheme for the oxidative metabolism of BHT and BHT-OH. Abbreviations used in the text are included below each compound.

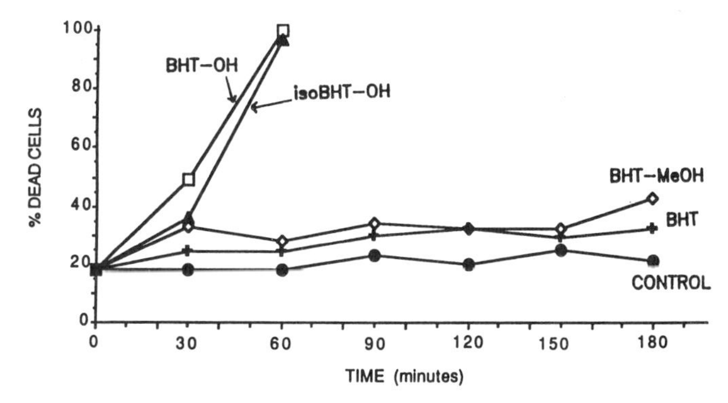

Figure 2. Effects of BHT and hydroxylated analogues on the viability of freshly isolated rat hepatocytes as a function of incubation time. The concentrations of the phenols were: BHT, 1.0 mM; BHT-MeOH, 1.0 mM; BHT-OH, 0.20 mM; *iso*BHT-OH, 0.30 mM.

The conversion of several phenols to their respective quinone methides by hepatic microsomes from phenobarbital-treated rats is shown in Figure 3. Clearly, quinone methides are important metabolites of the two most toxic phenols. BHT-QM and BHT-OH-QM were synthesized and their reactivities measured by monitoring the disappearance of the quinone methide chromophore at 287 nM (Figure 4) with the nucleophiles methanol and glutathione. The pseudo-first-order rate constants for reaction with methanol were 0.12×10^{-3} and 2.9×10^{-3} s^{-1} for BHT-QM and BHT-OH-QM, respectively, and 0.35×10^{-2} and 2.0×10^{-2} s^{-1} for reaction with glutathione. These results demonstrate that the hydroxylated quinone methide is several-fold more reactive as an electrophile than BHT-QM. The high reactivity has been rationalized by intramolecular hydrogen bonding between the side-chain hydroxyl and the ring oxygen, which stabilizes the aromatic resonance form resulting in increased positive charge density on the exocyclic methylene group relative to BHT-QM (Bolton et al., 1990). Direct spectral evidence for intramolecular hydrogen bonding in BHT-OH-QM has been obtained also. The data in Table 1 demonstrate that 0.2 mM BHT-OH depletes hepatic

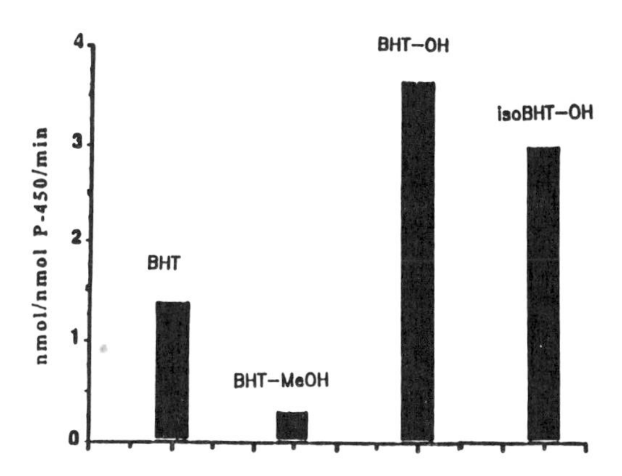

Figure 3. Oxidation of BHT and hydroxylated analogues (indicated above each bar) to quinone methides by rat liver microsomes. The products were analyzed by trapping with ^{3}H-glutathione and isolation of the adducts from incubation mixtures by HPLC methods.

glutathione more efficiently than does 3.0 mM BHT, an expected result based on the formation of a more reactive quinone methide from the former phenol.

Oxidation of BHT-OH to the reactive quinone methide BHT-OH-QM could explain the high toxicity of this phenol both in mouse lung and in rat hepatocytes. To test this hypothesis further, the structural analog *iso*BHT-OH (Figure 5) was prepared. This compound possesses the same relationship as BHT-OH between the side-chain hydroxyl and the ring oxygen favoring intramolecular hydrogen bonding, but it is a secondary rather than a primary alcohol so oxidation to an aldehyde cannot occur. *iso*BHT-OH also was oxidized to a quinone methide in microsomes (Figure 3) and produced similar cytotoxicity as BHT-OH (Figure 2). These results confirm the importance of a side-chain hydroxyl in quinone-methide-induced toxicity and, furthermore, indicate that oxidation of the hydroxyl of BHT-OH to an aldehydic group (Figure 5) is unlikely to be involved in its bioactivation.

In order to test the role of peroxyquinol formation in cytotoxicity, peroxyquinols derived from BHT, BHT-OH and *iso*BHT-OH were prepared and incubated with rat hepatocytes. The data in Table 2 demonstrate that these hydroperoxides cause a similar degree of cytotoxicity. BHT is substantially less toxic than the peroxyquinols, but in contrast, BHT-OH is more toxic than the peroxy compounds. These results indicate that quinone methide formation rather than peroxyquinol formation is the

Table 1. Depletion of reduced glutathione (GSH) in hepatocytes by BHT and hydroxylated analogues

Compound	Concentration (mM)	Time required to deplete GSH (min)
BHT	3.0	120
BHT-OH	0.20	60
*iso*BHT-OH	0.30	65

Compounds were incubated with isolated rat hepatocytes (10^6 cells/ml), and GSH concentrations were measured at 0, 5, 15, 30, 60, 90 and 120 min. GSH levels did not decline in the control incubation over 120 min.

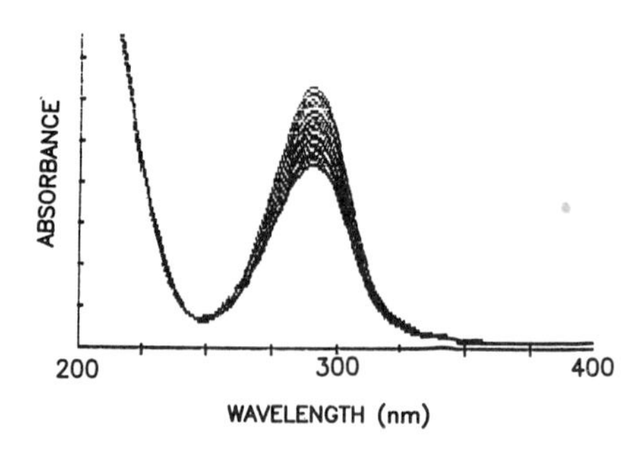

Figure 4. Ultraviolet absorbance of BHT-QM in 1:1 methanol:water. The rate of disappearance of BHT-QM was determined by recording repetative scans (1 s) at 60 s intervals with a diode array spectrophotometer.

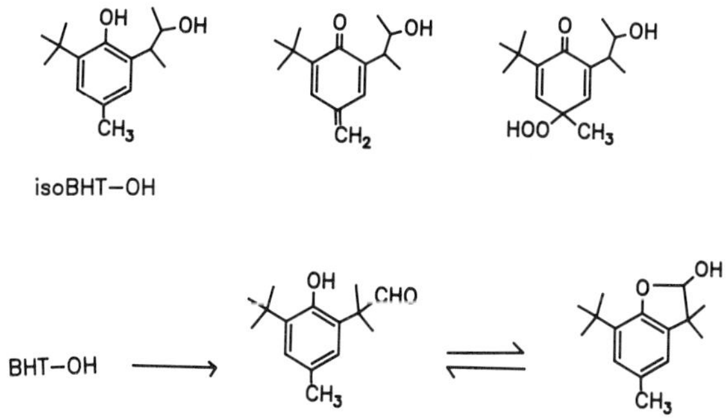

Figure 5. Structures of *iso*BHT and its quinoid products, and the aldehyde formed by oxidation of the primary hydroxyl group of BHT-OH. The latter would be expected to form an intramolecular hemiacetal.

Table 2. Cytotoxicity of phenols and peroxyquinols in isolated hepatocytes

Compound	Concentration (mM)	% Dead cells in 60 min
Control	0	18
phenols		
BHT	3.0	32
BHT-OH	0.20	100
*iso*BHT-OH	0.30	97
peroxyquinols		
BHT-PQ	0.25	84
BHT-OH-PQ	0.40	60
*iso*BHT-OH-PQ	0.50	70

Compounds were incubated with isolated rat hepatocytes (10^6 cells/ml), and cell viability measured by the ability of cells to exclude trypan blue after 60 min. BHT-OH killed 80% of the cells in 30 min.

principal route of toxicity for BHT-OH and *iso*BHT-OH. The low toxicity of BHT in rat hepatocytes could be due to conversion to the less reactive quinone methide BHT-QM.

In summary, the toxicity of BHT in mouse lung appears to be due to hydroxylation of a tert-butyl substituent producing BHT-OH. We have shown previously that BHT-OH is substantially more toxic than BHT to mouse lung (Malkinson et al., 1989; Thompson et al., 1989), and have now extended this comparison to rat hepatocytes. The hydroxylation pathway is nearly absent in rats, thereby explaining the facts that BHT does not cause pulmonary toxicity in this species, and that BHT is only mildly toxic to rat hepatocytes. Cytochrome P-450's from both species, however, oxidize BHT-OH to a quinone methide which is more electrophilic than BHT-QM, thereby providing an explanation for the high toxicity of BHT-OH (Bolton et al., 1990). Evidence was obtained that the enhanced reactivity of BHT-OH-QM is due to intramolecular hydrogen-bonding between the side-chain hydroxyl and the ring oxygen. An analogue of BHT-OH with a secondary hydroxyl group, *iso*BHT-OH, also forms a quinone methide that is activated by intramolecular hydrogen bonding, and also is effective at killing hepatocytes. This result demonstrates further that the toxicity of BHT-OH is due to formation of the activated quinone methide, and that oxidation of BHT-OH to an aldehyde-hemiacetal mixture (Figure 5) is not an important pathway of bioactivation. Also, evidence was obtained that the peroxyquinols are not the primary toxic metabolites of BHT-OH and *iso*BHT-OH, as these hydroperoxy derivatives were less toxic to rat hepatocytes than the parent phenols. These results further substantiate the proposal that quinone methide formation is the route of bioactivation leading to the cytotoxicity of BHT-OH both in mouse lung and rat hepatocytes.

ACKNOWLEDGEMENTS

We thank Dr. David Ross for assistance with the hepatocyte preparations. This research was supported by USPHS grant CA33497.

REFERENCES

Adam, W. and Lupon, P. (1988). Quinol epoxides from p-cresol and estrone by photooxygenation and titanium(IV)- or vanadium(V)-catalyzed oxygen transfer. *Chem. Ber.* **121**, 21-25.

Bolton, J.L., Sevestre, H., Ibe, B. and Thompson, J.A. (1990). Formation and reactivity of alternative quinone methides from butylated hydroxytoluene: possible explanation for species-specific pneumotoxicity. *Chem. Res. Toxicol.* **3**, in press.

Malkinson, A.M. (1983). Putative mutagens and carcinogens in foods. III. Butylated Hydroxytoluene (BHT). *Environ. Mutagen.* **5**, 353-362.

Malkinson, A.M., Thaete, L.G., Blumenthal, E.J. and Thompson, J.A. (1989). Evidence for a role of tert-butyl hydroxylation in the induction of pneumotoxicity in mice by butylated hydroxytoluene. *Toxicol. Appl. Pharmacol.* **101**, 196-2042.

Mizutani, T., Nomura, H., Nakanishi, K. and Fujita, S. (1987). Hepatotoxicity of butylated hydroxytoluene and its analogs in mice depleted of hepatic glutathione. *Toxicol. Appl. Pharmacol.* **87**, 166-176.

Moldeus, P., Hogberg, J. and Orrenius, S. (1978). Isolation and use of liver cells. In *Methods in Enzymology* (S. Fleischer and L. Packer, Eds.), Vol. 52, pp. 60-71. Academic Press, New York.

Reed D.J., Babson, J.R., Beatty, P.W., Brodie, A.E., Ellis, W.W. and Potter, D.W. (1980). High-performance liquid chromatography analysis of nanomole levels of glutathione, glutathione disulfide, and related thiols and disulfides. *Anal. Biochem.* **106**, 55-62.

Thompson, J.A. and Wand, M.D. (1985). Interaction of cytochrome P-450 with a hydroperoxide derived from butylated hydroxytoluene: mechanism of isomerization. *J. Biol. Chem.* **260**, 10637-10644.

Thompson, J.A., Malkinson, A.M., Wand, M.D., Mastovich, S.L., Mead, E.W., Schullek, K.M. and Laudenschlager, W.G. (1987). Oxidative metabolism of butylated hydroxytoluene by hepatic and pulmonary microsomes from rats and mice. *Drug Metab. Dispos.* **15**, 833-840.

Thompson, J.A. and Ross, D. (1988). Metabolism and toxicity of 2,6-di-t-butyl-4-hydroperoxy-4-methylcyclohexadienone (BHT-OOH) in isolated rat hepatocytes. *FASEB J.* **2**, A800.

Thompson, J.A., Schullek, K.M., Fernandez, C.A. and Malkinson, A.M. (1989). A metabolite of butylated hydroxytoluene with potent tumor-promoting activity in mouse lung. *Carcinogenesis* **10**, 773-775.

Wand, M.D. and Thompson, J.A. (1986). Cytochrome P-450-catalyzed rearrangement of a peroxyquinol derived from butylated hydroxytoluene. Involvement of radical and cationic intermediates. *J. Biol. Chem.* **261**, 14049-14056.

Witschi, H.P., Malkinson, A.M. and Thompson, J.A. (1989). Metabolism and pulmonary toxicity of butylated hydroxytoluene (BHT). *Pharmacol. Ther.* **42**, 89-113.

FURTHER EVIDENCE FOR THE ROLE OF MYELOPEROXIDASE IN THE ACTIVATION OF BENZO[A]PYRENE-7,8-DIHYDRODIOL BY POLYMORPHONUCLEAR LEUKOCYTESM

A. Trush, R. L. Esterline, W. G. Mallet, D. R. Mosebrook, and L. E. Twerdok

Department of Environmental Health Sciences
Division of Toxicological Sciences
Johns Hopkins School of Hygiene and Public Health
615 N. Wolfe St.
Baltimore, MD 21205

The carcinogenic initiating activity of benzo(a)pyrene has been attributed to the generation of a stereospecific bay region diolepoxide of benzo(a)pyrene-7,8-dihydrodiol (BP)diol), anti-diolepoxide 2 (Conney, 1982). Through a myeloperoxidase (MPO)-mediated reaction, polymorphonuclear leukocytes (PMNs) can activate BP-diol to a chemiluminescent dioxetane derivative and a genotoxic electrophilic intermediate, presumably BP-diolepoxide (Trush et al., 1985; Kensler et al., 1987). This study examines the ability of PMNs or MPO to generate diolepoxides from (±)BP-diol as well as the relationship between the oxidative capability of PMNs and their ability to activate BP-diol.

METHODS

PMNs were isolated from either human blood or the peritoneal cavity of rats or mice as previously described (Dix, T.A. et al., 1982; Esterline, R.L. et al., 1989). Human MPO was purchased from CalBiochem. The methods for assessing superoxide (0_2^-) generation, MPO activity, BP-diol chemiluminescence and covalent binding have been described previously (Trush et al., 1985; Kensler et al., 1987; Esterline et al., 1989). Diolepoxide formation was assessed indirectly through the formation of tetraols using a Waters HPLC system and a methanol/H_20 gradient (Dix and Marnett, 1983). Tetraols are stable products resulting from diolepoxide hydrolysis.

RESULTS AND DISCUSSION

As shown in Figure 1, the interaction of (±) BP-diol (5μM) with TPA-stimulated human PMNs ($1x10^7$) or a human MPO system (2.1 units MPO, 400 μM H_20_2) resulted in the generation of tetraols derived primarily from anti-diolepoxides (BPDE). The anti/syn ratio with PMNs or the MPO system was 6.5 and 5, respectively. Such a ratio is indicative of the epoxidation of BP-diol via a peroxidative reaction (Dix and Marnett, 1983), and in this particular case probably MPO. Accordingly, addition of azide, a MPO inhibitor, was effective in decreasing tetraol fromation from BP-diol with both PMNs and the MPO system (data not shown).

			TETRAOL FORMATION (pmol) PMNs	MPO
anti-BPDE	7,10/8,9-BaP-tetraol	+ 7/8,9,10-BaP-tetraol	116	146
syn-BPDE	7,9/8,10-BaP-tetraol	+ 7,9,10/8-BaP-tetraol	19	26

Figure 1. Formation of PB-7,8-diolepoxide-derived (BPDE) tetraols resulting from the interaction of BP-7,8-diol with human PMNs or a human myeloperoxidase (MPO) system.

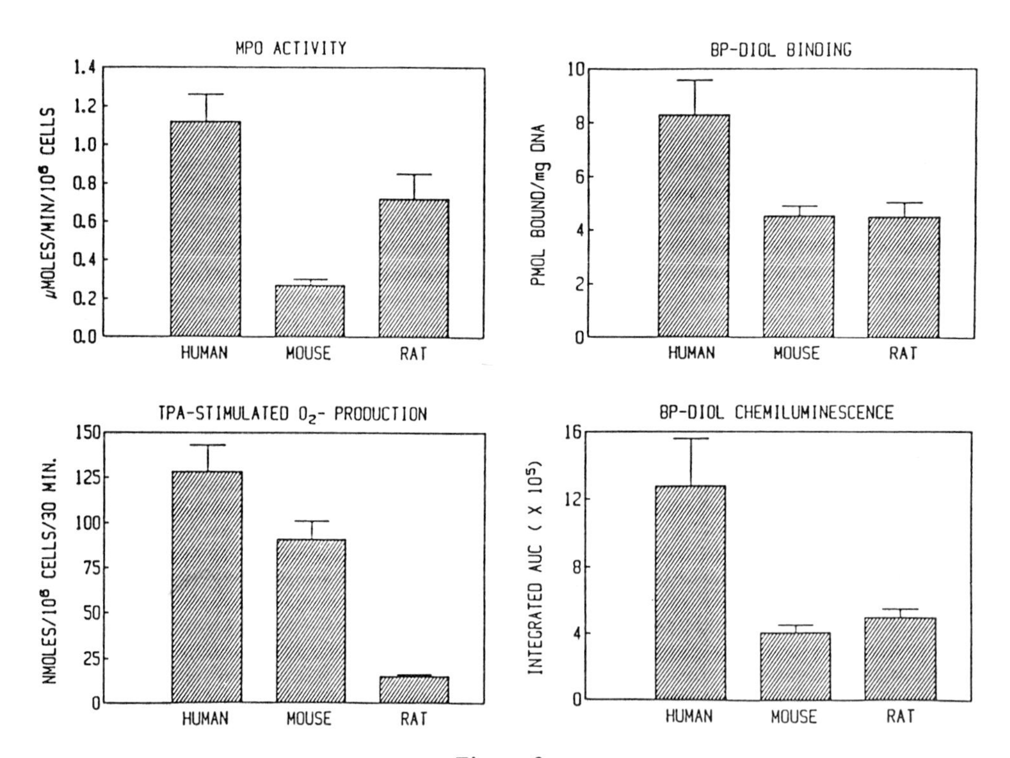

Figure 2.

Comparison of MPO activity of PMNs from human, rat or mouse revealed that human PMNs had the higher MPO activity (Figure 2). In addition, human PMNs generated superoxide, the precursor to H_2O_2, at a faster rate than cells from the other two species. Consistent, with a role for MPO in the activation of BP-diol with PMNs, human PMNs activated BP-diol to the greatest extent as indicated by chemiluminescence from it dioxetane intermediate (Seliger et al., 1982; Trush et al., 1985) and by its covalent binding to DNA. Rat and mouse PMNs exhibited an equivalent ability to activate BP-diol although they exhibited significant differences in MPO activity and superoxide generation.

The studies described here demonstrate that genotoxic diolepoxides are products of the activation of BP-diol by PMNs via a MPO-dependent reaction. Moreover, these data emphasize that at sites of PMN occurrence or accumulation (Twerdok and Trush, 1988; Esterline et al., 1989), the activation of BP-diol or other xenobiotics could create a highly localized genotoxic environment which could impact on human health. Humans may be particularly at risk from such a process due to the oxidative capability of their PMNs.

ACKNOWLEDGEMENTS

This research was supported by NIH grant ES-03760, ES-07141 American Cancer Soc. SIG-3 and Johns Hopkins CAAT.

REFERENCES

Conney, A.H. (1982). Induction of microsonal enzymes by foreign chemicals and carcinogenesis by polycyclic aromatic hydrocarbons: G.H.A. Clowes Memorial Lecture. *Cancer Res.* **42**, 4875-4917.

Dix, T.A. and Marnett, L.J. (1983). Metabolism of polycylic aromatic hydrocarbon derivatives to ultimate carcinogens during lipid peroxidation. *Science* **221**, 77-79.

Esterline, R.L., Bassett, D.J.P. and Trush, M.A. (1989). Characterization of oxidant generation by inflammatory cells lavaged from rat lungs following acute exposure to ozone. *Toxicol. Appl. Pharmacol.* **99**, 229-239.

Kensler, T.W., Egner, P.A., Moore, K.G., Taffe, B.G., Twerdok, L.E. and Trush, M.A. (1987). Role of inflammatory cells in the metabolic activation of polycyclic aromatic hydrocarbons in mouse skin. *Toxicol Appl. Pharmacol.* **90**, 337-346.

Seliger, H.H., Thompson, A., Hamman, J.P. and Posner, G.H. (1982). Chemiluminescence of benzo(a)pyrene-7,8-diol. *Photochem. Photobiol.* **36**, 359-365.

Trush, M.A., Seed, J.L. and Kensler, T.W. (1985). Oxidant dependent metabolic activation of polycyclic aromatic hydrocarbons by phorbol ester-stimulated human polymorphonuclear leukocytes: Possible link between inflammation and cancer. *Proc. Natl. Acad. Sci. USA* **82**, 5194-5198.

Twerdok, L.E. and Trush, M.A. (1988). Neutrophil-derived oxidants as mediators of chemical activation in bone marrow. *Chem.-Biol. Interactions* **65**, 261-273.

QUINONES AND THEIR GLUTATHIONE CONJUGATES AS IRREVERSIBLE INHIBITORS OF GLUTATHIONE S-TRANSFERASES

Ben van Ommen, Jan J.P. Bogaards, Jan Peter Ploemen, J. van der Greef* and Peter J. van Bladeren

TNO-CIVO Toxicology and Nutrition Institute
Department of Biological Toxicology and Department of Instrumental Analysis*
P.O. Box 360, 3700 AJ Zeist
The Netherlands

Quinones are known for their reactivity towards sulfhydryl groups, and thus are potential protein alkylating compounds. Human exposure to these compounds arises either from naturally occurring quinones or from quinones formed as metabolites from aromatic compounds. Normally, quinones are detoxified by conjugation with glutathione, producing the hydroquinone conjugates. Certain quinones, however, retain their oxidized nature after conjugation (e.g. halogen substituted quinones), while the hydroquinone conjugates have been shown to undergo intracellular oxidation. Thus, glutathione conjugates of quinones are formed, which display a special type of reactivity, i.e. selective inhibition of glutathione S-transferase isoenzymes (GST). This paper describes the development and use of glutathione conjugates of chlorinated benzoquinones as inhibitors of glutathione S-transferases.

MATERIALS AND METHODS

Time curves of enzyme inhibition were obtained as described previously (Van Ommen, 1988). Human and rat GST were purified by S-hexylglutathione affinity chromatography and chromatofocusing as described previously (Van Ommen, 1988; 1989). Covalent binding of tetrachloro-1,4-benzoquinone (TCBQ) to GST 4-4 was either quantified by spectropic methods or by using radioactive TCBQ (Van Ommen, 1989). For the localisation of the cysteine involved in inhibition of GST 4-4, 8 nmoles of ^{14}C-TCBQ were incubated with 16 nmoles of GST 4-4 (subunit concentration). After 10 minutes of incubation, the peptides were treated with cyanogen bromide (80 µg, volume 430 ul, 70% formic acid [Alin, 1986]) and the fragments were separated by reversed phase HPLC (water and acetronitril, both containing 0.1% trifluoro acetic acid). Fractions were collected and assayed by plasma desorption mass spectrometry.

RESULTS AND DISCUSSION

An irreversible inhibitor of the GST was developed, which combines a high chemical reactivity with a high degree of specificity for the active site of GST, the glutathione conjugate of tetrachloro-1,4-benzoquinone, 2-S- glutathionyl-3,5,6-trichloro-1,4-benzoquinone (GS-TCBQ). The benzoquinone moiety effectively reacts with a sulfhydryl group in or in the vicinity of the GST active site, while the glutathione moiety acts as a "targeting" component. GS-TCBQ irreversibly inhibits the catalytic activity of rat and human GST isoenzymes for 80% at equimolar concentrations of inhibitor and enzyme, independent of the enzyme concentration. At 0°C, the inhibition is achieved within 10-100 seconds, depending on the isoenzyme used. Competitive inhibitors of GST slow down but ultimately do not prevent inactivation by GS-TCBQ. The mercaptoethanol conjugate of TCBQ, which chemically behaves identically to GS-TCBQ, inhibits GST at a more than 20-fold lower rate, stressing the strong targeting effect of the glutathione part.

Quantification of the chemical modification of the cysteine residues by TCBQ showed all three cysteine residues to be modified, but already after the modification of one residue, the inhibition was complete. This suggests that modification of a single residue is responsible for inactivation. This cysteine residue was identified to be at position 114 by selective labeling, CNBR cleavage, and plasma desorption mass spectrometry of the HPLC eluate. The cyteine residue itself is not involved in the catalytic process, since the use of different substrates after modification resulted in a large variation in residual activity. Rather, the inhibition is caused by steric hindrance.

Structure activity relationships using various chlorinated benzoquinones and their glutathione conjugates were investigated with rat and human isoenzymes. Using rat GST 1-1, no inhibition was observed for 1,4-benzoquionone or its glutathione conjugate. chloro-1,4-benzoquinone did not inhibit GST 1-1, but the 2-chloro-6-glutathionyl-1,4-benzoquinone showed a very moderate inhibition. The three dichlorobenzoquinones displayed a considerable rate of inhibition towards rat GST 1-1: 50% inhibition was obtained within 300 seconds (Table 1). Conjugation of these quinones with glutathione increased the rate of inhibition dramatically: for the 2,3-isomer, a 2-fold increase was observed, while the 2,6-dichloro-3-glutathionyl-1,4-benzoquinone was 41 times faster in inhibiting rat GST 1-1, as compared to 2,6-dichloro-1,4-benzoquinone (Table 1). The extent of the "targeting effect" (the increase due to glutathione conjugation) observed for the 2,5-isomer was in between the effect found for the two previously mentioned dichlorobenzoquinones. The rates of inhibition by trichlorobenzoquinone was too high to measure a pronounced targeting effect.

Table 1. Inhibition of rat GST 1-1 by dichloro-benzoquinones and their glutathione conjugates

quinone	time at 50% inhibition (s)
2,3-diCl-BQ	150
2,3-diCl-GS-BQ	75
2,5-diCl-BQ	260
2,5-diCl-GS-BQ	20
2,6-diCl-BQ	290
2,6-diCl-GS-BQ	7

0.5 uM of rat GST 1-1 was incubated with the quinones (5 uM) and the time curve of the inhibition was determined (see Figure 1).

Looking at various human GST isoenzymes, both the rate of inhibition by 2,5-dichloro-1,4-benzoquinone and the targeting effect of the glutathione conjugate varied extensively: the mu class GST's mu and psi, and the alpha class GST B1-B1 were inhibited moderately by the quinone, while the glutathione conjugate very strongly increased the rate of inhibition (Figure 1). The closely related alpha GST B2-B2, however, was not inhibited by the quinone, nor by its glutathione conjugate. The explanation for this observation lies in the fact that B1-B1 contains one cysteine residue, while in B2-B2, this cysteine residue is absent. However, the acidic isoenzyme pi, which contains 4 cysteine residues, is also not inhibited by 2,5-dichlorobenoquinone or its glutathione conjugate. Apparently, the cysteines of this isoenzyme are not located in a position where chemical modification interferes with the catalytic properties of the enzyme.

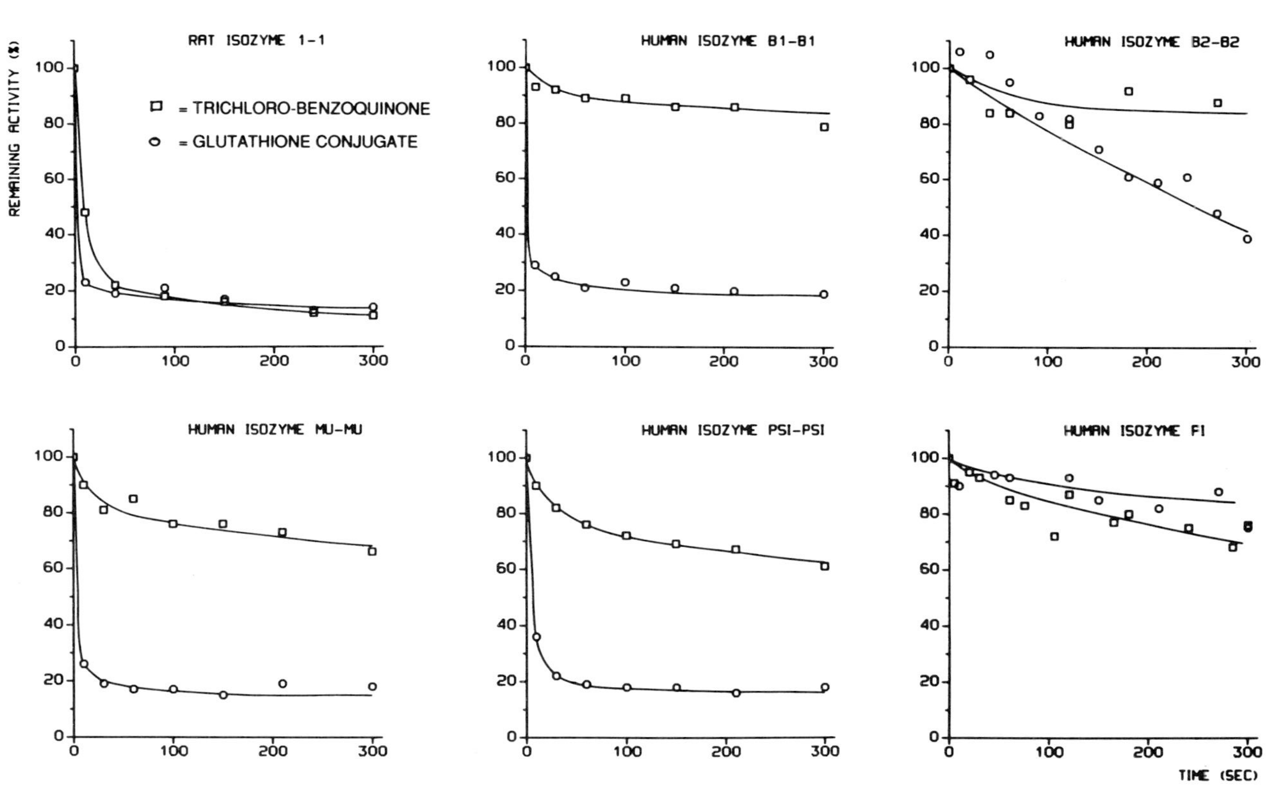

Figure 1. Inhibition of various human and rat glutathione S-transferases by 2,5-dichloro-1,4-benzoquinone and its glutathione conjugate. 0.5 uM of GST was incubated with the quinone (5 uM) at 0°C, and the enzyme activity was measured at various time intervals.

In conclusion, benzoquinones and their glutathione conjugates show to be promising inhibitors of GST. The targeting effect of the glutathione moiety can be exploited to increase the selectivity towards tranaferases. Use of these compounds in probing the active site of the GST is possible. Isoenzyme selectivity due to differences in cysteine content and location occurs and may be used in the development of isoenzyme-specific inhibitors.

REFERENCES

Alin, P., Mannervik, B., and Jörnvall, H. (1986). *Eur. J. Biochem.* **156**, 343.

van Ommen, B., den Besten, C., Rutten, A.L.M., Ploemen, J.H.T.M. Vos, R.M.E., Müller, F.,and van Bladeren, P.J. (1988). *J. Biol. Chem.* **263**, 12939.

van Ommen, B., Ploemen, J.H.T.M., Ruven, H.J., Vos, R.M.E., Bogaards, J.J.P., van Berkel, W.H.M and van Bladeren, P.J. (1989). *Eur. J. Biochem.* **181**, 423.

CYTOCHROME P450 IA2 ACTIVITY IN MAN MEASURED BY CAFFEINE METABOLISM: EFFECT OF SMOKING, BROCCOLI AND EXERCISE

K. Vistisen, S. Loft and H.E. Poulsen

Department of Pharmacology, University of Copenhagen
Juliane Mariesvej 20, DK-2100 Copenhagen, Denmark

Caffeine is sequentially metabolized by cytochrome P450IA2, N-acetyl transferase (NAT) and/or xanthine oxidase (XO). After ingestion of caffeine, equivalent to the content of a cup of coffee, the activity of these three enzymes can be estimated from the ratios of the formed metabolites excreted into the urine (Grant, D.M. et al., 1983; Grant, D.M. et al., 1986; Campbell, M.E. et al., 1987; Figure 1). Cytochrome P450IA2 and NAT are considered the most important enzymes in the metabolic activation of foreign com,pound, such as arylamines, into carcinogens (Butler, M.A. et al., 1989; Hein, D.W., 1988), whereas XO may be important in tissue damage after ischaemia and in relation to infections (Simpson, P.J. et al., 1987; Oda, T. et al., 1989).

In the present study we investigated factors influencing these enzyme activities measured by the metabolite ratios of dietary caffeine.

MATERIALS AND METHODS

In one, cross-sectional, experiment spot urine samples were collected from 335 healthy volunteers who gave informations regarding age, height, weight and the consumption of tobacco, alcohol, coffee, tea, coca cola, protein and cruciferous vegetables during the preceding two weeks. 171 of the subjects were women, twelve of whom were pregnant and 28 used oral contraceptives.

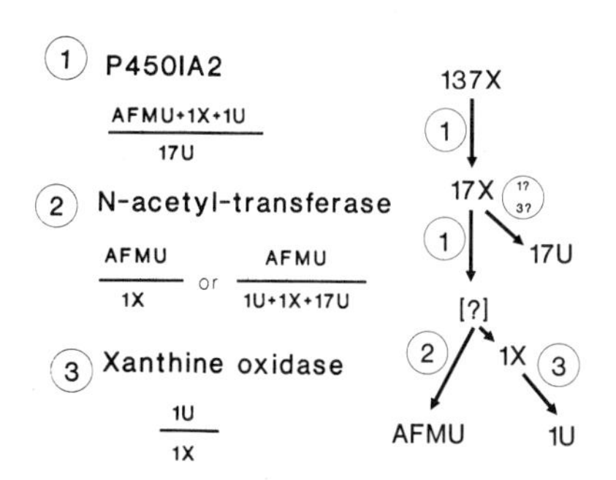

Figure 1. Metabolic ratios of caffeine

Biological Reactive Intermediates IV, Edited by C.M. Witmer *et al.*
Plenum Press, New York, 1990

In a second, longitudinal, experiment spot urine samp,les were collected from 23 healthy men before and after 30 days with vigourous exercise, 8 hr per day.

In a third, longitudinal, experiment spot urine samples were collected from 9 healthy subjects after two 10 day periods with a diet supplemented with 500 g green beans or 500 g broccoli in random order.

In all three experiments sampling of urine was preceded by ingestion of at least one cup of coffee, or equivalent, within 2-6 hr. The samples were acidified with HCl to pH 3.5 and stored at 20°C for subsequent HPLC analysis (Campbell, M.E. et al., 1987).

RESULTS

In 331 subjects the AFMU/1X ratio measuring NAT activity showed a typically bimodal distribution with 47% fast acetylators and 53% slow acetylators divided by an antimode of 0.5 (Figure 2). This is in agreement with previous reports on acetylator frequency in Denmark by the use of conventional probes (Hein, D.W., 1988).

The ratio reflecting P450IA2 activity (IA2) was normally distributed (Figure 2). In male and female subjects smoking 10 cigarettes/day or more the IA2 ratio was 66% and 70% higher than in the corresponding non-smoking groups, demonstrating the expected induction of P450IA by tobacco, $p<0.05$, but minimal sex-related differences (Campbell, M.E. et al., 1987; Figure 3). In 12 non-smoking pregnant women and in 28 smoking and non-smokingwomen using oral contraceptives the average IA2-ratio was reduced by 29% and 20% compared to the appropriate control groups ($p<0.05$: Figure 3), respectively, demonstrating the expected inhibition of P450IA2 (Campbell, M.E. et al., 1987; Knutti, R. et al., 1981).

The ratio reflecting XO-activity was normally distributed (Figure 2). Amalgating the male and female groups, but excluding pregnant women and oral contraceptives user, the XO-ratio was 1.04 ± 0.53 in the 191 non-smoking sub,jects compared to 1.26 ± 0.61 ($p<0.05$) and 1.29 ± 0.61 ($p<0.05$) in the 48 and 56 subjects smoking 1-9 and 10 or more cigarettes/day, respectively. This suggests that even light smoking increases XO activity.

Thirty days of vigorous exercise increased the IA2)ratio by 58%, i.e. from 5.2 ± 2.0 to 8.2 ± 2.2 ($p<0.05$, n=23), the XO-ratio increased from 0.73 ± 0.3 to 1.53 ± 0.64 ($p<0.05$, n=23), whereas the NAT-ratio was unchanged. An inducing effect of physical activity on cytochrome P450 activity has previously been demonstrated with antipyrine and aminopyrine as model compounds (Boel, J. et al., 1984).

After a bean and broccoli supplemented diet the IA2 ratio was 3.7 ± 1.1 and 4.4 ± 1.6 ($p<0.05$, n=9), correspon,ding to a 19% induction of P450IA2 activity by broccoli. This is in accordance with the expected effect of cruciferous vegetables as previously demonstrated with phena,cetin and antipyrine as probes (Pantuck, E.J. et al., 1979). The NAT- and XO-ratios were not significantly changed by the diets.

CONCLUSION

The ratios of metabolites from dietary caffeine in spot urine offer simple estimates of P450IA2, NAT and XO activity. In the present study the reliability of these indices was demonstrated by the expected distribution of the NAT-ratio and effects of smoking, pregnancy, oral contra,ceptive use, exercise and a diet rich in cruciferous vege,tables on the IA2-ratio.

The inducing effects of smoking and exercise on XO activity are not yet explained but may be of importance in ischaemic tissue damage related to smoking and physical strain, respectively.

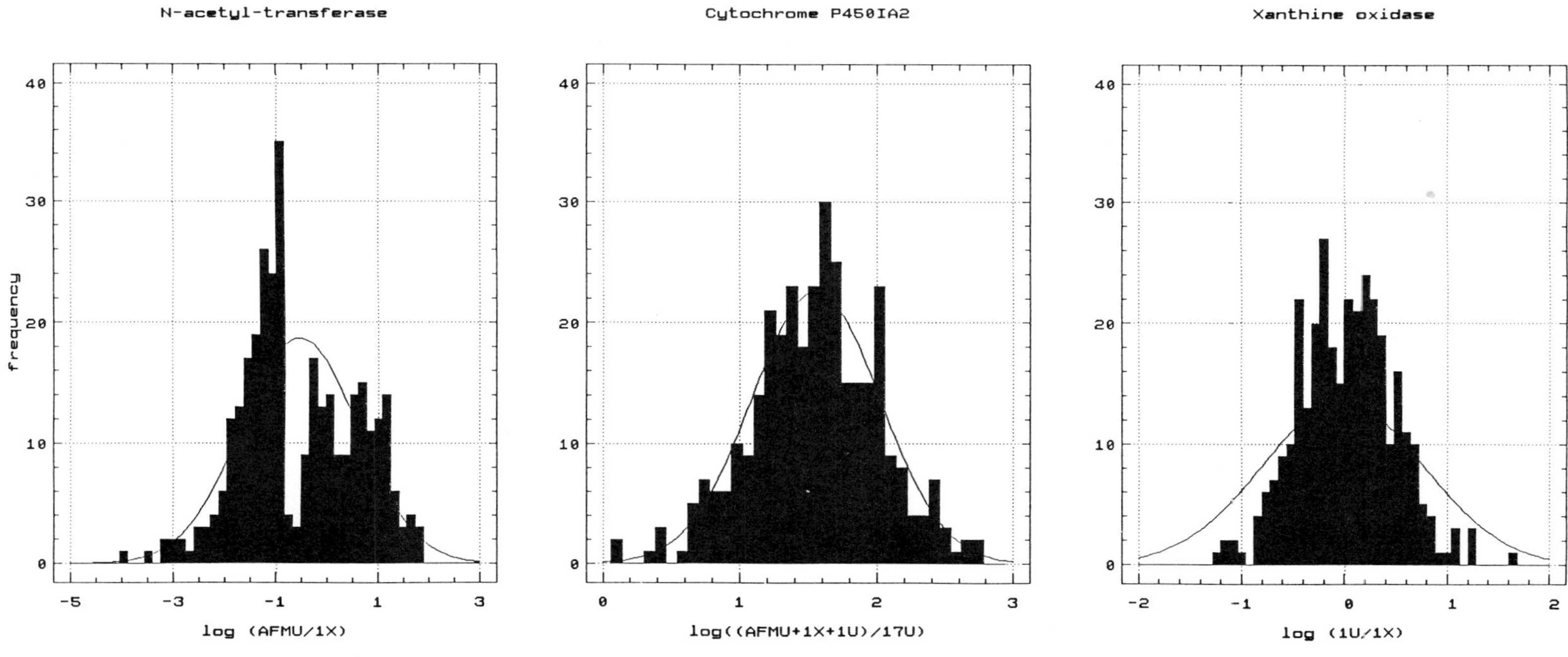

Figure 2. Distribution of the metabolic ratios of caffeine in 335 healthy subjects.

P450IA2

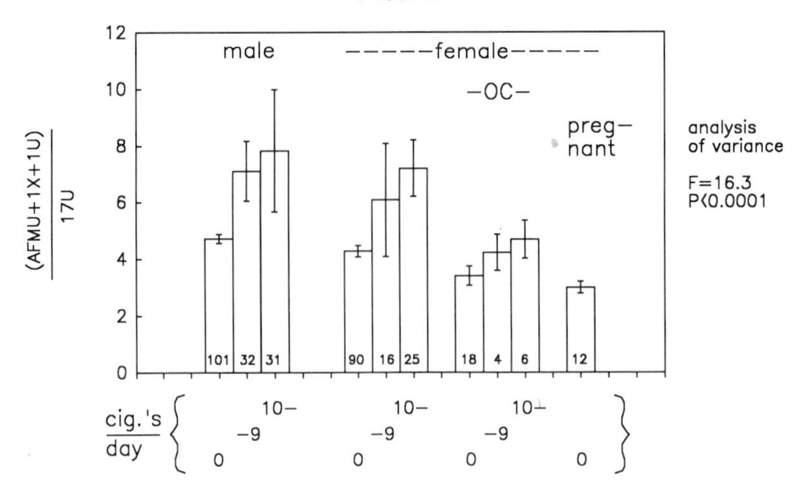

Figure 3. Effect of sex, smoking, pregnancy and oral contraceptive use on the IA2 metabolic ratio of caffeine.

The enzyme activities measurable from caffeine meta,bolism are especially relevant for the bioactivation of potentially toxic compounds and has retrospectively been linked to a variety of disease states, particularly cancer (Butler, M.A. et al., 1989; Hein, D.W., 1988; Guengerich, F.P., 1988). The assessment of caffeine metabolism by means of spot urine ratios of metabolites from dietarily ingested caffeine is applicable in large scale epidemiological studies. Thus, prospective testing of such relations between the enzyme activities and the development of disease is technically and practically feasible.

REFERENCES

Boel, J., Andersen, L.B., Rasmussen, B. and Hansen, S.H., Dossing M. (1984). Hepatic drug metabolism and physical fitness. In *Pharmacol. Ther.* **36,** 121-126.

Butler, M.A., Iwasaki, M., Guengerich, F.P. and Kadlubar, F.F. (1989) Human cytochrome P-450$_{PA}$ (P-450IA2), the phenacetin O-deethylase, is primarily responsible for the hepatic 3-demethylation of caffeine and N-oxidation of carcinogenic arylamines. *Proc. Natl. Acad. Sci.* **86,** 7696-7700.

Campbell, M.E., Spielberg, S.P. and Kalow, W. (1987). A urinary metabolite ratio that reflects systemic caffeine clearance. *Clin. Pharmacol. Ther.* **42,** 157-165.

Grant, D.M., Tang, B.K. and Kalow, W. (1983). Variability in caffeine metabolism. *Clin. Pharmacol. Ther.* **33,** 591-602.

Grant, D.M., Tang, B.K., Campbell, M.E. and Kalow, W. (1986). Effect of allo,purinol on caffeine disposition in man. *Br. J. Clin. Pharmacol.* **21,** 454-458.

Guengerich, F.P. (1988). Roles of cytochrome P-450 enzymes in chemical carcinogenesis and cancer chemotherapy. *Cancer Res.* **48,** 2946-2954.

Hein, D.W. (1988). Acetylator genotype and arylamine-induced carcinogenesis. *Biochem. et Biophysica. Acta.* **948,** 37-66.

Knutti, R., Rothweiler, H. and Sclatter, C.H. (1981). Effect of pregnancy on the pharmacokinetics of caffeine. *Eur. J. Clin. Pharmacol.* **21,** 121-126.

Oda, T., Akaike, T., Hamamoto, T., Suzuki, F., Hirano, T. and Maeda, H. (1989). Oxygen radicals in influenza-induced pathogenesis and treatment with pyran polymer-conjugated SOD. *Science* **244,** 974-976.

Pantuck, E.J., Pantuck, C.B., Garland, W.A., Min, B.H., Wattenberg, L.W., Anderson,

K.E. and Kappas, A. (1979). Stimulatory effect of brussel sprouts and cabbage on human drug metabolism. *Clin. Pharmacol. Ther.* 88-95.
Simpson, P.J., Lucchesi, B.R. (1987). Free radicals and myocardial ischemia and reperfusion injury. *J. Lab. Clin. Med.* **110**, 13-30.

TISSUE DIFFERENCE IN EXPRESSION OF CYTOCHROME P-450 BETWEEN LIVER AND LUNG OF SYRIAN GOLDEN HAMSTERS TREATED WITH 3-METHYLCHOLANTHRENE

Hiroshi Fujii, Ikuko Sagami, Tetsuo Ohmachi, Hideaki Kikuchi and Minro Watanabe[1]

Research Institute for Tuberculosis and Cancer
Tohoku University
Department of Cancer Chemotherapy and Prevention
Sendai 980, Japan

Abbreviations used: P-450, Cytochrome P-450; BP, benzo[a]pyrene: MC, 3-methylcholanthrene.

Most chemcicals formed in our environment require metabolic activation to exert their deleterious effects in target cells. The major activation pathway is oxidative metabolism mainly catalyzed by cytochrome P-450 (P-450)-dependent monooxygenase systems which have been resolved into at least two catalytic components, namely a species of P-450 and NADPH-cytochrome c (P-450) reductase. Furthermore, multiple forms of P-450 have been isolated and purified from some species of experimental animals including Syrian golden hamster and from human (Watanabe, M. et al., 1987; Guengerich, F.P. 1987; Mizokami, K. et al., (1983).

Most studies of xenobiotic metabolism have focues on the liver because this organ generally contains the highest concentration in the body of the enzymes involved in chemical biotransformation. Recently there have been increasing interest in the ability of extrahepatic tissues to metabolize exogeneous substrate, and interest has also focues on understanding the relationships between xenobiotic metabolism in the target organ and cell toxicity.

Because hamsters are more susceptible to polycyclic aromatic hydrocarbon-induced respiratory tract cancer than are rats (Schreiber, H. et al., 1975; Mass, M.J. et al.,1983), and has higher efficiency in metabolic activation in the Ames mutation tests (Phillipson, C.E. et al., 1989), we tried at first to purify pulmonary P-450 from Syrian golden hamsters treated with 3-methylcholanthrene (MC). It was demonstrated that the profile of benzo[a]pyren (BP) metabolites *in vitro* in a reconstituted enzyme system are different between rat and hamster lung, and that specific activities for the formation of 7,8- and 9,10-diols increase and those of 4,5-diol, 1,6- and 3,6-diones decrease in the hamster lung P-450MC system, compared with in the rat lung P-450MC system (Sagami, I. et al., 1987; Watanabe, M. et al., 1985).

Some reports (Burk, M.D. et al., 1976; Chiang, J.Y.L. et al., 1983) indicate that MC increases BP hydroxylation activity three-fold in lung microsomes but not in liver microsomes of the hamsters. Therefore, we tried to identify the immunochemical properties and N-terminal amino acid sequences of the hepatic P-450s purified from MC-treated hamster.

---

1. To whom correspondence should be addressed.

Biological Reactive Intermediates IV, Edited by C.M. Witmer *et al.*
Plenum Press, New York, 1990

MATERIALS AND METHODS

Male Syrian golden hamsters were obtained from Shizuoka Laboratory Animal Center, Shizuoka, Japan. MC was dissolved in corm oil and injected intraperitoneally at a dose of 25 µg per kg body weight every 24 h for 3 days. Twenty-four hours after the final injection of MC< the animals were killed by decapitation. Hepatic P-450MC I and P-450MC II, and pulmonary P-450MC were separated and purified from MC-treated hamsters, as previously described (Watanabe, M. et al., 1987). NADPH-cytochrome P-450 (cytochrome c) reductase was solubilized from the liver microsomes of MC-treated hamsters and purified by the modified method of Yasukochi and Masters (Yasukochi, Y. et al., 1976).

The purified rat liver P-450c and hamster liver P-450MC II were injected intradermally into female New Zealand white rabbits to obtain the antibody fractions, respectively. The antibody was partially purified according to the method previouslly reported (Sagami, I. et al., 1983). Western blots were performed by the method of Towbin et al. (1979) except that methanol was eliminated from the transfer buffer. After incubation with anti-P-450c or anti-P-450MC II, the nitrocellulose paper was developed with anti-rabbit IgG to which alkaline phosphatase was conjugated. Bands were visualized upon addition of 5-bromo-4-chloro-3-indelyl phosphate and nitroblue tetrazolium.

Sequencing was performed using an Applied Biosystems 470A gas-phase sequenator equipped with an on line model 120A phenyl thiohydantoin-analyzer. The purified P-450s were further purified by SDS-acrylamide gel electrophoresis. The Protein bands of P-450 were cut out and then electrophoretically eluted from gels using Atto Maxyield protein concentrator. The proteins recovered from the gel were lyophylized and then dissolved in 62.5% trifluoroacetic acids solution prior to sequencing.

RESULTS AND DISCUSSIONS

Generally two major forms of P-450 were purified from liver microsomes of MC-treated experimental animals. From liver of MC-treated hamsters the two forms of P450, P450MC I and P450MC II, were also purified, as reported previously (Watanabe, M. et al., 1987).

Figure 1 shows that Western blot analysis using the antibodies to rat P-450c and hamster P-450MC II provides evidence for the induction of both forms of P-450MC I and P-450MC II by MC treatment, respectively. It was confirmed that P-450MC I certainly cross-reacted with the antibody to rat P-450c after Western blot analysis, indicating the similar properties of the immunoreactivity. However, P450MC II does not react with P-450c on the Ouchterlony double diffusion plate and on the Western blot analysis (data not shown here).

The N-terminal amino acid sequences of the P-450MC I and P-450MC II are compared with those of MC-inducible P-450s obtained from the other species of experimental animals, as shown in Figure 2. P-450MC I lacks an N-terminal methionine, as shown in isosafrol-inducible rat P-450d, and is shown 75% similarity to rat P-450d and mouse P_3-450 over the first 16 amino acids, indicating a higher homology with P-450IA2 subfamily. On the other hand, P-450MC II carrying methionine as an N-terminal amino acid expressed to be low homology with P450IA1 subfamily, such as rat P-450c, mouse P_1-450 and rabbit form 6, which are major MC-inducible P-450 in the rodent liver. Fukuhara et al. (1989) reported that there are three forms of P-450 inducible by MC in hamster liver. P-450MC I and P-450MC II might be corresponded respectively to P-450 II and P-450AFB by Fukuhara et al. on spectrophotometrical spin states of oxidized form and the Soret maxima of co-reduced form of the purified P-450s and on catalytic properties in the reconstituted systems, but the molecular weights obtained on SDS-polyacrylamide gel electrophoresis are somewhat different from those of the purified P-450 herein reported.

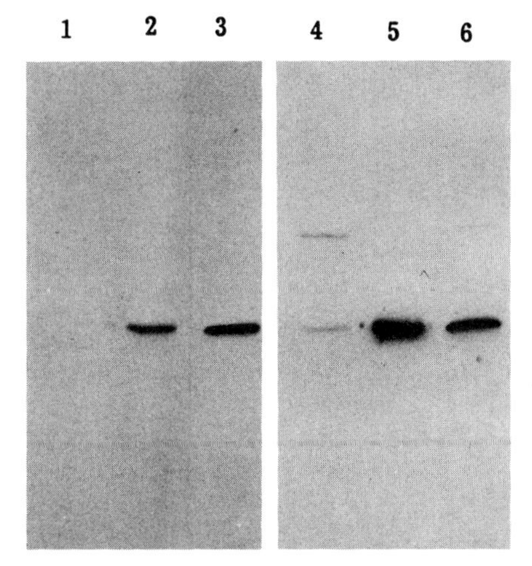

Anti-Rat Liver P-450c — Anti-Hamster Liver P-450MCII

Figure 1. Western blot analyses of hamster liver P-450MC I and P-450MC II. Liver microsomes of untreated hamster (1,4) and of MC-treated hamsters (2,5), and purified hamster liver P-450MC I (3) and P-450MC II (6) were reacted with antibodies against rat liver P-450c (1, 2, 3) and hamster liver P-450MC II, respectively.

The characterization of two major hepatic P-450s, P-450MC I and P-450MC II, and of one major pulmonary P-450MC (Sagami, et al., 1986) from MC-treated hanster are compared, as shown in Table 1. At present we are doing to obtain the cDNA of hepatic P-450MC I and P-450MC II and of pulmonary P-450MC from the respective cDNA library, and marked inductions of the three P-450 mRNA mentioned above were demonstrated by Northern blot analyses, respectively.

		1				5					10					15	
Hamster	MCI		A	L	S	Q	Y	T	S	L	S	T	E	L	V	L	A
Rat	d	M	A	F	S	Q	Y	I	S	L	A	P	E	L	L	L	A
Mouse	P_3	M	A	F	S	Q	Y	I	S	L	A	P	E	L	L	L	A
Rabbit	LM_4	A	M	S	P	A	A	P	L	S	V	T	E	L	L	L	V
Hamster	MCII	M	L	V	S	G	M	L	L	V	V	V	L	T	X	L	
Rat	c	M	P	S	V	Y	G	F	P	A	F	T	S	A	T	E	
Mouse	P_1	M	P	S	M	Y	G	L	P	A	F	V	S	A	T	E	
Rabbit	LM_6	M	V	S	D	F	G	L	P	T	F	I	S	A	T	E	

Figure 2. Comparison of the N-terminal amino acid sequences of P-450MC I and P-450MC II with those of cytochrome P-450IA family. Residues identical to the amino acid sequence of P-450MC I and P-450MC II are boxed, respectively. The amino acid sequences were obtained from the sources given in parentheses: Rat d (17), Mouse P_3 (18), Rabbit LM_4 (19), Rat c (20), Mouse P_1 (18), and Rabbit LM_6 (21).

Table 1. Comparison in Characteristics of Hepatic P-450MC I, P-450MC II and Pulmonary P-450MC from 3-Methylcholanthrene-treated Hamsters

	Liver		Lung
	P-450MC I	P-450MC II	P-450MC
Specific content: microsomes	1.90		0.086
(nmol/mg protein): purified forms	9.59	8.30	14.2
Molecular weight on SDS-PAGW (Mr.)	56,000	53,500	56,000
Maximum of reduced CO-complex (nm)	446.5	446.5	446.5
Spin states of heme iron	high	low	low
Catalytic activity in a reconstituted system (mol/min/mol P-450)			
Benzo[a]pyrene 3-hydroxylation	0.49	0.54	11.4
7-Ethoxycoumarin O-deethylation	3.27	2.20	0.93
Immunological analysis with antibody			
against rat hepatic P-450MC	++	-	+
against hamster liver P-450MC II	-	++	-

In summary the properties of the above three P-450s are mutually dissimilar. Hamster pulmonary P-450MC has properties similar to those of rat P-450MC (P-450c), showing higher catalytic activity in BP hydroxylation. It is of interest for us that there is an apparent tissue difference between liver and lung in the expression of hamster P-450 capable of catalysis of BP hydroxylation.

REFERENCES

Botelho, L.H., Ryan, D.E., Yuan, P.-M., Kutry, R., Shively, J.E. and Levin, W. (1982). Amino-terminal and carboxy-terminal sequence ofhepatic microsomal cytochrome P-450d, an unique hemeprotein from rats treated with isosafrole. *Biochemistry* **21**, 1152-1155.

Burk, M.D., and Prough, R.A. (1976). some characteristics of hamster liver and lung microsomal aryl hydrocarbon (biphenyl and benzo[a]pyrene) hydroxylation reactions. *Biochem. Pharmacol.* **25**, 2187-2195.

Chiange, J.Y. L. and Steggles, A.W. (1983). Identification and partial purification of hamster microsomal cytochrome P-450 isozymes. *Biochem. Pharmacol.* **32**, 1389-1397.

Coon, M.J., Black, S.D., Fujita, V.S., Koop, D.R. and Tarr, G.E. (1985). Structural comparison of multiple forms of cytochrome P-450. In *Microsomes and Drug Oxidations* (A.R. Boobis, J. Caldwell, F. DeMatteis and C.R. Elcombe, Eds.) pp. 42-51. Taylor and Francis, London.

Fujita, V.S., Black, S.D., Tarr, G.E., Koop, D.R. and Coon, M.J. (1985). On the amino acid sequence of cytochrome P-450 isozyme 4 from rabbit liver microsomes. *Proc. Natl. Acad. Sci. USA* **81**, 4260-4264.

Fukuhara, M., Nohmi, T., Mizokami, K., Sunouchi, M., Ishidate, M., Jr. and Takanaka, A. (1989). Characterization of three forms of cytochrome P-450 inducible by 3-methylcholanthrene in golden hamster livers with special reference to aflatoxin B_1 activation. *J. Biochem.* **106**, 253-258.

Gonzalez, F.J., Kimura, S. and Nebert, D.W. (1985). comparison of the flanking regions and introns of the mouse 2,3,7,8-tetrachlorodibenzo-p-dioxin-inducible cytochrome P_1-450 and P_3-450 genes. *J. Biol. Chem.* **260**, 5040-5049.

Guengerich, F.P. (1987). Enzymology of rat liver cytochrome P-450. In *Mammalian*

Cytochrome P-450 (F.P. Guengerich, Ed.), vol. 1 pp. 1-54. CRC, Boca Raton, Florida.

Guengerich, F.P. (1987). Characterization of human microsomal cytochrome P-450 enzymes. *Annu. Rev. Pharmacol. Toxicol.* **29**, 241-264.

Haniu, M., Ryan, D.E., Iida, S., Lieber, C.S., Levin, W. and Shively, J.E. (1984). NH_2-terminal sequence analyses of four rat hepatic microsomal cytochrome P-450. *Arch. Biochem. Biophys.* **235**, 304-311.

Mass, M.J., and Kaufman, D.G. (1983). A comparison between the activation of benzo[a]pyrene in organ cultures and microsomes from the tracheal epithelium of rats and hamsters. *Carcinogenesis* **4**, 297-303.

Mizokami, K., Nohmi, T., Fukuhara, M., Takanaka, A. and Omori, Y. (1983). Purification and characterization of a form of cytochrome P-450 with high specificity for aflatoxin B_1 from 3-methylcholanthrene-treated hamster liver. *Biochem. biophys. Res. Commun.* **139**, 1388-1397.

Phillipson, C.E. and Ioannides, C. (1989). Metabolic activation of polycyclic aromatic hydrocarbons to mutagens in the Ames test by various animal species including man. *Mut. Res.* **211**, 147-151.

Sagami, I., Ohmachi, T., Fujii, H. and Watanabe, M. (1986). Pulmonary microsomal cytochrome P-450 from 3-methylcholanthrene-treated hamster: Purification, characterization and metabolism of benzo[a]pyrene. *J. Biochem.* **100**, 449-457.

Sagami, I., Ohmachi, T., Fujii, H. and Watanabe, M. (1987). Benzo[a]pyrene metabolism by purified cytochrome P-450 from 3-methylcholanthrene-treated rats. *Xenobiotica* **17**, 189-198.

Sagami, I. and Watanabe, M. (1983). Purification and characterization of pulmonary cytochrome P-450 from 3-methylcholanthrene-treated rats. *J. Biochem.* **93**, 1499-1508.

Schreiber, H., Martin, D.H. and Pazmino, N. (1975). Species differences in the effect of benzo[a]pyrene-ferric oxide on the respiratory tract of rats and hamsters. *Cancer. Res.* **35**, 1654-1661.

Towbin, H., Staehelin, T. and Gordon, J. (1979). Electrophoretic transfer of proteins from polyacrylamide gels to nitrocellulose sheets: Procedures and some applications. *Proc. Natl. Acad. Sci. USA* **76**, 4350-4354.

Watanabe, M., Fujii, H., Sagami, I. and Tanno, M. (1987). Characterization of hepatic and pulmonary cytochrome P-450 in 3-methylcholanthrene-treated hamsters. *Arch. Toxicol.* **60**, 52-58.

Watanabe, M., Sagami, I., Ohmachi, T. and Fujii, H. (1985). Comparison in the character of purified pulmonary cytochrome P-450 between 3-methylcholanthrene-treated rat and hamster. In *Cytochrome P-450. Biochemistry, Biophysics and Induction* (L. Vereczkey and K. Magyar, Eds.), pp. 379-382. Elsevier Science, Amsterdam.

Yasukochi, Y. and Masters, B.S.S. (1976). Some properties of a detergent-solubilized NADPH-cytochrome c reductase purified by biospecific affinity chromatography. *J. Biol. Chem.* **251**, 5337-5344.

ROLE OF CALCIUM IN TOXIC AND PROGRAMMED CELL DEATH

Sten Orrenius, David J. McConkey and Pierluigi Nicotera

Department of Toxicology, Karolinska Institutet
P.O. Box 60400, S-104 01 Stockholm, Sweden

ABSTRACT

An uncontrolled and sustained increase in cytosolic Ca^{2+} concentration has been implicated as an early event in the developement of anoxic or toxic cell injury.

More recently it has become clear that an elevation of cytosolic Ca^{2+} is also involved in programmed cell death in the immune system. Here, we review some of our recent studies and provide further evidence for the role of Ca^{2+} in cell killing.

INTRODUCTION

Changes in the cytosolic free Ca^{2+} concentration function as an important intracellular signalling mechanism whereby growth factors, mitogens and hormones regulate many vital cell functions, including cell growth and differentiation, intracellular transport and secretion, and cell contraction and motility. Thus, it is apparent that perturbations of intracellular Ca^{2+} homeostasis may affect cellular functions and ultimately compromise cell survival. We now know of several potential mechanisms by which increases in cytosolic Ca^{2+}, elicited by chemical toxins, can activate lethal processes. In addition, it has recently become clear that elevations of cytosolic Ca^{2+} are involved in cell killing during "apoptosis" or programmed cell death. Studies in our laboratory have suggested that toxic cell killing and programmed cell death share similar features, including a sustained elevation of the cytosolic Ca^{2+} level, membrane blebbing, and the activation of Ca^{2+}-dependent catabolic enzymes (Orrenius et al., 1989).

In the following sections we shall briefly discuss the physiological regulation of intracellular Ca^{2+} homeostasis and the mechanisms responsible for the increases in cytosolic Ca^{2+} and the subsequent activation of Ca^{2+}-dependent lethal events during chemical toxicity and programmed cell death.

Regulation of the intracellular Ca^{2+} concentration

Mammalian cells have developed transport systems to maintain the intracellular Ca^{2+} concentration at a low level (Fig. 1). The very high concentration of Ca^{2+} in the extracellular space (10^{-3} M), as compared to the cytosolic Ca^{2+} level (10^{-7} M), generates a considerable electrochemical force which is balanced by the active extrusion of Ca^{2+} from the cell and by intracellular compartmentation processes (Carafoli, 1987). Although a substantial amount of Ca^{2+} is normally bound to cell constituents, such as phospholipids and Ca^{2+}-binding proteins, the buffering effect that they provide is limited by their total amount. Thus, it is generally accepted that the regulation of the cytosolic Ca^{2+} concentration is mainly dependent on membrane-associated transport systems, including distinct Ca^{2+}, Mg^{2+}-ATPases in the endoplasmic reticulum and plasma membrane and an energy-dependent uniport system in the mitochondrial inner membrane.

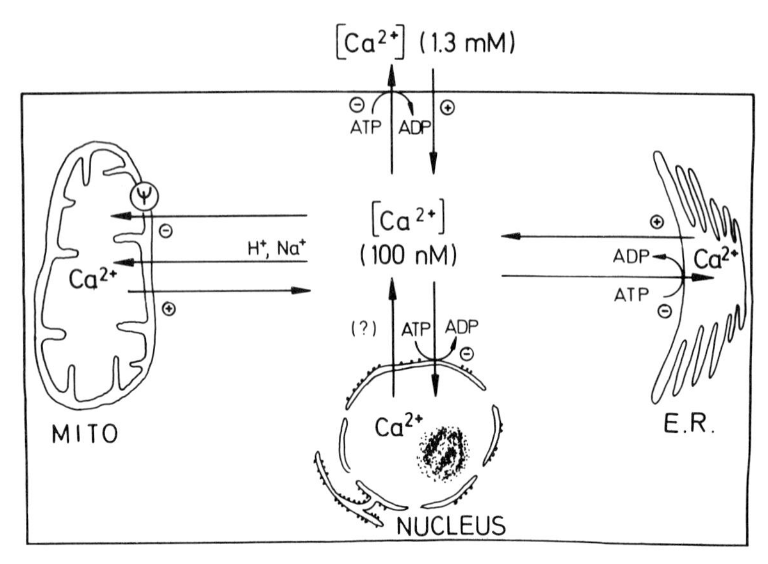

Figure 1. Schematic illustration of the regulation of intracellular Ca^{2+} compartmentation in mammalian cells.

Changes in intranuclear free Ca^{2+} can affect the behaviour of the nucleus during the cell cycle, activate nuclear processes such as DNA fragmentation and repair, and modulate gene transcription. The presence of pores in the nuclear membrane has been taken to indicate that ions and small molecules are freely diffusible in and out of the nucleus. On the other hand, it has long been known that the ion composition of the nucleus differs from that of the cytoplasm, and Ca^{2+} concentration gradients have recently been shown to exist between the nuclear matrix and the cytosol in individual muscle cells (Williams et al., 1985). Recent work in our laboratory has focussed on the role of Ca^{2+} in modulating the activity of nuclear enzymes involved in DNA fragmentation and repair. Particular attention has therefore been devoted to the mechanisms involved in Ca^{2+} transport across the nuclear membrane. We have been able to show that liver nuclei possess an active Ca^{2+} sequestration system, which is stimulated by ATP and appears to be calmodulin-dependent (Nicotera et al., 1989). This system operates at Ca^{2+} concentrations normally present in the cytosol of mammalian cells, and its stimulation results in an increased free Ca^{2+} concentration in the nuclear matrix.

Intracellular Ca^{2+} homeostasis in toxic injury and programmed cell death

Previous studies have shown that the function of Ca^{2+} transport systems is often impaired by toxic chemicals or by their reactive metabolites (Orrenius and Bellomo, 1986). Thus, the metabolism of quinones or hydroperoxides can result in the inhibition of Ca^{2+} transport systems in the endoplasmic reticulum and mitochondria. This causes the release of Ca^{2+} from the intracellular stores and a loss of their buffering capacity. In addition, several reactive species have been shown to inhibit Ca^{2+} efflux from cells by reacting with thiol groups required for plasma membrane Ca^{2+}- ATPase activity, or by depleting intracellular ATP. This, invariably results in a sustained increase in cytosolic free Ca^{2+} concentration.

Less clear are at present the effects of reactive species on intranuclear free Ca^{2+} concentration and on Ca^{2+}-dependent enzyme activities in the nuclear matrix. Although it appears that the generation of active oxygen species by xanthine/xanthine oxidase markedly reduces the extent of ATP-supported Ca^{2+} uptake by isolated liver nuclei, the rate of the increase in intranuclear free Ca^{2+} concentration is unaffected. Therefore it seems that, even under conditions of oxidative stress, a gradient between intranuclear and extranuclear Ca^{2+} concentrations can be mantained and Ca^{2+}-dependent nuclear enzymes (see below) can be stimulated.

A sustained elevation of cytosolic Ca^{2+} is now being recognized as a critical event in the initiation of apoptosis or programmed cell death in the immune system. Thus, changes in cytosolic Ca^{2+} are involved in the activation of cytotoxic T lymphocytes (CTL), in the delivery of the lethal hit to the target and, finally, in the killing of the target cell (Poenie et al., 1987). In addition, work in our laboratory has demonstrated that glucocorticoid hormones and the environmental contaminant 2,3,7,8-tetrachlorodibenso-p-dioxin (TCDD) can elicit lethal events in immature thymocytes by inducing a sustained increase in cytosolic Ca^{2+} (McConkey et al, 1988; 1989b). From these studies it has become apparent that an influx of extracellular Ca^{2+} is responsible for the increase in cytosolic Ca^{2+}. At least two mechanisms have been indicated whereby immune reactions can stimulate Ca^{2+} influx in cells undergoing programmed cell death: (i) the synthesis and delivery of pore-forming proteins, a family of proteins which polymerize to tubular lesions in target cell membranes (Young et al., 1986) and (ii) the opening of receptor-activated Ca^{2+} channels.

Increase in cytosolic free Ca^{2+} concentration

An imbalance between the influx and efflux of Ca^{2+} can result in the accumulation of intracellular Ca^{2+} and lead to an increased cytosolic Ca^{2+} concentration which is often more pronounced and sustained than the hormone-induced Ca^{2+} transients. Such sustained increases in Ca^{2+} have been observed in mammalian cells exposed to toxic agents or chemical anoxia and in cells of the immune system undergoing programmed cell death. Evidence for the lethal effect of such Ca^{2+} increases has come from experiments where intracellular Ca^{2+} chelators were used to buffer the changes in cytosolic Ca^{2+} produced by cytotoxic agents. Thus, protective effects of the Ca^{2+} chelators BAPTA and Quin-2 have been observed in hepatoma cells exposed to oxidative stress (Dypbukt-Källman et al., 1990), or to chemical anoxia (Nicotera et al., 1988), as well as in thymocytes treated with glucocorticoids or TCDD (McConkey et al., 1988, 1989b). The mechanisms by which a sustained elevation of cytosolic Ca^{2+} can cause cell killing during oxidative stress seem to involve both a disruption of the cytoskeletal network and an uncontrolled activation of Ca^{2+}-dependent catabolic enzymes (Figure 2).

On the other hand, it is becoming increasingly clear that elevations of cytosolic Ca^{2+} do not always activate cytotoxic mechanisms. Cells have developed control systems to protect themselves from the activation of lethal processes, and such mechanisms seem to be involved in positive cell selection and proliferation in the immune system. A brief description of Ca^{2+}-activated processes involved in cell killing and of possible cytoprotective mechanisms is given in the next sections.

Ca^{2+}-mediated cytoskeletal alterations

Disruption of cytoskeletal organization during toxic injury and programmed cell death can result in the appearance of surface protrusions known as blebs (Jewell et al., 1982). Blebs can rupture, thereby possibly precipitating cell death. The involvement of Ca^{2+} in bleb formation has been suggested by the observation that cells treated with Ca^{2+} ionophore A23187, or with toxic agents which cause sustained increases in cytosolic Ca^{2+}, present surface blebs. At least two distinct Ca^{2+}-dependent mechanisms have been shown to be involved in toxin-induced cytoskeletal alterations which lead to plasma membrane blebbing: (i) An increase in cytosolic Ca^{2+} can cause the dissociation of actin microfilaments from alpha-actinin, a protein that serves as an intermediate in the association of microfilaments with actin-binding proteins in the

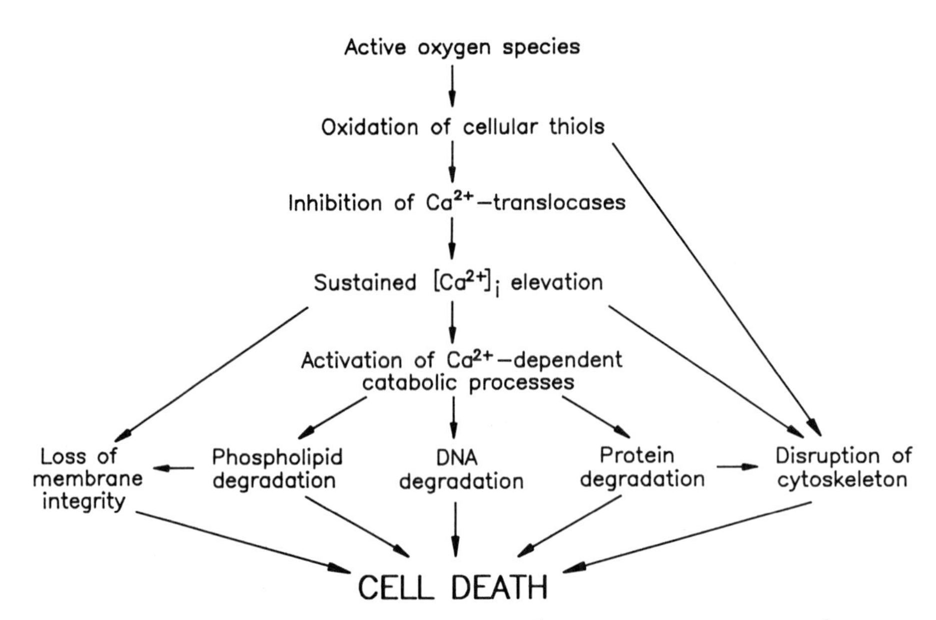

Figure 2. Sequence of events involved in the activation of Ca^{2+}-dependent degradation reactions during oxidative stress in liver cells.

plasma membrane; (ii) in addition, Ca^{2+} can activate proteases that cleave actin-binding proteins, removing the plasma membrane anchor for the cytoskeleton (Mirabelli et al., 1989). Further, Ca^{2+}-dependent transglutaminase activation has recently been proposed to cause cytoskeletal alterations in liver cells undergoing programmed cell death (Fesus et al., 1989). The latter mechanism may also be responsible for the increased membrane rigidity in apoptotic cells which are to be removed from the living tissue.

Ca^{2+}-activated phospholipases

Phospholipase activation has been proposed to cause cell damage through direct membrane lipid breakdown or by generating toxic metabolites. Stimulation of phospholipase A_2 produces lysophospholipids and arachidonic acid. The products of arachidonic acid metabolism, i.e. prostaglandins, thromboxane and leukotrienes, are known to be mediators of inflammatory reactions, while lysophospholipids are themselves cytotoxic agents. Support for a role of Ca^{2+}-dependent phospholipase activation in cell killing has come from studies demonstrating an accelerated phospholipid turnover in anoxic hepatocytes and myocardial cells (Farber and Young, 1981), and from the observation that phospholipase inhibitors could prevent ischemic injury in liver and heart (Chien et al., 1979) and protect from CCl_4 toxicity in liver cells (Glende and Pushpendran, 1986).

Ca^{2+}-activated neutral proteases

Non-lysosomal proteases include the ATP- and ubiquitin-dependent proteases, the metalloproteases, and the Ca^{2+}-activated proteases (calpains) which are present in virtually all mammalian cells and appear to be largely associated with cell membranes. The extralysosomal localization of this proteolytic system allows the proteases to participate in several specialized cell functions, including cytoskeletal and cell membrane remodelling, receptor cleavage and turnover, enzyme activation, and modulation of cell mitosis.

Ca^{2+} does not activate these enzymes by inducing allosterical changes, rather it triggers an autolytic process that results in increased activity at Ca^{2+} concentrations attained under physiological conditions. Substrate proteins for these enzymes include cytoskeletal elements and membrane integral proteins. Activation of Ca^{2+} proteases during cell injury has been described by our and other laboratories, and evidence for their involvement in cell killing has come from studies where inhibitors of these enzymes have been found to prevent, or delay, the onset of cytotoxicity (Nicotera et al., 1986). Although the substrates for these Ca^{2+}-activated neutral proteases during cell injury are largely unknown, recent studies have identified a cytoskeletal constituent as a target for their catabolic activity during oxidative injury (Mirabelli et al., 1989).

Ca^{2+}-activated endonucleases

It is now well established that programmed cell death in many systems is associated with the activation of an endogenous endonuclease that typically cleaves nuclear DNA in oligonucleosome-length fragments (Wyllie, 1980). Thus, endonuclease activation is involved in the killing of thymocytes and lymphocytes exposed to numerous stimuli. Several studies suggest that Ca^{2+} activates the endonuclease. Ionophore A23187 stimulates DNA fragmentation and apoptosis in thymocytes, and endonuclease activity in isolated nuclei is dependent on Ca^{2+} (Jones et al., 1989). Ca^{2+}-mediated DNA fragmentation appears also to be responsible for the killing of thymocytes exposed to the environmental contaminant TCDD. TCDD was found to induce endonuclease activation characteristic of thymocyte apoptosis *in vitro* and this effect was preceded by a sustained increase in cytosolic Ca^{2+} and could be prevented by agents which inhibited this Ca^{2+} increase (McConkey et al., 1988). More recently, we have shown that an organotin compound, tributyltin (TBT), which is immunotoxic *in vivo*, also stimulates DNA fragmentation and apoptosis in thymocytes. TBT was found to promote a rapid and sustained increase in the cytosolic Ca^{2+} concentration, which was caused by Ca^{2+} influx through a non-voltage dependent Ca^{2+} channel sensitive to Ni^{2+}. Both DNA fragmentation and cell death were prevented by buffering the TBT-induced increase in cytosolic Ca^{2+} with intracellular Ca^{2+} chelators (T.Y. Aw, P. Nicotera, L. Manzo, and S. Orrenius, unpublished results). A schematic representation of the events involved in thymocyte apoptosis is presented in Figure 3.

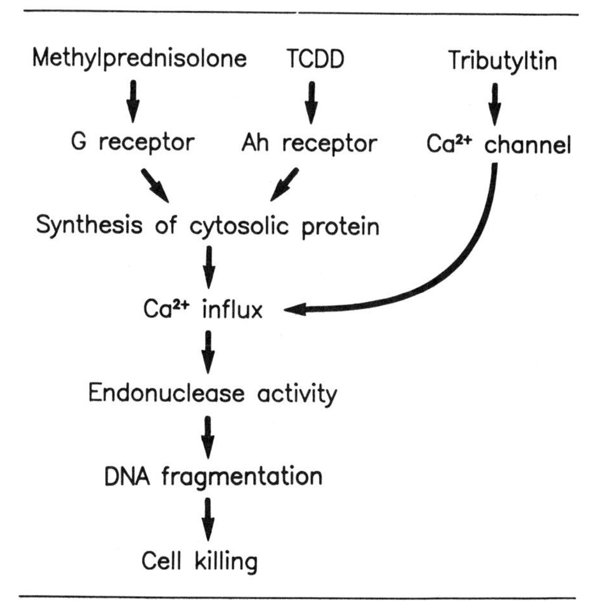

Figure 3. Schematic illustration of the mechanisms involved in glucocorticoid-activated thymocyte apoptosis and in TCDD- and tributyltin-induced cytotoxicity in thymocytes.

Possible cytoprotective mechanisms

It is paradoxal that sustained elevations of the cytosolic Ca^{2+} concentration, while compromising cell survival in some systems, can mediate the effects of growth hormones and other proliferative stimuli in others. Thus, in lymphocytes an elevation of cytosolic Ca^{2+} can stimulate proliferation, suggesting that mechanisms are activated to protect dividing cells from the activation of endonucleases or other Ca^{2+}-dependent catabolic processes. A clue to this apparent contradiction is found in the observation that Ca^{2+} signals alone are ineffective in promoting proliferation in the absence of additional stimuli, usually generated by growth factors. This requirement can be bypassed with phorbol esters, known as potent activators of protein kinase C. In our recent work we have found phorbol esters to be extremely potent in blocking endonuclease activation in thymocytes stimulated by antibodies to the CD3/T cell receptor complex, or by ionophore A23187 (McConkey et al., 1989a). Phorbol esters have also been shown to protect leukemia cells from DNA cleavage caused by topoisomerase II-reactive antineoplastic drugs, probably by inducing chromatin structural changes (Zwelling et al., 1988). Further work is required to identify the mechanism(s) by which these agents can prevent Ca^{2+}-dependent DNA cleavage.

CONCLUSIONS

A sustained elevation of cytosolic Ca^{2+} appears to be a common feature of both toxic cell killing and programmed cell death. Several biochemical mechanisms have been identified that can respond to such Ca^{2+} increases and mediate cell killing. Although different Ca^{2+}-dependent processes appear to be preferentially activated during toxic injury and apoptosis, there is now growing evidence that some of these mechanisms can be very similar. One such mechanism involves Ca^{2+}-dependent cytoskeletal alterations resulting in plasma membrane blebbing in cells exposed to chemical toxins as well as in cells undergoing apoptosis. Another common feature is the Ca^{2+}-dependent DNA damage. Whether similar mechanisms are responsible for DNA cleavage during chemical toxicity and programmed cell death remains, however, to be established.

REFERENCES

Carafoli, E. (1987). Intracellular Ca^{2+} homeostasis. *Ann. Rev. Biochem.* **56**, 395-433.

Chien, K.R., Pfau, R.G. and Farber, J.L. (1979). Ischemic myocardiac cell injury. Prevention by chlorpromazine of an accelerated phospholipid degradation and associated membrane dysfunction. *Am. J. Pathol.* **97**, 505-530.

Dypbukt-Källman, J., Thor, H. and Nicotera, P. (1990). Intracellular Ca^{2+} chelators prevent DNA damage and protect hepatoma 1c1c7 cells from killing by a redox cycling quinone. *Free Rad. Res. Comm.* in press.

Farber, J.L. and Young, E.E. (1981). Accelerated phospholipid degradation in anoxic rat hepatocytes. *Arch. Biochem. Biophys.* **221**, 312-320.

Fesus, L., Thomazy, V., Autuori, F., Ceru, M.P., Tarcsa, E. and Piacentini, M. (1989). Apoptotic hepatocytes become insoluble in detergents and chaotropic agents as a result of transglutaminase action. *FEBS Lett.* **245**, 150-154.

Glende, E.A. Jr and Pushpendran, K.C. (1986). Activation of phospholipase A_2 by carbon tetrachloride in isolated rat hepatocytes. *Biochem. Pharmacol.* **35**, 3301-3307.

Jewell, S.A., Bellomo, G., Thor, H., Orrenius, S. and Smith, M.T. (1982). Bleb formation in hepatocytes during drug metabolism is caused by disturbances in thiol and calcium ion homeostasis. *Science* **217**, 1257-1259.

Jones, D.P., McConkey, D.J., Nicotera, P. and Orrenius, S. (1989). Calcium-activated DNA fragmentation in rat liver nuclei. *J. Biol. Chem.* **264**, 6398-6403.

McConkey, D.J., Hartzell, P., Duddy, S.K., Håkansson, H. and Orrenius, S. (1988). 2,3,7,8,-Tetrachlorodibenzo-p-dioxin kills immature thymocytes by Ca^{2+}-mediated endonuclease activation. *Science* **242**, 256-259.

McConkey, D.J., Hartzell, P., Jondal, M. and Orrenius, S. (1989a). Inhibition of DNA

fragmentation in thymocytes and isolated thymocyte nuclei by agents that stimulate protein kinase C. *J. Biol. Chem.* **264**, 13399-13402.
McConkey, D.J., Nicotera, P., Hartzell, P., Bellomo, G., Wyllie, A.H. and Orrenius, S. (1989b). Glucocorticoids activate a suicide process in thymocytes through an elevation of cytosolic Ca^{2+} concentration. *Arch. Biochem. Biophys.* **269**, 365-370.
Mirabelli, F., Salis, A., Vairetti, M., Bellomo, G., Thor, H. and Orrenius, S. (1989). Cytoskeletal alterations in human platelets exposed to oxidative stress are mediated by oxidative and Ca^{2+}-dependent mechanisms. *Arch. Biochem. Biophys.* **270**, 478-488.
Nicotera, P., Hartzell, P., Baldi, C., Svensson, S.-Å., Bellomo, G. and Orrenius, S. (1986). Cystamine induces toxicity in hepatocytes through the elevation of cytosolic Ca^{2+} and the stimulation of a nonlysosomal proteolytic system. *J. Biol. Chem.* **261**, 14628-14635.
Nicotera, P., McConkey, D.J., Jones, D. and Orrenius, S. (1989). ATP simulates Ca^{2+} uptake and increases the free Ca^{2+} concentration in isolated rat liver nuclei. *Proc. Natl. Acad. Sci. USA* **86**, 453-457.
Nicotera, P., Thor, H. and Orrenius, S. (1988). Cytosolic free Ca^{2+} and cell killing in hepatoma 1c1c7 cells exposed to chemical anoxia. *FASEB J.* **3**, 59-64.
Orrenius, S. and Bellomo, G. (1986). Toxicological implications of perturbation of Ca^{2+} homeostasis in hepatocytes. In *Calcium and Cell Function* (W.Y. Cheung, Ed.), Vol. VI, pp. 185-208. Academic Press, Orlando, Florida, USA.
Orrenius, S., McConkey, D.J., Bellomo, G. and Nicotera, P. (1989). Role of Ca^{2+} in toxic cell killing. *Trends Pharmacol. Sci.* **10**, 281-285.
Poenie, M., Tsien, R.Y. and Schmitt-Verhulst, A.M. (1987). Sequential activation and lethal hit measured by $[Ca^{2+}]_i$ in individual cytolytic T cells and targets. *EMBO J.* **6**, 2223-2232.
Williams, D.A., Fogarty, K.E., Tsien, R.Y. and Fay, F. (1985). Calcium gradients in single smooth muscle cells revealed by the digital imaging microscope using Fura-2. *Nature* **318**, 558-561.
Wyllie, A.H. (1980). Glucocorticoid-induced thymocyte apoptosis is associated with endogenous endonuclease activation. *Nature* **284**, 555-556.
Young, J.D., Cohn, Z.A. and Podack, E.R. (1986). The ninth component of complement and the pore-forming protein (perforin 1) from cytotoxic T cells: structural, immunological and functional similarities. *Science* **233**, 184-190.
Zwelling, L.A., Chan, D., Hinds, M., Mayes, J., Silberman, L.E. and Blick, M. (1988). Effect of phorbol ester treatment on drug-induced topoisomerase II-mediated DNA cleavage in human leukemia cells. *Cancer Res.* **48**, 6625-6633.

MOLECULAR MECHANISMS OF γ-DIKETONE NEUROPATHY

Doyle G. Graham, Mary Beth Genter St. Clair, V. Amarnath and Douglas C. Anthony

Duke University
Durham, North Carolina 27710

We know from human and animal exposure data that inhalation of high concentrations of *n*-hexane are required, for prolonged periods of time, before a peripheral neuropathy will develop, a product of the rates of production of its ultimate toxic metabolite and of the reactions which occur within the axon (Yamamura, 1969; Herskowitz, et al., 1971). In both human and animal species, the greater vulnerability of long over short axons is a manifestation of the greater number of molecular targets for reaction with the toxic metabolite in the longer axons and a reflection of the stability of those targets. Looking back on the 20 years that we have been aware of this neurotoxicant, it is clear that these conclusions, so important to our understanding of the molecular pathogenesis of *n*-hexane neuropathy, might have been drawn from the clinical observation of these patients, and, subsequently, of exposed laboratory animals.

When *n*-hexane is inhaled, hepatic ω-1 oxidation results in the ultimate toxic metabolite, 2,5-hexanedione. The critical feature of this molecule is the γ-spacing of the two carbonyls, as shown by the observations that only γ-diketones are neurotoxic (Spencer et al., 1978; O'Donoghue and Krasavage, 1979), and that the potency of a γ-diketone precursor as a neurotoxicant depends upon the quantity of γ-diketones produced in metabolism (Krasavage et al., 1980).

The molecular target of γ-diketones is the amino group. Indeed, lysyl residues of proteins all over the body react with γ-diketones to yield pyrrolyl derivatives (DeCaprio and Weber, 1981; Graham, et al., 1982). Why, then, are the toxic effects limited to the nervous system and, in particular, the axon? Two anatomical facts stand out. One is that the intermediate filament of neurons, the neurofilament, is both extremely stable and the slowest moving component of axoplasm (Baitinger, et al., 1982). The three subunits of the neurofilament are synthesized and assembled in the neuronal perikaryon and phosphorylated as they enter the axon. The newly synthesized segment of the neurofilament then moves at 1mm/day without dissociation into component subunits, to the synapse where the neurofilament is degraded by a calcium-activated protease (Hoffman and Lasek, 1975; Schlaepfer and Zimmerman, 1981). Thus the stable, slowly-transported neurofilament is available for progressive covalent derivatization during chronic intoxication.

The second point of anatomy has to do with the reduction in the diameter of the axon which occurs when myelin is interrupted at nodes of Ranvier. Jones and Cavanagh (1983) have presented excellent pictorial evidence that these axonal constrictions present obstructions to the anterograde transport of neurofilament aggregates which develop during chronic γ-diketone intoxication. The masses of

Biological Reactive Intermediates IV, Edited by C.M. Witmer *et al.*
Plenum Press, New York, 1990

neurofilaments distort the anatomy at the node of Ranvier in a manner that suggests that covalent crosslinking of neurofilaments has occurred. With continued transport of neurofilaments up to this point of occlusion, large swellings of the axon form proximal to the nodes of Ranvier; distal to the swellings, axonal degeneration sometimes occurs.

Long before we understood the series of chemical reactions occurring within the axon, we appreciated that proteins could be crosslinked by γ-diketones *in vitro* (Graham et al., 1982, Graham et al., 1984). Though we and others (Carden et al., 1986; Lapadula et al., 1986; Genter St. Clair et al., 1990) have demonstrated neurofilament crosslinking after *in vivo* exposure, there has always been concern that the actual crosslinking reactions may have occurred during isolation of the neurofilaments or that neurofilament crosslinking was an epiphenomenon, i.e., an event not in the pathogenetic sequence leading to either the neurofilament-filled axonal swellings, or to the degeneration of the distal axon.

The use of γ-diketones with particular properties has allowed establishment of the following points: pyrrole formation is an actual step in the pathogenetic sequence; pyrrole oxidation, followed by nucleophilic attack and leading to covalent crosslinking of neurofilaments, are necessary for a given γ-diketone to be neurotoxic.

That pyrrole formation is the initial reaction in the axon leading to neurotoxicity has been defined by several types of experiments in which the predictable reaction of a given γ-diketone has allowed this hypothesis to be tested *in vivo*. Sayre and colleagues (Sayre et al., 1986) showed that 3,3-dimethyl-2,5-hexanedione, a γ-diketone incapable of pyrrole formation, is not neurotoxic. Studies in our laboratory which tested the hypothesis that pyrrole formation begins the neurotoxic sequence were initiated by Anthony, et al. (1983a,b). These studies showed that 3,4-dimethyl substitution of 2,5-hexanedione (DMHD) resulted in a γ-diketone which more rapidly formed pyrroles and was more neurotoxic than 2,5-hexanedione. These experiments predicted that the presence of two methyl groups would result in enhancement of both the rate of pyrrole formation and the rate of oxidation of the resulting pyrrolyl derivatives. Additional studies showed that di-substitution with ethyl, isopropyl, or phenyl groups resulted in marked steric interference with pyrrole formation and in loss of neurotoxic potential; thus, not all γ-diketones are neurotoxic, and a certain rate of pyrrole formation is required before a γ-diketone will be neurotoxic (Szakál-Quin, et al., 1986; Genter et al., 1987).

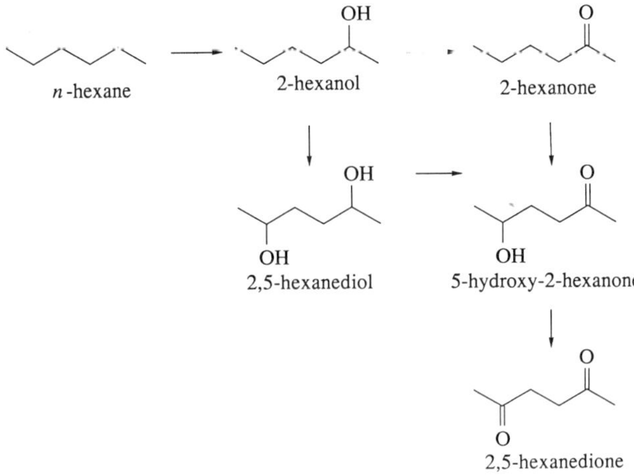

Figure 1. Metabolism of hexane. Both *n*-hexane and 2-hexanone (methyl n-butyl ketone) are neurotoxic, and both are activated through ω-1 oxidation to the ultimate toxic metabolite, 2,5-hexanedione (Anthony, 1990).

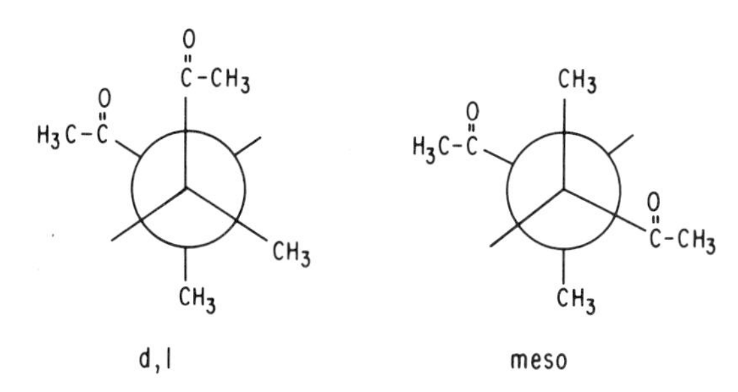

Figure 2. Newman projections showing the preferred conformations for *dl* and *meso* diastereomers of DMHD and other γ-diketones with 3,4-dialkyl substituents. 2,5-HD, whose 3 and 4 carbons are not chiral carbons, would preferentially be found with its oxygens in an antiperiplanar conformation, as shown for *meso* DMHD above (Genter et al., 1987).

The critical test of the role of pyrrole formation came in the synthesis and separation of the *d,l* and *meso* diastereomers of DMHD (Genter et al., 1987). After the demonstration that both ^{14}C-labeled diastereomers reach the nervous system at the same rate (Rosenberg, et al., 1987), the correlation between the faster rate of pyrrole formation and neurotoxic potency of the *d,l* over the *meso* diastereomer established the role of pyrrole formation unequivocally (Genter, et al., 1987).

More controversy has arisen over the issue of whether pyrrole derivatization is itself the event which leads to the neurofilament-filled axonal swellings. Sayre et al. (1985) and DeCaprio et al. (1985) have postulated that the conversion of hydrophilic lysyl amino groups to hydrophobic pyrrolyl derivatives leads to an effect on the interaction of neurofilaments with other axoplasmic components and altered transport. We have held the view that pyrrole oxidation with subsequent nucleophilic attack and neurofilament crosslinking were required for a γ-diketone to be neurotoxic (Graham, et al., 1982; 1984).

The idea that we needed a γ-diketone which could form pyrroles rapidly, but whose resulting pyrrole would be inert *in vivo*, and realization of such a compound were separated by four years of exploration of a number of blind alleys. After we had found several fluorine-containing γ-diketones either impossible to make or incapable of pyrrole formation, the ideal γ-diketone was synthesized. This γ-diketone, 3-acetyl-2,5-hexanedione (AcHD), forms pyrroles even faster than *meso* DMHD and results in massive pyrrole derivatization *in vitro* and *in vivo*. Yet the resulting pyrrolyl derivatives have very high oxidation potentials, such that pyrrole oxidation and protein crosslinking are not demonstrable. Thus, AcHD is the ideal γ-diketone to test the two opposing hypotheses. If pyrrole derivatization of lysyl amino groups itself results in neurotoxicity, then AcHD should be a potent neurotoxicant. If, on the other hand, pyrrole oxidation and neurofilament crosslinking are required for neurotoxicity, then AcHD should not be neurotoxic. When AcHD is administered to rats, extensive pyrrole derivatization of proteins occurrs, but there is neither clinical nor morphological evidence for neurofilament aggregation, axonal swellings, or axonal degeneration. Thus, these experiments establish that the neurotoxicity of γ-diketones requires protein crosslinking (Genter St. Clair et al., 1988).

Thus, over the past 10 years this laboratory has sought to test a multiple step hypothesis to explain the molecular pathogenesis of the axonopathy caused by *n*-hexane and methyl *n*-butylketone. Progress has come from increased understanding of the biology of axonal transport as well as the use of synthetic chemistry to create analogs to test specific steps in the pathogenetic sequence.

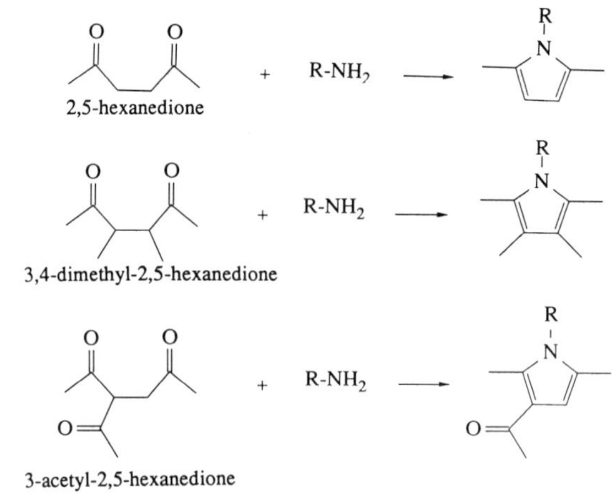

2,5-hexanedione + R-NH$_2$ →

3,4-dimethyl-2,5-hexanedione + R-NH$_2$ →

3-acetyl-2,5-hexanedione + R-NH$_2$ →

Figure 3. Substituted γ-diketones. Whereas 3,4-dimethyl substitution of 2,5-hexanedione (HD) yields a γ-diketone (DMHD) with accelerated rates of pyrrole formation and oxidation of the resulting tetramethylpyrrolyl ring, the 3-acetyl substitution of HD yields a γ-diketone (AcHD) with rapid rates of pyrrole formation but greatly reduced rates of oxidation of the pyrrole ring and of protein crosslinking (Anthony, 1990).

REFERENCES

Anthony, D. C. (1990). A molecular view of neurotoxicology. In Szabo, S. (ed.): *Chemical Pathology and Toxicology*, Volume 1: Why are Chemicals Toxic, Lewis Publishers and the American Chemical Society, NY, in press.

Anthony, D. C., Boekelheide, K., Anderson, C. W., and Graham, D. G. (1983a). The effect of 3,4-dimethyl substitution on the neurotoxicity of 2,5-hexanedione. II. Dimethyl substitution accelerates pyrrole formation and protein crosslinking. *Toxicol. Appl. Pharmacol.* **71**, 372-382.

Anthony, D. C., Boekelheide, K., and Graham, D. G. (1983b). The effect of 3,4-dimethyl substitution on the neurotoxicity of 2,5-hexanedione. I. Accelerated clinical neuropathy is accompanied by more proximal swellings. *Toxicol. Appl. Pharmacol.*, **71**, 362-371.

Baitinger, C., Levine, J., Lorenz, T., Simon, C., Skene, P., and Willard, M. (1982). Characteristics of axonally transported proteins. In Weiss, D.G. (ed.): *Axoplasmic Transport*. Springer-Verlag, Berlin, pp. 110-120.

Carden, M. J., Lee, V. M.-Y., and Schlaepfer, W. W. (1986). 2,5-Hexanedione neuropathy is associated with covalent crosslinking of neurofilament proteins. *Neurochem. Pathol.* **5**, 25-35.

Decaprio, A. P., and Weber, P. (1981). Conversion of lysine groups to substituted pyrrole derivatives by 2,5-hexanedione: A possible mechanism of protein binding. *Toxicologist* **1**, 134.

Decaprio, A. P., O'Neill, E. A. (1985). Alterations in rat axonal cytoskeletal proteins induced by *in vitro* and *in vivo* 2,5-hexanedione exposure. *Toxicol. Appl. Pharmacol.* **78**, 235-247.

Genter, M. B., Szákal-Quin, G.Y., Anderson, C. W., Anthony, D. C., and Graham, D. G. (1987). Evidence that pyrrole formation is a pathogenetic step in γ-diketone neuropathy. *Toxicol. Appl. Pharmacol.* **87**, 351-362.

Genter St. Clair, M. B., Amarnath, V., Moody, M. A., Anthony, D. C., Anderson, C. W., and Graham, D. G. (1988). Pyrrole oxidation and protein cross-linking are necessary steps in the development of γ-diketone neuropathy. *Chem. Res. Toxicol.* **1**, 179-185.

Genter St. Clair, M. B., Anthony, D. C., Wikstrand, C. J. and Graham, D. G. (1990). Neurofilament protein crosslinking in γ-diketone neuropathy: *In vitro* and *in vivo* studies using the seaworm Myxicola infundibulum. *Neurotoxicol.* **10**, in press.

Graham, D. G., Anthony, D. C., Boekelheide, K., Maschmann, N. A., Richards, R. G., Wolfram, J. W., and Shaw, B. R. (1982). Studies of the molecular pathogenesis of hexane neuropathy. II. Evidence that pyrrole derivatization of lysyl residues leads to protein crosslinking.

Graham, D. G., Szakál-Quin, GY., Priest, J.W., and Anthony, D.C. (1984). *In vitro* evidence that covalent crosslinking of neurofilaments occurs in γ-diketone neuropathy. *Proc. Natl. Acad. Sci. U.S.A.* **81**, 4979-4982.

Herskowitz, A., Ishii, N., and Schaumburg, H. (1971). *n*-hexane neuropathy. *N. Engl. J. Med.* **285**, 82-85.

Hoffman, P. N., and Lasek, R. J. (1975). The slow component of axonal transport: Identification of major structural polypeptides of the axon and their generality among mammalian neurons. *J. Cell Bio.* **66**, 351-366.

Jones, H. B. and Cavanagh, J. B. (1983). Distortions of the nodes of Ranvier from axonal distention by filamentous masses in hexacarbon intoxication. *J. Neurocytol.* **12**, 439-458.

Krasavage, W. J., O'Donoghue, J. L., Divincenzo, G. D., and Terhaar, C. J. (1980). The relative neurotoxicity of MnBK, *n*-hexane, and their metabolites. *Toxicol. Appl. Pharmacol.* **52**, 433-441.

Lapadula, D. M., Irwin, R. D., Suwita, E., and Abou-Donia, M. B. (1986). Cross-linking of neurofilament proteins of rat spinal cord *in vivo* after administration of 2,5-hexanedione. *J. Neurochem.* **46**, 1843-1850.

O'Donoghue, J. L., and Krasavage, W. J. (1979). Hexacarbon neuropathy: A γ-diketone neuropathy? *J. Neuropathol. Exp. Neurol.* **38**, 333.

Rosenberg C. K., Genter, M. B., Szakál-Quin, GY., Anthony, D. C., and Graham, D. G. (1987). *d,l* Versus *meso* 3,4-dimethyl-2,5-hexanedione: A morphometric study of the proximo-distal and temporal distribution of axonal swellings in the anterior root of the rat. *Toxicol. Appl. Pharmacol.* **87**, 363-373.

Sayre, L. M., Autilio-Gambetti, L., and Gambetti, P. (1985). Pathogenesis of experimental giant neurofilamentous axonopathies: A unified hypothesis based on chemical modification of neurofilaments. *Brain Res. Rev.* **10**, 69-83.

Sayre, L. M., Shearson, C. M., Wongmongkolrit, T., Medori, R., and Gambetti, P. (1986). Structural basis of γ-diketone neurotoxicity: non-neurotoxicity of 3,3-dimethyl-2,5-hexanedione, a γ-diketone incapable of pyrrole formation. *Toxicol. Appl. Pharmacol.* **84**, 36-44.

Schlaepfer, W. W., and Zimmerman, V.-J. P. (1981). Calcium-mediated breakdown of glial filaments and neurofilaments and in rat optic nerve and spinal cord. *Neuro-Chem. Res.* **6**, 243-255.

Szakál-Quin, GY., Graham, D. G., Millington, D. S., Maltby, D. A., and McPhail, A. T. (1986) Stereoisomer effects on the Paal-Knorr synthesis of pyrroles. *J. Org. Chem.* **51**, 621-624.

Spencer, P. S., Bischoff, M. C., and Schaumburg, H. H. (1978). On the specific molecular configuration of neurotoxic aliphatic hexacarbon compounds causing central-peripheral distal axonopathy. *Toxicol. Appl. Pharmacol.* **44**, 17-28.

Yamamura, Y. (1969). *n*-hexane polyneuropathy. *Folia Psychiat. Neurol. Jpn.* **23**, 45-47.

MICROTUBULE ASSEMBLY IS ALTERED FOLLOWING COVALENT MODIFICATION BY THE *n*-HEXANE METABOLITE 2,5-HEXANEDIONE[1]

Kim Boekelheide, Julia Eveleth, M. Diana Neely and Tracy M. Sioussat

Department of Pathology and Laboratory Medicine
Brown University,
Providence, RI 02912

INTRODUCTION

2,5-Hexanedione (2,5-HD)[1] is the ultimate toxic metabolite of aliphatic hexacarbon precursors such as *n*-hexane and methyl *n*-butyl ketone (Krasavage et al., 1980). 2,5-HD reacts with protein lysyl ε-amines to form pyrroles (DeCaprio et al., 1982; Graham et al., 1982; Anthony et al., 1983a,b; Genter et al., 1987). Pyrroles are unstable intermediates and undergo crosslinking reactions; the formation of protein crosslinks appears necessary for the development of testicular and nervous system toxicity (Boekelheide et al., 1988; St. Clair et al., 1988).

The Sertoli cell has been identified as one target cell of 2,5-HD-induced testicular injury (Chapin et al., 1982, 1983; Boekelheide, 1988a). We have proposed that a 2,5-HD-induced disruption of Sertoli cell microtubule function might result in the impairment of the important roles this cytoskeletal component performs in the structural, nutritional and hormonal support of germ cells (Russell et al., 1981; Vogl et al., 1983a,b; Neely and Boekelheide, 1988; Boekelheide et al., 1989). The data which implicate microtubule dysfunction in 2,5-HD-induced testicular injury arise from both *in vivo* (Boekelheide, 1987a; Boekelheide, 1988a,b; Boekelheide and Eveleth, 1988) and *in vitro* studies (Boekelheide, 1987b; Sioussat and Boekelheide, 1989).

The microtubule dysfunction induced by 2,5-HD is most apparent as an abnormality in microtubule assembly (Boekelheide, 1987a,b). Microtubules resemble hollow, cylindrical tubes and are polymers assembled from a core protein called tubulin. Tubulin is a 100 kDa dimer composed of two non-covalently linked subunits, the α and β tubulin monomers. Microtubule assembly is a complex process initiated by the formation of a stable nucleating element, followed by elongation of the growing microtubule and a final phase of balanced microtubule polymerization and depolymerization (Voter and Erickson, 1984). Both 2,5-HD exposure *in vivo* and 2,5-HD reaction with purified tubulin *in vitro* produce the same qualitative changes in microtubule assembly behavior (Boekelheide, 1987a,b).

2,5-HD modification enhances microtubule nucleation, producing a marked reduction in the time required for the assembly reaction to reach its maximal velocity (tVmax) and an increase in the maximal velocity of assembly (Vmax). These assembly

[1]. Abbreviations used: AcHD, 3-acetyl-2,5-hexanedione; bisANS, bis[8-anilinonaphthalene-1-sulfonate]; GTP, guanosine 5'-triphosphate; 2,5-HD, 2,5-hexanedione; Vmax, maximal velocity of assembly; tVmax, time required to achieve maximal velocity of assembly; glutamate assembly buffer, 1 M sodium glutamate, pH 6.60; Mes assembly buffer, 0.1 M 2-[N-morpholino]ethanesulfonic acid, 1 mM EGTA, 1 mM $MgCl_2$, pH 6.75.

Biological Reactive Intermediates IV, Edited by C.M. Witmer *et al.*
Plenum Press, New York, 1990

alterations are the result of a decrease in the kinetic constant for tubulin subunits dissociating from the microtubule polymer (Sioussat and Boekelheide, 1989). As a result of the decreased off rate, the critical concentration of assembly (the minimal tubulin concentration necessary for microtubule assembly) is lower in 2,5-HD-modified tubulin.

The altered assembly behavior of 2,5-HD-modified tubulin is associated with a conformational change in the α-tubulin subunit (Sioussat and Boekelheide, 1989). This conformational change is readily detected by the *de novo* appearance of a chymotryptic limited proteolytic site. This chymotryptic cleavage site has been localized to phenylalanine-169 of α-tubulin and an altered circular dichroism spectrum has indicated the presence of generalized secondary structural alterations in 2,5-HD-modified tubulin (Sioussat and Boekelheide, manuscript in preparation).

In recent experiments, sea urchin zygotes have been microinjected with 2,5-HD-treated tubulin (Sioussat et al., manuscript in preparation). Sea urchin zygotes are a useful system for the study of "*in vivo*" microtubule dynamics because of the ease of mitotic spindle visualization. Following 2,5-HD-treated tubulin microinjection, mitotic spindles were small with poor birefringence and poor astral development. Though good chromosome alignment was achieved at metaphase, little or no anaphase movement of chromosomes occurred and cleavage attempts frequently failed. These spindle effects can be explained by the non-dissociating properties of 2,5-HD-treated tubulin: the microinjected 2,5-HD-treated tubulin likely sequestered tubulin away from the mitotic spindle, resulting in small spindles, and stabilized spindle microtubules, contributing to poor anaphase movement.

In this article, we compare the effects of buffer, temperature, and tubulin concentration on the assembly of control tubulin and 2,5-HD-modified tubulin. The 2,5-HD congener, 3-acetyl-2,5-hexanedione (AcHD), is used to determine whether a crosslinking reaction is required for the appearance of the microtubule assembly alteration. Additional crosslinking agents, glutaraldehyde and dimethylsuberimidate, are examined for their ability to induce the same microtubule assembly alteration. Finally, the assembly of control tubulin and 2,5-HD-modified tubulin is compared in the presence of well-studied inhibitors of assembly, colchicine and bis[8-anilinonaphthalene-1-sulfonate] (bisANS).

RESULTS

All of these experiments used bovine brain tubulin freed of microtubule associated proteins and purified as previously described (Boekelheide, 1987b). The standard procedure for preparation of 2,5-HD-treated tubulin entailed a 16 hr incubation of once cycled purified bovine brain tubulin with 100 mM 2,5-HD in 1 M sodium glutamate, 1 mM guanosine 5'-triphosphate (GTP), pH 6.6, at 37°C followed by a cycle of temperature-dependent assembly and disassembly (Boekelheide, 1987b). Microtubule assembly was monitored by observing the change in optical density at 350 nm which occurred as tubulin polymerized (Boekelheide, 1987b). Protein concentrations were determined by the method of Lowry et al. (1951) or by using a dye binding assay (Bio-Rad, Rockville Centre, NY).

Assembly Alteration Requires Stabilized, Largely Polymeric Tubulin

Tubulin assembly behavior is dependent upon solvent characteristics (Lee and Timasheff, 1977). Solvent components which organize water induce rigid caged structures around hydrophobic regions of tubulin in solution (Gekko and Timasheff, 1981). T he presence of these caged structures favors polymerization due to entropy considerations since, in the microtubule, hydrophobic regions which produce solvent structuring form protein - protein contacts and are removed from solution (Arakawa and Timasheff, 1984).

Typical microtubule assembly buffers which favor assembly are 1 M sodium glutamate (Hamel and Lin, 1981) or buffers containing 4 M glycerol (Shelanski et al.,

1973). Buffers less supportive of microtubule assembly include 0.1 M 2-[*N*-morpholino]ethanesulfonic acid (Weisenberg, 1972) and 0.1 M piperazine-*N*,*N*'-bis[2-ethanesulfonic acid] (Lee and Timasheff, 1977). The microtubule assembly behavior of control tubulin and 2,5-HD-treated tubulin was compared in glutamate assembly buffer (1 M sodium glutamate, pH 6.60), and Mes assembly buffer (0.1 M 2-[*N*-morpholino]ethanesulfonic acid, 1 mM EGTA, 1 mM $MgCl_2$, pH 6.75) at 37°C (Fig. 1). Considering the difference in tubulin concentration in the two buffer systems, glutamate assembly buffer clearly supported assembly of both control and 2,5-HD-treated tubulin to a much greater extent than Mes assembly buffer. Indeed, control tubulin at 2.0 mg/ml did not polymerize in Mes assembly buffer while 2,5-HD-treated tubulin readily assembled under these conditions.

The differential support of assembly by glutamate assembly buffer and Mes assembly buffer was used to determine the optimal reaction conditions which result in the 2,5-HD-induced assembly alteration. In glutamate assembly buffer at 37°C, a 1.0 mg/ml solution of tubulin will be 93% in the polymer form since the critical concentration of assembly is known to be 0.07 mg/ml (Boekelheide, 1987b). On the other hand, a 2.3 mg/ml solution of tubulin in Mes assembly buffer will have no tubulin in the polymer form since this concentration is below the critical concentration for assembly (Siousssat and Boekelheide, 1989). The yield of purified tubulin following a 16 hr incubation with 100 mM 2,5-HD and 1 mM GTP at 37°C was compared in glutamate and Mes assembly buffers. 2,5-HD incubation and purification by a cycle of assembly and disassembly (a warm ultracentrifugation to pellet microtubules, resolubilization in cold buffer, and a cold ultracentrifugation) in glutamate buffer resulted in a 50% yield of 2,5-HD-treated tubulin with altered assembly behavior. However, no assembly competent tubulin could be recovered after 2,5-HD incubation in Mes assembly buffer following purification by a cycle of assembly and disassembly in the presence of 4 M glycerol added to lower the critical concentration of assembly. These results indicate that induction of the assembly alteration by 2,5-HD requires a stabilized form (presumably the polymeric microtubule form) of tubulin as a substrate.

Crosslinking Is Required for the Assembly Alteration

AcHD, a non-crosslinking 2,5-HD congener, readily reacts to form protein-bound pyrrole adducts, but does not produce the typical hexacarbon nervous system injury following *in vivo* exposure (St. Clair et al., 1988). In addition, AcHD had no effect upon subsequent microtubule assembly behavior when incubated *in vitro* with tubulin under conditions in which HD produces markedly abnormal assembly and crosslinked tubulin (Table 1). These results indicate that a crosslinking reaction is required for the occurrence of the microtubule assembly abnormality.

Additional crosslinking agents were studied to further characterize the chemical requirements for production of a 2,5-HD-like microtubule assembly alteration. Two commonly used amine crosslinkers, glutaraldehyde and dimethylsuberimidate, were capable of inducing a similar microtubule assembly abnormality; however, neither of these agents produced as dramatic a change in assembly characteristics as could be achieved with high concentrations of 2,5-HD (Table 1).

Temperature and Tubulin Concentration Affect Assembly

The microtubule assembly behavior of control tubulin and 2,5-HD-treated tubulin was examined over a range of concentrations (0.31 - 2.05 mg/ml) and at two different temperatures (32.5°C and 37°C). The most dramatically changed parameter of microtubule assembly of 2,5-HD-modified tubulin was the tVmax, a reflection of the nucleation time required for assembly. The tVmax of 2,5-HD-treated tubulin was greatly reduced at all tubulin concentrations and at both temperatures examined in comparison with control tubulin. For both control tubulin and 2,5-HD-treated tubulin at either temperature, the tVmax decreased at higher tubulin concentration; however, the extent of this change was much more pronounced in control tubulin. Similarly, for both control tubulin and 2,5-HD-treated tubulin at all concentrations, the tVmax was decreased at the higher temperature of assembly, an effect which was more pronounced in control tubulin.

Table 1. Effects of Various Chemicals upon Nucleation of Microtubule Assembly

Chemical	tVmax (% control)
3-acetyl-2,5-hexanedione (20 mM)[a]	100
2,5-hexanedione (100 mM)[a]	20
2,5-hexanedione (20 mM)[a]	36
glutaraldehyde (0.1%)[b]	39
dimethylsuberimidate (62 mM)[c]	28

Note. Shown is the tVmax (reflecting the nucleation time) for assemblies of chemically modified tubulin expressed as the percent of the tVmax of a comparable control assembly performed at the same time under similar conditions. In each case, purified bovine brain tubulin was incubated with the chemical (the concentration is shown in parenthesis) in glutamate assembly buffer with GTP at 37°C and then purified by a cycle of assembly and disassembly prior to analysis of the assembly behavior in glutamate assembly buffer. Listed below are the specific incubation and assembly conditions for each chemical.

a Incubated at 2.0 mg/ml for 16 hrs and assembled at 0.40 mg/ml at 30°C.

b Incubated at 3.1 mg/ml for 2 hrs and assembled at 0.55 mg/ml at 37°C.

c Incubated at 2.4 mg/ml for 3 hrs and assembled at 0.37 mg/ml at 37°C.

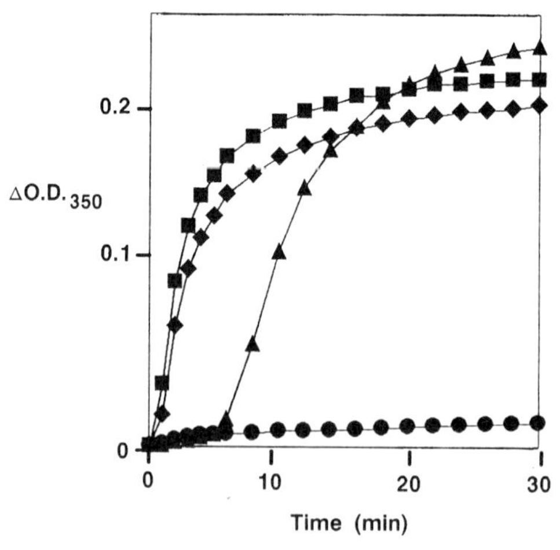

Figure 1. Comparison of 2,5-HD-treated and control tubulin assemblies in different buffer systems. Microtubule assembly was intitiated by raising the temperature to 37°C in the presence of GTP and was observed spectrophotometrically as a change in optical density at 350 nm. Tubulin was present at 0.34 mg/ml in glutamate assembly buffer (control tubulin, ▲; 2,5-HD-treated tubulin, ♦) and at 2.0 mg/ml in Mes assembly buffer (control tubulin, ●; 2,5-HD-treated tubulin, ■).

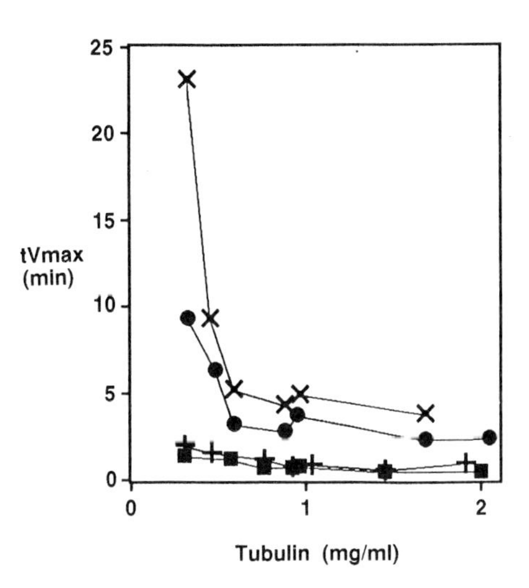

Figure 2. Comparison of the temperature and concentration dependence of 2,5-HD-treated and control tubulin polymerization. Tubulin at a variety of concentrations was analyzed for the nucleation time (tVmax) of assembly in glutamate assembly buffer in the presence of GTP at 32.5°C or 37°C (control tubulin at 37°C, •; control tubulin at 32.5°C, X; 2,5-HD-treated tubulin at 37°C, ▪; 2,5-HD-treated tubulin at 32.5°C, +).

Inhibitors of Tubulin Assembly

Microtubule assembly is inhibited by compounds, such as colchicine (Bergen and Borisy, 1983) and bisANS (Horowitz et al., 1984), which bind to free tubulin dimer in solution. Control and 2,5-HD-treated tubulin were compared for their assembly behavior in the presence of these two inhibitors.

Control and 2,5-HD-treated tubulin were assembled in the presence of colchicine without pre-incubation or pre-incubated with colchicine in the disassembled state at 4°C for 30 minutes prior to assembly (Fig. 3). Under both conditions of colchicine exposure, control tubulin was more sensitive than 2,5-HD-treated tubulin to colchicine inhibition of assembly, though preincubation increased colchicine inhibition of assembly of 2,5-HD-treated tubulin to a greater extent than that of control tubulin. One factor in the apparent resistance of 2,5-HD-treated tubulin to colchicine inhibition of assembly could be the rapid polymerization of this altered tubulin which removes the substrate, free tubulin in solution, from colchicine binding. The irreversible binding of colchicine to tubulin is known to require an extensive period of time, 10 - 20 minutes at 37°C, (Garland, 1978). These results are consistent with the following interpretation: 1) precocious assembly of 2,5-HD-treated tubulin decreased colchicine inhibition of its assembly, 2) preincubation at 4°C for 30 minutes prior to initiating assembly enhanced colchicine sensitivity, and 3) this time of preincubation at this low temperature was insufficient to fully inactivate 2,5-HD-treated tubulin since it remains less sensitive to colchicine inhibition of assembly than control tubulin.

BisANS binds to a site on tubulin distinct from the colchicine binding site (Prasad et al., 1986). Control and 2,5-HD-treated tubulin were assembled in the presence of bisANS (Fig. 4). When present in low concentrations, bisANS increased the nucleation time and decreased the maximal velocity and total turbidity for the assembly of control tubulin, as previously observed (Horowitz et al., 1984). In contrast, bisANS had no effect upon the nucleation time or maximal velocity for the assembly of 2,5-HD-treated tubulin assembly while increasing the final turbidity of the polymer solution. Examination of negatively stained preparations of 2,5-HD-treated microtubules indicated that bisANS induced an increased proportion of "sheet-like" open polymers as opposed to closed microtubules. This explains the higher turbidity observed, since open polymeric forms scatter more light than closed microtubules (Detrich et al., 1985). At these low concentrations, bisANS had little effect upon the final polymer mass formed, as determined by measuring the protein concentration in supernatants following ultracentrifugation of the assembled solutions. These experiments demonstrate that bisANS alters microtubule assembly of both control and 2,5-HD-treated tubulin; however, the nature of the bisANS-induced microtubule assembly abnormality may differ.

DISCUSSION

As shown by the experiments comparing 2,5-HD incubation in glutamate and Mes assembly buffers, tubulin must be in a stable, presumably polymeric, form to produce the alteration in assembly behavior. There are three factors which are likely to contribute to this requirement. First, a stabilizing agent such as 1 M sodium glutamate, which is known to slow the rate of tubulin denaturation (Arakawa and Timasheff, 1984), will increase the yield of active tubulin following a 16 hour

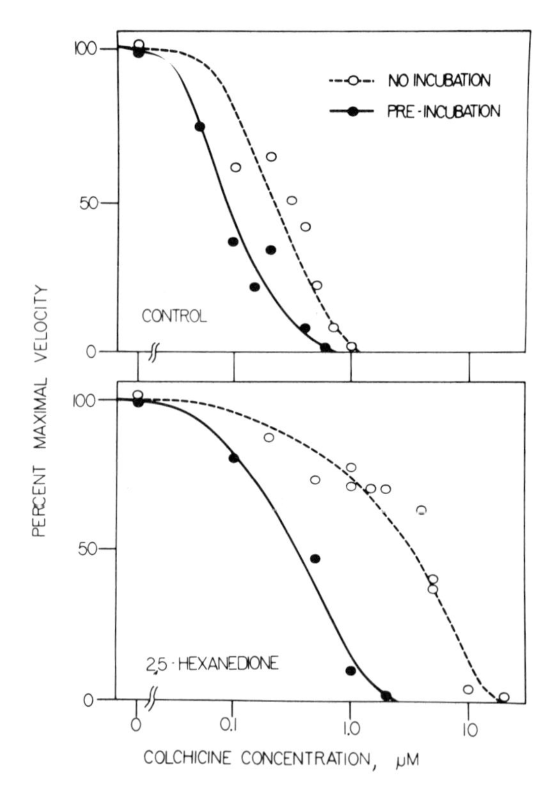

Figure 3. Colchicine inhibition of control and 2,5-HD-treated tubulin assembly. Tubulin at 0.4 mg/ml was assembled in glutamate assembly buffer at 37°C in the presence of GTP and various amounts of colchicine. Data are expressed as the percent of Vmax of assembly at each colchicine concentration with the Vmax without colchicine as 100%. Top, untreated tubulin (control); bottom, 2,5-HD-treated tubulin (2,5-hexanedione). Dashed lines, colchicine was added at the time of initiation of assembly; solid lines, tubulin was pre-incubated with colchicine for 30 minutes at 4°C before the initiation of assembly.

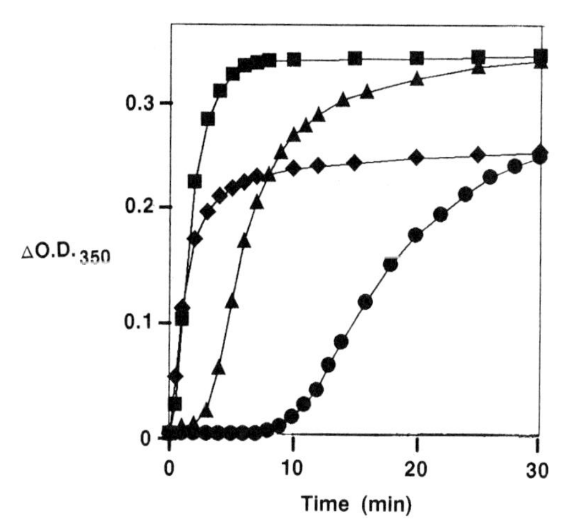

Figure 4. Effects of bisANS on control and 2,5-HD-treated tubulin polymerization. Tubulin at 0.4 mg/ml in glutamate assembly buffer was assembled at 37°C in the presense of GTP with and without added bis-ANS (control tubulin without bisANS, ▲; control tubulin with 20 mM bisANS, ●; 2,5-HD-treated tubulin without bisANS, ♦; 2,5-HD-treated tubulin with 20 mM bisANS, ■).

incubation at 37°C. Second, previous studies have shown that modification of one or two highly reactive tubulin lysine e-amines by reductive methylation results in inhibition of microtubule assembly (Szasz et al., 1982). These highly reactive amines are likely sites of pyrrole formation by 2,5-HD and are protected from modification when tubulin is polymerized in the microtubule form (Szasz et al., 1982). Finally, the requirement that crosslinking take place for the induction of the microtubule assembly abnormality, as shown in this study comparing the effects of 2,5-HD and AcHD and previous work (Boekelheide, 1987b), argues that a pro-assembly microtubule conformation of tubulin is generated by chemical fixation of tubulin in its microtubule form (Sioussat and Boekelheide, 1989).

This study demonstrates that a variety of crosslinking agents may alter microtubule assembly behavior. Both glutaraldehyde and dimethylsuberimidate produced microtubule assembly abnormalities (a decreased nucleation time and an increased maximal velocity of assembly) which were qualitatively similar to those produced by 2,5-HD. Both glutaraldehyde (Hardy et al., 1979) and dimethylsuberimidate (Galella and Smith, 1982) produce crosslinks between amine groups on proteins. Though the chemical nature of the 2,5-HD-induced crosslink in tubulin has not been determined, g-diketones are known to produce crosslinks which involve protein amines (Boekelheide et al., 1988; St. Clair et al., 1988). Thus, a common feature of these agents which induce microtubule assembly enhancement is their amine reactivity and crosslinking ability.

The temperature dependence of microtubule assembly may be relevant to the etiology of the testicular injury produced by experimental cryptorchidism. In experimental cryptorchidism, testicular injury is produced by raising the gonadal temperature from that of the scrotum (approximately 32.5°C) to body temperature (Nelson, 1951). The rise in gonadal temperature leads to germ cell loss within days by an as yet unexplained mechanism. Abnormalities have been observed in germ cell DNA synthesis (Nishimune and Aizawa, 1978), Leydig cell function (Demura et al., 1987) and in the ultrastructure and secretory capability of Sertoli cells (Hagenas and Ritzen, 1976; Kerr et al., 1979).

As show in this study, microtubule assembly at body temperature, as opposed to scrotal temperature, resulted in earlier and more rapid tubulin polymerization, an effect similar to that produced by 2,5-HD modification of tubulin. Raising the temperature of control tubulin assembly from 32.5°C to 37°C produced a 32% decrease in tVmax at 0.42 mg/ml. This extent of assembly alteration can be compared with that observed in previous *in vitro* studies of microtubule assembly following *in vivo* 2,5-HD exposure. Intoxication with 1% 2,5-HD in the drinking water produced a 27% - 45% decrease in the tVmax for assembly of purified testis tubulin at concentrations of 0.34 - 0.55 mg/ml (Boekelheide, 1987a; Boekelheide, 1988b; Boekelheide and Eveleth, 1988). Thus, the extent of the decrease in control tubulin tVmax produced by raising the temperature from 32.5°C to 37°C is similar to the decrease in tVmax produced by *in vivo* 2,5-HD exposure. Clearly, the *in vivo* environment is complex and accessory proteins may be capable of ameliorating temperature-induced alterations in microtubule assembly; nonetheless, it is interesting to speculate that a temperature-related microtubule assembly abnormality underlies germ cell loss in experimental cryptorchidism.

Colchicine is a well studied inhibitor of microtubule assembly. In previous experiments, we have shown that 2,5-HD-modified tubulin binds colchicine to the same extent as control tubulin when incubated at 37°C for 1 hr in a buffer which does not support polymerization (Boekelheide, 1987a). Our current results indicate that 2,5-HD-treated tubulin is protected from inactivation by low concentrations of colchicine due to its enhanced polymerization.

The assembly alterations produced by bisANS were distinctly different for unmodified and 2,5-HD-treated tubulin: while unmodified tubulin showed an increased nucleation time and a slowed elongation rate, 2,5-HD-treated-tubulin showed only an increase in the final turbidity of the assembled solution. These results suggest the presence of differences in the binding of bisANS to control and 2,5-HD-treated tubulin. However, these studies were performed in the highly supportive glutamate assembly buffer which may mask subtle alterations in the assembly behavior of 2,5-HD-treated tubulin. Additional experiments in a less supportive buffer, such as Mes assembly buffer, will be required to appreciate the extent of bisANS inhibition of 2,5-HD-treated tubulin assembly.

Biological systems are clearly sensitive to alterations in the dynamic properties of microtubules. Tubulin purified from brains and testes of 2,5-HD-exposed animals manifests altered microtubule assembly behavior and microinjection of 2,5-HD-treated tubulin into sea urchin zygotes perturbs the mitotic spindle. The microtubule assembly abnormality resulting from *in vivo* 2,5-HD exposure can be mimicked by *in vitro* incubation of purified tubulin with 2,5-HD. As shown in these experiments, *in vitro* co-incubation requires both a stable, presumably polymeric, form of tubulin as a substrate and chemical crosslinking of the protein. 2,5-HD-treated tubulin is strikingly altered in its temperature and concentration dependence of assembly and its response to common microtubule assembly inhibitors. The profound effects of 2,5-HD treatment upon microtubule dynamics indicate the potential sensitivity of polymeric biologic structures to xenobiotic disruption.

ACKNOWLEDGEMENTS

This work was supported by NIH grants K04 ES00193 and R01 ES05033 and the International Life Sciences Institute Research Foundation.

REFERENCES

Anthony, D.C., Boekelheide, K., Graham, D.G. (1983a). The effect of 3,4-dimethyl substitution on the neurotoxicity of 2,5-hexanedione. I. Accelerated clinical neuropathy is accompanied by more proximal axonal swellings. *Toxicol. Appl. Pharmacol.* 71, 362-371.

Anthony, D.C., Boekelheide, K., Anderson, C.W., Graham, D.G. (1983b). The effect of

3,4-dimethyl substitution on the neurotoxicity of 2,5-hexanedione. II. Dimethyl substitution accelerates pyrrole formation and protein crosslinking. *Toxicol. Appl. Pharmacol.* **71**, 372-382.
Arakawa, T., Timasheff, S.N. (1984). The mechanism of action of Na glutamate, lysine HCl, and piperazine-N,Ni-bis[2-ethanesulfonic acid] in the stabilization of tubulin and microtubule formation. *J. Biol. Chem.* **259**, 4979-4986.
Bergen, L.G., Borisy, G.G. (1983). Tubulin-colchicine complex inhibits microtubule elongation at both plus and minus ends. *J. Biol. Chem.* **258**, 4190-4194.
Boekelheide, K. (1987a). 2,5-Hexanedione alters microtubule assembly. I. Testicular atrophy, not nervous system toxicity, correlates with enhanced tubulin polymerization. *Toxicol. Appl. Pharmacol.* **88**, 370-382.
Boekelheide, K. (1987b). 2,5-Hexanedione alters microtubule assembly. II. Enhanced polymerization of crosslinked tubulin. *Toxicol. Appl. Pharmacol.* **88**, 383-396.
Boekelheide, K. (1988a). Rat testis during 2,5-hexanedione intoxication and recovery. I. Dose response and the reversibility of germ cell loss. *Toxicol. Appl. Pharmacol.* **92**, 18-27.
Boekelheide, K. (1988b). Rat testis during 2,5-hexanedione intoxication and recovery. II. Dynamics of pyrrole reactivity, tubulin content and microtubule assembly. *Toxicol. Appl. Pharmacol.* **92**, 28-33.
Boekelheide, K., Anthony, D.C., Giangaspero, F., Gottfried, M.R., Graham, D.G. (1988). Aliphatic diketones: influence of dicarbonyl spacing on amine reactivity and toxicity. *Chem. Res. Toxicol.* **1**, 200-203.
Boekelheide, K., Eveleth, J. (1988). The rate of 2,5-hexanedione intoxication, not total dose, determines the extent of testicular injury and altered microtubule assembly in the rat. *Toxicol. Appl. Pharmacol.* **94**, 76-83.
Boekelheide, K., Neely, M.D., Sioussat, T. (1989). The Sertoli cell cytoskeleton: A target for toxicant-induced germ cell loss. *Toxicol. Appl. Pharmacol.* **101**, 373-389.
Chapin, R.E., Morgan, K.T., Bus, J.S. (1983). The morphogenesis of testicular degeneration induced in rats by orally administered 2,5-hexanedione. *Exptl. Mol. Pathol.* **38**, 149-169.
Chapin, R.E., Norton, R.M., Popp, J.A., and Bus, J.S. (1982). The effects of 2,5-hexanedione on reproductive hormones and testicular enzyme activities in the F-344 rat. *Toxicol. Appl. Pharmacol.* **62**, 262-272.
DeCaprio, A.P., Olajos, E.J., Weber, P. (1982). Covalent binding of a neurotoxic n-hexane metabolite: conversion of primary amines to substituted pyrrole adducts by 2,5-hexanedione. *Toxicol. Appl. Pharmacol.* **65**, 440-450.
Demura, R., Suzuki, T., Nakamura, S., Komatsu, H., Jibiki, K., Odagiri, E., Demura, H., Shizume, K. (1987). Effect of uni- and bilateral cryptorchidism on testicular inhibin and testosterone secretion in rats. *Endocricol. Japon.* **34**, 911-917.
Detrich, H.W., III, Jordan, M.A., Wilson, L., Williams, R.C., Jr. (1985). Mechanism of microtubule assembly. Changes in polymer structure and organization during assembly of sea urchin egg tubulin. *J. Biol. Chem.* **260**, 9479-9490.
Galella, G., Smith, D.B. (1982). The cross-linking of tubulin with imidoesters. *Can. J. Biochem.* **60**, 71-80.
Garland, D.L. (1978). Kinetics and mechanism of colchicine binding to tubulin. Evidence for ligand-induced conformational change. *Biochemistry* **17**, 4266-4272.
Gekko, K., Timasheff, S.N. (1981). Mechanism of protein stabilization by glycerol. Preferential hydration in glycerol-water mixtures. *Biochemistry* **20**, 4667-4676.
Genter, M.B., Szakal-Quin, Gy., Anderson, C.W., Anthony, D.C., Graham, D.G. (1987). Evidence that pyrrole formation is a pathogenetic step in g-diketone neuropathy. *Toxicol. Appl. Pharmacol.* **87**, 351-362.
Graham, D.G., Anthony, D.C., Boekelheide, K., Maschmann, N.A., Richards, R.G., Wolfram, J.W., Shaw, B.R. (1982). Studies of the molecular pathogenesis of hexane neuropathy. II. Evidence that pyrrole derivatization of lysyl residues leads to protein crosslinking. *Toxicol. Appl. Pharmacol.* **64**, 415-422.
Hagenas, L., Ritzen, E.M. (1976). Impaired Sertoli cell function in experimental cryptorchidism in the rat. *Mol. Cell. Endocrinol.* **4**, 25-34.
Hamel, E., Lin, C.M. (1981). Glutamate-induced polymerization of tubulin. Characterization of the reaction and application to the large-scale purification of tubulin. *Arch. Biochem. Biophys.* **209**, 29-40.

Hardy, P.M., Hughes, G.J., Rydon, H.N. (1979). The nature of the cross-linking of proteins by glutaraldehyde. Part 2. The formation of quaternary pyridinium compounds by the action of glutaraldehyde on proteins and the identification of a 3-(2-piperidyl)-pyridinium derivative, anabilysine, as a cross-linking entity. *J. Chem. Soc. Perkin* I, 2282-2288.

Horowitz, P., Prasad, V., Luduena, R.F. (1984). Bis[1,8-anilinonaphthalenesulfonate]. A novel and potent inhibitor of microtubule assembly. *J. Biol. Chem.* **259**, 14647-14650.

Kerr, J.B., Rich, K.A., de Kretser, D.M. (1979). Effects of experimental cryptorchidism on the ultrastructure and function of the Sertoli cell and peritubular tissue of the rat testis. *Biol. Reprod.* **21**, 823-838.

Krasavage, W.J., O'Donoghue, J.L., DiVincenzo, G., Terhaar, C.J. (1980). The relative neurotoxicity of methyl n-butyl ketone, n-hexane and their metabolites. *Toxicol. Appl. Pharmacol.* **52**, 433-441.

Lee, J.C., Timasheff, S.N. (1977). *In vitro* reconstitution of calf brain microtubules. Effects of solution variables. *Biochemistry* **16**, 1754-1764.

Neely, M.D., Boekelheide, K. (1988). Sertoli cell processes have axoplasmic features: an ordered microtubule distribution and an abundant high molecular weight microtubule associated protein (cytoplasmic dynein). *J. Cell Biol.* **107**, 1767-1776.

Nelson, W.O. (1951). Mammalian spermatogenesis: effects of experimental cryptorchidism in the rat and non-descent of the testis in man. *Recent Prog. Horm. Res.* **6**, 29-62.

Nishimune, Y., Aizawa, S. (1978). Temperature sensitivity of DNA synthesis in mouse testicular germ cells *in vitro*. *Exp. Cell Res.* **113**, 403-408.

Prasad, A.R.S., Luduena, R.F., Horowitz, P.M. (1986). Bis[8-anilinonaphthalene-1-sulfonate] as a probe for tubulin decay. *Biochemistry* **25**, 739-742.

Russell, L.D., Malone, J.P., MacCurdy, D.S. (1981). Effect of the microtubule disrupting agents, colchicine, and vinblastine, on seminiferous tubule structure in the rat. *Tissue Cell* **13**, 349-367.

St. Clair, M.B.G., Amarnath, V., Moody, A., Anthony, D.C., Anderson, C.W., Graham, D.G. (1988). Pyrrole oxidation and protein cross-linking as necessary steps in the development of g-diketone neuropathy. *Chem. Res. Toxicol.* **1**, 179-185.

Shelanski, M.L., Gaskin, F., Cantor, C.R. (1973). Microtubule assembly in the absence of added nucleotides. *Proc. Nat. Acad. Sci. USA* **70**, 765-768.

Sioussat, T., Boekelheide, K. (1989). Selection of a nucleation-promoting element following chemical modification of tubulin. *Biochemistry* **28**, 4435-4443.

Szasz, J., Burns, R., Sternlicht, H. (1982). Effects of reductive methylation in microtubule assembly. *J. Biol. Chem.* **257**, 3697-3704.

Vogl, A.W., Lin, Y.C., Dym, M., Fawcett, D.W. (1983a). Sertoli cells of the golden-mantled ground squirrel (*Spermophilus lateralis*): a model system for the study of shape change. *Am. J. Anat.* **168**, 83-98.

Vogl, A.W., Linck, R.W., Dym, M. (1983b). Colchicine-induced changes in the cytoskeleton of the golden-mantled ground squirrel (*Spermophilus lateralis*) Sertoli cells. *Am. J. Anat.* **168**, 99-108.

Voter, W.A., Erickson, H.P. (1984). The kinetics of microtubule assembly. Evidence for a two-stage nucleation mechanism. *J. Biol. Chem.* **259**, 10430-10438.

Weisenberg, R.C. (1972). Microtubule formation *in vitro* in solutions containing low calcium concentrations. *Science* **177**, 1104-1105.

THE ROLE OF HEPATIC METABOLITES OF BENZENE IN BONE MARROW PEROXIDASE-MEDIATED MYELO- AND GENOTOXICITY

George Kalf, Robert Shurina, John Renz, and Michael Schlosser

The Department of Biochemistry and Molecular Biology
Jefferson Medical College of Thomas Jefferson University
Philadelphia, Pennsylvania, 19107

INTRODUCTION

Chronic exposure of humans to benzene causes bone marrow depression leading to pancytopenia and aplastic anemia (Goldstein, B.D., 1983). Benzene also causes genotoxic effects such as structural chromosome aberrations and DNA strand breaks (Dean, B.J., 1985) that might be related to the increased incidence of acute myelogenous leukemia that is associated with chronic exposure (Infante, P.F., White, M.C., 1983; Aksoy, M., 1985; Arp, E.W., et al., 1983).

Benzene metabolism, which is required for toxicity occurs predominantly in the liver via cytochrome P-450 (Sammett, D., et al., 1979; Tunek, A., 1980). Phenol (P), hydroquinone (HQ) and catechol are hepatic metabolites of benzene, and their production appears necessary for benzene-induced myelotoxicity (Arp, E.W., et al., 1983; Sammett, D., et al., 1979) but they show no overt toxicity to the liver. These metabolites are transported from the liver to the bone marrow (Rickert, D.E., et al., 1979; Greenlee, W.F., et al., 1981) where they are bioactivated in a peroxidase-mediated reaction (Sawahata, T., et al., 1985; Irons, R.D., 1985) to biological reactive compounds that bind to macromolecules and which have been implicated in mediating the toxic effects of benzene (Sammett, D., et al., 1979; Rickert, D.E., et al., 1979).

In bone marrow, benzene appears to have a cytotoxic affect on hematopoietic progenitor cells in intermediate stages of differentiation regardless of cell lineage (Irons, R.D., et al., 1979; Lee, E.W., et al., 1974; Tunek, A., et al., 1981). Injury to marrow stromal cells appears to be an important factor in benzene-induced myelosuppression (Frash, V.N., et al., 1976; Gaido, K., et al., 1986). The stromal macrophage, a cell essential in the regulation of hematopoiesis, has been implicated as a target of benzene-induced hematotoxicity (Lewis, J.G., et al., 1988; Thomas, D.J., et al., 1989; MacEachern, L., et al., 1988). Lewis et al. (1988) demonstrated *in vitro* a selective and pronounced inhibition of macrophage function following the addition of various benzene metabolites to the culture medium. Recently, Thomas and Wierda (1989) reported that bone marrow-derived macrophages, exposed to HQ or its oxidation product 1,4-benzoquinone, secreted less interleukin-1, a monokine capable of regulating the synthesis of several hematopoietic factors (Lee, M., et al., 1987). In addition, MacEachern et al. (1988) reported an activation of resident bone marrow macrophages in mice receiving benzene or P and HQ.

Macrophages contain considerable amounts of prostaglandin H synthase (PHS) an enzyme with both cyclooxygenase and peroxidase activities (Scott, W.A., et al., 1980; Ohki, S., et al., 1979). Since phenol and hydroquinone serve as reducing co-substrates

Biological Reactive Intermediates IV, Edited by C.M. Witmer *et al.*
Plenum Press, New York, 1990

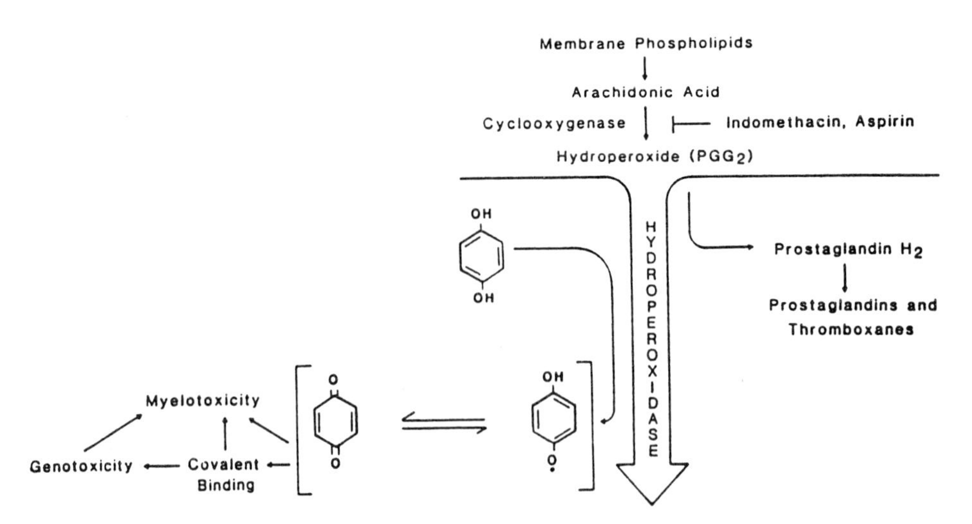

Figure 1. Postulated mechanism for the role of PHS in mediating benzene-induced bone marrow damage. Chronic exposure to benzene causes the release of arachidonic acid from the membrane lipid of marrow stromal cells particularly the resident stromal macrophage resulting in constitutive synthesis of prostaglandins. The oxidation of the co-substrate hydroquinone by PHS-peroxidase converts it to a reactive intermediate which can bind to macromolecules and cause toxicity. Inhibition of PHS-cyclooxygenase with indomethacin inhibits prostaglandin synthesis and prevents bioactivation of hydroquinone.

for the peroxidase of PHS (Markey, C.M., et al., 1987), macrophages have the capacity to oxidize these benzene metabolites to compounds capable of reacting with cellular macromolecules. Indeed, Post et al. have demonstrated that the macrophage can metabolize phenol to protein-binding species (Post, G., et al., 1986).

The role of PHS in benzene-induced myelotoxicity seems particularly relevant, as benzene administration elevates bone marrow levels of prostaglandin E_2 (Gaido, K.W., et al., 1987; Kalf, G.F., et al., 1989), a negative regulator of myelopoiesis (Gentile, P.S., et al., 1987). Nonsteroidal anti-inflammatory drugs, known inhibitors of cyclooxygenase, have been reported not only to inhibit this rise in prostaglandin, but to prevent benzene-induced myelotoxicity as well (Gaido, K.W., et al., 1987; Kalf, G.F., et al., 1989). Because nonsteroidal anti-inflammatory drugs inhibit the cyclooxygenase-catalyzed formation of prostaglandin G_2 (PGG_2), treatment with these agents would eliminate the peroxidase-catalyzed reduction of PGG_2, thereby avoiding the oxidation of phenolic co-substrates, if present, to reactive compounds.

The research to be described was designed to determine whether PHS produces benzene-induced myelo- and genotoxicity by the following proposed mechanism (Figure 1). Benzene is metabolized in the liver and bone marrow to P and HQ. Benzene per se may act on bone marrow cells to effect the constitutive release of arachidonic acid from membrane phospholipids via the activation of protein kinase C which is known to activate the arachidonic acid cascade in macrophages (Pfankuche, H.J., et al., 1986). Arachidonic acid would be converted by the cyclooxygenase component of PHS to the hydroperoxide (PGG_2). The hydroperoxide in turn would drive the endoperoxidase activity of the enzyme. The cooxidation of HQ or P during endoperoxidase conversion of PGG_2 to PGH_2, the immediate precursor molecule for prostaglandins, would result in increased levels of prostaglandins. In the case of HQ, cooxidation would generate p-benzoquinone which is known to cause genotoxic damage in the form of adducts, strand breaks, SCE and micronucleus formation (Dean, B.J., 1985). The inability of remaining viable stem and/or progenitor cells to proliferate due to the constitutive production of high levels of prostaglandins, known to be negative regulators of hematopoiesis, coupled with genotoxic damage from reactive metabolites such as p-benzoquinone might explain the benzene-induced myelotoxicity.

The experiments described herein demonstrate the arachidonic acid-dependent, indomethacin-sensitive, bioactivation of P and HQ to genotoxic compounds by macrophage peroxidase and purified PHS. Furthermore, we show that inhibition of PHS with indomethacin prevents benzene-induced myelo- and genotoxicity in mice when indomethacin is coadministered with benzene.

RESULTS

Activation of Phenol and Hydroquinone to Biological Reactive Intermediates in Macrophages

The inhibition of the toxic effects of benzene by indomethacin (Kalf, G.F., et al., 1989) suggested that metabolites of benzene such as P and/or HQ might be converted to reactive compounds by PHS-peroxidase. Consequently, we tested the ability of macrophages to effect the arachidonic acid-dependent metabolism of P and HQ to reactive species that bind to macromolecules. In these experiments mouse peritoneal macrophages or $P388D_1$ cells, which morphologically and functionally resemble macrophages, (Koren, H.S., et al., 1975) were used. Similar results can be obtained with both cell types.

Purified peritoneal macrophages incubated with $[^{14}C]P$ in the presence and absence of arachidonic acid metabolize P to reactive compounds that irreversibly bind macromolecules in a reaction that is dependent on the concentration of arachidonic acid (Figure 2). These results suggest that PHS-peroxidase in macrophages might be responsible for the cooxidation of P and/or HQ to reactive species during the benzene-induced formation of prostaglandins. As can be seen in Figure 3, $[^{14}C]HQ$ is also bioactivated by $P388D_1$ cells to species which bind to macromolecules. The addition of TPA causes the further activation of the arachidonic acid cascade which stimulates the oxidation of HQ to reactive species via PHS and this is prevented by indomethacin, a cyclooxygenase inhibitor. Benzene, which has been shown to activate protein kinase C (Roghani, M., et al., 1987) and thus the arachidonic acid cascade (Pfankuche, H.J., et al., 1986), also stimulates the bioactivation of HQ in an indomethacin-sensitive reaction.

Bioactivation of P or HQ to species which bind to macromolecules was shown to occur in a macrophage lysate. The bioactivation of P or HQ occurred in a time and concentration-dependent manner (data not shown). The binding of P or HQ equivalents to protein was decreased significantly, compared to the complete system, for incubations containing the peroxidase inhibitor aminotriazole and for reactions carried out in the absence of either the macrophage lysate or H_2O_2 (Figure 4). The effect of hydroxyl radical scavengers was investigated to determine if the macrophage lysate indirectly activated P and HQ through the generation of hydroxyl radicals. The addition of dimethylsulfoxide (DMSO) or mannitol to standard incubation mixtures unexpectedly resulted in a slight but significant increase in both P and HQ equivalents bound to TCA-precipitable material (Figure 4). The addition of the cytochrome P-450 inhibitors, metyrapone and SKF 525A to the macrophage lysate had no affect on the bioactivation of P or HQ (data not shown). These results suggested that adherent macrophages and a macrophage lysate were capable of the bioactivation of P or HQ to covalent binding species and that this was most probably catalyzed by PHS-peroxidase. We, therefore, investigated the PHS-catalyzed oxidation of HQ and examined the DNA-damaging effects of the reactive metabolite generated during the reaction.

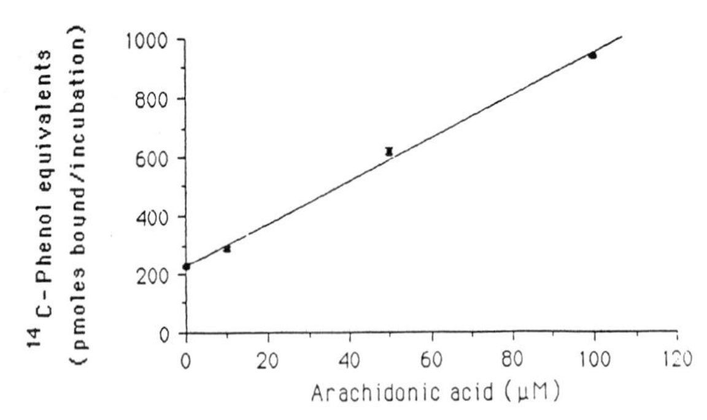

Figure 2. Aracidonic acid-dependent activation of phenol by macrophages. Incubation mixtures contained 2.3 x 10^6 macrophages, 0.25 mM [^{14}C] phenol (8000 dpm/nmole), 0.1 mM arachidonic acid diluted to a final volume of 1 mL with RPMI-1640 buffered to pH 7.3 with 5 mM HEPES. Reactions were terminated after 30 min by the addition of trichloracetic acid. Irreversible binding of [^{14}C] phenol equivalents was assessed by liquid scintillation counting after extensive washes with acetone/hexane/methanol. Reprinted from reference 26.

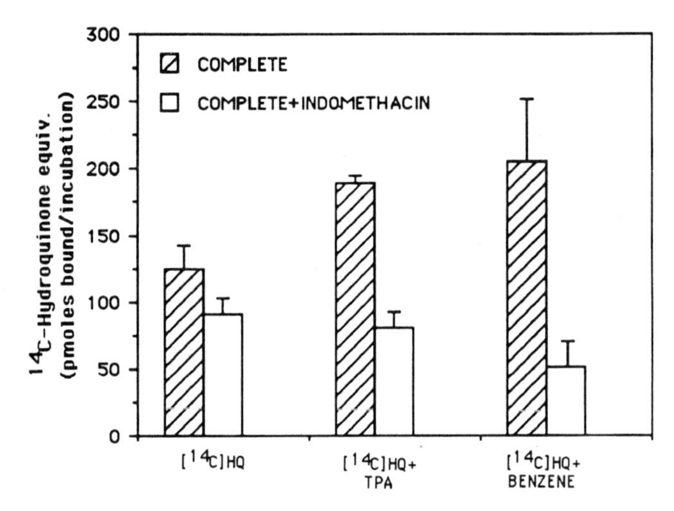

Figure 3. Covalent binding of [^{14}C] hydroquinone to macromolecules in P388D1 cells. P-388 cells (2x10^6 per incubation) were preincubated for 15 min. with either 100 mM indomethacin in ethanol or ethanol alone (1% final concentration), then treated with (50 µM) [^{14}C]HQ. Some incubations included 100 ng TPA or 170 mM benzene coadministered with the hydroquinone. After a 30 min. incubation at 37°C, 5% CO_2, cells were lysed by freezing and thawing in 10 mM Tris, 1 mM EDTA buffer pH 8.0 containing 2 mm PMSF, 0.1 mM Leupeptin, 0.1 mM pepstatin A and 0.01% Triton X-100. Macromolecules were precipitated in 10% TCA, washed 3 times with acetone followed by two washes in acetone/methanol (1:1, V:V), and radioactive HQ covalently bound to macromolecules was determined by scintillation spectrophotometry. Values represent the mean ± S.D. of three experiments.

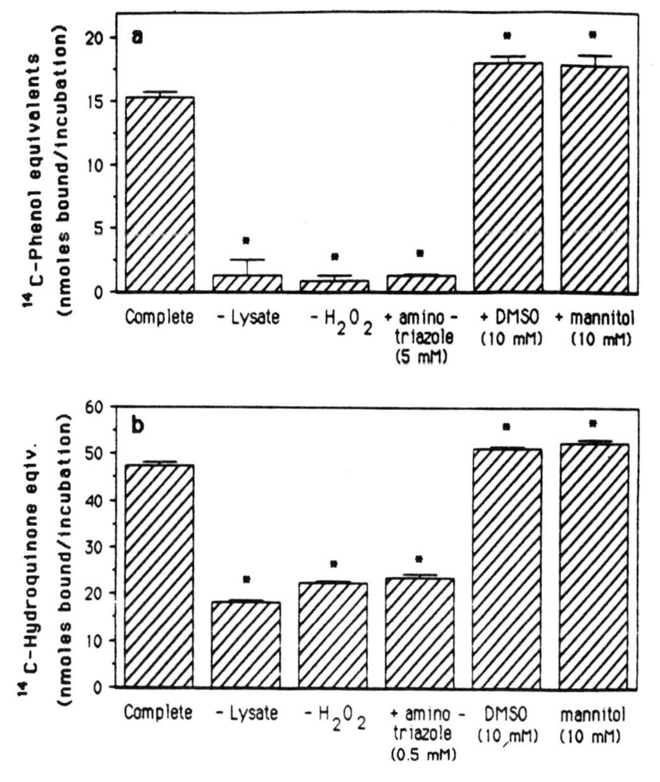

Figure 4. The effect of aminotriazole, DMSO, and mannitol on the H_2O_2-dependent activation of (a) ^{14}C-phenol and (b) ^{14}C-hydroquinone to protein-binding metabolites catalyzed by the macrophage lysate. Reprinted from reference 32.

Activation of Hydroquinone to p-Benzoquinone by PHS-Peroxidase

Purified PHS catalyzed the arachidonic acid-dependent oxidation of HQ to p-benzoquinone which was measured directly by HPLC (Table 1). p-Benzoquinone formation was dependent on the presence of enzyme and arachidonate. Addition of indomethacin significantly inhibited the formation of p-benzoquinone as did the addition of cysteine. The use of H_2O_2 in place of arachidonate also resulted in the formation of p-benzoquinone (data not shown). As expected, indomethacin had no affect on the H_2O_2-dependent oxidation of HQ to p-benzoquinone. When cysteine was added to standard reaction mixtures, p-benzoquinone was not detected (Table 1). A cysteine conjugate resulting from the PHS-catalyzed oxidation of HQ was detected which co-eluted on HPLC with an authentic monocysteine-hydroquinone conjugate synthesized as previously described (Schlosser, M.J., et al., 1989). As with the formation of p-benzoquinone, monocysteine-hydroquinone formation was dependent on the presence of PHS, arachidonate and cysteine (Table 2). Indomethacin inhibited the arachidonate-dependent but not the H_2O_2-dependent formation of monocysteine-hydroquinone.

In order to determine whether PHS can oxidize HQ to a DNA-binding species, calf thymus DNA was incubated with PHS and arachidonic acid for 10 min (Figure 5). Radioactivity bound to calf thymus DNA increased over the 10 min incubation period, whereas reaction mixtures containing heat-inactivated PHS remained constant over

Table 1. Archidonic acid-dependent Formation of p-Benzoquinone by Prostaglandin H Synthase°

System	nmol/incubation
complete [a]	17.3 ± 3.6
- PHS	0.9 ± 0.2
- arachidonate	1.0 ± 0.1
+ indomethacin (10 μM)	4.3 ± 0.7
+ cysteine (100 μM)	<0.1

[a] Complete system at 37°C contained hydroquinone (100 μM), PHS (5 μg/mL), hematin (1 μM), and arachidonic acid (100 μM) in 0.1 M Na^+/K^+ phosphate buffer, pH 7.0 in a final volume of 0.5 mL. Reactions were preincubated with indomethacin or 1% ethanol for 10 min. p-Benzoquinone was measured using HPLC with electrochemical detection. Data is expressed as the means ± standard deviations.

time. Similar results were obtained when mtDNA was used in place of calf thymus DNA (Table 3). mtDNA-binding was dependent on enzyme and was inhibited by indomethacin. Thus [^{14}C]HQ, functioning as a co-substrate for PHS-peroxidase, is cooxidized in an arachidonic acid- and time-dependent reaction to compound(s) that bind to mtDNA. [^{14}C]HQ-derived mtDNA adducts were isolated as deoxyribonucleoside adducts by hydrophobic chromatography on an LH-20 column following sequential digestion of radiolabeled mtDNA with nucleases (data not shown). A ^{14}C-labeled adduct was found to co-migrate on a thin layer cellulose chromatographic plate (Figure 6) with a 2'-deoxyguanosine (2'-dG) adduct standard. The structure of the 2'dG adduct, (3'OH) benzetheno (1,N2) deoxyguanosine has been reported (Snyder, R., et al., 1987) (Figure 7).

Table 2. Arachidonic acid-dependent Formation ofMonocysteine-hydroquinone by Prostaglandin H Synthase

System	nmol/incubation
complete a	14.6 ± 0.2
- PHS	0.8 ± 0.1
- arachidonate	0.6 ± 0.1
- cysteine	<0.1
+ indomethacin (10 μM)	6.3 ± 0.3

[a] Complete system at 37°C contained hydroquinone (100 μM), PHS (5 μg/mL), hematin (1 μM), cysteine (100 μM) and arachidonic acid (100 μM) in 0.1 M Na^+/K^+ phosphate buffer, pH 7.0 in a final volume of 0.5 mL. Reactions were preincubated with indomethacin or 1% ethanol for 10 min. Monocysteine-hydroquinone was measured using HPLC with electrochemical detection. Data is expressed as the means ± standard deviations.

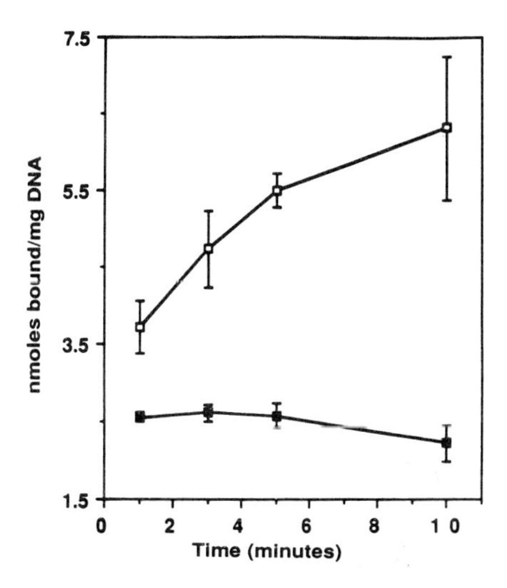

Figure 5. Time course for DNA binding by hydroquinone during PHS-catalyzed reactions: (?) contained calf thymus DNA (100 mg/mL), [^{14}C]hydroquinone (100 mM; 10,000 dpm/nmol), PHS (5 mg/mL) with hematin (1 mM), and arachidonic acid (10 mM); (?) as above except PHS was heat-inactivated. Data points represent the means ± standard deviations of triplicate determinations.

HQ has been reported to produce strand breaks in DNA (Sawahata, T., et al., 1985). Therefore, we tested for the ability of benzoquinone formed from HQ by PHS-peroxidase to cause strand breakage by determining its ability to nick supercoiled Bluescript plasmid DNA, thereby converting it to its open circle form. Bluescript is a small (3.9 Kb) circular plasmid that migrates slower on an agarose gel in its nicked open circular form than its supercoiled form. DNA damage was detected by incubating supercoiled Bluescript plasmid with PHS, arachidonic acid and HQ (Fig. 8). The oxidation of HQ by PHS produced product(s) capable of causing single strand breaks in the plasmid, thereby converting it to a nicked circle that could be isolated on an agarose gel. These PHS-catalyzed DNA strand breaks were found to be dependent on native enzyme and HQ for plasmid nicking, and were inhibited by indomethacin. When the autoradiogram prepared from a blot hybridization of the gel was scanned using an LKB laser scan densitometer, it was found that the intensity of the open circle band in the complete system containing active enzyme (lane 3) was 88% above the complete system that had been inhibited by indomethacin (lane 6).

The Role of PHS in the Induction of Myelo- and Genotoxicity in Mice by Benzene: The Effect of Indomethacin on Benzene-Induced Bone Marrow Depression and Micronucleus Formation in C57Bl/6 Mice

The oxidation of HQ to p-benzoquinone, the formation of a [^{14}C] HQ-derived adduct in DNA and the ability of p-benzoquinone to nick supercoiled plasmid DNA were inhibited by indomethacin. These results suggest that PHS-peroxidase plays a role in the oxidation of HQ to p-benzoquinone, a putative genotoxic metabolite of benzene and that, *in vivo*, toxicity to marrow progenitor cells induced by benzene might result from the production by PHS of biological reactive species from P and HQ acting as co-substrates for PHS-peroxidase. If this is indeed the case, then the myelotoxic effects of benzene should be prevented by the coadministration of PHS inhibitors such as indomethacin or other nonsteroidal anti-inflammatory drugs.

Table 3. Prostaglandin H Synthase-Catalyzed Activation of Hydroquinone to DNA-binding Metabolite(s)°

System	nmol/incubation
complete [a]	5.11 ± 0.55
- PHS	1.98 ± 0.46
+ indomethacin (10 mM)	2.07 ± 0.45

[a] Complete system contained [^{14}C]hydroquinone (100 μM; 10,000 dpm/nmol), PHS (5 μg/mL), hematin (1 μM), arachidonic acid (10 μM), and mtDNA (100 μg/mL) in 0.1 M K^+ phosphate buffer, pH 7.0 in a final volume of 0.5 mL. Reactions were preincubated with indomethacin or 1% ethanol for 1 min. Radioactivity bound to mtDNA was measured by liquid scintillation spectrometry. Data is expressed as the means ± standard deviations.

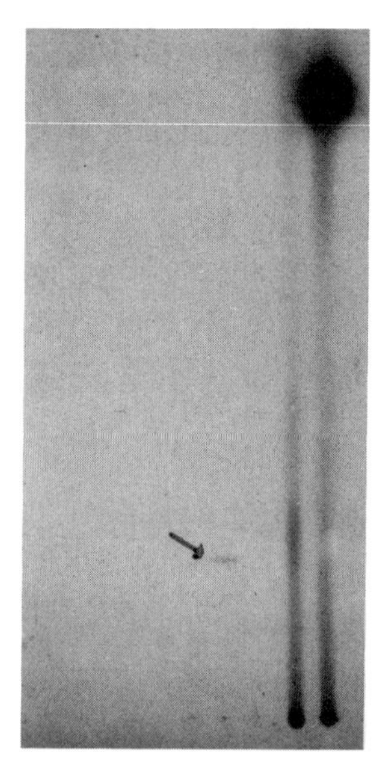

Figure 6. Autoradiogram of the TLC plate. Arrow indicates the position of the radiolabeled [^{14}C] p-benzoquinone-mtDNA deoxynucleoside adduct (Rf=0.22). The [^{14}C] p-benzoquinone-deoxyguanosine adduct standard appears in lane 2 at a Rf of 0.22. Lane 1 represents [^{14}C] hydroquinone alone.

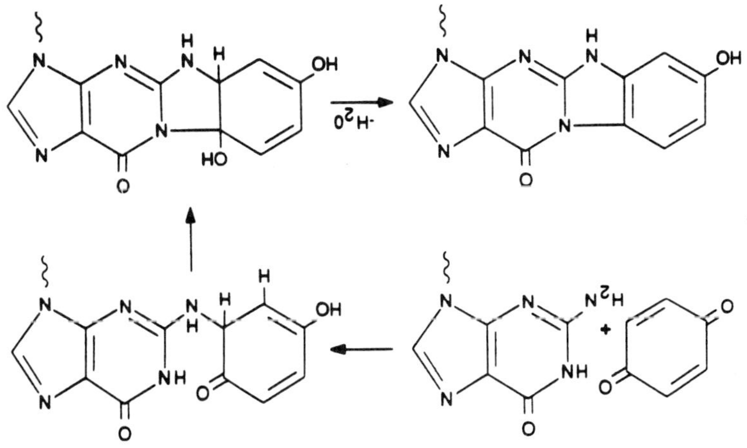

Figure 7. Proposed mechanism for the formation of adduct 2.

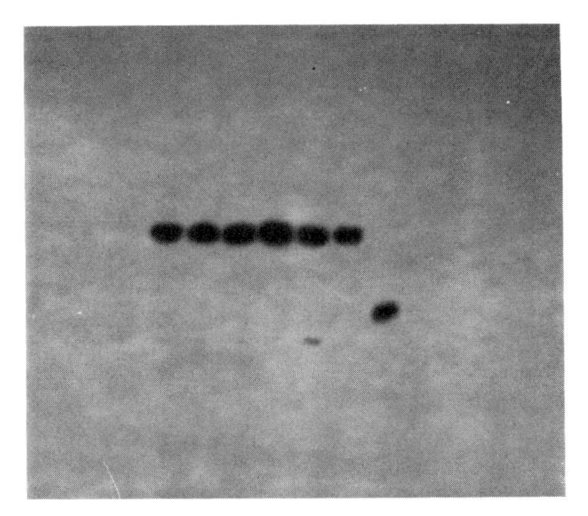

Figure 8. Blot hybridization of 75 ng Bluescript plasmid DNA incubated with the PHS complete system or with heat inactivated PHS in place of active enzyme, in the absence and presence of 10 mM indomethacin. Enzyme incubations and blot hybridizations were carried out routinely done in the lab. Lane 1: Linear Bluescript plasmid, digested with ECO R1. Lane 2: Supercoiled Bluescript plasmid. Lane 3: Supercoiled Bluescript plasmid incubated with the PHS complete system. Lane 4: Supercoiled Bluescript plasmid incubated with the PHS complete system, omitting HQ. Lane 5: Supercoiled Bluescript plasmid and heat inactivated enzyme, incubated with the PHS complete system. Lane 6: Supercoiled Bluescript plasmid and the PHS complete system in the presence of 10 mM indomethacin. Lane 7: Supercoiled Bluescript plasmid and heat inactivated enzyme incubated with the PHS complete system and 10 mM indomethacin.

Depression of bone marrow cellularity is a commonly used parameter to determine benzene toxicity. Administration of benzene (600 mg/kg body weight) twice a day for two days caused a significant decrease in bone marrow cellularity measured as a decrease in nucleated cells in the marrow (Table 4). The effect was dose-dependent and there was no change in body weight or viability of the cells flushed from the femur (data not presented). Coadministration of indomethacin (2 mg/kg body weight) with benzene completely prevented the depression of bone marrow cellularity. Indomethacin alone had no effect. Meclofenamate (4 mg/kg) or aspirin (50 mg/kg), which inhibit cycloxygenase activity of PHS by mechanisms different from that of indomethacin, also significantly prevented benzene-induced bone marrow depression (data not presented).

Under conditions where benzene causes significant myelotoxicity in the form of decreased bone marrow cellularity, it also causes genotoxicity measured by an increase in the frequency of micronucleus formation in peripheral blood polychromatic erythrocytes (PCE). It can be seen in Table IV that benzene at 600 mg/kg body weight also caused a 5.7-fold increase in the frequency of micronucleus formation. Coadministration of indomethacin with benzene prevented the increased frequency of micronucleus formation in PCE without affecting the division or maturation of nucleated erythroid precursors (NCE) as indicated by no change in the PCE/NCE. The administration of indomethacin alone had no affect on micronucleus formation.

Table 4. Prevention of Benzene-Induced Bone-Marrow Depression and Micronucleus Formation in Mice by Indomethacin

Group [a]	Nucleated bone marrow cells x 10^6/femur	Micronuclei/10^3		PCE/NCE x 10^3
		PCE	NCE	
control [b]	11.4 + 1.6	4.5 + 1.0	1.0 + 0.9	34.0 + 4.5
+ benzene [c]	6.2 + 1.6*	25.5 + 9.2*	0.5 + 0.4	24.0 + 10.4
+ benzene + indo [d]	9.4 + 0.7	10.0 + 5.0	0.75 + 0.3	32.4 + 3.0
+ indo [d]	12.4 + 1.8	5.5 + 3.0	1.7 + 0.8	32.0 + 7.6

[a] Each group consisted of 4 C57Bl/6 male mice.
[b] Control animals received corn oil and/or a solution of 4.2% ethanolic PBSA.
[c] Benzene (600 mg/kg in corn oil) was administred ip twice daily for two days.
[d] Indomethacin (2 mg/kg in 4.2% ethanolic PBSA was administered ip twice daily for two days.
Data are expressed as means ± SD
* $p \leq 0.01$ compared to control and indomethacin-treated groups.

DISCUSSION

Adherent macrophages bioactivate P in an arachidonic acid-dependent reaction to species capable of covalent binding to macromolecules (Figure 2). Similar results are obtained with HQ. The addition of TPA or benzene, which cause the release of arachidonic acid from plasma membrane lipids, stimulate the oxidation of [^{14}C] HQ to reactive species (Figure 3). This is prevented by indomethacin, a cyclooxygenase inhibitor. Similar results can be obtained with a lysate of adherent macrophages. The H_2O_2-dependent activation of P or HQ to biological reactive species is inhibited by aminotriazole, a peroxidase inhibitor (Figure 4), but not by inhibitors of cytochrome P-450 such as metyrapone or SKF 525A, or by hydroxyl radical scavengers (Figure 4). The arachidonic acid-dependent activation of HQ or P by the macrophage lysate was not attempted since the concentration of CTAB used to solubilize the membrane-bound peroxidase also inhibits the cyclooxygenase activity of PHS (data not shown). These results strongly implicate PHS-peroxidase in the bioactivation of these benzene metabolites to biological reactive compounds in macrophages. Further studies using purified PHS confirmed that the enzyme was capable of the arachidonic acid-dependent and indomethacin-sensitive oxidation of HQ to p-benzoquinone (Table 1) which could be measured directly by HPLC or trapped with cysteine as a monocysteine-hydroquinone conjugate (Schlosser, M.J., et al., 1989) (Table 2). The oxidation product(s) of HQ were also able to covalently bind to DNA (Figure 5).

[^{14}C] HQ, functioning as a cosubstrate for PHS-peroxidase, was also shown to be oxidized in an arachidonic acid- and time-dependent reaction to compound(s) that interact with DNA to form adducts and strand breaks. A ^{14}C-labeled adduct was isolated from mtDNA which was found to co-migrate on a thin layer cellulose chromatographic plate (Fig. 6) with a known 2'-deoxyguanosine adduct standard. The structure of the 2'dG adduct was identified as an adduct of p-benzoquinone with guanine to form (3'OH) benzetheno (1,N_2) deoxyguanosine (adduct 2; Fig. 7). Interaction of HQ with PHS-peroxidase and arachidonic acid in the presence of supercoiled Bluescript plasmid resulted in single strand breaks in the supercoiled DNA converting it to an open circle form. The PHS-catalyzed strand breaks were dependent on native enzyme, HQ, and arachidonic acid and were prevented by indomethacin (Fig. 8).

Taken together, these results implicate PHS-peroxidase in the bioactivation of HQ to p-benzoquinone, a genotoxic metabolite of benzene, and suggest that *in vivo* toxicity to marrow progenitor cells might result from the production of reactive species by PHS. That this may be the case is evidenced by the facts that benzene-induced depression of bone marrow cellularity and increased frequency of micronucleus formation in mice can be prevented by the coadministration of PHS inhibitors such as indomethacin (Table 4). Indomethacin does not appear to be modulating an alteration of bone marrow cellularity related to an initial benzene-induced inflammatory reaction since data obtained in a subchronic exposure study (to be published elsewhere) indicate that indomethacin can prevent the myelo- and genotoxic effects induced by benzene over a 3 week period. The amount of indomethacin used in our protocols had no effect on hepatic microsomal P-450 content or benzene hydroxylase activity (Pirozzi, S., et al., 1989) and the effective level of indomethacin was not sufficient to affect dihydrodiol dihydrogenase, phospholipase A_2 or myeloperoxidase, enzymes which might be expected to be involved in benzene metabolism.

REFERENCES

Aksoy, M. (1985). Malignancies due to occupational exposure to benzene. *Am. J. Ind.Med.*, **7**, 395-402.

Arp, E.W., Wolf, P.H., Checkoway, H. (1983). Lymphocyte leukemia and exposures to benzene and other solvents in the rubber industry. *J. Occup. Med.*, **25**, 598-602.

Dean, B.J. (1985). Recent findings on the genetic toxicology of benzene, toluene, xylenes and phenols. *Mutat. Res.*,**154**, 153-181.

Frash, V.N., Yushkov, B.G., Karalulov, A.V., and Suratov, V.L. (1976). Mechanism of action of benzene on hematopoiesis. Investigation of hematopoietic stem cells. *Bull. Exp. Biol. Med.*, **83**, 985-987.

Gaido, K., and Wierda, D. (1986). Hydroquinone suppression of bone marrow stromal cell supported hematopoiesis *in vitro* is associated with prostaglandin E_2 production. *Toxicologist.*, **6**, 286.

Gaido, K.W., and Wierda, D. (1987). Suppression of bone marrow stromal cell function by benzene and hydroquinone is ameliorated by indomethacin. *Toxicol. Appl. Pharmacol.*, **89** 378-390.

Gentile, P.S., and Pelus, L.M. (1987). *In vivo* modulation of myelopoiesis by prostaglandin E_2. II. Inhibition of granulocyte-monocyte progenitor cell (CFU-GM) cell-cycle rate. *Exp. Hematol.*, **15**, 119-126.

Goldstein, B.D. (1983). Clinical hematotoxicity of benzene. *Adv. Mod. Environ. Toxicol.*, **4**, 51-61.

Greenlee, W.F., Gross, E.A., Irons, R.D. (1981). A study on the disposition of ^{14}C-labeled, phenol, catechol and hydroquinone in rat during whole-body autoradiography. *Chem.-Biol. Interact.*, **33**, 285-299.

Infante, P.F., White, M.C. (1983). Benzene: epidemiologic observations of leukemia by cell type and adverse health effects associated with low-level exposure. *Environ. Health. Perspect.*, **52**, 75-82.

Irons, R.D., Heck, H. d'a, Moore, B.J., and Muirhead, K.A. (1979). Effects of short-term benzene administration on bone marrow cell cycle kinetics in the rat. *Toxicol. Appl. Pharmacol.*, **51**, 399-409.

Irons, R.D. (1985). Quinones as toxic metabolites of benzene. *J. Toxicol. Environ. Health.*, **16**, 673-678.

Kalf, G.F., Schlosser, M.J., Renz, J.F., and Pirozzi, S.J. (1989). Prevention of benzene-induced myelotoxicity by nonsteroidal anti-inflammatory drugs. *Environ. Health. Perspect.*, **82**, 57-64.

Koren, H.S., Handwerger, B.S., and Wunderlick (1975). Identification of macrophage-like characteristics in a cultured murine tumor cell line. *J. Immunol.*, **114**, 894-897.

Lee, E.W., Kocsis, J.J., and Snyder, R. (1974). Acute effects of benzene on ^{59}Fe incorporation into circulating erythrocytes. *Toxicol. Appl. Pharmacol.*, **27**, 431-436.

Lee, M., Segal, G.M., and Bagby, G.C. (1987). Interleukin-1 induces human bone marrow-derived fibroblasts to produce multilineage hematopoietic growth factors. *Exp. Hematol.*, **15**, 983-988.

Lewis, J.G., Odom, B., and Adams, D.O. (1988). Toxic effects of benzene and benzene metabolites on mononuclear phagocytes. *Toxicol. Appl. Pharmacol.*, **92**, 246-254.

MacEachern, L., Snyder, R., and Laskin, D. (1988). Activation of bone marrow macrophages and PMN following benzene treatment of mice. *Toxicologist.*, **8**, 72.

Markey, C.M., Alward, A., Weller, P.E., and Marnett, L.J. (1987). Quantitative studies of hydroperoxide reduction by prostaglandin H synthase. Reducing substrate specificity and the relationship of peroxidase to cyclooxygenase activities. *J. Biol. Chem.*, **262**, 6266-6279.

Ohki, S., Ogino, N., Yamamoto, S., and Hayaishi, O. (1979). Prostaglandin hydroperoxidase, an integral part of prostaglandin endoperoxide synthetase from bovine vesicular gland microsomes. *J. Biol. Chem.*, **254**, 829-836.

Pfankuche, H.J., Kaever, V., and Resch, K. (1986). A possible role of protein kinase C in regulating prostaglandin synthesis of mouse peritoneal macrophages. *Biochem. Biophys. Res. Commun.*, **139**, 604-611.

Pirozzi, S., Schlosser, M., and Kalf, G.F. (1989). Prevention of benzene-induced myelotoxicity and prostaglandin synthesis in bone marrow of mice by inhibitors of prostaglandin H synthase. *Immunopharmacol.*, **18**, 39-58.

Post, G., Snyder, R., and Kalf, G.F. (1986). Metabolism of benzene and phenol in macrophages *in vitro* and the inhibition of RNA synthesis by benzene metabolites. *Cell Biol. Toxicol.*, **2**, 231-246.

Rickert, D.E., Baker, T.S., Bus, J.S., Barrow, C.S., Irons, R.D. (1979). Benzene disposition in the rat after exposure by inhalation. *Toxicol. Appl. Pharmacol.*, **49**, 417-423.

Roghani, M., DaSilva, C., Guvelli, D., and Castagna, M. (1987). Benzene and toluene activate protein kinase C. *Carcinogenesis*, **8**, 1105-1107.

Sammett, D., Lee, E.W, Kocsis, J.J, Snyder, R. (1979). Partial hepatectomy reduces both metabolism and toxicity of benzene. *J. Toxicol. Environ. Health.*, **5**, 785-792.

Sawahata, T., Ricker, D.E., Greenlee, W.F. (1985) . Metabolism of benzene and its metabolites in bone marrow. In *Toxicology of the Blood and Bone Marrow*, Irons, R.D., ed. New York: Raven Press, 141-148.

Schlosser, M.J., and Kalf, G.F. (1989). Metabolic activation of hydroquinone by macrophage peroxidase. *Chem.-Biol. Interact.*, **72**, 191-207.

Schlosser, M. Shurina, R., and Kalf, G.F. (1989). Metabolism of phenol and hydroquinone to reactive products by macrophage peroxidase or purified prostaglandin H synthase. *Environ. Health. Perspect.*, **82**, 229-237.

Scott, W.A., Zrike, J.M., Hamill, A.L., Kempe, J., and Cohn, Z.C. (1980). Regulation of arachidonic acid metabolites in macrophages. *J. Exp. Med.*, **152**, 324-335.

Snyder, R., Jowa, L., Witz, G., Kalf, G.F., and Rushmore, T. (1987). Formation of reactive metabolites of benzene. *Arch. Toxicol.*, **60**, 61-64.

Thomas, D.J., Reasor, M.J., and Wierda, D. (1989). Macrophage regulation is altered by exposure to the benzene metabolite hydroquinone. *Toxicol. Appl. Pharmacol.*, **97**, 440-453.

Tunek, A., Platt, K.L., Przybylski, M., Oesch, F. (1980). Multi-step metabolic activation of benzene. Effect of superoxide dismutase on covalent binding to microsomal macromolecules, and identification of glutathione conjugates using high pressure liquid chromatography and field desorption mass spectrometry. *Chem.-Biol. Interact.*, **33**, 1-17.

Tunek, A., Olofsson, T., Berlin, M. (1981). Toxic effects of benzene and benzene metabolites on granulopoietic stem cells and bone marrow cellularity in mice. *Toxicol. Appl. Pharmacol.*, **59**, 149-156.

GLUTATHIONE CONJUGATION AS A MECHANISM OF TARGETING LATENT QUINONES TO THE KIDNEY

Serrine S. Lau and Terrence J. Monks

Division of Pharmacology and Toxicology
College of Pharmacy
University of Texas at Austin
Austin, Texas 78712

INTRODUCTION

One of the major difficulties in studying the toxicity of chemicals, in particular the factors contributing to target organ toxicity, is determining whether or not the ultimate (or penultimate) metabolites responsible for causing such toxicities are formed in the organ of susceptibility. This is both an interesting and challenging toxicological problem. Chemically reactive metabolites are capable of leaving the active site of the enzyme at which they are formed, of leaving the cell in which they are formed, and of leaving the organ in which they are formed. Thus, such metabolites are clearly capable of producing toxicity at sites distal to their site of formation. It is also important to note that not all chemically reactive metabolites are biologically reactive. That is, we know certain reactive metabolites are capable of interacting with cellular macromolecules without producing any discernable alteration in cellular structure or function.

Since the liver is the major organ contributing to the metabolism of chemicals, it seems likely that this organ is the major source of non-target organ derived metabolites capable of causing tissue damage. Indeed, this rational assumption provides the basis for this particular session on *Biological Reactive Intermediates*. Our own interest in this topic originated during our tenure at the National Institutes of Health, in the Laboratory of Chemical Pharmacology with Jim Gillette. At that time, convincing evidence had been presented that the nature of the reactive metabolite responsible for bromobenzene mediated liver necrosis was the 3,4-epoxide. Although the *mechanism* of bromobenzene mediated necrosis remains a contentious issue, the 3,4-epoxide remains the most likely culprit.

In contrast to brombenzene mediated hepatotoxicity, the metabolite(s) responsible for the nephrotoxicity caused by this compound remained unclear. However, the work of Reid (1973) suggested that bromobenzene nephrotoxicity might be caused by a metabolite formed in the liver and subsequently transported to the kidney. The nature of this liver derived circulating metabolite was unclear. A prime candidate was bromobenzene-3,4-oxide, which might be sufficiently stable to escape the liver and produce toxicity at extrahepatic sites. We were able to demonstrate that the 3,4-epoxide was indeed capable of diffusing out of hepatocytes and that as much as 65% of this metabolite was capable of leaving the cells in which it was formed (Monks et al., 1984; Gillette et al., 1984). Moreover, bromobenzene-3,4-oxide could be identified in retroorbital sinus blood of rats treated with bromobenzene (Lau et al., 1984a). However, the inability of the kidney to catalyse the formation of the 3,4-

Biological Reactive Intermediates IV, Edited by C.M. Witmer *et al.*
Plenum Press, New York, 1990

epoxide, in conjunction with the relatively high renal activity of the glutathione (GSH) S-transferase required for detoxication of the 3,4-epoxide (Monks and Lau, 1984) suggested that any of the epoxide reaching the kidney from the liver could be efficiently detoxified there.

The finding that a major phenolic metabolite of bromobenzene, *ortho*-bromophenol, was capable of causing severe GSH depletion in isolated hepatocytes (Thor et al., 1982) prompted us to investigate the *in vivo* effects of this metabolite. In contrast to its effects *in vitro*, *ortho*-bromophenol was not hepatotoxic *in vivo*. However, *ortho*-bromophenol was capable of reproducing the nephrotoxicity seen with bromobenzene and at much lower doses (Lau et al., 1984b). The data also indicated that the toxicity was caused by a metabolite of *ortho*-bromophenol, formed in the liver and transported by the circulation to the kidney. Subsequently, 2-bromohydroquinone (2-BrHQ) was identified as a major *in vivo* and *in vitro* metabolite of both brombenzene and *ortho*-bromophenol (Lau et al., 1984c) and was shown to cause nephrotoxicity at a dose much lower than that of its precursors. Although 2-BrHQ was found to be nephrotoxic, it had little effect on the liver. In contrast, whereas liver microsomes were capable of forming 2-BrHQ from *ortho*-bromophenol, kidney microsomes could not. This raised the question of how, and in what form might 2-BrHQ be transported to the kidney in order to elicit its selective toxicity to renal proximal tubules? Our most recent studies therefore focussed on the role played by GSH in selectively targeting 2-BrHQ, and other quinones, to renal proximal tubule cells and on the factors regulating the toxicity of these compounds. In particular, the effects of GSH conjugation, γ-glutamyl transpeptidase (γ-GT) and metabolism through the mercapturic acid pathway on 2-BrHQ toxicity have been investigated.

RESULTS AND DISCUSSION

Administration of 2-BrHQ to rats caused decreases in both liver and kidney GSH concentrations and the addition of GSH to liver microsomal incubations of [^{14}C]-*ortho*-bromophenol caused a decrease in the recovery of 2-BrHQ with a concomitant increase in the formation of water soluble metabolites (Lau *et al.*, 1984c). Oxidation of 2-BrHQ in the presence of glutathione (GSH) gave rise to isomeric and multi-substituted GSH conjugates. Adminstration of these conjugates to rats produced varying degrees of renal toxicity. In particular, 2-Br-(diGSyl)HQ was a potent and selective renal proximal tubular toxicant (Figure 1). Indeed, the dose of 2-Br-(diGSyl)HQ (20-30 μmol/kg) required to produce toxicity was approximately 300-450-fold less than that required of bromobenzene (9.3 mmol/kg). In contrast to the potent nephrotoxicity of 2-Br-(diGSyl)HQ, it had no apparent adverse effects on the liver. The reason(s) for the tissue selectivity and differential toxicity of the mono- and di-substituted GSH conjugates were subsequently investigated.

The tissue selectivity of 2-Br-(diGSyl)HQ appears to be a consequence of its targeting to renal proximal tubule cells by brush border γ-GT. In support of this

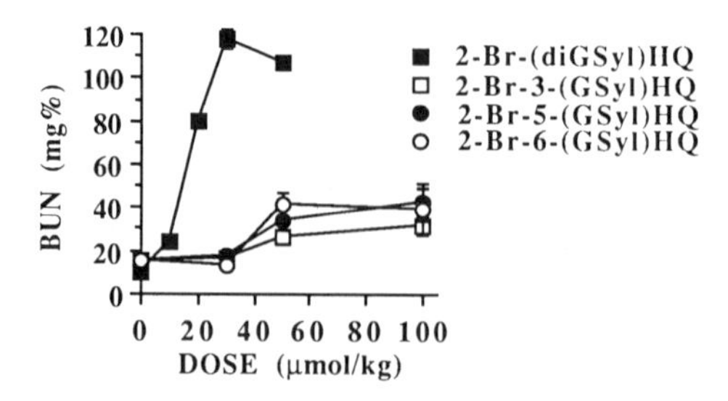

Figure 1. Relative nephrotoxicity of 2-bromo-(glutathion-S-yl)hydroquinones.

suggestion, inhibition of γ-GT by pretreatment of animals with AT-125 protected them against both 2-Br-(diGSyl)HQ (Monks *et al.*, 1988a) and 2,3,5-(triGSyl)HQ (Lau *et al.*, 1988a) mediated nephrotoxicity. The activity of γ-GT my be required to facilitate the transport of 2-Br-(GSyl)HQ conjugates into renal proximal tubule cells as their corresponding cysteinylglycine (CYSGLY) and/or cysteine (CYS) conjugates since the accumulation of 2-Br-(diGSyl)HQ into isolated renal slices was also inhibited by AT-125 (Lau *et al.*, 1988b). In addition, the uptake of 2-Br-(diGSyl)HQ into renal slices was much faster than that of the 2-Br-(monoGSyl)HQ conjugates, which may partially contribute to the enhanced toxicity of the di-conjugate.

Oxidation of 2-Br-(GSyl)HQ conjugates to their corresponding quinones, rather than metabolism to reactive thiols catalysed by cysteine conjugate β-lyase (Stevens and Jakoby, 1983) appears to be responsible for the formation of the majority of the reactive metabolites formed from these conjugates. Inhibition of β-lyase with aminooxyacetic acid, had only minor effects on either 2-Br-(GSyl)HQ covalent binding (10-26%) or 2-Br-(diGSyl)HQ nephrotoxicity (Monks et al., 1988a). Moreover, the toxicity of 6-bromo-2,5-dihydroxy-thiophenol, a putative β-lyase catalysed metabolite of 2-Br-3-(GSyl)HQ, was shown to be a function of the quinone moiety; thiophenols lacking the quinone group were not nephrotoxic (Monks et al., 1988b). In contrast to aminooxyacetic acid, ascorbic acid substantially (62-87%) inhibited the covalent binding of 2-Br-(monoGSyl)HQ metabolites to renal homogenates. Interestingly however, the covalent binding of 2-Br-(diGSyl)HQ to renal homogenates was only 30-45% of that seen with the 2-Br-(monoGSyl)HQ conjugates and ascorbic acid had only a minor effect (28%) on 2-Br-(diGSyl)HQ binding.

Differences in the electrochemical properties of the 2-Br-(GSyl)HQ conjugates appear to determine their relative reactivity since the ability of ascorbic acid to inhibit the covalent binding of the various isomers correlated with their oxidation potentials (Monks et al., 1988a). Interestingly, the most toxic conjugate, 2-Br-(diGSyl)HQ was the most stable to oxidation at pH 7.4 (Table 1). The conjugation of 2-BrHQ with GSH is therefore a detoxication reaction since the resulting conjugates are more difficult to oxidize than

Table 1. Half-wave Oxidation Potentials of Isomeric 2-Bromo-(Glutathion-S-yl)Hydroquinones and 3-Substituted Metabolites.

Compound	E1/2 (Volts)
2-Bromohydroquinone	+ 0.14
2-Bromo-(diglutathion-S-yl)hydroquinone	> + 1.10
2-Bromo-6-(glutathion-S-yl)hydroquinone	+ 0.37
2-Bromo-5-(glutathion-S-yl)hydroquinone	+ 0.41
2-Bromo-3-(glutathion-S-yl)hydroquinone	+ 0.43
2-Bromo-3-(cystein-S-yl)hydroquinone	+ 0.09
2-Bromo-3-N-acetyl(cystein-S-yl)hydroquinone	+ 0.45

Oxidation potentials were determined by hydrodynamic voltammetry at pH 7.4.

2-BrHQ. However, metabolism of the conjugates through the mercapturic acid pathway has significant effects on the reactivity of the intermediates. Hydrolysis of 2-Br-3-(GSyl)HQ by γ-GT and formation of the corresponding cysteine (CYS) conjugate results in the formation of a compound that is more readily oxidized than 2-BrHQ whereas N-acetylation of the CYS conjugate to give the mercapturate regenerates a compound that is more stable to oxidation than 2-BrHQ (Table 1). Thus the oxidation of 2-BrHQ is exquisitely regulated by its passage through the mercapturic acid pathway.

The ability of γ-GT to catalyse the formation of potentially reactive cysteinyl-quinones from relatively stable quinol-GSH conjugates may have important toxicological implications. For example, other cells expressing relatively high γ-GT activity, such as pancreatic acinar and ductile epithelial cells and the epithelial cells of jejenum, bile-duct, epididymis, seminal vesicles, bronchioles thyroid folicles, choroid plexus, ciliary body and retinal epithelium (Meister et al., 1976) may be exposed to higher concentrations of potentially toxic quinone-thioethers than other tissues. The possibility therefore exists for an interaction between hepatic and extra-hepatic cells capable of quinol oxidation and quinone-thioether formation, and those cells capable of selectively accumulating and activating these conjugates (Monks and Lau, 1989). In addition, γ-GT has been widely used as a marker for preneoplastic lesions in the liver during carcinogenesis. Abnormally high levels of γ-GT are also observed in tumors from a variety of tissues. If the renal toxicity of quinone-thioethers could be circumvented then such agents might prove useful directed against γ-GT in neoplastic cells. In this respect, preneoplastic liver nodules might be good targets for these compounds since normal liver contains negligible γ-GT and the quinone-thioethers have no apparent adverse effect on liver tissue. This might be an area worthy of future consideration (Monks and Lau, 1989).

Factors other than γ-GT probably contribute to the susceptibility to 2-Br-(diGSyl)HQ toxicity. For example, although species differences exist in the activity of renal γ-GT, susceptibility to 2-Br-(diGSyl)HQ nephrotoxicity did not correlate with this variability (Lau et al., 1990). The relative activities of cysteine conjugate N-acetyl transferase (Duffel and Jakoby, 1982) and the corresponding N-deacetylase (Suzuki and Tateishi, 1981) and the relative rates of cysteine conjugate oxidation and reduction, will contribute to the relative susceptibility of γ-GT containing cells to the toxicity of quinone-thioethers. Moreover, the presence of enzymatic and non-enzymatic (NAD[P])H, GSH, ascorbate, etc.) reductants may protect certain γ-GT containing tissues from the toxic effects of quinone-thioethers. In this respect, NAD(P)H quinone oxidoreductase (DT-diaphorase [EC 1.6.99.2]) activity is lower in rat renal cortex than in the papillae (Monks et al., 1988) and naphthoquinone-GSH conjugates have been shown to be substrates for this enzyme (Buffinton et al., 1989). However, pretreatment of rats with dicumarol which inhibited renal DT-diaphorase activity in a dose dependant manner (Figure 2) did not significantly affect the renal toxicity of 2-Br-(diGSyl)HQ (20 μmol/kg) or 2-Br-(monoGSyl)HQ conjugates (100 μmol/kg) (Figure 3). The data suggest that DT-diaphorase does not play an important role in the renal detoxication of quinone-thioethers. The lack of effect of dicumarol and the reactivity of the products of the γ-GT catalysed hydrolysis of the quinol-GSH conjugates may be pertinent to both the mechanism and site of action of these compounds.

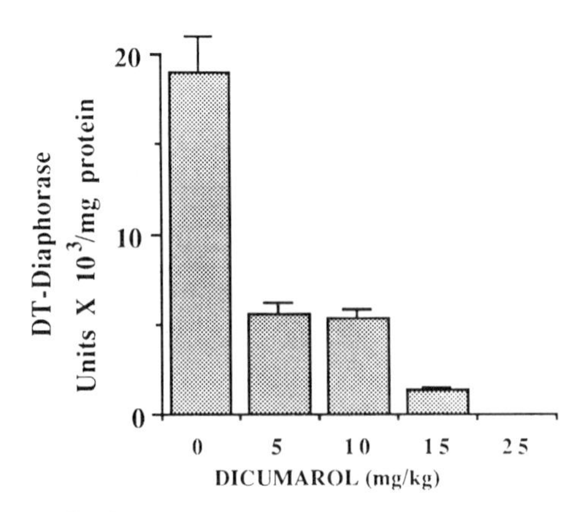

Figure 2. Effect of dicumarol on the *in vivo* activity of renal quinone reductase.

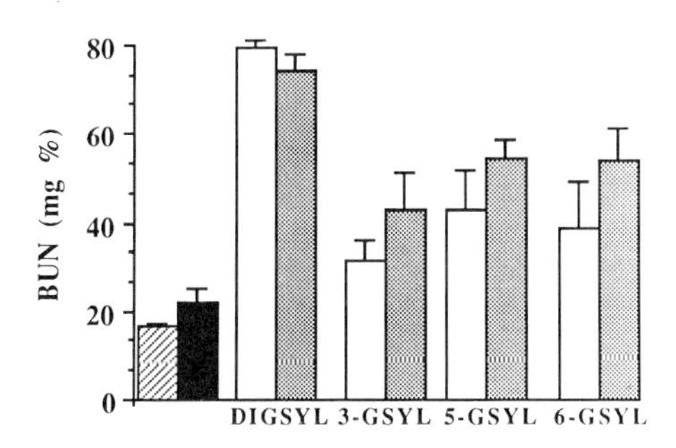

Figure 3. Lack of effect of dicumarol (15mg/kg) on the *in vivo* toxicity of 2-bromo-(glutathion-S-yl)hydroquinones ( ) Control; ( ) ontrol + dicumarol; ( ) 2-Br-(CGSyl)HQ: ( ) 2-Br-(GSyl)HQ + dicumarol.

The activities of the brush border membrane enzymes responsible for GSH conjugate metabolism result in the formation of products which can be readily oxidized and which may arylate membrane associated proteins. Thus, the CYS and/or CYSGLY conjugated quinols may be incapable of surviving passage through the plasma membrane. As a consequence, intracellular concentrations of these conjugates may be relatively low. Moreover, we have previously noted that the leakage of γ-GT into urine appears to be the most sensitive indicator of the toxicity of these conjugates suggesting that damage to the brush border membrane of renal proximal tubules may contribute to and possibly initiate the toxicity. Of course, a cause/effect relationship will be difficult to demonstrate.

In addition to the production of reactive cysteinyl-quinones, we have identified a novel pathway of quinol-GSH metabolism that diverges from the classical route of mercapturic acid synthesis. Thus the γ-GT mediated hydrolysis of 2-Br-3-(GSyl)HQ results in the subsequent oxidative cyclization of the CYSGLY and CYS conjugates and 1,4-benzothiazine formation (Figure 4) (Monks et al., 1990). The intramolecular cyclization reaction removes the reactive quinone function from the molecule and can therefore be considered an intramolecular detoxication reaction. The toxicity of quinol-GSH conjugates will therefore be dependent upon the relative rates of cysteine conjugate cyclization and the rate of cysteine conjugate arylation. The 1,4-benzothiazines undergo further oxidative coupling to give rise to dimeric and polymeric products analagous to the reactions observed during phaeomelanin synthesis from cystein-S-yl-3,4-dihydroxy-phenylalanine (Prota, 1988). These reactions may explain our inability to identify either cystein-S-yl or N-acetylcystein-S-yl conjugates as major *in vivo* metabolites of 2-BrHQ, despite the formation of the corresponding GSH conjugates (Lau et al., 1990).

The polymerization of the 1,4-benzothiazines results in the formation of insoluble pigments. The deposition of such pigments within proximal tubules could conceivably contribute to the toxicity of these compounds. To examine this possibility, we synthesised the L-homocysteine (HCyS) conjugates of 2-BrHQ. The additional methylene group in this compound should hinder the cyclization reaction. Consequently, if polymerization is a necessary step in the toxicity of the 2-Br-(GSyl)HQ conjugates, then the HCyS conjugates should exhibit little if any toxicity. However, administration of 2-BrHQ-HCyS to rats (100mg/kg) caused significant elevations in BUN (Figure 5) and histopathological alterations to the kidney consistent with those seen following treatment of rats with the 2-Br-(GSyl)HQ conjugates. The HCys analog of S-(1,2-dichlorovinyl)-L-cysteine is also nephrotoxic (Elfarra et al., 1986) but the toxicity of this compound is dependent upon its processing by β-lyase. Thus, HCyS conjugates can undergo a pyridoxal phosphate dependent γ-elimination

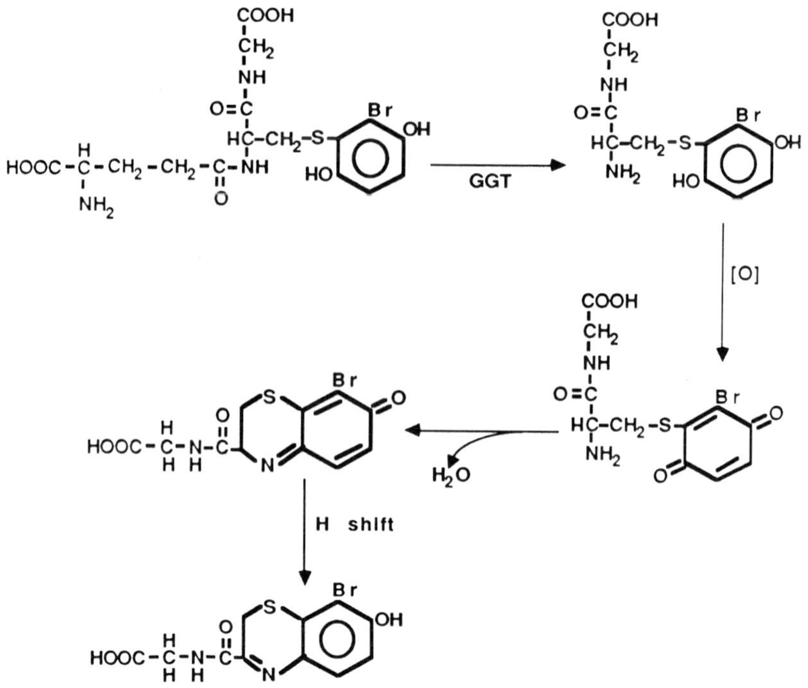

Figure 4. Oxidative cyclization and 1,4-benzothiazine formation from 2-bromo-3-(glutathion-S-yl)hydroquinone

reaction to yield a potentially reactive thiol and 2-oxo-3-butenoic acid. This reaction can be inhibited by aminooxyacetic acid (AOA). However, pretreatment of rats with AOA (55mg/kg) did not protect them against the nephrotoxicity of 2-BrHQ-HCys (Figure 5). These data suggest that 1,4-benzothiazine formation from 2-Br-(GSyl)HQ conjugates, by removing the reactive quinone function from the molecule, is most likely a detoxication reaction.

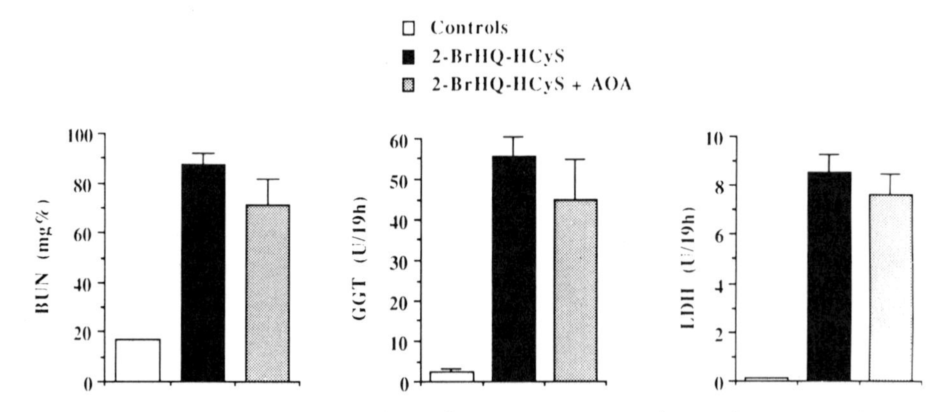

Figure 5. The nephrotoxicity of 2-bromohydroquinone-homocysteine conjugates and the effects of aminooxyacetic acid (55mg/kg).

There is now ample evidence attesting to the biological reactivity of quinone thioethers. Thus, in contrast to the generally accepted role of GSH conjugation serving as a detoxication mechanism conjugation of quinones with GSH results in the formation of a variety of biologically (re)active metabolites. For example, the GSH conjugate of menadione can itself redox cycle, with the concomitant formation of reactive oxygen species (Wefers and Sies, 1983). In addition, Ross *et al.* (1985) have isolated three GSH conjugates from the peroxidase-catalysed oxidation of *p*-phenetidine, which exist in both oxidized and reduced forms and which are readily interconverted by redox processes. Potter et al., (1986) have also demonstrated that the GSH conjugate of acetaminophen is readily oxidized to a free radical intermediate. It has been suggested that the redox activity of quinone-GSH conjugates may play a role in the oxidative damage of cataract (Wolff and Spector, 1987). Formation of a stable 2,6-dimethoxyquinone-GSH free radical has been demonstrated in bovine lens epithelial cells which may contribute to the cytotoxicity caused by this dietary quinone. Finally, the GSH conjugate(s) of N-(4-ethoxyphenyl)-*p*-benzoquinone imine has been shown to bind to DNA (Larsson et al., 1988) the GSH conjugates of menadione and toluquinone have been shown to be substrates for NADP-linked 15-hydroxyprostaglandin dehydrogenase and are mixed-type inhibitors of prostaglandin B_1 oxidation (Chung et al., 1987) and the GSH conjugate of tetra-chloro-1,4-benzoquinone is an effective inhibitor of the GSH S-transferases (van Ommen et al., 1988). Thus in light of the emerging evidence demonstrating the reactivity of a variety of quinone-thioethers studies on thioether conjugates of physiologically important endogenous quinones may be a fruitful area for future research.

ACKNOWLEDGEMENTS

Supported in part by USPHS awards ES 04662 and GM 39338.

REFERENCES

Duffel, M.W. and W.B. Jakoby (1982). Cysteine S-conjugate N-acetyltransferase from rat kidney microsomes. *Mol. Pharmacol.* **21**, 444-448.

Elfarra, A.A., Lash, L.H. and Anders, M.W. (1986). Metabolic activation and detoxication of nephrotoxic cysteine and homocysteine S-conjugates. *Proc. Natl. Acad. Sci. USA.* **83**, 2667-2671.

Gillette, J.R., Lau, S.S. and Monks, T.J. (1984). Intra and extracellular formation of metabolites from chemically reactive species. *Biochem. Soc. Trans.* **12**, 4-79.

Larsson, R., Boutin, J. and Moldeus, P. (1988). Peroxidase-catalysed metabolic activation of xenobiotics. In *Metabolism of Xenobiotics* (Gorrod, Oelschlager and Caldwell, eds.), pp 43-50, Taylor and Francis, New York.

Lau, S.S., Hill, B.A., Pinon, R.K. and Monks, T.J. (1989). Species differences in renal γ-glutamyl transpeptidase activity and susceptibility to 2-bromohydroquinone mediated nephrotoxicity. *Toxicologist*, **9**, 153.

Lau, S.S., Monks, T.J., Greene, K.E. and Gillette, J.R. (1984a). Detection and half-life of bromobenzene-3,4-oxide in blood. *Xenobiotica* **14**, 539-543.

Lau, S.S., Monks, T.J., Greene, K.E. and Gillette, J.R. (1984b). The role of *ortho*-bromophenol in the nephrotoxicity of bromobenzene. *Toxicol. Appl. Pharm.* **72**, 539-549.

Lau, S.S., Monks, T.J. and Gillette, J.R. (1984c). Identification of 2-bromohydroquinone as a metabolite of bromobenzene: Implications for bromobenzene induced nephrotoxicity. *J. Pharm. Exp. Ther.* **230**, 360-366.

Lau, S.S., Hill, B.A., Highet, R.J., and Monks, T.J. (1988a). Sequential oxidation and glutathione addition to 1,4-benzoquinone: Correlation of toxicity with increased glutathione substitution. *Molec. Pharmacol.* **34**, 829-836.

Lau, S.S., M.G. McMenamin, and T.J. Monks. (1988b). Differential uptake of isomeric 2-bromohydroquinone glutathione conjugates into rat kidney slices. *Biochem. Biophys. Res. Commun.*,**152**, 223-230.

Lau, S.S., and Monks, T.J. (1990). The *in vivo* disposition of 2-bromo-[^{14}C]-hydroquinone and the effect of γ-glutamyl transpeptidase inhibition. *Toxicol. Appl. Pharmacol.* **103**, in press.

Meister, A., Tate, S.S. and Ross, L.L. (1976). Membrane-bound γ-glutamyl transpeptidase. In *The Enzymes of Biological Membranes* (A. Martonosi, ed.), pp. 315-347, Plenum Publishing Corp., New York.

Monks, T.J., Lau, S.S. and Gillette, J.R. (1984). Diffusion of reactive metabolites out of hepatocytes: Studies with bromobenzene. *J. Pharmacol. Exp. Ther.* **228**, 393-399.

Monks, T.J., Highet, R.J. and Lau, S.S. (1988a) . 2-Bromo-(diglutathion-S-yl)-hydroquinone nephrotoxicity: Physiological, biochemical and electrochemical determinants. *Molec. Pharmacol.* **34**, 492-500.

Monks, T.J., Highet, R.J. and Lau, S.S. (1990). Oxidative cyclization, 1,4-benzothiazine formation and dimerization of 2-bromo-3-(glutathion-S-yl)hydroquinone. *Molec. Pharmacol.* In press.

Monks, T.J., Highet, R.J., Chu, P.S. and Lau, S.S. (1988b). Synthesis and nephrotoxicity of 6-bromo-2,5-dihydroxythiophenol. *Molec. Pharmacol.* **34**, 15-22.

Monks, T.J., Lau, S.S., Highet, R.J. and Gillette, J.R. (1985). Glutathione conjugates of 2-bromohydroquinone are nephrotoxic. *Drug Metab. Dispos.* **13**, 553-559.

Monks, T.J., and Lau, S.S. (1984). Activation and detoxification of bromobenzene in extrahepatic tissues. *Life Sci.* **35**, 561-568.

Monks, T.J., and Lau, S.S. (1987). Commentary: Renal transport processes and glutathione conjugate-mediated nephrotoxicity. *Drug Metab. Dispos.* **15**, 437-441.

Monks, T.J., and Lau, S.S. (1989). Sulfur-conjugate mediated toxicities. *Rev. Biochem. Toxicol.*, **10**, 41-90.

Potter, D.W., Miller, D.W. and Hinson, J.A. (1986). Horseradish-peroxidase catalysed oxidation of acetaminophen to intermediates that form polymers or conjugate with glutathione. *Molec. Pharmacol.* **29**, 155-162.

Prota, G. (1988) . Progress in the chemistry of melanins and related metabolites. *Med. Res. Rev.* **8**, 525-556.

Ross, D., Larsson, R., Norbeck, K. Ryhage, R. and Moldeus, P. (1985). Characterization and mechanism of formation of reactive products formed during peroxidase-catalysed oxidation of *p*-phenetidine. *Molec. Pharmacol.* **27**, 277-286.

Stevens, J. and Jakoby, W.B. (1983). Cysteine conjugate β-lyase. *Molec. Pharmacol.*, **23**, 761-765.

Suzuki,S. and Tateishi, M. (1981). Purification and characterization of a rat liver enzyme catalysing N-deacetylation of mercapturic acid conjugates. *Drug Metab. Dispos.* **9**, 573-577.

Wefers, H. and Sies, H. (1983). Hepatic low-level chemiuminesence during redox cycling of menadione and the menadione-glutathione conjugate: Relation to glutathione and NAP(P)H: quinone reductase (DT-diaphorase) activity. *Arch. Biochem. Biophys.* **224**, 568-578.

Wolff, S.P. and Spector, A. (1987). Pro-oxidant activation of ocular reductants. 2 Lens epithelial cell cytotoxicity of a dietary quinone is associated with a stable free radical formed with glutathione *in vitro*. *Exp. Eye Res.* **45**, 791-803.

HEPATIC BIOACTIVATION OF 4-VINYLCYCLOHEXENE TO OVOTOXIC EPOXIDES

Bill J. Smith, Donald R. Mattison*, and I. Glenn Sipes

Department of Pharmacology and Toxicology
College of Pharmacy
University of Arizona
Tucson, AZ USA 85721 and
*National Center for Toxicological Research
Jefferson, AR USA

INTRODUCTION

4-Vinylcyclohexene (VCH) is produced by a dimerization reaction of 1,3-butadiene (Rappaport and Fraser, 1976). The curing process of synthetic rubber production results in butadiene dimerization, discharge of VCH, and subsequent exposure of workers to VCH by inhalation (Rappaport and Fraser, 1977). Chronic exposure of humans to VCH may be of concern since a 2 yr bioassay indicated that VCH was carcinogenic to mice (Collins et al., 1987). Chronic gavage of female B6C3F1 mice caused the induction of rare ovarian tumors. These tumors were not observed in VCH-treated Fischer 344 rats (NTP, 1986). Studies in our laboratory have centered on determining the basis for the species difference in VCH-induced ovarian toxicity. Such studies have provided information which may aid in determining which species best predicts the response of humans exposed to VCH. In addition, clues have been obtained pertaining to the mechanism by which VCH produces ovarian injury.

The mechanism(s) by which chemicals induce ovarian cancers is/are unknown. However, it is likely that the sequence of events is similar to that of other chemically induced cancers. This would follow the pattern of initiation-promotion-progression. Investigation of the pathogenesis of polycyclic aromatic hydrocarbon-induced ovarian tumors revealed that destruction of the primordial or small oocyte occurs soon after carcinogen treatment and is probably associated with the initiation phase (Jull, 1973). Therefore, loss of oocytes, as determined by counting of serial sections of ovaries, was used as an indicator of toxicity in our studies with VCH. Most compounds which destroy oocytes are mutagenic (Dobson and Felton, 1983). Although VCH is not mutagenic, epoxide metabolites of VCH produced by hepatic microsomal cytochrome(s) P450 are mutagenic. These data suggest that hepatic metabolism of VCH may be involved in toxicity carcinogenecity (NTP, 1986, Watabe et al, 1981, and Simmon and Baden, 1980).

We proposed an activation scheme for VCH involving metabolism of VCH to a long lived biologically reactive intermediate(s) by the liver, delivery of reactive epoxide metabolite(s) to the ovary through the systemic circulation, and the subsequent destruction of oocytes by these reactive intermediates. In order to prove this hypothesis several experiments were performed. First, the ovotoxic potency of VCH was compared to that of the VCH epoxides in both species. Second, following ip administration of VCH the blood concentration of VCH epoxides was compared in the

Biological Reactive Intermediates IV, Edited by C.M. Witmer *et al.*
Plenum Press, New York, 1990

sensitive (B6C3F1 mouse) and resistant (F-344 rat) species. Third, the rate of VCH epoxidation by mouse and rat hepatic microsomes was compared. Finally, inhibition of VCH epoxidation was examined as a means of protecting animals from VCH-induced ovarian toxicity.

The species difference in VCH-induced ovarian toxicity may include differences in activation and detoxication pathways. Epoxide hydrolases are enzymes which catalyze the hydrolysis of alkene epoxides, a detoxication reaction for most of these compounds (Sipes and Gandolfi, 1986). Therefore, the rate of hydrolysis of VCH-1,2-epoxide by mouse and rat hepatic microsomes was compared.

METHODS

The methods used in the ovarian toxicity studies were reported previously by Smith et al. 1990a. The methods used to investigate the metabolism of VCH to VCH-1,2-epoxide *in vitro* and *in vivo* are reported by Smith et al. 1990b.

Preparation of hepatic cytosol and microsomes. Hepatic cytosol and microsomes were prepared by differential centrifugation using the method of Halpert et al. (1983). The supernatant of the first 100,000 x g centrifugation (cytosolic fraction) was dialyzed at 4°C against 3 changes of buffer containing 0.05 M Tris HCl, 0.1 M KCl, 0.001 M EDTA, and 2 x 10^{-5} M BHT (2 L) and stored at -70°C until use. Microsomal and cytosolic protein content was determined by the method of Lowry et al. (1951).

Determination of microsomal and cytosolic VCH-1,2-epoxide hydrolysis. The rate of VCH-1,2-epoxide hydrolysis was determined using a modified method of Watabe et al. (1981). Microsomal and cytosolic incubations were performed in 0.1 M Tris at pH 9.0 and 7.4, respectively. These are the optimal pH values for microsomal and cytosolic epoxide hydrolase (Ota and Hammock, 1980). Incubations with microsomes contained 0.5 mg/ml protein and cytosolic activity was assayed at a range of protein concentrations from 0.25 to 2.0 mg/ml. The assay conditions were optimized so that metabolite production was linear with the protein concentration (0.25 to 2 mg/ml) and incubation time (5 to 15 min). Blank incubations consisted of microsomal or cytosolic protein which had been boiled for 10 min. Incubation mixtures were preincubated for 3 min at 37°C and the reaction started by the addition of VCH-1,2-epoxide to a final concentration of 1 mM in tetrahydrofuran (THF was 2% i.e. 20 ml/ml of the entire incubation volume). The reaction was terminated after 10 minutes with 2 ml of ice cold ethyl acetate. Biphenyl (200 nmol) was then added as the internal standard. The aqueous phase was extracted twice with 2 ml of ethyl acetate, the extracts were combined, dried with anhydrous sodium sulfate, and evaporated to a small volume (approx. 0.2 ml) at 40°C under a gentle stream of nitrogen. The extracts were analyzed for VCH-1,2-diol by capillary gas-liquid chromatography.

Capillary Gas-Liquid Chromatographic Conditions. Analyses were performed on a Hewlett-Packard HP 5890 A gas chromatograph equipped with a 0.32 mm X 25 M RSL-300 capillary column (Alltec Associates, Dearfield, IL) and a flame ionization detector. The nitrogen carrier gas flow rate was 2 ml/min with a split flow rate of 30 ml/min. The detector gas flow rates for hydrogen and air were 30 and 240 ml/min, respectively. Nitrogen was used as the make up gas to adjust the total flow rate to 300 ml/min. The injector, oven, and detector temperatures were 250°C, 150°C, and 300°C, respectively. One µl of the extract was injected. VCH-1,2-diol and biphenyl eluted at 4.2 and 6.1 min, respectively. VCH-1,2-diol was quantified by comparing the peak area of VCH-1,2-diol and biphenyl to a standard curve prepared by extracting known amounts of VCH-1,2-diol (5-200 nmol) from blank incubations.

Chemicals. Vinylcyclohexene-1,2-epoxide and trichloropropene oxide were obtained from Aldrich Chemical Co. (Milwaukee, WI). Biphenyl was purchased from Sigma Chemical Co. (St. Louis, MO). 4-Vinylcyclohexene-1,2-dihydrodiol was synthesized by hydrolyzing the epoxide under acidic conditions using the method of Watabe et al. (1981). The structure of VCH-1,2-diol was consistent with data obtained by mass spectrometry, nuclear magnetic resonance spectroscopy (NMR), and infrared spectroscopy.

RESULTS

The ability of VCH or VCH epoxides to destroy oocytes was investigated in mice and rats (Smith et al., 1990a). VCH, VCH-1,2- epoxide, VCH-7,8-epoxide (mice only) and VCH diepoxide were administered to B6C3F1 mice and Fischer 344 rats ip for 30 days (Table 1). The results of this study are summarized in Table I. VCH produced a dose-dependent reduction in the number of small oocytes of mice while no detectable effect was observed in rats. Furthermore, the epoxides were much more potent that the parent compound in mice and destroyed small oocytes of rats with comparable but slightly lower potency compared with mice. These results suggest that mice are susceptible and rats resistant to VCH-induced ovarian tumors because of a species difference in the ability of VCH to destroy oocytes.

Table 1. ED_{50}[a] Values for the Reduction in Small Oocyte Counts in Mice and Rats Administered 4-Vinylcyclohexene (VCH) and VCH epoxides ip for 30 Days

Species	VCH	VCH-1,2-epoxide	VCH-7,8-epoxide	VCD
Mouse	2.7	0.5	0.7	0.2
Rat	> 7.4[b]	1.4	ND[c]	0.4

[a] Dose in mmol/kg/day which reduces the small oocyte count to 50% of that observed in control animals (data from Smith et al, 1990a).
[b] Highest dose given.
[c] Not done.

The destruction of oocytes by the epoxides of VCH in both the mouse and rat suggested that a species difference in VCH epoxidation could be responsible for the species difference in the ovarian toxicity of VCH. This was investigated by measuring VCH epoxidation *in vivo* and in *in vitro* incubations with hepatic microsomes (Smith et al., 1990b) The *in vivo* studies were performed by administering VCH (800 mg/kg, ip) to mice and rats and measuring the appearance of VCH monoepoxides in the blood. Only VCH-1,2-epoxide was detected and only in VCH- treated mice. This experiment demonstrated that a dramatic species difference exists between VCH-treated mice and rats in the blood concentration of an ovotoxic epoxide metabolite of VCH. At least part of this difference can be explained by differences in the rate of VCH epoxidation to the 1,2-epoxide by the liver since hepatic microsomes obtained from female mice catalyze this reaction about 6-7 fold more rapidly than hepatic microsomes from female rats. These data also suggest that the liver is the site of VCH epoxidation. The biochemical basis for this species difference in hepatic microsomal VCH epoxidation is due to differences in constitutive expression of the cytochrome P450 forms which are responsible for VCH epoxidation (Smith et al., 1990c). In female mice P450IIA and P450IIB forms catalyze about 80% of microsomal VCH epoxidation. Female F-344 rats lack a P450 form immunochemically related to mouse P450IIA and constitutively express P450IIB forms at very low levels.

Taken together these experiments indicate that VCH may be producing ovarian toxicity through a mechanism shown schematically in Figure 1. As shown, VCH is bioactivated in the liver to a reactive intermediate which is delivered to the ovary by the blood. The subsequent reactions which occur in the ovary resulting in oocyte destruction are unknown but may involve additional activation by the ovary and/or covalent binding of VCH epoxides to proteins or nucleic acids which are important for normal ovarian function. An additional experiment which supports this mechanism is

Relationship of VCH Epoxidation with Ovarian Toxicity

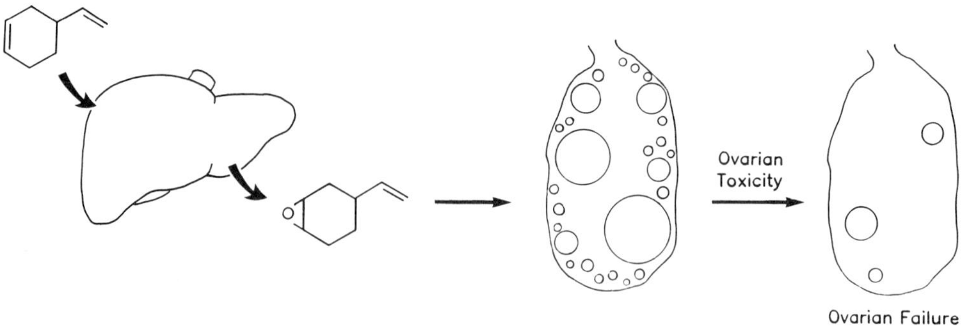

Ovarian Toxicity

Ovarian Failure

Figure 1. Proposed mechanism of VCH-induced ovarian toxicity in mice.

the inhibition of hepatic VCH epoxidation and partial protection from VCH-induced ovarian toxicity in chloramphenicol-pretreated mice (Smith et al., 1990a). Chloramphenicol inhibits hepatic cytochromes P450 because it is metabolized to a reactive intermediate which covalently binds to the apoprotein and inactivates the enzyme (Halpert et al., 1983). Administration of chloramphenicol (200 mg/kg) to mice prior to preparation of hepatic microsome resulted in a 69% reduction in hepatic microsomal VCH epoxidation. Furthermore, pretreatment of mice with chloramphenicol reduced the blood concentration VCH-1,2-epoxide by 50 % and partially prevented small oocyte loss in VCH-treated mice. This presumably occurred because of the decrease in the amount of systemically delivered epoxide to the ovary.

The amount of VCH-epoxide which reaches the systemic circulation from the liver in VCH-treated animals is a function of the rates of VCH epoxide formation and degradation. Reactions which are important in epoxide degradation include hydrolysis by epoxide hydrolases and/or conjugation with glutathione catalyzed by glutathione S-transferases (Sipes and Gandolfi, 1986). We have determined the rate of VCH-1,2-epoxide hydrolysis to its corresponding dihydrodiol by hepatic cytosol or microsomes to assay the activity of epoxide hydrolases toward VCH-1,2-epoxide. Essentially no enzymatic activity was present in hepatic cytosol from either species. However, hepatic microsomal hydrolysis of VCH-1,2-epoxide correlated well with protein concentration and incubation time in both species (Figure 2). The rate of hydrolysis was greater at pH 9 compared to pH 7.4 and was almost completely inhibited by the epoxide hydrolase inhibitor 3,3,3-trichloropropene oxide (data not shown). Shown in Table 2 is a comparison of the rate of hydrolysis of VCH-1,2-epoxide using female mouse and rat hepatic microsomes. Under the conditions of the assay rat microsomes catalyzed the hydrolysis of VCH-1,2-epoxide at a 2 fold greater rate compared to mouse microsomes.

DISCUSSION

The studies investigating the basis for the respective susceptibility and resistance of mice and rats to VCH-induced ovarian toxicity led us to propose a mechanism by which this might occur. We propose that VCH is bioactivated in the liver to VCH- 1,2-epoxide, a relatively stable reactive intermediate which is delivered to the ovary through the systemic circulation. The reason for the sensitivity of the ovary to VCH epoxides is unknown at this time and is currently under investigation. The production of reactive intermediates by the liver which can reach extrahepatic tissues and produce damage has been reported for a number of tissues including the kidney, central nervous system, testes, and bone marrow (see appropriate chapters in this book). Our studies with VCH show that the ovary can also be a target tissue for reactive intermediates generated by the liver. We believe that the ovary is probably a

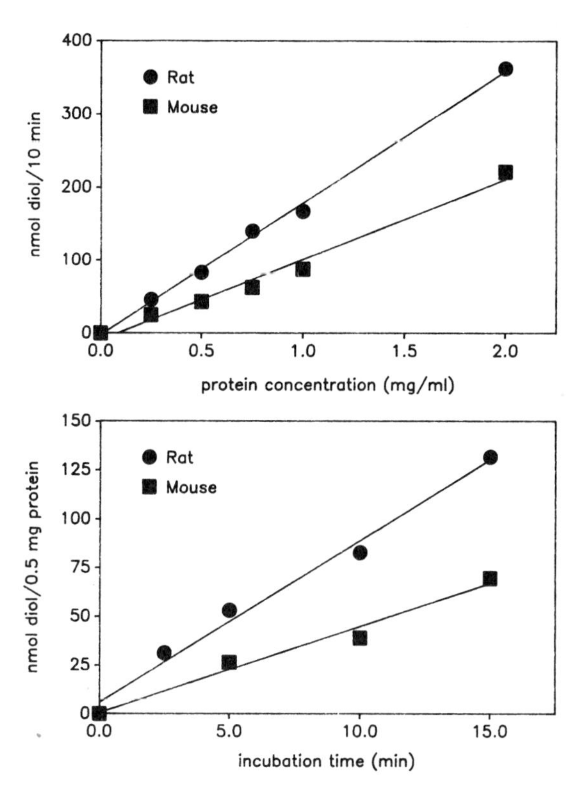

Figure 2. The effect of protein concentration and incubation time on hepatic microsomal VCH-1,2-epoxide hydrolysis. The points represent the mean of duplicate determinations.

Table 2. Comparison of Rates of Rat and Mouse Microsomal VCH-1,2-epoxide Hydrolysis

Species	nmol VCH-1,2-diol formed/min/mg microsomal protein[a]
Rat	18.7 ± 2.9[b,c]
Mouse	8.1 ± 1.2

[a] Reactions were preformed using 1 mM epoxide, 0.5 mg/ml protein, at 37°C, for 10 min.
[b] Data presented as mean and standard deviation (n=5).
[c] Significantly higher in rats $P < 0.002$.

target for other aliphatic olefinic hydrocarbons which can form stable epoxides. Candidates include 1,3-butadiene, cyclohexene, and isoprene among others. The ovary may also be involved in metabolizing VCH to reactive epoxides. Shiromizu and Mattison (1984) have shown that mouse ovaries contain all of the necessary enzymes to bioactivate benzo(a)pyrene to the ultimate ovotoxic metabolite benzo(a)pyrene-7,8-diol-9,10-epoxide. The experiments which examined the hydrolysis of VCH-1,2-epoxide highlight the importance of detoxication reactions in the susceptibility of a species to VCH-induced ovarian toxicity. Studies with hepatic microsomes show that female rats which are resistant to ovarian injury induced by VCH hydrolyzed the proposed ovotoxic metabolite, VCH-1,2-epoxide, at a greater rate than female mice which are susceptible to VCH-induced ovarian toxicity. Therefore, female rat liver performs VCH epoxidation more slowly yet hydrolyzes the ovotoxic epoxide more rapidly than female mouse liver. These findings further support our hypothesis and indicate why it is difficult to detect VCH-1,2-epoxide in the blood of VCH- treated female rats.

REFERENCES

Collins, J.J., Montali, R.J., and Manus, A.G. (1987). Toxicological evaluation of 4-vinylcyclohexene. II. Induction of ovarian tumors in female B6C3F$_1$ mice by 4-vinylcyclohexene. *J. Toxicol. Environ. Health* **21**, 507-524.

Dobson, R.L. and Felton, J.S. (1983). Female germ cell loss from radiation and chemical exposures. *Am. J. Indust. Med.* **4**, 175-190.

Halpert, J.R., Naslund, R.B., and Betner, I. (1983). Suicide inactivation of rat liver cytochrome P-450 by chloramphenicol in vivo and in vitro. *Mol. Pharmacol.* **23**, 445-452.

Jull, J.W. (1973). Ovarian tumorigenesis. *Methods Cancer Res.* **7**, 131-186.

Lowry, O.H. Rosebrough, N.J., Farr, A.L., Randall, R.J. (1951). Protein measurement with the folin phenol reagent. *J. Biol. Chem.* **193**, 265-275.

National Toxicology Program (1986). *Toxicology and carcinogenesis studies of 4-vinylcyclohexene in F344/N rats and B6C3F1 mice. NTP technical report no 303.* U.S. Department of Health and Human Services, Public Health Service, National Institutes of Health, Public Information, National Toxicology Program, P.O. box 12333, Research Triangle Park, N.C.

Ota, K. and Hammock, B.D. (1980). Cytosolic and microsomal epoxide hydrolases: Differential properties in mammalian liver. *Science* **207**, 1479-1480.

Rappaport, S.M. and Fraser, D.A. (1976). Gas chromatographic-mass spectrometric identification of volatiles released from rubber stock during simulated vulcanization. *Anal. Chem.* **48**, 476-481.

Rappaport, S.M. and Fraser, D.A. (1977). Air sampling and analysis in a rubber vulcanization area. *Am. Ind. Hyg. Assoc. J.* **38**, 205-209.

Shiromizu, K. and Mattison, D.R. (1984). The effect of intraovarian injection of benzo(a)pyrene on primordial oocyte number and ovarian aryl hydrocarbon [benzo(a)pyrene] hydroxylase activity. *Toxicol. Appl. Pharmacol.* **76**, 18-25.

Simmon, V.F. and Baden J.M. (1980). Mutagenic activity of vinyl compounds and derived epoxides. *Mutat. Res.* **78**, 227-231.

Sipes, I.G. and Gandolfi, A.J. (1986). Biotransformation of toxicants. In *Toxicology. The Basic Science of Poisons* (Klaassen, C., Amdur, M., and Doull, J. eds.), pp. 64-98, New York, NY.

Smith, B.J., Mattison, D.R., and Sipes, I.G. (1990a). The role of epoxidation in 4-vinylcyclohexene-induced ovarian toxicity. *Toxicol. Appl. Pharmacol.*, manuscript submitted.

Smith, B.J., Carter, D.E., and Sipes, I.G. (1990b). The disposition and *in vitro* metabolism of 4-vinylcyclohexene in the female mouse and rat. *Toxicol. Appl. Pharmacol.*, manuscript submitted.

Smith, B.J., Sipes, I.G., Stevens, J.C., and Halpert, J.R. (1990c). The biochemical basis for the species difference in hepatic microsomal 4-vinylcyclohexene epoxidation between female mice and rats. *Carcinogenesis*, manuscript submitted.

Watabe, T., Hiratsuka, A., Ozawa, N., and Isobe, M. (1981). A comparative study on the metabolism of d-limonene and 4-vinylcylohex-1-ene by hepatic microsomes. *Xenobiotica* **11**, 333-344.

TESTICULAR METABOLISM AND TOXICITY OF HALOGENATED PROPANES

E. Dybing, E.J. Soderlund, M. Låg, G. Brunborg, J.A. Holme, J.G. Omichinski*, P.G. Pearson** and S.D. Nelson**^

Department of Environmental Medicine, National Institute of Public Health, Geitmyrsveien 75, 0462 OSLO 4, Norway.
*NIDDK, National Institutes of Health, Bethesda, MD 20835, USA.
**Department of Medicinal Chemistry, University of Washington, Seattle, WA 98195, USA.

INTRODUCTION

A number of cases of infertility were discovered in 1977 among men working in a factory formulating the nematocide 1,2-dibromo-3-chloropropane (DBCP) (Whorton et al., 1977). These workers were found to have azoospermia or oligospermia. In the severely affected men, testicular biopsies revealed that the seminiferous tubules were devoid of spermatogenic cells (Biava et al., 1978). A similar testicular lesion had already been reported in rats, guinea pigs, rabbits, and monkeys after repeated inhalation exposure to DBCP (Torkelson et al., 1961). Later, the rat has been demonstrated to be a sensitive species towards the acute testicular necrogenic effects of DBCP (Kluwe, 1981). In addition to testicular necrosis and atrophy, also testicular DNA damage in the rat is known to be induced by DBCP (Bradley and Dysart 1985; Soderlund et al., 1988).

The present communication addresses the relationship between DBCP testicular necrosis and DNA damage; the role of metabolism in DBCP testicular toxicity; and a mechanism by which initial DBCP-induced DNA damage may ultimately lead to testicular necrosis.

METHODS

DBCP and selectively methylated and deuterated DBCP analogs were synthesized according to Omichinski et al. (1987a) and Omichinski and Nelson (1988). Other halogenated propanes were synthesized by published methods. The extent and severity of testicular necrosis and atrophy were determined as described by Kluwe (1983). Renal DNA damage was measured by the alkaline elution technique of Kohn et al. (1981) as modified by Brunborg et al. (1988). Testicular nuclear preparations were isolated using a modified procedure of Parodi et al. (1983) (Soderlund et al., 1988). Glutathione conjugates isolated from testicular cell incubations with DBCP were extracted, purified, derivatized and compared to synthetic standards using FAB tandem mass spectrometry (Pearson et al., submitted).

Biological Reactive Intermediates IV, Edited by C.M. Witmer *et al.*
Plenum Press, New York, 1990

RESULTS AND DISCUSSION

A number of halogenated propanes were studied for testicular necrogenic effects in the rat, and correlated with their ability to induce *in vivo* and *in vitro* testicular DNA damage (Table 1). There was a good quantitative relationship between these toxic endpoints, except for 1-bromo-2,3-dichloropro,pane. Both the type, the number and the position of the halogens affected activity. The vicinal dibromo-trihalogenated compounds were the most potent, whereas primary bromines resulted in a more potent compound than that with primary chlorines. The 1,2-dibrominated compound without the third halogen was devoid of activity.

Table 1. Relative potencies of halogenated propanes in causing testicular necrosis and *in vivo* and *in vitro* DNA damage

HALOGENATED PROPANE		RELATIVE POTENCY		
		Necrosis	In vivo DNA damage	In vitro DNA damage
1,2-Dibromo-3-chloropropane (DBCP)	Br Br Cl	+++	++++	+++
1,2,3-Tribromopropane	Br Br Br	+++	++++	+++
1,3-Dibromo-2-chloropropane	Br Cl Br	++	+++	+++
1-Bromo-2,3-dichloropropane	Br Cl Cl	0	++	++
1,3-Dichloro-2-bromopropane	Cl Br Cl	+	+	++
1,2-Dibromopropane	Br Br	0	0	0

Table 2. Testicular necrosis and DNA damage caused by DBCP, perdeutero-DBCP and methylated DBCP analogs in the rat

TREATMENT	NECROSIS[a] Mean grade ± S.D.	DNA DAMAGE Elution rate constant $x10^{-3}hr^{-1}$
Control	0.0 ± 0.0	6[b]
DBCP	3.0 ± 1.5	71
D_5-DBCP	3.1 ± 1.1	67
Control	0.0 ± 0.0	13[c]
DBCP	3.8 ± 0.5	133
C_1-Methyl-DBCP	0.0 ± 0.0	22
C_3-Methyl-DBCP	2.2 ± 1.7	47
1,2-Dibromo-4-chlorobutane	0.0 ± 0.0	16

a Determined 10 days after i.p. doses of 340 μmol/kg (n=8)
b Determined 3 hr after i.p. doses of 85 μmol/kg (n=3)
c Determined 3 hr after i.p. doses of 340 μmol/kg (n=3)
Adapted from (Soderlund et al., 1988).

Administration of DBCP and perdeutero-DBCP to rats demonstrated that the two compounds showed comparable testicular necrogenic as well as DNA damaging potencies (Table 2) (Soderlund et al., 1988). This makes a role for P450 metabolism unlikely in the *in vivo* activation of DBCP to intermediates involved in cell death and DNA damage. The effect of methylation on DBCP testicular toxicity and DNA damage was studied with three methyl,ated analogs (Table 2). Only the C=3-methyl analog caused necrosis as well as appreciable DNA damage. The close correlation between testicular necrosis and DNA damage suggests that for the halogenated propanes both toxic endpoints may be caused by the same chemical species and that they may be mechanistically interrelated.

Isolated testicular cells are able to convert DBCP to a DNA damaging species (Table 1, Fig. 1) (Omichinski et al., 1988), showing that metabolic activation of DBCP can occur in situ. This activation can be totally blocked by the addition of diethyl maleate (Fig. 1). Incubation of testicular cells in the presence of DBCP leads predominantly to the formation of 1,3-bis(S-glutathionyl)-propan-2-ol and S-(2,3-dihydroxypropyl) glutathione, as identified by FAB tandem mass spectrometry. Smaller amounts of S-(3-chloro-2-hydroxypropyl) glutathione are also formed. The bis-GSH metabolite is the major DBCP glutathione conjugate, and it shows full deuterium retention when testicular cells are incubated with perdeutero-DBCP. These findings are consistent with the in situ conversion of DBCP to a reactive episulfonium ion inter,mediate (Fig. 2) (Pearson et al., submitted).

An indispensable role of poly(ADP-ribosyl)ation in DNA excision repair has been proposed (Ueda and Hayashi, 1985). The nuclear poly(ADP-ribosyl)transferase regulates cellular metabolism in response to DNA damage through modulation of NAD^+ pools. Under conditions of extreme DNA damage this may lead to cell death (Berger, 1985). Administration of the poly(ADP-ribosyl)transferase inhibitor 3-aminobenzamide (3-ABA) to rats was found to block the histomorpho-logical evidence of testicular necrosis and atrophy (Table 3). 3-ABA has also been found to block DBCP-induced renal tubular necrosis in the rat (Dybing et al., 1989a) as well as the NAD^+ depletion and cytotoxicity of DBCP in isolated rat hepatocytes (Dybing et al., 1989b).

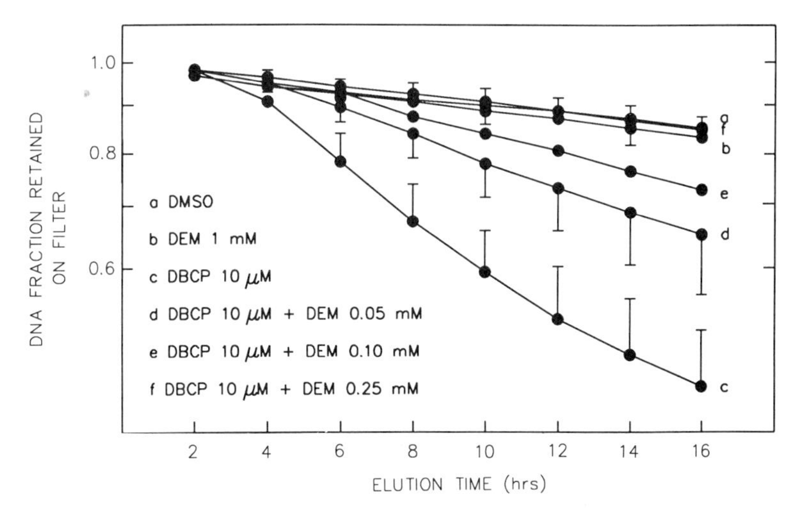

Figure 1. Effect of diethyl maleate (DEM) on DBCP-induced DNA damage in isolated rat testicular cells. 4×10^6 cells were incubated with test chemicals for 60 min at 33°C. Adapted from Omichinski et al. (1988).

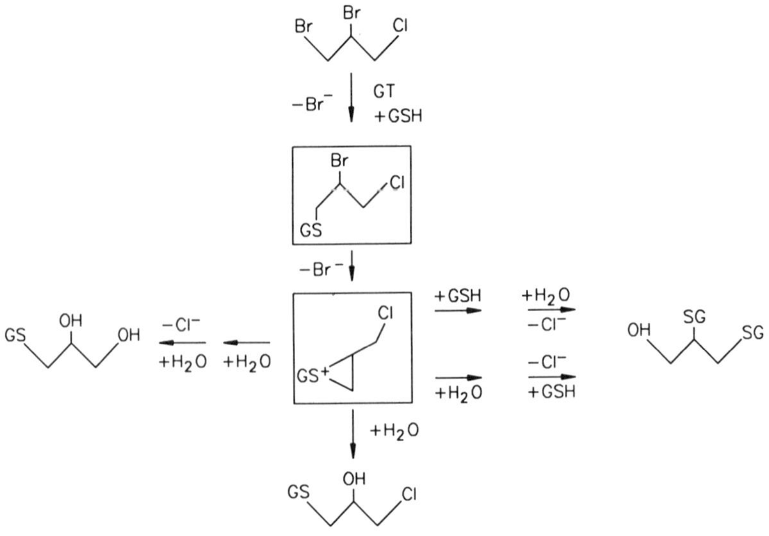

Figure 2. Proposed scheme for the formation of glutathione conjugates from DBCP in isolated rat testicular cells.

Table 3. Effect of 3-aminobenzamide (3-ABA) on DBCP-induced testicular necrosis and atrophy in the rat.

	TESTICULAR NECROSIS AND ATROPHY					
TREATMENT	0	1+	2+	3+	4+	Mean grade ± S.D.
DMSO	5	0	0	0	0	0.0 ± 0.0
3-ABA	5	0	0	0	0	0.0 ± 0.0
DBCP	1	0	0	3	1	2.6 ± 1.5
DBCP + 3-ABA	5	0	0	0	0	0.0 ± 0.0

Determined 10 days after 170 umol DBCP/kg i.p. without or with 4.4 nmol 3-ABA/kg 30 min before and 3.5 and 7.5 hr after DBCP.

Based on the results presented, the following sequence of events is proposed for the mechanism of DBCP testicular necrosis (Fig. 3): 1. Metabolic activation of DBCP in the testis by glutathione S-transferases to a reactive episulfonium ion; 2. interaction of this intermediate with DNA; 3. formation of DNA strand breaks; 4. activation of poly(ADP-ribosyl)transferase; 5. depletion of cellular NAD^+; and 6. further cellular steps ultimately leading to cell death.

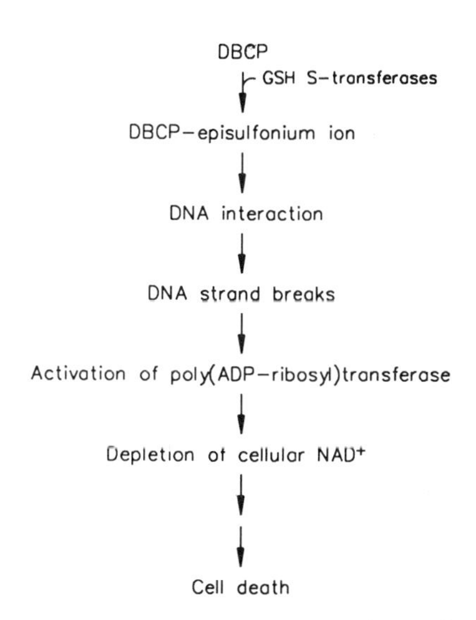

Figure 3. Involvement of poly(ADP-ribosyl)transferase in DBCP-induced organ damage.

ACKNOWLEDGEMENT

This research was supported by NIH Grant ES02728 and a grant from the Norwegian Council for Science and the Humanities.

REFERENCES

Berger, N.A. (1985). Poly(ADP-ribose) in the cellular response to DNA damage. *Radiation Res.* **101**, 4-15.

Biava, C.G., Smuckler, E.A. and Whorton, D. (1978). The testicular morphology of individuals exposed to dibromochloropropane. *Exp. Mol. Pathol.* **29**, 448-458.

Bradley, M.O. and Dysart, G. (1985). DNA single-strand breaks, doublestrand breaks and crosslinks in rat testicular germ cells: measurement of their formation and repair by alkaline and neutral filter elution. *Cell Biol. Toxicol.* **1**, 181-195.

Brunborg, G., Holme, J.A., Soderlund, E.J., Omichinski, J.G. and Dybing, E. (1988). An automated alkaline elution system. DNA damage induced by 1,2-dibromo-3-chloropropane *in vivo* and *in vitro*. *Anal. Biochem.* **174**, 522-536.

Dybing, E., Soderlund, E.J., Låg, M., Brunborg, G., Holme, J.A., Pearson, P.G. and Nelson, S.D. (1989a). Mechanism of 1,2-dibromo-3-chloropropane (DBCP) nephrotoxicity. Abstract, Int. Symp. Mechanisms of Tissue Specific Toxicity, Waarzburg.

Dybing, E., Soderlund, E.J., Låg, M., Brunborg, G., Omichinski, J.G., Dahl, J.E., Pearson, P.G., Holme, J.A. and Nelson, S.D. (1989b). Cytotoxicity via interactions with DNA. In *Proceedings of the 5th International Congress of Toxicology*, in press.

Kluwe, W.M. (1981). Acute toxicity of 1,2-dibromo-3-chloropropane in the F344 male rat. I. Dose-response relationships and differences in routes of exposure. *Toxicol. Appl. Pharmacol.* **59**, 71-83.

Kluwe, W.M.(1983). Chemical modulation of 1,2-dibromo-3-chloropropane toxicity. *Toxicology* **27**, 287-299.

Kohn, K.W., Erickson, L.C., Ewig, R.A.G. and Qwelling (1981). Measurement of strand breaks and crosslinks by alkaline elution. In: *DNA Repair: A Laboratory Manual of Recent Procedures* (E.C. Friedberg and P.C. Hanawalt, eds.), pp. 379-401, Marcel Dekker, New York, USA.

Omichinski, J.G. and Nelson, S.D. (1988). Synthesis of six specifically deuterated

analogs of 1,2-dibromo-3-chloropropane. *J. Labelled Comp. Radiopharmaceut.* **25**, 263-265.
Omichinski, J.G., Soderlund, E.J., Bausano, J., Dybing, E. and Nelson, S.D. (1987). Synthesis and mutagenicity of selectively methylated analogs of tris(2,3-dibromopropyl)phosphate and 1,2-dibromo-3-chloropropane. *Mutagenesis* **2**, 287-292.
Omichinski, J.G., Brunborg, G., Holme, J.A., Soderlund, E.J., Nelson, S.D. and Dybing, E. (1988). The role of oxidative and conjugative pathways in the activation of 1,2-dibromo-3-chloropropane to DNA-damaging products in rat testicular cells. *Mol. Pharmacol.* **34**, 74-79.
Parodi, S., Pala, M., Russo, P., Balbi, C., Abelmoschi, M.L., Taningher, M., Zunino, O., Ottagio, L., De Ferrari, M., Carbone, A. and Santi, L. (1983). Alkaline DNA fragmentation, DNA distanglement evaluated viscosimetrically and sister chromatid exchanges, after treatment *in vivo* with nitrofurantoin. *Chem. Biol. Interact.* **45**, 77-94.
Pearson, P.G., Soderlund, E.J., Dybing, E. and Nelson, S.D. Metabolic activation of 1,2-dibromo-3-chloropropane: evidence for the formation of reactive episulfonium intermediates. *Biochemistry*, submitted.
Soderlund, E.J., Brunborg, G., Omichinski, J.G., Holme, J.A., Dahl, J.E., Nelson, S.D. and Dybing, E. (1988). Testicular necrosis and DNA damage caused by deuterated and methylated analogs of 1,2-dibromo-3-chloropropane in the rat. *Toxicol. Appl. Pharmacol.* **94**, 437-447.
Torkelson, T.R., Sadek, S.E., Rowe, U.K., Kodama, J.K., Anderson, H.H., Loquvam, G.S. and Hine, C.H. (1961). Toxicological investigations of 1,2-dibromo-3-chloropropane. *Toxicol. Appl. Pharmacol.* **3**, 545-559.
Ueda, K., and Hayashi, O. (1985). ADP-ribosylation. *Ann. Rev. Biochem.* **54**, 73-100.
Whorton, D., Krauss, R.M., Marshall, S. and Milby, T.H. (1977). Infertility in male pesticide workers, *Lancet* **1**, 1259-1261.

LUNG VASCULAR INJURY FROM MONOCROTALINE PYRROLE, A PUTATIVE HEPATIC METABOLITE

Robert A. Roth* and James F. Reindel[†]

*Departments of Pharmacology and Toxicology, and of Pathology[†]
Michigan State University
East Lansing, Michigan 48824

ABSTRACT

The pyrrolizidine alkaloid, monocrotaline (MCT), is a plant toxin that causes injury to the vasculature of the lungs and pulmonary hypertension in animals. To produce lung injury, MCT is bioactivated in the liver by cytochrome P450 monooxygenases to pyrrolic metabolites which travel via the circulation to the lungs, where they cause injury by unknown mechanisms. One putative metabolite of MCT is monocrotaline pyrrole (dehydromonocrotaline, MCTP), a moderately reactive, bifunctional alkylating agent. A single, iv injection of chemically synthesized MCTP into rats causes delayed and progressive lung vascular injury and pulmonary hypertension similar to that caused by MCT itself.

Since pulmonary vascular endothelium is likely an important target of MCTP *in vivo*, the effects of MCTP on cultured endothelium were studied. A single application of MCTP to confluent monolayers of cultured endothelium from bovine pulmonary artery results in release of lactate dehydrogenase, some cell detachment from the growth surface and markedly altered morphology of remaining viable cells. These effects are dose-dependent and, as *in vivo*, are delayed in onset (1-2 days) and progressive. In endothelial cells of porcine origin, these particular responses to MCTP are also apparent but much less pronounced. Inhibition of proliferation of cells plated at low density occurred in both cell types at nominal MCTP concentrations (0.5 μg/ml) that were not overtly cytotoxic. These results indicate that MCTP causes a direct, dose-dependent injury to pulmonary vascular endothelium in culture that is delayed and progressive and suggest a mechanism by which MCT may act *in vivo* to cause lung injury and pulmonary hypertension.

INTRODUCTION

Pyrrolizidine alkaloids (PAs) are botanical toxins found in hundreds of plant species (Bull et al., 1968). They are of interest to the medical and scientific community in that many of them have been associated with intoxications of humans who have ingested cereal products contaminated with PA-containing seeds or herbal teas made with PA-containing plants (Huxtable, 1980; McLean, 1970). In addition, ingestion of PA-containing plants by grazing animals is a significant cause of animal disease in many parts of the world and causes substantial economic losses to the animal industry (Sippel, 1964; Piercy and Rusoff, 1946; McLean, 1970). Although toxic PAs are probably best known for their hepatotoxic effects, some PAs such as monocrotaline (MCT) cause pulmonary vascular disease and pulmonary hypertension in

Biological Reactive Intermediates IV, Edited by C.M. Witmer *et al.*
Plenum Press, New York, 1990

experimental animals (Turner and Lalich, 1965; Kay and Heath, 1969). MCT-induced pulmonary vascular disease in rats has been used as a model for certain forms of human chronic pulmonary hypertension (Voelkel and Reeves, 1979; Kay and Heath, 1969; Snow et al., 1982; White and Roth, 1989).

MCT is a toxic PA that occurs in the seeds and leaves of several plants of the *Crotalaria* genus, including the plant *C. spectabilis* (Bull et al., 1968; Neal et al., 1935; Adams and Rogers, 1939). It is a heterocyclic compound consisting of a retronecine nucleus esterified to the carboxyl groups of monocrotalic acid, a branched chained dicarboxylic acid (Figure 1). When high doses of MCT are given to rats, acute liver injury and death occurs within several days of administration (Schoental and Head, 1955). However, lower doses that produce only mild, transient liver injury result in lung injury that is associated with pulmonary hypertension and death several within weeks after administration (Hilliker et al., 1982). In this commentary, we will discuss the bioactivation of MCT to pneumotoxic metabolite(s), describe the pulmonary lesion, and summarize recent results using cultured pulmonary endothelium that have suggested an hypothesis to explain how MCT causes vascular injury and pulmonary hypertension.

RESULTS AND DISCUSSION

Metabolism of MCT

Initial metabolism of MCT and other PAs occurs in the liver by 3 major pathways (Mattocks and White, 1971; Allen et al., 1972; Tuchweber et al., 1974). Hydrolysis of the ester bonds and N-oxidation result in metabolites which, like the parent alkaloids, are relatively non-toxic. The third pathway leads to pyrrolic oxidation products which are generally considered to be responsible for the toxicity (Mattocks, 1968).

The PA pyrroles arise from the action of cytochrome P450 monooxygenases in the liver (Figure 1). Their formation occurs in the microsomal fraction, requires NADPH and molecular oxygen, and is inhibited by carbon monoxide, SKF 525A and other inhibitors of cytochrome P450-mediated reactions. Pretreatment of rats with phenobarbital increases the rate of pyrrole formation by hepatic microsomes. It has been proposed that PA pyrroles arise from initial hydroxylation of the PA at position 8 followed by leaving of the hydroxyl group and subsequent rearrangement of the product to a pyrrole (Mattocks, 1986). A recent study (Williams et al., 1989) indicates that cytochrome P450 PCN-E is the isozyme responsible for formation of at least one PA pyrrole (i.e., from senecionine) in rat liver.

Many PA pyrroles are unstable in aqueous solution and react readily with nucleophiles. Many have two electrophilic centers (Fig. 1) and are thus capable of acting as bifunctional alkylating agents (Kedzierski and Buhler, 1986). Indeed, DNA-DNA and DNA-protein crosslinking occurs in livers of MCT-treated rats (Petry et al., 1984). It is their capacity to react with tissue nucleophiles that likely renders the PA pyrroles responsible for the hepatotoxicity of the PAs.

Pyrroles are detected in lung after MCT administration (Mattocks, 1972). However, the lung and other extrahepatic tissues of the rat are apparently incapable of producing pyrrolic metabolites from the parent PAs (Hilliker et al., 1983; Mattocks, 1968; Mattocks and White, 1970, 1971). The pneumotoxicity that occurs from MCT is likely due to one or more pyrrolic metabolites produced in the liver. The chemical nature of monocrotaline pyrrole (MCTP, dehydromonocrotaline, Fig. 1) suggests that it is an important pneumotoxic metabolite. Primary pyrrolic metabolites of many PAs are quite unstable in aqueous milieu and may react before leaving hepatocytes in which they are formed. However, MCTP is among the more stable of the toxic, primary PA pyrroles (Karchesy and Deinzer, 1981; Karchesy et al., 1987). It has a half-life in rat serum of about five seconds (Bruner et al., 1986). This means that more than 15 seconds are required before 90% of MCTP is degraded. The circulation time of plasma in a rat is three seconds or less (Hanwell and Linzell, 1972). Thus, even if the survival of MCTP were somewhat shorter in hepatic intracellular milieu and blood than in serum, it seems likely that a fraction of MCTP formed *in vivo* in liver cells could escape and survive passage in the circulation to the next microvascular bed (i.e., that of the lung), where it could bind covalently to tissue components and initiate toxic changes.

MCTP that is synthesized chemically and given to rats intravenously (2-5 mg/kg) causes lung toxicity that is virtually identical to that produced by doses of MCT that are 10-20 times greater (Bruner et al., 1986). The route of administration is important in determining the toxic effects of MCTP. Injection of this unstable pyrrole into the hepatic portal vein produces only hepatic lesions (Butler et al., 1970), and subcutaneous administration leads only to local skin lesions (Hooson and Grasso, 1976). Intravenous administration alone leads to lung injury, a finding consistent with the idea that MCTP, once formed in the liver from MCT, is just stable enough to reach the next downstream capillary bed wherein it produces injury. Accordingly, the lungs may be an important target organ for MCT simply because of their position in the circulation relative to the liver (Figure 2).

Although the moderate chemical reactivity of MCTP and its biologic effects lend support to the contention that MCTP is an important pneumotoxic metabolite of MCT, it remains a putative proximate toxicant because it has not yet been isolated as such from target tissue. Indeed, Huxtable and coworkers have discovered a pyrrolic glutathione conjugate formed in the liver from MCT that is quite stable and that is also pneumotoxic (described elsewhere in this volume). Its greater stability and the need for doses approaching that of MCT itself to produce toxicity raise questions about whether and why it selectively targets the lung and about its importance *in vivo* as a pneumotoxic metabolite of MCT. Clearly, the roles of specific pyrrolic metabolites in MCT pneumotoxicity require additional study. It may be that several pyrrolic metabolites play important roles.

Nature of the Lung Injury

Since the requirement for bioactivation of MCT can limit experimental approaches and the interpretation of results of studies designed to explore biologic mechanisms via pharmacologic modulation, we have employed chemically synthesized MCTP in our studies. MCTP apparently does not require further bioactivation *in vivo* (Bruner et al., 1986), and the pathophysiology of MCTP pneumotoxicosis resembles closely that produced by MCT itself (Chesney et al., 1974; Butler, 1970; Butler et al., 1970; Lalich et al., 1977; Bruner et al., 1983, 1986; Reindel and Roth, unpublished

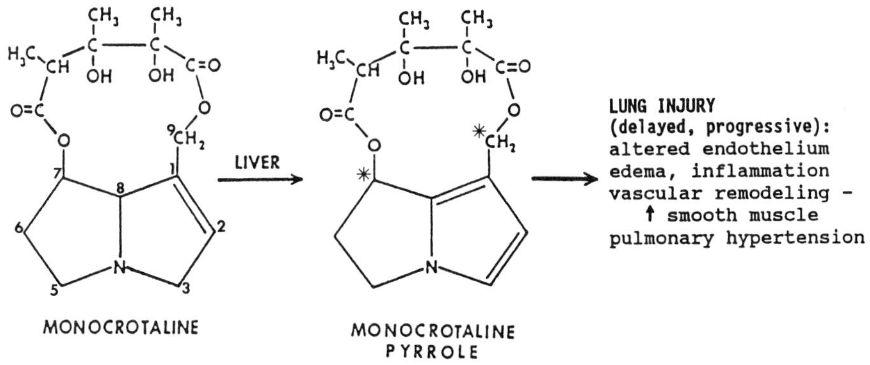

LIVER

MONOCROTALINE

MONOCROTALINE PYRROLE

LUNG INJURY
(delayed, progressive):
altered endothelium
edema, inflammation
vascular remodeling -
↑ smooth muscle
pulmonary hypertension

Figure 1. Metabolism and pneumotoxicity of monocrotaline. Monocrotaline is metabolized by liver to pyrrolic metabolites that have electrophilic centers (asterisks) and that cause delayed and progressive lung injury.

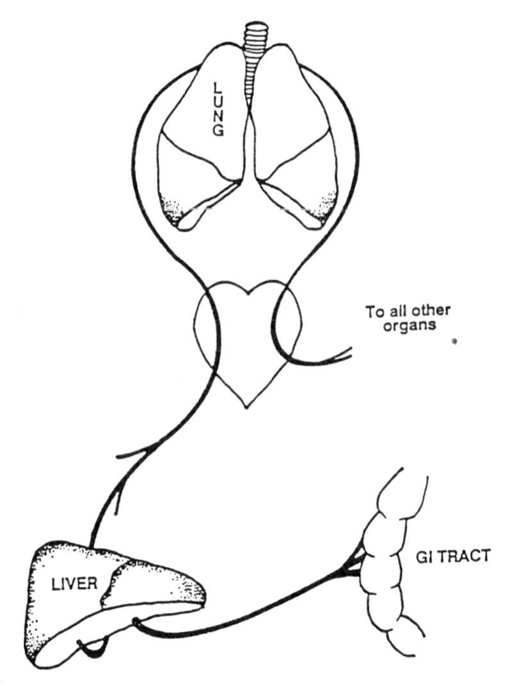

Figure 2. Liver-lung circulatory relationship. Xenobiotic agents consumed in food are absorbed in the gastrointestinal tract and enter the liver, where they may be bioactivated to toxic metabolites that are released into the circulation. The lungs may be a target for these because they represent the first capillary bed that is downstream from the liver.

observations). The onset of some of the major morphologic changes is summarized in the top panel of Figure 3. Lungs of rats given a single iv injection of a low dose (3.5 mg/kg) of MCTP appear normal for several days. By day 3, minimal edema around large airways and blood vessels and dilated lymphatic vessels are seen. By day 5, blebbing of occasional endothelial cell profiles is seen ultrastructurally, and this is accompanied by more pronounced edema in the perivascular and peribronchiolar interstitium and in alveolar septal walls. There is a slight increase in cellularity in interstitial areas, which is predominantly mononuclear in nature. Degeneration and necrosis of Type I cells are also evident. These morphologic alterations are accompanied by changes in quantifiable markers of lung injury and vessel leak. Instead of resolving, as happens with many other pneumotoxicants, the injury progressively worsens over the next week, so that by 14 days there is severe alveolar and interstitial edema with consolidation of large portions of lung parenchyma. Vascular remodeling takes place that involves thickening of the medial layer of pulmonary arteries and edema and inflammation of adventitia. Smooth muscle is apparent in small arterioles of the lung that are normally non-muscular. Increased numbers of alveolar macrophages are seen. Type II cells become enlarged and bizarre in appearance, and interstitial cells continue to increase in number and become extremely hypertrophic. The pronounced vascular structural changes are associated with sustained elevation in pulmonary arterial pressure and, by 2 weeks, hypertrophy of the right ventricle of the heart.

These toxic sequelae may be viewed as two phases: a lung injury phase that is characterized by delayed and progressive vessel leak and other manifestations of tissue damage which evolves into a pulmonary hypertensive phase characterized by vessel remodeling and elevated vascular pressure. This second phase likely reflects a response of the pulmonary vasculature to the initial, sustained lung injury.

Role of the Endothelial Cell in MCT-induced Pulmonary Hypertension

Several lines of evidence indicate that the endothelial cell may play a critical role in the development of MCT- and MCTP-induced lung disease and may in fact be the major target for the toxic effects of pyrrolic metabolites. In numerous studies in which rats were given MCT or MCTP, alterations in endothelial cell morphology and function were reported (Vincic et al., 1989; Kay et al., 1969; Plestina and Stoner, 1972; Butler, 1970; Meyrick and Reid, 1982; Rosenberg and Rabinovitch, 1988; Valdivia et al., 1967; Allen and Carstens, 1970). Alterations in morphology included vesicular blebbing of endothelial cells, increased prominence of pinocytotic vesicles, focal rarefaction of cell cytoplasm, marked enlargement of nuclei and cell hypertrophy. When the development of vascular alterations has been examined morphologically, many changes were slow to develop and progressive in that they were initially subtle or appeared only several days after treatment and worsened with time (Valdivia et al., 1967; Merkow and Kleinerman, 1966; Butler, 1970; Plestina and Stoner, 1972).

Alterations suggesting pulmonary endothelial cell dysfunction have also been reported. These include progressive decreases in uptake into lung of circulating 5-hydroxytryptamine and norepinephrine and altered plasminogen activator activity (Hilliker et al., 1983, 1984b; Molteni et al., 1984). Again, these changes were not apparent shortly after treatment but emerged slowly and gradually worsened days to weeks after exposure. The barrier function of the endothelium progressively deteriorates after exposure as indicated by a delayed but progressive enhancement in accumulation in lungs of circulating ^{125}I-albumin (Sugita et al., 1983; Bruner et al., 1986; Reindel et al., unpublished observations), exudation of fibrin and red blood cells into lung interstitium and alveoli and pulmonary platelet sequestration (White and Roth, 1988).

The delayed and progressive nature of the injury to endothelial cells seemed enigmatic, since reactive pyrroles such as MCTP bind rapidly to tissue macromolecules or are rapidly inactivated in aqueous environments. In fact, this suggested to some investigators that PA pyrroles, such as MCTP, may not be directly toxic to lung cells but may trigger indirect mechanisms, such as production and release of endogenous mediators of injury (e.g., leukotrienes, cytokines, etc.) (Roth and Ganey, 1987; Bruner et al., 1988; Langleben and Reid, 1985; Kay and Heath, 1966).

The effects of MCTP on endothelial cells in culture indicate that it is indeed directly toxic to cultured bovine and porcine pulmonary arterial endothelial cells (Roth and Reindel, 1988; Reindel and Roth, 1989; Roth et al., 1989a). The injury to these cells and cell monolayers was delayed in development and progressive, much like the injury *in vivo* (Fig. 3). Injury to monolayers of bovine pulmonary arterial endothelial cells was not apparent at 24 hours post- treatment, but thereafter 6-keto-prostaglandin $F_{1\alpha}$, the stable metabolite of prostacyclin, and lactate dehydrogenase activity gradually accumulated in the incubation medium over the subsequent 2-3 days. These changes were associated with a gradual increase in cell detachment from the monolayer surface. Surviving cells underwent pronounced hypertrophy, and they spread to cover up to 10 times the surface area of vehicle-treated control cells (Figure 4). Nuclei enlarged proportionally and had prominent, irregularly shaped nucleoli. The cytoplasm of some cells had perinuclear vacuoles and cytoplasmic filamentous structures radiating from the perinuclear region. For porcine cells, signs of cell injury developed even more slowly. Importantly, cell proliferation was blocked by very low concentrations of MCTP (0.5 μg/ml), and this hampered effective monolayer repair. Compensatory hypertrophy and spreading of cells remaining in the monolayer apparently maintained monolayer integrity for a while after the onset of injury: visible gaps appeared between cells only after several days to weeks following exposure.

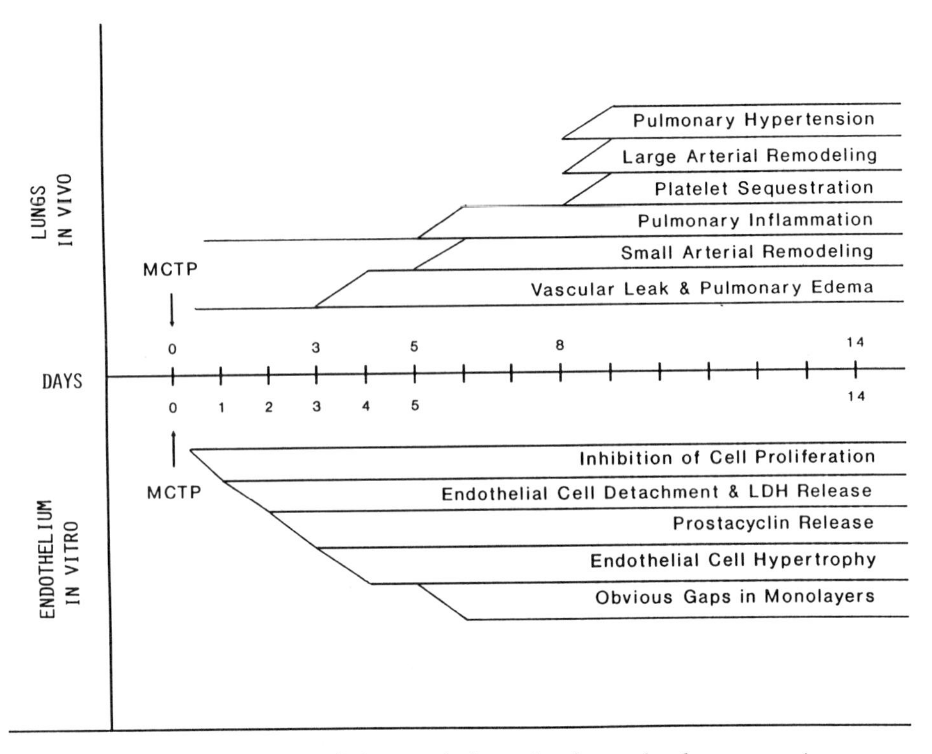

Figure 3. Summary of changes in lungs *in vivo* and pulmonary artery endothelium in vitro from MCTP. A single i.v. administration of 3-5 mg/kg MCTP (day 0) to rats results in lung injury. Although very subtle changes are detected early, the major manifestations of injury are delayed in appearance and progressive. MCTP applied to endothelial cell monolayers in culture also results in numerous changes that are delayed and progressive.

If comparable cellular responses occur *in vivo*, they could explain the delayed and progressive nature of MCT- and MCTP-induced pulmonary injury. If endothelial cell death does not occur until 24 hours or more after treatment, vascular injury might be delayed relative to that which occurs after exposure to other endothelial toxicants. The spreading of surviving cells, which can rapidly cover areas where endothelial cell detachment occurs without exposing sizable areas of the subendothelial matrix (i.e., a non-denuding injury), could also slow development and progression of vascular leak. The block in cell proliferation and, consequently, in the reparative response would ensure that the vascular leak is progressive and prolonged, since restoration of the endothelium by healthy endothelial cells would not readily occur until the cells incapable of reproduction were removed.

The persistant toxic effects of MCTP on the endothelium could contribute to the occurrence and maintenance of pulmonary hypertension in several ways. For example, gradual disruption of the endothelial barrier could facilitate the progressive accumulation of interstitial exudate. The mere presence of this exudate might limit vascular compliance. In addition, unresolved vascular leak leads to prolonged presence of large blood proteins in the lung interstitium which can invoke an inflammatory response and stimulate release of vasoactive molecules from activated interstitial inflammatory cells and mast cells. A site of accumulation of this proteinaceous fluid is in the perivascular and peribronchiolar interstitium, where small numbers of

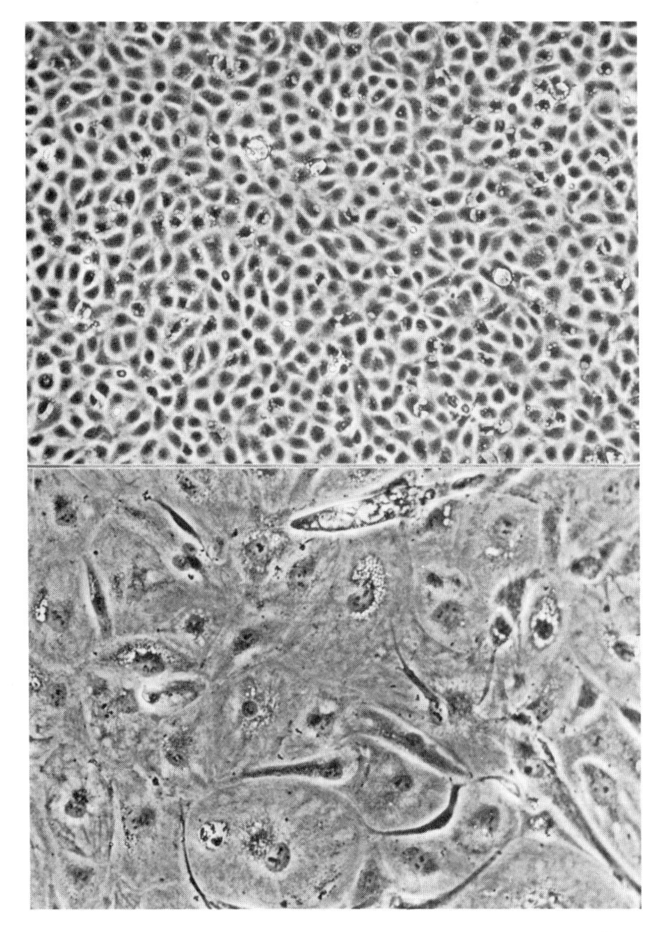

Figure 4. Photomicrographs of cultured bovine pulmonary arterial endothelial cells 14 days after a single exposure to DMF vehicle (top) or 50 µg/ml MCTP (bottom). (Phase contrast, 180X.)

inflammatory cells and mast cells exist. Upon activation, these cells are capable of releasing mitogens and vasoactive agents such as growth factors, oxygen radicals, leukotrienes and biogenic amines that influence the activity of smooth muscle cells in the media of blood vessels.

Damaged endothelial cells may also have a more direct influence on pulmonary vascular resistance (PVR) and/or vascular remodeling. Endothelial structural changes such as cell swelling, blebbing or hypertrophy could contribute to increased PVR by reducing the lumen diameter of capillaries and small arteries (Valdivia et al., 1967; Raczniak et al., 1979). Such morphologic changes have been identified *in vivo* (see above). In addition, endothelial cells are known to produce a variety of molecules that can influence smooth muscle activity. These include growth factors, vasoconstrictors such as endothelin, and vasodilators such as endothelium-derived relaxing factor and prostacyclin (Hammersen and Hammersen, 1985; Vane et al., 1987; Furchgott, 1983). Cultured bovine and porcine endothelial cells perturbed by MCTP release enhanced amounts of the stable metabolite of prostacyclin. Although prostaglandin is a vasodilator, vasoconstrictive factors may also be produced. Additional studies are needed to determine the effects of MCTP on the wide range of functions of endothelial cells.

The deterioration of the endothelial cell barrier may influence circulating blood components which, in turn, could influence PVR and pulmonary vascular remodeling. Platelets may adhere to the subendothelium when spreading and hypertrophy of endothelial cells can no longer maintain the integrity of the endothelial barrier. Indeed, platelet thrombi occur in MCT- and MCTP-treated animals (Lalich et al., 1977; Turner and Lalich, 1965; Valdivia et al., 1967), and platelet sequestration occurs in the lungs several days after MCTP administration, i.e., at about the time when major manifestations of lung injury become apparent (White and Roth, 1988). Platelet thrombi can themselves cause microvascular occlusion and can release vasoconstrictors and mitogens. That platelets are important in the development of pulmonary hypertension was shown in studies by Hilliker et al. (1984a) and Ganey et al. (1988). Depletion of blood platelets with antiplatelet antibodies diminished the pulmonary hypertension and right ventricular hypertrophy caused by MCTP. Which platelet factors contribute to the development of pulmonary hypertension in this disease are not known. The vasoconstrictor, thromboxane A_2 does not appear to be involved, since inhibition of TxA_2 production or receptor antagonism did not alter pulmonary hypertension (Ganey and Roth, 1986). Similarly, antagonists of 5HT receptors are without effect (Ganey et al., 1986). Other factors released by platelets, such as platelet-derived growth factor (PDGF), transforming growth factor β (TGFβ) or epidermal growth factor (EGF), may contribute to the development of pulmonary hypertension. PDGF is released from activated platelets and can serve as a smooth muscle cell mitogen and a vasoconstrictor (Ross and Vogel, 1978; Ross et al., 1986). TGFβ and EGF have hypertrophic effects on vascular smooth muscle that may be important in vascular remodeling. TGFβ- and EGF-like peptides are present in elevated amounts in lungs of MCT-treated rats (Orlinska et al., 1989; Rippetoe et al., 1989). Further study is needed to identify mediators of vascular remodeling and pulmonary hypertension in this model.

In summary, MCT is metabolized in the liver to a pneumotoxic pyrrolic metabolite, probably MCTP. The selectivity for lung is likely determined by the site of bioactivation, which is the liver, the moderately reactive nature of the toxic metabolite, and the relative positions of the liver and the lung in the circulation. The vascular bed of the lung would be the first to be encountered by a toxic metabolite that is produced in and escapes from the liver. Like MCT, MCTP causes injury *in vivo* that is delayed and progressive and that includes damaged endothelial cells, leaky vessels and persistent pulmonary hypertension. MCTP applied to cultures of pulmonary endothelial cells results similarly in delayed and progressive injury that is characterized by enhanced cell lysis and, in surviving cells, increased size and diminished capacity to proliferate (Fig. 3). Inasmuch as proliferation of endothelial cells is important in maintaining vessel homeostasis and repair, the inhibition of proliferation would be expected to have profound consequences. Persistent injury to vascular endothelium and failure to repair may result in processes that lead to vessel remodeling and pulmonary hypertension.

ACKNOWLEDEMENTS

The authors thank Diane Hummel for preparation of the manuscript and Gwen Suchanek and Susan Parkinson for preparation of figures. Supported by NIH grant ES02581.

REFERENCES

Adams, R., and Rogers, E. F. (1939). The structure of monocrotaline, the alkaloid in Crotalaria spectabilis and Crotalaria retusa. *I. J. Am. Chem. Soc.* **61**, 2815-2819.

Allen, J. R., and Carstens, L. A. (1970). Pulmonary vascular occlusions initiated by endothelial lysis in monocrotaline-intoxicated rats. *Exp. Mol. Pathol.* **13,** 159-171.

Allen, J. R., Chesney, C. F., and Frazee, W. J. (1972). Modifications of pyrrolizidine alkaloid intoxication resulting from altered hepatic microsomal enzymes. *Toxicol. Appl. Pharmacol.* **23,** 470-479.

Bruner, L. H., Carpenter, L. J., Hamlow, P., and Roth , R. A. (1986). Effect of a mixed function oxidase inducer and inhibitor on monocrotaline pyrrole pneumotoxicity. *Toxicol. Appl. Pharmacol.* **85,** 416-427.

Bruner, L. H., Hilliker, K. S., and Roth, R. A. (1983). Pulmonary hypertension and EKG changes from monocrotaline pyrrole in the rat. *Amer. J. Physiol.* **245,** H300-H306.

Bruner, L. H., Johnson, K. J., Till, G. O., and Roth, R. A. (1988). Complement is not involved in monocrotaline pyrrole-induced pulmonary injury. *Am. J. Physiol.* **254,** H258-H264.

Bull, L. B., Culvenor, C. C. J., and Dick, A. T. (1968). The Pyrrolizidine Alkaloids, North-Holland, Amsterdam.

Butler, W. H. (1970). An ultrastructural study of the pulmonary lesion induced by pyrrole derivatives of the pyrrolizidine alkaloids. *J. Pathol.* **102,** 15-19.

Butler, W. H., Mattocks, A. R., and Barnes, J. M. (1970). Lesions in the liver and lungs of rats given pyrrole derivatives of pyrrolizidine alkaloids. *J. Pathol.* **100,** 169-175.

Chesney, C. F., Allen, J. R., and Hsu, I. V. (1974). Right ventricular hypertrophy in monocrotaline pyrrole treated rats. *Exp. Mol. Pathol.* **20,** 257-268.

Furchgott, R. F. (1983). Role of the endothelium in responses of vascular smooth muscle. *Circ. Res.* **53,** 553-557.

Ganey, P. E., Hilliker-Sprugel, K., White, S. M., Wagner, J. G., and Roth, R. A. (1988). Pulmonary hypertension due to monocrotaline pyrrole is reduced by moderate thrombocytopenia. *Am. J. Physiol.* **255,** H1165-H1172.

Ganey, P. E., and Roth, R. A. (1986). Thromboxane does not mediate pulmonary vascular response to monocrotaline pyrrole. *Am. J. Physiol.* **252,** H743-H748.

Ganey, P. E., Sprugel, K. H., Boner, K. E., and Roth, R. A. (1986). Monocrotaline pyrrole induced cardiopulmonary toxicity is not altered by metergoline or ketanserin. *J. Pharmacol. Exp. Ther.* **237,** 226-231.

Hammersen, F., and Hammersen, E. (1985). Some structural and functional aspects of endothelial cells. *Basic Res. Cardiol.* **80,** 491-501.

Hanwell, A., and Linzell, J. L. (1972). Validation of the thermodilution technique for the estimation of cardiac output in the rat. *Comp. Biochem. Physiol.* **41,** 647-657.

Hilliker, K. S., Bell, T. G., and Roth, R. A. (1982). Pneumotoxicity and thrombocytopenia after a single injection of monocrotaline. *Amer. J. Physiol.* **242,** H573-H579.

Hilliker, K. S., Bell, T. G., Lorimer, D., and Roth, R. A. (1984a). Effects of thrombocytopenia on monocrotaline pyrrole-induced pulmonary hypertension. *Amer. J. Physiol.* **246,** H747-H753.

Hilliker, K. S., Garcia, C.M., and Roth, R. A. (1983). Effects of monocrotaline and monocrotaline pyrrole on 5-hydroxytryptamine and paraquat uptake by lung slices. *Res. Commun. Chem. Pathol. Pharmacol.* **40,** 179-197.

Hilliker, K. S., Imlay, M., and Roth, R. A. (1984b). Effects of monocrotaline treatment on norepinephrine removal by isolated, perfused rat lungs. *Biochem. Pharmacol.* **33,** 2692-2695.

Hooson, J., and Grasso, P. (1976). Cytotoxic and carcinogenic response to monocrotaline pyrrole. *J. Pathol.* **118,** 121-129.

Huxtable, R. J. (1980). Problems with pyrrolizidine alkaloids. *Trends in Pharmacol.* **1,** 299-303.

Karchesy, J. J., Arbogast, B., and Deinzer, M. L. (1987). Kinetics of alkylation reaction of pyrrolizidine alkaloid pyrroles. *J. Org. Chem.* **52,** 3867-3872.

Karchesy, J. J., and Deinzer, M. L. (1981). Kinetics of alkylation reactions of pyrrolizidine alkaloid derivatives. *Heterocycles* **16,** 631-635.

Kay, J. M., and Heath, D. (1966). Observations on the pulmonary arteries and heart weight of rats fed on Crotalaria spectabilis seeds. *J. Pathol. Bacteriol.* **92,** 385-394.

Kay, J. M., and Heath, D. (1969). Crotalaria spectabilis. The Pulmonary Hypertension

Plant. Charles C. Thomas, Springfield, IL.
Kay, J. M., Smith, P., and Heath, D. (1969). Electron microscopy of Crotalaria pulmonary hypertension. *Thorax* **24**, 511-526.
Kedzierski, B., and Buhler, D. R. (1986). The formation of 6,7-dihydro-7- hydroxyl-1-hydroxymethyl-5H-pyrrolizine, a metabolite pf pyrrolizidine alkaloids. *Chem.-Biol. Interactions* **57**, 217-222.
Lalich, J. J., Johnson, W. D., Raczniak, T. J., and Shumaker, R. C. (1977). Fibrin thrombosis in monocrotaline pyrrole-induced cor pulmonale in rats. *Arch. Pathol. Lab. Med.* **101**, 69-73.
Langleben, D., and Reid, L. M. (1985). Effect of methylprednisolone on monocrotaline-induced pulmonary vascular disease and right ventricular hypertrophy. *Lab. Invest.* **52**, 298-303.
Mattocks, A. R. (1968). Toxicity of pyrrolizidine alkaloids. *Nature* (London) **217**, 723-728.
Mattocks, A. R. (1972). Acute hepatotoxicity of pyrrolic metabolites in rats dosed with pyrrolizidine alkaloids. *Chem.-Biol. Interactions* **5**, 227-242.
Mattocks, A. R. (1986). Metabolism, distribution and excretion of pyrrolizidine alkaloids. In: Chemistry and Toxicology of Pyrrolizidine Alkaloids (A.R. Mattocks, ed.), pp. 167-168. Academic Press, London.
Mattocks, A. R., and White, I. N. H. (1970). Estimation of metabolites of pyrrolizidine alkaloids in animal tissues. *Anal. Biochem.* **38**, 529-535.
Mattocks, A. R., and White, I. N. H. (1971). The conversion of pyrrolizidine alkaloids to N-oxides and to dihydropyrrolizidine derivatives by rat liver microsomes *in vitro*. *Chem.-Biol. Interact.* **3**, 383-396.
McLean, E. K. (1970). The toxic action of pyrrolizidine (Senecio) alkaloids. *Pharmacol. Rev.* **22**, 429-483.
Merkow, L., and Kleinerman, J. (1966). An electron microscopic study of pulmonary vasculitis induced by monocrotaline. *Lab. Invest.* **15**, 547-564.
Meyrick, B., and Reid, L. (1982). Crotalaria-induced pulmonary hypertension: Uptake of ^{3}H-thymidine by the cells of the pulmonary circulation and alveolar walls. *Am. J. Pathol.* **106**, 84-94.
Molteni, A., Ward, W. F., Tsao, C-H., Port, C. D., and Solliday, N. H. (1984). Monocrotaline-induced pulmonary endothelial dysfunction in rats. *Proc. Soc. Exp. Biol. Med.* **176**, 88-94.
Neal, W. M., Rusoff, L., and Ahmann, C. F. (1935). The isolation and some properties of an alkaloid from Crotalaria spectabilis. *J. Am. Chem. Soc.* **57**, 2560-2561.
Orlinska, U., Shaio, R.-T., Olson, J. W., and Gillespie, M. N. (1989). Detection of transforming growth factor-β in lungs and platelets from monocrotaline- treated rats. *FASEB J.* **3**, A903.
Petry, T. W., Bowden, G. T., Huxtable, R. J., and Sipes, I. G. (1984). Characterization of hepatic DNA damage induced in rats by the pyrrolizidine alkaloid monocrotaline. *Cancer Res.* **44**, 1505-1509.
Piercy, P. L., and Rusoff, L. L. (1946). Crotalaria spectabilis poisoning in Louisiana livestock. *J. Am. Vet. Med. Assoc.* **108**, 69-73.
Plestina, R., and Stoner, H. B. (1972). Pulmonary oedema in rats given monocrotaline pyrrole. *J. Pathol.* **106**, 235-249.
Raczniak, T. J., Shumaker, R. C., Allen, J. R., Will, J. A., and Lalich, J. J. (1979). Pathophysiology of dehydromonocrotaline-induced pulmonary fibrosis in the beagle. *Respiration* **37**, 252-260.
Reindel, J. F., and Roth, R. A. (1989). Pulmonary arterial endothelial cells and smooth muscle cells differ in their response to monocrotaline pyrrole (MCTP). *FASEB J.* **3**, A1227.
Rippetoe, P. E., Olson, J. W., Maley, B. E., and Gillespie, M. N. (1989). Epidermal growth factor-like immunoreactivity in lungs from monocrotaline- treated rats. *FASEB J.* **3**, A903.
Rosenberg, H. C., and Rabinovitch, M. (1988). Endothelial injury and vascular reactivity in monocrotaline pulmonary hypertension. *Am. J. Physiol.* **255**, H1484-H1491.
Ross, R., Raines, E. W., and Bowne-Pope, D. F. (1986). The biology of platelet-derived growth factor. *Cell* **46**, 155-169.
Ross, R., and Vogel, A. (1978). The platelet-derived growth factor. *Cell* **14**, 203-210.
Roth, R. A., and Ganey, P. E. (1987). Arachidonic acid metabolites and the

mechanisms of monocrotaline pneumotoxicity. *Am. Rev. Respir. Dis.* **136**, 762-765.
Roth, R. A., and Reindel, J. F. (1988). Response of cultured endothelial cells to monocrotaline pyrrole. *FASEB J.* **2**, A1578.
Roth, R. A., and Reindel, J. F., and Hoorn, C. M. (1989). Differences in sensitivity to monocrotaline pyrrole of cultured endothelium and smooth muscle cells in culture. Abstracts of the Fifth International Congress of Toxicology, Taylor and Francis, London, p. 93.
Schoental, R., and Head, M. A. (1955). Pathologic changes in rats as a result of treatment with monocrotaline. *Brit. J. Cancer* **9**, 229-237.
Sippe, W. L. (1964). Crotalaria spectabilis in livestock and poultry. *Ann. N.Y. Acad. Sci.* **111**, 562-570.
Snow, R. L., Davies, P., Pontoppidan, H., Zapol, W. M., and Reid, L. (1982). Pulmonary vascular remodelling in adult respiratory distress syndrome. *Amer. Rev. Resp. Dis.* **126**, 887-892.
Sugita, T., Hyers, T. M., Dauber, I. M., Wagner, W. W., McMurtry, I. F., and Reeves, T. J. (1983). Lung vessel leak precedes right ventricular hypertrophy in monocrotaline-treated rats. *J. Appl. Physiol.* **54**, 371-374.
Tuchweber, B., Kovaks, K., Jago, M. V., and Beaulieu, T. (1974). Effect of steroidal and non-steroidal microsomal enzyme inducers on the hepatotoxicity of the pyrrolizidine alkaloids in rats. *Res. Commun. Chem. Pathol. Pharmacol.* **7**, 459-480.
Turner, J. H., and Lalich , J. J. (1965). Experimental cor pulmonale in the rat. *Arch. Pathol.* **79**, 409-418.
Valdivia, E., Lalich, J. J., Hayashi, Y., and Songrad, J. (1967). Alterations in pulmonary alveoli after a single injection of monocrotaline. *Arch. Pathol.* **84**, 64-76.
Vane, J. R., Gryglewski, R. J., and Botting, R. M. (1987). The endothelial cell as a metabolic and endocrine organ. *Trends in Pharmacol.* **8**, 491-496.
Vincic, L., Orr, F. W., Warner, J. A., Suyama, K. L., and Kay, J. M. (1989). Enhanced cancer metastasis after monocrotaline-induced lung injury. *Toxicol. Appl. Pharmacol.* **100**, 259-270.
Voelkel, N., and Reeves, J. T. (1979). Primary pulmonary hypertension. In Pulmonary Vascular Diseases (K. M. Mosea, ed.), pp. 573-628. Dekker, New York.
White, S. M., and Roth, R. A. (1988). Pulmonary platelet sequestration is increased following monocrotaline pyrrole treatment of rats. *Toxicol. Appl. Pharmacol.* **96**, 465-475.
White, S. M., and Roth, R. A. (1989). Progressive lung injury and pulmonary hypertension from monocrotaline. In: CRC Handbook of Animal Models of Pulmonary Disease, Vol. II (J. O. Cantor, ed.). CRC Press: Boca Raton, FL, pp. 75-91.
Williams, D. E., Reed, R. L., Kedzierski, B., Dannan, G. A., Guengerich, F. P., and Buhler, D. R. (1989). Bioactivation and detoxication of the pyrrolizidine alkaloid senecionine by cytochrome P-450 enzymes in rat liver. *Drug Metab. Dispos.* **17**, 387-392.

REACTIVE OXYGEN SPECIES IN THE PROGRESSION OF CCl_4-INDUCED LIVER INJURY

I.G. Sipes[1], A.E. El Sisi[1], W.W. Sim[2], S.A. Mobley[1], and D.L. Earnest[2]

Departments of Pharmacology and Toxicology[1]
College of Pharmacy and
Department of Internal Medicine[2]
College of Medicine
University of Arizona,
Tucson, Arizona 85721

ABSTRACT

Pretreatment of rats with large doses of vitamin A (retinol) dramatically increased the hepatotoxicity of carbon tetrachloride (CCl_4). Experiments were performed to elucidate the mechanism of this potentiation. Hypervitaminosis A was produced by oral administration of retinol, 250,000 IU/kg for seven days. CCl_4 was then administered at a dose of 0.15 ml/kg, ip.

This large dose of vitamin A did not enhance the biotransformation of CCl_4, but did produce a 4-fold increase in CCl_4-induced lipid peroxidation, as assessed by ethane exhalation. Because vitamin A has been shown to activate macrophages, it was hypothesized that this increased lipid peroxidation and liver injury resulted from the release of reactive oxygen species from activated Kupffer cells. By using a chemiluminescence assay, an enhanced release of free radicals was detected in Kupffer cells isolated from vitamin A pretreated rats. In addition, Kupffer cells from vitamin A pretreated rats displayed enhanced phagocytic activity *in vitro*, towards sheep red blood cells. *In vivo*, vitamin A pretreated rats cleared carbon particles from the blood 2-3 times faster than non-pretreated rats.

In vivo administration of superoxide dismutase (SOD) 2 hr after CCl_4 exposure did not influence CCl_4 toxicity in control rats but did block the enhanced ethane exhalation and also the potentiation of CCl_4 liver injury in vitamin A treated rats. Administration of methyl palmitate, an inhibitor of Kupffer cell function, did not inhibit CCl_4 toxicity in control rats, but did effectively block enhanced ethane exhalation and potentiation of CCl_4 injury in vitamin A treated rats. We conclude that potentiation of CCl_4 hepatotoxicity by hypervitaminosis A is mediated in part by reactive oxygen species released from activated Kupffer cells.

INTRODUCTION

Chronic ingestion of large doses of vitamin A has been reported to cause liver injury (Russell, et al., 1974; Howard and Willhite, 1986). In some cases, vitamin A was consumed as part of dietary faddism or used in a medically unsupervised manner for general health promotion or for treatment of nonspecific illness such as the common cold. Pharmacologic dosages of vitamin A and other carotenoids have been tested for

Biological Reactive Intermediates IV, Edited by C.M. Witmer *et al.*
Plenum Press, New York, 1990

therapeutic activity as adjuvants in cancer chemotherapy (Newton and Sporn, 1979; Bollag, 1983). The risk of developing significant liver disease from chronic ingestion of excess vitamin A is unknown. In some reports, liver disease has developed with as little as 50,000 IU consumed for 2-4 years (Farris and Erdman, 1982; Rubin et al., 1970). In other reports, ingestion of 50,000 IU per day for 17 years was without apparent harm (Krause, 1965). In those persons who have developed vitamin A hepatotoxicity, a wide spectrum of liver pathology has been described varying from a simple increase in fat to extensive fibrosis and cirrhosis (Ishak, 1987). The mechanism whereby vitamin A induces liver injury is unknown. Why some individuals and not others develop hepatotoxicity is also unclear. Recently, hypervitaminosis A was reported to potentiate the hepatotoxicity of a common toxicant, namely ethanol (Leo and Lieber, 1983). This observation has focused attention on the possibility that excessive vitamin A may also potentiate, or modify, the effects of other substances known to cause liver injury. If true, the sporadic development of liver disease in persons taking excess vitamin A might be explained, in part, by potentiation of liver injury from exposure to other drugs or environmental and occupational chemicals that normally would not cause clinical illness. To explore this possibility, we measured the effects of subacute hypervitaminosis A in liver injury produced by minimally hepatotoxic doses of CCl_4. The results showed that subacute hypervitaminosis A itself did not cause liver injury as assessed by serum enzymes or morphological changes. In contrast, vitamin A significantly potentiated liver injury by CCl_4. The mechanism of this potentiation appears to involve release of reactive species of oxygen from Kupffer cells activated by retinol.

MATERIALS AND METHODS

Male Sprague-Dawley rats (180-200g, Harlan) received Vitamin A (retinol, Aquasol-A, Armour Pharmaceutical) by oral gavage in a dose of 250,000 IU (i.e. 75 mg or 262 umole)/kg body weight daily for 7 days. At 24 hr after the last dose of vitamin A, carbon tetrachloride (CCl_4) was administered at a dose of 0.15 ml/kg in corn oil. The rats were killed by cervical dislocation 24 hr after administration of CCl_4. Blood and liver samples were collected for analysis. Plasma was isolated from the heparinized blood and analyzed for alanine aminotransferase activity (ALT) using a Sigma Diagnostic Kit (procedure No. 59 UV).

To determine the effect of vitamin A on CCl_4-induced lipid peroxidation, rats were housed in glass metabolism cages. Exhaled air was drawn through coconut charcoal that was contained in 13 ml glass tubes surrounded by dry ice. The ethane that was trapped by the charcoal was eluted with heat (230°C for 2 min) and a 5 ml sample of the headspace was withdrawn into a gas tight syringe. The 5 ml sample was compressed to 1 ml and injected into the gas chromatograph for ethane determination. Ethane retention was 6 min on a carboseive G column (80/100 mesh, Supelco) maintained at an oven temperature of 180°C.

To determine the effect of vitamin A on CCl_4 metabolism, $^{14}CCl_4$ (0.15 ml/kg, 21.4 µCi/mmole) was administered ip. Rats were then placed in glass metabolism cages for 24 hr for collection of the exhaled metabolites, $^{14}CO_2$ and $CHCl_3$, at numerous time points following CCl_4 administration. Exhaled air was passed through ice cold absolute ethanol (for collection of $CHCl_3$) and through ice cold Carbosorb (Packard) for collection of $^{14}CO_2$. The amount of $^{14}CO_2$ was determined by liquid scintillation spectroscopy and $CHCl_3$ by gas chromatography. Some rats receiving $^{14}CCl_4$ were killed at 0.5, 1, 2, or 4 hr to determine if vitamin A pretreatment altered the covalent binding of CCl_4-reactive metabolites to hepatic lipids and proteins. For this, 1 gm of liver was homogenized in cold Tris HCl buffer (pH 7.4). Cold ethanol was added to the homogenates. The pellet that formed upon centrifugation at 3000 x g was then treated with chloroform-methanol (3:1) to extract lipids. The pellet that contained protein was exhaustively extracted with solvents and the lipid extract concentrated by continuous evaporation with N_2. This process, as described by Sipes and Gandolfi (1982), results in lipid and protein extracts that are free from non-covalently bound $^{14}CCl_4$ equivalents.

To determine if vitamin A activated Kupffer cells, these cells were sterilely isolated by recirculation perfusion of a collagenase/pronase enzyme solution and purified by centrifugal elutriation (Knook and Sleyster, 1976). Briefly, the portal vein was cannulated and perfused *in situ* first with Gey's Balanced Salt Solution to remove red blood cells. This was followed by perfusion with 0.2% Pronase E during which time the liver was removed from the body cavity with the cannula in place. Pronase E selectively destroys parenchymal cells but does not permanently damage non-parenchymal cells. A solution composed of 0.05% collagenase and 0.05% Pronase E was then recirculated through the liver to degrade the liver's supporting matrix and destroy parenchymal cells. Following perfusion, the liver was minced with a razor blade and the resulting liver paste was incubated with slow stirring in a solution of 0.05% collagenase and 0.05% Pronase E to further dissociate the liver. The resulting suspension was filtered through gauze to remove any portion of undissociated liver, and then centrifuged to yield a sinusoidal cell-enriched pellet. This preparation was then purified by centrifugation on a one-step discontinuous Metrizamide gradient. At this point liver preparations are composed almost entirely of sinusoidal cells. Kupffer cells were separated from other sinusoidal cells by a centrifugal elutriation system explained in detail by Knook and Sleyster (1976). Kupffer cell identification and purity were established by electron microscopy and peroxidase staining. Kupffer cells maintained in culture for 24 hr were incubated with ^{51}Cr-labeled sheep red blood cells for 60 min to assess their phagocytic activity.

Release of reactive oxygen species by the cultured cells (as measured by free radical release) was assessed by chemiluminescence (Lee, P. et al., 1987). Briefly, media was decanted from vials containing attached Kupffer cells. The attached cells were then incubated for 15 min at 37°C in 4 ml of Hank's Balanced Salt Solution containing 15 mM luminol. 200 µl of zymosan solution (240 µg/ml final concentration) was then added to each vial and chemiluminescence was monitored using a liquid scintillation counter which was switched to the out-of-coincidence mode.

To assess the activity of Kupffer cells *in vivo*, the rate of disappearance of colloidal carbon was determined in control and vitamin A pretreated rats (Triarhou and Cerro, 1985). A solution of colloidal carbon (37.5 mg/ml, 100 mg/kg) was injected into the external jugular vein of anesthetized rats. Blood samples (50 µl) were obtained every 5 min for 30 min from the tail vein. After mixing with 1 ml of 0.1% Na_2CO_3, the absorbance at 640 nm was determined against a blank (blood obtained before carbon injection). To deactivate Kupffer cells, methyl palmitate (2 gm/kg;IV) was given 24 hr prior to CCl_4 in a solution of 0.2% Tween 20, 5% dextrose.

Since activated Kupffer cells are known to release active species of oxygen (i.e. superoxide anion, H_2O_2), superoxide dismutase (SOD) was administered IV 2 hr after CCl_4 to attenuate any hepatotoxicity dependent upon the presence of superoxide anion. SOD, which was conjugated with polyethylene glycol (PEG-SOD) to prolong its plasma half-life (Davis et al., 1980), was obtained from Enzon, Inc. The dose of PEG-SOD was 10,000 IU/kg administered in phosphate buffered saline.

Statistical analysis: Analysis of variance and the student t-test were used for statistical evaluation. Data were considered significantly different at $p<0.05$.

RESULTS

Seven day pretreatment with vitamin A dramatically increased the hepatotoxicity of CCl_4 as assessed by plasma ALT activity (Table 1). Plasma ALT activity was 17 times that observed in CCl_4 rats not pretreated with vitamin A. Histological evaluation of liver sections confirmed the presence of extensive necrosis in vitamin A pretreated rats administered CCl_4. Only minimal necrosis (only in cells adjacent to the terminal hepatic venules) was observed in control rats administered CCl_4. Vitamin A itself did not produce hepatic lesions.

Because of the critical role of bioactivation in CCl_4-induced liver injury, the effect of vitamin A pretreatment on CCl_4 biotransformation was determined.

Table 1. Potentiation of CCl_4-Induced Liver Injury by Vitamin A as Assessed by Plasma ALT Activity

	Plasma ALT Activity IU/L
Control	23 ± 3
Vitamin A	26 ± 4
CCl_4	210 ± 43
Vitamin A + CCl_4	3516 ± 814*

Vitamin A was administered daily for 7 days at a dose of 250,000 IU/kg by gavage. On day 8, CCl_4 was administered (0.15 ml/kg, ip). ALT activity was assessed at 24 hr after CCl_4. *$p < 0.01$ as compared to CCl_4 alone.

Vitamin A did not increase the rate of exhalation of two metabolites of CCl_4, $^{14}CO_2$ or $CHCl_3$ (data not presented). Vitamin A also did not increase the covalent binding of $^{14}CCl_4$-equivalents to hepatic lipids and proteins (Table 2).

Vitamin A pretreatment did increase the exhalation of ethane following administration of CCl_4 (Figure 1). Depending on the time of collection the amount of ethane exhaled was 4-7 times greater in vitamin A treated rats given CCl_4. Control rats and those receiving vitamin A alone exhaled equivalent amounts of ethane, which were well below these produced by CCl_4 in control animals.

Kupffer cells isolated from vitamin A pretreated rats and cultured for 24 hr had enhanced phagocytic activity towards opsonized sheep red blood cells. Activity was 50% higher than in Kupffer cells obtained from control rats (Figure 2). When incubated with zymosan (238 μg/ml), Kupffer cells isolated from vitamin A pretreated rats released more free radicals than Kupffer cells isolated from control rats (Figure 3). At maximum, the increase was over two-fold greater.

In vivo evidence also suggested that Kupffer cells were activated (more phagocytic) following vitamin A treatment. The rate of disappearance of colloidal carbon from the blood of vitamin A treated rats was 2-3 times more rapid than its disappearance from the blood of control rats. The half-life of carbon particles in blood was 6.5 and 18 min for vitamin A pretreated and control rats, respectively.

Table 2. Lack of Effect of Vitamin A on the Covalent Binding of $^{14}CCl_4$-Equivalents to Hepatic Proteins and Lipids

	Control + CCl_4	Vitamin A + CCl_4
Binding to Protein	pmol/mg protein	
0.5 hr	300 ± 100	400 ± 50
2.0 hr	295 ± 20	390 ± 25
Binding to Lipids	pmol/mg lipid phosphorus	
0.5 hr	60 ± 40	50 ± 30
2.0 hr	50 ± 10	30 ± 12

$^{14}CCl_4$ was administered (0.15 ml/kg, 21.4 μCi/mmole) to control or vitamin A pretreated rats (250,000 IU/kg, daily for 7 days). Livers were removed and processed for covalent binding of $^{14}CCl_4$ equivalents at 0.5 and 2 hr after administration.

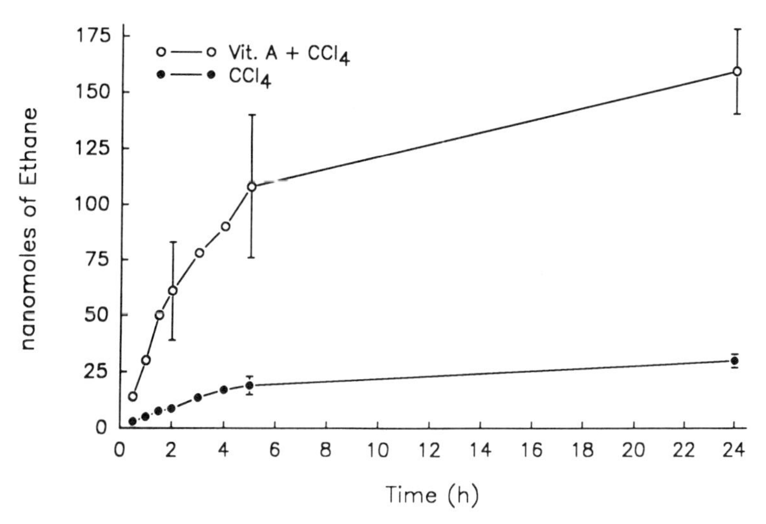

Figure 1. Enhancement of cumulative CCl_4-induced ethane exhalation by vitamin A pretreatment. Vitamin A was administered to male rats by oral gavage (250,000 IU/Kg/day for 7 days). $^{14}CCl_4$ was administered (0.15 ml/kg, ip), 24 hr after the last dose of vitamin A. Ethane was collected immediately after CCl_4 treatment. Data points are the mean of 4 rats per group. For clarity, S.D. are presented only for certain points.

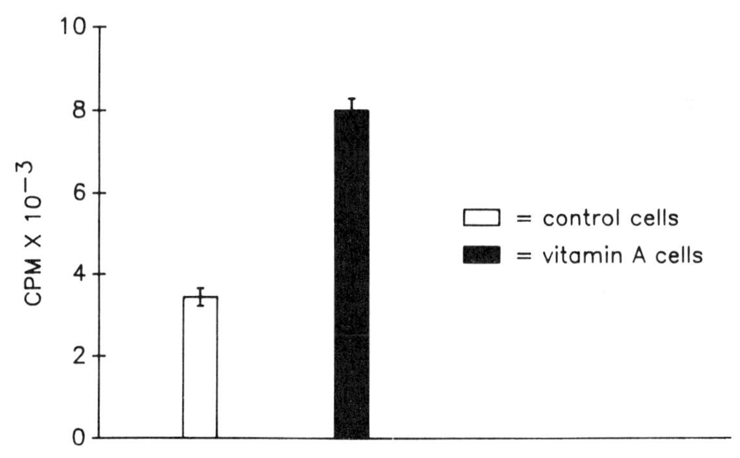

Figure 2. Phagocytosis of ^{51}Cr-labeled sheep red blood cells (sRBC) by Kupffer cells isolated from control rats or rats treated with vitamin A for 7 days. Kupffer cells were allowed to attach to culture plates for 24 hr and then incubated for 60 min at 37°C with opsonized sRBC. Kupffer cells were washed three times with phosphate buffered saline to remove unphagocytosed sRBC and then lysed with a 1% solution of NP-40 detergent and the radioactivity of the lysate determined.

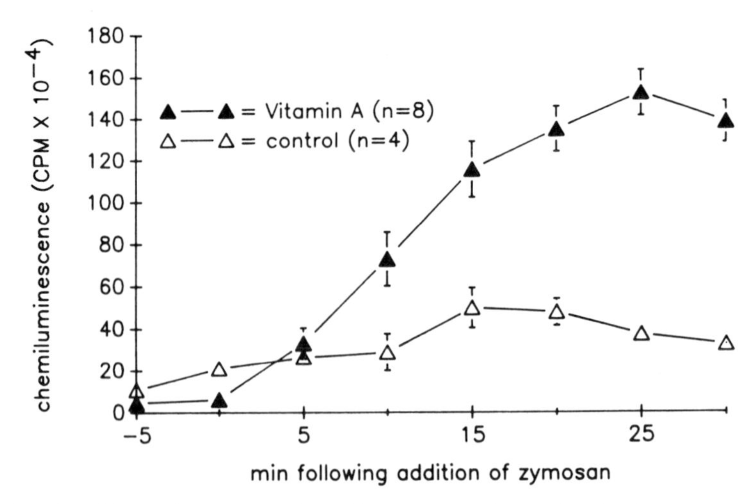

Figure 3. Kupffer cells were isolated from control or vitamin A treated rats and allowed to attach to plastic scintillation vials for 24 hr at 37°C. Free radical release was measured as CPM by luminol-enhanced chemiluminescence following addition of zymosan (238 μg/ml) to cell cultures.

These findings led to the hypothesis that reactive species of oxygen released from activated Kupffer cells are responsible for the increase in CCl_4-induced lipid peroxidation in vitamin A treated rats, and thus for the potentiation of CCl_4 liver injury. This hypothesis was tested by deactivation of Kupffer cells with methyl palmitate to decrease the release of reactive oxygen species, or by administration of PEG-SOD to consume superoxide anions that are released from Kupffer cells into the extracellular space.

As illustrated in Table 3, administration of PEG-SOD or methyl palmitate blocked the potentiation of CCl_4 hepatotoxicity by vitamin A pretreatment. The plasma ALT values were not significantly different between the CCl_4 alone (no vitamin A pretreatment) group and the vitamin A groups receiving CCl_4 and either methyl palmitate or PEG-SOD. Histological examination of liver sections confirmed this dramatic reduction in liver injury (data not presented). Importantly, PEG-SOD and methyl palmitate also inhibited the enhanced ethane exhalation observed in vitamin A pretreated rats administered CCl_4 (data not presented).

DISCUSSION

The results presented here demonstrate a novel biological interaction between a nutrient/drug (vitamin A) and a model hepatotoxicant (CCl_4). Pretreatment of rats with large doses of vitamin A greatly potentiates CCl_4-induced liver injury. Although it could be argued that CCl_4 was exacerbating the vitamin A effects of the liver, this appears unlikely for the following reasons:

1) Vitamin A alone, even in these high doses, did not cause apparent liver injury.
2) Vitamin A did not promote lipid peroxidation, as assessed by ethane exhalation.
3) The extensive liver injury observed in the vitamin A rats given CCl_4 was localized in the centrilobular region of the liver. It was characteristic of the liver injury produced by larger doses of CCl_4 (i.e. 1 to 2 ml/kg).
4) The severity and appearance of injury was similar to that produced by CCl_4 in phenobarbital or ethanol - induced rats.

Table 3. Inhibition of Vitamin A Potentiation of CCl_4-Induced Liver Injury by IV Administration of Superoxide Dismutase or Methyl Palmitate

	Plasma ALT activity IU/L
Treatment Group	
Control (No treatment or Vitamin A)	28 ± 4
CCl_4	168 ± 105
Vitamin A + CCl_4	2240 ± 522*
Vitamin A + CCl_4 + PEG-SOD	238 ± 127
Vitamin A + CCl_4 + methyl palmitate	87 ± 14

Vitamin A pretreated rats (250,000 IU/kg per day for 7 days) were administered CCl_4. PEG-SOD (10,000 IU/kg iv) was administered 2 hr post CCl_4. Methyl palmitate (2 gm/kg iv) was administered 24 hr before CCl_4. *$p < 0.05$ versus all other groups.

The assumption that vitamin A was modulating the response of the liver to CCl_4 directed the type of studies performed to determine the mechanism of this potentiation.

Interestingly, it was found that vitamin A had no effect on the rate of biotransformation of CCl_4, as assessed by exhalation of $^{14}CO_2$ and $CHCl_3$, or by the covalent binding of $^{14}CCl_4$ equivalents to hepatic lipids and proteins. Similarly, we found that vitamin A treatment did not increase hepatic microsomal cytochrome P-450 concentration or the *in vivo* metabolism of ^{14}C-aminopyrine (data not presented). These findings seemed to eliminate the possibility that vitamin A increased the biotransformation of CCl_4 to reactive metabolite(s) as a mechanism of potentiation of CCl_4-induced liver injury. Furthermore, vitamin A had no effect on the hepatic concentration of the key hepatoprotective agents, vitamin E and glutathione (data not presented). These results were surprising because CCl_4-induced lipid peroxidation, as measured by ethane exhalation, was greatly increased in vitamin A treated rats. Therefore, it was expected that vitamin A pretreatment would enhance the bioactivation of CCl_4 or deplete hepatoprotective factors. To us, the results suggested that the initiator of lipid peroxidation was not the trichloromethyl radical, but some other stimulus, probably not related to the CCl_4-molecule.

The finding that vitamin A pretreatment activated Kupffer cells (enhanced phagocytosis and enhanced clearance of carbon particles) was a critical observation. Kupffer cells, like other macrophages, are known to release a number of cytotoxic products, including reactive species of oxygen (O_2^- and H_2O_2) (Matuso et al., 1985). These reactive species of oxygen could serve as another stimulus for lipid peroxidation. This stimulus would not be derived from CCl_4 and would actually be formed outside of the hepatocyte. Therefore, it was necessary to show if Kupffer cells from vitamin A treated rats released more oxygen radicals upon stimulation and to determine a role for oxygen radicals in the potentiation of CCl_4-induced liver injury. The increased chemiluminescence observed in Kupffer cells isolated from vitamin A pretreated rats is evidence for an enhanced release of reactive oxygen species (Fujiwara et al., 1989). We also obtained evidence for a role of these reactive oxygen species in the vitamin A potentiation of CCl_4-induced hepatotoxicity. Administration of PEG-SOD effectively blocked the potentiation of injury observed in the vitamin A pretreated rats. Since the PEG-SOD should distribute into the extracellular space (as opposed to intracellular uptake), it appears to consume superoxide anion released from Kupffer cells.

The data obtained in the methyl palmitate studies support the hypothesis that the *in vivo* source of these reactive species of oxygen is the Kupffer cells. Methyl palmitate is known to deactivate Kupffer cells (Al-Tuwaijri et al., 1981). We did show that methyl palmitate blocked the enhanced clearance of carbon particles from the blood of vitamin A pretreated rats (data to shown). Importantly, methyl palmitate completely inhibited the enhanced exhalation of ethane and the potentiation of CCl_4 liver injury in vitamin A pretreated rats.

Taken together, the results suggest a novel mechanism of interaction between CCl_4 and vitamin A. One effect of administration of large doses of vitamin A is activation of the resident macrophages of the liver, the Kupffer cells. In response to a second chemical, CCl_4, the activated Kupffer cells are stimulated to release reactive species of oxygen. These reactive oxygen species can initiate peroxidation of lipids present in the hepatocellular membrane, an event associated with increased cell killing. The net result is a dramatic potentiation of CCl_4-induced liver injury. It appears that vitamin A does not affect the initial events following CCl_4 intoxication, but acts to promote the progression of cell injury.

Many other factors are known to modulate Kupffer cell activity - exposure to bacteria, endotoxin, stress, etc. These factors may, through their effects on Kupffer cells, alter the progression of chemical-induced liver injury. Alterations in Kupffer cell activity may be another of those unknown or unanticipated factors that explain animal (and human ?) variability in the response of the liver to hepatotoxicants. It should be stressed that the doses used in these studies far exceed those expected for humans. However, these doses were only administered to rats for one week, as opposed to humans, who may ingest smaller doses (i.e., 300,000 IU per person) but for prolonged periods of time. It is the concept of Kupffer cell activation in modulation of chemical induced liver injury which is most important. Knowledge of what other chemicals share this phenomenon will be important in elucidation of mechanisms of chemical induced hepatotoxicity.

REFERENCES

Al-Tuwaijri, A., Akdamar, K. and DiLuzio, R. (1981). Modification of galactosamine-induced liver injury in rats by reticuloendothelial stimulation or depression. *Hepatology* **1**(2), 107-113.

Bollag, W. (1983). Vitamin A and retinoids: From nutrition to pharmacotherapy in dermatology and oncology. *Lancet* **1** (8329), 860-863.

Davis, F., Pyatak, P. and Abuchowshi, A. (1980). Preparation of PEG-SOD adduct and an examination of its blood circulating life and anti-inflammatory activity. *Res. Comm. Chem. Pathol. Pharmacol.* **29**(1), 113-127.

Farris, W.A. and Erdman, J.W. (1982). Protracted hypervitaminosis A following long-term, low-level intake. *JAMA* **247**, 1317-1318.

Fujiwara, K., Ogata, I. and Mochida, S. (1989). *In situ* evaluation of stimulatory state of hepatic macrophage based on ability to produce superoxide in rats. In, *Cells of the Hepatic Sinusoid*, Vol. 2 (E. Wisse, D.L. Knook and K. Decker, eds.), pp. 204-205, Kupffer Cell Foundation, The Netherlands.

Howard, W.B. and Willhite, C. (1986). Toxicity of retinoids in humans and animals. *J. Toxicol. - Toxin Reviews* **5**, 55-94.

Ishak, K.G. (1987). New development in diagnostic liver pathology. In, *Pathogenesis of Liver Disease* (E. Farber, M.J. Phillips and N. Kaufman, eds.), pp. 223-373, Williams and Wilkins, Baltimore.

Knook, D. and Sleyster, C. (1976). Separation of Kupffer and endothelial cells of rat liver by centrifugal elutriation. *Exp. Cell. Res.* **99**, 445-449.

Krause, R.F. (1965). Liver lipids in a case of hypervitaminosis A. *Am. J. Clin. Nutr.* **16**, 455-457.

Lee, P., Walker, E.R., Miles, P.R. and Castranova, V. (1987). Differential generation of chemiluminescence from various cellular fractions obtained by dog lung lavage. In, *Cellular Chemiluminescence*, Vol. III. (K. Van Dyke and V. Castranova, eds.), pp. 53-60, Boca Raton, Florida.

Leo, M.A. and Lieber, C.S. (1983). Hepatic fibrosis after long-term administration of ethanol and moderate vitamin A supplementation in the rat. *Hepatology* **3**, 1-11.

Matuso, S., Nakagawara, A., Ikeda, K., Mitsuyama, M. and Nomoto, K. (1985). Enhanced release of reactive oxygen intermediates by immunologically activated Kupffer cells. *Clin. Exp. Immunol.* **59**, 203-209.

Newton, D. and Sporn, M. (1979). Chemoprevention of cancer with retinoids. *Fed. Proc.* **58**, 2528-2534.

Rubin, E., Florman, A.L., Degnan, T. and Diaz, J. (1970). Hepatic injury in chronic hypervitaminosis A. *Amer. J. Dis. Child.* **119**, 132-138.

Russell, R.M., Boyer, J.L., Bagheri, S.A. and Hruban, Z. (1974). Hepatic injury from chronic hypervitaminosis A resulting in portal hypertension and ascites. *N. Eng. J. Med.* **291**, 435-440.

Sipes, I.G. and Gandolfi, A.J. (1982). Bioactivation of aliphatic organohalogens: Formation, detection and relevance. In, *Toxicology of the Liver* (G. Plaa and W. Hewitt, eds.) pp. 181-212, Raven Press, New York.

Triarhou, L. and Cerro, M. (1985). Colloidal carbon as a multilevel marker for experimental lesions. *Experientia* **41**, 620-621.

PARENCHYMAL AND NONPARENCHYMAL CELL INTERACTIONS IN HEPATOTOXICITY

Debra L. Laskin

Joint Graduate Program in Toxicology
Rutgers University
Piscataway, NJ 08854

INTRODUCTION

Hepatic nonparenchymal cells represent approximately 30-35% of the cells in the liver. The majority of these cells, which include Kupffer cells, endothelial cells, fat storing cells and pit cells reside within the hepatic sinusoids. Because of their small size relative to hepatocytes and the small volume that they occupy in the liver, nonparenchymal liver cells have largely been ignored in studies aimed at elucidating cellular mechanisms of heptotoxicity. However, following exposure to hepatotoxic chemicals, nonparenchymal cells become "activated". They release large quantities of highly reactive mediators such as superoxide anion and hydrogen peroxide, eicosinoids and proteolytic enzymes which have the capacity to damage hepatic tissue. Thus these cells may contribute to pathophysiologic processes leading to toxicity.

Our laboratory has been interested in studying the potential role of activated inflammatory macrophages and Kupffer cells in hepatotoxicity. For our studies we have used acetaminophen and lipopolysaccharide (LPS) as model hepatotoxicants. Treatment of rats with acetaminophen or LPS results in a rapid accumulation of macrophages in the liver in the absence of cellular necrosis (Laskin and Pilaro, 1986; Pilaro and Laskin, 1986). When the macrophages were isolated from these animals, they were found to display properties of activated mononuclear phagocytes including altered morphology and enhanced functional responsiveness (Laskin and Pilaro, 1986; Pilaro and Laskin, 1986). Based on these data, we proposed a model for the role of activated macrophages in hepatotoxicity (Laskin et al., 1986). In the present studies, we extend these observations and our model to include hepatic endothelial cells. We specifically focused on these nonparenchymal cells since they are the predominant sinusoidal cell type in the liver.

MATERIALS AND METHODS

Isolation of Liver Macrophages and Endothelial Cells

Macrophages and endothelial cells were isolated from livers of female Sprague-Dawley rats (200-250 g) by combined collagenase/pronase perfusion followed by centrifugal elutriation and differential centrifugation on a metrizamide gradient (Pilaro and Laskin, 1986). Cells were identified morphologically and histochemically by nonspecific esterase and peroxidase staining and were found to be greater than 85% pure. In some experiments, rats were pretreated for 24 hr with 1.2 g/kg acetaminophen (PO) or for 48 hr with 5 mg/kg of Escherichia coli LPS 0128:B12 (IV).

Biological Reactive Intermediates IV, Edited by C.M. Witmer *et al.*
Plenum Press, New York, 1990

Measurement of Chemotaxis

Chemotaxis of macrophages through millipore filters was measured using the modified Boyden chamber technique (Boyden, 1962). For our studies, we used a 48-well microchemotaxis chamber (Laskin, and Pilaro, 1986). The chemotactic agent used was C5a which was prepared from endotoxin activated rat serum (Laskin et al., 1981).

Measurement of Phagocytosis and Oxidative Metabolism

Phagocytosis of ^{51}Cr-labeled, opsonized sheep red blood cells (sRBC) by liver macrophages was quantified as described previously (Laskin and Pilaro, 1986; Pilaro and Laskin, 1986). Superoxide anion release by macrophages in response to C5a was measured spectrophotometrically as the superoxide dismutase inhibitable reduction of ferricytochrome C (Laskin and Pilaro, 1986; Pilaro and Laskin, 1986). Hydrogen peroxide production by was measured by flow cytometry using 2',7-dichlorofluorescin diacetate (DCFH-DA) (Laskin et al., 1988a; Laskin et al., 1988b).

Measurement of Interleukin-1 (IL-1) and Interleukin-6 (IL-6) Production

Following isolation, macrophages and endothelial cells were inoculated into 35 mm culture dishes and supernatants collected 1-24 hr later. IL-1 and IL- 6 activity in macrophage and endothelial cell culture supernatants were determined by bioassay using the D10.G4.1 and the B9 cell lines which proliferate in the presence of IL-1 and Il-6, respectively (Kurt-Jones et al., 1985; Helle et al., 1988).

RESULTS

Activation of Liver Macrophages and Endothelial Cells by Hepatotoxicants

Treatment of rats with acetaminophen or LPS resulted in extensive infiltration of mononuclear cells into the liver in the absence of necrosis (Laskin and Pilaro, 1986; Pilaro and Laskin, 1986). This was associated with a significant increase in the number of both macrophages and endothelial cells recovered from livers of these rats when compared to controls (Table 1). Macrophages isolated from treated rats displayed morphologic characteristics of "activated" mononuclear phagocytes. These cells were larger and more vacuolated than resident Kupffer cells and exhibited an increased cytoplasmic:nuclear ratio (Laskin and Pilaro, 1986; Pilaro and Laskin, 1986). In general, endothelial cells isolated from treated rats were similar in morphology to resident endothelial cells, although they appeared larger and more granular (now shown). To determine if macrophages from treated rats were functionally "activated", we analyzed their chemotactic, phagocytic and respiratory burst activity. We found that macrophages from LPS (Table 2) and acetaminophen (not shown) treated rats phagocytized three to five times more opsonized ^{51}Cr-sRBC than did resident Kupffer cells (Table 2). These macrophages also displayed enhanced chemotaxis and released elevated levels of superoxide anion (Table 2) and hydrogen peroxide (Table 3). These responses are all characteristics of "activated" inflammatory macrophages (Adams and Hamilton, 1984).

We next analyzed the effects of LPS treatment of rats on the functional activity of endothelial cells. Using DCFH-DA and flow cytometry, we found that these cells produced significantly more hydrogen peroxide than resident endothelial cells (Table 3). In fact, following stimulation with phorbol myristate acetate, endothelial cells from LPS-treated rats produced levels of hydrogen peroxide that were comparable to levels produced by activated macrophages in the absence of stimulation (Table 3).

In further experiments we analyzed the effects of hepatotoxicant exposure on IL-1 and Il-6 production by Kupffer cells and endothelial cells. These mediators are potent immunoregulatory molecules and are known to have significant effects on hepatocyte functioning (Dinarello, 1989). We found that while Kupffer cells and endothelial cells from untreated rats produced similar amounts of IL-6, resident endothelial cells produced significantly more IL-1 than resident Kupffer cells (not shown). Treatment of rats with LPS had no effect on the amount or kinetics of IL-6 production by either macrophages or endothelial cells. In contrast, both cell types produced significantly more IL-1 following LPS treatment. In addition, the endothelial cells appeared to be more sensitive to the effects of LPS treatment than were the macrophages.

Table 1. Alterations in Macrophage and Endothelial Cell Number Following Hepatotoxicant Treatment[a]

Treatment	Cells/gr Perfused Liver X 10^6	
	Macrophages	Endothelial Cells
None	2.3 ± 0.3	12.7 ± 1.8
LPS	6.2 ± 0.8*	19.9 ± 2.8*
Acetaminophen	5.7 ± 0.8*	16.0 ± 2.0*

[a]Macrophages and endothelial cells were isolated from female rats treated with 5 mg/kg LPS (IV, 48 hr) or 1.2 g/kg acetaminophen (PO, 24 hr) prior to nonparenchymal cell isolation.
*Significant difference ($p \leq 0.05$) between treated and untreated rats.

Table 2. Comparison of the Functional Activity of Macrophages from Livers of Control and LPS-treated Rats

	Control	LPS-treated
Phagocytosis (cpm/10^5 cells)	41450.2 ± 384.6	92370.6 ± 422.9*
Chemotaxis (cells/10 oil fields)	58.2 ± 9.3	1568.5 ± 213.4*
Superoxide Anion (nmoles/10^6 cells)	5.9 ± 0.4	8.6 ± 1.2*

[a]Each point is the mean of three to six samples ± SEM from one representative experiment.
*Significant differences ($p \leq 0.05$) between control and LPS-treated rats (Student's t test).

Table 3. Production of Hydrogen Peroxide by Resident and LPS-Activated Nonparenchymal Cells[a]

	Control	LPS-treated
Macrophages		
unstimulated	88	92
stimulated	116	125
Endothelial cells		
unstimulated	55	69
stimulated	59	86

[a]Macrophages and endothelial cells from control or LPS-treated rats were incubated with 20 uM DCFH-DA and 170 nM 12-O-tetradecanoyl phorbol-13- acetate or control as a stimulus. After 45 min at 37°C, fluorescence intensity, which is directly proportional to hydrogen peroxide production, was measured by flow cytometry on a Coulter EPICS ELITE flow cytometer. The data are presented as mean fluorescence channel number. On our flow cytometer, each histogram is presented on a three-decade log scale which is divided into 256 channels.

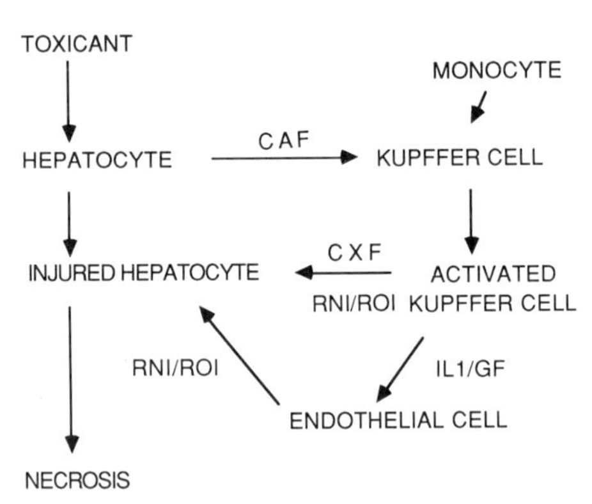

Figure 1. Model for the potential role of macrophages and endothelial cells in hepatotoxicity. Toxic doses of drugs or chemicals injure hepatocytes. These injured hepatocytes release factors that attract Kupffer cells to specific regions of the liver. Additional mononuclear phagocytes are also recruited from blood and bone marrow precursors. Once localized in the liver, the macrophages become activated by hepatocyte-derived factors and release mediators that induce proliferation and activation of endothelial cells. Activated macrophages and endothelial cells also release medaitors that contribute to the damage initiated by toxicants. This eventually leads to cell death and necrosis. CAF, chemotactic/activating factors; CXF, cytotoxic factors; RNI, reactive nitrogen intermediates; ROI, reactive oxygen intermediates; IL1, interleukin 1; GF, growth factors.

Effects of Modifying Nonparenchymal Cell Function on Hepatotoxicity

If activated inflammatory macrophages and endothelial cells play a role in liver injury, then modulating the function of these cells should modify the hepatotoxicity of agents like acetaminophen. To test this possibility, rats were pretreated with 5 mg/kg LPS, which activates liver macrophages and endothelial cells (see above) and then with acetaminophen. We found that LPS enhanced the toxicity of acetaminophen as evidenced morphologically by the rapid development of liver necrosis. In contrast, accumulation of macrophages in the liver and subsequent toxicity of acetaminophen was blocked by pretreatment of the rats with 5 mg/kg dextran sulfate or 7 mg/kg gadolinium chloride, two inhibitors of macrophage function (Souhami and Bradfield, 1974; Husztik et al., 1980).

DISCUSSION

Activated macrophages are characterized by altered morphology, increased functional capacity and the release of reactive mediators. Although it is generally assumed that this results in more rapid removal and efficient destruction of foreign antigens, recent evidence suggests that release of mediators by activated macrophages may also contribute to tissue injury (Tanner et al., 1980; Arthur et al., 1985). In this paper we present evidence that macrophages as well as endothelial cells isolated from the livers of rats treated with hepatotoxicants such as acetaminophen or LPS are "activated". They display altered morphology and enhanced release of reactive oxygen intermediates and cytokines. Furthermore modification of the functioning of these cells modifies hepatotoxicity. Taken together these data suggest that activated macrophages and endothelial cells contribute to hepatotoxicity.

Tissue damage typically results in a rapid accumulation of monocytes and neutrophils at the injured site. These cells are derived from precursor cells in the blood and bone marrow (Diesselhoff et al., 1979). Once localized in the tissue, the monocytes mature into macrophages and become "activated". Twenty-four hr following treatment of rats with acetaminophen or LPS we obtained a two to three-fold increase in the number of macrophages and a 30-40% increase in the number of endothelial cells isolated from the livers. Macrophages isolated from treated rats displayed morphological and functional characteristics of "activated" mononuclear phagocytes. A similar infiltration of activated macrophages into the liver has been reported previously following treatment of rats with hepatotoxicants such as phenobarbital and carbon tetrachloride (Laskin et al., 1988a; Lautenschlager et al., 1982). Furthermore, our data indicate that Kupffer cell accumulation and activation is mediated by factors released from injured hepatocytes (Laskin et al., 1986). These results suggest that accumulation of activated macrophages in response to injured hepatocytes may be an important event in the development of hepatotoxicity induced by drugs and chemicals.

The data presented in this paper also suggest that hepatic endothelial cells may contribute to toxicity induced by chemicals. Endothelial cells recovered from rats treated with LPS, like macrophages, display morphologic and functional characteristics of activated cells including increased production of IL-1 and hydrogen peroxide. The capacity of endothelial cells to produce these mediators may represent an important mechanism by which these cells participate in inflammatory and immune reactions associated with hepatotoxicity. Vascular endothelial cells have been reported to proliferate and to produce increased amounts of superoxide anion, as well as IL-1 and IL-6 in response to macrophage-derived cytokines (Ooi et al., 1983; Matsubara and Ziff, 1986; Sironi et al., 1989). Endothelial cells activated by inflammatory mediators are also known to release eicosinoids, reactive nitrogen intermediates, and lysososmal enzymes (Knook et al., 1976; Cotran, 1987; Salvemini et al., 1989). These data indicate that inflammatory cells and their products may play a regulatory role in the growth and integrity of the microvasculature. Similiarly, in the liver it is possible that cytokines released by inflammatory macrophages and activated Kupffer cells promote the production of reactive oxygen intermediates or hydrolytic enzymes by endothelial cells. Reactive mediators released by stimulated endothelial cells may directly injure the matrix of the vasculature as well as the surrounding hepatic tissue. In this regard, in preliminary studies we found that Kupffer cells activated by LPS release a factor that stimulates proliferation and activation of hepatic endothelial cells.

The precise role of hepatic macrophages and endothelial cells in the pathophysiology of liver injury is unknown. It has been suggested that macrophages recruited to the liver can promote hepatic damage through the secretion of toxic products (Tanner et al., 1980; Arthur et al., 1985). A similar role may exist for endothelial cells. Our findings that macrophages and endothelial cells from treated rats produce elevated levels of reactive oxygen intermediates and IL-1 support this hypothesis as do our findings that modulation of nonparenchymal cell activity modifies the hepatotoxicity of acetaminophen. Based on these results, we propose an expanded model for the potential role of macrophages and endothelial cells in chemical and drug-induced hepatotoxicity (Figure 1). According to this model, hepatocytes, damaged by toxicants like acetaminophen or LPS, release factors that attract Kupffer cells to the injured area. Additional mononuclear phagocytes are also recruited from blood and bone marrow precursors. Once localized in the injured area, the macrophages become activated by hepatocyte-derived factors and release mediators that induce proliferation and activation of endothelial cells. Activated macrophages and endothelial cells also release mediators that contribute to damage initiated by toxicants. This eventually leads to cell death and necrosis. The data from our laboratory and those of other investigators support this model of hepatotoxicity. Additional studies on the nature of the mediators released from nonparenchymal cells and their effects on hepatocytes will be particularly relevant for understanding mechanisms of liver injury.

REFERENCES

Adams, D.O. and Hamilton, T.A. (1984). The cell biology of macrophage activation. *Annu. Rev. Immunol.* **2**, 283-318.

Authur, M.J.P., Bentley, I.S., Tanner, A.R., Kowalski Saunders, P., Millward-Sadler, G.H., and Wright, R. (1985). Oxygen derived free radicals promote hepatic injury in the rat. *Gastroenterology* **89**, 1114-1122.

Boyden, S.V. (1962). The chemotactic effect of mixtures of antibody and antigen on polymorphonuclear leukocytes. *J. Exp. Med.* **115**, 453-466.

Cotran, R.S. (1987). New roles for the endothelium in inflammation and immunity. *Amer. J. Pathol.* **129**, 407-413.

Diesselholf-Den Dulk, M.M., Crofton, R.W., and van Furth, R. (1979). Origin and kinetics of Kupffer cells during an acute inflammatory response. *Immunol.* **377**, 7-14.

Dinarello, C.A. (1989). Interleukin-1 and its related cytokines. In *Macrophage-Derived Cell Regualtory Factors.* Cytokines (C. Sorg, ed.), pp. 105-154, Karger, Basel.

Helle, M, Boeije, L., and Aarden, L.A. (1988). Functional discrimination between interleukin 6 and interleukin 1. *Eur. J. Immunol.* **18**, 1535-1540.

Husztik, E., Lazar, G., Parducz, A. (1980). Electron microscopic study of Kupffer cell phagocytosis blockade induced by gadolinium chloride. *Br. J. Exp. Pathol.* **61**, 624-630.

Knook, D.L., Blansjaar, N., Sleyster, ECh. (1976). Isolation and characterization of Kupffer and endothelial cells from the rat liver. *Exp. Cell Res.* **99**, 444-449.

Kurt-Jones, E.A., Beller, D.I., Mizel, S.B., and Unanue, E.R. (1985). Identification of a membrane-associated interleukin 1 in macrophages. *Proc. Natl. Acad. Sci. USA* **82**, 1204-1208.

Laskin, D.L. and Pilaro, A.M. (1986). Potential role of activated macrophages in acetaminophen hepatotoxicity. I. Isolation and characterization of activated macrophages from rat liver. *Toxicol. Appl. Pharmacol.* **86**, 204-215.

Laskin, D.L., Laskin, J.D., Weinstein, I.B., and Carchman, R.A. (1981). Induction of chemotaxis in mouse peritoneal macrophages by phorbol ester tumor promoters. *Cancer Res.* **41**, 1923-1928.

Laskin, D.L., Pilaro, A.M. and Ji, S. (1986). Potential role of activated macrophages in

acetaminophen hepatotoxicity. II. Mechanism of macrophage accumulation and activation. Toxicol. *Appl. Pharmacol.* **86**, 216-226.
Laskin, D.L., Robertson, F.M., Pilaro, A.M. and Laskin, J.D. (1988a). Activation of liver macrohages following phenobarbital treatment of rats. *Hepatology* **8**, 1051-1055.
Laskin, D.L., Sirak, A.A., Pilaro, A.M., and Laskin, J.D. (1988b). Functional and biochemical properties of rat Kupffer cells and peritoneal macrophages. *J. Leuk. Biol.* **44**, 71-78.
Lautenschlager, I., Vaananen, H., and Kulonea, E. (1982). Quantitative study on the Kupffer cells in the liver of ethanol- and carbon tetrachloride-treated rats. *Acta. Pathol. Microbiol. Immunol. Scand.* **90**, 347-351.
Matsubara T. and Ziff, M. (1986). Increased superoxide anion release from human endothelial cells in response to cytokines. *J. Immunol.* **137**, 3295-3298.
Ooi, B.S., MacCarthy, E.P., Hsu, A., Ooi, Y.M. (1983). Human mononuclear cell modulation of endothelial cell proliferation. *J. Lab. Clin. Med.* **102**, 428-433.
Pilaro, A.M. and Laskin, D.L. (1986). Accumulation of activated mononuclear phagocytes in the liver following lipopolysaccharide treatment of rats. *J. Leuk. Biol.* **40**, 29-41.
Salvemini, D., Korbut, R., Anggard, E., and Vane, J.R. (1989). Lipopolysaccharide increases release of nitric oxide-like factor from endothelial cells. *Eur. J. Pharmacol.* **171**, 135-136.
Sironi, M. Breviario, F., Proserpio, P., Biondi, A., Vecchi, A., Van Damme, J., Dejana, E., and Mantovani, A. (1989). IL-1 stimulates IL-6 production in endothelial cells. *J. Immunol.* **142**, 549-553.
Souhami, R.L. and Bradfield, J.W. The recovery of hepatic phagocytosis after blockade of Kupffer cells. *J. Reticuloendothel. Soc.* **16**, 75-86.
Tanner, A., Kehani, A., Reiner, R., Holdstock, G., and Wright, R. (1981). Proteolytic enzymes released by liver macrohages may promote hepatic injury in a rat model of hepatic damage. *Gastroenterology* **80**, 647-654.

SIGNAL PATHS AND REGULATION OF SUPEROXIDE, EICOSANOID AND CYTOKINE FORMATION IN MACROPHAGES OF RAT LIVER

Karl Decker

Institute of Biochemistry
Albert-Ludwigs-University
Freiburg, Federal Republic of Germany

INTRODUCTION

Biological signal molecules mediate information between different cells. They are usually elicited by stimuli that include toxins, stress factors, nerve stimulation and metabolic deficiences. Like neurotransmitters they are in most cases shortlived and narrow-(or medium-)-ranged; unlike neurotransmitters and some hormones they are not stored within the producer cells. Some of them also differ from neurotransmitters and hormones in that they are autostimulatory, i.e. eliciting the same cells which produce them. Signal molecules show a high to medium specificity with respect to the elicitation of synthesis and secretion, to producer and target cells and to the mechanism of action.

A great number of cell types is able to produce signal molecules of one or another kind. Within the liver, the sessile macrophages, the Kupffer cells, appear to be the most active and broad-ranged producers of mediators (Table 1). They belong to the mononuclear phagocyte system and are thus part of the body's defensive machinery. Their major functions include phagocytosis, immunomodulation, presentation of antigens and biochemical attack. Kupffer cells are the first macrophages to meet foreign and noxious material coming from the gut. Immunocomplexes, viruses, unicellular organisms and their components, e.g. endotoxins from Gram-negative intestinal bacteria, are known to get internalized by Kupffer cells. The responses elicited by these agents depend on the kind of contact with the cell; surface-bound receptors or intracellular binding sites trigger different signal paths and regulatory phenomena.

METHODS

Kupffer cells were obtained from the livers of Wistar rats; they were prepared basically by the method of Brouwer et al. (1984) with some modifications. Primary cultures were maintained and checked for purity and viability as described previously (Eyhorn et al., 1988). Usually, experiments were performed 72 h after seeding. Eicosanoids were qualitatively and quantitatively analyzed by radio-immunoassays or by radio-HPLC if the material was derived from radioactive arachidonic acid (Tran-Thi et al., 1987). The isolated perfused liver was used in a way given in detail already (Tran-Thi et al., 1988). TNF-α of human or murine origin was generously supplied by Dr. G. Adolf, Ernst-Boehringer-Institut, Vienna (Austria), IL-6 by Dr. L. Aarden, Central Laboratory of the Netherland Red Cross, Amsterdam (The Netherlands). Established bioassays were used for the quantitation of IL-6 (Aarden et al.,1985) and

Table 1

Peptides:	
	Interleukin-1
	Interleukin-6
	Interferon α/β
	Tumor necrosis factor-α (Cachectin)
	Transforming growth factor-β
Lipids:	
	Prostaglandin D_2
	E_2
	$F_{2\alpha}$
	Thromboxane A_2
	Prostacyclin
	Leukotriene B_4
	Platelet-activating factor
Inorganic:	
	Nitric oxide (NO), or Endothelium-derived relaxing factor(EDRF)
	Superoxide (O_2^-)

TNF (Zacharchuk et al., 1983). O_2- was measured in the media as the SOD-inhibitable cytochrome c reduction (Nakagawara and Minakami, 1975).

RESULTS

Signal paths operating in Kupffer cells (Dieter et al., 1987b, 1988) are schematically outlined in Figure 1. The sequence involving phospholipase C and proteinkinase C operates in phagocytosis-stimulated cells and leads to the production both of eicosanoids and superoxide (Dieter et al., 1987b). Phorbol esters (phorbol 12-myristate 13-acetate) are thought to intervene in this pathway by interaction with proteinkinase C.

F_c and glycoconjugate receptors trigger an excess oxygen consumption, superoxide release (Bhatnagar et al., 1981, Latocha et al., 1989), Ca^{2+} flux (Birmelin and Decker, 1983) and eicosanoid synthesis (Latocha et al., 1989, Birmelin and Decker, 1984). The signal transduction in these cases appears to involve G proteins (sensitive to pertussis toxin), the activation of phospholipase C (enhanced turnover of phosphatidyl inositol), and proteinkinase C (inhibited by staurosporin) (Dieter, P. et al., unpublished). Superoxide production accompanied by oxidative burst and enhanced NADPH generation (Bhatnagar et al., 1981) involves proteinkinase C but does not depend on Ca^{2+} movement (Dieter et al., 1988).

Prostaglandin E_2 (PGE_2) leads to an unusually long-lasting (up to 24 h) enhancement of the activity of adenylate cyclase and a concomitant increase of the intracellular cAMP level in rat Kupffer cells (Bhatnagar et al., 1982). The evidence for the presence of specific PGE_2 receptors as well as the involvement of G proteins in the cAMP- and the phospholipase C-dependent signal paths is still indirect only.

Also, the primary biochemical events that participate in the interaction of Kupffer cells with lipopolysaccharides (endotoxins of Gram-negative intestinal bacteria) are not yet elucidated. Activation of macrophages requires in most cases the combined action of an immunomodulator (biological response modifier) and a cytokine (Adams and Hamilton, 1984). These inflammation-associated lipopolysaccharides seem to be the most important modulators of Kupffer cell activity. Interferon-gamma is the most potent and probably universal activator of macrophages (Pestka and Langer, 1987). It is a peptidic mediator and thus can be expected to interact with plasma membrane-located binding sites or receptors.

Inorganic mediators

Among the mediators known to be released by rat Kupffer cells two are of inorganic nature. One of these is **nitric oxide (NO)**, the endothelium-derived relaxing factor (EDRF). It is produced by a variety of cells from L-arginine (Palmer et al., 1987) and increases the activity of the soluble guanylate cyclase. The lipopolysaccharide-triggered NO production by rat Kupffer cells is regulated by PGE_2 and cAMP (T. Gaillard et al., unpublished).

Superoxide is the major reactive oxygen species produced by stimulated rat Kupffer cells. It is questionable whether superoxide must be considered as a mediator. It has been claimed to function as a chemoattractant; whether this is of any biological significance remains to be established. But superoxide is an important reactant of phagocytosis- or phorbol ester- stimulated Kupffer cells and most likely involved in the destruction of phagocytosed particles. O_2^- appears to be most efficient if this material is quickly engulfed in coated pits or phagosomes so that the superoxide produced in the periplasmic space remains concentrated in a narrow area. Superoxide accumulates in the medium of phagocytosing Kupffer cells only if the formation of vesicles is prevented by prior addition of cytochalasin B (Latocha et al., 1989) (Figure 2).

Superoxide production is intrinsically accompanied by the accumulation of protons in the cytosol. Thus, phagocytosis would lead to intracellular acidification followed by inactivation of processes such as eicosanoid synthesis (Dieter et al., 1987b). However, in intact Kupffer cells superoxide formation is accompanied by the activation of the Na^+/H^+ antiporter that exchanges intracellular protons for extracellular sodium ions (Decker and Dieter, 1987). Superoxide production differs

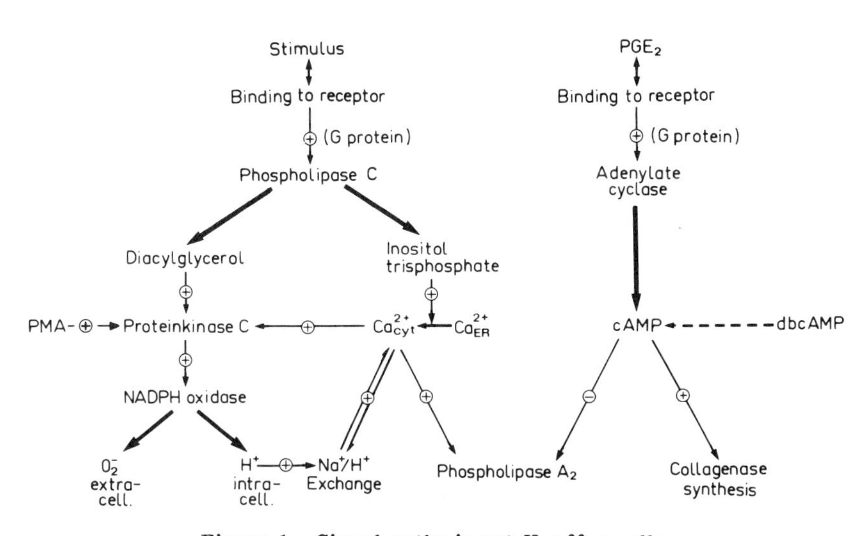

Figure 1. Signal paths in rat Kupffer cells

from the synthesis of eicosanoids in that it does not require an influx of calcium ions (Dieter et al., 1988). Furthermore, O_2- is not produced by lipopolysaccharide-stimulated rat Kupffer cells.

Eicosanoids

All responses of stimulated Kupffer cells so far examined lead to the production of **arachidonic acid** metabolites. They are formed from the free polyunsaturated fatty acid involving oxygen- requiring enzymatic processes. Arachidonic acid, however, is almost absent from the intracellular milieu as free acid (Irvine, 1982). It is integrated mainly in phospholipids occupying the sn- 2 position of phosphatides. In the extracellular space arachidonic acid may be transported together with other free fatty acids attached to albumin or other fatty acid-binding proteins. Arachidonic acid in serum is said to reach levels of up to 100 µM during severe general inflammations. When this happens it may have important consequences for the production of eicosanoids because this level is sufficient to elicit under culture conditions maximal rates of synthesis of most (but not all!) eicosanoids in potentially eicosanoid-producing cells (Dieter et al., 1989a,b).

Many cells are equipped with a fatty acid transport protein that picks up extracellular free fatty acids and facilitates their transfer into the cytoplasmic space. The existence of such a protein has not yet been confirmed for rat Kupffer cells. However, arachidonic acid added in the absence of albumin to primary cultures of Kupffer cells is very rapidly converted to eicosanoids as well as integrated into membrane phospholipids (Dieter et al., 1989b).

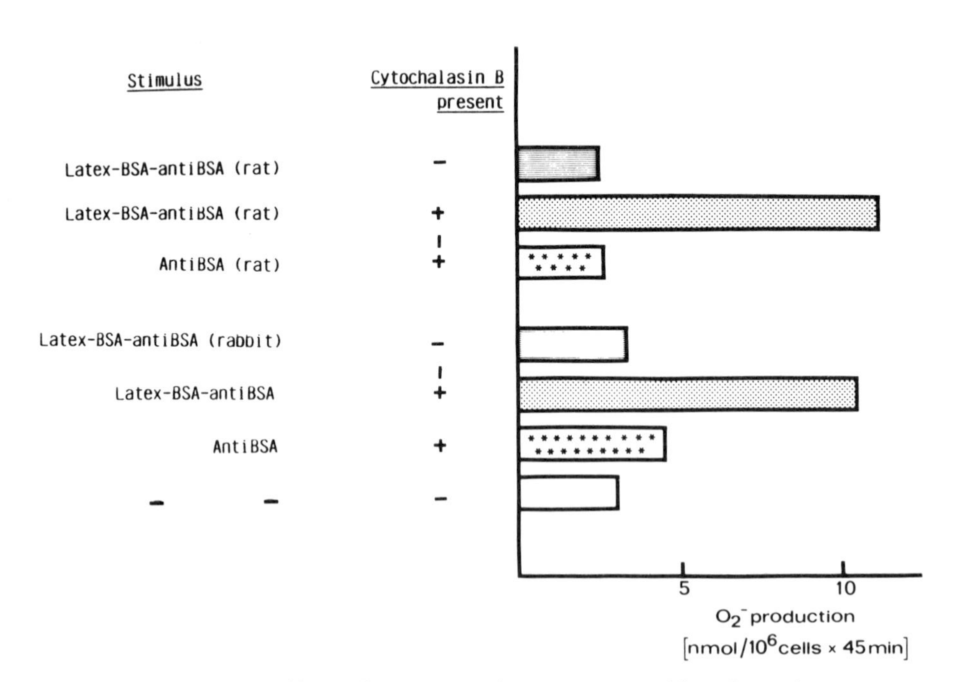

Figure 2. Effect of Cytochalasin B on superoxide release by phagocytosing rat Kupffer cells. Latex beads of 0.885 µ diameter were covalently covered with bovine serum albumin and reacted with antisera against BSA. The cells were exposed to cytochalasin B (5 µg/ml) 5 min prior to the immunocomplex-coated particles.

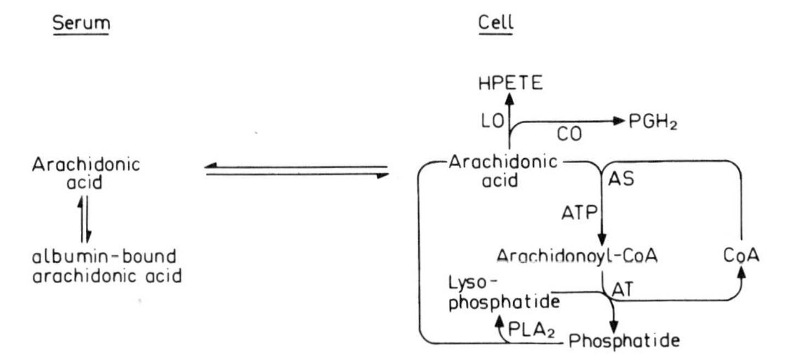

Figure 3. **Arachidonic acid metabolism in rat Kupffer cells. AS, acyl-CoA synthetase; AT, lysolecithin acyltransferase; CO, cyclo-oxygenase (prostaglandin H synthase); HPETE, hydroper-oxyeicosatetraenoic acid; LO, 5-lipoxygenase; PLA_2, phospholipase A_2.**

Albumin-bound arachidonic acid appears to be a rather poor donor for intracellular arachidonate at least in Kupffer cells. Even more, albumin present in the culture medium of Kupffer cells partially suppresses the formation of eicosanoids from *endogenous*, i.e. phosphatide-bound arachidonic acid (Dieter et al., 1990). This observation appears to have implications for our understanding of the topology of eicosanoid formation as it requires the transient presence of free arachidonic acid in a location that is accessible to extracellular albumin. The fact that activated phospholipase A_2 as well as many of the arachidonic acid-metabolizing enzymes, e.g. prostaglandin H synthase and activated 5-lipoxygenase are closely attached to the plasma membrane, is to be seen in this context.

Several enzymatic reactions are known to release fatty acids from the sn-2 position of phosphatides. In rat Kupffer cells the major route is via **phospholipase A_2** (Fig. 3). This enzyme is found in high activity in cell-free extracts of Kupffer cells. In the presence of divalent cations such as Ca^{2+} in physiological concentrations, the activity becomes attached to the membrane fraction of the cell (Krause et al., 1989). As the stimulation of eicosanoid formation requires the mobilization of Ca^{2+} it is most likely that the membrane-attached phospholipase A_2 is the species able to release arachidonic acid from phosphatides of the plasma membrane. In addition to phospholipase A_2 the combination of phospholipase C and diacylglycerol lipase appears to participate in the arachidonic acid liberation from phosphatidyl inositol (Dieter et al., 1989b).

The major pathway of arachidonic acid metabolism in rat Kupffer cells is initiated by **prostaglandin H synthase,** a two-headed enzyme that includes the O_2-requiring cyclo-oxygenase and the glutathione-dependent hydroperoxidase activities. The prostaglandin H_2 formed may be converted by specific enzymes to thromboxane A_2, prostacyclin and the various prostaglandins.

Prostaglandins and thromboxane are the predominant eicosanoids synthesized by activated rat Kupffer cells (Table 2). The pattern of arachidonic acid derivatives depends to some extent on the nature of the stimulating agent: The contact with phagocytosable particles, phorbol ester or Ca^{2+} ionophore yields very similar eicosanoid profiles with PGD_2 as the major metabolite followed by thromboxane, PGE_2, $PGF_{2\alpha}$ and tiny amounts of prostacyclin (Decker, 1987). The synthesis of these compounds starts almost immediately and comes to an end about 1 h after stimulation. The response to the inflammation-related agents, lipopolysaccharide (endotoxin), tumor necrosis factor-α (TNF) and some viruses, e.g. Sendai or Newcastle Disease Virus, differs in two important aspects: a) The rate of prostanoid synthesis is much slower than that elicited by the above mentioned stimuli, but continues for more than 24 hours. b) The pattern of accumulating eicosanoids shows PGE_2 (and thromboxane) as the major prostaglandin (Peters and Decker, 1989). As far as synthetic rates are concerned, PGE_2 synthesis is increased while PGD_2 production is about the same as after zymosan stimulation. Due to the longer incubation periods necessary to quantitate the eicosanoid synthesis triggered by immunomodulators, however, the greater lability of PGD_2 in aqueous solution (as compared to the other prostanoids present) influences the profile of the accumulating prostaglandins significantly; the degradation of PGD_2 leads mainly to 9-deoxy-Δ^9-PGD_2 and Δ^{12}-PGJ_2. An enhanced rate of PGE_2 synthesis can also be measured in cell-free preparations of LPS-treated vs. untreated Kupffer cells (Grewe et al., 1989). It appears that the conversion of PGH_2 to PGE_2 is specifically increased.

Arachidonic acid must be considered a substrate rather than an elicitor for eicosanoid synthesis in rat Kupffer cells. Its addition to the medium allows to bypass the phospholipase A_2 step and, in most instances, to realize the potential of a given cell for eicosanoid synthesis. There is one notable exception, however: **Thromboxane A_2** (detectable as its stable degradation product, thromboxane B_2) is not formed by rat Kupffer cells from added arachidonic acid, but from endogenous phospholipid-associated arachidonic acid only (Dieter et al., 1989a) (Fig. 4). The cause of this phenomenon has not yet been elucidated, but it suggests topological peculiarities of thromboxane synthase.

The presence and activity of **5-lipoxygenase** in rat Kupffer cells is still a matter of some controversy (Sakagami et al., 1988). We find products of the 5-lipoxygenase pathway in mouse Kupffer cells and in rat peritoneal macrophages (Decker et al., 1989), but could only demonstrate a tiny formation of LTB_4 in cultures and no significant 5-lipoxygenase activity in cell-free extracts from Kupffer cells (Rothacher et al., unpublished).

Table 2

STIMULUS	EICOSANOID found in the medium (pmol * 10^{-6}cells * h^{-1})			
	PGD_2	PGE_2	$PGF_{2\alpha}$	Tx
None	260	9	< 5	< 5
Zymosan	2100	145	65	170
Phorbol ester	2350	134	70	165
PAF	n.d.	45	n.d.	n.d.
TNF*	12	14	< 10	15
LPS*	13	16	< 10	15
A 23187	1780	132	60	145
Arachidonate	2420	152	75	< 20

* Average of the 24 h value. The rates are not linear over the 24 h period.

PGD_2 is much less stable in aqueous media than PGE_2.
Small amounts of material co-eluting with HETEs are also formed; they are probably side products of prostaglandin H synthase. 5-HETE could not be detected as product of stimulated rat Kupffer cells.
The data represent the production of an average batch of rat Kupffer cells after 72 h in primary culture. The yield of total eicosanoids may vary between cell batches, but the relative contribution of the different compounds is rather constant

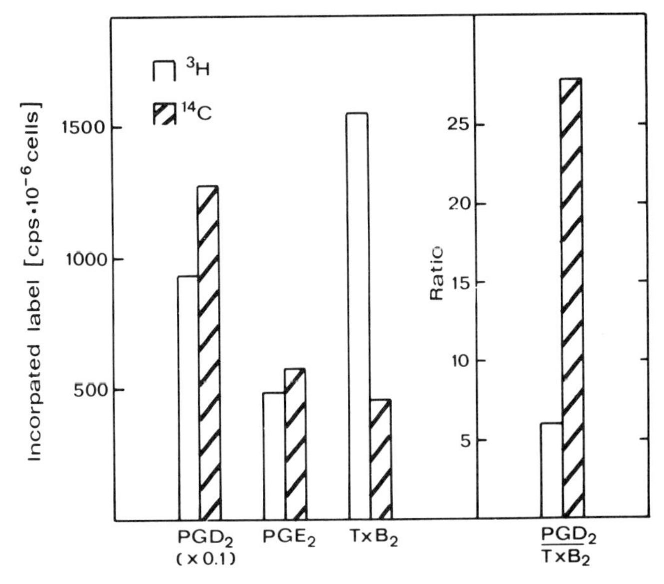

Figure 4. Different incorporation of free and membrane-bound arachidonic acid in prostaglandin D_2 and thromboxane, respectively. Rat Kupffer cells were incubated with ^3H-arachidonic acid for 24 h. The media were replaced and free ^{14}C-arachidonic acid was added together with zymosan. 60 min later, the media were removed and processed for HPLC and isotope analysis. Open bars, ^3H; hatched bars, ^{14}C content.

Kupffer cells are the major producers of eicosanoids in rat liver (Decker, 1987). The data reported so far were mostly obtained from work with primary cultures of rat Kupffer cells. Therefore, it was of great importance to prove that the mediator production in the cultured cells reflects the response in the intact organ both qualitatively and quantitatively. Zymosan (Dieter et al., 1987a) as well as phorbol ester (Häussinger et al., 1987, Tran-Thi et al., 1987,1988) elicited the Kupffer cell-specific reactions in the **isolated perfused liver.** The kinetics of formation and the amounts produced (sum of unchanged and partially degraded eicosanoids in the effluent and the bile) agreed remarkably well with the values calculated from the cell culture data.

Much less is known, however, about the radius of action of the signals produced by these macrophages and the targets of their message. Hepatocytes do not produce significant amounts of eicosanoids (Tran-Thi et al., 1987). But they participate in the signal system in two ways: They take up these substances, inactivate and partially release them into bile and they respond to them in various ways including stimulation of glycogenolysis and ion fluxes (Häussinger et al., 1987, Tran-Thi et al., 1987,1988). The topological aspects of the liver sinusoid, in particular the preferentially periportal location of the Kupffer cells, are important for the efficiency of signal transmission. Studies using the isolated, phorbol ester-stimulated rat liver perfused either in the ortho- or retrograde direction have shown this aspect very convincingly (Tran-Thi et al.,1988). Thromboxane exerts a surprisingly strong effect on the perfusion pressure. The point of attack is not yet known; it might be a sinusoidal cell, e.g. the Ito cell, the proposed sinusoidal sphingters (Häussinger, 1989) or the smooth muscle cells of hepatic vessels.

PGE_2 is an autostimulant; it activates adenylate cyclase, raises the intracellular cAMP level and elicits collagenase production in Kupffer cells (Bhatnagar et al., 1982). It also plays an important regulatory role in the release of some cytokines from these cells.

Cytokines

In addition to arachidonic acid metabolites, stimulated Kupffer cells may also release signals of polypeptide nature.

Interleukin-6 (IL-6) is well known as the mediator responsible for the elicitation of acute phase protein synthesis in hepatocytes (Andus et al., 1988). Thus, a production of this cytokine in close proximity to the parenchymal cells appears to be of particular importance. IL-6 as well as other cytokines is not elicited by phagocytotic stimuli, phorbol ester or calcium ionophore. Endotoxin, interferon-gamma and some viruses were found to elicit Kupffer cells to release IL-6 (Busam et al., 1989) (Figure 5).

In rats exposed to lipopolysaccharide Kupffer cells are the major source of **tumor necrosis factor-α** (TNF). After a short lag phase of about 30-60 min, TNF increases in the medium of cultured Kupffer cells for up to 3 hours; then the production stops and cells become refractory to repeated stimulation (Karck et al., 1988).

Regulation of mediator synthesis

The kinetics of TNF release are quite different from those of PGE_2 (Fig. 6), although a very close and intricate interaction between TNF and PGE_2 appears to exist: TNF itself is able to elicit prostanoid synthesis in Kupffer cells. The rate of production and the pattern of products are almost the same as those obtained after LPS treatment (Peters et al., 1989). On the other hand, PGE_2 was shown to inhibit specifically and in a dose-dependent manner the LPS-triggered TNF release from Kupffer cells (Karck et al.,1988) (Figure 7).

Furthermore, the inhibitory action is quite rapid; given simultaneously with LPS, PGE_2 leads to a >80% suppression of the TNF release (Karck et al., 1988). The same is also true for the dexamethasone blockade of TNF release. This almost instantaneous inhibitory action is in apparent contrast to the time requirement for its suppressive effect on eicosanoid production (Dieter et al.,1986) and suggests a different mechanism of action; it may also explain clinical findings on the rather rapid glucocorticoid effects seen in severe inflammatory or shock states.

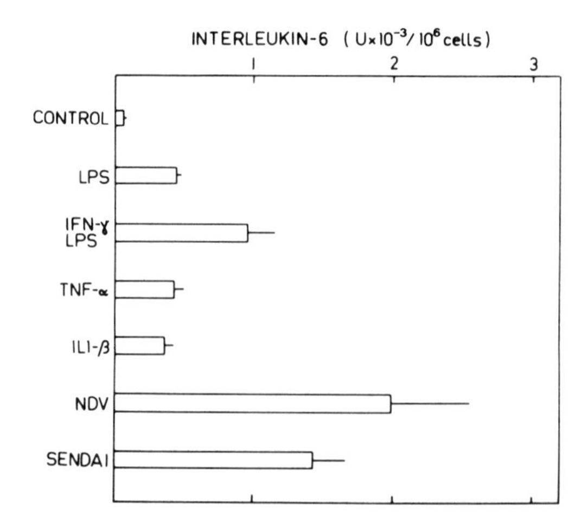

Figure 5. Interleukin-6 production by stimulated rat Kupffer cells. NDV, Newcastle Disease Virus.

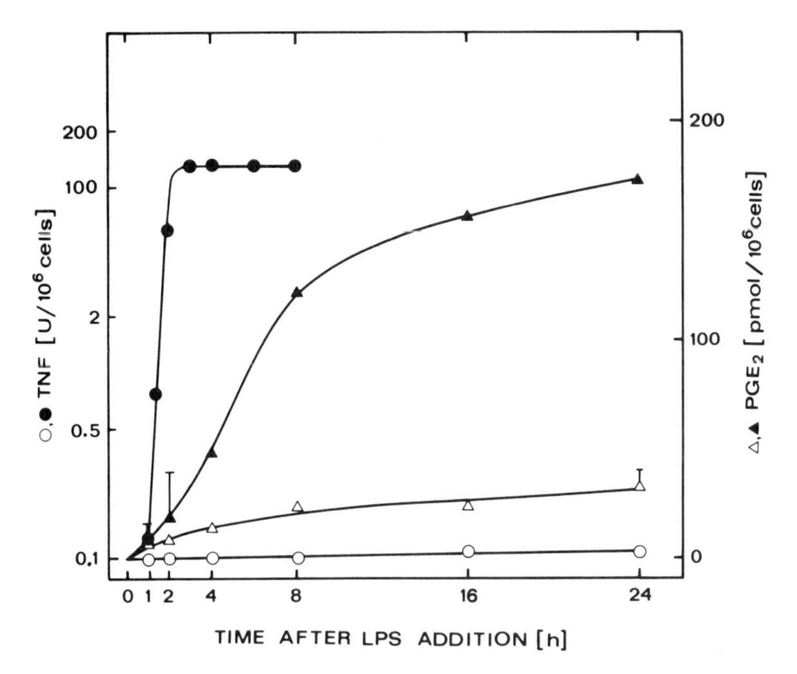

Figure 6. Time course of the syntheses of TNF-α and PGE_2 in LPS-stimulated rat Kupffer cells. Full symbols, LPS-treated cells; open symbols untreated controls.

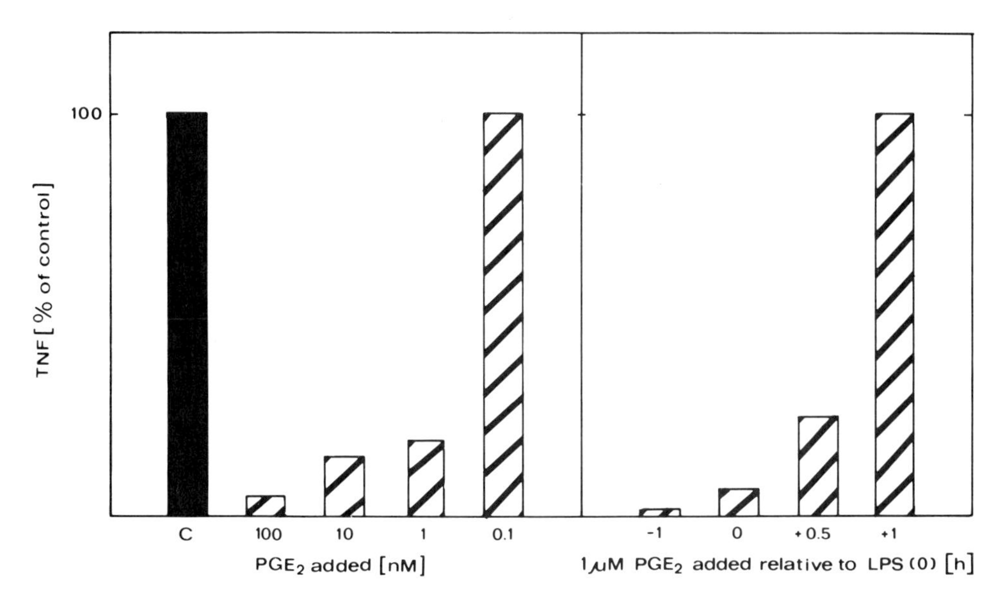

Figure 7. Dose- and time-dependent inhibition of TNF-α synthesis by PGE_2.

Table 3

Addition	Cytokine formed		
	TNF	IL-6	IL-1
	($U*10^{-6}$cells*4h)	($U*10^{-6}$cells*24h)	(% of control)
None	< 20	44 ± 12	n.d.
LPS	680 ± 256	364 ± 44	100
PGE_2	< 20	52 ± 16	n.d.
LPS + PGE_2	64 ± 16	332 ± 38	15

Rat Kupffer cells in primary culture were exposed to LPS (100 ng/ml) and 100 nM PGE_2, respectively and TNF and IL-6 measured. With regard to IL-6, the same results were obtained using 1 ng LPS/ml. For IL-1 synthesis, 10 µg LPS and 1 µM PGE_2 were used. The data are taken from Busam et al [1989] and Shirahama et al [1988]

An inhibitory effect of PGE_2 is also reported for the LPS-elicited **interleukin-1** (IL-1) production by rat Kupffer cells (Shirahama et al.,1988). In contrast to its suppression of TNF and IL-1 production, PGE_2 does not affect the synthesis and release of IL-6 by LPS- or virus-stimulated rat Kupffer cells (Table 3). As PGE_2 is assumed to trigger the cAMP-mediated signal transduction it will be of interest to pinpoint the expected bifurcation in the signal paths leading to the synthesis of IL-1 on the one and IL-6 and TNF on the other hand.

The regulatory role of PGE_2 in Kupffer cells appears to be even more complicated. In contrast to the PGE_2 synthesis triggered by zymosan, LPS-elicited release of PGE_2 displays specific feedback inhibition (Table 4); PGE_2 suppresses its own synthesis in a dose-dependent manner. This effect is not obtained if TNF instead of LPS is used as a stimulant of rat Kupffer cells; rather, a stimulatory effect on prostanoid synthesis is observed (Table 5). Again, this different behavior suggests peculiarities in the transduction of inflammatory signals in macrophages that are not yet understood.

As a consequence of these findings a **self-limiting regulatory cycle** emerges that involves both PGE_2 and TNF (Fig. 8). Exposure of Kupffer cells to lipopolysaccharide leads to the rather fast release of TNF. This cytokine as well as LPS itself is able to induce eicosanoid synthesis with a clear preference for PGE_2. A certain independence of the effect of LPS and TNF follows from the fact that the presence of anti-TNF antibodies can only partially suppress the LPS-triggered PGE_2 formation. PGE_2 thus accumulating is able to inhibit any further TNF production and also to slow down the synthesis of more PGE_2. The time courses of PGE_2 and TNF release and the refractoriness for the latter may be a consequence of this regulatory self-limitation.

Figure 8 includes the effects of TNF on sinusoidal endothelial cells. These cells are able to bind TNF rather firmly on their surfaces (Schlayer et al., 1987); TNF induces neutrophilic granulocytes to stick to the endothelial cells with the ensuing release of enzymes such as elastase and myeloperoxidase. This effect of the TNF-bearing endothelial cells can be completely abolished by addition of anti-TNF antibodies. It is evident that TNF interaction with sinusoidal endothelial cells - a situation which seems to occur in liver sinusoids early after LPS + D-galactosamine treatment (Schlayer et al.,1988) - will lead to destructive processes in the

Table 4

Stimulus	PGE_2	TxB_2	n
	released into the medium ^3H content (arbitrary units)		
None	21 ± 18	34 ± 18	5
LPS	395 ± 117	257 ± 79	5
LPS + 1 µM PGD_2	446 ± 109	247 ± 45	3
LPS + 10 nM PGE_2	269 ± 30	203 ± 69	2
LPS + 100 nM PGE_2	195 ± 31	137 ± 44	2
LPS + 1 µM PGE_2	131 ± 39	120 ± 37	5

Kupffer cells were prelabeled with ^3H-arachidonic acid, the washed cells were exposed to LPS (400 ng/ml) and unlabeled prostanoids. Analysis by radio-HPLC. n = number of independent experiments.

Table 5

Stimulus	PGE_2	TxB_2	n
	released into the medium ^3H content (arbitrary units)		
None	21 ± 18	34 ± 19	5
TNF	203 ± 63	207 ± 40	3
TNF + PGE_2	382 ± 33	354 ± 109	3

Kupffer cells were prelabeled with ^3H-arachidonic acid, the washed cells were exposed to TNF (8000 U/ml) and unlabeled PGE_2 (1µM). Analysis by radio-HPLC. n = number of independent experiments.

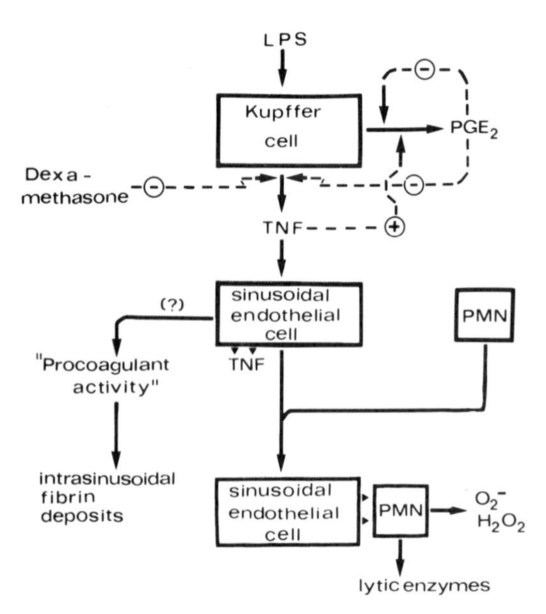

Figure 8. Regulatory circuit in LPS-treated rat Kupffer cells. PMN, polymorphonuclear granulocytes.

neighborhood of these degranulated neutrophils. In line with this deduction is the observation that prior injection of LPS-treated mice with anti-TNF antibodies abolished the hepatotoxic response (Pober et al.,1986).

REFERENCES

Aarden, L., Lansdorp, P. and De Groot, E. (1985). A growth factor for B cell hybridomas produced by human monocytes. *Lymphokines* **10**, 175-185.

Adams, D.O. and Hamilton, T.A. (1984). The cell biology of macrophage activation. *Ann. Rev. Immunol.* **2**, 283-318.

Andus, T., Geiger, T., Hirano, T., Kishimoto, T., Tran-Thi, T.A., Decker, K. and Heinrich, P.C. (1988). Regulation of synthesis and secretion of major rat acute-phase proteins by recombinant human interleukin-6 (BSF-2/IL-6) in hepatocyte primary cultures. *Eur. J. Biochem.* **173**, 287-293.

Bhatnagar, R., Schirmer, R., Ernst, M. and Decker, K. (1981). Superoxide release by zymosan-stimulated rat Kupffer cells in vitro. *Eur. J. Biochem.* **119**, 171-175.

Bhatnagar, R., Schade, U., Rietschel, T.E. and Decker, K. (1982). Involvement of prostaglandin E and cyclic adenosine 3',5'-monophosphate in lipopolysaccharide-stimulated collagenase release by rat Kupffer cells. *Eur. J. Biochem.* **125**, 125-130.

Birmelin, M. and Decker, K. (1983). Ca^{2+} flux as an initial event in phagocytosis by rat Kupffer cells. *Eur. J. Biochem.* **131**, 539- 543.

Birmelin, M. and Decker, K.(1984). Synthesis of prostanoids and cyclic nucleotides by phagocytosing rat Kupffer cells. *Eur. J. Biochem.* **142**, 219-225.

Brouwer, A., Barelds, R.J. and Knook, D.L. (1984). Centrifugal separation of mammalian cells. In *Centrifugation, a Practical Approach* (D.Rickwood, Ed.) pp.183-218. IRL Press, Oxford, UK.

Busam,K ., Bauer, T., Bauer, J., Gerok, W. and Decker, K. (1989). Regulation of interleukin-6 secretion by rat Kupffer cells. *J.Hepatol.* **9** (suppl.1), S 13

Decker, K. (1987). Eicosanoids as signal molecules between hepatocytes and sinusoidal cells. In *Modulation of liver cell expression* (W. Reutter, H. Popper, I.M. Arias, P.C. Heinrich, D. Keppler and L. Landmann, Eds.) pp.397-409. MTP Press, Lancaster, UK.

Decker, K. and Dieter, P. (1987). The stimulus-activated Na^+/H^+ exchange in macrophages, neutrophils and platelets. In *pH Homeostasis, Mechanism and Control* (D. Häussinger, Ed.) pp. 79- 96, Academic Press, London, UK.
Decker, T., Lohmann-Matthes, M.-L., Karck, U., Peters, T. and Decker, K. (1989). Cytotoxicity, tumor necrosis factor and prostaglandin release after stimulation: an interspecies comparison of rat Kupffer cells, murine Kupffer cells and murine inflammatory liver macrophages. *J. Leukocyte Biol.* **45**, 139-147.
Dieter, P., Schulze-Specking, A. and Decker, K. (1986). Differential inhibition of prostaglandin and superoxide production by dexamethasone in primary cultures of rat Kupffer cells. *Eur. J. Biochem.* **159**, 451-457.
Dieter, P., Altin, J.G., Decker, K. and Bygrave, F.L. (1987a). Possible involvement of eicosanoids in the zymosan- and arachidonic acid-induced oxygen uptake, glycogenolysis and Ca^{2+} mobilisation in the perfused rat liver. *Eur. J. Biochem.* **165**, 455- 460.
Dieter, P., Schulze-Specking, A., Karck, U. and Decker, K. (1987b). Prostaglandin release but not superoxide production by rat Kupffer cells in vitro depends on Na^+/H^+ exchange. *Eur. J. Biochem.* **170**, 201-206.
Dieter, P., Schulze-Specking, A. and Decker, K. (1988). Ca^{2+} requirement of prostanoid but not of superoxide production by rat Kupffer cells. *Eur. J. Biochem.* **177**, 61-67.
Dieter, P., Peters, T., Schulze-Specking, A. and Decker, K. (1989a). Independent regulation of thromboxane and prostaglandin synthesis in liver macrophages. *Biochem. Pharmacol.* **38**, 1577-1581.
Dieter, P., Schulze-Specking, A. and Decker, K. (1989b). Signal transduction during stimulus-induced synthesis of prostanoids and superoxide in liver macrophages. In *Cells of the Hepatic Sinusoid_* (E. Wisse, D.L. Knook and K. Decker, Eds.) Vol. II,pp. 190-193, Kupffer Cell Foundation, Rijswijk, The Netherlands.
Dieter, P., Krause, H. and Schulze-Specking, A. (1990). Arachidonate metabolism in macrophages is affected by albumin. *Eicosanoids* in press
Eyhorn, S., Schlayer, H.-J., Henninger, H.P., Dieter, P., Hermann, R., Woort-Menker, M., Becker, H., Schaefer, H.E. and Decker, K. (1988). Rat hepatic sinusoidal endothelial cells in monolayer culture. Biochemical and ultrastructural characteristics. *J. Hepatol.* **6**, 23-35.
Grewe, M., Schulze-Specking, A. and Decker, K. (1989). Prostaglandin H_2 synthase of rat Kupffer cells. In *Cells of the Hepatic Sinusoid* (E. Wisse, D.L. Knook and K. Decker, Eds.) Vol. II, pp. 206-207, Kupffer Cell Foundation, Rijswijk, The Netherlands.
Häussinger, D., Stehle, T., Tran-Thi, T.-A., Decker, K. and Gerok, W. (1987). Prostaglandin responses in isolated perfused rat liver: Ca^{2+} and K^+ fluxes, hemodynamics and metabolic effects. *Biol. Chem. Hoppe-Seyler* **368**, 1509-1513.
Häussinger, D. (1989). Regulation of hepatic metabolism by extracellular nucleotides and eicosanoids. The role of cell heterogeneity. *J. Hepatol.* **8**, 259-266.
Irvine, R.F. (1982). How is the level of free arachidonic acid controlled in mammalian cells? *Biochem. J.* **204**, 3-16.
Karck, U., Peters, T. and Decker, K. (1988). The release of tumor necrosis factor from endotoxin-stimulated rat Kupffer cells is regulated by prostaglandin E_2 and dexamethasone. *J. Hepatol.* **7**, 352-361.
Krause, H., Dieter, P., Schulze-Specking, A. and Decker, K. (1989). Regulation of phospholipase A_2 in rat Kupffer cells. *J. Hepatol.* **9**, Suppl. 1, S 178.
Latocha, G., Dieter, P.,Schulze-Specking, A. and Decker, K. (1989). F_c receptors mediate prostaglandin and superoxide synthesis in cultured rat Kupffer cells. *Biol. Chem. Hoppe-Seyler* **370**, 1055-1061.
Nakagawara, A. and Minakami, S. (1975). Generation of superoxide by leukocytes treated with cytochalasin E. *Biochem. Biophys. Res. Comm.* **64**, 760-767.
Palmer, R.M., Ferrige, A.G. and Moncada, S. (1987). Nitric oxide accounts for the biological activity of endothelium-derived relaxing factor. *Nature* (London) **327**, 524-526.
Pestka, S. and Langer, J.A. (1987). Interferons and their action. *Ann. Rev. Biochem.* **15**, 727-777.

Peters, T. and Decker, K. (1989). Recombinant mouse tumor necrosis factor alpha induces prostaglandin E_2 synthesis in rat Kupffer cells. In *Cells of the Hepatic Sinusoid* (E. Wisse, D.L. Knook and K. Decker, Eds.) Vol. II, pp. 182-185, Kupffer Cell Foundation, Rijswijk, The Netherlands.

Pober, J.S., Gimbrone, Jr. M.A., Lapierre, L.A., Mendrick, D.L., Fiers, W., Rothlein, R. and Springer, T.A. (1986). Overlapping patterns of activation of human endothelial cells by interleukin 1, tumor necrosis factor and immune interferon. *J. Immunol.* **137**, 1893-1896.

Sakagami, Y., Mizoguchi, Y., Seki, S., Kobayashi, K., Morisawa, S.and Yamamoto, S. (1988). Release of peptide leukotrienes from rat Kupffer cells. *Biochem. Biophys. Res. Commun.* **156**, 217-221.

Schlayer ,H.-J., Karck, U., Ganter, U., Hermann, R. and Decker, K. (1987). Enhancement of neutrophil adherence to isolated rat liver sinusoidal endothelial cells by supernatants of lipopolysaccharide-activated monocytes - role of tumor necrosis factor. *J. Hepatol.* **5**, 311-321.

Schlayer, H.-J., Laaf, H., Peters, T., Woort-Menker, M., Estler, C., Karck, U., Schaefer, H.E. and Decker, K. (1988). Involvement of tumor necrosis factor in endotoxin-triggered neutrophil adherence to sinusoidal endothelial cells of mouse liver and its modulation in acute phase. *J. Hepatol.* **7**, 239-249.

Shirahama, M., Ishibashi, H., Tsuchiya, Y., Kurokawa, S., Hayashida, K., Okumura, Y. and Niho, Y. (1988). Kupffer cells may autoregulate interleukin 1 production by producing interleukin 1 inhibitor and prostaglandin E_2. *Scand. J. Immunol.* **28**, 719-725.

Tran-Thi, T.-A., Gyufko, K., Henninger, H., Busse, R. and Decker, K. (1987). Studies on synthesis and degradation of eicosanoids by rat hepatocytes in primary culture. *J. Hepatol.***5**, 322-331.

Tran-Thi, T.-A., Gyufko, K., Häussinger, D. and Decker, K. (1988). Net prostaglandin release by perfused rat liver after stimulation with phorbol 12-myristate 13-acetate. *J. Hepatol.* **6**, 151-157.

Zacharchuk, C.M., Drysdale, B.E., Mayer, M.M. & Shin, H.S. (1983). Macrophage-mediated cytotoxicity: role of a soluble macrophage cytotoxic factor similar to lymphotoxin and tumor necrosis factor. *Proc. Natl. Acad. Sci. USA* **80**, 6341-6345.

REACTIVE METABOLITES FROM N-NITROSAMINES

Michael C. Archer, Jamie R. Milligan, Sandra Skotnicki and Shi-Jiang Lu

Department of Medical Biophysics
University of Toronto
Ontario Cancer Institute
500 Sherbourne Street
Toronto, Canada M4X 1K9

INTRODUCTION

N-Nitrosodialkylamines are metabolically activated by hydroxylation at the carbon atom alpha to the *N*-nitroso group. Spontaneous cleavage of the C-N bond in the α-hydroxynitrosamine leads to production of an alkydiazohydroxide that decomposes to an alkydiazonium ion. In principle, the diazonium ion can lose nitrogen to form a carbonium ion, although we have presented evidence that carbonium ions are not involved in DNA alkylation (Park et al., 1980). It is not known which of the activated species first interacts with the DNA. The α-hydroxynitrosamine has a half life of the order of seconds at physiological pH (Mochizuki et al., 1980), and may be the transport form of the activated carcinogen (Gold and Linda, 1979) which diffuses from its site of synthesis into the nucleus. The α-hydroxynitrosamine may also be the species which first interacts with the DNA.

In the present study, we have determined whether DNA alkylation by *N*-nitroso compounds displays any sequence specificity that may be indicative of interaction of the α-hydroxy intermediate with the DNA. We have, furthermore, determined whether there is selectivity in the alkylation of DNA with respect to DNA conformation or particular genes within the genome.

RESULTS AND DISCUSSION

We first investigated the formation in vitro of 7-methylguanine in various DNA substrates including calf thymus DNA, a negatively supercoiled plasmid, and two synthetic polynucleotides of defined sequence following their reaction with *N*-nitroso(acetoxymethyl)methylamine (AcONDMA), *N*-nitroso(acetoxybenzyl)-methylamine (AcONMBzA), and methylnitrosurea (MNU). α-Acetoxy-*N*-nitrosamines have been used extensively as stable derivatives of activated *N*-nitrosamines (Archer and Labuc, 1985), since they are readily hydrolyzed to produce the corresponding α-hydroxy-*N*-nitrosamines in the presence of non-specific esterase. MNU does not form an α-hydroxy derivative, but undergoes non-enzymatic, base-catalyzed decomposition to yield methyldiazotate.

Figure 1 shows that AcONDMA and AcONMBzA in the presence of esterase, and MNU, produce the same pattern of methylation as determined by 7-methylguanine and

Biological Reactive Intermediates IV, Edited by C.M. Witmer *et al.*
Plenum Press, New York, 1990

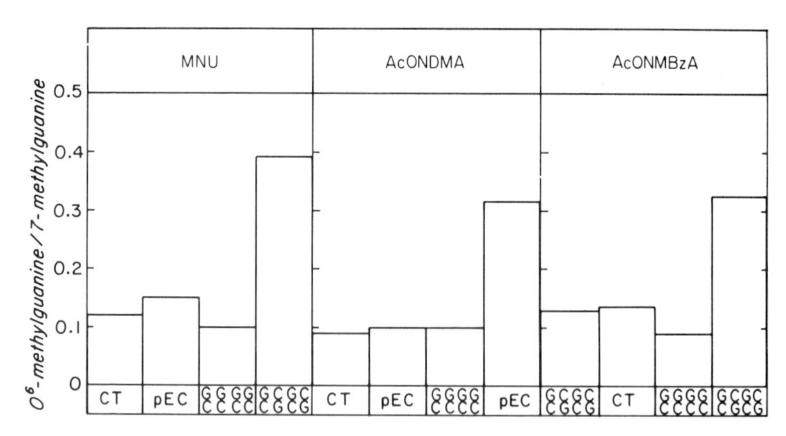

Figure 1. O6-Methylguanine/7-methylguanine ratios in DNA substrates reacted with methylating agents. (CT = calf thymus DNA; pEC = supercoiled plasmid; GGGG/CCCC = poly(dG).poly(dC); GCGC/CGCG = poly(dGdC).poly(dGdC).

O6-methylguanine formation, when they react with various DNA substrates (Milligan et al. 1989). These data suggest that the three N-nitroso compounds methylate DNA via a common intermediate, probably the methyl diazonium ion. As expected, for calf thymus DNA, a supercoiled plasmid, and poly(dG).poly(dC), 7-methylguanine was formed in about ten times the yield of O6-methylguanine. For poly(dGdC).poly(dGdC), however, the yield of O6-methylguanine was the same as the other DNA substrates, but the 7-methylguanine yield was only about one third of what we expected. This result is similar to that observed by Briscoe and Cotter (1984, 1985) for MNU.

To determine whether sequence selectivity played a role in these observations, we methylated the plasmid pBR322, then isolated a 345 bp BamHI-HindIII restriction fragment. 7-Methylguanine and 3-methyladenine residues were removed by heating to 65° at neutral pH. The samples were then end-labeled with ^{32}P and treated with spermidine to cleave the DNA at a purinic sites. Mixtures of small fragments were finally separated by gel electrophoresis.

No differences were observed between lanes in which the DNA was reacted with the three N-nitroso compounds, confirming our observations described above. There seemed, however, to be some preference for reaction at particular guanines with the restriction fragment. In order to quantify this observation, the sequencing gel autroradiograph was analyzed by densitometry which is shown in Figure 2. Integration of the peak areas indicated that individual guanines can differ by about 5-fold in their reactivity with the nitroso compounds. Dimethyl sulfate showed a similar variation in the reactivity of guanines, but these was no correlation between the reactivity of the nitroso compounds and dimethyl sulfate. It appears, therefore, that sequence selectivity depends on both the stereoelectronic environment of the guanines and the nature of the alkylating agent.

Although our results provide evidence for sequence selectivity in the methylatin of guanines in DNA, it is possible that the effect of dimethylsulfoxide that we used to aid the solubility of the carcinogens may have promoted the B to Z transition in poly(dGdC).poly(dGdC) (Van de Sande and Jovin, 1982; Saenger et al., 1986), which in turn may have affected the reactions. We therefore examined the reaction of MNU with B- and Z-DNA, and extended the studies to include DNA in the cruciform conformation (Wells, 1988), and H-DNA, a structure in which dA:dT Watson-Crick base pairs alternate with Hoogsteen (syn) dG:dC pairs (Wells, 1988). Dr. David Pulleyblank (University of Toronto) generously supplied us with plasmids in which inserts in the non-B topological forms could be prepared by topoisomerase I catalyzed relaxation in the presence of ethidium (Peck et al., 1982). Following methylation, a fragment of the plasmid containing the insert was isolated by restriction endonuclease digestion, and methylation sites were determined as described above. In this way, we were able to compare directly methylation of the same sequence in the B and non-B conformations.

The sequencing gels showed that DNA sequences in the Z, cruciform and H conformations are methylated by MNU in a manner that is indistinguishable from the reaction of MNU with the same sequences in the B conformation. Electronic factors, therefore, rather than steric factors appear to dominate the methylation reaction. Furthermore, any conformational changes induced in poly(dGdC).poly(dGdC) by the co-solvent in our first study, had no effect on the methylation reaction. Our results may have important implications for chemical carcinogenesis, since some repair enzymes are unable to act on methylated guanines in non-B DNA, thereby increasing the carcinogenic potential of the lesions.

We have recently described a method for quantifying alkylated bases in defined gene sequences of genomic DNA following treatment of an animal with a carcinogen (Milligan and Archer, 1988). First, the DNA is digested with a restriction endonuclease to generate a fragment containing the gene of interest. Following depurination (pH8 at 65°), reaction with spermidine generates strand breaks at the apurinic sites. Finally, gel electrophoresis followed by Southern transfer enables sequences of interest to be visualized using specific probes. Strand breaks in methylated restriction fragments reduces their intensity compared to unmethylated fragments. Using this method, we have shown that a 10kb fragment containing the Ha-ras proto-oncogene is extensively methylated in DNA isolated from the breast tissue of female rats 6h following treatment with MNU (Fig 3). In contrast, fragments of similar size containing both the Ki- and N-ras genes are considerably less methylated. Preliminary data on the 5-methylcytosine content (MspI versus HpaII digestion) of these genes suggests that the Ha-ras gene is transcriptionally active in this tissue, while the Ki- and N-ras genes are either inactive or at least much less active than the Ha-ras gene. In this animal model, the Ha-ras gene has been shown to be activated by point mutation to a transforming gene by a single dose of MNU, while neither the Ki- nor the N-ras gene is mutated (Zarbl et al., 1985). It appears, therefore, that the extent of alkylation of a given gene, which depends on its transcriptional activity, may be a factor in determining whether the gene is mutated by a chemical carcinogen.

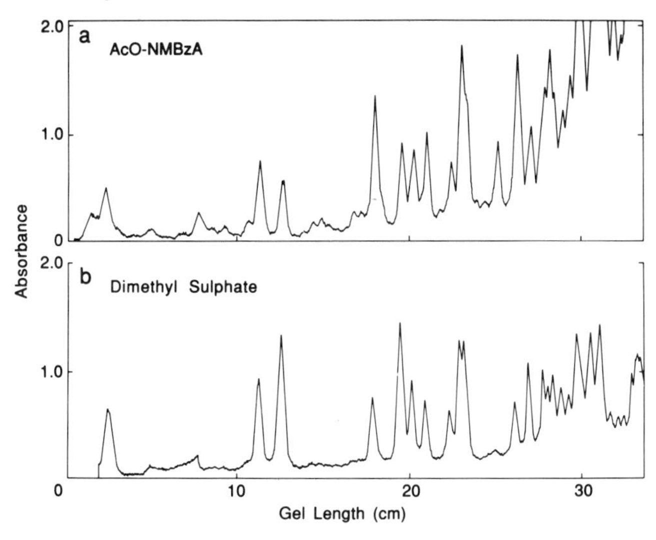

Figure 2 Densitometric scans of lanes in polyacrylamide sequencing gel in which 345bp BamHI-Hind III restriction fragments of pBR322 DNA treated with AcONMBzA or dimethyl sulfate were end labeled with ^{32}P, then cleaved by reaction with piperidine (Maxam and Gilbert, 1980).

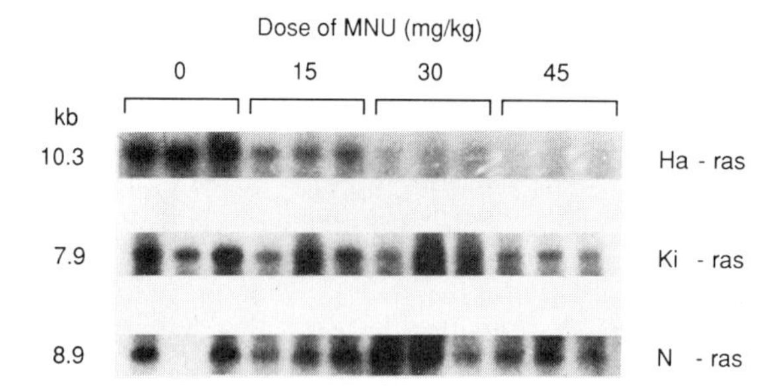

Figure 3 Single strand breaks in restriction fragments of genomic DNA containing the Ha, Ki, or *N*-ras genes of female rat breast 6h following treatment with various doses of MNU. Each lane utilized DNA from a separate animal. Experimental conditions were as described by Milligan and Archer (1988).

ACKNOWLEDGEMENTS

This work was supported by Grants MT-7025 and MT-10491 from the Medical Research Council of Canada.

REFERENCES

Archer, M.C. and Labuc, G.E. (1985). Nitrosamines. *In Bioactivation of Foreign Compounds* (M.W. Anders, Ed.) pp. 403-429, Academic Press, New York.
Briscoe, W.T. and Cotter, L.E. (1984). The effects of neighboring bases on *N*-methyl-*N*-nitrosourea alkylation of DNA. *Chem.-Biol. Interact.* **52,** 103-110.
Briscoe W.T. and Cotter L.E. (1985). DNA sequence has an effect on the extent and kinds of alkylation of DNA by a potent carcinogen. *Chem.-Biol. Interact.* **56,** 321-331.
Gold, B. and Linder, W. B. (1979). α-Hydroxynitrosamines: transportable metabolites of dialkylnitrosamines. *J. Am. Chem. Soc.* **101,** 6772-6773.
Milligan, J.R. and Archer, M.C. (1988). Alkylation of individual genes in rat liver by the carcinogen *N*-nitrosodimethylamine. *Biochem. Biophys. Res. Comm.* **155,** 14-17.
Mochizuki M., Anjo, T. and Okada, M. (1980). Isolation and characterization of N-alkyl-*N*-(hydroxymethyl)nitrosamines from *N*-alkyl-*N*-(hydroperoxymethyl)-nitrosamines by deoxygenation. *Tetrahedron Lett.* **21,** 3693-3696.
Park, K.K., Archer, M.C. and Wishnok, J.S. (1980). Alkylation of nucleic acids by *N*-nitroso-di-*N*-propylamine: evidence that carbonium ions are not significantly involved. *Chem.-Biol. Interact.* **29,** 139-144.
Peck, L.J. Nordheim, A., Rich, A. and Wang, J.C. (1982). Flipping of cloned d(pCpG)n.d(pCpG)n DNA sequences from right to left-handed helical structure by salt, Co (III), or negative supercoiling. *Proc. Natl. Acd. Sci. USA* **79,** 4560-4564.
Saenger, W., Hunter, W.N. and Kennard, O. (1986). DNA conformation is determined by economics in the hydration of phosphate groups. *Nature* **324,** 385-388.
Van de Sande, J.H. and Jovin, T.M. (1987). Z DNA, the left-handed helical form of poly[d(G-C)] in $MgCl_2$-ethanol, is biologically active. *EMBO J.* **1** 115-120.
Wells, R.D. (1988). Unusal DNA structures. *J. Biol. Chem.* **263,** 1095-1098.
Zarbl, H., Sukumar, S., Arthur, A.V., Martin-Zanca, D. and Barbacid, M. Direct mutagenesis of Ha-ras-1 oncogenes by *N*-nitroso-*N*-methylurea during initiation of mammary carcinogenesis in rats. *Nature* **315,** 382-385.

BISFURANOID MYCOTOXINS: THEIR GENOTOXICITY AND CARCINOGENICITY

Dennis P. H. Hsieh and David N. Atkinson

Department of Environmental Toxicology,
University of California at Davis
Davis, CA 95616

INTRODUCTION

Bisfuranoid mycotoxins are a family of fungal metabolites that contain in their molecules a characteristic dihydrobisfuran ring structure. Some representative bisfuranoid mycotoxins are aflatoxins B_1, G_1, M_1, sterigmatocystin, O-methylsterigmatocystin, and versicolorin A (See Figure 1 & 2).

Among these, Aflatoxin B_1 (AFB_1) and sterigmatocystin (ST) are well known potent hepatocarcinogens in some laboratory animals (IARC,1987a). AFB_1 was designated by IARC as a human carcinogen in 1987 (IARC, 1987b) and is considered by the agency as an etiological factor of primary liver cancer in human populations. Primary liver cancer is the major form of cancer in some African and Asian countries, such as Mozambique, Swaziland, Kenya, Thailand, Taiwan, and some parts of China (Van Rensburg et al., 1985; Peers et al., 1987; Beasley and Hwang, 1984; Sun and Chu, 1984). The occurrence of aflatoxin in food commodities has been a worldwide concern and made headlines in the media repeatedly.

OCCURRENCE

The bisfuranoid mycotoxins are biogenetically related to the biosynthesis of AFB_1 in fungi. The pathway of aflatoxin biosynthesis is shown in Figure 1. Bisfuranoid mycotoxins are decaketides, derived from 10 acetate units (Hsieh et al., 1989). One acetyl CoA primer and 9 malonyl CoA units undergo a head-to-tail assembly to form an anthraquinone with a C6 side chain. The C6 side chain undergoes deacetylation and cyclization to form the characteristic dihydrobisfuran ring of versicolorin A (VA). The anthraquinone moiety of VA is then converted to xanthone to form sterigmatocystin. The xanthone is then converted to coumarin to form AFB_1.

This entire pathway is only possessed by two fungal species, Aspergillus flavus and A. parasiticus. The pathway terminating at ST is possessed by at least 5 fungal species, and the pathway terminating at VA is possessed by still more fungal species (Cole and Cox, 1981). This means that the occurrence of VA may be more widespread than ST and the occurrence of ST may be more widespread than AFB_1. AFB_1 has been found to occur in corn, peanut, cottonseed, and other calory-rich commodities (Pohland and Wood, 1987).

Biological Reactive Intermediates IV, Edited by C.M. Witmer *et al.*
Plenum Press, New York, 1990

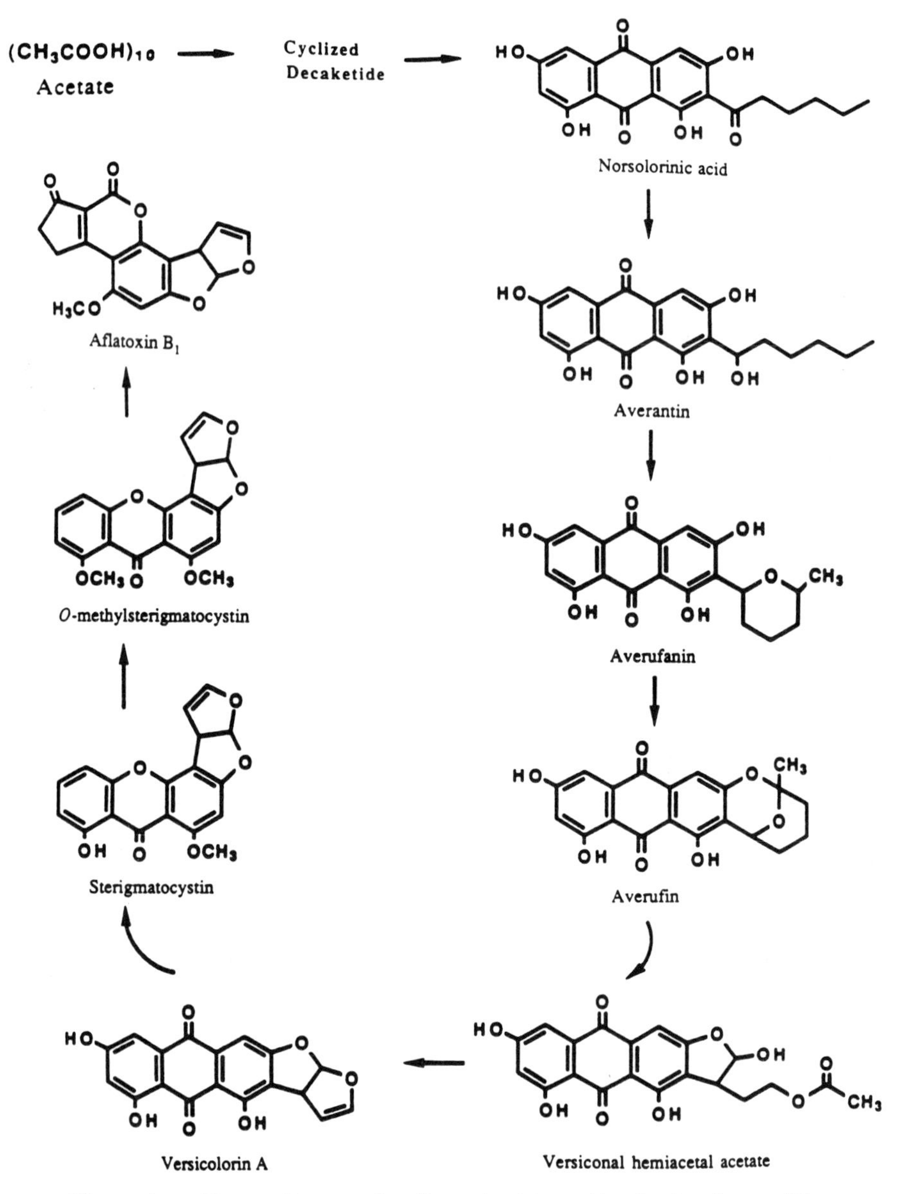

Figure 1. The pathway of aflatoxin biosynthesis in *Aspergillus parasiticus.*

MUTAGENICITY AND CARCINOGENICITY

Ames Salmonella typhymurium microsomal/mutagenicity assays on the intermediates of the aflatoxin pathway have revealed that only the compounds containing the dihydrobisfuran ring are mutagens (Wong et al., 1977). The structure required for the mutagenicity has been determined to be the vinyl ether double bond of the terminal furan ring. The derivatives of the bisfuranoid mycotoxins with this double bond saturated, such as AFB_2 , G_2, and versicolorin C, are not mutagenic in the Ames test.

The carcinogenicity of AFB_1 and ST to laboratory animals have been well established as mentioned earlier. VA has also been shown to be carcinogenic in rainbow trout (Hendricks et al., 1980).

METABOLIC ACTIVATION

The mechanism of toxic action of AFB_1 has been extensively studied. It is now established that AFB_1 is a procarcinogen requiring metabolic activation to covalently bind to DNA, RNA, and proteins to exert toxic effects (Swenson et al., 1975; Lin et al., 1977). The active form of AFB_1 has been identified as the 8,9-oxide of AFB_1, or AFB_1 epoxide (Essigmann et al., 1977; Lin et al., 1977). The requirement of metabolic activation for AFB_1 to modify DNA has been well demonstrated by the in vitro enzymatic and chemical syntheses of AFB_1-DNA adducts using the hepatic cytochrome P-450 associated monooxygenase and 3-chloroperbenzoic acid as activation agents, respectively (Lin et al., 1977; Martin and Garner, 1977). Using dimethyldioxirane as the oxidizing agent, the AFB_1-oxide has been synthesized, isolated, and characterized in vitro (Baertschi et al., 1988). In addition to epoxidation, the primary metabolism of AFB_1 also includes hydroxylation to AFM_1 and AFQ_1, demethylation to AFP_1, and reduction to AFL (Hsieh et al., 1977). These primary metabolites, whose structures are shown in Figure 2, all still contain the vinyl ether double bond in the dihydrobisfuran ring, and therefore can be epoxidated and activated to form mutagenic species (Wong and Hsieh, 1976). When a susceptible animal such as rat is given AFB_1, the hepatocellular DNA would be modified to a large extent by the epoxide of AFB_1 and to a small extent by the AFB_1 metabolites (Essigmann et al., 1982). The principal adduct, 8,9-dihydro-9-hydroxy-(N7-guanyl) AFB_1 (ADA), accounts for more than 80% of the total aflatoxin-DNA adducts.

However, the mutagenic potency of AFB_1 is much greater than that of its metabolites, indicating that all the primary metabolic pathways, except epoxidation, are detoxication pathways and that the coumarin structure in AFB_1 is the optimal structure for the biological activity of bisfuranoid compounds (Wong and Hsieh, 1976).

The appropriate enzymes that catalyze the formation of the various AFB_1 metabolites have been largely isolated and characterized. The epoxidase that activates AFB_1 is localized both in the endoplasmic reticulum and the nuclear membrane (Ueno et al., 1983; Shimada et al., 1987). The principal enzyme involved in the bioactivation of aflatoxins in human liver has been identified as cytochrome $P\text{-}450_{NF}$, the nifedipine oxidase (Shimada et al., 1989).

METABOLIC FATES OF AFB_1-OXIDE

The AFB_1 epoxide is an electrophile that readily reacts with the nucleophiles in DNA, RNA, protein, and reduced glutathione. The covalent binding of AFB_1 epoxide with DNA has been extensively characterized. Due to steric restrictions of the DNA backbone structure, AFB_1-oxide only forms adducts with guanines in DNA at the N-7 position (Loechler et al., 1988). Under physiological conditions, the epoxide can be enzymatically or spontaneously hydrolyzed to form a diol which in turn can undergo ring opening to form a dialdehyde intermediate to form Schiff bases with the primary amino groups of proteins (Sabbioni et al., 1987). The plasma albumin adducts of AFB_1 have been isolated and characterized recently as a potential marker of molecular dosimetry of AFB_1 in humans (Gan et al., 1988).

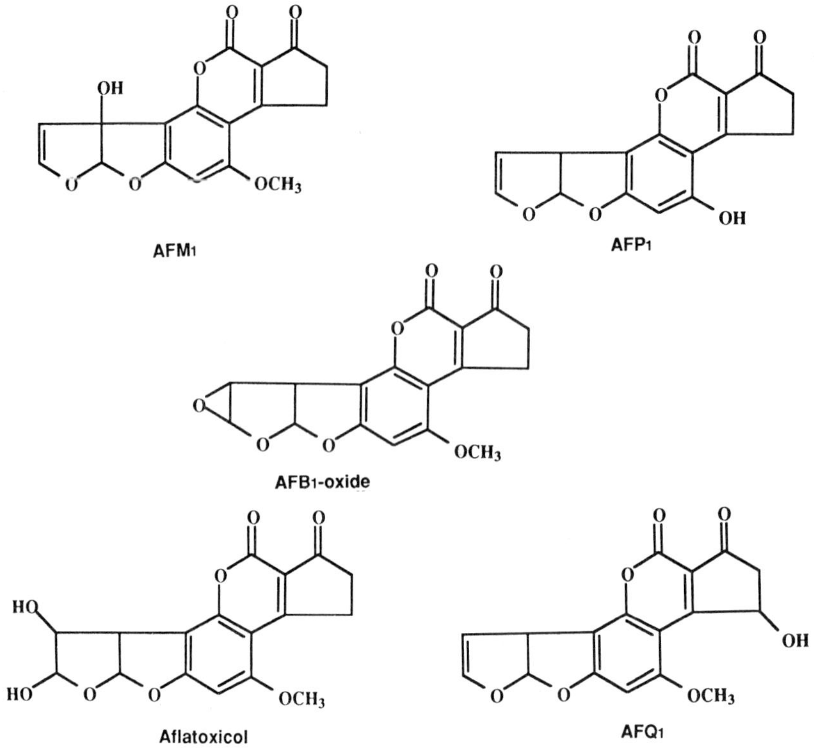

AFM1 AFP1

AFB1-oxide

Aflatoxicol AFQ1

Figure 2. Primary metabolites of AFB_1

FATES AND CONSEQUENCES OF AFB_1 ADDUCTS

There are three major metabolic fates for the AFB_1-guanyl-DNA adducts (ADA) in the rat liver. First, they undergo normal excision repair to have the lesion on the DNA removed. Second, the adduct may be removed to leave an apurinic site on the DNA as a lesion. And third, the adduct may cause the opening of the imidazole ring to form a formamidopyrimidyl derivative (AFB_1-FAPy-DNA), which upon acid hydrolysis yields 8,9-dihydro-8-(N5-formyl-2',5',6'-triamino-4-oxo-N5-pyrimidyl)-9-hydroxy AFB_1 (AFB_1-FAPy) (Lin et al., 1977). The structures of AFB_1-guanyl-DNA and AFB_1-FAPy-DNA adducts are shown in Figure 3.

The apurinic site tends to pair preferentially with adenine, resulting in a guanine to thymine transversion, a significant point mutation (Foster et al., 1983; Schaaper et al., 1983). The FAPy adducts, being resistant to repair and removal of the adduct, becomes a persistent lesion in the modified DNA that may possibly represent a non-informational site, also preferentially paired by adenine during DNA replication (Wogan, 1989). It may possibly also impair DNA replication and chromosomal aberrations. The FAPy adducts in RNA may conceivably impair transcription and translation processes and alter gene expression.

The adducts formed with RNA and critical proteins are obvious chemical lesions involved in the acute toxicity of AFB_1. Adduct formation with structural or other non-critical proteins, on the other hand, may reduce the toxicity by sequestering the toxin.

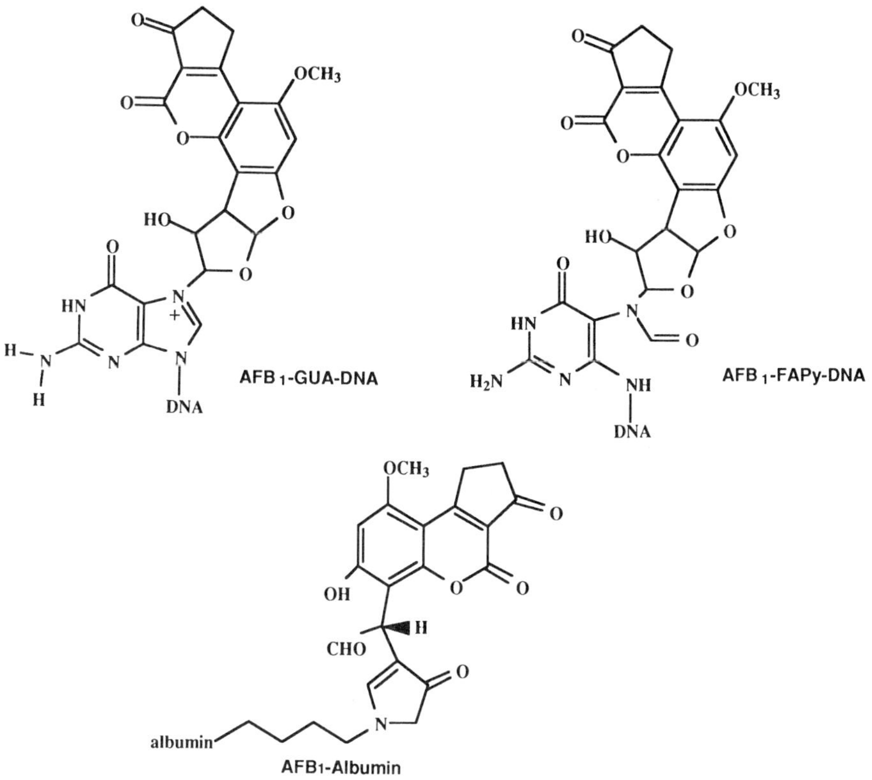

Figure 3. The structures of AFB_1-DNA adducts

DNA MODIFICATION AND CARCINOGENICITY

In the liver of a rat that has received a single dose of AFB_1, the primary ADA in the liver is formed rapidly after dosing and then disappears exponentially with time, with part of it being transformed into AFB_1-FAPy-DNA (Croy and Wogan. 1981). The latter therefore would increase with time after dosing and then persist in the liver. When animals are exposed continuously to AFB_1 at different doses, the FAPy adducts will reach different steady state levels after a few weeks of exposure (Buss and Lutz, 1988; Lutz, 1987). It has been observed that the steady state levels of the FAPy adducts are linearly correlated with the incidence of liver tumors in a similar manner in the rat and rainbow trout (Bechtel, 1989). The level corresponding to 100% tumor incidence is around 2 adducts per 10^6 nucleotides.

Using immunoassay techniques, the AFB_1-FAPy-DNA adducts have been found in most of the liver specimens taken from primary liver cancer patients in Czechoslovakia (Garner et al., 1988), Taiwan (Hsieh et al., 1988; Takayama et al., 1989), and Japan (Takayama et al., 1989). Seven out of 8 liver specimens from Czechoslovakia, 24 out of 26 liver specimens from Taiwan, and 15 out of 17 liver specimens from Japan were found to contain FAPy adducts at the levels ranging from 0.2 to 3.6 adducts per million nucleotides. The consistent occurrence of the AFB_1-DNA adducts in the liver specimens taken from liver cancer patients suggests that these patients had been definitely exposed to AFB_1 and that AFB_1-FAPy-DNA adducts are indeed useful biomarkers for assessing the human exposure to AFB_1.

MOLECULAR SITES OF DNA MODIFICATION

The exact role of AFB_1-FAPy adducts in hepatocarcinogenesis is not clear. But the occurrence of AFB_1 adducts in hepatic DNA is not evenly distributed. There are regions in rat liver DNA that are especially susceptible to attack by AFB_1-oxide. Examples include rat liver mitochondrial DNA (Niranjan et al., 1982), ribosomal RNA gene sequences of rat liver DNA (Irvin and Wogan, 1984), and transcriptionally active regions of rat liver nucleolar chromatin (Yu, 1983). The susceptibility of these binding sites is most likely associated with their accessibility to the carcinogen due to less protection by histones. Other factors that may contribute to preferential binding are specific neighboring base composition and nucleotide sequence. Identification of "hot spots" of AFB_1 adduct formation in DNA has been an interesting subject of investigation. Limited studies have indicated that the preferred guanine to be attacked by the AFB_1-oxide is the one followed by another guanine or by cytosine in the nucleotide sequence (Modali and Yang, 1986; Yang et al., 1985).

Studies on the DNA binding of sterigmatocystin have revealed that it is also activated by epoxidation at the vinyl ether double bond of the dihydrobisfuran ring and forms N-7 guanine and FAPy adducts with DNA (Reddy et al., 1985). It is likely that the mode of action of all bisfuranoid compounds is similar to that of AFB_1 and that they are all genetic carcinogens. The difference in the relative mutagenic potency of various bisfuranoid mycotoxins makes this family of compounds useful in the study of quantitative structure-activity relationships.

SUMMARY

Based on the mode of action of AFB_1 and the activities of its biologically active intermediates, one may conclude that:

1. The mode of toxic action of the bisfuranoid mycotoxin is through epoxidation of the vinyl ether double bond of their dihydrobisfuran functionality.

2. The DNA and plasma albumin adducts formed *in vivo* may be useful in the molecular dosimetry of these environmental carcinogens.

3. There appears to be a linear correlation between the steady state levels of AFB_1-FAPy-DNA adducts and the carcinogenicity of AFB_1. Elucidation of the molecular basis of this correlation may shed light on the mechanism of AFB_1-induced carcinogenesis.

4. Consistent appearance of AFB_1-DNA adducts in the livers of liver cancer patients tested is supportive of the IARC conclusion that AFB_1 is a human carcinogen involved in human primary liver cancer.

ACKNOWLEDGMENT

The authors' studies on the genotoxicity and carcinogenicity of bisfuranoid mycotoxins have been supported by NIEHS training grant ES07059 and the Western Regional Research Project on Food Safety, W-122.

REFERENCES

Baertschi, S. W., Raney, K. D., Stone, M. P., and Harris, T. M. (1988). Preparation of the 8,9-epoxide of the mycotoxin aflatoxin B_1: the ultimate carcinogenic species. *J. Am. Chem. Soc.* **110,** 7929-7931.

Beasley, R. P., and Hwang, L. (1984). Epidemiology of hepatocellular carcinoma. In *Viral Hepatitis and Liver Disease* (G. N. Vyas, J. L. Dienstag, and J. H. Hoffnagle, eds.), pp. 209-224. New York: Grune & Stratton.

Bechtel, D. H. (1989). Molecular dosimetry of hepatic aflatoxin B_1-DNA adducts:

linear correlation with hepatic cancer risk. *Regul. Toxicol. Pharmacol.* **10**, 74-81.
Buss, P. and Lutz, W. K. (1988). Steady-state DNA adduct level in rat liver after chronic exposure to low doses of aflatoxin B_1 and 2-acetylaminofluorine. Abstract. *Proc. Amer. Assoc. Cancer Res.* **29**, 380.
Cole, R. J., and Cox, R. H. (1981). Handbook of Toxic Fungal Metabolites. New York: Academic Press.
Croy, R. G. and Wogan, G. N. (1981). Temporal patterns of covalent DNA adducts in rat liver after single and multiple doses of aflatoxin B_1. *Cancer Res.* **41**, 197-203.
Essigmann, J. M., Croy, R. G., Bennett, R. A., and Wogan, G. N. (1982). Metabolic activation of aflatoxin B_1: patterns of DNA adduct formation, removal, and excretions in relation to carcinogenesis. *Drug Metab. Rev.* **13**, 581-602.
Essigmann, J. M., Croy, R. G., Nadzan, A. M., Busby, W. F., Reinhold, V. N., Buchi, G., and Wogan, G. N. (1977). Structural identification of the major DNA adduct formed by aflatoxin B_1 *in vitro*. *Proc. Natl. Acad. Sci. USA* **74**, 1870-1874.
Foster, P. L., Eisenstadt, E., and Miller, J. H. (1983). Base substitution mutations induced by metabolically activated AFB_1. *Proc. Natl. Acad. Sci. USA* **80**, 2695-2698.
Gan, S., Skipper, P. L., Peng, X., Groopman, J. D., Chen, J., Wogan, G. N., and Tannenbaum, S. R. (1988). Serum albumin adducts in the molecular epidemiology of aflatoxin carcinogenesis: correlation with aflatoxin B_1 intake and urinary excretion of aflatoxin M_1. *Carcinogenesis* **9**, 1323-1325.
Garner, R. C., Dvorackova, I., and Tursi, F. (1988). Immunoassay procedures to detect exposure to aflatoxin B_1 and benzo(a)pyrene in animals and man at the DNA level. *Int. Arch. Occup. Environ. Health* **60**, 145-150.
Hendricks J.D., Sinnhuber R.O., Wales J.H., Stack M.E., and Hsieh D.P.H. (1980). Hepatocarcinogenicity of sterigmatocystin and versicolorin A to rainbow trout (Salmo gairdneri) embryos. *J. Natl. Cancer Inst.* **64**, 1503-1509.
Hsieh, D. P. H., Wan, C. C., and Billington, J. A. (1989). A versiconal hemiacetal acetate converting enzyme in aflatoxin biosynthesis. *Mycopathologia* **107**, 121-126.
Hsieh, D. P. H., Wong, Z. A., Wong, J. J., Michas, C., and Ruebner, B. H. (1977). Comparative metabolism of aflatoxin. In *Mycotoxins in Human and Animal Health* (J. V. Rodricks, C. W. Hesseltine, and M. A. Mehlman, Ed.). pp. 37-50. Park Forest South, IL: Pathotox Publishers.
Hsieh, L., Hsu, S., Chen, D., and Santella, R. M. (1988). Immunological detection of aflatoxin B_1-DNA adducts formed *in vivo*. *Cancer Res.* **48**, 6328-6331.
IARC. (1987a). IARC monographs on the evaluation of carcinogenic risks to humans - overall evaluations of carcinogenicity: an updating of IARC Monographs volumes to 42. *Suppl.* **7**, 56-75.
IARC. (1987b). IARC monographs on the evaluation of carcinogenic risks to humans - overall evaluations of carcinogenicity: an updating of IARC Monographs volumes 1 to 42. *Suppl.* **7**, 82-87.
Irvin, T. R., and Wogan, G. N. (1984). Quantitation of aflatoxin B_1 adduction within the ribosomal RNA gene sequences of rat liver DNA. *Proc. Natl. Acad. Sci. USA* **81**, 664-668.
Lin, J., Miller, J. A., and Miller, E. C. (1977). 2, 3-Dihydro-2-(guan-7-yl)-3-hydroxy-aflatoxin B_1, a major acid hydrolysis product of aflatoxin B_1-DNA or -ribosomal RNA adducts formed in hepatic microsome-mediated reactions and in rat liver *in vivo*. *Cancer Res.* **37**, 4430-4438.
Loechler, E. L., Teeter, M. M., and Whitlow, M. D. (1988). Mapping the binding site of aflatoxin B_1 in DNA: molecular modeling of the binding sites for the N(7)-guanine adduct of aflatoxin B_1 in different DNA sequences. *J. Biomol. Struct. Dyn.* **5**, 1237-1257.
Lutz, W. K. (1987). Quantitative evaluation of DNA-binding data in vivo for low-dose extrapolations. *Arch. Toxicol. Suppl.* **11**, 66-74.
Martin, C. N., and Garner, R. C. (1977). Aflatoxin B-oxide generated by chemical or enzymic oxidation of aflatoxin B_1 causes guanine substitution in nucleic acids. *Nature* **267**, 863-865.
Modali, R. and Yang, S. S. (1986). Specificity of aflatoxin B_1 binding on human proto-

oncogene nucleotide sequence. In *Monitoring of Occupational Genotoxicants* (M. Sorsa and H. Norppa, ed.), pp. 147-158. Alan R. Liss.

Niranjan, B. G., Bhat, N. K., and Avadhani, N. G. (1982). Preferential attack of mitochondrial DNA by aflatoxin B_1 during hepato-carcinogenesis. *Science* **215**, 73-75.

Peers, F., Bosch, X., Kaldor, J., Linsell, A., and Pluumen, M. (1987). Aflatoxin exposure, hepatitis B virus infection and liver cancer in Swaziland. *Int. J. Cancer* **39**, 545-553.

Pohland, A. E. and Wood, G. E. (1987). Occurrence of mycotoxins in food. In *Mycotoxins in Food* (P. Krogh, ed.), pp. 35-64. London: Academic Press.

Reddy, M. B., Irvin, T. R., and Randerath, K. (1985). Formation and persistence of sterigmatocystin-DNA adducts in rat liver determined via 32P-postlabeling analysis. *Mutat. Res.* **152**, 85-96.

Sabbioni, G., Skipper, P. L., Buchi, G., and Tannenbaum, S. R. (1987). Isolation and characterization of the major serum albumin adduct formed by aflatoxin B_1 *in vivo* in rats. *Carcinogenesis* **8**, 819-824.

Schaaper, R. M., Kunkel, T. A., and Loeb, L. A. (1983). Infidelity of DNA synthesis associated with bypass of apurinic sites. *Proc. Natl. Acad. Sci. USA* **80**, 487-491.

Shimada, T., and Guengerich, F. P. (1989). Evidence for cytochrome $P\text{-}450_{NF}$, the nifedipine oxidase, being the principal enzyme involved in the bioactivation of aflatoxins in human liver. *Proc. Natl. Acad. Sci. USA* **86**, 462-465.

Shimada, T., Nakamura, S., Imaoka, S., and Funae, Y. (1987). Genotoxic and mutagenic activation of aflatoxin B_1 by constitutive forms of cytochrome P-450 in rat liver microsomes. *Toxicol. Appl. Pharmacol.* **91**, 13-21.

Sun, T. and Chu, Y. (1984). Carcinogenesis and prevention strategy of liver cancer in areas of prevalence. *J. Cell Physiol. Suppl.* **3**, 39-44.

Swenson D.H., Miller J.A., and Miller E.C. (1975). The reactivity and carcinogenicity of aflatoxin B_1-2,3-dichloride, a model for the putative 2,3-oxide metabolite of aflatoxin B_1. *Cancer Res.* **35**, 3811-3823.

Takayama, K., Wakabayashi, K. L., Hsieh, D. P. H., Sugimura, T., and Nagao, M. (1989). Detection of aflatoxin-DNA adducts in liver of humans in Taiwan and Japan. Abstract. The 1989 International Chemical Congress of Pacific Basin Societies, Honolulu, Hawaii, December, 1989.

Ueno Y., Ishii K., Omata Y., Kamataki T., and Kato R. (1983). Specificity of hepatic cytochrome P-450 isoenzymes from PCB-treated rats and participation of cytochrome b5 in the activation of aflatoxin B_1. *Carcinogenesis* **4**, 1071-1077.

Van Rensburg, S. J., Cook-Mozaffari, P., Van Schakkwyk, D. J., Van Der Watt, J. J., Vincent, T. J., and Purchase, I. F. (1985). Hepatocellular carcinoma and dietary aflatoxin in Mozambique and Transkei. *Br. J. Cancer* **51**, 713-726.

Wogan, G. N. (1989). Molecular and cellular events associated with aflatoxin-induced hepatocarcinogenesis. *Pure Appl. Chem.* **61**, 1-6.

Wong J.J. and Hsieh D.P.H. (1976). Mutagenicity of aflatoxins related to their metabolism and carcinogenic potential. *Proc. Natl. Acad. Sci. USA* **73**, 2241-2244.

Wong J.J., Singh R., and Hsieh D.P.H. (1977). Mutagenicity of fungal metabolites related to aflatoxin biosynthesis. *Mutat. Res.* **44**, 447-450.

Yang, S. S., Taub, J. V., Modali, R., Vieira, W., Yasei, P., and Yang, G. C. (1985). Dose dependency of aflatoxin B_1 binding on human high molecular weight DNA in the activation of Proto-Oncogene. *Environ. Health Perspect.* **62**, 231-238.

Yu, F. (1983). Preferential binding of aflatoxin B_1 to the transcriptionally active regions of rat liver nucleolar chromatin *in vivo* and *in vitro*. *Carcinogenesis* **4**, 889-893.

COVALENT BONDING OF BAY-REGION DIOL EPOXIDES TO NUCLEIC ACIDS

Donald M. Jerina*, Anju Chadha*, Albert M. Cheh*, Mark E. Schurdak+, Alexander W. Wood+ and Jane M. Sayer*

Laboratory of Bioorganic Chemistry* and Department of Oncology+
The National Institutes of Health
NIDDK, Bethesda, MD 20892, and
Roche Research Center
Nutley, NJ 07110

INTRODUCTION

Fifteen years ago, we proposed bay-region diol epoxides as the reactive metabolites responsible for the tumorigenic activity of polycyclic aromatic hydrocarbons (Jerina and Daly, 1976; Jerina et al., 1976; Jerina and Lehr, 1977). In the intervening period, extensive studies from several laboratories have provided substantial evidence that over a dozen hydrocarbons are activated by this pathway (Jerina et al., 1984; Lehr et al., 1985; Thakker et al., 1985). Examples have been forthcoming from a variety of structural types including benz- and dibenzanthracenes and acridines, benz- and dibenzpyrenes, chrysenes, and benzo[c]phenanthrene. To date, there are no known examples of carcinogenic alternant polycyclic aromatic hydrocarbons for which a bay-region diol epoxide is not an ultimate carcinogen. Although a number of mechanistically attractive alternatives to bay-region diol epoxides exist (Watabe et al., 1989; Cavalieri and Rogan, 1985; Miller et al., this volume), the extent of their contribution to the overall carcinogenicity of the hydrocarbons remains to be established.

In mammals, the hydrocarbons are metabolized to an enantiomeric pair of trans benzo-ring (R,R)- and (S,S)-dihydrodiols, each of which is further epoxidized at either face of the bay-region double bond to a pair of diastereomerically related bay-region diol epoxides. Thus enantiomeric pairs of diol epoxide-1 (DE-1, in which the benzylic hydroxyl group and epoxide oxygen are cis) and diol epoxide-2 diastereomers (DE-2, in which these groups are trans) are produced (Figure 1). In rat liver microsomes, particularly those from 3-methylcholanthrene-treated animals, this process is associated with a very high degree of stereoselectivity such that the (R,S,S,R) DE-2 enantiomer greatly predominates due to high stereoselectivity of cytochrome P450c (P450IA1) and regioselectivity of epoxide hydrolase (Jerina et al., 1985). As a result of direct chemical synthesis, it became possible to evaluate the relative mutagenic and tumorigenic properties of these diol epoxides, either as pairs of racemic diastereomers or as sets of all four optical antipodes. Two very important conclusions have emerged from these studies. First, conformation and absolute configuration are major determinants in the expression of tumorigenic activity by bay-region diol epoxides. For example, only the DE-2 diastereomers from benzo[a]pyrene, benz[a]anthracene and chrysene elicit strong tumorigenic responses while their DE-1 diastereomers are weak or inactive by comparison (Jerina et al., 1984). These differences in activity have been related to altered conformation for the two types of diastereomers. The DE-1

Biological Reactive Intermediates IV, Edited by C.M. Witmer *et al.*
Plenum Press, New York, 1990

diastereomers show a preference for the conformation in which the hydroxyl groups are axial whereas the much more tumorigenic DE-2 diastereomers prefer to have their hydroxyl groups equatorial. Alteration of these conformational preferences has been shown to modulate tumorigenic response. For example, introduction of a 6-fluoro group into benzo[a]pyrene DE-2 causes the diastereomer to prefer the conformation with axial hydroxyl groups. Although there is little change in solvolytic reactivity due to this substitution (Yagi et al., 1987), the fluoro analog lacked tumorigenic activity (Chang et al., 1987). The reverse situation, high tumorigenic response from a DE-1 diastereomer that has a preference for the conformation with equatorial hydroxyl groups, has also been demonstrated for the hydrocarbon benzo[c]phenanthrene. Due to severe crowding in the bay region ("fjord region") of this hydrocarbon, both diastereomers prefer the conformation with equatorial hydroxyl groups, and both diastereomers are highly tumorigenic (Levin et al., 1980). In terms of absolute configuration, studies of five hydrocarbons have shown that the (R,S,S,R) DE-2 diastereomer of the four possible isomers has from most to nearly all of the tumorigenic activity for each set (Jerina et al., 1986). These results are suggestive that an epoxide with a reactive (R)-benzylic center is important to the expression of tumorigenicity. Second, attempts to correlate chemical reactivity, either by rates of hydrolysis or by extent of covalent bonding to purified calf thymus DNA versus tetraol hydrolysis products with tumorigenicity have been unsuccessful (Jerina et al., 1986). This lack of correlation between chemical reactivity and tumorigenic activity may be more a reflection of attempts to correlate the wrong chemical reactions with biological response than it is the actual absence of a correlation. In summary, it is clear that a workable chemical model for the reaction(s) that leads to cell transformation has yet to be identified and that either the initial chemical damage or some subsequent cellular processing of this damage is subject to remarkably rigid stereochemical control.

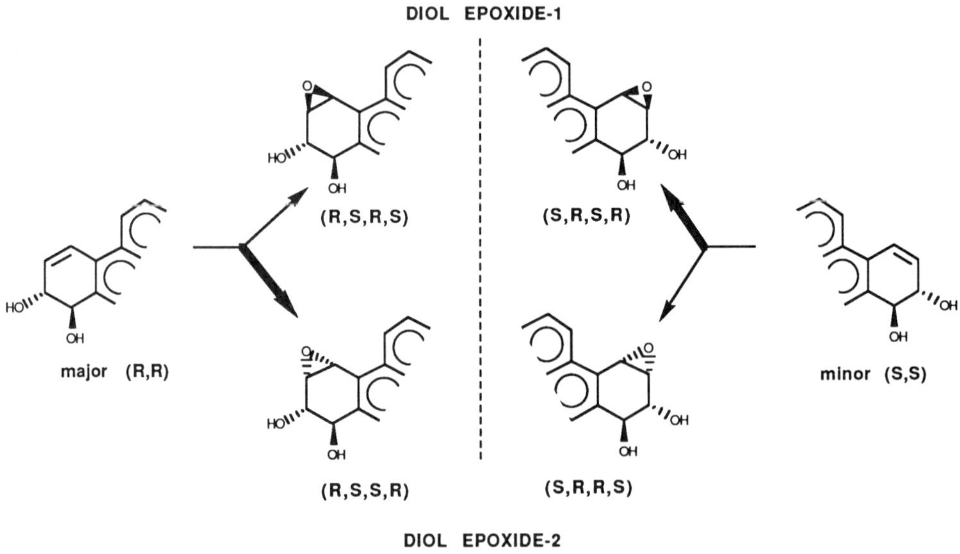

Figure 1. Generalized structures of the metabolically formed bay-region diol epoxides. With liver microsomes from 3-methylcholanthrene-treated rats, the (R,S,S,R) isomer (absolute configurations specified from the benzylic hydroxyl carbon around the ring to the benzylic epoxide carbon) greatly predominates due to the stereospecificity of cytochrome P450c and the regiospecificity of epoxide hydrolase.

Although transformation of a cell from a normal to a neoplastic state may well be the result of an accumulation of several genetic and epigenetic events, it is generally observed that most chemical carcinogens are also mutagens, and that their mutagenic activity is related to their ability to induce neoplasia. Thus these closely related sets of four isomeric bay-region diol epoxides represent unique probes, the use of which has the potential for dissection of the mutagenic from the tumorigenic response since all four isomers from a given hydrocarbon are potent mutagens in mammalian and/or bacterial test systems, but only the (R,S,S,R)-enantiomer has high tumorigenic activity. Current efforts have been directed toward establishing a data base of the types of adducts which are formed when these sets of four isomers react with purified DNA in aqueous solution. Based on studies with bay-region diol epoxides from five hydrocarbons, it is now clear that the exocyclic amino groups of the purine bases are the predominant targets for alkylation, although other very minor adducts at other sites have been detected with specific optically active diol epoxide isomers.

MECHANISM OF DNA ADDUCT FORMATION

The epoxide group of bay-region diol epoxides is a potentially reactive electrophilic group which can undergo either unimolecular ring cleavage, to give a resonance-stabilized benzylic carbocation, or direct attack at the benzylic carbon by solvent or other nucleophiles. Epoxide ring opening in the solvolyses of diol epoxides occurs by both an acid catalyzed (k_H) and a pH-independent (k_o) pathway (Whalen et al., 1977; Whalen et al., 1978; Sayer et al., 1981). The products of diol epoxide solvolysis in aqueous media are two tetraols derived from attack of water on the benzylic carbon atom cis and trans to the oxirane oxygen. Under conditions where the k_o pathway predominates, a keto diol that results from a 1,2-hydride migration to the benzylic carbon can also be formed from DE-1 isomers (Yagi et al., 1977; Whalen et al, 1977; Whalen et al., 1978). As a consequence of these two mechanistic pathways, the pH-rate profiles for diol epoxide solvolyses are biphasic, and follow the rate law shown below where k_{obsd} is the observed pseudo first order rate constant for diol epoxide disappearance at any given pH. At low pH values, where the k_H term dominates, the

$$k_{obsd} = k_H[H^+] + k_o$$

observed rate constant is directly proportional to the hydronium ion activity, whereas at pH values above neutrality, k_o dominates and k_{obsd} becomes independent of pH. For either diastereomer, a fairly good correlation exists between log k_o and the ease of carbocation formation as predicted by the quantity $\Delta E_{deloc}/\beta$ derived from Hückel molecular orbital calculations. The steep slopes of these linear free energy relationships (estimated to be 60-90% of the maximal value expected for full carbocation formation) indicate that substantial positive charge is developed at carbon in the transition state(s) for the k_o reaction (Sayer et al., 1982a). This observation may be consistent with either an S_N1 mechanism in which a carbocation intermediate is formed or an S_N2 mechanism in which there is extensive benzylic C-O bond breaking but little bond formation to the incoming solvent molecule. Recent evidence (Islam et al., 1990) strongly suggests that benzo[a]pyrene DE-1 and DE-2 solvolyze by different mechanisms. In the case of DE-1, rate determining epoxide ring opening gives a carbocation intermediate that can be trapped by nucleophiles in a subsequent fast (in some cases probably diffusion-controlled) step. However, the formation of such an intermediate from DE-2 could not be demonstrated, and the predominant mechanistic pathway for the k_o solvolysis of this diol epoxide may involve nucleophilic attack by water.

Solvolysis reactions of diol epoxides are subject to general acid catalysis by buffers and other acidic species (Whalen et al., 1979; Whalen et al., 1990). In addition, a number of molecules with aromatic or heterocyclic ring systems that are capable of complexation with the aromatic hydrocarbon moiety also exhibit markedly enhanced reactivity with diol epoxides, either as catalysts for hydrolysis or as nucleophilic trapping agents that form covalent adducts at the benzylic carbon. At a concentration of 0.1 mM, riboflavin-5'-phosphate, whose phosphate group acts as a general acid catalyst for the hydrolysis of benzo[a]pyrene DE-2, accelerates the hydrolysis of this diol epoxide to tetraols at pH 7 by a factor of >10, whereas ribose-5-phosphate, a compound of comparable pK_a that lacks the isoalloxazine ring moiety, requires an approximately 200-fold higher concentration to produce a similar catalytic effect (Wood et al., 1982). Similarly, rate constants for the catalysis of solvolysis of this diol epoxide by nucleoside monophosphates are 5-50 times larger than the corresponding rate constants for ribose-5-phosphate (Gupta et al., 1987). Ellagic acid, a tetracyclic tetraphenol, as its dianion, reacts more than 3000 times faster than phenol itself with benzo[a]pyrene DE-2 to give an adduct derived from attack of one of its phenolic hydroxyl groups at the benzylic C-10 of the diol epoxide (Sayer et al., 1982b). It should be noted that there are several mechanisms by which complexation may relate to reactivity or catalytic effectiveness in these systems. 1) Complexation can enforce physical proximity between the epoxide substrate and a catalytic group (such as a phosphate in riboflavin-5'-phosphate) or a reactant group (such as a hydroxyl in ellagic acid) thus providing an entropic advantage for reaction in dilute solution. 2) The complex itself can either stabilize or destabilize the positively charged transition state for epoxide ring opening (e.g., by charge-transfer interactions), depending on whether the complexing agent is an electron donor or acceptor. 3) In some complexes, solvolysis could be retarded if complexation in a hydrophobic environment sequesters the diol epoxide from solvent molecules.

The reaction of diol epoxides with DNA is a multistep process which involves initial, rapid physical interaction between the nucleic acid and the diol epoxide to form a complex that is analogous to those observed or postulated for small molecules (Geacintov, 1988). This complexation, which occurs with a time constant on the order of 5 msec or less (Geacintov et al., 1981), is followed by slower chemical reactions that lead to tetraol hydrolysis products as well as covalently bound diol epoxide-DNA adducts. For all diol epoxides studied to date, with the exception of those derived from benzo[c]phenanthrene, the predominant reaction of the complex involves DNA-catalyzed hydrolysis to tetraols, with stable covalent adducts comprising only a minor fraction (2-30%) of the total product yield. The simplest model for such a reaction process is given in Figure 2. Evidence for physical complex formation between the diol epoxides and DNA is provided both by spectroscopic shifts seen upon complexation and by the observation of saturation kinetics (equation below), such that at high concentrations of DNA, the observed pseudo first order rate constant, k_{obsd}, for the

$$k_{obsd} = (k_u + k_{cat}K_e[DNA])/(1 + K_e[DNA])$$

slow step at a given pH approaches a value, k_{cat}, that is independent of DNA concentration and corresponds to the total rate of disappearance of the complexed diol epoxide (k_t + k_c). This type of kinetic behavior is general for those diol epoxides whose reactions with DNA have been investigated (cf. Table 1), and is also observed for reactions of diol epoxides with several polynucleotides (Michaud et al., 1983; Islam et al., 1987; Geacintov et al., 1988).

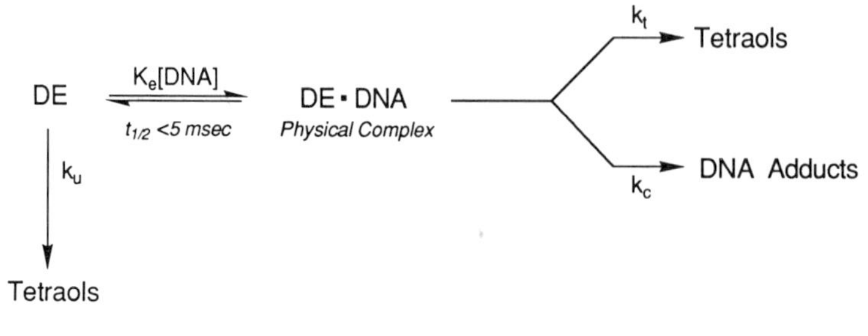

$$k_{obsd} = \frac{k_u + (k_c + k_t)K_e[DNA]}{1 + K_e[DNA]}$$

Figure 2. Reaction mechanism of diol epoxides with DNA.

Table 1. Selected Kinetic and Equilibrium Constants for Reactions of Racemic Diol Epoxides and Related Compounds with Calf Thymus DNA at pH 7.0, 23-25°C[a]

Compd	$K_e(M^{-1})$	$10^3\ k_u$ (s^{-1})	$10^2\ k_{cat}$ (s^{-1})	f_{cov}	References
			[NaCl] = 0		
BaP DE-2	12000[b] 14000[d]	0.65	6.7[b] 45[d]	0.1[b,c]	Geacintov et al., 1982b; 1984; Michaud et al., 1983
5-MeChr DE-2[e]	2810	0.87	0.87	0.07	Kim et al., 1985
3-MC DE-2[b]	4600	0.7	3.9	0.22	Kim et al., 1986
BA 1,2-DE-2[b]	4000	0.17	2.0	0.23	Carberry et al., 1988
BA 1,2-DE-1[b]	600	1.1	1.0	0.08	Carberry et al., 1988
BA 10,11-DE-2[b]	850	0.01	0.5	0.13	Carberry et al., 1989
BA 10,11-DE-1[b]	400	0.02	0.2	0.03	Carberry et al., 1989
			[NaCl] = 0.1 M		
BaP DE-2	4000[e] 2400[d]	3.4	1.5[e] 1.0[d]	0.08[e]	Kim et al., 1984 Michaud et al., 1983
BaP DE-1[f]	1070	~10	3.6	g	Islam et al., 1987
1-OP[e]	4000	0.46	0.58	0.08	Kim et al., 1984
5-MeChr DE-2	450	0.2	0.18	0.06	Kim et al., 1985

[a] Abbreviations are: BaP DE, benzo[a]pyrene 7,8-diol 9,10-epoxide; 5-MeChr DE, 5-methylchrysene 1,2-diol 3,4-epoxide; BA 1,2-DE, benz[a]anthracene 3,4-diol 1,2-epoxide; BA 10,11-DE, benz[a]anthracene 8,9-diol 10,11-epoxide; 3-MC DE, 3-methylcholanthrene 9,10-diol 7,8-epoxide; 1-OP, 1-oxiranylpyrene.

[b] In 5 mM sodium cacodylate buffer.

[c] Upon reaction with salmon DNA, which has the same (42% G + C) composition as calf thymus DNA, the (+)- and (-)-enantiomers of BaP DE-2 exhibit rate and equilibrium constants that are identical to each other, but different values of f_{cov}: f_{cov} [(+)-enantiomer], 0.14; f_{cov} [(-)-enantiomer], 0.03 (MacLeod and Zachary, 1985b). These values are consistent with the extents of covalent bonding to DNA observed by us under different conditions (see following section).

[d] In 1 mM sodium cacodylate buffer.

[e] In 5 mM sodium phosphate buffer.

[f] In 0.2 mM buffers.

[g] The fraction of covalent adduct formation determined by us under different conditions (37°C, 10% acetonitrile, 10 mM Tris buffer, pH 7.4) is 0.04-0.05 (see following section).

These reactions can be viewed as proceeding via competitive pathways, namely, solvolysis of the free diol epoxide in solution (k_u) to give tetraols (and in some cases ketone), and reaction of diol epoxide physically bound to DNA to give tetraols and covalent adducts. The observed distribution of products is a consequence of this competition. Mathematical expressions for the product distribution as a function of DNA concentration have been derived by Geacintov (1986) for several mechanisms analogous to that shown in Figure 2. At low DNA concentrations the k_u pathway is competitive with the DNA catalyzed processes, and both pathways contribute to product formation. Thus, the fraction of diol epoxide converted to covalent adducts, f_{cov}, under these conditions increases with increasing DNA concentration and is proportional to K_e[DNA]. At higher DNA concentrations f_{cov} approaches a constant value that corresponds to $k_c/(k_c + k_t)$ and is characteristic of the reaction of physically bound diol epoxide. The DNA concentration at which this occurs depends on both the equilibrium constant K_e and the rate enhancement, k_{cat}/k_u, upon noncovalent binding of the diol epoxide to DNA, such that the limiting value of f_{cov} will be observed when the DNA concentration is $>>(k_u/k_{cat})(1/K_e)$. For typical K_e values of 10^3 -10^4 M, and a rate enhancement of tenfold or greater for the bound diol epoxide (Table 1), the half maximal value for f_{cov} should be reached at or below a DNA concentration of 0.1 Mm, or 0.03 mg/ml, assuming an average molecular weight of 300 per nucleotide unit. Thus, at DNA concentrations of 0.8 mg/ml, as used in our bonding studies (see below), the observed efficiencies of covalent bonding relative to solvolysis should correspond to their limiting values.

Observed values of K_e and k_{cat} at constant pH are sensitive to ionic strength, and decrease markedly at high salt concentrations. These salt effects are probably a combined result of conformational factors such as changes in the tightness of the DNA helix as well as charge neutralization of phosphodiester groups. The latter effect should diminish the electrostatic stabilization of the positively charged transition state for epoxide ring opening by the polyanionic phosphodiester backbone. The pH dependence of k_{cat} for benzo[a]pyrene diol epoxides indicates that this macroscopic constant contains pH-independent and hydronium ion-dependent terms that are analogous to k_H and k_o for the uncomplexed diol epoxides in aqueous solution (Michaud et al., 1983; Geacintov et al., 1984; Islam et al., 1987). Hydronium ion dependence of k_{cat} may result either from specific hydronium ion catalysis of reaction of the DNA-bound diol epoxide or from general acid catalysis involving the small fraction of the phosphodiester groups that are protonated at pH values near neutrality. The suggestion (MacLeod and Zachary, 1985a) that protonated exocyclic amino groups of deoxyguanosine may be involved catalytically seems less likely, based on the very low microscopic pk_a values (ca.-2 to -3) expected for such groups (Abrams and Kallen, 1976), which would make the fraction of these groups that are protonated at neutrality even smaller than the corresponding fraction of protonated phosphodiesters. The fraction of complexed diol epoxide that is converted to covalent adducts is independent of pH between pH 7 and 9.5 for benzo[a]pyrene DE-2, indicative of similar pH dependencies for both product-forming pathways (k_c and k_t) that comprise k_{cat}. Similarly, this partitioning ratio exhibits little sensitivity to ionic strength over a range of 0 to 1.5 M NaCl (Geacintov et al., 1984).

DE-2 isomers appear to be somewhat better noncovalent binders to DNA (K_e) than their DE-1 diastereomers (Table 1), as well as to partition more efficiently (f_{cov}) to adducts relative to tetraols. Within the group of compounds studied, diol epoxides and analogs containing a pyrene moiety exhibit the highest values of K_e, but there is no obvious correlation between structures of the parent hydrocarbons and the fraction of DNA-bound diol epoxide that is eventually converted to covalent adducts. For DE-2 diastereomers derived from different hydrocarbons, rates (k_{cat}) of DNA catalyzed diol epoxide disappearance follow the same order as those observed under the same conditions in the absence of DNA (k_u). Solvolytic reactions of DE-1 isomers exhibit substantially greater contributions of uncatalyzed, relative to acid-catalyzed reaction pathways at pH values near neutrality (see for example, Whalen et al., 1977). Thus, the higher reactivity of benzo[a]pyrene DE-1 relative to DE-2 in both the k_u and k_{cat} processes results from the large contribution (ca. 90% of k_{cat} at pH 7) of pH-independent processes to these rates.

The mechanism shown in Figure 2 represents a gross oversimplification of the actual, microscopic processes that occur upon physical interaction and covalent bonding of the diol epoxides to DNA. Spectroscopic investigations (Geacintov, 1988) have suggested that the physical process represented by the equilibrium constant K_e corresponds predominantly to the formation of species (Site I complexes) that resemble intercalative complexes. For example, substantial (~10 nm) red shifts in the UV spectra of these complexes are observed, suggestive of stacking between the hydrocarbon and the nucleic acid bases. Linear dichroism (LD) measurements have been used to estimate the angles between the axes of the hydrocarbon chromophore and the helix axis of DNA when the DNA is oriented in an electric field or a flow system; negative values of the LD parameter ΔA, observed in virtually all physical complexes of PAH epoxide derivatives with DNA that have been investigated to date, are characteristic of large values ($> 55°$) of this angle and consistent with intercalation.

Upon covalent adduct formation, the conformations of the initial complexes appear to undergo substantial reorganization in some but not all cases (Geacintov, 1988; Gräslund and Jernström, 1989; Roche et al., 1989). Shifts to positive LD values and/or considerable diminution of the red shifts of the hydrocarbon chromophore occur in the major species formed upon covalent bonding to DNA of benzo[a]pyrene (+)-(R,S,S,R) DE-2, as well as racemic benz[a]anthracene 3,4-diol 1,2-epoxide-2 (Carberry et al., 1988) and 5-methylchrysene 1,2-diol 3,4-epoxide. These spectral features are consistent with orientation of the hydrocarbon chromophore at an external site (Site II) in the major or minor groove rather than intercalated in the DNA. It has been pointed out that Site II adducts could represent a variety of possible conformations in addition to location in a groove, including structures in which the DNA is bent or distorted at the adduct bonding site (Hogan et al., 1981).

In the case of a number of other bay-region diol epoxides and related compounds, including benz[a]anthracene 3,4-diol 1,2-epoxide-1, benzo[a]pyrene 7,8-diol 9,10-epoxide-1, the 7,8,9,10-tetrahydro-9,10-epoxide of benzo[a]pyrene, 1-oxiranylpyrene, the 9,10,11,12-tetrahydro 9,10-epoxide of benzo[e]pyrene, and benzo[e]pyrene 7,8-diol 9,10-epoxide-2, all as the racemic compounds, observed negative LD spectra are most consistent with a predominance of Site I covalent adducts, which bear a closer resemblance to the corresponding noncovalent complexes, although small proportions of Site II adducts may be formed as well (Geacintov, 1988; Gräslund and Jernström, 1989; Carberry et al., 1988). Unlike its (+)-(R,S,S,R)-enantiomer, benzo[a]pyrene (-)-(S,R,R,S)-DE-2 also appears to form both Site I and Site II adducts, with Site I adducts predominating. Values for the angle between the DNA and hydrocarbon axes that are large but less than 90°, as determined by LD studies, as well as only a small red shift in many of these Site I covalent adducts, make their detailed structural assignments ambiguous (Gagliano et al., 1982; Geacintov et al., 1982a; Kim et al., 1984). Local distortion of the DNA structure is possible, and true intercalation of the hydrocarbon chromophore into the DNA is probably not involved. In any event, the classification of adducts into two types is somewhat arbitrary, since even within a general type (Site I or Site II), substantial heterogeneity may exist (Eriksson et al., 1988). For example, the covalent DNA adducts of benzo[a]pyrene (R,S,S,R) DE-2, which are thought to be predominantly of the Site II type, comprise at least two components with different lifetimes for fluorescence decay (Jernström et al., 1984). Characterization of the adducts at the nucleoside level (see below) has shown that several chemically distinct products may arise from a given optically pure diol epoxide, namely, those derived upon cis or trans attack of the exocyclic amino group of adenine or guanine upon the benzylic carbon atom of the epoxide. This fact, in addition to the sequence heterogeneity of natural DNA molecules, provides an obvious rationale for the apparent spectral heterogeneity of many covalent DNA-diol epoxide adducts.

Based on the foregoing considerations, a somewhat more realistic but still oversimplified version of the molecular events that occur upon reaction of a diol epoxide with DNA is presented in Figure 3, which resembles the "two-site models" described by Meehan and Bond (1984) and Geacintov (1985). Such a scheme raises the obvious question of whether one or both of the chemical reactions that lead to adducts and tetraols occur(s) from the predominant Site I physical complex or from a minor, perhaps undetectable, fraction of complexes that more closely resemble non-intercalated or groove-bound species. It has recently been pointed out by Geacintov (1988) and by MacLeod (1990) that, so long as intercalated and externally complexed diol epoxide molecules are in *rapid equilibrium* with each other, the mathematical form of the rate equation will be identical to that given above, except that the experimental rate and equilibrium constants will be functions of several microscopic constants. Likewise, the observed similarity (Geacintov et al., 1984) of effects of varying reaction conditions such as temperature, ionic strength and pH on the partitioning of DNA-diol epoxide complexes to tetraols and adducts, although consistent with similar transition-state structures for these two chemical processes, does not necessarily require that they occur at the same noncovalent binding site. Thus, kinetics do not readily provide an answer to this question.

STRUCTURES OF DIOL EPOXIDE-DNA ADDUCTS

We have recently undertaken a program to identify and characterize, at the nucleoside level, the major adducts isolated after covalent modification of DNA by polycyclic aromatic diol epoxides, and to develop a structural and spectroscopic data base for these adducts (Jerina et al., 1988; Chadha et al., 1990). Initial impetus for this program was provided by the startling observation that, upon reaction with DNA, diol epoxides derived from benzo[c]phenanthrene partitioned exceptionally efficiently to covalent adducts relative to tetraols: in the presence of 0.8 mg/ml (ca. 2.5 mM) DNA, 55-75% of the total diol epoxide was converted to covalent adducts (Jerina et al., 1986). This contrasts with the values of f_{cov} shown in Table 1 for several other diol epoxides under different conditions, as well as with observations made by us with other diol epoxides (see Figure 6 later). Diol epoxides derived from 5- methylchrysene (Reardon et al., 1987), which, like the benzo[c]phenanthrene derivatives, possess a highly hindered bay region, also exhibit high efficiencies of covalent bonding to DNA, although not so high as the benzo[c]phenanthrene derivatives, and preliminary studies also indicate similar behavior for the fjord-region benzo[g]chrysene diol epoxides.

In typical experiments, DNA adducts are formed by reaction of the appropriate optically active or racemic diol epoxide (dissolved in acetonitrile, acetone or tetrahydrofuran) with 10 volumes of calf thymus DNA (0.8 mg/ml in Tris/HCl buffer, pH 7.0-7.4) at 37°C. Upon completion of reaction, solvolysis products of the diol epoxide are extracted with organic solvent, and the modified DNA in the aqueous solution is then subjected to enzymatic hydrolysis to the nucleoside level. Although

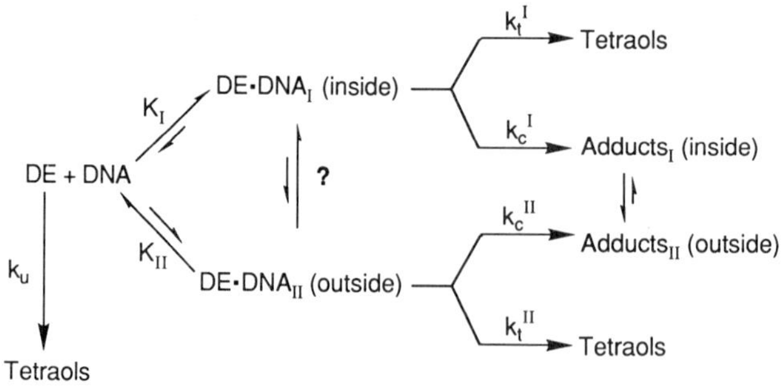

$$k_{obsd} = \frac{k_u + K_I(k^I_t + k^I_c)[DNA] + K_{II}(k^{II}_t + k^{II}_c)[DNA]}{1 + (K_I + K_{II})[DNA]}$$

Figure 3. Reaction mechanism for the formation of adducts and tetraols via multiple DNA complexes. Note that the mathematical form of the rate law will be unchanged, even if more than two nonequivalent complexes are involved.

separation of the multiple nucleoside adducts thus formed (typically two dG and two dA adducts from each optically active diol epoxide) is not always a trivial problem, and chromatographic methods must be developed individually for each parent hydrocarbon, the use of reverse phase HPLC has proven highly effective for the separation of these compounds (Dipple et al., 1987; Chadha et al., 1989).

Major isolated adducts derived from the parent hydrocarbons studied in detail thus far have involved nucleophilic attack of the exocyclic amino group of the purine bases at the benzylic carbon of the diol epoxides (Figure 4) (Agarwal et al., 1987; Reardon et al., 1987; Cheng et al., 1988a; Chadha et al., 1989). A dC adduct at the exocyclic amino group has also been identified from (-)- (S,R,R,S)-DE-2 of dibenz[a,j]anthracene (Chadha et al., 1989). Chemical characterization of these adducts has relied heavily on the physical techniques of circular dichroism (CD), ^{1}H NMR and mass spectrometry as well as the determination of pH-titration curves. Identification of dC vs dA adducts can sometimes be accomplished by UV spectra, especially when combined with the observation that the dA adducts elute later than, and are generally quite well separated from, the corresponding dG and dC adducts on reverse phase HPLC. Confirmation of the identity of the nucleoside base can be accomplished by chemical ionization mass spectrometry (NH_3 gas) using a direct exposure probe. In our experience the fully acetylated adducts afford $(M+1)^+$ peaks, whereas the unacetylated adducts are less suitable for mass spectrometry since they undergo loss of the sugar under the mass spectral conditions.

That the position of attachment of the diol epoxide moiety is at the exocyclic amino group of the base has been deduced by a combination of pH titration, ^{1}H NMR spectroscopy and chemical stability considerations. For the determination of titration curves, CD spectroscopy is particularly useful as a method for monitoring subtle structural changes induced by protonation of the bases in cases where little or no change in the UV absorption spectrum is observed because of the overpowering contribution of the hydrocarbon chromophore (Kasai et al., 1978). Typical pK_a values (in 1:9 methanol-water determined by this method for adducts of dG or G are 1.4-2.4 (for protonation of the 5-membered ring) and 9.1- 9.8 (for loss of H-1) (Moore et al., 1977; Kasai et al., 1978; Agarwal et al., 1987; Chadha et al., 1989). The observation of a dissociable proton on N-1 indicates a lack of substitution at N-1 or O-6, and is probably also inconsistent with substitution at N-3, which should lead to rearrangement of double bonds in the 6-membered ring and loss of the hydrogen at

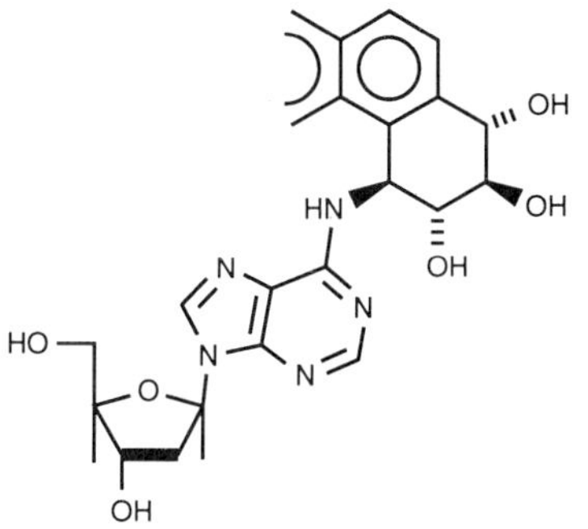

Figure 4. For the four optically active diol epoxides from a given hydrocarbon, sixteen major adducts are observed. These arise by both cis and trans opening of the epoxide groups at their benzylic positions by the exocyclic amino groups of the two purine bases. The structure shown illustrates trans opening of a (S,R,S,R) DE-1 isomer by dG. Note that the benzylic (R)-epoxide carbon of the diol epoxide has changed to an (S)-center on trans opening by the amino group. On cis opening, the absolute configuration at this benzylic center would remain unchanged.

N-1. Typical dA or A adducts exhibit a single pK_a at 2.2-2.8 (Agarwal et al., 1987; Chadha et al., 1989). This excludes substitution at N-1 (expected pK_a ca. 8.5). The single dC adduct whose structure has been determined has a pK_a of 2.6, which is inconsistent with substitution at N-3 or O-2 (expected pK_a values > 8). For purine adducts containing intact base and sugar rings, considerations of chemical stability exclude substitution at N-7 of dG or at N-7 or N-3 of dA, since facile ring opening or deglycosylation of such adducts would be expected under the conditions used for adduct formation and enzymatic hydrolysis of the DNA. Finally, the appropriate C-H protons of the purine and pyrimidine bases can be identified in NMR spectra; thus, substitution at a purine or pyrimidine carbon is ruled out.

Table 2. ^1H NMR Coupling Constants for the Methine Protons of the Tetrahydro Benzo-ring of Diol Epoxide Adducts (as the Peracetates) Formed by Alkylation of the Exocyclic Nitrogen of Purine Nucleosides[a]

Diol Epoxide	Stereochemistry of Addition		$J_{1,2}$	$J_{2,3}$ (Hertz)	$J_{3,4}$
DE-1	trans	(normal)	3.8-4.3	6.3-7.1	4.7-6.1
		(hindered bay)	2.5-3.5	2.8-3.5	7.8-8.2
DE-1	cis	(normal)	4.2-4.7	10.9-11.9	7.9-8.2
		(hindered bay)	3.2-3.5	8.5-10.2	2.0-3.0 (dA)
					4.3-4.6 (dG)
DE-2	trans	(normal)	2.0-4.0	1.9-2.8	8.8-9.4
		(hindered bay)	3.0-4.5	2.6-3.0	8.3-9.4
		(5-MeChr)	4.0-7(?)		
DE-2	cis	(normal)	5.2-5.7	2.1-2.7	2.8-4.0
		(hindered bay)	3.9-4.7	1.8-2.1	6.0-9.3

[a] In acetone-d_6. For consistency, ring positions are numbered in order, beginning with the N-substituted benzylic position as H-1, regardless of the numbering conventions used for individual hydrocarbons. Data for normal bay-region adducts are based upon the parent hydrocarbons dibenz[a,j]anthracene, benz[a]anthracene and benzo[a]pyrene. Data for hindered bay-region adducts are based upon the deoxyribonucleoside adducts from the parent hydrocarbons benzo[c]phenanthrene, 7,12-dimethylbenz[a]anthracene (cis-dA adducts of DE-1 only: cf. Cheng et al., 1988a) and 5-methylchrysene (DE-2 adducts only: Reardon et al., 1987). In the case of the cis DE-2 derivatives where a significant conformational difference exists between the bay- and fjord-region derivatives, the 5-MeChr derivatives exhibit intermediate values of $J_{3,4}$ (6-7 Hz).

Use of the fully acetylated derivatives in acetone-d_6 has proven to be the method of choice for NMR spectral studies because of the superior resolution and stability of the adducts in this solvent. NMR spectra in Me_2SO-d_6 of the unacetylated adducts derived from the benzo[c]phenanthrene diol epoxides were unsatisfactory because of extreme line broadening (Agarwal et al., 1987). For acetylated adducts of bay-region diol epoxides, coupling constants (Table 2) and chemical shifts (Table 3) of the methine protons on the tetrahydro benzo-ring exhibit patterns that are characteristic of the structure of the adduct and the stereochemistry (cis vs. trans) of opening of the epoxide, and are relatively insensitive to the nature of the aromatic moiety. In the absence of steric hindrance in the bay region, the following features of the NMR spectra are of particular diagnostic value. (1) For adducts derived from DE-1, large values (10.9-11.9 and ca. 8 Hz, respectively) for the coupling constants $J_{2,3}$ and $J_{3,4}$ (where H-1 represents the bay-region benzylic proton and H-4 the non-bay-region benzylic proton) are diagnostic of cis opening of the epoxide ring. In the corresponding trans adducts, these coupling constants are 6.3-7.1 and 4.7-6.1 Hz,

Table 3. ^{1}H NMR Chemical Shifts for the Methine Protons of the Tetrahydro Benzo-ring of Diol Epoxide Adducts (as the Peracetates) Formed by Alkylation of the Exocyclic Nitrogen of Purine Nucleosides[a]

Compd	Stereochemistry of Addition	δ_1	δ_2	δ_3	δ_4
DE-1/G	trans	6.49-6.64	5.80-5.93	5.58-5.63	6.32-6.68
	(BcPh)	6.22		5.17	
	cis	6.78-7.03	5.55-5.70	5.87-5.98	6.32-6.66
	(BcPh)		5.05		
DE-1/A	trans	6.66-7.05	5.72-5.89	5.55-5.63	6.41-6.79
	(BcPh)			5.25	6.87
	cis	7.12-7.35	5.65-5.74	6.21-6.27	6.32-6.59
	(BcPh)		4.85	5.85	6.20
	(DMBA)		5.13	5.92	6.13
DE-2/G	trans	6.12-6.52	5.84-6.21	5.73-5.88	6.50-6.81
	(BcPh)		5.90		
	(5-MeChr)		6.38 (?)		
	cis	6.80-6.93	5.83-6.05	5.61-5.72	6.31-6.66
	(BcPh)	7.0-7.3	6.14-6.30		6.71
DE-2/A	trans	6.60-6.95	5.92-6.09	5.95-6.03	6.52-6.81
	(BcPh)		6.25		6.43
	(5-MeChr)		6.25	5.80	6.50
	cis	7.02-7.23	5.80-5.99	5.63-5.76	6.31-6.67
	(BcPh)	7.40	6.10		6.71
	(5-MeChr)	7.16	6.12		6.59

[a] In acetone-d_6. For consistency, ring positions are numbered in order, beginning with the N-substituted benzylic position as H-1. Data for bay-region adducts are based on available chemical shifts for adducts derived from the parent hydrocarbons dibenz[a,j]anthracene, benz[a]anthracene and benzo[a]pyrene. Chemical shifts for the compounds with hindered bay regions, benzo[c]phenanthrene (BcPh), 7,12- dimethylbenz[a]anthracene (DMBA) and 5-methylchrysene (5-MeChr) diol epoxide derivatives, are given only when they differ significantly from the range observed for the less hindered bay-region derivatives.

respectively. (2) In the DE-2 series, adducts derived from cis ring opening exhibit coupling constants, $J_{3,4}$, of ca. 3 Hz, whereas for the corresponding trans adducts these coupling constants are much larger: ca. 9 Hz. These differences are a consequence of the conformations of the substituted cyclohexene ring. Thus, H-2, H-3 and H-4 are all pseudoaxial in the cis DE-1 adducts and pseudoequatorial in the trans DE-1 adducts. For DE-2, H-3 and H-4 of the cis adducts are pseudoequatorial, whereas these protons are pseudoaxial in the trans adducts.

Conformational differences between these bay-region adducts derived from benzo[a]pyrene (Cheng et al., 1989; Sayer et al., 1990), benz[a]anthracene or dibenz[a,j]anthracene (Chadha et al., 1989) and adducts with a highly hindered fjord region or methyl-substituted bay region result in markedly different coupling constants for the two classes of adducts (Table 2). For example, in the case of benzo[c]phenanthrene derivatives in the DE-1 series $J_{2,3}$ is large (8.5- 10.2 Hz) in the cis adducts and small (2.8-3.5 Hz) in the trans adducts, whereas $J_{3,4}$ is small in the cis adducts and large in the trans adducts. It is not possible to distinguish between the cis and trans DE-2 adducts from this parent hydrocarbon on the basis of coupling constants; however, an extreme downfield shift (ca. 0.7-1.0 ppm) for H-1, observed in the cis relative to the trans adducts, makes possible the assignment of stereochemistry in this case. Similar, but smaller, downfield shifts (cf. Table 3) are observed in the adducts that have less hindered bay regions.

Circular dichroism spectroscopy also provides an exceptionally useful tool for the assignment of stereochemistry to the diol epoxide-nucleoside adducts, and because of its sensitivity is of particular value in cases where only trace amounts of adducts are available. The intense CD spectra of the adducts derived from purine nucleosides result from exciton interactions between the purine and the polycyclic aromatic ring system. Thus, the signs of the CD bands are independent of the orientation of the non-chromophoric hydroxyl groups on the tetrahydro benzo-ring and the sugar, and depend only on the skew sense between the aromatic chromophores, which is determined by the absolute configuration at the N-substituted benzylic carbon of the diol epoxide moiety. In the absence of additional information concerning the conformations of the adducts, theoretical prediction of the signs of the CD bands is problematic, since the relative orientations of the electric dipole transition moments of the chromophoric

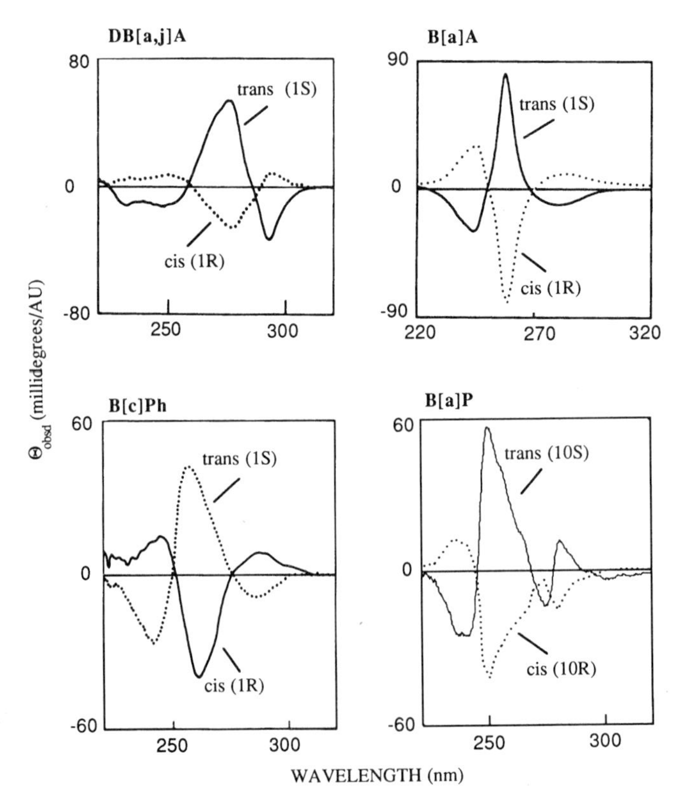

Figure 5. Circular dichroism spectra of cis/trans pairs of dG adducts derived from (R,S,S,R) diol epoxide-2 isomers of dibenz[a,j]anthracene, benz[a]anthracene, benzo[c]phenanthrene and benzo[a]pyrene.

groups preferred in these adducts are unknown. However, an empirical relationship has been established between observed CD spectra and the absolute configuration at the N-substituted benzylic carbon for adducts whose absolute and relative configurations are known independently from the absolute configuration of the parent diol epoxide together with NMR data for the adduct (Sayer et al., 1990). For example, in the case of N-2 dG adducts derived from the four parent hydrocarbons studied by us as well as from 5-methylchrysene (Reardon et al., 1987), the intense central CD band at 260-280 nm (Figure 5) is positive when the N-substituted benzylic carbon has S absolute configuration, and negative when this carbon has R absolute configuration. Thus, CD spectra permit the assignment of absolute configurations to minor adducts that cannot be obtained in quantities sufficient for NMR spectroscopy. If the absolute configuration of the parent diol epoxides is also known for such adducts, their relative stereochemistry (cis vs trans) of attack at the benzylic carbon of the epoxide can be assigned. An analogous correlation of CD spectra with absolute configuration at the N- substituted benzylic carbon also exists in the case of dA adducts; in this case the shorter-wavelength band of the two intense bands observed in the 250-280 nm region is positive for adducts with S absolute configuration and negative for adducts with R absolute configuration at this carbon (cf. Chadha et al., 1989; Cheng et al., 1988b; Vericat et al., 1989).

EXTENT AND DISTRIBUTION OF DIOL EPOXIDE-DNA ADDUCTS

Although as yet somewhat incomplete, a large body of data now exists on the relative tumorigenicity and the ability to bond covalently to DNA for the four optically active bay-region diol epoxides from five different hydrocarbons. A comparison of the efficiency with which these sets of four isomers bond to DNA relative to the extent to which they undergo solvolysis is shown in Figure 6. Notably, the (R,S,S,R) DE-2 isomer shows the highest percentage of bonding relative to solvolysis for each set. However, it is our view that the differences in extent of bonding between isomers within each set is not sufficiently large to account for their

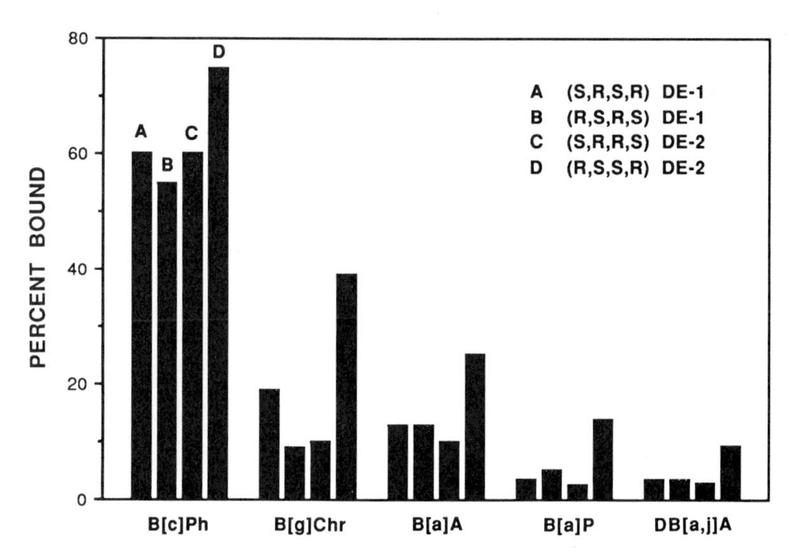

Figure 6. Distribution between tetraols and adducts on calf thymus DNA for the four optically active bay-region diol epoxides from benzo[c]phenanthrene, benzo[g]chrysene (11,12-diol-13,14-epoxide positional isomers only), benz[a]anthracene, benzo[a]pyrene and dibenz[a,j]anthracene.

much larger differences in tumorigenic response. Notably, the (R,S,R,S) DE-1 isomer of benzo[c]phenanthrene is very effective in its ability to bond to DNA yet is inactive as a carcinogen. Comparison of the sets of isomers between different hydrocarbons also fails to provide a correlation. Although the (R,S,S,R) DE-2 isomer of benzo[c]phenanthrene is the most potent known diol epoxide carcinogen and is the most efficient of these sixteen diol epoxides in its ability to covalently bond to DNA, the same isomer from benzo[a]pyrene is relatively inefficient in its ability to bond despite the fact that it is highly tumorigenic. In addition, the bonding efficiency of the benzo[a]pyrene diol epoxides is similar to that for the diol epoxides from dibenz[a,j]anthracene, yet the latter diol epoxides are thought to be weak or inactive as carcinogens. The percentage of bonding to dG relative to total dG + dA adducts (Figure 7) also fails to provide a clear correlation with tumorigenic response. For three of the hydrocarbons, each of their four diol epoxide isomers prefer bonding to dG, yet there are both weak and potent carcinogens within each set. Although all of the benzo[c]phenanthrene diol epoxide isomers exhibit preferential bonding to dA, there is no correlation with their tumorigenic activity. For example, the (R,S,R,S) DE-1 and (R,S,S,R) DE-2 isomers both bind extensively to dA yet the former is inactive as a carcinogen in two tumor models. Examination of the ratio of cis to trans opening of these diol epoxides by the exocyclic amino groups of dA and dG has also failed to provide a correlation with tumorigenic response. At present, the data obtained in studies of the reactions of these diol epoxides with purified DNA in aqueous solution indicate that the covalent bonding site(s) critical to cell transformation is obscured by noncritical bonding.

SEQUENCE SPECIFICITY OF COVALENT BONDING

In view of the above, it has become necessary to determine whether there is sequence specificity in the reactions of bay-region diol epoxides with DNA. Chemical activation of ras proto-oncogenes (Bowden, this volume: Anderson et al., this volume) whose altered gene products are found in high incidence in tumor cells, including those induced by polycyclic aromatic hydrocarbons (Bizub et al., 1986; You et al., 1989),

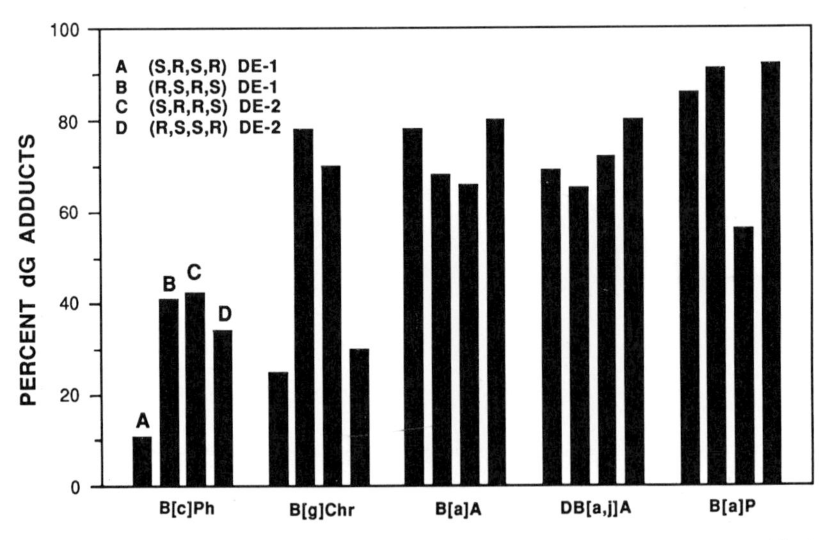

Figure 7. Percent dG adducts relative to total dG + dA adducts on calf thymus DNA for benzo[c]phenanthrene, benzo[g]chrysene (11,12-diol-13,14-epoxide positional isomers only), benz[a]anthracene, dibenz[a,j]anthracene and benzo[a]pyrene.

could be the result of a high degree of modification of specific codons within such genes. Since polycyclic aromatic hydrocarbons were known to cause mutations at adenine residues in the CAA codon 61 of the rodent H-ras oncogene, Reardon et al. (1989) sought to determine if this and codon 12 (GGA) represented "hot spots" for the chemical reaction of diol epoxides with DNA. The plasmid pAL-7, which contains rat c-H-ras sequences, was allowed to react to similar extents with optically active diol epoxides from benzo[c]phenanthrene and benzo[a]pyrene. The modified template was annealed with labeled primers adjacent to either codon 12 or 61 and replicated with Sequenase. Since Sequenase is partially blocked by the presence of bulky hydrocarbon adducts (cf. Reardon et al., 1990; Strauss et al., this volume), examining the products on denaturing polyacrylamide gels established points at which synthesis had stopped. The four optically active diol epoxides from benzo[c]phenanthrene, which all show a preference for dA residues in calf thymus DNA (Figure 7), showed reaction at each of the dA residues in codon 61, but these were not the most intense bands on the gel. Although the (R,S,S,R) DE-2 isomer of benzo[a]pyrene very effectively binds to dG relative to dA residues in calf thymus DNA (Figure 7), bands corresponding to binding at either dG residue in codon 12 were again not the most intense bands on the gel. Through the use of specific base labeling and ^{32}P-postlabeling analysis, Schurdak et al. (1990) have examined the base-specific bonding of the four optically active diol epoxides from benzo[a]pyrene and benzo[g]chrysene to double stranded 19-mers containing H-ras codons 12 and 61, respectively. For diol epoxides from both hydrocarbons, the (R,S,S,R) DE-2 enantiomer was the most efficient in bonding to the respective 19-mer as was the case with calf thymus DNA (Figure 6). In the case of the benzo[a]pyrene isomers, only the nontumorigenic (S,R,R,S) DE-2 isomer showed a preference for bonding to both guanines in codon 12 that was in excess of that expected for all of the guanines in the oligomer. For the benzo[g]chrysene diol epoxides, none of the isomers preferentially bound to the central dA of codon 61 (a known site of activation of H-ras by benzo[g]chrysene) relative to the other adenines in the sequence. Thus neither of these codons represent hot spots for bonding that correlate with tumorigenicity for the tested diol epoxides *in vitro*. Boles and Hogan (1986) developed a laser scission technique to map covalent bonding sites of racemic benzo[a]pyrene DE-2 on a DNA fragment containing the chicken β-globin gene. They observed preferential formation of adducts within runs of dG. Koostra, et al.(1989) applied the laser scission procedure and as well as a T7 RNA polymerase stop assay to determine the location of racemic benzo[a]pyrene DE-2 adducts in a DNA fragment containing the hamster aprt gene plus its flanking regions. Both methods indicated a preferred formation of adducts in runs of dG, including the upstream GC box consensus sequences that serve to activate aprt expression.

MUTAGENIC SPECIFICITY OF DIOL EPOXIDES

An alternative approach in the search for sequence specificity in the reactions of bay-region diol epoxides with DNA has been evaluation of the spectrum of mutagenic events which occur after DNA has been exposed to these metabolites. Deciphering the origin of chemically induced mutations is complex. The spectrum of observed mutations reflects a series of chemical and biochemical steps: initial selectivity on adduct formation, potential selective elimination of certain adducts by DNA repair systems, and differential responses that DNA polymerases might exhibit toward selected adducts. Deoxyadenosine, for example, is often inserted across from a bulky, non-instructional adduct. In addition, some mutations, in particular those occurring in the third bases of codons, may be silent and thus will not be scored. Although initial experiments were in bacteria (Eisenstadt et al., 1982), recent studies have focused on adducted DNA which has been processed in mammalian cells. Yang et al. (1987) have examined mutations produced in the supF gene, which codes for a tyrosine suppressor tRNA, following exposure to racemic benzo[a]pyrene DE-2. After replication in a human embryonic kidney cell line, the supF shuttle vector was rescued from the mammalian cells, amplified in a strain of E. coli, and sequenced. The mutational spectrum from racemic benzo[a]pyrene DE-2 at the dihydrofolate reductase (dhfr) locus in Chinese hamster ovary cells has also been evaluated with the aid of the polymerase chain reaction technique (Carothers and Grunberger, 1990). In all of the experiments, most of the mutations were the result of base substitutions via specific transversions of GC→TA. Since the (R,S,S,R) DE-2 enantiomer has a marked specificity toward guanine relative to adenine and since the (S,R,R,S) DE- 2 enantiomer is relatively unreactive toward DNA (cf. Figure 6 & 7), selective modification of guanine is easily rationalized. The *supF* gene has also been targeted with the optically active (R,S,S,R) enantiomer of benzo[c]phenanthrene DE-2 (Bigger et al., 1989). Since this diol epoxide shows a marked specificity (~2:1) toward adenine relative to guanine (Figure 7), a much higher percentage of point mutations involving adenine relative to guanine was anticipated and was found (~50%). In all of the studies involving racemic benzo[a]pyrene DE-2, the central bases of 5'-AGG-3' and of 5'-GGA-3' were major targets. For the (R,S,S,R) DE-2 enantiomer of benzo[c]phenanthrene DE-2, the central bases of 5'- AGA-3', 5'-AAC-3' and of 5'-GAG-3' were the most frequent sites of mutation within trinucleotide sequences. Although attempts to identify longer target nucleotide sequences common to all studies have not been rewarding, it does appear that purine rich runs of bases provide the best targets for the diol epoxides.

SUMMARY

Although the solution chemistry of diol epoxides is now fairly well understood, a great deal remains to be elucidated regarding their reaction in the presence of DNA. Not only DNA but also small molecules are capable of sequestering diol epoxides in aqueous solutions with equilibrium constants on the order of 10^2-10^4 M^{-1}. In the case of DNA, at least two major families of complexes are presently recognized, possibly the result of groove binding vs. intercalation. As is the case for diol epoxides free in solution, the complexed diol epoxides undergo solvolysis to tetraols and in some cases possibly to keto diols as well. Fractionation between covalent bonding and solvolysis from within the complex(s) is determined more by the nature of the parent hydrocarbon from which the diol epoxide is derived than any other factor. Studies of a wide variety of alkylating and arylating agents have show that practically every potentially nucleophilic site on DNA can serve as a target for modification. In the case of the diol epoxides, practically all of the modification occurs at the exocyclic amino groups of the purine bases. In contrast to the diol epoxides, other epoxides such as those derived from aflatoxin B_1, vinyl chloride, propylene, 9-vinylanthracene, and styrene preferentially bind to the aromatic ring nitrogens N-7 in guanine and N-3 in adenine (cf. Chadha et al., 1989). Molecular modeling as well as the spectroscopic evidence suggests that the hydrocarbon portion of the diol epoxides lies in the minor groove of DNA when bound to the exocyclic 2-amino group of guanine and in the major groove when bound to the exocyclic 6-amino group of adenine. Detailed conformational analysis of adducted DNA should prove to be extremely valuable in developing mechanistic models for the enzymatic processing of chemically altered DNA. At present, the critical lesion or lesions responsible for induction of neoplasia remains obscured by the large number of apparently noncritical adducts which form when polycyclic hydrocarbon diol epoxides bond to DNA.

REFERENCES

Abrams, W. R. and Kallen, R. G. (1976). Estimates of microscopic ionization constants for heteroaromatic exocyclic amines including purine and pyrimidine nucleotides and amides based upon a reactivity-basicity correlation for N-hydroxymethylation reactions with formaldehyde. *J. Am. Chem. Soc.* **98**, 7789-7792.

Anderson, M. W., You, M., Maronpot, R. R. and Reynolds, S. (1991). Proto-oncogene activation in rodent and human tumors. This volume.

Agarwal, S. K., Sayer, J. M., Yeh, H. J. C., Pannell, L. K., Hilton, B. D., Pigott, M. A., Dipple, A., Yagi, H. and Jerina, D. M. (1987). Chemical characterization of DNA adducts derived from the configurationally isomeric benzo[c]phenanthrene-3,4-diol 1,2-epoxides. *J. Am. Chem. Soc.* **109**, 2497-2504.

Bigger, C. A. H., Strandberg, J., Yagi, H., Jerina, D. M. and Dipple, A. (1989).

Mutagenic specificity of a potent carcinogen, benzo[c]phenanthrene (4R,3S)-diol (2S,1R)-epoxide, which reacts with adenine and guanine in DNA. *Proc. Natl. Acad. Sci USA.* **86**, 2291-2295.

Bizub, D., Wood, A. W. and Skalka, A. M. (1986). Mutagenesis of the Ha-ras oncogene in mouse skin tumors induced by polycyclic aromatic hydrocarbons. *Proc. Natl. Acad. Sci. USA* **83**, 6048-6052.

Boles, T. C. and Hogan, M. E. (1986). High-resolution maping of carcinogen binding sites on DNA. *Biochemistry* **25**, 3039-3043.

Bowden, G. T. (1991). Oncogene activation and differential gene expression during multistage skin carcinogenesis. This volume.

Carberry, S. E., Shahbaz, M., Geacintov, N. E. and Harvey, R. G. (1988). Reactions of stereoisomeric and structurally related bay region diol epoxide derivatives of benz[a]anthracene with DNA. Conformations of non-covalent complexes and covalent carcinogen-DNA adducts. *Chem.-Biol. Interactions* **66**, 121- 145.

Carberry, S. E., Geacintov, N. E. and Harvey, R. G. (1989). Reactions of stereoisomeric non-bay-region benz[a]anthracene diol epoxides with DNA and conformations of non-covalent complexes and covalent adducts. *Carcinogenesis* **10**, 97-103.

Carothers, A. M. and Grunberger, D. (1990). DNA base changes in benzo[a]pyrene diol epoxide-induced dihydrofolate reductase mutants of Chinese hamster ovary cells. *Carcinogenesis* **11**, 189-192.

Cavalieri, E. L. and Rogan, E. G. (1985). One-electron oxidation in aromatic hydrocarbon carcinogenesis. In *Polycyclic Hydrocarbons and Carcinogenesis* (ACS Symposium Series No. 283) (R. G. Harvey, Ed.), pp. 289-305. American Chemical Society, Washington, DC.

Chadha, A., Sayer, J. M., Yeh, H. J. C., Yagi, H., Cheh, A. M., Pannell, L. K. and Jerina, D. M. (1989). Structures of covalent nucleoside adducts formed from adenine, guanine and cytosine bases of DNA and the optically active bay-region 3,4-diol 1,2-epoxides derived from dibenz[a,j]anthracene. *J. Am. Chem. Soc.* **111**, 5456-5463.

Chadha, A., Sayer, J. M., Agarwal, S, K., Cheh, A. M., Yagi, H., Yeh, H. J. C. and Jerina, D. M. (1990). Formation of covalent adducts between DNA and optically active bay-region diol epoxides of dibenz[a,j]anthracene. In *Polynuclear Aromatic Hydrocarbons: Measurements, Means and Metabolism* (M. Cooke, K. Loening, and J. Merritt, Eds.), pp. 179-194. Battelle Press, Columbus, OH.

Chang, R. L., Wood, A. W., Conney, A. H., Yagi, H., Sayer, J. M., Thakker, D. R., Jerina, D. M. and Levin, W. (1987). Role of diaxial versus diequatorial hydroxyl groups in the tumorigenic activity of a benzo[a]pyrene bay-region diol epoxide. *Proc. Natl. Acad. Sci. USA* **84**, 8633-8636.

Cheng, S. C., Prakash, A. S., Pigott, M. A., Hilton, B. D., Roman, J. M., Lee, H., Harvey, R. G. and Dipple, A. (1988a). Characterization of 7,12-dimethylbenz[a]anthracene-adenine nucleoside adducts. *Chem. Res. Toxicol.* **1**, 216-221.

Cheng, S. C., Prakash, A. S., Pigott, M. A., Hilton, B. D., Lee, H., Harvey, R. G., and Dipple, A. (1988b). A metabolite of the carcinogen 7,12-dimethylbenz[a]anthracene that reacts predominantly with adenine residues in DNA. *Carcinogenesis* **9**, 1721-1723.

Cheng, S. C., Hilton, B. D., Roman, J. M., and Dipple, A. (1989). DNA adducts from carcinogenic and noncarcinogenic enantiomers of benzo[a]pyrene dihydrodiol epoxide. *Chem. Res. Toxicol.* **2**, 334-340.

Dipple, A., Pigott, M. A., Agarwal, S. K., Yagi, H., Sayer, J. M. and Jerina, D. M. (1987). Optically active benzo[c]phenanthrene diol epoxides bind extensively to adenine in DNA. *Nature* **327**, 535-536.

Eisenstadt, E., Warren, A. J., Porter, J., Atkins, D. and Miller, J. H. (1982). Carcinogenic epoxides of benzo[a]pyrene and cyclopenta[c,d]pyrene induce base substitutions via specific transversions. *Proc. Natl. Acad. Sci. USA* **79**, 1945-1949.

Eriksson, M., Nordén, B., Jernström, B. and Gräslund, A. (1988). Binding geometries of benzo[a]pyrene diol epoxide isomers covalently bound to DNA. Orientational distribution. *Biochemistry* **27**, 1213-1221.

Gagliano, A. G., Geacintov, N. E., Ibanez, V., Harvey, R. G. and Lee, H. M. (1982).

Application of fluorescence and linear dichroism techniques to the characterization of the covalent adducts derived from interaction of (±)-trans-9,10-dihydroxy-anti-11,12-epoxy-9,10,11,12-tetrahydro-benzo[e]pyrene with DNA. *Carcinogenesis* **3**, 969-976.

Geacintov, N. E. (1985). Mechanisms of interaction of polycyclic aromatic diol epoxides with DNA and structures of the adducts. In Polycyclic Hydrocarbons and Carcinogenesis (ACS Symposium Series No. 283) (R. G. Harvey, Ed.), pp. 107-124. American Chemical Society, Washington, DC.

Geacintov, N. E. (1986). Is intercalation a critical factor in the covalent binding of mutagenic and tumorigenic polycyclic aromatic diol epoxides to DNA? *Carcinogenesis* **7**, 759-766.

Geacintov, N. E. (1988). Mechanisms of reaction of polycyclic aromatic epoxide derivatives with nucleic acids. In *Polycyclic Aromatic Hydrocarbon Carcinogenesis: Structure-Activity Relationships*, Volume II (S. K. Yang and B. D. Silverman, Eds.), pp. 181-206. CRC Press, Boca Raton, FL.

Geacintov, N. E., Yoshida, H., Ibanez, V. and Harvey, R. G. (1981). Non-covalent intercalative binding of 7,8-dihydroxy-9,10-epoxy-benzo[a]pyrene to DNA. *Biochem. Biophys. Res. Commun.* **100**, 1569-1577.

Geacintov, N. E., Gagliano, A. G., Ibanez, V. and Harvey, R. G. (1982a). Spectroscopic characterizations and comparisons of the structures of the covalent adducts derived from the reactions of 7,8-dihydroxy-7,8,9,10- tetrahydrobenzo[a]pyrene-9,10-oxide, and the 9,10-epoxides of 7,8,9,10-tetrahydrobenzo[a]pyrene and 9,10,11,12-tetrahydrobenzo[e]pyrene with DNA. *Carcinogenesis* **3**, 247-253.

Geacintov, N. E., Yoshida, H., Ibanez, V. and Harvey, R. G. (1982b). Noncovalent binding of 7β,8α-dihydroxy-9α,10α-epoxytetrahydrobenzo[a]pyrene to deoxyribonucleic acid and its catalytic effect on the hydrolysis of the diol epoxide to tetrol. *Biochemistry* **21**, 1864-1869.

Geacintov, N. E., Hibshoosh, H., Ibanez, V., Benjamin, M. J. and Harvey, R. G. (1984). Mechanisms of reaction of benzo[a]pyrene-7,8-diol-9,10-epoxide with DNA in aqueous solutions. *Biophysical Chemistry* **20**, 121-133.

Geacintov, N. E., Shahbaz, M., Ibanez, V., Moussaoui, K., and Harvey, R. G. (1988). Base-sequence dependence of noncovalent complex formation and reactivity of benzo[a]pyrene diol epoxide with polynucleotides. *Biochemistry* **27**, 8380-8387.

Gräslund, A. and Jernström, B. (1989). DNA-carcinogen interaction: covalent DNA-adducts of benzo[a]pyrene 7,8-dihydrodiol 9,10-epoxides studied by biochemical and biophysical techniques. *Quart. Rev. Biophys.* **22**, 1-37.

Gupta, S. C., Islam, N. B., Whalen, D. L., Yagi, H. and Jerina, D. M. (1987). Bifunctional catalysis in the nucleotide-catalyzed hydrolysis of (±)-7β,8α-dihydroxy-9α,10α-epoxy-7,8,9,10-tetrahydrobenzo[a]pyrene. *J. Org. Chem.* **52**, 3812-3815.

Hogan, M. E., Dattagupta, N. and Whitlock, J. P., Jr. (1981). Carcinogen-induced alteration of DNA structure. *J. Biol. Chem.* **256**, 4504-4513.

Islam, N. B., Whalen, D. L., Yagi, H. and Jerina, D. M. (1987). pH Dependence of the mechanism of hydrolysis of benzo[a]pyrene-cis-7,8-diol 9,10-epoxide catalyzed by DNA, poly(G), and poly(A). *J. Am. Chem. Soc.* **109**, 2108-2111.

Islam, N. B., Whalen, D. L., Yagi, H. and Jerina, D. M. (1990). Differences in the mechanisms of the spontaneous hydrolysis reactions of bay region benzo[a]pyrene 7,8-diol 9,10-epoxides: Trapping of a carbocationic intermediate. *J. Am. Chem. Soc.* **112**, in press.

Jerina, D. M. and Daly, J. W. (1976). Oxidation at Carbon. In *Drug Metabolism - from Microbe to Man* (D. V. Parke and R. L. Smith, Eds.), pp. 13-32. Taylor and Francis Ltd., London, UK.

Jerina, D. M., Lehr, R. E., Yagi, H., Hernandez, O., Dansette, P. M., Wislocki, P. G., Wood, A. W., Chang, R. L., Levin, W. and Conney, A. H. (1976). Mutagenicity of benzo[a]pyrene derivatives and the description of a quantum mechanical model which predicts the ease of carbonium ion formation from diol epoxides. In *In Vitro Metabolic Activation In Mutagenesis Testing* (F. J. de Serres, J. R. Fouts, J. R. Bend, and R. M. Philpot, Eds.), pp. 159-177. Elsevier/North-Holland Biomedical Press, Amsterdam.

Jerina, D. M. and Lehr, R. E. (1977). The bay-region theory: A quantum mechanical approach to aromatic hydrocarbon-induced carcinogenicity. In *Microsomes and Drug Oxidations* (Proceedings of the 3rd International Symposium) (V. Ullrich, I.

Roots, A. G. Hildebrandt, R. W. Estabrook, and A. H. Conney, Eds.), pp. 709-720. Pergamon Press, Oxford, UK.

Jerina, D. M., Yagi, H., Thakker, D. R., Sayer, J. M., van Bladeren, P. J., Lehr, R. E., Whalen, D. L., Levin, W., Chang, R. L., Wood, A. W. and Conney, A. H. (1984). Identification of the ultimate carcinogenic metabolites of the polycyclic aromatic hydrocarbons: Bay-region (R,S)-diol-(S,R)-epoxides. In *Foreign Compound Metabolism* (J. Caldwell and G. D. Paulson, Eds.), pp. 257-266. Taylor and Francis Ltd., London, UK.

Jerina, D. M., Sayer, J. M., Yagi, H., van Bladeren, P. J., Thakker, D. R., Levin, W., Chang, R. L., Wood, A. W. and Conney, A. H. (1985). Stereoselective metabolism of polycyclic aromatic hydrocarbons to carcinogenic metabolites. In *Microsomes and Drug Oxidations* (A. R. Boobis, J. Caldwell, F. De Matteis, and C. R. Elcombe, Eds.), pp. 310-319. Taylor and Francis Ltd., London, UK.

Jerina, D. M., Sayer, J. M., Agarwal, S. K., Yagi, H., Levin, W., Wood, A. W., Conney, A. H., Pruess-Schwartz, D., Baird, W. M., Pigott, M. A. and Dipple, A. (1986). Reactivity and tumorigenicity of bay-region diol epoxides derived from polycyclic aromatic hydrocarbons. In *Biological Reactive Intermediates III.* (J. J. Kocsis, D. J. Jollow, C. M. Witmer, J. O. Nelson, and R. Snyder, Eds.), pp. 11-30. Plenum Press, New York.

Jerina, D. M., Cheh, A. M., Chadha, A., Yagi, H. and Sayer, J. M. (1988). Binding of metabolically formed bay-region diol epoxides to DNA. In *Microsomes and Drug Oxidations: Proceedings of the 7th International Symposium* (J. O. Miners, D. J. Birkett, R. Drew, B. K. May, and M. E. McManus, Eds.), pp. 354-362. Taylor and Francis, London, UK.

Jernström, B., Lycksell, P.-O., Gräslund, A. and Nordén, B. (1984). Spectroscopic studies of DNA complexes formed after reaction with anti- benzo[a]pyrene-7,8-dihydrodiol-9,10-oxide enantiomers of different carcinogenic potency. *Carcinogenesis* 5, 1129-1135.

Kasai, H., Nakanishi, K. and Traiman, S. (1978). Two micromethods for determining the linkage of adducts formed between polyaromatic hydrocarbons and nucleic acid bases. *J. Chem. Soc. Chem. Commun.*, 798-800.

Kim, M.-H., Geacintov, N. E., Pope, M. and Harvey, R. G. (1984). Structural effects in reactivity and adduct formation of polycyclic aromatic epoxide and diol epoxide derivatives with DNA: Comparisons between 1-oxiranylpyrene and benzo[a]pyrenediol epoxide. *Biochemistry* **23**, 5433-5439.

Kim, M.-H., Geacintov, N. E., Pope, M., Pataki, J. and Harvey, R. G. (1985). Reaction mechanisms of trans-1,2-dihydroxy-anti-3,4-epoxy-1,2,3,4-tetrahydro-5-methylchrysene with DNA in aqueous solutions. *Carcinogenesis* **6**, 121-126.

Kim, M.-H., Geacintov, N. E., McQuillen, D. G., Pope, M., and Harvey, R. G. (1986). Reaction mechanisms of trans-1,2-dihydroxy-anti-7,8-epoxy-7,8,9,10-tetrahydro-3-methylcholanthrene with DNA in aqueous solutions. Conformation of adducts. *Carcinogenesis* **7**, 41-47.

Koostra, A., Lew, L. K., Nairn, R. S. and MacLeod, M.C. (1989). Preferential modification of GC boxes by benzo[a]pyrene-7,8-diol-9,10-epoxide. *Mol. Carcinogenesis* 1, 239-244.

Lehr, R. E., Kumar, S., Levin, W., Wood, A. W., Chang, R. L., Conney, A. H., Yagi, H., Sayer, J. M. and Jerina, D. M. (1985). The bay-region theory of polycyclic aromatic hydrocarbon carcinogenesis. In *Polycyclic Hydrocarbons and Carcinogenesis* (ACS Symposium Series No. 283) (R. G. Harvey, Ed.), pp. 63-84. American Chemical Society, Washington, DC.

Levin, W., Wood, A. W., Chang, R. L., Ittah, Y., Croisy-Delcey, M. Yagi, H., Jerina, D. M. and Conney, A. H. (1980). Exceptionally high tumor-initiating activity of benzo[c]phenanthrene bay-region diol-epoxides on mouse skin. *Cancer Res.* **40**, 3910-3914.

MacLeod, M. C. (1990). The importance of intercalation in the covalent binding of benzo[a]pyrene diol epoxide to DNA. *J. Theor. Biol.* **142**, 113-132.

MacLeod, M. C. and Zachary, K. L. (1985a). Involvement of the exocyclic amino group of deoxyguanosine in DNA-catalysed carcinogen detoxification. *Carcinogenesis* **6**, 147-149.

MacLeod, M. C. and Zachary, K. (1985b). Catalysis of carcinogen-detoxification by DNA: comparison of enantiomeric diol epoxides. *Chem.-Biol. Interactions* **54**, 45-55.

Meehan, T. and Bond, D. M. (1984). Hydrolysis of benzo[a]pyrene diol epoxide and its covalent binding to DNA proceed through similar rate-determining steps. *Proc. Natl. Acad. Sci. USA* **81**, 2635-2639.

Michaud, D. P., Gupta, S. C., Whalen, D. L., Sayer, J. M. and Jerina , D. M. (1983). Effects of pH and salt concentration on the hydrolysis of a benzo[a]pyrene 7,8-diol-9,10-epoxide catalyzed by DNA and polyadenylic acid. *Chem.-Biol. Interactions* **44**, 41-52.

Miller, J. A., Surh, Y.-J. and Miller, E. C. (1991). Electrophylic sulfuric acid ester metabolites of hydroxymethyl aromatic hydrocarbons as precursors of hepatic benzylic DNA adducts *in vivo*. This volume.

Moore, P. D., Koreeda, M., Wislocki, P. G., Levin, W., Conney, A. H., Yagi, H. and Jerina, D. M. (1977). *In vitro* reactions of the diastereomeric 9,10-epoxides of (+)- and (-)-trans-7,8-dihydroxy-7,8-dihydrobenzo[a]pyrene with polyguanylic acid and evidence for formation of an enantiomer of each diastereomeric 9,10-epoxide from benzo[a]pyrene in mouse skin. In *Drug Metabolism Concepts* (ACS Symposium Series No. 44) (D. M. Jerina, Ed.), pp. 127-154. American Chemical Society, Washington, DC.

Reardon, D.B., Prakash, A. S., Hilton, B. D., Roman, J. M., Pataki, J., Harvey, R. G. and Dipple, A. (1987). Characterization of 5-methylchrysene-1,2-dihydrodiol-3,4-epoxide-DNA adducts. *Carcinogenesis* **8**, 1317-1322.

Reardon, D. B., Bigger, C. A., Strandberg, J., Yagi, H., Jerina, D. M. and Dipple, A. (1989). Sequence selectivity in the reaction of optically active hydrocarbon dihydrodiol epoxides with H-ras DNA. *Chem. Res. Toxicol.* **2**, 12-14.

Reardon, D. B,. Bigger, C. A. and Dipple, A. (1990). DNA polymerase action on bulky deoxyguanosine and deoxyadenosine adducts. Chem. Res. Toxicol. **3**, 165-168.

Roche, C. J., Geacintov, N. E., Ibanez, V., and Harvey, R. G. (1989). Linear dichroism properties and orientations of different ultraviolet transition moments of benzo[a]pyrene derivatives bound noncovalently and covalently to DNA. *Biophysical Chemistry* **33**, 277-288.

Sayer, J. M., Yagi, H., Croisy-Delcey, M. and Jerina, D. M. (1981) Novel bay-region diol epoxides from benzo[c]phenanthrene. *J. Am. Chem. Soc.* **103**, 4970-4972.

Sayer, J. M., Lehr, R. E., Whalen, D. L., Yagi, H. and Jerina, D. M. (1982a). Structure-reactivity indices for the hydrolysis of diol epoxides of polycyclic aromatic hydrocarbons. *Tetrahedron Lett.* **23**, 4431-4434.

Sayer, J. M., Yagi, H., Wood, A. W., Conney, A. H. and Jerina, D. M. (1982b). Extremely facile reaction between the ultimate carcinogen benzo[a]pyrene-7,8-diol 9,10-epoxide and ellagic acid. *J. Am. Chem. Soc.* **104**, 5562-5564.

Sayer, J. M., Chadha, A., Agarwal, S. K., Yeh, H. J. C., Yagi, H., and Jerina, D. M. (1990). Covalent nucleoside adducts of benzo[a]pyrene 7,8-diol 9,10- epoxides: Structural reinvestigation and characterization of a novel adenosine adduct on the ribose moiety. *J. Org. Chem.* **55**, in press.

Strauss, B., Turkington, E., Sahm, J. and Wang, J. (1991). Mutagenic consequences of the alteration of DNA by chemicals and radiation. This volume.

Schurdak M. E., Bekesi, E., Jerina, D. M., Yagi, H., Bushman, D. R., Lehr, R. E. and Wood, A. W. (1990). *Proc. Am. Assoc. Cancer Res.* **31**, 90.

Thakker, D. R., Yagi, H., Levin, W., Wood, A. W., Conney, A. H. and Jerina, D. M. (1985). Polycyclic aromatic hydrocarbons: Metabolic activation to ultimate carcinogens. In *Bioactivation of Foreign Compounds* (M. W. Anders, Ed.), pp. 177- 242. Academic Press, New York.

Vericat, J. A., Cheng, S. C., and Dipple, A. (1989). Absolute stereochemistry of the major 7,12-dimethylbenz[a]anthracene-DNA adducts formed in mouse cells. *Carcinogenesis* **10**, 567-570.

Watabe, T., Ogura, K., Okuda, H., and Hiratsuka, A. (1989). Hydroxymethyl sulfate esters as reactive metabolites of the carcinogens 7-methyl- and 7,12-dimethylbenz[a]anthracenes and 5-methylchrysene. In *Xenobiotic Metabolism and Disposition* (R. Kato, R. W. Estabrook, and M. N. Cayen, Eds.), pp. 393-400. Taylor and Francis, London, UK.

Whalen, D. L., Montemarano, J. A., Thakker, D. R., Yagi, H. and Jerina, D. M. (1977). Changes of mechanism and product distributions in the hydrolysis of benzo[a]pyrene 7,8-diol 9,10-epoxide metabolites induced by changes in pH. *J. Am. Chem. Soc.* **99**, 5522-5524.

Whalen, D. L., Ross, A. M., Yagi, H., Karle, J. M. and Jerina, D. M. (1978).

Stereoelectronic factors in the solvolysis of bay region diol epoxides of polycyclic aromatic hydrocarbons. *J. Am. Chem. Soc.* **100**, 5218-5221.

Whalen, D. L., Ross, A. M., Montemarano, J. A., Thakker, D. R., Yagi, H. and Jerina, D. M. (1979). General acid catalysis in the hydrolysis of benzo[a]pyrene 7,8-diol 9,10-epoxides. *J. Am. Chem. Soc.* **101**, 5086-5088.

Whalen, D. L., Islam, N. B., Gupta, S. C., Sayer, J. M. and Jerina, D. M. (1990). Kinetic studies of the reactions of benzo[a]pyrene 7,8-diol 9,10-epoxides with general acids, nucleotides, DNA and micelles. In *Polynuclear Aromatic Hydrocarbons: Measurements, Means and Metabolism* (M. Cooke, K. Loening and J. Merritt, Eds.), pp. 1017-1032. Battelle Press, Columbus, OH.

Wood, A. W., Sayer, J. M., Newmark, H. L., Yagi, H., Michaud, D. P., Jerina, D. M. and Conney, A. H. (1982). Mechanism of the inhibition of mutagenicity of a benzo[a]pyrene 7,8-diol 9,10-epoxide by riboflavin 5'-phosphate. *Proc. Natl. Acad. Sci. USA* **79**, 5122-5126.

Yagi, H., Thakker, D. R., Hernandez, O., Koreeda, M. and Jerina, D. M. (1977). Synthesis and reactions of the highly mutagenic 7,8-diol 9,10-epoxides of the carcinogen benzo[a]pyrene. *J. Am. Chem. Soc.* **99**, 1604-1611.

Yagi, H., Sayer, J. M., Thakker, D. R., Levin, W. and Jerina, D. M. (1987). Effects of a 6-fluoro substituent on the solvolytic properties of the diastereomeric 7,8-diol 9,10-epoxides of the carcinogen benzo[a]pyrene. *J. Am. Chem. Soc.* **109**, 838-846.

Yang, J.-L., Maher, V. M. and McCormick, J. J. (1987). Kinds of mutations formed when a shuttle vector containing adducts of (±)-7β,8α-dihydroxy-9α,10α-epoxy-7,8,9,10-tetrahydrobenzo[a]pyrene replicates in human cells. *Proc. Natl. Acad. Sci. USA* **84**, 3787-3791.

You, M., Candrian, U., Maronpot, R. R., Stoner, G. D. and Anderson, M. W. (1989). Activation of the Ki-ras protooncogene in spontaneously occurring and chemically induced lung tumors of the strain A mouse. *Proc. Natl. Acad. Sci. USA* **86**, 3070- 3074.

ELECTROPHILIC SULFURIC ACID ESTER METABOLITES OF HYDROXY-METHYL AROMATIC HYDROCARBONS AS PRECURSORS OF HEPATIC BENZYLIC DNA ADDUCTS IN VIVO

James A. Miller, Young-Jooh Surh, Amy Liem, and
Elizabeth C. Miller (deceased)

McArdle Laboratory for Cancer Research and Environmental
Toxicology Center
University of Wisconsin Medical School
Madison, WI 53706

INTRODUCTION

In recent years work in our laboratory has identified the principal ultimate electrophilic and carcinogenic metabolites formed *in vivo* in mouse liver from several classes of chemical carcinogens, especially in relation to DNA adduct and hepatoma formation. These carcinogens comprised several naturally occuring alkenylbenzenes and their proximate carcinogenic 1'-hydroxy metabolites (Boberg et al., 1983; Wiseman et al., 1987), the synthetic alkyne, 1'-hydroxy-2',3-'dehydroestragole (Fennell et al., 1985); 4-aminoazobenzene and its proximate carcinogenic metabolite, N-hydroxy-4-aminoazobenzene (Delclos et al., 1986), and 2-acetylamino-fluorene (Lai et al., 1985, 1987, 1988). In each of these cases, the ultimate reactive carcinogen leading to DNA adduct formation and hepatoma formation appears to be an electrophilic sulfuric acid ester generated by one or more 3'-phospho-adenosine-5'-phosphosulfate (PAPS)-dependent sulfotransferase activities in the liver cytosol. It appears possible from the work of others that the carcinogenic activity of 2,6-dinitrotoluene (Kedderis et al., 1984; Chism and Rickert, 1989), certain nitrosamine derivatives (Sterzel and Eisenbrand, 1986; Kokkinakis et al., 1986; Kroeger-Koepke et al., 1989), and certain drugs with β-aminoalcohol moiety such as pronethalol (Bicker and Fischer, 1974) may also involve sulfuric acid ester metabolites. The oncogenic purine derivative, 3-hydroxyxanthine has also been thought to be activated by sulfotransferase activity (Stöhrer et al., 1972; Anderson et al., 1978).

This report concerns the PAPS-plus sulfotransferase-dependent formation in rodent liver *in vivo* of benzylic DNA adducts of several hydroxymethyl polycyclic aromatic hydrocarbons. The concept that these adducts might be formed from certain methyl-substituted polycyclic aromatic hydrocarbons is relatively old and is due largely to Flesher, as noted in Figure 1, who has offered supporting data from carcinogenicity tests and metabolic studies *in vitro* (Flesher and Sydnor, 1971, 1973). Further supporting carcinogenicity and mutagenicity data were provided by Cavalieri et al. (1979). More definitive metabolic data *in vitro* have been publsihed in the last few years by Watabe and his associates for several hydroxymethyl polycyclic aromatic hydrocarbons (reviewed recently by Watabe et al., 1989). Those studies began with the demonstration that an electrophilic sulfuric acid ester was formed from 7-hydroxymethyl-12-methylbenz[a]anthracene (HMBA) (Figure 2) by PAPS-dependent sulfotransferase activity in rat liver cytosol (Watabe et al., 1982). This ester reacted with the amino groups of deoxyguanosine and deoxyadenosine residues in added calf thymus DNA to form stable benzylic DNA adducts. It was further shown that the

Biological Reactive Intermediates IV, Edited by C.M. Witmer *et al.*
Plenum Press, New York, 1990

metabolically formed or chemically synthesized sulfuric acid ester was directly mutagenic in *Salmonella typhimurium* TA98. These basic observations were also noted for several other hydroxymethyl hydrocarbons in similar studies *in vitro* (Watabe et al., 1982, 1985, 1986, 1987; Okuda et al., 1986, 1988, 1989).

Our studies started with a full confirmation of the basic findings of Watabe et al. (1982) for the metabolic activation of HMBA by rat liver cytosolic PAPS-dependent sulfotransferase activity. These studies employed the radiomatic and fluorometric assays of sulfotransferase activity as outlined in Figure 3. Greater cytosolic sulfotransferase activity for HMBA was noted for rat liver as compared to mouse liver (Table 1). As noted in Table 2, the rat hepatic sulfotransferase activity was only poorly inhibited by pentachlorophenol, a strong inhibitor of phenol sulfotransferase activity (Mulder and Scholtens, 1977), but more strongly inhibited by dehydroepiandrosterone (DHEA) which is a typical substrate for hydroxysteroid sulfotransferases (Lyon and Jakoby, 1980). These findings suggest that the sulfotransferase activity that utilizes HMPA as a substrate may be a hydroxysteroid sulfotransferase. This possibility is supported by the age- and sex-dependent differences noted in Figure 4 for the sulfotransferase activities of rat liver cytosols for HMPA. These activities diminished with age much faster in male rats than in females. Adult males had no detectable activity in their hepatic cytosols. Treatment of female and male rats with testosterone propionate considerably lowered their cytosolic sulfotransferase activities while estradiol benzoate markedly increased the activity in each sex (Figure 5). Evidence for the endocrine control of rat hepatic sulfotransferase activity for HPBA was further provided by comparing the sulfotransferase activities in cytosols from gonadectomized rats with those in control animals (data not shown). These patterns are consistent with age- and sex-differences reported previously for the rat hepatic sulfotransferase activities that catalyze the sulfonation of hydroxysteroids (Carlstedt-Duke and Gustafsson, 1973; Singer and Sylvester, 1976; Singer, 1985; Iwasaki et al., 1986).

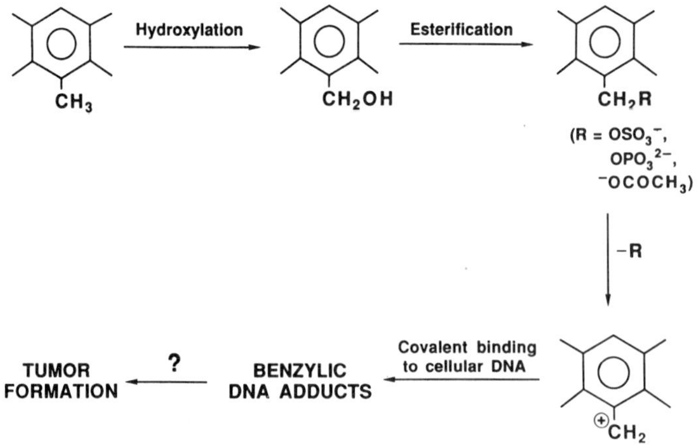

Figure 1. A hydroxylation-esterification pathway for the metabolic activation of methyl-substituted polycyclic aromatic hydrocarbons to form benzylic DNA adducts; proposed by Flesher, *Cancer Res.*, **31**, 1951 (1971) and *Int. J. Cancer*, **11**, 433 (1973).

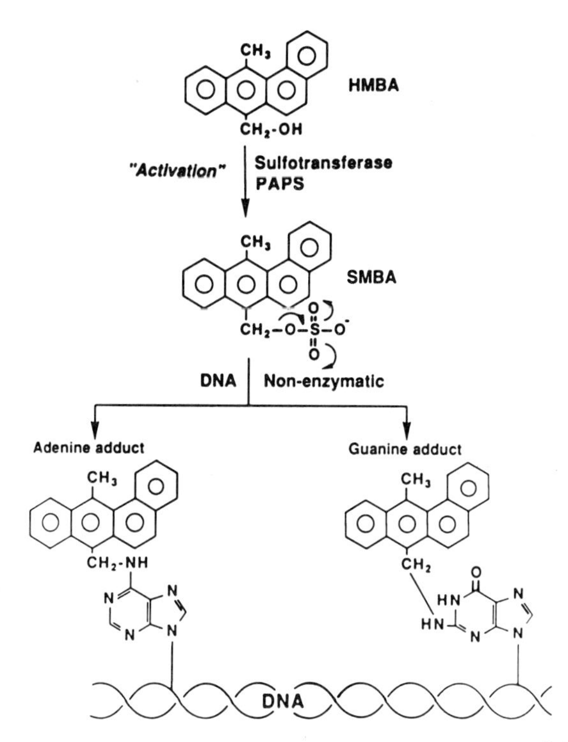

Figure 2. Activation of 7-hydroxymethyl-12-methylbenz[a]anthracene (HMBA) to 7-sulfooxymethyl-12-methylbenz[a]anthracene (SMBA) by sulfotransferase activity in rat liver cytosol; adapted from Watabe et al., *Science*, **215**, 403 (1982) and *Biochem. Pharmacol.*, **34**, 3002 (1985).

Table 1. The sulfotransferase activities for HMBA in rat and mouse liver cytosols[a]

Species	Sulfotransferase activity (pmol DNA adducts/30 min/mg cytosolic protein)
Mouse	57 ± 27[b]
Rat	2290 ± 990

a Liver cytosols from 12-day-old male rats and mice were assayed for sulfotransferase activity toward HMBA using calf thymus DNA as a nucleophilic acceptor. The details of the assay condition are as described in the previous report by Surh et al. (1987).

b Values are mean ± S.D. of determinations on 3 separate livers.

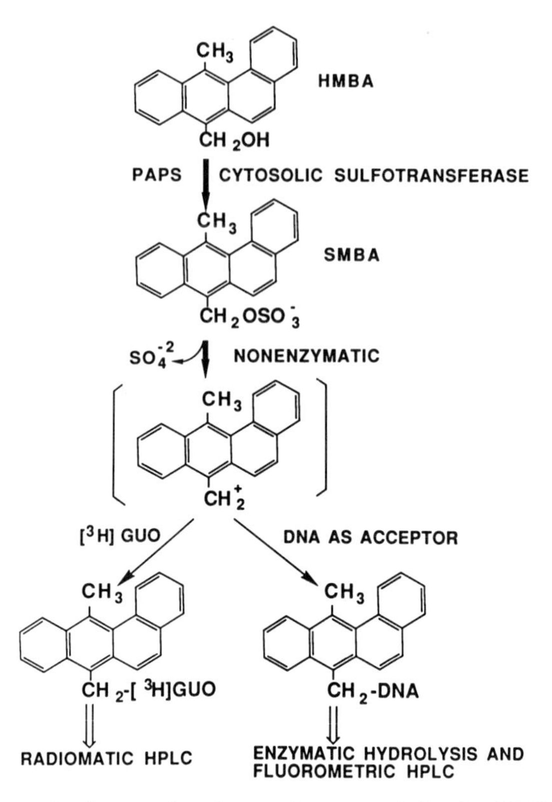

Figure 3. Assay of sulfotransferase activity for HMBA.

Table 2. The inhibition of sulfotransferase activity for HMPA in rat liver cytosols by pentachlorophenol (PCP) and dehydroepiandrosterone (DHEA)[a]

Inhibitor	Conc. (μmol)	Sulfotransferase activity[b]	% inhibition
None	-	342 ± 41	-
PCP	100	291 ± 77	15
DHEA	100	113 ± 23[c]	67

[a] The sulfotransferase activities for HMBA in infant male rats were determined in the presence or absence of sulfotransferase inhibitors. [^{3}H]guanosine was added to the incubation mixtures to trap the labile sulfuric acid ester of HMBA.

[b] Enzyme activity is expressed as pmol of [^{3}H]guanosine adduct formed/mg cytosolic protein/20 min.

[c] Statistically different from the value obtained in the absence of an inhibitor ($P < 0.01$).

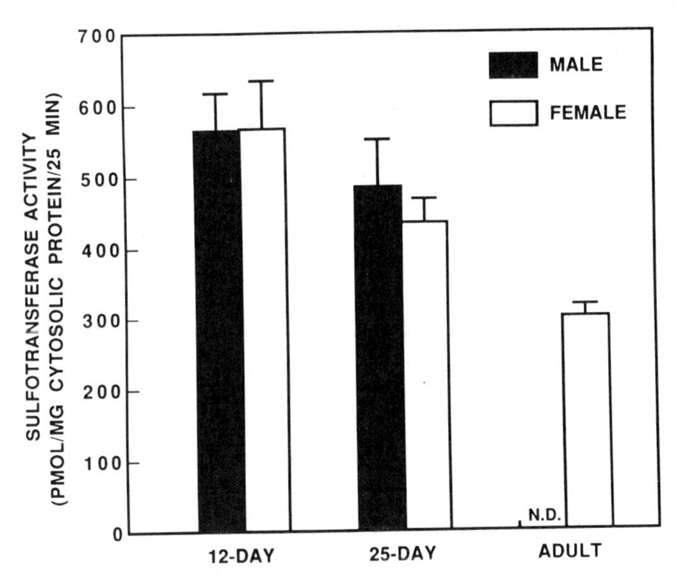

Figure 4. Sex- and age-dependent sulfotransferase activities for HMPA in rat liver cytosols. The activities are given in pmol of [^{3}H]guanosine adduct formed/mg cytosolic protein/25 min. N.D.; not detectable.

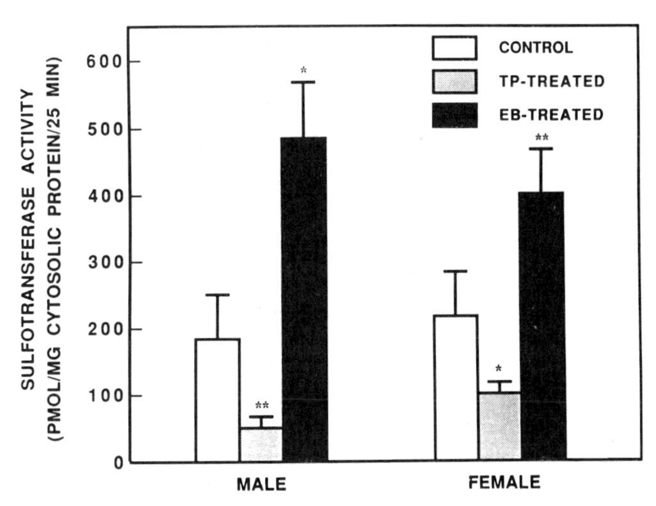

Figure 5. Effect of administration of gonadal hormones on the hepatic cytosolic sulfotransferase activities for HMBA in rats. Sprague-Dawley rats (28-day-old) were given daily subcutaneous injections of 1.0 mg of testosterone propionate (TP) or 0.2 mg of estradiol benzoate (EB) in 0.2 ml trioctanoin for 12 days. Control animals were treated with vehicle only. The unit of enzyme activity is same as in Figure 4. Asterisks indicate statistically significant differences from the respective controls ($^*P < 0.05$; $^{**}P < 0.01$).

Because of our interest in the sulfonation in vivo of hydroxy metabolites of several classes of chemical carcinogens and the lack of studies on the formation of benzylic DNA adducts in vivo from hydroxymethyl aromatic hydrocarbons, we have recently investigated this subject. In our first report (Surh et al., 1987) we noted the formation of the benzylic adducts of (deoxy)guanosine and (deoxy)-adenosine in the hepatic DNA and RNA of preweanling rats and mice given i.p. injections of HMBA and its electrophilic sulfuric acid ester. Figure 6 shows the fluorometric analysis by HPLC of these adducts in the enzymatic hydrolysates of rat hepatic DNA after administration of the solvent (dimethylsulfoxide) or 0.25 μmol per gram body weight of HMBA or its synthetic sulfuric acid ester, sodium 7-sulfooxymethyl-12-methylbenz[a]anthracene (SMBA). The peaks labeled as G and A denote the benzylic adducts of deoxyguanosine and deoxyadenosine, respectively. These adducts formed *in vivo* were identical chromatographically with the synthetic adducts obtained chemically from the reaction of SMBA with the appropriate deoxyribonucleoside. Before chromatography, the benzylic adducts were extracted from the DNA hydrolysate with chloroform:*n*-propanol (7:3, v/v/). This extraction separates the benzylic adducts from more polar adducts of HMBA, presumably including those derived from the dihydrodiol epoxide pathway described by Dr. Jerina elsewhere in this volume. As noted in Figure 6, injection of HMBA and SMBA in rats gave rise to more of the G adduct than the A adduct. A more polar unidentified adduct (U) was also observed. It seems likely that this adduct may be a ring-hydroxy derivative of the G adduct. This unidentified adduct is found only in the liver DNA exposed to HMBA and SMBA *in vivo* and was not found in DNA exposed to synthetic SMBA *in vitro* or to this ester when it was generated by incubation of HMBA with rat liver cytosol fortified with PAPS.

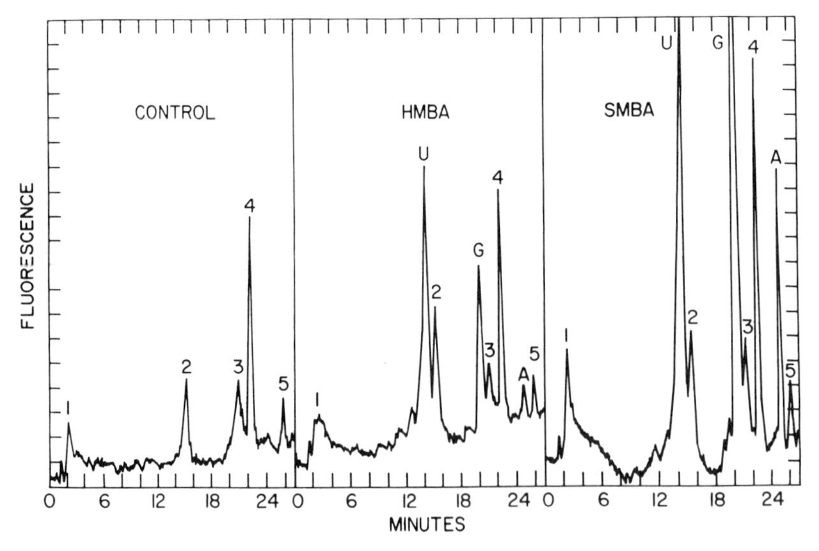

Figure 6. HPLC profiles of benzylic nucleoside adducts from hydrolysates of hepatic DNA isolated from preweanling male rats 6 hours after i.p. injections of DMSO (solvent control), 0.25 μmol HMBA or 0.25 μmol SMBA. G and A are the benzylic adducts of deoxyguanosine and deoxyadenosine, respectively. U denotes an unidentified adduct. The benzylic adducts were selectively extracted from the hydrolysates by chloroform:*n*-propanol (7:3, v/v) and analyzed by reverse-phase HPLC with a 25-min linear gradient of 60-100% methanol in water. Adducts were detected by their fluorescence (excitation, 361 nm; emission, 418 nm).

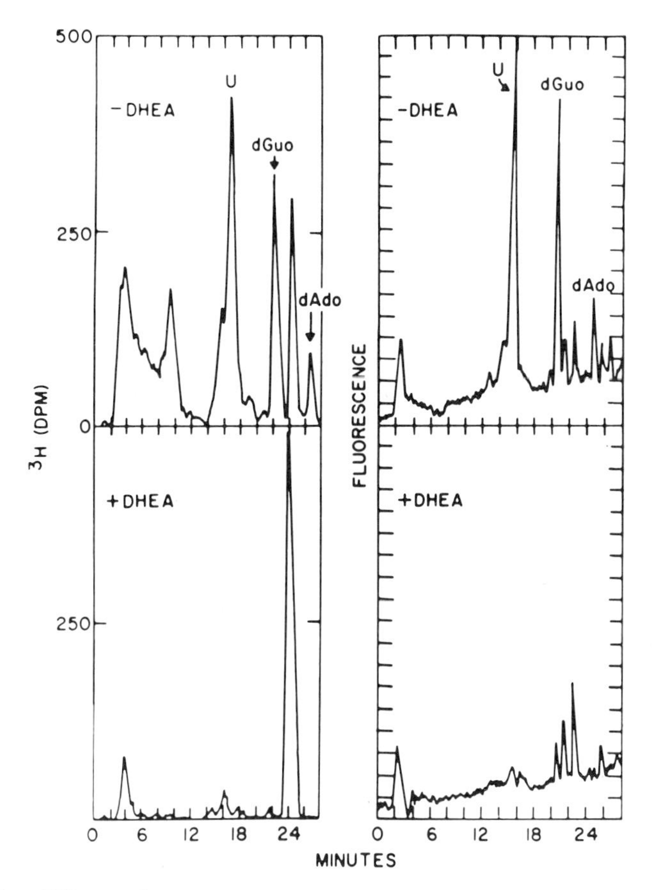

Figure 7. Effect of DHEA pretreatment on the formation of benzylic DNA adducts from [^{3}H]HMPA in rat liver. Preweanling male Sprague-Dawley rats were injected i.p. with DHEA (0.5 μmol/10 μl trioctanoin/g body weight) or vehicle only one hour before a dose of [^{3}H]HMPA (0.25 μmol/g body weight). Hepatic benzylic DNA adducts were analyzed as described in Figure 6.

Since SMBA is labile in aqueous media with a half-life of less than 1 minute (Watabe et al., 1987), we were surprised that i.p. injections of this ester in infant rats gave rise to much higher amounts of the U, G and A adducts than did equimolar amounts of HMBA (Figure 6). It seems probable that this labile ester is transported to the liver and then to the hepatic DNA in a manner that protects some of the dose from solvolysis to form HMBA. HMBA is presumably metabolized to SMBA in the liver cells before adduction to the DNA. This pathway is strongly indicated by the ability of pretreatment of the rats with DHEA (0.5 μmol per gram body weight) before the dose of HMBA (Figure 7) to greatly decrease the formation of the benzylic adducts in the hepatic DNA. As noted above, DHEA is a strong inhibitor of the hepatic sulfotransferase activity for HMBA. The amounts of the G and A adducts were calculated from the fluorescence of standard adducts of known specific radioactivity prepared from the enzymatic reaction of [^{3}H]HMBA with the deoxyribonucleosides. [^{3}H]HMBA of known specific activity was also used to show that the sum of the benzylic G and a adducts (approximately 10 pmol per mg DNA at 6 hour following a dose of 0.25 μmol HMBA/g body weight in preweanling male S.D. rats) in the hepatic DNA accounted for approximately 20-30% of the total radioactivity bound to the DNA in the liver. The remainder presumably includes the adducts formed by the dihydrodiol-epoxide pathway of metabolism of HMBA.

Figure 8 shows the structures of HMBA and those of three other hydroxymethyl hydrocarbons that we have also studied for their ability to form hepatic benzylic DNA adducts in the rat. HMPA and HMBP contain bay-regions from which dihydrodiol-epoxide derived DNA adducts may be formed *in vivo*. HMA and HMP do not contain bay-regions, but adducts derived from non-bay-region metabolites may be formed *in vivo*. We have found that all of these hydrocarbons form benzylic DNA adducts *in vivo* and *in vitro* in rat liver. Our data on HMBP have been published (Surh et al., 1989) and studies on HMA and HMP are in progress in our laboratory.

Until recently we have regarded the metabolically formed sulfuric acid esters of these hydroxymethyl hydrocarbons as the final reactive electrophiles in the formation of the benzylic adducts in the liver DNA of the rat. However, an important finding has been recently reported on the effect of chloride ion on the mutagenicity of 1-sulfooxymethylpyrene (SMP) (R. Henschler et al., 1989a, b). Chloride ion at physiological concentrations was found to greatly enhance the mutagenicity of this ester for *Salmonella typhimurium* TA98 by over 50-fold. Furthermore, these investigators showed that this sulfuric acid ester reacted with chloride ion to form 1-chloromethylpyrene (CMP) as noted in Figure 9. This product was much more mutagenic than the ester. We have confirmed these interesting results and have studied CMP and SMP in two ways. As shown in Table 3, the intrinsic reactivities of CMP and SMP in forming benzylic adducts in DNA *in vitro* in non-enzymatic chemical reactions are not very different. However, as noted in Table 4, the abilities of these two electrophiles to form benzylic DNA adducts in the livers of rats *in vivo* after i.p. injection are quite different. In contrast to the bacterial mutagenicity data the sulfuric acid ester is about 6 times more effective than the chloromethyl hydrocarbon in the rat system. Several factors may account for this difference. A likely factor may be the probable greater solubility of the less polar chloromethyl derivatives in fat depots as compared to that of the polar sulfooxymethyl derivative. Further studies will be required to determine the electrophilic nature of the final reactive metabolites of HMP and similar hydroxymethyl polycyclic aromatic hydrocarbons during the formation of the benzylic DNA adducts in the rat liver.

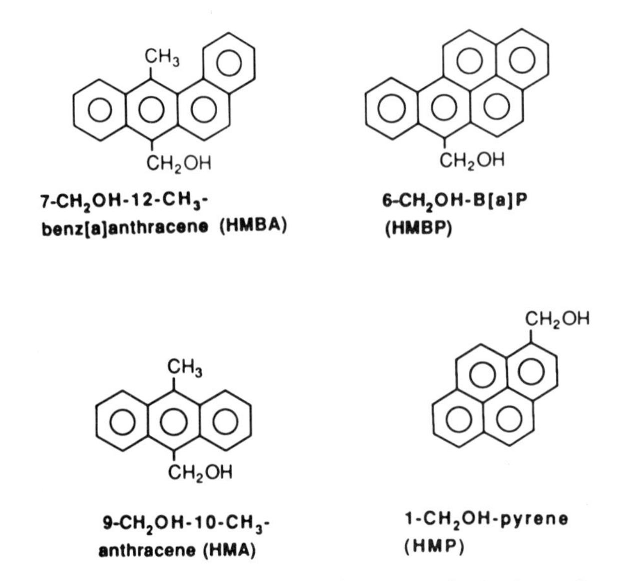

Figure 8. The hydroxymethyl polycyclic aromatic hydrocarbons included in this study.

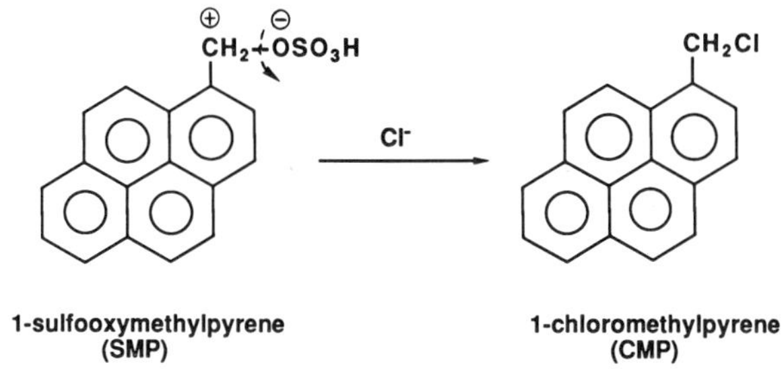

1-sulfooxymethylpyrene (SMP) $\xrightarrow{Cl^-}$ 1-chloromethylpyrene (CMP)

Figure 9. The reaction of chloride ion with 1-sulfooxymethylpyrene (SMP); Henschler et al., *Naunyn-Schmiedeberg's Arch. Pharmacol.* **339**, (Suppl.), R26 (1989) and *Mutagenesis*, **4**, 309 (1989).

Table 3. Benzylic DNA ddduct formation from 1-hydroxymethylpyrene (HMP), 1-sulfooxymethylpyrene (SMP) and 1-chloro-methylpyrene (CMP) in reactions *in vitro*[a]

Hydrocarbon	Experimental conditions	Benzylic dGuo adduct
HMP	Liver cytosol + PAPS	644
SMP	Non-enzymatic	3246
CMP	Non-enzymatic	3781

[a] Hydrocarbons (0.1 µmol) dissolved in 50 µl DMSO were incubated at 37°C for 30 min with ~ 2 mg calf thymus DNA in a final volume of bis-Tris buffer, pH 7.4 (50 µM) containing $Mg(OAc)_2$ (5 µM) in the presence (HMP) or absence (SMP and CMP) of sulfotransferase system consisting of liver cytosol and PAPS (100 µM). DNA was isolated and enzymatically digested as previously (Surh et al., 1989). Benzylic adducts were analyzed by reverse-phase HPLC with a 30-min linear gradient of 60-100% MeOH. The major benzylic adduct (dGuo adduct) was quantitated by comparison with the fluorescence (excitation, 275 nm; emission, 389 nm) of the standard adduct of known specific radioactivity.

Table 4. The formation of benzylic DNA adducts in the livers of preweanling rats injected i.p. with 1-hydroxymethyl-, 1-sulfooxymethyl- and 1-chloromethylpyrene[a]

Hydrocarbon	Dose (µmol/g body wt.)	Benzylic dGuo adduct (pmol/mg hepatic DNA)
HMP	0.25	94 ± 20
SMP	0.25	1458 ± 495
CMP	0.25	222 ± 82

[a] Animals were killed 6 hour after treatment of hydrocarbons and hepatic DNA was isolated and digested. Benzylic DNA adducts were analyzed by HPLC as described in the foot-note of Table 3.

Of special interest to us is the role of the metabolically derived sulfuric acid esters of these hydroxymethyl hydrocarbons in carcinogenesis by these agents. Unfortunately, it has proved difficult to test more than very low levels of these esters, since they are considerably more toxic and lethal than the parent hydroxymethyl hydrocarbons. For example, in the peritoneal cavity of mice they produce severe adhesions. Data on the initiation of liver tumors in male B6C3F$_1$ infant mice by DMBA, HMBA and SMBA are shown in Table 5. These tumors were produced by single i.p. doses at the levels indicated. DMBA proved to be very active in this system and HMBA was appreciably active. However, SMBA was no more active than the comparable low level of HMBA. In other studies, we have found SMBA to be quickly and completely hydrolyzed in water to HMBA. It was possible to apply higher levels of SMBA to mouse skin and the data in Table 6 show that SMBA had about the same carcinogenicity in this tissue as did HMBA at the three dose levles tested. Again, it is likely that SMBA is rapidly hydrolyzed to form HMBA. Thus, the toxicity and the hydrolysis of SMBA back to HMBA make it very questionable if the carcinogenicities of SMBA and HMBA can be adequately compared in direct tests in the animal. Inside the liver cell, for example, a reactive form of HMBA (presumably SMBA or the chloromethyl hydrocaron, or both) is required to form the benzylic DNA adducts observed. The test animal thus imposes a water barrier to the direct introduction of a water-sensitive compound such as SMBA into cells and may invalidate direct carcinogenicity tests. To circumvent this problem, we have also tested the effect of the sulfotransferase inhibitor DHEA on the carcinogenicity of HMBA in the liver, skin and lung of mice. The administration of DHEA one hour before the application of the carcinogen did not alter the carcinogenicity of HMBA in these tissues (data not shown). These results do not indicate a significant role of the benzylic DNA adducts of HMBA in carcinogenesis by this hydrocarbon in these tissues in the mouse, and other DNA adducts of HMBA, including those derived from the dihydrodiol-epoxide metabolic pathway, may be more important in this regard. Further data on the metabolism of HMBA in these tissues of the mouse will be needed, however, to support this conclusion.

Table 5. The comparative hepatocarcinogenicities of single i.p. doses of 7,12-dimethylbenz[a]anthracene (DMBA), 7-hydroxymethyl-12-methylbenz-[a]-anthracene (HMBA) and 7-sulfooxymethyl-12-methylbenz-[a]anthracene (SMBA) in 12-day-old male B6C3F$_1$ mice

Compound	Dose (μmol/g body weight)	No. of mice autopsied[a]	% hepatoma-bearing mice	Average No. of tumors/mouse
DMBA	0.01	33	100	25.7 ± 7.0
HMBA	0.03	30	100	9.9 ± 3.5
	0.01	32	100	4.5 ± 2.1
	0.003	36	42	0.5 ± 0.6
SMBA	0.003	33	39	0.4 ± 0.7
Vehicle only (10 μl trioctanoin/g body weight)		34	15	0.2 ± 0.4

a The tumors were counted at 10 months

Table 6. The formation of skin tumors in female CD-1 mice by a single topical applications of DMBA, HMBA and SMBA[a]

Initiator	Dose (μmol)	No. of mice at 22 wk	% tumor-bearing mice	No. of papillomas/ mouse
DMBA	0.03	30	100	13.0 ± 6.4
HMBA	0.3	30	73	3.4 ± 3.0
	0.1	30	77	1.6 ± 1.3
	0.03	30	50	1.0 ± 1.1
SMBA	0.3	29	76	3.1 ± 6.2
	0.1	30	67	1.1 ± 1.1
	0.03	30	27	0.3 ± 0.6
Solvent only (acetone: DMSO, 95:5, v/v)		30	13	0.1 ± 0.3

[a] One week after the initiation dose, promotion was begun by application twice weekly of 2.5 μg of tetradecanoyl phorbol acetate.

REFERENCES

Anderson, L.M., McDonald, J.J., Budinger, J.M., Mountain, I.M. and Brown, G.B. (1978). 3-Hydroxyxanthine: transplacental effects and ontogeny of related sulfate metabolism in rats and mice. *J. Natl. Cancer Inst.* **61**, 1405-1410.

Bicker, U. and Fischer, W. (1974). Enzymatic aziridine synthesis from β-amino-alcohols; a new example of endogenous carcinogen formation. *Nature* **249**, 344-345.

Boberg, E.W., Miller, E.C., Miller, J.A., Poland, A. and Liem, A. (1983). Strong evidence from studies with brachymorphic mice and pentachlorophenol that 1'-sulfooxysafrole is the major ultimate electrophilic and carcinogenic metabolite of 1'-hydroxysafrole in mouse liver. *Cancer Res.* **43**, 5163-5173.

Carlstedt-Duke, J. and Gustafsson, J.-A. (1973). Sexual differences in hepatic sulphurylation of deoxycorticosterone in rats. *Eur. J. Biochem.* **36**, 172-177.

Cavalieri, E.L., Roth, R.W. and Rogan, E.G. (1979). Hydroxylation and conjugation at the benzylic carbon atom: a possible mechanism of carcinogenic activation for some methyl-substituted aromatic hydrocarbons. In *Polynuclear Aromatic Hydrocarbons* (P.W. Jones and P. Leber, Eds.), pp. 517-529. Ann Arbor Science, ann Arbor, MI.

Chism, J.P. and Rickert, D.E. (1989). *In vitro* activation of 2-aminobenzyl alcohol and 2-amino-6-nitrobenzyl alcohol, metabolites of 2-nitrotoluene and 2,6-dinitrotoluene. *Chem. Res. Toxicol.* **2**, 150-156.

Delclos, K.B., Miller, E.C., Miller, J.A. and Liem, A. (1986). Sulfuric acid esters as major ultimate electrophilic and hepatocarcinogenic metabolites of 4-aminoazobenzene and its N-methyl derivatives in infant male C57BL/6JxC3H/HeJ F_1 (B6C3F_1) mice. *Carcinogenesis*, **7**, 277-287.

Fennell, T.R., Wiseman, R.W., Miller, J.A. and Miller, E.C. (1985). Major role of hepatic sulfotransferase activity in the metabolic activation, DNA adduct formation and carcinogenicity of 1'-hydroxy-2',3'-dehydroestragole in infant male C57BL/6JxC3H/HeJ F_1 mice. *Cancer Res.* **45**, 5310-5320.

Flesher, J.W. and Sydnor, K.L. (1971). Carcinogenicity of derivatives of 7,12-dimethylbenz[a]anthracene. *Cancer Res.* **31**, 1951-1954.

Flesher, J.W. and Sydnor, K.L. (1973). Possible role of 6-hydroxymethylbenzo[a]pyrene as a proximate carcinogen of benzo[a]pyrene and 6-methylbenzo[a]pyrene. *Int. J. Cancer* **11**, 433-437.

Henschler, R., Seidel, A. and Glatt, H.R. (1989a). Phase-II-metabolite genotoxicity of polycyclic aromatic hydrocarbons: a chloromethyl derivative as a mutagenic intermediate from a benzylic sulfate ester. *Naunyn-Schmiedeberg's Arch. Pharmacol.* **339** (Suppl.), R26.

Henschler, R., Pauly, K., Godtel, U., Seidel, A., Oesch, F. and Glatt, H.R. (1989b). The mutagenicity of sulphate esters in *Salmonella typhimurium* is strongly affected by the ions present in the exposure medium. *Mutagenesis* **4**, 309 (GUM abstracts).

Iwasaki, K., Shiraga, T., Tada, K., Noda, K. and Noguchi, H. (1986). Age- and sex-related changes in amine sulphoconjugation in Sprague-Dawley strain rats. Comparison with phenol and alcohol sulphoconjugations. *Xenobiotica* **16**, 717-723.

Kedderis, G.L., Dyroff, M.C. and Rickert, D.E. (1984). Hepatic macromolecular covalent binding of the hepatocarcinogen 2,6-dinitrotoluene and its 2,4-isomer *in vivo*: modulation by the sulfotransferase inhibitors pentachlorophenol and 2,6-dichloro-4-nitrophenol. *Carcinogenesis* **5**, 1199-1204.

Kokkinakis, D.M., Hollenberg, P.F. and Scarpelli, D.G. (1986). The role of hepatic sulfotransferases in the activation of β-hydroxynitrosamines to potentially mutagenic agents. *Proc. Am. Assoc. Cancer Res.* **30**, 167.

Lai, C.-C., Miller, J.A., Miller, E.C. and Liem, A. (1985). N-Sulfooxy-2-aminofluorene is the major ultimate electrophilic and carcinogenic metabolite of N-hydroxy--acetylaminofluorene in the livers of infant male C57BL/6JxC3H/HeJ F_1 (B6C3F_1) mice. *Carcinogenesis* **6**, 1037-1045.

Lai, C.-C., Miller, E.C., Miller, J.A. and Liem, A. (1987). Initiation of hepato-carcinogenesis in infant male B6C3F1 mice by N-hydroxy-2-aminofluorene or N-hydroxy-2-acetylaminofluorene depends primarily on metabolism to N-sulfooxy-2-aminofluorene and formation of DNA-(deoxyguanosine-8-yl)-2-aminofluorene adducts. *Carcinogenesis* **8**, 471-478.

Lai, C.-C., Miller, E.C., Miller, J.A. and Liem, A. (1988). the essential role of microsomal deacetylase activity in the metabolic activation, DNA-(deoxyguanosine-8-yl)-2-aminofluorene adduct formation and initiation of liver tumors by N-hydroxy-2-acetylaminofluorene in the livers of infant male B6C3F_1 mice. *Carcinogenesis* **9**, 1295-1302.

Lyon, E.S. and Jakoby, W.B. (1980). The identity of alcohol sulfotransferase with hydroxysteroid sulfotransferases. *Arch. biochem. biophys.* **202**, 474-481.

Mulder, G.J. and Scholtens, E. (1977). Phenol sulfotransferase and uridine diphosphate glucuronyltransferase from rat liver *in vivo* and *in vitro*. 2,6-dichloro-4-nitrophenol as a selective inhibitor of sulphation. *Biochem. J.* **165**, 553-559.

Okuda, H., Hiratsuka, A., Nojima, H. and Watabe, T. (1986). A hydroxymethyl sulphate ester as an active metabolite of the carcinogen, 5-hydroxymethylchrycene. *Biochem. Pharmacol.* **35**, 535-538.

Okuda, H., Nojima, H., Watanabe, N., Miwa, K. and Watabe, T. (1988). Activation of the carcinogen, 5-hydroxymethylchrycene, to the mutagenic sulphate ester by mouse skin sulphotransferase. *Biochem. Pharmacol.* **37**, 970-973.

Okuda, H., Nojima, H., Watanabe, N. and Watabe, T. (1989). Sulphotransferase-mediated activation of the carcinogen 5-hydroxymethyl-chrycene. Species and sex differences in tissue distribution of the enzyme activity and a possible participation of hydroxysteroid sulphotransferases. *Biochem. Pharmacol.* **38**, 3003-3009.

Singer, S.S. and Sylvester, S. (1976). Enzymatic sulfation of steroids II. The control of the hepatic cortisol sulfotransferase activity and of the individual hepatic steroid sulfotransferases of rats by goands and gonadal hormones. *Endocrinology*, **99**, 1346-1352.

Singer, S.S. (1985). Preparation and characterization of the different kinds of sulfotransferases. In *Biochemical Pharmacology and Toxicology* (D. Zakim Ed.), Vol. I, pp. 95-159. Academic Press, New York.

Sterzel, W. and Eisenbrand, G. (1986). N-Nitrosodiethanolamine is activated in the rat to an ultimate genotoxic metabolite by sulfotransferase. *J. Cancer Res. Clin. Oncol.* **111**, 20-24.

Stohrer, G., Corbin, E. and Brown, G.B. (1972). Enzymatic activation of the oncogen 3-hydroxyxanthine. *Cancer Res.* **32**, 637-642.

Surh, Y.-J., Lai, C.-C., Miller, J.A. and Miller, E.C. (1987). Hepatic DNA and RNA adduct formation from the carcinogen 7-hydroxymethyl-12-methylbenz[a]anthracene and its electrophilic sulfuric acid ester metabolite in preweanling rats and mice. *Biochem. Biophys. Res. Commun.* **144**, 576-582.

Surh, Y.-J., Liem, A., Miller, E.C. and Miller, J.A. (1989). Metabolic activation of the carcinogen 6-hydroxymethylbenzo[a]pyrene: formation of an electrophilic sulfuric acid ester and benzylic DNA adducts in rat liver *in vivo* and in reactions *in vitro*. *Carcinogenesis* **10**, 1519-1528.

Watabe, T., Ishizuka, T., Isobe, M. and Ozawa, N. (1982). A 7-hydroxymethyl sulfate ester as an active metabolite of 7,12-dimethylbenz[a]anthracene. *Science* **215**, 403-405.

Watabe, T., Hiratsuka, A., Ogura, K. and Endoh, K. (1985). A reactive hydroxymethyl sulfate ester formed regioselectively from the carcinogen, 7,12-dihydroxymethylbenz[a]anthracene, by rat liver sulfotransferase. *Biochem. Biophys. Res. Commun.* **131**, 694-699.

Watabe, T., Hakamata, Y., Hiratsuka, A. and Ogura, K. (1986). A 7-hydroxymethyl sulphate ester as an active metabolite of the carcinogen, 7-hydroxymethylbenz[a]anthracene. *Carcinogenesis* **7**, 207-214.

Watabe, T., Hiratsuka, A. and Ogura, K. (1987). Sulphotransferase-mediated covalent binding of the carcinogen 7,12-dihydroxymethylbenz[a]anthracene to calf thymus DNA and its inhibition by glutathione transferase. *Carcinogenesis* **8**, 445-453.

Watabe, T., Ogura, K. Okuda, H. and Hiraszuka, A. (1989). Hydroxymethyl sulfate esters as reactive metabolites of the carcinogens, 7-methyl- and 7,12-dihydroxymethylbenz[a]anthracenes and 5-methylchrycene. In *Xenobiotic Metabolism and Disposition (Proc. 2nd Int. ISSX Meeting)* (R. Kato, R.W. Estabrook and M.N. Cayen, Eds.) pp. 393-400. Taylor and Francis, London.

Wiseman, R.W., Miller, E.C., Miller, J.A. and Liem, A. (1987). Structure-activity studies of the hepatocarcinogenicities of alkenylbenzene derivatives related to estragole and safrole on administration to preweanling male C57BL/6JxC3H/HeJ F_1 mice. *Cancer Res.* **47**, 2275-2283.

HETEROCYCLIC AMINES: NEW MUTAGENS AND CARCINOGENES IN COOKED FOODS

Takashi Sugimura and Keiji Wakabayashi

National Cancer Center
1-1, Tsukiji 5-Chome, Chuo-ku
Tokyo 104, Japan

OCCURRENCE OF HETEROCYCLIC AMINES IN COOKED FOOD

Studies on heterocyclic amines (HAs), mutagens and carcinogens were initiated by the simple idea that smoke produced with broiling fish and meat should contain carcinogenic compounds as cigarette smoke does. Soon it was found that not only smoke produced by cooking fish and met, but also charred parts on fish and meat contained high mutagenic activity (Sugimura, T., et al., 1977). Among pyrolysates of food components, only that of protein showed strong mutagenic activity (Nago, M., et al., 1977). This led to the isolations of mutagens from pyrolysates of amino acids and proteins. Microbial assay using *Salmonella typhimurium* TA98 with metabolic activation was used for montoring the purifications of mutagens, and HAs such as Trp-P-1, Trp-P-2, Glu-P-1-2, A α C and MeA α C (see Table 1 for full names) were isolated as new mutagens (Sugimura, T., et al., 1982; Sugimura, T., et al. 1986; Sugimura T., et al., 1988; Yoshida, D., et al., 1978).

Our attempts to isolate mutagens directly from broiled fish and fried meat were also successful and a different class of HAs, IQ, MeIQ and MeIQx were isolated (see Table 1 for full names) (Sugimura, T., et al., 1982; Sugimura, T., et al. 1986; Sugimura T., et al., 1988). Later, a Swedish group and then ours found that heating a mixture of creatinine, sugars and amino acids resulted in the formations of IQ, MeIQx, 4,8-DiMeIQx and 7,8-DiMeIQx (Jägerstad, M., et al., 1984; Negishi, C., et al., 1984; Negishi, C., et al., 1985; Grivas, S., et al., 1985; Grivas, S., et al., 1986). Creatinine probably serves as a precursor for the amino-imidazole moieties of IQ, MeIQx and DiMeIQxs, while their quinoline and quinoxaline moieties could be derived from reaction products of sugars and amino acids.

The structures of HAs were determined by X-ray crystallography and NMR, mass and UV spectroscopies and confirmed by comparisons with the compounds synthesized chemically (Sugimura, T., et al., 1982; Sugimura, T., et al., 1986; Sugimura, T., et al., 1988; Yoshida, D., et al., 1978). The structures of these compounds are shown in Figure 1. Recently, PhIP was isolatd from fried beef by scientists in the Lawrence Livermore National Laboratory, California (Felton, J.S., et al., 1986).

Most of these HAs are newly registered chemicals. HAs of the IQ type, such as IQ, MeIQ, MeIQx, 4,8-DiMeIQx and PhIP, which are also called aminoimidazoazaarenes, have been found in various kinds of cooked fish and meat (Sugimura, T., et al., 1988; Felton, J.S., et al., 1984; Yamaizumi, Z., et al., 1986; Becher, G., et al., 1988; Zhang, X-M., et al., 1988; Murray, S.,, et al., 1988; Gross, G.A. et al., 1989), fried egg patties (Grose, K.R., et al., 1986) and beef extract

Biological Reactive Intermediates IV, Edited by C.M. Witmer *et al.*
Plenum Press, New York, 1990

(Sugimura, T., et al., 1988; Hargraves, W.A., et al., 1983; Hayatsu, H., et al., 1983; Turesky, R.J., et al., 1983). The non-IQ type HAs isolated from pyrolysates of amino acids and proteins, Trp-P-1, Trp-P-2, Glu-P-2, A α C and MeA α C were also found in cooked protein foods (Sugimura, T., et al., 1988; Yamaizumi, Z., et al., 1980; Yamaguchi, K., et al., 1980; Matsumoto, T., et al., 1981). Furthermore, roasted coffee beans were demonstrated to contain MeIQ (Kikugawa, K., et al., 1989). As examples of substances with comparably strong mutagenicity, only 2-(2-furyl)-3-(5-nitro-2-furyl)acrylamide, AF-2, a food additive as a preservative (Kondo, S., et al, 1973; Yahagi, T., et al., 1974) and aflatoxin B1, a fungus toxin (Int. Agency Res. Cancer, 1987) can be sited. Therefore, we were surprised to find potent mutagens in ordinary cooked foods.

CHARACTERISTICS OF HAs AS GENOTOXIC SUBSTANCES

HAs must be metabolically activated to exert mutagenicity. Among many molecular species of chytochrome P-450s, P-448-H induced by polychlorinated biphenyls or 3-methylcholanthrene is the most effective for converting HAs to *N*-hydroxy derivatives (Kato, R., et al., 1987; Yamazoe, Y., et al. 1988). Cultured mammalian cells infected with vaccinia virus bearing cloned c-DNA of a singel molecular species of cytochrome P-450 have been useful for identifying the particular molecular species of P-450 responsible for metabolic activation of HAs (Aoyama, T., et al., 1989). HAs were specifically activated by P_3450 (Aoyama, T., et al., 1989; Snyderwine, E.G. et al., 1989). *N*-hydroxy derivatives were further activated by enzyme esterifications such as *O*-acetylation, *O*-sulfonylation and amino acid *O*-acylation (Kato, R., et al., 1987; Snyderwine, E.G. et al., 1988). The ultimate forms of Trp-P-2, Glu-P-1 and IQ obtained in this way were shown to bind mainly to position C-8 of guanine bases in DNA (Hashimoto, Y., et al., 1980; Hashimoto, Y. et al., 1980; Snyderwine, E.G., et al., 1988). The structures of the adducts of Glu-P-1 and IQ with guanine are shown in Figure 2 (Hashimoto, Y., et al., 1980; Snyderwine, E.G., et al., 1988). In addition, activation of HAs by prostaglandin hydroperoxidase has also been reported (Nemoto, N., et al., 1984; Wild, D., et al., 1987; Petry, T.W., et al., 1989).

Table 1. Abbreviations and Chemical Names of HAs

Abbreviation	Chemical Name
Trp-P-1	3-Amino-1,4-dimethyl-5*H*-pyrido[4,3-*b*]indole
Trp-P-2	3-Amino-1-methyl-5*H*-pyrido[4,3-*b*]indole
Glu-P-1	2-Amino-6-methyldipyrido[1,2-*a*:3',2'-*d*]imidazole
Glu-P-2	2-Aminodipyrido[1,2-*a*:3',2'-*d*]imidazole
A α C	2-Amino-9*H*-pyrido[2,3-*b*]indole
MeA α C	2-Amino-3-methyl-9*H*-pyrido[2,3-*b*]indole
IQ	2-Amino-3-methylimidazo[4,5-*f*]quinoline
MeIQ	2-Amino-3,4-dimethylimidazo[4,5-*f*]quinoline
MeIQx	2-Amino-3,8-dimethylimidazo[4,5-*f*]quinoxaline
4,8-DiMeIQx	2-Amino-3,4,8-trimethylimidazo[4,5-*f*]quinoxaline
7,8-DiMeIQx	2-Amino-3,7,8-trimethylimidazo[4,5-*f*]quinoxaline
PhIP	2-Amino-1-methyl-6-phenylimidazo[4,5-*b*]pyridine

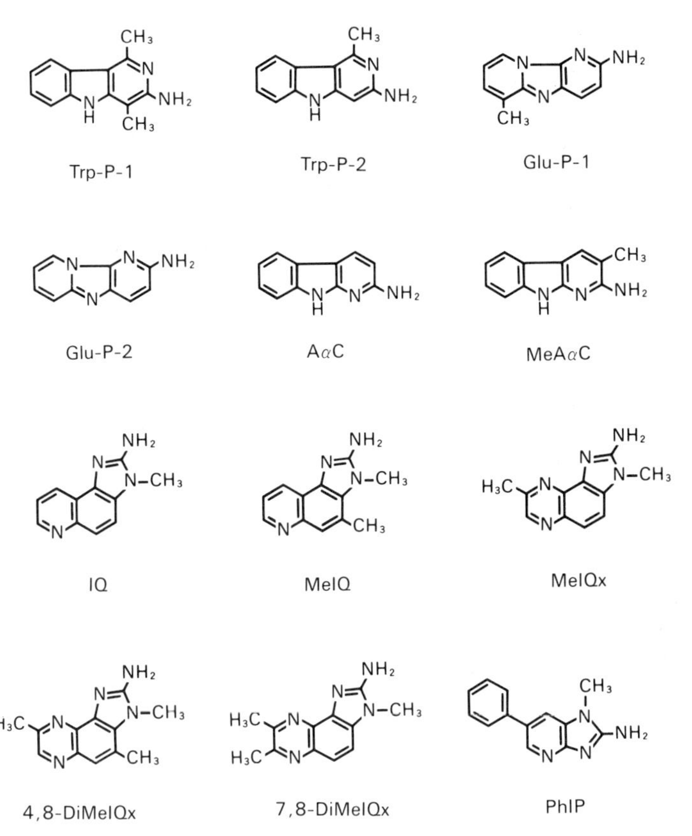

Figure 1. Structures of HAs insolated from pyrolysates of amino acids and a protein, and from cooked fish and meat.

Figure 2. Structures of adducts of Glu-P-1 and IQ with guanine.

In the presence of a metabolic activation system, HAs have been shown to induce sister-chromatide exchanges: Trp-P-1, Trp-P-2, glu-P-1 and A α C in a permanent human lymphoblastoid cell line, NL3 (Tohda, H. et al., 1980), IQ in Chinese hamster ovary cells (Thompson, L.H., et al., 1983) and PhIP in Chinese hamster V79 cells (Holme, J.A., et al., 1989). HAs such as Glu-P-1, Glu-P-2, A α C and MeA α C are reported to cuase chromosomal aberrations in cultured Chinese hamster lung cells (Ishidate, M., 1983). IQ induces single-strand DNA breaks in mouse leukemia cells and rat hepatocytes (Caderni, G., et al., 1983) and unscheduled DNA synthesis in rat hepatocytes (Barnes, W.S. et al., 1985). Most HAs have been shown to induce diphtheria toxin-resistant mutants of cultured Chinese hamster lung cells (Nakayasu, M., et al., 1983). In addition, some HAs give positive results in the wing spot test in *Drosophila melanogaster* (Yoo, M.A., et al., 1983).

CARCINOGENIC CHARACTERISTICS OF HAs

Trp-P-1, Trp-P-2, and Glu-P-1 transformed Syrian golden hamster embryo cells *in vitro* (Takayama, S., et al., 1977; Takayama, S., et al., 1979). IQs also induced transformed type III foci in a mouse embryo fibroblast cell line (Cortesi, E., et al., 1983). Subcutaneous injection of Trp-P-1 into F344 rats and Syrian golden hamsters resulted in local production of fibrosarcomas (Ishikawa, T., et al., 1979). IQ induced tumors in the mammary gland, liver and ear duct in rats when given by intragastric intubation (Tanaka, T., et al., 1985). Painting the skin of the back of mice with some HAs and then 12-O-tetradecanoylphorbol-13-acetate (TPA) resulted in development of papillomas and squamous cell carcinomas (Takahashi, M., et al., 1986; Sato, H., et al., 1987).

Long term feeding experiments on the carcinogenic effects of HAs are crucial, because HAs are present in foods. All long-term feeding experiments on HAs gave positive results in rats and mice (Sugimura, T., 1986; Sugimura, T., et al., 1988; Esumi, H., et al., 1989). In rats, tumors developed in the liver, small and large intestines, Zymbal gland, clitoral gland, skin, oral cavity and mammary gland. In mice, tumors developed in the liver, forestomach, lung, blood vessels, hematopoietic system and lymphoid tissue. These results are summarized in Table 2.

Studies on the effects of HAs on non-human primates are of special interest from the view point of involvement of HAs in development of cancer in humans. ^{32}P-postlabeling analysis demonstrated that the pattern of DNA adducts formed in the liver of rats fed IQ was the same as that of those formed in the liver of monkeys given IQ (Snyderwine, E.G., et al., 1988). The intensities of spots, namely the levels of total DNA adducts in the two were similar at comparable doses of IQ. This means that the susceptibilities of rats and monkeys to IQ are similar. Actually, the administration of IQ to monkeys resulted in the development of hepatocellular carcinomas as in rats (Adamson, R.H., et al., 1990). Furthermore, the capacities of liver enzymes from rats, monkeys and humans to activate HAs are within the same range (Ishida, Y., et al., 1987; Yamazoe, Y., et al., 1988; McManus, M.E., et al., 1989). These results suggest that HAs may play some role in the development of cancer in humans.

DISCUSSION

HAs are newly identified mutagens/carcinogens produced by heating amino acids, proteins, fish and meat and are actually present in foods. Their quantification can now be achieved by a combination of methods for their efficient partial purification and sensitive detection (Yamaizumi, Z., et al., 1986; Murray, S., et al., 1988; Gross, G.A. et al., 1989; Takahashi, M., et al., 1985; Turesky, R.J., et al., 1988). Their partial purification involves XAD-2 column chromatography, acid-base liquid partition and blue cotton adsorption (Hayatsu, H., et al., 1983; Hayatsu, H., 1990). HAs in partially purified samples are analyzed by liquid chromatography in a system equipped with apparatus for UV-, fluorescence- and electrochemical-detection, and mass spectrometry and also by gas chromatography-mass spectrometry. The most abundant HA is PhIP, its level being 27.0 - 69.2 ng/g cooked food. The level of MeIQx is second

Table 2. Carcinogenicities of HAs

Chemical	Species	Concentration (%)	Target Organs
Trp-P-1	Rats	0.015	Liver
	Mice	0.02	Liver
Trp P-2	Mice	0.02	Liver
Glu-P-1	Rats	0.05	Liver, small & large intestines, Zymbal gland, clitoral gland
	Mice	0.05	Liver, blood vessels
Glu-P-2	Rats	0.05	Liver, small & large intestines, Zymbal gland, clitoral gland
	Mice	0.05	Liver, blood vessels
A α C	Mice	0.08	Liver, blood vessels
MeA α C	Mice	0.08	Liver, blood vessels
IQ	Rats	0.03	Liver, small & large intestines, Zymbal gland, clitoral gland, skin
	Mice	0.03	Liver, forestomach, lung
MeIQ	Rats	0.03	Large intestines, Zymbal gland, skin, oral cavity, mammary gland
	Mice	0.03	Liver, forestomach
MeIQx	Rats	0.04	Liver, Zymbal gland, clitoral gland, skin
	Mice	0.06	Liver, lung, hematopoietic system
PhIP	Mice	0.04	Lymphoid tissue

highest, being of 0.64 - 6.44 ng/g cooked food (Sugimura, T., et al., 1988; Zhang, X-M., et al., 1988). On HPLC of urine from human subjects who ate fried ground beef, mutagenicity was found in positions corresponding to possible metabolites of MeIQx (Hayatsu, H., et al., 1985). This finding provides important information on the adsorption, metabolism and excretion of HAs in humans.

Most experiments in rodents on the long-term effects of HAs have been carried out at a single dose close to the maximum tolerable dose. Data obtained in long-term feeding experiments in mice with various doses of MeIQx (0.002, 0.006, 0.02 and 0.06% in the diet) indicated that the dose-response for carcinogenicity was not linear: no liver tumors developed in mice fed 0.002 or 0.006 % MeIQx and liver tumors were induced in mice given 0.02 and 0.06 % of MeIQx at incidences of 3 and 82%, respectively. On the other hand, a linear relation was obtained between the dose of MeIQx and the DNA-adduct level in the livers of rats given diet containing 0.0004 to 0.04 % MeIQx (Yamashita, K., et al., 1990), suggesting that there may be no threshold for the formation of DNA adducts by HAs.

The extent of human exposure to HAs is certainly insufficient alone to explain the development of cancer in humans. However the effects of HAs on mouse skin carcinogenesis are greatly enhanced by TPA (Takahashi, M., et al., 1986; Sato, H., et al., 1987) and on rat hepatocarcinogenesis by phenobarbital (Tsuda, H., et al., 1988).

Human cancers show various genetic alterations (Yamada, H., et al., 1986; Yokota, J., et al., 1987; Vogelstein, B., et al., 1988; Yokota, J., et al., 1989; Mori, N., et al., 1989). The frequency of familial retinoblastoma, which arises from cells with abnormality in one allele of the *RB* gene is at least 10^5 times that of sporadic retinoblastoma, which arises from cells with two intact alleles of the *RB* gene (Knudson, A.G., 1985). Cells with some of the multiple genetic alterations found in malignant tumors may exist in the human body. Those cells may be caused by exposure to vast numbers of various types of environmental mutgens, but each of those at very minute dose levels, and then by non-specific cell proliferation, such as tissue damage/regeneration with trauma, infection and autoimmune conditions. Cells with genetic alterations may grow monoclonally to produce preneoplastic lesions that are more susceptible than normal cells to HAs. Thus in considering cancer prevention, estimation of the risks of HAs to humans should not be based simply on a comparison of doses producing tumors in long-term experiments in rodents with the levels of exposure of humans. It is still premature to evaluate the real risk of the contribution of HAs to development of human cancer. At present, the most realistic method for reducing the risk of HAs is to reduce the formation and intake of HAs as far as possible by pragmatic measures without affecting the taste of foods or normal daily life.

REFERENCES

Adamson, R.H., Thorgeirsson, U.P., Snyderwine, E.G., Thorgeirsson, S.S., Reeves, J., Dalgard, D.W., Takayama, S. and Sugimura, T. (1990). Carcinogenicity of 2-amino-3-methylimidazo[4,5-f]-quinoline in nonhuman primates: Induction of tumors in three Macaques. *Jpn. J. Cancer Res.* **81**, 10-14.

Aoyama, T., Gonzalez, F.J. and Gelboin, H.V. (1989). Mutagen activation by cDNA-expressed P_1450, P_3450, and P450a. *Mol. Carcinog.* **1**, 253-259.

Barnes, W.S., Lovelette, C.A., tong, C., Wiliams, G.M. and Weisburger, J.H. (1985). Genotoxicity of the food mutagen 2-amino-3-methylimidazo[4,5-f]quinoline (IQ) and analogs. *Carcinogenesis* **6**, 441-444.

Becher, G., Knize, M.G., Nes, I.F. and Felton, J.S. (1988). Isolation and identification of mutagens from a fried Norwegian meat product. *Carcinogenesis* **9**, 247-253.

Caderni, G., Kreamer, B.L. and Dolara, P. (1983). DNA damage of mammalian cells by the beef extract mutagen 2-amino-3-methylimidaxo[4,5-f]quinoline. *Food. Chem. Toxicol.* **21**, 641-643.

Cortesi, E. and Dolara, P. (1983). Neoplastic transformation of BALB 3T3 mouse embryo fibroblasts by the beef extract mutagen 2-amino-3-methylimidazo[4,5-f]quinoline. *Cancer Lett.* **20**, 43-47.

Esumi, H., Ohgaki, H., Kohzen, E., Takayama, S., and Sugimura, T. (1989). Induction of lymphoma in CDF1 mice by the food mutagen, 2-amino-1-methyl-6-phenylimidazo[4,5-b]pyridine. *Jpn. J. Cancer Res.* **80**, 1176-1178.

Felton, J.S., Knize, M.G., Shen, N.H., Lewis, P.R., Andreson, B.D., Happe, J. and Hatch, F.T. (1986). The isolation and identification of a new mutagen from fried ground beef: 2-amino-1-methyl-6-phenyl-imidazo[4,5-b]pyridine (PhIP). *Carcinogenesis* **7**, 1081-1086.

Felton, J.S., Knize, M.G., Wood, C., Wuebbles, B.J., Healy, S.K., Stuermer, D.H., Bjeldanes, L.F., Kimble, B.H. and Hatch, F.T. (1984). Isolation and characterization of new mutagens from fried ground beef. *Carcinogenesis* **5**, 95-102.

Grivas, S., Nyhammar, T., Olsson, K. and Jägerstad, M. (1985). Formatin of a new mutagenic DiMeIQx compound in a model system by heating creatinine, alanine and fructose. *Mutat. Res.* **151**, 177-183.

Grivas, S., Nyhammar, T., Olsson, K. and Jägerstad, M. (1986). Isolation and identification of the food mutagens IQ and MeIQx from a heated model system of creatinine, glycine and fructose. *Food Chem.* **20**, 127-136.

Grose, K.R., Grant, J.L., Bjeldanes, L.F., Andreson, B.D., Heales, S.K., Lewis, P.R., Felton, J.S. and Hatch, F.T. (1986). Isolation of the carcinogen IQ from fried egg patties. *J. Agric. Food. Cehm.* **34**, 201-202.

Gross, G.A., Philippossian, G. and Aeschbacher, H.U. (1989). An efficient and convenient method for the purification of mutagenic heterocyclic amines in heated meat products. *Carcinogenesis* **10**, 1175-1182.

Hargraves, W.A. and Pariza, M.W. (1983). Pruification and mass spectral characterization of bacterial mutagens from commercial beef extract. *Cancer Res.* **43**, 1467-1472.

Hashimoto, Y., Shudo, K. and Okamoto, T. (1980). Metabolic activation of a mutagen, 2-amino-6-methyldipyrido[1,2-a:3',2'-d]imidazole. Identification of 2-hydroxyamino-6-methyldipyrido[1,2-a:3',2'-d]imidazole and its reaction with DNA. *Biochem. Biophys. Res. Commun.* **92**, 971-976.

Hashimoto, Y., Shudo, K. and Okamoto, T. (1980). Activation of a mutagen, 3-amino-1-methyl-5H-pyrido[4,3-b]indole. Identification of 3-hydroxyamino-1-methyl-5H-pyrido[4,3-b]indole and its reaction with DNA. *Biochem. Biophys. Res. Commun.* **96**, 355-362.

Hayatsu, H., Matsui, Y., Ohara, Y., Oka, T. and Hayatsu, T. (1983). Characterization of mutagenic fractions in beef extract and in cooked ground beef. Use of blue-cotton for efficient extration. *Gann.* **74**, 472-482.

Hayatsu, H., Oka, T., Wakata, A., Ohara, Y., Hayatsu, T., Kobayashi, H., and Arimoto, S. (1983). Adsorption of mutagens to cotton bearing covalently bound trisulfocopper-phthalocyanine. *Mutat. Res.*, **119**, 233-238.

Hayatsu, H., Hayatsu, T., and Ohara, Y. (1985). Mutagenicity of human urine caused by ingestion of fried ground beef. *Jpn. J. Cancer. Res.* (Gann) **76**, 445-448.

Hayatsu, H. (1990). Blue cotton-broad possibility in assessing mutagens/carcinogens in the environment. In, *Advances in Mutagenesis Research-1*, G. Obe, ed., pp. 1-26, Springer-Verlag, Berlin.

Holme, J.A., Wallin, H., Brunborg, G., Soderlund, E.J., Hongslo, J.K. and Alexander, J. (1989). Genotoxicity of the food mutagen 2-amino-1-methyl-6-phenylimidazo[4,5-b]pyridine (PhIP): formation of 2-hydroxamino-PhIP, a directly acting genotoxic metabolite. *Carcinogenesis* **10**, 1389-1396.

Int. Agency Res. Cancer (1987), "IARC Monographs on the Evaluation of Carcinogenic Risks to Humans" Supplement 6, Genetic and related effects: an updating of selected IARC Monographs from Volumes 1 to 42, Int. Agency Res. Cancer, Lyon, pp. 40-45.

Ishida, Y., Negishi, C., Umemoto, A., Fujita, Y., Sato, S., Sugimura, T., Thorgeirsson, S.S. and Adamson, R.H. (1987). Activation of mutagenic and carcinogenic heterocyclic amines by S-9 from the liver of a rhesus monkey. *Toxic. In Vitro* **1**, 45-48.

Ishidate, M. (1983). Chromosomal aberration test *in vitro* (M. Ishidate, ed.) p. 256, Realize Inc., Tokyo.

Ishikawa, T., Takayama, S., Kitagawa, T., Kawachi, T., Kinebuchi, M., Matsukura, N., Uchida, E. and Sugimura, T. (1979). *In vivo* experiments on tryptophan pyrolysis products. In, *Naturally Occurring Carcinogens-Mutagens and Modulators of Carcinogenesis.* E.C. Miller, J.A. Miller, I. Hirono, T. Sugimura, and S. Takayama eds., pp. 159-167, Japan Scientific Societies Press/University Park Press, Tokyo/Baltimore.

Jägerstad, M., Olsson, K., Grivas, S., Negishi, C., Wakabayashi, K., Tsuda, M., Sato, S. and Sugimura, T. (1984). Formation of 2-amino-3,8-dimethylimidazo[4,5-f]quinoxaline in a model system by heating creatinine, glycine and glucose. *Mutat. Res.* **126**, 239-244.

Kato, R., and Yamazoe, Y. (1987). Metabolic activation and covalent binding to nucleic acids of carcinogenic heterocyclic amines from cooked foods and amino acid pyrolysates. *Jpn. J. Cancer Res.* (Gann), **78**, 297-311.

Kikugawa, K., Kato, T. and Takahashi, S. (1989). Possible presence of 2-amino-3,4-dimethylimidazo[4,5-f]quinoline and other heterocyclic amine-like mutagens in roasted coffee beans. *J. Agric. Food. Chem.* **37**, 881-886.

Knudson, A.G. (1985). Hereditary cancer, oncogenes, and antioncogenes. *Cancer Res.* **45**, 1437-1443.

Kondo, S., and Ichikawa-Ryo, H. (1973). Testing and classification of mutagenicity of furylfuramide in *Escherichia coli*. *Jpn. J. Genet.*, **48**, 295-300.

Matsumoto, T., Yoshida, D. and Tomita, H. (1981). Determination of mutagens, amino-carbolines in grilled foods and cigarette smoke condensate. *Cancer Lett.* **12**, 105-110.

McManus, M.E., Felton, J.S., Knize, M.G., Burgess, W.M., roberts-Thomson, S., Pond, S.M., Stupans, I. and Veronese, M.E. (1989). Activation of the food-derived mutagen 2-amino-1-methyl-6-phenylimidazo-[4,5-b]pyridine by rabbit and human liver microsomes and purified forms of cytochrome P-450. *Carcinogenesis* **10**, 357-363.

Mori, N., Yokota, J., Oshimura, M., Cavenee, W.K., Mizoguchi, H., Noguchi, M., Shimosato, Y., Sugimura, T. and Terada, M. (1989). Concordant deletions of chromosome 3p and loss of heterozygosity for chromosomes 13 and 17 in small cell lung carcinoma. *Cancer Res.* **49**, 5130-5135.

Murray, S., Gooderham, N.J., Boobis, A.R. and Davies, D.S. (1988). Measurement of MeIQx and DiMeIQx in fried beef by capillary column gas chromatography electron capture negative ion chemical ionisation mass spectrometry. *Carcinogenesis* **9**, 321-325.

Nagao, M., Honda, M., Seino, Y., Yahagi, T., Kawachi, T. and Sugimura, T. (1977). Mutagenicities of protein pyrolysates. *Cancer Lett.* **2**, 335-340.

Nakayasu, M., Nakasato, F., Sakamoto, H., Terada, M. and Sugimura, T. (1983). Mutagenic activity of heterocyclic amines in Chinese hamster lung cells with diphtheria toxin resistance as a mrker. *Mutat. Res.* **118**, 91-102.

Negishi, C., Wakabayashi, K., Tsuda, M., Sato, S., Sugimura, T., Saito, H., Maeda, M. and Jägerstad, M. (1984). Formation of 2-amino-3,7,8-trimethylimidazo[4,5-f]quinoxaline, a new mutagen, by heating a mixture of creatinine, glucose and glycine. *Mutat. Res.*, **140**, 55-59.

Negishi, C., Wakabayashi, K., Yamaizumi, Z., Saito, H., Sato, S., Sugimura, T. and Jägerstad, M. (1985). Identification of 4,8-DiMeIQx, a new mutagen,. Selected Abstracts of Papers presented at the 13th Annual Meeting of the Environemntal Mutagen Society of Japan, 12-13 Oct. 1984, Yokyo (Japan). *Mutat. Res.* **147**, 267-268.

Nemoto, N. and Takayama, S. (1984). Activation of 2-amino-6-methyl-dipyrido[1,2-a:3',2'-d]imidazole, a mutagenic pyrolysis product of glutamic acid, to bind to microsomal protein by NADPH-dependent and -independent enzyme systems. *Carcinogenesis* **5**, 653-656.

Petry, T.W., Josephy, P.D., Pagano, D.A., Zeiger, E., Knecht, K.T. and Eling, T.E. (1989). Prostaglandin hydroperoxidase-dependent activation of heterocyclic aromatic amines. *Carcinogenesis* **10**, 2201-2207.

Sato, H., Takahashi, M., Furukawa, F., Miyakawa, Y., Hasegawa, R., Toyada, K., Hayashi, Y. (1987). Initiating activity in a two-stage mouse skin model of nine mutagenic pyrolysates of amino acids, soybean globulin and proteinaceous food. *Carcinogenesis* **8**, 1231-1234.

Snyderwine, E.G., Yamashita, K., Adamson, R.H., Sato, S., Nagao, M., Sugimura, T. and Thorgeirsson, S.S. (1988). Use of the ^{32}P-postlabeling method to detect DNA adducts of 2-amino-3-methylimidazo[4,5-f]quinoline (IQ) in monkeys fed IQ: identification of the N-(deoxyguanosin-8-yl)-IQ adduct. *Carcinogenesis* **9**, 1739-1743.

Snyderwine, E.G., Wirth, P.J., Roller, P.P., Adamson, R.H., Sato, S. and Thorgeirsson, S.S. (1988). Mutagenicity and *in vitro* covalent DNA binding of 2-hydroxyamino-3-methylimidazolo[4,5-f]quinoline. *Carcinogenesis* **9**, 411-418.

Snyderwine, E.G., Roller, P.P., Adamson, R.H., Sato, S. and Thorgeirsson, S.S. (1988). Reaction of *N*-hydroxylamine and *N*-acetoxy derivatives of 2-amino-3-methylimidazo[4,5-f]quinoline with DNA. Synthesis and identification of *N*-(deoxyguanosin-8-yl)-IQ. *Carcinogenesis* **9**, 1061-1065.

Snyderwine, E.G. and Battula, N. (1989). Selective mutagenic activation by cytochrome P_3-450 of carcinogenic arylamines found in foods. *J. Natl. Cancer Inst.* **81**, 223-227.

Sugimura, T., Nagao, M., Kawachi, T., Honda, M., Yahagi, T., Seino, Y., Sato, S., Matsukura, N., Matsushima, T., Shirai, A., Sawamura, M. and Matsumoto, H. (1977) Mutagen-carcinogens in food, with special reference to highly mutagenic pyrolytic products in broiled foods. In, *Origins of Human Cancer* (H.H. Hiatt, J.D. Watson, and J.A. Winsten, eds.), pp. 1561-1577. Cold Spring Harbor Laboratory, Cold Spring Harbor, NY.

Sugimura, T. (1982). Mutagens, carcinogens, and tumor promoters in our daily food. *Cancer* **49**, 1970-1984.

Sugimura, T. (1986). Studies on environmental chemical carcinogenesis in Japan. *Science*, **233**, 312-318.

Sugimura, T., Sato, S. and Wakabayashi, K. (1988). Mutagens/carcinogens in pyrolysates of amino acids and proteins and in cooked foods: heterocyclic aromatic amines. In *Chemical Induction of Cancer, Structural Bases and Biological Mechanisms* (Y-T. Woo, D.Y. Lai, J.C. Arcos, and M.F. Argus eds.), Vol. IIIC, pp, 681-710, Academic Press, San Deigo.

Takahashi, M., Wakabayashi, K., Nagao, M., Yamamoto, M., Masui, T., Goto, T., Kinae, N., Tomita, I. and Sugimura T. (1985). Quantification of 2-amino-3-methylimidazo[4,5-f]quinoline (IQ) and 2-amino-3-dimethylimidazo[4,5-f]quinoxaline (MeIQx) in beef extracts by liquid chromatography with electrochemical detection (LCEC). *Carcinogenesis* **6**,1195-1199.

Takahashi, M., Furukawa, F., Miyakawa, Y., Sato, H., Hasegawa, R. and Hayashi, Y. (1986). 3-Amino-1-methyl-5H-pyrido[4,3-b]indole initiates two-stage carcinogenesis in mouse skin but is not a complete carcinogen. *Jpn. J. Cancer. Res.* (Gann) **77**, 509-513.

Takayama, S., Katoh, Y., Tanaka, M., Nagao, M., Wakabayashi, K., and Sugimura, T. (1977). *In vitro* transformation of hamster embryo cells with tryptophan pyrolysis products. *Proc. Jpn. Acad.* **53**B, 126-129.

Takayama, S., Hirakawa, T., Tanaka, M., Kawachi, T. and Sugimura, T. (1979). *In vitro* transformation of hamster embryo cells with a glutamic acid pyrolysis product. *Toxicol. Lett.* **4**, 281-284.

Tanaka, T., Barnes, W.S., Williams, G.M. and Weisburger, J.H. (1985). Multipotential carcinogenicity of the fried food mutgen 2-amino-3-methylimidazo[4,5-f]quinoline in rats. *Jpn. J. Cancer Res.* (Gann) **76**, 570-576.

Thompson, L.H., Carrano, A.V., Salazar, E., Felton, J.S. and Hatch, F.T. (1983). Comparative genotoxic effects of the cooked-food-related mutagens Trp-P-2 and IQ in bacteria and cultured mammalian cells. *Mutat. Res.* **117**, 243-257.

Tohda, H., Oikawa, A., Kawachi, T. and Sugimura, T. (1980). Induction of sister-chromatid exchanges by mutgens from amino acid and protein pyrolysates. *Mutat. Res.* **77**, 65-69.

Tsuda, H., Asamoto, M., Ogiso, T., Inoue, T., Ito, N. and Nagao, M. (1988). Dose-dependent induction of liver and thyroid neoplastic lesions by short-term administration of 2-amino-3-methylimidazo[4,5-f]quinoline combined with partial hepatectomy followed by phenobarbital or low dose 3'-methyl-4-dimethylaminoazobenzene promotion. *Jpn. J. Cancer. Res.* (Gann) **79**, 691-697.

Turesky, R.J., Wishnok, J.S., Tannenbaum, S.R., Pfund, R.A. and Buchi, G.H. (1983). Qualitative and quantitative characterization of mutagens in commercial beef extract. *Carcinogenesis* **4**, 863-866.

Turesky, R.J., Bur, H., Huynh-Ba, T., Aeschbacher, H.U. and Milon, H. (1988). Analysis of mutagenic heterocyclic amines in cooked beef products by high-performance liquid chromatography in combination with mass spectrometry. *Fd. Chem. Toxic.* **26**, 501-509.

Vogelstein, B., Fearon, E.R., Hamilton, S.R., Kern, S.E., Preisinger, A.C., Leppert, M., Nakamura, Y., White, R., Smits, A.M.M. and bos, J.L. (1988). Genetic alterations during colorectal-tumor development. *N. Engl. J. Med.* **319**, 525-532.

Wild, D. and Degen, G.H. (1987). Prostaglandin H synthase-dependent mutagenic activation of heterocyclic aromatic amines of the IQ-type. *Carcinogenesis* **8**, 541-545.

Yahagi, T., Nagao, M., Hara, K., Matsushima, T., Sugimura, T. and Bryan, G.T. (1974). Relationships between the carcinogenic and mutagenic or DNA-modifying effects of nitrofuran derivatives, including 2-(2-furyl)-3-(5-nitro-2-furyl)acrylamide, a food additive. *Cancer Res.* **34**, 2266-2273.

Yamada, H., Yoshida, T., Sakamoto, H., Terada, M. and Sugimura, T. (1986). Establishment of a human pancreatic adenocarcinoma cell line (PSN-1) with amplifications of both c-*myc* and activated c-Ki-*ras* by a point mutation. *Biochem. Biophys. Res. Commun.* **140**, 167-173.

Yamaguchi, K., Shudo, K., Okamoto, T., Sugimura, T. and Kosuge, T. (1980). Presence of 2-aminodipyrido[1,2-a:3',2'-d]imidazole in broiled cuttlefish. *Gann.* **71**, 743-744.

Yamaizumi, Z., Shiomi, T., Kasai, H., Nishimura, S., Takahashi, Y., Nagao, M. and Sugimura, T. (1980). Detection of potent mutagens, Trp-P-1 and Trp-P-2, in broiled fish. *Cancer Lett.* **9**, 75-83.

Yamaizumi, Z., Kasai, H., Nishimura, S., Edmonds, C.G. and McCloskey, J.A. (1986).

Stable isotope dilution quantification of mutagens in cooked foods by combined liquid chromatography-thermospray mass spectrometry. *Mutat. Res.* **173**, 1-7.

Yamashita, K., Adachi, M., Kato, S., Nakagama, H., Ochiai, M., Wakabayashi, K., Sato, S., Nagao, M. and Sugimura, T. (1990). DNA adducts formed by 2-amino-3,8-dimethylimidazo[4,5-f]quinoxaline in rat liver: dose-response on chronic administration. *Jpn. J. Cancer. Res.* in press.

Yamazoe, Y., Abu-Zeid, M., Yamaguchi, K., and Kato, R. (1988). Metabolic activation of pyrolysate arylamines by human liver microsomes: possible involvement of a P-448-H type cytochrome P-450. *Jpn. J. Cancer Res.* (Gann) **79**, 1159-1167.

Yamazoe, Y., Abu-Zeid, M., Manabe, S., Toyama, S. and Kato, R. (1988). Metabolic activation of a protein pyrolysate promutagen 2-amino-3,8-dimethylimidazo[4,5-f]quinoxaline by rat liver microsomes and purified cytochrome P450. *Carcinogenesis* **9**, 05-109.

Yokota, J., Wada, M., Shimosato, Y., Terada, M. and Sugimura, T. (1987). Loss of heterozygosity on chromosomes 3, 13, and 17 in small-cell carcinoma and on chromosome 3 in adenocarcinoma of the lung. *Proc. Natl. Acad. Sci. USA* **84**, 9252-9256.

Yokota, J., Tsukada, Y., Nakajima, T., Gotoh, M., Shimosato, Y., Mori, N., Tsunokawa, Y., Sugimura, T. and Terada, M. (1989). Loss of heterozygosity on the short arm of chromosome 3 in carcinoma of the uterine cervix. *Cancer Res.* **49**, 3598-3601.

Yoo, M.A., Ryo, H., Todo, T. and Kondo, S. (1985). Mutagenic potency of heterocyclic amines in the *Drosophila* wing spot test and its correlation to carcinogenic potency. *Jpn. J. Cancer Res.* (Gann), **76**, 468-473.

Yoshida, D., Matsumoto, T., Yoshimura, R., and Matsuzaki, T. (1978). Mutagenicity of amino-α-carbolines in pyrolysis products of soybean globulin. *Biochem. Biophys. Res. Commun.* **83**, 915-920.

Zhang, X-M., Wakabayashi, K., Liu, Z-C., Sugimura, T. and Nagao, M. (1988). Mutagenic and carcinogenic heterocyclic amines in Chinese cooked foods. *Mutat. Res.* **201**, 181-188.

ACETAMINOPHEN AND PROTEIN THIOL MODIFICATION

Sidney D. Nelson, Mark A. Tirmenstein, Mohamed S. Rashed and Timothy G. Myers

Department of Medicinal Chemistry, BG-20
University of Washington
Seattle, WA USA 98195

INTRODUCTION

Acetaminophen (4'-hydroxyacetanilide, APAP) is a widely-used analgesic and antipyretic drug which, while considered to be safe at therapeutic doses, can cause acute hepatic centrilobular necrosis in both humans and experimental animals when consumed in large doses (Boyd and Bereczky, 1966; Prescott et al., 1971; for a review see Hinson, 1980). In a series of classic studies (Mitchell et al., 1973a,b; Jollow et al., 1973, 1974; Potter et al., 1973, 1974), protein thiol group arylation by a reactive quinone imine metabolite of APAP was implicated in the pathogenesis of hepatotoxicity. Indirect evidence to support the hypothesis that cysteinyl thiol groups were arylated was provided by mass spectral characterization of thioether metabolites of acetaminophen (Jollow et al., 1974; Knox and Jurand, 1977; Nelson et al., 1981), and the 3-position of the aromatic ring was determined by ^{1}H and ^{13}C-NMR to be the site of conjugation with glutathione (Hinson et al., 1982). A few years later (Streeter et al., 1984; Hoffman et al., 1985), cysteinyl thioether conjugates at the same position of the aromatic ring were characterized as the major protein bound residues of acetaminophen.

Recently, immunological detection of acetaminophen bound to proteins has corroborated the results of the physicochemical analyses reported above (Bartolone et al., 1987; Roberts et al., 1987). Based on mass spectral analysis (Hoffman et al., 1985; Axworthy et al., 1988) and epitope characterization (Potter et al., 1989) of acetaminophen in mouse liver and serum protein adducts, most of the adduct is the 3-cysteine conjugate of APAP itself. Lesser amounts of the adduct (<30%) have lost the acetyl group and correspond to conjugates of p-benzoquinone (Axworthy et al., 1988). This agrees with earlier studies of the binding of selectively radiolabeled analogs of APAP (Nelson et al., 1981) which showed that approximately 20% of the acetyl group is lost as acetamide, most likely by hydrolysis of the initially formed reactive metabolite, *N*-acetyl-*p*-benzoquinone imine (NAPQI).

The development of antibodies to adducts of acetaminophen provides an exciting approach to the isolation and identification of target proteins. This process has already begun, and molecular weights of a few arylated proteins have been determined (Bartolone et al., 1988; 1989). The major drawback to this approach is that other target proteins may not be identified. The antibodies may not access some bound adducts, and will not recognize other types of protein damage caused by reactive metabolites of acetaminophen. For example, it is now known that the major reactive metabolite of APAP is NAPQI which, like most quinones, can act as both an

Biological Reactive Intermediates IV, Edited by C.M. Witmer *et al.*
Plenum Press, New York, 1990

electrophile and an oxidant (Blair et al., 1980; Dahlin et al., 1984). Moreover, acute toxicity to hepatocytes caused by either NAPQI (Albano et al., 1985; Moore et al., 1985) or APAP (Moore et al., 1985; Tee et al., 1986) is significantly attenuated by the late addition of the disulfide reducing agent, dithiothreitol (DTT). This would indicate that oxidation of protein thiols may be important in the pathogenesis of hepatoxicity caused by APAP, a kind of protein damage not recognized by the antibodies to arylated proteins.

In order to further delineate the mechanisms of protein thiol modification by APAP and to further examine how this may relate to its mechanism of hepatoxicity, we have carried out studies on the mechanism of thiol oxidation by NAPQI, and we have compared the effects of APAP and its non-hepatotoxic regioisomer, 3'-hydroxyacetanilide (AMAP) on thiol status in mouse liver.

RESULTS

Glutathione Oxidation by NAPQI

We found that the reaction of NAPQI with GSH was first order in the quinone imine and GSH, and that over a wide range of reactant concentrations and pH, the reaction yields the three products (GS-APAP, APAP and GSSG) in an invariant 2:2:1 ratio (Coles et al., 1988). There are two likely explanations for these results, and both involve the formation of an ipso adduct (Meisenheimer-type complex) of GSH with NAPQI. A role for such adducts has been proposed previously (Fernando et al., 1980; Smith and Mitchell, 1985; Rundgren et al., 1988), and a carbinolamide (hydrated ipso adduct) of NAPQI is formed in aqueous media (Novak et al., 1986).

We studied the reaction of NAPQI with GSH by thermospray LC/MS. Immediately after mixing an equimolar mixture of the two in phosphate buffer at pH 7.4, a sample was introduced into the LC/MS system and the total ion current (TIC) plus scans of eluting ionizable products were recorded (Figure 1). At least six different products were observed including APAP (6 min) and GSSG (2.6 min). A major product appeared at approximately 5 min in the TIC that had the same retention time and spectrum as the 3-glutathionyl adduct of APAP. A second glutathionyl metabolite of APAP appeared at approximately 3.5 min in the TIC, based on ions in the scan at *m/z* 130 for the glutamate moiety, *m/z* 152 for APAP, *m/z* 184 for a thiol derivative of APAP and *m/z* 308 for glutathione. The latter three ions are for protonated species and indicate decomposition of the adduct in the ion source. Work is underway to better characterize this presumed ipso adduct by both MS and NMR techniques.

Effects of APAP and AMAP on Hepatic Thiols

AMAP is the meta positional isomer of APAP that was found several years ago to bind to hepatic proteins of hamsters and yet not cause hepatotoxicity (Roberts and Jollow, 1979). We have carried out studies in mice on the metabolism of AMAP and found that it is metabolized to dihydroxyacetanilides that are sequentially oxidized to quinones and react with GSH to form GSH conjugates (Rashed and Nelson, 1989a,b).

Although the quinone metabolites of AMAP are similar to NAPQI and bind to about the same overall extent in mouse liver, they are apparently more reactive and bind to proteins within the cell that are nearer to their site of formation in the endoplasmic reticulum. In fact, we have found significant differences in the intracellular distribution of covalent binding to mouse liver at a time after administration when total covalent binding is approximately equal (Tirmenstein and Nelson, 1989). For example, AMAP binds more extensively to proteins in the microsomal fraction, whereas APAP binds more extensively to proteins in the mitochondrial fraction (Figure 2). Moreover, APAP depletes mitochondrial GSH to a greater extent than AMAP, and at 6 h post-administration, about one-third of the GSH recovered is in the oxidized form (GSSG) which is consistent with the known oxidative properties of NAPQI (Blair et al., 1980; Dahlin et al., 1984; Albana et al., 1985; Coles et al., 1988).

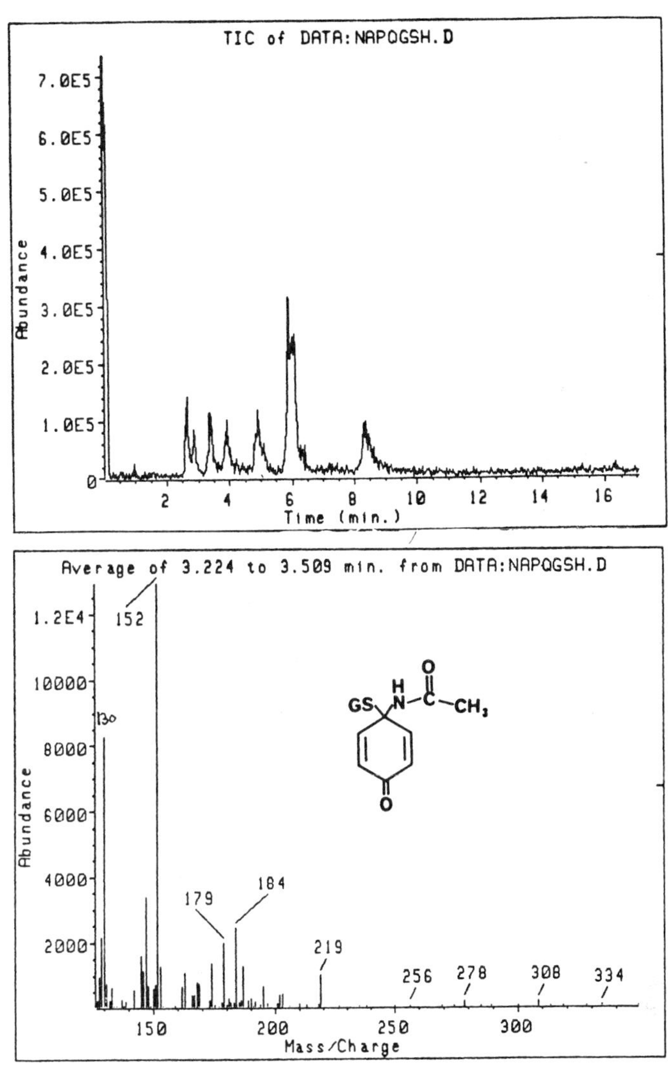

Figure 1. HPLC/MS of an equimolar mixture of NAPQI and GSH. The top panel is the total ion chromatogram (TIC), and the bottom panel is the mass spectrum of the region of the TIC that contains the suspected glutathione ipso adduct of NAPQI.

These studies indicate that thiols in the mitochondria may be an important target in the pathogenesis of acetaminophen hepatotoxicity. We examined this by the administration of L-buthionine sulfoximine (BSO) to deplete mitochondrial glutathione. Although AMAP depletes cytosolic GSH, it has little effect (unlike APAP) on mitochondrial GSH (Figure 3). However, when administered after BSO which depletes both cytosolic and mitochondrial GSH, a marked depletion occurs to about 10% of control levels. In contrast, diethylmaleate (DEM) primarily depletes cytosolic GSH and only depletes mitochondrial GSH, in conjunction with AMAP, to about 40% of control values (Figure 3). Subsequently, we have found that only the pretreatment with BSO causes AMAP to become hepatotoxic. This would indicate the importance of mitochondrial thiols in the toxic effects of APAP and AMAP.

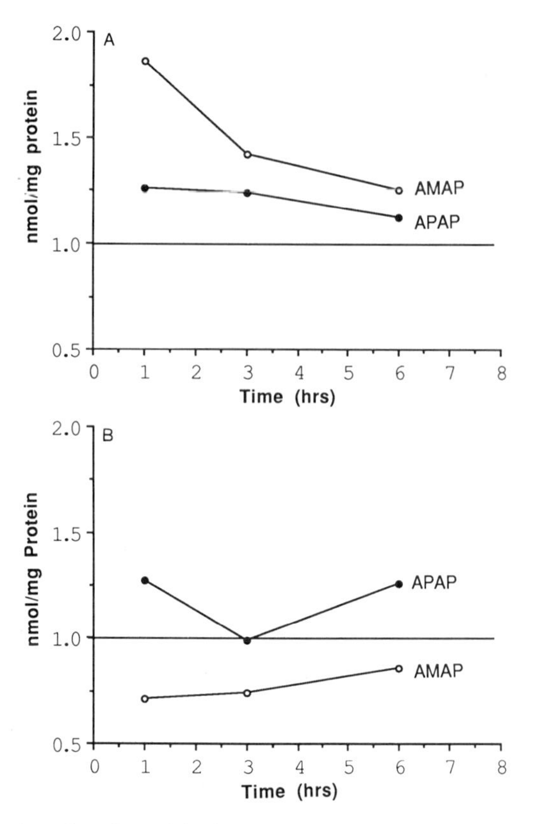

Figure 2. A. Covalent binding to microsomal protein (relative to homogenate); B. Covalent binding to mitochondrial protein (relative to homogenate). Covalent binding of radiolabel from [^{14}C-ring]-APAP and AMAP to microsomal and mitochondrial proteins relative to total homogenate at various times after the administration of each drug to mice.

Overall, hepatocellular protein thiols also are modified in different ways by AMAP and APAP. Although reactive metabolites of both AMAP and APAP covalently bind to about the same extent to hepatocellular protein thiols, APAP causes almost three times as much noncovalent loss of protein thiols 1 h after treatment of mice than does AMAP (Table 1). Based on the pronounced effect of DTT on restoration of the protein thiols, we believe that the noncovalent loss of protein thiols caused by reactive metabolites of APAP is primarily caused by the formation of disulfides though this remains to be demonstrated. Furthermore, the studies on ipso adduct formation suggests to us that, at least in part, the oxidation reactions to produce the disulfides occur through bimolecular reactions of the ipso adducts with thiol groups in amino acids (e.g., cysteine), peptides (GSH), or other proteins. Previously, we have found that NAPQI and related quinoid compounds will cause cross-linking of protein molecules (Streeter et al., 1986).

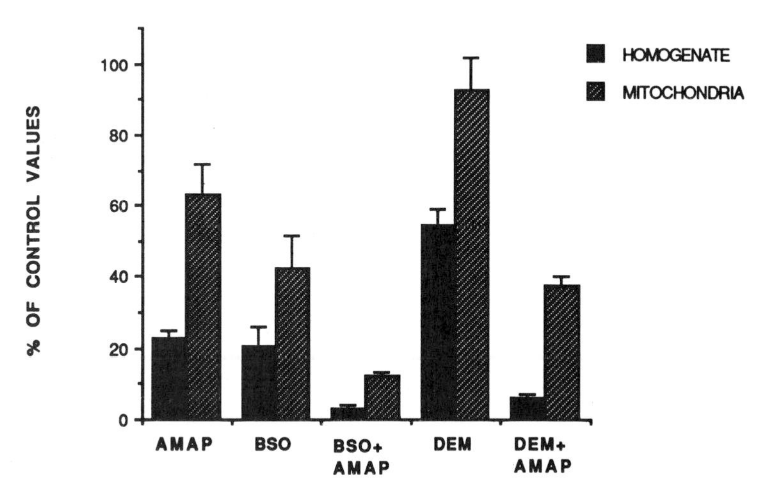

Figure 3. Effects of BSO and DEM on liver homogenate and mitochondrial glutathione levels in mice 2 h after the administration of either vehicle or AMAP.

Table 1. Hepatic Covalent Binding Levels and the Loss of Protein Thiols Induced by Acetaminophen and 3'-Hydroxyacetanilide One Hour after Administration.

	Loss of Protein Thiols* (nmol/mg prot)	Covalently Bound (nmol/mg prot)	Noncovalent Loss of Protein Thiols
AMAP	6.0	1.0	5.0
APAP	14.9	0.9	14.0
AMAP+DTT	2.1	1.0	1.1
APAP+DTT	0.1	0.9	---

*Determined by subtracting values from control levels.

Additional studies were carried out on the effects of AMAP and APAP on functional activities of some enzymes known to contain thiol groups important for their activities, and whose activities provide protection against cellular insult. Both cytosolic GSH peroxidase and thioltransferase activities were decreased 30-40% 1 h after APAP treatment, and GSH peroxidase was still inhibited 6 h post-treatment (Figure 4). Glutathione reductase, on the other hand, was not found to be significantly affected (data not shown). This may provide a partial explanation for the observed loss of protein thiols (mostly by disulfide formation) without a large increase in GSSG concentrations.

The most dramatic effect that we observed was on the enzyme xanthine dehydrogenase. In livers from mice treated with either vehicle or AMAP this enzyme was present mostly (about 90%) in its dehydrogenase form. However, 1 h after APAP treatment about 40% of it was present (representing a 4-fold increase) in its oxidase form (Figure 5). The conversion of xanthine dehydrogenase to xanthine oxidase may occur by oxidation of thiols, by thiol adduct formation, or by proteolysis (Della Corte and Stirpe, 1972). We are now attempting to determine if reactive metabolites of NAPQI mediate any of the above processes.

The Effects of Acetaminophen and 3'-Hydroxyacetanilide on Protein Thiol Levels and On Glutathione Peroxidase and Thioltransferase Activity in Mouse Liver

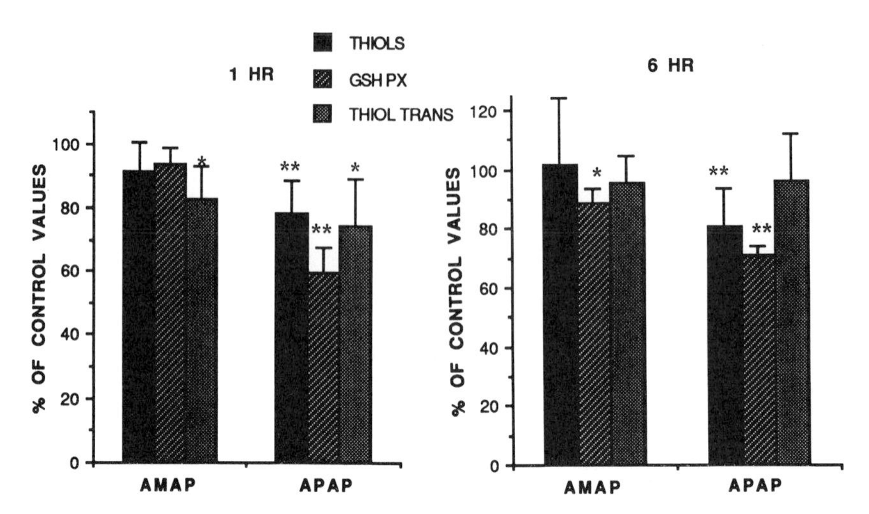

Figure 4. Effects of AMAP and APAP on the activities of liver cytosolic glutathione peroxidase and thiol transferase activities 1 h and 6 h after administration of the drugs to mice.

Because the xanthine oxidase form of the enzyme is known to generate toxic oxygen species, we determined the effect of the xanthine oxidase inhibitor, allopurinol, on APAP hepatotoxicity in mice. Allopurinol was found to decrease the severity of hepatotoxicity (based on visual inspection of the livers and serum ALT level decreases) 24 h after administration of APAP. Such an effect has been previously observed without an effect on reactive metabolite formation from APAP (Jaeschke and Mitchell, 1989).

The Effects of Acetaminophen and 3'-Hydroxyacetanilide on the Conversion of Xanthine Dehydrogenase to the Oxidase Form

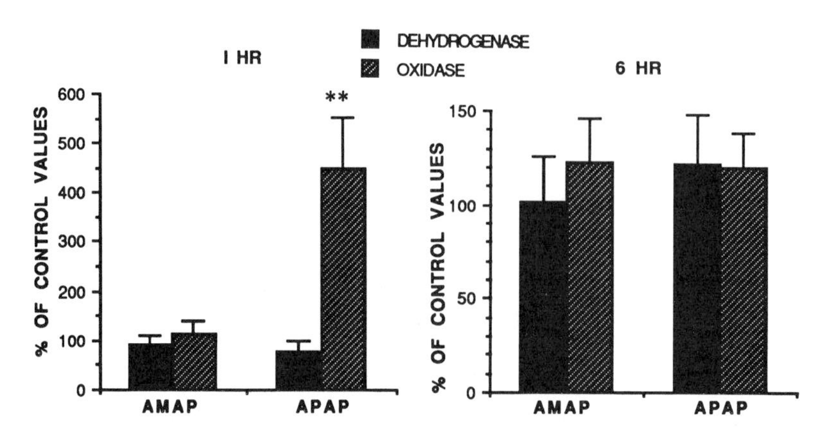

Figure 5. Effects of AMAP and APAP on the conversion of xanthine dehydrogenase to xanthine oxidase in liver cytosol 1 h and 6 h after administration of the drugs to mice.

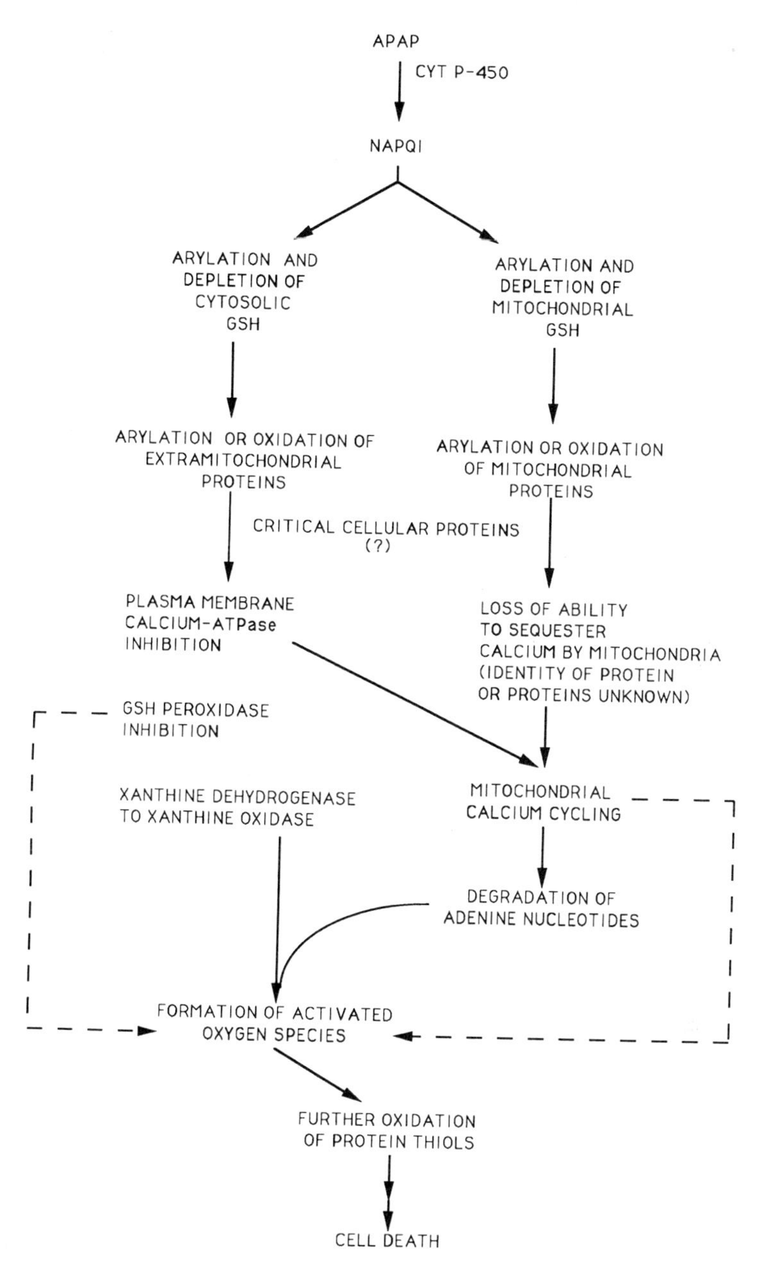

Figure 6. A scheme proposed for the pathogenesis of hepatotoxicity caused by APAP in mice.

DISCUSSION

It is clear that APAP modifies protein thiols in the target organ for toxicity by more than one mechanism. What is not clear is which of the mechanisms is responsible for the pathogenesis of hepatotoxicity. Based on the results presented here, as well as results reported by several other laboratories, we believe that both covalent and noncovalent modifications of protein thiols by reactive metabolites of APAP are important in the initiation of a biochemical cascade of events that leads to hepatotoxicity.

Our proposed mechanism (Figure 6) is based on the fact that both chemically and biochemically, the major reactive metabolite of APAP, NAPQI, is a quinone imine that is both a soft electrophile and an oxidant. Therefore, it can arylate and oxidize both non-protein and protein thiols. Studies with the regioisomer, AMAP, suggest that modification of mitochondrial thiols may be an important early event in the pathogenetic process. Some critical cellular proteins may be affected, something that is being actively pursued by immunological and molecular biological methods, at least in the identification of adducted proteins.

The arylating and oxidizing events apparently affect the function of several cellular proteins. Plasma membrane Ca^{2+}-ATPase is inhibited and the mitochondria lose their ability to sequester Ca^{2+}, which may lead to increases in cytosolic Ca^{2+} levels, mitochondrial Ca^{2+} cycling, and activation of Ca^{2+} proteases. This latter process may be responsible for the conversion of xanthine dehydrogenase to its oxidase form, or arylation and/or oxidation of critical thiol groups in the enzyme may be involved. Adenine nucleotides, particularly ATP, may then be degraded with the production of reactive oxygen species that may lead to additional protein thiol oxidation, particularly because enzymes such as GSH peroxidase and thioltransferase have compromised activities after APAP administration. Finally, macrophages and neutrophils may be attracted to the regions of the liver that are being damaged, and may lead to additional protein thiol modification by releasing oxidants (Laskin and Pilaro, 1986; Jaeschke and Mitchell, 1989).

In summary, the covalent and noncovalent modification of protein thiols by APAP reactive metabolites may both contribute to cell death, or, depending on the state of the hepatocyte and the time of its demise, either one or the other of these thiol group modifications may play the most significant role.

REFERENCES

Albano, E., Rundgren, M., Harvison, P.J., Nelson, S.D. and Moldeus, P. (1985). Mechanisms of N-acetyl-p-benzoquinone imine cytotoxicity. *Mol. Pharmacol.* **28**, 306-311.

Axworthy, D.B., Hoffman, K.-J., Streeter, A.J., Calleman, C.J., Pascoe, G.A. and Baillie, T.A. (1988). Covalent binding of acetaminophen to mouse hemoglobin. Identification of major and minor adducts formed in vivo and implications for the nature of arylating metabolites. *Chem.-Biol. Interactions* **68**, 99-116.

Bartolone, J.B., Sparks, K., Cohen, S.D. and Khairallah, E.A. (1987). Immunochemical detection of acetaminophen-bound liver proteins. *Biochem. Pharmacol.* **36**, 1193-1196.

Bartolone, J.B., Birge, R.B., Sparks, K., Cohen, S.D. and Khairallah, E.A. (1988). Immunochemical analysis of acetaminophen covalent binding to proteins. *Biochem. Pharmacol.* **37**, 4763-4774.

Bartolone, J.B., Beierschmitt, W.P., Birge, R.B., Hart, S.G.E., Wyand, S., Cohen, S.D. and Khairallah, E.A. (1989). Selective acetaminophen metabolite binding to hepatic and extrahepatic proteins: An *in vivo* and *in vitro* analysis. *Toxicol. Appl. Pharmacol.* **99**, 249-249.

Blair, I.A., Boobis, A.R., Davies, D.S. and Cresp, T.M. (1980). Paracetamol oxidation: Synthesis and reactivity of N-acetyl-p-benzoquinone imine. *Tetrahedron Lett.* **21**, 4947-4950.

Boyd, E.M. and Bereczky, G.M. (1966). Liver necrosis from paracetamol. *Br. J. Pharmacol. Ther.* **26**, 606-614.

Coles, B., Wilson, I., Wardman, P., Hinson, J.A., Nelson, S.D. and Ketterer, B. (1988). The spontaneous and enzymatic reaction of N-acetyl-p-benzoquinonimine with glutathione: a stopped-flow kinetic study. *Arch. Biochem. Biophys.* **264**, 253-260.

Dahlin, D.C., Miwa, G.T., Lu, A.Y.H. and Nelson, S.D. (1984). N-Acetyl-p-

benzoquinone imine: A cytochrome P-450-mediated oxidation product of acetaminophen. *Proc. Natl. Acad. Sci. USA*, **81**, 1327-1331.
Della Corte, E. and Stirpe, F. (1972). The regulation of rat liver xanthine oxidase. *Biochem. J.* **126**, 739-745.
Fernando, C.R., Calder, I.C. and Ham, K.N. (1980). Studies on the mechanism of toxicity of acetaminophen: Synthesis and reactions of N-acetyl-2,6-dimethyl- and N-acetyl-3,5-dimethyl-p-benzoquinone imines. *J. Med. Chem.* **23**, 1153-1158.
Hinson, J.A. (1980). Biochemical Toxicology of acetaminophen. *Rev. Biochem. Toxicol.* **2**, 103-129.
Hinson, J.A. Monks, T.J., Hong, R.S. Highet, R.J. and Pohl, L.R. (1982). 3-(Glutathion-S-yl)acetaminophen: A biliary metabolite of acetaminophen. *Drug Metab. Dispos.* **10**, 47-50.
Hoffman, K.-J., Streeter, A.J., Axworthy, D.B. and Baillie, T.A. (1985). Identification of the major covalent adduct formed *in vitro* and *in vivo* between acetaminophen and mouse liver proteins. *Mol. Pharmacol.* **27**, 566-573.
Jaeschke, H. and Mitchell, J.R. (1989). Neutrophil accumulation exacerbates acetaminophen-induced liver injury. *FASEB J.* **3**, A920.
Jollow, D.J., Mitchell, J.R., Potter, W.Z., Davis, D.C., Gillette, J.R. and Brodie, B.B. (1973). Acetaminophen-induced hepatic necrosis. II. Role of covalent binding *in vivo*. *J. Pharmacol. Exp. Ther.* **187**, 195-202.
Jollow, D.J., Thorgeirsson, S.S., Potter, W.Z., Hashimoto, M. and Mitchell, J.R. (1974). Acetaminophen-induced hepatic necrosis. VI. Metabolic disposition of toxic and non-toxic doses of acetaminophen. *Pharmacology* **12**, 251-271.
Knox, J.H. and Jurand, J. (1977). Determination of paracetamol and its metabolites in urine by high-performance liquid chromatography using reversed-phase bonded supports. *J. Chromatog.* **142**, 651-670.
Laskin, D.L. and Pilaro, A.M. (1986). Potential role of activated macrophages in acetaminophen hepatotoxicity. *Toxicol. Appl. Pharmacol.* **86**, 204-215.
Mitchell J.R., Jollow, D.J., Potter, W.Z., Davis, D.C., Gillette, J.F. and Brodie, B.B. (1973a). Acetaminophen-induced hepatic necrosis. I. Role of drug metabolism. *J. Pharmacol. Exp. Ther.* **187**, 185-194.
Mitchell, J.R., Jollow, D.J., Potter W.Z., Gillette, J.R. and Brodie, B.B. (1973b). Acetaminophen-induced hepatic necrosis. IV. Protective role of glutathione. *J. Pharmacol. Exp. Ther.* **187**, 211-217.
Moore, M., Thor, H., Moore, G., Nelson, S., Moldeus, P. and Orrenius, S. (1985). The toxicity of acetaminophen and N-acetyl-p-benzoquinone imine in isolated hepatocytes is associated with thiol depletion and increased cytosolic Ca^{2+}. *J. Biol. Chem.* **260**, 13035-13040.
Nelson, S.D., Vaishnav, Y., Kambara, H. and Baillie, T.A. (1981). Comparative electron impact, chemical ionization and field desorption mass spectra of some thioether metabolites of acetaminophen. *Biomed. Mass Spectrom.* **8**, 244-251.
Nelson, S.D., Forte, A.J., Vaishnav, Y., Mitchell, J.R., Gillette, J.R. and Hinson, J.A. (1981). The formation of arylating and alkylating metabolites of phenacetin in hamsters and hamster liver microsomes. *Mol. Pharmacol.* **19**, 140-145.
Novak, M., Pelecanou, M. and Pollack, L. (1986). Hydrolysis of the model carcinogen N-(pivaloyloxy)-4-methoxyacetanilide: Involvement of N-acetyl-p-benzoquinone imine. *J. Amer. Chem. Soc.* **108**, 112-120.
Potter, W.Z., Davis, D.C., Mitchell, J.R., Jollow, D.J., Gillette, J.R. and Brodie, B.B. (1973). Acetaminophen-induced hepatic necrosis. III. Cytochrome P-450-mediated covalent binding *in vitro*. *J. Pharmacol. Exp. Ther.* **187**, 203-210.
Potter, W.Z., Thorgeirsson, S.S., Jollow, D.J. and Mitchell, J.R. (1974). Acetaminophen-induced hepatic necrosis. V. Correlation of hepatic necrosis, covalent binding, and glutathione depletion in hamsters. *Pharmacology* **12**, 129-143.
Potter, D.W., Pumford, N.R., Hinson, J.A., Benson, R.W. and Roberts, D.W. (1989). *J. Pharmacol. Exp. Ther.* **248**, 182-189.
Prescott, L.F., Wright, N., Roscoe, P. and Brown, S.S. (1971). Plasma paracetamol half-life and hepatic necrosis in patients with paracetamol overdosage. *Lancet* **1**, 519-522.
Rashed, M.S. and Nelson, S.D. (1989a). Characterization of glutathione conjugates of

reactive metabolites of 3'-hydroxyacetanilide, a non-hepatotoxic positional isomer of acetaminophen. *Chem. Res. Toxicol.* **2**, 41-45.

Rashed, M.S. and Nelson, S.D. (1989b). Use of thermospray liquid chromatography-mass spectrometry for characterization of reactive metabolites of 3'-hydroxyacetanilide, a non-hepatotoxic regioisomer of acetaminophen. *J. Chromatog.* **474**, 209-222.

Roberts, S.A. and Jollow, D.J. (1979). Acetaminophen structure-toxicity studies: In vivo covalent binding of a nonhepatotoxic analog, 3-hydroxyacetanilide. *Fed. Proc.* **38**, 462 (abstract).

Roberts, D.W., Pumford, N.R., Potter, D.W., Benson, R.W. and Hinson, J.A. (1987). A sensitive immunochemical assay for acetaminophen-protein adducts. *J. Pharmacol. Exp. Ther.* **241**, 527-533.

Rundgren, M., Porubek, D.J., Harvison, P.J., Cotgreave, I.A., Moldeus, P. and Nelson, S.D. (1988). Comparative cytotoxic effects of N-acetyl-p-benzoquinone imine and two dimethylated analogues. *Mol. Pharmacol.* **34**, 566-572.

Smith, C.V. and Mitchell, J.R. (1985). Acetaminophen hepatotoxicity *in vivo* is not accompanied by oxidant stress. *Biochem. Biophys. Res. Commun.* **133**, 329-336.

Streeter, A.J., Dahlin, D.C., Nelson, S.D. and Baillie, T.A. (1984). The covalent binding of acetaminophen to protein. Evidence for cysteine residues as major sites of arylation *in vitro*. *Chem.-Biol. Interactions* **48**, 349-366.

Streeter, A.J., Harvison, P.J., Nelson, S.D. and Baillie, T.A. (1986). Cross-linking of protein molecules by the reactive metabolite of acetaminophen, N-acetyl-p-benzoquinone imine, and related quinoid compounds. In *Biological Reactive Intermediates* III (J.J. Kocsis, D.J. Jollow, C.M. Witmer, J.O. Nelson and R. Snyder, eds.) pp. 727-737. Plenum Press, New York.

Tee, L.B.G., Boobis, A.R., Hugett, A.C. and Davies, D.S. (1986). Reversal of acetaminophen toxicity in isolated hepatocytes by dithiothreitol. *Toxicol. Appl. Pharmacol.* **83**, 294-314.

Tirmenstein, M.A. and Nelson, S.D. (1989). Subcellular binding and effects on calcium homeostasis produced by acetaminophen and a nonhepatotoxic regioisomer, 3'-hydroxyacetanilide, in mouse liver. *J. Biol. Chem.* **264**, 9814-9819.

FORMATION AND REACTIVITY OF A QUINONE METHIDE IN BIOLOGICAL SYSTEMS

David Thompson[1] and Peter Moldéus

Department of Toxicology, Karolinska Institutet,
S-104 01 Stockholm, Sweden

INTRODUCTION

The formation of reactive electrophilic intermediates during xenobiotic metabolism which interact with cellular macromolecules is the foundation of current hypotheses on the mechanism of toxicity and carcinogenicity of many chemicals. Aromatic compounds which contain one or more oxygen atoms in functional groups attached to the ring have been extensively studied. These compounds include phenols, hydroquinones and catechols which are oxidized to quinones. Quinones are cytotoxic by virtue of their ability to redox cycle and generate reactive oxygen species and also by their ability to alkylate cellular macromolecules. More recently, compounds such as acetaminophen and phenetidine have been shown to form quinoneimines which are also highly cytotoxic (Dahlin, D.C. et al., 1984; Ross, D. et al., 1985). It is now becoming apparent that other derivatives of the quinonoid moiety may also be highly reactive; for example, quinone methides and imine methides (Mizutani, T. et al., 1982; Yost, G.S., 1989). For comparison, the chemical structures of a quinone, quinoneimine and quinone methide are shown in Figure 1.

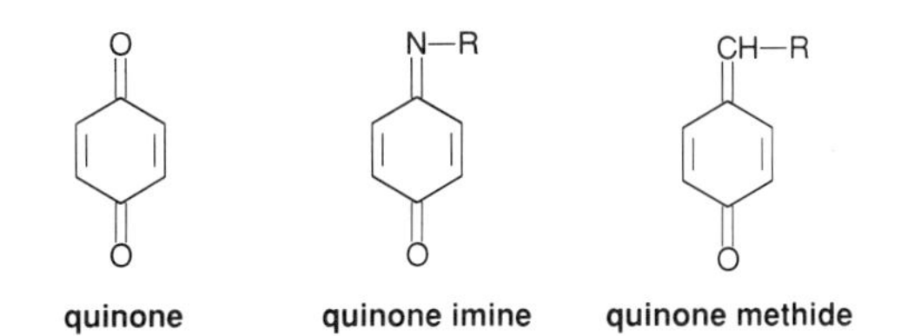

Figure 1. Comparative chemical structures of a quinone, quinone imine and quinone methide.

Present address: National Institute of Environmental Health Sciences, Research Triangle Park, NC 27709

Quinone methides are suspected to be unstable intermediates in many chemical and biochemical phenol oxidation reactions (Turner,A.B., 1964; Wagner, H.-U. et al., 1972; Filar, L.J. et al., 1961; Becker, H.-D., 1965; Moore, H.W. et al., 1986; Dyall, L.K. et al., 1971; Brown, Jr., K.S. et al., 1971; Jurd, L., 1977). Quinone methides are formed from the two-electron oxidation of phenols with *ortho* or *para* methyl or alkyl groups. The reactivity of unsubstituted quinone methides is generally so high that they cannot be isolated under normal conditions; only polymeric products result. The quinone methide can be isolated, however, if the methylene carbon is substituted or if the benzenoid character of the ring is weak, such as with methyleneanthrone. Quinone methides display high reactivity toward nucleophiles, undergoing Michael additions with many nucleophilic groups. *Para* quinone methides undergo coupling reactions which are distinct from *ortho* compounds, the latter forming cyclic ether derivatives.

Eugenol (4-allyl-2-methoxyphenol) possesses structural requirements necessary for forming a quinone methide metabolite. Eugenol is present in a large number of foods and medicines and is used extensively in dentistry (IARC 1985). Therefore we studied the possible formation of a quinone methide metabolite from eugenol in enzymatic and cellular systems (Thompson, D. et al., 1989; Thompson, D. et al., 1989; Thompson, D. et al., 1990; Thompson, D. et al., 1990). The quinone methide appears to be formed from both peroxidase and microsomal (cytochrome P-450) enzyme systems. The quinone methide is also formed in intact cellular systems (human polymorphonuclear leukocytes and rat hepatocytes) where it is highly reactive and cytotoxic.

RESULTS

The quinone methide of eugenol can be chemically synthesized in a relatively pure form using silver(I) oxide in carbon tetrachloride. Zanarotti (Zanarotti, A., 1985) studied the reactivity of this compound and related compounds with various nucleophiles. Generally two adducts were obtained, one at the C-1' and one at the C-3' carbon of the allyl chain. We found that the quinone methide was brightly yellow colored and had an absorbance maximum at 346 nm in methanol. In aqueous solutions it decayed rapidly and the addition of glutathione caused an immediate decrease in absorbance. Leary (1972) reported that eugenol quinone methide, formed from flash photolysis of coniferyl alcohol, has a half life in aqueous solutions of 5.78 min. We further studied the reactivity of the quinone methide with glutathione and observed the formation of three distinct adducts on analysis by HPLC. Mass spectroscopy confirmed that each product was a eugenol-glutathione conjugate and two dimensional COSY proton NMR suggested that all of the adducts were formed on the allyl portion of the eugenol molecule. One of the conjugates had a much larger extinction at 280 nm than the other two, and this conjugate had glutathione bound to C-3' carbon. With this knowledge we could either detect the quinone methide directly in biological incubations (by measuring absorbance changes at 350 nm) or trap it with glutathione. We wanted to see if this metabolite was formed in enzymatic reactions, particularly using peroxidase and microsomal enzyme systems.

Peroxidase enzymes

Eugenol has been known as a substrate for peroxidases for many years, primarily because of research into the structure and formation of lignin (Siegel, S.M., 1956). The peroxidation of eugenol leads to the formation of a polymer which resembles lignin. However, other products of eugenol oxidation by peroxidases have not been studied. We used several peroxidase enzymes for our studies of eugenol metabolism, including horseradish peroxidase, myeloperoxidase and prostaglandin H synthase (Thompson, D. et al., 1989). In incubations with each peroxidase eugenol was oxidized to at least three products: a phenoxyl radical, a transient, yellow-colored intermediate which rapidly disappeared (absorbance maximum 350 nm, which appeared to be the quinone methide), and (at high concentrations of eugenol) a polymer which precipitated out of solution. The phenoxyl radical was detected by fast flow ESR techniques in incubations containing horseradish peroxidase.

The transient yellow-colored intermediate was tentatively identified as a quinone methide by the following criteria: (1) the absorbance maximum was similar to synthetically prepared eugenol quinone methide (350 nm), and (2) the compound reacted directly with glutathione, forming conjugates which were identical (spectrally and chromatographically) with those formed using synthetically prepared quinone methide. Conventional chromatographic methods (reverse and normal phase HPLC) were not successful in isolating the quinone methide for unequivocal identification. This is apparently due to the high reactivity of this compound with itself and column packing material.

In addition to reacting directly with the quinone methide, glutathione prevented the formation of this compound by reducing the phenoxyl radical back to eugenol. This resulted in the formation of the glutathione thiyl radical, extensive oxygen uptake and the formation of oxidized glutathione (GSSG). These results suggested two possible mechanisms for depletion of glutathione by the peroxidative oxidation of eugenol - (1) oxidation of glutathione by phenoxyl radicals and (2) conjugation of glutathione with the quinone methide.

The toxicity of eugenol quinone methide could not be tested directly because of low solubility and instability in aqueous solutions. Therefore, the cytotoxicity of the peroxidase-generated metabolites of eugenol was tested using isolated rat hepatocytes as a target cell. Eugenol was metabolized extracellularly by horseradish peroxidase and hydrogen peroxide (continuously generated by the reaction of glucose and glucose oxidase) and the toxic effects of these metabolites on cell viability were measured using trypan dye exclusion. We observed a concentration-dependent toxicity of eugenol on hepatocytes. Complete cell death was seen after 90 minutes with a concentration of 1 mM eugenol. The cytotoxic effects were peroxidase- and hydrogen peroxide-dependent. Although we could not attribute the toxicity to the quinone methide metabolite alone, these observations demonstrate that peroxidase generated metabolites of eugenol, including the phenoxyl radical and quinone methide, are cytotoxic.

Microsomal enzymes

Incubations were also performed using microsomal fractions isolated from rat liver and lung (Thompson, D. et al., 1990). The incubations contained eugenol, microsomes, glutathione and an NADPH generating system. We attempted to establish that quinone methide was being formed by trapping it with glutathione. Upon HPLC analysis we found three peaks which contained both eugenol and glutathione. These were the same three peaks obtained when synthetic quinone methide was reacted with glutathione and in the peroxidase reactions described above. Quantitation of the metabolites formed in microsomal incubations using radiolabeled eugenol and radiolabeled glutathione showed that each of the metabolites were formed in equivalent amounts.

The formation of glutathione conjugates in microsomal incubations was found to be inhibited by cytochrome P-450 inhibitors, including metyrapone, SKF 525- A, piperonyl butoxide and α-naphthoflavone. The reaction was also oxygen- dependent. Pretreatment of rats with cytochrome P-450 inducers revealed that the enzyme responsible for eugenol oxidation is inducible by 3-methylcholanthrene.

Radiolabeled metabolites of eugenol were bound to microsomal protein and this was inhibited by glutathione, also suggesting that quinone methide was being formed in these incubations. The addition of glutathione-depleted cytosol, which contained active glutathione S-transferase, had no effect on the rate of formation of glutathione conjugates in these incubations. This indicated that conjugation occurs nonenzymatically, as would be expected with quinone methide. Finally, although we could not measure formation of quinone methide directly because of interference by NADPH at 350 nm, we did observe the formation of a metabolite absorbing at 350 nm in the presence of cumene hydroperoxide which was sensitive to glutathione added either before or after the start of the reaction. Cumene hydroperoxide catalyzed the formation of the same eugenol-glutathione conjugates with microsomes as did NADPH.

It therefore appeared that both peroxidase and cytochrome P-450 catalyzed the formation of a reactive metabolite from eugenol which was capable of covalently binding to cellular macromolecules, forming conjugates with glutathione and which was most likely a quinone methide. We next turned our attention to study intact cellular systems (one containing peroxidase and one with cytochrome P-450) which would likely form the quinone methide intracellularly and study the potential cytotoxic effects.

Human polymorphonuclear leukocytes (PMNs)

PMNs are capable of activating an increasing number of xenobiotics to reactive intermediates, usually through the action of myeloperoxidase. When stimulated, PMNs generate superoxide and subsequently hydrogen peroxide via the activity of a membrane enzyme, NADPH oxidase. Myeloperoxidase metabolizes this hydrogen peroxide in the presence of halide to form cytotoxic products which are used to kill invading microorganisms. We observed that myeloperoxidase, isolated and purified from human PMNs, oxidized eugenol to the quinone methide similar to horseradish peroxidase (Thompson, D. et al., 1989). In intact PMNs eugenol had several deleterious effects. Eugenol interfered with the oxidative burst as indicated by decreased formation of superoxide and hydrogen peroxide and also by inhibiting oxygen uptake in whole cells. Eugenol was cytotoxic to PMNs in the presence, but not absence, of the stimulator phorbol ester. Similarly, eugenol decreased intracellular glutathione in PMNs when stimulated by phorbol ester, but not when quiescent. Covalent binding of eugenol metabolites was detected in stimulated cells and this binding was partially inhibitable by sodium azide, which inhibits myeloperoxidase. Thus, our evidence suggested that eugenol is oxidized in intact PMNs, that this oxidation leads to glutathione depletion and formation of reactive metabolites, and that the cytotoxicity seen in PMNs may be due to the quinone methide of eugenol. In other studies eugenol has been reported to inhibit cell migration and chemiluminescence in human PMNs (Fotos, P.G. et al., 1987) and also to be cytotoxic to guinea pig PMNs (Suzuki, Y. et al., 1985).

Isolated rat hepatocytes

The metabolism and toxicity of eugenol was also investigated in isolated rat hepatocytes (Thompson, D. et al., 1990). In this system eugenol was actively metabolized to form conjugates with glucuronic acid, sulfate and glutathione. The glucuronide was the major metabolite observed. Only one glutathione conjugate was observed, but it was identical to one of the conjugates formed with peroxidases. At concentrations between 0.5 and 1.5 mM eugenol, we observed a rapid decrease in intracellular glutathione. After incubations of about 2 hr, glutathione levels in eugenol-treated cells were less than 30% of control cells. Cellular blebbing became apparent at this time. At about 3 hr significant cytoxicity was observed in eugenol-treated cells. If thiol levels were depleted with diethylmaleate, cytoxicity was observed at earlier time points, whereas in the presence of N- acetylcysteine, which provides cysteine for glutathione synthesis, cytotoxicity was prevented. With radiolabeled eugenol covalent binding to cellular macromolecules was observed, demonstrating that reactive intermediates were being formed. Covalent binding was prevented in the presence of N-acetylcysteine or metyrapone, a cytochrome P-450 inhibitor. Our experiments with hepatocytes indicated that eugenol is cytotoxic in this system and that metabolism to a reactive intermediate, possibly a quinone methide, may mediate these effects.

DISCUSSION

We investigated the oxidative metabolism of eugenol in several biological systems. Both peroxidase and cytochrome P-450 enzymes metabolized eugenol to a reactive quinone methide metabolite (see Figure 2). This intermediate is capable of conjugating with glutathione and other cellular macromolecules. In intact cellular systems, human PMNs and rat hepatocytes, generation of the quinone methide led to cytotoxicity while prevention of its formation preserved cellular integrity. These

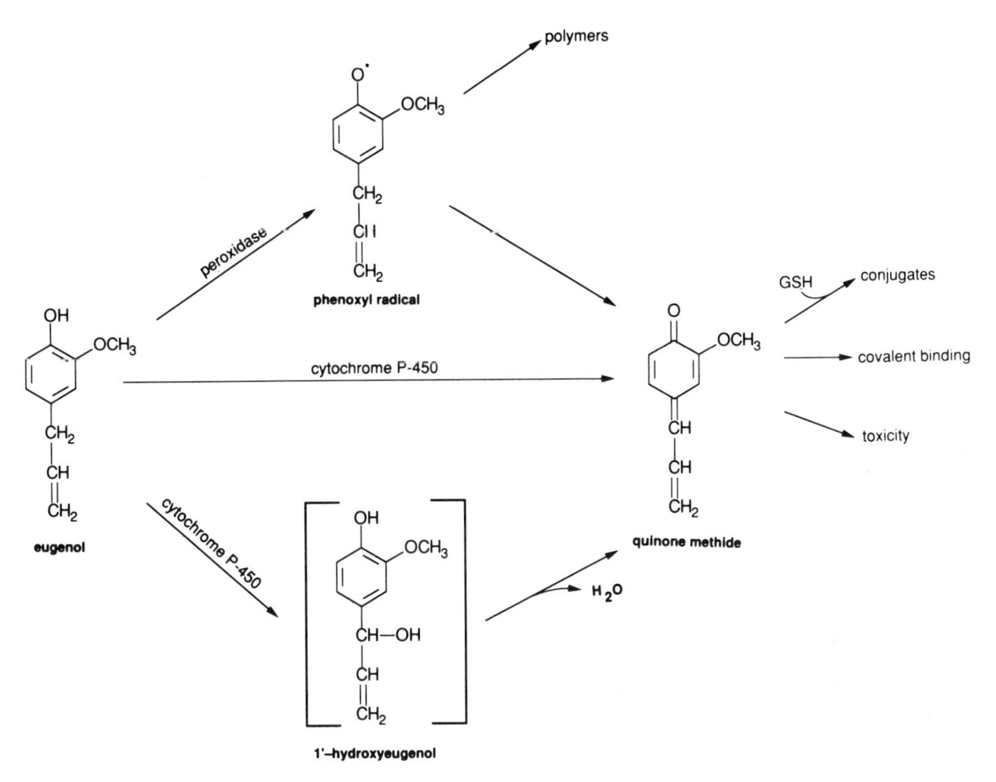

Figure 2. Proposed metabolic pathways for the formation of quinone methide from eugenol with both peroxidase and cytochrome P-450 enzyme systems.

observations suggest that eugenol may be cytotoxic in other cell types which have these activating systems present. Also, these observations may help explain, or at least suggest a mechanism for, the toxic effects of eugenol discussed below.

Eugenol is the major component in oil of cloves and is present in high concentrations in clove cigarettes. A few years ago the Center for Disease Control began receiving reports of acute pulmonary illness in persons who had recently smoked clove cigarettes (1985). The major symptoms included pulmonary edema, bronchospasm and hemoptysis. In some of the more severe cases, concurrent or preceeding respiratory infections were noted. La Voie et al. (LaVoie, E.J. et al., 1986) reported an acute toxic effect of eugenol on rat lungs following intratracheal injection. Together with reports of toxic effects of eugenol on PMNs discussed above, these observations suggest that eugenol may be contributing to these respiratory symptoms by both suppressing normal PMN function in fighting off infection, and by toxic effects due to formation of quinone methide in pulmonary cells as well as PMNs. It should be noted, however, that association between clove cigarettes and lung damage has not been conclusively demonstrated, especially since the concentration of eugenol reached in the lung via inhalation is probably very low (Clark, G.C., 1988). In addition, immunologic reactions may be involved since eugenol is a weak contact allergen and only a small proportion of persons who smoke clove cigarettes experience its toxic effects. Data which which would be very useful would be the extent of pulmonary illnesses in Indonesia, where clove cigarettes are manufactured and widely used.

Another area where toxic effects from eugenol might be apparent is in dentistry where eugenol is used as an analgesic and as a component of several dental materials, such as anodyne dressings, pulp capping agents and in root canal materials. High local concentrations of eugenol invoke an inflammatory response which may be related to a cytotoxic effect on cells immediately surrounding the application area (Webb, Jr., J.G. et al., 1981). In addition, migration of PMNs into the area of inflammation might be impeded as well as their functional capacity. Eugenol has also been described as possessing neurotoxic efects and inhibiting nerve transmission in dental tissue (Trowbridge, H. et al., 1982; Kozam, G., 1977). These effects are not well characterized, and their relationship to eugenol metabolism is not clear but deserves further study.

The metabolism of eugenol to a quinone methide suggests a possible difference between eugenol, which is noncarcinogenic, and other carcinogenic allylbenzenes such as methyleugenol, safrole and estragole (Miller, E.C. et al., 1983). These latter compounds form 1'-hydroxy metabolites which are then sulfated to form electrophilic sulfate esters, which are the purported carcinogenic metabolites. However, if eugenol forms a 1'-hydroxy metabolite (see Figure 2), it may lose water to form the quinone methide, which the other carcinogenic metabolites cannot do because their structures prohibit quinone methide formation. Thus, eugenol may form smaller amounts of potentially carcinogenic metabolites, instead forming the quinone methide which, under normal conditions, would be detoxified by the cell. Thus the formation of quinone methide may actually preclude the formation of carcinogenic metaboites from eugenol.

Besides eugenol, other xenobiotics have been reported to form toxic quinone methides. The best studied example is butylated hydroxytoluene which causes pulmonary damage in mice (Marino, A.A. et al., 1972). This compound also causes liver damage in mice and rats and hemorrhagic death in rats (Nakagawa, Y. et al., 1984; Mizutani, T. et al., 1987; Takahashi, O. et al., 1978). Some or all of these effects may be related to the formation of a quinone methide metabolite. Other compounds whose toxicity may involve quinone methides include adriamycin and daunomycin (Sinha, B.K., H. et al., 1981) as well as a series of phenolic insect sterilants which Jurd (Jurd, L. et al., 1979) suggested were toxic because of their ability to form quinone methides *in vivo*.

In summary, we have suggested that eugenol is metabolized by both peroxidase and cytochrome P-450 to a quinone methide which is highly reactive and elicits cytotoxic effects in cells in which it is formed. There is accumulating evidence implicating quinone methides as reactive intermediates which may play a significant role in the toxicity of certain phenolic compounds. Since there are many xenobiotics which are structurally capable of being oxidized to quinone methides, this route of biotransformation should be considered in the study of metabolism and toxicity of these chemicals.

REFERENCES

Becker, H.-D. (1965). Quinone dehydrogenation. I. The oxidation of monohydric phenols. *J. Org. Chem.* **30,** 982-989.

Brown, Jr., K.S. and Baker, P.N. (1971). Stable *ortho*- and *para*-naphthaquinone methides. *Tetrahedron Lett.* **No. 38,** 3505-3508.

Clark, G.C. (1988). Acute inhalation toxicity of eugenol in rats. *Arch. Toxicol.* **62,** 381-386.

Dahlin, D.C., Miwa, G.T., Lu, A.Y.H. and Nelson, S.D. (1984). N-acetyl-p-benzoquinone imine: a cytochrome P-450-mediated oxidation product of acetaminophen. *Proc. Natl. Acad. Sci. USA* **81,** 1327-1331.

Dyall, L.K. and Winstein, S. (1971). Nuclear magnetic resonance spectra and characterization of some quinone methides. *J. Am. Chem. Soc.* **94,** 2196-2199.

Filar, L.J. and Winstein, S. (1961). Preparation and behavior of simple quinone methides. *Tetrahedron Lett.* **No. 25,** 9-16.

Fotos, P.G., Woolverton, C.J., Van Dyke, K. and Powell, R.L. (1987). Effects of eugenol on polymorphonuclear cell migration and chemi-luminescence. *J. Dent. Res.* **66,** 774-777.

IARC (1985). Eugenol. IARC Monogr. Eval. Carcinog. *Risk Chem. Hum.* **36,** 75-97.

Jurd, L. (1977). Quinones and quinone-methides. I. Cyclization and dimerisation of crystalline *ortho*-quinone methides from phenol oxidation reactions. *Tetrahedron* **33,** 163-168.

Jurd, L. Fye, R.L. and Morgan, Jr., J. (1979). New types of insect chemo-sterilants. Benzylphenols and benzyl-1,3-benzodioxole derivatives as additives to housefly diet. *J. Agr. Food Chem.* **27,** 1007-1016.

Kozam, G. (1977). The effect of eugenol on nerve transmission. *Oral Surg.* **44,** 799-805.

LaVoie, E.J., Adams, J.D., Reinhardt, J., Rivenson, A. and Hoffman, D. (1986). Toxicity studies on clove cigarette smoke and constituents of clove: determination of the LD_{50} of eugenol by intratracheal installation. *Arch. Toxicol.* **59,** 78-81.

Leary, G. (1972). Chemistry of reactive lignin intermediates. I. Transients in coniferyl alcohol photolysis. *J. Chem. Soc. Perkin Trans. 2* **5,** 640-642.

Marino, A.A. and Mitchell, J.T. (1972). Lung damage in mice following intraperitoneal injection of butylated hydroxytoluene. *Proc. Soc. Exp. Biol. Med.* **140,** 122-125.

Miller, E.C., Swanson, A.B., Phillips, D.H., Fletcher, T.L., Liem, A. and Miller, J.A. (1983). Structure-activity studies of the carcinogenicities in the mouse and rat of some naturally occurring and synthetic alkenyl-benzene derivatives related to safrole and estragole. *Cancer Res.* **43,** 1124-1134.

Mizutani, T., Ishida, I., Yamamoto, K. and Tajima, K. (1982). Pulmonary toxicity of butylated hydroxytoluene and related alkylphenols: Structural requirements for toxic potency in mice. *Toxicol. Appl. Pharmacol.* **62,** 273-281.

Mizutani, T. Nomura, H., Nakanishi, K. and Fujita, S. (1987). Hepatotoxicity of butylated hydroxytoluene and its analogs in mice depleted of hepatic glutathione. *Toxicol. Appl. Toxicol.* **87,** 166-176.

Moore, H.W., Czerniak, R. and Hamdan, A. (1986). Natural quinones as quinonemethide precursors - ideas in rational drug design. *Drugs Exptl. Clin. Res.* **12,** 475-494.

Morbidity and Mortality Weekly Report (1985). Illnesses possibly associated with smoking clove cigarettes. Vol. **34,** pp. 297-299.

Nakagawa, Y., Tayama, K, Nakao, T. and Hiraga, K. (1984). On the mechanism of butylated hydroxytoluene-induced hepatic toxicity in rats. *Biochem. Pharmacol.* **33,** 2669-2674.

Ross, D., Larsson, R., Norbeck, K., Ryhage, R. and Moldéus, P. (1985). Characterization and mechanism of formation of reactive products formed during peroxidase-catalyzed oxidation of p-phenetidine. *Mol. Pharmacol.* **27,** 277-286.

Siegel, S.M. (1956). The biosynthesis of lignin: Evidence for the participation of celluloses as sites for oxidative polymerization of eugenol. *J. Am. Chem. Soc.* **78,** 1753-1755.

Sinha, B.K. and Gregory, J.L. (1981). Role of one-electron and two-electron reduction products of adriamycin and daunomycin in deoxyribonucleic acid binding. *Biochem. Pharmacol.* **30,** 2626-2629.

Suzuki, Y., Sugiyama, K. and Furuta, H. (1985). Eugenol-mediated super-oxide generation and cytotoxicity in guinea pig neutrophils. *Japan. J. Pharmacol.* **39,** 381-386.

Takahashi, O. and Hiraga, K. (1978). Dose-response study of hemorrhagic death by dietary butylated hydroxytoluene (BHT) in male rats. *Toxicol. Appl. Pharmacol.* **43,** 399-406.

Thompson, D., Norbeck, K., Olsson, L.-I., Constantin-Teodosiu, D., Van der Zee, J. and Moldéus, P. (1989). Peroxidase-catalyzed oxidation of eugenol: Formation of a cytotoxic metabolite(s). *J. Biol. Chem.* **264,** 1016-1021.

Thompson, D., Constantin-Teodosiu, D., Norbeck, K., Svensson, B. and Moldéus, P. (1989). Metabolic activation of eugenol by myeloperoxidase and polymorphonuclear leukocytes. *Chem. Res. Toxicol.* **2,** 186-192.

Thompson, D., Constantin-Teodosiu, D., Egestad, B., Mickos, H. and Moldéus, P. (1990). Formation of glutathione conjugates during oxidation of eugenol by microsomal fractions of rat liver and lung. *Biochem. Pharmacol.* (in press).

Thompson, D., Constantin-Teodosiu, D. and Moldéus, P. (1990) . Metabolism and cytotoxicity of eugenol in isolated rat hepatocytes. *Chem-Biol. Interactions* (manuscript submitted).
Trowbridge, H., Edwall, L. and Panopoulos, P. (1982). Effect of zinc oxide-eugenol and calcium hydroxide on intradental nerve activity. *J.Endodontics* **8,** 403-406.
Turner,A.B. (1964). Quinone Methides. *Q. Rev. Chem. Soc. Lond.* **18,** 347-360.
Wagner, H.-U. and Gompper, R. (1972). Quinone methides. In: *The Chemistry of Quinonoid Compounds*, S. Patai (ed.), pp. 1145-1178, Wiley, NY.
Webb, Jr., J.G. and Bussell, N.E. (1981). Comparison of the inflammatory response produced by commercial eugenol and purified eugenol. *J. Dent. Res.* **60,** 1724-1728.
Yost, G.S. (1989). Mechanisms of 3-methylindole pneumotoxicity. *Chem. Res. Toxicol.* **2,** 273-279.
Zanarotti, A. (1985). Synthesis and reactivity of vinyl quinone methides. *J. Org. Chem.* **50,** 941-945.

MECHANISMS FOR PYRROLIZIDINE ALKALOID ACTIVATION AND DETOXIFICATION

Donald R. Buhler[1,2], Cristobal L. Miranda[2], Bogdan Kedzierski[2], and Ralph L. Reed[2]

Toxicology Program[1] and Department of Agricultural Chemistry[2]
Oregon State University
Corvallis, OR 97331

INTRODUCTION

The pyrrolizidine alkaloids (PAs) constitute a large group of hepatotoxic and carcinogenic plant constituents of wide geographic and botanical distribution. These alkaloids are responsible for the death of livestock throughout the world and for occasional human poisonings following the consumption of contaminated foods or the injudicious use of herbal medicines (Bull et al., 1968; Mattocks, 1986; Hirano, 1981; Huxtable, 1980; Peterson and Culvenor, 1983). PAs are relatively nontoxic but are bioactivated *in vivo* primarily via the liver, through enzymatic dehydrogenation to form highly reactive pyrrole- type metabolites. It is thought that PAs are initially converted to the corresponding dehydropyrrolizidine alkaloids (PA pyrroles) which then can either alkylate protein, DNA or other cellular nucleophiles (Hsu et al. 1975; Reed et al. 1988; Wickramanayake et al., 1985) or be hydrolyzed to the more stable pyrrolic alcohol, such as (R)-6,7-dihydro-7-hydroxy-1-hydroxymethyl-5H-pyrrolizidine (DHP) in the case of retronecine or heliotridine based PAs (Jago et al., 1979; Kedzierski and Buhler, 1985, 1986; Mattocks, 1986; Mattocks and White, 1971). PAs also can be oxidized *in vivo* to relatively nontoxic PA N-oxides and hydrolyzed to the corresponding amino alcohol (Kedzierski and Buhler, 1986; Mattocks, 1986; Ramsdell et al., 1987).

In vitro experiments (Guengerich, 1977; Hirono, 1981; Mattocks, 1986; Ramsdell et al., 1987) have shown that conversion of the PAs to toxic pyrrolic metabolite(s) and to PA N-oxides is primarily mediated by cytochrome P-450s. Rats displayed a marked sex difference in the oxidation of PAs, especially with respect to N-oxidation (Williams et al., 1989a). However, dexamethasone markedly increased the oxidation of PAs to pyrrolic metabolites and to PA N- oxides with almost complete elimination of the large differences in rates of metabolism between the sexes. Studies performed in reconstituted systems and with anti-cytochrome P-450 antibodies have shown that the male-specific cytochrome P-450 IIC11 (UT-A) was primarily responsible for the N-oxidation of the PAs while cytochrome IIIA-2 (PCN-E) catalyzed most of the conversion of PAs to reactive pyrrolic metabolites (Williams et al., 1989a).

In addition to cytochrome P-450, N-oxidation of PAs also could be catalyzed by microsomal flavin-containing monooxygenase (FMO). We have recently shown that purified pig liver FMO but not purified rabbit lung FMO could readily N-oxidize the PAs senecionine and lasiocarpine (Williams et al., 1989b). Studies with rat liver microsomes, however, suggest that cytochrome P-450 and not FMO catalyzes most of the PA N-oxidation in that species.

Biological Reactive Intermediates IV, Edited by C.M. Witmer *et al.*
Plenum Press, New York, 1990

In the present study we have investigated the role of glutathione (GSH) as a trapping agent for the reactive pyrrole metabolites of the PA senecionine. Additional experiments with the glutathione-pyrrole adduct have provided evidence that initial cytochrome P-450 oxidation of PAs results in formation of a pyrrole, such as dehydrosenecionine (senecionine pyrrole), that is more reactive than DHP.

MATERIALS AND METHODS

The PA senecionine was isolated from Senecio jacobaea extracts by preparative HPLC (Ramsdell et al., 1987) and DHP and senecionine N-oxide standards were prepared as previously described (Kedzierski and Buhler, 1986). [^{3}H]-Senecionine was prepared biosynthetically as described by Reed and Buhler (1988) and [^{35}S]-glutathione was purchased from NEN Research Products (Boston, MA). Monocrotaline (Trans World Chemical Co.) was converted to the DHP enantiomer dehydroretronecine (DHR) by standard methods (Culvenor et al., 1970). The conjugate of DHR with glutathione (GSH) was synthesized chemically (Robertson et al., 1977). Chromatography of microsomal incubation mixtures was performed as described previously (Kedzierski and Buhler, 1986) using a PRP-1 reversed-phase styrene-divinyl benzene column (Hamilton Co., Reno, NV) and an acetonitrile-0.1 NH_4OH gradient with detection at 220 or 235 nm by a Hewlett-Packard (Palo Alto, CA) Model 1040A diode array detector. Radioactive peaks were detected with a Radiomatic (Tampa, FL) Flo-One Beta detector. HPLC analysis of incubation mixtures containing the GSH-DHP conjugate or GSH-DHR was conducted on a PRP-1 column eluted with 0.1 M ammonium formate (pH 5.6) and a 0-25% acetonitrile gradient. For mass spectrometry, the GSH-DHP conjugate was further purified on a PRP-1 column eluted with 3.6 mM ammonium acetate (pH 6.2) and 0-75% acetonitrile. The isolated GSH-DHP peak and the synthetic GSH-DHR standard were characterized by negative-ion fast atom bombardment (FAB) mass spectrometry on a Kratos MS-50 instrument. Quantitation of radioactivity was by liquid scintillation spectrometry with a Packard (Downers Grove, IL) Model 4530 spectrometer.

Rats (male Sprague-Dawley, 155-180 g) were injected ip with either 80 mg/kg sodium phenobarbital, 50 mg/kg dexamethasone for 3 days or with saline vehicle. Animals were fasted overnight and killed by cervical dislocation, the livers removed and homogenized in ice-cold 10 mM potassium phosphate (pH 7.4) containing 1.15% KCl in a Potter-Elvehjem homogenizer. Washed microsomes then were prepared as previously described (Ramsdell et al., 1987). Microsomal incubations were conducted with senecionine (1-2 mM) or [^{3}H]- senecionine incubated at pH 7.4 for 15 min or longer at 37° with or without added GSH as described previously by Kedzierski and Buhler (1986). The cooled reaction mixture was centrifuged at 46,000 g and aliquots injected directly onto the PRP-1 column (Kedzierski and Buhler, 1986) as described above.

RESULTS AND DISCUSSION

Incubation of senecionine in the presence of rat hepatic microsomes resulted in the formation of two major metabolites, the reactive pyrrolic alcohol DHP and senecionine N-oxide (Buhler and Kedzierski, 1986; Williams et al., 1989a). Pretreatment of the animals with either phenobarbital or dexamethasone but not β-naphthoflavone enhanced microsomal oxidation of senecionine to DHP but either decreased or had no effect on senecionine N-oxide production.

When incubations run with 0.15 M GSH were analyzed by HPLC, the DHP peak was no longer present and a new UV absorbing peak, eluting one minute earlier, was detected (Figure 1). This latter peak was radioactive when [^{3}H]- senecionine or [^{35}S]-GSH was included in the incubation mixture and chemically synthesized GSH-DHR coeluted with the microsomally produced conjugate on HPLC analysis.

The conjugate was isolated and purified by preparative HPLC and then compared with the synthetic GSH conjugate of DHR. The FAB spectrum for both conjugates

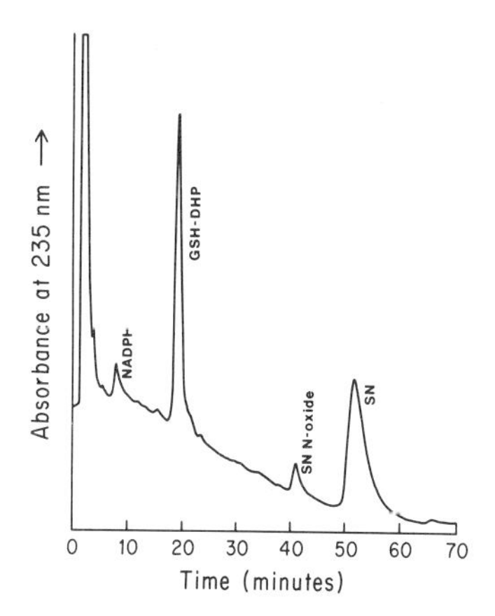

Figure 1. HPLC analysis of senecionine metabolites after incubation with rat hepatic microsomes in the presence of GSH. Liver microsomes (1 mg) from adult PB-pretreated male rats were incubated with senecionine (2 mM) and GSH (100 mM) and analyzed by HPLC or a PRP-1 column as described in Materials and Methods.

were identical with major peaks at m/e 441 and 306, corresponding to the M-1 and the GSH-minus-1 fragment ions, respectively (Figure 2). Peaks also occurred at m/e 479 and 344, resulting from the addition of potassium ion to each of the two peaks. The microsomal product also gave a m/e 463 peak from sodium addition to the M-1 fragment. These findings are consistent with the addition of GSH to either the C-7 or C-9 position of DHP or dehydrosenecionine (senecionine pyrrole). Previous studies by Robertson et al. (1977), Karchesy and Deinzer (1981) and Wickramanayake et al. (1985), however, have indicated that the C-7 position of DHR and the PA pyrroles is favored for attack by nucleophiles in chemical synthesis.

Production of the GSH conjugate and senecionine N-oxide was essentially linear over a two hour incubation period (data not shown). Microsomal conversion of senecionine to DHP, in the absence of added GSH, and to the GSH-DHP conjugate when GSH was added was increased by phenobarbital pretreatment of the animals (Table 1). Addition of GSH enhanced total conversion of senecionine to soluble pyrrole perhaps by trapping the reactive pyrrolic metabolites of the PA and thus protecting cytochrome P-450 against inactivation. PAs are known to inhibit cytochrome P-450 dependent oxidations (Miranda et al., 1980) possibly through the binding of pyrrolic metabolites to protein.

Formation of the GSH adduct from senecionine was directly related to the amount of GSH added to the incubation mixture. As GSH levels increased, concentration of free DHP decreased concomitant with increasing GSH-DHP conjugate formation (Figure 3). Incubation with only 1 mM GSH reduced final DHP levels by about one-half while with 25 mM GSH only a small amount of DHP was detected. As GSH concentrations increased, senecionine N-oxide levels slightly increased and those of senecionine decreased. To quantitate conjugate formation, incubations also were performed with [^{3}H]-senecionine, with and without added GSH (Figure 4). At concentrations of 10 and 50 mM GSH, formation of the pyrrole was enhanced over that seen in the absence of GSH with virtually all of the pyrrole present in the microsomal incubation mixture being bound to GSH as the GSH-DHP conjugate. These results provide additional evidence for formation of a GSH-DHP conjugate during microsomal incubation of the PA senecionine in the presence of GSH and confirm the action of GSH in enhancing senecionine metabolism apparently by protecting against the pyrrolic metabolite inactivation of cytochrome P-450.

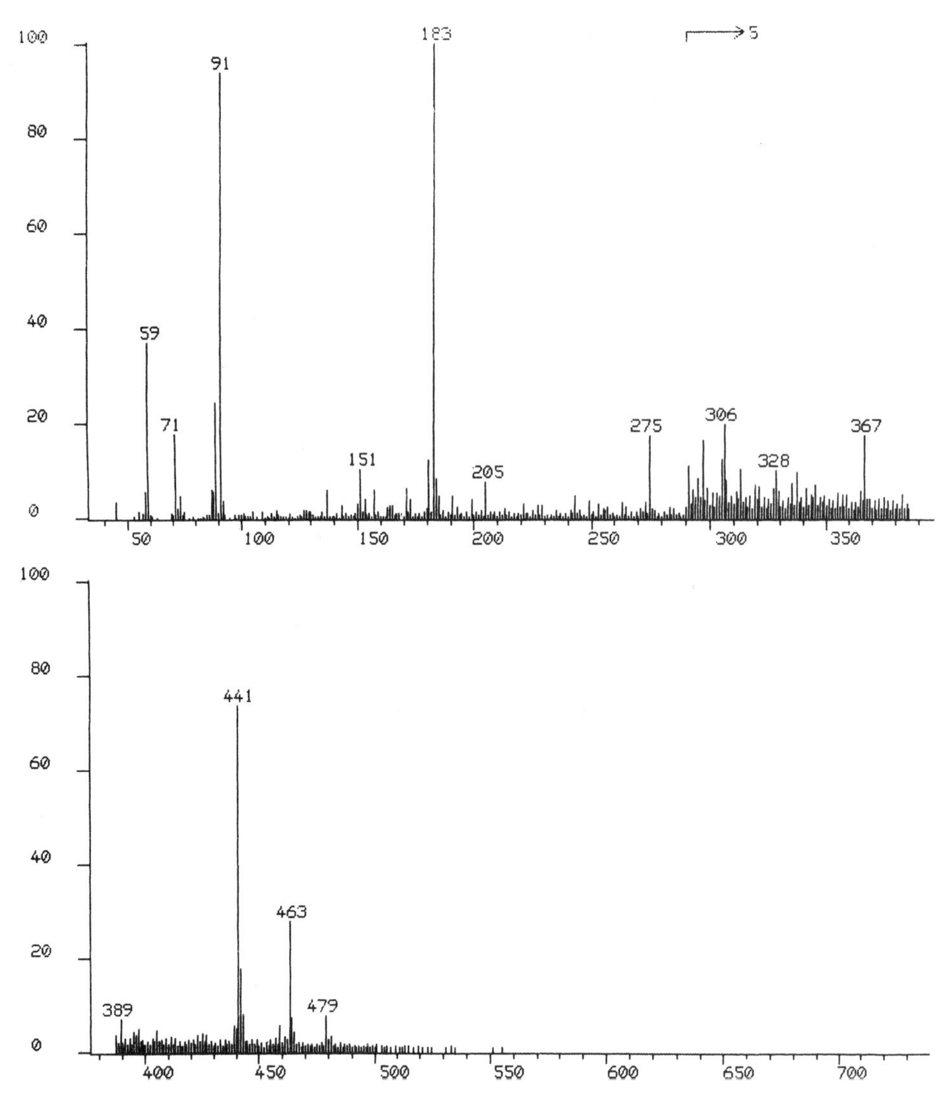

Figure 2. Negative ion FAB mass spectrometry of microsomally produced GSH-DHP conjugate.

Table 1. Rat Liver Microsomal Formation of DHP-GSH Conjugate (in the presence of 50 mM GSH) and DHP (without added GSH).*

Group	DHP	DHP-GSH
Control	3.30 ± 0.19	4.27 ± 0.28
PB pretreated	9.23 ± 0.24†	11.56 ± 0.61†

* Liver microsomes (0.2 nmol P-450/ml) from adult male rats were incubated with senecionine (1 mM) with and without GSH (50 mM) and analyzed by HPLC as described in Materials and Methods.
† Significantly different from controls at $P < 0.005$.

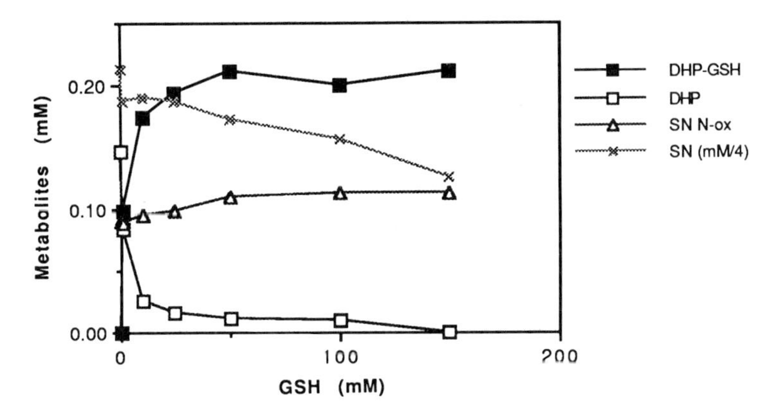

Figure 3. Liver microsomes (0.2 nmol cytochrome P-450) from adult PB-pretreated male rats were incubated for 1 hour with senecionine (1 mM) and various concentrations of GSH and analyzed by HPLC as described in Materials and Methods.

The possible pathways for formation of a GSH-DHP conjugate are summarized in Figure 5. Senecionine is oxidized initially to a reactive metabolite such as dehydrosenecionine (senecionine pyrrole). GSH could then form the GSH-DHP conjugate either by direct reaction with dehydrosenecionine or by combining with the dehydrosenecionine hydrolysis product, DHP. Experiments, therefore, were designed to determine the relative importance of these two pathways in conjugate production.

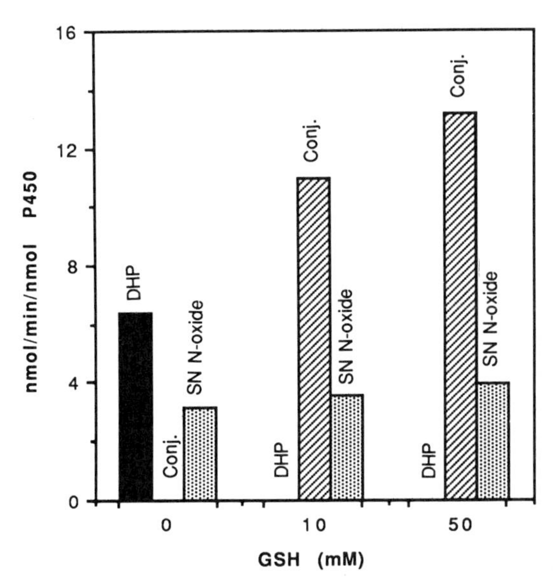

Figure 4. Radiochemical quantitation of metabolites from the incubation of [^{3}H]-senecionine (1 mM) with (10 and 50 mM) and without GSH and then analyzed by HPLC and liquid scintillation counting as described in Materials and Methods.

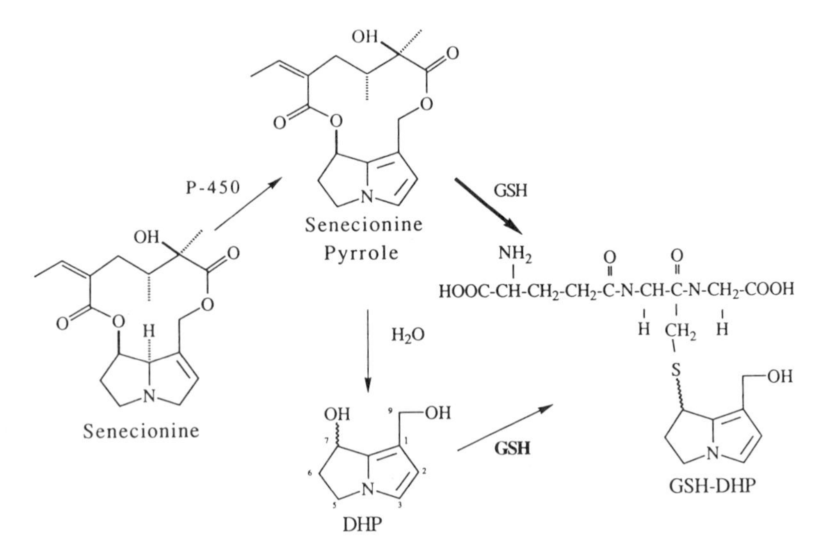

Figure 5. Proposed structure and possible routes of microsomal formation of the GSH-DHP conjugate.

When 200 mM GSH was added to the supernatant from a centrifuged microsomal incubation mixture, only a small fraction (<5%) of the DHP present was converted to the conjugate. When DHR (1 mM) was added to a microsomal incubation mixture in the presence of 50 mM GSH but without adding senecionine, less than 10% of the DHR reacted with GSH in one hour to form the GSH-DHR conjugate. The failure of GSH to yield appreciable amounts of the conjugate when it was added to a centrifuged senecionine microsomal incubation mixture together with the low rate of conjugate formation when DHR was added to a microsomal bioactivation containing only GSH, all provides strong evidence that GSH-DHP conjugate formation occurs only as a result of GSH reaction with a more reactive pyrrole than DHR.

It seems likely, therefore, that GSH preferentially reacts with the strong electrophile (Karchesy and Deinzer, 1981) dehydrosenecionine as rapidly as the latter is formed by the cytochrome P-450 catalyzed oxidation of senecionine. GSH at concentrations as low as 1 mM decreases DHP levels by over 50% in senecionine microsomal incubation mixtures suggesting that *in vivo* the normal physiological concentrations of 5-10 mM GSH in the liver serves as an important modulator of reactive pyrrole concentrations and hence significantly influence PA toxicity. Further studies are needed, however, to better define the role of GSH in the toxicity of the PAs, especially the toxicity and fate of the GSH-DHP conjugate.

ACKNOWLEDGEMENTS

The authors would like to thank Ms. Marilyn Henderson for her excellent technical assistance and Mr. Brian Arbogast for performing the mass spectrometry.

REFERENCES

Buhler, D.R. and Kedzierski, B. (1986). Biological reactive intermediates of pyrrolizidine alkaloids. In *Biological Reactive Intermediates* (J.J. Kocsis, D.J. Jollow, C.M. Witmer, J.O. Nelson and R. Snyder, eds.), pp. 611-620, Plenum Publishing Corp., NY.

Bull, L.B., Culvenor, C.C.J. and Dick, A.T. (1968). *The Pyrrolizidine Alkaloids*, North-

Holland, Amsterdam.
Culvenor, C.C.J., Edgar J.A., Smith, L.W. and Tweddale, H.J. (1970). Dihydropyrrolizidines. III. Preparation and reactions of derivatives related to pyrrolizidine alkaloids, *Aust. J. Chem.* **23**, 1853.
Guengerich, F.P. (1977). Separation and purification of multiple forms of microsomal cytochrome P-450. *J. Biol. Chem.* **252**, 3970-3979.
Hirono, I. (1981). Natural carcinogenic products of plant origin. *C.R.C. Crit. Rev. Toxicol.* **8**, 235.
Huxtable, R.J. (1980). Herbal teas and toxins: Novel aspects of pyrrolizidine poisoning in the United States. *Persp. Biol. Med.* **24**, 1.
Jago, M.V., Edgar, J.A., Smith, L.W. and Culvenor, C.C.J. (1979). Metabolic conversion of heliotridine-based pyrrolizidine alkaloids to dehydroheliotridine. *Mol. Pharmacol.* **6**, 402.
Karchesy, J.J. and Deinzer, M.L. (1981). Kinetics of alkylation reactions of pyrrolizidine alkaloid derivatives. *Heterocycles*, **16**, 631.
Kedzierski, B. and Buhler, D.R. (1986). Method for determination of pyrrolizidine alkaloids and their metabolites by high-performance liquid chromatography. *Anal. Biochem.* **152**, 59.
Kedzierski, B. and Buhler, D.R. (1985). Configuration of necine pyrroles-toxic metabolites of pyrrolizidine alkaloids. *Toxicol. Lett.* **25**, 115.
Mattocks, A.R. and White, I.N.H. (1971). The conversion of pyrrolizidine alkaloids to N-oxides and to dihydropyrrolizine derivations by rat-liver microsomes *in vitro*. *Chem.-Biol. Inter.* **3**, 383.
Mattocks, A.R. (1986). *Chemistry and Toxicology of Pyrrolizidine Alkaloids*, Academic Press, New York, NY. 393 p.
Miranda, C.L., Cheeke, P.R. and Buhler, D.R. (1980). Comparative effects of the pyrrolizidine alkaloids jacobine and monocrotaline on hepatic drug metabolizing enzymes in the rat. *Res. Commun. Chem. Pharmcol.* **29**, 573.
Peterson, J.E. and Culvenor, C.C.J. (1983). Hepatotoxic pyrrolizidine alkaloids. In: *Handbook of Natural Toxins. Vol. 1, Plant and Fungal Toxins*, R.F. Keeler, K.R. Van Kampen and L.F. James, eds., Academic Press, New York.
Ramsdell, H.S., Kedzierski, B. and Buhler, D.R. (1987). Microsomal metabolism of pyrrolizidine alkaloids from Senecio jacobaea. Isolation and quantification of 6,7-dihydro-7-hydroxy-1-hydroxymethyl-5H-pyrrolizidine and N-oxides by high performance liquid chromatography. *Drug Metab. Dispos.* **15**, 32.
Reed, R.L. and Buhler, D.R. (1988). The synthesis of ^{3}H-putrescine and subsequent biosynthesis of ^{3}H-jacobine, a pyrrolizidine alkaloid from *Senecio jacobaea*. *J. Labelled Cmpds. and Radiopharm.* **25**, 1041-1047.
Reed, R.L., Ahern, K.G., Pearson, G.D. and Buhler, D.R. (1988). Crosslinking of DNA by dehydroretronecine, a metabolite of pyrrolizidine alkaloids. *Carcinogenesis* **9**, 1355-1361.
Robertson, K.A., Seymour, J.L., Hsia, M.T. and Allen, J.R. (1977). Covalent interaction of dehydroretronecine, a carcinogenic metabolite of the pyrrolizidine alkaloid monocrotaline with cysteine and glutathione. *Cancer Res.* **37**, 3141.
Shu, I.C., Robertson, A.A., Shumaker, R.C. and Allen, J.R. (1975). Binding of tritiated dehydroretrocine to macromolecules. *Res. Commmun. Chem. Pathol. Pharmcol.* **11**, 99-106.
Wickramanayake, P.P., Arbogast, B.L., Buhler, D.R., Deinzer, M.L. and Burlingame, A.L. (1985). Alkylation of nucleosides and nucleotides by dehydroretronecine: Characterization of adducts by liquid secondary ion mass spectrometry. *J. Am. Chem. Soc.* **107**, 2485-2488.
Williams, D.E., Reed, R.L., Kedzierski, B., Guengerich, F.P. and Buhler, D.R. (1989a). Bioactivation and detoxication of the pyrrolizidine alkaloid senecionine by cytochrome P-450 enzymes in rat liver. *Drug Metab. Dispos.* **17**, 387-392.
Williams, D.E., Reed, R.L., Kedzierski, B., Ziegler, D.M. and Buhler, D.R. (1989b). The role of flavin-containing monooxygenase in the N-oxidation of the pyrrolizidine alkaloid senecionine. *Drug. Metab. Dispos.* **17**, 380-386.

SULFUR CONJUGATES AS PUTATIVE PNEUMOTOXIC METABOLITES OF THE PYRROLIZIDINE ALKALOID, MONOCROTALINE

R. J. Huxtable, R. Bowers, *A. R. Mattocks and M. Michnicka

Department of Pharmacology, College of Medicine
University of Arizona
Tucson, Arizona 85724
*MRC Toxicology Unit, Medical Research Council Laboratories, Carshalton, Surrey, England

Well over 200 pyrrolizidine alkaloids are known, distributed in a number of plant families (Huxtable, 1989; Mattocks, 1986). Important pyrrolizidine-containing families include Crotalaria, Symphytum, Heliotropium and Senecio. Alkaloids containing a 1,2-double bond (Fig. 1) are hepatotoxic, producing veno-occlusive disease in experimental animals, livestock and humans. A relatively small number of alkaloids are also pneumotoxic, producing pulmonary arterial hypertension, arterial medial hyperplasia, endothelial proliferation, and right ventricular hypertrophy in selected species (Huxtable, 1990). Pneumotoxic alkaloids include the Crotalaria lkaloids, monocrotaline and fulvine, and the Senecio alkaloids, senecionine and seneciphylline. Pulmonary hyperplasia and right ventricular hypertrophy are produced in rats by monocrotaline following a single injection of 0.12-0.36 mmole/kg, or a total dose of 0.04 mmole/kg given subacutely (Shubat, Hubbard and Huxtable, 1989). However, pneumotoxicity is delayed in appearance. Following a single injection, pulmonary hyperplasia is first detectable after seven days, pulmonary arterial blood pressure is raised after nine days, and right ventricular hypertrophy develops after fourteen days. These effects are caused by hepatic metabolites of monocrotaline. Monocrotaline itself has no effect on the isolated, perfused lung, and lung tissue appears to be incapable of metabolizing the alkaloid (Gillis, Huxtable and Roth, 1978).

The Nature of the pneumotoxic Metabolite

It was proposed by Mattocks in 1968 that 1,2-unsaturated pyrrolizidines were metabolized to pyrroles, or dehydroalkaloids (Fig. 1) (Mattocks, 1968). Pyrroles are alkylating agents, and it has been proposed that they are responsible for the toxicity of pyrrolizidines. Mattocks' basic premise, that pyrrolizidines are metabolized to pyrroles, has been well substantiated in the intervening years, and it is generally accepted that pyrrolic metabolites are responsible for the toxicity of these alkaloids. However, the nature of the metabolite responsible for the pneumotoxicity is problematic.

The primary pyrrole formed from monocrotaline, dehydromonocrotaline (Figure 1), produces pulmonary hyperplasia and right ventricular hypertrophy on intravenous administration to rats at doses of 0.015 mmole/kg or higher (Bruner, Hilliker and Roth, 1983). However, dehydromonocrotaline is not pneumotoxic when given subcutaneously, because its high reactivity leads to its destruction at the site of injection. Dehydromonocrotaline has a half-live in water of 3.5 seconds (Mattocks, 1989). Dehydromonocrotaline has not been detected as a metabolite of monocrotaline leaving

Biological Reactive Intermediates IV, Edited by C.M. Witmer *et al.*
Plenum Press, New York, 1990

the liver, although recent studies indicate a small amount of a highly reactive pyrrole does exit the liver following perfusion with monocrotaline (Mattocks, 1989). Given the reactivity of dehydromonocrotaline, however, it is difficult to see how much could be delivered to the lungs following hepatic formation.

The secondary pyrrole, dehydroretronecine (Figure 1) has been detected in urine of monocrotaline-exposed rats (Hsu, Allen and Chesney, 1973). It has also been found as a microsomal metabolite of Senecio alkaloids (Ramsdell, Kedzierski and Buhler, 1987; Kedzierski and Buhler, 1986). However, nonester pyrroles are much less active as alkylating agents than are the ester pyrroles. Single injections of 1.31 mmole/kg of dehydroretronecine by the intravenous, intraperitoneal or subcutaneous routes produce a lower degree of right ventricular hypertrophy than is achieved by 1/7 this dose of monocrotaline (Huxtable, unpublished observations). Subacutely, total doses of 0.35 mmole/kg of dehydroretronecine are required to produce right ventricular hypertrophy, as compared to 0.04 mmole/kg of monocrotaline. These findings indicate that dehydroretronecine does not have the requisite potency to be responsible for the pneumotoxicity of monocrotaline.

Following perfusion of the isolated liver with monocrotaline, we have reported the appearance of a novel pyrrole in the perfusate which was provisionally and unimaginatively named Compound A (Figure 1) (Lafranconi and Huxtable, 1984). This pyrrole, on perfusion through the isolated lung, inhibited serotonin transport in a manner earlier reported for lungs from monocrotaline or dehydroretronecine-exposed rats. Compound A was also found to be excreted in bile in remarkably high concentrations (Lafranconi, Ohkuma and Huxtable, 1985). Thus, perfusion of 300 μM monocrotaline resulted in 5 mmole Compound A in the bile. This highly polar metabolite has now been identified as 7-glutathionyldehyroretronecine (Figure 1). The position of thiolation is based on the conversion of compound A in buffered ethanolic silver nitrate to an ethyl ether with a mass spectrum and chromatographic behavior identical to that of authentic 7-ethoxydehydroretronecine.

Bile also contains substantial concentrations of glutathione, up to 5 mM in the rat (Ballatori et al., 1986). However, the formation of 7-glutathionyl-dehydroretronecine appears to occur in the liver cell, as incubation of bile with synthetically prepared dehydromonocrotaline does not result in formation of 7-glutathionyldehydroretronecine as detectable by TLC.

Synthesis and Pneumotoxicity of 7-Glutathionyl-dehydroretronecine

7-glutathionyldehydroretronecine was synthesized in a two-step procedure from monocrotaline. Monocrotaline was converted to dehydromonocrotaline by oxidation with o-chloranil, as previously described (Mattocks, Jukes and Brown, 1989). Dehydromonocrotaline (0.55 nmole) in dimethoxyethane is added to a solution of glutathione (0.81 mmole) in phosphate buffer (0.5 M; pH 7.4). After 5 min, ethanol is added and the precipitated salts centrifuged out. The supernatant is freeze-dried, and the glutathionyldehydroretronecine purified by silica gel chromatography.

7-Glutathionyldehydroretronecine mimicks the pneumotoxicity of monocrotaline in rats (Figure 2). Doses of 0.23 mmole/kg i.v., or 0.109 mmole/kg given subacutely by subcutaneous injection over six days, are equitoxic with monocrotaline (20 mg/l in drinking water for 6 days; total dose 0.07 mmole/kg) as shown by increase in lung mass, right ventricular hypertrophy (right ventricular to body weight ratio and right ventricular to left ventricular weight ratio) and decreased body weight gain. A lower dose of 0.11 mmole/kg i.v. of glutathionyldehydroretronecine produces a somewhat lower degree of right ventricular hypertrophy than is seen with monocrotaline (Figure 2). This level of glutathionyldehydroretronecine did not result in pulmonary hyperplasia 19 days after the injection. However, we have shown that pulmonary hyperplasia always precedes right ventricular hypertrophy, but that at low acute levels of monocrotaline, the pulmonary hyperplasia will regress over a 3-week period (Shubat, Hubbard and Huxtable, 1989).

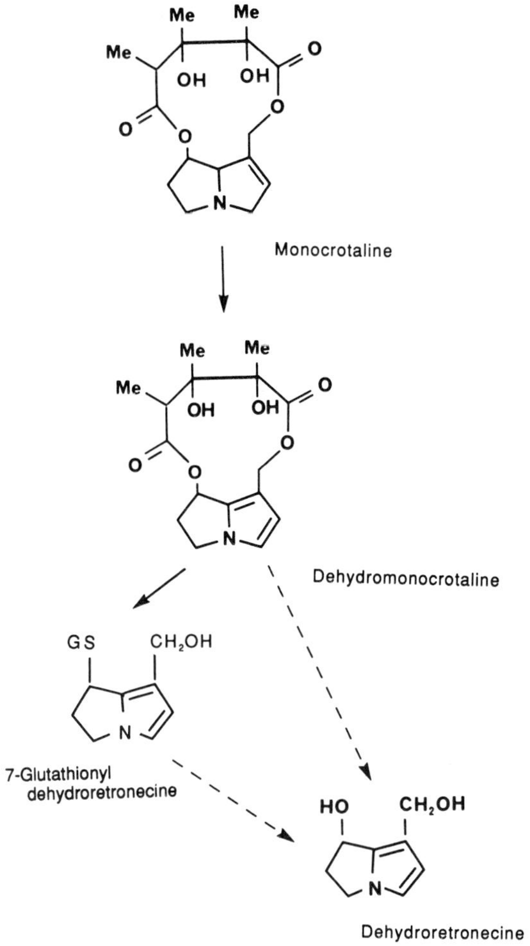

Monocrotaline

Dehydromonocrotaline

7-Glutathionyl dehydroretronecine

Dehydroretronecine

Figure 1. A proposed scheme for monocrotaline metabolism. the pneumotoxic alkaloid, monocrotaline, is oxidized by a P450 mixed function oxidase system in the liver to the pyrrole, dehydromonocrotaline. Dehydromonocrotaline either reacts with macromolecular constituents of the cell, producing hepatotoxicity, or undergoes hydrolysis to the relatively inocuous secondary pyrrole, dehydroretronecine. A thrid pathway, conjugation with glutathione to produce glutathionyldehydroretronecine (Compound A) may be responsible for the pneumotoxicity. The differing pneumotoxicity of the various pyrrolizidines may be a function of the proportion of the primary pyrrole diverted into the conjugation pathway.

Similar toxicity is produced by 0.10 mmole/kg of glutathionyldehydroretronecine delivered over a 1-week period via Alzet minipumps implanted subcutaneously or intraperitoneally. However, the total dose actually delivered in this experiment is uncertain, as glutathionyldehydroretronecine slowly breaks down on storage in aqueous solution.

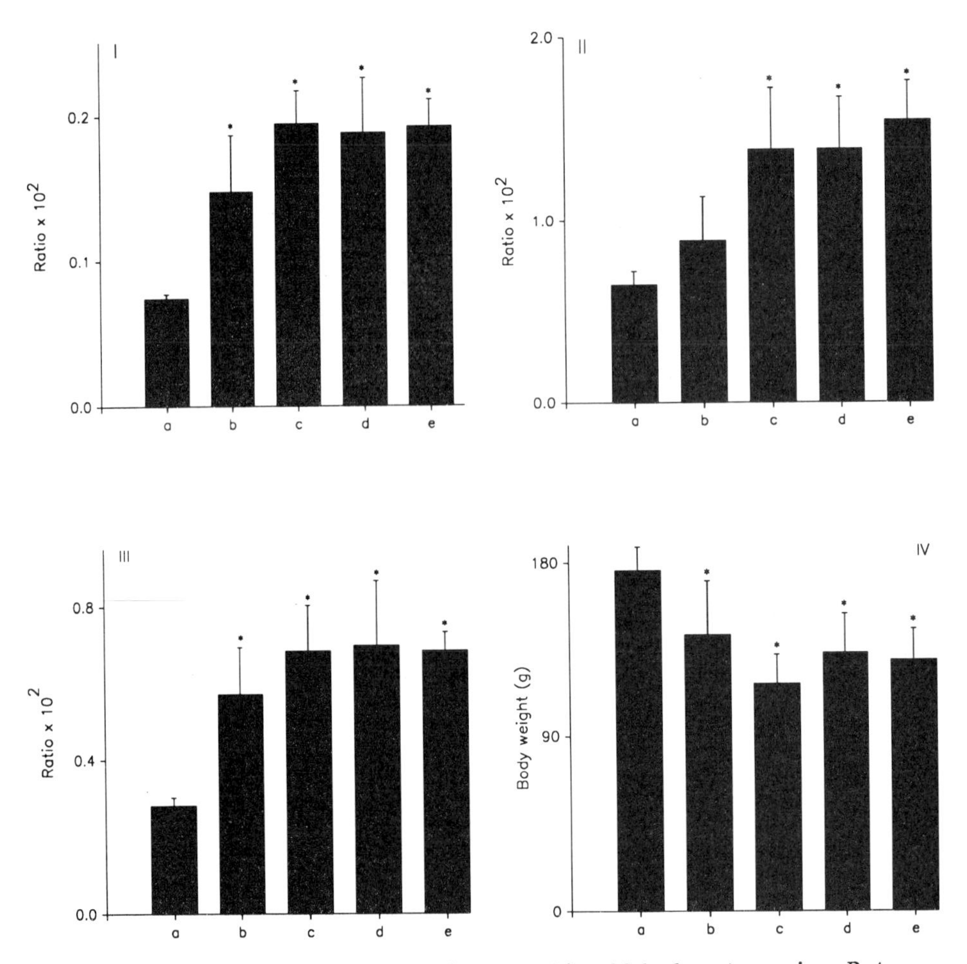

Figure 2. The pneumotoxicity of 7-glutathionyldehydroretronecine: Rats of initial body weight 50 g were assigned to one of the following treatments: (a) control; (b) 0.11 mmoles/kg i.v. glutathionyldehydroretronecine; (c) 0.23 mmoles/kg i.v. glutathionyldehydroretronecine; (d) 0.011 mmoles/kg subcutaneously per day for 6 days of glutathionyldehydroretronecine; or (e) monocrotaline in drinking water (20 mg/l) for 6 days. Animals were killed 20 days after beginning of treatment. Organ weight changes at death are shown on the panels for (I) right ventricular to body weight ratio; (II) lung to body weight ratio; (III) right ventricular to left ventricular weight ratio; and (IV) body weight. Each group contained 7 animals. *$p < 0.05$ by analysis of variance.

In monocrotaline-exposed rats, there is a progressive decrease in the ability of the lungs to sequester and metabolize serotonin (Huxtable, Ciaramitaro and Eisenstein, 1978; Roth, Dotzlaf, Baranyi, Kuo and Hook, 1981). This is due to an inhibition of the endothelial transport process, rather than to a deficiency in monoamine oxidase activity. In rats injected with 0.11 mmole/kg i.v. of glutathionyldehydroretronecine,

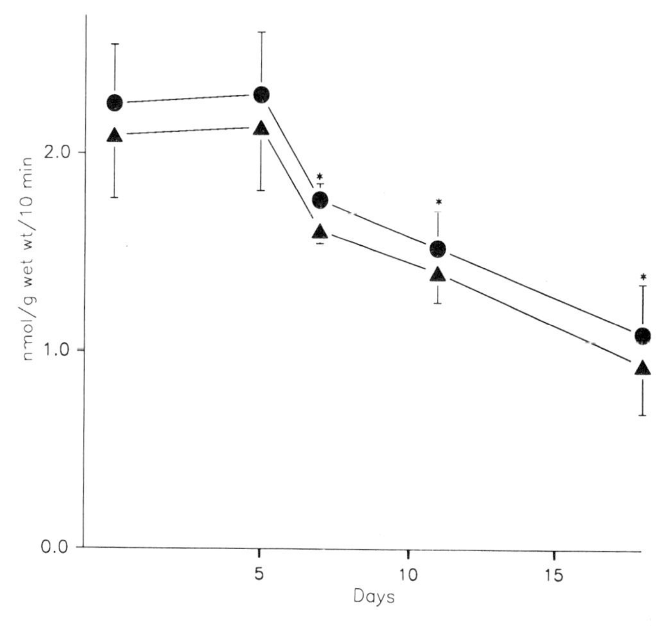

Figure 3. The effect of treatment with glutathionyldehydroretronecine *in vivo* on lung serotonin transport *in vitro*. Animals received a single injection of 0.11 mmoles/kg i.v. of glutathionyldehydroretronecine. They were killed at the times indicated, and lungs isolated and serotonin transport and metabolizing capabilities determined as previously described (Huxtable et al., 1978; Lafranconi and Huxtable, 1984). Each group contained 5-6 animals. *p <0.05 compared to control group, as determined by analysis of variance.

there is a similar progressive decrease in the serotonin-transporting capabilities of the lung (Fig. 3).

Perfusion of 1 mM glutathionyldehydroretronecine through isolated rat lungs also results in the inhibition of serotonin transport (Fig. 4).

The Pneumotoxicity of 7-Cysteinyldehydroretronecine

7-glutathionyldehydroretronecine can potentially be metabolized further by the mercapturic acid (Bakke and Gustafsson, 1984) and the β-lyase pathways (Dekant, Lash and Anders, 1988). These could yield as further metabolites various thiolated derivatives of dehydroretronecine, including glycinylcysteinyl, cysteinyl, N-acetylcysteinyl, thio- and thiomethyl. We have synthesized 7-cysteinyldehydroretronecine and find that at a dose level of 0.195 mmole/kg i.v. it mimicks all aspects of the pneumotoxicity of monocrotaline, and 7-glutathionyldehydroretronecine.

CONCLUSIONS

When monocrotaline is perfused through the isolated liver, 7-glutathionyldehydroretronecine is the only detectable pyrrole released into bile and perfusate. The quantities formed account for most of the monocrotaline metabolized. 7-Glutathionyldehydroretronecine and its analog 7-cysteinyldehydroretronecine

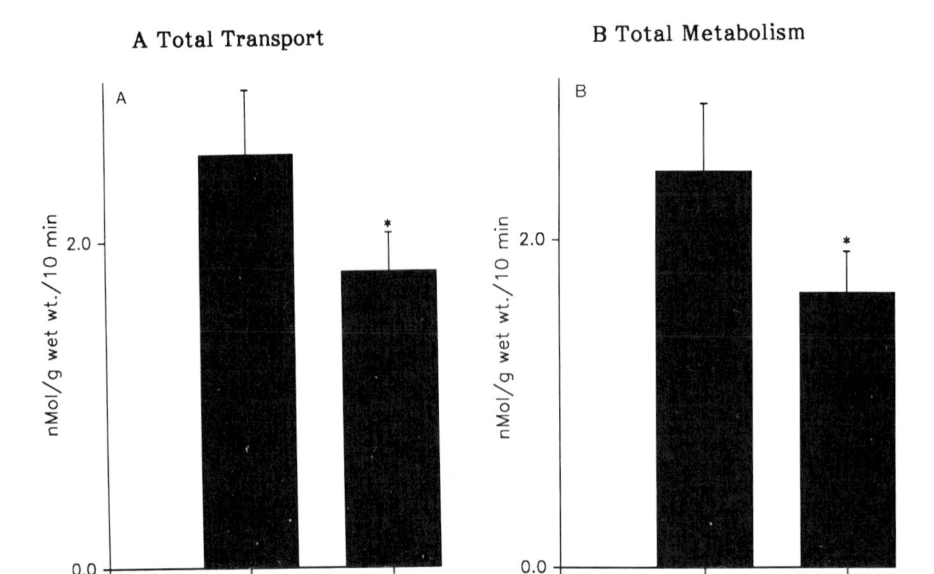

Figure 4. Effect of glutathionyldehydroretronecine *in vitro* on serotonin transport and metabolism by the isolated lung. Glutathionyldehydroretronecine (1 mM) was co-perfused with 0.1 mM serotonin for a 10 min period. The transport and metabolism of serotonin was determined as previously described (Huxtable et al., 1978; Lafranconi and Huxtable, 1984). Each group contained 6 animals. *p <0.01 according to the Student unpaired t-test.

reproduce all aspects of the pneumotoxicity of monocrotaline, with potencies approximately the same as that of monocrotaline.

We conclude from this that 7-thioethers of dehydroretronecine are effective delivery systems for alkylating pyrroles. Glutathionyldehydroretronecine must be considered as a candidate for the pneumotoxic metabolite of monocrotaline. In support of this, preliminary work shows that the pneumotoxic alkaloid, retrorsine (Mattocks, 1972), also results in the production of glutathionyldehydroretronecine in the isolated liver, whereas perfusion with the non-pneumotoxic alkaloid, heliotrine, does not.

We propose that the primary metabolite of monocrotaline in the liver is dehydromonocrotaline. This highly reactive substance has three pathways of further reaction; it can react with cell macromolecular constituents producing hepatotoxicity, it can be hydrolyzed to inert products such as dehydroretronecine, or it can be non-altruistically conjugated with glutathione, and exported to produce toxic consequences elsewhere. It is also possible that a small quantity of dehydromonocrotaline may survive passage from the liver, and reach the lungs. However, to sustain the suggestion that dehydromonocrotaline is the pneumotoxic metabolite, it will be necessary to demonstrate that sufficient quantities of dehydromonocrotaline to cause pneumotoxicity do, indeed, reach the lungs.

A further point of interest that glutathionyldehydroretronecine is unreactive in transferring pyrrole groups to a thiosepharose acceptor in a chemical assay. This suggests that alkylation to produce toxicity is not simply a chemical reaction, but involves an enzyme. Candidate enzymes include a glutathione transferase, or possibly a flavin monoxygenase. If the latter, the sulfoxide intermediate formed provides a much readier leaving group than the thiolate grouping of a thioester.

ACKNOWLEDGEMENT

This work was supported by Public Health Service Grant HL-25258.

REFERENCES

Bakke, J. and Gustafsson, J.-A. (1984). Mercapturic acid pathway metabolites of xenobiotics: generation of potentially toxic metabolites during enterohepatic circulation. *TIPS* **5**, 517-521.

Bruner, L.H., Hilliker, K.S. and Roth, R.A. (1983). Pulmonary hypertension and ECG changes from monocrotaline pyrrole in the rat. *Am. J. Physiol.* **245**, H300-H306.

Dekant, W., Lash, L.H. and Anders, M.W. (1988). Fate of glutathione conjugates and bioactivation of cysteine S-conjugates by cysteine conjugate beta-lyase. In *Glutathione conjugation: Mechanisms and Biological Significance*, edited by Sies, H. and Ketterer, B. pp. 415-447. Academic Press Limited, London.

Gillis, C.N., Huxtable, R.J. and Roth, R.A. (1978). Effect of monocrotaline pretreatment of rats on removal of 5-hydroxytryptamine and norepinephrine by perfused lung. *Brit. J. Pharmacol.* **63**, 435-443.

Hsu, I.C., Allen, J.R. and Chesney, C.F. (1973). Identification and toxicological effects of dehydroretronecine, a metabolite of monocrotaline. *Proc. Soc. Exp. Biol. Med.* **144**, 834-838.

Huxtable, R.J. (1989). Human health implications of pyrrolizidine alkaloids and herbs containing them. In Toxicants of Plant Origin, Vol I: Alkaloids edited by Cheeke, P.R. pp. 41-86. CRC Press, Boca Raton, FL.

Huxtable, R.J. (1990). Activation and pulmonary toxicity of pyrrolizidine alkaloids. *Pharmacology and Therapeutics*, in press.

Huxtable, R.J., Ciaramitaro, D. and Eisenstein, D. (1978). The effect of a pyrrolizidine alkaloid, monocrotaline, and a pyrrole, dehydroretronecine, on the biochemical functions of the pulmonary endothelium. *Mol. Pharmacol.* **14**, 1189-1203.

Kedzierski, B. and Buhler, D.R. (1986). The formation of 6,7-dihydro-7-hydroxy-1-hydroxymethyl-5H-pyrrolizine, a metabolite of pyrrolizidine alkalods. *Chem-Biol. Inter.* **57**, 217-222.

Lafranconi, W.M. and Huxtable, R.J. (1984). Hepatic metabolism and pulmonary toxicity of monocrotaline using isolated perfused liver and lung. *Biochem. Pharm.* **33**, 2479-2484.

Lafranconi, W.M., Ohkuma, S. and Huxtable, R.J. (1985). Biliary excretion of novel pneumotoxic metabolites of the pyrrolizidine alkaloid, monocrotaline. *Toxicon* **23**, 983-992.

Mattocks, A.R. (1968). Toxicity of pyrrolizidine alkaloids. *Nature* (London) **217**, 723-728.

Mattocks, A.R. (1972). Acute hepatotoxicity and pyrrolic metabolites in rats dosed with pyrrolizidine alkaloids. *Chem. -Biol. Interactions* **5**, 227-242.

Mattocks, A.R. (1986). Chemistry and Toxicology of Pyrrolizidine Alkaloids, London: Academic Press, pp. 1-393.

Mattocks, A.R. (1990). Pyrrolizidine alkaloids: what metabolites are responsible for extrahepatic tissue damage in animals? Proceedings of the Third International Symposium on Poisonous Plants, edited by James, L.F., in press.

Mattocks, A.R., Jukes, R. and Brown, J. (1989). Simple procedures for preparing putative toxic metabolites of pyrrolizidine alkaloids. *Toxicon* **27**, 561-567.

Ramsdell, H.S., Kedzierski, B. and Buhler, D.R. (1987). Microsomal metabolism of pyrrolizidine alkaloids from Senecio jacobaea: isolation and quantification of 6,7-dihydro-7-hydroxy-1-hydroxymethyl-5H-pyrrolizine and N-oxides by high performance liquid chromatography. *Drug Metab. Dispos.* **15**, 32-36.

Roth, R.A., Dotzlaf, L.A., Baranyi, B., Kuo, C.H. and Hook, J.B. (1981). Effect of monocrotaline ingestion on liver, kidney, and lung of rats. *Toxicol. Appl. Pharmacol.* **60**, 193-203.

Shubat, P.J., Hubbard, A.K. and Huxtable, R.J. (1989). Dose-response relationship in intoxication by the pyrrolizidine alkaloid monocrotaline. *J. Toxicol. Environ. Health* **28**, 445-460.

THE METABOLISM OF BENZENE TO MUCONIC ACID, A POTENTIAL BIOLOGICAL MARKER OF BENZENE EXPOSURE

G. Witz, T.A. Kirley, W.M. Maniara, V.J. Mylavarapu, and B.D. Goldstein

UMDNJ-Robert Wood Johnson Medical School
Dept. of Environmental & Community Medicine and EOHSI
Piscataway, NJ

INTRODUCTION

Benzene, the simplest aromatic hydrocarbon, is a ubiquitous environmental pollutant. It is a volatile compound, extensively used in industry as an important synthetic intermediate. It is also an additive in gasoline. In addition to occupational exposure, a major human exposure results from cigarette smoke (Wallace, 1989). The toxicity of benzene as a leukemogen in humans (Goldstein, 1977) and a hematotoxic and carcinogenic agent in experimental animals (Snyder and Kocsis, 1977; Maltoni et al., 1989) has been well established. It is generally accepted that benzene toxicity is mediated by metabolites (Snyder et al., 1981). The nature of the metabolite(s) responsible for benzene toxicity is at present not known.

The metabolism of benzene is complex and results in the formation of a variety of ring)hydroxylated compounds, quinones, glucuronide and sulfate conjugates (Kalf, 1987), as well as the ring)opened metabolites trans,trans)muconaldehyde and trans,trans)muconic acid (Latriano et al., 1986; Parke and Williams, 1953). Until recently it was thought that metabolism leading to hydroquinone/benzoquinone represents a pathway for the formation of toxic benzene metabolites. With the indentification of trans,trans)muconaldehyde (muconaldehyde, MUC) as a microsomal hematotoxic metabolite of benzene in our laboratory (Latriano et al., 1986; Witz et al., 1985), the ring)opening pathway leading to reactive metabolic products is being considered to represent a second pathway for the formation of toxic benzene metabolites.

The formation of trans,trans-muconic acid (muconic acid, MA) as a urinary metabolite of benzene was firmly established by Parke and Williams (1953) who showed that oral administration of ^{14}C-benzene to rabbits leads to urinary excretion of -)1.8% of the administered dose (0.4-0.34 g/kg) as muconic acid. Subsequently, benzene was shown to be metabolized to urinary muconic acid in mice and rats (Gad-El Karim et al., 1985; Sabourin et al., 1988). At present the pathway for muconic acid formation is not known. Since we originally postulated that muconaldehyde may be a likely precursor of muconic acid and a potential toxic metabolite of benzene (Goldstein et al., 1982), we addressed the question of muconaldehyde metabolism to determine its relationship to muconic acid formation including the formation of other toxic ring-opened intermediates. These studies, summarized below, indicate that a possible pathway for muconic acid formation from benzene involves metabolism of muconaldehyde to the mixed aldehyde-carboxylic acid derivative of MUC, followed by a second oxidation to muconic acid (Figure 1).

Biological Reactive Intermediates IV, Edited by C.M. Witmer *et al.*
Plenum Press, New York, 1990

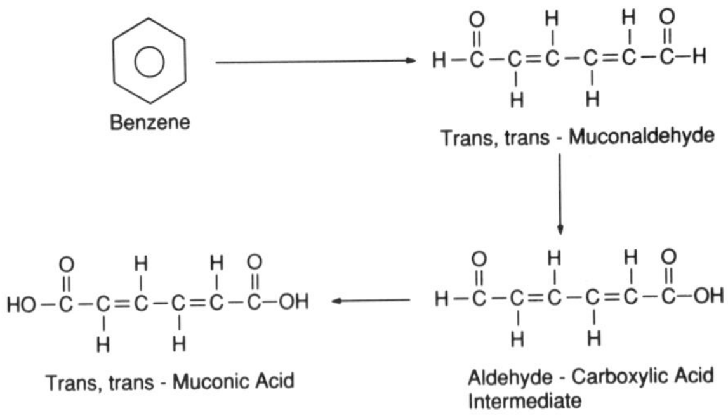

Benzene

H–C–C=C–C=C–C–H

Trans, trans - Muconaldehyde

H–C–C=C–C=C–C–OH

Aldehyde - Carboxylic Acid Intermediate

HO–C–C=C–C=C–C–OH

Trans, trans - Muconic Acid

Figure 1. Potential metabolic pathway for the formation of trans,trans-muconic acid from benzene via trans,trans-muconaldehye. (From Kirley et al., 1989; reprinted with permission from Academic Press, Inc.).

Among the benzene metabolites, muconic acid is unique since its sole known source at present is benzene. Thus, unlike phenol which may be derived from a variety of sources including benzene, muconic acid has the potential for serving as a biological marker for benzene exposure. To determine the dose response relationship for muconic acid formation from benzene, urinary muconic acid formation was studied in mice administered benzene doses ranging over three orders of magnitude, i.e. high hematotoxic doses and low doses in keeping with known levels of human benzene exposure.

In order to further assess our hypothesis concerning the potential involvement of muconaldehyde in benzene toxicity, the dose response relationship for urinary muconic acid formation from benzene was studied in the benzene sensitive DBA/2N mice and the less benzene sensitive C57BL/6 mice (Longacre et al., 1981). Since we established that muconaldehyde may be a precursor of muconic acid, the latter might serve as an indicator of muconaldehyde formation. If muconaldehyde should paly a role in benzene hematotoxicity we would expect to find at least one of the following three possibilities: 1) evidence of increased metabolism of benzene to muconic acid in the more sensitive strain, 2) a difference in the metabolism of muconaldehyde to muconic acid between the sensitive and less sensitive strain, or 3) a greater response of the target organ to muconaldehyde in keeping with its response to benzene. Should none of these be observed it would provide less support for the hypothesis that muconaldehyde or other ring-opened products play an important role in benzene hematotoxicity.

The results of our studies described below indicate that benzene sensitive DBA/2N mice excrete significantly more muconic acid at high hematotoxic benzene doses compared with the less benzene sensitive C57BL/6 mice, suggesting that metabolism to ring-opened products might play a role in benzene toxicity.

RESULTS AND DISCUSSION

Metabolism of Muconaldehyde to a Monocarboxylic Acid Derivative and to Muconic Acid

The *in vitro* metabolism of muconaldehyde was investigated using purified yeast aldehyde dehydrogenase (ALDH) and a mouse liver soluble fraction. Yeast ALDH has been reported to have varying specificities to a wide variety of aldehyde substrates including crotonaldehyde, another alpha,beta-unsaturated aldehyde (Lundquist, 1958). A mouse liver soluble fraction of liver cytosol containing cytosolic and mitochondrial ALDHs was used for MUC metabolism studies to approximate likely metabolic conditions in mouse liver, a major site of metabolism of benzene. The results of these studies described in detail by Kirley et al. (1989) indicate that purified yeast ALDH supplemented with NAD^+ oxidizes MUC in a biphasic manner with apparent K_m values of 0.48 and 3.2 uM and corresponding V_{max} values of 604 and 1227 nmol/min/mg protein. Using high pressure liquid chromatography, a product "X" with polarity intermediate between that of the dialdehyde trans,trans-muconaldehyde and the diacid trans,trans-muconic acid was detected in the ALDH incubation mixtures (Figure 2). The same product was also detected in a DBA/2J mouse liver soluble fraction supplemented with NAD^+ and incubated with MUC (Figure 2). This product was isolated from a scaled-up incubation mixture containing MUC, purified yeast ALDH, and NAD^+. Mass spectral analysis of this product and functional group spot tests were consistent with a structure in which one of the aldehydic groups of MUC has been oxidized to a carboxylic acid group. Thus the unknown product "X" is the monocarboxylic acid derivative of MUC. In the presence of the mouse liver soluble fraction supplemented with NAD^+, the monocarboxylic acid derivative of MUC was metabolized to trans,trans-muconic acid.

The results described above suggest that one mechanism for the formation of urinary muconic acid may involve a two step oxidation of the aldehyde functional groups. This process leads to the formation of an intermediate mixed aldehyde-carboxylic acid MUC derivative which is oxidized to muconic acid. In comparison with MUC, the monocarboxylic acid derivative of MUC is expected to be less reactive towards Michael addition reactions since the carboxylic acid functional group deactivates the adjacent double bond towards nucleophilic addition. However, the double bond adjacent to the remaining aldehydic group is still activated and may participate in Michael additions involving thiols or primary amines. If muconaldehyde is formed in the liver, the formation of the mixed aldehyde-carboxylic acid derivative might provide an intermediate with greater potential for reaching the bone marrow. The reduced reactivity of the mixed aldehyde-carboxylic acid derivative relative to

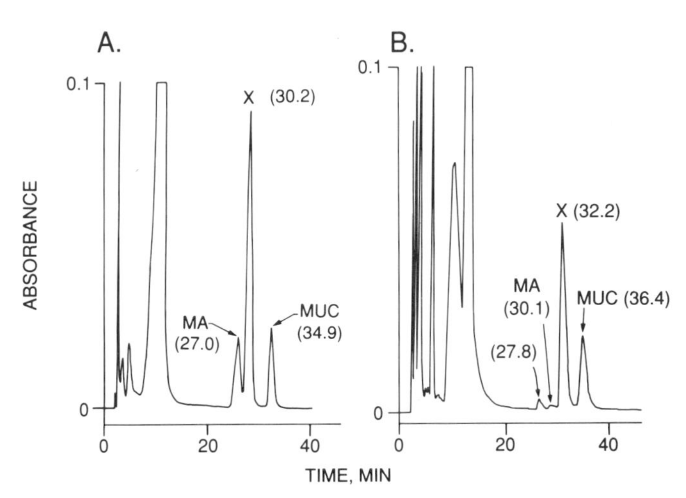

Figure 2. (A) High-pressure liquid chromatogram of the methanolic sample from the complete yeast aldehyde dehydrogenase mixture incubated with trans,trans-muconaldehyde. This methanolic sample was spiked with trans,trans-muconic acid (MA) and trans,trans-muconaldehyde (MUC).(X, intermediate monoacid derivative of trans,trans-muconaldehyde). (B) High-pressure liquid chromatogram of the methanolic sample from the mouse liver soluble fraction incubated with MUC. (From Kirley et al., 1989; reprinted with permission from Academic Press, Inc.).

that of MUC may also facilitate its detection *in vivo* relative to that of MUC. The possible *in vivo* formation of the monocarboxylic acid derivative of MUC and its toxicity in relation to MUC as well as benzene are subjects of future investigation in our laboratory.

Dose-Response Studies on the Formation of Urinary Muconic Acid from Benzene in DBA/2N and C57BL/6 Mice

A sensitive assay for urinary muconic acid was developed and used for the measurement of urinary MA in DBA/2N and C57BL/6 mice administered benzene doses ranging over three orders of magnitude, i.e. 0.5 to 880 mg/kg. Benzene was dissolved in corn oil and administered intraperitoneally (i.p.). The mice, 2 per cage, were housed in metabolic cages and urine was collected into ascorbic acid for 24 hr following benzene administration. Urinary muconic acid was also measured for mice treated i.p. with 0.5-5 mg/kg MUC dissolved in corn oil and 0.7 to 7.1 mg/kg muconic acid dissolved in phosphate buffered saline, pH 7.4. The results of these studies summarized below are described in detail in Witz et al. (1990).

In both strains, the percent of benzene dose excreted as urinary MA decreased with an increase in benzene dose, i.e. from 9.8 to 0.4% in DBA/2N mice and from 17.6 to 0.2% in C57BL/6 mice treated with 0.5 to 880 mg/kg benzene. Significant differences in the percent dose excreted as urinary muconic acid within the first 24 hr after treatment were observed between DBA/2N and C57BL/6 mice at very high and very low doses of benzene. At 880, 440 and 220 mg/kg benzene, the percent dose excreted as urinary muconic acid was significantly higher in DBA/2N mice compared with C57BL/6 mice. As shown in Figure 3 at 440 and 880 mg/kg benzene DBA/2N mice excrete about twice as much urinary muconic acid compared C57BL/6 mice. The dose response curve (Figure 3) also indicates that a plateau is reached for urinary muconic acid excretion at 220 mg/kg benzene in C57BL/6 mice and at 440 mg/kg benzene in DBA/2N mice.

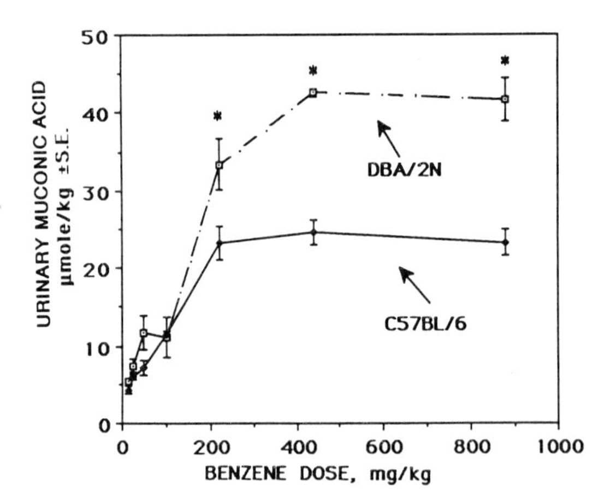

Figure 3. Excretion of urinary muconic acid by DBA/2N and C57BL/6 mice treated with benzene. Mice were housed in metabolic cages (2 per cage) and administered 12.5 to 880 mg/kg benzene intraperitoneally. Urines were collected for 24 hr, extracted and analyzed by HPLC as described in Materials and Methods. Each point is the mean ± S.E. of 3-5 urine samples. *p ▪ 0.05. (From Witz et al., 1990; reprinted with permission from Pergamon Journals Ltd).

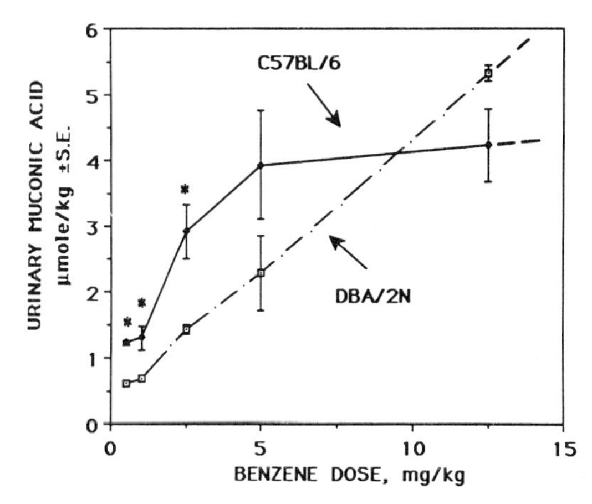

Figure 4. Excretion of urinary muconic acid by DBA/2N and C57BL/6 mice treated with low doses of benzene. Mice were housed in metabolic cages (2 per cage) and administered 0.5 to 12.5 mg/kg benzene intraperitoneally. Urines were collected for 24 hr, extracted and analyzed by HPLC as described in Materials and Methods. Each point is the mean ± S.E. of 3-5 urine samples. *p ≤ 0.05. (From Witz et al.,1990; reprinted with permission from Pergamon Journals Ltd).

In contrast to the results obtained at high doses of benzene, the percent dose excreted as urinary muconic acid was significantly less in DBA/2N mice than in C57BL/6 mice after treatment with 2.5, 1.0, and 0.5 mg/kg benzene. In C57BL/6 mice, 1.23 to 2.92 umole/kg muconic acid is excreted compared with 0.63 to 1.44 umole/kg muconic excreted by DBA/2N mice after administration of 0.5 to 2.5 mg/kg benzene, respectively, (Figure 4).

The results described above indicate (i) that the percent dose excreted as urinary muconic acid increases in both strains with a decrease in administered benzene dose, and (ii) that significant differences between DBA/2N and C57BL/6 mice exist in urinary muconic acid excretion at the low dose and the high dose end of the dose response curve for benzene doses ranging from 0.5 to 880 mg/kg. The observed differences between DBA/2N and C57BL/6 mice in the percent dose metabolized to urinary muconic acid after administration of hematotoxic benzene doses, i.e. 220-880 mg/kg, cannot be ascribed to differences in the metabolism of MUC to MA if MUC is indeed a precursor of MA, since DBA/2N and C57BL/6 mice excrete similar amounts of MA after treatment with MUC (Witz et al., 1990). These two strains also exhibited no differences in the disposition of MA (Witz et al., 1990). The observed differences in the percent benzene dose excreted as urinary MA therefore reflect real differences in the percent benzene dose metabolized to muconic acid. The present data, which indicate twice as much muconic acid excretion in benzene sensitive DBA/2N mice compared with less benzene sensitive C57BL/6 mice after administration of 440 and 880 mg/kg benzene, are in keeping with, but do not prove, the hypothesis that metabolism of benzene to ring-opened compounds may play a role in benzene toxicity.

These findings may be pertinent to the development of a biomarker of benzene exposure and effect. If the dose response curve for the metabolism of benzene to urinary muconic acid is similar in mice and humans, the increased metabolism to urinary muconic acid at low benzene exposures coupled with the great sensitivity of the assay suggests that urinary muconic acid may be a valid biomarker of benzene exposure.

ACKNOWLEDGEMENTS

We thank Mrs. Toni Myers for the preparation of this manuscript. This research was supported by NIH Grant ES02558.

REFERENCES

Gad-El Karim, M.M., Sadagopa Ramanujam, V.M., and Legator, M.S. (1985). Trans, trans-muconic acid, an open-chain urinary metabolite of benzene in mice. Quantification by high-pressure liquid chromatography. *Xenobiotica* **15**, 211-220.

Goldstein, B.D. (1977). Hematotoxicity in humans. In: Benzene Toxicity: A Critical Evaluation. (Eds., J. Laskin and B.D. Goldstein). *J. Toxicol. Environ. Health*, Supplement 2, pp. 69-105.

Goldstein, B.D., Witz, G., Javid, J., Amoruso, M., Rossman, T., and Wolder, B. (1982). Muconaldehyde, a potential toxic intermediate of benzene metabolism. In *Biological Reactive Intermediates* (R. Snyder, D.V. Parke, J.J. Kocsis, D. Jallow, G.G. Gibson, and C.M. Witmer, Eds.), Vol. 2 pp. 331-339. Plenum, New York.

Kalf, G.F. (1987). Recent advances in the metabolism and toxicity of benzene. *Critical Reviews in Toxicology*, **18**, 141-159.

Kirley, T.A., Goldstein, B.D., Maniara, W.M., and Witz, G. (1989). Metabolism of trans,trans-muconaldehyde, a microsomal hematotoxic metabolite of benzene, by purified yeast aldehyde dehydrogenase and a mouse liver soluble fraction. *Toxicol. Appl. Pharmacol.* **100**, 360-367.

Latriano, L., Goldstein, B.D., and Witz, G. (1986). Formation of muconaldehyde, an open-ring metabolite of benzene in mouse liver microsomes: An additional pathway for toxic metabolites. *Proc. Natl. Acad. Sci.* **83**, 8356-8360.

Longacre, L.S., Kocsis, J.J., and Snyder, R. (1981). Influence of strain differences in mice on the metabolism and toxicity of benzene. *Toxicol. Appl. Pharmacol.* **60**, 398-409.

Lundquist, F. (1958). Enzymic determination of acetaldehyde in blood. *Biochemistry* **68**, 172-177.

Maltoni, C., Ciliberti, G., Cotti, G., Conti, B., and Belpoggi, F. (1989). Benzene, an experimental multipotential carcinogen: Results of the long)term bioassays performed at the Bologna Institute of Oncology. *Environ. Health Perspect.* **82**, 109-124.

Parke, D.V., and Williams, R.T. (1953). The metabolism of benzene. (a) The formation of phenylglucuronide and phenylsulfuric acid from (^{14}C)benzene. (b) The metabolism of (^{14}C)phenol. *Biochem. J.* **55**, 337-340.

Sabourin, P.J., Bechtold, W.E., Birnbaum, L.S., Lucier, G., and Henderson, R.F. (1988). Differences in the metabolism of inhaled [^{3}H]benzene by F344/N rats and B6C3F1 mice. *Toxicol. Appl. Pharmacol.* **94**, 128-140.

Snyder, R., and Kocsis, J.J. (1975). Current concepts of chronic benzene toxicity. *CRC Critical Rev. Toxicol.* **3**, 265-287.

Snyder, R., Lee, L.S., and Witmer, C.M. (1981). Biochemical Toxicology of benzene. In Reviews in Biochemical Toxicology (E. Hodgson, J.R. Bend, and R.M. Philpot, Eds.,) Elsevier/North Holland, Amsterdam, pp. 123-153.

Wallace, L.A. (1989). Major sources of benzene exposure. *Environ. Health Perspect.* **82**, 165-169.

Witz, G., Rao, G.S. and Goldstein, B.D., (1985). Short-term toxicity of trans,trans-muconaldehyde. *Toxicol. Appl. Pharmacol.* **80**, 511-516.

Witz, G., Maniara, W., Mylavarapu, V., and Goldstein, B.D. (1990). Comparative metabolism of benzene and trans,trans-muconaldehyde to trans,trans-muconic acid in DBA/2N and C57BL/6 mice. Biochem. Pharmacol. In Press.

LESSONS ON THE SECOND CANCERS RESULTING FROM CANCER CHEMOTHERAPY

Bernard D. Goldstein, M.D.

Department of Environmental and Community Medicine
UMDNJ-Robert Wood Johnson Medical School
675 Hoes Lane
Piscataway, NJ 08854 and
Director
Environmental and Occupational Health Sciences Institute

INTRODUCTION

In this paper we explore some of the insights into the mechanisms of human carcinogenesis and into the assessment of cancer risk that could be obtained by the study of individuals at risk of development of a second cancer due to treatment with chemotherapeutic agents. We urge a concerted effort to study the processes by which cancer chemotherapeutic agents result in the production of a second cancer.

Chemicals are of major importance in causing cancer, whether these chemicals come from outside the body in the form of pollutants, food, or personal habits, or from within the body. Thus studies of the mechanisms of chemical carcinogenesis have received much attention in recent years with the hope that some of the mysteries that govern this important group of human diseases will be unraveled. The two major study approaches are to apply known amounts of potential cancer-causing chemicals *in vitro* or to laboratory animals, and then to extrapolate the consequences to humans, or to evaluate the outcome in humans previously exposed in an uncontrolled and unfortunate manner to a poorly defined amount of a potential carcinogen. The former approach is one in which the carcinogen dose can be clearly identified but findings must be extrapolated to humans. In the latter, actual carcinogen dose is usually difficult to estimate accurately. In the present paper, we will discuss the potential value of studying patients who for beneficial therapeutic reasons are given known doses of chemical carcinogens thereby permitting, with consent, the ethical study of the development of cancer in humans.

Cancer as a consequence of common forms of chemotherapy has been well documented in recent years. Radiation therapy has long been known to lead to an increased risk of tumors, with local radiation leading to tumors at the local site, and whole body radiation leading to systemic tumors (Kohn and Fry, 1984). Many chemotherapeutic agents were originally designed and tested on the basis of radiomimetic characteristics. Alkylating agents capable of producing DNA damage, either directly or through metabolism to electrophilic agents, have been the mainstay of cancer chemotherapy since the 1950s.

The studies that document second malignancies in cancer survivors have largely been retrospective. Acute leukemias and myelodysplastic syndromes, and less commonly chronic leukemias, lymphomas, sarcomas, and a wide range of solid tumors

Biological Reactive Intermediates IV, Edited by C.M. Witmer *et al.*
Plenum Press, New York, 1990

have all been reported. These have been exhaustively reviewed in a recent paper by Loescher, et al (1989). There are well-documented associations between treatment and second malignancies for primary sites including prostate, testis, ovary, uterine cervix, breast, small cell lung cancer, and most notably Hodgkin's disease. The alkylating agents are most commonly implicated and include melphalan, cyclophosphamide, procarbazine, chlorambucil, thiotepa, and treosulfan. Risks increase with increasing dose, but not always in a linear fashion (Kaldor et al., 1990a; Kaldor et al., 1990b). Acute leukemias and myelodysplastic syndromes have been reported following treatment for Hodgkin's disease, as well as ovarian, lung, and breast cancers. Recent data from Hodgkin's disease and ovarian cancer survivors confirm that the risk after high dose radiation alone (2 to 11 fold) is much lower than after alkylating agent chemotherapy (57-136 fold) (Kaldor et al., 1990a; Kaldor et al., 1990b). Peak incidences of leukemia occur at approximately 6 years, and although this seems to decline by 10 years, most cohorts have not been observed long enough to define a point at which the risk becomes insignificant.

The risks of secondary cancers are unfortunately relatively high, often ranging from 2-10%. In a subset of children with acute lymphoblastic leukemia, a 19% incidence of AML as a second tumor was noted, perhaps reflecting treatment with epipodophyllotoxins (Pui et al., 1989). This high level of an adverse outcome makes prospective study of cohorts receiving cancer chemotherapy likely to provide valuable information about chemical carcinogenesis. Such an approach may also provide insight as to how best to minimize this risk.

MECHANISM OF ACTION

There is a reasonably large body of information concerning the mechanism of action of those cancer chemotherapeutic agents that have been identified as increasing the risk of secondary tumors (Calabresi and Parks, 1985). Certain of these would be classified as primary or direct-acting carcinogens due to their ability to engage in an electrophilic attack on DNA (Colvin, 1982). Alkylating agents are a diverse group of compounds that are capable of adding alkyl groups to DNA. The mustard agents might be considered prototypic and are certainly among the most well studied in terms of structure activity relationships. Nitrogen mustards have in common the presence of two bis-(2-chloroethyl) groups attached to the nitrogen, with the third substituent varying depending upon the compound. For mechlorethamine, a component of the frequently used MOPP regime for Hodgkin's disease, the substituent is a methyl group. Mechlorethamine is a rapidly direct acting compound requiring intravenous administration. Other commonly used nitrogen mustards are melphalan, chlorambucil, and cyclo-phosphamide. The latter is of note because it appears to require cytochrome P-450 dependent metabolic activation in the liver with transport of the active metabolite to target sites. In this respect it may be similar to benzene, and, as with any agent from which hepatic metabolism produces a reactive intermediate toxic to a distant organ, poses questions as to why the liver seems unaffected. There are differences in the stability and reactivity of alkylating chemotherapeutic agents as well as often poorly understood differences among the types of tumors that are most responsive to each individual mustard agent. The demonstration of differences in the responsiveness of individual types of tumors has almost completely resulted from clinical trial and error rather than through a predictive mechanistic approach.

The action of nitrogen mustards on the genome includes alkylation of DNA, intra and interchain crosslinking, depurination, formation of abnormal base pairs and DNA chain scission. Which of these effects predominates or is responsible for second tumors is unknown. Alkylating agents may also have antitumor properties through reaction with integral proteins such as DNA polymerases, repair enzymes, etc.

Other classes of chemotherapeutic agents also seem to produce their toxicity through alkylation alone or in combination with other effects. Nitrosoureas appear to produce toxicity through both methylation and carbamoylation. Carbazine derivatives also appear to be alkylating agents as well as antimetabolites, and busulfan, an alkanesulfonic acid used in the treatment of chronic granulocytic leukemia is an alkylating agent that appears particularly capable of crosslinking guanine residues. The action of cis-platinum derivatives, implicated as carcinogenic in patients treated for ovarian cancer, seems to include direct reaction with DNA through chelation of the O6 and N7 positions of guanine. Free radicals and lipid peroxidation have also been suggested to be of importance in the genesis of secondary tumors, particularly from chemotherapeutic agents such as adriamycin and bleomycin in which active states of oxygen are involved (Sangeetha et al., 1990).

Studies of chemical carcinogenesis have in the past often focussed on a single mutation believed necessary for cancer. In evaluating human cancer, such as in colorectal carcinogenesis (Vogelstein et al., 1989), it has become evident that there are multiple mutation and non-mutational steps, and more than one oncogene that is activated or repressed. Studies of humans treated with chemical carcinogenesis should permit us to unravel this process in a way that may be particularly valuable to assessing the risks of chemical carcinogens (see below) and perhaps might allow identification of that portion of human cancer which is due to exogenous chemicals.

TYPES OF CANCER

An intriguing area for toxicologists is the relationship between chemical structure and biological effect. Evaluation of secondary tumors permits determination of the susceptibility of different organs to a systemic carcinogen, and allows exploration of the basis for the observed differences in susceptibility.

Two patterns appear reasonably clear in patients treated with systemic carcinogens. The first, described in more detail below, is that certain cancers of the hematopoietic system occur more rapidly than do cancers of other organ systems (Coleman et al., 1982; Tucker et al., 1988). The second pattern is that the cluster of tumor types that occurs with agents primarily causing immune suppression is different from the cluster occurring with classic alkylating agents (Penn, 1981; Hoover, 1977).

The reason for the prominence of the hematopoietic system in toxicity and tumor formation following exposure to radiation and systemic carcinogens is often ascribed to the rapid rate of turnover of this tissue. However, other rapidly proliferating tissue, such as the villous crypts of the small intestine, are highly resistant to cancer formation. Other explanations include the relatively loose architecture of the hematopoietic system. Not only do mature cells circulate, but precursor cells can be cloned from peripheral blood suggesting that the pluripotential bone marrow precursor cell can wander as a single cell, perhaps thereby having less of a barrier to serum concentrations of an alkylating agent than do precursor cells anchored in a more clearly defined organ architecture. A more prosaic explanation is that cancer occurs predominately in the hematopoietic system because it is the system that is the limiting factor in the dosage selected for chemotherapy. It may be that the choice of a dose that destroys perhaps 80-90% of hematopoietic progenitor cells is about the right dose level to optimize carcinogenesis in surviving cells.

Therapy aimed primarily at immunosuppression results in a completely different pattern of tumors than does chemotherapy for cancer. The production of tumors following immunosuppression, primarily non-Hodgkin's lymphoma and some cases of Kaposi's sarcoma, was first noted in association with attempts to prevent the rejection of transplanted kidneys. This is the same pattern that one sees in patients who have immunosuppression as a result of infection with the AIDS virus. Non-Hodgkins's lymphomas secondary to immunotherapy have the shortest latency period of any tumors known to occur as a result of chemicals, some cases being reported in under one year (Penn, 1981; Hoover, 1977). In contrast, it is rare that a case of acute myelogenous leukemia occurs less than two years following the onset of chemotherapy (Kaldor et al., 1990a; Kaldor et al., 1990b), or following exposure to known leukemogens such as benzene (Goldstein, 1989a) or ionizing radiation (Linet, 1985; Brill et al., 1982).

LATENCY PERIOD

Data on human cancers can be used to ask the questions of the extent to which the latency period between onset of chemotherapy and the development of cancer is determined by the type of the second tumor and whether time to tumor depends upon dose.

That intrinsic tumor biology is the key determinant in latency period is apparent from comparing the onset of acute myelogenous leukemia with that of adenocarcinoma following radiation or therapy with alkylating agents. Acute myelogenous leukemia is noted perhaps as early as two years following onset of chemotherapy or radiation, with a peak incidence following a latency period of perhaps five to ten years, and, as discussed below, an apparent decrease in risk after that time (Blayney et al., 1987; Pedersen-Bjergaard et al., 1987). In contrast, solid tumors such as adenocarcinoma seem to appear decades after the original insult (Penn, 1981; Kleinerman et al., 1982; Tucker et al., 1988). As is well known, the average latency period for mesotheliomas is in the range of three decades following initial exposure to asbestos (Selikoff et al., 1979).

Among the questions that remain to be answered are whether, for a given tumor and dose, there is a different latency period at different ages. A recent reanalysis of the Doll and Peto British doctor lung cancer data suggests that age influences lung cancer risk for smokers independent of the duration of smoking (Moolgavkar et al., 1989). Increasing age was also a strong risk factor for treatment-related AML after alkylating agent therapy for Hodgkin's disease (Pedersen-Bjergaard et al., 1987). Whether age also affects latency period is unclear.

DECREASE IN RISK OVER TIME

As noted above one of the more intriguing observations in recent years has been that the risk of acute myelogenous leukemia following chemotherapy for a second malignancy decreases after about 10 years (Blayney et al., 1987; Kaldor et al., 1990a; Kaldor et al., 1990b; Pedersen-Bjergaard et al., 1987). It is perhaps not surprising that a similar rise and then fall in myelogenous leukemia is seen in atom bomb survivors (Brill et al., 1962). In addition, one might infer that such an effect is occurring in Turkey among the benzene-exposed shoe workers whom Aksoy and his colleagues have followed for many years (Aksoy, 1989; Goldstein, 1989b).

The mechanism by which there is a decrease in AML risk over these many years is unclear. It may represent repair, or perhaps recruitment of the originally mutated stem cells into a differentiated state leading to cell death. Documentation of this effect, including understanding of its mechanism, will have important consequences for the estimation of lifetime risk from exposure to a leukemogen. This could be particularly important for a compound such as benzene.

BIOLOGICAL MARKERS

There has been increasing attention to the development of biological markers of exposure, of effect, and of susceptibility to xenobiotics, in part reflecting the availability of new tools capable of discerning relatively miniscule chemical and biological effects in human tissues (National Academy of Sciences, 1989). Of particular interest have been techniques able to detect small amounts of adducts bound to DNA and other macromolecules. Techniques such as P32 post-labeling and monoclonal antibodies promise to find one altered nucleotide in 10^{10} or more DNA bases. Initially, these approaches will be of great value in developing markers of exposure in that, once the turnover time of these adducts is known, they should provide an integration of the internal dose reaching the genome. This will be particularly valuable in studying the relationship between external and internal dose where the external dose is known, such as in humans receiving chemotherapeutic alkylating agents.

As compared to markers of exposure, the ability of such adducts to serve as biological markers of effect will present greater problems. One can predict that it will be difficult to relate observed adducts to the eventual biological effects as it is unlikely that all adducts are equal in terms of their potential for carcinogenesis. For example, Ludlum has suggested that crosslinking of DNA strands may be responsible for cytotoxicity of certain alkylating agents while adducts leading to point mutations might be responsible for cancer initiation (Ludlum and Tong, 1985; Ludlum et al., 1988). It would be of value to follow the formation of various adducts to the eventual development of a secondary malignancy, including relating the observations to genomic alteration such as the activation of oncogenes. This could lead to establishing biomarkers of effect predictive of cancer due to environmental and occupational chemicals and could help in designing therapy to avoid secondary malignancies.

One of the major unanswered questions about human cancer is why only certain individuals develop cancer while others do not, despite similar exposure. Study of the recipients of chemotherapy with alkylating agents, comparing those who do or do not develop cancer or molecular markers predictive of cancer, should also allow explanation of the molecular basis of susceptibility, including development of susceptibility markers. For example, a current area of interest is the role of glutathione-S-transferase in protecting against electrophiles. Different subtypes of this enzyme are known, and there appears to be substantial variability in the activity of such subtypes in human tissue. Exploration of the possibility that the activity of GST subtypes is a valid susceptibility marker may be of particular value in patients receiving chemotherapy in that it might help modify therapy for "susceptible" individuals so as to decrease the likelihood of second malignancies. "Non-susceptibles" could receive higher doses with less concern about secondary tumors.

PERTINENCE TO RISK ASSESSMENT

Information about the relationship between dose and response is at the core of accurate risk assessment for human cancer. Appropriate mathematical models of this relationship have been a subject of debate. What is clear is that the best mathematical model is the one that is most firmly grounded in the biology of chemical carcinogenesis. Study of chemical carcinogenesis in humans given known amounts of alkylating agents presents an excellent opportunity to test many of the fundamental assumptions of these models. For example, Moolgavkar and his colleagues have examined the fit to human epidemiology data of a recessive oncogenesis model which suggests two initiations are necessary for cancer (Moolgavkar et al., 1989). Assessment of the development of specific mutations in oncogenes from the tissue of humans receiving known doses of chemical carcinogens provides a firmer basis for such modelling activities. Models of tumor promotion, so-called non-genotoxic effects, will be more sound when the role of such effects as gene replication on the progression and clinical manifestation of human tumors can be unravelled.

Development of biomarkers of exposure and effect that can be linked together is also pertinent to risk assessment. Exploring the relation of a biomarker that is predictive of cancer with the external dose of a carcinogen may permit direct measurement of the lower end of the dose response curve. This, in essence, can lead to the replacement of mathematical extrapolation by direct biological observations at the lower levels of exposure of concern in the general environment or controlled workplace.

In summary, a set of essential questions could be addressed through study of human chemical carcinogenesis resulting from cancer chemotherapy (Table 1). Pursuit of information related to types of cancer, latency period, intervals of decreased risk, biological markers, and application to risk assessment is likely to contribute substantially to our understanding of exposure, mechanisms of action, and effective treatment of dosage.

Table I. Questions That Can Be Approached Through the Study of Iatrogenic Chemical Carcinogenesis

---

1. Are the types of cancers observed specific for the chemotherapeutic agents(s) received?

2. To what extent is latency period from time of onset of chemotherapy to diagnosis of cancer determined by the dose of the chemical or the biology of the cancer?

3. Is there ever a time period following cessation of chemotherapy after which the risk of cancer substantially decreases?

4. Can evaluation of DNA adducts or other changes in the blood of individuals receiving cancer chemotherapy be a useful approach for the development of biological markers of carcinogenesis?

5. What is the pertinence of iatrogenic chemical carcinogenesis to risk assessment for environmental chemical carcinogenesis?

REFERENCES

Aksoy, M. (1989). Hematotoxicity and carcinogenicity of benzene. *Env. Health Perspectives* **82**, 193-197.

Blayney, D.W., Longo, D.L., Young R.C., Greene, M.H., Hubbard, S.M., Postal, M.G., Duffey, P.L. and Devita, JR., V.T. (1987). Decreasing risk of leukemia with prolonged follow-up after chemotherapy and radiotherapy for Hodgkin's disease. *NEJM* **316**, 710-714.

Brill, B., Tomonaga, M. and Heyssel, R.M. (1962). Leukemia in man following exposure to ionizing radiation. *Ann. Intern. Med.* **56**, 590-609.

Calabresi, P. and Parks Jr., R.E. (1985). Antiproliferative agents and drugs used for immunosuppression. In *Goodman and Gilman's The Pharmacological Basis of Therapeutics* (L.S. Goodman, T.W. Rall and F. Murad, Eds.). pp. 1247-1306. MacMillan Publishing Company, New York.

Coleman, C.N., Kaplan, H.S., Cox, R., Varghese, A., Butterfield, P. and Rosenberg, S.A. (1982). Leukemias, non-Hodgkin's lymphomas and solid tumors in patients treated for Hodgkin's disease. *Cancer Surv.* 734-744.

Colvin, M. (1982). The alkylating agents. In *Pharmacologic Principles of Cancer Treatment* (B.A. Chabner, Ed.). pp. 276-308. W.B. Saunders Co., Philadelphia.

Goldstein, B.D. (1989a). Occam's razor is dull. *Environ. Health Perspectives* **82**, 199-206.

Goldstein, B.D. (1989b). Clinical hematotoxicity of benzene. In *Benzene: Occupational and Environmental Hazards - Scientific Update* (M.A. Mehlman, Ed.). pp. 55-65. Princeton Scientific Publishing Co., Inc., Princeton, NJ.

Hoover, R. (1977). Effects of drugs: Immunosuppression. In Origins of Human Cancer (H.H. Hiatt, J.D. Wetson and J.A. Winsten, Eds.). pp. 369-379. Cold Spring Harbor Conferences on Cell Proliferation 4, Cold Spring Harbor, NY.

Kaldor, J.M., Day, N.E., Pettersson, F., Clarke, E.A., et al. (1990a). Leukemia following chemotherapy for ovarian cancer. *NEJM* **322**, 1-6.

Kaldor, J.M., Day, N.E., Clarke, E.A., Van Leeuwen, F.E., et al. (1990b). Leukemia following Hodgkin's disease. *NEJM* **322**, 7-13.

Kleinerman, R.A., Curtis, R.E., Boice Jr., J.D., Flannery, J.T. and Fraumeni Jr., J.F. (1982). Second cancers following radiotherapy for cervical cancer. *J. Natl. Cancer Inst.* 69, 1027-1033.

Kohn, H.I. and Fry, R.J. (1984). Radiation carcinogenesis. *NEJM* **310**, 504-511.

Linet, M.S. (1985). The leukemias: Epidemiologic Aspects. In *Monographs in Epidemiology and Biostatistics*, Volume 6, Oxford University Press, New York.

Loescher, L.J., Welch-McCaffrey, D., Leigh, S.A., Hoffman, B. and Meyskens Jr., F.L. (1989). Surviving adult cancers. Part 1: Physiologic effects. *Annals of Internal Medicine* **111**, 411-432.

Ludlum, D.B. and Tong, W.P. (1985). DNA modification by the nitrosoureas: Chemical nature and cellular repair. In *Cancer Chemotherapy*, Vol. II (E.M. Muggia and M. Nishoff, Eds.). pp. 141-154. Martinus Nijhoff, Boston.

Ludlum, D.B., Colinas, R.J., Kirk, M.C. and Mehta, J.R. (1988). Reaction of reduced metronidazole with guanosine to form an unstable adduct. *Carcinogenesis* **9**, 593-596.

Moolgavkar, S.H., Dewanji, A. and Luebeck, G. (1989). Cigarette smoking and lung cancer: Reanalysis of the British Doctor's Data. *J. Natl. Cancer Inst.* **81**, 415-420.

National Academy of Sciences (1989). Biologic Markers in Reproductive Toxicology. National Academy Press, Washington D.C.

Pedersen-Bjergaard J., Specht, L., Larsen, S.O. and Ersboll J. (1987). Risk of therapy-related leukaemia and preleukaemia after Hodgkin's diseases. Relation to age, cumulative dose of alkylating agents, and time from chemotherapy. *Lancet* **2**, 83-88.

Penn, I. (1981). Second malignancies following radiotherapy or chemotherapy for cancer. In *The Immunopharmacologic Effects of Radiation Therapy* (B. Serrou, C. Rosenfeld and J.B. Dubois, Eds.). pp. 415-437. Raven Press, New York.

Pui, C-H., Behm, F.G., Raimondi, S.C., Dodge, R.K., George, S.L., Rivera, G.K., Mirro Jr., J., Kalwinsky, D.K., Dahl, G.V., Murphy, S.B., Crist, W.M. and Williams, D.L. (1989). Secondary acute myeloid leukemia in children treated for acute lymphoid leukemia. *NEJM* **321**, 136-142.

Sangeetha, P., Das, U.N., Koratkar, R. and Suryaprabha, P. (1990). Increase in free radical generation and lipid peroxidation following chemotherapy in patients with cancer. *Free Radical Biol. & Med.* **8**, 15-19.

Selikoff, I.J., Hammond, E.L. and Seidman, H. (1979). Mortality experience of insulation workers in the USA and Canada 1943-1976. *Ann. NY Acad. Sci.* **330**, 91-116.

Tucker, M.A., Coleman, C.N., Cox, R.S., Varghese, A. and Rosenberg, S.A. (1988). Risk of second cancers after treatment for Hodgkin's disease. *NEJM* **318**, 76-81.

Vogelstein B., et al. (1989). Genetic alterations accumulate during colorectal tumorigenesis. In *Recessive Oncogenes and Tumor Suppression* (W. Cavenee, N. Hastle, E. Stanbridge, Eds.). Cold Spring Harbor Laboratory, Cold Spring Harbor, NY.

GENETIC POLYMORPHISM OF DRUG METABOLISM IN HUMANS

A.S. Gross, H.K. Kroemer and M. Eichelbaum[1]

Dr. Margarete Fischer-Bosch-Institut für Klinische Pharmakologie
Auerbachstrasse 112
D-7000 Stuttgart 50
Federal Republic of Germany

INTRODUCTION

Drug metabolizing enzymes are of paramount importance in drug detoxification as well as chemical mutagenesis, carcinogenesis and toxicity mediated via metabolic activation. Thus genetically determined differences in the activity of these enzymes can influence individual susceptibility to adverse drug reactions, drug induced diseases and certain types of chemically induced cancers. The genetic polymorphisms of three human drug metabolising enzymes, namely N-acetyltransferase and two cytochrome P-450 isozymes (P-450IID6: debrisoquine / sparteine polymorphism, P-450IIC10: mephenytoin polymorphism) have been firmly established. Based on the metabolic handling of certain probe drugs the population can be divided into two phenotypes: the rapid acetylator / extensive metabolizer and slow acetylator / poor metabolizer. These polymorphisms have provided useful tools for the study of the relationship between genetically determined differences in the activity of drug metabolizing enzymes and the risk of adverse drug reactions and certain types of chemically induced diseases and cancers.

A genetic polymorphism is a monogenic trait which exists as at least two distinct phenotypes in the general population, each occurring with a frequency of not less than 1 or 2%. Therefore, rare metabolic deficiencies in drug metabolism are not covered by this definition and are not dealt with in this article. Genetic polymorphisms in drug metabolism have been observed for both Phase I and Phase II reactions and will be outlined in detail. In addition the possible scenarios for the influence of genetic polymophisms on drug toxicity will be discussed.

POLYMORPHISM OF N-ACETYLATION

Historical background

Isoniazid has been used for the treatment of tuberculosis since 1952 (Robitzek et al., 1952). As early as 1953 Bönicke and Reif reported that a population of 86 patients could be divided into two subgroups. In one group most of the dose administered was recovered in urine as isoniazid conjugate whereas the other group excreted only a small fraction of the dose as the conjugated parent compound. These findings were

1. Address for correspondence: Prof. M. Eichelbaum, Dr. Margarete Fischer-Bosch-Institut fur Klinische Pharmakologie, Auerbachstrasse 112, D-7000 Stuttgart 50, FRG.

Biological Reactive Intermediates IV, Edited by C.M. Witmer *et al.*
Plenum Press, New York, 1990

expanded upon by Biehl (1957) and Mitchell and Bell (1957) who described a clearcut bimodality in parameters describing isoniazid disposition. The major metabolites excreted were identified as acetylisoniazid and nicotinic acid / isonicotinoylglycine (Hughes et al., 1955). The patients which excreted most of the drug as conjugated parent compound were referred to as slow acetylators as compared to the remaining patients who were described as rapid acetylators (Price Evans et al., 1960). In addition to isoniazid (Price Evans & White, 1964) a variety of probe drugs including sulphadimidine (Rao et al., 1970), dapsone (Drayer & Reidenberg, 1977) and caffeine (Grant et al., 1984) have also been used to establish the acetylator phenotype.

Frequency of Impaired Acetylation and Interethnic Differences

In Caucasians the incidence of the rapid metabolizer phenotype for acetylation is approximately 50% (McQueen, 1980). In Oriental populations, a higher incidence of rapid acetylators has been reported (88-90 % and 78-85 % in Japanese and Chinese, respectively). Family studies have revealed that variability in acetylation is genetically determined by at least two alleles at a single gene locus (Iselius & Price Evans, 1983). Moreover, acetylation capacity is influenced by factors such as age, sex and weight.

Molecular Aspects

Until recently, the cause of the deficiency in acetylation observed in slow acetylators was not known. Most information has been obtained using animal models. Liver samples obtained from New Zealand rabbits which could be phenotyped as slow or rapid acetylators contained equal quantities of N-acetyltransferases (Andres & Weber, 1986). In contrast, human liver samples from slow and rapid acetylators exhibited differences in kinetic experiments which suggested that the variation in acetylation capacity could be attributed to different quantities of N-acetyltransferases (Jenne, 1965). More recently, Grant et al. (1989a) were able to isolate two kinetically different N-acetyltransferases designated NAT 1 and NAT 2 from human livers. Further studies compared the *in vivo* acetylation capacity in patients who donated liver wedge biopsy specimens with the *in vitro* activity of N-acetyl transferase observed in these specimens (Grant et al., 1989b). A significant correlation between *in vivo* and *in vitro* activity was noted. Moreover, Western blots from liver samples of slow acetylators showed a decrease or absence of immunoreactive protein. These data suggest that defective drug acetylation in the slow acetylator phenotype is due to a decrease in the amount of the enzyme present. Isolation of a human gene encoding arylamine N-acetyltransferases has recently been reported (Grant et al., 1989c) and this tool will be used to elucidate whether the interphenotype variability originates on a pre- or post-translational level.

Clinical Consequences

The metabolism of a variety of frequently used drugs is dependent on acetylator phenotype (Table 1). The relationship between acetylator phenotype and development of disease has been investigated and reviewed recently by Price Evans (1989). A high incidence of slow acetylators has been reported in patients with chemically induced bladder cancer (Price Evans et al. (1983). More recently, Horai et al. (1989) reported no association between acetylator phenotype and the development of spontaneous bladder cancer in a population of 51 Japanese patients who were not exposed to occupational risk factors. Drug induced lupus has been reported to be more frequent in slow acetylators. This subgroup of patients developed antinuclear antibodies at lower doses of procainamide than fast acetylators (Woosley et al., 1978). In addition, a higher incidence of slow acetylators was noted in a group of patients with Gilberts syndrome (Platzer et al., 1978) which is an inherited conjugation defect leading to hyperbilirubinaemia. The data presently available on the possible association between diabetes mellitus and acetylator phenotype are also reviewed by Price Evans (1989). In European patients the incidence of diabetes is higher among rapid acetylators. The opposite has been reported in a Saudi Arabian population (Price-Evans et al., 1985). Therefore, some diseases appear to be associated with acetylator phenotype. In the case of chemically induced bladder cancer such an association may be anticipated since xenobiotic aromatic amines which have been held responsible for the development of the disease are polymorphically acetylated. The association of acetylator phenotype with Gilberts syndrome, diabetes and other disorders is based on the the higher incidence of slow acetylators in the populations studied and the underlying causes are not yet understood.

Table 1. Compounds which are polymorphically acetylated (from Price-Evans, 1989)

Aminoglutethimide	Dapsone	Prizidilol
7-Amino nitrazepam	Dipyrone	Procainamide
Amrinone	Endralazine	Sulfamerazine
Caffeine	Hydralazine	Sulfamethazine
Clonazepam metabolites	Isoniazid	Sulfapyridine

POLYMORPHIC OXIDATION OF SPARTEINE / DEBRISOQUINE

Historical Background

The polymorphic oxidative metabolisms of sparteine and of debrisoquine were discovered independently (Eichelbaum, 1975; Eichelbaum et al., 1979; Mahgoub et al., 1977). During routine pharmacokinetic studies of the two drugs exaggerated pharmacological responses were observed in particular individuals. Further studies indicated that the individuals affected were not able to metabolize the probe drugs in the same manner as the other subjects. Excessively high plasma concentrations due to accumulation of the parent drug occurred and in turn produced in the side effects reported. Familial studies demonstrated that the impaired ability to metabolize these drugs was inherited in a Mendelian manner as an autosomal recessive trait (Price Evans et al., 1980). Later studies established that the two defects in metabolism cosegregate in Caucasian subjects and individuals who are poor metabolizers (PMs) of sparteine are also poor metabolizers of debrisoquine (Eichelbaum et al., 1982a). The remainder of the population is referred to as extensive metabolizers (EMs).

The oxidizer phenotype is assigned by administering a small subtherapeutic dose of one of the probe drugs. The parent drug and metabolites formed by the pathway affected by the genetic polymorphism excreted in the urine over 6 or 12 hours post dose are assayed and the metabolic ratio, or amount of unchanged drug divided by the total amount of the metabolites, is calculated. The metabolic ratio is bimodally distributed. Extensive metabolizers have low metabolic ratios and poor metabolizers are individuals with a metabolic ratio greater than the antimode. The drugs whose metabolism cosegregates with that of sparteine / debrisoquine are listed in Table 2 and include widely used compounds such as ß-adrenoceptor antagonists, antiarrhythmics, opiates and antidepressants. The evidence that the metabolism of these drugs cosegregates with that of sparteine / debrisoquine has been obtained in familial studies and in investigations of panels of previously phenotyped poor and extensive metabolizers (phenotyped panel approach) where different pharmacokinetic parameters have been observed in the two groups. Evidence has also been obtained from *in vitro* investigations using human hepatic microsomes. Competitive inhibition of metabolism of a probe drug by a compound being investigated is indicative of affinity for the same enzyme and thus further studies to establish whether the disposition is polymorphic are warranted. For example autoantibodies directed against cytochrome P-450IID6 from patients with autoimmune hepatitis (LKM1) can be used *in vitro* to specifically inhibit the metabolic pathway catalyzed by this P-450 isozyme. An animal model of this genetic polymorphism using the male Sprague Dawley and the female Dark Agouti rat has been developed (Al-Dabbagh et al., 1981). The interested reader is referred to recent reviews on the genetic polymorphism in sparteine / debrisoquine metabolism (Brosen & Gram, 1989; Eichelbaum & Gross, 1990).

Table 2. Compounds whose metabolism cosegregates with that of sparteine / debrisoquine (from Eichelbaum and Gross, 1990).

Alprenolol	Encainide	Perhexiline
Amiflamine	Flecainide	Perphenazine
Amitryptiline	Guanoxan	Phenformin
Bufuralol	Indoramin	Propafenone
CGP 15210G	Imipramine	Propranolol
Clomipramine	Methoxyamphetamine	Timolol
Codeine	Metoprolol	Tomoxetine
Desipramine	Nortriptyline	
Dextromethorphan	N-propylajmaline	

Frequency of Impaired Sparteine / Debrisoquine Hydroxylation and Interethnic Differences

The incidence of poor and extensive metabolizers has been investigated using a number of probe drugs and in different ethnic groups (Eichelbaum & Gross, 1990). The incidence of poor metabolizers varies from 18% in the San Bushmen of South Africa (Sommers et al., 1988) to 1% or less in Oriental populations (Nakamura et al., 1985). The incidence of poor metabolizers in the Caucasian population lies between 5 and 10%. The frequency of the PM phenotype, however, depends on where the antimode is set when the poor metabolizer population is defined. In each population studied it must be initially established that the distribution of the metabolic ratio is bimodal and the antimode should be set according to the data observed, rather than by applying the antimode for the Caucasian population to different ethnic groups. Studies in Africa indicate that in these populations the metabolism of the probe drugs sparteine, debrisoquine and metoprolol does not cosegregate. Further studies to elucidate the discrepancy between the African and Caucasian data are required.

Molecular Aspects

The poor metabolizer phenotype is characterized by a lack of an isozyme of cytochrome P-450 termed P-450IID6 (also P-450db1; P-450IID1) (Meyer et al., 1990). The gene encoding for cytochrome P-450IID6 is located on the long arm of chromosome 22 (Eichelbaum et al., 1987; Gonzalez et al., 1988a). P-450 IID6 has been isolated from both animal and human livers and the N-terminal sequence matches that of a cDNA containing the full protein coding sequence isolated from a human liver λ gtII library using a polyclonal antibody against the equivalent rat protein (Gonzalez et al., 1988a). The cDNA isolated was functionally expressed in an SV-40-based COS cell system and the expressed enzyme has the same functional characteristics as the human microsomes or the purified protein (Gonzalez et al., 1988b). At least three different mutational events cause aberrant splicing of pre-mRNA in livers from poor metabolizers (Gonzalez et al., 1988b). These abnormal RNAs lead to the synthesis of unstable, truncated or prematurely terminated proteins. Restriction fragment length polymorphism analysis of genomic DNA harvested from leukocytes of both poor and extensive metabolizers has also been investigated (Skoda et al., 1988). Using the restriction endonuclease Xbal and the cDNA previously isolated, it is possible to identify 33% of poor metabolizers of Caucasian origin on the basis of RFLP alone (Skoda et al., 1988). Using present techniques it is not possible to identify all poor metabolizers, however, studies are continuing to establish the exact nature of the DNA mutations which lead to the poor metabolizer phenotype. It is hoped that in the future genotype assignment from a single blood sample will be in all cases an unequivocal predictor of phenotype and therefore the administration of probe drugs can be avoided. It has been speculated that P-450 IID6 evolved in order to metabolize xenobiotics, possibly plant alkaloids, in the diet (Fonne-Pfister & Meyer, 1988).

Clinical Consequences

The clinical consequences of the impaired oxidation of substrates of cytochrome

P-450IID6 depends upon the type of compound, whether the pathway affected is of major importance to the elimination of the drug and whether the activity is also dependent on the active metabolites which may be present (Eichelbaum & Gross, 1990; Brosen & Gram, 1989). Therefore, each drug must be evaluated individually and no generalizations can be made. The genetically determined interindividual variation in cytochrome P-450IID6 metabolism is associated with considerable interindividual variation in the metabolism of the drugs used as probes to establish phenotype. For example sparteine is a potent oxytocic, however during trials of its efficacy a substantial number of tetanic uterine contractions with adverse fetal outcome occurred (Newton et al., 1966). The patients affected were in all probability poor metabolizers of sparteine, who would attain very high plasma concentrations. These concentrations were associated with an exaggerated pharmacological effect and the adverse outcomes observed.

When therapeutic efficacy is dependent on an active metabolite, the formation of which is impaired in PMs, therapeutic failure may result. For example the analgesic activity of codeine is thought to be attributable to the active metabolite morphine. This metabolic step is impaired in PMs (Dayer et al., 1988; Chen et al., 1989) and therefore in PM patients no analgesia is anticipated after codeine administration (Desmeules et al., 1989). In contrast, both the parent drug and an active metabolite contribute to the antiarrhythmic efficacy of propafenone. Formation of the active metabolite 5-hydroxypropafenone is catalyzed by P-450 IID6 (Kroemer et al., 1989) and therefore impaired in PMs (Siddoway et al., 1987). At the same dose substantially higher plasma concentrations of propafenone are observed in poor relative to extensive metabolizers. The antiarrhythmic efficacy of propafenone in EMs depends both on the parent compound and 5-hydroxypropafenone. In PMs the lack of active metabolite is counterbalanced by higher concentrations of parent drug and thus similar antiarrhythmic efficacy is observed in patients of both phenotypes at the same dose. In contrast, propafenone mediated ß-adrenoceptor blockade is elicited only by the parent compound and therefore PMs are at greater risk of developing ß-adrenoceptor mediated adverse effects as higher concentrations of the parent drug are observed in this phenotype.

POLYMORPHIC OXIDATION OF MEPHENYTOIN

Historical Background

Mephenytoin (3-methyl-5 phenyl-5 ethylhydantoin) has been used as an anticonvulsive agent since the early 1940s. The major metabolic pathways are formation of 4-hydroxymephenytoin and N-desmethylmephenytoin (Nirvanol), the latter metabolite contributing to the anticonvulsant activity of mephenytoin. During pharmacokinetic studies a subject with diminished urinary excretion of 5-hydroxymephenytoin was observed (Küpfer et al., 1984). Moreover, stereoselective mephenytoin disposition, which is due to preferential 4-hydroxylation of S-mephenytoin and which leads to predominance of the R-enantiomer in plasma, was absent in this volunteer. Further studies revealed that this defect was familial.

Frequency of Impaired Mephenytoin Hydroxylation and Interethnic Differences

Mephenytoin phenotype has been assigned on the basis of the hydroxylation index, the molar ratio of the test dose of S-mephenytoin administered and the amount of 4-hydroxymephenytoin eliminated in the urine over 0 to 8 hours (Küpfer & Preisig, 1984). An alternative basis for phenotype assignment is assessment of the ratio of R- to S-mephenytoin in urine as pronounced enantioselectivity in disposition is noted in metabolizers of mephenytoin and diminished enantioselectivity is observed in poor metabolizers (Wedlund et al., 1985). The use of a metabolic ratio is hampered by the fact that urinary excretion of unchanged mephenytoin is very low. The hydroxylation index and the 8 hr urinary enantiomeric ratio have been used to phenotype 156 unrelated Caucasians, the incidence of slow metabolizers being 2.6% (Wedlund et al., 1984). All studies performed to date have identified 33 PMs among 834 Caucasians, which represents a frequency of 3.7% (Wilkinson et al., 1989). The defect in mephenytoin metabolism is inherited independently of the defect in sparteine / debrisoquine metabolism (Küpfer & Preisig, 1984). The frequency of PMs appears to be considerably higher in patients of Oriental origin, with an incidence of up to 18% in a Japanese population (Nakamura et al., 1985). In contrast, no PMs for mephenytoin were detected among 90 healthy Cuna Amerindians in Panama (Inaba et al., 1988). Inaba and coworkers (1986) established that the deficiency in mephenytoin 4-hydroxylation is an autosomal recessive trait which involves at least two alleles at a single gene locus and follows Mendelian laws.

Molecular Aspects

Studies using conventional biochemical techniques indicate that mephenytoin metabolism is mediated by P-450 isozymes referred to as P-450 MP. P-450 MP (P-450 IIC10) is a member of the IIC subfamily which is characterized by inducibility by phenobarbital (Gonzalez, 1989). The gene for P-450 IIC10 is located on the long arm of chromosome 10 (Meehan et al., 1988). Higher Km and lower Vmax values for 4-hydroxylation were reported in *in vitro* studies using biopsy liver samples from two patients phenotyped *in vivo* as PMs of mephenytoin as compared to the Vmax and Km values in EMs. These findings are in agreement with *in vivo* data (Meier et al., 1985). At least two enzymes which 4-hydroxylate mephenytoin have been isolated and are designated P-450 MP1 and P-450 MP2 (Shimada et al., 1986). No detailed insight into the molecular basis of the mephenytoin polymorphism has been obtained so far despite extensive efforts. P-450 IIC10 appears to be present in poor metabolizers of mephenytoin and the cause of the metabolic defect may be due to formation of a structural variant with altered functional properties (Gut et al., 1986).

A new approach has recently been taken using antibodies obtained from sera of patients suffering from tielinic acid induced hepatitis (Meier & Meyer, 1987). These antibodies, designated LKM2, are strong inhibitors of P-450 IIC10. Isolation of P-450 IIC10 using LKM2 imunoaffinity columns from EM and PM liver microsomes revealed that similar proteins are present in the livers of both phenotypes. Catalytic activity, however, is impaired in the protein isolated from PM livers. If indeed a minor structural change is responsible for altered P-450 IIC10 affinity leading to a deficiency in mephenytoin 4-hydroxylation, molecular biological techniques are likely to detect this alteration. Consequently, expression libraries have been screened and several clones of P-450 IIC10 have been isolated. Two clones named MP8 and MP4 differ by only two bases in the coding region (Umbenauer et al., 1987; Ged et al., 1988). Both are expressed in human liver samples. Attempts have been made to establish a restriction fragment length polymorphism using MP8 as a probe to test human genomic DNA after digestion with different restriction enzymes (Ged et al., 1988). However, RFLP patterns do not match *in vivo* phenotypes and this probe therefore appears to be an unsuitable tool for assigning phenotype.

Clinical Consequences of the Mephenytoin Hydroxylation Polymorphism

The *in vivo* metabolism of a variety of drugs appears to be corregulated with that of mephenytoin. Küpfer and Branch (1985) reported that mephobarbital 4-hydroxylation is essentially absent in PMs of mephenytoin. Similarly, the metabolism of hexobarbital was found to be impaired in a poor metabolizer of mephenytoin (Knodell et al., 1988). These substrates display close structural similarity to mephenytoin. The relationship between structural changes and affinity to P-450 IIC10 has been studied (Hall et al., 1987). The minimal requirement for binding to P-450 IIC10 appears to be an aryl or cyclohexenyl ring positioned alpha to the carbonyl carbon of an N-alkyllactam in a 5 or 6 membered ring.

The metabolism of propranolol to naphthoxylactic acid cosegregates with mephenytoin 4-hydroxylation (Ward et al., 1989). Propranolol 4-hydroxylation cosegregates with sparteine / debrisoquine metabolic polymorphism (Raghuram et al., 1984). Thus, propranolol appears to be the first substrate metabolized by both P-450 IID6 and P-450 IIC10. More recently, the plasma clearance of both diazepam and desmethyldiazepam has been reported to be reduced in poor metabolizers of mephenytoin (Sanz et al., 1989). In virtue of the fact that benzodiazepines are among the most commonly prescribed drugs these findings, if confirmed, would yield further insight into interindividual differences in response to benzodiazepines.

TOXICOLOGICAL IMPLICATIONS

In contrast to the therapeutic implications of these polymorphisms for the drugs affected, where it has been documented that phenotype is important for therapeutic outcome and drug concentration related side effects, the consequences of genetic polymorphisms for drug toxicity are less well established. Relationships between certain drug induced diseases or cancers and phenotype point to the possible involvement of these polymorphic enzymes in the generation of disease. With regard to the susceptibility of individuals of the two phenotypes of the genetic polymorphisms in drug metabolism for drug mediated toxicity, at least three scenarios can be anticipated.

1. *The toxicity of the drug is attributable to the parent compound and the metabolism of the drug proceeds exclusively via the polymorphic enzyme.* No alternative pathways of biotransformation are available. Thus the slow acetylator / poor metabolizer phenotype will be more prone to such toxicity since at the same level of exposure this phenotype will accumulate the drug to a greater extent as a result of impaired metabolism.

a. Isoniazid induced polyneuropathy due to depletion of vitamin B6 stores occurs as a result of the formation of an isoniazid-pyridoxal phosphate complex. Since slow acetylators, due to diminished metabolizing capacity, attain higher plasma concentrations of isoniazid than rapid metabolizers at the same dose, a greater depletion of Vitamin B6 can be expected to occur in slow acetylators (Devadatta et al., 1960). It is now common practice to administer vitamin B6 concurrently with isoniazid and consequently polyneuropathy is no longer observed in either phenotype.

b. Perhexiline is a highly lipophilic basic drug which is almost entirely eliminated by metabolism. Perhexiline metabolism cosegregates with that of sparteine / debrisoquine (Cooper et al., 1984). Therapy with this antianginal drug is occasionally associated with disabling peripheral neuropathy. A retrospective study has demonstrated that 50% of twenty patients who developed irreversible peripheral neuropathy while being treated with perhexiline were poor metabolizers of debrisoquine (Shah et al., 1982). In these poor metabolizer patients higher perhexiline plasma concentrations were attained at the routine doses administered, which resulted in the higher incidence of the serious adverse effect observed.

c. Chemically induced diseases may also be more frequent in individuals unable to metabolize these drugs. *In vitro* data suggest that the neurotoxin MPTP (N-methyl-4-phenyl-1,2,3,6 tetrahydropyridine) and several analogues are metabolized by cytochrome P-450 IID6 (Fonne -Pfister et al., 1987). This compound is related to the pesticide Paraquat and epidemiological data suggest that there is a relationship between pesticide use and the incidence of Parkinson's disease (Barbeau et al., 1986). In addition it has been reported that poor metabolizers of sparteine / debrisoquine are at increased risk of suffering from Parkinson's disease (Barbaeu et al., 1986). One may have some reservations concerning the data as at the time of determination of phenotype the patients were receiving drugs which may interfere with phenotype assignment. However, it may also be postulated that poor metabolizers are at a higher risk of chemically induced Parkinson's disease as they have a genetically determined impairment in the ability to metabolize environmental neurotoxins.

2. *Metabolism via the polymorphic enzyme constitutes the major route of elimination and this pathway leads to the formation of a nontoxic metabolite. Impairment in this pathway shifts metabolism to an alternative pathway via which a reactive intermediate is being formed. In such a situation the slow / poor metabolizer phenotype is at a much greater risk of drug related toxicity.*

Therapy with the tuberculostatic drug isoniazid is associated with a high incidence of hepatotoxicity. Animal experiments demonstrated that isoniazid itself is not hepatotoxic but rather that the hydrolysis products acetylhydrazine and hydrazine are toxic. Acetylhydrazine is in turn further metabolized via two different routes (Figure 1). One pathway involves acetylation to nontoxic diacetylhydrazine and the other pathway oxidative metabolism catalyzed by cytochrome P-450. It is the latter pathway which leads to the formation of a reactive intermediate which is responsible for hepatotoxicity. Therefore the susceptibility of a patient to isoniazid hepatotoxicity is dependent on the activity of several enzymes. The majority of studies indicate that slow acetylators have a much higher risk of developing hepatotoxicity than rapid acetylators. Rapid acetylators, however, form more acetylhydrazine, the precursor of the toxic metabolite, than slow acetylators and, based on this observation, it has been proposed that rapid acetylators should be more susceptible to isoniazid hepatotoxicity (Mitchell et al., 1975). However, this concept does not take into account the fact that N-acetylation of acetylhydrazine leading to formation of the nontoxic metabolite is also polymorphic. Thus the proportion of N-acetylhydrazine which is available for oxidative metabolism via which the toxic metabolite is generated, very much depends on the proportion which is metabolized via acetylation. Although rapid metabolizers form more acetylhydrazine than slow acetylators, the proportion of acetylhydrazine metabolized to diacetylhydrazine is higher in this phenotype and hence less acetylhydrazine is available for oxidative metabolism. In support of this assumption are data from Musch et al. (1982) who demonstrated that severe hepatotoxicity is observed only in slow acetylators. Detailed studies on the metabolic disposition of isoniazid and acetylhydrazine in patients on isoniazid therapy showed that plasma levels of acetylhydrazine are higher in slow acetylators with hepatitis (Table 3). By contrast, rapid metabolizers have higher plasma concentrations of diacetlyhydrazine (Table 3). The percentage of the isoniazid dose converted to acetylhydrazine in rapid acetylators (40%) is greater than that observed in slow acetylators (25%) (Table 4). However, if one evaluates the further disposition of acetlyhydrazine then striking differences in relation to phenotype in the two pathways involved in acetylhydrazine metabolism become apparent (Table 4). In slow acetylators formation of diacetylhydrazine constitutes a minor route of elimination and most acetylhydrazine is metabolized oxidatively. A significant correlation between the amount of isoniazid excreted in urine and the amount of acetylhydrazine oxidatively metabolized has been observed (Figure 2). This

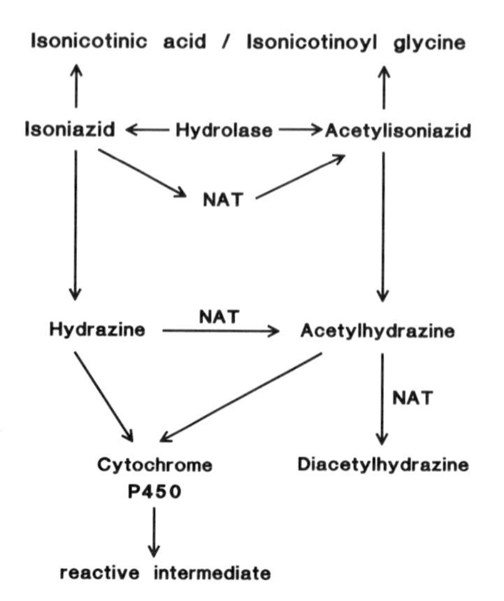

Figure 1. Metabolic pathways of isoniazid (NAT = N-acetyltransferase).

indicates that the slower the rate of acetylation, the greater the proportion of acetylhydrazine metabolism which is shifted to the metabolic pathway which forms the toxic metabolite (Eichelbaum et al., 1982b). These data are in agreement with clinical studies which clearly demonstrate that during combined tuberculosis therapy, isoniazid hepatitis is observed almost exclusively in slow acetylator patients.

Table 3. Morning trough levels of isoniazid and metabolites in five patient groups (dose 73 umol/kg) in relation to acetylator phenotype (R: rapid acetylator; S: slow acetylator).

	Patient Group				
Acetylator phenotype	R	R	S	S	S
Transaminases [U/L]	< 30	> 50	< 30	< 150	> 150
n	12	2	12	10	11
Isoniazid [umol/L]	0.7±0.3	0.9	4.1±2.3	5.8±5.0	6.6±3.6
Acetylisoniazid [umol/L]	2.8±1.8	1.2	1.2	n.d.	n.d.
Acetylhydrazine [umol/L]	6.5±4.2	8.6	8.8±2.6	9.5±4.2	11.0±3.7
Diacetylhydrazine [umol/L]	5.9±1.8	6.5	5.6±5.1	3.5±1.6	3.3±2.9

n.d. = not detectable

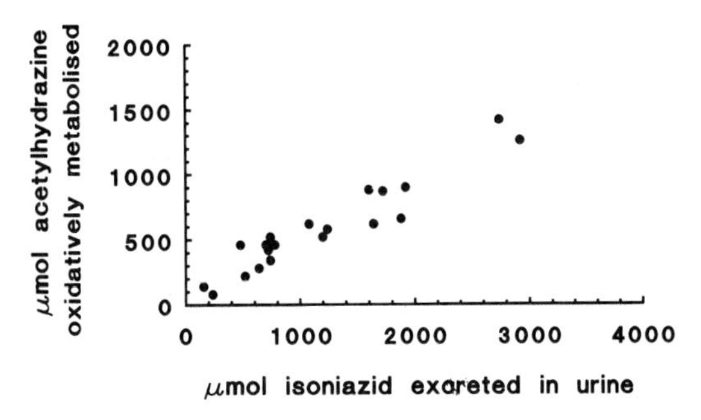

Figure 2. Relationship between the amount of isoniazid excreted in the urine and the amount of acetylhydrazine / hydrazine formed not accounted for by urinary excretion of acetylhydrazine and diacetylhydrazine and which is therefore oxidatively metabolized ($r^2 = 0.84$).

Table 4. Percent of isoniazid dose converted to acetylhydrazine / hydrazine and the disposition of acetylhydrazine in relation to acetylator phenotype (Eichelbaum et al., 1982).

	Rapid (n=26)	Slow (n=37)
Total acetylhydrazine	40.1	25.8
Renal excretion	2.1	3.3
Acetylation	25.9	4.9
Oxidation	10.8	17.4

3. *The toxicity is mediated by a reactive intermediate generated by a polymorphic enzyme.* In this case extensive metabolizers are at a much higher risk than poor metabolizers of developing toxicity or cancer. Most carcinogens require metabolic activation to reactive intermediates in order to induce neoplasia. Thus, if an enzyme which exhibits a genetic polymorphism is involved in metabolic activation of a carcinogen, high activity of such an enzyme could render the extensive metabolizer phenotype more susceptible to cancer. Although no experimental data are presently available to indicate that cytochrome P-450 IID6 participates in metabolic activation of carcinogens, epidemiological data support this contention. Cigarette smokers of the EM phenotype for debrisoquine have a four fold greater risk of lung cancer than poor metabolizers. If in addition to cigarette smoking extensive metabolizers are occupationally exposed to lung carcinogens such as polycyclic aromatic hydrocarbons and/or asbestos the relative risk for lung cancer increases to 18-fold (Caporaso et al., 1989). Similarly extensive metabolizers of debrisoquine have a higher risk of aggressive nonindustrially related bladder cancer (Kaisary et al., 1987). These data indicate that a synergism in risk between extensive metabolizer phenotype and the exposure to pre-carcinogens exists.

ACKNOWLEDGEMENTS

Supported by the Robert Bosch Foundation and the Deutschen Forschungsgemeinschaft Schwerpunktprogramm Klinische Pharmakologie Ei/157/1.

REFERENCES

Al-Dabbagh, S.A., Idle, J.R. and Smith, R.L. (1981). Animal modelling of human polymorphic drug oxidation -the metabolism of debrisoquine and phenacetin in rat inbred strains. J. *Pharm. Pharmacol.* **33:** 161-164.

Anders, H.H. and Webe,r W.W. (1986). N-Acetylation pharmacogenetics: Michaelis Menten constants for arylamine drugs as predictors of their N-acetylation rates *in vivo. Drug Metab. Dispos.* **14:** 382-385.

Barbeau, A., Cloutier, T., Roy M., Plasse, L., Paris, S. and Poirier, J. (1985). Ecogenetics of Parkinson's disease: 4-hydroxylation of debrisoquine. *Lancet* **ii:** 1213-1215.

Barbeau, A., Roy, M., Cloutier, T., Plasse, L. and Paris, S. (1986). Environmental and genetic factors in the etiology of Parkinson's disease. *Adv. Neurol.* **45:** 299-306.

Biehl, J.P. (1957). Emergence of drug resistance as related to the dosage and metabolism of isoniazid. Trans 16th Conf Chemother Tuberc Washington D.C. US Veterans Adm. Army Navy, 108-113.

Bönicke, R. and Reif, W. (1953). Enzymatische Inaktivierung von Isonicotinsäurehydrazid im menschlichen und tierischen Organismus. *Arch. Exp. Pathol. Pharmakol.* **220:** 321-333.

Brosen, K. and Gram, L.F. (1989). Clinical significance of the sparteine / debrisoquine oxidation polymorphism. *Eur. J. Clin. Pharmacol.* **36:** 537-547.

Caporaso, N., Hayes, R.B., Dosemeci, M., Hoover, R., Ayesh, R., Hetzel, M. nad Idle, J. (1989). Lung cancer risk, occupational exposure, and the debrisoquine metabolic phenotype. *Can. Res.* **49:** 3675-3679.

Chen, Z.R., Somogyi, A.A. and Bochner, F. (1988). Polymorphic O-demethylation of codeine. *Lancet* **ii:** 914-915.

Cooper, R.G., Evans, D.A.P. and Whibley, E.J. (1984). Polymorphic hydroxylation of perhexiline maleate in man. *J. Med. Genet.* **21:** 27-33.

Dayer, P., Desmeules, J., Leemann, T. and Striberni, R. (1988). Bioactivation of the narcotic drug codeine in human liver is mediated by the polymorphic monooxygenase catalyzing debrisoquine 4-hydroxylation (cytochrome P-450 dbl/bufl). *Biochem. Biophys. Res. Commun.* **152:** 411-416.

Desmeules, J., Dayer, P., Gascon, M.-P. and Magistris, M. (1989). Impact of genetic and environmental factors on codeine analgesia. *Clin. Pharmacol. Ther.* **45:** 122.

Devadatta, S., Gangadharan, P.R.J., Andrews, R.H., Fox, W., Ramakrishnan, C.V., Selhon, J.B. and Veru, S. (1960). Peripheral neuritis due to isoniazid. *Bull World Health Org.* **23:** 587-598.

Drayer, D.E. and Reidenberg, D.M. (1977). Clinical consequences of polymorphic acetylation of basic drugs. *Clin. Pharmacol. Ther.* **22:** 251-258.

Eichelbaum, M. (1975). Ein neuentdeckter Defekt im Arzneimittelstoffwechsel des Menschen: Die fehlende N-Oxidation des Spartein. Habilitationsschrift. Medizinische Fakultät Rheinischen Friedrich-Wilhelms-Universität Bonn.

Eichelbaum, M. and Gross, A.S. (1990). The genetic polymorphism of debrisoquine/sparteine - clinical aspects. *Pharmacol. Ther.* In press.

Eichelbaum, M., Spannbrucker, N., Steincke, B. and Dengler, H.J. (1979). Defective N-oxidation of sparteine in man: a new pharmacogenetic defect. *Eur. J. Clin. Pharmacol.* **16:** 183-187.

Eichelbaum, M., Bertilsson, L., Säwe, J. and Zekorn, C. (1982a). Polymorphic oxidation of sparteine and debrisoquine: Related pharmacogenetic entities. *Clin. Pharmacol. Ther.* **31:** 184-186.

Eichelbaum, M., Musch, E., Castro-Parra, M. and v. Sassen, W. (1982b). Isoniazid hepatotoxicity in relation to acetylator phenotype and isoniazid metabolism. *Br J. Clin. Pharmacol.* **14:** 575P-576P.

Eichelbaum, M., Baur, M.P., Dengler, H.J., Osikowska-Evers, B.O., Tieves, G., Zekorn, C. and Rittner, C. (1987). Chromosomal assignment of human cytochrome P-450 (debrisoquine/sparteine type) to chromosome 22. *Br J. Clin. Pharmacol.* **23:** 455-458.

Fonne-Pfister, R., Bargetzi, M.J. and Meyer, U.A. (1987). MPTP, the neurotoxin inducing Parkinson's disease, is a potent competitive inhibitor of human and rat cytochrome P450 isozymes (P450bufl, P450dbl) catalyzing debrisoquine 4-hydroxylation. *Biochem. Biophys. Res. Comm.* **148:** 1144-1150.

Fonne-Pfister, R. and Meyer, U.A. (1988). Xenobiotic and endobiotic inhibitors of cytochrome P-450dbl function, the target of the debrisoquine/sparteine type polymorphism. *Biochem. Pharmacol.* **37:** 3829- 3835.

Ged, C., Umbenauer, D.R., Bellew, T.M., Bork, R.W., Srivastava, P.K., Shinriki, N., Lloyd ,R.S. and Guengerich, F.P. (1988). Characterization of cDNAs, mRNAs and proteins related to human liver microsomal cytochrome P-450 S-mephenytoin 4-hydroxylase. *Biochemistry* **27:** 6929-6940.

Gonzalez, F.J. (1989). The molecular biology of cytochrome P-450s. *Pharmacol. Rev.* **40:** 243-288.

Gonzalez, F.J., Vilbois, F., Hardwick, J.P., McBride, O.W., Nebert, D.W., Gelboin, H.V. and Meyer, U.A. (1988a). Human debrisoquine 4-hydroxylase (P450IID1): cDNA and deduced amino acid sequence and assignment of the CYP2D locus to chromosome 22. *Genomics* **2:** 174-179.

Gonzalez, F.J., Skoda, R.C., Kimura, S., Umeno, M., Zanger, U.M., Nebert, D.W., Gelboin, H.V., Hardwick, J.P. and Meyer, U.A. (1988b). Characterization of the common genetic defect in humans deficient in debrisoquine metabolism. *Nature* (Lond) **331:** 442-446.

Grant, D.M., Tang, B.K. and Kalow, W. (1984). Polymorphic N-acetylation of a caffeine metabolite. *Clin. Pharmacol. Ther.* **33:** 355-359.

Grant, D.M., Lottspeich, F. and Meyer, U.A. (1989a). Evidence for two closely related isozymes of arylamine N-acetyltransferase in human liver. *FEBS Letters* **244:** 203-207.

Grant, D.M., Eichelbaum, M. and Meyer, U.A. (1989b). Genetic polymorphism of N-acetyltransferase: enzyme activity and content in liver biopsies correlates with acetylator phenotype determined with caffeine. *Eur. J. Clin. Pharmacol.* **36:** 199.

Grant, D.M., Blum, M., Demierre, A. and Meyer, U.A. (1989c). Nucleotide sequence for an intronless gene for a human arylamine N-acetyltransferases related to polymorphic drug acetylation. *Nucleic Acids Res.* **17:** 3978.

Gut, J., Meyer, U.T., Catin, T. and Meyer, U.A. (1986). Mephenytoin type polymorphism in drug oxidation: purification and characterization of a human liver cytochrome P-450 isozyme catalyzing microsomal mephenytoin hydroxylation. *Biochem. Biophys. Acta.* **884:** 435-447.

Hall, S.D., Guengerich, F.P., Branch, R.A. and Wilkinson, G.R. (1987). Characterization and inhibition of mephenytoin 4-hydroxylase activity in human liver microsomes. *J. Pharmacol. Exp. Ther.* **240:** 216-222.

Horai, Y., Fujita, K. and Ishizaki, T. (1989). Genetically determined N-acetylation and oxidation capacities in Japanese patients with non occupational urinary bladder cancer. *Eur. J. Clin. Pharmacol.* **37:** 581-587.

Hughes, H.B., Schmidt, L.H. and Biehl, J.P. (1955). The metabolism of isoniazid, its implications in therapeutic use. Trans. 14th Conf. Chemother. Tuberc. Washington D.C., U.S. Veterans Adm Army Navy, 217-222.

Inaba, T., Jurima, M. and Kalow, W. (1986). Family studies of mephenytoin hydroxylation deficiency. *Am. J. Hum. Genet.* **38:** 768-772.

Inaba, T., Jorge, L.F. and Arias, T.D. (1988). Mephenytoin hydroxylation in the Cuna Indians of Panama. *Brit. J. Clin. Pharmacol.* **25:** 75-79.

Iselius, L. and Price Evans, D.A.P. (1983). Formal genetics of isoniazid metabolism in man. *Clin. Pharmacokin.* **8:** 541-544.

Jenne, J.W. (1965). Partial purification and properties of the isoniazid transacetylase in human liver: Its relationship to the acetylation of p-amino-salicylic acid. *J. Clin. Invest.* **44:** 1992-2002.

Kaisary, A., Smith, P., Jaczq, E., McAllister, B., Wilkinson ,G.R., Ray, W.A. and Branch, R.A. (1987). Genetic predisposition to bladder cancer: ability to hydroxylate debrisoquine and mephenytoin as risk factors. *Can. Res.* **47:** 5488-5493.

Knodell, R.G., Dubey, R.K., Wilkinson, G.R. and Guengerich, F.P. (1988). Oxidative metabolism in human liver: relationship to polymorphic S-mephenytoin 4-hydroxylation. *J. Pharamcol. Exp. Ther.* **245:** 845-849.

Kroemer, H.K., Mikus, G., Kronbach, T., Meyer, U.A. and Eichelbaum, M. (1989). *In vitro* characterization of the human cytochrome P-450 involved in polymorphic oxidation of propafenone. *Clin. Pharmacol. Ther.* **45:** 28-33.

Küpfer, A. and Preisig, R. (1984). Pharmacogenetics of mephenytoin: A new drug hydroxylation polymorphism in man. *Eur. J. Clin. Pharmacol.* **26:** 753-759.

Küpfer, A. and Branch, R.A. (1985). Stereoselective mephobarbital hydroxylation cosegregates with mephenytoin hydroxylation. *Clin. Pharmacol. Ther.* **38:** 414-418.

Küpfer, A., Desmond, P.V., Schenker, S. and Branch, R.A. (1984). Mephenytoin hydroxylation deficiency: kinetics after repeated doses. *Clin Pharmacol Ther* **35:** 33-39.

Mahgoub, A., Idle, J.R., Dring, L.G., Lancaster, R. and Smith, R.L. (1977). Polymorphic hydroxylation of debrisoquine in man. *Lancet* **ii,** 584-586.

McQueen, E.G. (1980). Pharmacological bases of adverse drug reactions. In *Drug Treatment*, ed Avery G.S., 2nd edn, pp 202-235 Auckland: Adis Press.

Meehan, R.R., Gosden, J.R., Rout, D., Hastie, N.D., Friedberg, T., Adesnik, M., Buckland, R., van Heyningen, V., Fletcher, J., Spurr, N.K., Sweeney, J. and Wolf, C.R. (1988). Human cytochrome P450 PB- 1: a multigene family involved in mephenytoin and steroid oxidation that maps to chromosome 10. *Am. J. Hum. Genet.* **42:** 26-37.

Meier, U.T., Dayer, P., Male, P.J., Kronbach, T. and Meyer, U.A. (1985). Mephenytoin hydroxylation polymorphism: characterization of the enzymatic deficiency in liver microsomes of poor metabolizers phenotyped *in vivo*. *Clin. Pharmacol. Ther.* **38:** 488-494.

Meier, U.T. and Meyer, U.A. (1987). Genetic polymorphism of human cytochrome P-

450 (S) mephenytoin 4-hydroxylase. Studies with human autoantibodies suggest a functionally altered P-450 isozyme as cause of genetic deficiency. *Biochemistry* **26:** 8466-8474.
Meyer, U.A., Skoda, R.C. & Zanger, U.M. (1990). The genetic polymorphism of debrisoquine / sparteine metabolism - molecular mechanisms. *Pharmacol Ther,* In press.
Mitchell, R.S. & Bell, J.C. (1957). Clinical implications of isoniazidc PAS and streptomycin blood levels in pulmonary tuberculosis. *Trans Am Clin Clim Ass* **69:** 98-105.
Mitchell, J.R., Thorgeirsson, U.P., Black, M., Timbrell, J.A., Snodgrass, W.R., Potter, W.Z., Jollow, D.J. & Keiser, H.R. (1975). Increased incidence of isoniazid hepatitis in rapid acetylators: Possible relation to hydrazine metabolites. *Clin Pharmacol Ther* **18:** 70-79.
Musch, E., Eichelbaum, M., Wang, J.K.V., Sassen, W., Castro-Parra, M. & Dengler, H.J. (1982). Die Häufigkeit hepatotoxischer Nebenwirkungen der tuberkulostatischen Kombinationstherapie (INH, RMP, EMB) in Abhängigkeit vom Acetyliererphänotyp. *Klin Wochenschr* **60:** 513-519.
Nakamura, K., Goto, F., Ray W.A., McAllister, C.B., Jacqz, E. Wilkinson, G.R. & Branch, R.A. (1985). Interethnic differences in genetic polymorphism of debrisoquin and mephenytoin hydroxylation between Japanese and Caucasian populations. *Clin Pharmacol Ther* **38:** 402-408.
Newton, B.W., Benson, R.C. & McCarriston, C.C. (1966). Sparteine sulphate: A potent capricious oxytocic. *Am J Obstet Gynecol* **94:** 234-241.
Platzer, R., Küpfer, A., Bircher, J. & Preisig, R. (1978). Polymorphic acetylation and aminopyrine demethylation in Gilbert's syndrome. *Eur J Clin Invest* **8:** 219-223.
Price Evans, D.A. (1989). N-Acetyltransferase. *Pharmacol Ther* **42:** 157-234.
Price Evans, D.A. & White, T.A. (1964). Human acetylation polymorphism. *J Lab Clin Med* **63:** 394- 403.
Price Evans, D.A., Manley, K.A. & Mc Kusick, V.A. (1960). Genetic control of isoniazid metabolism in man. *Br Med J* **2:** 485-461.
Price Evans ,D.A., Mahgoub, A., Sloan, T.P., Idle, J.R. & Smith, R.L. (1980). A family and population study of the genetic polymorphism of debrisoquine oxidation in a white British population. *J Med Genet* **17:** 102-105.
Price Evans, D.A., Eze, L.C. & Whibley, E.J. (1983). The association of the slow acetylator phenotype with bladder cancer. *J Med Gen* **20:** 321-329.
Price Evans, D.A., Paterson, S., Francisco, P. & Alvarez, G. (1985). The acetylator phenotypes of Saudi Arabian diabetics. *J Med Genet* **22:** 479-483.
Raghuram, T.C., Koshakji, R.P., Wilkinson, G.R. & Wood, A.J.J. (1984). Polymorphic ability to metabolize propranolol alters 4-hydroxypropranolol levels but not beta blockade. *Clin Pharmacol Ther* **36:** 51-56.
Rao, K.V.N., Mitchison, D.A., Nair, N.G.K., Prema, K. & Tripathy, S.P. (1970). Sulfadimidine acetylation test for classification of patients as slow or rapid inactivators of isoniazide. *Br Med J* **3:** 495-497.
Robitzek, E.H., Selikoff, I.J. & Ornstein, G.G. (1952). Chemotherapy of human tuberculosis with hydrazine derivatives of isonicotinic acid. *Q Bull Sea View Hosp N.Y.* **13:** 27-51.
Sanz, E.J., Villen, T., Alm, C. & Bertilsson, L. (1989). S-mephenytoin hydroxylation phenotypes in a Swedish population determined after coadministration with debrisoquine. *Clin Pharmacol Ther* **45:** 495- 499.
Shah, R.R., Oates, N., Idle, J.R., Smith, R.L. & Lockhart, J.D. (1982). Impaired oxidation of debrisoquine in patients with perhexiline neuropathy. *Br Med J* **284:** 295-299.
Shimada, T., Misono, K.S. & Guengerich, F.P. (1986). Human liver microsomal cytochrome P-450 mephenytoin 4-hydroxylase, a prototype of genetic polymorphism in oxidative drug metabolism: purification and characterization of two similar forms involved in the reaction. *J Biol Chem* **261:** 909- 921.
Siddoway, L.A., Thompson, K.A., McAllister, C.B., Wang, T., Wilkinson, G.R., Roden, D.M. & Woosley, R.L. (1987). Polymorphism of propafenone metabolism and disposition in man: clinical and pharmacokinetic consequences. *Circulation* **75:** 785-791.
Skoda, R., Gonzalez, F.J., Demierre, A. & Meyer, U.A. (1988). Two mutant alleles of

the human cytochrome P 450 db1 gene (P 450 IID1) associated with genetically deficient metabolism of debrisoquine and other drugs. *Proc Natl Acad Sci* **85:** 5240-5243.

Sommers, De K., Moncrieff, J. & Avenant, J. (1988). Polymorphism of the 4-hydroxylation of debrisoquine in the San Bushmen of Southern Africa. *Human Toxicol* **7:** 273-276.

Umbenauer, D.R., Martin, M.V., Lloyd, R.S. & Guengerich, F.P. (1987). Cloning and sequence determination of a complementary DNA related to human liver microsomal cytochrome P-450S- mephenytoin 4-hydroxylase. *Biochemistry* **26:** 1094-1099.

Ward, S.A., Walle, T., Walle, U.K., Wilkinson, G.R. & Branch, R.A. (1989). Propranolol's metabolism is determined by both mephenytoin and debrisoquin hydroxylase activities. *Clin Pharmacol Ther* **45:** 72- 79.

Wedlund, P.J., Aslanian, W.S., McAllister, C.B., Wilkinson, G.R. & Branch, R.A. (1984). Mephenytoin hydroxylation deficiency in Caucasians: frequency of a new oxidative drug metabolism polymorphism. *Clin Pharmacol Ther* **36:** 773-780.

Wedlund, P.J., Aslanian, W.S., Jacqz, E., McAllister, C.B., Branch, R.A. & Wilkinson, G.R. (1985). Phenotypic differences in mephenytoin pharmacokinetics in normal subjects. *J Pharmacol Exp Ther* **234:** 662-669.

Wilkinson, G.R., Guengerich, F.P. & Branch, R.A. (1989). Genetic polymorphism of S-mephenytoin hydroxylation. *Pharmacol Ther* **43:** 53-76.

Woosley, R.L., Drayer, D.E., Reidenberg, M.M., Nies, A.S., Carr, K. & Oates, J.A. (1978). Effect of acetylator phenotype on the rate at which procainamide induces antinuclear antibodies and the lupus syndrome. *New Engl J Med* **298:** 1157-1159.

HUMAN HEALTH RISK ASSESSMENT AND BIOLOGICAL REACTIVE INTERMEDIATES: HEMOGLOBIN BINDING

L. Ehrenberg and Margareta Tornqvist

Department of Radiobiology
University of Stockholm
S-106 91 Stockholm, Sweden

ABSTRACT

There is ample evidence to show that the demonstration of adducts to hemoglobin and other proteins of electrophilically reactive compounds or biological reactive intermediates (BRI) is a relevant indication of the formation of the corresponding DNA adducts, and also that the rates of formation of protein- and DNA adducts are proportional. Measurement of hemoglobin and DNA adducts are therefore complementary. The former has so far been used mainly in the monitoring of low-mol.wt alkylators and BRI whereas DNA adduct monitoring has been applicable mostly to bulky compounds.

Since dose-response curves are presumably linear, demonstration of adducts should be taken as identification of genotoxic risk factors. The fast development of analytical methods renders quantification of associated risks increasingly important: For instance, using adduct analysis in the search for a priori unknown carcinogens/mutagens, analytical procedures should be developed towards a power permitting detection of unacceptable risks, at the same time as unnecessary banning of factors originating from beneficial procedures should be avoided.

Adduct level as a measure of dose

In early kinetic studies of monofunctional alkylating agents (Osterman-Golkar et al., 1970) it was observed that the mutation frequency was proportional to the rate of reaction with nucleophiles of a low strength, corresponding to oxygens of DNA (Turtoczky and Ehrenberg, 1969; Ehrenberg, 1979). The agents were then compared at equal dose, dose (D) being defined as concentration (C) integrated over time.

$$D = \int_t C(t)dt \qquad (1)$$

It was further found that the alkylators produced a frequency of forward mutation equal to that induced by 1 rad of gamma-radiation at the doses which cause alkylation of one out of 10 million of (hypothetical) DNA oxygens. Since this equivalence was approximately valid in widely different species (bacteria, plants, mammalian cells) it appeared that the same ratio would apply also for humans and could form a basis for cancer risk estimation, once doses in target cells - target doses (Ehrenberg et al., 1983) - could be measured. Generally it is not possible to measure doses by repeated determinations of the often very low concentrations of reactive

Biological Reactive Intermediates IV, Edited by C.M. Witmer *et al.*
Plenum Press, New York, 1990

chemicals or biological reactive intermediates (BRI). Doses are better monitored through levels of products of reaction with ("adducts to") nucleophilic atoms in sufficiently stable macromolecules. A reactive compound absorbed as such, or a BRI, here denoted RX, formed from a precursor, A, reacts with nucleophiles, Y, according to

$$RX + Y \xrightarrow{k2} RY + X \qquad (2)$$

The degree of alkylation or other chemical change, will be related to dose according to

$$[RY]/[Y] = k_2 \int_t [RX](t)dt = k_2 D \qquad (3)$$

where k_2 is the second-order rate constant for reaction (2). If k_2 is known - when its value is not extreme k_2 may be determined in vitro (Segerbäck, 1983, 1990) - the dose following acute exposure to A (or RX) may be solved from equation (3), using the analytically measured level [RY]/[Y].

Hemoglobin (Hb) and DNA as dose monitors

Most environmental exposures (including formation of BRI associated with living habits, in the sense of Higginson and Muir, 1979) are chronic or intermittent, leading to steady-state (SS) adduct levels, which are inversely proportional to the rates of elimination, k_-

$$[RY]_{SS}/[Y] = \frac{k_2 \, [\hat{RX}]}{k_-} \qquad (4)$$

where $[\hat{RX}]$ is the average concentration of RX. If $[\hat{RX}]$, calculated from (4), is expressed in mM, the annual dose would then be

$$D_{an} = [\hat{RX}] \cdot 8760 \quad mMh/y$$

In blood samples, the samples most conveniently obtained from humans, DNA in WBC, Hb in RBC and plasma proteins offer possibilities for dose monitoring. Although the measurement of DNA adducts would seem most relevant, considering that DNA is the target structure for genotoxic action of chemicals, some properties such as availability in small amounts per sample and varying, partly high rates of repair, the main determinant of k_- in equation (4) (Ehrenberg et al., 1986) seemed to render DNA less useful as a sensitive and precise dosimeter. In contrast, Hb is available in 1000 times larger amounts in a sample and has, through the constant life-span of erythrocytes, ca 126 d in healthy persons, a well-defined value of k_- (Osterman-Golkar et al., 1976). When dosimetric work began some 15 years ago it also appeared that available analytical techniques were more promising with regard to both identification and quantification of amino-acid adducts than of DNA-base adducts. The decision to study the suitability of Hb as a dose monitor was further based on the fact that alkylating agents and other electrophilic reagents are not specifically reactive towards one or the other type of macromolecule. In a number of studies it has also been confirmed for several compounds that rates of formation of Hb and DNA adducts are proportional (Farmer et al., 1987). Both rates are also proportional to the applied dose of reagents - if there is no low-capacity sink, as in the case of dichlorvos (Segerbäck and Ehrenberg, 1981) - or dose of precursor (with due regard to saturation effects at higher concentrations). It is concluded that the demonstration of Hb adducts (and also adducts to plasma proteins) is a relevant measure of theformation of the corresponding DNA adducts.

Adducts to Hb (or plasma proteins) and to DNA in blood cells give information on the dose in blood (D_{blood}), which is equal to the target dose (D_{targ}) only when a one-compartment model applies, as is the case with ethylene oxide in rodents (Segerbäck,

1983, 1985). In other cases such as short-lived BRI formed predominantly in the liver (Ehrenberg and Osterman-Golkar, 1980), gradients of D_{targ}/D_{blood} have to be assessed by measurement in animal models or, preferently, in human biopsy and autopsy materials.

Analytical procedures, detection levels

Mass-spectrometric (MS) methods directly permit both chemical identification and quantification of adducts; for low-mol.wt amino-acid adducts preseparation is suitably obtained by gas chromatography (GC). At an early stage of the work on alkylating agents or BRI, cysteine-S and histidine-ring-N adducts were determined after total acid or enzymatic hydrolysis of the globin. Although this method permitted dose monitoring in persons occupationally exposed to ethylene oxide (Calleman et al., 1978) and propylene oxide (Osterman-Golkar et al., 1984) it was tedious, expensive, insufficiently sensitive and unsuitable in search for a priori unknown adducts. In order to overcome these drawbacks the development of methods was directed towards specific splitting off of adducts to particular sites. A break-through was obtained by a modification of the Edman sequencing method - the "N-alkyl Edman method" - permitting specific determination of alkylated N-termini (valines in Hb) (Törnqvist et al., 1986a). All the mentioned drawbacks were overcome by this technique. The fact that no pre-fractionation is involved, the Edman derivatives (pentafluorophenylthiohydantoins) of substituted valines being extracted in one fraction, is advantageous to the analysis of unknowns. The mildness of the procedure, operating in neutral solution, counteracts artefact formation. Alkylvalines are further mostly chemically stable in vivo, in contrast to methionine and carboxyl adducts, and they are not misincorporated in regular protein synthesis, as are amino acids substituted in the residues (Kautiainen et al., 1986).

In the development of the method it was set as a goal that monitoring should be possible into the region of adduct levels where the associated risks are becoming acceptably low (Törnqvist et al., 1986a). On the basis of a review of available information related to risks generally acceptable in everyday life, ICRP (1977) concluded that a fatal risk in the range of 10^{-6} - 10^{-5} per year would be likely to be acceptable to individual members of the public. The steady-state Hb-adduct level associated with these risks, resulting from chronic or intermittent exposure to ethylene oxide and compounds with similar reaction pattern, was estimated, from the radiation-dose equivalence of chemical dose (cf. below), to be 1-10 pmol alkylvaline per g globin. The N-alkyl Edman method permits, in its present state, GC-MS measurement of adducts in this range of levels (Törnqvist, 1990). However, due to the occurrence of variable background levels of several low-mol.wt adducts (cf. below) increments due to specific exposures have to be determined as differences, with a reduced sensitivity in consequence.

The measurement of DNA adducts at steady-state levels occurring concomitantly with 1-10 pmol alkylvaline in Hb is at present hardly possible due to repair or chemical instability (of adducts to guanine-N-7) (Ehrenberg et al., 1986). The fact that bulky adducts to leukocyte DNA can be measured by the ^{32}P-postlabelling technique (Gupta and Randerath, 1988) in workers exposed to PAH (Phillips et al., 1988a) and in placenta (Everson et al., 1986) and lung cells (Phillips et al., 1988b) from smokers seems, however, to signify that this method is at present preferable in the case of high-mol.wt BRI. The fact that relatively high adduct levels are observed may also indicate the possible existence of a population of long-lived cells with a very low rate of DNA repair. For a dosimetry useful for risk estimation the kinetics of cell renewal and overall DNA repair should be clarified.

Applications of Hb dosimetry

The N-alkyl Edman method has been used for monitoring occupational exposures to ethylene oxide, EO (Duus et al. 1989), propylene oxide, PO (Högstedt et al., 1990), ethene (which is metabolically converted to EO) (Törnqvist et al., 1989a) and styrene (a predominating BRI of which is styrene oxide) (Christakopoulos et al., 1990). Increments of 2-hydroxyethyl- and 2-hydroxypropylvaline in globin from smokers originate probably from ethene and propene, converted to EO and PO, respectively (Törnqvist et al., 1986b; Törnqvist, 1989). Adducts originating from ethene and propene in urban air could so far not be measured above the background levels of the adducts (Törnqvist, 1988), but were safely determined in animals experimentally exposed to gasoline or diesel exhausts (Törnqvist et al., 1988a). In smokers it has also been possible to measure Hb adducts from aromatic amines which, via the nitrosoaromatic metabolites, give rise to sulfinamides with cysteine (Bryant et al., 1987).

Exposure to EO 40 h/week gives rise to valine adduct levels ca 2,400 pmol per g globin per ppm, and exposures to ethene to ca 15 times lower levels, indicating that some 6% of inhaled ethene is converted to EO (Törnqvist 1989; Törnqvist et al., 1990). Exposure to propene and, particularly, styrene give rise to still lower adduct levels, partly due to faster detoxification of the BRIs.

Background adduct levels

Background levels of adducts identical with those produced by low-mol.wt alkylating agents are regularly encountered in knowingly unexposed persons. These background adducts certainly tend to reduce the power of dose monitoring at low exposure levels. The occurrence of such adducts should, however, be taken as a demonstration of the potential of a search for a priori unknown adducts as a method to map humans for total loads of BRIs as possible causes of diseases.

Among background adducts measured, methyl is found at levels compatible with S-adenosylmethionine in red cells being the alkylator (Törnqvist et al., 1988b; Törnqvist, 1989). Hydroxyethyl seems to originate, via EO, from spontaneous ethene production, with an involvement of intestinal bacteria and lipid peroxidation (Törnqvist et al., 1989b). The source of background 2-hydroxypropyl is at present unknown.

A number of Schiff bases and oxoalkyl adducts are determined after reduction with NaBH4 to hydroxyalkylvalines. A high level of the 2-hydroxyethyl adduct, presumably from glycolaldehyde, a natural BRI, prevents determination of doses of the BRI from vinyl chloride, chloroethylene oxide, by this method. Adducts from formaldehyde, acetaldehyde, glyoxal and malonaldehyde (associated with peroxidation) are effectively measured as alkylvalines by this technique (Kautiainen et al., 1989).

A number of Schiff bases and oxoalkyl adducts are determined after reduction with $NaBH_4$ to alkylvalines or hydroxyalkylvalines. A high level of the 2-hydroxyethyl adduct, presumably from glycolaldehyde, a natural BRI, prevents determination of doses of the BRI from vinyl chloride, chloroethylene oxide, by this method (Svensson and Osterman-Golkar, 1987). Adducts from formaldehyde, acetaldehyde, glyoxal and malonaldehyde (associated with peroxidation) are effectively measured as alkylvalines by this technique (Kautiainen et al., 1989).

Need for a quantitative aspect

The management of genotoxic factors meets the dilemma that, in contrast to poisons in a classical sense (Paracelsus, 1538), a no-effect threshold dose cannot be defined and most probably does not exist (Ehrenberg et al., 1983). Therefore, as first developed for protection against ionizing radiation (ICRP, 1977), a risk has to be taken that is judged to be acceptable considering the benefits of the activity leading to exposure. One consequence of the absence of safe thresholds, in addition to the randomness (with probabilities determined by kinetic laws) of nucleophilic substitution reactions, is that the demonstration of an adduct - and its causative reagent or precursor - has to be taken as the identification of a risk factor (Ehrenberg and Osterman-Golkar, 1980). Analytical methods are at present undergoing a fast development towards lower detection levels, i.e. increased resolving power. Although the soundness of such a development of the analytical tools, with increased possibilities of unveiling the initiators in today's cancer, cannot be questioned, one may at the same time fear that unless the risks associated with findings are quantified and properly evaluated, a number of chemicals of potential value to human health and well being may be improperly banned (Törnqvist et al., 1990). It is at the same time essential to be aware of the meaning, in terms of risk, of the detection level of an analytical method, in order to avoid acceptance of false hypotheses of zero risk when no adducts of the kind looked for are seen (Ehrenberg, 1984).

At least provisionally a way of solving these problems is offered by the radiation-dose equivalence ("rad-equivalence") of chemical target dose (Ehrenberg, 1979, 1980), defined according to equation (1) or based on the level of DNA adducts associated with this dose. Further, some rules for what can be considered an acceptable risk have to be developed, with due regard to the distribution of the factor considered in the population, i.e. as to whether the collective risk or the idividual risk becomes limiting. At present the standards vary considerably.

ACKNOWLEDGEMENT

Financial support from the National Swedish Environment Protection Borad and Shell Internationale Research Maatschappij B.V., The Hague, is acknowledged.

REFERENCES

Bryant, M.S., Skipper, P.L., Tannenbaum, S.R. and Maclure, M. (1987). Hemoglobin adducts of 4-aminobiphenyl in smokers and non-smokers. *Cancer Res.* **47**, 602-608.

Calleman, C.-J., Ehrenberg, L., Jansson, B., Osterman-Golkar, S., Segerbäck, D., Svensson, K. and Wachtmeister, C.A. (1978). Monitoring and risk assessment by means of alkyl groups in hemoglobin in persons occupationally exposed to ethylene oxide. *J. Environ. Pathol. Toxicol.* **2**, 427-442.

Christakopoulos, A., Svensson, K., Bergmark, E. and Osterman-Golkar S. (1989). Monitoring of styrene exposure through hemoglobin adducts by the modified Edman procedure. To be published.

Duus, U., Osterman-Golkar, S., Törnqvist, M., Mowrer, J., Holm, S. and Ehrenberg, L. 1989). Studies of determinants of tissue dose and cancer risk from ethylene oxide exposure. In *Proc. Symp. Management of Risk from Genotoxic Substances in the Environment* (L. Freij, Ed.), pp. 141-153. Swedish National Chemicals Inspectorate, Solna, Sweden.

Ehrenberg, L. (1979). Risk assessment of ethylene oxide and other compounds. In *Assessing Chemical Mutagens: The Risk to Humans* (V.K. McElheny and S. Abrahamson, Eds) Banbury Rep., 1, pp. 157-190. Cold Spring Harbor Laboratory.

Ehrenberg, L. (1980). Purposes and methods of comparing effects of radiation and chemicals. In *Radiobiological Equivalents of Chemical Pollutants*, pp. 11-22, 23-36. IAEA, Vienna.

Ehrenberg, L. (1984). Aspects of statistical inference in testing for genetic toxicity. In *Handbook of Mutagenicity Test Procedures, 2nd Edition* (B.J. Kilbey et al. Eds) pp. 775-822. Elsevier, Amsterdam.

Ehrenberg, L. and Osterman-Golkar, S. (1980). Alkylation of macromolecules for detecting mutagenic agents. Teratog. *Carcinog. Mutag.* **1**, 105-127.

Ehrenberg, Moustacchi, E. and Osterman-Golkar, S. (1983). Dosimetry of genotoxic agents and dose-response relationships of their effects. Mutat. *Res.* **123**, 121-182.

Ehrenberg, L., Osterman-Golkar, S., Segerbäck, D. and Törnqvist, M. (1986). Power of methods for monitoring exposure to genotoxic chemicals by covalently bound adducts to macromolecules. In *Environmental Mutagenesis and Carcinogenesis* (N.K. Notani and P.S. Chauhan Eds), pp. 155-166. Bhabha Atomic Research Centre, Bombay.

Everson, R.B., Randerath, E., Santella, R.M. et al. (1986). Detection of smoking-related covalent DNA adducts in human placenta. *Science* **231**, 54-57.

Farmer, P.B., Neumann, H.-G. and Henschler, D. (1987). Estimation of exposure of man to substances reacting covalently with macromolecules. *Arch. Toxicol.* **60**, 251-260.

Gupta, R.C. and Randerath, K. (1988). Analysis of DNA adducts by 32P labelling and thin layer chromatography. In *DNA Repair* (E.C. Friedberg and P.C. Hanawalt, eds) Marcel Dekker, New York.

Higginson, J. and Muir, C.S. (1979). Environmental carcinogenesis: Misconceptions and limitations to cancer control. *J. Natl. Cancer Inst.* **63**, 1291-1298.

Hgstedt, B., Bergmark, E., Törnqvist, M. and Osterman-Golkar, S. (1990). Chromosomal aberrations and micronuclei in lymphocytes in relation to alkylation of hemoglobin in workers exposed to ethylene oxide and propylene oxide (manuscript).

ICRP Publication No. 26 (1977). Recommendations of the International Commission on Radiological Protection. *Ann. ICRP*, Vol. 1, No 3, Pergamon Press, Oxford.

Kautiainen, A., Törnqvist, M., Svensson, K. and Osterman-Golkar, S. (1989). Adducts of malonaldehyde and a few other aldehydes to hemoglobin. *Carcinogenesis* **10**, 2123-2130.

Kautiainen, A., Osterman-Golkar, S. and Ehrenberg, L. (1986). Misincorporation of alkylated amino acids into hemoglobin - a possible source of background alkylations. *Acta Chem. Scand.* B, **40**, 453-456.

Osterman-Golkar, S., Bailey, E., Farmer, P.B., Gorf, S.M. and Lamb, J.H. (1984). Monitoring exposure to propylene oxide through the determination of hemoglobin alkylation. *Scand. J. Work Environ. Health* **10**, 99-102.

Osterman-Golkar, S., Ehrenberg, L. and Wachtmeister, C.A. (1970). Reaction kinetics and biological action in barley of monofunctional methanesulfonic esters. *Radiat. Botany* **10**, 303-327.

Osterman-Golkar, S., Ehrenberg, L., Segerbäck, D. and Hällström, I. (1976). Evaluation of genetic risks of alkylating agents. II. Haemoglobin as a dose monitor. *Mutat. Res.* **34**, 1-10.

Paracelsus (1538) See Deichmann, W.B., Henschler, D., Holmstedt, B. and Keil, G. (1986). What is there that is not poison? A study of the *Third Defence* by Paracelsus. *Arch. Toxicol.* **58**, 207-213.

Phillips, D.H., Hemminki, K., Alhonen, A., Hewer, A. and Grover, P.L. (1988a). Monitoring occupational exposure to carcinogens: detection by ^{32}P-postlabelling of aromatic DNA adducts in white blood cells from iron foundry workers. *Mutat. Res.* **204**, 531-541.

Phillips, D.H., Hewer, A., Martin, C.N., Garner, C.G. and King, M.M. (1988b). Correlation of DNA adduct levels in human lung with cigarette smoking. *Nature* **336**, 790-792.

Segerbäck, D. (1983). Alkylation of DNA and hemoglobin in the mouse after exposure to ethene and ethene oxide. *Chem.-Biol. Interactions* **45**, 139-151.

Segerbäck, D. (1985). In vivo dosimetry of some alkylating agents as a basis for risk estimation. Ph.D. Thesis, University of Stockholm, Stockholm.

Segerbäck, D. (1990). Reaction product in hemoglobin and DNA after *in vitro* treatment with ethylene oxide and N-(2-hydroxyethyl)-N-nitrosourea. *Carcinogenesis* 11. (In press.)

Segerbäck, D. and Ehrenberg, L. (1981). Alkylating properties of dichlorvos (DDVP). *Acta Pharmacol. Toxicol.* **49**, Suppl. 5, 56-66.

Svensson, K. and Osterman-Golkar, S. (1987). *In vivo* 2-oxoethyl adducts in hemoglobin and their possible origin. In *Application of Short-Term Bioassays in the Analysis of Complex Environmental Mixtures V* (S.S. Sandhu et al., Eds), pp. 49-66. Plenum Press, New York.

Turtóczky, I. and Ehrenberg, L. (1969). Reaction rates and biological action of alkylating agents preliminary report on bactericidal and mutagenic action in E.coli. *Mutat. Res.* **8**, 229-238.

Törnqvist, M. (1988). Search for unknown adducts: increase of sensitivity through preselection by biochemical parameters. *IARC Sci. Publ.* **89**, 378-383.

Törnqvist, M. (1989). Monitoring and cancer risk assessment of carcinogens, particularly alkenes in urban air. Ph.D. Thesis, University of Stockholm, Stockholm.

Törnqvist, M. (1990). The N-alkyl Edman method for hemoglobin adduct measurement: Updating and applications to humans. (In press.)

Törnqvist, M., Segerbäck, D. and Ehrenberg, L. (1990). The "rad-equivalence approach" for ssessment and evaluation of cancer risks, exemplified by studies ofethylene oxide and ethene. (In press.)

Törnqvist, M., Mowrer, J., Jensen, S. and Ehrenberg, L., (1986a). Monitoring of environmental cancer initiators through hemoglobin adducts by a modified Edman degradation method. *Anal. Biochem.* **154**, 255-266.

Törnqvist, M., Osterman-Golkar, S., Kautiainen, A., Jensen, S., Farmer, P.B. and Ehrenberg, L. (1986b). Tissue doses of ethylene oxide in cigarette smokers determined from adduct levels in hemoglobin. *Carcinogenesis* **7**, 1519-1521.

Törnqvist, M., Kautiainen, A., Gatz, R.N. and Ehrenberg, L. (1988a). Hemoglobin adducts in animals exposed to gasoline and diesel exhausts. 1. Alkenes. *J. Appl. Toxicol.* **8**, 159-170.

Törnqvist, M., Osterman-Golkar, S., Kautiainen, A., N£slund, M., Calleman, C.J. and Ehrenberg, L. (1988b). Methylations in human hemoglobin. *Mutat. Res.* **204**, 521-529.

Törnqvist, M., Almberg, J., Nilsson, S. and Osterman-Golkar, S. (1989a). Ethene oxide doses in ethene exposed fruit store workers. *Scand. J. Work Environ. Health* **15**, 436-438.

Törnqvist, M., Gustafsson, B., Kautiainen, A., Harms-Ringdahl, M., Granath, F. and Ehrenberg, L. (1989b). Unsaturated lipids and intestinal bacteria as sources of endogenous production of ethene and ethylene oxide. *Carcinogenesis* **10**, 39-41.

QUANTITATING THE PRODUCTION OF BIOLOGICAL REACTIVE INTERMEDIATES IN TARGET TISSUES: EXAMPLE, DICHLOROMETHANE[1]

Richard H. Reitz

Chemical Industry Institute of Toxicology
6 Davis Drive
Research Triangle Park, NC 27709

INTRODUCTION

Dichloromethane (CH_2Cl_2, DCM) is a widely used industrial solvent. In addition to industrial applications such as film processing, it has been used for the preparation of consumer products such as cosmetics, decaffeinated coffee, and paint strippers. With such a wide potential for human exposure to DCM, the need for a good toxicology database is obvious, and a number of acute, subchronic, and chronic studies have been conducted with this material. In general the results from these studies have been reassuring until the National Toxicology Program completed a long term inhalation bioassay of methylene chloride in rats and mice (NTP, 1985).

In the NTP study, high incidences of malignant lung and liver tumors were observed in $B_6C_3F_1$ mice inhaling either 2000 or 4000 ppm of DCM throughout their lifetime. The incidences of lung and liver tumors in treated and control mice in the NTP study are summarized in Table 1.

The results observed in the NTP inhalation study are inconsistent with those observed in another bioassay of DCM in $B_6C_3F_1$ mice reported by Serota et al. (1984). In the Serota study, DCM was administered in the drinking water and the incidences of lung and liver tumors in treated animals were not significantly different than control groups (Table 2, lung tumors not shown).

The obvious question raised by the NTP studies is whether human populations exposed to much lower concentrations of DCM vapor (or to low doses of DCM through other routes, such as drinking water) are likely to develop tumors similar to those seen in $B_6C_3F_1$ mice. While there are many biological factors which must be considered in answering this question, one of the primary considerations involves the delivery of toxic species to the target tissues in the animals. We have recently developed a physiologically-based pharmacokinetic (PB-PK) model capable of quantitatively describing the delivery of DCM and its metabolites to target organs (Andersen et al., 1987), and have suggested that this model would be useful for the preparation of quantitative risk estimations with DCM.

---

1. Portions of this material have been previously published in Andersen et al., (1987) *Toxicol. Appl. Pharmacol.*, **87**, 185-205 and Reitz et al. (1988) *Toxicol. Lett.*, **43**, 97-116.

Biological Reactive Intermediates IV, Edited by C.M. Witmer *et al.*
Plenum Press, New York, 1990

RESULTS AND DISCUSSION

Before a realistic quantitative risk estimation can be performed, it is necessary to consider information on the mechanism(s) by which DCM influences the tumorigenic process in the $B_6C_3F_1$ mouse.

Dihalomethanes, including DCM, are metabolized by two major pathways: (1) an oxidative pathway (Kubic et al., 1974) that appears to yield CO as well as considerable amounts of CO_2, (Gargas et al., 1986) and (2) a glutathione-dependent pathway (Ahmed and Anders, 1978) which produces formaldehyde and CO_2 but no CO (Figure 1). Potentially reactive intermediates are formed in each of the metabolic pathways for DCM: formyl chloride in the oxidative pathway, and formaldehyde and chloromethyl glutathione in the conjugative pathway.

Distribution of DCM metabolism between these two pathways is dose-dependent. The oxidative (MFO) pathway is a high-affinity limited-capacity pathway which saturates at relatively low atmospheric concentrations (200-500 ppm). In contrast, the conjugative (GSH) pathway has lower affinity for DCM, but does not appear to saturate at experimentally accessible concentrations (<5000 ppm). Thus at low concentrations, the MFO pathway accounts for most of the DCM metabolized, but as exposure concentrations are increased above the MFO saturation level, disproportionate increases in the amount of DCM metabolized by the secondary GSH pathway are seen (Andersen et al., 1987).

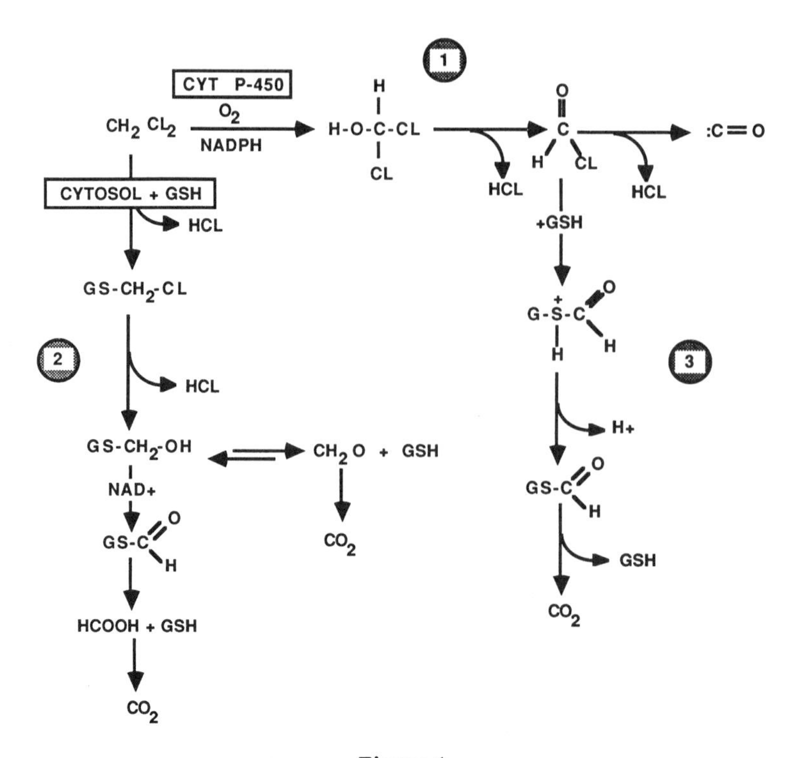

Figure 1

With this knowledge of DCM metabolism, three different hypotheses can be formulated to account for the tumorigenicity of DCM in the $B_6C_3F_1$ mouse:

(I) Tumorigenicity results from production of biologically reactive intermediates (BRI's) by the conjugative (GSH) pathway in target tissues.

(II) Tumorigenicity results from production of BRI's by the oxidative (MFO) pathway in target tissues.

(III) Tumorigenicity results from the presence of parent chemical in the target tissues.

Since the chemical reactivity of DCM is very low, it is unlikely that parent material would bond directly to biological macromolecules. Thus it is unlikely that the parent chemical could cause genetic alterations (through covalent binding to DNA) similar to those produced by many other chemical carcinogens (Miller and Miller, 1966).

This raises the question as to whether some physical property of the parent material (e.g. alteration of membrane permeability) might be responsible for the tumorigenicity of DCM. We consider this to be unlikely, because another solvent with physical properties similar to DCM (1,1,1-trichloroethane) has recently been tested in a long term inhalation bioassay in $B_6C_3F_1$ mice and it failed to induce either lung or liver tumors (Quast et al., 1984). The primary difference between DCM and 1,1,1-trichloroethane is that 1,1,1-trichloroethane is very poorly metabolized by mammalian species.

Collectively these observations suggest that the tumorigenicity of DCM probably resulted from the production of reactive metabolites of DCM rather than from a physical effect of the solvent itself. Consequently, we have concluded that DCM is unlikely to directly influence tumor frequencies in the lung and liver of $B_6C_3F_1$ mice and have rejected hypothesis (III). This leaves the task of determining whether BRI's produced by MFO (Hypothesis II) or GSH transferase (Hypothesis I) are involved in the tumorigenic action of DCM.

The availability of a mathematical model capable of quantitatively describing the production of metabolites in target tissues by either the MFO or GSH pathway (Andersen et al., 1987) provides an opportunity to test these hypotheses against the results obtained in the two long term bioassays of DCM. For this purpose, predicted amounts of DCM metabolism produced by each pathway (mg equivalents of DCM metabolized per liter of liver tissue per day) under the various bioassay conditions are listed in Table 3.

The MFO pathway saturates between 200 and 500 ppm of DCM in inhaled air. Consequently the predicted amounts of MFO metabolites formed in lung and liver tissue are nearly identical for the two dose levels in the inhalation study (2000 & 4000 ppm; Table 3). This prediction is inconsistent with the observation that liver and lung tumors are consistently higher in the 4000 ppm exposure group than the 2000 ppm exposure group in the inhalation study (Table 1).

The PB-PK model also predicts that the levels of MFO metabolites produced by administration of 250 mg/kg/day of DCM to $B_6C_3F_1$ mice in drinking water would be similar to the amounts of MFO metabolites produced in the inhalation study (Table 3). This prediction is also inconsistent with the hypothesis that MFO metabolites are responsible for the tumorigenicity of DCM, because no statistically significant increases in tumors were seen in the drinking water study (Table 2).

Predicted rates of metabolism of DCM by the GSH pathway correlate much better with the induction of lung and liver tumors in these two chronic studies. Predicted levels of GSH metabolites at 4000 ppm are higher than the predicted levels of GSH metabolites at 2000 ppm (Table 3). In contrast, the predicted levels of GSH metabolites formed in target tissues during administration of DCM in drinking water are very low. Thus the pattern of tumor induction in both studies shows a good correlation with the rates of metabolism of DCM by the GSH pathway.

Table 1. Tumor incidences (combined adenomas and carcinomas) in male and female $B_6C_3F_1$ mice as reported by the National Toxicology Program (NTP, 1985) in a chronic inhalation study (6 hr/day, 5 days/week) of DCM for chronic toxicity and/or carcinogenicity.

	PPM in Chamber Atmosphere		
	0	2000	4000
MALE MICE			
Liver Tumors	44%	49%	67%
Lung Tumors	10%	54%	80%
FEMALE MICE			
Liver Tumors	6%	33%	83%
Lung Tumors	6%	63%	85%

Table 2. Liver tumor incidences (combined adenomas and carcinomas) in $B_6C_3F_1$ mice as reported by Serota et al. (1984) in a chronic drinking water study of DCM for chronic toxicity and/or carcinogenicity (lung tumor incidences in treated groups were not significantly different from controls; data not shown).

	Mg/kg/day from Water					
	0	0	60	125	185	250
MALE MICE						
Liver Tumors	18%	20%	28%	30%	31%	28%
FEMALE MICE						
Liver Tumors	6%	6%	4%	4%	10%	6%

Table 3. Predicted amounts of DCM metabolized by the oxidative (MFO) pathway or the conjugative pathway (GSH) in lung and liver tissue under bioassay conditions. Values given are mg equivalents of DCM metabolized by the indicated pathway per liter of tissue per day. Inhalation values are multiplied by 5/7 to correct for exposure 5 days/week.

	Control	Inhal (2000)	Inhal (4000)	Water (250)
Liver				
MFO	0	3573	3701	5197
GSH	0	851	1811	15
Lung				
MFO	0	1531	1583	1227
GSH	0	123	256	1

Another indication that the GSH pathway plays a critical role in the induction of lung and liver tumors comes from long term inhalation bioassays of DCM in the Syrian Golden hamster. Burek et al. (1984) reported that exposure of hamsters to DCM concentrations up to 3500 ppm failed to affect the incidences of tumors in the lung and liver (or any other site). The relative insensitivity of the hamster correlates well with *in vitro* measurements of the levels of MFO and GSH transferase in tissue preparations from the lung and livers of various species (Reitz et al., 1988).

Homogenates of lung and liver from $B_6C_3F_1$ mice contained high levels of both MFO and GSH-transferase (Table 4). In contrast, tissue homogenates from the lung and livers of hamsters contained very low levels of GSH-transferase, although the levels of MFO activity in the livers of hamsters were slightly higher than in the livers of $B_6C_3F_1$ mice (Table 4). Again, the relative levels of GSH-transferase activity in the two species correlates well with the sensitivity of these species to the tumorigenicity of DCM, but the relative levels of MFO are inconsistent with the bioassay results. It is noteworthy that the levels of GSH-transferase in homogenates of human tissue are similar to those present in hamster tissues, but much lower than those in tissues of the mouse.

Having accepted the hypothesis that BRIs from the GSH pathway are involved in the tumorigenicity of DCM, it is possible to incorporate quantitative information from the PB-PK model into a quantitative risk assessment for DCM (Reitz et al., 1988; 1989). In the absence of a PB-PK model, Singh et al. (1985) assumed that the amount of DCM-derived material (the "delivered dose") arriving at the target sites was directly proportional to the concentration of DCM in inhaled air from very high concentrations (4000 ppm) to very low concentrations (<1 ppm) as shown by the dashed line in Figure 2.

This assumption is a bad approximation of the dose-dependency of GSH metabolites in the target tissues. The solid line in Figure 2 shows the dose-dependency of GSH metabolites predicted by the PB-PK model. Levels of GSH metabolites predicted by the PB-PK model are close to those obtained by linear extrapolation at concentrations above 1000 ppm. However, a marked nonlinearity in the curve is apparent in the region where MFO becomes saturated in mice (200-500 ppm). At low concentrations (1 ppm) where MFO is below saturation, the levels of GSH metabolite predicted by the PB-PK model are about 20 fold lower than predicted by linear extrapolation from 4000 ppm. This suggests that linear extrapolation of tumorigenicity results obtained at 4000 ppm down to low levels of DCM exposure (such as might be encountered by human populations) may significantly overestimate hazard.

Singh et al. (1985) also employed another assumption to extrapolate between species (i.e. from $B_6C_3F_1$ mice to humans). They assumed that on a mg/kg/day basis, humans were approximately 13 times more sensitive to DCM toxicity (including mice carcinogenicity) than humans (the "surface area rule"). This rule of thumb is probably approximately correct when parent chemical is directly responsible for toxicity, because the smaller species have higher levels of detoxifying enzymes and hence are better able to protect themselves by removing the chemical from the body. However, in the case of DCM, metabolism by the GSH pathway serves to activate DCM, so this assumption is inappropriate.

The heavy solid line in Figure 2 shows the dose-dependency of GSH metabolism on inhaled concentration for humans as calculated by the PB-PK model. It is note-

Table 4. Enzyme activities measured *in vitro* in tissue preparations from $B_6C_3F_1$ mice, F344 rats, Syrian Golden Hamsters, and healthy human accident victims. Details of the enzyme preparation and enzyme assays may be found in Reitz et al. (1988). MFO enzymes were assayed at a substrate concentration of 5 mM DCM, while GSH-Transferase enzymes were assayed at a substrate concentration of 40 mM DCM. Activities are given in nmoles/min/mg protein.

	MFO		GSH	
	Liver	Lung	Liver	Lung
Mouse	11	5	26	7
Hamster	14	1	1	<0.2
Rat	4	0.2	7	1
Human	5	<0.1	2	0.4

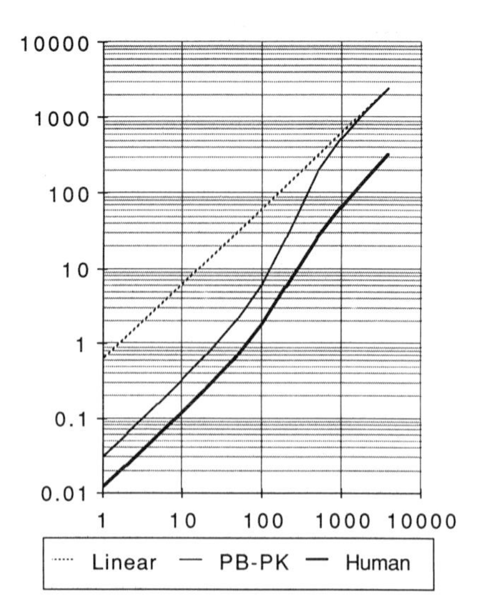

Figure 2

worthy that throughout the entire range of exposure concentrations, the PB-PK predicts that exposure of humans and $B_6C_3F_1$ mice to equivalent concentrations of DCM will result in lower values of the "delivered dose" for humans. Consequently, interspecies extrapolations of DCM tumorigenicity from $B_6C_3F_1$ mice to humans with the "surface area" rule will significantly overestimate the actual risk to humans.

The net effect of replacing the two default assumptions of Singh et al. with quantitative information based on mechanistic information and PB-PK modeling has been summarized elsewhere (Reitz et al., 1988). Using the linearized multistage model of Howe and Crump (1982), Singh et al. calculated that the lifetime risk for humans continuously exposed to 1 microgram/cubic meter of DCM was 4.1 x 10-6 (Singh et al., 1985). Incorporation of PB-PK principles into this same procedure reduced the calculated upper boundary on risk to 3.7 x 10-8 (Reitz et al., 1988).

SUMMARY

Development of quantitative mathematical models for the production of BRIs in target tissues can provide valuable insights into the mechanisms of toxicity for specific chemicals. Furthermore, use of mechanistic information and mathematical modeling in the hazard evaluation process should significantly reduce the uncertainty inherent in extrapolating the results of animal toxicity tests to man.

REFERENCES

Ahmed, A.E., and Anders, M.W. (1976). Metabolism of dihalomethanes to formaldehyde and inorganic halide. I. *In vitro* studies. *Drug. Metab. Dispos.*, **4**, 357-361.

Andersen, M.E., Clewell, H.J., Gargas, M.L., Smith, F.A., and Reitz, R.H. (1987). Physiologically-based pharmacokinetics and the risk assessment process for methylene chloride. *Toxicol. Appl. Pharmacol.* **87**, 185-205.

Burek, J.D., Nitschke, K.D., Bell, T.J., Wackerle, D.L., Childs, R.C., Beyer, J.D., Dittenber, D.A., Rampy, L.W., and McKenna, M.J. (1984). Methylene chloride: A 2 year inhalation toxicity and oncogenicity study in rats and hamsters. *Fundam. Appl. Toxicol.*, **4**, 30-47.

Gargas, M.L., Clewell, H.J., and Andersen, M.E., (1986). Metabolism of inhaled dihalomethanes *in vivo*: differentiation of kinetic constants for two independent pathways. *Toxicol. Appl. Pharmacol.*, **82**, 211-223.

Howe, R.B., and Crump, K.S., (1982). GLOBAL83: A computer program to extrapolate quantal animal toxicity data to low doses (May, 1982). OSHA Contract No. 41USC252C3.

Kubic, V.L., Anders, M.W., Engel, R.R., Barlow, C.H., and Caughey, W.S. (1974). Metabolism of dihalomethanes to carbon monoxide. I. *In vivo* studies. *Drug. Metab. Dispos.*, **2**, 53-57.

Miller, E.C., and Miller, J.A., (1966). Mechanisms of chemical carcinogenesis: Nature of proximate carcinogens and interactions with macromolecules. *Pharmacol. Rev.*, **18**, 805.

National Toxicology Program (NTP) (1985). NTP Technical Report on the Toxicology and Carcinogenesis Studies of Dichloromethane in F-344/N Rats and $B_6C_3F_1$ Mice (Inhalation Studies). NTP-TR-306 (board draft).

Reitz, R.H., Mendrala, A.L., Park, C.N., Andersen, M.E., and Guengerich, F.P. (1988). Incorporation of *in vitro* enzyme data into the physiologically-based pharmacokinetic (PB-PK) model for methylene chloride: Implications for risk assessment. *Toxicology Letters*, **43**, 97-116.

Reitz, R.H., Mendrala, A.L. and Guengerich, F.P. (1989). *In Vitro* Metabolism of Methylene Chloride in Human and Animal Tissues: Use in Physiologically-Based Pharmacokinetic Models. *Toxicol. Appl. Pharmacol.* **97**, 230-246.

Serota, D., Ulland, B., and Carlborg, F. (1984). Hazelton Chronic Oral Study in Mice. Food Solvents Workshop I: Methylene Chloride, March 8-9, Bethesda, Maryland.

Singh, D.V., Spitzer, H.L., and White, P.D. (1985). Addendum to the Health Assessment Document for Dichloromethane (Methylene Chloride). Updated carcinogenicity assessment of dichloromethane. EPA/600/8-82/004F.

TRAPPING OF REACTIVE INTERMEDIATES BY INCORPORATION OF ^{14}C-SODIUM CYANIDE DURING MICROSOMAL OXIDATION

John W. Gorrod, Catherine M.C. Whittlesea, and Siu Ping Lam[1]

Chelsea Department of Pharmacy, King's College London
University of London, Manresa Road
London, SW3 6LX U.K.

INTRODUCTION

It has long been known that nicotine is metabolized to the corresponding amide cotinine (McKennis et al., 1957). The metabolism of nicotine has been investigated extensively (see Gorrod and Jenner, 1975; Nakayama, 1988). It has been suggested that the enzymatic formation of cotinine occurred via an intermediate (Hucker et al., 1960), which was then metabolized by "aldehyde" oxidase or "iminium oxidase" to form cotinine (Brandange and Lindblom, 1979; Hibberd and Gorrod, 1980). Evidence of a reactive intermediate probably from the cytochrome P-450 dependent oxidation of nicotine, was achieved when a stable product, identified as 5'-cyanonicotine was formed when cyanide was present during the metabolism (Murphy, 1973). This observation led Murphy (1973) to propose nicotine $\Delta^{1'(5')}$ iminium ion as the reactive intermediate. To determine the site of enzyme activity responsible for the production of this reactive intermediate, experiments were carried out as in the original work by Hucker et al. (1960), and Booth and Boyland, (1971). Incorporation of radioactivity from ^{14}C-cyanide by the substrate was examined by various cell fractions, i.e. isolated microsomes, and 140,000g soluble fraction. The metabolism of nicotine has been studied in many species. Miller and Larson, (1953) observed its metabolism by tissue slices from mouse, rabbit, rat, and dog. Enzymic oxidation of nicotine has also been observed in guinea pig (Booth and Boyland, 1971), and rabbit (Hucker et al., 1960). It was therefore decided to investigate the *in vitro* metabolism of nicotine to the cyanonicotine adduct by different mammalian laboratory species.

The technique of trapping metabolically generated electrophilic species using radioactive cyanide has been used to investigate the metabolism of other compounds (Chiba et al., 1985; Ho and Castagnoli, 1980; Hoag et al., 1988; Singer et al., 1987).

A large number of drugs in clinical practice are alicyclic amines. These compounds include those containing three to seven membered ring systems; others incorporate a further nitrogen or oxygen hetero atom in the ring. These compounds would also be expected to be metabolized to a product containing a carbonyl group alpha to the nitrogen atom. In order to investigate the metabolic activation of a large number of drugs a screening procedure has been developed to determine those alicyclic amine drugs that may be converted to an electrophilic species (iminium ion?) and therefore would be expected to be metabolized to the corresponding amide. Stability of the iminium ion, pKa of the base, lipophilicity and steric factors may all be related

---

[1]. Present address for Siu Ping Lam is Medicines Control Agency, Dept. of Health, Market Towers, 1, Nine Elms Lane, London, SW8 5NQ U.K.

Biological Reactive Intermediates IV, Edited by C.M. Witmer *et al.*
Plenum Press, New York, 1990

to amide metabolite formation. Studies to investigate controlling physico-chemical properties for the generation of iminium intermediates from a series of N-substituted phenylpiperidines are thus necessary.

MATERIALS AND METHODS

Glucose-6-phosphate dehydrogenase was purchased from the Boehringer Mannheim Corporation (London) Ltd. Nicotinamide adenine dinucleotide phosphate ($NADP^+$), and ^{14}C labelled sodium cyanide [$Na^{14}CN$], specific activity 8.0 mCi/mmol, dimethyl sulfoxide and nicotine were obtained from Sigma Uk Ltd. The therapeutic drugs and N-substituted phenylpiperidine compounds were donated by various pharmaceutical companies. All compounds were used as supplied without further purification. Glucose-6-phosphate, Scintran Cocktail T and all other chemcials and solvents used were obtained from British Drug House, Dorset (UK). The animals used in the investigations were male rats, mice, hamsters, guinea pigs, dog and monkey.

Male rat hepatic microsomes and soluble fraction (140,000g supernatant) (0.5 g/ml of original liver weight) were prepared using a standard sedimentation procedure at pH 7.4 (Gorrod et al., 1975), in experiments to determine the site of enzyme activity. Hepatic microsomes used in all other experiments were prepared at 0°C using the calcium chloride precipitation method (Schenkman and Cinti, 1978). For the experiments to study the incorporation of ^{14}C-sodium cyanide by alicyclic amine drugs and N-substituted phenylpiperidine compounds, microsomes were suspended in Sorensen's buffer pH 8.0 at the same concentration.

Protein concentrations were determined by the method of Lowry et al. (1951), as modified by Miller (1959). Cytochrome P-450 content was determined by method of Omura and Sato (1962).

Radiolabelled sodium cyanide [$Na^{14}CN$] (0.25 ml, 10^{-3}M aq. solution) was added to 25 ml Erlenmeyer flasks at 37°C. The corresponding buffer either Sorensen's or phosphate buffer (2 ml) containing standard cofactor solution was then added 5 minutes before addition of hepatic microsomes (1.0 ml), followed immediately by addition of the substrate (2 μmol). In experiments to determine the site of enzyme activity in some incubation flasks hepatic microsomes were replaced with an equal volume of the soluble fraction. Nicotine was metabolized as a positive control in experiments to determine incorporation of cyanide by alicyclic amine drugs. The incubation time was 5 mins, the reaction was terminated and the reaction mixture extracted by addition of dichloromethane (3x5ml). The combined organic extracts were washed with double distilled water (3x2.5 ml). Anhydrous K_2CO_3 was added. Organic extracts were evaporated to dryness under a stream of nitrogen at 25-30°C. The residue was dissolved in 5 mls of Scintran Cocktail T scintillation fluid, and the radioactivity counted for 2 mins. Ionization constatns (pKa) were measured potentiometrically by titration with HCl solution (0.1M) (Albert and Serjeant, 1971).

RESULTS AND DISSCUSSION

Determination of the Site of Iminium Ion Formation from Nicotine Metabolism in the Presence of ^{14}C-sodium Cyanide Using Hepatic Microsomes and 140,000g Soluble Fraction

The results are shown in Table 1 and show that no radioactivity was detected in the extracts obtained from incubations of the 140,000g soluble fraction. Showing that the reactive intermediate was not formed by the cytosolic enzymes alone. Radioactivity was detected in the microsomal fraction. This observation agrees with the proposal that the first stage in nicotine metabolism involving the reactive intermediate occurs in microsomes. Analysis of control flasks containing no microsomes, or no 140,000g soluble fraction showed no retention of radioactivity. The possibility that a synergestic action may occur when the microsomal and 140,000g soluble fraction are both present is being investigated.

Table 1. Site of the Metabolims of Nicotine to Cyanonicotine by Rat

CELL FRACTION	RADIOACTIVITY (dps/mg protein)
140,000g soluble fraction	No activity
Washed microsomes	200

Results are the mean of duplicate experiments differing by not more than ± 10%.

Studies on the In Vitro Metabolism of Nicotine in the Presence of ^{14}C-sodium Cyanide with Hepatic Microsomes from Various Species

The metabolism of nicotine was carried out using hepatic microsomes from rats, mice, hamsters, guinea pigs, monkey, dog and rabbits. The results of ^{14}C-cyanide incorporation are shown in Figure 1. It was found that incorporation of radioactivity was greatest using microsomes obtained from monkey followed by rabbit, guinea pig > dog > hamster > mouse > rat. It is interesting to observe that similar activity for cotinine formation in various species was rabbit and hamster > guinea-pig and mouse > rat, although guinea-pig showed marked inter-animal variation (Jenner et al,. 1973). The inter-species differences in cyanonicotine formation observed could be a result of different amounts of the enzyme responsible for the rective intermeidate formation or differences in specific enzyme activities. Due to difficulty in obtaining monkey tissue, further experiments were carried out using rabbit preparations. It is of particular interest to note that the primate gave the highest incorporation of radioactivity, and would therefore be of interest to undertake similar experiments using human liver as enzyme source.

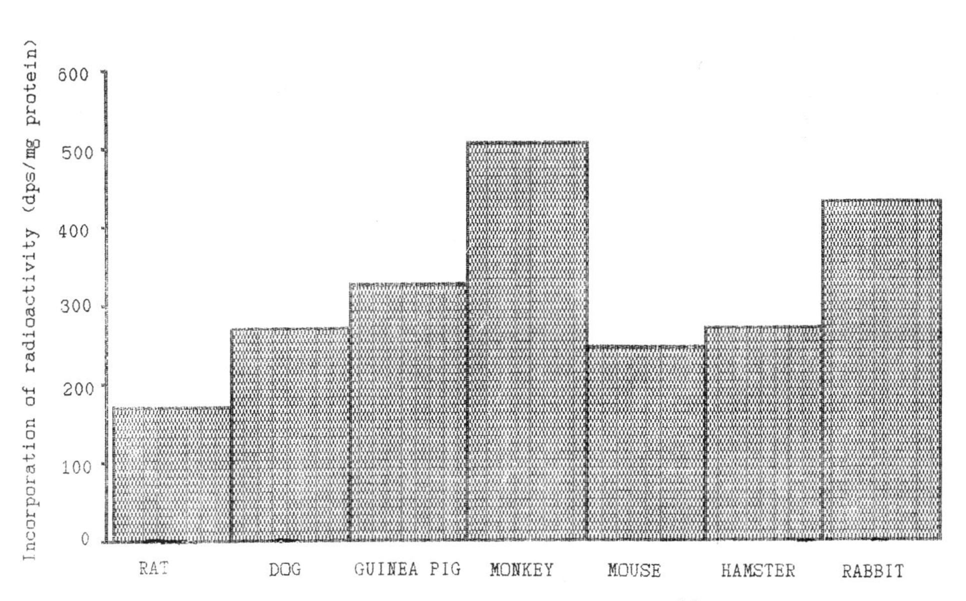

Figure 1. Metabolism of nicotine in the presence of ^{14}C-sodium cyanide by different mammalian laboratory species. Results are the mean of duplicate experiments differing by not more than ± 10%.

Studies on the In Vitro Metabolism of Alicyclic Amine Durgs in the Presence of ^{14}C-sodium Cyanide

Results in Table 2 show that under the conditions used extracts obtained following the metabolism of clemastine, prolintane, thioridazine, triprolidine and dipipanone had a greater incorporation of radioactivity from ^{14}C-syanide than did nicotine.

It has been proposed that the metabolic transformation of nicotine to cotinine occurred via an intermediate iminium ion (Murphy, 1973). Prolintiane is metabolized to oxoprolintane (Yoshihara and Yoshimura, 1972), which is similar to the amide formation from nicotine and tremorine (Hucker et al., 1972). In the present study metabolites from prolintane incorporated over eight times the radioactivity from ^{14}C-syanide compared with that from nicotine. Clemastine showed incorporation similar to nicotine. Metabolic oxidation at the pyrrolidine ring has been suggested for this compound (Tham et al., 1978). Our work would suggest that metabolism to the corresponding amide metabolite of clemastine could take place. There is no information on the metabolism of triprolidine or dipipanone, although this study suggests that amide metabolites may be formed. Procyclidine was found to produce metabolites with incorporation of cyanide at a level similar to nicotine. Metabolsim of this compound *in vitro* and *in vivo* in rat is by hydroxylation of the cyclohexane ring; no oxo-pyrrolidine metabolites have been reported (Paeme et al., 1984). Our experiments suggest that formation of a reactive intermediate through a process similar to nicotine and prolintane may occur when procyclidine is metabolized in the rabbit. Thiroidazine was also found to produce higher levels of radioactive products when metabolized under our conditions than nicotine. Thioridazine is metabolized by rat to a large number of compounds, but no piperidone has been reported (Dinovo et al., 1976). Flecainide was found to be only half as active as nicotine in the incorporatin of ^{14}C-cyanide. One of the major metabolites in hamn was the meta-o-dealkylated lactam of flecainide (McQuinn et al., 1984). Our results suggest that the rabbit may also form an amide metabolite via an iminium intermediate which was trapped by the radioactive cyanide present in the incubation medium.

In vivo metabolism of doxapram in the dog produces several metabolites, one an amide in the morpholino ring (Pitts et al., 1973). In this investigation, metabolic extracts from doxapram whilst less radioactive than those from nicotine, suggest that amide formation involving an iminium ion intermediate frm doxapram could occur in the rabbit.

Table 2. Metabolism of Alicyclic Amine Drugs in the Presence of Radiolabelled ^{14}C-Sodium Cyanide, by Rabbit Hepatic Microsomes, and Their Activity Compared With That of Nicotine

DRUG COMPOUND	RADIOACTIVITY (dps/mg protein)	% COMPARED WITH NICOTINE METABOLISM
Triprolidine	5854	1270
Prolintane	3983	864
Thioridazine	1263	274
Dipipanone	1009	219
Clemastine	484	105
Nicotine	461	100
Procyclidine	429	93
Viloxazine	354	77
Flecainide	237	52
Doxapram	198	43
Dextromoramide	159	34

The results are the mean of two separate determinations.

In man some viloxazine metabolites contain a α-carbonyl group adjacent to the morpholino-ring (Case and Reeves, 1975). Viloxazine was found to produce a radioactive metabolite, but the activity was less than that from nicotine. This again suggests that the metabolic pathway of viloxazine in the rabbit may involve the formation of a heterocyclic amide. A pathway for dextromoramide metabolism has been proposed (Caddy et al., 1980), as similarities in structure with doxapram suggested it would be metabolized to form an amide metabolite in the morpholino ring (Pitts et al., 1973). Dextromoramide also contains a pyrrolidine ring, and metabolism similar to nicotine also might occur. Radioactivity incorporation results obtained in this study supports the proposal for metabolic amide formation from dextromoramide. A number of other compounds used in clinical practice failed to show any significant incorporation of ^{14}C-cyanide under our experimental conditions.

Whilst our results have indicated that ^{14}C-cyanide incorporation can take place the site of incorporation has not been determined in any case. It is known that nicotine in addition to forming an iminium ion as reactive intermediate to cotinine, it also forms an iminium ion which is involved in N-dealkylation (Nguyen et al., 1979). So that incorporation may be carrried out at sites in the molecule other than the alicyclic ring system. Conversely it is known that metabolite formation of a pyrrolidone from a pyrrolidine, N-(5-pyrrolidinopent-3-ynyl)-succinimide, does not always proceed via production of an intermediate capable of reacting with cyanide (Hallstrom et al., 1981). Thus our results whilst being indicative of reactive intermediate formation and amide production do not preclude amide formation via an alternative mechanism.

Studies on the Incorporation of ^{14}C-sodium Cyanide Using a Series of N-substituted Phenylpiperidines

Results are shown in Table 3. the order of incorporation was methyl and allyl substituted phenylpiperidine > butylphenylpiperidine > phenylpiperidine and phenylpropylphenylpiperidine > ethylphenylpiperidine. Differences in the lipophilicity, pKa and steric properties may affect the biotransformation of compounds. Determination of these parameters may enable a structure activity relationship to be established for generation of iminium ions from N-substituted phenylpiperidine compounds. Beckett and Shenoy (1973) found a direct relationship between partition co-efficient and total metabolism for a number of N-alkylamphetamines *in vivo*. A mathematical correlation using hydrophobic fragmental constant (f) was developed (Rekker, 1977). This method of calculation has been used to determine the apparent partition co-efficients for the N-substituted phenylpiperidine series used (Table 3). all the substituted compounds were calculated to be more hydrophobic than the parent compound, phenylpiperidine. the lipophilicity of these compounds does not correlate with the incorporation of radioactivity when metabolized by rabbit hepatic microsomes in the presence of ^{14}C-sodium cyanide. The most lipophilic compound, N-phenylpropylphenylpiperidine showed greater incorporation of radioactivity than the ethyl substituted compound, and a similar activity to phenylpiperidine the least lipophilic compound. Determination of the partition co-efficients using the octanol/water system may allow a better correlation. Ionization constants for the N-substituted phenylpiperidine compounds showed a great difference between phenylpiperidine and the other compounds of the series (Table 3). Distribution of a compound in a system is not only governed by its partition co-efficient but also its degree of ionization. Phenylpiperidine would be ionized at the pH of the metabolism mixture (pH=8), however the other compounds would be virtually unionized. Phenylpiperidine appears to be an anomoly in this series, nevertheless the ionization constants of the other compounds do not correlate with the observed differences in incorporation of radioactivity.

CONCLUSIONS

The metabolism of nicotine in the presence of radiolabelled ^{14}C-sodium cyanide to a reactive intermediate trapped by cyanide, and measured as radioactivity incorporated by the sample, was found to be formed in microsomes but not in the soluble fraction. Microsomes from different species were able to form this radio-

Table 3. Incorporation of ^{14}C-sodium Cyanide and Some Physico-chemical Properties of a Series of N-substituted Phenylpiperidine Compounds

Compound	Radioactivity (dps/mg protein)	Ionization constant (pKa)	Calculated partition coefficient (lpg P)
Phenylpiperidine	120	0.7 ± 0.04	2.561
N-methylphenyl-piperidine	212	3.40 ± 0.04	2.801
N-ethylphenyl-piperidine	88	3.37 ± 0.05	3.331
N-butylphenyl-piperidine	182	3.49 ± 0.03	4.391
N-allylphenyl-piperidine	210	3.38 ± 0.04	3.564
N-phenylpropyl-phenylpiperidine	107	3.43 ± 0.03	5.575

Results of the metabolism experiment indicated by incorporation of radioactivity are the mean of two experiments differing by not more than ± 10%. Ionization constants were calculated at 17°C. The compounds were dissolved in dimethylsulfoxide (3% of the aq. solution). These results are expressed as mean ± S.D. (n=8). Calculated partition co-efficients based on log P (piperidine), determined using the hydrophobic fragmental constant.

active adduct but to different extents. Monkey microsomes were found to have the highest activity but beacuse of the expense and difficulty in obtaining the liver, rabbit microsomes was therefore used in further work on the metabolism of alicyclic compounds. Five compounds were found to have a greater incorporation of radioactivity compared with nicotine. These results from metabolism of the alicyclic amine drugs represent a preliminary study of their potential for amide formation from these compounds, involving an unstable iminium intermediate step. the intermediate formed, although transient, may be reactive enough to bind to proximal nucleophilic centres of cellular macromolecules resulting in tissue damage. The importance of this remains unknown.

No clear structure-activity relationship for the incorporation of radioactivity by N-substituted phenylpiperidine compounds could be determined by the calculated partition co-efficients and the ionization constants.

ACKNOWLEDGEMENTS

C. Whittlesea would like to thank the Royal Pharmaceutical Society of Great Britain for a grant to undertake this work. This study was carried out during a Maplethorpe fellowship to S.P.Lam. We would like to thank all the pharmaceutical companies for donating drug samples and information on the metabolism of their compounds. In particular Dr. P. Murphy and Lilly Research Laboratories, Indiana for the N-substituted phenylpiperidine compounds.

REFERENCES

Albert, A., Serjeant, E.P. (1971). The determination of ionization constants: A Laboraotry Manual, 2nd Ed. Chapman and Hall.

Beckett, A.H., Shenoy, E.V.R. (1973). The effect of N-alkyl chain length and stereochemistry on the absorption, metabolism and urinary excretion of N-alkylamphetamines in man. *J. Pharm. Pharmacol.* **25**, 793.

Booth, J., Boyland, C. (1971). Enzymic oxidation of (-)nicotine by guinea pig tissues *in vitro*. *Biochem. Pharmacol.* **20**, 407-415.

Brandange, S., Lindblom, L. (1979). The enzyme 'aldehyde oxidase' is an iminium oxidase. reaction with nicotine Δ^{1} '(5') iminium ion. *Biochem. Biophys. Res. Commun.* **91**, 991-996.

Caddy, B., Idowu, R., Tilstone, W.J., and Thomson, N.C. (1980). Analysis and disposition of dextromoramide in body fluids. Forensic toxicology. Proceedings of the European meeting of the International Association of Forensic Toxicologists. Ed. J. Oliver. Croom Helm, London, p 126.

Case, D.E. and Reeves, P.R. (1975). The dispisition and metabolism of I.C.I. 58,834 (viloxazine) in humans. *Xenobiotica* **5**, 113-129.

Chiba, K., Peterson, L.A., Castagnoli, K.P., Trevor, A.J. and Castagnoli, Jr., N. (1985). Studies on the molecular mechanism of bioactivation of the selective nigrostriatal toxin 1-methyl-4-phenyl-1,2,3,6-tetrahydropyridine. *Durg. Metab. Dispos.* **13**, 342-347.

Dinovo, E.G., Gottschalk, L.A., Noindi, B.R. and Geddes, P.D. (1976). GLC analysis of thioridazine, mesoridazine and their metabolites. *J. Pharma. Sci.* **65**, 667-669.

Gorrod, J.W. and Jenner, P. (1975). The metabolism of tobacco alkaloids. In *Essays in Toxicology*, Ed. Hayes, Jr., W.J., Academic Press, New York, Vol. 16, pp 35.

Gorrod, J.W., Temple, D.J. and Beckett, A.H. (1975). The metabolism of N-ethyl-methylaniline by rabbit liver microsomes. The measurement of metabolites by gas-liquid chromatography. *Xenobiotica* **5**, 435-464.

Hallstrom, G., Lindeke, B. and anderson, E (1981). Metabolism of N-(5-pyrrolidinopent-3-ynyl)-succinimide (BL 14) in rat liver preparations. Characterization of four oxidative reactions. *Xenobiotica.* **11**, 459-471.

Hibberd, A.R. and Gorrod, J.W. (1980). Nicotine Δ^{1} '(5') iminium ion: a reactive intermediate in nicotine metabolism. In *Biological Reactive Intermediates-II, Chemical Mechanisms and Biological Effects*, Part B. Eds. Synder, R., Parke, D.V., Kocsis, J.J., Jollow, D.J., Gibson, G.G. and Witmer, C.M. Plenum Press, New York, pp 1121.

Ho, B. and Castagnoli, Jr., N. (1980). Trapping of metabolically generated electrophilic species with cyanide ion: metabolism of 1-benzylpyrrolidine. *J. Med. Chem.* **23**, 133-139.

Joag, M., Schmidt-Peetz, M., Lampen, P., Trevor, A. and Castagnoli, Jr., N. (1988). Metabolic studies on phencyclidine: characterization of a phencyclidine iminium ion metabolite. *Chem. Res. Toxicol.* **1**, 128-131.

Hucker, H.B., Gillette, J.R. and Brodie, B.B. (1960). Enzymic pathway for the formation of cotinine, a major metabolite of nicotine in rabbit liver. *J. Pharmacol. Exptl. Ther.* **129**, 94-100.

Hucker, H.B., Stauffer, S.C. and Rhodes, R.E. (1972). Metabolism of a pharmacologically active pyrrolidine derivative (prolintane) by lactam formation. *Experientia.* **28**, 430-431.

Jenner, P., Gorrod, J.W. and Beckett, A.H. (1973). Species variation in the metabolism of R-(+)- and S-(-)-nicotine by α-C- and N-oxidation *in vitro*. *Xenobiotica.* **3**, 573-580.

Lowry, D., Rosenbrough, H.J., Farr, A.L. and Randall, R.J. (1951). Protein measurement with the Folin phenol reagent. *J. Biol. Chem.* **193**, 265-275.

McKennis, Jr, H., Turnbull, L.B. and Bowman, E.R. (1957). γ-3(3-pyridyl)-γ-methylaminobutyric acid as a urinary metabolite of nicotine. *J. Am. Chem. Soc.* **79**, 6342.

McQuinn, R.L., Quarfoth, G.J., Johnson, J.D., Banitt, E.H., Pathre, S.V., Chang, S.F., Ober, R.E. and Conard, G.J. (1984). Biotransformation and elimination of ^{14}C-flecanide acetate in humans. *Durg. Metab. Dispos.* **12**, 414.

Millar, G.L. (1959). Protein determination for large numbers of samples. *Anal. Chem.* **31**, 964.

Millar, A.L. and Larson, P.J. (1953). Observations on the metabolism of nicotine by tissue slices. *J. Pharm. Exp. therap.* **109**, 2218-2222.

Murphy, P. (1973). Enzymatic oxidation of nicotine to nicotine Δ^{1} '(5') iminium ion, a newly discovered intermediate in the metabolism of nicotine. *J. Biol. Chem.* **248**, 2796-2800.

Nakayama, H. (1988). Nicotine metabolism in mammals. *Drug. Metab. Drug. Interac.* **6**, 95-122.

Nguyen, T., Gruenke, L.D. and Castagnoli, N. Jr. (1979). Metabolic oxidation of nicotine chemically reactive intermediates. *J. Med. Chem.* **22**, 259-263.

Omura, T. and Sato, R. (1962). A new cytochrome in liver microsomes. *J. Biol. Chem.* **237**, 1375-1376.

Paeme, G., Grimee, R. and Vercruysse, A. (1984). *In vitro* metabolism of procyclidine in the rat. *ERur. J. Metab. Pharmaco.* **9**, 311-313.

Pitts, J.E., Bruce, R.B. and Forehand, J.B. (1973). Identification of doxapram metabolites using high pressure ion exchange chromatography and mass spectroscopy. *Xenobiotica.* **3**, 73-83.

Rekker, R.F. (1977). The hydrophobic fragmental constant: its derivation and application: a means of characterizing membrane systems. Pharmacochemistry Libray, Vol. 1. Elservir Scientific Publishing Co., Amsterdam.

Schenkman, J.B. and Cinti, D.L. (1975). Preparation of microsomes with calcium. *Method in Enzymology* **52**, 83-89.

Singer, S.S., Lijinsky, V., Kratz, L.E., Castagnoli, Jr., N. and Rose, J.E. (1987). A comparison of *in vivo* and *in vitro* metabolites of the H1-antagonist N,N-dimethyl-N-2-pyridyl-N-(2-thienylmethyl)-1,2-ethane-diamine (methapyrilene) in the rat. *Xenobiotica* **17**, 1279-1291.

Tham, R., Norlander, B., Hagermark, O. and Fransson, L. (1978). Gas chromatography of clemastine. A study of plasma kinetics and biological effect. *Arzneim. Forsch.* **28**, 1017-1020.

Yoshihara, S. and Yoshimura, H. (1972). Metabolism of drugs-LXXVIII the formatin *in vitro* of oxoprolintane by rabbit liver. *Biochem. Pharmacol.* **21**, 3205-3211.

MULTIPLE BIOACTIVATION OF CHLOROFORM: A COMPARISON BETWEEN MAN AND EXPERIMENTAL ANIMALS

Luciano Vittozzi, Emanuela Testai, Angelo De Biasi

Istituto Superiore di Sanità
Department of Comparative Toxicology and Ecotoxicology
Rome, Italy

Chloroform carcinogenic action is species-specific and the target organ depends on the animal species (Bull, R.J., 1985). Several experimental studies have shown that B6C3F1 mice develop hepatic tumours. Sprague-Dawley (SD) rats are among the resistant animal species. Epidemiology studies have shown a correlation between human exposure to $CHCl_3$ in drinking water and colon cancer (Gottlieb, M.S. et al., 1981). Results of human studies, however, may be affected by confounding factors, and warrant further scientific support. Therefore, we have undertaken this study to find a metabolic feature able to explain the different susceptibility of animals and man to $CHCl_3$ toxicity.

Male Sprague-Dawley rats and B6C3F1 mice were from Charles River; human liver, colon and ileum specimens were kindly supplied by Prof. G.M. Pacifici, Pisa (Italy). Microsomes from colon and small intestine mucosa were prepared essentially according to Fang and Strobel (Fang, W.F., et al., 1978) and Stohs et al. (1976), respectively. Chloroform (radiochemical purity >98%) was from NEN and was used in a range of specific radioactivity of 0.2-3 nCi/nmole, as required.

The standard incubation mixture, prepared in stoppered vessels, contained: microsomal protein (2 mg/ml), G6P (2mM), $MgCl_2$ (2 mM) G6PDH (1 U/ml), NADP (0.2 nM), EDTA (1 mM) and $CHCl_3$ in 50 mM Tris-HCl buffer (pH 7.4). To obtained hypoxic conditions, the mixture was flushed with nitrogen for 20 min before injection of NADP and $^{14}CHCl_3$. When anaerobic conditions were required, the mixture contained also D-glucose (60 mM), glucose oxidase (12.5 U/ml) and catalase (3000 U/ml) and was flushed with nitrogen for 20 min before injection of NADP and $^{14}CHCl_3$. Covalent binding of chloroform metabolites to lipid and protein was measured on aliquots of the incubation mixtures as described previously (Testai E. et al., 1986). To measure the radioactivity associated to the polar head and to the fatty acyl chain of microsomal phospholipids, extracted lipids were transesfterified with MeOH and H_2SO_4 and fractions separated following standard methods.

As shown in Table 1, $CHCl_3$ is bioactivated *in vitro* by hepatic microsomes of B6C3F1 mice through different processes evidenced by the effects of varying oxygen and chloroform concentration and the GSH addition. At low $CHCl_3$ concentrations (0.1 mM) the decrease of pO_2 caused a strong inhibition of chloroform biotransformation; furthermore the scavenging action of 3 mM GSH was nearly total, independent of the oxygenation conditions. On the contrary, at 5 mM $CHCl_3$, the lipid covalent binding was slightly decreased on decreasing the oxygen partial pressure from 20% to 1%, but is stimulated when the oxygen is further scavenged to complete anoxia. The $CHCl_3$ metabolites produced in anaerobiosis were almost completely

Biological Reactive Intermediates IV, Edited by C.M. Witmer *et al.*
Plenum Press, New York, 1990

scavenged by GSH whereas the protective action of reduced glutathione was less and less effective on lowering oxygen partial pressure.

Therefore, it can be supposed that at low chloroform concentrations an oxidative process only takes place, whereas at high concentration, chloroform bioactivation is due to two different processes, showing opposite dependences on oxygen concentration.

The metabolites produced through the oxygen-inhibited mechanism are the only ones which can bind microsomal lipids even in the presence of GSH. The metabolites produced by the different metabolic pathways show a clear regioselectivity of binding to microsomal phospholipids: the oxidation metabolites bind preferentially to the phospholipid polar head (PH) whereas the other intermediates, likely radicals produced with a reductive mechanism, prefer binding to the fatty acyl chain (FAC). This feature is evidenced by the ratio between radioactivity associated to PH and FAC (PH/FAC ratio), which is 10.8 in room air-equilibrated incubations and 0.09 in complete anaerobiosis.

Table 1. Covalent Binding of $^{14}CHCl_3$ Metabolites to Lipid in B6C3F1 Liver Microsomes

	$CHCl_3$ 0.1 mM		$CHCl_3$ 5 mM	
	- GSH	+ GSH	- GSH	+ GSH
pO_2				
20%	7.90 ± 0.43	0.11 ± 0.51	14.03 ± 0.53	0.17 ± 0.01
1%	1.05 ± 0.01	0.17 ± 0.04	10.03 ± 0.65	4.01 ± 0.67
0%	0.54 ± 0.07	0.09 ± 0.01	18.3 ± 0.51	10.60 ± 1.1

Results are expressed as nmol/mg lipid. They represent means ± S.E. of at least three experiments performed with different microsomal preparations.

Table 2. Covalent Binding of $^{14}CHCl_3$ Metabolites to Liver Microsomal Lipid of Different Species

	20%pO2	1% pO2	0% pO2
B6C3F1 mice	7.9 ± 0.4	1.0 ± 0.01	0.54 ± 0.07
SD - rats	1.3 ± 0.09	0.4 ± 0.005	0.09 ± 0.005
Human Samples			
1	0.2	0.3	
2	0.5	0.3	
3	0.2	0.3	
4			0.1
5			0.0

$^{14}CHCl_3$ concentration was 0.1 mM. Results are expressed as nmol/mg lipid. They represent means ± S.E. of at least three experiments performed with different microsomal preparations.

As bioactivation of $CHCl_3$ in hypoxic conditions is nearly absent in liver microsomes from untreated S.D. rats (Testai, E. et al., 1986), it seems that this metabolic pathway is species-specific, whereas the oxidative process occurs in both species. The comparison between experimental animals and human liver microsomes at 0.1 mM $CHCl_3$ indicated that human samples bioactivated $CHCl_3$ at levels comparable to the levels found with rat liver microsomes, but of limited statistical significance (Table 2).

On the contrary, with colon mucosa microsomes from human biopsies in reductive conditions only substantial levels (3.2 and 5.8 nmol/mg lipid) of covalent binding to lipid were reached in 2 out of 4 samples.

In comparison, colon mucosa microsomes obtained from either control and β-NF-pretreated rats were not able to bioactivate chloroform. In conclusion, the reductive activation should receive more attention as a possible determinant in the carcinogenic action of chloroform in physiologically hypoxic tissues. In fact, it appears to be present in different species in the organs which are targets of the action of chloroform and where the physiological conditions of oxygenation and GSH concentration do not protect against the production of damage.

REFERENCES

Bull, R.J. (1985). Carcinogenic and mutagenic properties of chemicals in drinking water. *Sic. Total Environm.* **47**, 385-413.

Fang, W.F. and Strobel, H.J. (1978). The drug and carcinogen metabolic system of rat colon microsomes. *Arch. Biochem. Biophys.* **186**, 128-138.

Gottlieb, M.S., Carr, J.K., Morris, D.T. (1981). Cancer and drinking water in Louisiana: colon and rectum. *Int. J. Epidemiol.* **10**, 117-125.

Stohs, S.J., Grafstrom, R.C., Burke, M.D., Moldeus, P.W. and Orrenius, S.G. (1976). The isolation of rat intestinal microsomes with stable cytochrome P-450 and their metabolism of benzo(a)pyrene. *Arch. Biochem. Biophys.* **177**, 105-116.

Testai, E. and Vittozzi, L. (1986). Biochemical alterations elicited in rat liver microsomes by oxidation and reduction products of chloroform metabolism. *Chem. Biol. Interact.* **59**, 157-171.

CCl_4-INDUCED CYTOCHROME P-450 LOSS AND LIPID PEROXIDATION IN RAT LIVER SLICES

Shana Azri, Heriberto P. Mata, A. Jay Gandolfi, and Klaus Brendel

Department of Anesthesiology
University of Arizona
College of Medicine
Tucson, Arizona 85724

Previous studies from this laboratory have shown a concentration-time dependent, site specific cytotoxicity of CCl_4 in slices. The present study examined radical formation and lipid peroxidation as a potential mechanism of toxicity. Liver slices from male Sprague-Dawley rats (220- 250 g) were exposed to 0.57 mM CCl_4 by vaporization using a roller incubation system. Cytochrome P-450 is responsible for the bioactivation of CCl_4 generating the CCl_3 radical, which also acts as a suicide inhibitor. In slices exposed to CCl_4, cytochrome P-450 loss occurred in a time-dependent manner relative to controls (↓ 58% at 9 hr). A 48 hr fast prior to sacrifice both enhanced and accelerated both cytochrome P-450 loss as well as CCl_4 toxicity in slices. Unlike cytochrome P-450, glutathione levels were not altered over the course of the experiment. These studies suggest that centrilobular hepatocytes are more susceptible to CCl_4 induced injury. Covalent binding studies using $^{14}CCl_4$ confirmed CCl_3 radical formation by cytochrome P-450. Binding to slice proteins plateaued as early as 30 min following CCl_4 administration whereas lipid binding was saturated by 60 min. Covalent binding of the CCl_3 radical was increased two-fold 60 min following phenobarbital pretreatment whereas allylisopropylacetamide caused the converse (50% at 60 min). Conjugated diene formation, an index of lipid peroxidation as detected by 15 min following

CCl_4 exposure

These results demonstrate that CCl_4-induced radical formation and subsequent lipid peroxidation can be measured in slices and that this system may allow future mechanistic and site specific studies.

INTRODUCTION

Previous studies from this laboratory have demonstrated a concentration- and time-dependent, site-specific cytotoxicity of CCl_4 in slices which could be accelerated following phenobarbital pretreatment and prevented by allylisopropylacetamide, a cytochrome P-450 inhibitor (Azri et al, 1989). However, actual biotransformation as well as potential mechanisms involved in CCl_4 toxicity have not been examined using this slice system.

Carbon tetrachloride is both a suicide inhibitor of cytochrome P-450 and a site-specific hepatotoxicant. A specific isozyme of cytochrome P-450 (cytochrome P-450IIE1), induced by fasting, is believed to be responsible for the homolytic cleavage of CCl_4, which results in the formation of the trichloromethyl radical (Hong et al., 1987). Subsequent binding of this radical in the centrilobular region is believed to initiate lipid peroxidation and subsequent cell death (Recknagel and Glende, 1973).

The present study examines the loss of cytochrome P-450 in slices from both fed and fasted rats, covalent binding of the trichloromethyl radical, the presence of lipid peroxidation in slices exposed to CCl_4, and lastly if glutathione content is altered as a result of CCl_4 exposure, demonstrating increased susceptibility of centrilobular hepatocytes.

METHODS

Liver Slice Preparation

Male Sprague-Dawley rats (220-250 g) were used in these studies. Animals were fed Teklab rat chow and water *ad libitum* except in experiments where animals were fasted for 48 hr prior to sacrifice. Animals were killed by cervical dislocation and their livers excised. Cylindrical cores (1 cm) were made and slices were cut (250-300 µm) using a Krumdieck slicer (Azri et al, 1989).

Incubation System

The slices were incubated in 1.7 ml Waymouth's MB 752/1 media (pH = 7.4) containing 10 mg/ml gentamycin which maintains slice viability for at least 24 hr. Slices were loaded onto wire cradles which fit into conventional scintillation counting vials (2 slices/vial). Slices were incubated for up to 9 hr in a temperature controlled roller incubator. CCl_4 was vaporized from a wick suspended from a septa which fit into the caps forming an air tight seal. Slices were preincubated for 1 hr at 37°C prior to CCl_4 exposure (Azri et al, 1989).

Analytical Assays

The primary indicator for viability was intracellular K^+ leakage using flame photometry (DiRenzo et al, 1984). To determine the amount of covalent binding to slice macromolecules, slices were exposed to 0.57 mM $^{14}CCl_4$ (53 mCi/mmol) for 90 min under 95% O_2. Covalent binding in slices from animals pretreated with 80 mg/kg phenobarbital or 100 mg/kg. AIA was also examined to determine the relationship between cytochrome P-450 status and radical formation. Lipid and protein fractions were isolated using exhaustive extraction techniques (Gandolfi et al, 1980).

Conjugated diene levels in slices were determined using a modified method of Sell and Reynolds (1969). Total cytochrome P-450 content in slices was determined using a modified method of Estabrook et al. (1972). Glutathione content was determined via a modified spectrophotometric method using Ellman's reagent (Waters, personal communication).

Statistics

Each data point reflects 3-5 experiments (individual animals) with 6 slices used for each condition/experiment. The N values are per slice taken from different experiments. Comparisons were made using the ANOVA and Student Newman Keuls analysis. $p < 0.05$ was considered significant.

RESULTS

Covalent binding to slice proteins plateaued 30 min following administration whereas lipid binding continued to increase steadily over the 60 min exposure time (Figure 1). The amount of $^{14}CCl_4$ bound to slice lipids was increased two-fold following pretreatment with phenobarbital relative to that seen in nonpretreated animals (Table 1). Protein binding was unchanged. AIA pretreatment caused a

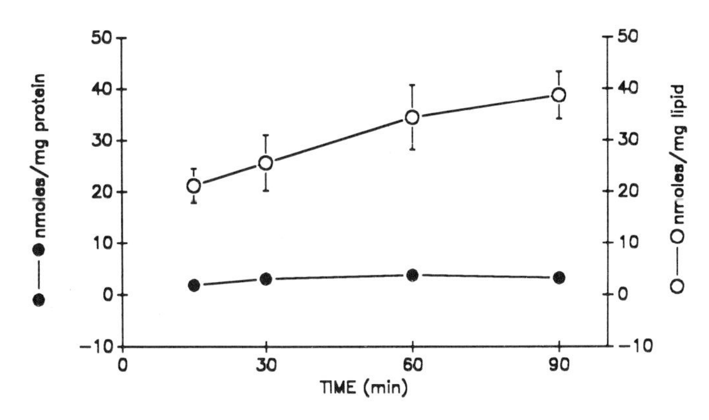

Figure 1. Covalent binding of $^{14}CCl_4$ to slice proteins and lipids. Slices were exposed to 0.57 mM $^{14}CCl_4$ at 37R C, 95% O_2. Results are expressed as the mean + SEM of 3-6 experiments consisting of 6 slices each.

significant decrease in both lipid and protein binding due to the inhibition of CCl_4 bioactivation.

Conjugated diene formation increased over time in slices exposed to 0.57 mM CCl_4 which is a direct consequence of attack of fatty acids by CCl_3 (Figure 2). No conjugated dienes were detected in control slices.

A 48 hr fast prior to sacrifice enhanced intracellular K^+ loss (CCl_4 toxicity) in slices exposed to CCl_4 relative to that seen in slices from fed animals (Figure 3). Similarly, loss of cytochrome P-450 occurred in a time-dependent manner and was enhanced in slices from fasted animals (Figure 4). 6 and 9 hr subsequent to CCl_4 administration total cytochrome P-450 content in slices from fed animals was only 41% and 53% of controls, respectively. Cytochrome P-450 content in fasted animals was 40% of controls at 6 hr and could not be detected at 9 hr following CCl_4 administration. Glutathione levels remained constant and identical to that seen in respective controls, over the duration of the experiment.

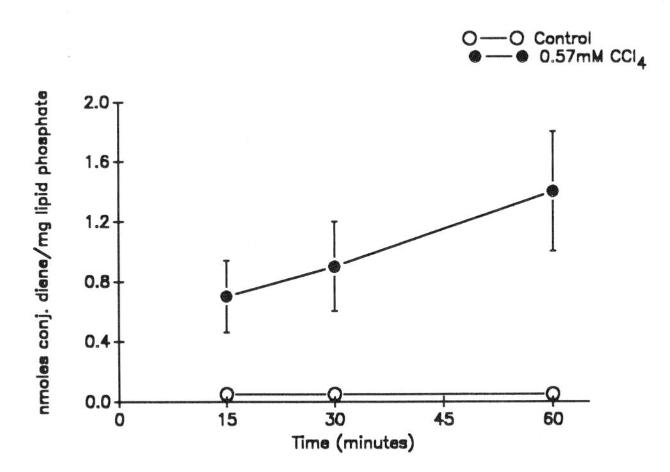

Figure 2. Conjugated diene formation in rat liver slices exposed to CCl_4. Rat liver slices were incubated with 0.57 mM CCl_4 for 60 min. Each point represents the mean + SEM of 3-4 experiments of 6 slices for each experimental point.

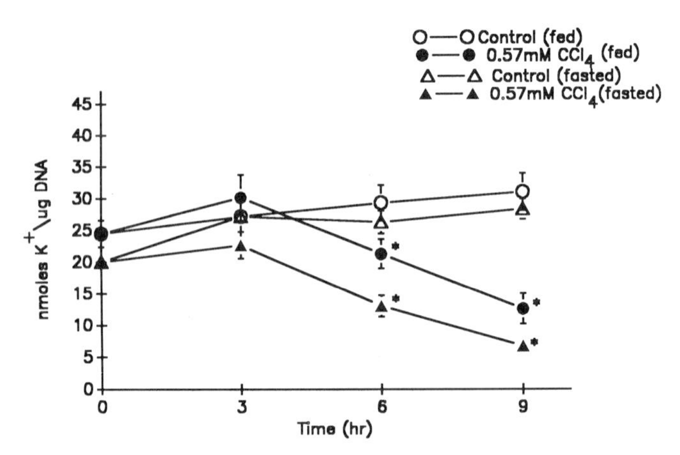

Figure 3. Intracellular K^+ levels in rat liver slices from fed and fasted animals. Animals were fasted for 48 hr unless otherwise indicated. Slices were exposed to 0.57 mM CCl_4 at 37R C under 95% O_2. Results are expressed as the mean + SEM of 3-6 experiments consisting of 4-6 slices each. $*p < 0.05$.

Table 1. Covalent binding of $^{14}CCl_4$ to rat liver slice lipids and proteins.

Pretreatment	Covalent Binding Protein (nmoles/mg protein)	 Lipid (nmoles/mg lipid)
Control	3.8 ± 0.8	34.4 ± 6.3
Phenobarbital	4.0 ± 0.3	70.6 ± 4.6*
AIA	2.1 ± 0.2*	11.7 ± 3.6*

Slices were exposed to 0.57 mM $^{14}CCl_4$ for 60 min at 37°C under 95% O_2. Results are expressed as the mean = SEM of 3-6 experiments consisting of 6 slices each. $*p < 0.05$.

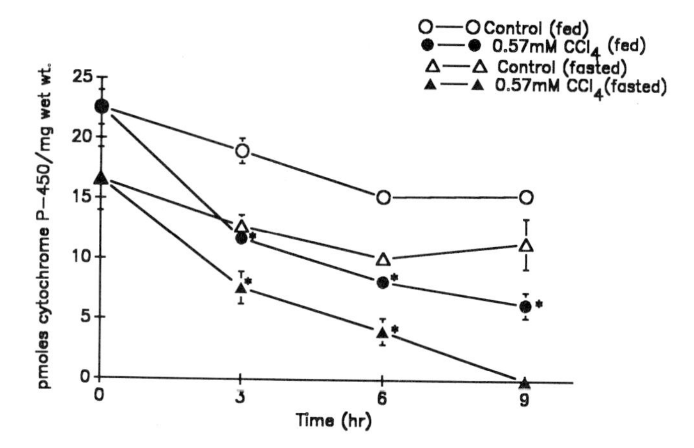

Figure 4. Cytochrome P-450 content in rat liver slices from fed and fasted animals. Rat liver slices were exposed to 0.57 mM CCl_4 at 37°C in 95% O_2, for a total of 9 hr. Each point represents the mean + SEM of 3-4 experiments of 6 slices for each experimental point. *$p < 0.05$.

DISCUSSION

CCl_4 toxicity is dependent upon cytochrome P-450 bioactivation where the CCl_3-Cl bond is homolytically cleaved forming the trichloromethyl radical. Formation and subsequent binding of this radical mediates peroxidative hepatocellular damage which occurs in centrilobular hepatocytes (Recknagel and Glende, 1973). Using precision-cut rat liver slices, these studies focused on actual biotransformation of CCl_4 as well as the initial cellular events which occur following exposure to CCl_4.

The amount of free radical formation was reflected by the degree of covalent binding of $^{14}CCl_4$ to slice lipids and proteins. Since radical formation results from cytochrome P-450 dependent bioactivation of CCl_4, induction or inhibition of mixed function oxidase activity was found to directly affect the amount of binding to slice macromolecules, ie. phenobarbital increased the amount of binding whereas AIA caused the converse to occur. A similar time course indicated that along with covalent binding, lipid peroxidation ensued (the index of lipid peroxidation being conjugated diene formation). These studies demonstrated the detection of initial peroxidative events mediating CCl_4 toxicity in slices.

CCl_4 is a well-documented suicide substrate of cytochrome P-450 (Glende, 1972). A specific isozyme of cytochrome P-450 (cytochrome P-450IIE1), induced by ethanol, fasting, or acetone is believed to be responsible for the bioactivation/metabolism of CCl_4 (Hong et al., 1987; Johansson et al, 1988). In these studies, CCl_4 exposure not only caused a significant loss of cytochrome P-450 content in slices but enzyme loss was enhanced in animals fasted 48 hr prior to sacrifice. CCl_4 toxicity (intracellular K^+ loss) also increased in slices from fasted animals relative to that seen in slices from fed animals. In addition, intracellular K^+ loss was augmented in slices from fasted animals exposed to CCl_4 relative to that seen in liver slices from fed animals.

Lastly, CCl_4 induced cytochrome P-450 loss reflected the site-specificity of this hepatotoxicant, since this enzyme is located exclusively in centrilobular hepatocytes. This was confirmed by the maintenance of glutathione levels in slices exposed to CCl_4 since glutathione is present predominantly in periportal hepatocytes (Jungermann and Katz, 1989).

Precision-cut liver slices are particularly suitable for toxicological analysis for site-specific toxicants such as CCl_4 since the architecture of the organ is maintained whilst extraneous physiological influences such as blood flow are removed. Experimental results using this organ culture system concur with those found *in vivo* and in other *in vitro* systems. Using precision-cut liver slices, further elucidation of the nature and mechanisms of CCl_4 toxicity as well as the actual characterization of specific cellular responses may be examined.

ACKNOWLEDGEMENTS

This study was supported by an NIH grant GM 38290 and National Research Service Award F32ES 05474-01 BI-4. The authors wish to express their thanks to Patricia Kime for her assistance in the preparation of this manuscript.

REFERENCES

Azri, S., Gandolfi, A.J., and Brendel, K. (1989). Carbon tetrachloride toxicity in precision-cut rat liver slices. *In Vitro Toxicol.* in press.

Chen, P.S., Toribara, T.Y., and Warner, H. (1956). Microdetermination of phosphorus. *Anal. Chem.* **28,** 1756-1758.

Direnzo, A.B., Gandolfi, A.J., Brooks, S.D., and Brendel K. (1985). Toxicity and biotransformation of volatile halogenated anesthetics in rat hepatocytes. *Drug Chem. Toxicol.* **8,** 207-218.

Estabrook, R.W., Peterson, J., Baron, J., and Hildebrant, A. (1972). The spectrophotometric measurement of turbid suspensions associated with drug metabolism. *Methods Pharmacol.* **2,** 303-350.

Gandolfi, A.J., White, R.D., Sipes, I.G., and Pohl, L.R. (1980). Bioactivation and binding of halothane *in vitro*: Studies with 3H and 14C halothane. *J. Pharm. Exp. Ther.* **214,** 721-725.

Glende, E.A. (1972). Carbon tetrachloride-induced protection against carbon tetrachloride toxicity. The role of the liver microsomal drug-metabolizing system. *Biochem. Pharmacol.* **21,** 1697-1702.

Hong, J., Pan, J., Gonzales, F.J., Gelboin, H.V., and Yang, C.S. (1987). The induction of a specific form of cytochrome P-450 (P450j) by fasting. *Biochem. Biophys. Res. Commun.* **142,** 1077-1083.

Johansson, I., Ekstrom, G., Scholte, B., Puzycki, D.,Jornvall, H., and Ingelman-Sundberg, M. (1988). Ethanol-, fasting- and acetone-inducible cytochrome P-450 in rat liver: regulation and characteristics of enzymes belonging to the IIB and IIE gene subfamilies. *Biochemistry* **27,** 1925-1934.

Jungermann, K., and Katz, N. (1989). Functional specialization of different hepatocyte populations. *Physiol. Rev.* **69,** 708-764.

Recknagel, R.O., and Glende, E.A. (1973). Carbon tetrachloride hepatotoxicity: An example of lethal cleavage. *CRC Crit. Rev. Toxicol.* **27,** 263-295.

Sell, D.A., and Reynolds, E.S. (1969). Liver parenchymal cell injury. VII lesions of membranous cellular components following iodoform. *J. Cell. Biol.* **41,** 736-752.

ALTERATION OF BENZO(A)PYRENE-DNA ADDUCT FORMATION BY RATS EXPOSED TO SIMPLE MIXTURES

S. Stephen Bentivegna and Charlotte M. Witmer

Joint Graduate Program in Toxicology/UMDNJ
Robert Wood Johnson Medical School
Piscataway, NJ

Testing the potential toxicity of mixtures is a difficult task compounded by the fact that very little is generally known about the constituents of most complex mixtures. New applications of existing methods are needed to investigate the potential adverse health effects of both simple and complex mixtures. Our laboratory has been investigating the effects of mixtures on the formation and persistence of adducts on DNA by a known toxicant along with resultant changes in metabolic enzymes as a means of studying the effects of mixtures. This approach is based on the fact that most multi-component and simple environmental mixtures have many components in common, and common effects may thus allow prediction of actions of mixtures.

We chose benzo(a)pyrene (BP) as the known toxin in applying the above approach. BP is a widespread environmental pollutant that forms adducts with DNA which can be measured at very low levels by the 32P-postlabeling method of Reddy and Randerath (1986). To study the effects a simple mixture might have on BP-DNA adduct formation in lung we exposed Fischer 344 rats to BP alone (5 mg/kg) and to BP simultaneously with a mixture (Mix) of carbon tetrachloride (0.5 ml/kg), monochlorobenzene (300 mg/kg) and lead acetate (55 mg/kg) for two days and because of resultant toxicity droppped the doses to one half these levels for another five days. The animals were sacrificed at 48, 72 and 168 hours after the last dosing. Lungs were immediately excised and portions were used to measure levels of DNA adducts and activities of aryl hydrocarbon hydroxylase (AHH) and glutathione transferase (GST).

Treatment with both BP and BP+Mix resulted in the formation of 3 major and 5 minor adducts. At 48 and 72 hours after the last dose no differences in the number of BP-DNA adducts or in the levels of the major adducts were seen between the two treatment groups. By 168 hours the levels of adducts in the BP+Mix group as measured by the RAL values were significantly higher than the BP only group. AHH activity in the BP+Mix group was increased over the BP only group at 48 hours with no difference at 72 hours, while by 168 hours the activity in the BP only group was significantly higher than BP+Mix group. GST values were not different in the two groups at 48 and 72 hours but by 168 hours the BP+Mix group activity was significantly higher than BP only. Results so far do not show a direct correlation between adduct levels and the levels of enzyme activity.

Biological Reactive Intermediates IV, Edited by C.M. Witmer *et al.*
Plenum Press, New York, 1990

MATERIALS AND METHODS

Benzo(a)pyrene, Tris, micrococcal nuclease, Tricaprylin, LiCl, LiOH, and CHES were from Sigma Chemical Co. Nuclease P1 and spleen phosphodiesterase were from Boehringer-Mannheim Biochemicals. Polynucleotide kinase was from United States Biochemicals. Monochlorobenzene and carbon tetrachloride were from Aldrich Chemical Co. All other chemicals were from Fisher, Inc. ^{32}P (200 uCi/ul) in HCl free water was from ICN Biochemicals. PEI cellulose thin-layer chromatography sheets by Machery-Nagel were from Brinkman Instruments.

Male Fischer 344 rats (Taconic Farms) were treated i.p. with either 5 mg/kg BP alone in Tricaprylin or 5 mg/kg BP plus an oral dose of a mixture of carbon tetrachloride (0.5 ml/kg), monochlorobenzene (300 mg/kg) both in corn oil and lead acetate in water (55 mg/kg) given simultaneously. After two days the mixture dose was reduced by one half to reduce the toxicity seen at the original dose. Control animals recieved dose vehicles only. All dosing was for seven consecutive days. Animals were sacrificed 48, 72 and 168 hours after the last dose by ether anesthesia and exsanguination. Lungs were removed and frozen at -80° until processed.

DNA was extracted from 0.3-0.5 gm of lung tissue by enzyme treatment and solvent extraction as described by Reddy and Randerath (1987). Samples of DNA (10 ug) were digested with a mixture of spleen phosphodiesterase and micrococcal nuclease for 3 hr at 37°. These digests were then treated with nuclease P1 for 45 min at 37° to dephosphorylate the unmodified nucleotides. Adducts were then labeled by incubation with ^{32}P-ATP and 4 U. polynucleotide kinase. This mix was spotted on a PEI cellulose sheet and developed with 2.3 M sodium phosphate, pH-5.8. The origin was cut out and tranferred to another PEI sheet and chromatography was continued with 5.3 M lithium formate, 8.5 M urea, pH-3.5, followed by 4 N ammonium hydroxide/isopropanol (1:1), and finally with 1.0 M sodium phosphate, pH-6.0. Adduct spots were located by autoradiography at -70°, cut out and quantitated by Cerenkov counting.

Microsomes were prepared from the remaining lung tissue by homogenization followed by differential centrifugation. The pellet from the 105,000g centrifugation step was resuspended in phosphate buffer and used for the AHH assay (Van Cantfort 1977). Microsomal protein was incubated with ^{3}H-BP mixed with cold BP, NADH and NADPH. The reaction mix was extracted and an aliquot of the aqueous phase was counted by liquid scintillation counting.

Cytosol from the 105,000g centrifugation step was used to measure glutathione transferase activity (Habig 1974). Cytosolic protein was incubated with glutathione and 1-chloro-2,4-dinitrobenzene at 30° and the increase in absorbance at 340 nm was measured.

RESULTS AND DISCUSSION

There were no DNA adducts found in lungs from animals treated with dosing vehicles in this postlabeling/chromatographic procedure, while in lungs from animals treated with BP only (for 7 days) there were three major adducts and generally five minor ones. The locations and numbers of adducts were the same for the two treatment groups (Figure 1), only the amounts of the adducts differed. Adduct B co-chromatographed with a known standard of pdGp-N^2-(C10-BPT) prepared from (+)-anti-benzo(a)pyrene diol epoxide and calf thymus DNA. Adduct A has not been characterized. Quantification of the adducts by calculation of the relative adduct labeling (RAL) values resulted in the values shown in Table 1. Levels of the main adducts (A and B) did not differ between treatment groups at the 48 and 72 hour periods, nor was there a significant difference between the two time periods for the individual adducts. However, the 168 hour samples from the BP+Mix group had significantly higher RAL values than the samples from the BP only group. The maximum RAL values for the BP only group were at 72 hours and were declining at 168 hours. The RAL values for adducts A and B in the BP+Mix group were still rising at 168 hours. The minor adducts in both treatment groups were decreasing to the limit of detection (one adduct in 10^9 nucleotides pairs) at the last time point.

Table 1. Mean Relative Adduct Labeling Values from Rat Lung DNA

GROUP	48 HR	72 HR	168 HR
ADDUCT A ($\times 10^{-8}$)			
BP ONLY	1.26 ± 0.63	1.67 ± 1.19	1.08 ± 0.28
BP + MIX	2.24	1.21 ± 0.50	1.73 ± 0.31*
ADDUCT B ($\times 10^{-8}$)			
BP ONLY	3.30 ± 1.70	5.71 ± 2.35	4.67 ± 0.65
BP + MIX	4.08	5.49 ± 1.35	6.39 ± 0.89*

A: ACTIVITY PRESENTED AS MEAN ± S.D., N=3 OR 4
*: SIGNIFICANTLY DIFFERENT FROM BP ONLY, $P < 0.05$

Table 2. ARYL Hydrocarbon Hydroxylase Values from Rat Lung[A]

GROUP	48 HR	72 HR	168 HR
CONTROL[B]	10.8 ± 6.0	10.8 ± 6.0	10.8 ± 6.0
BP ONLY	64.8 ± 8.1	63.5 ± 11.7	89.0 ± 10.1
BP + MIX	87.1 ± 9.1*	77.7 ± 14.3	62.0 ± 12.5*

A: ACTIVITY PRESENTED AS MEAN + S.D., IN PMOLES/MIN/MG PROT
B: ALL CONTROL ANIMALS POOLED, N=6, ALL OTHERS N=3 OR 4.
*: SIGNIFICANTLY DIFFERENT FROM BP ONLY, $P < 0.05$.

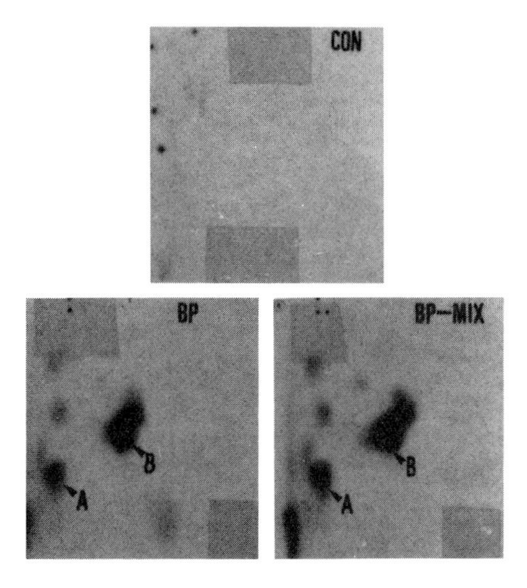

Figure 1. Autoradiograms of Lung DNA from Control and Treated Rats (168 hr).

Table 2 presents the results of the AHH activity measurements in lung from these animals. Both treatment groups showed an induction of activity over vehicle controls. At 48 hours the BP+Mix group activity was significantly increased over the BP only group while at 72 hours there was no difference in the activities of the two groups. At 168 hours after dosing, the the reverse was true i.e.- the activity of the BP only group was markedly increased over the BP+Mix treatment.

Table 3 summarizes the results of the GST assay in lung from the same animals. At 48 and 72 hours there was no difference between any of the treatment groups. At 168 hours the BP+Mix group showed increased activity over both the control and BP only groups.

Table 3. Glutathione Transferase Values from Rat Lung[A]

GROUP	48 HR	72 HR	168 HR
CONTROL[B]	145 ± 27	145 ± 27	145 ± 27
BP ONLY	171 ± 35	162 ± 18	140 ± 16
BP + MIX	143 ± 5	178 ± 11	170 ± 14*

A: ACTIVITY PRESENTED AS MEAN + S.D., IN NMOLES/MIN/MG PROT
B: ALL CONTROL ANIMALS POOLED, N=6, ALL OTHERS N=3 OR 4.
*: SIGNIFICANTLY DIFFERENT FROM BP ONLY, $P < 0.05$

The results of the postlabeling assay indicate that there are increased levels of BP-DNA adducts in animals given the mixture and BP simultaneously. While this was not evident at the early time points it was significant by one week after the last dose. Previous workers using dermal dosing (Springer et. al. 1989) have reported that some complex mixtures alter not only total BP binding to skin DNA but also the ratio of the different enantiomers, i.e. the ratio of anti/syn BPDE-dG adducts. The presumed syn BPDE-dG adduct can be resolved in the chromatography system used in this study and can be seen in Figure 1 as the spot migrating just above and to the right of adduct B. Preliminary results indicate that the alteration of ratio might be occuring with this simple mixture.

In this study the activities of the AHH enzyme system do not directly correlate with the differences in adduct levels. The possibility exists that the increased AHH activity at the late time points in the BP only group is due to retention of unmetabolized BP in an unidentified organ. The levels of GST activity reflect the differences in adduct level. This is an unexpected result as the GST system conjugates BP metabolites to water soluble mercapturic acid derivatives and increased activity of GST would thus be expected to lead to lower rather than higher levels of adducts. However, GST exists as a number of isozymes and the increased level seen in this total activity assay could be due to an increase in an isoform that is not efficient in the removal of BP metabolites (Robertson 1986). Studies are being carried out to determine if any one of the components in the mixture has the major effect on the changes seen in GST.

REFERENCES

Habig, W.H., Pabst, M.J., Jakoby, W.B. (1974). Glutathione S-transferases: The first step in mercapturic acid formation. *Jour. Biol. Chem.* **249**, 7130.

Reddy, M.V., and Randerath, K. (1986). Nuclease P1-mediated enhancement of sensitivity of ^{32}P-postlabeling test for structurally diverse DNA adducts. *Carcinogenesis* **7**, 1543.

Reddy, M.V., and Randerath, K. (1987). ^{32}P analysis of DNA adducts in somatic and reproductive tissues of rats treated with the anticancer antibiotic, mitomycin. *Mutat. Res.* **179**, 75.

Robertson, I.G.C., Jensson, H., Mannervik, B., Jernstrom, B. (1986). Glutathione transferases in rat lung: the presence of transferase 7-7, highly efficient in the conjugation of glutathione with the carcinogenic (+)-benzo(a)pyrenediolepoxide. *Carcinogenesis* **7**, 295.

Springer, D.L., Mann, D.B., Dankovic, D.A., Thomas, B.L., Wright, C.W., Mahlum, D.D. (1989). Influences of complex organic mixtures on tumor-initiating activity, DNA binding and adducts of benzo(a)pyrene. *Carcinogenesis* **10**, 131.

Van Cantfort, J., De Graeve, J., Gielen, J.E. (1977). Radioactive assay for aryl hydrocarbon hydroxylase. Improved method and biological importance. *Biochem. Biophys. Res. Comm.* **79**, 505.

COMPARISON OF THE TOXICITY OF NAPHTHALENE AND NAPHTHALENE-1,2-DIHYDRODIOL (DIOL)

R.E. Billings, N.E. Miller, J.E. Dabbs, S.E. LeValley, J.R. Hill and C.E. Green

Dept. of Biochemical Toxicology, Life Sciences Div.
SRI International
Menlo Park, CA
Depts. of Surgery and Pharmacology
Univ. of Nevada Medical School
Reno, NV

Naphthalene metabolism and toxicity have been extensively investigated and these data indicate that the lung and eye toxicity are mediated by reactive intermediates. The elegant studies conducted in Buckpitt's laboratory implicate a 1,2-epoxide as one reactive intermediate which is extensively conjugated with glutathione (Buckpitt et al., 1987). Data from this laboratory also suggest that the liver may form a precursor metabolite or reactive metabolite which is transported to the lung (Buckpitt and Warren, 1983). Studies of the ocular effects of naphthalene suggest that the well known liver metabolite, naphthalene-1,2-dihydrodiol (DIOL), may be transported to the eye where it is oxidized to the catechol metabolite, 1,2-dihydroxynaphthalene (VanHeyningen and Pirie, 1967). The catechol is easily oxidized non-enzymatically presumably to potentially reactive quinones and semiquinones. A more recent report also suggests the involvement of quinones in the ocular toxicity of naphthalene (Wells et al., 1989) and an *in vitro* study indicates that multiple quinone-derived metabolites are formed from DIOL when its oxidation is catalyzed by hepatic dihydrodiol dehydrogenase (Smithgall et al., 1988). Naphthalene is extensively metabolized and at least two epoxides as well as the the catechol/quinone metabolites are potentially reactive, toxic metabolites (Horning et al., 1980).

The present studies were conducted to characterize the toxicity of DIOL and to compare it with naphthalene toxicity.

MATERIALS AND METHODS

DIOL was synthesized by standard methods. ^{14}C-labeled DIOL was synthesized from ^{14}C-labeled naphthalene using isolated rat hepatocytes.

Swiss Webster mice (20-30 gm) were used for all experiments. Hepatocytes were prepared by whole liver collagenase perfusion and they were incubated in modified Waymouth's medium at a concentration of 2×10^6 cells/ml (Green et al., 1983). Lactate dehydrogenase activity (LDH) in the incubation medium was determined by standard methods and compared with total cellular LDH measured following lysis of the cells with Triton X100. Intracellular reduced glutathione (GSH) concentrations were determined by measuring the fluorescence of the o-phthaldehyde adduct and total glutathione (GSH + GSSG) concentrations were measured with glutathione reductase

Biological Reactive Intermediates IV, Edited by C.M. Witmer *et al.*
Plenum Press, New York, 1990

(Ku and Billings, 1986). Covalent protein binding was determined in incubations with ^{14}C-labeled substrates. Proteins were precipitated with trichloroacetic acid or acetone and then extracted 6-8 times with methanol. Radioactivity which remained associated with the proteins was assumed to be covalently bound. After removal of protein, the incubations were analyzed for metabolites by HPLC using a ^{18}C-column (Beckman Ultrasphere) and a methanol/water/acetic acid gradient essentially as described by Buckpitt et al. (1984). Separate samples were analyzed with and without prior treatment with Glusulase in order to hydrolyze glucuronide/sulfate conjugates.

For the *in vivo* toxicity experiments, the mice were fasted overnight and then injected i.p. with either naphthalene dissolved in corn oil (0.1 ml/10 gm body weight) or DIOL dissolved in phosphate-buffered saline ((0.1 ml/10 gm body weight). GSH concentrations in liver and lung were determined 4 hr after dosing. Lung and liver samples were obtained from mice 24 hr after injections. Histological specimen were prepared and examined by light microscopy.

RESULTS AND DISCUSSION

Metabolism and Toxicity of Naphthalene and Diol In Mouse Hepatocytes

When naphthalene was incubated with mouse hepatocytes it was observed that DIOL was the major metabolite. These data are shown in Table 1. Based upon the amount of metabolite detected with and without hydrolysis of conjugates with Glusulase, the DIOL was not conjugated whereas the phenolic metabolite, 1-naphthol, was extensively conjugated with glucuronide and/or sulfate. This result is similar to previously reported results with bromobenzene metabolites in that the dihydrodiols are much less extensively conjugated than the phenols (Dankovic et al., 1985); this was postulated to account for the lack of further hydroxylation of the phenols to catechol metabolites (Billings, 1985).

Table 1. Metabolism of Naphthalene and Naphthalene-1,2-Dihydrodiol in Mouse Hepatocytes

Substrate	Minutes Incubated	Percent Substrate Incubated			
		Covalent Binding*	Polar Metab.	Diol	1-Naphthol
Naphthalene (0.50 Mm)	60 (Heat-killed)**	0.2	--	--	--
	15	0.7	9	5	2
	30	1.3	12	11	6
	60	2.0	14	16	8
DIOL (0.25 Mm)	60 (Heat-killed)**	1.5	--	81	--
	15	13	--	58	--
	30	19	--	47	--
	60	22	--	40	--

Data are shown for samples assayed after Glusulase hydrolysis.

* Amount of radioactivity bound to protein after methanol extraction. No correction for recovery has been made.
** Cells were heated at 50° C. for 30 min. prior to incubation.

It is apparent that DIOL is extensively bound to cellular proteins and this binding is about 10-fold greater than observed with naphthalene. At a DIOL concentration of 0.25 mM, covalent protein binding was 5.5 nmol/mg protein whereas at a naphthalene concentration of 0.5 mM, protein binding was 0.5 nmol/mg protein. At these concentrations, both DIOL and naphthalene decreased the hepatocyte concentration of GSH to approximately 10% of control values. Neither GSH nor GSSG accumulated in the medium and GSSG was not observed in the cells. These data suggest that both naphthalene and DIOL are converted to reactive metabolite(s) which form GSH conjugates and are covalently bound to proteins. It was also observed that DIOL was more rapidly toxic to the liver cells. Thus, after a 1 hr incubation with DIOL (0.75 mM), 90% of the cellular LDH activity was in the medium. In contrast, there was no naphthalene-related leakage of LDH after a 1 hr incubation with naphthalene (1.0 mM), although 50% of the cellular LDH activity was in the medium after 3 hr of incubation.

These data with isolated hepatocytes show that naphthalene is extensively converted to DIOL in isolated liver cells. The diol is not conjugated and, *in vivo*, may be transported to target tissues such as the lung and eye. In hepatocyte suspensions, DIOL readily renters the liver cells and is extensively converted to reactive metabolites which are conjugated with glutathione and covalently bind to cellular proteins. It is likely that these reactive metabolites are responsible for the cellular toxicity of DIOL and contribute to the toxicity of naphthalene in isolated hepatocytes.

Toxicity In Vivo of Naphthalene and Diol

As previously reported (Warren et al, 1982), naphthalene caused severe bronchial/bronchiolar necrosis in the lungs of mice given doses of 200-400 mg/kg. The high dose was lethal to 60% of the mice. DIOL was found to be more toxic than naphthalene and was lethal to 3 of 6 mice given 100 mg/kg. Death of these animals occurred rapidly (2-4 hr after dosing). In the surviving animals, a moderate degree of lung necrosis, similar to that observed with naphthalene, was observed. Liver toxicity was not observed with either DIOL or naphthalene. Both DIOL and naphthalene decreased lung and liver concentrations of GSH. The two compounds were approximately equally effective on a dosage basis.

These data show that DIOL is more toxic than naphthalene. However, while lung toxicity is observed with DIOL, death of the animals is apparently unrelated to the necrosis since this occurred rapidly before the lung lesion developed and at doses where the lesion was only moderate in the surviving mice. Nevertheless, it is clear that the dynamics are very different when DIOL is formed *in situ* from naphthalene in the liver and transported to the lung, the first organ to receive blood via the hepatic vein. Overall the data obtained in isolated hepatocytes and *in vivo* support the concept that DIOL metabolites contribute to the toxic effects of naphthalene. The nature of these reactive metabolites (e.g. quinones derived from the catechol and/or epoxides) remains to be investigated.

ACKNOWLEDGEMENTS

These studies were supported by NIEHS grant ES4336. Dr. Charles Tyson provided encouragement and helpful discussions.

REFERENCES

Billings, R.E. (1985). Mechanisms of catechol formation from aromatic compounds in isolated rat hepatocytes. *Drug Metab. Disp.* **13**, 287-290.

Buckpitt, A.R. and Warren, D.L., 1983. Evidence for hepatic formation, export and covalent binding of reactive naphthalene metabolites in extrahepatic tissues *in vivo*. *J.Pharmacol. Exp. Ther.* **225**, 8-16.

Buckpitt, A.R., Bahnson, L.S. and Franklin, R.B., 1984. Hepatic and pulmonary microsomal metabolism of naphthalene to glutathione adducts: Factors affecting the relative rates of conjugate formation. *J. Pharmacol. Exp. Ther.* **232**, 291-300.

Buckpitt, A.R., Castagnoli, N. Nelson, S.D., Jones, A.D. and Bahnson, L.S. (1987). *Drug. Metab. Disp.* **15**, 491-498.

Dankovic, D., Billings, R.E., Seifert, W. and Stillwell, W.G. (1985). Bromobenzene metabolism in isolated rat hepatocytes: incorporation studies. *Mol. Pharmacol.* **27**, 287-295.

Green, C.E. , Dabbs, J.E. and Tyson, C. (1983). Functional integrity of isolated rat hepatocytes prepared by whole liver vs. biopsy perfusion. *Anal. Biochem.* **129**, 269-276.

Horning, M.G., Stillwell, W.G., Griffin, G.W. and Tsang, W.-S. (1980). Epoxide intermediates in the metabolism of naphthalene by the rat. *Drug Metab. Disp.* **8**, 404-414.

Ku, R.H. and Billings, R.E. (1986). The role of mitochondrial glutathione and cellular protein sulfhydryls in formaldehyde toxicity in glutathione toxicity in glutathione-depleted rat hepatocytes. *Arch. Biochem. Biophys.* **247**, 183-189.

Smithgall, T.E., Harvey, R.G., and Penning, T.M. (1988). Spectroscopic identification of quinones as the products of polycyclic aromatic trans-dihydrodiol oxidation catalyzed by dihydrodiol dehydrogenase. A potential route of proximate carcinogen metabolism. *J.Biol. Chem.* **263**, 1814-20.

VanHeyningen, R. and Pirie, A. (1967). The metabolism of naphthalene and its toxic effect on the eye. *Biochem. J.* **102**, 842-852.

Warren, D.L., Brown, D.L. and Buckpitt, A.R. (1982). Evidence for cytochrome P-450 mediated metabolism in the bronchiolar damage by naphthalene. *Chem.-Biol. Interactions* **40**, 287-303.

Wells, P.G., Wilson, B. and Lubek, B.M. (1989). *In vivo* murine studies on the biochemical mechanism of naphthalene cataractogenesis. *Toxicol. Appl. Pharmacol.* **99**, 466-473.

SELECTIVE BINDING OF ACETAMINOPHEN (APAP) TO LIVER PROTEINS IN MICE AND MEN

Raymond B. Birge*,**, John B. Bartolone**[1], Charles A. Tyson***, Susan G. Emeigh Hart,*,*** Steven D. Cohen* and Edward A. Khairallah**

SRI International***
Menlo Park, CA 94025
Toxicology Program; Departments of Pharmacology and Toxicology*, Molecular and Cell Biology** and Pathobiology***
University of Connecticut
Storrs, CT 06269

Acetaminophen (N-acetyl-p-aminophenol, APAP) toxicity is dependent upon APAP activation by cytochrome P450 to a metabolite which binds covalently to macromolecules as cellular glutathione becomes depleted (Potter, et al., 1973; Mitchell, et al., 1973; Jollow, et al., 1973; Dahlin, et al., 1984). Studies with an affinity purified anti-APAP antibody have demonstrated that, in mice, the covalent binding is not random, but is highly selective, and that binding to cytosolic protein(s) of approximately 58kD was most closely associated with APAP toxicity (Bartolone, et al., 1987; 1988; 1989a; Birge, et al., 1988; 1989; Beierschmitt, et al., 1989). While human liver has been demonstrated to activate APAP *in vitro* (Dybing, 1977), covalent binding has not been demonstrated during human APAP poisoning. The present study demonstrates for the first time that a similar, selective arylation of proteins also occurs in humans during APAP hepatotoxicity.

MATERIALS AND METHODS

^{125}I - conjugated goat anti-rabbit IgG was purchased from Dupont New England Nuclear (Boston, MA). Ultra pure electrophoretic grade Tris-HCl, acrylamide, and N,N'-methylene bis-acrylamide were obtained from Boehringer Mannheim Biochemicals (Indianapolis, IN). Nitrocellulose membranes (0.2 µm) were purchased from Schleicher and Schuell (Keene, N.H.) and all other reagents were purchased from Sigma Chemical Co. (St. Louis, MO).

A frozen (-80°C) liver specimen, obtained within 1 hour after death, was provided by the Liver Tissue Procurement and Distribution Center (University of Minnesota Hospital, St. Paul, MN) from a 5 yr old white, female, APAP fatality. Plasma APAP was 52 µg/ml approximately 24 hr after ingestion, and death occurred 44 hr later. Sections of healthy liver from a 17 yr old white male were obtained from the National Disease Research Interchange (Philadelphia, PA) as a control. At the time of assay samples were thawed and immediately homogenized 1:40 (w/v) in STM buffer (0.25 M sucrose, 10 mM Tris-HCl, and 1 mM $MgCl_2$, pH 7.4) and centrifuged at 8,500g. The

[1] Present address, Richardson Vicks, Inc. 1 Far Mill Crossing, Shelton, CT 06484.

Biological Reactive Intermediates IV, Edited by C.M. Witmer *et al.*
Plenum Press, New York, 1990

supernatant was then centrifuged at 105,000g at 4°C to separate microsomal and cytosolic fractions for subsequent immunochemical analysis.

Three-month old male mice [Crl:CD-1(ICR)BR (Charles River Laboratories), Wilmington, MA], were also used. They were housed in stainless steel cages with free access to food (Purina Rodent Chow) and water in an animal facility with controlled light and temperature. They were fasted for 18 - 20 hr and then dosed with 600 mg APAP/kg, po, in 50% propylene glycol / water (injection volume: 10 ml/kg) and killed by decapitation 4 hr later. Controls were given vehicle only. Tissues were homogenized and separated into microsomal and cytosolic fractions as described above.

APAP covalent binding was assayed by SDS-PAGE and Western blotting as previously described (Bartolone, et al., 1988). Briefly, proteins (30 μg/lane) were resolved according to molecular weight (Laemmli, 1970) on a discontinuous 10% SDS-PAGE slab gel system with a 3% stacking layer at a constant current of 20 mA/slab. Protein content was determined by the method of Lowry, et al., (1951) using bovine serum albumin (BSA) as the standard. Resolved proteins were electroblotted to nitrocellulose membranes at 60 V for 6 hr and, after rinsing and blocking with 1% BSA, were incubated with affinity purified anti APAP antisera for subsequent reaction with ^{125}I-conjugated goat anti-rabbit IgG.

RESULTS AND DISCUSSION

Figure 1 provides a comparison of the selective protein arylation which occurred in mouse and human livers after APAP exposure, *in vivo*. In the human, covalent binding was primarily detected in the cytosolic fraction with the most prominent immunostaining on bands of approximately 58 and 130 kD with lesser arylation detected at the 38 and 62 kD bands. By contrast, there was no apparent selective binding in the microsomal fraction. In the mouse the major adducts were on proteins in the 44, 58 and 130 kD bands in cytosol and 44 kD in microsomes. This clearly demonstrates that selective APAP binding also occurs in human liver. Furthermore, the 130 kD and the very prominent 58 kD protein bands were targeted in the cytosol of both mouse and human. The prominence of the 58 kD binding in human liver after APAP poisoning and its frequent association with hepatotoxicity in mice (Bartolone, et al., 1987; 1988; 1989a; Birge, et al., 1988; 1989; Beierschmitt, et al., 1989) suggests that it may also be important in human APAP hepatotoxicity. However, the cytosolic 38 and 62 kD proteins which were arylated in human liver were not major targets in mouse liver. It is possible that these differences may reflect different molecular weights of similar proteins in the two species. In addition, the microsomal protein at 44 kD in mouse liver was not a prominent target in the human liver. This difference may merely reflect the relatively rapid turnover of the arylated 44 kD protein (Bartolone, et al., 1989b), or it may be due to differences in proteins associated with the specific isoforms of cytochrome P-450 involved in APAP activation in each species (Gonzalez, 1989).

Immunohistochemical analysis of liver from the APAP fatality with the anti-APAP antibody (not presented here) revealed that APAP binding was localized to the centrilobular regions with no staining in the periportal areas. This is consistent with previous reports which similarly localized binding and damage after APAP (Jollow, et al., 1973; Placke, et al., 1987; Bartolone, et al., 1989b; Roberts, et al., 1989)

The results of this study directly link covalent binding with APAP-induced hepatotoxicity for the first time in humans. The observations that in both human and mouse liver there was similar centrilobular and protein selective covalent binding suggests that further study of the prominent, commonly targeted 58 kD protein will provide valuable information about the mechanistic importance of such binding in APAP hepatotoxicity.

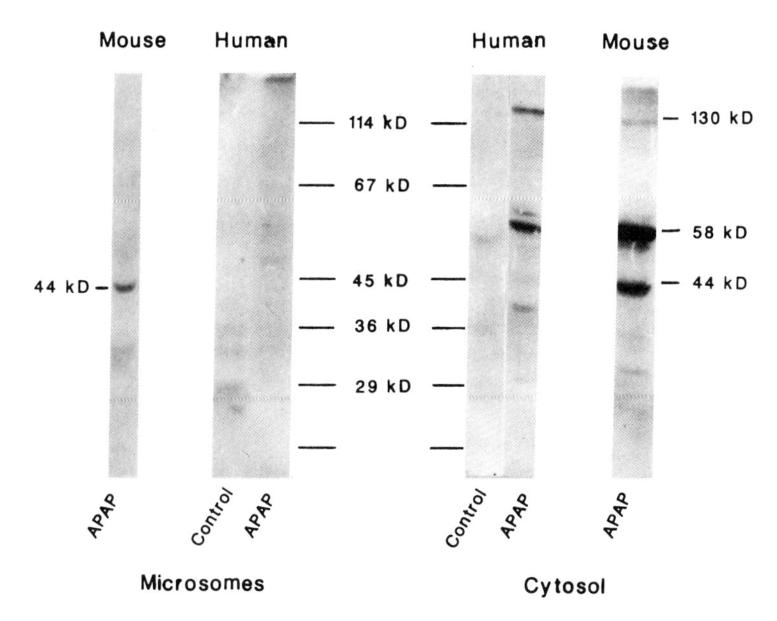

Figure 1. Selective protein arylation after *in vivo* exposure to APAP. Western blot analysis of hepatic cytosol and microsomes from a mouse given an hepatotoxic dose of APAP (600 mg/kg, po) and a human APAP fatality were analyzed immunochemically with affinity purified anti-APAP antibody. Protein standards and the major arylated proteins are indicated.

ACKNOWLEDGEMENTS

Supported in part by the Center for Biochemical Toxicology and PHS Grants ES-07163, GM-31460 and ES-55109.

REFERENCES

Bartolone, J.B., Sparks, K., Cohen, S.D. and Khairallah, E.A. (1987). Immunochemical detection of acetaminophen bound liver proteins. *Biochem. Pharmacol.* **36**, 1193-1196.

Bartolone, J.B., Birge, R.B., Sparks, K., Cohen, S.D. and Khairallah, E.A. (1988). Immunochemical analysis of acetaminophen covalent binding to proteins: Partial characterization of the major acetaminophen-binding liver proteins. *Biochem. Pharmacol.* **37**, 4763-4774.

Bartolone, J.B., Beierschmitt, W.P., Birge, R.B., Emeigh Hart, S.G., Wyand, S., Cohen, S.D. and Khairallah, E.A. (1989a). Selective acetaminophen metabolite binding to extrahepatic proteins: An *in vivo* and *in vitro* analysis. *Toxicol. Appl. Pharmacol.* **99**, 240-249.

Bartolone, J.B., Cohen, S.D., and Khairallah, E.A. (1989b). Immunochemical localization of acetaminophen-bound liver proteins. *Fund. Applied. Toxicol.* **13**, 859-862.

Beierschmitt, W.P., Brady, J.T., Bartolone, J.B., Wyand, D.S., Khairallah, E.A. and Cohen, S.D. (1989). Selective protein arylation and the age-dependency of acetaminophen hepatotoxicity in mice. *Toxicol. Appl. Pharmacol.* **98**, 517-529.

Birge, R.B., Bartolone, J.B., Nishanian, E.V., Bruno, M.K., Mangold, J.B., Cohen, S.D.

and Khairallah, E.A. (1988). Dissociation of covalent binding from the oxidative effects of acetaminophen: studies using dimethylated acetaminophen derivatives. *Biochem. Pharmacol.* **37**, 3383-3393.

Birge, R.B., Bartolone, J.B., McCann, D.J., Mangold, J.B., Cohen, S.D. and Khairallah, E.A. (1989). Selective protein arylation by acetaminophen and 2,6-dimethyl-acetaminophen in cultured hepatocytes from phenobarbital-induced and uninduced mice. *Biochem. Pharmacol.* **38**, 4429-4438.

Dahlin, D.C., Miwa, G.T., Lu, A.Y.H. and Nelson, S.D. (1984). N-acetyl-p-benzoquinone imine: a cytochrome P-450-mediated oxidation product of acetaminophen. *Proc. Nat. Acad. ScI.* **81**, 1327-1331.

Dybing, E. (1977). Activation of α-methyldopa, paracetamol and furosemide by human liver microsomes. *Acta Pharmacol.* et toxicol. **41**, 89-93.

Gonzalez, F.J. (1988). The molecular biology of the cytochrome P-450s. *Pharmacol. Rev.* **40**, 243-288.

Jollow, D.J., Mitchell, J.R., Potter, W.Z., Davis, D.C., Gillette, J.R. and Brodie, B.B. (1973). Acetaminophen-induced hepatic necrosis. II. Role of covalent binding *in vivo*. *J.Pharmacol. Exp. Ther.* **187**, 185-194.

Laemmli, U.K. (1970). Cleavage of structural proteins during the assembly of the head bacteriophage T4. *Nature* (London) **227**, 680-685.

Lowry, O.H., Rosebrough, N.J., Farr, A.L. and Randall, R.J. (1951). Protein measurement with the Folin phenol reagent. *J. Biol. Chem.* **193**, 265-275.

Mitchell, J.R., Jollow D.J., Potter, W.Z., Davis, D.C., Gillette, J.R. and Brodie, B.B. (1973). Acetaminophen-induced hepatic necrosis. I. Role of drug metabolism. *J. Pharmacol. Exp. Therap.* **187**, 185-194.

Placke, M.E., Ginsberg, G.L., Wyand, D.S. and Cohen, S.D. (1987). Ultrastructural and biochemical changes during acute acetaminophen-induced hepatotoxicity in the mouse. A time and dose study. *Toxicologic Pathol.* **15**, 431-438.

Potter W.Z., Davis D.C., Mitchell J.R., Jollow, D.J., Gillette, J.R. and Brodie, B.B. (1973). Acetaminophen-induced hepatic necrosis. III. Cytochrome-P450 mediated covalent binding *in vitro*. *J. Pharmacol. Exp. Ther.* **187**, 203-210.

Roberts, D.W., Hinson, J.A., Benson, R.W., Pumford, N.R., Warbritton, A.R., Crowell, J.A. and Bucci, T.J. (1989). Immunohistochemical localization of 3-(cystein-S-yl) acetaminophen protein adducts in livers of mice treated with acetaminophen. *The Toxicologist.* **9**, 47.

POST-TREATMENT PROTECTION WITH PIPERONYL BUTOXIDE AGAINST ACETAMINOPHEN HEPATOTOXICITY IS ASSOCIATED WITH CHANGES IN SELECTIVE BUT NOT TOTAL COVALENT BINDING

Joseph T. Brady[1], Raymond B. Birge[1,2], Edward A. Khairallah[2] and Steven D. Cohen[1]

Toxicology Program: Departments of Pharmacology & Toxicology[1] and Molecular and Cell Biology[2]
University of Connecticut,
Storrs, CT 06269

Acetaminophen (APAP) induced hepatic centrilobular necrosis has been associated with cytochrome P-450-mediated generation of an electrophilic, reactive metabolite which covalently binds to cellular macromolecules as glutathione becomes depleted (Jollow, et al., 1973; Mitchell, et al., 1973a; 1973b; Potter, et al., 1973; 1974). Covalent binding has been well-correlated with the incidence and severity of liver necrosis and prior cytochrome P450 inhibition blocks both covalent binding and the atotoxicity (Jollow, et al., 1973; Potter, et al., 1973, 1974; Mitchell, et al., 1973a). In addition, administration of the cytochrome P-450 inhibitor, piperonyl butoxide (Pip B), 2 hrs after APAP, the time of maximal covalent binding (Jollow,et al., 1973, Ginsberg and Cohen, 1985) reduced the severity of liver damage (Brady, et al., 1988). The present study demonstrates that Pip B's post-treatment protection is associated with alterations in selective protein arylation by APAP without a change in total covalent binding.

METHODS

Three-month old male mice [Crl:CD-1(ICR)BR (Charles River Laboratories), Wilmington, MA], were housed in stainless steel cages with free access to food (Ralston Purina Company, St. Louis, MO) and water in an animal facility with controlled light and temperature. They were fasted for 18 - 20 hr and then dosed with 600 mg APAP/kg, po, in 50% propylene glycol / water (injection volume: 10 ml/kg). Two hr later, either Pip B (600 mg/kg) or corn oil vehicle (5 ml/kg) was administered, ip, and mice were killed by decapitation after an additional 2 hr (i.e., 4 hr after APAP). For assessment of total covalent binding, tritiated APAP was diluted with unlabeled APAP such that each mouse received approximately 0.1 mCi of tritiated APAP in the 600 mg/kg dose. Controls were given vehicle only.

Total and selective APAP covalent binding were measured in liver 4 hr after APAP dosing. The radiometric method of Jollow, et al. (1973) was used for determination of total liver covalent binding in whole liver homogenates. Electrophoresis (1-D, SDS-PAGE), western blotting and autoradiography techniques for the immunochemical assessment of selective APAP covalent binding using an affinity-purified anti-APAP antibody were performed as described by Bartolone, et al. (1988). The Student's t-test was used for statistical analysis with $p \leq 0.05$ accepted as significant.

Biological Reactive Intermediates IV, Edited by C.M. Witmer *et al.*
Plenum Press, New York, 1990

RESULTS AND DISCUSSION

In mice, APAP covalent binding to liver macromolecules was maximal at approximately 2 hr after dosing (Jollow, et al., 1973; Ginsberg and Cohen, 1985). Additionally, post-treatment with the MFO inhibitor, Pip B, at this time significantly reduced the severity of APAP-induced liver damage (Brady, et al., 1988). In the present studies, Pip B post-treatment did not change total liver covalent binding which was 0.417 ± 0.043 and 0.457 ± 0.062 nmols of APAP bound/mg homogenate protein (mean ± SE, n=6) for control and Pip B post-treated animals, respectively. Figure 1 shows the selective APAP protein binding in liver microsomal and cytosolic fractions from the individual control and Pip B post-treated mice. The immunochemical analysis revealed that the Pip B post-treatment altered the pattern of APAP binding to liver proteins. In agreement with our previous findings (Bartolone, et al., 1987; 1988), the two most heavily targeted protein bands were at 58 kD in cytosol and at 44 kD in both microsomes and cytosol. Pip B post-treatment greatly reduced APAP binding to the 58 kD protein in the cytosol. Additionally, the Pip B post-treatment caused an alteration in the distribution of the arylated 44 kD protein with a decrease in the microsomes and an increase in the cytosol.

The present findings provide evidence for the earlier suggestion (Gillette, 1974) that a critical determinant of hepatotoxicity may be covalent binding to specific protein targets which represent only a small portion of total protein arylation. Similarly, we have previously shown that the age-dependent difference in APAP hepatotoxicity between two and three month old mice was associated with differences in selective protein arylation without detectable differences in total covalent binding (Beierschmitt, et al., 1989). In that report we cautioned against questioning the role of covalent binding in toxicity simply on the basis of dissociation of total binding from the extent of liver damage. The present study reinforces that caution.

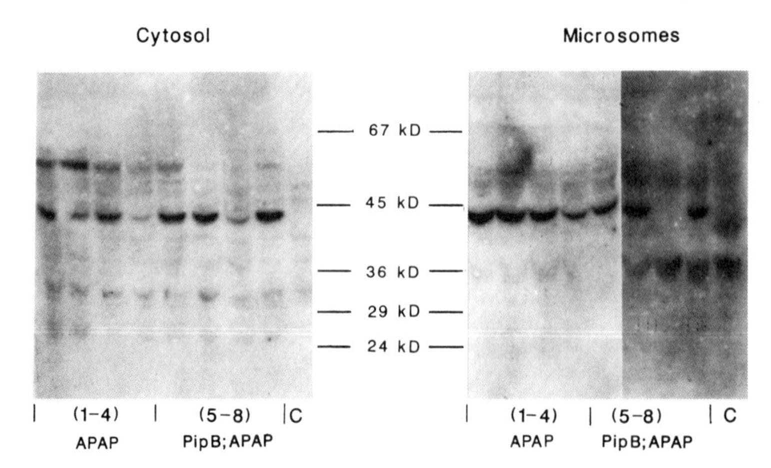

Figure 1. Mice were given 600 mg APAP/kg, po, followed by 600 mg Pip B/kg or corn oil vehicle, 5 ml/kg, ip, 2 hr later. Mice were killed 4 hr after APAP dosing and livers collected for 1-D electrophoresis, western blotting and immunochemical determination of covalent binding in microsomes and cytosol. Lanes 1-4 represent 4 individual mice that received APAP and corn oil vehicle. Lanes 5-8 represent 4 individual animals that received APAP and Pip B.

The present findings also suggest that Pip B's post-treatment protection was due to interruption of APAP covalent binding to "critical" cellular targets which may be found in the 58 kD band. When considered along with other evidence associating selective 58 kD binding with toxicity, it seems likely that the binding pattern changes observed with the protection may be mechanistically important. For example, the age dependancy for APAP hepatotoxicity in mice (Beierschmitt, et al., 1989) was associated with more 58 kD arylation in the sensitive mice, and that 58 kD binding was greater in mice given an hepatotoxic dose of APAP than in those given a less than hepatotoxic dose (Bartolone, et al., 1987). Similarly, hepatocytes in culture were resistant to 2,6-dimethyl APAP where only minimal 58 kD binding was detected. However, phenobarbital induction sensitized the cells and the resultant toxicity was accompanied by 58 kD binding (Birge, et al., 1989). In the same study, the induction increased APAP toxicity with an associated increase in 58 kD arylation. Lastly, APAP-induced necrosis has been observed in the lungs, liver and kidneys of mice and the only major APAP target arylated in all 3 tissues at organ-toxic doses was at 58 kD (Placke, et al., 1987; Bartolone, et al., 1989).

In summary, the diminution in APAP hepatotoxicity caused by Pip B post-treatment was associated with changes in selective protein arylation by APAP but there were no detectable changes in total covalent binding. Existing evidence suggests that a critical target which is mechanistically involved in APAP hepatotoxicity may be located in the 58 kD protein band of mouse liver cytosol.

ACKNOWLEDGEMENTS

Supported in part by the Center for Biochemical Toxicology and NIH grants ESO7163 and GM31460 and an ICI Americas Predoctoral Fellowship in Toxicology to Joseph T. Brady.

REFERENCES

Bartolone, J.B., Sparks, K., Cohen, S.D. and Khairallah, E.A. (1987). Immunochemical detection of acetaminophen bound liver proteins. *Biochem. Pharmacol.* **36**, 1193-1196.

Bartolone, J.B., Birge, R.B., Sparks, K., Cohen, S.D. and Khairallah, E.A. (1988). Immunochemical analysis of acetaminophen covalent binding to proteins: Partial characterization of the major acetaminophen-binding liver proteins. *Biochem. Pharmacol.* **37**, 4763-4774.

Bartolone, J.B., Beierschmitt, W.P., Birge, R.B., Emeigh, Hart, S.G., Wyand, D.S., Cohen, S.D. and Khairallah, E.A. (1989). Selective acetaminophen metabolite binding to extrahepatic proteins: An in vivo and in vitro analysis. *Toxicol. Appl. Pharmacol.* **99**, 240-249.

Beierschmitt, W.P., Brady, J.T., Bartolone, J.B., Wyand, D.S., Khairallah, E.A. and Cohen, S.D. (1989). Selective protein arylation and the age-dependency of acetaminophen hepatotoxicity in mice. Toxicol. Appl. Pharmacol. **98**, 517-529.

Birge, R.B., Bartolone, J.B., McCann, D.J., Mangold, J.B., Cohen, S.D. and Khairallah, E.A. (1989). Selective protein arylation and 2,6-dimethyl acetaminophen in cultured mouse hepatocytes from phenobarbital-induced and uninduced mice: relationship to cytotoxicity. *Biochem. Pharmacol.* **31**, 3745-3749.

Birge, R.B., Bartolone, J.B., Nishanian, E.V., Bruno, M.K., Mangold, J.B., Cohen, S.D. and Khairallah, E.A. (1988). Dissociation of covalent binding from the oxidative effects of acetaminophen: studies using dimethylated acetaminophen derivatives. *Biochem. Pharmacol.* **38**, 4429-4439.

Brady, J.T., Montelius, D.A., Beierschmitt, W.P., Wyand, D.S., Khairallah, E.A. and Cohen, S.D. (1988). The effect of piperonyl butoxide post-treatment on acetaminophen hepatotoxicity. *Biochem. Pharmacol.* **37**, 2097-2099.

Dahlin, D.C., Miwa, G.T., Lu, A.Y.H. and Nelson, S.D. (1984). N-acetyl-p-benzoquinone imine: a cytochrome P-450-mediated oxidation product of acetaminophen. *Proc. Nat. Acad. Sci.* **81**, 1327-1331.

Gillette, J.R. (1974). A perspective on the role of chemically reactive metabolites of foreign compounds in toxicity: I. Correlation of changes in covalent binding of reactive metabolites with changes in the incidence and severity of toxicity. *Biochem. Pharmacol.* **23** 2785-2794.

Ginsberg, G.L. and Cohen, .SD. (1985). Plasma membrane alterations andcovalent binding to hepatic organelles after an hepatotoxicdose of acetaminophen (APAP). *The Toxicologist* **5**, 154.

Jollow, D.J., Mitchell, J.R., Potter, W.Z., Davis, D.C., Gillette, J.R. and Brodie, B.B. (1973). Acetaminophen-induced hepatic necrosis. II. Role of covalent binding *in vivo*. *J. Pharmacol. Exp. Ther.* **187**, 185-194.

Mitchell, J.R., Jollow, D.J., Potter, W.Z., Davis, D.C., Gillette, J.R., and Brodie, B.B. (1973a). Acetaminophen-induced hepatic necrosis. I. Role of drug metabolism. *J. Pharmacol. Exp. Therap.* **187**, 185-194.

Mitchell, J.R., Jollow, D.J., Potter, W.Z., Gillette, J.R. and Brodie, B.B. (1973b). Acetaminophen-induced hepatic necrosis. IV. Protective role of glutathione. *J. Pharmacol. Exper. Therap.* **187**, 211-217.

Placke, M.E., Wyand, D.S. and Cohen, S.D. (1987). Extrahepatic lesions induced by acetaminophen in the mouse. *Toxicologic Pathol.* **15**, 381-387.

Potter, W.Z., Davis, D.C., Mitchell, J.R., Jollow, D.J., Gillette, J.R. and Brodie, B.B. (1973). Acetaminophen-induced hepatic necrosis. III. Cytochrome-P450 mediated covalent binding *in vitro*. *J. Pharmacol. Exp. Ther.* **187**, 203-210.

Potter, W.Z., Thorgeirsson, S.S., Jollow, D.J. and Mitchell, J.R. (1974). Acetaminophen-induced hepatic necrosis. V. Correlation of hepatic necrosis, covalent binding and glutathione depletion in hamsters. *Pharmacol.* **12**, 129-143.

COVALENT BINDING OF A HALOTHANE METABOLITE AND NEOANTIGEN PRODUCTION IN GUINEA PIG LIVER SLICES

Alan P. Brown, Kenneth L. Hastings, A. Jay Gandolfi, and Klaus Brendel

Department of Anesthesiology
University of Arizona
College of Medicine
Tucson, Arizona 85724

Halothane hepatitis is a rare and potentially fatal consequence to the use of this anesthetic. The physiological basis of the disease appears to be an immune response to neoantigens formed by the covalent binding of halothane metabolites to liver protein. Liver slices were used to study the condition for halothane associated neoantigen formation *in vitro*. Liver slices, (1 cm diameter, 300 μm thick) from male Hartley guinea pigs (600 g) were exposed to either 1.0 or 1.7 mM halothane (media concentration) in 95% O_2/5% CO_2 for 12 hr. Covalent binding was determined using ^{14}C-halothane. Neoantigens were detected by western immunoblot assay using rabbit anti-trifluoroacetylated albumin antiserum. Covalent binding was detected by 1 hr of incubation and increased linearly through 12 hr. Covalent binding preceeded and correlated with the appearance of neoantigen. By 12 hr of incubation, 5 neoantigens were seen with molecular weights ranging from 51-97 kD. These neoantigens have molecular weights similar to those seen *in vivo*. This *in vitro* model system can be used to examine the mechanism for covalent binding and neoantigen production in the hepatocyte.

INTRODUCTION

The anesthetic halothane can be oxidatively biotransformed by the liver cytochrome P-450 system to produce a highly reactive intermediate, trifluoroacetyl chloride, capable of covalently binding to liver protein. This intermediate can react with lysine to produce trifluoroacetyl-N-e-amino lysine (TFA-lysine) which can act as an epitope to alter the immunogenicity of native liver proteins. This results in production of neoantigens capable of eliciting an immune response. A hypersensitivity reaction may follow leading to an allergic hepatitis, which is potentially fatal (Pohl et al, 1988).

Previous studies, involving the use of human liver samples as well as rabbits and rats exposed to halothane *in vivo* have demonstrated the presence of liver neoantigens reactive with anti-TFA antibodies. Studies using rat liver microsomes exposed *in vitro* have also produced similar neoantigens (Kenna et al, 1987; Kenna et al, 1988; Kenna, Neuberger, Williams, 1988; Roth et al, 1988).

An *in vitro* liver slice system was used to exposed viable guinea pig liver cells to halothane. Halothane biotransformation has previously been demonstrated in this system (Ghantous et al, 1989a; Ghantous et al; 1989b). Studies were done to determine if covalent binding of a reactive intermediate would occur, leading to neoantigen

Biological Reactive Intermediates IV, Edited by C.M. Witmer *et al.*
Plenum Press, New York, 1990

production. Western immunoblot analysis using anti-TFA antibodies were used to screen for neoantigen presentation. Liver proteins were assayed for covalently bound halothane metabolites by ^{14}C-radiolabeling. The role of the oxidative pathway in neoantigen presentation was studied using deuterated halothane. Deuterated halothane, which is oxidatively metabolized to a much smaller extent, should inhibit the formation of neoantigens.

MATERIALS AND METHODS

Animals

Adult male outbred Hartley guinea pigs, 600-650 g, were used (Sasco, Inc.).

In Vitro Liver Slice Exposure

Cores were taken (1 cm) from various areas of the liver lobes and slices (30-35 mg wet weight, 250-300 mm thick) were prepared using a Krumdieck tissue slicer. Liver slices were incubated in glass scintillation vials (3 per vial) on stainless steel mesh cylinders circumscribed with 2 wheels. Liver slices were incubated at 37R C in Krebs-Henseleit buffer (supplemented) in a 95:5 O_2:CO_2 atmosphere. After a 1 hr pre-incubation, either halothane (Abbott Laboratories), deuterated halothane (8) or 1-^{14}C-halothane (New England Nuclear) were injected into the vials and allowed to vaporize to produce a media concentration of 1.7 mM. Control and exposed slices were taken at various time points (1, 3, 6, and 12 hr) and sonicated in water to produce a whole liver cell homogenate (6, 7).

Radiolabeled Covalent Binding Assay

^{14}-C-halothane (0.5 mCi) was injected into the vials to produce a media concentration of either 1.0 or 1.7 mM. Liver proteins were precipitated, washed extensively with ethanol followed by trichloroacetic acid, and then dissolved in 1 N NaOH. Aliquots were analyzed for radioactivity and protein content.

Western Immunoblot Analysis

Proteins were precipitated with ethanol and separated on 12% SDS polyacrylamide mini gels with a 4% stacking gel (BIO RAD). Proteins were transferred to polyvinylidene difluoride membranes (Millipore) and incubated with rabbit anti-trifluoroacetylated-rabbit serum albumin antibody. Specifically bound antibody was detected by incubating the membrane with biotinylated anti-rabbit antiserum followed by an avidin-biotin-peroxidase complex (Vector Laboratories). The membranes were developed using the peroxidase substrate diaminobenzidine/$NiCl_2$. Molecular weight determinations were performed using biotinylated molecular weight standards transferred onto the blotting membrane (Della-Penna, Christoffersen, Bennett, 1986).

RESULTS

Studies using ^{14}C-halothane demonstrate that covalent binding of a halothane intermediate occurs by 1 hr of incubation and increase linearly to 12 hr. Covalent binding was seen to be dose dependent (Figure 1). Western immunoblot analysis of halothane exposed liver slice show formation of 5 neoantigens developing progressively from 1 to 12 hr of incubation where presentation is greatest (Figure 2). These neoantigens have molecular weights of 97 kD, 62 kD, 57 kD, 54 kD, and 51 kD. These results correlate with the linear increase in covalent binding over the same time period. Exposure to deuterated halothane did not produce neoantigens over a 12 hr incubation (Figure 3). This should be expected due to the decrease in total oxidative metabolism of the compound.

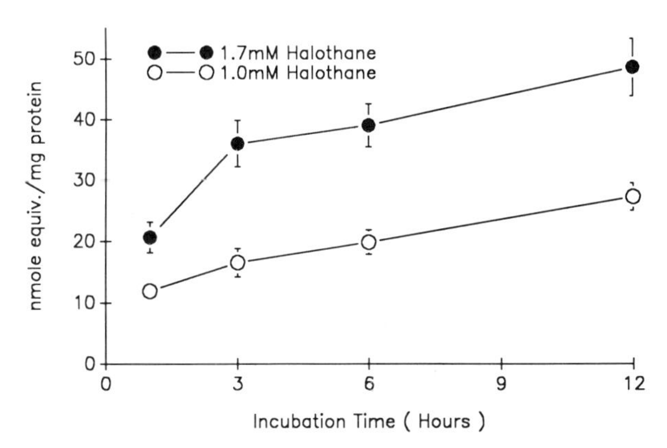

Figure 1. Covalent binding of halothane to guinea pig liver proteins. Guinea pig liver slices were exposed to 1.0 or 1.7 mM ^{14}C-halothane (0.5 mCi) in 95% O_2. Covalent binding is expressed in nanomole equivalents/mg protein. N=24-30 slices from 3 animals. Values are means ± standard error of the mean.

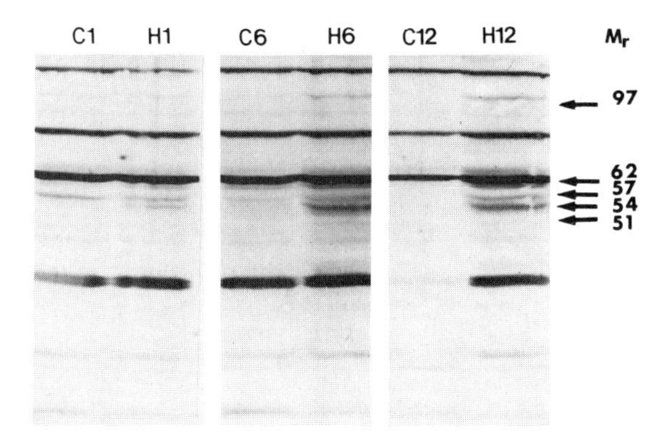

Figure 2. Neoantigen formation in guinea pig liver slices exposed to halothane. Guinea pig liver slices were incubated for 1, 6, and 12 hr with 1.7 mM halothane and taken for western immunoblot analysis. C=control.

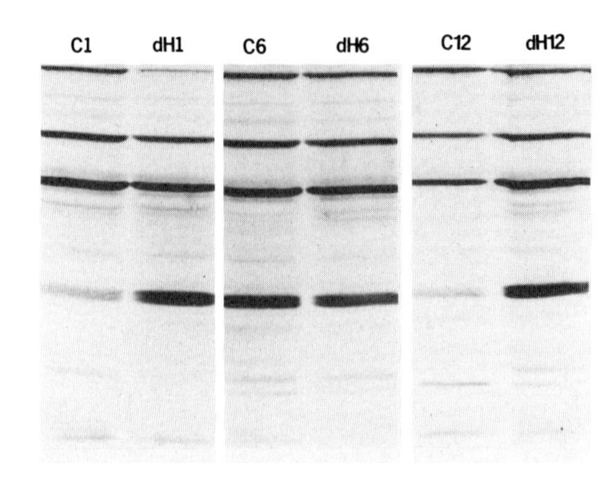

Figure 3. Neoantigen formation in guinea pig liver slices exposed to deuterated halothane. Guinea pig liver slices were incubated for 1, 6, and 12 hr with 1.7 mM d-halothane and taken for western immunoblot analysis. C=control.

DISCUSSION

In this *in vitro* model system, halothane is oxidatively metabolized to produce a reactive intermediate, presumably trifluoroacetyl chloride. This leads to covalent covalent binding to guinea pig liver proteins to produce a hapten capable of altering protein antigenicity. Covalent binding appears to precede the formation of neoantigens reactive with anti TFA-RSA rabbit antibodies. Oxidative metabolism is required for neoantigen presentation. The neoantigens produced in this *in vitro* system correspond to those seen in other *in vivo* studies (2, 3, 4, 5). This model system can be used for mechanistic studies of covalent binding and neoantigen production in liver tissue upon exposure to the anesthetic halothane. Future studies can include other compounds suspected of eliciting an idiosyncratic hypersensitivity response in the liver.

REFERENCES

Della-Penna, D., Christoffersen, R.E., Bennett, A.B. (1986). Biotinylated proteins as molecular weight standards on western blots. *Analytical Biochem.* **152,** 329-332.

Ghantous, H.N., Fernando, J., Gandolfi, A.J., Brendel, K. (1989). Toxicity of halothane in guinea pig liver slices. *Toxicology*, accepted.

Ghantous, H.N., Fernando, J., Gandolfi, A.J., Brendel, K. (1989). Biotransformation of halothane in guinea pig liver slices. *Drug Metab.*, accepted.

Kenna, J.G., Neuberger, J., Williams, R. (1987). Identification by immunoblotting of three halothane-induced liver microsomal polypeptide antigens recognized by antibodies in sera from patients with halothane associated hepatitis. *J. Pharm. Exp. Ther.* **242,** 733-740.

Kenna, J.G., Satoh, H., Christ, D.D., Pohl, L.R. (1988). Metabolic basis for a drug hypersensitivity: Antibodies in sera from patients with halothane hepatitis recognize liver neoantigen that contain the trifluoroacetyl group derived from halothane. *J. Pharm. Exp. Ther.* **245,** 1103-1109.

Kenna, J.G., Neuberger, J., Williams, R. (1988). Evidence for expression in human liver of halothane-induced neoantigens recognized by antibodies in sera from patients with halothane hepatitis. *Hepatology* **8,** 1635-1641.

Lind, R.C., Gandolfi, A.J., Hall, P.M. (1990). Covalent binding of oxidative biotransformation intermediates is associated with halothane hepatotoxicity in guinea pigs. Submitted, *Anesthesiology*.

Pohl, L.R., Satoh, H., Christ, D.D., Kenna, J.G. (1988). The immunological and

metabolic basis of drug hypersensitivities. *Ann. Rev. Pharmacol.* **28**, 367-387.
Roth, T.P., Hubbard, A.K., Gandolfi, A.J., Brown, B.R. (1988). Chronology of halothane-induced antigen expression in halothane exposed rabbits. *Clin. Exp. Immunol.* **72**, 330-336.

EXTENSIVE ALTERATION OF GENOMIC DNA AND RISE IN NUCLEAR Ca^{2+} *IN VIVO* EARLY AFTER HEPATOTOXIC ACETAMINOPHEN OVERDOSE IN MICE[1]

Sidhartha D. Ray, Christopher L. Sorge, Asadollah Tavacoli, Judy L. Raucy, and George B. Corcoran

Toxicology Program, College of Pharmacy
University of New Mexico
Albuquerque, NM 87131

Much attention has fallen upon Ca^{2+} and the critical role it appears to play in the process of lethal cell injury (Schanne et al., 1979; Moore, 1980; Jewell et al., 1982). Ca^{2+} appears to serve as the principal intracellular messenger that conveys initial damage arising from alkylation or peroxidation to discrete secondary sites that are essential to cell viability. Nonetheless, several aspects of the Ca^{2+} hypothesis of cell death remain elusive, including the location of initially damaged Ca^{2+} regulatory sites important to cell death, and the actual identity of the vital function or functions that deteriorate under excessive Ca^{2+} activity. Cell death is believed to occur via two distinct processes; apoptosis, or the programmed cell death seen during physiological events such as organ development and cell renewal, and necrosis, or the unprogrammed cell death that follows substantive pathologic insult (Wyllie, 1980; Duvall and Wyllie, 1986). A number of the steps leading to apoptosis in immature thymocytes are well defined. These include Ca^{2+} influx into the cell, endonuclease activation in the nucleus, and DNA degradation into periodic fragments not seen in necrosis (Cohen and Duke, 1984; McConkey et al., 1988a). The present study assesses whether acetaminophen-induced liver necrosis may share certain key steps in common with the process of apoptosis. Toxic doses of acetaminophen are known to cause Ca^{2+} to accumulate in liver and cytosolic Ca^{2+} activity to rise within 2 hr (Corcoran et al., 1987, 1988). We now examine the nucleus for lethal actions of unchecked Ca^{2+} activity in acetaminophen-induced necrosis, and we specifically monitor genomic DNA as a critical secondary target in this form of cell death.

MATERIALS AND METHODS

NIH Swiss mice (25-35g, Harlan S-D, Indianapolis, IN) received chow and tap water *ad libitum* in a controlled environment for one week before study. Acetaminophen, alanine aminotransferase kit (#59-UV), diphenylamine, proteinase-K, RNase 'A', sarkosyl, sodium secobarbital, tris-HCl and Triton X-100 were from Sigma (St. Louis, MO), Ca^{2+} determination kit #351130 and sucrose from Boehringer Mannheim (Indianapolis, IN) and agarose from Bio-Rad (Richmond, CA). Mice given 600 mg/kg acetaminophen ip were sacrificed to collect plasma for alanine aminotransferase (EC 2.6.1.2) determination based upon Wroblewski and LaDue (1956), and to remove livers for immediate liquid N_2 freezing. Hepatocyte nuclei were isolated by sucrose density centrifugation and verified for purity with trypan blue (Chauveau et al., 1956; Hewish and Burgoyne, 1973). Nuclear pellets ultrasonicated in 1% HCl were held 18 hr at 22°C and centrifuged. Ca^{2+} measured in supernatant fractions was confirmed by atomic absorption (Ray Sarkar and Chouhan, 1973; Baginski

Biological Reactive Intermediates IV, Edited by C.M. Witmer *et al.*
Plenum Press, New York, 1990

et al., 1973). Intact and fragmented DNA in nuclei or liver homogenates were separated by differential sedimentation (Wyllie, 1980) and quantitated colorimetrically (Burton, 1956). DNA isolation for electrophoresis followed the same procedure except the homogenization medium contained 5 mM EDTA. Nuclear pellets were treated with 25 mM sodium citrate buffer pH 7 containing 0.5% sarkosyl, 4 M guanidinium thiocyanate, and 0.1 M 2-mercaptoethanol for 15-20 min at 0-4°C. Some nuclei and homogenates were treated with proteinase-K (45 min, 37°C) and later RNase 'A.' DNA extracted with phenol-chloroform and precipitated with ethanol/sodium acetate (Sambrook et al., 1989) was quantified fluorometrically (Hoechst dye B-33258, standard calf thymus DNA), and 4-7 μg DNA/lane were separated by electrophoresis (1.6% agarose gels, 0.4 μg/ml ethidium bromide, Hoefer Instruments, San Francisco, CA). Results are Mean ± SEM. Analysis included ANOVA with Fisher PLSD comparison or Mann-Whitney nonparametric comparison, and linear regression and correlation (Zar, 1984). Differences were attributed to treatment rather than chance variation when $p < 0.05$.

RESULTS AND DISCUSSION

This study of acetaminophen hepatotoxicity examined whether some aspects of the highly integrated process of drug-induced cell necrosis *in vivo* might resemble steps in cell death by apoptosis *in vitro*. A 600 mg/kg dose of acetaminophen produced >100 fold increases in alanine aminotransferase activity from 4 hr onward, significant accumulation of total Ca^{2+} (bound plus free) in the nucleus between 2-3 hr, and sustained increases in the fragmentation of nuclear DNA from 2 hr onward (Table 1). Ca^{2+} accumulation, DNA fragmentation, and decreased nuclear DNA recovery all appeared prior to extensive hepatocellular damage. The timing of these events suggests that faltering Ca^{2+} regulation throughout the cell, but particularly in the nucleus, resulting in substantial DNA changes. One component appears to be activation of an endonuclease very early during necrosis, analogous to what has been described for apoptosis *in vitro* (Wyllie, 1980; Duvall and Wyllie, 1986; Cohen and Duke, 1984; McConkey et al., 1988a). Regression analysis examining the relatedness of these events showed strong linear correlations between nuclear Ca^{2+} accumulation and liver damage (Figure 1A), between DNA fragmentation and liver damage (Figure 1B), and between DNA fragmentation and nuclear Ca^{2+} accumulation (Figure 1C). Results of DNA separation by electrophoresis gave a picture of changes that was qualitatively similar to sedimentation analysis but quantitatively quite different from results reported in Table 1. By 2 hr, slowly migrating large DNA had diminished substantially in isolated nuclei (Figure 2A) and whole liver homogenates (Figure 2B) while the DNA fragments that accumulated often appeared in a ladder-like pattern. These findings suggest activation of a constitutive endonuclease in mouse liver nuclei by supraphysiologic Ca^{2+} concentrations prior to attack upon intranucleosomal regions of genomic DNA. McConkey and colleagues (1988b) describe a similar stimulation of endogenous endonuclease activity in isolated hepatocytes killed by oxidative stress. Specific knowledge of nuclear Ca^{2+} activity would be more informative than our measurement of total nuclear Ca^{2+} concentration during acetaminophen injury. Nonetheless, reproducible early increases in total nuclear Ca^{2+} coincided with evidence of endonuclease activation, reflected in DNA fragmentation (Table 1; Figure 1,2). Ca^{2+} homeostasis seems to be under intensive regulation in the nucleus as elsewhere throughout the cell. The concentration gradient between nucleus and cytoplasm (Williams et al., 1988) implies the presence of structures that transport Ca^{2+} out of the nuclear compartment. Skeletal muscle nuclei exhibit Ca^{2+}-stimulated ATPase activity (Kulikova et al., 1982). Isolated liver nuclei accumulate Ca^{2+} *in vitro* via an ATP and calmodulin dependent process, and contain a constitutive endonuclease that cleaves DNA into intranucleosomal fragments in response to submicromolar Ca^{2+} concentrations (Jones et al., 1989). In order for us to have observed significant accumulation of Ca^{2+} in the nucleus and fragmentation of DNA by acetaminophen, it may have been necessary for acetaminophen to impair both plasma membrane and nuclear membrane pumps which provide for Ca^{2+} extrusion.

Table 1. Time Course of Nuclear Ca^{2+} Accumulation, DNA Fragmentation, and Liver Damage Following Administration of 600 mg/kg Acetaminophen to Mice

Time Hr	Total Nuclear Ca^{2+} μg/g liver	Fragmented DNA (%)	Plasma ALT U/L	Nuclear DNA Recovery[a] mg/g liver
0	7.95 ± 1.84	7.62 ± 0.90	58 ± 3.3	2.94 ± 0.23
1	9.75 ± 2.34	8.82 ± 0.76	56 ± 5.25	2.70 ± 0.23
2	11.6 ± 2.00 ¶	9.82 ± 0.70 *	93 ± 19 *	2.64 ± 0.08
3	11.3 ± 1.69 *	8.50 ± 0.60	356 ± 106 *	2.01 ± 0.33
4	22.9 ± 3.22 *	8.87 ± 0.48 *	4914 ± 1257 *	1.87 ± 0.23 **
6	29.0 ± 6.75 *	9.35 ± 0.69 *	13884 ± 2110 *	1.57 ± 0.30 **
12	32.1 ± 4.28 *	20.5 ± 2.15 *	25348 ± 2317 *	1.34 ± 0.28 **
24	48.9 ± 10.1	14.0 ± 3.78 *	20396 ± 3623 *	1.32 ± 0.49 **

[a] DNA recovery from mice not treated with acetaminophen was 3.42 ± 0.14 mg/g liver.
¶ $p \leq 0.07$ vs time zero by Mann-Whitney nonparametric comparison.
* $p \leq 0.05$ vs time zero by Mann-Whitney nonparametric comparison.
** $p \leq 0.05$ vs time zero by Fisher PLSD test.

In our experiments, overall DNA fragmentation was not extensive at 2-4 hours after acetaminophen overdose. This probably reflects several factors: 1) the number of target cells undergoing DNA fragmentation was small relative to uninjured nontarget cells; 2) the declining yield of DNA from 2 hr onward biased downward the estimates of fragmentation due to loss of the most severely damaged nuclei, and 3) the relatively insensitive DNA sedimentation assay failed to disclose the full nature of changes in large genomic DNA. The latter 2 limitations were obviated in large part by isolating DNA from whole liver homogenates and evaluating it by agarose gel electrophoresis (Figure 2).

During acute chemical injury, sustained increases in cytosolic Ca^{2+} activity can inflict a variety of potentially lethal cellular lesions. These include cytoskeletal alterations leading to plasma membrane blebbing (Jewell et al., 1982), protein-thiol oxidation and interference with mitochondrial function (Moore et al., 1985), activation of phospholipases with resultant membrane damage (Chien et al., 1979), stimulation of Ca^{2+}-dependent neutral proteases (Nicotera et al., 1986), and activation of a constitutive endonuclease in the nucleus (Hewish and Burgoyne, 1973; Chien et al., 1979; Vanderbilt et al., 1982). Which of these lesions or ones yet undiscovered is critical to lethal cellular damage *in vivo* remains unclear. Orrenius and colleagues report that DNA fragmentation is the lesion responsible for thymocyte killing *in vitro* by glucocorticoids (McConkey et al., 1989). Interestingly, Long et al. (1989) find no early breakdown of hepatic DNA *in vivo* or *in vitro* during carbon tetrachloride or dichloroethylene toxicity, prompting their proposal that Ca^{2+} concentration probably increases in the cytosol of liver cells but not sufficiently in the nucleus to activate endonucleases. The present results with acetaminophen in mice differ from those observed with halocarbons in rats. It is certainly possible that DNA fragmentation represents a bystander event during lethal cell injury *in vivo*. However, these contrasting data may also indicate that certain hepatotoxins activate discrete cellular events that may be held in common by the processes of apoptosis and necrosis, particularly if sites of Ca^{2+} control in the nucleus become damaged.

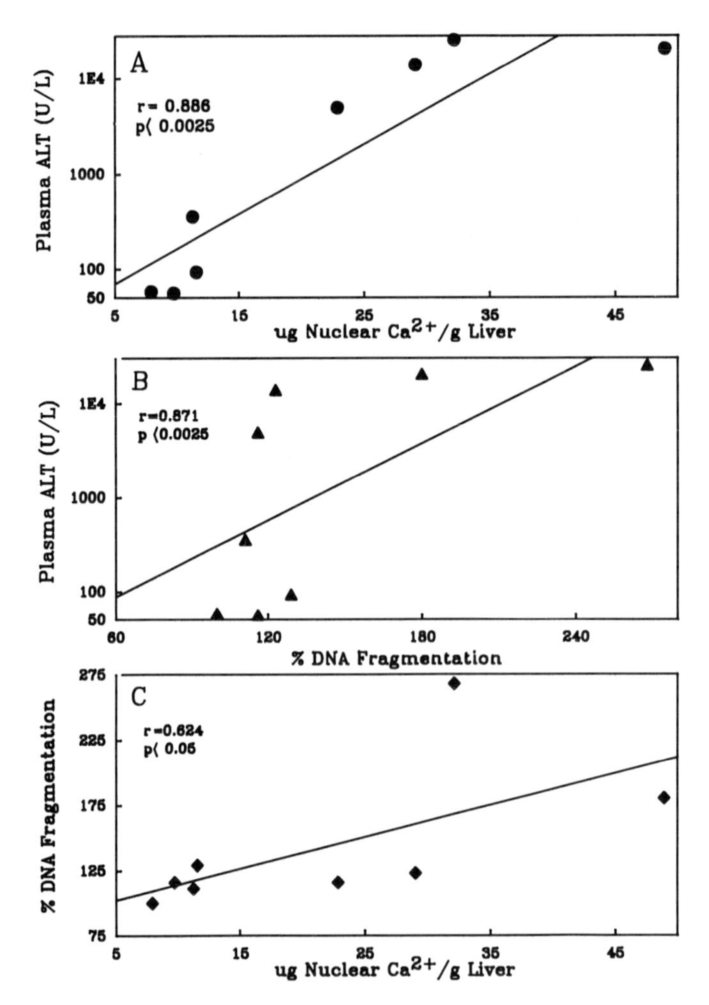

Figure 1. Linear regression analyses of variables thought to be involved in acetaminophen-induced liver necrosis. Parameter estimates for nuclear Ca^{2+} accumulation, DNA fragmentation expressed as a percentage of time zero fragmentation, and liver injury expressed as ALT activity, as reported in Table 1, were regressed against one another ignoring time as a variable. Panel A) Plasma ALT vs Nuclear Ca^{2+} Accumulation; Panel B) Plasma ALT vs. % of Control DNA Fragmentation; Panel C) % of Control DNA Fragmentation vs Nuclear Ca^{2+} Accumulation.

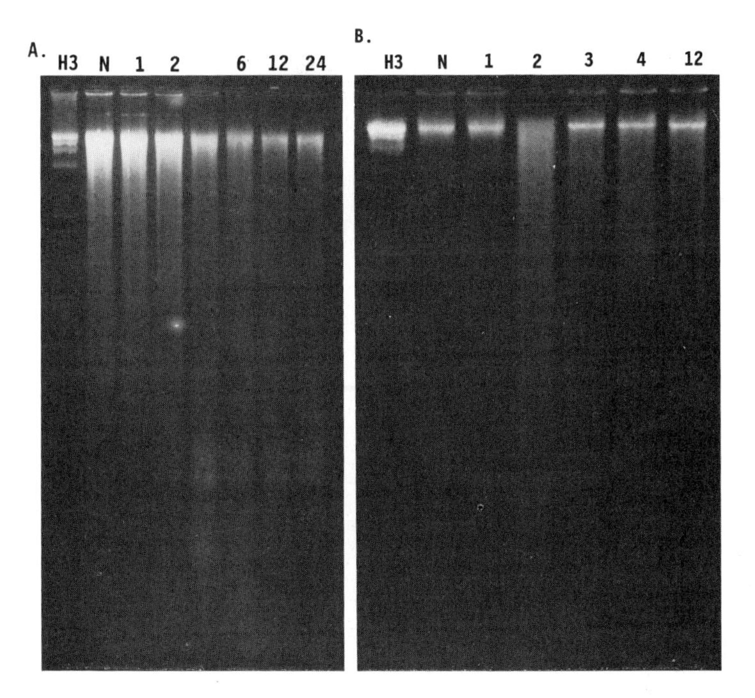

Figure 2. Electrophoretograms demonstrate the time course of acetaminophen-induced damage to hepatocellular genomic DNA. Ethidium bromide stained agarose gels are representative of 4 independent experiments. Each lane contains DNA from one animal at one time point. Loss of large genomic DNA with concomitant appearance of a ladder-like fragmentation pattern typical of apoptosis was observed by 2 hours in isolated nuclei (Panel A, 7 μg DNA/lane) and in total liver homogenates after proteinase-K and RNase 'A' treatment (Panel B, 4 μg DNA/lane). LEGENDS: H3 (molecular weight markers from Hind-III digested λ DNA), N (naive control), and 0, 1, 2, etc. (hours after acetaminophen treatment).

SUMMARY

Hepatotoxic doses of acetaminophen cause early impairment of Ca^{2+} homeostasis. In this *in vivo* study, 600 mg/kg acetaminophen caused total nuclear Ca^{2+} and % fragmented DNA to rise in parallel from 2-6 hr, followed by large later increases mirroring frank liver injury. Agarose gel electrophoresis revealed substantial loss of large genomic DNA from 2 hours onward, with accumulation of DNA fragments in a ladder-like pattern resembling apoptosis. Extensive late cleavage of DNA probably resulted from cell death, whereas degradative loss of large genomic DNA at 2 hours arose at an early enough point to contribute to acetaminophen-induced liver necrosis in mice.

ACKNOWLEDGEMENTS

We thank Ms. Michelle Wood for skilled lab assistance, Dr. Peter C. Simons for sound scientific guidance, and Drs. William Hadley and Daniel Salazar for valuable suggestions to improve the manuscript. This work is supported in part by Grants # GM 41564 and # AA 08139 from the Department of Health and Human Services.

REFERENCES

Baginski, E.S., Marie, S.S., Clark, W.L., and Zak, B. (1973). Direct microdetermination of serum calcium. *Clin. Chim. Acta* **46**, 49-54.

Burton, K. (1956). A study of the conditions and mechanism of the diphenylamine reaction for the colorimetric estimation of deoxyribonucleic acid. *Biochem. J.* **62**, 315-322.

Chauveau, Y., Moule, Y., and Rouiller, C. (1956). Isolation of pure and unaltered liver nuclei: morphology and biochemical composition. *Exp. Cell Res.* **11**, 317-320.

Chien, K.R., Pfau, R.G., and Farber, J.L. (1979). Ischemic myocardial cell injury: prevention by chlorpromazine of an accelerated phospholipid degradation and associated membrane dysfunction. *Am. J. Pathol.* **97**, 505-530.

Cohen, J.J., and Duke, R.C. (1984). Glucocorticoid activation of a calcium)dependent endonuclease in thymocyte nuclei leads to cell death. *J. Immunol.* **132**, 38-42.

Corcoran, G.B., Wong, B.K., and Neese, B.L. (1987). Early sustained rise in total liver calcium during acetaminophen hepatotoxicity in mice. *Res. Comm. Chem. Pathol. Pharmacol.* **58**, 291-305.

Corcoran, G.B., Bauer, J.A., and Lau, D.T. (1988). Immediate rise in intracellular calcium and glycogen phosphorylase a activities upon acetaminophen covalent binding leading to hepatotoxicity in mice. *Toxicology* **50**, 157-167.

Duvall, E., and Wyllie, A. (1986). Death and the cell. *Immunol. Today* **7**, 115-119.

Hewish, D.R., and Burgoyne, L.A. (1973). The calcium dependent endonuclease activity of isolated nuclear preparations. Relationships between its occurrence and the occurrence of other classes of enzymes found in nuclear preparations. *Biochem. Biophys. Res. Comm.* **52**, 475-481.

Jewell, S.A., Bellomo, G., Thor, H., Orrenius, S., and Smith, M.T. (1982). Bleb formation in hepatocytes during drug metabolism is caused by disturbances in thiol and calcium homeostasis. *Science* **217**, 1257-1259.

Jones, D.P., McConkey, D.J., Nicotera, P., and Orrenius, S. (1989). Calcium-activated DNA fragmentation in rat liver nuclei. *J. Biol. Chem.* **264**, 6398-6403.

Kulikova, O.G., Savostianov, G.A., Beliavsteva, L.M., and Razumov,skaia, N.I. (1982). ATPase activity and ATP)dependent accumulation of Ca^{2+} in skeletal muscle nuclei. Effects of denervation and electrical stimulation. *Biokhimiia* **47**, 1216-1221.

Long, R.M., Moore, L., and Schoenberg, R. (1989). Halocarbon hepatotoxicity is not initiated by CaC?C2+D?Dstimulated endonuclease activation. *Toxicol. Appl. Pharmacol.* **97**, 350-359.

McConkey, D.J., Hartzell, P., Duddy, S.K., Hakansson, H., and Orrenius, S. (1988a). 2,3,7,8) Tetrachlorodibenzo-p-dioxin (TCDD) kills immature thymocytes by Ca^{2+} mediated endonuclease activation. *Science* **242**, 256-259.

McConkey, D.J., Hartzell, P., Nicotera, P., Wyllie, A.H., and Orrenius, S. (1988b). Stimulation of endogenous endonuclease activity in hepatocytes exposed to oxidative stress. *Toxicol. Lett.* **42**, 123-130.

McConkey, D.J., Hartzel, P., Nicotera, P., and Orrenius, S. (1989). Calcium-activated DNA fragmentation kills immature thymocytes. *FASEB J.* **3**, 1843-1849.

Moore, L. (1980). Inhibition of liver microsomal calcium pump by *in vivo* administration of carbon tetrachloride, chloroform, and 1,1-dichloroethylene. *Biochem. Pharmacol.* **29**, 2505-2511.

Moore, M., Thor, H., Moore, G., Nelson, S.D., Moldeus, P., and Orrenius, S. (1985). The toxicity of acetaminophen and N-acetyl-*p*-benzoquinone imine in isolated hepatocytes is associated with thiol depletion and increased cytosolic Ca^{2+}. *J. Biol. Chem.* **260**, 13035-13040.

Nicotera, P., Hartzell, P., Baldi, C., Svenson, S.)A., Bellomo, G., and Orrenius, S. (1986). Cystamine induces toxicity in hepatocytes through the elevation of cytosolic Ca^{2+} and the stimulation of a non-lysosomal proteolytic system. *J. Biol. Chem.* **261**, 14628-14635.

Ray Sarkar, B.C., and Chouhan, U.P.S. (1967). A new method for determining micro quantities of calcium in biological materials. *Anal. Biochem.* **20**, 155-166.

Sambrook, J., Fritsch, E.F., and Maniatis, T. (1989). Molecular Cloning, A Laboratory Manual, 2nd ed, Cold Spring Harbor Laboratory, New York.
Schanne, F.A.X., Kane, A.B., Young, E.E., and Farber, J.L. (1979). Calcium dependence of toxic cell death: a final common pathway. *Science* **206**, 700-702.
Vanderbilt, J.N., Bloom, K.S., and Anderson, J.N. (1982). Endogenous nuclease. Properties and effects on transcribed genes in chromatin. *J. Biol. Chem.* **257**, 13009-13017.
Williams, D.A., Becker, P.L., and Fay, F.S. (1988). Regional changes in calcium underlying contraction of single smooth muscle cells. *Science* **235**, 1644-1648.
Wroblewski, F., and LaDue, J. (1956). Serum glutamic)pyruvic transaminase in cardiac and hepatic disease. *Proc. Soc. Exp. Biol. Med.* **91**, 569-571.
Wyllie, A.H. (1980). Glucocorticoid)induced thymocyte apoptosis is associated with endogenous endonuclease activation. *Nature* **284**, 555-556.
Zar, J.H. (1984). *Biostatistical Analysis*, 2nd ed, Prentice-Hall, Englewood Cliffs, New Jersey.

THE POSSIBLE ROLE OF GLUTATHIONE ON THE HEPATOTOXIC EFFECT OF PAPAVERINE HYDROCHLORIDE *IN VITRO*

Julio C. Davila*, Daniel Acosta*, and Patrick J. Davis**

Department of Pharmacology and Toxicology* and Department of Medicinal Chemistry**, College of Pharmacy
The University of Texas
Austin, Texas 78712

Papaverine hydrochloride (papaver) is an isoquinoline alkaloid widely used as a smooth muscle relaxant agent. It is metabolized mainly in the liver by the P-450 system (O-demethylation) to yield several phenolic metabolites (4'-,6-, 7-, 3'-, and 4',6-demethylated phenols) (Belpaire et al., 1975). Papaver induces hepatotoxicity by an idiosyncratic immune effect (Kiaer et al., 1974) or a direct metabolite-related hepatocellular toxic effect (Acosta et al., 1980; Davila et al., 1989). At the present time, however, there is no direct evidence which suggests that toxic metabolites derived from papaver are responsible for liver cell injury. We had demonstrated previously that a time lag of 6 hr was necessary before papaver injured cultured hepatocytes and that the papaver-phenolic metabolites were considerably less toxic than the parent compound. They were ranked in toxicity as follows: papaver>6OH>4'OH>3'OH (Acosta et al., 1980; Davila et al., 1989). According to these results, it seems that the metabolite(s) is (are) not the major cause of liver injury. It is interesting to note, however, that certain principal metabolites of papaver are nearly as toxic as the parent compound, while others clearly represent detoxification pathways; therefore, we can not rule out the possibility that papaver-derived intermediate metabolite(s) is (are) involved in the toxic insult. Some mechanistic alternatives are suggested: 1) papaver itself may be directly responsible for cellular toxicity or be metabolized to toxic products by a pathway not involving O-dealkylation; 2) phenol formation (e.g. at 4'-position) may require an additional O-dealkylation or aromatic hydroxylation to yield a catechol which may undergo oxidation to a reactive ortho-quinone; 3) quinone-methides may be formed following O-dealkylation (Figure 1); and 4) the time lag of 6 hr may represent the time required for bioconversion and accumulation of sufficient quantities of toxic metabolites and/or saturate and deplete intracellular glutathione stores.

In view of these results, it was of interest to investigate the role of glutathione and lipid peroxidation in explaining the hepatotoxicity by papaver in cultured cells. As indicated in Figure 2, hepatocytes demonstrated a concentration- and time-dependent decrease in reduced glutathione (GSH) levels. GSH was significantly reduced after 8 hr treatment with papaver at 10^{-3} M and 12 hr at 10^{-4} M. In addition, the administration of buthionine sulphoximine (BSO) (glutathione synthesis inhibitor) to the cultured cells, enhanced the toxicity of papaver as early as 2 hr and damaged most of the cells by 4 hr (Table 1). We had previously shown that papaver was toxic to the hepatocytes at 10^{-4} M for 8 hr (Davila et al., 1989). Furthermore, the malondialdehyde formation did not show significant differences from control, indicating that lipid peroxidation may not be involved in the toxic insult by papaver (data not shown). In conclusion, changes in intracellular GSH levels may have a role in

Biological Reactive Intermediates IV, Edited by C.M. Witmer *et al.*
Plenum Press, New York, 1990

Table 1. Liver cell cultures pretreated with buthionine sulphoximine (BSO) for 24 hr and then treated with papaverine hydrochloride (papaver) at 10^{-5} M, 10^{-4} M and 10^{-3} M for 2 and 4 hr. Data are represented as mean ± S.E. (n=4). Asterisks indicate significant differences from control, ($P<0.05$). Assays were performed in duplicate.

Compound	Conc (M)	LDH release units/mg protein 2 hr	LDH release units/mg protein 4 hr
BSO	control	50 ± 1	51 ± 1
	10^{-5}	50 ± 1	56 ± 1
	10^{-4}	155 ± 2*	190 ± 2*
	10^{-3}	290 ± 2*	385 ± 3*

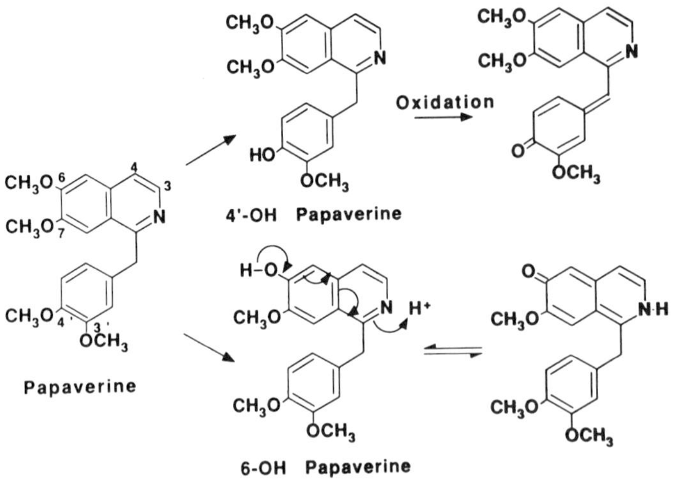

Figure 1. Possible formation of quinone methides from O-dealkylated metabolites of papaverine.

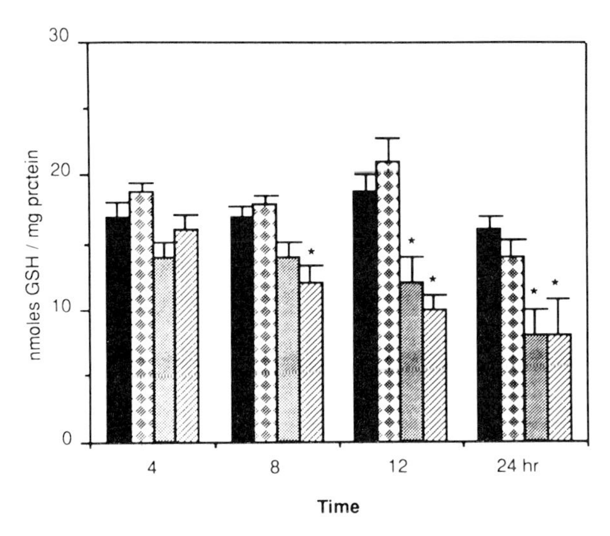

Figure 2. Reduced glutathione (GSH) measured in cultured hepatocytes after 4, 8, 12 and 24 hr exposures to papaverine hydrochloride (papaver) at 10^{-5} M, ▨; 10^{-4} M, ▨ and 10^{-3} M, ▨. Each bar represent the mean ± S.E. (n=4). Asterisks show significant differences when compared to control, ■, ($P<0.05$). Assays were performed in duplicate.

papaver-induced hepatotoxicity. The inhibition of glutathione synthesis by BSO and the increase of cell injury provide evidence for a crucial role of GSH in the mechanism of toxicity. Depletion of intracellular GSH may result in a loss of the balance between metabolic activation and inactivation leading to oxidative stress and then cell damage.

REFERENCES

Belpaire, F.M., Bogaert, M.G., and Rossel, M.T. (1975). Metabolism of papaverine I. Identification of metabolites in rat liver. *Xenobiotica* **5**, 413-420.

Kiaer, H.W., Olsen, S., and Ronnov-Jenssen, V. (1974). Hepatotoxicity of papaverine. *Arch. Pathol.* **98**, 292-296.

Acosta, D., Anuforo, D.C., and Smith, R.V. (1980). Cytotoxicity of acetaminophen and papaverine in primary cultures of rat hepatocytes. *Toxicol. Appl. Pharmacol.*, **53**, 306-314.

Davila, J.C., Hsieh, G.C., Reddy, C.G., Acosta, D., and Davis, P.J. (1989). Toxicity assessment of papaverine hydrochloride and papaverine-derived metabolites *in vitro*. *The Pharmacologist* **31**, 179.

IN VIVO AND *IN VITRO* EVIDENCE FOR *IN SITU* ACTIVATION AND SELECTIVE COVALENT BINDING OF ACETAMINOPHEN (APAP) IN MOUSE KIDNEY

Susan G. Emeigh Hart*,†, Raymond B. Birge‡,†, Richard W. Cartun*, Charles A. Tysonƒ, Jack E. Dabbsƒ, Ervant V. Nishanian†, D. Stuart Wyand*, Edward A. Khairallah‡, and Steven D. Cohen†.

SRI International, Menlo Park, CAƒ, and Toxicology Program: Departments of Pathobiology*, Pharmacology and Toxicology†, and Molecular and Cell Biology‡,
University of Connecticut
Storrs, CT

Acetaminophen (APAP, N-acetyl-p-aminophenol) is a widely used analgesic and antipyretic which, at high doses, causes acute hepatic centrilobular necrosis in man and a variety of laboratory animals (Proudfoot and Wright, 1970; Boyer and Rouff, 1971; Mitchell et al., 1973). In addition, acute renal proximal tubular necrosis following APAP has been reported in man (Kleinman et al., 1980; Cobden et al., 1982; Kher and Makker, 1987; Davenport and Finn, 1988). A similar lesion has been described in the Fischer rat (McMurtry et al., 1978; Newton et al., 1983) andthe CD-1 mouse (Placke et al, 1987) but the APAP metabolite responsible for the toxicity is different between species. In the rat, APAP is enzymatically deacetylated to p-aminophenol, a potent nephrotoxicant whose activation is independent of cytochrome P450 (Crowe et al., 1979; Calder et al., 1979; Newton et al., 1982; 1985a; 1985b). By contrast, enzymatic deacetylation of APAP is not required in the mouse but instead activation of intact APAP by cytochrome P450 appears to mediate nephrotoxicity (Bartolone et al, 1989; Emeigh Hart et al., 1989a; 1989b). Since hepatic metabolism of APAP is similarly dependent on cytochrome P450 (Mitchell et al, 1973), nephrotoxicity and renal adduct accumulation could arise from transport of a liver-generated metabolite or adduct to the kidney and not from *in situ* metabolism of APAP. The purpose of the present study was to determine, using renal proximal tubule (RPT) cell suspensions and immunohistochemistry, if the mouse kidney could generate such adducts *in situ*.

MATERIALS AND METHODS

Three month old, male Crl:CD-1(ICR)BR mice (Charles River Laboratories, Wilmington, MA) were used for all studies. Animals were housed in a temperature and humidity controlled animal facility with a 12 hour light-dark cycle and were acclimatized for two weeks prior to use. Food (Purina Rodent Chow #5001, Ralston Purina Company, St. Louis, MO) and tap water were available *ad libitum* except that all animals were fasted for 18 hours prior to their use in any experiments.

Ammonium persulfate, Coomassie brilliant blue R-250, and 2-mercaptoethanol were from BioRad Laboratories (Richmond, CA); acrylamide and N, N'-methylene-*bis*-acrylamide from Boehringer Mannheim Biochemicals (Indianapolis, IN); nitrocellulose (0.2 μm) from Schleicher and Schuell (Keene, NH); bovine serum albumin (BSA) from Miles Scientific (Naperville, IL); and deferoxamine from the Ciba

Biological Reactive Intermediates IV, Edited by C.M. Witmer *et al.*
Plenum Press, New York, 1990

Pharmaceutical Company (Summit, NJ). All other reagents were from the Sigma Chemical Company (St. Louis, MO) and were of reagent grade.

Mice were given APAP (600 mg/kg, 10 ml/kg, po) in alkaline deionized distilled water or vehicle only and were killed by cervical dislocation 4 hours later. Kidneys were excised and sections were trimmed into plastic cassettes, immersed in 0.9% NaCl at room temperature, and fixed by microwave irradiation at 60° C for 3 minutes. The cassettes were immersed in 50% ethanol for storage until routine processing. Paraffin-embedded sections (5Mm) were placed onto poly-L-lysine coated glass slides and routinely rehydrated. Sections for anti-APAP antibody incubation were digested for 15 minutes in 0.5% pepsin in 0.01 N HCl. All sections were then incubated overnight at 4° C in diluted primary antibody (1:10 for the affinity-purified anti-APAP antibody (Bartolone et al., 1987), 1:250 for the anti-58 kD antibody (Nishanian, 1989) and 1:500 for the anti-mouse liver cytochrome P450 antibody (gift of Dr. J. Crivello, University of Connecticut). Visualization of the bound anti-APAP was by the AS/AP Universal Rabbit Detection System #1035 (BioCan Scientific, Portland, ME); the other antibodies were visualized by peroxidase-antiperoxidase staining (Sternberger et al., 1970) using commercially available reagents (Dako Corporation, Santa Barbara, CA).

RPT cells were isolated from mice by the method of Jones et al. (1979) as modified by Green et al. (1990). After preincubation as described (Tyson et al., 1990b), RPT cells were exposed to either 0, 10 or 25 mM APAP for 4 hr. Incubations were then stopped and cytotoxicity was assessed by net lactate dehydrogenase (LDH) release and measurement of basal and nystatin-stimulated O_2 consumption (Tyson et al., 1990b).

Particulate and soluble fractions were prepared from replicate RPT suspensions and subjected to SDS-PAGE and Western blot analysis as described (Bartolone et al, 1987). Similar fractions were prepared from the kidneys of mice treated with APAP *in vivo* for comparison.

RESULTS AND DISCUSSION

Dose-dependent APAP toxicity was detected in the mouse RPT cell suspensions. Net LDH release increased to 20.2 and 38.8% after a 4 hr incubation with 10 or 25 mM APAP, respectively. By contrast, LDH release in controls after the 4 hr incubation was only 14.3%. Basal O_2 consumption decreased from 58.2 nmol/min-mg protein at the start of the incubation, to 31.2, 23.2 and 6.0 nmol/min- mg protein after a 4 hr exposure to 0, 10 and 25 mM APAP, respectively. Nystatin-stimulated O_2 consumption was 142% of the basal level in control cells but was decreased to 114% and 103% by 10 or 25 mM APAP, respectively.

Western blot analysis of proteins from the RPT suspensions demonstrated selective arylation of proteins with approximate molecular weights of 130, 58, 44 and 33 kD in the soluble fraction, and of 44 kD in the particulate fraction. The selectivity was similar to that seen in similar fractions from kidneys of mice treated with APAP *in vivo*, and the intensity of the arylated bands increased as a function of APAP concentration in the RPT suspensions (Figure 1).

Immunohistochemical staining of kidneys from APAP-treated mice also provided evidence of APAP binding which was restricted to the proximal tubule cells (Figure 2), and cytochrome P450 staining was also most intense in the renal proximal tubule cells, with no detectable staining of the renal medulla (not shown). By contrast, staining for the 58 kD protein demonstrated that it was uniformly distributed throughout all epithelial cells of the kidney (not shown).

The results of this study indicate that the response of mouse RPT cells to APAP differs markedly from that reported for rat RPT cells. In this study, exposure of mouse RPT cells to 10 and 25 mM APAP resulted in a dose-dependent decrease in viability and function, whereas previously no toxicity was observed in rat RPT suspensions exposed to similar APAP concentrations (Jones et al., 1979; Tyson et al.,

1990a). This suggests that *in situ* metabolism of APAP may contribute to nephrotoxicity in mice. This is further supported by the observation that the same kidney proteins were arylated by APAP *in vivo*, and in the RPT suspensions *in vitro*, and that binding increased with increasing APAP concentration as did cytotoxicity.

APAP binding was primarily located in the convoluted portion of the proximal tubule, the segment of the nephron which later undergoes necrosis in the CD-1 mouse Placke et al., 1987; Emeigh Hart et al. 1989a). This distribution is most similar to the distribution of cytochrome P450 both as observed with immunohistochemistry in this study and also as reported by others (Zenser et al., 1978; Foster et al., 1986). The selectivity of adduct formation does not appear to be related to the content of the 58 kD target protein in tissues, since it was homogeneously distributed and also was evident in cell types which had no detectable APAP adducts.

Previous studies of APAP-induced nephrotoxicity in mice have demonstrated that both adduct formation and tubular necrosis can be decreased by inhibition of the activity of cytochrome P450 (Bartolone et al., 1989; Emeigh Hart et al, 1989b), suggesting its importance to nephrotoxicity in this species. These studies further support this hypothesis by demonstrating that adducts are preferentially formed in the cell types with the highest quantity of cytochrome P450, and that the cell types which contain adduct are in the areas which later undergo necrosis. The present observation that APAP is toxic to mouse RPTs in suspension and results in the formation of adducts of identical selectivity to those formed in mouse kidney *in vivo* further suggests that toxicity results from the ability of renal cytochrome P450 to activate APAP *in situ*, and that adducts found in the kidney are the result of such *in situ* activation.

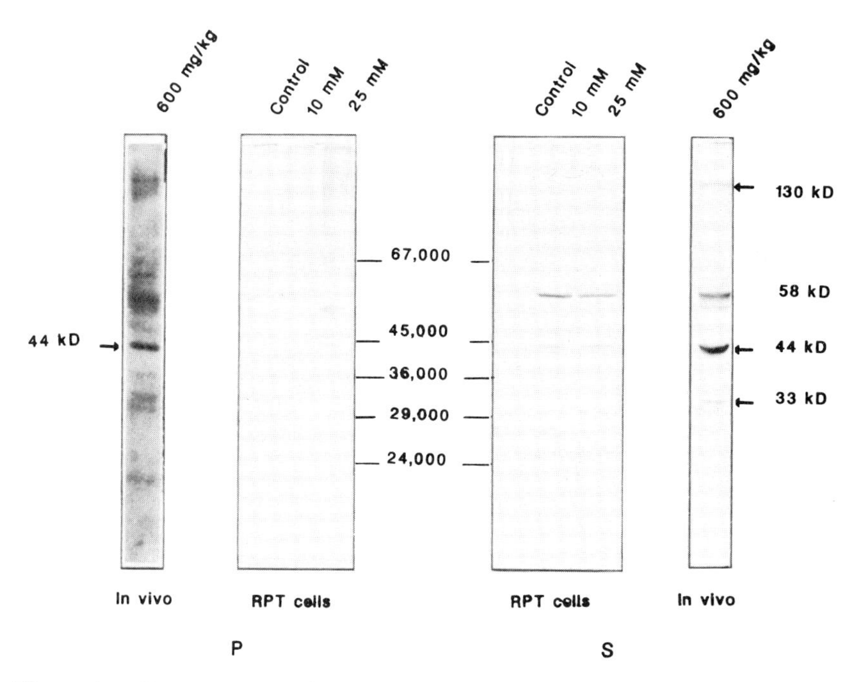

Figure 1. Western blot of soluble (S) and particulate (P) fractions from cultured RPT cells incubated in medium containing APAP (10 mM and 25 mM) or no APAP (control). Lanes marked in vivo are similar fractions from the kidney cortex of animals dosed with 600 mg/kg APAP, p.o., and killed 4 hours later. Arrows indicate the major target proteins of the mouse kidney. Molecular weight standards are also indicated.

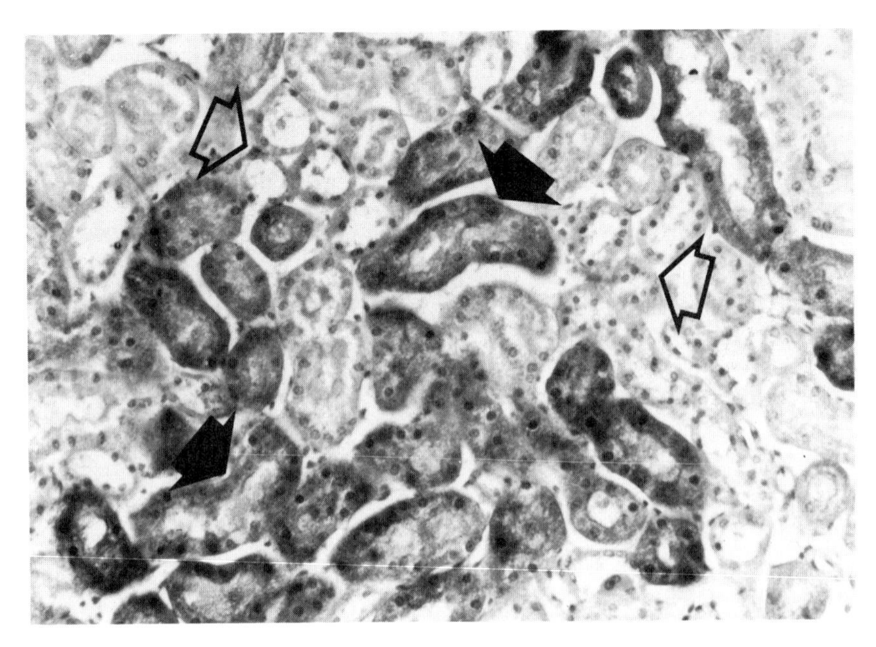

Figure 2. Immunohistochemical detection of bound APAP in mouse kidney. Mice were given APAP (600 mg/kg, po) and killed four hours later. Mouse kidney (x160) stained with anti-APAP antibody as described in text. The darkly staining proximal tubule cells (filled arrows) contain APAP adducts while no APAP is detectable in distal tubules (open arrows).

ACKNOWLEDGEMENTS

Supported in part by the Center for Biochemical Toxicology and NIH grants ES07163, GM31460 and ES55109.

REFERENCES

Bartolone, J.B., Sparks, K., Cohen, S.D., and Khairallah, E.A. (1987). Imunochemical detection of acetaminophen-bound liver proteins. *Biochem Pharmacol* **36**, 1193-1196.

Bartolone, J.B., Beierschmitt, W.P., Birge, R.B., Emeigh, Hart S.G., Wyand, S., Cohen, S.D. and Khairallah, E.A. (1989). Selective acetaminophen metabolite binding to hepatic and extrahepatic proteins: an *in vivo* and *in vitro* analysis. *Toxicol Appl Pharmacol* **99**, 240-249.

Björk, S., Svalander, C.T., and Aurell, M. (1988). Acute renal failure after analgesic drugs including paracetamol (acetaminophen). *Nephron* **49**, 45-53.

Boyer, T.D., and Rouff, S.L. (1971). Acetaminophen-induced hepatic necrosis and renal failure. *J Am Med Assoc.* **218**, 440-441.

Calder, I.C., Yong, A.C., Woods, R.A., Crowe, C.A., Ham, K.N., and Tange, J.D. (1982). The nephrotoxicity of p-aminophenol. II. The effect of metabolic inhibitors and inducers. *Chem-Biol Interact.* **27**, 245-254.

Cobden, I., Record, C.O., Ward, M.K., and Kerr, D.N.S. (1982). Paracetamol-induced acute renal failure in the absence of fulminant liver damage. *B. Med. J.* **284**, 21-22.

Crowe, C.A., Yong, A.C., Calder, I.C., Ham, K.N. and Tange, J.D. (1982). The nephrotoxicity of p-aminophenol. I. The effect on microsomal cytochromes, glutathione and covalent binding in kidney and liver. *Chem-Biol. Interact.* **27**, 235-243.

Davenport, A., and Finn, R. (1988). Paracetamol (acetaminophen) poisoning resulting in acute renal failure without hepatic coma. *Nephron* **50**, 55-56.

Emeigh, Hart S.G., Meyers, L.L., Beierschmitt, W.P., Wyand, D.S., Khairallah, E.A., and Cohen, S.D. (1989). Deacetylation of acetaminophen to p-aminophenol is not required for nephrotoxicity in the CD-1 mouse. *Toxicologist* **9**, 176.

Emeigh Hart, S.G., Beierschmitt, W.P., Wyand, D.S., Khairallah, E.A., and Cohen, S.D. (1989). Acetaminophen nephrotoxicity in the CD-1 mouse: the possible role of cytochrome P450 oxidation. *Pharmacologist* **31**, 177.

Foster, J.R., Elcombe, C.R., Boobis, A.R., Davies, D.S., Sesardic, D., McQuaide, J., Robson, R.T., Hayward, C., and Lock, E.A. (1986). Immunocytochemical localization of cytochrome P-450 in hepatic and extra-hepatic tissues of the rat with a monoclonal antibody against cytochrome P-450 c. *Biochem Pharmacol.* **35**, 4543-4554.

Green, C.E., Dabbs, J.E., Tyson, C.A., and Rauckman, E.J. (1990). Oxidative stress initiated during isolation of rat renal proximal tubules limits *in vitro* survival. Renal Failure, in press.

Jones, D.P., Sundby, G.-B., Ormstad, K., and Orrenius, S. (1979). Use of isolated kidney cells for study of drug metabolism. *Biochem. Pharmacol.* **28**, 929-935.

Keaton, M.R. (1988)., Acute renal failure in an alcoholic during therapeutic acetaminophen ingestion. *Southern Med. J.* **81**, 1163-1166.

Kher, K., and Makker, S. (1987)., Acute renal failure due to acetaminophen ingestion without concurrent hepatotoxicity. *Am. J. Med.* **82**, 1280-1281.

Kleinman, J.G., Breitenfield, R.V., and Roth, D.A. (1980)., Acute renal failure associated with acetaminophen ingestion: report of a case and review of the literature. *Clin. Nephrol.* **14**, 201-205.

McMurtry, R.J., Snodgrass, W.R., and Mitchell, J.R. (1978). Renal necrosis, glutathione depletion and covalent binding after acetaminophen. *Toxicol. Appl. Pharmacol.* **46**, 87-100.

Mitchell, J.R., Jollow, D.J., Potter, W.Z., Davis, D.C., Gilette, J.R., and Brodie, B.B. (1973). Acetaminophen-induced hepatic necrosis. I. Role of drug metabolism. *J. Pharm. Exper. Ther.* **187**, 185-194.

Newton, J.F., Kuo, C.-H., Gemborys, M.W., Mudge, G.H., and Hook, J.B. (1982). Nephrotoxicity of p-aminophenol, a metabolite of acetaminophen, in the Fischer 344 rat. *Toxicol. Appl. Pharmacol.* **65**, 336-344.

Newton, J.F., Yoshimoto, M., Bernstein, J., Rush, G.F., and Hook, J.B. (1983a). Acetaminophen nephrotoxicity in the rat. I. Strain differences in nephrotoxicity and metabolism. *Toxicol. Appl. Pharmacol.* **69**, 291-306.

Newton, J.F., Yoshimoto, M., Bernstein, J., Rush, G.F., and Hook, J.B. (1983b). Acetaminophen nephrotoxicity in the rat. II. Strain differences in nephrotoxicity and metabolism of p-aminophenol, a metabolite of acetaminophen. *Toxicol. Appl. Pharmacol.* **69**, 307-318.

Newton, J.F., Pasino, D.A., and Hook, J.B. (1985a). Acetaminophen nephrotoxicity in the rat: quantitation of renal metabolic activation *in vivo*. *Toxicol. Appl. Pharmacol.* **78**, 39-46.

Newton, J.F., Kuo, C.-H., DeShone, G.M., Hoefle, D., Bernstein, J. and Hook, J.B. (1985b). The role of p-aminophenol in acetaminophen-induced nephrotoxicity: effect of bis(p-nitrophenyl)phosphate on acetaminophen and p-aminophenol nephrotoxicity and metabolism in Fischer 344 rats. *Toxicol. Appl. Pharmacol.* **81**, 416-430.

Nishanian, E.V. (1989). Isolation and partial characterization of a major acetaminophen binding protein. PhD Dissertation, University of Connecticut.

Placke, M.E., Wyand, D.S., and Cohen, S.D. (1987). Extrahepatic lesions induced by acetaminophen in the mouse. *Toxicol. Pathol.* **15**, 381-387.

Proudfoot, A.T., and Wright, N. (1970). Acute paracetamol poisoning. *B. Med. J.* **3**, 557-558.

Sternberger, L.A., Hardy, P.H., Cuculis, J.J., and Meyer, H. (1970). The unlabeled antibody-enzyme method of immunohistochemistry. Preparation and properties of soluble antigen-antibody complex (horseradish peroxidase-antihorseradish peroxidase) and its use in the identification of spirochetes. *J. Histochem. Cytochem.* **18**, 315-333.

Tyson, C.A., Dabbs, J.E., Birge, R.B., Emeigh Hart, S.G., Bartolone, J.B., Cohen, S.D., and Khairallah, E.A. (1990a). Selective acetaminophen (APAP) binding to renal proximal tubule (RPT) proteins in CD-1 mice does not involve deacetylation to p-aminophenol as in the F-344 rat. *Toxicologist* **10**, 229.

Tyson, C.A., Dabbs, J.E., Cohen, P.M., Green, C.E., and Melnick, R.L. (1990b). Studies of nephrotoxic agents in an improved renal proximal tubule system. *Toxicology In Vitro*, in press.

Zenser, T.V., Mattammal, M.B., and Davis, B.B. (1978). Differential distribution of the mixed-function oxidase activities in rabbit kidney. *J. Pharm. Exper. Ther.* **207**, 719-725.

METABOLISM OF DICHLOROBENZENES IN ORGAN CULTURED LIVER SLICES

R. Fisher, S. McCarthy, I.G. Sipes, R.P. Hanzlik* and K. Brendel

University of Arizona
Departments of Pharmacology and Toxicology
Tucson, Arizona 85724 and
*University of Kansas
Department of Medicinal Chemistry
Lawrence, Kansas 66045

ABSTRACT

A novel *in vitro* system was used to evaluate tissue specific toxicity. This system utilizes precision cut organ slices in dynamic organ culture and is viable for up to 24 hrs. The three isomers of dichlorobenzene were added to liver slices prepared from Sprague Dawley rats or human donors. The precursor dichlorobenzenes were radiolabelled and metabolites were separated by classes (i.e. glucuronides, sulfates and glutathione and cysteine conjugates). Covalent Binding of the dichlorobenzenes was also determined after extensive extraction of the tissue. The total amount of metabolism of the dichlorobenzenes varied depending on the isomer and the type of tissue. For example, the Sprague-Dawley rat liver slices metabolized 1,2-DCB and 1,3-DCB at approximately the same rate while 1,4-DCB was metabolized at a slower rate. This metabolism profile was also seen in the majority of the adult human liver slices. However, the fetal human slices showed that 1,4-DCB was metabolized to a greater extent than 1,3-DCB or 1,2-DCB while 1,3-DCB was metabolized at a faster rate than 1,2-DCB. Our results show that liver slices in organ culture are a suitable system for species comparisons and of structure/activity relationships in xenobiotic metabolism with an emphasis on the fate of reactive intermediates. In addition, this system is suitable for evaluation of hepatotoxic potency.

INTRODUCTION

The availability of *in vitro* test systems for human tissue toxicity studies are limited and inadequate. The problems in utilizing humans for *in vivo* toxicity studies and the use of whole human liver perfusions are quite obvious. Problems with human isolated hepatocytes for *in vitro* studies include the difficulty with which they are obtained and the absence of the functional heterogeneity of the intact liver. A system which bridges the gap between isolated hepatocytes, organ perfusions and intact animals is desirable. The establishment of a novel dynamic organ culture system brought about such a system which is directly applicable to human liver slices. In the past we have reported on the relative hepatotoxicity of 1,2; 1.3 and 1,4 dichlorobenzene in human liver slices. Previous work with the dichlorobenzenes in Sprague-Dawley rat liver slices, maintained in our culture system, shows that there are differences in toxicity between the three isomers (Fisher et al., 1990). Since one can correlate the rat *in vitro* data with similar data found *in vivo* for Sprague-Dawley

Biological Reactive Intermediates IV, Edited by C.M. Witmer *et al.*
Plenum Press, New York, 1990

rats, one should be able to extrapolate from the human liver slices toxicity data to humans. In this work we have investigated the relative metabolism of the three isomers in rat and human tissue.

METHODS

Incubation Procedure

Liver slices (200-300 μM) were prepared in a modified version of the Krumdieck (Krumdieck et al (1980) tissue slicer in cold Krebs bicarbonate buffer. Within 20 min after slicing, individual slices were floated onto stainless steel mesh cylinders which were then placed in glass scintillation vials containing 1.7 ml of Krebs-HEPES buffer supplemented with gentamycin (84 μg/ml) and glucose (450 mg/100 ml). Vials were capped with aluminum lined caps and placed horizontally on a heated (37°C) vial rotator and incubated at 1 RPM for the indicated time periods. ^{14}C labelled dichlorobenzenes were added at a concentration of 0.1 mM (Smith et al., 1987). The stock solutions of radioactive halobenzenes were adjusted to equal specific activity.

Extraction Procedure

Slices are retrieved at the end of the incubation period and are homogenized in their respective supernatants. Aliquots of these homogenates are placed on dry pretreated filter papers which are then extracted. The basic idea behind wet filter extraction is similar to other procedures based on spreading the aqueous phase on a large surface (i.e. Feansie or absorbent cotton packed into small columns). Extraction then proceeds by equilibrating this absorbed aqueous phase with an organic solvent not miscible with water. Aliquots of 100 μl of fluid to be extracted are spotted on pencil numbered 3MM Whatman filter paper disks of 2.3 cm diameter. These disks are then inserted into individual chambers of an extraction basket made from a histological slice carrier and the basket placed into a glass staining dish with the lid filled with extractant to 0.5 cm above the filter paper disks. Magnetic stir bars inside the staining dishes and a 6-position stirr plate are used to circulate the extractant through the individual chambers of the extraction baskets. The baskets are removed from the staining dishes after 15 minutes of agitation and the extractant is collected in a radioactive waste container. This procedure is repeated for a total of 5 times. In addition and analogous to the above procedure methanol is used for extraction of parent and all metabolites. Protein precipitates on the filters and only protein bound label is retained on the filters. After the filter paper disks have gone through the extraction procedure the baskets are dried in the fume hood and the dry filter papers placed onto the bottom of counting vials (pencil numbers facing up) and soaked overnight with 1 ml of 0.25 M H_3PO_4. On the next morning counting vials are filled with 15 ml Universol (ICN) scintillation counting cocktail and counted in a scintillation counter with quench correction capabilities and filters for selective ^{14}C counting.

Statistical Analysis

The data are presented as the mean ± SEM for values compiled from 3-6 tissues in which three slices were used in each individual experiments. The averaged values were evaluated by a one-way ANOVA program and differences between each dichlorobenzene isomer were determined using Fisher's test for multiple comparisons (Sokal et al., 1981).

RESULTS

In Sprague-Dawley rat liver slices the total metabolism of 1,2-DCB and 1,3-DCB isomers are not significantly different from each other, however, 1,2-DCB metabolism is slightly higher than that of the other isomers. 1,4-DCB is metabolized to a significantly lower level than the other two isomers. The amount of covalent binding, glucuronide and glutathione-cysteine conjugates formed does not differ between the

Table 1. Extraction Modalities

1. Extract with hexane from alkaline papers[b]

    Remaining on paper: sulfate
    glucuronide
    GSH/cysteine conjugates
    phenols

    Extract will remove: parent

2. Extract with hexane from acid papers[a]

    Remaining on paper: sulfate
    glucuronide
    GSH/cysteine conjugates
    bound

    Extract will remove: parent and phenol

3. Extract with methanol from acid papers[a]

    Remaining on paper: bound

    Extract will remove: parent and also water soluble metabolites

4. Incubate with glucuronidase then extract with hexane from acid papers[a]

    Remaining on paper: sulfate
    GSH/cysteine conjugates
    bound

    Extract will remove: parent and hydrolyzed glucuronides

5. Incubate with glucuronidase and sulfatase then extract with hexane from acid papers[a]

    Remaining on paper: GSH/cysteine conjugates
    bound

    Extract will remove: parent and hydrolyzed sulfates and glucuronides

6. Incubate with sulfatase and glucuronidase inhibitor then extract with hexane from acid papers[a]

    Remaining on paper: glucuronide
    GSH/cysteine conjugates
    bound

    Extract will remove: parent and hydrolyzed sulfates

[a]Acid filters are produced by freshly soaking filter papers in 1 M phosphate buffer pH3 then drying them.
[b]Alkaline filters are produced by freshly soaking filter papers as supplied in a 1 M phosphate buffer of pH11 then drying them.

Table 2. The Metabolism of the Dichlorobenzene Isomers in Sprague-Dawley Rat Liver Slices[a]

Compound[b]	Tissue #[c]	Total Metabolism[d]	Covalent Binding[e]	Glucuronide[f]	Sulfate[f]	(GSH, Cysteine-conjugate)[f]
1,2	1	14903 ± 1146	12.4 ± 0.7	0.9 ± 0.9	0.0 ± 0.0	81.3 ± 3.3
	2	10748 ± 588	12.7 ± 0.8	3.7 ± 1.6	1.9 ± 1.2	74.2 ± 2.5
	3	14677 ± 1490	11.3 ± 1.4	9.1 ± 3.8	5.0 ± 1.1	62.7 ± 6.8
	Average	13442 ± 1348	12.1 ± 0.4	4.6 ± 2.4	2.3 ± 1.5	72.7 ± 5.4
1,3	1	14383 ± 3594	8.9 ± 0.5	2.6 ± 1.1	2.4 ± 0.4	76.6 ± 0.8
	2	15792 ± 1510	7.4 ± 0.8	5.2 ± 2.6	4.9 ± 2.7	66.4 ± 6.4
	3	22944 ± 215	10.9 ± 0.7	8.1 ± 0.8	2.1 ± 1.1	66.0 ± 1.3
	Average	17706 ± 2650	9.1 ± 1.0	5.3 ± 1.6	3.1 ± 0.9	69.7 ± 3.5
1,4	1	4069 ± 196	9.8 ± 0.1	0.3 ± 0.3	7.9 ± 1.3	67.4 ± 1.5
	2	10040 ± 1944	8.7 ± 1.4	5.5 ± 1.9	16.0 ± 3.3	55.1 ± 8.6
	3	9076 ± 321	14.7 ± 0.3	0.0 ± 0.0	6.9 ± 3.8	64.6 ± 2.5
	Average	7728 ± 1850	11.1 ± 1.8	1.9 ± 1.8	10.3 ± 2.9	62.4 ± 3.7

a) Sprague-Dawley rats were obtained from Harlan Sprague-Dawely
b) Each compound was dissolved in DMSO at a concentration of 0.1 mM
c) A total of three rat livers were used with three slices per condition
d) Values are presented as DPM/mg protein ± SEM at 4 hrs
e) Values are presented as % of total metabolism at 4 hrs
f) Significant difference between isomers at the $P > 0.05$ level (ANOVA)

Table 3. The Metabolism of the Dichlorobenzene Isomers in Adult Human Liver Slices[a]

Compound[b]	Tissue #[c]	Total Metabolism[d]	Covalent Binding[e]	Glucuronide[f]	Sulfate[f]	(GSH, Cysteine-conjugate)[f]
	1	20069 ± 1911	8.2 ± 4.2	32.1 ± 2.4	12.9 ± 4.6	54.9 ± 4.5
	2	18128 ± 1422	5.5 ± 0.2	52.7 ± 1.3	13.9 ± 1.5	33.5 ± 2.1
1,2	3	5831 ± 165	9.9 ± 0.3	45.4 ± 0.9	0 ± 0	54.6 ± 0.9
	4	7878 ± 1328	6.2 ± 0.7	38.0 ± 2.1	2.0 ± 1.3	60.0 ± 3.3
	5	11823 ± 209	12.8 ± 0.5	57.9 ± 2.1	3.4 ± 0.1	38.8 ± 2.1
	Average	12745 ± 2783	8.5 ± 1.3	45.2 ± 4.7	6.4 ± 2.9*	48.4 ± 5.1
	1	30266 ± 4437	4.8 ± 1.1	31.1 ± 2.8	26.0 ± 4.8	42.8 ± 2.9
	2	33277 ± 1191	5.2 ± 0.3	46.1 ± 1.1	25.4 ± 2.2	28.4 ± 1.1
1,3	3	13024 ± 529	5.2 ± 0.3	48.9 ± 3.4	12.2 ± 3.0	38.9 ± 0.4
	4	11387 ± 419	6.1 ± 0.9	27.9 ± 7.4	17.2 ± 3.4	54.8 ± 4.1
	5	19477 ± 1557	10.4 ± 0.6	47.7 ± 2.3	14.0 ± 2.1	38.3 ± 3.4
	Average	21486 ± 4437	6.3 ± 1.0	40.3 ± 4.5	19.0 ± 2.9	40.6 ± 4.3
	1	15828 ± 1247	4.6 ± 0.4	11.3 ± 6.5	16.9 ± 1.1	71.8 ± 6.1
	2	13184 ± 310	6.0 ± 0.3	20.7 ± 0.8	32.6 ± 1.2	46.7 ± 0.5
1,4	3	11365 ± 331	6.1 ± 0.4	33.6 ± 1.0	16.1 ± 0.4	55.2 ± 0.6
	4	12906 ± 526	6.4 ± 0.1	27.0 ± 2.6	21.3 ± 0.6	51.7 ± 2.1
	5	13790 ± 569	7.2 ± 0.1	31.2 ± 2.9	22.2 ± 2.7	46.6 ± 0.2
	Average	13414 ± 723	6.1 ± 0.4	24.8 ± 4.0	21.8 ± 2.9	54.4 ± 4.6

a) Human adult tissue was obtained from the Arizona Organ Bank, the International Institute For the Advancement of Medicine and the National Disease Research Institute
b) Each compound was dissolved in DMSO at a concentration of 0.1 mM
c) A total of five human adult livers were used with three slices per condition
d) Values are presented as DPM/mg protein ± SEM at 4 hrs
e) Covalent binding is presented at % of total metabolism ± SEM at 4 hrs
f) These values are presented as % of soluble metabolites in the media ± SEM at 4 hrs
*) Significant difference between isomers at the $P > 0.05$ level (ANOVA)

Table 4. The Metabolism of the Dichlorobenzene Isomers in Fetal Human Liver Slices[a]

Compound[b]	Tissue #[c]	Total Metabolism[d]	Covalent Binding[e]	Glucuronide[f]	Sulfate[f]	(GSH, Cysteine-conjugate)[f]
1,2	1	1809 ± 90	18.7 ± 1.1	6.4 ± 1.4	8.4 ± 1.4	85.2 ± 1.6
	2	1203 ± 61	14.4 ± 1.0	0 ± 0	1.0 ± 1.0	99.0 ± 1.0
	3	1330 ± 27	14.8 ± 0.6	0 ± 0	6.2 ± 3.4	93.8 ± 3.4
	4	1010 ± 42	16.0 ± 0.4	0 ± 0	2.5 ± 0.8	97.5 ± 0.8
	5	805 ± 61	17.5 ± 1.0	0 ± 0	2.0 ± 2.0	98.0 ± 2.0
	6	1502 ± 243	15.9 ± 0.7	9.5 ± 2.9	11.8 ± 9.0	78.9 ± 6.2
	Average	1276 ± 145*	16.2 ± 0.7*	2.7 ± 1.7*	5.3 ± 1.7*	92.1 ± 3.3*
1,3	1	6447 ± 551	9.4 ± 0.9	16.5 ± 3.4	13.5 ± 3.3	70.1 ± 5.8
	2	4803 ± 610	4.9 ± 0.4	10.7 ± 2.6	36.6 ± 2.2	52.6 ± 1.2
	3	6376 ± 316	4.8 ± 0.1	7.0 ± 3.1	46.2 ± 3.0	46.8 ± 0.4
	4	4933 ± 349	4.6 ± 0.4	6.1 ± 1.6	46.8 ± 0.8	47.1 ± 1.3
	5	3800 ± 266	6.5 ± 0.2	10.0 ± 1.8	37.4 ± 2.1	52.7 ± 0.9
	6	7127 ± 150	6.0 ± 0.7	16.7 ± 0.4	27.2 ± 4.1	56.2 ± 3.7
	Average	5581 ± 515*	6.0 ± 0.7	11.2 ± 1.9	34.6 ± 5.1	54.3 ± 3.5
1,4	1	8595 ± 476	6.8 ± 0.1	15.1 ± 1.3	18.8 ± 5.6	66.1 ± 4.3
	2	7267 ± 539	4.1 ± 0.3	8.4 ± 1.8	30.4 ± 1.5	61.1 ± 1.7
	3	8663 ± 238	4.8 ± 0.2	9.1 ± 4.9	33.6 ± 3.0	57.3 ± 3.9
	4	7378 ± 287	5.1 ± 0.3	12.9 ± 1.7	29.8 ± 0.9	57.3 ± 2.3
	5	5248 ± 479	8.6 ± 0.4	6.6 ± 0.7	26.0 ± 2.7	67.4 ± 1.9
	6	9438 ± 108	6.9 ± 1.0	12.8 ± 0.9	21.3 ± 1.6	65.9 ± 0.7
	Average	7764 ± 606*	6.1 ± 0.7	10.8 ± 1.3	26.7 ± 2.3	62.5 ± 1.9

a) Human fetal tissue was obtained from the International Institute For the Advancement of Medicine
b) Each compound was dissolved in DMSO at a concentration of 0.1 mM
c) A total of six human fetal livers were used with three slices per condition
d) Values are presented as DPM/mg protein ± SEM at 4 hrs
e) Covalent binding is presented at % of total metabolism ± SEM at 4 hrs
f) These values are presented as % of soluble metabolites in the media ± SEM at 4 hrs
*) Significant difference between isomers at the $P > 0.05$ level (ANOVA)

three isomers. However, the amount of sulfate formed is higher when 1,2-DCB is metabolized than when 1,2-DCB or 1,2-DCB is metabolized.

In adult human liver slices the total metabolism of the three isomers of dichlorobenzene is not significantly different, however, 1,3-DCB metabolism is slightly higher than 1,2-DCB or 1,4-DCB metabolism. 1,4-DCB significantly forms less glucuronide than 1,2- DCB or 1,3-DCB while the 1,2-DCB significantly forms less sulfate than 1,3-DCB or 1,4-DCB. The covalent binding and the glutathione and cysteine conjugate values are not significantly different between the three isomers.

In fetal human liver slices all three isomers are significantly different from each other. The 1,4-DCB isomer is metabolized to a greater extent than 1,3-DCB or 1,2-DCB. While the 1,3-DCB isomer is metabolized to a greater extent than 1,2-DCB. The covalent binding and glutathione-cysteine conjugate values were significantly higher for the 1,2-DCB isomer than that of 1,3-DCB and 1,4-DCB. The amount of glucuronides and sulfates formed were significantly lower for 1,2-DCB than for 1,3-DCB or 1,4-DCB.

CONCLUSION

The overall metabolism of 1,2-DCB and 1,3-DCB isomers is more extensive than that of 1,4-DCB in Sprague-Dawley rat liver slices. This was not the case in adult human liver slices or fetal human liver slices. All three isomers were metabolized to the same extent in adult human liver slices while 1,4-DCB was metabolized to a greater extent than the other two isomers in fetal human liver slices. There were no differences in the covalent binding or amount of gluthione-cysteine conjugates formed between the isomers for either rat or adult liver slices. The covalent binding and the gluthione-cysteine conjugates were higher for fetal human liver slices which metabolized the 1,2-DCB isomer. There were no differences in the amount of glucuronides formed between the isomers in rat slices, however, a significantly smaller amount of glucuronides were formed with the 1,2-DCB isomer in fetal human slices and with the 1,4-DCB isomer in adult human slices. Fetal and adult human slices produced a significantly lesser amount of sulfate with the 1,2-DCB isomer than the other two while rat slices showed an increase in the production of sulfate with the 1,4-DCB isomer.

ACKNOWLEDGEMENT

We are grateful for the editorial assistance of Mrs. Anita Finnell. We thank Dr. Thomas Burka, the project officer at NIEHS, for helpful discussions. Human tissue was acquired from the Arizona Organ Bank and through the services of the International Institute for the Advancement of Medicine and the National Disease Research Exchange. We are indebted to the personnel at these institutions for their extreme cooperativeness.

REFERENCES

Fisher, R., Smith, P.F., Sipes, I.G., Gandolfi, A.J., Krumdieck, C.L. and Brendel, K. (1990). Toxicity of chlorobenzenes in cultured rat liver slices. *In Vitro Tox.* **3**, 2.

Krumdieck, C.L., Dos Santos, J.E. and Ho, K.J. (1980). A new instrument for the rapid preparation of tissue slices. *Analyt. Biochem.* **104**, 118-123.

Smith, P.F., Fisher, R., Shubat, P.J., Gandolfi, A.J., Krumdieck, C.L. and Brendel, K. (1987). *In vitro* cytotoxicity of allyl alcohol and bromobenzene in a novel organ culture system. *Toxicol. Appl. Pharmacol.* **87**, 509-522.

Sokal, R.R. and Rolf, F.J. (1981). *Biometry*, W.H. Freeman, New York.

INHIBITION OF PROTEIN SYNTHESIS AND SECRETION BY VOLATILE ANESTHETICS IN GUINEA PIG LIVER SLICES

Hanan Ghantous, Jeannie Fernando, A. Jay Gandolfi, Klaus Brendel

Department of Anesthesiology
University of Arizona
College of Medicine
Tucson, Arizona 85724

ABSTRACT

The decrease in protein synthesis and secretion caused by volatile anesthetics was investigated using Hartley male guinea pig liver slices. Precision-cut liver slices (250-300 mM thick) were incubated in sealed roller vials (3 slices/vial) containing Krebs-Hensleit buffer at 37°C under 95% O_2 atmosphere. Volatile anesthetics were injected through a teflon septa cap on a filter paper wick and vaporized to produce constant concentration in the medium. A concentration (1-2.1 mM) and time related (0-24) decrease in protein synthesis (^{3}H-leucine incorporation) and secretion by halothane and d-halothane was observed. d-Halothane was less inhibiting than halothane. Inhibition was not on the uptake of the ^{3}H-leucine but with its incorporation in the nascent peptide. The effects of enflurane (2.2 mM), isoflurane (2.2 mM), and sevoflurane (1.3 mM) on protein synthesis and secretion were also studied. The rank order of decrease in protein synthesis caused by the volatile anesthetics studied was halothane > isoflurane > enflurane > sevofluane > d-halothane. Enflurane, isoflurane, and sevoflurane increased the protein secretion while halothane and d-halothane caused a pronounced decrease. Alterations in protein synthesis and secretion appears to be an early and sensitive indicator of cytotoxin injury.

INTRODUCTION

Previous studies have shown that the dynamic organ culture of precision guinea pig liver slices is a suitable tool for studying the biotransformation and toxicity of volatile anesthetics (Ghantous et al, 1989a; Ghantous et al, 1989b). 1,2). Guinea pig liver slices remained viable for over 12 hr of incubation and were able to biotransform halothane to either oxidative or reductive metabolites depending on the O_2 atmosphere (Ghantous et al, 1989a) Volatile anesthetics have also been shown to be cytotoxic to guinea pig liver slices as indicated by the leakage of K^+ ion across the membrane (Ghantous et al, 1989b). However, for relatively weak toxins such as volatile anesthetics, high concentrations were required for long periods of time to produce damage to cellular membranes. Thus, a more sensitive indicator is needed to monitor the intracellular changes leading to the toxicological damage. Volatile anesthetics are known to disrupt many aspects of cellular biochemical processes including that of protein synthesis and secretions. Several reports have described the effects of anesthetics on protein synthesis and secretion in different cellular systems (Bedows, Law, Knight et al, 1988; Horber et al, 1988). However, most of these studies used higher concentrations of volatile anesthetics than used in this study.

Biological Reactive Intermediates IV, Edited by C.M. Witmer *et al.*
Plenum Press, New York, 1990

The object of this study is (1) to determine the effect of halothane on protein synthesis and secretion, and to examine the role of its oxidative pathway of biotransformation on toxicity by studying the effect of d-halothane (which is resistant to oxidative metabolism) on protein synthesis and secretion; (2) to compare the effects of halothane, d-Halothane, enflurane, isoflurane, and sevoflurane on protein synthesis and secretion using guinea pig liver slices.

METHODS

Liver Slice Preparations

Male Hartley guinea pigs (600-650 g) were terminated and the liver was excised. Cores (1 cm) were taken and slices (250-300 mm) prepared using a Krumdieck tissue slicer (Ghantous et al, 1989). Slices were maintained on ice in oxygenated (95/5 O_2/CO_2) Krebs-Hensleit buffer (pH 7.4) until used.

Incubation

Slices were incubated in sealed roller vials (3 slices/vial) containing 1.6 ml Krebs-Hensleit buffer (plus vitamins, amino acids, glutamine, gentamycin) at 37R C (Ghantous et al, 1989). Vials were gassed with 95% O_2/CO_2 prior to exposure and incubation. After a 1 hr preincubation, volatile anesthetics were injected through a Teflon septa cap onto a filter paper wick and vaporized. Due to differences in partition coefficients, the following concentrations of anesthetic in the media were obtained: 1.7 mM halothane and d-halothane, 1.1 mM enflurane and isoflurane, and 1.3 mM sevoflurane.

Protein Synthesis and Secretion Analysis

Protein synthesis was measured by the amount of ^3H-leucine incorporated into slice proteins. Leucine-4,5-^3H (0.3 mCi/ml) was added to the incubating media. After incubation the slices were sonicated in 1 ml ice cold 1 N KOH. Aliquots (50 ml) of the homogenates were taken for protein determination. An equal volume of 1.5 N acetic acid was added to the rest of the homogenate and centrifuged at 3000 xg for 10 min. The pellet was then suspended in 1 ml of 0.5 N NAOH and neutralized by the addition of 250 ml of 2 N HCl. The radioactivity was quantified by liquid scintillation counting. Results were expressed as dpm ^3H incorporated/mg protein.

Protein secretion was measured in the culture medium. Aliquots of 1 ml of medium was mixed with 0.2 ml ice cold 10% perchloric acid. The denatured extra cellular protein were precipitated by centrifugation at 3000 xg for 10 min and the pellet was washed 3 times by resuspending in 2 ml ice cold 20% perchloric acid. Final pellet was dissolved in 1 ml 0.5 N NaOH and neutralized by 200 ml 2 N HCl. The radioactivity was quantified by liquid scintillation counting. Results were expressed as dpm ^3H in secreted protein/ml of culture media.

RESULTS

Halothane produced a concentration related decrease in protein synthesis (Figure 1) and protein secretion (Figure 2) in the liver slices when incubated under 95% O_2 atmosphere. The inhibition in protein secretion was more pronounced than the inhibition of protein synthesis. d-Halothane which is relatively resistant to oxidative biotransformation, produced less inhibition of protein synthesis and secretion than did halothane (Figure 3). When liver slices were exposed to volatile anesthetics, the rank order of decrease in protein synthesis produced was halothane > isoflurane > enflurane > sevoflurane > d-halothane (Figure 4). Halothane and d-halothane caused a pronounced inhibition of protein secretion while isoflurane, enflurane, and sevoflurane did not.

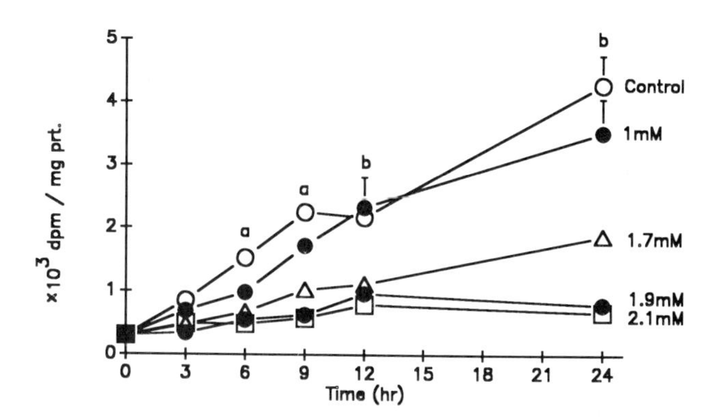

Figure 1. Effect of halothane on protein synthesis in the exposed to 1.5-2.1 mM halothane under 95% O_2 incorporated/mg protein ± SE of the mean. N = 12-15 slices from 3-4 separate animals. [a]Indicates all values significantly different from control ($p < 0.05$). [b]Indicates that all values are significantly different from control except after 1 mM halothane ($p < 0.05$).

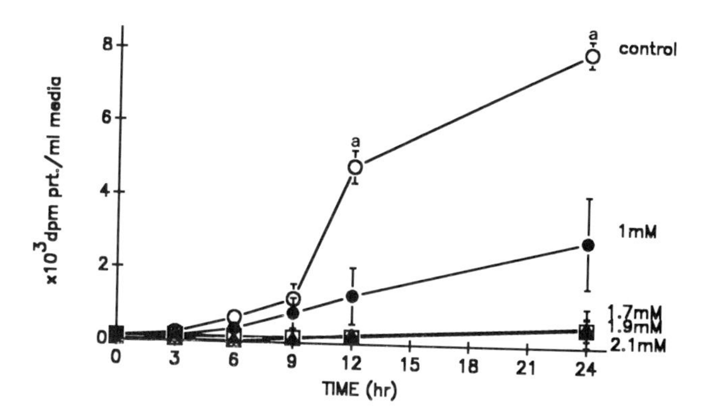

Figure 2. Effect of halothane on protein secretion by guinea pig liver slices. Liver slices were exposed to 1.5-2.1 mM halothane under 95% O_2 atmosphere. Values and means of dpm ^3H in secreted proteins/ml culture media ± SE of the mean. N = 6-8 media samples. [a]Indicates all values significantly different from control ($p < 0.05$).

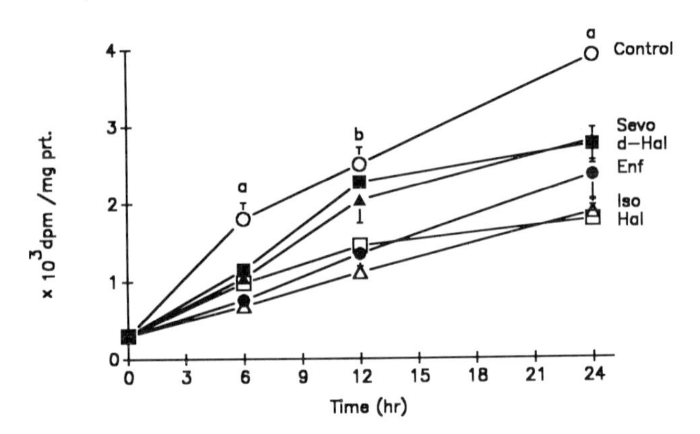

Figure 3. Effect of halothane and d-halothane on protein synthesis and secretion. Liver slices were exposed to 1.7 mM of halothane and d-halothane separately. Values are means of dpm ^{3}H-incorporated protein ± SE of the mean and dpm ^{3}H in secreted protein/ml culture media ± SE of the mean. N = 12-15 slices and 6-8 media samples from 3-4 separate animals.

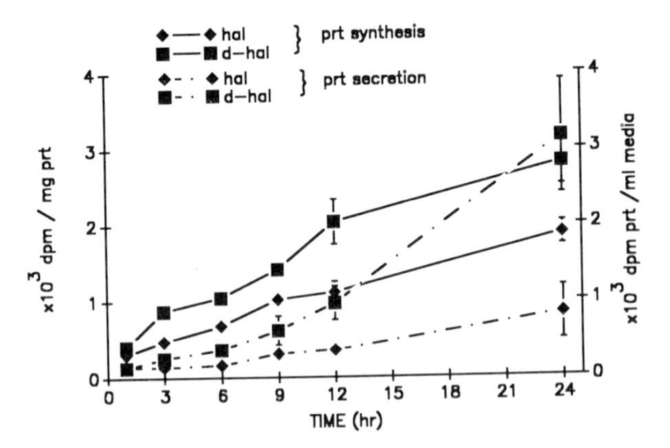

Figure 4. Effect of volatile anesthetics on protein synthesis in the guinea pig liver slices. Values are means of dpm ^{3}H-incorporating/mg protein + SE of the mean. N = 12-15 slices from 3-4 separate animals. [a]Indicates values significantly different ($p < 0.05$) from control. [b]Indicates that only halothane, enflurane and isoflurane are significantly different ($p < 0.05$) from control.

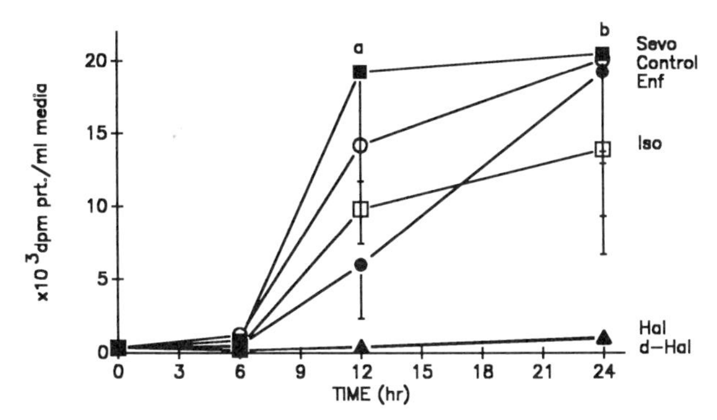

Figure 5. Effect of volatile anesthetics on protein secretion by guinea pig liver slices. Values are means of dpm ^{3}H in secreted proteins/ml culture media ± SE of the mean. N = 6-8 media samples. [a]Indicates all values significantly different from control and each other except halothane and d-halothane ($p < 0.05$). [b]Indicates that all values significantly different from control and each other except the following pairs; halothane-d-halothane and control-sevoflurane ($p < 0.05$).

DISCUSSION

This study examined the effect of volatile anesthetics on protein synthesis and secretion in liver slices as a sensitive indicator to assess toxicological damage. Halothane inhibited protein synthesis and secretion in a concentration-response manner in the liver slices. d-Halothane, which is relatively resistent to oxidative biotransformation, produced less inhibition of protein synthesis and secretion than halothane, indicating the importance of oxidative metabolism in halothane toxicity. These results are consistent with the report of Lind et al. (1989) that oxidative biotransformation is an important toxic mechanism *in vivo*. Using protein synthesis as an indicator of toxicity, halothane, enflurane and isoflurane were more toxic to the guinea pig liver slices than sevoflurane and d-halothane. This is in contrast to the *in vivo* situation where under certain conditions halothane is hepatotoxic while enflurane and isoflurane are not (Lind et al., 1985; Lind et al., 1988).

Inhibition of protein synthesis and secretion appears to be a valid indicator of early cellular toxicity to subtle toxicants such as volatile anesthetics. Low concentrations caused significant inhibition without the risk of a generalized injury response due to a "solvent effect".

ACKNOWLEDGEMENTS

This research was supported by grant DK 16715.

REFERENCES

Bedows, E., Law, S.T., and Knight, P. (1988). Inhibition of cellular protein synthesis by volatile anesthetics. *In Vitro Tox.* **2**, 19-29.

Ghantous, H.N., Fernando, J., Gandolfi, A.J. and Brendel, K. (1989). Biotransformation of halothane in guinea pig liver slices. *In Vitro Tox.*, accepted.

Ghantous, H.N., Fernando, J., Gandolfi, A.J. and Brendel, K. (1989). Toxicity of halothane in guinea pig liver slices. *Toxicology*, accepted.

Horber, F.F., Krayer, S., Rehder, K. and Hayward, M.W. (1988). Anesthesia with halothane and nitrous oxide alters protein and amino acid metabolism in dogs. *Anesthesiology* **69**, 319-326.

Lind, R.C., Gandolfi, A.J., Sipes, I.G. and Brown, B.R. (1985). Comparison of the requirements for hepatic injury with halothane and enflurane in rats. *Anesth. Analg.* **64**, 955.

Lind, R.C. and Gandolfi, A.J. (1988). Hypoxia and anesthetic-associated liver injury in guinea pigs. *Anesthesiology* **68.**

Lind, R.C., Gandolfi, A.J., Hall, P de la M. (1989). The role of oxidative biotransformation of halothane in the guinea pig model of halothane-associated hepatotoxicity. *Anesthesiology* **70**, 649-653.

DICHLOROBENZENE HEPATOTOXICITY STRAIN DIFFERENCES AND STRUCTURE ACTIVITY RELATIONSHIPS

Lhanoo Gunawardhana and I. Glenn Sipes[1]

Department of Pharmacology and Toxicology
College of Pharmacy
University of Arizona
Tucson, Arizona 85721

Environmentally significant halogenated compounds include the three isomers of dichlorobenzene (DCB), i.e. ortho (1,2-DCB), meta (1,3-DCB) and para (1,4-DCB), which have been identified as contaminants in air and water at sites throughout the United States (United States Environmental Protection Agency, 1985). This environmental contamination by the dichlorobenzenes reflects their wide-spread use in industry. These compounds are extensively used as solvents, fumigants and intermediates in the production of pesticides and dyes (Hawley, 1971).

Among the toxicities caused by dichlorobenzene isomers in animals, hepatotoxicity of 1,2-DCB and nephrotoxicity of 1,4-DCB are well documented (Brodie et. al., 1971 and Charbonneau et. al., 1989). The role of structure activity relationships in the differential hepatotoxicity of dichlorobenzene isomers has been reported previously (Stine et al., 1986). A further investigation has been undertaken to compare the hepatotoxicity of dichlorobenzene isomers in two different strains of rats frequently used in toxicological research i.e. Fischer-344 (F-344) and Sprague Dawley (SD) rats.

MATERIALS AND METHODS

Animals

Male, F-344 and SD rats, 9-10 week old, were purchased from Harlan Sprague-Dawley Inc., Indianapolis, IN.

Chemicals

Phenobarbital was purchased from Mallinckrodt Inc. (Paris, KY). 1,2-dichlorobenzene, 1,3-dichlorobenzene and 1,4-dichlorobenzene were purchased from Aldrich Chemical Company Inc. (Milwaukee, WI). The purity of all three isomers of dichlorobenzene as determined by NMR was > 99%.

---

[1]To whom all correspondence should be sent: Dr. I. Glenn Sipes, Department of Pharmacology and Toxicology, College of Pharmacy, University of Arizona, Tucson, Arizona 85721.

Biological Reactive Intermediates IV, Edited by C.M. Witmer *et al.*
Plenum Press, New York, 1990

Phenobarbital Pretreatment

Phenobarbital, dissolved in 0.9% NaCl, was administered i.p. at a dose of 80 mg/kg/day for 3 days. The injection volume was 2 ml/kg. Dichlorobenzenes were administered 24 hr after the last dose of phenobarbital (as described below).

Intraperitoneal Administration of Dichlorobenzenes

The dosing solutions were prepared on the day they were used. Appropriate amounts of the dichlorobenzene isomers were dissolved in corn oil (Mazola^{R}) to prepare 5.4 mmol/kg dosing solutions. Each animal received 2ml/kg of the respective dosing solutions via intraperitoneal injection.

Plasma Alanine Aminotransferase (ALT) Activity

Animals were killed 24 hr after the administration of dichlorobenzenes and blood samples were removed from the inferior vena cava. Plasma was prepared by centrifugation at 1500 g for 15 min. ALT activity in the plasma was assayed using a commercial kit purchased from Sigma Chemical Company (St. Louis, MO). Results are reported as U/L plasma.

Statistical analysis

Statistical differences were determined by analysis of variance (ANOVA) followed by Newman - Keuls test. Differences were considered significant at $p < 0.05$.

RESULTS

The most interesting finding of this work was the remarkable difference in the plasma ALT activities of F-344 and SD rats exposed to 1,2-DCB (Table 1). F-344 rats administered 1,2-DCB had 75 fold greater plasma ALT activities than the SD rats. Morphological examination of liver obtained from F-344 and SD rats confirmed the dramatic difference in hepatotoxicity induced by 1,2-DCB (data not shown). Pretreatment of animals with phenobarbital potentiated the hepatotoxicity of 1,2-DCB in both F-344 and SD rats administered a non hepatotoxic dose of (0.9 mmol/kg) 1,2-DCB (Figure 1). The plasma ALT activities in F-344 rats treated with 1,3-DCB was also significantly higher than in SD rats ($p< 0.05$). 1,4-DCB did not result in elevations of plasma ALT activities in either strain of rats. Furthermore, the hepatotoxic potential of the three isomers of dichlorobenzene in F-344 rats was different. The rank order of hepatotoxicity was 1,2-DCB >> 1,3-DCB >>> 1,4-DCB (Table 1).

DISCUSSION

The responses to certain environmental chemicals in man and animals exhibit considerable intraspecies variation. Therefore, generalization of findings from one strain of animal to a whole species or extrapolation of data from one strain of animals to humans could be misleading.

The results of this study show that large differences exist between F-344 and SD rats in the hepatotoxicity of 1,2- and 1,3-DCB (Table 1). These differences, as indicated by plasma ALT activity, were not due to differences in hepatic ALT content in these two strains (data not shown). Moreover, the centrilobular necrosis observed in the F-344 rats subsequent to 1,2-DCB treatment and the lack of such changes in the SD rats confirm that differences in plasma ALT activity reflect differences in the extent of tissue damage. Dramatic differences in the hepatotoxic potential of the three dichlorobenzene isomers (1,2-DCB >> 1,3-DCB >>> 1,4-DCB) were also observed in the F-344 rats as reported by Stine et al. (1986).

The enhancement of the hepatotoxicity of 1,2-DCB by phenobarbital pretreatment indicates that microsomal enzymes may be involved in the bioactivation of these compounds. In addition the differences in the enhancement of toxicity by phenobarbital in F-344 and SD rats may reflect the differences in certain microsomal enzymes in these two strains (Figure 1). It appears that differences in the biotransformation of 1,2-DCB may be important for the differential hepatotoxicity of this compound in F-344 and SD rats.

Table 1. Comparison of Dichlorobenzene-Induced Elevations of Plasma ALT (U/L) Activity in Male Fischer-344 and Sprague-Dawley Rats

	SD	F-344
Control	52 ± 1.6 (6)	38 ± 0.4 (6)
1,2-DCB (o-DCB) (5.4 mmol/kg)	194 ± 127 (6)	14716[a,c] ± 1207 (6)
1,3-DCB (m-DCB) (5.4 mmol/kg)	71 ± 11 (6)	848[b,d] ± 213 (6)
1,4-DCB (p-DCB) (5.4 mmol/kg)	26 ± 2 (4)	39 ± 3 (4)

Data expressed as the mean ± SEM (n)
[a] $p < 0.05$ (F-344>SD)
[b] $p < 0.05$ (F-344>SD)
[c] $p < 0.05$ (1,2-DCB>1,3-DCB)
[d] $p < 0.05$ (1,3-DCB>1,4-DCB)
Statistical analysis by ANOVA followed by Newman-Keuls test. ALT activity was determined at 25°C, but was corrected to 30°C as described by Sigma (No. 59-UV).

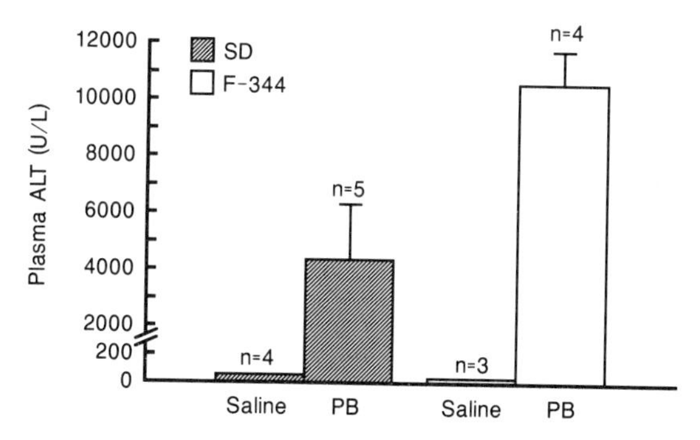

Figure 1. 1,2-DCB (0.9 mmol/kg) induced elevations of plasma ALT (U/L) activity in phenobarbital (PB) pretreated, 9-10 week old, male, F-344 and SD rats. See text for phenobarbital pretreatment. Data are expressed as the mean plasma ALT (U/L) ± SEM. ALT activity was determined at 25°C, but was corrected to 30°C as described by Sigma (No. 59-UV)

This differential hepatotoxicity of DCB in F-344 and SD rats underscores the importance of comparing the toxicity of a compound among different strains of a single species. Such a comparison might facilitate the search for a suitable animal model to assess the potential for human toxicity produced by a particular chemical. Understanding the molecular basis for such dramatic differences among strains of animals will help identify factors that may predispose certain humans to chemical induced toxicity.

ACKNOWLEDGEMENTS

This work was supported by NIEHS ES - 85320.

REFERENCES

Brodie, B. B., Reid, W. D., Cho, A. K., Sipes, I. G. Krishna, G. and Gillete, J. R. (1971). Possible mechanism of liver necrosis caused by aromatic organic compounds. *Proc. Nat. Acad. Sci.* **68,** 160-164.

Charbonneau, M., Strasser, J. Jr., Lock, E. A., Turner, M. J. Jr. and Swenberg, J. A. (1989). Involvement of reversible binding to 2u-globulin in 1,4-dichlorobenzene induced nephrotoxicity. *Toxicol. Appl. Pharmacol.* **99,** 122-132.

Hawley, G. G. (1971). *The Condensed Chemical Dictionary*, 8th ed., pp. 283-284. Van Nostrand Reinhold, New York/London.

Stine, E. R., Barr, J., Carter, D. E. and Sipes, I. G. (1986). Hepatotoxicity of the three isomers of dichlorobenzene as a function of chlorine position. *Toxicologist* **6,** pp. 119.

U.S.A. (1985). Health assessment document for chlorinated benzenes. Final Report, January, Washington, DC. EPA 600/8-84-0015F.

THE USE OF ^{19}F NMR IN THE STUDY OF PROTEIN ALKYLATION BY FLUORINATED REACTIVE INTERMEDIATES

James W. Harris and M. W. Anders

Department of Pharmacology and Environmental Health Sciences Center
University of Rochester
Rochester, NY 14642, U.S.A.

Cysteine *S*-conjugates are formed as a result of enzymatic conjugation of xenobiotics with the tripeptide glutathione and subsequent peptidase cleavage. Enzymatic *N*-acetylation of cysteine *S*-conjugates yields the corresponding mercapturic acids, which are excreted in the urine. Thus glutathione and cysteine *S*-conjugate formation are associated with detoxication and excretion of xenobiotics. However, cysteine *S*-conjugate β-lyase, which is identical to glutamine transaminase K, catalyzes β-elimination reactions of cysteine *S*-conjugates, yielding a thiolate, pyruvate, and ammonia.

Cysteine *S*-conjugates of many industrially important halogenated alkenes are nephrotoxic (Anders, 1988). Metabolism of such conjugates by β-lyase yields thiolates that are precursors of reactive, electrophilic species which may bind covalently to cellular proteins. However, neither the adducts formed with proteins nor the target amino acids have been identified. In the present research, ^{19}F nuclear magnetic resonance (NMR) has been used to study the renal protein adducts resulting from metabolism of the cysteine conjugates of chlorotrifluoroethene and tetrafluoroethene.

METHODS

S-(2-Chloro-1,1,2-trifluoroethyl)-*L*-cysteine (CTFC) and *S*-(1,1,2,2-tetrafluoroethyl)-*L*-cysteine (TFEC) were prepared by stirring an alkaline, ethanolic solution of cysteine under an atmosphere of chlorotrifluoroethene or tetrafluoroethene, respectively. After recrystallization from 5:2 ethanol:water, pH 5, the conjugates were characterized by ^{1}H and ^{19}F NMR and by fast atom bombardment mass spectrometry. The proreactive intermediate 2-chloro-1,1,2-trifluoroethyl 2-nitrophenyl disulfide (CTFC disulfide) was prepared by reacting chlorotrifluoroethene with 2- methyl-2-propane thiol. The resulting *tert*-butyl sulfide was reacted with 2-nitrophenylsulfenyl chloride to yield CTFC disulfide.

Male Fisher 344 rats were anesthetized with ketamine/xylazine (90 mg/kg and 8 mg/kg, respectively) and given 1.0 mmol/kg body weight of either CTFC or TFEC by intraperitoneal injection. One hour later the anesthetized animals were killed and the kidneys were excised. Mitochondrial, microsomal, and cytosolic fractions were prepared by differential centrifugation; special attention was paid to insure that each fraction was free of contamination by other fractions. The fractions were suspended in water and placed in dialysis tubing (Spectrapor 3, Spectrum Medical Industries, 3500 M.W. cutoff). Equilibrium dialysis in 10 mM phosphate buffer, pH 7, with 0.1% (w/v)

Biological Reactive Intermediates IV, Edited by C.M. Witmer *et al.*
Plenum Press, New York, 1990

SDS was employed to remove any metabolites that were not covalently bound (Sun and Dent, 1980). The dialysis was carried out at 4°C with constant stirring; the buffer was changed several times during at least 48 hr of SDS dialysis. After dialysis, the protein was lyophilized and then dissolved in D_2O for ^{19}F NMR analysis. Protein concentrations in the NMR tube were kept as high as possible while still maintaining a liquid, albeit viscous, solution. A minimum of 40 mg protein per ml of D_2O was used. NMR spectra were recorded with a Bruker WP-270 spectrometer operating at 254.18 MHz for fluorine and equipped with a dedicated 5 mm ^{19}F probe. Spectra were acquired at room temperature with the sample spinning. A pulse width of 3 μsec and a total acquisition time of 0.9 sec per transient were used. Exponential multiplication of the FID was not employed. Depending on signal strength and desired signal-to-noise ratio, 2000 or more transients were required. All spectra were referenced to a dilute solution of trifluoroacetamide in D_2O (δ=0 ppm) in a sealed coaxial tube. The chemical shift of this reference standard is less pH sensitive than the commonly used trifluoroacetate.

Proteins were alkylated with CTFC disulfide as follows: Renal protein from untreated rats was fractionated and shown to contain no fluorine resonances. To an aqueous suspension of this protein (pH 7.4) was added neat CTFC disulfide and Triton X-100 to aid in the dissolution of the hydrophobic disulfide. The mixture was stirred overnight and dialyzed as described above. Alkylated proteins were proteolyzed by incubating the treated and dialyzed protein with proteinase K at 37°C. The reaction was followed by placing the mixture in a NMR tube and acquiring ^{19}F spectra periodically. Lipid-containing subcellular fractions (e.g., mitochondria or microsomes) may not undergo complete proteolysis unless the lipid is removed with organic solvent.

RESULTS AND DISCUSSION

A single, broad ^{19}F NMR signal remained after SDS dialysis of renal subcellular fractions from rats given CTFC or TFEC. The CTFC or TFEC alkylation product resonated near 56 or 40 ppm, respectively, upfield from trifluoroacetamide. The fluorine atoms of synthetic chlorofluoroacetyl and difluoroacetyl thioamides resonate in these same chemical shift regions (Table 1). Single, broad signals representing alkylated proteins were observed in mitochondrial, microsomal, and cytosolic fractions of rat kidney. Although quantitative information was difficult to obtain by NMR, mitochondria appeared to be more highly alkylated than the other fractions studied.

Incubation of CTFC disulfide with protein from untreated rats yielded the same alkylation pattern found with *in vivo* administration of CTFC. Reduction of CTFC disulfide produces the same thiol that is released by β-lyase metabolism of CTFC; this thiol may rearrange to a reactive thionoacyl fluoride, the putative alkylating species (Dekant et al., 1987). The spectra in Figures 1A and 1C show the single resonance found after preparation and dialysis of synthetically alkylated mitochondrial and cytosolic protein, respectively; proteolysis of these preparations converted the broad NMR resonance, attributed to anisotropic motion, to a sharp doublet with little change in chemical shift and J_{HF} = 49 Hz (see Figures 1B and 1D). These results indicate that a single amino acid is alkylated. *In vitro* incubation of CTFC disulfide with N-α-blocked amino acids and various organic amines indicates that lysine is the target of alkylation (Table 1).

The spectroscopic evidence indicates that amino groups (i.e., lysine) in protein are targets for alkylation by the putative chlorofluorothionoacetyl fluoride generated by CTFC metabolism. Similar behavior would be expected from the difluorothionoacetyl fluoride presumably generated by TFEC metabolism, although experiments to verify this point have not been conducted (Table 1). Less stable amino acid adducts may also be formed but not detected because of loss during sample handling.

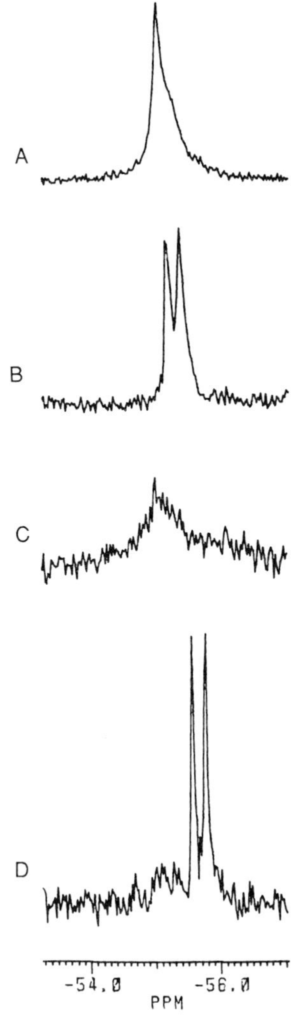

Figure 1. ^{19}F NMR spectra of alkylated proteins (ppm upfield of trifluoroacetamide standard). Mitochondrial (A) and cytosolic (C) renal protein from untreated rats after incubation with CTFC disulfide and dialysis; after proteolysis of the treated mitochondrial (B) and cytosolic (D) protein.

Table 1. Summary of NMR Data[a]

Experiment	TFEC	CTFC
in vivo protein	~ 40 ppm, broad	~ 56 ppm, broad
in vitro protein	not done	~ 56 ppm, broad
proteolysis of *in vitro* protein	not done	55.7 ppm, sharp doublet J_{HF} = 49 Hz
in vitro lysine	not done	56.4 ppm, sharp doublet J_{HF} = 49 Hz
in vitro, absence of nucleophiles	not done	no resonances near 56 ppm
thioamide of benzylamine	41.6 ppm, doublet J_{HF} = 56 Hz[b,c]	58.5 ppm, doublet J_{HF} = 49 Hz[c,d]
thioamide of aniline	39.7 ppm, doublet J_{HF} = 57 Hz[b,c]	not done

[a]Upfield of trifluoroacetamide with D_2O as solvent, unless noted; *in vitro* experiments use the proreactive intermediate CTFC disulfide.
[b]Data taken from Commandeur et al., 1989.
[c]Gives interpretable mass spectrum.
[d]Solvent was $CDCl_3$.

In summary, ^{19}F NMR was used to study the alkylation of proteins in renal tissue of rats given the nephrotoxic cysteine *S*-conjugates CTFC and TFEC. Proteolysis of alkylated protein converts the broad NMR signal to a single doublet with clearly resolved, characteristic geminal H-F coupling. These results indicate that lysine is the only amino acid that forms stable adducts with the thionoacyl fluoride reactive intermediates generated in these experiments. Analysis of tissue by ^{19}F NMR may prove useful in characterizing cellular targets of reactive intermediates for several classes of compounds.

ACKNOWLEDGEMENT

This research was supported by NIEHS grant ES03127 to M.W.A.

REFERENCES

Anders, M. W. (1988). Glutathione-dependent toxicity: Biosynthesis and bioactivation of cytotoxic *S*-conjugates. ISI Atlas of Science: *Pharmacology* **2**, 99-104.

Commandeur, J. N. M., De Kanter, F. J. J., and Vermeulen, N. P. E. (1989). Bioactivation of the cysteine *S*-conjugate and mercapturic acid of tetrafluoroethylene to acylating reactive intermediates in the rat: Dependence of activation and deactivation activities on acetyl coenzyme A availability. *Mol. Pharm.* **36**, 654-663.

Dekant, W., Lash, L. H., and Anders, M. W. (1987). Bioactivation mechanism of the cytotoxic and nephrotoxic *S*-conjugate S-(2-chloro-1,1,2-trifluoroethyl)-L-cysteine. *Proc. Natl. Acad. Sci. USA* **84**, 7443-7447.

Sun, J. D., and Dent, J. G. (1980). A new method for measuring covalent binding of chemicals to cellular macromolecules. *Chem.-Biol. Interact.* **32**, 41-61.

S-ETHYLTHIOTRIFLUOROACETATE ENHANCEMENT OF THE IMMUNE RESPONSE TO HALOTHANE IN THE GUINEA PIG

Kenneth L. Hastings, Susan Schuman, Alan P. Brown, Cindy Thomas, and A. Jay Gandolfi

Department of Anesthesiology
University of Arizona
College of Medicine
Tucson, Arizona 85724

Halothane hepatitis appears to be a hypersensitivity reaction to neoantigens formed when halothane reactive intermediates covalently bind to liver proteins. Production of an animal model for halothane hepatitis has been largely unsuccessful, possibly due to inadequate cytochrome P-450 mediated biotransformation of halothane to reactive intermediates. S-ethylthiotrifluoroacetate (SETFA) reacts rapidly and efficiently with protein lysine groups to form trifluoroacetylated proteins. The advantage to this procedure is that cytochrome P-450 mediated activation is not required. Male Hartley guinea pigs (600-700 g) were injected (ip) with SETFA (50 ml) emulsified in an adjuvant-mycobacterial protein mixture. Subsequent exposure to halothane produced a 10x increase in anti-trifluoro acetylated albumin antibody response compared to controls, as well as centrilobular lesions consistent with halothane toxicity. Western-type immunoblot analysis demonstrated three unique antigen bands in microsomal fractions from SETFA-treated guinea pigs. The adjuvant-mycobacterial mixture alone caused only minor liver lesions and minimal immune response. SETFA may be effective for producing the reactive intermediate of halothane metabolism, bypassing the need to induce atypical halothane metabolism to produce an immunopathological response.

INTRODUCTION

Many of the features of acute halothane toxicity can be produced in the guinea pig, but the immunopathological events presumed to be necessary for the induction of halothane hepatitis can only be partially produced (Lunam et al., 1985). Although guinea pigs exposed to halothane produce liver neoantigens and antibodies to these neoantigens, production of these immunological markers has not resulted in symptoms indicative of fulminant hepatitis (Siadat-Pajouh et al, 1987). In order to produce such a model, two biochemical events must occur: the animal must be able to metabolize halothane to form the necessary neoantigen and the animal's immune system should respond to this neoantigen as it would to a foreign protein.

In this study, we have investigated the effect of S-ethylthiotrifluoroacetate (SETFA) treatment on the immune response to halothane in the guinea pig. SETFA is used to produce the artificial neoantigen used as a mimic for the naturally occurring halothane-associated neoantigen. When reacted with protein, SETFA forms trifluoroacetylated amino acid adducts, of which trifluoroacetylated lysine appears to be the target epitope for the immunopathological response (Kenna et al., 1988; Satoh

Biological Reactive Intermediates IV, Edited by C.M. Witmer *et al.*
Plenum Press, New York, 1990

et al., 1987). The objective of this study was to determine if SETFA can act as an appropriate mimic for the oxidative biotransformation products of halothane, reacting *in vivo* with liver proteins to produce neoantigens that would induce an immunopathological response.

MATERIALS AND METHODS

Animals. Male Sasco Hartley albino guinea pigs were used throughout.

SETFA Treatment. For *in vivo* exposure, SETFA was emulsified with killed, desiccated *Mycobacterium butyricum* in complete Freund's adjuvant.

Trifluoroacetylated Biological Material. Trifluoroacetylated white blood cells (TFA-WBC) were prepared by incubating guinea pig lymphocytes with SETFA. After reaction, the lymphocytes were collected, washed, and injected back into the guinea pigs from which the cells were obtained. Trifluoroacetylated mycobacterial protein (TFA-MB) was formulated by reacting *M. butyricum* with SETFA (Goldberger and Anfinsen, 1962).

Methods of Exposure. Thirty-five guinea pigs were divided into 5 experimental groups.

SEFTA + HALOTHANE: A.) Ten guinea pigs were each injected ip with 50 ml SETFA and 20 mg *M. butyricum* emulsified in 450 ml adjuvant. Two control animals received *M. butyricum* in adjuvant, ip. Two SETFA treated animals were killed two weeks after treatment for histopathological evaluation. Two weeks after treatment, four SETFA treated animals and one control were exposed to 1% halothane for 4 hr by inhalation. These animals were exposed in the same way to halothane two weeks after the first exposure. All of the guinea pigs were killed at 3 or 7 days after the second exposure.

B.) Six guinea pigs were treated with SETFA as described above. The animals were exposed to halothane 4 weeks after treatment in the same manner as above. Two weeks after the first exposure the animals were exposed again to halothane as for group A. The animals were killed at 3 or 7 days after the second exposure.

TFA-WBC: Five guinea pigs were given TFA-WBC preparations ip. Four weeks after treatment, the animals were exposed twice over a two week period to halothane as described in "A". They were killed at 3 or 7 days after the second exposure.

TFA-MB: Six guinea pigs were given TFA-MB, 20 mg in 500 ml adjuvant each, ip. Four weeks after treatment the animals were exposed twice to halothane and then killed as described in "A".

HALOTHANE: Six guinea pigs were kept untreated for four weeks, then exposed twice to halothane and killed as described in "A".

ELISA

Antibodies reactive with trifluoroacetylated albumin were quantitated by an enzyme-linked immunosorbent assay (ELISA) (Callis et al., 1987). Microtiter plates were coated with either trifluoroacetylated guinea pig serum albumin (TFA-GPA) or trifluoroacetylated rabbit serum albumin (TFA-RSA). Diluted guinea pig serum samples were added to the plates followed by overnight incubation at 4R C. Bound antibodies were detected with peroxidase-labeled goat anti-guinea pig antibodies using o-phenylenediamine as the substrate. Antibody titers were determined by regression analysis.

Immunoblot Analysis

Proteins in liver tissue homogenates were separated by SDS-PAGE using a 4.5% stacking gel and a 12% resolving gel. The proteins were transferred to a nitrocellulose paper followed by incubation with rabbit anti-trifluoroacetylated albumin antiserum (Hubbard et al., 1989). Bound antibody was detected by incubation with biotinylated anti-rabbit IgG followed by incubation with avidin-biotin-peroxidase complex. The immunoblots were developed with diamino-benzidine and nickel chloride.

RESULTS

Treatment of guinea pigs with SETFA *in vivo* followed by exposure to halothane resulted in the production of anti-halothane metabolite antibody titers comparable to titers seen in animals that were actively immunized with trifluoroacetylated albumin [Table 1] (Siadat-Pajouh et al., 1987). Increasing the time between SETFA treatment and halothane exposure from two weeks to four weeks resulted in a higher mean antibody titer but also greater variability in antibody response (Figure 1).

Three out of five guinea pigs treated with TFA-WBC and then exposed to halothane produced anti-halothane metabolite antibody titers comparable to SETFA treated animals. Animals treated with TFA-MB produced antibody titers comparable to SETFA treated guinea pigs. Unlike the other experimental groups, these animals also produced relatively high antibody titers before halothane exposures, indicating active immunization by TFA-MB. Untreated guinea pigs exposed to halothane produced very low titers of anti-halothane metabolite antibodies.

Western-type immunoblot analysis demonstrated unique neoantigens (apparent MW = 51k, 43k, and 32k) in liver homogenates from animals treated with SETFA and then exposed to halothane.

Histopathological examination of liver tissue demonstrated leukocytic infiltrates in the SETFA treated animals, both before and after subsequent halothane exposure. Granu-lomatous lesions seen in livers from these animals were also seen in animals receiving only mycobacterial protein. Other pathological features were typical of acute halothane toxicity.

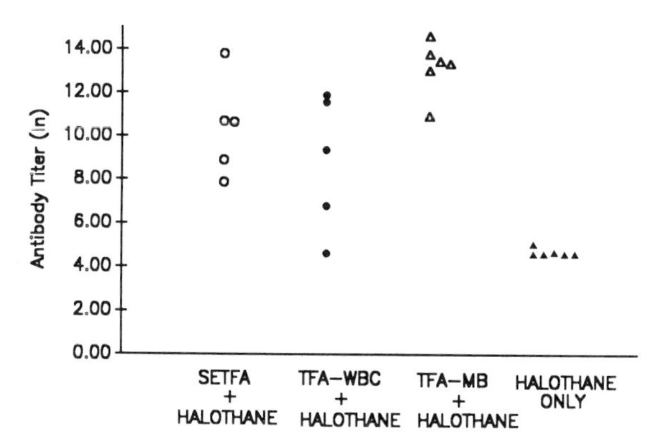

Figure 1. Anti-halothane metabolite antibody titers of animals treated with either SETFA, TFA-WBC, or TFA-MB, and then after one month exposed twice to halothane. Untreated control guinea pigs were likewise exposed twice to halothane. The animals were killed at 3 or 7 days after the second halothane exposure and antibody titers were determined by ELISA on serum samples collected then.

Table 1. Anti-halothane metabolite antibody titers of guinea pigs treated with SETFA

Group	Pre-Treatment	Day 2*	Day 7*	Day 14*	I Day 2+	I Day 7+	II Day 3++	II Day 7++ (Term)
SETFA + HALOTHANE (N=4)	287 ± 1	253 ± 1	236 ± 1	999 ± 1	5131 ± 2	22,115 ± 6	6180	19,990
							53,691	48,098
SETFA (N=4)	235 ± 2	190 ± 1	220 ± 1	888 ± 4	ND	ND	ND	825 ± 3
Myobacterial Control	176	189	213	379	ND	ND	ND	507
Mycobacteria + HALOTHANE	237	181	242	597	ND	966	ND	816

All data expressed as the reciprocal of the titer as determined by ELISA, ± SD.

*Days after SETFA treatment. +Days after first halothane exposure. ++Days after second halothane exposure. Titers on individual animals given for SETFA + halothane group.

DISCUSSION

The role of the immune system in the development of halothane hepatitis remains controversial (Farrell, 1988). Although both halothane-associated liver neoantigens and anti-halothane metabolite antibodies can be produced by exposing guinea pigs to halothane, fulminant hepatitis is not produced in this animal model (Siadat-Pajouh et al., 1987; Hubbard et al., 1989; Brown et al., 1990; Lunan, Hall, and Cousins, 1989; Hastings et al., 1990). One possible reason for the failure to produce an immunopathologic response to halothane exposure is inadequate biotransformation of the anesthetic to protein reactive intermediates. Thus, although neoantigen formation appears to occur in the guinea pig, the amount of halothane metabolite-protein adduct might not be sufficient to induce an immune response adequate for immunopathy.

SETFA reacts preferentially at pH 10 with free amino groups to form trifluoroacetyl adducts, thought to be identical in structure to the epitopes formed when halothane metabolites bind to proteins (Kenna et al., 1988). This reaction does not require biotransformation in order to produce neoantigen; thus, if producible *in vivo* could serve as a way to overcome inadequate metabolism to induce an immune response to halothane.

There are two potential problems to using SETFA: 1.) it may not adequately react at physiological pH to form the required protein adducts, 2.) it would be too toxic to be used in the guinea pig. Histopathologic examination of liver tissue from two guinea pigs killed two weeks after dosing with 50 ml of SETFA demonstrated leukocytic infiltrates, but no extensive pathology.

The effect of SETFA treatment appears to be to prime the immune system to react to the biotransformation products of halothane. Although a slight increase in anti-halothane metabolite antibody titer appears before exposure to halothane, significant titers were observed only after SETFA treatment and subsequent halothane exposure. The possibility that SETFA reacts in vivo with mycobacterial protein to form complete antigen was investigated by treating animals with TFA-MB. Animals treated with TFA-MB developed high titers of anti-halothane metabolite antibody titers both before and after halothane exposure, indicating active immunization as opposed to the immune priming that appears to be the result of SETFA treatment.

To investigate whether trifluoroacetylation of self-protein could induce an immune response to halothane, white blood cells from guinea pigs were trifluoroacetylated and injected back into the donor animals. Four of five animals responded to subsequent halothane exposure by producing anti-halothane metabolite antibody titers comparable to SETFA treated animals, indicating that self-protein trifluoroacetylation could, in fact, be the basis of immune priming by SETFA.

Immunoblot analysis demonstrated that SETFA-treated, halothane exposed guinea pigs produced detectable halothane-associated neoantigens not seen in any of the other experimental animals. In what way SETFA facilitates the appearance of these neoantigens requires further investigation. Finally, none of the animals developed symptoms of fulminant hepatitis, either before or after halothane exposure. Thus, the experimental demonstration of immune-mediated halothane hepatotoxicity remains to be realized.

ACKNOWLEDGEMENTS

This research was supported by NIH grant GM 34788.

REFERENCES

Brown, A.P., Hastings, K.L., Gandolfi, A.J. and Brendel, K. Covalent binding of halothane metabolite and neoantigen production in guinea pig liver slices. IV International Symposium on Biological Reactive Intermediates, Tucson, January, 1990.

Callis, A.H., Brooks, S.D., Roth, T.P., Gandolfi, A.J. and Brown, B.R. (1987). Characterization of a halothane-induced humoral immune response in rabbits. *Clin. Exp. Immunol.* **67**, 343-351.

Farrell, G.C. (1988). Mechanism of halothane-induced liver injury: Is it immune or metabolic idiosyncrasy. *J. Gastroenterol. Hepatol.* **3**, 465-482.

Goldberger, R. and Anfinsen, C. (1962). The reversible masking of amino groups in ribonuclease and its possible usefulness in the synthesis of the protein. *Biochemistry* **1**, 401-405.

Hastings, K.L., Schuman, S., Brown, A.P., Hubbard, A.K. and Gandolfi, A.J. (1990). Examination of guinea pig liver for halothane-associated neoantigen production in animals exposed to halothane, enflurane, or isoflurane. *Anesthesiology*, submitted.

Hubbard, A.K., Roth, T.P., Schuman, S. and Gandolfi, A.J. (1989). Localization of halothane-induced antigen *in situ* by specific anti-halothane metabolite antibodies. *Clin. Exp. Immunol.* **76**, 422-427.

Kenna, J.G., Satoh, H., Christ, D.D. and Pohl, L.R. (1988). Metabolic basis for a drug hypersensitivity: Antibodies in sera from patients with halothane hepatitis recognize liver neoantigens that contain the trifluoroacetyl group derived from halothane. *J. Pharmacol. Exp. Therap.* **245**, 1103-1109.

Lunam, C.A., Cousins, M.S. and Hall, P.M. (1985). Guinea pig model of halothane-associated hepatotoxicity in the absence of enzyme induction and hypoxia. *J. Pharmacol. Exp. Therap.* **232**, 802-809.

Lunam, C.A., Hall, P.M. and Cousins, M.J. (1989). The pathology of halothane hepatotoxicity in a guinea pig model: A comparison with human halothane hepatitis. *Br. J. Exp. Path.* **70**, 533-541.

Siadat-Pajouh, M., Hubbard, A.K., Roth, T.P. and Gandolfi, A.J. (1987). Generation of halothane-induced immune response in a guinea pig model of halothane hepatitis. *Anesth. Analg.* **66**, 1209-1214.

Roth, T.P., Hubbard, A.K., Gandolfi, A.J. and Brown, B.R. (1988). Chronology of halothane-induced antigen expression in halothane-exposed rabbits. *Clin. Exp. Immunol.* **72**, 330-336.

Satoh, H., Gillette, J.R., Takemura, T., Ferrans, V.J., Jelenich, S.E., Kenna, J.G., Neuberger, J. and Pohl, L.R. (1986). Investigation of the immunological basis of halothane-induced hepatotoxicity. In, *Biological Reactive Intermediates-III*, (Ed. J.J. Kocsis, D.J. Jollow, C.M. Witmer, J.O. Nelson, and R. Snyder), pp. 657-673. Plenum Press, New York.

Satoh, H., Fukuda, Y., Anderson, D.K., Ferrans, V.J., Gillette, J.R. and Pohl, L.R. (1987). Immunological studies on the mechanism of halothane-induced hepatotoxicity: Immunohistochemical evidence of trifluoroacetylated hepatocytes. *J. Pharmacol. Exp. Therap.* **233**, 857-862.

BONE MARROW DNA ADDUCTS AND BONE MARROW CELLULARITY FOLLOWING TREATMENT WITH BENZENE METABOLITES *IN VIVO*

Christine C. Hedli, Robert Snyder and Charlotte M. Witmer

Joint Graduate Program in Toxicology
Rutgers University/UMDNJ
Robert Wood Johnson Medical School
Piscataway, NJ

Benzene[1] (BZ) is a widely used industrial solvent, exposure to which has been linked to the development of aplastic anemia and leukemia. BZ is an animal carcinogen (Cronkite et al., 1984), but the ultimate carcinogenic metabolite still has not been determined. Recent work in our laboratory involving the mechanism of toxicity of BZ has been concerned with the capacity of BZ or its metabolites to interact with DNA. Previous studies have identified an adduct formed by the reaction of p-benzoquinone and deoxyguanosine *in vitro* (Jowa et al., 1986). Our current efforts have focused on the detection of adducts between BZ metabolites and DNA *in vivo*.

In this study, we have investigated the *in vivo* patterns of DNA adduct formation by the BZ metabolites phenol (P) and hydroquinone (HQ) using the [32P] postlabeling technique. In comparison to vehicle treated control animals, three additional adducts were detected following intraperitoneal dosing of the animals with a mixture of 50 mg/kg each of P and HQ, for four consecutive days. Similar patterns of adduct formation were detected in pooled bone marrow DNA isolated from the femurs of rats treated as above with doses of up to 100 mg/kg each of P and HQ. To determine whether there is a correlation between adduct formation and toxicity, we measured the number of nucleated bone marrow cells in the femurs of the treated animals (bone marrow cellularity). Measurement of the number of nucleated bone marrow cells in the femurs of these animals showed that treatment of rats with 100 mg/kg each of P and HQ caused significant depression of bone marrow cellularity, while the two lower doses (50 mg/kg and 75 mg/kg of each) caused no such depression. These results demonstrate that reactive metabolites of BZ form adducts *in vivo* with the DNA of rat bone marrow, the target organ of BZ toxicity. These adducts can be detected at doses of P plus HQ that do not cause significant depression of nucleated bone marrow cells. This suggests that a more sensitive assay is required for measurement of the effect of adducts on cellular physiology.

MATERIALS AND METHODS

Nuclease P1, T4 polynucleotide kinase, spleen phosphodiesterase, and molecular biology grade phenol were all obtained from Boeringer Mannheim Biochemicals. Micrococcal nuclease, Tris, dAMP, HEPES, Ches, and DTT were obtained from Sigma. Carrier-free [32P] phosphate (about 200 mCi/ml) was purchased from ICN Chemical

---

[1]Abbreviations used are: Ches = 2[N-cyclohexylamino]ethane sulfonic acid; and HEPES = N-2-hydroxyethylpiperazine-N''-2-ethane sulfonic acid.

and Radioisotope Division. RPMI 1640 medium (Cellgro) was obtained from Fisher. PEI-cellulose thin layer chromatography sheets (Machery-Nagle) were purchased from Brinkmann. Hydroquinone was of the highest grade possible and was obtained from Aldrich. Male Sprague-Dawley rats (Taconic Farms) were treated intraperitoneally with either 0.9 % NaCl or with 50 mg/kg, 75 mg/kg or 100 mg/kg each of P and HQ for 4 consecutive days. Animals were sacrificed 24 hours after the last treatment, and their femurs were removed and cleaned (of muscle). One femur per animal was used for measurement of bone marrow cellularity. The opposite femur was used for DNA isolation. The epiphysial plate on each end of the femur was removed. Bone marrow cells were flushed into a tissue culture tube by forcing a solution of RPMI medium containing 15 u/ml heparin and 10 mM HEPES, pH 7.4 through the femur using a syringe. Cells were collected by centrifugation at 630g at 4 degrees for 5 minutes. Following removal of red blood cells by hypotonic lysis, nucleated cells were again collected by centrifugation. For determination of bone marrow cellularity, the number of viable cells were counted using the trypan blue exclusion method. Viability was 98% or greater in all cases.

DNA was isolated from nucleated bone marrow cells using a solvent extraction technique as described by Reddy and Randerath (1987). The nucleated bone marrow cells were pooled for each treatment group prior to DNA isolation. 10 ug samples of DNA were digested and radiolabeled with [32P] ATP according to the postlabeling procedure of Reddy and Randerath (1986) except that Ches, pH 9.2 was used in the labeling buffer in place of bicine. Adducts were separated by multidimensional chromatography as described by Reddy et al (1989). Briefly, the following solvents were used : D1=2.3 M sodium phosphate, pH 5.8; D3=2.5 M lithium formate, 4.7 M urea, pH 3.5; D4=0.4 M sodium phosphate, 0.25 M Tris-HCl, 4.25 M urea, pH 8.0; D5=1.7 M sodium phosphate, pH 6.0. Labeled adducts were detected by screen intensified autoradiography. The amount of radioactivity incorporated into adduct spots was determined by Cerenkov counting of adduct spots. Adduct levels were expressed as Relative Adduct Labeling (RAL).

RESULTS AND DISCUSSION

Autoradiograms of thin layer chromatograms of postlabeled bone marrow DNA are shown in Figures 1 and 2. As shown on Figure 1 several adducts were detected in rat bone marrow following vehicle treatment. Three additional adduct spots (indicated by arrows), which were absent in postlabeled DNA isolated from vehicle control treated rats, were present on chromatograms of post-labeled DNA isolated from rats that were treated with 50 mg/kg each of P and HQ (Figure 1). As illustrated in Figure 2, those adducts could also be detected after treatment of the animals with 75 mg/kg and 100 mg/kg each of P and HQ. RAL values for these adducts were calculated to range between 1 and 8 X10-9.

Table 1 shows the effect of P plus HQ on rat bone marrow cellularity. In comparison to vehicle control treated animals, treatment with 50 mg/kg each of P and HQ had no significant effect on bone marrow cellularity. Significant depression of bone marrow cellularity was not detectable until 100 mg/kg of each metabolite was used. This dose was, however, lethal to 3 out of 5 animals. Adducts were thus detected in rat bone marrow DNA following treatment of the animals with doses of P and HQ that did not significantly effect bone marrow cellularity. Thus, at this time of observation, patterns of adduct formation did not correlate with this index of toxicity.

The *in vivo* detection of benzene metabolite/DNA adducts in bone marrow has not been reported before in rats, and is particularly significant since bone marrow is the target organ of BZ toxicity. The detection of bone marrow DNA adducts that were absent in control treated rats in animals treated with three different combinations of P/HQ suggests that adduct formation can be correlated with exposure to P/HQ at this range of exposure. Due to the low amounts of DNA that can be isolated from a single rat femur, these experiments were performed with pooled

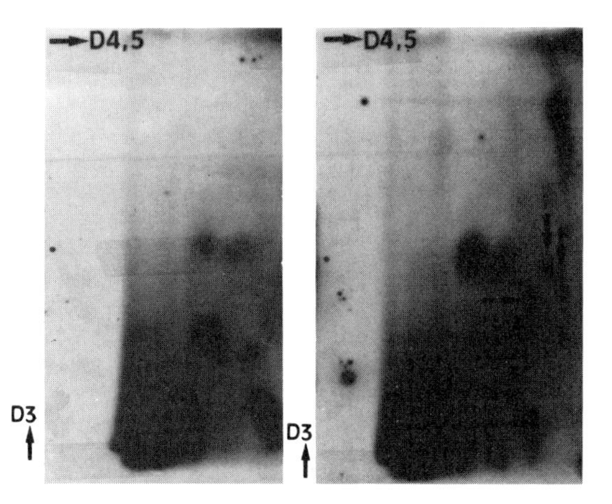

Figure 1.

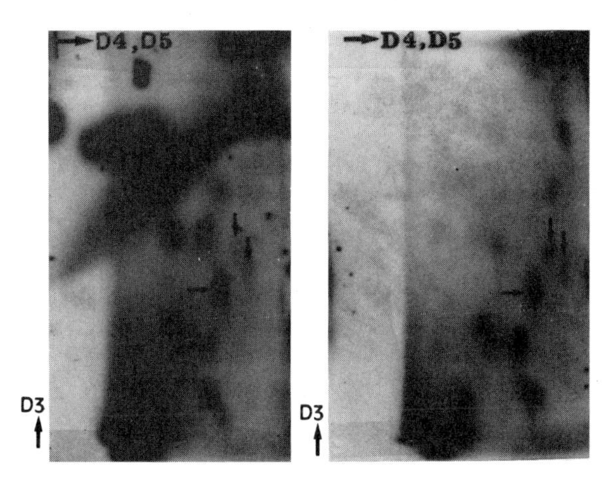

Figure 2.

samples of rat bone marrow DNA. Additional experiments with more groups of treated animals will allow us to accurately quantify and statistically analyze the levels of adducts measured.

Table 1. Effect of Phenol (P) and Hydroquinone (HQ) on Rat Bone Marrow Cellularity

Treatment	Nucleated Bone Marrow Cells X 10^6/Femur[C]	% of Control
Vehicle	30.6 + 1.2[A] (5)	100
50mg/kg each P+HQ	29.9 +9.4[A] (5)	97.7
75mg/kg each P+HQ	27.3[B] (2)	89.2
100mg/kg each P+HQ	18.0[B] (2)	58.8

[A]The data are expressed as mean ± the standard deviation.
[B]The data are expressed as mean only.
[C]The number in parentheses indicates the number of animals per treatment.

The detection of adducts rat bone marrow following treatment of the animals with doses of P plus HQ that have no significant effect on bone marrow cellularity at the time of observation suggests that a more sensitive index is required to assess the effect of adducts on bone marrow cell physiology. The effect of BZ/BZ metabolite treatment on the production of bone marrow growth factors or enzymes involved in benzene metabolism, such as cytochrome P450, are examples of factors that might be investigated.

REFERENCES

Cronkite, E.R., Bullis, J.E., Inoue, T. and Drew, R. (1984). Benzene inhalation produces leukemia in mice. *Toxicol. Appl. Pharmacol.* **75**, 358.

Jowa, L., Winkle, S., Kalf, G.F., Witz, G., and Snyder, R. (1986). Deoxyguanosine adducts from benzoquinone and hydroquinone. In: *Advances in Experimental Medicine and Biology: Biological Intermediates III* (J. Kocsis, D.J. Jollow, C.M. Witmer, J.O. Nelson, and R. Snyder, Eds.), Plenum Publishing, pp. 825-832.

Reddy, M.V., Blackburn, G.R., Irwin, S.E.,Kommineni, C., Mackerer, C.R., and Mehlman, M.A. (1989). A method for *in vitro* culture of rat Zymbal gland:use in mechanistic studies of benzene carcinogenesis in combination with 32P-postlabeling. *Environ. Health Perspect.* **82**, 239.

Reddy, M.V., Blackburn, G.R., Schreiner, C.A., Mehlman, M.A. and Mackerer, C.R. (1989). 32P analysis of DNA adducts in tissues of benzene-treated rats. *Environ. Health Perspect.* **82**, 253.

Reddy, M.V., and Randerath, K. (1986). Nuclease P1-mediated enhancement of sensitivity of 32P-postlabeling test for structurally diverse DNA adducts. *Carcinogenesis* **7**, 1543.

Reddy, M.V., and Randerath, K. (1987). 32P analysis of DNA adducts in somatic and reproductive tissues of rats treated with the anticancer antibiotic, mitomycin C. *Mutat. Res.* **179**, 75.

THE ROLE OF γ-GLUTAMYL TRANSPEPTIDASE IN HYDROQUINONE-GLUTATHIONE CONJUGATE MEDIATED NEPHROTOXICITY

Barbara A. Hill, Herng-Hsiang Lo, Terrence J. Monks and Serrine S. Lau

Division of Pharmacology and Toxicology
College of Pharmacy
The University of Texas at Austin
Austin, Texas

We have previously shown that oxidation of 2-bromohydroquinone (2-BrHQ) in the presence of glutathione (GSH) gives rise to a mixture of mono- and di-substituted GSH conjugates. Administration of the di-substituted conjugates to rats (30 μmol/kg, intravenously) caused extensive necrosis of renal proximal tubular cells, while the three mono-substituted GSH conjugates exhibited significantly less toxicity (Monks, et al., 1985, 1988). We subsequently investigated the toxicological properties of the products formed by the interaction of GSH with 1,4-benzoquinone (BQ). The chemical reaction of BQ with GSH resulted in the formation of GSH adducts that exhibited increasing degrees of GSH substitution (Lau, et al., 1988). Administration of these conjugates (50 μmol/kg, i.v.) to male Sprague-Dawley rats caused varying degrees of renal proximal tubular necrosis (Table 1). 2,3,5-(tri-GSyl)HQ was the most potent nephrotoxicant, demonstrating a 10 fold elevation in blood urea nitrogen (BUN) as compared to control animals. 2,3-, 2,5- and 2,6-(di-GSyl)HQ exhibited similar degrees of nephrotoxicity, eliciting an approximately 3 fold elevation in BUN over control levels. 2-(GSyl)HQ and 2,3,5,6-(tetra-GSyl)HQ were not toxic at the 50 μmol/kg dose (Table 1). However, 2-(GSyl)HQ produced signs of toxicity at a dose of 250 μmol/kg (BUN = 43.6 ± 4.3). None of these conjugates caused any elevations in serum pyruvate glutamate transaminase (SGPT) (Table 1) or histological alterations in the liver, suggesting that quinol-GSH conjugates may be selective nephrotoxicants.

Although the reason for the differential nephrotoxicity exhibited by the various isomers is unclear, the activity of γ-glutamyltranspeptidase (GGT) appears to be essential for the expression of toxicity. For example, pretreatment of animals with AT-125, (Acivicin; L-[αS-5S]-α-amino-3-chloro-4,5-dihydro-5-isoxazoleacetic acid, 10 mg/kg) an inhibitor of several glutamine utilizing enzymes (Weber, 1983), including GGT (Monks et al., 1985) protected rats from 2,3,5-(triGSyl)HQ mediated nephrotoxicity (Lau et al., 1988). Additional evidence for the importance of GGT in mediating 2,3,5-(tri-GSyl)HQ nephrotoxicity has also now been demonstrated in vitro using LLC-PK1 cells (a renal proximal tubular derived cell line). LLC-PK1 cells were grown in Costar Transwell Cell Culture Chambers which permits the cells to be treated either apically or basolaterally with 2,3,5-(tri-GSyl)HQ. Apical treatment of LLC-PK1 cells with 0.5 mM 2,3,5-(tri-GSyl)HQ caused a maximal LDH leakage (68.0 ± 5.9%) after 15 hr. However, basolateral treatment with 0.5 mM 2,3,5-(tri-GSyl)HQ caused only 17.5 ± 0.1% LDH leakage after 15 hr. These results correlated directly with GGT activity measured at confluency. The activity of GGT within the basolateral membrane represented only 17% of that present on the apical brush border membrane. Interestingly, in addition to causing LDH leakage, 2,3,5-(tri-GSyl)HQ also caused the dose-dependent appearance in the culture medium of the membrane bound

Biological Reactive Intermediates IV, Edited by C.M. Witmer *et al.*
Plenum Press, New York, 1990

enzyme GGT, with a concomitant decrease in the total activity of GGT in LLC-PK1 cells (56-62% inhibition between 0.5-1.0 mM; Table 2). Whether the disruption of the brush border membrane and the apparent inhibition of GGT activity contributes to the cytotoxicity of 2,3,5-(triGSyl)HQ, or whether these events occur as a consequence of the cytotoxicity remain to be determined. However, the data do indicate that the brush border membrane may be a potential target for these conjugates.

Table 1. Comparative nephrotoxicity of HQ-GSH conjugates

Compound	BUN (mg%)	SPGT (U/L)
2-(GSyl)HQ	18.7 ± 1.9	23.1 ± 2.6
2,3-(DiGSyl)HQ	46.8 ± 12.1	30.1 ± 3.0
2,5-(DiGSyl)HQ	45.0 ± 5.9	28.9 ± 4.6
2,6-(DiGSyl)HQ	51.8 ± 2.0	29.4 ± 5.7
2,3,5-(TriGSyl)HQ	171.6 ± 13.7	27.9 ± 1.9
2,3,5,6-(TetraGSyl)HQ	16.4 ± 1.4	34.2 ± 5.8
Control	16.4 ± 1.4	29.1 ± 3.8

Male Sprague-Dawley rats received 50 μmol/kg i.v. of the HQ-GSH conjugates. Blood samples were obtained from the retro-orbital sinus 24 hr after dosing and blood urea nitrogen (BUN) and serum glutamate pyruvate transaminase (SGPT) were determined. Control animals received saline. Values represent the mean ± S.E. (n=4).

Table 2. The Effect of 2,3,5-(Tri-glutathion-S-yl)hydroquinone on γ-Glutamyl Transpeptidase in LLC-PK1 Cells

Dose (mM)	GGT units in the media	GGT units in the cells	Total GGT activity/well
0	50.4 ± 5.1 (8.0%)	629 ± 61.2	679 ± 66.1
0.25	48.4 ± 9.6 (11.7%)	370 ± 61.8	418 ± 71.0
0.50	133 ± 13.0 (51.2%)	127 ± 75.9	260 ± 85.2
1.0	180 ± 51.5 (60.6%)	117 ± 52.2	297 ± 93.6

Unit Definition: 1 unit = 1 nmol of p-nitroanaline formed per minute at 25°C. Confluent cell cultures grown on 6 well culture dishes (9.5 cm^2/well) were washed twice with 20 mM HEPES/Tris buffer, pH 7.3, containing 116.5 mM NaCl, 6.4 mM KCl, 1.8 mM $CaCl_2/2H_20$, 0.8 mM $MgSO_4/7\ H_2O$ and 5.5 mM D-glucose (HEPES/Tris-EBSS) (Schaeffer and Stevens, 1986) followed by the addition of 0.5 mM 2,3,5-(triGSyl)HQ dissolved in the same buffer. The cells were then incubated at 37°C in an atmosphere of 5% CO_2/95% air for 15 hr. The activity of GGT in the medium and triton (0.5%)-treated lysates was determined according to Sigma Bulletin 545. Values represent mean ± S.D. (n=3). Numbers in parentheses represent % of total GGT that leaked into the medium.

The important role that GGT plays in mediating 2,3,5-(triGSyl)HQ nephrotoxicity suggested that differences in the metabolism of the various HQ-GSH conjugates might contribute to the differential nephrotoxicity. We therefore determined the kinetics of the GGT-mediated hydrolysis (in the absence of glycylglycine) and transpeptidation (in the presence of glycylglycine) of the HQ-GSH conjugates. There was a decrease in the efficiency of the GGT-mediated hydrolysis and transpeptidation (Vmax/Km) with increased GSH substitution (Table 3). Thus, the most nephrotoxic isomer, 2,3,5-(triGSyl)HQ was the poorest substrate for GGT. However, since the capacity of renal GGT (approx. 4 x 10^5 units/kidney) to metabolize the HQ-GSH conjugates appears to far exceed the dose required to elicit toxicity (10-50 μmol/kg), it is unlikely that the differential nephrotoxicity observed after HQ-GSH administration *in vivo* is a consequence of their differential metabolism by GGT. The finding that 2,3,5,6-(tetraGSyl)HQ does not appear to be a substrate for GGT, at least under the present experimental conditions, probably contributes to the lack of toxicity of this conjugate.

Table 3. The kinetic parameters for the GGT mediated hydrolysis (- glycylglycine) and transpeptidation (+ glycylglycine) of HQ-GSH conjugates

HQ-GSH	- glycylglycine		+ glycylglycine		Vmax/Km	
Conjugate	Km	Vmax	Km	Vmax	-glygly	+glygly
2-(GSyl)HQ	37	18	68	123	0.49	1.80
2,3-(diGSyl)HQ	62	13	139	88	0.20	0.63
2,5-(diGSyl)HQ	222	45	91	95	0.20	1.04
2,6-(diGSyl)HQ	75	19	127	127	0.26	1.00
2,3,5-(triGSyl)HQ	36	5	89	12	0.14	0.14
2,3,5,6-(tetraGSyl)HQ	--	--	--	--	--	--

The rate of hydrolysis (-glygly) and transpeptidation (+glygly) was determined by measuring the disappearance of each substrate at different concentrations (2 μM - 160 μM) by HPLC. Chromatographic conditions consisted of a linear gradient of methanol : water : acetic acid (5:94:1) to methanol : water : acetic acid (20:79:1) at a flow rate of 1 ml/min over 20 min. Km and Vmax were obtained from the Lineweaver-Burk equation.

In summary, the selective nephrotoxicity of HQ-GSH conjugates is probably a consequence of their selective uptake into renal cells, as the corresponding cysteine and/or cysteinlyglycine conjugate, mediated by GGT, followed by oxidation to the quinone. The differential nephrotoxicity of the HQ-GSH conjugates is probably due to a combination of physiological, biochemical and electrochemical factors. The *mechanism* of toxicity of HQ-GSH conjugates remains unclear and is currently under investigation. (Supported by NIH grants ES 04662 and GM 39338).

REFERENCES

Lau, S.S., Hill, B.A., Highet R.J. and Monks, T.J. (1988). Sequential oxidation and glutathione addition to 1,4-benzoquinone: Correlation of toxicity with increased glutathione substitution. *Molec. Pharmacol.* **34**, 829-836.

Monks, T.J. Lau, S.S., Highet, R.J. and Gillette, J.R. (1985). Glutathione conjugates of 2-bromohydroquinone are nephrotoxic. *Drug Metab. Dispos.* **13**, 553-559.

Monks, T.J., Highet, R.J. and Lau, S.S. (1988). 2-Bromo-(diglutathion-S-yl)hydroquinone nephrotoxicity: Physiological biochemical and electrochemical determinants. *Molec. Pharmacol.* **34**, 492-500.

Schaeffer, V.H. and Stevens, J.L. (1986). The transport of S-cysteine conjugates in LLC-PK1 cells and its role in toxicity. *Molec. Pharmacol.* **31**, 506-512.

Weber, G (1983). Biochemical strategy of cancer cells and the design of chemotherapy. *Cancer Res.*, **43**, 3466-3492.

CYTOCHROME P450IIE1 METABOLISM OF PYRIDINES: EVIDENCE FOR PRODUCTION OF A REACTIVE INTERMEDIATE WHICH EXHIBITS REDOX-CYCLING ACTIVITY AND CAUSES DNA DAMAGE

Sang Geon Kim and Raymond F. Novak

The Institute of Chemical Toxicology
Wayne State University
Detroit, MI 48201

To date some thirty pyridine derivatives have been identified as being present in cigarette smoke and smoke condensate as revealed by chromatographic analysis of smoke composition (Schumacher, et al., 1977; Moree-Testa, et al.; 1984). 3-Hydroxypyridine has been shown to be present in tobacco smoke and its content was estimated to be as high as 0.13 mg per cigar smoked (Schmelt and Stedman, 1962; Osman et al., 1963). Although studies on the metabolism of 3- and 2-hydroxypyridine in microorganisms have been reported (Houghton and Cain, 1972), information concerning the cytochrome P-450-dependent metabolism of these agents or the specificity of individual forms of P-450 for these substrates is lacking.

Recent research has shown the solvent pyridine to be rapid and efficacious inducer of cytochromes P-450 in rabbits and rats (Kaul and Novak, 1987; Kim et al., 1988; Novak et al., 1989). Moreover, immunochemical studies revealed that P450IIE1 was a major form of P-450 induced in rabbit and rat hypatocytes and that this form exhibited a unique substrate specificity for the conversion of pyridine to pyridine N-oxide (Kaul and Novak, 1987; Kim et al., 1988; Novak et al., 1989).

In view of the specificity of P450IIE1 for pyridine, experiments were initiated to examine the scope of this activity towards other pyridine derivatives. The oxidation of hydroxylated pyridine derivatives such as 2- or 3-hydroxypyridine could result in the formation of catechols or quinones capable of producing cytotoxicity. The conversion of 2- or 3-hydroxypyridine to 2,5-dihydroxypyridine by cytochrome P450IIE1 and the resultant effects of the redox cycling activity of this metabolite on DNA integrity was examined (Kim and Novak, 1990). The specificity of P450IIE1 for oxidation of 3-hydroxypyridine to 2,5-dihydroxypyridine is reviewed in concert with the ability of this metabolite to stimulate cofactor consumption and superoxide production. Data showing that reduced glutathione diminishes both the rate of NADPH oxidation and superoxide production is also presented. These results are of particular significance given the strong correlation of increased incidence of cancer associated with alcohol and tobacco consumption (McCoy and Wynder, 1979; Lieber, et al., 1979; Blot, et al., 1988).

MATERIALS AND METHODS

Materials

2-Hydroxypyridine and 3-hydroxypyridine were obtained from Aldrich Chemical

Company, Milwaukee, WI. 2,5-Dihydroxypyridine was kindly supplied by Dr. E. J. Behrman, Ohio State University, Columbus, Ohio.

Preparation of Microsomes

Male New Zealand white rabbits (2.0-2.5 kg) were injected daily for 5 days with pyridine (100 mg/kg, i.p.), isosafrole (150 mg/kg, i.p., 4d in corn oil) or given 0.1% phenobarbital in drinking water *ad libitum* and fasted 18 hr prior to sacrifice. Microsomes were prepared by differential centrifugation (Kaul and Novak, 1987). Protein was assayed by the method of Lowry (Lowry et al., 1951) and cytochrome P-450 content was determined according to the method of Omura and Sato (Omura and Sato, 1964).

3-Hydroxypyridine Hydroxylase Assay

2,5-Dihydroxypyridine formation was assayed by HPLC according to a modified method of Blaauboer and Paine (Blaauboer and Paine, 1980) as has been described previously (Kaul and Novak, 1987; Kim et al, 1988).

Microsomal NADPH Oxidation

NADPH oxidation was monitored at 37°C in an incubation mixture containing 0.1 M potassium phosphate buffer (pH 7.5), 0.1 mg phenobarbital-induced microsomal protein, 0.1 mM NADPH in the presence or absence of 2,5-dihydroxypyridine (0.5 mM) in a total volume of 1 ml. Following a 3 min preincubation period, the reaction was initiated by NADPH addition. NADPH oxidation was determined by the absorbance at 340 nm using an extinction coefficient of 6.33 $mM^{-1}cm^{-1}$.

SOD-inhibitable Cytochrome c Reduction

The SOD-inhibitable rate of acetylated cytochrome c reduction was employed as an index of $O_2\cdot^-$ production. The incubation mixture consisted of 0.1 M potassium phosphate buffer (pH 7.5), 0.1 mg phenobarbital-induced microsomal protein, 0.1 mM NADPH, 0.1 mM Na2EDTA and 26 μM acetylated cytochrome c and incubations were performed in the presence and absence of (0.5 mM) 3-hydroxypyridine or 2,5-dihydroxypyridine. The incubations were performed at 37°C and the reactions initiated by NADPH addition to the preequilibrated reaction mixture. The rate of reduction was monitored by following the increase in the absorbance at 550 nm (extinction coefficient, 19.6 $mM^{-1}cm^{-1}$). The rate of $O_2\cdot^-$ production was calculated from the difference in the rate of acetylated cytochrome c reduction in the absence and presence of 350 μg/ml SOD.

RESULTS AND DISCUSSION

Pyridine-induced rabbit hepatic microsomes, which have been shown to contain an elevated level of P450IIE1, catalyzed a 9-fold increase in the rate of hydroxylation of 2,5-dihydroxypyridine production, relative to uninduced microsomes, with saturation occurring at 0.5 mM substrate (Kim and Novak, 1990). Although isosafrole-induced microsomes also showed a high affinity monooxygenase (K_M= 134 μM) for the production of 2,5-dihydroxypyridine, only a 1.6-fold increase in V_{max}, relative to control microsomes was monitored (Kim and Novak, 1990). Phenobarbital-induced microsomes exhibited a higher K_M value, 590 μM, and an intermediate V_{max} (2.5 nmol/min/mg protein) which represented a 4-fold increase over control. When metabolic rates were normalized to account for the different P-450 levels present in the various microsomal suspensions, a pronounced substrate specificity was evident with a V_{max} value of 3.0 nmol/min/nmol P-450 for pyridine-induced microsomes; a value ~3- to 4-fold greater than that associated with either control or phenobarbital-induced microsomal suspensions (Kim and Novak, 1990).

That P450IIE1 was the primary catalyst of 3-hydroxypyridine oxidation to 2,5-dihydroxypyridine was confirmed through competitive inhibition with p-nitrophenol, a substrate specific for P450IIE1, and through studies using the reconstituted enzyme system (Kim and Novak, 1990).

Interestingly, the rapid rate of the metabolism of 3-hydroxypyridine in pyridine-induced microsomes resulted in a ~25% decrease in total P-450 content suggesting that the production of 2,5-dihydroxypyridine caused significant damage to microsomal P-450 (P45OIIE1).

Because the production of 2,5-dihydroxypyridine affected P-450 content and somewhat unusual kinetics were observed at saturating substrate concentrations, the redox cycling activity of this metabolite was examined. 2,5-Dihydroxypyridine (0.5 mM) stimulated the rate of cofactor consumption over 4-fold, to ~69 nmol/min/mg protein, while 3-hydroxypyridine failed to increase the basal rate of cofactor oxidation in phenobarbital-induced hepatic microsomes (Figure 1). Since a reduced semiquinone free radical was likely the reactive intermediate, studies employing reduced glutathione (GSH) (0.5 mM) in the incubation medium were performed to examine the effect of GSH on the rate of cofactor oxidation stimulated by 2,5-dihydroxypyridine (Kim and Novak, 1990).

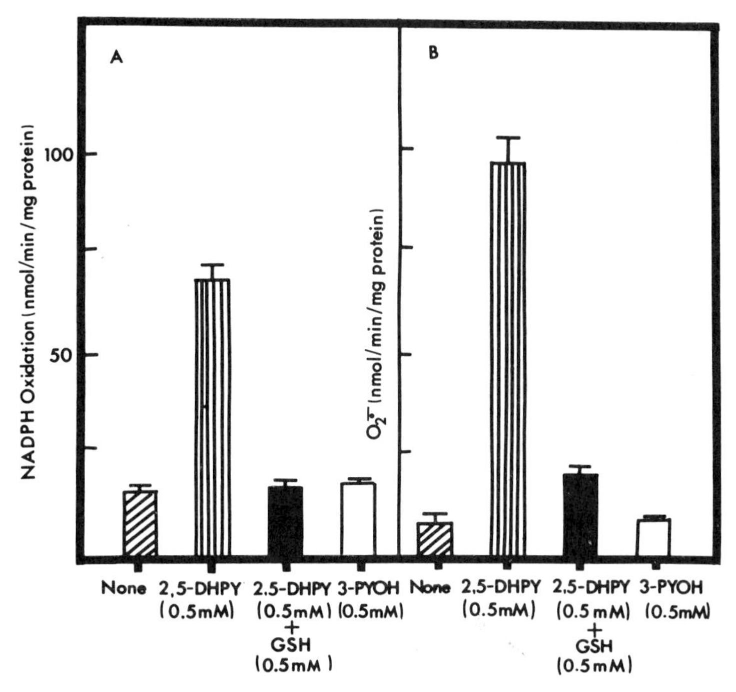

Figure 1. Redox-cycling activity of 3-hydroxypyridine and 2,5-dihydroxypyridine and the relative effects of reduced glutathione on this activity in rabbit hepatic microsomal suspensions. Panel A: The effects of 2,5-dihydroxypyridine (2,5-DHPY), 2,5-dihydroxypyridine and glutathione (2,5-DHPY + GSH) and 3-hydroxypyridine (3-PYOH) on cofactor oxidation as compared to control. Panel B: The effects of 2,5-dihydroxypyridine (2,5-DHPY), 2,5-dihydroxypyridine and glutathione (2,5-DHPY + GSH) and 3-hydroxypyridine (3-PYOH) on superoxide production relative to control.

GSH (0.5 mM) effectively decreased the rate of NADPH oxidation to the basal level (Figure 1, Panel A). In association with increased cofactor oxidation, 2,5-dihydroxypyridine at 0.5 mM increased $O_2 \cdot^-$ production by ~12-fold, while 3-hydroxypyridine (0.5 mM) alone failed to affect the rate of $O_2 \cdot^-$ generation (Figure 1, Panel B). GSH addition (0.5 mM) effectively decreased the rate of $O_2 \cdot^-$ production by ~80%, (Figure 1, Panel B), consistent with the ability of GSH to trap radical intermediate(s) and thereby diminish $O_2 \cdot^-$ production.

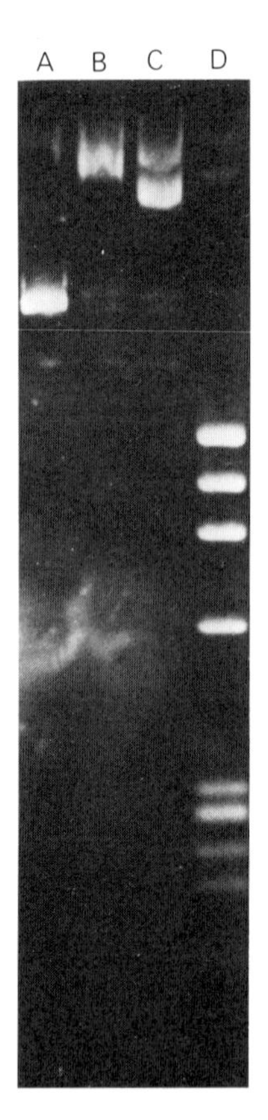

Figure 2. Strand scission of ∅X-174 supercoiled DNA by 2,5-dihydroxypyridine. ∅X-174 DNA (2.0 μg) was incubated with 5 mM 2,5-dihydroxypyridine for 60 min at 37°C (pH 7.4). The open circular, supercoiled and linear forms of DNA were separated on a 1.2% agarose gel. Lane A contains the band associated with the linear form of ∅X-174 DNA (0.3 μg) which was digested with the restriction endonuclease PstI. Lanes B and C contain 0.3 μg of the ∅X-174 DNA which had been incubated in the presence and absence of 2,5-dihydroxypyridine, respectively. Lane D is the HaeIII digest of ∅X-174 DNA showing the different molecular weight fragments. These results indicate that the DNA damage produced by 2,5-dihydroxypyridine was primarily single strand breakage of ∅X-174 (i.e. conversion of the supercoiled form to the open circular form).

The ability of 2,5-dihydroxypyridine to damage DNA was examined. 2,5-Dihydroxypyridine converted supercoiled ⌀X-174 DNA to the open circular form which was shown by a decreased intensity of the supercoiled ⌀X-174 DNA band and an increase in intensity of the band associated with the open circular form (Figure 2). The change in ⌀X-174 DNA topology produced by 2,5-dihydroxypyridine was concentration-dependent (10 µM to 1 mM at pH 7.4) with an estimated EC_{50} of ~60 µM (Kim and Novak, 1990). Moreover, neither pyridine (1 mM) nor 3-hydroxypyridine (1 mM) when incubated with ⌀X-174 DNA produced DNA damage. Catalase (0.5-1.0 µg) inhibited the DNA strand scission produced by 1 mM 2,5-dihydroxypyridine; in contrast, superoxide dismutase (1 µg) failed to protect DNA from damage. DNA damage also resulted from conversion of 3-hydroxypyridine to 2,5-dihydroxypyridine in a reconstituted enzyme system (Kim and Novak, 1990).

In summary, the alcohol-inducible form of P-450 (P450IIE1) catalyzes the conversion of 3-hydroxypyridine, a significant constituent of tobacco smoke, to 2,5-dihydroxypyridine. This metabolite stimulates the rate of cofactor oxidation and superoxide generation and produces DNA damage in vitro. Elevation of P450IIE1 levels through alcohol consumption, in conjunction with exposure to pyridine or other nitrogen heterocycles present in tobacco smoke, may result in deleterious effects which contribute to cellular toxicity and to an increased risk of cancer. Moreover, the ability of glutathione to conjugate the reactive intermediate formed may result in conjugates which are toxic to different organs or cell types as has been reported for the glutathione conjugates of bromohydroquinone or benzoquinone (Monks, et al., 1985; Lau, et al., 1988).

ACKNOWLEDGEMENTS

The research was supported by USPHS Grant ES03656 from the National Institutes of Environmental Health Sciences.

REFERENCES

Blaauboer, B.J. and Paine, A.J. (1980). Reversed phase high performance liquid chromatography of heterocyclic aromatic N-oxides: Application to measurement of N-oxidation by microsomes and rat hepatocytes in primary culture. *Xenobiotica* **10**, 655-660.

Blot, W.J., McLaughlin, J.K., Winn, D.M., Austin, D.F., Greenberg, R.S., Preston-Martin S., Bernstein, L., Schoenberg, J.B., Stemhagen, A. and Fraumeni Jr., J.F. (1988). Smoking and drinking in relation to oral and pharyngeal cancer. *Cancer Res.* **48**, 3282-3287.

Houghton, C. and Cain, R.B. (1972). Microbial metabolism of the pyridine ring. *Biochem. J.* **130**, 879-893.

Kaul, K. L. and Novak, R.F. (1987). Inhibition and induction of rabbit liver microsomal cytochrome P-450 by pyridine. *J. Pharmacol. Exp. Ther.* **243**, 384-390.

Kim, S.G. and Novak, R.F. (1990). Role of P450IIE1 in the metabolism of 3-hydroxypyridine a constituent of tobacco smoke: Redox cycling and DNA strand scission by the metabolite 2,5-dihydroxypyridine. *Cancer Res.* in press.

Kim, S.G., Williams, D.E., Scheutz, E.G., Guzelian, P.S. and Novak, R.F. (1988). Pyridine induction of cytochrome P-450 in the rat: Role of P-450j (alcohol-inducible form) in pyridine N-oxidation. *J. Pharmacol. Exp. Ther.* **246**, 1175-1182.

Lau, S.S., Hill, B.A., Highet, R.J., and Monks, T.J. (1988). Sequential oxidation and glutathione addition to 14-benzoquinone: Correlation of toxicity with increased glutathione substitution. *Mol. Pharmacol.* **34**, 829-836.

Lieber, C.S., Seitz, H.K., Garro, A.J. and Worner, T.M. (1979). Alcohol-related diseases and carcinogenesis. *Cancer Res.* **39**, 2863-2886.

Lowry, O.H., Rosebrough, N.J., Farr, A.L. and Randall, R.T. (1951). Protein measurement with the Folin phenol agent. *J. Biol. Chem.* **193**, 265-275.

McCoy, G.D. and Wynder, E.L. (1979). Etiological and preventive implications in alcohol carcinogenesis. *Cancer Res.* **39**, 2844-2850.

Monks, T.J., Lau, S.S., Highet, R.J. and Gillette, J.R. (1985). Glutathione conjugates of 2-bromohydroquinone are nephrotoxic. *Drug Metab. Dispos.* **13**, 553-559.
Moree-Testa, P., Saint-Jalm, Y. and Testa, A. (1984). Identification and determination of imidazole derivatives in cigarette smoke. *J. Chromatogr.* **290**, 263-274.
Novak, R.F., Kaul, K.L. and Kim, S.G. (1989). Induction of the alcohol-inducible form of cytochrome P-450 by nitrogen-containing heterocycles: Effects on pyridine N-oxide production. *Drug Metab. Rev.* **20**, 781-792.
Omura, T. and Sato, R. (1964) The carbon monoxide-binding pigment of liver microsomes. I. Evidence for its hemoprotein nature. *J. Biol. Chem.* **239**, 2370-2378.
Osman, S., Schmeltz, I., Higman, H.C. and Stedman, R.L. (1963). Volatile Phenols of Cigar Smoke. *Tobacco Sci.* **1117**, 144-146.
Schmelt, I. and Stedman, R.L. (1962). Pyridin-3-ol in Cigar Smoke. Chem. Ind. (London), 1244-1255.
Schumacher, J.N., Green, C.R., Best, F.W. and Newell, M.P. (1977). Smoke Composition. An extensive investigation of the water-soluble portion of cigarette smoke. *J. Agric. Food Chem.* **25**, 310-320.

MORPHOLOGICAL CELL TRANSFORMATION AND DNA ADDUCTION BY BENZ(J)ACEANTHRYLENE AND ITS PRESUMPTIVE REACTIVE METABOLITES IN C3H10T1/2CL8 CELLS[1]

Jessica Lasley*, Susan Curti*, Jeffrey Ross**, Garret Nelson*, Ramiah Sangaiah***, Avram Gold*** and Stephen Nesnow[1]**

Environmental Health Research* and Testing and Carcinogenesis and Metabolism Branch**, U.S. Environmental Protection Agency
Research Triangle Park, NC 27709 (U.S.A.)
Department of Environmental Sciences and Engineering***
University of North Carolina at Chapel Hill
Chapel Hill, North Carolina 27514 (U.S.A.)

Abbreviations: B(j)A, benz(j)aceanthrylene; C3H10T1/2, C3H10T1/2CL8; B(j)A-diol-epoxide, 9,10-dihydro-9,10-dihydroxy-B(j)A-7,8-oxide.

Benz(j)aceanthrylene [B(j)A], a cyclopenta-fused polycyclic aromatic hydrocarbon found in coal combustion emissions [Grimmer et al., 1985; Schmidt et al., 1986], is a strong inducer of morphological transforming activity in C3H10T1/2CL8 (C3H10T1/2) mouse embryo fibroblasts [Mohapatra et al., 1987]. The metabolism of B(j)A by these cells has been studied and the results differ from those found with Aroclor-1254 induced rat liver S9. In C3H10T1/2 cells the major metabolite is 9,10-dihydro-9,10-dihydroxy-B(j)A with minor amounts of 1,2-dihydro-1,2-dihydroxy-B(j)A [Mohapatra et al., 1987], while in rat liver S9 the major metabolite is 1,2-dihydro-1,2-dihydroxy-B(j)A [Nesnow et al., 1988]. These results suggest that while the major route of metabolic activation by Aroclor-1254 induced rat liver S9 is through B(j)A-1,2-oxide, the arene oxide at the cyclopenta-ring, metabolic activation of B(j)A in C3H10T1/2 cells might be via two routes: botharene oxide [B(j)A-1,2-oxide] formation and bay-region diol-epoxide formation [9,10-dihydro-9,10-dihydroxy-B(j)A-7,8-oxide]. This study was undertaken to clarify the role of the putative reactive intermediates, 9,10-dihydro-9,10-dihydroxy-B(j)A-7,8-oxide [B(j)A-diol-epoxide], and B(j)A-1,2-oxide in the metabolic activation of B(j)A in C3H10T1/2 cells.

MATERIALS AND METHODS

Chemicals

Chemicals were either purchased from commercial sources or prepared according to previously reported procedures. The details of the preparation of B(j)A-diol-epoxide will be reported elsewhere (A. Gold and R. Sangaiah,unpublished data). Biochemicals and medium components were obtained from Grand Island Biological Co. (Grand Island, NY).

[1] Correspondence: Dr. Stephen Nesnow, Carcinogenesis and Metabolism Branch, MD-68, U.S. Environmental Protection Agency, Research Triangle Park, NC 27711 (U.S.A.).

Morphological Transformation Assays

The mouse embryo fibroblast cell line C3H10T1/2 (passage 7) was employed using the morphological transformation and cytotoxicity procedures of Nesnow *et al.* (1982) without any modification. C3H10T1/2 cells were seeded for cytotoxicity studies at 200 per dish (6 dishes/concentration) and transformation studies at 1000 per dish (24 dishes/concentration) in 60-mm plastic petri dishes in 5 ml of medium. Cells were treated with PAH dissolved in acetone either one day [for cytotoxicity assays] or 5 days [for transformation assays] after seeding. After a 24-hr exposure, the cells were fed with fresh complete medium containing 25 μg/ml Garamycin (Schering Corp., Kenilworth, NJ). One week after the treatment, the dishes treated for cytotoxicity determinations were fixed with methanol and stained with Giemsa. The medium in the dishes treated for cell transformation was changed weekly, and at the end of 6 weeks, the cells were fixed, stained, andscored for morphological transformation [Reznikoff et al., 1973].

DNA Adducts

C3H10T1/2 fibroblasts were seeded at 50,000 cells per flask for DNA adduct analysis [Nesnow et al., 1989]. Twenty flasks with cells in mid-log growth (approximately 70% confluent) were treated and after a 24-hour exposure, the flasks were washed and the cells were trypsinized. DNA was isolated from C3H10T1/2 cells by spermine extraction [Ross et al., unpublished] and DNA adducts were analyzed by the ^{32}P-postlabeling assay [Gupta et al., 1985] after enhancement of the adduct detection limit using the P1 nuclease method [Reddy and Randerath, 1986].

RESULTS AND DISCUSSION

C3H10T1/2 cells became morphologically transformed after treatment with B(j)A, B(j)A-1,2-oxide [Bartczak et al., 1987], or B(j)A-diol-epoxide. B(j)A and B(j)A-diol-epoxide displayed similar concentration-response curves from 0 to 0.5 μg/ml (Figure 1). B(j)A produced Type II and III foci in 75% of dishes treated at 0.5 μg/ml and induced 1.38 foci/dish. The number of foci per dish increased more than 3 fold as the concentration of chemical was increased from 1.0 to 2.5 μg/ml. All of the dishes treated with 2.5 μg/ml of B(j)A contained morphologically transformed foci. B(j)A was relatively non-cytotoxic over the concentration range used, maximally decreasing survival by 15%. B(j)A-diol-epoxide was highly active inproducing foci: 71% of the dishes contained at least one Type II or III focus at 0.5 μg/ml with 1.33 foci/dish (based on all treated dishes). At higher concentrations the transformation response decreased, probably as a function of the increased cytotoxicity which resulted in 0% survival at 1.0 μg/ml. Forty-six per cent of dishes treated with B(j)A-1,2-oxide at 0.5 μg/ml contained at least one morphologically transformed focus with a mean of 1.08 foci/dish (based on alltreated dishes). B(j)A-1,2-oxide induced approximately the same cytotoxicity response as B(j)A.

B(j)A DNA adducts in C3H10T1/2 cells were isolated, separated, identified, and quantitated using the ^{32}P-postlabeling method. B(j)A forms two major groups of adducts: a group of B(j)A-1,2-oxide-2'-deoxyguanosine adducts and a group ofB(j)A-diol-epoxide-2'-deoxyguanosine adducts. These adducts were identified by cochromatography with the products of reaction of B(j)A-diol-epoxide and B(j)A-1,2-oxide with polydeoxyguanylic acid. Furthermore, the profile of B(j)A-diol-epoxide adducts in C3H10T1/2 cells was identical with that obtained with polydeoxyguanylic acid. Similar results were obtained in C3H10T1/2 cells treated with B(j)A-1,2-oxide. Minor amounts of B(j)A-diol-epoxide and B(j)A-1,2-oxide adducted to 2'-deoxyadenosine and 2'-deoxycytosine were also observed with the homopolymer reactions. DNA adduction in C3H10T1/2 cells via the B(j)A-diol-epoxide pathway occurred to a much greater extent than via the B(j)A-1,2-oxide pathway.

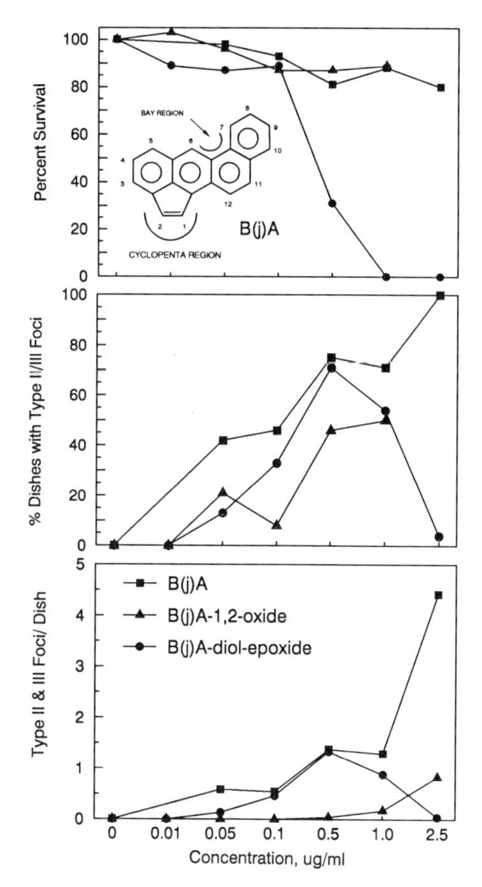

Figure 1. Cytotoxicity and morphological transforming activity of B(j)A and its presumptive reactive intermediates in C3H10T1/2 cells.

In the present study B(j)A and B(j)A-diol-epoxide were equipotent morphological transforming agents in C3H10T1/2 cells. B(j)A-1,2-oxide also induced transformation but was approximately 30% less active. Although evidence supports activation of B(j)A both at the cyclopenta-ring and at the bay region in C3H10T1/2 cells, the quantitative DNA adduct and transformation data suggest that the B(j)A-diol-epoxide pathway may predominate.

ACKNOWLEDGEMENTS

The research described in this article has been reviewed by the Health Effects Research Laboratory, U.S. Environmental Protection Agency and approved for publication. Approval does not signify that the contents necessarily reflect the views of the Agency nor does mention of trade names or commercial products constitute endorsement or recommendation for use. This work was supported in part by the U.S. Environmental Protection Agency, Grant CR-811817 (UNC), Contract 68-02-4031 (EHRT), and USPHS Grant No. ES03433 (UNC).

REFERENCES

Bartczak, A.W., Sangaiah, R., Ball, L.M., Warren, S.H. and Gold, A. (1987). Synthesis and bacterial mutagenicity of the cyclopenta oxides of the four cyclopenta-fused isomers of benzanthracene. *Mutagenesis* **2**, 101-105.

Grimmer, G., Jacob, J., Dettbarn, G. and Naujack, K.W. (1985). Determination of polycyclic aromatic hydrocarbons, azaarenes, and thiaarenes emitted from coal-fired residential furnaces by gas chromatographic/mass spectrometry. *Fresenius Z. Anal. Chem.* **322**, 595-602.

Gupta, RC. (1985). Enhanced sensitivity of ^{32}P-postlabeling analysis of aromatic carcinogen:DNA adducts. *Cancer Res.* **45**, 5656-5662.

Mohapatra, N., MacNair, P., Bryant, B.J., Ellis, S., Rudo, K., Sangaiah, R., Gold, A. and Nesnow, S. (1987). Morphological transforming activity and metabolism of cyclopenta-fused isomers of benz[a]anthracene in mammalian cells. *Mutation Res.* **188**, 323-334.

Nesnow, S., Garland, H. and Curtis ,G. (1982). Improved transformation of C3H10T1/2CL8 cells by direct- and indirect-acting carcinogens. *Carcinogenesis* **3**, 377-380.

Nesnow, S., Easterling, R.E., Ellis, S., Watts, R. and Ross, J. (1988). Metabolism of benz(j)aceanthrylene (cholanthrylene) and benz(l)aceanthrylene by induced rat liver S9. *Cancer Lett.* **39**, 19-27.

Nesnow, S., Ross, J., Mohapatra, N., Gold, A., Sangaiah, R. and Gupta, R. (1989). DNA adduct formation, metabolism, and morphological transforming activity of aceanthrylene in C3H10T1/2CL8 cells. *Mutation Res.* **222**, 223-235.

Reddy, M.V. and Randerath, K. (1986). Nuclease P1-mediated enhancement of sensitivity of ^{32}P-postlabeling test for structurally diverse DNA adducts. *Carcinogenesis* **7**, 1542-1551.

Reznikoff, C.A., Bertram, J.S., Brankow, D.W. and Heidelberger, C. (1973). Quantitative and qualitative studies of chemical transformation of cloned C3H mouse embryo cells sensitive to post confluence inhibition of cell division. *Cancer Res.* **33**, 3239-3249.

Ross, J., Nelson, G., Kligerman, A., Erexson, G., Bryant, M., Earley, K., Gupta, R. and Nesnow, S. Formation and persistence of novel benzo(a)pyrene adducts in rat lung, liver and peripheral blood lymphocyte DNA. Submitted.

Schmidt, W., Grimmer, G., Jacob, J. and Dettbarn, G. (1986). Polycyclic aromatic hydrocarbons and thiaarenes in the emission from hard coal combustion. *Toxicol. Environ. Chem.* **13**, 1-16.

COVALENT BINDING OF OXIDATIVE BIOTRANSFORMATION REACTIVE INTERMEDIATES TO PROTEIN INFLUENCES HALOTHANE-ASSOCIATED HEPATOTOXICITY IN GUINEA PIGS

Richard C. Lind and A. Jay Gandolfi

Department of Anesthesiology
University of Arizona
College of Medicine
Tucson, Arizona 85724

ABSTRACT

Halothane ($CF_3CBrClH$; H) biotransformation by cyt P-450 produces reactive intermediates along both oxidative (acyl chloride) and reductive (free radical) pathways that ultimately generate the metabolites trifluoroacetic acid and F^-, respectively. Inhibiting oxidative metabolism with deuterated halothane (d-H) reduces resultant injury in our guinea pig model of acute H hepatotoxicity. To elucidate whether covalent binding of rective intermediates to proteins (oxidative pathway) or lipids (reductive pathway) is a mechanism of necrosis, male outbred Hartley guinea pigs (600-725 g), N=8, were exposed to either 1% (v/v) H or d-H at either 40% or 10% O_2 for 4 hr. One-half of the animals were killed immediately after exposure for binding studies; the remainder at 96 hr post for evaluation of hepatotoxicity. Covalent binding of halothane intermediates to liver protein or lipid was determined by measuring the fluoride content of the bound moieties. The use of d-H and/or 10% O_2 during exposure led to 63-88% reductions ($p < 0.01$) in plasma trifluoroacetic acid concentrations (H-40% O_2 = 546 ; 73 mM, N=8) which were accompanied by 33-60% decreases ($p < 0.01$) in binding to liver proteins (H-40% O_2 = 1.36; 0.26 nmoles bound F^-/mg protein, N=4), 78-84% decreases ($p < 0.05$) in 48 hr plasma ALT levels (H-40% O_2 = 308; 219, control = 23 + 3, N=4) and a total amelioration of centilobular necrosis. Deuteration of H did not alter plasma F^-concentrations or binding to lipid (H-40% O_2 = 6.7; 1.9 M F^-, N=8 and 5.6;1.8 nmoles bound F^-/mg lipid phosphate, N=4). Exposure to H or d-H under 10% O_2 increased ($p < 0.05$) plasma F^- by almost 2 x and covalent binding to lipid by 5-6 x ($p < 0.01$) but ameliorated the centrilobular lesion. Covalent binding of oxidative pathway generated H reactive intermediates to protein is thus indicated as the mechanism of H-necrosis in guinea pigs.

INTRODUCTION

Halothane biotransformation by the hepatic cytochrome P-450 enzyme system occurs along two pathways. The oxidative route produces a trifluoroacyl acid chloride intermediate that can either covalently bind to free amino groups on subcellular proteins or undergo hydrolysis to produce the metabolite, trifluoroacetic acid (TFA) (Satoh et al, 1986). When there is a lack of sufficient oxygen to maintain oxidative biotransformation, electrons are inserted into halothane (reductive pathway) generating free radical intermediates (DeGroot and Noll, 1983). These radicals can react with the double bonds of subcellular lipids or breakdown further releasing fluoride ion (F^-) (DeGroot and Noll, 1983). Selective inhibition of oxidative

Biological Reactive Intermediates IV, Edited by C.M. Witmer *et al.*
Plenum Press, New York, 1990

biotransformation with deuterated halothane (d-halothane) significantly reduces halothane-associated acute centrilobular necrosis in guinea pigs (Lind, Gandolfi, Hall, 1989). Thus, the role of covalent binding of reactive halothane metabolic intermediates to cellular proteins and lipids in producing hepatotoxicity was determined. In order to alter flux along the two metabolic pathways, d-halothane (inhibits oxidative metabolism) and/or very low oxygen concentrations during exposure (inhibits oxidative promotes reductive metabolism) were used. By removing the livers of guinea pigs immediately after exposure, isolating the protein and lipid, and determining the amount of bound residues containing fluorine, a measure of the covalent binding of halothane intermediates could be made and compared with resultant necrosis in animals allowed to survive for 96 hr after exposure.

MATERIALS AND METHODS

Exposure conditions. Groups (N=8) of male, outbred Hartley guinea pigs (600-700 g) from Sasco Inc. (Omaha, NE) were exposed to either 1% (v/v) halothane (Abbott Laboratories, North Chicago, IL) or d-halothane (Lind, Gandolfi, Hall, 1989), F_IO_2 = 0.40 or 0.10 (balance N2), flow rate = 3 l/min, for 4 hr in a 180 l plexiglass chamber. Control animals (N=4) were untreated.

Sample collection. One half of the animals were killed immediately after exposure. Blood was drawn, and their livers were removed and frozen on dry ice for bound organic fluorine analysis. Plasma samples were obtained from the remaining guinea pigs by toenail bleedings immediately after exposure (0 hr) and at 24, 48, 72, and 96 hr. These animals were killed at 96 hr and liver tissue sections collected for subsequent histopathologic analysis. The presence of focal to confluent centrilobular necrosis was used to classify an animal as being responsive to halothane (Lind, Gandolfi, Hall, 1989).

Plasma enzyme and metabolite analysis. Plasma alanine aminotransferase (ALT) levels were measured spectrophotometrically (Procedure 59-UV; Sigma Chemical Company, St. Louis, MO). Values at 48 hr were used for comparing degrees of hepatic injury (Lind, Gandolfi, Hall, 1987). Plasma concentrations of F^- and TFA were determined in 0 hr samples with specific ion electrodes (Lind, Gandolfi, Hall, 1987) (Orion Research, Boston, MA) and by gas chromatography (Maiorino, Gandolfi, Sipes, 1980), respectively.

Protein and lipid isolation. Quadruplicate samples of liver tissue (100 mg) were sonicated in 1 ml H_2O. Protein was precipitated and lipid extracted with addition of 5 ml of -20° C ethanol. Following centrifugation, protein pellets were washed twice with 1 ml 0.6 mM TCA and twice with 5 ml 2:1 chloroform:methanol (C/M), and dried. The ethanol and C/M extracts were combined as the lipid fraction, dried, redissolved in 5 ml C/M and transferred to another tube. This extract was washed twice with 1 ml C/M-saturated H_2O, and dried. One of the quadruplicate samples was analyzed for protein or lipid phosphate content. Average protein content was 11.1 ± 0.86 mg protein/sample. Lipids averaged 1.07 ± 0.05 mg lipid phosphate/sample. These values were used to normalize bound organic fluorine values.

Bound organic fluorine analysis. Using a modification of our sodium fusion technique (Gandolfi, Sipes, Brown, 1981), sodium metal (15-20 mg) was added to each sample and the test tubes intensely heated in a burner flame. One ml H_2O and 50 ml glacial acetic acid were then added. Samples were directly analyzed for F^- with specific ion electrodes.

Statistical Analysis. Values are mean ± SD. Analysis of ALT, halothane metabolites, and bound organic fluorine were made using ANOVA with a Newman-Keuls multiple comparison test. Due to increasing standard deviations with increasing means, log of ALT values was used for analysis. Incidences of centrilobular necrosis were compared by chi-square. A $p < 0.05$ was considered significant.

Table 1. Effect of Oxygen Concentration and d-Halothane on Halothane Biotransformation, Covalent Binding, and Hepatotoxicity

F_IO_2	Treatment	Metabolite Concentrations		Bound Organic Fluorine		Hepatotoxicity	
		TFA (mM)	Fluoride ion (mM)	Protein (nmoles F/ mg protein)	Lipid (nmoles F/mg lipid phosphorous)	ALT (Units/ml)	Necrosis (#/N)
0.40	Halothane	546 ± 73^b	6.7 ± 1.9^a	1.36 ± 0.26^b	5.7 ± 1.8^a	308 ± 219^d	$3/4^d$
0.40	d-Halothane	203 ± 40^a	7.7 ± 2.7^a	0.91 ± 0.09^c	6.3 ± 1.7	47 ± 50	0/4
0.10	Halothane	88 ± 27	10.8 ± 2.3	0.81 ± 0.05	33.7 ± 2.6	69 ± 31	0/4
0.10	d-Halothane	65 ± 27	15.5 ± 2.9	0.55 ± 0.19	31.4 ± 15.0	49 ± 27	0/4

Values are x ± SD. N=8 for 0 hr plasma metabolites. N=4 for all other values. $^a p<0.05$ vs $F_IO_2 = 0.10$ values. $^b p<0.01$ vs all other values. $^c p<0.05$ vs $F_IO_2 = 0.10$, d-halothane values. $^d p<0.05$ vs all other values and control values (23 ± 3 units/ml, 0/4 with necrosis).

RESULTS

The use of d-halothane and/or low oxygen concentrations (F_IO_2 = 0.10) significantly decreased oxidative biotransformation, as indicated by 0 hr plasma TFA concentrations, and covalent binding of halothane metabolic intermediates to hepatic proteins as indicated by bound organic fluorine (Table 1). These decreases were associated with an amelioration of halothane hepatotoxicity as indicated by significant decreases in ALT levels and incidences in of centrilobular necrosis (Table 1). Significant increases in reductive biotransformation, as indicated by 0 hr plasma F^- concentrations, and covalent binding to lipids occurred with the use of low oxygen concentrations (FI02 = 0.10) during exposure to halothane or d-halothane (Table 1). These were not associated with the development of centrilobular necrosis (Table 1).

DISCUSSION

Halothane-associated hepatotoxicity in guinea pigs appears to be the result of covalent binding to hepatic proteins by the trifluoroacyl acid chloride intermediate generated during oxidative biotransformation. Binding to subcellular lipids by free radicals produced via reductive halothane biotransformation does not appear to be involved in producing the centrilobular lesion that is characteristic of this animal model (Lind, Gandolfi, Hall, 1987; 1989). Identification of the specific hepatic proteins (enzymes?) which show differential degrees of bound halothane intermediates under the conditions of this study could well elucidate the specific mechanism(s) by which halothane can cause acute hepatic necrosis (Kenna et al, 1987).

ACKNOWLEDGEMENTS

This research was supported by grant DK 16715.

REFERENCES

DeGroot, H.E. and Noll, T. (1983). Halothane hepatotoxicity: Relation between metabolic activation, hypoxia, covalent binding, lipid peroxidation and liver cell damage. *Hepatology* **3**, 601-606.

Gandolfi, A.J., Sipes, I.G. and Brown, B.R. (1981). Detection of covalently bound halothane metabolites in the hypoxic rat model for halothane hepatotoxicity. *Fund. Appl. Toxicol.* **1**, 255-259.

Kenna, J.G., Satoh, H., Christ, D.D. and Pohl, L.R. (1987). Metabolic basis for a drug hypersensitivity: Antibodies in sera from patients with halothane hepatitis recognize liver neoantigens that contain the trifluoroacetyl group derived from halothane. *J. Pharmacol. Exp. Therap.* **245**, 1103-1109.

Lind, R.C., Gandolfi, A.J. and Hall, P de la M. (1987). Halothane hepatotoxicity in guinea pigs. *Anesth. Analg.* **66**, 222-228.

Lind, R.C., Gandolfi, A.J. and Hall, P de la M. (1989). The role of oxidative biotransformation of halothane in the guinea pig model of halothane-associated hepatotoxicity. *Anesthesiology* **70**, 649-653.

Maiorino, R.M., Gandolfi, A.J. and Sipes, I.G. (1980). Gas chromatographic method for the halothane metabolites, trifluoroacetic acid and bromide in biological fluids. *J. Anal. Toxicol.* **4**, 250- 254.

Satoh, H., Gillette, J.R., Takemura, T., Ferrans, V.J., Jelenich, S.E., Kenna, J.G., Neuberger, J. and Pohl, L.R. Investigation of the immunological basis of halothane-induced hepatotoxicity. In, *Biological Reactive Intermediates III* (Eds. Kocis, J.J., Jollow, D.J., Witmer, C.M., Nelson, J.O. and Snyder, R., pp. 657-673, Plenum Publishing Corp., New York, 1986.

THE NEPHROTOXICITY OF 2,5-DICHLORO-3-(GLUTATHION-S-YL)-1,4-BENZOQUINONE, AND 2,5,6-TRICHLORO-3-(GLUTATHION-S-YL)-1,4-BENZOQUINONE IS POTENTIATED BY ASCORBIC ACID AND AT-125

Terrence J. Monks*, Serrine S. Lau*, Jos. J.W. Mertens, Johan H.M. Temmink and Peter J. van Bladeren

*Division of Pharmacology and Toxicology
College of Pharmacy
The University of Texas at Austin, Austin, TX
and Department of Toxicology
Agricultural University
Wageningen, The Netherlands

We have previously shown that oxidation of either 2-bromohydroquinone (2-BrHQ) or hydroquinone (HQ) in the presence of glutathione (GSH) results in the formation of several isomeric and multi-substituted GSH conjugates. Administration of 2-Br-(diGSyl)HQ and 2,3,5-(tri-GSyl)HQ (10-30 μmol, i.v.) to male Sprague Dawley rats caused selective necrosis of renal proximal tubules (Monks et al., 1985, 1988; Lau et al., 1988). The tissue selectivity of these conjugates appears to be a consequence of their metabolism and selective accumulation by renal proximal tubular cells, mediated by γ-glutamyl transpeptidase (γ-GT). Pretreatment of animals with AT-125 (Acivicin; L-[aS-5S]-a-amino-3-chloro-4,5-dihydro-5-isoxazoleacetic acid, 10 mg/kg) an inhibitor of several glutamine (Weber, 1983) utilizing enzymes, including γ-GT (Monks et al., 1985) protected rats from both 2-Br-(diGSyl)HQ and 2,3,5-(triGSyl)HQ mediated nephrotoxicity (Monks et al., 1988; Lau et al., 1988). The cytotoxicity appears to be a consequence of oxidation of the corresponding cysteinylglycine(CYSGLY) and/or cysteine conjugates to their equivalent quinones. We have therefore investigated the toxicity of the GSH conjugates of 2,5-dichloro-1,4-benzoquinone (DCBQ) and 2,3,5-trichloro-1,4-benzoquinone (TCBQ). These compounds have certain similarities to the 2-BrHQ and HQ conjugates, the major difference being that reaction of the chloroquinones with GSH results in the formation of a conjugate that resides in the quinone form when reaction with GSH occurs at a chlorine-substituted carbon atom (Figure 1).

Administration of both DC(GSyl)BQ and TC(GSyl)BQ to rats caused elevations in blood urea nitrogen (BUN) (Table 1) and in the urinary excretion of glucose, lactate dehydrogenase (LDH) and γ-GT (Mertens et al., 1990). However, the dose required to produce toxicity (150-200 μmol/kg) was approximately ten-fold higher than that required of either 2-Br-(diGSyl)HQ or 2,3,5-(triGSyl)HQ (10-20 μmol/kg). It seems likely that the higher dose is required to overcome the interaction of the quinone conjugates with tissue macromolecules prior to their reaching the kidney. In support of this interpretation, coadministration of ascorbic acid (0.6 or 0.3 mmol/kg) with either DC(GSyl)BQ (200 μmol/kg) or TC(GSyl)BQ (100 μmol/kg) respectively, resulted in a significant increase in nephrotoxicity as evidenced by histological examination of kidney slices and by elevations in BUN ($p<0.01$) (Table 1). This protocol may result in the increased delivery of DC(GSyl)BQ and TC(GSyl)HQ to the kidney by maintaining the conjugates in the reduced form, thereby preventing their interaction with nucleophilic sites on plasma proteins and/or with other extra-renal macromolecules.

Biological Reactive Intermediates IV, Edited by C.M. Witmer *et al.*
Plenum Press, New York, 1990

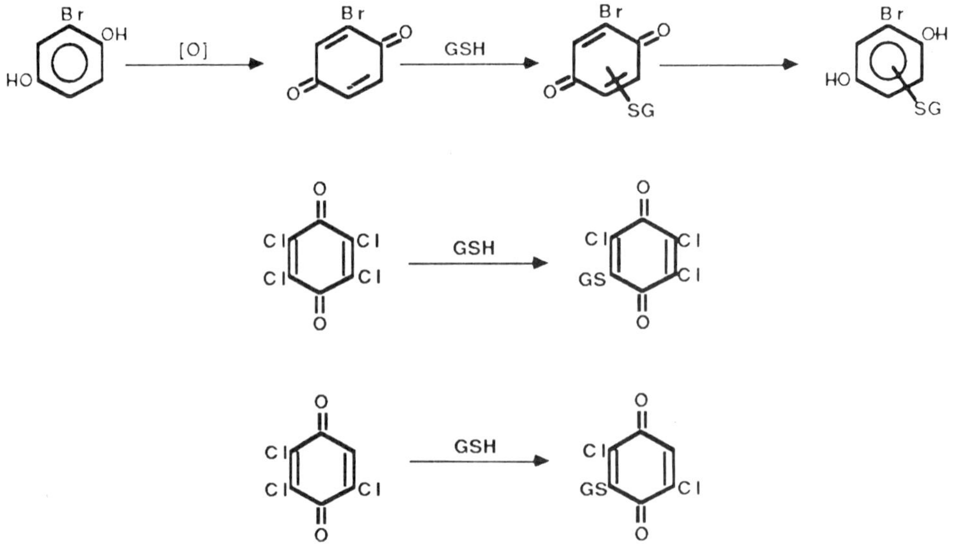

Figure 1. Glutathione addition/substitution to quinones resulting in the formation of conjugates that reside in either the reduced or oxidized forms.

Pretreatment of rats with AT-125 further potentiated the toxicity of both DC(GSyl)HQ and TC(GSyl)HQ, as indicated by increases in BUN (Table 1). In addition, although AT-125 caused a decrease in the urinary excretion of glucose, this was probably a consequence of decreased glomerular filtration rates concomitant with the enhanced toxicity (Mertens et al., 1990). T he effects of AT-125 on both DC(GSyl)HQ and TC(GSyl)BQ nephrotoxicity are in contrast to its protective effect on 2-Br-(diGSyl)HQ and 2,3,5-(tri-GSyl)HQ mediated nephrotoxicity. Although the mechanism of this potentiation is unclear it may be related to effects subsequent to the γ-GT catalysed formation of the CYSGLY conjugates. We have recently shown that the γ-GT catalysed hydrolysis of 2-Br-3-(GSyl)HQ results in the subsequent oxidative cyclization of the CYSGLY conjugate and 1,4-benzothiazine formation (see Lau and Monks, this volume). This intramolecular cyclization reaction removes the reactive quinone moiety from the molecule and can be considered an intramolecular detoxication reaction. Differences in the rate at which the various quinone-thioethers undergo this cyclization reaction may determine their response to the inhibition of γ-GT. Differences in the lipophilicity of the quinone/quinol-GSH conjugates may also play a

Table 1. Effects of Ascorbic Acid and AT-125 on DC(GSyl)BQ and TC(GSyl)BQ Nephrotoxicity

Treatment	DC(GSyl)BQ	TC(GSyl)BQ
	BUN (mg%)	
GSH conjugate alone	47.7 ± 7.6	24.9 ± 3.5
Conjugate + ascorbate	119.1 ± 18.1	56.6 ± 0.9
Conjugate + ascorbate + AT-125	208.6 ± 10.4	83.8 ± 4.5

Blood samples were obtained from the retro-orbital sinus 19h after dosing and BUN determined. Control animals received DMSO (BUN = 18.0 ± 1.1). AT-125 (10 mg/kg) was administered 1h prior to i.v. injection of the conjugates and ascorbic acid. Values represent the mean ± S.E. (n = 4).

role in their differential response to AT-125. The data also raise questions on the mechanism of transport of the quinone-GSH conjugates into renal proximal tubule cells under conditions of decreased γ-GT activity. The extracellular (tubular lumen) oxidation of these thioethers and subsequent arylation of brush border membrane proteins, may be sufficient to initiate toxicity. (Supported by NIH grants ES 04662, GM 39338, and NATO 542/87).

REFERENCES

Lau, S.S., Hill, B.A., Highet, R.J., and Monks, T.J. (1988). Sequential oxidation and glutathione addition to 1,4-benzoquinone: Correlation of toxicity with increased glutathione substitution. *Molec. Pharmacol.* **34**, 829-836.

Mertens, J.J.W.M., Temmink, J.H.M., Van Bladeren, P.J., Lau, S.S. and Monks,T.J. (1990). Inhibition of g-glutamyl transpeptidase potentiates the nephrotoxicity of glutathione conjugated chlorohydroquinones. In preparation.

Monks, T.J., Lau, S.S., Highet, R.J. and Gillette, J.R. (1985). Glutathione conjugates of 2-bromohydroquinone are nephrotoxic. *Drug Metab. Dispos.* **13**, 553-559.

Monks, T.J., Highet, R.J. and Lau, S.S. (1988). 2-Bromo-(diglutathion-S-yl)hydroquinone nephrotoxicity: Physiological biochemical and electrochemical determinants. *Molec. Pharmacol.* **34**, 492-500.

Weber, G. (1983). Biochemical strategy of cancer cells and the design of chemotherapy. *Cancer Res.*, **43**, 3466-3492.

GENERATION OF FREE RADICALS RESULTS IN INCREASED RATES OF PROTEIN DEGRADATION IN HUMAN ERYTHROCYTES

Anne M. Mortensen, Melissa Runge-Morris and Raymond F. Novak

The Institute of Chemical Toxicology
Wayne State University
Detroit, MI 48201 (AMM, RFN) and
The Department of Molecular Biology
Northwestern University Medical School
Chicago, Il 60611

The hydrazines represent a class of agents well recognized for their potent hemolytic activity in red blood cells. Phenylhydrazine, one of the most potent hemolyzing agents known, is one of the few agents capable of producing hemolysis in vitro. Phenylhydrazine reacts with the heme of oxyhemoglobin to generate free radical species which alkylate the porphyrin ring system (Saito and Itano, 1981; Augusto et al., 1982; Winterbourn and French, 1977; Ortiz de Montellano et al., 1983). Similar oxidation reactions have been reported for other substituted hydrazines and alkyl hydrazines (Winterbourn and French, 1977; Ortiz de Montellano et al., 1983). The organic hydroperoxides (cumene hydroperoxide, t-butyl hydroperoxide) also decompose to yield organic free radicals capable of producing protein damage (Trotta et al., 1983; Thornalley et al., 1983; Taffe et al., 1987; Maples et al., 1990).

Previous research has shown that the hydrazines stimulate the rate of protein degradation in red cells (Goldberg and Boches, 1982; Fagan. et al., 1986). Subsequent research suggested that reactive oxygen species may be involved in damaging protein and stimulating the degradation rate of damaged protein (Davies and Goldberg, 1987, 1988). More recently, we have presented evidence which suggests that reactive oxygen species are not the primary initiators of protein damage following treatment with substituted hydrazines or with organic hydroperoxides (Runge-Morris, et al., 1988, 1989). Rather, organic free radicals generated from decomposition of these agents appear to be the more potent and proximate initiators of protein damage (Runge-Morris, et al., 1988, 1989).

The red cell contains a number of proteolytic enzymes capable of degrading protein which is damaged as a consequence of either biosynthetic error, senescent degeneration, oxidant stress or xenobiotic insult. The calcium-dependent neutral protease system (CANP; calpain/calpastatin) in human erythrocytes has recently been implicated in the degradation of protein damaged by divicine, the toxic quinoid compound present in fava beans (Morelli, et al., 1986).

Previous studies, conducted primarily using fluorescence or metabolic assays for tyrosine or alanine content, respectively, explored the effects of free radical scavengers, antioxidants and various inhibitors on the rate of xenobiotic-stimulated protein degradation (Goldberg and Boches, 1982; Fagan et al., 1986, Davies and Goldberg, 1987, 1988; Runge-Morris, 1988). Detailed mechanistic studies however, were more difficult to conduct owing to a significant potential for interference of

Biological Reactive Intermediates IV, Edited by C.M. Witmer *et al.*
Plenum Press, New York, 1990

inhibitors or other reagents in the fluorometric or metabolic assays. Moreover, such experiments were generally limited in scope because of sample size, preparation time, specificity and sensitivity. Thus, a direct analysis of amino acid release from red cells was initiated using HPLC employing dabsyl derivatization according to previously published methodology (Knecht and Chang, 1986.) Although some 16 individual amino acids were observed in the samples from control and treated cells, the data presented will focus on quantifying changes in lysine, histidine and tyrosine levels since these peaks have been unambiguously assigned in the observed HPLC chromatograms, are readily resolved, and the standard curve of peak area versus concentration is linear over the concentration range of interest.

SDS-PAGE has been employed to assess the relative effects of the hydrazines and hydroperoxides on cellular proteins which may serve as targets for these free radicals and the effects of various scavengers of free radicals on proteolysis have been examined.

MATERIALS AND METHODS

Blood Sample Preparation

Units of freshly isolated human red blood cells were obtained from the blood bank and prepared as described previously (Dershwitz and Novak, 1982). Aliquots of packed red cells were diluted to yield a final hematocrit of 33% with buffer A which contained 115 mM NaCl, 4 mM KCl, 1 mM $MgCl_2$, 0.3 mM Na_2SO_4, 20 mM monobasic sodium phosphate and 10 mM glucose, adjusted to pH 7.45 with NaOH (Dershwitz and Novak, 1982).

Incubations

Incubations were performed in 50cc glass vials in a 37°C shaking water bath for 2 hrs. Red blood cell suspensions were incubated in the absence or presence of 4mM phenylhydrazine or 2 or 4 mM cumene hydroperoxide and contained cycloheximide (50 mM) as described previously (Runge-Morris, et al., 1988, 1989; Novak, et al., 1988). Free radical scavengers were pre-incubated with red cell suspensions for 30 minutes at 37°C prior to hydrazine or hydroperoxide addition. An aliquot of the incubation was frozen for HPLC analysis and red cell ghosts were isolated using the following protocol. One ml of the incubation mixture was diluted with 2 ml of normal saline, centrifuged at 3000 x g for 5 minutes, and the red cell pellet was washed once with 2 ml of normal saline. Red cells were then lysed with 5 ml of Nanopure water, and ghosts were pelleted at 6000 x g for 10 minutes. Ghosts were washed three times with Nanopure water and then diluted in buffer A. Protein determinations were carried out as described by Lowry et al. (Lowry, et al., 1951); SDS-PAGE was performed according to Laemmeli (Laemmeli, et al., 1970) using 12 and 7% gels.

HPLC Amino Acid Analysis

A 0.1 ml volume of red cells was added drop-wise to an equal volume of acetonitrile and vortexed vigorously. Following centrifugation for 15 min in a clinical centrifuge, the supernatant was removed and the amino acids derivatized with dabsyl chloride using a modification of a method described previously (Knecht and Chang, 1986).

HPLC analysis was performed on a model 400 BioRad HPLC system. A 5 micron C_{18} Column (Spheri-5, Applied Biosystems) fitted with a guard column was used. A flow rate of 1.9 ml/min was employed and sample detection accomplished at 436 nm using 0.005 AUFS. The gradient consisted of solvent A (25 mM sodium acetate in 4% dimethylformamide) and solvent B (100% acetonitrile) with an increase from 10 to 35% solvent B occurring over 20 min. followed by an increase from 35 to 55% solvent B in 12 min.(Runge-Morris, et al., 1989). The retention times of the individual amino acids in the HPLC chromatograms were confirmed using derivatized standards and the

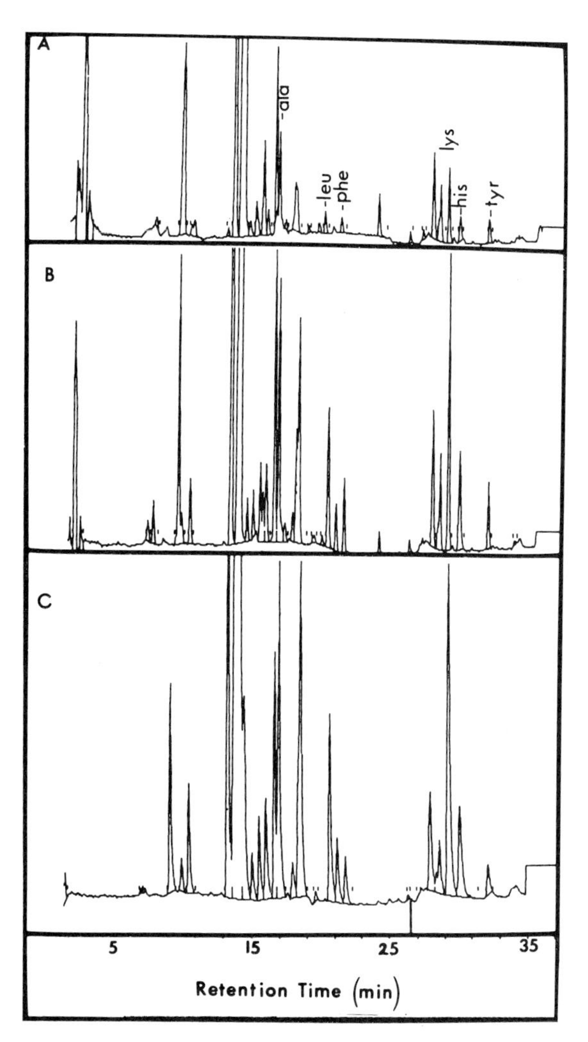

Figure 1. HPLC analysis of amino acid release from a control red cell suspension following a 2 hr incubation at 37°C (panel A), from red cells treated with 2 mM cumene hydroperoxide following a 2hr incubation at 37°C (panel B) and from a red cell suspension following incubation with 4 mM phenylhydrazine for 2 hr at 37°C (panel C).

Table 1. The Effects of Free Radical Scavengers on Phenylhydrazine and Cumene Hydroperoxide-stimulated Proteolysis in Human Red Blood Cells

Addition	Lysine	Amino Acid Release[a] Histidine (nmol/hr/ml RBC)	Tyrosine
None	16.7 ± 3.0 (100)[c]	8.3 ± 1.5 (100)	7.0 ± 1.8 (100)
+ NAC (20mM)[b]	12.8 ± 2.4 (77)	8.0 ± 2.6 (97)	4.0 ± 3.3 (57)
+ DTT (50mM)[b]	19.2 ± 5.1 (115)	8.3 ± 2.3 (100)	6.0 ± 2.0 (86)
phenylhydrazine (4mM)	121.7 ± 7.7 (730)	48.9 ± 11.1 (589)	19.2 ± 2.2 (274)
+ NAC (20mM)	67.1 ± 11.7 (400)	20.3 ± 7.1 (245)	5.6 ± 2.2 (80)
+ DTT (50mM)	78.3 ± 6.9 (469)	33.9 ± 8.7 (408)	12.1 ± 2.6 (172)
cumene hydroperoxide (2 mM)	50.3 ± 2.1 (300)	30.0 ± 1.9 (361)	15.1 ± 2.4 (216)
+ NAC (20mM)	27.4 ± 6.7 (164)	14.5 ± 5.3 (175)	6.0 ± 2.2 (86)
+ DTT (50mM)	17.3 ± 8.7 (104)	9.4 ± 5.4 (113)	5.7 ± 3.3 (81)

[a] Data represent the mean ± S.D. of three determinations.
[b] Abbreviations: NAC, N-acetylcysteine; DTT = Dithiothreitol
[c] % control

linearity of peak area with concentration from 20 to 150 pmoles confirmed using a standard curve.

RESULTS AND DISCUSSION

SDS-PAGE analysis of membrane proteins from red cells treated with either phenylhydrazine or cumene hydroperoxide revealed a substantial loss of protein bands migrating in the region of band 2.1 (ankyrin), bands 2.2, 3, and 6 (glyceraldehyde-3-phosphate dehydrogenase) and a decrease in intensity of bands 1 and 2 (spectrin), bands 4.1, 4.2 and 5 (actin). In addition, a noticeable smearing of a protein band appearing in the region of 30 kda also occurred. These observations also correlate with the appearance of new protein spots in the 2D SDS-PAGE of phenylhydrazine- or cumene hydroperoxide-treated samples as compared to control and the disappearance of protein spots which existed in control red cell preparations. Thus, incubation of red cells with agents which form free radical species results in damage to cytoskeletal proteins as well as to hemoglobin and other proteins present in the cytosol.

In association with protein damage, a significant stimulation of the rate of protein degradation occurred as was demonstrated using HPLC analysis (Figure 1). Both a concentration- and time-dependence for release of amino acids could be demonstrated for phenylhydrazine- and cumene hydroperoxide-treated red cells. Following a 2 hr incubation, phenylhydrazine (4 mM) produced an 8-, 6- and 3-fold stimulation in the rate of Lys, His, and Tyr release, respectively, whereas cumene hydroperoxide (2 mM) increased Lys, His, and Tyr release 3-, 3.6- and 2-fold, respectively, relative to control (Table 1).

Inclusion of free radical trapping agents diminished significantly both the hydrazine- and hydroperoxide-stimulated rate of proteolysis suggesting that these agents trapped free radical species and thereby diminished protein damage. Inclusion of N-acetylcysteine or dithiothreitol in the incubation with hydrazine or hydroperoxide diminished proteolysis in red cells by 35 to 75% with only a slight effect monitored for control red cell suspensions (Table 1). The use of other inhibitors, such as methylsulfoxide or dimethylfuran to diminish reactive oxygen species such as hydroxyl radical, failed to affect the rate of hydrazine- or hydroperoxide-stimulated proteolysis.

In summary, these results show that free radical generation results in both cytosolic and membrane protein damage, that protein damage is reflected by increased rates of proteolysis and that free radical scavengers diminish xenobiotic-stimulated proteolysis likely reflecting their ability to trap reactive free radicals.

ACKNOWLEDGEMENTS

This work was supported by USPHS Grant ES 02521 from the National Institutes of Environmental Health Sciences.

REFERENCES

Augusto, O., Kunze K.L. and Ortiz de Montellano, P.R. (1982). N-phenylprotoporphyrin IX formation in hemoglobin-phenylhydrazine reaction. *J. Biol. Chem.* **257**, 6231-6241.

Davies, K.J.A, and Goldberg, A.L. (1987). Oxygen radicals stimulate intracellular proteolysis and lipid peroxidation by independent mechanisms in erythrocytes. *J. Biol. Chem.* **262**, 8220-8226.

Davies, K.J.A and Goldberg, A.L. (1988). Proteins damaged by oxygen radicals are rapidly degraded by extracts of red blood cells. *J. Biol. Chem.* **262**, 8227-78234.

Dershwitz, M. and Novak, R.F. (1982). Studies on the mechanism of nitrofurantoin-mediated red cell toxicity. *J. Pharmacol. Exp. Ther.* **222**, 430-434.

Fagan, J.M., Waxman, L. and Goldberg, A.L. (1986). Red blood cells contain a pathway of the degradation of oxidant-damaged hemoglobin that does not require ATP or ubiquitin. *J. Biol. Chem.* **261**, 5705-5713.

Goldberg, A.L. and Boches, F.S. (1982). Oxidized proteins in erythrocytes are rapidly degraded by the adenosine triphosphate-dependent proteolytic system. *Science* **215**, 404-418.

Knecht, R. and Chang, Y.J. (1986). Liquid chromatographic determination of amino acids after gas-phase hydrolysis and derivatization with (dimethylamino)azobenzenesulfonyl chloride. *Anal. Chem.* **58**, 2375-2379.

Laemmli, U.K. (1970). Cleavage of structural proteins during assembly of the head of the bacteriophage T_4. *Nature* (Lond.) **227**, 680-685.

Lowry, O.H., Rosebrough, N.J., Farr, A.L. and Randall, R.T. (1951). Protein measurement with the Folin phenol reagent. *J. Biol. Chem.* **193**, 265-275.

Maples, K.R., Kennedy, C.H., Jordan, S.J. and Mason, R.P. (1990). *In Vivo* thiyl free radical formation from hemoglobin following administration of hydroperoxides. *Arch. Biochem. Biophys.* 278, in press.

Morelli, A., Grasso, M., and DeFlora, A. (1986). Oxidative inactivation of the calcium-stimulated neutral proteinase from human red blood cells by divicine and intracelllular protection by reduced glutathione. *Arch. Biochem. Biophys.* **251**, 1-8.

Novak, R.F., Kharasch, E.D., and Wendell N. (1988). Nitrofurantoin-stimulated proteolysis in human erythrocytes: a novel index of toxic insult by nitroaromatics. *J. Pharmacol. Exp. Ther.* **247**, 439-444.

Ortiz de Montellano, P.R., Augusto, O., Violo F., and Kunze K.L. (1983). Carbon radicals in the metabolism of alkyl hydrazines. *J. Biol. Chem.* **258**, 8623-8629.

Runge-Morris, M., Iacob, S., and Novak, R.F. (1988). Characterization of hydrazine-stimulated proteolysis in human erythrocytes. *Toxicol. Appl. Pharmacol.* **94**, 414-426.

Runge-Morris, M., Frank, P., and Novak, R.F. (1989). Differential effects of organic hydroperoxides and hydrogen peroxide on proteolysis in human erythrocytes. *Chem. Res. Toxicol.* **2**, 76-83.

Saito, S. and Itano H.A. (1981). β-Mesophenylbiliverdin IXa and N-phenylprotoporphyrin IX products of the reaction of phenylhydrazine with oxyhemoproteins. *Proc. Natl. Acad. Sci., U.S.A.* **78**, 5508-5512.

Taffe, B.G., Takahashi, N., Kensler, T.W., and Mason, R.P. (1987). Generation of free radicals from organic hydroperoxide tumor promoters in isolated mouse keratinocytes. *J. Biol. Chem.* **262**, 12143-12149.

Thornalley, P.J. Trotta, R.J. and Stern A. (1983). Free radical involvement in the oxidative phenomena induced by t-butyl hydroperoxide in erythrocytes. *Biochem Biophys. Acta* **759**, 16-22.

Trotta, R.J. Sullivan, S.G. and Stern A. (1983). Lipid peroxidation and haemoglobin degradation in red blood cells exposed to t-butyl hydroperoxide. *Biochem. J.* **212**, 759-772.

Winterbourn, C.C. and French J.K. (1977). Free-radical production from acetylphenylhydrazine and haemoglobin. *Biochem. Soc. Trans.* **5**, 1480-1481.

FREE RADICALS GENERATED IN ETHANOL METABOLISM MAY BE RESPONSIBLE FOR TUMOR PROMOTING EFFECTS OF ETHANOL

Siraj I. Mufti

Department of Pharmacology and Toxicology
The University of Arizona Health Sciences Center
Tucson, Arizona 85721

Chronic alcohol consumption is considered a major risk factor for human cancers [for example, Hakulinen et al., 1974; Breslow and Enstrom, 1974; Feldman et al., 1975; Williams and Horm, 1977; Schottenfeld, 1979; Kono and Ikeda, 1979; Tuyns, 1979; Doll and Peto, 1981]. The main sites associated with alcohol-related cancers are the oral cavity, esophagus and liver [Hakulinen et al., 1974; Breslow and Enstrom, 1974; Feldman et al., 1975; Kono and Ikeda, 1979; Rothman et al., 1980; Tuyns, 1982; Vassallo et al., 1985] and 75% of esophageal cancers and 36% of hepatic cancers in the U.S.A. are attributable to excessive alcohol consumption [Rothman et al., 1980]. Despite the strong epidemiologic evidence, however, experimental evidence for the association is not clear and often contradictory results have been obtained [Schmahl et al., 1965; Gibel, 1967; Schmahl, 1976; Griciute et al., 1982; Gabrial et al., 1982; Habs and Schmahl, 1981; Teschke et al., 1983; McCoy et al., 1986]. In order to clarify the role of ethanol, recently we carried out studies where ethanol was administered either during the initiation of carcinogenesis or later as a tumor promoter after initiation with the carcinogen was completed [Mufti et al., 1989]. Our results showed that ethanol increased esophageal tumor incidence only when administered after treatment with esophagus specific carcinogen, N-nitrosomethylbenzylamine (NMB_ZA) was completed. However, in studies of carcinogenesis induced by N-nitrosodiethylamine (NDEA), reported in another article in these proceedings, we did not find an increase either in liver tumors or gamma glutamyltranspeptidase-altered hepatic islands that could be attributed to the tumor promoting effects of ethanol [Mufti and Sipes, 1990]. This paper is a preliminary report of our investigations into the mechanisms of alcohol-related carcinogenesis that could be used to explain the contradictory effects of ethanol obtained with NMB_ZA- and NDEA-induced carcinogenesis.

MATERIALS AND METHODS

Chemicals

NMB_ZA was purchased from Ash Stevens, Chicago, Illinois, and NDEA from Eastman Kodak Company. Purity of the compounds was 99% and verified by Dr. Karl Schram of our clinical chemistry department through mass spectrometric-gas chromatographic procedures. These carcinogens were kept in tightly sealed containers, stored under refrigeration and used only in small aliquots to make fresh solutions each time animals were treated. Corn oil was used as the vehicle for NMB_ZA and a 0.9% NaCl solution for NDEA. The laboratory safety guidelines issued for handling carcinogens and chemicals by the Office of Risk Management, University of Arizona were strictly followed.

Biological Reactive Intermediates IV, Edited by C.M. Witmer *et al.*
Plenum Press, New York, 1990

Other chemicals including dietary constituents were purchased either from ICN or Sigma chemical companies.

Animals and Their Diet

Male Sprague Dawley rats weighing 100-120 g were purchased from Harlan Laboratories, Indianapolis, IN. The animals were isolated for a week after arrival and then housed in suspended cages in the Division of Animal Resources in a room allotted for our experimental animals. This room had regulated temperature (23°C), humidity (30-40%) and a day (16h) and night (8h) cycle. The health of animals and their maintenance was under the care of the Division of Animal Resources, University of Arizona.

Rats were fed regular Purina chow before and during treatments with carcinogens. A week after the carcinogen treatment was completed, the rats were fed an isocaloric liquid diet formulated from basic ingredients in our laboratory according to the suggestions of Lieber and DeCarli [1982] and including minerals and vitamins for rats as described by the American Institute of Nutrition [1977]. Ethanol constituted 7% of the diet in gram weight and substituted for part of dextrin-maltose of the control diet. Animals were pair fed so that the animals on control diet received no more than their ethanol consuming counterparts. The animals received ethanol until the end of experiment or until their termination. The blood ethanol content was checked at random every two months through a gas chromatographic analysis [Taylor et al., 1984] and showed no significant variation throughout, range 0.08 to 0.18% with a mean of 0.09%.

Treatment of Animals

The rats received the following carcinogen treatments: 1. NMB_ZA, i.p., at a rate of 2.5 mg/kg bodyweight, 3 times a week for 3 weeks. NMB_ZA is a potent esophagus specific carcinogen [Druckrey et al., 1967]. 2. A single dose of 100 mg/kg of NDEA followed 2 weeks later by 2 acetylaminofluorene (2AAF) at a dose of 6 mg/rat/day for 7 days and on day 3 of 2AAF, CCl_4 at 2 ml/kg [Lans et al., 1983]. This regimen subjects hepatic cells to selective pressures for hepatocarcinogenesis.

Measurement of Ethane Exhalation

Ethane was measured as previously done by us [Szebeni et al., 1986]. In brief, the animals were placed in air-tight metabolic chambers with incoming and existing air passages. The air was allowed to pass at a flow rate of 200 ml/min. Any ethane from ambient air was absorbed by a carbon trap placed in the passage of incoming air. The air existing from the chamber paused through a trap of coconut charcoal that was cooled to dry ice temperature to absorb ethane. Ethane exhaled by animals was collected for three 30 min cycles, using fresh charcoal each time, for a total of 90 min. The charcoal was placed in a screw-top test tube and sealed with a teflon-lined silicon rubber-septum. The tubes and their contents were heated to 230-240°C for 10 min to desorb ethane. Ethane was determined by head space gas chromatography using a Hewlett-Packard gas chromatograph equipped with a flame ionization detector and a carbosive B column. Cumulative ethane exhaled was averaged to minimize fluctuations in production. Data was statistically analyzed by the unpaired Student t test.

Separation of hepatic nuclei and determination of DNA strand breaks

Hepatic nuclei were separated as previously done by us [White et al., 1981; Mufti and Sipes, 1988]. These nuclei were subjected to alkaline elution technique devised by Kohn et al. [1987] and modified by us [Mufti et al., 1988]. Accordingly, measured volumes of nuclear preparations were used and fractions collected at 30 min intervals for 3 hrs. The DNA eluting in the fractions was determined fluorimetrically as described by Kissane and Robins [1958] and calculated as fraction of total DNA placed on the filter. DNA-protein cross-links were determined by adding proteinase K (0.5

mg/ml) immediately after the nuclei were lysed on the filter and allowed to drain before the start of elution.

Measurement of hepatic non-protein sulfhydral (GSH) Content

Non-protein GSH was determined according to the procedure of Sedlak and Lindsay [1968]. For this, aliquots of the liver homogenate were diluted 1:2 with a Tris-KCl buffer at a pH of 8.9. Then, 0.2 ml of 5% sulfosalicylic acid was added and the samples centrifuged at 3,000 r.p.m. for 15 min to sediment denatured protein. Aliquots (0.5 ml) of the resulting supernatants were mixed with 0.1 ml of Ellman's reagent (5,5'-dithio-(bis)-nitrobenzoic acid) and absorbence at 412 nm recorded. The pH of the reaction mixture was kept between 8 and 9 to avoid variation in color development. The net absorbence (read out minus turbidity blank) was used to determined the concentration of GSH from a standard curve. Duplicate determinations were made for each sample and the results expressed as umoles/g of tissue.

RESULTS AND DISCUSSION

The increase in cancer risk observed with alcohol consumption must be explained in terms of ethanol effects on other carcinogens, since ethanol itself is not carcinogenic. Our recent observation that ethanol acted to enhance esophageal tumor incidence only after treatment with NMB_ZA was completed is consistent with a tumor promoting role for ethanol [Mufti et al., 1989]. A particularly attractive mechanism for the tumor promoting effects of ethanol involves generation of free radicals. This suggestion is based on the following observations. In a previous study, we observed that there was an increase in ethane exhalation by animals that were chronically fed ethanol [Szebeni et al., 1986] thus lending support to the hypothesis that ethanol intake is conducive to lipid peroxidation, i.e., generation of free radicals [for a review, Dianzani, 1985]. In a subsequent study, we also observed an increase other lipid peroxidation products such as diene-conjugates and fluorescent lipids with chronic ethanol consumption [Odeleye et al., 1990]. Furthermore, administration of vitamin E, an antioxidant inhibited the ethanol-induced increase in lipid peroxidation products. A number of investigators implicate lipid peroxidation in alcoholic liver disease [for example, Shaw et al., 1983; Ryle, 1984; Feher et al., 1987]. Furthermore, a National Institute of Health workshop has reviewed a large amount of experimental evidence that indicates that a variety of tumor promoters generate free radicals [Copeland, 1983].

In studies reported here, young Sprague Dawley rats were treated with either an esophagus specific carcinogen NMB_ZA or a liver NDEA-2AAF-CCl_4 carcinogensis protocol and then administered ethanol for 12 months. The levels of ethane exhaled by these animals were measured. The results presented in Table I confirm our previous data that chronic ethanol consumption leads to an increase in excretion of ethane. However, treating the animals with NMB_ZA or NDEA before administration of ethanol affected the results differently. While the increase in ethane exhalation consequent upon ethanol consumption was present in NMB_ZA treated rats, such an increase was absent in the NDEA treated rats. The results indicate that prior treatment with the NDEA carcinogen regimen leads to a cancellation of the ethanol effect on lipid peroxidation. NDEA is known to damage the hepatocellular endoplasmic reticulum and cause functional alterations in cell membranes [Garcea et al., 1984], the cell components that are involved in lipid peroxidation by ethanol [Lieber and Savolainen, 1984]. Thus our results suggest that ethanol is not capable of inducing lipid peroxidation in liver cells that are affected by the NDEA treatment. Therefore, if tumor promotion by ethanol is due to generation of free radicals, it does not occur with the NDEA treatment. This conclusion is further supported by our observation of a significant decrease in mixed function oxidases with the NDEA treatment protocol and its further reduction with ethanol consumption, reported in an accompanying paper [Mufti and Sipes, 1990].

Table 1. Ethane Exhalation by Rats Treated With Carcinogens and Administered Ethanol as a Tumor Promoter

Treatment		No. of Animals	Ethane Exhaled (n moles)
Untreated	Control	4	4 ± 0.9
	Ethanol	4	9 ± 0.8[a]
NMB_zA	Control	8	3.2 ± 0.4
	Ethanol	6	9.2 ± 2.6[b]
NDEA	Control	9	4 ± 0.6
	Ethanol	7	3.6 ± 0.6[c]

NDEA and NMB_zA were administered to rats as described in the text. The animals on these experiments were chronically fed 7% ethanol diet starting a week after treatment with the carcinogen was completed. Ethane exhaled was measured after 12 months of chronic ethanol consumption. Each ethane measurement represents cumulative ethane exhaled at 90 min after the start of experiment.

[a,b] Ethane exhaled is significantly higher compared to the controls ($p < 0.05$).
[c] Differences in ethane exhaled by NDEA-treated control and ethanol fed rats are not significant.

Free radicals cause damage to DNA. The DNA damage apparent as strand breaks may present discontinuities in the DNA backbone thus increasing the probability of an alteration in gene expression or of an aberrant rearrangement within the genome [Copeland, 1983]. These experiments were carried out before ethanol effects on NDEA-induced carcinogenesis (discussed in the other paper) were known and we wanted to determine whether or not strand breaks could be induced by administering ethanol as a promoter of NDEA-initiated carcinogenesis. Hepatic nuclei from rats, that were treated with the NDEA-2AAF-CCl_4 regimen and then fed ethanol, were isolated and subjected to the alkaline elution technique. The results (data not shown) did not indicate any strand breaks. Thus if free radicals act to promote carcinogenesis by inducing strand breaks, these were not produced when ethanol was administered after the NDEA treatment. We are interested in determining induction of strand breaks in esophageal DNA with ethanol administration in NMB_zA-induced carcinogenesis [Mufti et al., 1989] where we have found that ethanol increased free radicals as well as tumor incidence. Unfortunately the experiments on production of free radicals and DNA strand breaks in the esophagus could not be carried out in the present study since they involve a pooling of esophagi which requires more animals and expenditure.

Among antipromoters are included the availability of such endogenous compounds as glutathione which serve as scavengers of free radicals, thereby, protecting the cells against their ill effects [for example, Meister and Anderson, 1983]. For this reason, we wanted to determine the levels of glutathione in organ where it is stored i.e. liver, in animals treated with NMB_zA and NDEA. The results presented in Figure 1 show that NDEA treated ethanol-fed and their control rats had significantly higher levels of glutathione, umoles/g of liver, compared to NMB_zA-treated rats. Thus an increase in glutathione levels occurring with NDEA treatment could have obliterated free radicals and their tumor promoting effects. Furthermore, our preliminary data indicates that there is an increase in hepatic superoxide dismutase levels in the NDEA-treated rats.

Based on the above observations, we hypothesize that free radicals generated in metabolism of ethanol may be responsible for the tumor promoting effects of ethanol. This would provide an explanation for the increase in cancer risk that is observed with excessive alcohol consumption.

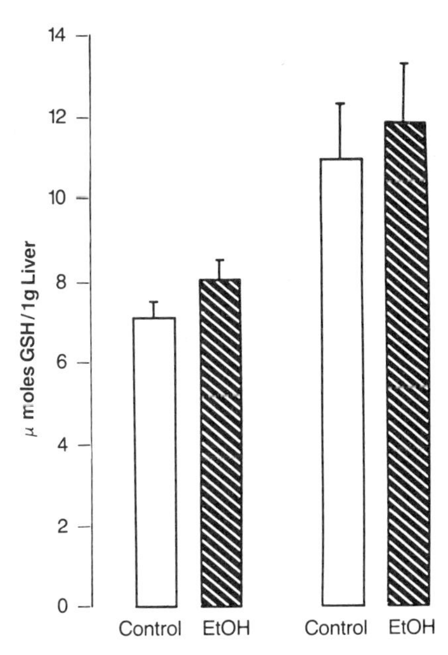

Figure 1. NMB_zA or NDEA was administered to rats as described in the text. The animals were fed 7% ethanol diet a week after the carcinogen treatment and continued on this diet until their termination at 12 months. Livers were excised and assayed for GSH as described. NDEA treated rats (data represented on right of the figure) have significantly higher GSH compared to NMB_zA treated rats ($n = 5$, $p < 0.05$).

ACKNOWLEDGMENTS

I appreciate the help and assistance of my friend and colleague Dr. Cleamond D. Eskelson in determining ethane. The research described here was supported by the U.S. National Institutes of Health grants, ES03438 and CA51088.

REFERENCES

American Institute of Nutrition (1977). Report of the American Institute of Nutrition Ad Hoc Committee on standards for nutritional studies. *J Nutr.* **107**, 1340-1348.

Breslow, N.E. and Enstrom, J.E. (1974). Geographic correlations between cancer mortality rates and alcohol-tobacco consumption in the United States. *J Natl. Cancer Inst.* **53**, 631-639,.

Copeland, E.S. (1983). A National Institutes of Health Workshops report, Free radicals in promotion - A chemical pathology study section workshop. *Cancer Res.* **43**, 5631-5637.

Dianzani, M.U. (1985). Lipid peroxidation in ethanol poisoning: A critical reconstruction. Alcohol Alcoholism 20:161-173.

Doll, R. and Peto, R. (1981). The causes of cancer: quantitative estimates of avoidable risk of cancer in the United States today. *J. Natl. Cancer Inst.* **66**, 1191-1308.

Druckrey, H., Preussmann, R., Ivankovic, S. and Schmahl, D. (1967). Organotrope carcinogene Wirkungen bei 65 verschiedenen N-nitroso-Verbindungen an BD-ratten. *Z. Krebsforsch* **69**, 103-201.

Feher, J., Csomos, G. and Vereckei, A. (Eds), *Free Radicals in Medicine*. Springer-Verlag, New York, 1987.

Feldman, J.G., Haan, M., Nagarajen, M. and Kissin, B. (1975). A case-control investigation of alcohol, tobacco, and diet in head and neck cancer. *Prev. Med.* **4**, 444-463.

Gabrial, G.N., Schrager, T.F. and Newberne, P.M. (1982). Zinc deficiency, alcohol and a retinoid: association with esophageal cancer in rat. *J. Natl. Cancer Inst.* **68**, 785-789.

Garcea, R., Canuto, R.A., Biocca, M.E., Muzio, G., Rossi, M.A. and Dianzani, M.U. (1984). Functional alterations of the endoplasmic reticulum and the detoxification systems during diethyl-nitrosamine carcinogenesis in rat liver. *Cell Biochem. Funct.* **2**, 177-181.

Gibel, Von W. (1967). Experimental studies on syncarcinogenesis of oesophageal cancer. *Arch. Geschwulstforsch* **30**, 181-189.

Griciute, L., Castegnaro, M. and Bereziat, J-C. (1982). Influence of ethyl alcohol in the carcinogenic activity of N-nitrosodi-n-propylamine. In Bartsch, H., Castegnaro, M, O'Neill, I.K., Okada, M. and Davis, W. (eds). N-Nitroso Compounds: Occurrence and Biological Effects. IARC Scientific Publication, Lyon, France, Vol. 41; pp. 643-648.

Habs, M. and Schmahl, D. (1981). Inhibition of the hepatocarcinogenic activity of diethylnitrosamine (DENA) by ethanol in rats. *Hepato-gastroenterol* **28**, 242-244.

Hakulinen, T., Lehtimaki, L., Lehtonen, M. and Teppo, L. (1974). Cancer morbidity among two male cohorts with increased alcohol consumption in Finland. *J. Natl. Cancer Inst.* **52**, 1711-1714.

Kissane, J.M. and Robins, E.J. (1958). The fluorimetric measurement of deoxyribonucleic acid in animal tissues with special reference to the central nervous system. *J. Biol. Chem.* **233**, 184-188.

Kohn, K.W., Erickson, L.C., Ewig, R.A.G. and Friedman, C.A. (1976). Fractionation of DNA from mammalian cells by alkaline elution. *Biochemistry* **15**, 4629-4637.

Kono, S. and Ikeda, M. (1979). Correlation between cancer mortality and alcoholic beverage in Japan. *Br. J. Cancer* **40**, 449-455.

Lans, M., DeGerlache, J., Taper, H.S., Preat, V. and Roberfroid, M.B. (1983). Phenobarbital as a promoter in the initiation/selection process of experimental rat hepatocarcinogenesis. *Carcinogenesis* **4**, 141-144.

Lieber, C.S. and DeCarli, L.M. (1982). The feeding of alcohol in liquid diets: two decades of applications and 1982 update. *Alcohol Clin. Exp. Res.* **6**, 523-531.

Lieber, C.S. and Savolainen, M. (1984). Ethanol and lipids. *Alcohol Clin. Exp. Res.* **8**, 409-423.

McCoy, C.D., Hecht, S.S. and Furuya, K. (1986). The effect of chronic ethanol consumption on the tumorigenicity of N-nitrosopyrrolidine in male Syrian golden hamsters. *Cancer Lett.* **33**, 151-159.

Meister, A. and Anderson, M.E. (1983). Glutathione. *Ann. Rev. Biochem.* **52**, 711-768.

Mufti, S.I. and Sipes, I.G. (1988). Differential induction of DNA strand breaks by nitrosamines in the rat liver and esophagus. *Cancer Lett.* **40**, 203-211.

Mufti, S.I., Becker, G. and Sipes, I.G. (1989). Effect of chronic dietary ethanol consumption on the initiation and promotion of chemically-inducedesophageal carcinogenesis in experimental rats. *Carcinogenesis* **10**, 303-309.

Mufti, S.I. and Sipes, I.G. (1990). A reduction in mixed function oxidases and in tumor promoting effects of ethanol in a NDEA-initiated hepatocarcinogenesis model. These proceedings.

Odeleye, O., Mufti, S.I., Eskelson, C.D. and Watson, R.R. Vitamin E reduces ethanol and cod liver oil induced *in vivo* lipid peroxidation in rats. Submitted for publication, 1990.

Rothman, K., Garfinkel, L., Keller, A.Z., Muir, C.S. and Schottenfeld, D.(1980). The proportion of cancer atributable to alcohol consumption. *Prev. Med.* **9**, 174-179.

Ryle, P.R. (1984). Free radicals, lipid peroxidation, and ethanol hepatotoxicity. *Lancet* **11**, 461.

Schmahl, D., Thomas, C., Sattler, W. and Scheld, G.F. (1965). Experimental studies of syncarcinogenesis: III. Attempts to induce cancer in rats by administering diethylnitrosamine and CCl_4 (or ethyl alcohol) simultaneously. In addition, an experimental contribution regarding 'alcoholic cirrhosis'. *Z. Krebsforsch* **66**, 526-532.

Schmahl, D. (1976). Investigations on esophageal carcinogenicity by methyl-phenyl-nitrosamine and ethyl alcohol in rats. *Cancer Lett.* **1**, 215-218.

Schottenfeld, D. (1979). Alcohol is a co-factor in the etiology of cancer. *Cancer* **43**, 1961-1966.

Sedlak, J. and Lindsay, R.H. (1968). Estimation of total, protein-bound and non-protein sulfhydryl groups in tissue with Ellman's reagent. *Anal. Biochem.* **25**, 192-205.

Shaw, S., Rubin, K.P. and Lieber, C.S. (1983). Depressed hepatic glutathione and increased diene conjugates in alcoholic liver disease. Evidence of lipid peroxidation. *Dig. Dis. Sci.* **28**, 585-589.

Szebeni, J., Eskelson, C.D., Mufti, S.I., Watson, R.R. and Sipes, I.G. (1986). Inhibition of ethanol induced ethane exhalation by carcinogenic pretreatment of rats 12 months earlier. *Life Sci.* **39**, 2587-2591.

Taylor, G.F., Turrill, G.H. and Carter, G. (1984). Blood alcohol analysis: a comparison of the gas chromatographic assay with an enzyme assay. *Pathology* **16**, 157-159.

Teschke, R., Minzlaff, M., Oldiges, H. and Frenzel, H. (1983). Effect of chronic alcohol consumption on tumor incidence due to dimethylnitrosamine administration. *J. Cancer Res. Clin. Oncol.* **106**, 58-64.

Tuyns, A.J. (1979). Epidemiology of alcohol and cancer. *Cancer Res.* **39**, 2840-2843.

Tuyns, A.J. (1983). Oesophageal cancer in non-smoking drinkers and in non-drinking smokers. *Int. J. Cancer* **32**, 443-444.

Vassallo, A., Correa, P., Stefani, E.D., Cendan, M., Zavala, D., Chen, V., Carzogolio, J. and Deneo-Pellegrini, H. (1985). In Uruguay: a case-control study. *J. Natl. Cancer Inst.* **75**, 1005-1009.

Williams, R.R., and Horm, J.W. (1977). Association of cancer sites with tobacco and alcohol consumption and socioeconomic status of patients: interview study from the Third National Cancer Survey. *J. Natl. Cancer Inst.* **58**, 525-547.

White, R.D., Sipes, I.G., Gandolfi, A.J. and Bowden, G.T. (1981). Characterization of hepatic DNA damage caused by 1,2-dibromoethane using the alkaline elution technique. *Carcinogenesis* **2**, 839-844.

COMPOSITION OF HEPATIC LIPIDS AFTER ETHANOL, COD LIVER OIL AND VITAMIN E FEEDING IN RATS

Olalekan E. Odeleye*, Ronald R. Watson*, Cleamond D. Eskelson** and Siraj I. Mufti***

Family and Community Medicine*, Surgery**, and Pharmacology and Toxicology***
University of Arizona
College of Medicine
Tucson Arizona 85724

One of the numerous metabolic effects of alcohol consumption is its altering the metabolism of fatty acids in tissues. This alterations lead to an increase in the ratio of linoleic to arachidonic acid in the phospholipid fraction of the tissues (Alling et al., 1984, Reitz 1979). The additional associated depletion of the alpha-tocopherol (Vitamin E) content of cell membranes following prolonged ethanol administration (Bjorneboe et al., 1987) suggests that increased lipid peroxidation may play a role in the pathogenesis of alcoholic fatty liver. If increased peroxidation of membrane lipids can lead to the disruption of the structural and functional properties of the membranes, then it might be expected that the fluidity and composition of the liver would be altered. Such an observation has been confirmed (Chin et al., 1977). It would, therefore, follow that such alterations may be prevented by a dietary antioxidant such as alpha tocophetol. In this study, the development of liver membrane proxidation and alcoholic fatty liver was promoted through chronic ethanol and cod liver oil. Supplementation of the diets with vitamin E was used to study the role of vitamin E in preventing hepatic fatty acid changes.

MATERIALS AND METHODS

Twenty adult male Sprague-Dawley were randomly divided into 4 groups of 5 animals per group. The rats were adjusted to a liquid diet (Lieber-DeCarli 1986) over a period of 5 days. Following the adjustment period, the animals then received one of the following diets: (A) control rats received the liquid diet in which dextrin maltose provided 36% of the calories in the diet; (B) ethanol fed rats received a diet in which ethanol isocalorically substituted for dextrin maltose and provide 36% of the caloric requirement; (C) the cod liver oil fed rats received a modified form of the control diet in which cod liver oil replaced the corn oil, olive oil and cotton seed oil and provides 36% of the calories or 39.6g/Kg of the diet; (D) the fourth group of rats received ethanol, cod liver oil and vitamin E diet. In this diet, ethanol substituted isocalorically for dextrin maltose, cod liver oil replaced all the oils in the diet, and the diet was supplemented with an additional 142 IU d alpha-tocopherol per Kg of diet. The animals were sacrificed by decapitation, the livers were removed, blotted on a filter paper, weighed, frozen in a sealed vial and maintained at -70°C until analyzed. Total hepatic liver was extracted as described (Folch et al., 1951). Triacylglycerols in the extract were determined by a colorimetric method (Biggs et al., 1977). Total cholesterol was estimated as outlined (Zak 1985). Phospholipid content of the extract was estimated by the method of Raheja et al., (Raheja et al., 1973). This method of phospholipid determination does not require predigestion of the phospholipid. The fatty acid profile in the extract was determined as described elsewhere (Eskelson et al., 1988). Briefly, to 1 ml of the extract, 200\mcg of diheptadecanoyl lecithin was added as an internal standard and the extract evaporated to dryness using a steady stream of nitrogen. 0.2\ml of chloroform and 2 ml of 12% boron trifluoride-methanol was added to the dry residue and the solution vortexed thoroughly and incubated at 80°C for 45 minutes. The dry residue was dissolved in chloroform and various amounts of this solution was used for analysis in a gas-liquid chromatograph. The chromatograph was fitted with a 6' x 1/8" glass column with 10% EGSS-X on chromophore-P. The fatty acid methylester peaks were identified by comparing individual peaks with standard methyl ester.

RESULTS

Table 1 shows changes in the major lipid classes in the treated animals. Hepatic triacylglycerol, phospholipid, cholesterol and the cholesterol/phospholipid molar ratio were significantly higher in rats fed ethanol than in controls ($p<0.05$). Supplementation of the alcohol diet with vitamin E decreased the triglyceride, and phospholipid in the ethanol and cod liver oil fed rats ($p<0.05$). Cod liver oil reduced hepatic cholesterol and cholesterol/phospholipid ratio, and these values are further increased with vitamin\E supplementation. Ethanol and cod liver oil increased hepatic triglyceride, cholesterol, and phospholipid but decreased the cholesterol/phospholipid ratio ($p<0.05$) by the concomitant increase in the phospholipid values observed. Table 2 shows that the ingestion of alcohol and/or cod liver oil altered the fatty acid composition of hepatic lipids. The relative amounts of 18:2/20:4 ratio increased in alcohol fed rats but decreased following supplementation with vitamin E. Cod liver oil fed rats showed increased 16:1 but decreased 18:2, 20:4, 18:0/18:1 and 18:2/20:4 ($p<0.05$). Rats fed cod liver oil supplemented with vitamin\E exhibited increased fatty acid levels of 14:0, 16:0, 18:0, 18:1, 18:2 and 18:2/20:4 and decreased 18:1/16:0 and ratio of unsaturated/saturated fatty acid ($p<0.05$).

Table 1: Effects of Cod Liver Oil, Ethanol and Vitamin E on Major Lipids in Rat Liver

Treatment	Cholesterol (mg/g)	Triglyceride (mg/g)	Phospholipid (mg/g)	Cholesterol/ Phospholipid Molar ratio
Control	4.37 + 0.37	13.97 + 0.82	26.26 + 3.16	0.17 + 0.004
Ethanol treated	5.68 + 0.32*	26.57 + 1.30*	31.54 + 1.65*	0.18 + 0.002
Ethanol + Cod Liver Oil	4.88 + 0.25	29.31 + 2.72*	38.17 + 2.79*	0.13 + 0.004*
Ethanol + Cold Liver Oil + Vitamin E	4.94 + 0.25	25.46 + 1.02*+	33.65 +0.90*,a	0.14 + 0.004*

Treatment of the animal groups is described in material and method. Results are mean + S.D. of five rats. *p\\0.05, statistically significant from control; +p\\0.05 from ethanol + cod liver oil.

Table 2: Fatty Acid Composition of Total Lipid from Livers of Rats Fed Ethanol, Cod Liver Supplemented with Vitamin E

Fatty Acid and Fatty Acid Ratio	DIET GROUP Control	Cod Liver Oil	Ethanol + Cod Liver Oil	Ethanol + Cod Liver Oil + Vitamin E
16:0	1.50 + 0.13	1.73 + 0.15	3.21 + 0.08*,+	3.40 + 0.25*
16:1	0.1	0.33 + 0.12*	0.43 + 0.09*	0.60 + 0.15*
18:0	2.33 + 0.24	2.45 + 0.19	4.26 + 0.15*,+	4.37 + 0.27*
18:1	1.84 + 0.14	2.18 + 0.18	2.54 + 0.17*	3.13 + 0.11*,+
18:2	2.71 + 0.29	1.78 + 0.13*	1.55 + 0.06*	1.74 + 0.28*
20:4	2.65 + 0.16	2.09 + 0.23*	1.51 + 0.17*,+	1.45 + 0.19*
16:0/16:1	15.21 + 0.24	5.26 + 0.22*	7.47 + 0.57*,+	5.66 + 0.23*,+
18:0/18:1	1.29 + 0.16	1.13 + 0.05	1.70 + 0.06	1.40 + 0.07+
18:1/16:0	1.21 + 0.14	1.25 + 0.10	0.77 + 0.14*,+	0.92 + 0.08*
18:2/20:4	1.05 + 0.15	0.88 + 0.09	1.07 + 0.09	1.19 + 0.10
Unsaturated/ Saturated Fatty acids	1.98 + 0.13	1.51 + 0.05	0.78 + 0.03*,	1.88 + 0.06*,+

Values are mean + SD\mg fatty acid/g total lipids
*Significantly different from control ($p<0.05$)
+Significantly different from ethanol + cod liver oil ($p<0.05$)
Significantly different from cod liver oil ($p<0.05$)

DISCUSSION

This data demonstrates that changes occur in the hepatic lipids following chronic ethanol administration. These changes are observed in the main lipid classes and the methyl esters of the fatty acids. Of significant interest is the role of vitamin E supplementation in modifying hepatic cholesterol. Cholesterol is an important determinant of membrane physical structure and it has an ordering effect on cell membrane. Increased cholesterol of the hepatic cells may explain the changed physical state of the cell, thus the partition of both lipid soluble and lipid insoluble substances into the liver cell, and thus, the development of a fatty liver. The increase in the saturation of membrane fatty acids in ethanol treated rats may also contribute to cell-fat accumulation. Our results showing an accumulation of large amounts of triglycerides, and that vitamin E has no effect on this accumulation is in consonance with earlier studies (Gavino et al., 1981). The alterations in the acyl chain indicating that ethanol consumption stimulated accumulation of C16:0, 18:0, 18:1 and 18:2 but depressed the desaturation step required for the elongation of C20:4 (arachidonic acid) implicate the depression of those enzymes required for this elongation step. This result is confirmed by the accumulation of palmitic acid in rats fed with the supplemental vitamin E.

In conclusion, the study indicates that there is a synergistic effect of ethanol and cod liver oil in peroxidative damage in alcoholic fatty liver, and that vitamin E plays a role in reducing these peroxidative damages and the saturation of hepatic membrane fatty liver which would otherwise contribute to hepatic fatty liver.

ACKNOWLEDGMENTS

This study was supported by Grant AA 08037 (R.R.W.) from the National Institute of Alcohol Abuse and Alcoholism and a National Cancer Institute Grant #51088 (S.I.M.).

REFERENCES

Alling, C.S., Gustavsson, L., Kristenson-Aas, A., and Wallerstedt, S. (1984). Changes in fatty acid composition of major glycerophospholipids in erythrocyte membrane from chronic alcoholics during withdrawal. *Scad. J. Clin. Lab. Invest.* **44**, 283-289.

Biggs, H.G., Erickson, J.M., and Morland, A. (1977). A manual colorimetric assay of triglyceride in serum. In: *Selected Methods of Clinical Chemistry*, **8**, 71-76. Cooper GR, ed., American Association of Clinical Chemist, Washington, D.C., .

Bjornrboe, G.A., Bjornrboe, A., Hagen, B.F., Morland, J., and Drevon, C.A. (1987). Reduced hepatic alpha tocopherol content of liver after long-term administration of ethanol to rats. *Biochem. Biophys. Acta.* **918**, 236-241.

Chin, J.H., Goldstein, D.B. (1977). Effects of low concentration of ethanol on the fluidity of spin labeled erythrocyte and brain membranes. *Mol. Pharmacol.* **13**, 435-441.

Eskelson, C.D., Stiffel, V., Owen, J.A., Chvapil, M., Vickers, A., and Brendel, K. (1988). Changes in the fatty acid profile in lung, liver and serum of rats given silica. *Life Sci.* **42**, 1455-1457.

Folch, J., Ascoli, I., Lees, M., Meath, J.A., and LeBaron, F.N. (1951). Preparation of lipid extract in brain tissues. *J. Biol. Chem.* **191**, 833-838.

Gavino, V.C., Miller, J.S., Dillman, J.M., Milo, G.E., and Cornwell, D.G. (1981). Polyunsaturated fatty acid accumulation in the lipids of cultured fibroblasts and smooth muscle cells. *J. Lipid Res.* **22**, 57-62.

Lieber, C.S. and DeCarli, L.M. (1986). The feeding of ethanol in liquid diets. 1986 update. *Alcohol Clin. Exp. Res.* **45**, 550-553.

Raheja, R.K., Kaur, C., Singh, A., and Bhatia, I.S. (1973). New Colorimetric methods for the quantitative estimation of phospholipid without acid digestion. *J. Lipid Res.* **14**, 695-697.

Reitz, R.C. (1979). The effects of ethanol administration on lipid metabolism. *Prog. Lipid Res.* **18**, 87-115.

Zak, B. (1985). Total and free cholesterol. In: *Standard Method of Clinical Chemistry*, **5**, 79-99, Mietis, S, ed., Academic Press, New York, N.Y.

DIETARY POLYUNSATURATED FATTY ACID PROMOTE PEROXIDATION AND ITS POSSIBLE ROLE IN THE PROMOTION OF CANCER

Olalekan E. Odeleye*, Ronald R. Watson*, Cleamond D. Eskelson** and Siraj I. Mufti***

Family and Community Medicine*, Surgery**, and Pharmacology and Toxicology***
University of Arizona
College of Medicine
Tucson Arizona 85724

It is known that an interplay between dietary, hereditary and environmental factors initiate and/or promote cancer. Several factors associated with high risk of the development of diseases including cancer have been identified. Included in these is the amount and type of dietary fat. Circumstantial evidence provided by epidemiological studies is further supported by laboratory models that provide strong correlation between dietary fat and cancer. These studies have consistently shown that high dietary fat level of 20 percent (w/w) or 40 percent by calories promote spontaneous or induced mammary tumor growth in rats and mice (Tinsley et al., 1981, Carrol et al., 1970). Polyunsaturated fatty acid containing several double bonds are unstable molecules and are readily oxidized by non-specific and specific lipoxygenases and cyclo-oxygenase enzymes systems to yield free radicals and peroxides, which are toxic to cells. Although the peroxidation of lipids have been involved in certain forms of tissue pathology and tumor growth, the mechanism and role of these fatty acids in the development of tumors has not been completely defined. Therefore, the effect of dietary cod liver oil, a polyunsaturated fatty acid, *in vivo* -1-lipid peroxidation as measured by ethane exhalation was investigated.

MATERIALS AND METHODS

Ten adult male Sprague-Dawley rats (120-150g) were housed in a controlled environment at 28° with a 12-hr light/dark cycle. Rats were divided into two groups. The control group received the Lieber-DeCarli diet (Lieber et al., 1986) during the 28-day experimental period. The cod liver oil (CLO) supplemented group received a modified form of the control diet in which CLO replaced corn seed oil, cotton seed oil, and olive oil. The CLO provided 35% of the total calories of the diet (3.96% w/v). This level of dietary CLO is a crucial threshold for tumor promotion (Begin et al., 1988).

Ethane exhalation was collected non-invasively according to a method previously described (Eskelson et al, 1987). Briefly, each rat was placed in a metabolic chamber consisting of the bottom of a glass dessicator and covered with a silicon sealed glass plate with a hole in the center. To prevent any contamination from external air, the inspirate entered the chamber via a rubber tube and then passed through a large volume of freshly prepared activated charcoal. The unidirectional flow of gas exited the chamber via a teflon tubing through a replaceable glass tube that contained the

Biological Reactive Intermediates IV, Edited by C.M. Witmer *et al.*
Plenum Press, New York, 1990

purified charcoal. The charcoal containing tubing was surrounded by dry ice maintained at -70°C to ensure the retention of the excluded ethane in the charcoal. The tubings were placed at 30-minutes intervals, and the used charcoal was poured into a 13.2 ml glass test tube and tightly sealed with a screw cap containing teflon septum. The charcoal was then heated on a heating block maintained at 240°C to desorb the ethane, and a 2 ml of gas head space was removed using an air-tight syringe. The sample was analyzed in a gas chromatograph (Model 402; Hewlett-Packard, Palo Alto, CA) containing a 6 foot long column packed with carbosphere of a surface area of about 1000m2/g and a pore size of about 13 A (Supelco, Bellefonte, PA). The oven temperature was maintained at 240°C and the injection port at 190°C. The ethane retention time and a standard curve were measured by injecting known volumes of pure ethane into the column.

RESULTS AND DISCUSSION

The weight gain were similar in both groups of animals. Administration of CLO significantly increased total ethane exhalation ($p<0.05$, Figure 1A) and ethane exhalation per unit body weight ($p<0.05$, Figure 1B). Expressed either as total ethane exhalation or ethane exhalation per unit of body weight, the increase is about 127-fold in the CLO group compared to the controls. Figure 2 show the cumulative ethane exhalation in the groups of rats. The CLO treated rats exhaled ethane at a linear rate with a correlation coefficient of 0.98. Dietary polyunsaturated fatty acids are potent substrate and/or inducers of in\vivo lipid peroxidation (Roehm et al., 1971), and its administration increases the peroxy radical load and activates the anti-oxidant system (Iritani et al, 1982).

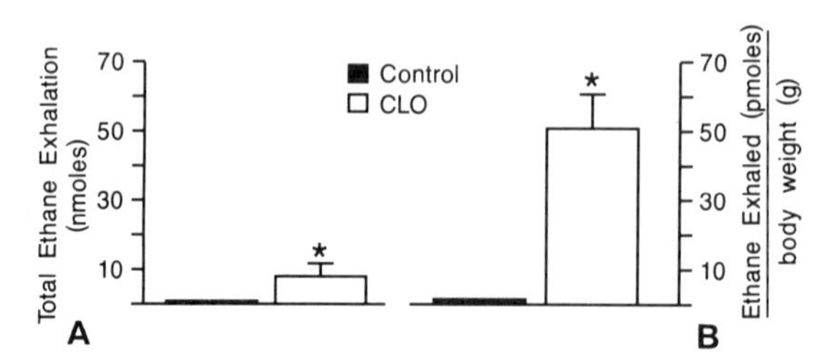

Figure 1A, 1B. Ethane exhalation expressed as (a) total; and (b) per gm body wt. Results are mean + S.D. of five animals. *$p<0.05$ statistically significant from controls.

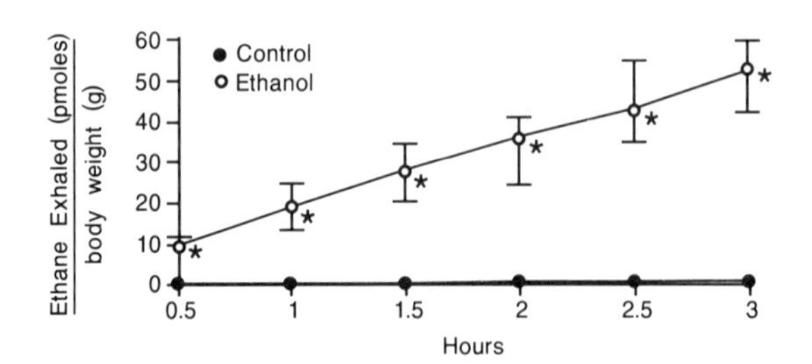

Figure 2. Effect of cod liver oil on cummulative ethane exhalation. Animals were treated as described under Materials and Methods. *$p<0.05$ statistically significant from control values.

In summary, dietary fatty acids in CLO are potent substrates for free radical attack and thus lipid peroxidation as measured by ethane exhalation. This increase in lipid peroxidation mediated by free radical attack of the cell membranes provides a possible hypothesis that partly explain the role of dietary polyunsaturated fatty acids in tumorigenesis.

ACKNOWLEDGMENTS

Supported by Grant AA-08037 (R.R.W.) from the National Institute on Alcohol Abuse and Alcoholism and a National Cancer Institute Grant. Grant #51088 (S.I.M.).

REFERENCES

Begin, E., Ellis, G., and Horribin, D.F. (1988). Polyunsaturated fatty acid-induced cytotoxicity against tumor cells and its relationship to lipid peroxidation. *J. Natl. Can. Inst.* **80**, 188-194.

Carrol, K.K. and Khor, H.T. (1970). Effects of high dietary fat and dose level of 7,12 dimethyl-benz-(alpha) anthracene on mammary tumor incidence in rats. *Cancer Res.* **30**, 2260-2264.

Eskelson, C.D., Chvapil, M., and Sipes, I.G. Effects of ultraviolet light irradiation, carbon tetrachloride, and trichloethylene on ethane exhalation in rats. In: *Prostaglandins and Lipid Metabolism in Radiation Injury*. Eds. Warden T.L., Hughes H.N. Plenum Press, N.Y. 1987, 393-401.

Iritani, N. and Ikeda, Y. (1982). Activation of catalase and other enzymes by corn oil intake. *J. Nutr.* **112**, 2235-2239.

Lieber, C.S. and DeCarli, L.M. (1986). The feeding of ethanol in liquid diets. 1986 Update. *Alcohol Clin. Exp. Res.* **45**, 550-553.

Roehm, J.N., Hadley, J.G., and Manzel, D.B. (1971). Oxidation of unsaturated fatty acids by oxone and nitrogen dioxide. *Arch. Envir. Health* **23**, 142-148.

Tinsley, I.J., Scmidt, J.A., and Pierce, D.A. (1981). Influence of dietary fatty acids on the incidence of mammary tumors in the C3H mouse cancer. *Cancer Res.* **41**, 1460-1465.

CHIRAL EPOXIDES, THEIR ENANTIOSELECTIVE REACTIVITY TOWARDS NUCLEIC ACIDS, AND A FIRST OUTLINE OF A QUANTUM CHEMICAL STRUCTURE-REACTIVITY CALCULATION

H. Peter, B. Marczynski, D. Wistuba*, L.V. Szentpály**, G. Csanády, H.M. Bolt

Inst. of Occup. Health at the University of Dortmund, Ardeystr. 67, D-4600 Dortmund
Inst. Org. Chem.*, University of Tübingen, Auf der Mogenstelle 18, D-7400 Tübingen, FRG.
Univ. of the West Indies**, Chem. Dept., Mona Campus, Kingston 7, Jamaica WI

INTRODUCTION

Liver microsomes of mice, rats and humans metabolize prochiral vinyl precursors to chiral epoxides with different ratios for the two enantiomers (Wistuba et al. 1989), which may interact in different ways with chiral structures in the cell. Thus, the pharmacological effects of chiral pharmaceutics (L-Dopa, D-Penicillamine) are known to depend on the configuration of the molecule. A similar enantioselectivity may therefore modulate genotoxicity of metabolically formed chiral epoxides when they react with chiral (in terms of their conformation) macromolecular nucleic acids.

METHODS

Equilibrium Dialysis

Pure enantiomers of either propylene oxide or styrene oxide were added to a 4 ml buffer solution of homobasic polynucleotides (100mers) or oligonucleotides (dimers) with 3-6 mmol base equivalents per litre in gas tight sealed head space vials. The buffer solution contained 0.2 M NaCl, 1 mM EDTA, 10 mM Tris-HCl, with a pH of 7.3. The ratio of the enantiomer added ranged from 0.1 to 1 with regard to the base equivalents. The samples were then incubated in a water bath at 37°C by shaking for 4 hours. To remove unreacted epoxide, the samples were subsequently shock frozen, lyophilized and the dried nucleic acids and buffer salts redissolved in 4 ml aq.bidist.. For equilibrium dialysis, 0.5 ml portions of these stock solutions were used. Eqilibrium dialysis was performed according to the method described by Uhlenbeck et al. (1970).

Melting Behavior

Synthetic homobasic polynucleotides when reacted as described above were hyybridized with their counterparts and melting behavior was investigated.

Circular Dichroism Spectra

PolyC and polyU reacted with either pure enantiomer of propylene oxide or styrene oxide were analyzed either single-stranded or after hybridization with their

Biological Reactive Intermediates IV, Edited by C.M. Witmer *et al.*
Plenum Press, New York, 1990

complementary polynucleotides in the double-stranded form. Circular dichroism spectra in the range of 220 - 320 nm were registered on a Dichrographe R.J. Mark III (Jobin Yvon Division d'Instruments S.A.).

HPLC-analysis of Nucleosides

After registration of circular dichroism spectra, aliquots of the pretreated polynucleotides were hydrolyzed enzymatically with RNase, phosphodiesterase and alkaline phosphatase, and the hydrolysate analyzed for binding products with HPLC. The conditions were:

LiChrospher 100 RP-18, 10 µm, 250 x 10 mm (Merck); gradient elution with solution A: 0.1 N triethylammonium acetate, pH 7.4; solution B: 10 % aqueous methanol; flow rate: 3 ml/min; UV detection at 254 nm.

From the resulting UV peak pattern, a semi-quantitative estimation was made for binding products.

Quantum Chemical Assessment of Enantioselective Reactivity

The reaction of cytosine with ethylene oxide and with R- and S-propylene oxide was chosen as a model to study a possible stereoselective reactivity of epoxides with nonchiral nucleobases. The method is described in detail by Peter and Csanády (1990).

Furthermore, a possible stereoselective reactivity of R- and S-propylene oxide towards chiral N-acetyl-L-cysteine was performed. 15 ml pre-equilibrated (pH 7.15, T 20°C) aqueous solutions of 0.05 M amino acid were mixed with 300 µl 1.4 mM R- and S-propylene oxide and incubated at 20°C. 100 µl samples were drawn at different time intervals and were reacted with 500 µl 0.1 M pentamethylene-dithiocarbamic acid sodium salt. The amount of dithiocarbamic acid-ester was determined by HPLC (column: 250 x 4 mm Lichrospher C_{18}, eluent: 50% methanol in 0.01 M Tris/$HClO_4$ pH 7.15, flow: 0.4 ml/min, detection at 280 nm). The second order kinetic constants were determined.

RESULTS

Equilibrium Dialysis

The influence of R- and S-propylene oxide on the base pairings C-G and C-A is shown in Figure 1 with pre-incubated polyC. An increasing equilibrium constant is indicative for stronger interactions between the homobasic polynucleotide with the hybridized dinucleotide and vice versa, thus reflecting influences of the epoxide. R- and S-propylene oxide shifted the molar equilibrium constants differently when preincubated polyC was used in the hybridization experiments with GpG, whereas the shift of the constant showed no stereoselective difference in the pairing with ApA (see Figure 1).

When polyC was pre-incubated with either enantiomer of styrene oxide and then hybridized with GpG or ApA, again the shift in the molar equilibrium constant was antagonistic for the pair C-G and uniform for the pair C-A.

When GpG was pre-incubated with either enantiomer of propylene oxide or styrene oxide and subsequently equilibrated with non-reacted polyC, the equilibrium constant shifted non-stereoselectively in the same direction for both chiral epoxides.

Pre-incubations of polyU with R- and S-propylene oxide and subsequent hybridization with GpG or ApA showed an antagonistic shift in the equilibrium constant for the base pairing U-A. When the non-complementary dinucleotide GpG was recombined with pre-incubated polyU an antagonistic shift in the equilibrium constant was also observed, but with an apparant increase for S-propylene oxide and an apparent decrease in the case of R-propylene oxide.

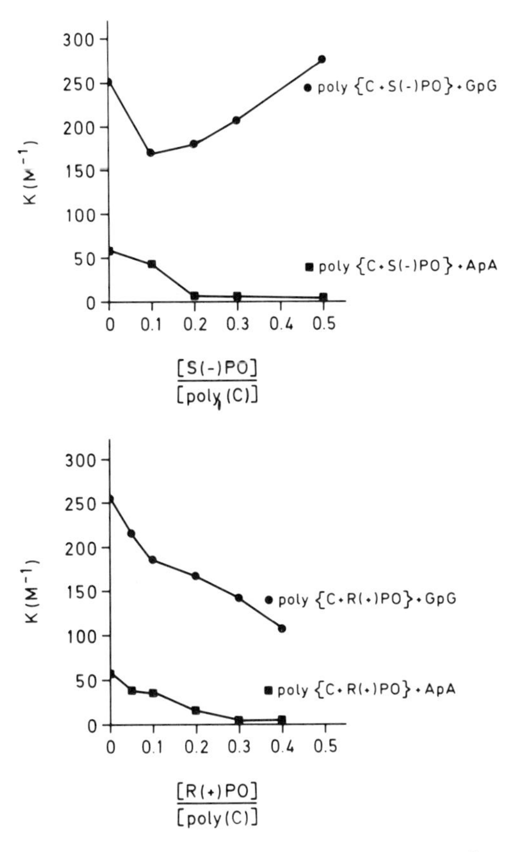

Figure 1. Shift in the molar association constant of polyC reacted with different concentrations (ratio of the epoxide to base equivalents) of the two enantiomers of propylene oxide after subsequent hybridization with GpG or ApA.

Pretreatment of polyU with either enantiomer of styrene oxide shifted the equilibrium constants in recombination with the dinucleotides ApA or GpG. Both enantiomers of styrene oxide reduced the affinity between the complementary bases U-A and enhanced it in the non-complementary pairing U-G.

When the dinucleotide GpG was incubated with either enantiomer of propylene oxide or styrene oxide before recombination with polyU, the equilibrium constant was generally elevated.

Melting Behavior

When the equilibrium constant is raised, an increase in melting temperature is to be expected, and vice versa. The melting behavior cannot be registered in the case of hybridized polyC·polyG because of the very high natural melting temperature of this double-stranded nucleic acid. When polyC or polyU was reacted with either enantiomer of propylene oxide or styrene oxide and subsequently recombined with polyA or polyG for registration of melting temperature, a change in melting behavior was observed (Figure 2 and 3), which was in concordance with the shift of the equilibrium constants in the corresponding equilibrium dialysis experiments (Figure 1).

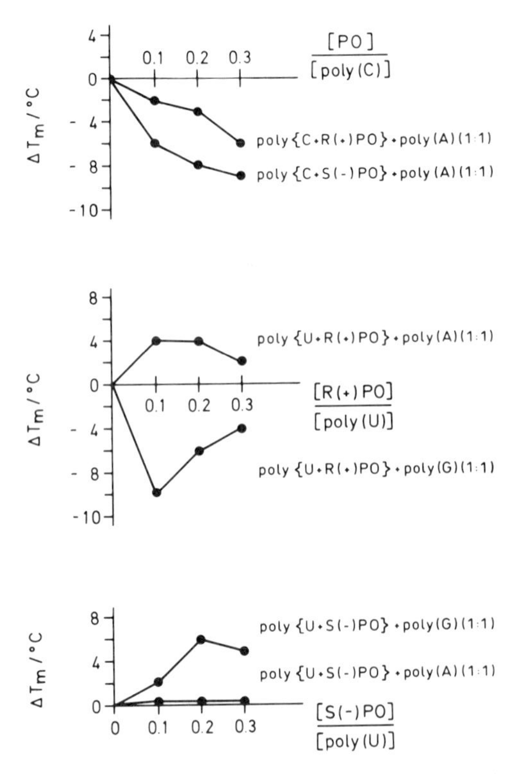

Figure 2. Shift in the melting temperature Tm of the double-strand when polyC or polyU are reacted with the two enantiomers of propylene oxide and subsequently recombined with complemetary or non-complementary polynucleotides.

Circular Dichroism (CD) Spectra

No substantial changes were observed in the spectra of singe-stranded polyC with R- or S-styrene oxide. In contrast, after hybridization with polyG, a dramatic change in the spectrum was seen in the particular case of pretreatment with R-styrene oxide (see Figure 4a+b). This coincides with a stronger mutagenicity of this enantiomer (Pagano et al. 1982).

PolyC reacted with either enantiomer of propylene oxide showed stereoselectively different CD-spectra of the single strand.

When polyU was reacted with either enantiomer of propylene oxide CD-spectra also differed stereoselectively. However, after hybridization of the reacted polyU with polyA, no significant differences were observed.

HPLC-analysis of Nucleosides

Obviously, in the case of polyC pretreated with the enantiomers of propylene oxide, an additional peak appears only when R-propylene oxide is incubated. This stereoselective binding product with cytosine residues has not yet been identified.

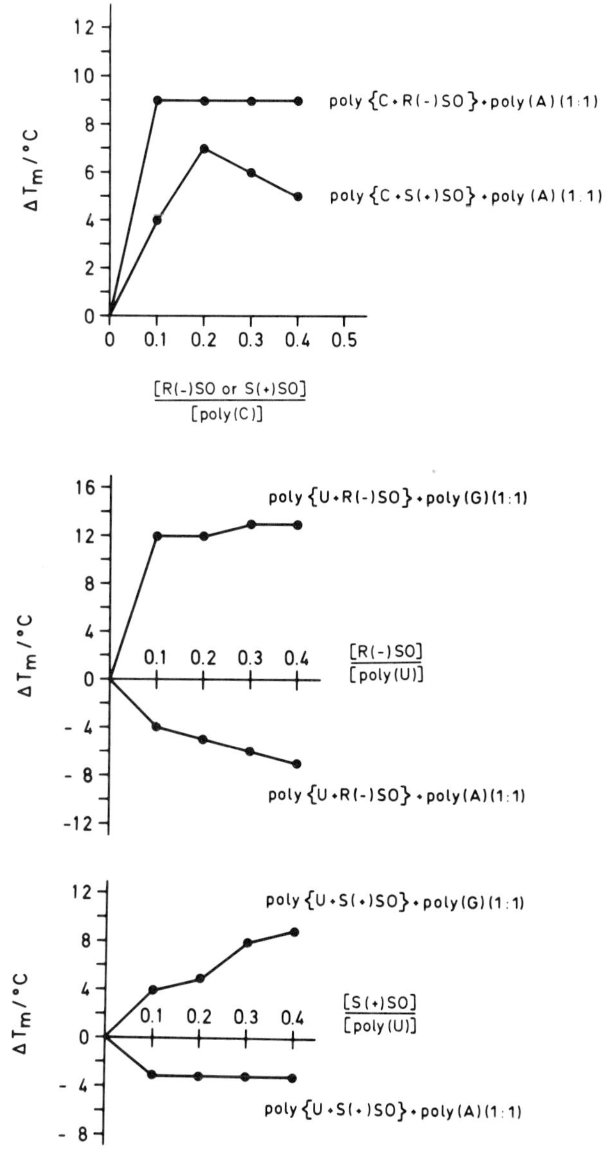

Figure 3. Shift in the melting temperature Tm of the double-strand when polyC or polyU are reacted with the two enantiomers of styrene oxide and subsequently recombined with complemetary or non-complementary polynucleotides.

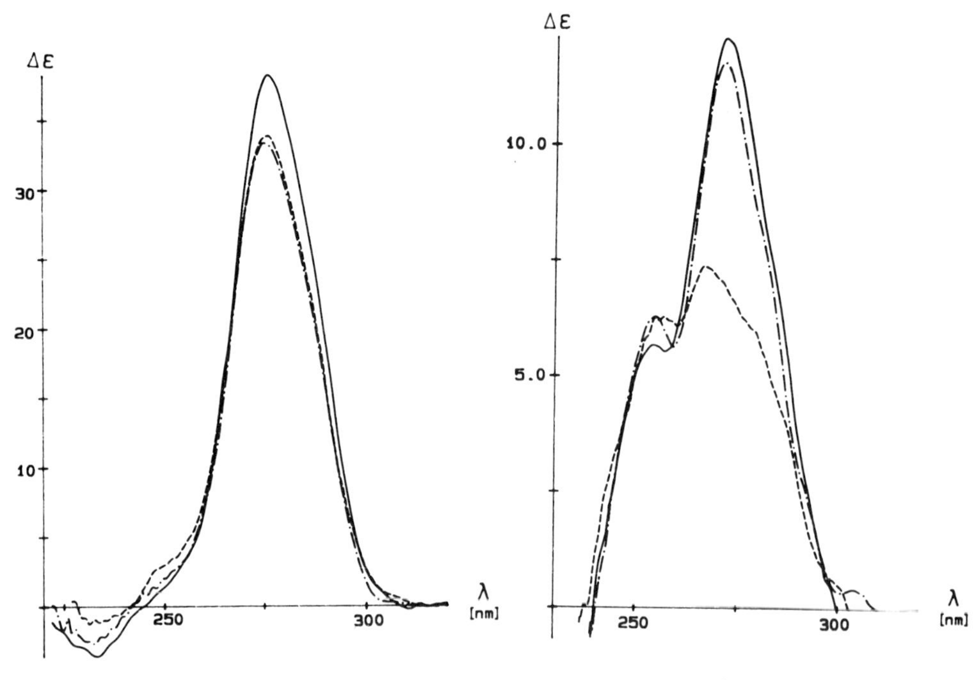

Figure 4a.
CD spectra of non-reacted single-strand polyC (_____), polyC reacted with R-styrene oxide (-----) and polyC reacted with S-styrene oxide (._._.).

Figure 4b.
CD spectra of non-reacted polyC.polyG (_____), polyC reacted with R-styrene oxide (-----) and polyC reacted with S-styrene oxide after hybridization with polyG (._._.).

Furthermore, we observed stereoselective differences in the peak pattern when we analyzed polyU reacted with either enantiomer of styrene oxide. This clearly demonstrate that the chiral epoxides investigated so far have a stereoselective reactivity towards chiral nucleic acids.

Quantum Chemical Assessment of Enantioselective Reactivity

In order to provide a quantum-chemical explanation for the above described results, calculations were performed according to the MNDO-method.

These calculations showed no enantioselective difference in reactivity of R- or S-propylene oxide towards the nonchiral nucleobase cytosine. In contrast, incubation experiments with the chiral compound N-acetyl-L-cysteine with and enantiomers of propylene oxide gave significantly different kinetic constants for the two forms (Table 1).

Table 1. Kinetic Constants for the Reaction of Either Enantiomer of Propylene Oxide or Its Racemate with N-Acetyl-L-Cysteine

Compound	k $[M^{-1}min^{-1}]$
R(+)-propylene oxide	$(1.02 \pm 0.09) \times 10^{-2}$
S(-)-propylene oxide	$(1.23 \pm 0.10) \times 10^{-2}$
racemic propylene oxide	$(1.12 \pm 0.11) \times 10^{-2}$

The difference between the reaction constants of the pure enantiomers was determined to be slight but significant ($p \geq 99\%$). This indicates that only *chiral* compounds or *chiral* macromolecular nucleic acids show a enantioselective reactivity with *chiral* epoxides.

DISCUSSION

The two enantiomers of both propylene oxide and styrene oxide were reacted with synthetic homobasic nucleic acids and subsequently subjected to equilibrium dialysis and an investigation of melting behaviour. An enantioselective difference in reactivity towards single-stranded nucleic acids subsequently hybridized to their double-strands was found. The reaction of the pure enantiomers of both propylene oxide and styrene oxide resulted in different changes of the conformation of the macromolecular nucleic acid, as could be verified by circular dichroism spectra. A subsequent hplc analysis of nucleosides enzymatically released from the reacted polynucleotides showed that the changes in conformation had been caused by an enantioselectively different covalent binding of the chiral epoxides to nucleobases.

REFERENCES

Pagano, D.A., Yagen, B., Hernandez, O., Bend, J.R., Zeiger, E. (1982). Mutagenicity of (R) and (S) styrene 7,8-oxide and the intermediary mercapturic acid metabolites formed from styrene 7,8-oxide. *Environ. Mutagenesis* **4**, 575-584.

Peter, H., Csanády, G. (1990). A note on the preference of aliphatic epoxides for the N-7 position of guanine in DNA. *Arch. Toxicol.*, in press

Uhlenbeck, O.C., Baller, J., Doty, P. (1970). Complementary oligonucleotide binding to the anticodon loop of fMet-transfer RNA. *Nature* **225**, 508-510.

Wistuba, D., Nowotny, H.P., Träger, O., Schurig, V. (1989). Cytochrome P-450-catalyzed asymmetric epoxidation of simple prochiral and chiral aliphatic alkenes: species dependence and effect of enzyme induction on enantioselective oxirane formation. *Chirality* **1**, 127-136.

COMPARISON OF 8-HYDROXYDEOXYGUANOSINE AND 5-HYDROXY-METHYLURACIL AS PRODUCTS OF OXIDATIVE DNA DAMAGE

David W. Potter* and Zora Djuric**

*Rohm and Haas Co.
Spring House, PA and
**Dept. of Internal Medicine
Wayne State University, Detroit, MI

In vitro experiments with calf thymus DNA demonstrated that both 8-hydroxydeoxyguanosine (8OHdG) and 5-hydroxymethyluracil (5OHmU) are formed by H_2O_2 and Fe^{2+} in a concentration dependent fashion. Mannitol completely inhibited 8OHdG while 5OHmU formation was only partially inhibited. Equimolar concentrations of EDTA and Fe^{2+} stimulated adduct formation. Singlet oxygen, generated in reaction mixtures containing methylene blue and light, resulted in formation of both 8OHdG and 5OHmU.

INTRODUCTION

Cellular oxidative stress may result in damage to critical macromolecules, such as DNA. Numerous DNA lesions are formed during exposure to reactive oxygen intermediates, including 8OHdG and 5OHmU. It has been suggested that reactive oxygen intermediates may be involved in mutagenesis, carcinogenesis and aging (Ames, 1983). Since different types of DNA damage may be formed by different reactive oxygen intermediates the mechanisms of 8OHdG and 5OHmU formation have been evaluated.

METHODS

Calf thymus DNA (1 mg/ml) was reacted at 37o with various combinations of the following components: 0-200 uM H_2O_2, 0-400 uM $FeCl_2$, 0-400 uM $FeCl_3$, 0-800 uM EDTA and 0 or 100 mM mannitolin 100 mM potassium phosphate, pH 7.4. In other incubation mixtures, calf thymus DNA was reacted at room temperature with 0-200 uM methylene blue and light from a 75 watt bulb in 100 mM potassium phosphate, pH 7.4. After incubation for 15 min, DNA was precipitated by addition of 1 M NaCl (0.1 ml/ml sample) and ice-cold isopropyl alcohol (1:1). DNA was collected by centrifugation at 16,000 x g for 5 min and was subsequently washed twice with 5 ml 70% ethyl alcohol and redissolved in 5 mM Bis Tris, 0.1 mM EDTA, pH 7.1. After enzymatic hydrolysis of the DNA, 8OHdG was quantified by HPLC with electrochemical detection (Floyd et al., 1986) after DNA was hydrolyzed with DNases (Beland et al., 1984). 5OHmU was quantified after acid hydrolysis of DNA and its trimethylsilyl derivative was detected by gas chromatography and mass spectromety based on the methods of Dizdaroglu and Bergtold (1986).

Biological Reactive Intermediates IV, Edited by C.M. Witmer *et al.*
Plenum Press, New York, 1990

RESULTS

In reaction mixtures containing DNA, 0-200 uM H_2O_2, 200 uM $FeCl_2$ and 0 or 100 mM mannitol, a H_2O_2-dependent increase of both 8OHdG and 5OHmU formation was observed. The ambient levels of 8OHdG and 5OHmU in calf thymus DNA were approximately 2 8OHdG/10^5 deoxyguanosine (dG) and 3 5OHmU/10^4 thymidine (dT). While Fe^{2+} alone caused a slight increase in DNA adducts, H_2O_2 alone had no effect. Equimolar concentrations of Fe^{2+} and H_2O_2 increased 8OHdG formation by about 40-fold and 5OHmU by about 9-fold. Mannitol completely decreased formation of 8OHdG; however, mannitol decreased 5OHmU formation by only 50%.

EDTA stimulated formation of 8OHdG and 5OHmU. Although equimolar concentrations of EDTA-Fe^{2+} stimulated oxidation of DNA by approximately 3-fold, a ratio of EDTA/Fe^{2+} > 1 decreased formation of 8OHdG and 5OHmU. With high concentrations of EDTA, formation of oxidized bases was approximately the same as formation without EDTA.

The role of Fe^{3+} in DNA oxidation reactions was also evaluated. In reaction mixtures containing 100 uM H_2O_2, 200 uM $FeCl_2$, 0 or 200 uM EDTA, 0-400 uM $FeCl_3$ and 100 mM potassium phosphate, pH 7.4, Fe^{3+} did not have an effect on 8OHdG formation. However, 5OHmU formation was stimulated at concentrations of Fe^{3+} above 100 uM.

It has previously been proposed by Floyd et al. (1989), that methylene blue and light forms 8OHdG via singlet oxygen; yet, oxidation of other bases was not reported. In the presence of various amounts of methylene blue, both 8OHdG and 5OHmU were formed in a concentration dependent manner.

DISCUSSION

The results suggest that both 8OHdG and 5OHmU are formed via reactive oxygen intermediates. However, the intermediates involved in the oxidation of dG and dT may be different. For example, in reaction mixtures containing H_2O_2 and Fe^{2+}, mannitol completely decreased 8OHdG formation; yet, mannitol only partially decreased 5OHmU formation. The results are consistent with 8OHdG formation by a hydroxyl radical addition reaction. Oxidation of dT appears to be more complex. Mannitol only partially decreased 5OHdG formation suggesting that oxidation may have occured via a different mechanism than hydroxyl radical addition. Oxidation of dT may have resulted from an intermediate similar to what has been suggested for lipid peroxidation reactions, namely a reactive oxygen-iron complex (Giorgio and Aust, 1989). Furthermore, the observed stimulation of dT oxidation by Fe^{3+} is consistent with involvement of an oxygen-iron complex as proposed for lipid peroxidation (Giorgio and Aust, 1989). In contrast, Fe^{3+} had no effect on 8OHdG formation.

Singlet oxygen generated from methylene blue and light has previously been shown to form 8OHdG (Floyd et al., 1989). It was suggested that oxidation of dT did not occur. Results presented here indicate that singlet oxygen is equally effective in formation of 5OHmU and 8OHdG.

The data presented in this paper are consistent with 8OHdG formation via hydroxyl radical addition, 5OHmU formation via electron abstraction of dT by an oxygen-iron complex followed by oxygen addition and both adducts also being formed from singlet oxygen by an unknown mechanism.

REFERENCES

Ames, B.N. (1983). Dietary carcinogens and anticarcinogens. Oxygen radicals and degenerative diseases. *Science* **221**, 1256-1264.

Beland, F.A., Fullerton, N.F., and Heflich, R.H. (1984). Rapid isolation, hydrolysis and chromatography of formaldehyde-modified DNA. *J. Chromatogr.* **308**, 121-131.

Dizdaroglu, M., and Bertold, D.S. (1986). Characterization of free radical-induced base damage in DNA at biologically relevant levels. *Anal. Biochem.* **156**, 182-188.

Floyd, R.A., Watson, J.J., Wong, P.K., Altmiller, D.H., and Rickard, R.C. (1986). Hydroxyl free radical adduct of deoxyguansine: sensitive detection and mechanisms of formation. *Free Rad. Res. Comms.* **1**, 163-172.

Floyd, R.A., West, M.S., Eneff, K.L., and Schneider, J.E. (1989). Methylene blue plus light mediates 8-hyroxyguanine formation in DNA. *Arch. Biochem. Biophys.* **273**, 106-111.

Giorgio, M., and Aust, S.D. (1989). The role of iron in oxygen radical mediated lipid peroxidation. *Chem.-Biol. Interactions* **71**, 1-19.

MODULATION OF AORTIC SMOOTH MUSCLE CELL PROLIFERTION BY DINITROTOLUENE

K. Ramos[1], K. McMahon, C. Alipui* and D. Demick**

Department of Veterinary Physiology and Pharmacology
Texas A&M University
College Station, TX 77843 and
Departments of Pathology** and Pharmacology*
Texas Tech University Health Sciences Center
Lubbock, TX 79430

Technical grade dinitrotoluene (DNT) is a mixture of approximately 76% 2,4-DNT, 19% 2,6-DNT and 5% other isomers. Because DNT is commonly used in the manufacture of explosives and several commercial products, concerns have been raised that this chemical might represent a significant occupational hazard (Levine, 1987). Toxicity studies in rodents have shown that prolonged administration of DNT causes cancers of the liver, gall bladder, and kidney and benign tumors of connective tissues. Bond et al. (1981) have proposed that the hepatocarcinogenic effect of DNT is mediated by a toxic metabolite formed by intestinal flora upon reduction of nitrobenzylalcohol, an oxidative metabolite of dinitrotoluene.

Retrospective cohort mortality studies have shown an increased incidence of atherosclerotic heart disease rather than cancer in workers exposed to DNT (Levine, 1987). This correlation suggests that DNT, or its metabolites, exert toxic effects within the vessel wall to initiate and/or promote the atherosclerotic process. This proposal is consistent with previous studies which show that chemical carcinogens cause atherosclerotic lesions in several animal species (Albert et al., 1977; Bond et al., 1981). Because changes of cellular proliferation characterize the initial stages of atherosclerosis (Campbell and Campbell, 1987), the present studies were conducted to assess the effects of DNT on the ^{3}H-thymidine incorporation in aortic smooth muscle cells obtained from animals exposed subchronically to DNT.

METHODS AND RESULTS

Male Sprague-Dawley rats (175-200g) were given daily IP injections of 2,4- or 2,6-DNT (0.5, 5, 10 mg/kg) or MCT oil for 8 weeks. At the end of the dosing regimen, aortae were excised and processed for the isolation of smooth muscle cells as described previously (Ramos and Cox, 1987). Cells were grown in Medium 199 supplemented with 10% fetal bovine serum and 2 mM glutamine. The incorporation of ^{3}H-thymidine into DNA was measured as an index of cellular proliferation in confluent primary cultures of aortic smooth muscle cells from control or DNT-treated animals. Cultures were labelled with 5 uCi/ml [methyl-^{3}H]-thymidine triphosphate for 1 hr and processed according to Palmberg et al. (1985). The results presented in Figure 1 show

---

1. Send correspondence to: Dr. Kenneth S. Ramos, Dept. of Physiology and Pharmacology, College of Veterinary Medicine, Texas A&M University, College Station, TX 77843

that exposure of rats to either 2,4- or 2,6-DNT is associated with a marked reduction in the incorporation of ^{3}H-thymidine by aortic smooth muscle cells in primary culture. This inhibitory response was observed in cells from animals at all doses tested.

In other studies, subcultured aortic smooth muscle cells obtained from naive animals were incubated with various concentrations of 2,4- or 2,6-DNT (1-100 mM) for 3 hr. Thymidine incorporation was measured during this period in cycling cultures or cultures previously synchronized by serum-deprivation and subsequent co-incubation with 10 uM hydroxyurea (Ramos, 1990). This approach was used to determine if DNT modulates scheduled and/or unscheduled DNA synthesis, respectively, in aortic smooth muscle cells. The results presented in Table 1 show that neither 2,4- nor 2,6-DNT altered the extent of ^{3}H-thymidine incorporation in cycling or growth-arrested cultures.

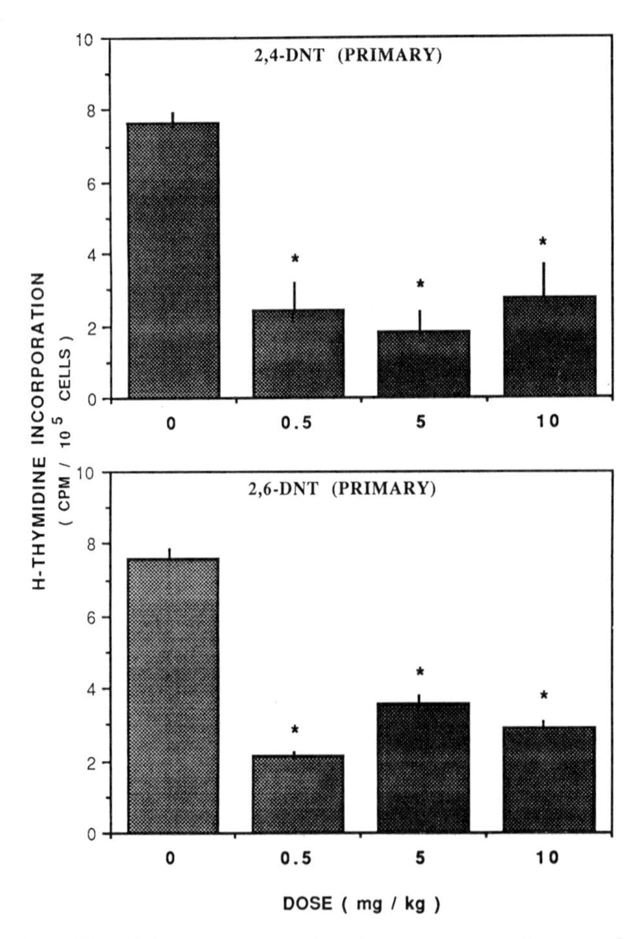

Figure 1. Thymidine incorporation in primary cultures of aortic smooth muscle cells obtained from animals treated with 2,4- or 2,6-dinitrotoluene for 8 weeks.

Table 1. DNA Synthesis in Cultured SMC upon Exposure to Dinitrotoluene

Treatment	Concentration	DNA Synthesis Scheduled[a]	DNA Synthesis Unscheduled[b]
Vehicle Control	-	218.71 ± 29.02	33.64 ± 3.21
2,4-DNT	1	213.22 ± 26.81	-
	10	248.30 ± 29.30	41.96 ± 2.18
	100	306.53 ± 55.64	34.21 ± 5.20
2,6-DNT	1	272.80 ± 24.12	-
	10	181.36 ± 12.09	32.32 ± 6.54
	100	299.69 ± 53.52	25.78 ± 8.42

[a] Scheduled DNA synthesis was measured as 3H-thymidine incorporation/mg cell protein for 3 hours in cycling cultures of aortic smooth muscle cells from naive animals upon exposure to DNT.

^{3}H-Thymidine incorporation in untreated controls was similar to that observed in cultures exposed to the vehicle.

[b] Unscheduled DNA synthesis was measured in serum-deprived cultures in the presence of 10 uM hydroxyurea, an inhibitor of scheduled DNA synthesis.

DISCUSSION

Our results suggest that the atherogenic effect of DNT may be due to alterations in the regulation of vascular smooth muscle cell growth. Dinitrotoluene, or toxic metabolites formed *in vivo*, may exert genotoxic effects in smooth muscle cells which lead to interference with cellular proliferation. The inability of DNT to alter DNA synthesis *in vitro* suggests that the inhibitory response observed is not due to direct actions of DNT. Generalized cytotoxicity was ruled out as a possible mechanism since exposure of aortic smooth muscle cells to DNT *in vitro* for 24 or 48 hr was not associated with cytophasmic enzyme leakage (results not shown). We have obtained preliminary evidence suggesting that diaminotoluene, a reduced metabolite of DNT, inhibits scheduled DNA synthesis in cultured aortic smooth muscle cells. This observation supports the concept that, as described for the hepatocarcinogenic effect of DNT, metabolic activation of the parent compound is essential for the modulation of smooth muscle cell growth. Studies are now in progress to determine if the reduced thymidine incorporation associated with DNT treatment *in vivo* reflects differences in the mitogenic responsiveness of smooth muscle cells.

The apparent discrepancy between the toxicologic profile of DNT in rodents and humans can be reconciled when one considers that atherogenesis and carcinogenesis are both associated with disturbances in the regulation of cellular differentiation and proliferation. Within this context, Benditt (1978) has proposed that smooth muscle cells within the atherosclerotic plaque are the progeny of a single, mutated cell and thus, represent benign neoplastic lesions. Although the significance of our observations within the context of atherogenesis remains to be established, our results support the concept that modulation of vascular smooth muscle cell proliferation by DNT may play an important role in the expression of cardiovascular toxicity.

ACKNOWLEDGEMENTS

This research was supported in part by grant RD-5-90 from Texas A&M University.

REFERENCES

Albert, R.E., Vanderlaan, M., Burns, F.J., and Nishuzumi, M. (1977). Effect of carcinogens on chicken atherosclerosis. *Can. Res.* **37**, 2232-2235.

Benditt, E.P. (1978). The monoclonal theory of atherogenesis. In *Atherosclerosis Reviews* (R. Paoletti and A.M. Gotto, Jr., Eds). Vol. 3, pp 77-85, Raven Press, New York.

Bond, J.A., Yang, H-Y.L., Majesky, M.W., Benditt, E.P., and Juchau, M.R. (1980). Metabolism of benzo(a)pyrene and 7,12-dimethylbenz(a)anthracene in chicken aortas: monooxygenation, bioactivation to mutagens, and covalent binding to DNA *in vitro*.

Bond, J.A., Medinsky, M.A., Dent, J.G., and Rickert, D.E. (1981). Sex-dependent metabolism and biliary excretion of [2,4-^{14}C]Dinitrotoluene in isolated perfused livers.

Campbell, G.R. and Campbell, J.H. (1987). Phenotypic modulation of smooth muscle cells in culture. In *Vascular Smmoth Muscle Cells in Culture* (J.H. Campbell and G.R. Campbell, Eds.), pp 39-56, CRC Press, Florida.

Levine, R.J. (1987). Dinitrotoluene: Human atherogen, carcinogen, neither or both? *CIIT Act.* **7**, 1-5.

Palmberg, L., Sjolund, M., and Thyberg, J. (1985). Phenotype modulation in primary cultures of arterial smooth muscle cells: Regulation of cytoskeleton and activation of synthetic activities. *Differentiation* **29**, 275-283.

Ramos, K., and Cox, L.R. (1987). Primary cultures of rat aortic endothelial and smooth muscle cells: An *in vitro* model to study xenobiotic-induced vascular cytotoxicity. *In Vitro Cell. Devel. Biol.* **23**, 288-296.

Ramos, K. (1990). Cellular and molecular basis of xenobiotic-induced cardiovascular toxicity: Application of cell culture systems. In *Focus on Cellular Molecular Toxicology and In Vitro Toxicology*, (D. Acosta, Ed.), in press.

REACTIVE POTENTIAL OF DIETHYLSTILBESTROL REACTIVE METABOLITES TOWARDS CELLULAR NUCLEAR PROTEINS: IMPLICATIONS FOR ESTROGEN*INDUCED CARCINOGENESIS

Deodutta Roy

Department of Pharmacology and Toxicology
University of Texas Medical Branch
Galveston, Texas 77550

Diethylstilbestrol (DES), a synthetic estrogen, has increasingly been associated with human cancer (IARC 1979). Over the years, DES has been shown to induce tumors in reproductive- and various other organs in animals (IARC 1979). The exact mechanism of estrogen-induced carcinogenesis is not understood. A role of reactive metabolites in estrogen-induced carcinogenesis has been postulated (Metzler, 1987), because hormonal potency could not be correlated with tumor incidence.

Chromosomal proteins, histones and nonhistones, play important roles in the regulation of gene expression and chromosome structure (van Holde, 1989). Histones influence the structure and assembly of nucleosomes, while nonhistones have critical roles in gene activation and transcription. DES reactive metabolites have been shown to bind irreversibly to microsomal proteins. Recently, direct evidences of oxidation of DES to DES-p-quinone and covalent binding of DES-p-quinone to DNA both *in vitro* and *in vivo* have been provided (Roy and Liehr, 1989, Liehr et al., 1989). It was, therefore, of interest to determine the nature of alkylation of histones and nonhistones by metabolically activated DES. In the present study synergistic covalent attack of DES reactive metabolites on the nucleophilic sites of nuclear proteins has been demonstrated.

MATERIALS AND METHODS

[Monoethyl-^{3}H]Diethylstilbestrol (specific activity, 70 mCi/mmol) was purchased from Amersham Corporation, Arlington Heights, IL. NADPH, DES, glutathione (GSH), and other chemicals were purchased from Sigma Chem. Co., St. Louis, MO. 3-OH-DES was a gift from Eli Lilly, Indianapolis. Syrian male hamsters were purchased from Sprague Dawley Inc., Houston, TX. Hamster liver microsomes and hamster kidney nuclei were prepared by the method of Dignam and Strobel (1977) and Yu (1975), respectively.

In Vitro Incubation System

Reaction mixtures consisted of 10 mM phosphate buffer, pH 7.5, 50 uM DES (containing 10 uCi ^{3}H-DES), 1 mM NADPH, 4 mg hamster liver microsomal protein and kidney nuclei equivalent to 10 mg nuclear protein in a final volume of 1 ml. The reactions were carried out for 30 min at 37°C. The incubations were terminated by chilling in ice and nuclei were pelleted by centrifuging at 1500 x g for 20 min.

Biological Reactive Intermediates IV, Edited by C.M. Witmer *et al.*
Plenum Press, New York, 1990

Extraction of Free DES and Its Metabolites

Histone and nonhistone proteins from nuclei were separated by selective acid extraction (Pezzuto et al., 1976). Free DES and its metabolites were extracted with organic solvents from histone and nonhistone fractions as described previously (Pezzuto et al., 1976). Radioactivities in both nonhistone and histone fractions were measured in LS-5000 Beckman liquid scintillation counter. Protein contents were determined by the method of Bradford (1976).

RESULTS AND DISCUSSION

Binding of DES to nuclear proteins in the presence of NADPH and microsomes were ten times higher than that observed in the absence of NADPH (15 pmol/mg protein/30 min). This binding was irreversible in nature because the radioactivity coprecipitated with nuclear proteins and cannot be extracted with organic solvents. Binding was inhibited by 7,8-benzoflavone, a potent inhibitor of cytochrome P-450 (Table 1). Binding of DES to nuclear proteins in the presence of nitrogen or CO gas atmosphere was lower than that of aerobic condition. Binding of DES to nuclear proteins in the absence of NADPH was not inhibited by 7,8-benzoflavone or CO. Therefore, binding of DES to nuclear proteins in the absence of NADPH was considered as a background. Thus, inhibition of binding of DES to nuclear proteins by 7,8-benzoflavone and CO suggests that DES reactive metabolites, generated by cytochrome P-450 system, irreversibly binds to nuclear proteins.

Table 1. Irreversible Binding of $[^{3}H]$DES To Nuclear Proteins

Conditions	pmol/mg protein/30 min	
	Histones	Nonhistones
No NADPH	2.5 + 0.17	7.8 + 1.2
NADPH	52.0 + 7.50	130.0 + 15.0
NADPH, 7,8-benzoflavone (500 uM)	15.0 + 3.0	47.0 + 6.0
NADPH, N_2	9.0 + 1.2	35.0 + 4.0
NADPH, CO	14.0 + 1.8	25.0 + 3.8
NADPH, Ascorbate (1 mM)	21.0 + 2.5	55.0 + 6.0

Reactions were carried as described in methods section. Each value is the mean of four experiments + standard deviation.

Addition of ascorbate in NADPH-dependent incubations resulted in a significant inhibition of binding of DES to nuclear proteins (Table 1). DES metabolites were extracted from this incubation by ether, dried under N_2, and the dried extract was derivatized with heptafluorobutyric acid anhydride. Analysis of derivatized products by gas chromatography using electron capture detector revealed that 20% of the DES was metabolized to 3-OH-DES during this incubation. Formation of 3-OH-DES (20%), a precursor of DES o-quinone (Kalyanaraman et al., 1989), in the presence of ascorbate was higher than that of observed in the absence of ascorbate (5%). Addition of ascorbate in incubation mixtures resulting in a decrease in binding of DES reactive metabolites to nuclear proteins and an increase in the formation of 3-OH-DES appear to suggest that metabolically activated 3-OH-DES might be involved in binding to nuclear proteins. Oxidation of DES to DES-p-quinone in a NADPH-mediated microsomal system has been previously demonstrated (Roy and Liehr, 1989). Addition of DT-diaphorase (equivalent to the substrate concentration, 50 nmol), known to reduce DES quinone to hydroquinone by exempting the formation of DES semiquinone, partially inhibited the binding to nuclear proteins (Table 1). Thus, these results suggests that both DES para and/or ortho semiquinone/quinone might be involved in covalent binding to nuclear proteins.

Thiols (GSH, cysteine, and Cysteamine) and thiol modifiers (diamide, N-ethylmaleimide, and $HgCl_2$) inhibited the covalent binding to nuclear proteins (Table 2). The inhibitory influence of thiols and thiol modifiers on binding suggests that nucleophilic sites of nuclear proteins might be susceptible to the covalent attack of DES reactive metabolites.

Table 2. Influence of Thiols and Thiol Modifying Agents on Irreversible Binding of [^{3}H]DES To Nuclear Proteins

Conditions	pmol [^{3}H]DES bound/mg nuclear proteins/30 min
No NADPH	15.0 + 2.0
NADPH (control)	145.0 + 16.0
Thiols:Glutathione (500 uM)	56.0 + 4.0
Cysteine (500 uM)	76.0 + 6.8
Cysteamine (500 uM)	36.0 + 3.2
Thiol modifiers:	
$HgCl_2$ (50 uM)	47.0 + 5.7
Diamide (1 mM)	62.0 + 7.0
N)ethylmaleimide (5 mM)	86.0 + 11.0

Reaction conditions were the same as described in methods section, except nuclei were preincubated with thiols or thiol modifying agents for 5 minutes as described previously (Roy and Snodgrass, 1990). Each value is the mean + standard deviation.

Covalent modification to nonhistone proteins by electrophilic metabolites of xenobiotic compounds has been implicated to participate in chemical carcinogenesis (Gronow, 1980). Histones as well as nonhistones irreversible modification caused by synergistic covalent attack of DES reactive metabolites, as shown here, may thus have significant effects in changing nucleosome assemble and ultimate function, which may play a role in DES-induced carcinogenesis.

REFERENCES

Bradford, M.M. (1976). A rapid and sensitive method for the quantification of microgram quantities of protein utilizing the principles of protein-dye-binding. *Anal. Biochem.* **72**, 248-254.

Dignam, J.D. and Strobel, H.W. (1977). NADPH-cytochrome P-450 reductase from rat liver: Purification by affinity chromatography and characterization. *Biochemistry* **16**, 1116-1122.

Grownow, M. (1980). Nuclear proteins and chemical carcinogenesis. *Chem. Biol. Interact.* **29**, 1-30.

van Holde, K.E. (1989). Chromatin structure and transcription. In: *Chromatin*, Alexender Rich ed. Dpringer Verlag, New York, pp 355-408.

IARC (1979). The evaluation of the carcinogenic risk of chemicals to hormone: Sex hormone. *IARC Monographs*, **21**, 173-221.

Kalyanaraman, B., Seally, R., and Liehr, J.G. (1989). Characterization of semiquinone free radicals formed from Astilbene catechol estrogens: an ESR spin stabilization and spin trapping study. *J. Biol. Chem.* **264**, 11014-11019.

Liehr, J.G., Roy, D., and Gladek, A. (1989). Mechanism of inhibition of renal carcinogenesis in male Syrian hamsters by vitamin C. *Carcinogenesis* **10**, 1983-1988.

Metzler, M (1987). Metabolic activation of xenobiotic stilbene estrogens. *Fed. Proc.* **46**, 1855-1857.

Pezzuto, J.M., Lea, M.A., and Yang, C.S. (1976). Binding of metabolically activated benzopyrene to nuclear macromolecules. *Cancer Res.* **36**,3647-3653.

Roy, D. and Liehr, J.G. (1989). Metabolic activation of diethylstilbestrol to diethylstilbestrol-4'4"-quinone in Syrian hamsters. *Carcinogenesis* **10**, 1241-1245, 1989.

Roy, D. and Snodgrass, W.R. (1990). Covalent binding of phenytoin to protein and modulation of phenytoin metabolism by thiols in A/J mouse liver microsome. *J. Pharmacol. Exp. Ther.*, In Press.

Yu, F. (1975). An improved method for the quantitative isolation of rat liver nuclear polymerase. *Biochim. Biophys. Acta.* **395**, 329-336.

BIOCHEMICAL AND MORPHOLOGIC RESPONSE OF NASAL EPITHELIA TO HYPEROXIA

Patrick J. Sabourin, Kristen J. Nikula, Amie J. Birdwhistell, Breton C. Freitag, and Jack R. Harkema

Lovelace Inhalation Toxicology Research Institute
P.O. Box 5890
Albuquerque, NM 87185

The nasal epithelium of people receiving oxygen therapy can be exposed to up to 100% oxygen by intranasal cannulae. The morphologic effect of hyperoxia on nasal epithelia has not been studied, but it is known that another oxidant, ozone, induces hyperplasia and metaplasia of the nasal epithelium. The purpose of our study was to examine the effects of acute and chronic hyperoxia on nasal epithelial morphology, and on antioxidant enzyme and cytochrome P450-dependent monooxygenase activities in nasal epithelium.

Male F344/N rats (8-12 weeks old) were exposed to ambient air or 85% oxygen, 24 hrs/day, for 1 or 11 weeks (total atmospheric pressure = 620 torr). The nonciliated, cuboidal to columnar (NCC) epithelium of the nasal turbinate, lateral wall, and maxilloturbinate; the respiratory epithelium of the nasal septum; and, the olfactory epithelium of the ethmoturbinates were examined.

Using fixed tissue sections from the anterior nasal cavities, we determined the number of epithelial nuclei/mm of basal lamina, the arithmetic mean thickness of the epithelium, and the number of bromodeoxyuridine (BrdU)-labeled (S-phase) nuclei/mm of basal lamina. For the ultrastructural studies, the NCC epithelium was removed, processed, and examined by electron microscopy.

For biochemical assays, rats were asphyxiated with CO_2, after which the nasal septum, the NCC and the ethmoturbinates were removed. Within each exposure group, we prepared four pooled samples (10 rats per pool) of each of the three tissue regions. Samples were homogenized, centrifuged (1000 x g and 9000 x g), and divided into aliquots for the enzyme assays. Superoxide dismutase (Misra, H.P., et al., 1972); catalase (Tolbert, N.E., et al., 1974); glutathione peroxidase (Paglia, D.E., et al., 1967; Lawrence, R.A., et al., 1976); glucose-6-phosphate dehydrogenase (Cohen, P., et al., 1975); cytochrome P-450 dependent monooxygenase, as measured by the O-deethylation of 3-cyano-7-ethoxycoumarin (White, I.N.H., 1988); and protein (Lowry, O.H., et al., 1951) were measured, as previously described.

There were no significant differences between the control and oxygen-exposed (1 and 11 weeks) epithelia (NCC and respiratory) in the number of epithelial nuclei/mm basal lamina. Oxygen exposure (1 week) resulted in a significant (25 - 50%) increase in the arithmetic mean thickness of the NCC epithelium (Figure 1). Continued oxygen exposure (11 weeks) resulted in a 65 - 110% increase in the NCC epithelial thickness.

Biological Reactive Intermediates IV, Edited by C.M. Witmer *et al.*
Plenum Press, New York, 1990

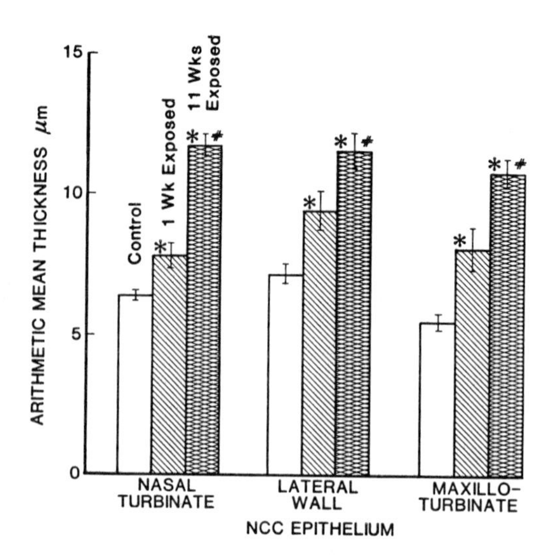

Figure 1. Oxygen Exposure Causes an Increased Mean Thickness of the Nasal Nonciliated to Columnar (NCC) Epithelium. There was no significant difference between the 1 wk and 11 wk controls in the mean thickness of the epithelium in each region, so the control values were pooled. *Significantly different from control by Student's t test ($p<0.05$). #Significantly different from 1 week exposed ($p<0.05$).

After one week of oxygen exposure, there were more BrdU-labeled nuclei in the epithelia of the nasal turbinate (NCC, 90 +/- 34%), maxilloturbinate (NCC, 206 +/- 80%), and nasal septum (respiratory; 350 +/- 83%) than in the same epithelia from control animals. Ultrastructurally, the main oxygen-induced alteration in the NCC epithelium was an apparent increase in number and size of mitochondria within the hypertrophic epithelial cells.

After one week of hyperoxic exposure, there were significant ($p<0.05$) increases in the specific activity of glucose-6-phosphate dehydrogenase in the NCC, the respiratory, and the olfactory epithelia, and in the specific activity of glutathione peroxidase in the NCC and olfactory epithelia (Figure 2). The specific activity of cytochrome P450-dependent monooxygenase was significantly decreased (62% of control) in the NCC epithelium. This decrease probably reflects an increase in non-microsomal protein.

Because the number of cells/surface area of the basal lamina did not change with oxygen exposure, but the volume of epithelium/surface area of the basal lamina increased in the NCC epithelium, the volume of the cells must have increased. Thus, oxygen exposure induced hypertrophy, but not hyperplasia of the NCC epithelium. Ultrastructural evaluation showed an increase in cytoplasmic organelles, particularly mitochondria, which is indicative of cellular hypertrophy and not swelling.

The increase in S-phase nuclei suggests that cell turnover was increased after one week of exposure in three of the four regions examined. Because there was no hyperplasia after either one or 11 weeks of exposure, the cell turnover most likely replaced cells damaged by the hyperoxia, but was not an additive proliferation.

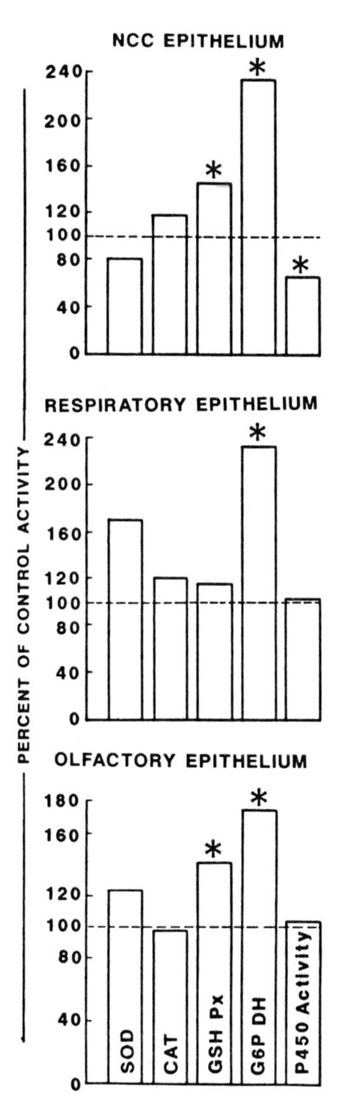

Figure 2. Specific Activity of Antioxidant Enzymes and P-450-Dependent Monooxygenase in Nasal Epithelia. Results are expressed as percent of control activity. *Significantly different from control by Student's t test ($p<0.05$).

In summary, hyperoxia induces hypertrophy, but not hyperplasia, of the NCC epithelium in rats. The ultrastructural correlate of the cellular hypertrophy is mitochondrial hypertrophy and hyperplasia. Similar mitochondrial hypertrophy and hyperplasia has been seen in type II cells and pulmonary endothelial cells after hyperoxic exposure. Antioxidant enzyme activities in rat nasal epithelia are increased by hyperoxic exposure. The increase in cell turnover seen after 1 week of hyperoxic exposure may reflect injury to sensitive epithelial cells and their subsequent replacement by more resistant cells, perhaps by hypertrophic cells with an increased antioxidant enzyme content. These results suggest that oxygen therapy may induce morphologic and biochemical alterations in the nasal epithelia. These alterations appear to be a local protective response of certain cells types to hyperoxia. The effects of these adaptations on air flow and clearance remain to be determined.

ACKNOWLEDGEMENTS

Research sponsored by the United States Department of Energy's Office of Health and Environmental Research under Contract No. DE-AC04-76EV01013.

REFERENCES

Cohen, P. and M.A. Rosemeyer (1975). *Meth. Enzymol.* **41**, 208.
Lawrence, R.A. and R.F. Burke (1976). *Biochem. Biophys. Res. Commun.* **71**, 952.
Lowry, O.H., et al. (1951). *J. Biol.* **193**, 265.
Misra, H.P. and I. Fridovich (1972). *J. Biol. Chem.* **247**, 3170.
Paglia, D.E. and W.N. Valentine (1967). *J. Lab. & Clin. Med.* **70**, 158.
Tolbert, N.E. (1974). *Meth. Enzymol.* **31**, 734.
White, I.N.H. (1988). *Analyt. Biochem.* **172**, 304.

MEMBRANE STABILIZATION AS A FUNDAMENTAL EVENT IN THE MECHANISM OF CHEMOPROTECTION AGAINST CHEMICAL INTOXICATION

*Howard G. Shertzer, **Malcolm Sainsbury and ***Marc L. Berger

University of Cincinnati Medical Center*
Cincinnati, OH USA 45267-0056
University of Bath**, Bath, UK BA2 7AY and
Jefferson Medical College***
Philadelphia, PA, USA 19107

INTRODUCTION

Elucidation of the mechanisms involved in chemical intoxication is difficult due to the concurrence of multiple cellular events. The major physiological consequences of toxicant-induced injury are organellar dysfunctions, including plasma membrane leakiness, loss of mitochondrial energy homeostasis, limited or pervasive autolysis due to release of lysozomal proteolytic enzymes, nucleic acid damage, and impairment of endoplasmic reticular functions including protein turnover, biotransformation, subcellular packaging and calcium homeostasis (Kaplowitz et al., 1986; Popper & Keppler, 1986). Often organellar associated events are correlated with biochemical events, such as alterations in thiol status, ion homeostasis, ATP levels or lipid peroxidation (Jones et al., 1986; Mitchell et al., 1982; Brattin et al., 1985; Comporti, 1989). In this report we have examined the physicochemical properties of 10 hydrophobic antioxidants that protect against chemical toxicity in isolated rat hepatocytes. Protection is shown to be correlated with both radical quenching and membrane stabilization properties of these compounds.

MATERIALS AND METHODS

Compounds were obtained from Sigma Chemical Co. (St. Louis, MO), Aldrich Chemical Co. (Milwaukee, WI) and Eastman Kodak Co. (Rochester, NY). Indenoindoles were synthesized as reported (Shertzer & Sainsbury, 1988). Hepatocytes were prepared from male Sprague-Dawley rats as described (Reitman et al.,1988). Regression and statistical evaluations were performed using Sigma-Plot (Jandel Scientific, Corte Madera, CA).

RESULTS AND DISCUSSION

During the past decade significant progress has been made in defining the intracellular events that occur during the process of chemical intoxication. Specific mechanisms have been defined involving lesions in ion, energy or redox homeostasis resulting in the impairment of organellar function and cell death. Recent studies have demonstrated that hydrophobic antioxidants protect against chemically induced hepatotoxicity (Shertzer et al., 1987a and 1987b; Shertzer et al., 1988; Shertzer and Sainsbury, 1988). In this study, ten hydrophobic antioxidants were examined for their

Biological Reactive Intermediates IV, Edited by C.M. Witmer *et al.*
Plenum Press, New York, 1990

capacity to inhibit toxicity of methylmethanesulfonate (MMS) and N-methyl-N'-nitro-N-nitrosoguanidine (MNNG) toxicity in isolated rat hepatocytes. These were butylated hydroxytoluene (BHT), 5,10-dihydroindeno[1,2-b]indole (DHII), N,N'-diphenyl-p-phenylenediamine (DPPD), 1,2,3,3a,4,8b-hexahydrocyclopenta[b]indole (HHCPI), indole-3-carbinol (I-3-C), indole-3-ethanol (I-3-E), iso-DHII, promethazine (PM), 1,2,3,4-tetrahydrocyclopenta[b]indole (THCPI), and D-alpha-tocopherol. The concentration of antioxidant required to produce 50% inhibition of liposomal iron/ascorbate-initiated lipid peroxidation (e.g. antioxidant efficacy), was plotted against the concentration that delayed by 1 hr the decline to 50% viability produced by MNNG and MMS in hepatocytes (e.g. chemoprotective efficacy), as shown in Figure 1.

Although a significant correlation is observed for MNNG, this is not the case for MMS. Still, all of the compounds were protective against MMS hepatotoxicity, suggesting that protection was related to an intrinsic chemical or physical property of the protectants other than antioxidation. That property appears to be related to membrane stabilization, as indicated by the significant second order correlations between the ability of the antioxidants to reduce the osmotic fragility of the red blood cell (RBC) membrane (expressed as % protection from osmolysis/uM compound) and the ability of antioxidants to act as chemoprotectants, as shown in Figure 2.

It is attractive to speculate that either the antioxidative or membrane stabilizing property of a protective compound may contribute to chemoprotection, depending upon the circumstances of intoxication. For example, if a toxicant acts via a free radical mechanism, then either membrane stabilizing or anti-oxidant properties of a protectant may be relevant. On the other hand, if a toxicant acts by a purely electrophilic mechanism (binding to a critical cellular target) leading to membrane destabilization and loss of organellar integrity, then only the membrane stabilizing property of a protectant would be operative. This does not rule out other mechanisms for protection, such as cytoplasmic calcium buffering. We have approached the topic of chemoprevention, by focussing on general chemical and physical perturbations, rather than specific biochemical or site-specific damage. In so doing, we hope to develop broad based strategies for intervention or arrest of intoxication resulting from exposure to chemical toxicants and ionizing radiation. We anticipate that definitive and predictive structure-activity relationships for chemoprotective agents will have a number of benefits that include the diminished use of animals in the early testing phases for new putative chemoprotective compounds, the design of studies to elucidate basic mechanisms for toxicity and chemoprotection, and the design of more potent chemoprotective compounds with therapeutic applications. (Supported by USPHS grant ES-03373 and the UK Cancer Research Campaign).

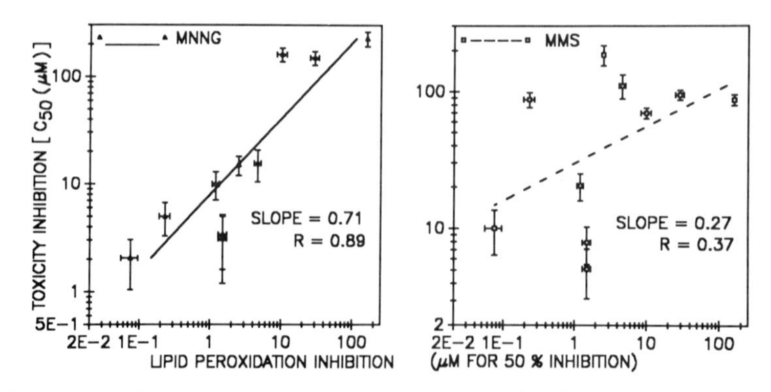

Figure 1. Correlations between inhibition of cell killing and inhibition of lipid peroxidation by various compounds. Each point represents the values obtained with a different compound. The left and right panels depict results obtained using the toxicants MNNG and MMS, respectively.

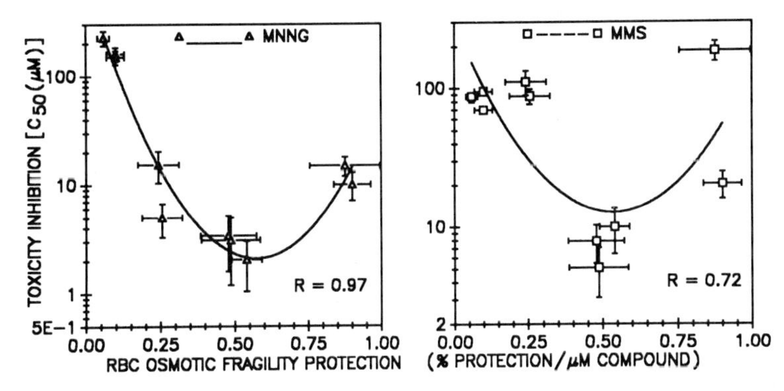

Figure 2. Correlations between inhibition of cell killing and RBC osmotic fragility protective index value by various compounds. The data obtained for the toxicants MNNG and MMS are shown in the left and right panels, respectively.

REFERENCES

Brattin, W.J., Glende, E.A. Jr., and Recknagel, R.O.(1985). Pathological mechanisms in carbon tetrachloride hepatotoxicity. *J. Free Radicals Biol. Med.* **1**, 27-38.

Comporti, M. (1989). Three models of free radical-induced cell injury. *Chem.-Biol. Interactions* **72**, 1-56.

Jones, T.W., Thor, H., and Orrenius, S.(1986). *In vitro* studies of mechanisms of cytotoxicity. *Fd. Chem. Toxicol.* **24**, 769-773.

Kaplowitz, N., Aw, T.Y., Simon, F.R., and Stolz, A.(1986). Drug-induced hepatotoxicity. *Ann. Int. Med.* **104**, 826-839.

Mitchell, J.R., Corcoran, G.B., Hughes, H., Lauterburg, B.H., and Smith, C.V.(1982). Acute lethal liver injury caused by chemically reactive metabolites. In *Organ-Directed Toxicity: Chemical Indices and Mechanisms* (S.S. Brown and D.S. Davies, Eds.), pp. 117-129. Pergamon Press, NY.

Popper, H., and Keppler, D.(1986). Networks of interacting mechanisms of hepatocellular degeneration and death. *Prog. Liv. Dis.* **8**, 209-235.

Shertzer, H.G., Niemi, M.P., Reitman, F.A., Berger, M.L., Myers, B.L., and Tabor, M.W. (1987a). Protection against carbon tetrachloride hepatotoxicity by pretreatment with indole-3-carbinol. *Exptl. Molec. Pathol.* **46**, 180-189.

Shertzer, H.G., Tabor, M.W., and Berger, M.L.(1987b). Protection from N-nitrosodimethylamine mediated liver damage by indole-3-carbinol. *Exptl. Molec. Pathol.* **47**, 211-218.

Shertzer, H.G., Tabor, M.W., and Berger, M.L.(1988). Intervention in free radical mediated hepatotoxicity and lipid peroxidation by indole-3-carbinol. *Biochem. Pharmacol.* **37**, 333-338.

Shertzer H.G., and Sainsbury, M.(1988). Protection against carbon tetrachloride hepatotoxicity by 5,10-dihydroindeno[1,2-b]indole, a potent inhibitor of lipid peroxidation. *Fd. Chem. Toxicol.* **26**, 517-522.

BIOCHEMICAL EFFECTS AND TOXICITY OF MITOXANTRONE IN CULTURED HEART CELLS

N.G. Shipp and R.T. Dorr

Dept. of Pharmacology and Toxicology and
The Arizona Cancer Center
University of Arizona
Tucson, AZ 85721

In response to the cumulative cardiotoxicity associated with the clinical use of the antitumor agent doxorubicin, many structurally related anthraquinone drugs have been developed. Mitoxantrone has been the most successful of these, and is used with increasing frequency in the treatment of leukemia, lymphoma, and breast cancer. Response rates are similar to those obtained with doxorubicin treatment, and notably, mitoxantrone is less cardiotoxic. However, while more courses of mitoxantrone therapy can be given, patients eventually develop a cumulative cardiotoxicity which is clinically similar to that observed following prolonged doxorubicin treatment (Henderson et al., 1989).

Since mitoxantrone is structurally similar to doxorubicin, it was originally thought that cardiotoxicity was caused by active oxygen species resulting from redox cycling of the drug. However, there is little evidence of mitoxantrone-induced free-radical formation *in vitro*. In fact, mitoxantrone has been shown to inhibit some lipid peroxidation processes under certain conditions *in vitro* (Kharasch and Novak, 1985). Furthermore, the one-electron reduction potential of mitoxantrone (-.527 V) is well below the range of intracellular reductases, making intracellular reduction and the initiation of redox cycling an unlikely event.

It has since been suggested that mitoxantrone may undergo intracellular oxidation to a reactive intermediate, or alternately to a cyclic metabolite which has been enzymatically produced and purified. Covalent binding of mitoxantrone to DNA has been shown to occur in combination with horseradish peroxidase and hydrogen peroxide (Reszka et al., 1989).

The aim of this study was to quantitate the toxicity of mitoxantrone and its purified cyclic metabolite in an *in vitro* heart cell culture system. Furthurmore, some of the biochemical effects of mitoxantrone in isolated heart cells are described.

MATERIALS AND METHODS

Chemicals

^{14}C-mitoxantrone was custom synthesized and kindly suppled by Dr. W. Murdock of American Cyanamid, Pearl River, NJ.

Biological Reactive Intermediates IV, Edited by C.M. Witmer *et al.*
Plenum Press, New York, 1990

Cell Culture. Hearts from 1-2 day old neonatal Sprague-Dawley rats were isolated under sterile conditions, minced, and dissociated in 0.2% crude trypsin using a serial digestion technique. Heart cells were plated in Primaria 35 mm tissue culture dishes, or 24 well plates, and grown in a supplemented Liebovitz's M3 media (Dorr et al., 1988).

Assay

ATP levels were determined photometrically following trichloroacetic acid extraction using the firefly luciferin-luciferase bioluminescence assay (Dorr et al., 1988). Malondialdehyde (MDA) was measured as thiobarbituric acid-reactive material using the spectrophotometric technique of Stacy et al. (Stacy et al, 1988). Covalent binding of radiolabeled mitoxantrone to heart cell protein was determined using the method of Wallin et al (Wallin et al., 1981). Heart cell post-nuclear supernatent fractions were prepared by homogenizing three day old cultured heart cells in Hanks Balanced Salt Solution (HBSS), and spinning 5 min in a microcentrifuge (12,000g).

RESULTS AND DISCUSSION

The chemical structure of mitoxantrone and its horseradish peroxidase/H_2O_2-derived metabolite are shown in Figure 1. Formation of the metabolite proceeds via a free radical intermediate which has been identified by EPR spectroscopy as an unstable cation radical (Reszka et al., 1989). The cytotoxicity of these compounds was assessed at 72 hours after a 3 hr drug exposure by measuring heart cell ATP levels (Figure 2). The cyclic metabolite was 15-fold less cardiotoxic than mitoxantrone based on IC_{50} values for intracellular ATP (1.8 μg/ml and 27 μg/ml for mitoxantrone and metabolite respectively).

Previous studies have suggested that mitoxantrone does not stimulate production of active oxygen species or lipid peroxidation (Kharasch and Novak, 1985). In this system malondialdehyde levels were not elevated at any time up to 24 hr following exposure to a cytotoxic concentration of mitoxantrone (results not shown). Malondialdehyde levels 24 hr after drug treatment are shown in Figure 3. In order to test whether mitoxantrone could inhibit peroxidation, a lipid hydroperoxide generating system was developed. Exposure of heart cells to a mixture containing ADP-Fe^{3+} chelate and H_2O_2 (250 μM ADP, 25 μM Fe^{3+} and 25 μM H_2O_2) resulted in a two fold elevation in heart cell MDA levels.

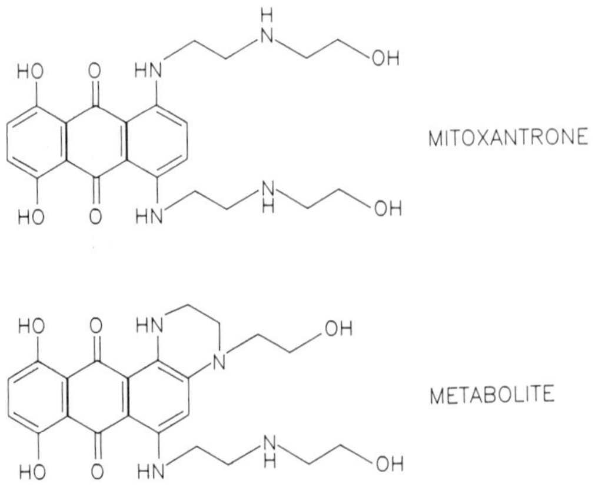

Figure 1. Structures of mitoxantrone and its cyclic metabolite. The metabolite is produced by oxidative metabolism of the parent compound (Reszka et al., 1989).

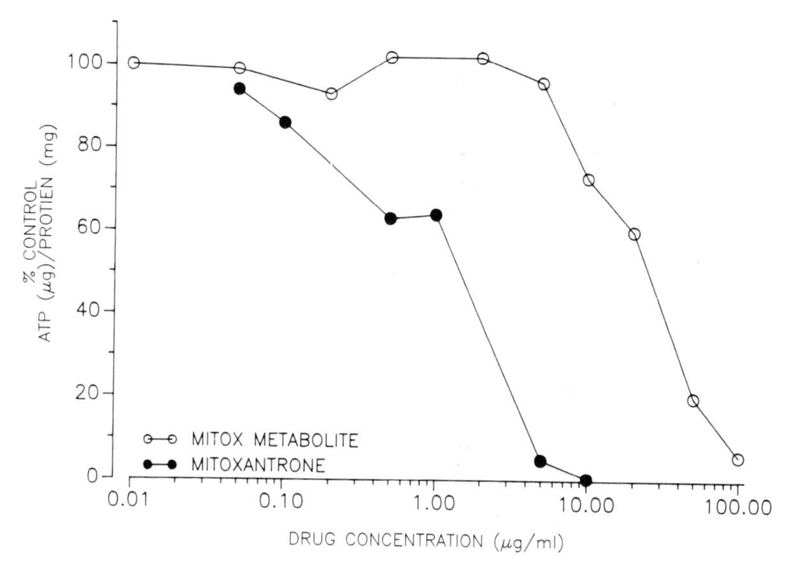

Figure 2. ATP levels in myocytes treated with varying concentrations of mitoxantrone or its cyclic metabolite. On day 3 myocytes were incubated with drug concentrations indicated for 3 hr. Cultures were then post-incubated in drug-free medium for 72 hr. Each point represents average of 3-6 experiments. Standard deviation was less than 10% of mean value.

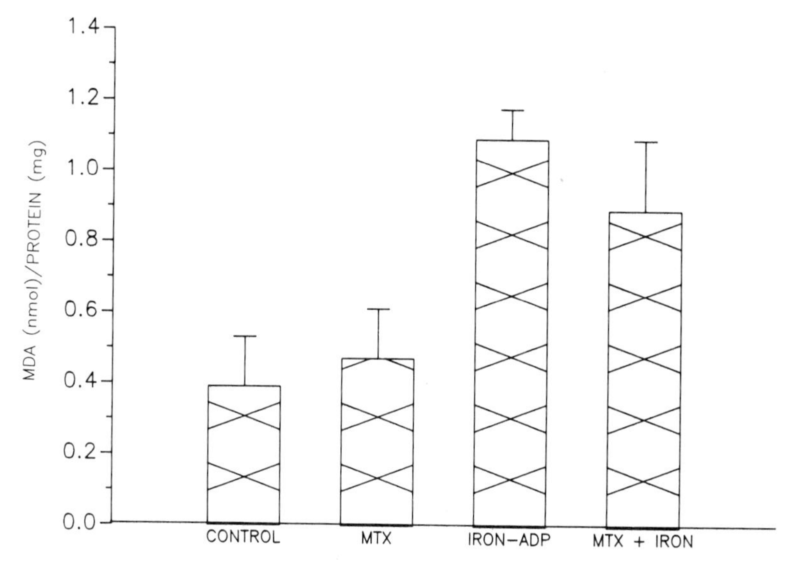

Figure 3. Malondialdehyde levels in myocytes under varying conditions. Cells were treated on day 3 with 1.5 μg/ml mitoxantrone, ADP-iron/H2O2 (see text) or a combination of both for 3 hr. MDA was measured 24 hr later. Bars represent the average of 5-7 experiments +/- standard deviation.

Thus, while mitoxantrone has been shown to terminate some lipid hydroperoxide-initiated free radical reactions, the combination of mitoxantrone with the ADP-Fe^{3+}/H_2O_2 system did not lessen MDA production significantly.

Iron is also postulated to play a role in doxorubicin cardiotoxicity. To assesss the potential involvement of iron in mitoxantrone cardiotoxicity, heart cells were treated with an iron chelating agent. ICRF-187 was added to heart cells alone (50 μg/ml, 72 hr continuous exposure) and in combination with mitoxantrone (1.5 μg/ml, 3 hr exposure). ATP levels were measured at 72 hr as an indicator of cardiac cell viability. The results show that ICRF-187 alone was non-cardiotoxic, and in combination with mitoxantrone, was able to protect heart cells from mitoxantrone cardiotoxicity (Figure 4).

ICRF-187 has also been shown to be effective in the prevention of doxorubicin-induced clinical cardiotoxicity (Herman and Ferrans, 1987). While the mechanism(s) of cardiotoxicity for mitoxantrone and doxorubicin appears to differ, ICRF-187 nonetheless exerts a similar protective effect on heart tissue. It thus may be useful clinically for the prevention of mitoxantrone cardiotoxicity.

Covalent binding of mitoxantrone to cell macromolecules would be expected if the drug can undergo oxidative activation as in the presence of horseradish peroxidase and H_2O_2. As an initial assessment of this possibility, heart cell post nuclear supernatent fractions were exposed to radiolabeled mitoxantrone for 3 hr at 4°C and 37°C. After this, the level of covalently bound drug was determined. Fractions incubated at 37°C bound significantly more drug (1.37 +/- 0.26 nmole mitoxantrone/mg protein) than fractions incubated a 4°C (0.94 +/- 0.13 nmole mitoxantrone/mg protein). This suggests that an enzymatic process may result in increased levels of covalently-bound mitoxantrone.

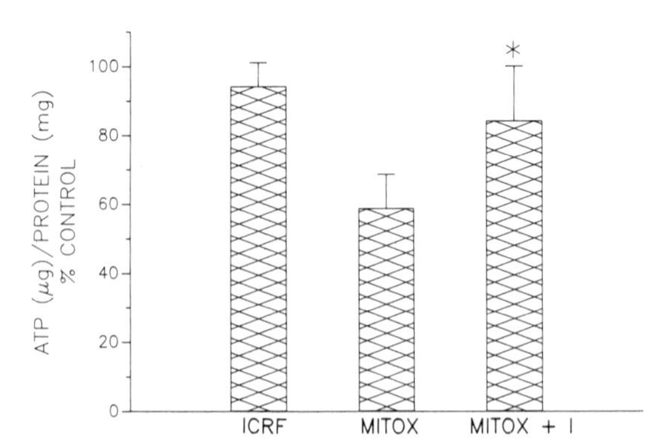

Figure 4. ATP levels in myocytes exposed to mitoxantrone, ICRF-187 or a combination of drugs. Myocytes were treated on day 3 with 1.5 μg/ml mitoxantrone for 3 hr, and post-incubated in drug-free medium for an additional 72 hr. Cells receiving ICRF-187 were treated 3 hr prior to, during and for 72 hr after mitoxantrone exposure, or a continuous exposure in the absense of mitoxantrone (see text). Bars represent the average of 6 experiments +/- standard deviation. * = significant difference from mitoxantrone alone, $p < 0.01$.

In summary, we have shown that mitoxantrone is toxic to cultured heart cells at concentrations that are clinically relevant. A purified cyclic metabolite of mitoxantrone produced by oxidative metabolism of the parent compound, was 15 fold less toxic to cultured heart cells. Heart cell post-nuclear supernatent fractions incubated with radiolabeled mitoxantrone at 37°C had significantly elevated levels of covalently bound drug over fractions incubated at 4°C. This suggests that enzymatic activation by a heart cell homogenate leads to covalent binding by mitoxantrone. However, mitoxantrone neither stimulates MDA formation nor inhibits MDA production initiated by a lipid hydroperoxide generating system. Finally, the chelating agent, ICRF-187, protects cultured heart cells from mitoxantrone-induced cytotoxicity and may be useful to prevent cardiotoxicity in cancer patients.

REFERENCES

Dorr, R. T., Bozak, K. A., Shipp, N. G., Hendrix, M., Alberts, D. S., and Ahmann, F. (1988). *In vitro* rat myocyte cardiotoxicity model for antitumor antibiotics using adenosine triphosphate/protein ratios. *Cancer Research* **48**, 5222-5227.

Henderson, I. C., Allegra, J. C., Woodcock, T., Wolff, S., Bryan, S., Cartwright, K., Dukart, G., and Henry, D. (1989). Randomized clinical trial comparing mitoxantrone with doxorubicin in previously treated patients with metastatic breast cancer. *J. Clin. Oncol.* **7**, 560-571.

Herman, E. H., and Ferrans, V. J. (1987). Amelioration of chronic anthracycline cardiotoxicity by ICRF-187 and other compounds. *Cancer Treat. Rev.* **14**, 225-229.

Kharasch E. D., and Novak, R. F. (1985). Mitoxantrone and ametantrone inhibit hydroperoxide-dependent initiation and propagation reactions in fatty acid peroxidation. J. Biol. Chem. **260**, 10645-10652.

Reszka, K., Hartley, J. A., Kolodziejczyk, P., and Lown, J. W. (1989). Interaction of the peroxidase-derived metabolite of mitoxantrone with nucleic acids. *Biochem. Pharm.* **38**, 4253-4260.

Stacy, N. H., Cantilena, L. R., and Klaassen, C. D. (1980). Cadmium toxicity and lipid peroxidation in sisolated hepatocytes. *Toxicol. Appl. Pharmacol.* **53**, 470-480.

Wallin, H., Schelin, C., Tunek, A., and Jergil, B. (1981). A rapid and sensitive method for determination of covalent binding of benzo(a)pyrene to proteins. *Chem.-Biol. Interactions.* **38**, 109-118.

EVIDENCE FOR THE INDUCTION OF AN OXIDATIVE STRESS IN RAT HEPATIC MITOCHONDRIA BY 2,3,7,8-TETRACHLORODIBENZO-P-DIOXIN (TCDD)

S.J. Stohs, N.Z. Alsharif, M.A. Shara, Z.A.F. Al-Bayati and Z.Z. Wahba

School of Pharmacology and Allied Health
Creighton University
Omaha, NE 68178

TCDD is prototypical of a large number of halogenated polycyclic hydrocarbons that occur as environmental contaminants and pollutants (Safe, 1986; Poland and Knutson, 1982; Poland et al., 1985). Following the acute exposure to a toxic dose of TCDD and its bioisosteres, a progressive weight loss with hypophagia and depletion of adipose tissue occurs (Safe, 1986; Kociba and Schwetz, 1982; Neal et al., 1982; Poland and Knutson, 1982). In addition, hepatic, gastric and epidermal lesions, thymic involution, and increased excretion of porphyrins occur (Poland and Knutson, 1982; Kociba and Schwetz, 1982). Evidence indicates that the mechanism of toxicity of TCDD and its bioisosteres involves binding to a specific TCDD (Ah) receptor, interaction of this complex with chromatin, and the ultimate production of a pleiotropic response (Cook et al., 1987; Safe, 1988; Poland and Knutson, 1982). the post-translational pathways and mechanisms involved in the expression of the toxic effects of TCDD are not known.

We have hypothesized that production of an oxidative stress plays an importnat role in expression of the toxic manifestations of TCDD, providing at least in part an explanation for the diverse types of toxic responses which occur following exposure to TCDD. Previous studies have shown that administration of TCDD to rats induces large increases in hepatic lipid peroxidation as determined by the production of malondialdehyde (MDA) by isolated microsomes (Hassan et al., 1983; Al-Bayati et al., 1987; Wahba et al., 1989a), and the increase in whole liver MDA content as determined by the formation of thiobarbituric acid reactive substances (TBARS) (Al-Bayati et al., 1987; Stohs et al.., 1984; Shara and Stohs, 1987; Hermansky et al., 1988; Wahba et al., 1989a).

Little information is available concerning the possible role of mitochondria in TCDD-induced oxidative stress and subsequent tissue damage. Previous studies have shown an increase in the iron and copper content of hepatic mitochondria following exposure of rats to TCDD (Wahba et al., 1988; Al-Bayati et al., 1987). Iron is involved in hepatic microsomal lipid peroxidation (Al-Bayati and Stohs, 1987), and iron is presumably involved in the production of reactive oxygen species and lipid peroxidation in mitochondria. The incubation of hepatic mitochondria from TCDD-treated rats with hepatic nuclei from control animals results in an increase in DNA single strand breaks in the nuclei (Wahba et al., 1989b). Thus, mitochondria may be involved in the production of the TCDD-induced oxidative stress.

Biological Reactive Intermediates IV, Edited by C.M. Witmer *et al.*
Plenum Press, New York, 1990

MATERIALS AND METHODS

Animals and Treatment

Female Sprague-Dawley rats weighing 160-180 g were treated orally with 100 μg TCDD/kg orally as a single dose in corn oil:acetone (10:1) or the vehicle. The animals were killed on days 0, 1, 3, 5, 7 or 9 and hepatic mitochondria were isolated by differential centrifugation in a 0.10 M phosphate buffer, pH 7.4.

Analytical Methods

Mitrochondrial lipid peroxidation was determined as the content of thiobarbituric acid reactive substances (TBARS) by the method of Uchiyama and Mihara (1978), and as the time-dependent formation of TBARS according to the method of Miles et al. (1980). Calcium content of mitochondria was assessed by flame ionization spectrometry (Goldstein et al., 1973). Membrane fluidity of mitochondria was evaluated utilizing steady state fluorescence spectroscopy with 1,6-diphenyl-1,3,5-hexatriene as the fluorescent probe (Stubbs et al., 1980). Non-protein sulfhydryl content of mitochondria was determined using Ellman's reagent following precipitation of the protein with 50% trichloroacetic acid according to the method of Sedlack and Lindsay (1967). Mitochondrial NADPH content was estimated by the enzymatic cycling spectrophotometric method of Giblin and Reddy (1980). Protein content was determined by the colorometric method of Lowry et al. (1951). Significant differences (P , 0.05) between pairs of mean values were determined by Student's t test.

RESULTS

Lipid Peroxidation and Calcium Content

The time-dependent effects following the administration of 100 μg TCDD/kg on hepatic mitochondrial lipid peroxidation and calcium content were examined (Table 1). Significant increases in lipid peroxidation, determined both as NADPH-dependent TBARS formation and TBARS content, and calcium content of mitochondria increased 3 days post-treatment. Maximum increases in lipid peroxidation of 225-255% were observed by day 5 and remained elevated through the remainder of the 9 day study. In contrast, the calcium content of mitochondria continued to gradually increase, with a 180% increase being observed by the 9th day.

Table 1. TCDD-Induced Increases in Hepatic Mitochondrial Lipid Peroxidation and Calcium content

Days after TCDD treatment	Lipid Peroxidation: TBARS Formation (nmoles/mg protein/min)	Lipid Peroxidation: TBARS Content (nmoles/mg protein)	Calcium Content (nmoles/mg protein)
0	0.24 ± 0.03	0.22 ± 0.04	11.1 ± 1.0
1	0.29 ± 0.02	0.26 ± 0.03	12.0 ± 0.9
3	0.38 ± 0.04*	0.35 ± 0.06*	13.4 ± 1.1*
5	0.51 ± 0.05*	0.50 ± 0.04*	15.1 ± 1.9*
7	0.54 ± 0.04*	0.56 ± 0.05*	17.4 ± 2.1*
9	0.50 ± 0.06*	0.55 ± 0.06*	20.5 ± 2.2*

Female rats were treated with 100 μg TCDD/kg orally as a single dose in corn oil:aceton (10:1) or the vehicle. The animals were killed on the days indicated above. Each value is the mean ± S.D. of 4-8 animals. *$P < 0.05$ with respect to the corresponding control (0 day) group.

Membrane Fluidity, and NADPH and Non-protein Sulfhydryl Contents

The time-dependent effect of TCDD administration on mitochondrial membrane fluidity as well as the mitochondrial ocontents of NADPH and non-protein sulfhydryls were determined (Table 2). Significant decreases in both membrane fluidity and non-protein sulfhydryl content of mitochondria were first observed 3 days after the administration of TCDD. Both parameters gradually decreased with time, and by day 9 a 20% decrease in membrane fluidity was observed while a 70% decrease in the non-protein sulfhydryl content was noted.

No change occurred in the hepatic mitochondrial NADPH content until 7 days after the administration of TCDD. By day 9, a 25% decrease in the NADPH content was observed. A gradual decrease in NADPH content continued after day 9 (data not shown).

Table 2. TCDD-Induced Decreases in Hepatic Mitochondrial Membrane Fluidity, Non-Protein Sulfhydryl Content and NADPH Content

Days after TCDD treatment	Membrane fluidity (r)	NADPH content (nmoles/mg protein)	Non-protein sulfhydryl content (nmoles/mg protein)
0	0.160 ± 0.006	0.94 ± 0.04	9.68 ± 0.68
1	0.153 ± 0.005	0.99 ± 0.08	8.31 ± 0.71
3	0.142 ± 0.005*	10.4 ± 0.11	4.94 ± 0.48*
5	0.136 ± 0.006*	1.05 ± 0.10	3.78 ± 0.44*
7	0.132 ± 0.003*	0.77 ± 0.07*	3.39 ± 0.32*
9	0.128 ± 0.004*	0.71 ± 0.06*	2.71 ± 0.23*

Female rats were treated with 100 μg TCDD/kg orally as a single dose in corn oil:acetone (10:1) or the vehicle. The animals were killed on the days indicated above. Each value is the mean ± S.D. of 4-8 animals. *P , 0.05 with respect to the corresponding control (0 day) group.

DISCUSSION

The results clearly indicate that a lag period exists in the hepatic mitochondrial responses to the oral administration of TCDD. No significant changes in any of the parameters which were measured occurred prior to the third day post-treatment. Three days after the administration of TCDD significant increases in lipid peroxidation and calcium content (Table 1) and significant decreases in membrane fluidity and non-protein sulfhydryl content were observed (Table 2). The only parameter that was measured which did not significantly change by the third day post-treatment was the NADPH content (Table 2).

Taken together, the above results clearly indicate that an oxidative stress occurs in hepatic mitochondria following the administration of TCDD. The delayed response with respect to NADPH content may reflect a compensatory mechanism on the part of the mitochondria to protect this organelle. Only after depletion of the components necessary for NADPH generation, as glucose-6-phosphate, or inactivation of enzymes such as glucose-6-phosphate dehydrogenase is a decrease in NADPH content observed.

The mitochondrial production of reactive oxygen species is well known (Nohl and Jordan, 1986; Luschen et al., 1974). Mitochondrial swelling is an indication of toxic insult to the mitochondria, and TCDD has been shown to increase the number of mitochondria as well as produce mitochondrial swelling (Rozman et al., 1986). Thus, hepatic mitochondria may be a target organelle of TCDD.

Several studies have shown that mitochondria may serve as a source of reactive oxygen species in response to TCDD. These reactive oxygen species may therefore contribute to the overall oxidative stress which is observed in animals following exposure to TCDD. Studies in our laboratories have demonstrated that hepatic mitochondria from TCDD-treated animals can produce reactive oxygen species resulting in DNA damage when incubated with hepatic nuclei from control animals (Wahba et al., 1989b). Studies by Noh. et al (1989) have shown that the addition of TCDD to beef heart mitochondria results in an inhibition of respiration with a partial uncoupling of oxidative phosphorolation, leading to the increased productin of reactive oxygen species. No similar studies have been conducted with hepatic mitochondria from rats or mice.

The precise role of the TCDD-induced oxidative stress in the overall toxic manifestations of this xenobiotic have not been clearly determined to date. However, reactive oxygen species are implicated in the sequence of events leading to TCDD-induced tissue damage and cell death. Further investigations are required to elucidate the precise role of reactive oxygen species and oxidative stress in the toxicity of TCDD.

ACKNOWLEDGEMENTS

These studies were supported in part by grant No. ES04367 from the National Institute of Environmental Health Sciences. The authors thank Ms. LuAnn Schwery for technical assistance.

REFERENCES

Al-Bayati, Z.A.F. and Stohs, S.J. (1987). The role of iron in 2,3,7,8-tetrachlorodibenzo-p-dioxin (TCDD)-induced lipid peroxidation by rat liver microsomes. *Toxicol. Lett.* **38**, 115-121.

Al-Bayati, Z.A.F., Murray, W.J., and Stohs, S.J. (1987). TCDD-induced lipid peroxidation in hepatic and extrahepatic tissues of male and female rats. *Arch. Environ. Contam. Toxicol.* **16**, 259-266.

Cook, J.C., Gaido, K.W>, and Greenlee, W.F. (1987). Ah receptor: relevance of mechanistic studies to human risk assessment. *Environ. Hlth. Persp.* **76**, 71-77.

Giblin, F.J. and Reddy, V.M. (1980). Pyridine nucleotides in ocular tissues as determined by the cycling assay. *Exp. Eye Res.* **31**, 601-609.

Goldstein, J.A., Hickman, P., Bergman, H., and vos, J.G. (1973). Hepatic porphyria induced by 2,3,7,8-tetrachlorodibenzo-p-dioxin in the mouse. *Res. comm. Chem. Path.* **6**, 919-928.

Hassan, M.Q., Stohs, S.J., and Murray, W.J. (1983). comparative ability of TCDD to induce lipid peroxidation in rats, guinea pigs and Syrian golden Hamsters. *Bull. Environ. Contam. Toxicol.* **31**, 649-657.

Hermansky, S.J., Holcslaw, T.L., Murray, W.J., Markin, R.S., and Stohs, S.J. (1988). Biochemical and functional effects of 2,3,7,8-tetrachlorodibenzo-p-dioxin (TCDD) on the heart of female rats. *Toxicol. Appl. Pharmacol.* **95**, 174-184.

Kociba, R.J. and Schwetz, B.A. (1982). Toxicity of 2,3,7,8-tetrachlorodibenzo-p-dioxin (TCDD). *Drug Metab. Rev.* **13**, 387-406.

Lowry, O.H., Rosebrough, N.J., Farr, W.L., and Randall, R.J. (1951). Protein measurement with folin phenol reagent. *J. Biol. Chem.* **193**, 265-275.

Luschen, G., Azzi, A., Richter, C., and Flohe, L. (1974). Superoxide radicals as precursors of mitochondrial hydrogen peroxide. *FEBS Lett.* **42**, 68-72.

Miles, P.R., Wright, J.R., Browman, L., and Colby, H.D. (1980). Inhibition of hepatic microsomal lipid peroxidation by drug substrates without drug metabolism. *Biochem. Pharmacol.* **29**, 565-570.

Neal, R.A., Olson, J.R., Gasiewicz, T.A. and Geiger, L.E. (1982). the toxicokinetics of 2,3,7,8-tetrachlorodibenzo-p-dioxin in mammalian systems. *Drug Met. Rev.* **13**, 355-385.

Nohl, H. and Jordan, W. (1986). The mitochondrial site of superoxide formation. *Biochem. Biophys. Res. Comm.* **138**, 533-539.

Nohl, H., Desilva, D., and Summer, K.H. (1989). 2,3,7,8-Tetrachlorodibenzo-p-dioxin induces oxygen activation associated with cell respiration. *Free Rad. Biol. Med.* **6**, 369-374.

Poland, A. and Knutson, J.C. (1982). 2,3,7,8-Tetrachlorodibenzo-p-dioxin and related halogenated aromatic hydrocarbons: Examination of the mechanism of toxicity. *Ann. Rev. Pharmacol. Toxicol.* **22**, 517-554.

Poland, A. and Knutson, J.C. and Gluver E. (1985). Studies on the mechanism of action of halogenated aromatic hydrocarbons. *Clin. Physiol. Biochem.* **3**, 147-154.

Rozman, K., Pereira, D., and Iatropoulos, M.J. (1986). Histopathology of interscapular brown adipose tissue, thyroid, and pancreas in 2,3,7,8-tetrachlorodibenzo-p-dioxin (TCDD)-treated rats. *Toxicol. Appl. Pharmacol.* **82**, 551-559.

Safe, S.H. (1986). Comparative toxicology and mechanism of action of polychlorinated dibenzo-p-dioxins and dibenzofurans. *Ann. Rev. Pharmacol. Toxicol.* **26**, 371-399.

Safe, S.H. (1988). the aryl hydrocarbon (Ah) receptor. *ISI Atlas Sci: Pharmacol.*, 78-83.

Sedlack, J. and Lindsay, R.H. (1968). Estimation of total, protein-bound and non-protein bound sulfhydryl groups in tissue with Ellman's Reagent. *Anal. Biochem.* **25**, 192-205.

Shara, M.A. and Stohs, S.J. (1987). Biochemical and toxicological effects of 2,3,7,8-tetrachlorodibenzo-p-dioxin (TCDD) congeners in female rats. *Arch. Environ. Contam. Toxicol.* **16**, 597-605.

Stohs, S.J., Hassan, M.Q., and Murray, W.J. (1984). Induction of lipid peroxidation and inhibition of glutathione peroxidase by TCDD. *Banbury Report* #18, Banbury Center, Cold Spring Harbor Laboratory, NY., pp. 241-253.

Stubbs, C.D., Tsang, W.M., Belin, J., Smith, A.D., and Johnson, S.M. (1980). Incubation of exogenous fatty acids with lymphocytes: changes in fatty acid composition and effects on the rotational relaxation time of 1,6-diphenyl-1,3,5-hexatriene. *Biochemistry* **19**, 2756-2762.

Uchiyama, M. and Mihara, M. (1978). Determination of malondialdehyde precursor in tissues by thiobarbituric acid test. *Anal. Biochem.* **86**, 271-278.

Wahba, Z.Z., Al-Bayati, Z.A.F., and Stohs, S.J. (1988). Effect of 2,3,7,8-tetrachlorodibenzo-p-dioxin (TCDD) on the hepatic distribution of iron, copper, zinc and magnesium in rats. *J. Biochem. toxicol.* **3**, 121-129.

Wahba, Z.Z., Murray, W.J., Hassan, M.Q., and Stohs, S.J. (1989a). comparative effects of pair-feeding and 2,3,7,8-tetrachlorodibenzo-p-dioxin (TCDD) on various biochemical parameters in female rats. *Toxicology* **59**, 311-323.

Wahba, Z.Z., Lawson, T.A., Murray, W.J., and Stohs, S.J. (1989b). Factors influencing the induction of DNA single strand breaks in rats by 2,3,7,8-tetrachlorodibenzo-p-dioxin (TCDD). *Toxicology* **58**, 57-69.

White, R.D., Sipes, I.G., Gandolfi, A.J. and Bowden, G.T. (1981). Characterization of the hepatic DNA damage caused by 1,2-dibromoethane using the alkaline elution technique. *Carcinogenesis* **2**, 839.

ANTIOXIDATION POTENTIAL OF INDOLE COMPOUNDS - STRUCTURE ACTIVITY STUDIES

M. Wilson Tabor[1], Eugene Coats[2], Malcolm Sainsbury[3], and Howard G. Shertzer[1]

Institute Environmental Health[1] and College of Pharmacy[2]
University of Cincinnati Medical Center
Cincinnati, OH 45267 and
University of Bath[3]
Bath, England

Many indole compounds, including dietary indoles such as indole-3-carbinol [I-3-C], have been shown to protect against chemical carcinogenesis (Wattenberg 1983) and chemically induced hepatotoxicity (Shertzer et al. 1987a, 1987b, 1988a). Such protection may be due to the ability of I-3-C to scavenge biologically reactive electrophiles (Shertzer et al. 1987b, 1988b) and free radicals (Shertzer et al. 1987a; Shertzer et al. 1988a). To develop better chemoprotective indoles with lower intrinsic toxicity, 28 structurally related indoles were investigated relative to antioxidant efficacies and physicochemical properties (Log P and electronic effects).

METHODS

Chemicals

Speciality chemicals were obtained from Sigma Chemical Co., St. Louis, MO and Aldrich Chemical Company, Milwaukee, WI. Non-commercially available indole derivatives were synthesized as described (Tabor et al., 1990) and structures were verified using both high resolution Fourier transformed nuclear magnetic resonance and mass spectrometry techniques. All other chemicals were reagent grade or better, and were used without further purification.

Assays

The two cell-free soybean phospholipid peroxidation assays utilized in these studies (Shertzer et al., 1988a) consisted of either a solution of lipid in chlorobenzene [CB system], or a mixture of phosopholipid vesicles in aqueous buffer [Fe system]. Reactions were initiated with either azobisisobutyronitrile (AIBN) or iron/ascorbate, respectively. For each lipid peroxidation assay system, 10 different concentrations of individual indole compounds were assessed for inhibition, with at least 4 different concentrations on each side of the concentration which gave 50% inhibition. Results were calculated as the average percent of control (no inhibitor) rates of lipid peroxidation versus concentration of inhibitor.

Correlation Analysis

Potential structure activity relationships between indole antioxidant effects and physiochemical properties were computed via multiparameter linear regression analysis using a program designed and made available to us by Corwin Hansch, Pomona College, Claremont, CA. Using methods of Hansch and Leo (1979), lipophilic properties were estimated by log P(octanol/ water partition coefficient) where log P values of the indoles were calculated from available measured partition coefficients via additivity principles, and electronic effects were parameterized with Hammett sigma constants. For these structure-activity relationship evaluations, the *in vitro* antioxidant activities were converted to the form: log $1/I_{50}$. Minimum neglect of differential overlap [MNDO] calculations were computed using a microcomputer version of MNDO (QCMO 002/QCPE 353) obtained from the Quantum Chemistry Program Exchange of Indiana University in Bloomington, IN.

RESULTS AND DISCUSSION

The micromolar concentrations of indole compounds required for 50% inhibition are summarized in Table 1 along with the values for 2,6-di-t-butyl-4-methylphenol [BHT] and alpha-tocopherol as standard anti-oxidants. Results from the two *in vitro* test systems are not parallel, reflecting the homogeneous nature of the chlorobenzene (CB) assay system as opposed to the heterogeneous aqueous Fe system.

Table 1. 50% Lipid Peroxidation Inhibition Constants for Indole Derivatives as Determined in Iron/ascorbate [Fe] and Azobisisobutyronitrile [CB] Initiated Assay Systems.

No.	Antioxidant	50%I Constants	
		[Fe] Assay	[CB] Assay
1	5-hydroxy-3-(4-methoxybenzyl)indole	0.25	**
2	2,6-di-t-butyl-4-methylphenol *	1.2	18
3	3-(4-N,N-dimethylaminobenzyl)indole	1.5	16
4	ethyl 2,2-(di-3-indolyl)acetate	5.5	250
5	alpha-tocopherol *	10	40
6	3-(2,4,6-trimethylbenzyl)indole	11	100
7	3-(4-hydroxybenzyl)indole	12	13
8	3[2-(3,4-dimethoxybenzyl)ethyl]indole	12	115
9	3-benzyl-1-methylindole	13	500
10	3,3-methylene-bisindole (di-indole)	15	64
11	3-(2-methyl-2-phenylethyl)indole	18	80
12	2,3-dimethylindole	20	56
13	3-(4-methoxybenzyl)indole	24	300
14	3-hydroxyethylindole(indole-3-ethanol)	29	**
15	3-(4-methoxybenzyl)-1-methylindole	31	400
16	3-benzylindole	36	185
17	3-methylindole (skatole)	100	**
18	1H-indol-3-yl-3-methoxyphenylmethanone	>250	>250
19	3-hydroxymethylindole(indole-3-carbinol)	160	120
20	3-(2-pyridylmethyl)indole	325	200
21	5-methoxy-3-(3-pyridylmethyl)indole	450	1100
22	3-(3-pyridylmethyl)indole	500	500
23	3-butyroindole(K+salt)	800	**
24	indole	800	1250
25	3-acetamidoindole	>1250	**
26	indole-3-acetonitrile	>1250	1250
27	5-methoxy-1H-indole-3yl-3-pyridinylmethanone	>1250	>1250
28	1-methyl-3-(2-methyl-2-phenylethyl)indole	>1250	>1250
29	indole-3-(3-lactic acid)	>1250	**
30	indole-3-acetic acid	>1250	**

* Standard Antioxidant
** Not Determined

In the CB system, antioxidants and lipid substrate are dispersed evenly, whereas in the aqueous assay, the indoles partition into the lipid vesicles membrane to produce much higher concentrations of lipid substrate and antioxidant.

While all molecules investigated do posses an indole nucleus, this in itself is not sufficient for significant antioxidant activity since unsubstituted indole exhibits only weak inhibition. Compounds 9, 15 and 28 all contain N-methyl modifications while compounds 15, 12, and 11 are the corresponding structures with a free indole N-H. The only significant difference in antioxidant activity occurs with compound 28 versus compound 11, while the other two pairs are essentially equivalent. Thus, it would seem that the nature of the substituent at the 3 position of the indole nucleus does exert a profound influence upon antioxidant activity. Futhermore, methylation of the pyrrole ring enhanced antioxidant potency. In both antioxidation assays, addition of each successive methyl group (compounds 24, 17 and 12, respectively) decreased the I_{50} by about 5-fold.

Structure Activity Relationship Regression Equations

EQ. 1	Log 1/[Fe] = 0.58 {0.454}	Log P - 3.82 {1.78}
	n = 10 s = 0.594	r = 0.729
EQ. 2	Log 1/[Fe] = -1.61 {0.47}	sigma - 1.65 {0.21}
	n = 10 s = 0.291	r = 0.942

Log 1/[Fe] is log $1/I_{50}$ determined in the iron/ascorbate assay; log P is the calculated partition coefficient; sigma is the Hammet sigma constant; numbers in brackets, { }, are the 95% confidence intervals associated with the coefficients; n is the number of data points used to derive the equation; s is the standard deviation; r is the correlation coefficient.

Another subset of the molecules in Table 1 was used to develop quantitative structure-activity relationships. This subset examined here consisted of the substituted benzylic indoles (compounds 1, 3, 6, 7, 9, 13, 15, and 16) and the very similar pyridylmethyl indoles (compounds 20-22) which facilitated the examination of electronic effects upon the 3-methylene of the indole through the Hammett sigma constant and estimated log P values as an indicator of relative lipophilicity. Regression analysis of these data produced the best correlation [r = 0.669] of the CB antioxidation activity with the electronic parameter, sigma, indicating the presence of other factors in this system that cannot be identified by simple correlation analysis. However, regression analysis of the Fe antioxidation activity in the Fe assay afforded equations 1 and 2 as the most meaningful relationships, which assess the effects of lipophilic character and electronic character, respectively, on antioxidant behavior. While Equation 1 suggests that partitioning equilibria may have some influence on the observed activity, Equation 2 is the most significant and indicates that the negative coefficient associated with the Hammett sigma parameter suggests that electron donation to the 3-methylene position increases antioxidant potency.

While many of the indoles are too large to be amenable to molecular orbital calculations, several of the smaller structures in Table 1 (compounds 14, 17, 19, 25, 26, and 30) have been examined by MNDO in an effort to gain further insight into structural features which may relate to antioxidant activity. The results, data not shown, of both atomic charge and ionization potential computations indicate significant contributions of the 2 and 3 positions of the indole nucleus, as well as the 3-methylene position, to antioxidant activity. Indole compounds containing substituents enhancing electron donation to the 3-methylene group show increased antioxidant potential.

In conclusion, this structure-activity assessment has provided a valid physicochemical basis for the development of better chemoprotective compounds.

ACKNOWLEDGEMENTS

Supported by NIEHS ES-03373 and UK Cancer Research Campaign.

REFERENCES

Hansch, C. and Leo, A.J. (1979). Substituent constants for correlation analysis in chemistry and biology, John Wiley & Sons, New York.

Shertzer, H.G., Niemi, M.P., Reitman, F.A., Berger, M.L. Byers, B.L. and Tabor, M.W. (1987a). Protection against carbon tetrachloride hepatotoxicity by pretreatment with indole-3-carbinol. *Exptl. Molec. Pathol.* **46**, 180-189.

Shertzer, H.G., Tabor, M.W. and Berger, M.L. (1987b). Protection from N-nitrosodimethylamine mediated liver damage by indole-3-carbinol. *Exptl. Molec. Pathol.* **4**, 211-218.

Shertzer, H.G., Berger, M.L. and Tabor, M.W. (1988a). Intervention in free radical mediated hepatotoxicity and lipid peroxidation by indole-3-carbinol. *Biochem. Pharmacol.* **37**, 333-338.

Shertzer, H.G. and Tabor, M.W. (1988b). Nucleophilic index value: implications for protection by indole-3-carbinol from N-nitrosodimethylamine cyto and genotoxicity in mouse liver. *J. Appl. Toxicol.* **8**, 105-110.

Tabor, M.W., Coats, E.C., Hogan, I.T.D., Sainsbury, M. and Shertzer, H.G. (1990). Structure-activity studies on the antioxidation potential of indole compounds, submitted.

Wattenberg, L.W. (1983). Inhibition of neoplasia by minor dietary constitutents. *Cancer Research* (Suppl) 43, 24485-24535.

MEASUREMENT OF STYRENE-OXIDE CYSTEINE ADDUCTS IN HEMOGLOBIN BY SELECTIVE CATALYTIC REDUCTION

David Ting, Martyn T. Smith, Penelope Doane-Setzer, Jeff Woodlee, and S. M. Rappaport[1]

Dept. of Biomedical and Environmental Health Sciences
School of Public Health
University of California
Berkeley, CA 94720

Alkylating agents, either direct-acting or generated *in vivo*, form an important class of environmental mutagens and carcinogens. These agents are characterized by their ability to bind covalently to nucleophilic sites in proteins and DNA. Due to the long life time of red blood cells and the stability of protein adducts, Ehrenberg and coworkers first (Calleman 1978; Osterman-Golkar 1976) proposed the use of hemoglobin (Hb) as a dosimeter for alkylation *in vivo*.

Since then, several methods have been devised to detect hemoglobin adducts (Bryant 1987; Calleman 1978; Farmer 1980; Neumann 1984). Here, we report an alternative method that is sensitive, specific and relatively simple. It is based on the well known desulfurization by Raney nickel (Ra-Ni) (Pettit 1962; Pizey 1974), in which the carbon-sulphur bond formed by the alkylating agent and cysteine is selectively cleaved and at least one new carbon-hydrogen bond is formed. Pachecka et al. (Pachecka 1979) demonstrated that when styrene-7,8-oxide-glutathione adducts were refluxed with Ra-Ni in ethanol, 1-phenylethanol (1-PE) and 2-phenylethanol (2-PE) were produced. This reaction between cysteine-bound styrene oxide and Ra-Ni is shown in Figure 1. We utilized this reaction to determine the amount of cysteine-bound styrene-7,8-oxide (SO) which reacted with Hb *in vitro* and *in vivo*.

MATERIALS AND METHODS

C-14 labeled SO was obtained from Amersham; it was diluted to a specific activity of 1.5 mCi/mmol with unlabeled SO before use. Styrene and SO were purchased from Aldrich and used without further purification. Human blood was incubated with SO *in vitro* at 37°C for 2 hr. In the *in vivo* experiment, Male Sprague-Dawley rats weighing 300g were given a single i.p. injection of styrene in corn oil. Blood was collected from each animal by cardiac puncture 20 hr after dosing. Globin (Gb), from both human and rat blood, was prepared by the following procedure. After lysing the red blood cells with water and spinning down the cell debris, the Hb solution was dialysed exhaustively at 5°C against distilled water. Then, the Hb solution was added dropwise to a cold 0.1% HCl, acetone solution to precipitate Gb (Anson 1930). The Gb was isolated by filtration, washed with acetone and hexane, and dried to constant weight in a desiccator. The radioactivity associated with the Gb samples was determined by liquid scintillation counting. The Gb was denatured by heating to 80°C

1. Author to whom correspondence should be sent, Tel# 415/642-7916, FAX# 415/642-5815.

Biological Reactive Intermediates IV, Edited by C.M. Witmer *et al.*
Plenum Press, New York, 1990

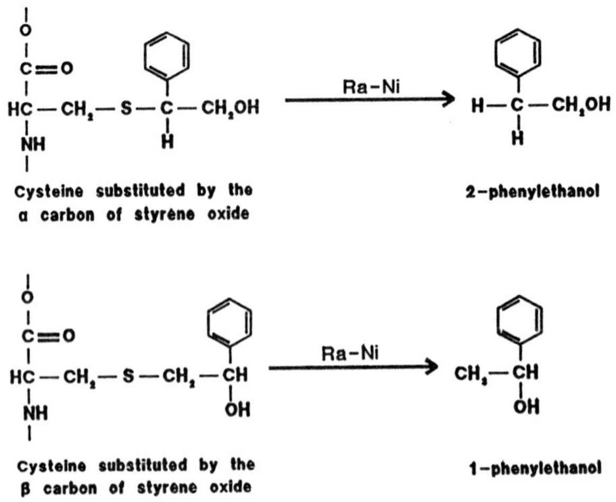

Ra-Ni

Cysteine substituted by the α carbon of styrene oxide

2-phenylethanol

Ra-Ni

Cysteine substituted by the β carbon of styrene oxide

1-phenylethanol

Figure 1. Cleavage of styrene-oxide-cysteine by Ra-Ni.

for 25 min and then digested with Protease XXV (from Sigma, about 7% W/W of Gb) at 37°C for 6 hr. After adjusting the pH to 12, the solution was spiked with an internal standard (100 ng of 3-phenyl-1-propanol). The sample was then shaken with Ra-Ni (50% slurry, from Aldrich, 5-6 g per g of Gb) at 5°C for 40 min to cleave the cysteine adduct. Upon completion of the reaction, the solution was extracted twice with 6 ml of ethyl ether. The ether fractions were combined, washed twice with 0.1 M HCl and then reduced almost to dryness under a stream of nitrogen. To derivatize the products of the Ra-Ni reaction, 0.5 ml of hexane, 3 ul of pyridine and 1.5 ul of pentafluorobenzyl chloride (PFB-chloride, from Aldrich) were added to the residue and the solution was warmed to 50°C for 20 min. After cooling to room temperature, the sample was dried under a stream of nitrogen, and the remaining residue was dissolved in 0.5 ml of 85% methanol and extracted with 0.5 ml hexane (Bogaert, 1978). A portion of this solution was injected into a Varian 3700 gas chromatograph equipped with an electron-capture detector and a capillary column (15-m x 0.32-mm i.d. DB-5, 1.0 um film thickness from J and W. Scientific). The injection volume was 2 ul with a split ratio of 20 to 1. The carrier gas was helium with a linear velocity of 29 cm/sec. Analysis was performed isothermally at 180°C.

RESULTS AND DISCUSSION

Our interest in monitoring the bioavailable dose of SO received by humans exposed to styrene led us to seek a simple method for monitoring SO-Gb adducts. In developing such a method we were intrigued with Pacheka's use of Ra-Ni to cleave SO-cysteine adducts from glutathione conjugates (Pachecka 1979). We applied a modification of that procedure to release adducts from Gb which had been derived from either human blood, modified with styrene oxide *in vitro*, or blood from rats to which styrene had been administered *in vivo*. We observed in both cases that reaction of Gb with Ra-Ni yielded 2-PE, one product which would be expected to result from cleavage of SO-cysteine.

Analysis of Human Blood Reacted with Styrene Oxide In Vitro

5-ml portions of fresh human whole blood were reacted *in vitro* with SO at concentrations ranging between 27 and 318 uM. Applying the Ra-Ni method to the Gb samples, a linear dose-response relationship was observed as shown in Figure 2. The amount of 2-PE detected was directly proportional to the amount of styrene oxide

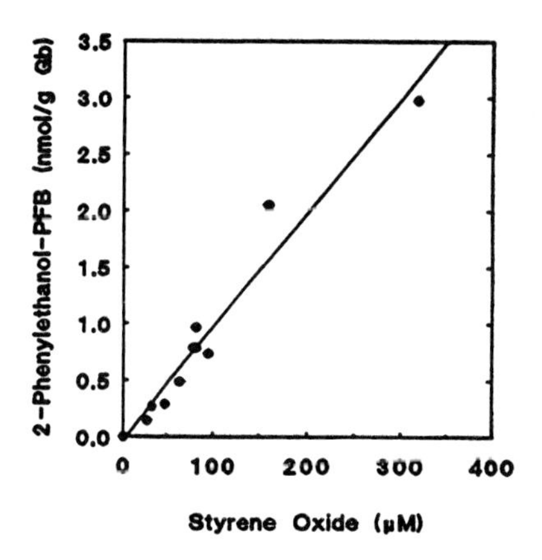

Figure 2. Dose-response curve showing globin adducts in human whole blood modified with styrene-oxide *in vitro*. (Corrected for 81.6% recovery).

added, with a slope of 9.83×10^{-3} nmol 2-PE/g Gb/uM styrene oxide (r=0.97). The detection limit of 2-PE-PFB was estimated to be 0.04 nmol/sample; assuming 5ml of blood per sample.

Analysis of Blood From Rats Treated with Styrene In Vivo

Since styrene is metabolized to SO in many animal species including humans (Leibman 1975; Ohtsuji 1971), we wished to demonstrate that this method can detect SO formed *in vivo*. Male Sprague-Dawley rats were treated with 0, 0.5, 1, 2 and 3 mmol styrene/kg body weight by i.p. injection. Upon analysis of the Gb isolated from these animals, a linear dose-response curve was observed (Figure 3), with a slope of 2.30 nmol 2-PE/g Gb/mmol styrene/kg body weight (r=0.90).

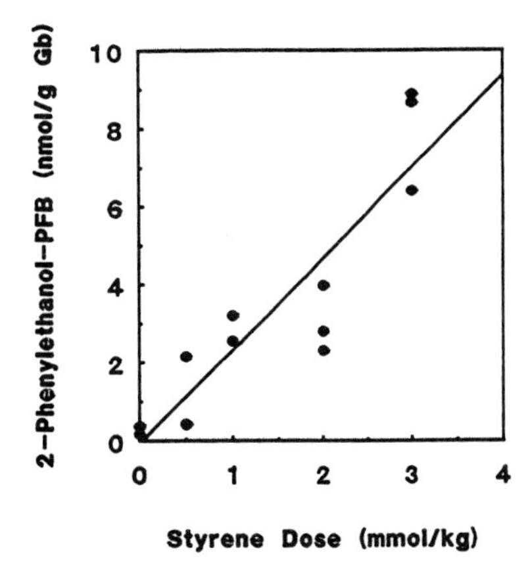

Figure 3. Dose-response curve showing globin adducts in rats following a single i.p. dose of styrene. (Corrected for 81.6% recovery).

Proportion of Cysteine Adducts

Although we showed that Ra-Ni could be successfully applied to cleave styrene-oxide adducts from Gb; our work indicated that only about 6% of the total binding was actually recovered as isomers of phenylethanol (Table 1) following the Ra-Ni reaction. This suggests either that the Ra-Ni is rather inefficient at cleaving cysteine adducts from Gb or (more likely) that other nucleophilic sites on the Gb chain account for the bulk of reaction with SO. Further experimentation is required to resolve this question.

Table 1. Styrene oxide-globin adducts detected by the Ra-Ni method* (Estimated mean ± std. dev. and number of replicates for each group)

Styrene-Oxide Conc. (uM)	% of total adducts detectable by the Ra-Ni method
33	5.8 ± 0.6 (2)
50	4.8 (1)
82	6.1 ± 0.9 (3)
Total	5.8 ± 0.8 (6)

* Corrected for 81.6% recovery and assuming 60% of styrene oxide-cysteine adducts are alpha-substituted (Pachecka, 1979). Total adducts were determined by counting Gb which had been reacted with 14C-labeled styrene oxide.

ACKNOWLEDGEMENT

This work was supported by grant RO1OH02221 from the National Institute for Occupational Safety and Health of the Centers of Disease Control, by grant P42ES04705 of the National Institute for Environmental Health Sciences, and by the Health Effects Component of the University of California Toxic Substances Research and Teaching Program.

REFERENCES

Anson, M. L. a. M., A.E. (1930). Protein coagulation and its reversal. *J. Gen. Physiol.* **13**, 469-475.

Bryant, M. S., Skipper, P. L., Tannenbaum, S. R. and Maclure, M. (1987). Hemoglobin adducts of 4-aminobiphenyl in smokers and nonsmokers. *Cancer Res.* **47**, 602-608.

Calleman, C. J., Ehrenberg, L., Jansson, B., Osterman-Golkar, S., Segerback, D., Svensson, K. and Wachtmeister, C.A. (1978). Monitoring and risk assessment by means of alkyl groups in hemoglobin in persons occupationally exposed to ethylene oxide. *J. Environ. Pathol. Toxicol.* **2**, 427-442.

Farmer, P. B., Bailey, E., Lamb, J. H. ,and Connors, T. A. (1980). Approach to the quantitation of alkylated amino acids in hemoglobin by gas chromatography mass spectrometry. *Biomed. Mass Spectrometry* **7**, 41-46.

Leibman, K. C. (1975). Metabolism and toxicity of styrene. *Environ. Health Perspect.* **11**, 115-119.

Neumann, H. G. (1984). Analysis of hemoglobin as a dose monitor for alkylating and arylating agents. *Arch. Toxicol.* **56**, 1-6.

Ohtsuji, H. a. I., M. (1971). The metabolism of styrene in the rat and the stimulatory effect of phenobarbital. *Toxicol. Appl. Pharmacol.* **18**, 321-328.

Osterman-Golkar, S., Ehrenberg, L., Segerback, D. and Hallstrom, I. (1976). Evaluation of genetic risks of alkylating agents. II. Hemoglobin as a dose monitor. *Mutat. Res.* **34**, 1-10.

Pachecka, J., Gariboldi, P., Cantoni, L., Belvedere, G. Mussini, E. and Salmona, M.

(1979). Isolation and structure determination of enzymatically formed styrene oxide glutathione conjugates. *Chem.-Biol. Interactions* **27**, 313-321.
Pettit, G.R.A.V.T., E.E. (1962). Desulfurization with Raney nickel. *Organic Reactions* **12**, 356-529.
Pizey, J. S. (1974). Raney nickel. *Synthetic Reagents* II, 177-288.

STUDIES ON BIOCHEMICAL DETERMINANTS OF QUINONE-INDUCED TOXICITY IN PRIMARY MURINE BONE MARROW STROMAL CELLS

Lorraine E. Twerdok and Michael A. Trush

Johns Hopkins University School of Hygiene and Public Health
Department of Environmental Health Sciences,
Division of Toxicological Sciences
Baltimore, MD 21205

Bone marrow is a target organ for toxicities induced by a spectrum of chemicals including the environmental pollutants benzene (Snyder et al., 1981) and benzo,[a]pyrene (BP) (Nebert and Jensen, 1979). Metabolism appears to be important for the expression of hematotoxicity by these two compounds, and metabolism of BP or benzene results in a number of metabolites including redox-active quinones. It has been demonstrated that metabolites of benzene, including hydroquinone (HQ) and benzoquinone (BZQ), concentrate in the marrow, and are believed to be the proximate or active metabolites responsible for the toxic effects of benzene. In support of this, Gaido and Wierda (1984, 1985) have identified bone marrow stromal cells as targets of toxicity of several benzene metabolites, particularly hydroquinone. We hypothesize that biochemical activities, such as quinone reductase (QR; NADPH:DT diaphorase) at the level of the bone marrow may contribute to target organ specificity in chemically-induced bone marrow toxicity. Quinone reductase is a widely distributed, cytosolic flavo-protein that has been shown to protect cells against the toxicity of quinones and their metabolic precursors (Kappus and Sies, 1981). Thus, the studies we report here have investigated susceptibility to quinone-generating metabolites of benzene and QR activity in cultured bone marrow stromal cells derived from DBA/2 mice.

MATERIALS AND METHODS

Animals and Cell Isolation from Bone Marrow

Male DBA/2 mice (25-30g) were obtained from Jackson Laboratories (Bar Harbor, ME) and housed in an air conditioned room with a light period from 6 a.m. to 6 p.m. Purina laboratory chow and water were available *ad libitum*, unless otherwise indicated.

Bone marrow cells were flushed from the femurs of mice according to the method of Oliver and Goldstein (1978). The procedure used to establish primary adherent stromal cell cultures was a modification of the method of Zipori and Bol (1979). Cells were used in experiments during the after being in primary culture for 12 to 18 days, unless otherwise specified.

Determination of Enzyme Activities

Dicoumarol-inhibitable QR activity was assayed by a modification of theprocedure developed by Prochaska and Santamaria (1988). Glutathione S-

Biological Reactive Intermediates IV, Edited by C.M. Witmer *et al.*
Plenum Press, New York, 1990

transferase (GST) activity was measured by the method of Habig et al. (1974), using 1-chloro-72,4-dinitrobenzene (CDNB) as substrate at pH 6.5. Basal QR and GST activities and activities following 24 h DTT treatment were determined on cells cultured in TC100 dishes.

Chemical Toxicity to Stromal Cells

Toxicity assays were performed as previously described (Twerdok and Trush,1990). Crystal violet staining, which correlates linearly with protein content and cell number, was used to assess cytotoxicity.

QR Protection Against HQ Toxicity

For QR protection studies, cells were treated with DTT 24 h prior to treatment with HQ, then assessed for survival or fixed and stained for esterase activity following 24 hr incubation with HQ. Esterase staining was performed using two different substrates, α-naphthyl acetate (nonspecific) and alpha-naphthyl butyrate (Sigma diagnostic kits 90-A1 and 181-B). Stromal cell-conditioned media (conditioned for 7 d) from control and HQ-treated cells (20 μM) was concentrated by centrifugation in "centri-cells" (Polysciences, Inc), and used as the source of colony-stimulating-factor for granulocyte/macrophage colony forming assays (Jones et al., 1989).

In Vivo Feeding of DTT

DBA/2 mice (9-10 week old males) were acclimated and maintained on an antioxidant-free powdered diet (AIN-76 purified diet with additional menadione and without ethoxyquin, TEKLAD DIETS) for one week, then the test group was switched to diet containing 0.1% DTT, a concentration which was well tolerated by the animal based on appearance, weight and activity levels. After 6 days on test or control diet, animals were sacrificed and bone marrow removed for enzyme analysis. QR activity was determined in wholebone marrow and 24 h primary cultures of bone marrow stromal cells.

RESULTS AND DISCUSSION

Bone marrow is the target organ of many diverse chemicals that humans are exposed to occupationally. We are presently investigating biochemical mechanisms at the level of bone marrow which could contribute to cellular toxicity resulting from biologically reactive intermediates. Bone marrow stromal cells, which consist of primarily fibroblasts and macrophages, were susceptible to toxicity induced by two redox-active metabolites of benzene, BZQ and HQ. LC50's for BZQ and HQ in primary cultures of DBA/2-derived bone marrow stromal cells were 6.98 ± 0.28 μM and 49.2 ± 6.40 μM, respectively. Non-cytotoxic doses of HQ impaired stromal cell ability to support granulocyte/macrophage colony formation *in vitro* by approximately 50%. 1,2-Dithiole-3-thione (DTT), a compound capable of inducing phase II enzymes *in vivo* in rat liver (Kensler et al., 1987), protected gross survival of stromal cells against HQ-induced toxicity (Table 1). In order to identify the underlying mechanisms of DTT chemoprotection, we assessed the effect of DTT on QR (NADPH:DT diaphorase) and GST activities in the stromal cells. DTT induced QR activity in primary stroma cells in a dose-dependent manner, however, there were no corresponding changes in GST activity, indicating that induction of QR activity may be important in DTT's chemoprotection (Figure 1). Basal QR activity and percent esterase-positive (macrophages) remained essentially constant in the first 2-3 wks of primary culture of bone marrow stroma. Treatment with HQ increased QR activity but reduced the percent esterase-positive cells, indicating that macrophages may be more sensitive to HQ-induced toxicity than the fibroblastic element of bone marrow stroma. Pretreatment of cells with DTT 24 h prior to administration of HQ restored the percent esterase-positive cells to control values. *In vivo* feeding studies with DTT induced QR activity within the bone marrow compartment. Thus, QR activity and its inducibility may prove to be important factors in chemically-induced bone marrow toxicity and carcinogenicity.

Table 1

Hydroquinone Concentration (μM)	Percent Survival of Stromal Cells - DTT	+ DTT (75 μM)
0	100	100
30	81 ± 1[b]	99 ± 2[c]
50	59 ± 5	92 ± 3[c]
75	28 ± 7	90 ± 4[c]
95	0.0	77 ± 3[c]

[a]cells were induced for 24 h where indicated, then treated for 24 h with the indicated concentration of HQ; survival was assessed 24 h later.
[b]all values represent mean ± SEM with an n of at least 3.
[c]significantly different from corresponding uninduced value by student's t-test, $p<0.05$.

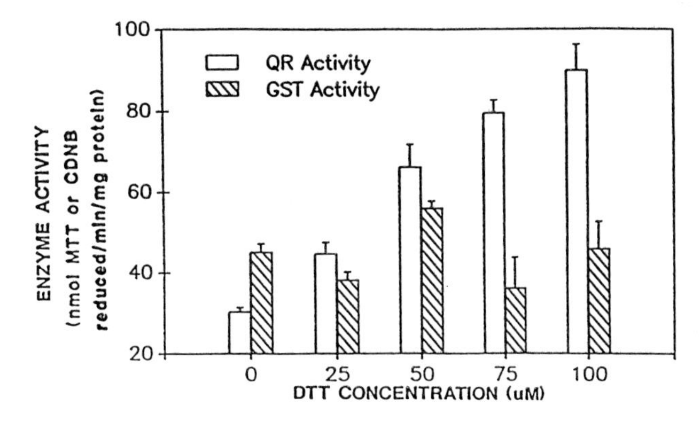

Figure 1. DTT effect on QR and GST activity in bone marrow stromal cells.

ACKNOWLEDGEMENTS

Supported by a Hazleton Graduate Student Fellowship, ES 03760, ES 07141, OH 02632 and American Cancer Society SIG-3. Dr. T.W. Kensler is gratefully acknowledged for the gift of DTT.

REFERENCES

Gaido, K.W. and Wierda, D. (1984). *In vitro* effects of benzene metabolites on mouse bone marrow stromal cells. *Toxicol. Appl. Pharmacol.* **76**, 45-55.

Gaido, K.W. and Wierda, D. (1985). Modulation of stromal cell function in DBA/2J and B6C3F1 mice exposed to benzene or phenol. *Toxicol. Appl. Pharmacol.* **1**, 469-475.

Habig, W.H., Pabst, M.J. and Jakoby, W.B. (1974). Glutathione S-transferases. The first enzymatic step in mercapturic acid formation. *J. Biol. Chem.* **249**, 7130-7139.

Jones, R.J., Celano, P., Sharkis, S.J. and Senesenbrenner, L.L. (1989). Two phases of engraphment established by serial bone marrow transplantation in mice. *Blood* **73**(2), 397-401.

Kappus, H. and Sies, H. (1981). Toxic effects associated with oxygen metabolism: redox cycling and lipid peroxidation. *Experientia* **37**, 1233-1241.

Kensler, T.W., Egner, P.A., Dolan, P.M., Groopman J.D. and Roebuck B.D. (1987).

Mechanism of protection against aflatoxin tumorigenicity in rats fed 5-(2-pyrazinyl)-4-methyl-1,2-dithiol-3-thione (Oltipraz) and related 1,2-dithiol-3-thiones and 1,2-dithiol-3-ones. *Cancer Res.* **47**, 4271-4277.

Nebert, D.W. and Jensen, N.M. (1979). Benzo[a]pyrene-initiated leukemia in mice association with allelic differences at the Ah locus. *Biochem. Pharm.* **7**, 149-151.

Oliver, J.P. and Goldstein, A.L. (1978). Rapid method for preparing bone marrow cells from small laboratory animals. *J. Immunol. Methods,* **19**, 289-292.

Prochaska, H.J. and Santamaria, A.B. (1988). Direct measure,ment of NAD(P)H:quinone reductase from cells cultured in microtiter wells: a screening assay for anticarcinogenic enzyme inducers. *Analytical Biochem.* **169**, 328-336.

Snyder, R., Longacre, S.L., Witmer, S., Kocsis, J.J., Andrews, L.S. and Lee, E.W. (1981). Biochemical toxicology of benzene. In: *Reviews in Biochemical Toxicology*, vol. 3, (Hodgeson E, Bend JR and Philpot RM). pp. 123-153. Elsevier/North Holland, New York.

Twerdok, L.E. and Trush, M.A. (1990). Differences in quinone reductase activity in primary bone marrow stromal cells derived from C57Bl/6 and DBA/2 mice. *Res. Comm. Chem. Path. Pharmacol.* in press.

Zipori, D. and Bol, S. (1979). The role of fibroblastoid cells and macrophages from mouse bone marrow in the *in vitro* growth promotion of hemopoietic tumour cells. *Exp. Hematol.* **7**, 206-218.

SELECTIVE ALTERATION OF CYTOKERATIN INTERMEDIATE FILAMENT BY CYCLOSPORINE A IS A LETHAL TOXICITY IN PTK2 CELL CULTURES

Lawrence A. Vernetti*, A. Jay Gandolfi* and Raymond B. Nagle**

Department of Anesthesiology*
Department of Pathology**
University of Arizona
Tucson, Arizona 85724

The cytoplasm of eukaryotic cells contain a series of three filamentous structures, microtubules, microfilaments, and intermediate filaments that are termed the cytoskeleton. Cytokeratin, one type of intermediate filament, has no known physiological function, yet, can comprise up to 30% of the total cytoplasmic protein content. As there are no selective toxins to cytokeratins, it is not known if alterations to these hydrophobic filaments is a lethal event. Cyclosporine A, a novel hydrophobic immunosuppressant compound used to prevent allograft rejection, may show a selective toxicity to the cytokeratin filaments. This effect is seen in PtK_2 cell cultures as a single large perinuclear aggregate of collapsed cytokeratin filaments (5 mM, 72 hr). Microtubules and microfilaments are not affected in PtK_2 cell cultures (5 mM, 72 hr). Increased LDH levels into cell culturing media occur soon after cyclosporine exposure to PtK_2 cell cultures (5 mM, 2 hr). Cytokeratin filaments show no changes at 12 hr exposure but show thickening, decreased plasma membrane attachments and some peri-nuclear ring formations at 24 hr (5 mM, 24 hr). Cyclosporine G, an analog of cyclosporine A, does not exhibit the cytokeratin filament collapse (5 mM, 72 hr). The effect of cyclosporine A on DNA binding protein (M_r 64 kd), believed to be a nuclear scaffolding protein related to intermediate filaments, exhibited an early invagination and folding of the nuclear membrane (5 mM, 4 hr). Due to a hydrophobic bonding potential between cyclosporine A and cytokeratin and cytokeratin-like intermediate filaments, cyclosporin A may be a selective cytokeratin toxin. Alteration of the cytokeratin filaments in PtK_2 cell cultures may be a lethal event.

INTRODUCTION

Cyclosporine A (CsA) has been shown to alter various intracellular functions. CsA induces lipid peroxidation in rat renal microsomes (Inselmann et al., 1988), inhibits protein kinase C activity in rat renal tubular epithelial cells (Walker et al., 1989), inhibits the activity of a cis-trans isomerase in pig kidney (Takahashi et al., 1989), and increases intracellular free calcium by binding directly to calmodulin (Colambani et al., 1987). Since CsA exposure produces an increase in intracellular vacuolation and nuclear lobulation, Colambani (Colambani et al., 1987) has suggested that CsA may alter the cytoskeleton of eukaryotic cells. These structural elements include the three components of the cytoskeleton, microtubules microfilaments, and intermediate filaments, and structural elements of the nuclear matrix.

Biological Reactive Intermediates IV, Edited by C.M. Witmer *et al.*
Plenum Press, New York, 1990

CsA is a cyclic peptide composed of 11 hydrophobic, neutral amino acids, and in seven of these amino acids the hydrophobicity is increased due to an amide hydrogen replacement by a methyl group. Further, three free amide hydrogens participate in intracyclic binding. The combination of the hydrophobic amino acids, intracyclic binding of amide hydrogens, and the methylamide hydrogen substitutions make CsA extremely lipophilic. Approximately 80% of cytokeratin protein monomers are of an alpha helical structure with a repeating heptad (abcdefg) amino acid sequence (Nagle, 1988). The first and fourth amino acid are always a hydrophobic residue. It is possible that a hydrophobic interaction between CsA and the cytokeratins may be a contributing factor in CsA associated nephrotoxicity.

METHODS

Cell Culture

Stock cultures of PtK_2 cells (ATC collection, Rockville, MD) were maintained at 37R C (5% CO_2 in air) in Hams F12/DME 1:1 media supplemented with 10% fetal calf serum and 1% penicillin/streptomycin (Gibco). Cells were then exposed to either vehicle (0.1% EtOH) or 5 mM CsA (Sandoz, LTD., Basel, Switzerland).

Cell Viability and LDH Leakage

At 0, 12, 24, 36, 48, and 72 hr, cell culturing media was decanted and saved for LDH leakage analysis, while the monolayer of cells was dislodged by a 0.25% trypsin treatment for 2 min. Cells were resuspended by viability counting by trypan blue stain or disrupted for total LDH enzymatic activity.

Histologic Evaluation

Cells dislodged from each time point by treatment with trypsin were allowed to settle on glass coverslips (2 hr), fixed, then incubated in primary antibody [anti-microtubule (Blose and Meltzer, 1981), anti-cytokeratin (Nagle et al., 1986), anti-nuclear scaffold protein - 2 (Cress and Kurath, 1988)] or rhodamine-conjugated phalloidin (Barak and Yocum, 1981). Coverslips then were incubated with fluorescein conjugated secondary antibody [anti-mouse IgG (Weir et al., 1979), anti-rabbit (Gown, 1986), 20 min]. Coverslips were mounted cell side down in a gelvatol mounting solution (16% gelvatol w/v in 33% glycerol-PBS, pH 7.4). All photomicrographs were taken with Tri-X 400 B & W film.

Two Dimensional Electrophoresis

Methodology of cytokeratin extraction according to Achtstaetter et al. (1986). Methodology of non equilibrium pH gel electrophoresis was according to O'Farrell et al. (1977) and DeRobertis et al. (1977). Protein spots were visualized with ammonical silver (10 mM $AgNO_3$, 25 mM NaOH, 100 mM NH_4OH).

Experimental Protocol

To examine the effect of CsA on the intact cytoskeleton and nuclear matrix, the cell line PtK_2 and nuclear matrix, the cell line PtK_2 was exposed to CsA through the cell culturing media. PtK_2 cell cultures were derived from a proximal tubular epithelial source and allow excellent immuno- histologic evaluation of cytoskeletal arrays. Murine antibodies raised to the cytokeratins and to microtubules were used for immnohistologic visualization at various time points during exposure. Rhodamine conjugated phalloidin was used to visualize microfilament arrays. A polyclonal antibody raised to a nuclear scaffolding protein-2 (Mr 64 kd) was used to visualize alterations in the nuclear matrix. Two dimensional electrophoresis and ammonical silver stain for proteins was used to describe decreased expression of three acidic cytokeratin protein monomers.

RESULTS

Examination of cell viability by trypan blue exclusion staining reveal a steady decrease in cells exposed to 5 mM CsA at 24-72 hr, then a abrupt decrease in viability at 72-96 hr (Figure 1). The LDH total/LDH leakage ratio into cell culturing media rise steadily from 12-48 hr (6-8 fold increase) then an abrupt increase at 72 hr (14 fold increase).

In PtK_2 cells exposed to 5 mM CsA for 72 hr there is the loss of cell to cell attachment and cytoplasmic constriction as the cells become elongated or rounded in appearance. At 72 hr, however, immunofluorescent staining exhibited no apparent alterations in the microtubule or microfilament arrays. Exposure to CsA did appear to decrease the number of actin microfilament stress fibers over controls, while leaving he microfilament network intact.

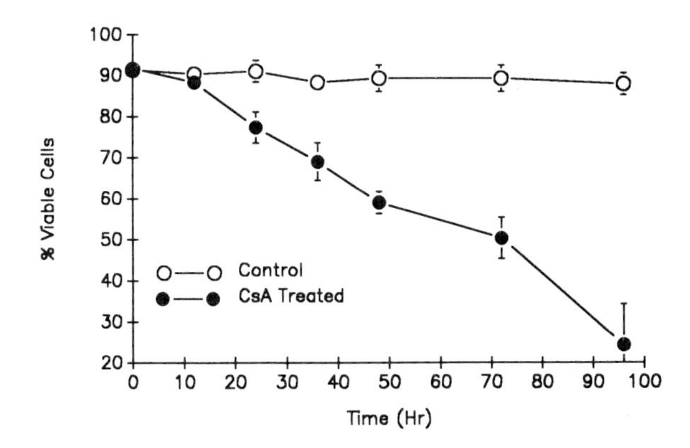

Figure 1. Effects of CsA of PtK_2 viability. Trypan blue exclusion staining depicts the decrease in cells exposed to 5 mM CsA.

The immunofluorescence staining patterns did reveal a selective alteration in the cytokeratin intermediate filament component of the cytoskeleton over controls at 72 hr (Figure 2 a,b). This pattern of altered cytokeratin was noticeable at 24 hr, with formation of peri-nuclear rings of cytokeratins, and continued at 36, 48 and 72 hr, revealing increasing numbers of peri-nuclear rings and eventual formation of a single aggregate clump of cytokeratins within the cytoplasm. Two dimensional electrophoretic gels of cytokeratin extractions reveal a decrease or elimination of a lessor triplet of acidic cytokeratins (human equivalents #15, 16, and 17).

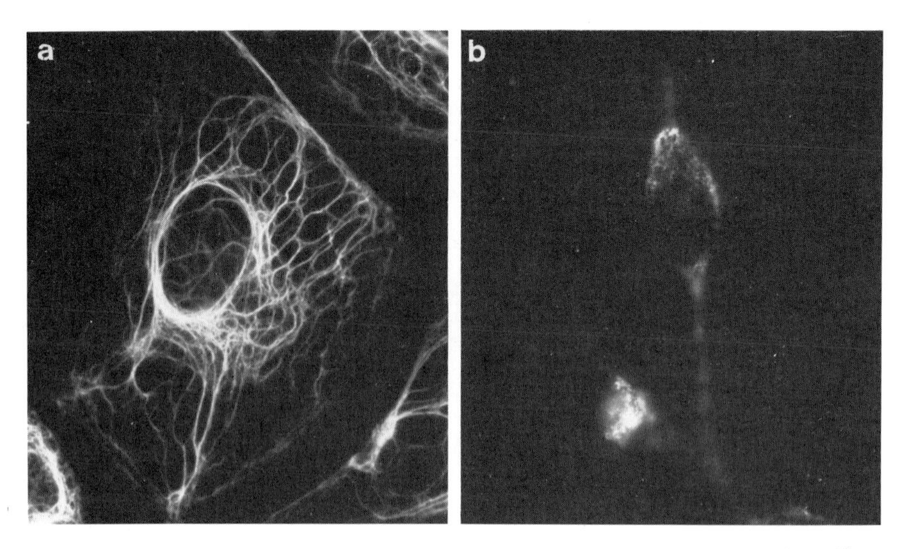

Figure 2. a. Cells exposed to vehicle for 72 hr illustrate cytokeratin intermediate filament arrays in normal relationships.

b. Cells exposed to 5 mM CsA for 72 hr illustrate collapse of cytokeratin arrays into a single aggregate clump located within the cytoplasm.

In the nucleus, immunofluorescence staining utilizing anti-protein 2 can demonstrate an early (4 hr) folding and buckling of the nuclear scaffold in cells exposed to CsA.

DISCUSSION

Although PtK_2 cells appeared elongated and spindle shaped or rounded up and condensed after 72 hr incubation in CsA, there was no discernable effect on the microtubule or microfilament arrays. There was a dramatic peri-nuclear ring formation of cytokeratin filaments at 24 hr and appearance of single aggregate of cytokeratins within the cell cytoplasm at 48 and 72 hr. Consistent with this visual alteration was a decrease in expression of 3 acidic cytokeratin proteins. Chen et al. (1988) has shown that aleration of cytokeratin protofilaments in yeast is a lethal event. These experiments suggest an association between CsA, the decline in cellular viability, the visual alteration in the immunofluorescent staining pattern of cytokeratins, and the decrease of certain cytokeratin proteins.

It is not known how CsA may interact with cytokeratin proteins. There may be a direct binding of drug to protein due to hydrophobic interactions, or there may be an interaction between drug and the regulatory system of the cytokeratins.

Little is known about the structural elements of the nuclear scaffolding, a group of detergent and salt insoluble proteins at the base of DNA loops (Cress and Kurath, 1988). Some evidence suggests these scaffolding proteins are related to intermediate filaments (Kackson and Cook, 1988. This study suggests an early (4 hr) CsA-associated folding or buckling of nuclear scaffold protein-2 (Mr 64kd), but the importance of this observation has yet to be determined.

ACKNOWLEDGEMENTS

This research was supported by the Arizona Kidney Foundation and by a University of Arizona Graduate School Fellowship.

REFERENCES

Achtstatter, T., Hatzfeld, M., Quinlan, R.A., Parmelee, D.C., Franke, W.W. (1986). Separation of cytokeratin polypeptides by gel electrophoretic and chromatographic techniques and their identification by immunoblotting. *Methods in Enzymology* **134**, 355-371.

Barak, L.S., Yocum, R.R. (1981). 7-Nitrobenz-2-oxa-1,3- diazole (NBD)-phallacidin; synthesis of a fluorescent actin probe. *Analyt. Biochem.* **31**, 110-117.

Blose, S.H., Meltzer, D.I. (1981). Visualization of the 10-nm filament vimentin rings in vascular endothelial cells *in situ*: close resemblence to vimentin cytoskeletons found in monolayer *in vitro*. *Exp. Cell. Res.* **135**, 299-300.

Chen, W. (1988). Yeast plasmid protein is a karyoskeleton component. *EMBO J.* **7**, 4323-4328.

Colambani, P.M., Donnenberg, A.D., Robb, A., Hess, A.D. (1987). Use of T lymphocyte clones to analyze cyclosporin binding. *Transplan. Proc.* **17**, 1413-1419.

Cress, A.E., Kurath, K. (1988). Identification of attachment proteins for DNA in Chinese hamster ovary cells. *J. Biol. Chem.* **263**, 19678-19684.

DeRobertis, E., Partington, G.A., Longthorne, R.F., Gurdon, J.B. (1977). Somatic nuclei in amphibian oocytes: evidence for selective gene expression. *J. Embryol. Ex. Morph.* **50**, 199-214.

Gown, A.M. (1986). Immunocytochemical analysis of the cellular composition of human atherosclerotic lesions. *Am. J. Path.* **125**, 191-207.

Inselmann, G., Blank, M., Bauman, K. (1988). Cyclosporine A induced lipid peroxidation in microsomes and effect on active and passive glucose transport by brush border membrane vesicle of rat kidney. *Re. Comm. Chem. Pathol. Pharmacol.* **62**, 207-220.

Jackson, D.A., Cook, P.R. (1988). Visualization of a filamentous nucleoskeleton with a 23 nm axial repeat. *EMBO J.* **7**, 3677-3687.

Moll, R. et al .(1982). The catalog of human cytokeratins: patterns and expression in normal epithelia, tumors, and cultured cells. *Cell* **31**, 11-24.

Nagle, R.B. (1988). Intermediate filaments: a review of basic biology. *Am. J. Surg. Path.* 12, 4-16.

Nagle, R.B., Bocker, W., Davis, J.R., Heid, H.W., Kaufman, M., Lucas, D.O., Jarasch, E.D. (1988). Characterization of breast carcinoma by two monoclonal antibodies distinguishing myoepithelial from luminal epithelial cells. *J. Hist. CytoChem.* **34**, 869-881.

O'Farrell, P.Z. et al. (1977). High resolution two dimensional electrophoresis of basic as well as acidic protein. *Cell* **12**, 1133-1142.

Takahashi, N., Hayano, T., Susuki, M. (1989). Petidyl-prolyl cis-trans isomerase is the cyclosporin A-binding protein cyclophilin. *Nature* **337**, 473-475.

Walker, R.J., Lazzaro, V.A., Duggin, G.G., Horvath, J.S., Tiller, D.J. (1989). Cyclosporin A inhibits protein kinase C activity: a contributing mechanism in the development of nephrotoxicity. *Biochem. Biophys. Re. Comm.* **160**, 409- 415.

Weir, D.M. (1979). Handbook of Experimental Immunology, Blackwell Scientific Publications, England 3rd edition.

RAT HEPATIC DNA DAMAGE INDUCED BY 1,2,3-TRICHLOROPROPANE

Gregory L. Weber and I. Glenn Sipes[1]

Department of Pharmacology and Toxicology
College of Pharmacy
University of Arizona
Tucson, Arizona 85721

1,2,3-Trichloropropane (TCP) is used as an industrial solvent and as an intermediate in pesticide and polysulfide rubber manufacturing. It is structurally similar to other halogenated alkanes with known carcinogenic potential (1,2-dibromoethane and 1,2-dibromo-3-chloropropane). Preliminary evidence from a chronic bioassay study conducted by the National Toxicology Program indicates that TCP administration results in an increased incidence of tumors in both male and female F-344 rats and B6C3F1 mice (Mahmood et al.,1988).

TCP requires bioactivation to become mutagenic. In the *Ames Salmonella* mutagenicity assay, TCP was found to be mutagenic for strains TA100 and TA1535 in the presence of rat hepatic S9 fraction (Ratpan and Plaumann, 1985) and mutagenic in TA100 in the presence but not in the absence of S9 or liver microsomes (Mahmood et al.,1988). We have previously shown that TCP binds covalently to rat hepatic protein, RNA and DNA *in vivo* (Weber and Sipes, 1990). The purpose of the present study was to determine if TCP could cause DNA damage in the form of DNA strand breaks, DNA-DNA or DNA-protein cross-links as determined by the alkaline elution procedure.

MATERIALS AND METHODS

Animals. Male F-344 rats (220-260g) were purchased from Harlan Sprague Dawley Inc. (Indianapolis, IN). The animals were kept in rooms with a 12-hour light/dark cycle, maintained at 22°C and were allowed food (Wayne Lab-Blox^R) and water *ad libitum*.

Dosing. All animals were dosed between 08:00 and 10:00. TCP (30, 100, 300mg/kg) was given in soybean oil at 2.5ml/kg by intraperitoneal injection. All animals were killed by CO_2 asphyxiation. Blood (3ml) and livers were quickly removed and placed on ice or in phosphate buffer saline containing 1mM Na_2EDTA (PBS) pH 7.2 at 4°C. Plasma alanine aminotransferase (ALT) activity was measured spectrophotometrically by the method of Bergmeyer et al. (1978).

Alkaline Elution. Rat hepatic nuclei were prepared by a modification of the procedure reported by Cox et al. (1973). Briefly, 2 grams of liver were minced in 2 volumes PBS and gently squashed through a 60 mesh wire screen. Nuclei were pelleted

---

1. To whom all correspondence should be sent: Dr. I. Glenn Sipes, Department of Pharmacology/Toxicology, College of Pharmacy, University of Arizona, Tucson, Arizona 85721

Biological Reactive Intermediates IV, Edited by C.M. Witmer *et al.*
Plenum Press, New York, 1990

by centrifugation at 50xg for 4 minutes at 4°C. Alkaline elution was performed by a modification of the method of Eastman and Bresnick (1978). Nuclei (50-100 μg DNA) were applied to 47 mm diameter polycarbonate 2 μm filters (Nucleopore, Corp., Pleasanton, CA) with 20 ml of PBS. The nuclei were lysed with 20 ml of lysing buffer containing 0.2% sodium lauroylsarcosinate (Sarkosyl), 2M NaCl and 40mM Na_2EDTA adjusted to pH 9.7 with NaOH. The nuclei were allowed to lyse for 1 hr then washed twice with 10 ml of buffer containing 20mM H_4EDTA adjusted to pH 10.0 with tetrapropylammonium hydroxide (RSA, Corp., Ardsley, NY). The DNA was eluted from the filters with a buffer containing 20mM H_4EDTA adjusted to pH 12.3 with tetrapropylammonium hydroxide at a flow rate of 0.1ml/min. Six 3ml fractions were collected and the DNA quantified fluorometrically using Hoechst 33258 (Murray et al., 1987). The alkaline elution results were plotted as the log of the fraction of DNA retained on the filter versus elution volume. The elution rate constant (k) was calculated as the slope of the linear portion of the elution curve, k=(-2.303) x slope.

Statistical analysis. Statistical differences were calculated with one-way analysis of variance (ANOVA) using the Duncan's multiple comparisons test. Differences were considered significant at $p<0.05$.

RESULTS AND DISCUSSION

TCP induced hepatic DNA damage (strand breaks) as soon as 1 hr after an intraperitoneal dose (Figure 1). The elution rate constants were significantly different at all time points except for 12 hr. The extent of DNA damage showed a continual decrease out to 12 hr with a transit rise at 24 hr. At 24 hr the animals had significantly elevated plasma ALT activity, demonstrating a cytotoxic response. The rise in the DNA damage at 24 hr is thus, most likely due to cell degeneration and death and not due to direct DNA damage caused by TCP. TCP induced DNA damage in a dose dependent manner (Figure 2). TCP at 0, 30, 100 or 300mg/kg produced elution rate constants of 0.010 ± 0.002, 0.027 ± 0.010, 0.041 ± 0.004, and 0.041 ± 0.006 (mean ± SE of at least 3 animals), respectively. The positive control in our single strand break system was the known carcinogen, 1,2-dibromoethane (EDB). Four hours after an equal molar dose of EDB (0.68mmol/kg), which was equal to 100mg/kg TCP, EDB had an elution rate constant of 0.081 or approximately twice that of TCP. When TCP was assayed for the generation of DNA cross-links, neither DNA-DNA or DNA-protein cross-links were detected.

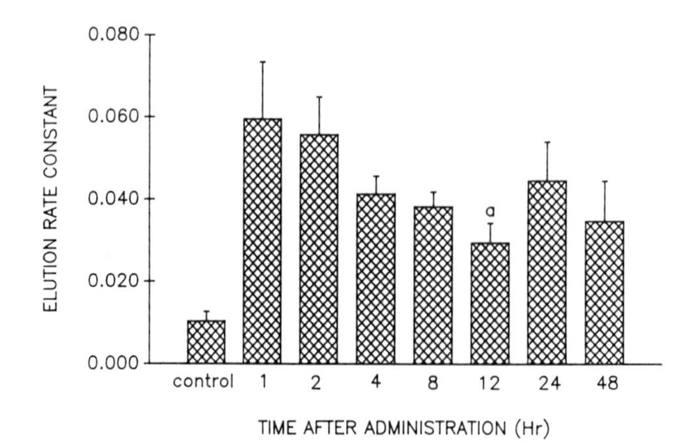

Figure 1. The effect of time on the induction of DNA strand breaks by 100mg/kg TCP. Data are mean elution rate constants (k) ± SE for at least 4 animals at each time. [a]Not significantly different from control; all other values significantly different from control.

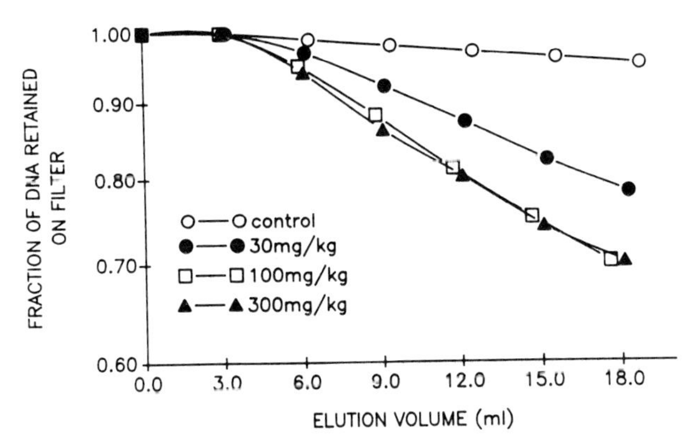

Figure 2. Effect of dose of TCP on the production of DNA strand breaks. Animals were dosed with either (0, 30, 100, or 300mg/kg TCP) and killed 4 hr later. Results are expressed as fraction of DNA retained on the filter versus elution volume.

The observations that TCP is mutagenic, can covalently bind to DNA and induces DNA single strand breaks as reported here, demonstrates that TCP is genotoxic and that DNA is the major target for TCP in relation to its tumorigenic affect.

ACKNOWLEDGMENTS

This work was supported by the National Toxicology Program (ES-85230).

REFERENCES

Bergmeyer, H.U., Scheibe, P., and Wahlefeld, A.W. (1978). Optimization of methods for aspartate aminotransferase and alanine aminotransferase. *Clin. Chem.* **24,** 58-73.

Cox, R., Damjanov, I., Abanobi, S.E., and Sarma, D.S.R. (1973). A method for measuring DNA damage and repair in the liver *in vivo*. *Cancer Res.* **33,** 2214.

Eastman, A., and Bresnick, E. (1978). A technique for the measurement of breakage and repair of DNA alkylated *in vivo*. *Chem.- Biol. Interactions* **23,** 369-377.

Mahmood, N.A., and Burka, L.T., and Cunningham, M.L. (1988). Metabolism and mutagenicity of 1,2,3-Trichloropropane. *Pharmacologist* **30,** A8.

Murray, D., VanAnkeren, S.C., and Meyn, R. (1987). Applicability of the alkaline elution procedure as modified for the measurement of DNA damage and its repair in nonradioactively labeled cells. *Anal. Biochem.* **160,** 149-159.

Ratpan, F., and Plaumann, H. (1985). Mutagenicity of halogenated three-carbon compounds and their methylated derivatives. *Environ Mutagenesis* **7,** (Suppl 3) 15.

Weber, G.L., and Sipes, I.G. (1990). Covalent interactions of 1,2,3-trichloropropane with hepatic macromolecules: Studies in the male F-344 rat. *Toxicol. Appl. Pharmacol.*, in press.

OXIDATION OF REDUCED PORPHYRINS BY THE MITOCHONDRIAL ELECTRON TRANSPORT CHAIN: STIMULATION BY IRON AND POTENTIAL ROLE OF REACTIVE OXYGEN SPECIES

James S. Woods and Karen M. Sommer

Department of Environmental Health
University of Washington
Seattle, Washington, 98195

Numerous studies have demonstrated a contributory role of iron in the pathogenesis of various inherited and chemically-induced disorders of porphyrin metabolism (porphyrias) (De Matteis et al, 1988, Smith and Francis, 1987, Lambrecht et al, 1988). However, the precise mechanism by which iron leads to excess tissue porphyrin accumulation and excretion in these disorders remains to be determined. Recent studies from these laboratories (Woods and Calas, 1989) have demonstrated that reduced porphyrins (porphyrinogens) are readily oxidized *in vitro* by reactive oxidizing species, such as superoxide (O_2^-) and hydroxyl ($OH\cdot$) radicals, and that this effect is dramatically stimulated by the presence of iron compounds in the reaction mixture. Inasmuch as mitochondria are a principal locus both of porphyrin metabolism (e.g. Tait, 1978) as well as of the generation of reduced oxygen species (O_2^-), H_2O_2) (Forman and Boveris, 1982) in cells, it is reasonable to postulate that, in the presence of excess iron, reduced porphyrins could be oxidized by reactive oxidants derived from the mitochondrial electron transport chain, contributing to the accumulation and excretion of oxidized porphyrins observed in iron-exacerbated porphyrinopathies. The present studies were conducted to test the hypothesis that reduced porphyrins are oxidized by reactive oxidants derived from the mitochondrial electron transport chain in the presence of iron.

MATERIALS AND METHODS

Materials

Male Sprague-Dawley rats (175-200 gms) were obtained from Tyler Laboratories, Inc., Bellevue, WA. Reduced β-nicotinamine adenine dinucleotide (NADH), sodium azide (NaN_3), sodium succinate, ethylenediaminetetraacetic acid (EDTA), reduced glutathione (GSH), superoxide dismutase (SOD) and catalase (CAT) were obtained from Sigma Chemical Co, St. Louis, MO. Uroporphyrin I was purchased from Porphyrin Products, Logan, UT. All other chemicals were obtained from standard commercial sources and were of the highest available purity. All solutions were prepared with metal-free deionized water.

Mitochondria Preparation

Mitochondria were prepared from rat liver or kidney cortex essentially as described by Johnson and Lardy (1967) using 0.25 M sucrose, 0.05 M Tris buffer, pH 7.5, and 3 mM EDTA. Pellets were washed twice with the same solution and finally

Biological Reactive Intermediates IV, Edited by C.M. Witmer *et al.*
Plenum Press, New York, 1990

suspended 1:1 with respect to original tissue weight in 140 mM KCl. Protein concentrations were determined by the method of Smith et al (1985).

Iron Chelates and Porphyrin Reduction

Iron-EDTA chelates were prepared by mixing a solution of 40 mM ferric chloride with an equal volume of 41 mM EDTA, and appropriate dilutions of this solution were made with deionized water. Uroporphyrinogen was prepared by reduction of uroporphyrin with freshly ground 3% sodium amalgam under N_2 and was neutralized with 0.05 M phosphate (Na_2HPO_4/NaH_2PO_4) buffer, pH 7.5, before use.

Porphyrinogen Oxidation

Oxidation of uroporphyrinogen at 37° was followed spectrofluorometrically using a Shimadzu RF-5000U recording spectrofluorometer by monitoring the increase in fluorescence of the oxidized porphyrin at the emission wavelength (620 nm) following excitation at 395 nm. Reaction mixtures contained 0.1 M HEPES buffer, pH 7.45, 600-800 ug mitochondrial protein, 0.4 mM NADH or 5 mM succinate, and 1 mM NaN_3 in a total volume of 3 ml. Fe-EDTA and other components were added as indicated. Uroporphyrinogen was added to the sample cuvette after a 5 minute incubation in a final concentration of 1 uM.

Statistical Analyses

Statistical differences between groups were determined by means of Student's test.

RESULTS AND DISCUSSION

Previous studies from this laboratory have demonstrated the oxidation of reduced porphyrins by reactive oxygen species *in vitro* (Woods, 1988, Woods and Calas, 1989). As summarized in Table 1, uroporphyrinogen was readily oxidized to the corresponding porphyrin in the presence of a O_2^- generating system comprised of 0.1 U/ml xanthine oxidase and 0.2 mM hypoxanthine in 0.1 M Tris buffer, pH 7.5, at 37°. This effect was increased 9.5-fold when 1 uM Fe^{+3}-EDTA was added to the reaction mixture, and by 17-fold by the addition of 10 uM Fe^{+3}-EDTA. Porphyrinogen oxidation was significantly suppressed by the addition of SOD or various OH· radical scavengers or GSH to the reaction mixture. Similarily, uroporphyrinogen was readily oxidized *in vitro* in the presence of a OH·-generating system comprised of 10 uM Fe^{+3}-EDTA and 2.5 mM H_2O_2 (Woods and Calas, 1989). This effect was also attenuated by OH· radical scavengers as well as GSH.

The capacity of mitochondria from mammalian tissues to generate reduced oxygen species (O_2^-, H_2O_2) in the presence of specific respiratory chain substrates and electron transport inhibitors is well established and has been described by various investigators (Forman and Boveris, 1982, Boveris and Chance, 1973, Loschen et al., 1971). When incubated in the presence of either NADH or succinate and NaN_3, mitochondria from either liver or kidney oxidize uroporphyrinogen at rate significantly faster than that observed in the absence of a respiratory chain substrate and inhibitor (Table 2). Addition of iron as 100 uM Fe^{+3}-EDTA to reaction mixtures dramatically increases the rate of porphyrinogen oxidation by mitochondria from either tissue. This response was observed when either NADH or succinate was used as the respiratory chain substrate (Table 2).

The oxidation of reduced porphyrins by the mitochodrial electron transport chain is stimulated in direct proportion to the concentration of iron in the reaction mixture. As depicted in Figure 1, the rate of uroporphyrinogen oxidation by hepatic or renal mitochondria increased in a dose-related fashion with respect to iron concentrations ranging from 0 to 50 uM in the reaction cuvette. Higher concentrations of iron produced no greater stimulation of porphyrinogen oxidation, suggesting the limiting capacity of the mitochondrial electron transport chain for generation of reduced oxygen species under the experimental conditions employed. In this regard, it was observed in other experiments that, in the presence of a fixed concentration of Fe-EDTA (100 uM), porphyrinogen oxidation increased in direct proportion to the amount of mitochondrial protein present in the reaction mixture in concentrations between 25 and 250 ug/ml (data not shown).

Table 1. Oxidation of Uroporphyrinogen *In Vitro* by O_2^- and $OH\cdot$ Generating Systems

Sample Composition	Porphyrinogen oxidation rate (pmoles/min)
1. Xanthine oxidase (0.1 U/ml) + hypoxanthine (0.2 mM) (Control-1)	150 ± 38
2. 1 + Fe^{+3}-EDTA (1 uM)	1425 ± 78*
3. 1 + Fe^{+3}-EDTA (10 uM)	2551 ± 132*
4. 1 + SOD (25 U/ml)	11 ± 10*
5. Fe^{+3}-EDTA (10 uM) + H_2O_2 (2.5 mM) (Control-2)	723 ± 61
6. 5 + mannitol (10 mM)	218 ± 21**
7. 5 + allopurinol (5 mM)	250 ± 40**

Reaction mixtures contained components indicated in 3 ml 0.1 M HEPES buffer, pH 7.45. Oxidation reactions were initiated after a 5 min. incubation period by adding uroporphyrinogen to the sample cuvette in a final concentration of 1 uM. Reactions were conducted at 37°, as described in Materials and Methods. Values in this and following tables and figure are expressed as the means ± SEM of at least 3 replicate experiments.

* $p < 0.05$ with respect to Control-1

** $p < 0.05$ with respect to Control-2

Table 2. Effects of Iron on Porphyrinogen Oxidation by Hepatic and Renal Mitochondrial Preparations.

Sample Composition	Porphyrinogen Oxidation rate (pmoles/min/mg protein) Liver	Kidney
Mitochondria only (Control)	33 ± 9	45 ± 17
+ NADH, NaN_3	65 ± 17*	86 ± 11*
+ NADH, NaN_3, Fe^{+3}-EDTA	260 ± 33*	351 ± 43*
+ succinate, NaN_3	61 ± 10*	78 ± 15*
+ succinate, NaN_3, Fe^{+3}-EDTA	312 ± 23*	382 ± 49*

Reaction mixtures contained 0.4 mM NADH or 5 mM succinate, 1 mM NaN_3, 600-800 ug mitochondrial protein and 0.1 M PO_4 buffer, pH 7.5, in a total volume of 3 ml. Fe as Fe^{+3}-EDTA (100 uM) was added where indicated. Uroporphyrinogen was added after a 5 minute incubation of the reaction mixture at 37° in a final concentration of 1 uM.

* $p < 0.05$ with respect to Control.

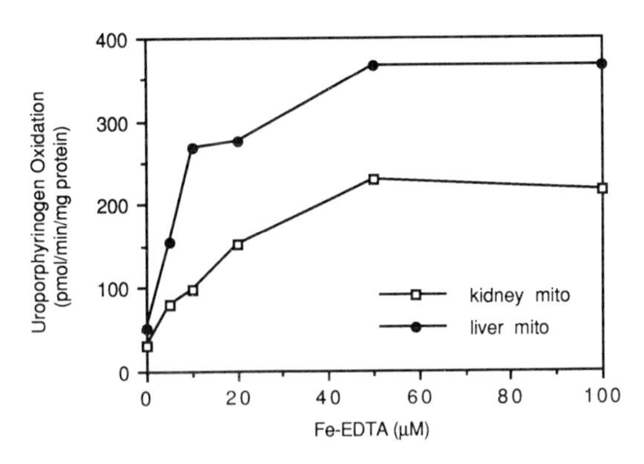

Figure 1. Uroporphyrinogen oxidation by the mitochondrial electron transport chain varies directly with added iron concentration. Reaction mixtures contained 5 mM succinate, 1 mM NaN_3, and 600-800 ug mitochondrial protein in 3 ml 0.1 M HEPES buffer, pH 7.45. Fe^{+3}-EDTA was added in concentrations indicated. Uroporphyrinogen was added after a 5 minute incubation of the reaction mixture in a final concentration of 1 uM.

Table 3. Effects of Reactive Oxidant Scavengers on Iron-Stimulated Mitochondrial Porphyrinogen Oxidation.

Sample Composition	Porphyrinogen oxidation rate (pmoles/min/mg protein)	
	Liver	Kidney
NADH, NaN_3, Fe-EDTA (Control)	201 ± 32	303 ± 43
+ SOD (120 U/ml)	234 ± 33	266 ± 43
+ CAT (5000 U/ml)	144 ± 48	226 ± 22*
+ mannitol (10 mM)	173 ± 21	211 ± 17*
+ GSH (5 mM)	48 ± 15*	70 ± 16*

Reaction mixtures contained 0.4 mM NADH, 1 mM NaN_3, 100 uM Fe^{+3}-EDTA, 600-800 ug mitochondrial protein and other components in concentrations indicated in a total volume of 3 ml 0.1 M PO_4 buffer, pH 7.5. Uroporphyrinogen was added after a 5 minute incubation in a final concentration of 1 uM.
* $p < 0.05$ with respect to Control.

The oxidation of reduced porphyrins by iron-supplemented mitochondria was attenuated by various antioxidants and free radical scavengers. As shown in Table 3, both mannitol and glutathione were moderately effective in attenuating the rate of porphyrinogen oxidation by either hepatic or renal mitochondria. Superoxide dismutase (250 U/ml), however, was relatively ineffective in this respect, suggesting, perhaps the inability of exogenous SOD to gain access to the site of reactive oxidant production within the mitochondrial matrix (Burkitt and Gilbert, 1989). The probability that free radical-mediated oxidation of cell constituents occurs within the immediate vicinity of reactive oxidant production has been suggested by various investigators (Borg and Schaich, 1987, Bucher et al, 1983, Halliwell and Guttridge, 1984).

In summary, the present studies demonstrate that reduced porphyrins are oxidized by the electron transport chain of respiring mitochondria *in vitro* and that this effect is dramatically stimulated by addition of iron to the reaction mixture. These observations are consistent with the hypothesis that iron catalyzes the production of reactive oxidants from mitochondrial-generated reduced oxygen species, which, in turn, promote porphyrinogen oxidation. Should these reactions occur *in vivo*, these results could suggest a plausible mechanism by which iron facilitates the production of excess porphyrins observed in various inherited or acquired disorders of porphyrin metabolism.

ACKNOWLEDGEMENT

This work was supported by the National Institutes of Health Grants ES03628 and ES04696.

REFERENCES

Borg, D.C. and Schaich, K.M. (1987). Iron and iron-derived radicals, In: *Oxygen Radicals and Tissue Injury*. Proceedings of an Upjohn Symposium. (Ed. Halliwell, B), pp. 20-26.

Boveris, A. and Chance, B. (1973). The mitochondrial generation of hydrogen peroxide. General properties and effect of hyperbaric oxygen. *Biochem. J.* **134**,707-716.

Bucher, J.R., Tein, M. and Aust, S.D. (1983). The requirement for ferric in the initiation of lipid peroxidation by chelated ferrous iron. *Biochem. Biophys. Res. Commun.* **111**,777-784.

Burkitt, M.J. and Gilbert, B.C. (1989). The control of iron-induced oxidative damage in isolated rat liver mitochondria by respiration state and ascorbate. *Free Rad. Res. Commun.* **5**,333-344.

De Matteis, F., Harvey, C., Reed, C. and Hempenius, R. (1988). Increased oxidation of uroporphyrinogen by an inducible liver microsomal system. *Biochem. J.* **250**, 161-169.

Forman, H.J. and Boveris, A. (1982). Superoxide radical and hydrogen peroxide in mitochondria. In: *Free Radicals in Biology*, Vol 5 (Ed. Pryor, WA), pp. 65-90, Academic Press, New York.

Halliwell, B. and Guttridge, J.M.C. (1984). Oxygen toxicity, oxygen radicals, transistion metals and disease. *Biochem. J.* **219**, 1-4.

Johnson, D. and Lardy, H. (1967). Isolation of liver and kidney mitochondria. *Methods Enzymol.* **10**, 94-96.

Lambrecht, R.W., Sinclair, P.R., Bement, W.J., Sinclair, J.F., Carpenter, H.M., Buhler, D.R., Urquhart, A.J. and Elder, E.H. (1988). Hepatic uroporphyrin accumulation and uroporphyrinogen decarboxylase activity in cultured chick-embyyo hepatocytes and in Japanese quail and mice treated with polyhalogenated and aromatic compounds. *Biochem. J.* **253**, 131-138.

Loschen, G.L., Flohe, L. and Chance, B. (1971). Respiratory chain linked H_2O_2 production in pigeon heart mitochondria. *FEBS. Lett.* **18**, 261-264.

Smith A.G. and Francis, J.E. (1987). Chemically-induced formation of an inhibitor of hepatic uroporphyrinogen decarboxylase in inbred mice with iron overload. *Biochem. J.* **246**, 221-226.

Smith, P.K., Krohn, R.I., Hermanson, G.T. (1985). Measurement of protein using bicinchoninic acid. *Analytic. Biochem.* **150**, 76-85.

Tait, G.H. (1978). The biosynthesis and degradation of heme. In: *Heme and Hemoproteins* (Eds. De Matteis, F. and Aldridge, W.N.), pp.1-48, Springer-Verlag, Berlin.

Woods, J.S. (1988). Attenuation of porphyrinogen oxidation by glutathione *in vitro* and reversal by porphyrinogenic trace metals. *Biochem. Biophys. Res. Commun.* **152**, 1428-1434.

Woods, J.S. amd Calas, C.A. (1989). Iron stimulation of free radical-mediated porphyrinogen oxidation by hepatic and renal mitochondria. *Biochem. Biophys. Res. Commun.* **160,** 101-08.

FUTURE RESEARCH NEEDS FOR THE APPLICATION OF MECHANISTIC DATA TO RISK ASSESSMENT

Donald J. Reed

Oregon State University
Department of Biochemistry and Biophysics
and Environmental Health Sciences Center
Corvallis, OR 97331

Risk assessment benefits enormously from the mechanistic data derived from research activity in toxicology which utilizes biological systems that range from whole animal, whole organ, organ slices, cells to the molecular level. New research techniques and instrumentation have vastly improved our ability to investigate complex toxicokinetic questions and to determine the roles for very small quantities of both low and high molecular weight molecules in toxic events. Some of these advances will be discussed relative to the requirements for making predictions for risk assessment associated with exposure to toxic chemicals.

Historically, chemical risk perception has relied on human beings using their senses of sight, smell, and taste to detect harmful or unsafe food, water and air, known as intuitive toxicology. Even today, the Food and Drug Administration utilizes smell as a major criteria for the quality of fish products. However, because of our increased ability to detect and quantitate chemical interactions, contemporary risk assessment in general has changed very dramatically in the last decade. Today, modern technologies have made our senses and intuitions inadequate to assess the dangers inherent in exposure to a chemical substance. We now depend on the sciences of toxicology and risk assessment, which involve extensive regulatory policies and bureaucracies, for the setting of standards and enforcement of these standards. Thus, risk assessment has become a complex process with major economic consequences as well as health considerations.

An extremely important component of risk assessment, as related to the acceptance of any exposure level, is the perceived risk to such exposure (Slovic, 1986, 1987). Perceived risk often involves strong and conflicting views about the nature and seriousness of the risks of modern life. An under appreciated aspect of risk assessment is that human beings view quite differently risk that they personally assume in their daily lives compared to those risks that they consider to be mandated upon them by private and/or governmental institutions. The acceptable voluntary risk by individuals has been estimated to be 1,000 times greater than that for involuntary risk (Starr, 1969). Moreover, these controversies elicit comments such as "How extraordinary! The richest, longest lived, best protected, most resourceful civilization, with the highest degrees of insight into its own technology, is on its way to becoming the most frightened", (Wildavsky, 1979).

Biological Reactive Intermediates IV, Edited by C.M. Witmer *et al.*
Plenum Press, New York, 1990

RISK ASSESSMENT BY PHYSIOLOGICALLY-BASED PHARMACOKINETIC MODELS

Physiologically-based pharmacokinetic (PB-PK) models, as described elsewhere in this book, are being used to provide an experimental method to reduce the uncertainty inherent in extrapolating the results of animal toxicity tests to man. Differences among animals and man are better understood with such models and thus advances are being made in many areas related to the data needed for risk assessment. A major step has been taken in the utilization of exposure dose and the extent of metabolism as a component in risk assessment. These experimental systems have required new methods of analysis accompanied by various assumptions. A few examples are given below.

Trichloroethylene risk assessment has involved the assumption that the amount of trichloroethylene metabolized per square meter of surface area was equivalent among species to facilitate the calculation of human equivalent doses from animal data (Bruckner et al., 1989). Drinking water exposures to trichloroethylene with rats has been quantitated by the amount of 14C-trichloroethylene metabolized. These data were used as an indicator of toxicity in a PB-PK model (Koizumi, 1989). In this model it was assumed that man drinks per day two liters of water containing 32 ppb trichloroethylene and 5 ppb tetrachloroethylene (Koizumi, 1989).

As more Pb-PK models are developed and with some based on data from *in vitro* systems, validation will require additional advanced analytical techniques. For example, risk assessment of exposure to methylene chloride has utilized both exposure dose and metabolism for estimation of *in vivo* rates of metabolism. *In vitro* incubations of lung and liver tissues from B6C3F1 mice, F344 rats, Syrian Golden hamsters, and humans indicate that metabolic rate constants support the PB-PK model for methylene chloride but that mouse data may greatly over estimate human tumor formation by methylene chloride (Reitz et al., 1989).

In the utilization of metabolism data as part of the risk assessment for any chemical in standard animal cancer assays, the administration of chemicals at maximally-tolerated doses (MTD) is postulated to increase cell division (mitogenesis). Since cell division increases also enhance the rates of mutagenesis and carcinogenesis, research is needed to understand better the influence of stimulated mitogenesis on the changes in the mechanism(s) of chemically induced toxicity (Review by Ames et al., 1990).

Ames and coworkers (1990) have analyzed existing pesticides data and calculated that 99.99% (by weight) of the pesticides in human diets are naturally)occurring chemicals that are produced by plants as defense agents. Support for the concept that many naturally-occurring substances are carcinogenic is provided by the ranking of possible carcinogenic hazards (Ames et al., 1988) (Table 1). Described in units defined as HERP (Human Exposure dose/Rodent Potency dose), the data show that rodent carcinogens vary in potency by more that 10 millionfold (Ames et al., 1987). Ames et al., also conclude that, if tested at MTD, a high proportion of natural chemicals would also be carcinogenic. Thus, much more research is needed on the relative contribution of synthetic and naturally occurring biological reactive intermediates and pesticides in general.

It would seem that risk assessment in the future will need to focus on research that addresses the possibility that natural and synthetic pesticides may have equal probability of being positive in high-dose animal cancer tests due to similarities in mechanism of toxicity. Further, Ames et al., (1990) suggest that at the low doses of most human exposures where mitogenesis due to high doses does not occur, the hazards to exposure may be lower that is commonly assumed from existing risk assessment analyses. Testing the long term effects of both natural and synthetic pesticides at such low doses in various model systems can benefit from new methodologies that will extend the current risk assessment practices. Some of those new technologies are indicated below.

Table I. Survey of Mass Spectrometric Techniques for the Analysis of Biopolymers

Technique*	Mass Range (Daltons)	Sensitivity (pmole)
Fast Atom Bombardment	20,000	1-100
Fast Ion Bombardment (M)	20,000	0.1-10
Continuous Flow FAB (M)	20,000	0.01-0.1
Plasma Desorption (T)	50,000	0.01-0.1
Pulsed Ion Desorption (T)	30,000	0.001-0.1
Pulsed Laser Desorption (T)	$>10^5$	0.001-1
Ekectrospray (Q)	$>10^5$	1-0.1

* Type of instrument: (M) = magnetic sector; (T) = time of flight; (Q) = quadrupole

ADVANCES IN ANALYTICAL TECHNIQUES

During the past 5-10 years the ability to analyze, with a high degree of sensitivity, metabolites and especially non-volatile metabolite conjugates has advanced to a remarkable degree. These new techniques including those of mass spectrometry, have permitted detailed analyses of metabolic processes in biological systems ranging from whole animals to *in vitro* cell incubations. Many of these advances have resulted from new ionization techniques of non-volatile compounds for mass spectrometry. Even more significant developments are being applied to the analysis of large proteins (Table 2). Proteins with molecular weights between 14,000 and 175,000 have been investigated by matrix-assisted laser desorption/ionization mass spectrometry (Karas et. al., Biomed. Environmental Mass Spectrometry 18, 641-643, 1989). High abundance molecular ions were obtained with absolute sample amounts ranging from 5 pmol down to 50 fmol. The volatization of high molecular weight DNA by pulsed laser ablation caused ejection of DNA molecules as large as 410 kilodaltons which could be further analyzed by polyacrylamide gel electrophoresis (Nelson et al., 1989). The potential for direct acquisition of sequence information in the mass spectrometer appears quite promising.

Additional improvements have been made on mass spectrometric analyses by the direct introduction of small quantities of samples with on-line coupling to separation systems such as HPLC and capillary-zone electrophoresis. The mass spectra of compounds with molecular weights in the range of 2,500-20,000 daltons have been obtained with a quadrupole mass spectrometer equipped with an atmospheric pressure ion source and an ion-spray interface to produce ions via the ion evaporation process (Covey et al., 1988). The extreme sensitivity of ion-spray allows the application of this technique to pmol levels compared to nmol levels for similar compounds by FAB and thermospray (Covey et al., 1988). Furthermore, the precision is claimed to be ± 1 dalton which is of importance to characterization of protein adducts.

From the advances in mass spectrometry it is possible to predict that future studies with biological reactive intermediates will be extended to the characterization of additional intermediates from xenobiotics. Even more significant will be the characterization of many more endogenous or naturally occurring biological reactive intermediates. With new technologies being developed rapidly, it appears possible to obtain mass spectral characterizations of specific proteins and presumably protein adducts of biological reactive intermediates in samples in which only limited separation or purification are required prior to analysis bylaser desorption/ionization mass spectrometry (Barofsky, personal communication, 1990).

Table II. Ranking Possible Carcinogenic Hazards by Ames et al., (1988)

Possible Hazard: HERP (%)	Daily Human Exposure	Carcinogen Dose per 70-kg Person	Potency of Carcinogen: TD50 (mg/kg)	
			Rats	Mice
		Pesticide and Other Residues		
0.0002*	PCBs: Daily dietary intake	PCBs, 0.2 μg (US average)	1.7	(9.6)
0.0004	EDB: Daily dietary intake (from grains and grain products)	Ethylene dibromide, 0.42 μg (US average)	1.5	(5.1)
		Natural Pesticides and Dietary Toxins		
0.003	Bacon, cooked (100 g)	Dimethylnitrosamine, 0.3 μg	(0.2)	0.2
0.006		Diethylnitrosamine, 0.1 μg	0.02	(+)
0.03	Comfrey herb tea, 1 cup	Symphytine, 38 μg (750 μg of pyrrolizidine alkaloids)	1.9	(?)
0.03	Peanut butter (32 g; one sandwich)	Aflatoxin, 64 ng (US average, 2ppb)	0.003	(+)
0.07	Brown mustard (5 g)	Allyl isothiocyanate, 4.6 mg	96	(-)
0.1	Basil (1 g of dried leaf)	Estragole, 3.8 mg	(?)	52
0.1	Mushroom, one raw (15 g) (*Agaricus bisporus*)	Mixture of hydrazines, and so forth	(?)	20,300
2.8*	Beer (12 ounces; 354 ml)	Ethyl alcohol, 1,8 ml	9110	(?)
		Environmental Pollution		
0.0004*	Well water, 1 liter contaminated, Woburn	Trichloroethylene, 267 μg	(-)	941
0.6	Conventional home air (14 hour/day)	Formaldehyde, 598 μg	1.5	(44)
0.004		Benzene, 155 μg	(157)	53
		Occupational Exposure		
5.8	Formaldehyde: Workers' average daily intake	Formaldehyde, 6.1 mg	1.5	(44)
140	EDB: Workers' daily intake (high exposure)	Ethylene dibromide, 150 mg	1.5	(5.1)
		Drugs		
[0.3]	Phenacetin pill (average dose)	Phenacetin, 300 mg	1246	(2137)

* HERP from carcinogens thought to be nongenotoxic.

MOLECULAR BIOLOGY

A major advance in molecular biology is the application of a thermally stable DNA polymerase to the synthesis of segments of DNA by a technique known as the polymerase chain reaction (PCR) (Guyer and Koshland Jr., 1989). Major features of the PCR includes the ability to amplify genetic information a millonfold within hours by the completion of 20 cycles of the PCR by automated procedures for repeated synthesis and heat denaturation cycles to anneal to the primers. The genetic information being amplified can be a gene or segment of DNA ranging from a single cell to large quantities of DNA. Because of the success of the PCR, it may replace gene cloning as the DNA amplification method of choice. The multitude of PCR applications include 1) direct sequencing of PCR amplified DNA; 2) engineering of DNA; 3) mutation detection; and 4) detection of gene expression with RNA from 10 to 1,000 cells.

PCR applications can involve several different strategies depending on the type of primer used for DNA synthesis. The standard PCR require primers of known sequences at both ends of a target sequence of DNA. An inverse PCR utilizes a circularized target sequence which is amplified at flanking unknown sequences away from the two primer sites. This occurs as the synthesis proceeds in the opposite direction taken by standard PCR primers and is possible by the use of reversed primer sequences. A third strategy is the use of only one defined primer to amplify genes that encode proteins for which partial sequences are known. Oligonucleotide complements to these sequences are prepared by automated chemical synthesis.

Variation on cloning and gene amplification in combination with the culture of human cells makes possible a virtually limitless range of experiments that can assess the degree to which the formation of biological reactive intermediates leads to alteration in gene expression and the formation of gene products that are important for risk assessment. Genetic susceptibility, as measured by alteration in gene expression, may well be correlated with the degree of formation of biological reactive intermediates and/or the degree to which such intermediates alter the function of both genes and gene products. Cell protective systems such as those that depend on glutathione for the detoxication of biological reactive intermediates, are also expected to be closely related to genetic susceptibility.

REFERENCES

Ames, B.N., Magaw, R. and Gold, L. S. (1987). Ranking Possible Carcinogenic Hazards. *Science*, **236**, 271-280.

Ames, B. M., Profet, M. and Gold, L. S. Dietary Pesticides (99.99% All Natural), Mitogenesis, Mutagenesis, and Carcinogenesis, Mutagens in Food: Detection and Prevention, H. Hayatsu, Ed. (CRC Press, Inc. Boca Raton, Fl) in press.

Bruckner, J.V., Davis, B.D. and Blancato, J.N. (1989). Metabolism, Toxicity, and Carcinogenicity of Trichlorethylene. *Crit. Rev. Toxicol.*, **20**, 31-50.

Covey, T. R., Bonner, R. F., Shushan, B. I., and Henion, J. (1988). The Determination of Protein, Oligonucleotide and Peptide Molecular Weights by Ion-spray Mass Spectrometry. *Rapid Communications in Mass Spectrometry*, **2**, 249-256.

Guyer, R. L. and Koshland, D. E. Jr. (1989). Perspective "The Molecule of the Year." *Science*, **246**, 1543-1546.

Karas, M., Ingendoh, A., Bahr, U., and Hillenkamp,F. (1989). Ultraviolet-Laser Desorption/Ionization Mass Spectrometry of Femtomolar Amounts of Large Proteins. *Biomedical Environmental Mass Spectrometry*, 641-643.

Koizumi, A. (1989). Potential of Physiologically Based Pharmacokinetics to Amalgamate Kinetic Data of Trichloroethylene and Tetrachloroethylene Obtained in Rats and Man. *Br. J. Ind. Med.* **46**, 239-249.

Nelson, R. W., Rainbow, J. J., Lohr, D. E., and Williams ,P. (1989). Volatilzation of High Molecular Weight DNA by Pulsed Laser Ablation of Frozen Aqueous Solutions. *Science*, **246**, 1585-1587.

Reitz, R.H., Mendrala, A.L., and Guengerich,F.P. (1989). *In vitro* Metabolism of

Methylene Chloride in Human and Animal Tissues: Use in Physiologically Based Pharmacokinetic Models. *Toxicol. Appl. Pharmacol.*, **97**, 230-246.

Slovic, P. (1987). Perception of Risk. *Science*, **236**, 280-285.

Slovic, P. (1986). Informing and Educating the Public About Risk. *Risk Analysis*, **6**, 403-415.

Starr, C. (1969). Social Benefit versus Technological Risk. *Science*, **165**, 1232-1238.

Wildavsky, A. (1979). No Risk Is the Highest Risk of All. *Am. Sci.*, **67**, 32-37.

CONTRIBUTORS

Anders, M. W., University of Rochester, Rochester, NY, USA
Anderson, M.W., National Institutes of Health, Research Triangle Park, NC, USA
Appleton, M.L., University of Utah, Salt Lake City, Utah, USA
Archer, M.C., University of Toronto, Toronto, Canada
Arndt, K., University of Illinois at Urbana-Champaign, Urbana, IL, USA
Azri, S., University of Arizona, Tucson, AZ, USA
Bentivegna, S., Robert Wood Johnson Medical School, Piscataway, NJ, USA
Billings, R.E., University of Nevada Medical School, Reno, NV, USA
Birge, R.B., University of Connecticut, Storrs, CT, USA
Boekelheide, K., Brown University, Providence, RI, USA
Bohr, V.A., National Cancer Institute, NIH, Bethesda, MD, USA
Bradshaw, T.P., Medical University of South Carolina, Charleston, SC, USA
Brady, J.T., University of Connecticut, Storrs, CT, USA
Brown, A.P., University of Arizona, Tucson, AZ, USA
Bruno, M.K., University of Connecticut, Storrs, CT, USA
Buhler, D.R., Oregon State University, Corvallis, OR, USA
Chepiga, T.A., Rutgers University/Robert Wood Johnson Medical School, Piscataway, NJ, USA
Chinje, E.C., University of Surrey, Guildford, Surrey, England, UK
Corcoran, G.B., University of New Mexico, Albuquerque, NM, USA
Davila, J.C., The University of Texas, Austin, TX, USA
Decker, K., Albert-Ludwigs-University, Freiburg, Fed. Rep. Germany
Dybing, E., National Institute of Public Health, Norway
Ehrenberg, L., University of Stockholm, Stockholm, Sweden
Eichelbaum, M., Dr. Margarete Fischer-Bosch-Institut für Klinische Pharmakologie, Fed. Rep. Germany
Elguindi, S., University of Toronto, Toronto, Ontario Canada
Faria, E.C., Rutgers, The State University of New Jersey, Piscataway, NJ, USA
Fisher, R., University of Arizona, Tucson, AZ, USA
Fitzsimmons, M.E., University of Rochester, Rochester, NY, USA
Gervasi, P.G., Istituto di Mutagenesi e Differenziamento, CNR, Pisa, Italy
Ghantous, H., University of Arizona, Tucson, AZ, USA
Gillette, J.R., National Heart, Lung, and Blood Institute, Bethesda, MD, USA
Goldstein, B.D., UMDNJ-Robert Wood Johnson Medical School, Piscataway, NJ, USA
Gorrod, J.W., University of London, London, UK
Graham, D.G., Duke University, Durham, NC, USA
Guengerich, F. Peter, Vanderbilt University School of Medicine, Nashville, TN, USA
Gunawardhana, L., University of Arizona, Tucson, AZ, USA
Halpert, J.R., University of Arizona, Tucson, AZ, USA
Hargus, S.J., University of Rochester, Rochester, NY, USA
Harris, J.W., University of Rochester, Rochester, NY, USA
Hart, S.G.E., University of Connecticut, Storrs, CT, USA
Hastings, K.L., University of Arizona, Tucson, AZ, USA
Hedli, C.C., Robert Wood Johnson Medical School, Piscataway, NJ, USA
Hill, B.A., The University of Texas at Austin, Austin, TX, USA
Hsieh, D.P.H., University of California at Davis, Davis, CA, USA
Huxtable, R.J., University of Arizona, Tucson, AZ, USA
Jaeschke, H., Baylor College of Medicine, Houston, TX, USA

Jerina, D.M., The National Institutes of Health, NIDDK, Bethesda, MD, USA
Kalf, G., Jefferson Medical College of Thomas Jefferson University, Philadelphia, PA, USA
Kehrer, J.P., The University of Texas at Austin, Austin, TX, USA
Kelner, M.J., University of California San Diego, San Diego, CA, USA
Kim, S.G., Wayne State University, Detroit, MI, USA
Laskin, D.L., Rutgers University, Piscataway, NJ, USA
Lasley, J., U.S. Environmental Protection Agency, Research Triangle Park, NC, USA
Lau, S.S., University of Texas at Austin, Austin, TX, USA
Lauriault, V.V., University of Toronto, Toronto, Ontario, Canada
Levine, W.G., Albert Einstein College of Medicine, Bronx, NY, USA
Lin, R.C., Indiana University School of Medicine, and the Veterans Affairs Medical Center, Indianapolis, IN, USA
Lind, R.C., University of Arizona, Tucson, AZ, USA
Lutz, W.K., University of Zurich, Schwerzenbach, Switzerland
Ma, Q., Rutgers University, Piscataway, NJ, USA
Maher, V.M., Michigan State University, East Lansing, MI, USA
Manno, M., University of Padua Medical School, Via J. Facciolati, Padova, Italy
Mansuy, D., Université René Descartes, Paris Cedex 06, France
Marnett, L.J., Vanderbilt University School of Medicine, Nashville, TN, USA
McCarthy, T.J., Rutgers University/UMDNJ-RW Johnson Medical School, Piscataway, NJ, USA
McGirr, L.G., University of Toronto, Toronto,Ontario, Canada
McMillan, D.C., Medical University of South Carolina, Charleston, SC, USA
Miller, J.A., University of Wisconsin Medical School, Madison, WI, USA
Moldéus, P., Karolinska Institutet, Stockholm, Sweden
Monks, T.J., The University of Texas at Austin, Austin, TX, USA
Mortensen, A.M., Wayne State University, Detroit, MI, USA
Mufti, S.I., The University of Arizona, Tucson, AZ, USA
Nelson, S.D., University of Washington, Seattle, WA, USA
Netter, K.J., Philipps-University, Lahnberge, Fed. Rep. Germany
O'Brien, P.J., University of Toronto, Toronto, Ontario, Canada
Odeleye, O.E., University of Arizona, Tucson, AZ, USA
Oesch, F., University of Mainz, Federal Republic of Germany
Orrenius, S., Karolinska Institutet, Stockholm, Sweden
Peter, H., University of Dortmund, Dortmund, Fed. Rep. Germany
Pohl, L.R., National Institutes of Health, Bethesda, MD, USA
Potter, D.W., Rohm and Haas Co., Spring House, PA, USA
Preston, B.D., College of Pharmacy, Piscataway, NJ, USA
Ramos, K., Texas A&M University, College Station, TX, USA
Reed, Donald J., Oregon State University, Corvallis, OR, USA
Reitz, R.H., Chemical Industry Institute of Toxicology, Research Triangle Park, NC, USA
Roth, R.A., Michigan State University, East Lansing, MI, USA
Roy, D., University of Texas Medical Branch, Galveston, TX, USA
Sabourin, P.J., Lovelace Inhalation Toxicology Research Institute, Albuquerque, NM, USA
Santella, R.M., Columbia University, New York, NY, USA
Shertzer, H.G., University of Cincinnati Medical Center, Cincinnati, OH, USA
Shipp, N.G., University of Arizona, Tucson, AZ, USA
Sies, H., University of Düsseldorf, West Germany
Silva, J.M., University of Toronto, Toronto, Ontario, Canada
Sipes, I.G., University of Arizona, Tucson, AZ, USA
Smith, B.J., University of Arizona, Tucson, AZ, USA
Smith, R.P., Dartmouth Medical School, Hanover, NH, USA
Stohs, S.J., Creighton University, Omaha, NE, USA
Strauss, B., The University of Chicago, Chicago, IL, USA
Streeter, A.J., National Cancer Institute, Frederick Cancer Research Facility, Frederick, MD, USA
Subrahmanyam, V.V., University of California at Berkeley, Berkeley, CA, USA
Sugimura, T., National Cancer Center, Tokyo, Japan
Surh, Y.-J., University of Wisconsin Medical School, Madison, WI, USA

Tabor, M.W., University of Cincinnati Medical Center, Cincinnati, OH, USA
Thomas, J.A., Iowa State University, Ames, IA, USA
Thompson, J.A., University of Colorado, Boulder, CO, USA
Tice, R.R., Integrated Laboratory Systems, Research Triangle Park, NC, USA
Ting, D., University of California, Berkeley, CA, USA
Trush, A., Johns Hopkins School of Hygiene and Public Health, Baltimore, MD, USA
Twerdok, L.E., Johns Hopkins University, Baltimore, MD, USA
Uetrecht, J., University of Toronto, Toronto, Canada
Ullrich, V., University of Konstanz, Federal Republic of Germany
van Ommen, B., TNO-CIVO Toxicology and Nutrition Institute, The Netherlands
Vernetti, L.A., University of Arizona, Tucson, AZ, USA
Vistisen, K., University of Copenhagen, Copenhagen, Denmark
Vittozzi, L., Istituto Superiore di Sanità, Rome, Italy
Watanabe, M., Tohoku University, Sendai, Japan
Weber, G.L., University of Arizona, Tucson, AZ, USA
Witz, G., UMDNJ-Robert Wood Johnson Medical School, Piscataway, NJ, USA
Woods, J.S., University of Washington, Seattle, WA, USA
Ziegler, D.M., The University of Texas at Austin, Austin, TX, USA

INDEX

THE LIBRARY
UNIVERISTY OF CALIFORNIA, SAN FRANCISCO
(415) 476-2335

THIS BOOK IS DUE ON THE LAST DATE STAMPED BELOW

Books not returned on time are subject to fines according to the Library Lending Code. A renewal may be made on certain materials. For details consult Lending Code.

14 DAY
MAY - 4 1994

14 DAY
MAY 18 1994

RETURNED
MAY 18 1994

Series 4128